The Nuclear Equation of State

Part A: Discovery of Nuclear Shock Waves and the EOS

NATO ASI Series

Advanced Science Institutes Series

A series presenting the results of activities sponsored by the NATO Science Committee, which aims at the dissemination of advanced scientific and technological knowledge, with a view to strengthening links between scientific communities.

The series is published by an international board of publishers in conjunction with the NATO Scientific Affairs Division

A	**Life Sciences**	Plenum Publishing Corporation
B	**Physics**	New York and London
C	**Mathematical and Physical Sciences**	Kluwer Academic Publishers Dordrecht, Boston, and London
D	**Behavioral and Social Sciences**	
E	**Applied Sciences**	
F	**Computer and Systems Sciences**	Springer-Verlag
G	**Ecological Sciences**	Berlin, Heidelberg, New York, London,
H	**Cell Biology**	Paris, and Tokyo

Recent Volumes in this Series

Volume 210—Crystal Growth in Science and Technology
edited by H. Arend and J. Hulliger

Volume 211—Phase Transitions in Soft Condensed Matter
edited by Tormod Riste and David Sherrington

Volume 212—Atoms in Strong Fields
edited by Cleanthes A. Nicolaides, Charles W. Clark, and Munir H. Nayfeh

Volume 213—Interacting Electrons in Reduced Dimensions
edited by Dionys Baeriswyl and David K. Campbell

Volume 214—Science and Engineering of One- and Zero-Dimensional Semiconductors
edited by Steven P. Beaumont and Clivia M. Sotomayor Torres

Volume 215—Soil Colloids and their Associations in Aggregates
edited by Marcel F. De Boodt, Michael H. B. Hayes, and Adrien Herbillon

Volume 216A—The Nuclear Equation of State, Part A:
Discovery of Nuclear Shock Waves and the EOS
edited by Walter Greiner and Horst Stöcker

Volume 216B—The Nuclear Equation of State, Part B:
QCD and the Formation of the Quark-Gluon Plasma
edited by Walter Greiner and Horst Stöcker

Series B: Physics

The Nuclear Equation of State

Part A: Discovery of Nuclear Shock Waves and the EOS

Edited by

Walter Greiner and Horst Stöcker

Johann Wolfgang Goethe University
Frankfurt am Main, Federal Republic of Germany

Springer Science+Business Media, LLC

Proceedings of a NATO Advanced Study Institute
on The Nuclear Equation of State,
held May 22-June 3, 1989,
in Peñiscola, Spain

Library of Congress Cataloging-in-Publication Data

NATO Advanced Study Institute on the Nuclear Equation of State (1989 Peñiscola, Spain)
The nuclear equation of state / edited by Walter Greiner and Horst Stöcker.
p. cm. -- (NATO ASI series. Series B, Physics ; vol. 216A-216B)
"Published in cooperation with NATO Scientific Affairs Division."
"Proceedings of a NATO Advanced Study Institute on the Nuclear Equation of State, held May 22-June 3, 1989, in Peñiscola, Spain"--T.p. verso.
Includes bibliographical references.
Contents: pt. A. Discovery of nuclear shock waves and the EoS -- pt. B. QCD and the formation of the quark-gluon plasma.
ISBN 978-1-4612-7877-1 ISBN 978-1-4613-0583-5 (eBook)
DOI 10.1007/978-1-4613-0583-5

1. Matter, Nuclear--Congresses. 2. Heavy ion collisions--Congresses. 3. Nuclear astrophysics--Congresses. I. Greiner, Walter, 1935- . II. Stöcker, Horst. III. North Atlantic Treaty Organization. Scientific Affairs Division. IV. Title. V. Series: NATO ASI series. Series B, Physics ; v. 216A-216B.
QC793.3.N8N38 1989
539.7--dc20 89-71006
CIP

Originally published by Plenum Press, New York in 1989
Softcover reprint of the hardcover 1st edition 1989

PREFACE

The NATO Advanced Study Institute on ***The Nuclear Equation of State*** was held at Peñiscola Spain from May 22 – June 3, 1989.

The school was devoted to the advances, theoretical and experimental, made during the past fifteen years in the physics of nuclear matter under extreme conditions, such as high compression and high temperature. More than 300 people had applied for participation – this demonstrates the tremendous interest in the various subjects presented at the school. Indeed, the topic of this school, namely

the Nuclear Equation of State,

- plays the central role in high energy heavy ion collisions;
- contains the intriguing possibilities of various phase transitions (gas – vapor, meson condensation, quark – gluon plasma);
- plays an important role in the static and dynamical behavior of stars, especially in supernova explosions and in neutron star stability.

The investigation on the nuclear equation of state can only be accomplished in the laboratory by compressing and heating up nuclear matter and the only mechanism known to date to achieve this goal is through shock compression and –heating in violent high energy heavy ion collisions. This key mechanism has been proposed and highly disputed in the early 70's. It plays a central role in the whole field of high energy heavy ion physics, and particularly in our discussions during the two weeks at Peñiscola.

The first week was mostly devoted to high energy heavy ion collisions, in particular to the experimental discovery of the collective flow, multifragmentation, particle production etc. and their dependence on the nuclear equation of state and other characteristic quantities of nuclear matter (e.g. its viscosity coefficients). This was followed by the presentation of a broad range of other areas of research where the nuclear equation of state is vital, reaching from nuclear physics (compressibility) and meson condensates to gravitation theory and astrophysics (neutron stars, supernova explosions). The second week focused on quantum chromodynamics and the statistical mechanics and kinetic theory of quarks and hadrons, in particular the search for a phase transition to a quark gluon plasma in ultra–relativistic heavy ion collisions. Under these extreme conditions one expects the nucleons in the compression shock zone to be so densely packed that considerable overlap between them will appear and, consequently, the individual nucleon bags may dissolve and the quark–gluon plasma (QGP) may be created. This is a theoretically conjectured new phase of nuclear matter. It is of enormous interest because quarks would exist relatively free within the QGP while they are ordinarily confined to small baryon and meson bags. First experimental data taken at Brookhaven's AGS and CERN's SPS by various collaborations have been presented. The formation of the QGP may eventually be detected by observing strangeness and charm, or, more promisingly, via exotic fragments like anti–nuclei or strange matter droplets. The strangelets are a particularly appealing theoretical prediction, because they represent a qualitative (not quantitative) signal for the formation of quark matter – in fact, they would be an exotic remnant of such a hypothetical state.

We were fortunate to be able to call on a body of lecturers who not only made considerable personal contributions to this research but who are also noted for their lecturing skills. Their enthusiasm and dedication for their work was readily transmitted to the students resulting in a very successful school. This enthusiasm is also reflected in the contributions to these Proceedings which, as I believe, will in time become a standard source of reference for future work on the physics of high energy heavy ion collisions and will help to spread the benefits of the school to a larger audience than those who were able to attend. I am happy that a number of East German, Hungarian, Polish and Soviet colleagues were able to participate. They have made important contributions to the field so that their presence has certainly further contributed to the atmosphere of the meeting.

The wide variety of topics discussed by the invited lecturers and by the seminar speakers made it necessary to publish these proceedings in two volumes: The first volume covers the physics of shock waves and the EoS in high energy heavy ion collisions, nuclear matter theory, astrophysics and special topics. The second volume is concerned with non–perturbative QCD, ultrarelativistic heavy ion collisions and the phase transition to the quark–gluon plasma.

Special highlights were three distinguished lectures. Erwin Schopper and Arthur Poskanzer lectured on "Early experiments on nuclear shock waves". Professor Schopper was personally involved in pioneering first experiments aimed at nuclear shock waves while Dr. Poskanzer together with Prof. Hans Gutbrod and their associates achieved the breakthrough in discovering the nuclear shock waves in all details predicted by the Frankfurt school. In their lectures they exhibited the experimental roots of the modern development of these ideas. Professor Edward Teller is a giant in physics and technology, who made many fundamental contributions to science. Hearing and seeing him during his lectures on "Ideas on high T_c–superconductors" and on "Smart Pebbles" was of great benefit to all members of the institute. No wonder that the lecture theater was overcrowded.

The School was sponsored by and had the support of the NATO Research Council, the Bundesministerium für Forschung und Technologie, Vanderbilt University (Nashville, -Tennessee), the Gesellschaft für Schwerionenforschung (GSI), the University of Valencia, and also the Spanish Government.

The meeting took place at the "Instituto de Estudios de Administracion Local" at Peñiscola, a center adjoint to the University of Valencia. With its beautiful scenery, its facilities and last but not least its management it contributed to the success of the school.

Finally I wish to acknowledge the contributions made by all those who took part in organizing and running the school. Above all, I am grateful to Prof. Elugio Oset (Valencia) who helped tremendously in the local organization of the school. He not only drew my attention to the beautiful ancient city of Peñiscola, but also helped actively to hire Spanish folklore–singers and dancers and organized an excursion to a very entertaining baby bull fight. Special thanks go also to my secretary Katja Matto, who handled the correspondence before, during, and after the school, managed the budget, and helped to make these Proceedings possible. I am grateful to the Organizing Committee for their help in preparing the school, in particular I wish to thank my colleague Horst Stöcker and many other members of the Institute of Theoretical Physics, Johann Wolfgang Goethe–University at Frankfurt am Main.

Walter Greiner

CONTENTS

THE NUCLEAR EQUATION OF STATE IN SUPERNOVAE AND NEUTRON STARS

Opening Remarks

Walter Greiner
Institut für Theoretische Physik
der Johann Wolfgang Goethe Universität
Frankfurt am Main, Germany

I would like to welcome you at Peniscola/Spain to our Nato Advanced Study Institute on

The Nuclear Equation of State

You came from many different parts of the world to attend this school, which deals with a novel field of physics with exciting new aspects and promising scientific adventures. Far too many have applied for participation(in fact more than 300 people). I had to turn down a great number and have hoped that still another great number would finally not show up, because I could not offer the support they wanted. As you can see by looking at the great many people who are here today , I have not been successful in this respect. Let us hope that we can manage. I apologize in advance for any inconvenience. Take it leasurely and be assured that we try our best to make these two weeks a very successful meeting. Before I try to describe our motivation for the physics we shall be discussing during this conference, let me thank a few institutions and persons, who have made this meeting possible. There is first the *Nato Scientific Council* who approved this NASI. I express my gratitude to the members of this council. Second, I would like to thank *Dr. D. Hartwig* and through him the *Bundesministerium für Forschung und Technology* of The Federal Republic of Germany, who gave additional support to our school. Third, it is a pleasure to especially thank *Prof. E. Oset* from the *University of Valencia*, who not only helped tremendously with the local organization and who drew my attention to this beautiful town of Peniscola, but also raised some funds from the Spanish Government in support of Spanish attendends. *Prof. Paul Kienle*, the director general of GSI(Darmstadt) helped also supporting the meeting, as did *Prof. J. Hamilton(Vanderbilt University*, Nashville, Tennessee), the director of the Joint Heavy Ion Institute at Oak Ridge/Tennessee. Finally I would explicitly like to thank my colleague and friend *Prof. Horst Stöcker* who helped me outstandingly with many useful advices and suggestions. He has been, and still is a permanent scientific consultant. Consider him a co-director of this NASI!
As you may have noticed, I used four different phrases to describe our gathering here; namely Advanced Study Institute, School, Meeting and Conference. This is my intention!Iwant us to have something of everything. We should have a school, in which advanced new research and ideas are represented in the most pedagogical way so that our numerous students and researchers can follow easily and get a smooth introduction into the new field. The lecturers should keep this in mind, as should the seminar speakers(who are mostly in action during the late afternoon session). I want also that the usual atmosphere of a scientific meeting or conference prevails here, since there are so many new, recent developments in our field which must be brought up, carefully discussed, with opposing views being friendly confronted. I use the word friendly, because there is no need for disharmony. The various aspects of the Nuclear Equation of State are so far reaching and so rich in content that there is enough for everybody in this area of research. Moreover, it is only harmony which can create new ideas for new adventures. This is certainly true for me, but I believe that it is also true for the most of you; if not all of us.

The Nuclear Equation of State, Part A
Edited by W. Greiner and H. Stöcker
Plenum Press, New York

For some time I have been thinking about the proper title of our meeting. There were various posibilities: should we call it "Physics of High Energy Heavy Ion Collisions" or "Nuclear Shock Waves" ?These titles would have been to narrow. I finally decided to call it "The Nuclear Equation of State". Indeed it is the nuclear equation of state which plays the central role in high energy heavy ion collisions, which contains the intriguing posibilities of various phase transitions (gas - vapor, meson condensation, quark - gluon plasma) and which plays an important role in the static and dynamical behaviour of stars, especially in supernova explosions. Of course, the investigation on the nuclear equation of state can only be accomplished in the laboratory by compressing and heating up nuclear matter. *The only mechanism known to achieve that is through shock wave compression and heating in high energy heavy ion collisions.* This key mechanism has been proposed in the early 70'ies. It plays a central role in the hole field of high energy heavy ion physics, and particularly in our discussions during the next two weeks here at Peniscola. It is therefore appropriate to outline the basic features of nuclear shock waves, to make you familiar with basic definitions and concepts(which are described and refined by other speakers later on). It is also important to describe to you the history of nuclear shock waves. They were highly disputed after they were proposed by Scheid, Müller and Greiner in 1973/1974. The arguments pro and contra nuclear shock waves, the difficult experimental steps, paved with doubts on one side and conviction for their existence on the other side, until finally all the theoretical predictions concerning nuclear shock waves were verified, deserve to be remembered. I have therefore made sure that two early-timers in the field, namely Prof. Erwin Schopper(Frankfurt) and Dr. Arthur Poskanzer(Berkeley) present parts of the history of their experimental envolvement in the great endeavor of the discovery of nuclear shock waves. I, of course, shall give you my theory-oriented point of view.

The Concept of The Nuclear Equation of State

Suppose there is nuclear matter consisting of nucleons only, of density $\rho(\vec{r},t)$ and temperature $T(\vec{r},t)$ extended over all space. If we cut out of that a volume V and ask for the energy content in that volume, it will be given by

$$E_V = \int_V \rho W(\rho,T) dV \tag{1}$$

The thus defined energy functional $W(\rho,T)$ is what we call in this simple situation of neutron matter only the nuclear equation of state. It contains all important global properties of nuclear matter. For example, the pressure p is given by(sec. eq. (10)below)

$$p = \rho^2 \left.\frac{\partial W}{\partial \rho}\right|_{T=const} \tag{2}$$

and hence it can be brought into the form

$$p = p(\rho,T) \tag{3}$$

Since $\rho = \frac{N}{V}$ where N is the total number of nucleons and V is their volume, we obtain finally

$$p = p(V,T) \tag{4}$$

and this is what is usually understood as the equation of state. We have learned it in school. In nuclear physics it is more convenient to use $W(\rho,T)$ instead of (4), because $W(\rho,T)$ exhibits a number of features in a more transparent form.

It was already in 1968 when Scheid and Greiner (Phys. Rev. Lett. 21(1968)1479 and Z. f. Physik 226(1969)364) used the nuclear equation of state to calculate interaction potentials for nucleus-nucleus collision and also the ground state properties of ordinary, finite nuclei(charge-, mass distribution, binding energy, radii, etc). In this classical paper the compression of nuclear matter and its side wards flow in central collisions were already discussed (see fig. 1)

They used an improved ansatz compared to (1) for the energy $E_V(\rho)$ of a piece of nuclear matter, namely

$$\begin{aligned} E_V(\rho) = & \int_V \rho(W(\rho) + m_N c^2) d\tau + \frac{e^2}{2}\int\int_V \frac{\rho_p(\vec{r}_1)\rho_p(\vec{r}_2)}{|\vec{r}_1 - \vec{r}_2|} d\tau_1 d\tau_2 \\ & + C_{ex}\int_V \rho_p^{4/3} d\tau + C_{sy}\int_V (\rho_p - \rho_n)^2 \rho^{\nu} d\, d\tau \\ & + \frac{V_0}{4\pi}\int\int_V \rho(\vec{r}_1)\frac{e^{-|\vec{r}_1 - \vec{r}_2|/\mu}}{|\vec{r}_1 - \vec{r}_2|}(\rho(\vec{r}_2) - \rho(\vec{r}_1)) d\tau_1 d\tau_2\,. \end{aligned} \tag{5}$$

The various terms have a very plausible interpretation: The first term takes over the role of the *volume energy* in the Bethe-Weizsäcker mass formula. $W(\rho)$ is the *energy per nucleon* (1) in infinitely extended nuclear matter, which is called the "*equation of state for cold nuclear matter*". It is indeed a quite fundamental functional which contains the basic global properties of nuclear matter. The form $W(\rho)$ will be discused in some detail below. $m_N c^2 \approx 923 MeV$ denotes the rest energy of a nucleon at equilibrium density. Since $\int \rho d\tau = A$, the number of nucleons, the term $\int \rho m_N c^2 d\tau = A m_N c^2$ contributes a constant to (5). It is often deleated because the total rest energy of the nucleons has no effect whatsoever on the determination of physical quantities, such as density distribution, *etc.*

VOLUME 21, NUMBER 21 PHYSICAL REVIEW LETTERS 18 NOVEMBER 1968

ION-ION POTENTIALS AND THE COMPRESSIBILITY OF NUCLEAR MATTER

Werner Scheid, Rainer Ligensa, and Walter Greiner
Department of Physics, University of Virginia, Charlottesville, Virginia,
and Institut fur Theoretische Physik der Universität, Frankfurt am Main, Germany
(Received 4 September 1968)

With a schematic model for the nuclear matter we give a unified treatment of the real and imaginary parts of the elastic O^{16}-O^{16} scattering potential. The model connects the parameters of the potential with the density and binding properties of the O^{16}-O^{16} system and reproduces the structure of the excitation function quite well. It is shown that the nuclear compressibility can be obtained from the scattering data, and in the case of the S^{32} compound system there results an effective compressibility (finite quenching of the nuclei) of about 200 MeV.

We propose a simplified and phenomenological model of the nuclear matter similar in the fundamental ideas to those of the recently proposed nuclear matter theories of Bethe[3] and Brueckner.[4] The following Ansatz for the total nuclear energy is constructed under the condition that we get a linear and analytically solvable equation for the nuclear density distribution ρ:

$$E = W_0 A + \frac{C}{2\rho_0}\int(\rho-\rho_0)^2 d\tau + \frac{V}{8\pi}\int \rho(\vec{r}_1)\frac{e^{-|\vec{r}_1-\vec{r}_2|/\mu}}{|\vec{r}_1-\vec{r}_2|}[\rho(\vec{r}_2)-\rho(\vec{r}_1)]d\tau_1 d\tau_2 + \frac{1}{2}\left(\frac{eZ}{A}\right)^2\int\frac{\rho(\vec{r}_1)\rho(\vec{r}_2)}{|\vec{r}_1-\vec{r}_2|}d\tau_1 d\tau_2. \quad (1)$$

The total energy (1) is thus composed of an energy proportional to the particle number A, an essentially repulsive energy, an attractive interaction energy ($V < 0$) of Yukawa type with the range μ, and the Coulomb energy in which we put the charge density proportional to the nuclear density.

Disregarding the Coulomb energy and assuming constant density in (1) we obtain the binding energy per nucleon in infinite nuclear matter,

$$\frac{E}{A} = W_0 + \tfrac{1}{2} C \frac{\rho_0}{\rho}\left(\frac{\rho}{\rho_0}-1\right)^2. \quad (2)$$

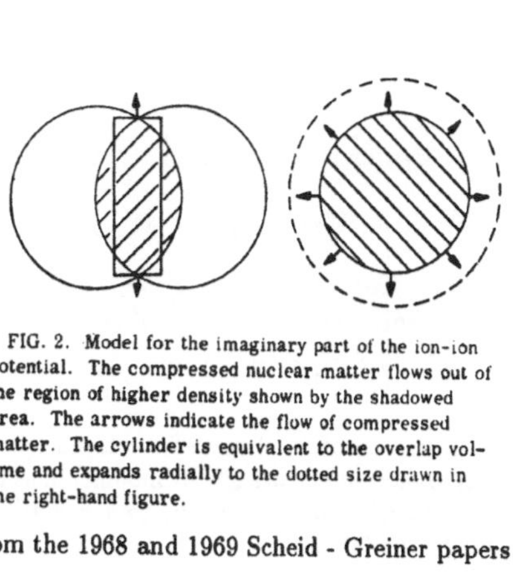

FIG. 2. Model for the imaginary part of the ion-ion potential. The compressed nuclear matter flows out of the region of higher density shown by the shadowed area. The arrows indicate the flow of compressed matter. The cylinder is equivalent to the overlap volume and expands radially to the dotted size drawn in the right-hand figure.

Fig. 1a. Excerpts from the 1968 and 1969 Scheid - Greiner papers quoted in the text.

The second term, involving only the proton density ($\rho = \rho_p + \rho_n$), is the *Coulomb energy*, followed by an *exchange correction* due to the Pauli principle acting between the nucleons. Its strength is characterized by the coefficient C_{ex} given by

$$C_{ex} = -\frac{3}{4}\left(\frac{3}{\pi}\right)^{1/3} = -1.0635 MeV fm. \tag{6}$$

The fourth term describes the *symmetry energy*; it is proportional to the square of the local deviation between proton and neutron densities. The power ν of the additional factor ρ^{ν} is not easily fixed theoretically; in practice different values of ν in the range $-1 \leq \nu \leq +\frac{1}{3}$ have been tested and have been shown to yield equally good fits for the ratio of $\frac{Z}{A}$ along the valley of stable isotopes - and even far beyond - for all known nuclei. The coefficient C_{sy} varies with the assumed value for ν and was found to be(J. Fink et. al. , Z. Physik A 323(1986)189).

$$C_{sy} = \gamma e^{\beta\nu}; \gamma = 215 MeV fm^3; e^{\beta} = 9.39 fm^3$$

The last term is the *surface energy*. For the surface interaction, which is dominated by the long-range components in the nucleon-nucleon interaction, a Yukawa form was assumed; because of the

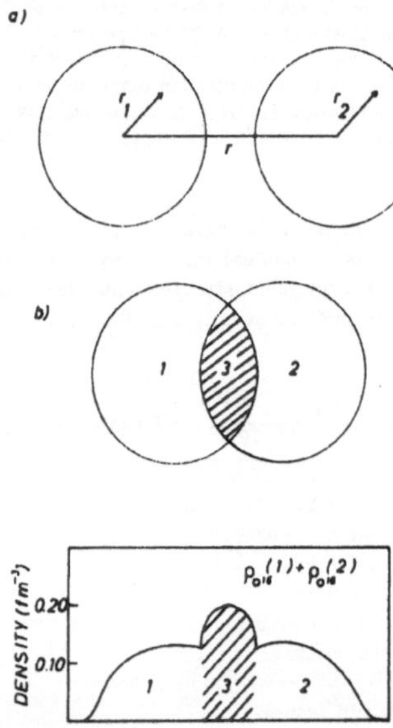

Fig. 1b. The superposition of the densities: (a)The coordinates $\vec{r}_1$ and $\vec{r}_2$ are measured from the center of the nuclei. The coordinate $\vec{r}$ denotes the relative distance of two centers. (b)The density distribution of nuclei 1 and 2 overlap in the region 3 shown by the shadowed area, in which the matter is compressed. The lower diagram contains the density distribution along an intercept through the centers of the two nuclei.

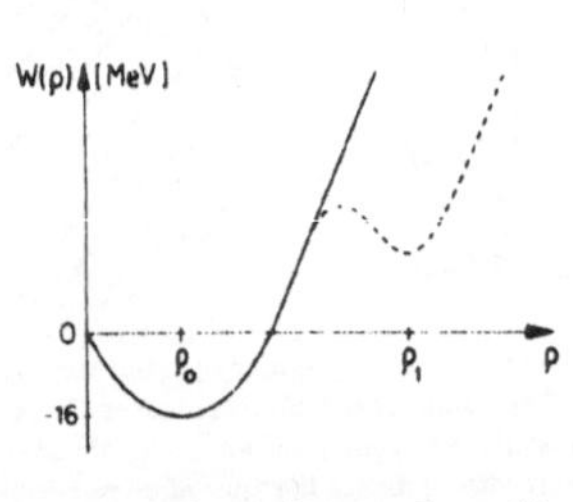

Fig. 2. Typical form of the nuclear equation of state for cold nuclear matter. ρ_0 is the nuclear matter density. The dashed curve shows schematically its modifications for a nuclear density isomer at density ρ_1. Such new phases of nuclear matter have up to now not been observed. Experiments indicate a functional as shown by the full curve.

factor $\rho(\vec{r}_2) - \rho(\vec{r}_1)$, it contributes only where density gradients are present. The range μ critically determines the surface thickness of the nuclear density distribution, as shown below. Not much is presently known about the nuclear equation of state $W(\rho)$ in general. The short-range nature of nuclear interactions implies $W(\rho = 0) = 0$, *i. e.* infinitely dilute nuclear matter is unbound. The saturation of nuclear forces leads to a minimum of $W(\rho)$ at the saturation or equilibrium density ρ_0 (see fig. 2). Fits to nuclear binding energies and density distribution yield (H. Stock, Nucl.

Phys. A 237(1975)365, J. P. Blaizot, Phys. Report 64C(1980)/7)

$$\rho_0 = 0.166 \pm 0.027 fm^{-3} \quad and \quad W_0 \equiv W(\rho_0) = -16 \pm 0.5 MeV. \tag{7}$$

At this point the pressure(which is the sum of the positive Pauli pressure from the nuclear zero-point motion and the negative pressure from the attractive long-range interactions between the nucleons) vanishes:

$$p(\rho_0) = \rho^2 \left. \frac{\partial W}{\partial \rho} \right|_{\rho=\rho_0} = 0. \tag{8a}$$

This formula is rather plausibile, since it describes the pressure as an energy density. The pressure is positive for $\partial W/\partial \rho > 0$, negative for $\partial W/\partial \rho < 0$, and vanishes at the equlibrium density $\rho = \rho_0$(see fig. 2). This can be understood as follows: In thermodynamics the pressure p of N particles is given in terms of their inner energy E and the volume V as $p = -\partial E/\partial V|_{T=const.}$. Now, $\partial/\partial V = \partial\rho/\partial V . \partial/\partial\rho = -N/V^2 \partial/\partial\rho$, because $\rho = N/V$ is the particle density. Hence it follows that

$$p = \left. \frac{N}{V^2} \frac{\partial E}{\partial \rho} \right|_{T=const} = \left. \frac{N^2}{V^2} \frac{\partial}{\partial \rho} \left(\frac{E}{N} \right) \right|_{T=const} = \left. \rho^2 \frac{\partial W}{\partial \rho}(\rho) \right|_{T=const.} \tag{8b}$$

The curvature of $W(\rho)$ near ρ_0 characterizes the compressibility of equilibrium nuclear matter: Larger curvature obviously implies that a greater energy is necessary for compression; no curvature at all would mean that nuclear matter at different density ρ always has the same energy per nucleon. The curvature can be determined from density oscillations(giant monopol vibrations)in spherical nuclei and was measured as(see the contributions of Blaizot and of Sharma):

$$K \equiv 9\rho^2 \left. \frac{\partial W}{\partial \rho^2} \right|_{\rho=\rho_0} = 210 \pm 30 MeV \tag{9}$$

Its value is also related to the speed of propagation of small density oscillations or *sound velocity* in nuclear matter,

$$v_s = \sqrt{\frac{K}{9m_N}}, \tag{10}$$

where m_N is the nucleon mass. To understand this expression let us make a short digression into sound waves in fluids and gases which depend sensitively on the compresibility. Naturally, the viscosity also plays a role and for the exact description of this phenomenon we would have to use the ***Navier-Stokes equation*** of fluid dynamics(see J. Maruhn's lecture). Here we are interested in a simplified description neglecting the influence of viscosity, *i. e.* we do not consider the damping of the collective motion we call sound. Also, external forces are not considered. The equations of motion are therefore ***Euler's equation***

$$m_N \rho \left(\frac{\partial \vec{v}}{\partial t} + (\vec{v}.\vec{\nabla})\vec{v}(\vec{x},t) \right) = \vec{\nabla} p(\vec{x},t) \tag{11}$$

and the ***continuity equation***

$$\frac{\partial \rho}{\partial t} + div(\rho \vec{v}) = 0. \tag{12}$$

Here m_N is the mass of the particles of the gas, *i. e.* in our case the nucleon mass. The characteristic feature of sound waves is that the velocity of the gas of fluid elements involved is small, but not their change with time and space. In other words, for sound waves one can neglect $\vec{v}$ as compared to $\partial\vec{v}/\partial t$ and $\vec{\nabla}.\vec{v}$. In particular one can neglect the nonlinear term $(\vec{v}.\vec{\nabla}).\vec{v}$ relative to $\partial\vec{v}/\partial t$ in (11). In view of this the continuity equation simplifies also by neglecting $\vec{v}.\nabla\rho$ relative to $\rho \, div \, \vec{v}$ and eqs. (11) and (12) become

$$m_N \frac{\partial \vec{v}}{\partial t} = -\frac{1}{\rho} \nabla p,$$

$$\frac{\partial \rho}{\partial t} = -\rho \, div \, \vec{v} \tag{13}$$

Now one needs a relation between density and pressure. If ρ_0 and p_0 denote the corresponding values at equilibrium one may write

$$\rho = \rho_0(1 + \kappa(p - p_0)) \tag{14}$$

where

$$\begin{aligned}
\kappa &= \frac{1}{\rho}\frac{\partial\rho}{\partial p}\Bigg|_{\substack{\rho=\rho_0 \\ T=const}} = \frac{V}{N}\frac{\partial N/V}{\partial p}\Bigg|_{\substack{\rho=\rho_0 \\ T=const}} = -\frac{1}{V}\frac{\partial V}{\partial p}\Bigg|_{\substack{\rho=\rho_0 \\ T=const}} \\
&= \frac{1}{\rho}\frac{1}{\frac{\partial p}{\partial \rho}}\Bigg|_{\substack{\rho=\rho_0 \\ T=const}} \\
&= \frac{1}{\rho}\frac{1}{2\rho\frac{\partial W}{\partial\rho} + \rho^2\frac{\partial^2 W}{\partial\rho^2}}\Bigg|_{\substack{\rho=\rho_0 \\ T=const}} \\
&= \frac{1}{\rho^3\frac{\partial^2 W}{\partial\rho^2}\Big|_{\substack{\rho=\rho_0 \\ T=const}}} = \frac{1}{\rho}\frac{9}{K}.
\end{aligned} \tag{15}$$

is the ***compressibility*** and K is the ***compression constant*** defined in eq. (9). In the penultimate step, eq. (7) has been used. From (14) it also follows that

$$\nabla p = \frac{1}{\kappa\rho_0}\nabla\rho. \tag{16}$$

Differentiating the second eq. (13) with respect to time and applying the divergence operator to the first one yields, after eliminating the term $\partial/\partial t div\vec{v}$ and neglecting second-order contributions,

$$m_N\frac{\partial^2\rho}{\partial t^2} = \Delta p$$

or, because of (16)

$$m_N\kappa\rho_0\frac{\partial^2\rho}{\partial t^2} - \Delta p = 0. \tag{17}$$

This can be cast in the form

$$\frac{1}{c_s^2}\frac{\partial^2\rho}{\partial t^2} - \Delta\rho = 0, \tag{18}$$

which is the wave equation for sound waves with the sound velocity

$$c_s^2 = \frac{1}{m_N\kappa\rho_0} = \frac{9m_N}{K}. \tag{19}$$

The constraints on $W(\rho)$ near $\rho = \rho_0$ leave a lot of freedom for extrapolation to larger densities. In order to select between different functions $W(\rho)$, experimental information about the nuclear equation of state at larger-than-equilibrium density is required. We shall see in the following chapter that heavy ion- collisions at high energy, where nuclear matter is stongly compressed, can provide such information.

Two commonly used parametrization of $W(\rho)$ originate from the extended liquid drop model by Scheid and Greiner(Z. Physik 226(1969) 364) (see figs. 2, 3). The so-called *"linear equation of state"* is defined by

$$W(\rho) = \begin{cases} \frac{K(\rho-\rho_0)^2}{18\rho\rho_0} + W_0; & \rho \geq \rho_c, \\ -a\rho; & \rho < \rho_c, \end{cases} \tag{20}$$

where $\rho_c = \rho_0/(1 - qW_0/K)$ *and* $a = K(1 - \rho_0^2/\rho_c^2)/18\rho_0$. Its name stems from its linear growth at large values of ρ. The *"quadratic equation of state"* is given by

$$W(\rho) = \frac{K(\rho - \rho_0)^2}{18\rho_0^2} + W_0 \tag{21}$$

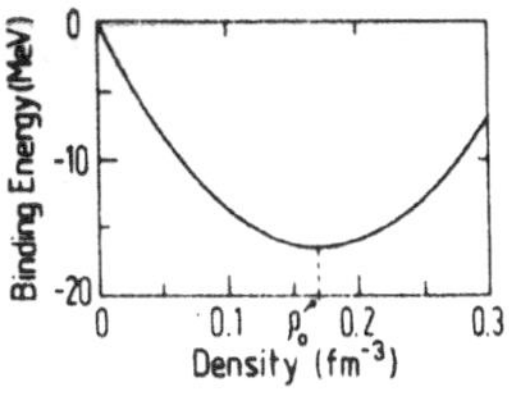

Fig. 3. The binding energy per nucleon in infinite nuclear matter with an equal number of protons and neutrons and without Coulomb energy. It has a minimum at $\rho_0 \approx 0.17 fm^{-3}$, which correspond to equilibrium nuclear matter. The functional has been obtained with the *ansatz* $W(\rho) = A\rho^2 + B\rho + C$ and the parameters were fixed by the conditions $W(0) = 0, W(\rho_0) = -16.5MeV, \partial W/\partial\rho|_{\rho=\rho_0} = 0$. This yields $A = 571 MeV fm^6, B = -194 MeV fm^3, C = 0$.

In this case the requirement $W(0) = 0$ and W_0 from (7) fixes K to be about 288 MeV, a value somewhat larger than the experimental one(9).

The Nuclear Equation of State and Finite Nuclei

The minimum of the energy functional under variation of $\rho_p(r)$ and $\rho_N(r)$ determines the optimal density distributions of the nuclei. The variation has to be performed in such a way that the total number of protons and neutrons is conserved:

$$\frac{\delta(E + \lambda_n N + \lambda_p Z)}{\delta\rho_p} = \frac{\delta(E + \lambda_n N + \lambda_p Z)}{\delta\rho_n} = 0 \tag{22}$$

For spherically symmetric nuclei one obtains the coupled equations (J. Fink et. al. , Z. Physik A 323 (1986) 189)

$$\rho_n \left[\frac{d(\rho W)}{d\rho} - 2C_{sy}(\rho_p - \rho_n)\rho^\nu + C_{sy}(\rho_p - \rho_n)^2 \nu\rho^{\nu-1} - 2V_0\mu^2\rho + \psi(r) + \frac{\lambda_n}{4\pi}\right] = 0. \tag{23a}$$

and

$$\rho_p \left[\frac{d(\rho W)}{d\rho} + 2C_{sy}(\rho_p - \rho_n)\rho^\nu + C_{sy}(\rho_p - \rho_n)^2 \nu\rho^{\nu-1} - 2V_0\mu^2\rho + \psi(r) + e\phi(r) + \frac{4}{3}C_{ex}\rho_p^{1/3} + \frac{\lambda_p}{4\pi}\right] = 0, \tag{23b}$$

with the *Coulomb and Yukawa potentials*

$$\Delta_r\phi(r) = -4\pi e\rho_p(r) \tag{24}$$

and

$$\Delta_r\psi(r) - \frac{1}{\mu^2}\psi(r) = -V_0\rho(r). \tag{25}$$

Here Δ_r is the radial Laplace operator. The constant total rest energy $A.m_N c^2$ of all nucleons does not, of course, appear in eqs. (23). The factors ρ_p and ρ_n in (23a, b) indicate that the equations are only valid as long as the respective densities are positive. Since in general the neutron densities reach out farther than the proton densities (because of neutron excess and the symmetry energy, which tries to keep the neutron density close to the proton density wherever the latter does not vanish), eq. (23a) still must be solved even when eq. (23b) has ceased to be relevant because $\rho_p = 0$.

The above system of equations has a unique solution for a stable equilibrium density. It can easily be seen that in the case of no Coulomb, no symmetry, no exchange, and no Yukawa force the density distribution has the shape of a square well with a constant value $\rho = \rho_0$. An attractive Yukawa force ($V_0 < 0$) causes surface effects; in particular the density profile now goes to zero smoothly. The surface thickness (the distance within which the density decreases from 9o to 1o%) and the mean square radius of the density distribution depends on the choice of the two parameters V_0 and μ in the Yukawa energy and, to a small extent, on the parameters of the volume energy

$\int \rho W(\rho)d\tau$. For $W(\rho)$ given by eq. (21), a good fit to the binding energies and the measured proton density distributions for a larger number of spherical nuclei is obtained by choosing(see the paper by J. Fink et. al.)

$$V_0 = -1429 MeV fm \; and \; \mu = 0.5 fm. \tag{26}$$

A Taylor expansion of the difference $\rho(\vec{r}_2) - \rho(\vec{r}_1)$ in the surface energy in (5) shows that the combination $-V_0\mu^4 \simeq 138 MeV fm^5$ should determine the binding energy and surface thickness of the density distribution. Numerical studies of the complete surface energy in (5) yield more or less equal results for all combinations of V_0 and μ related by $V_0\mu^{3.6} = -117 MeV fm^{4.6}$, with best results for the choice (26). The change in the power of μ between the exact calculations and the lowest-order Taylor expansion gives a measure of the importance of high-order density gradients for the nuclear binding energy. Some sample density distributions for light and heavy nuclei obtained with the parameter given above are are shown in fig. 4. With the set of parameters $C_{ex} = -\frac{3}{4}(3/\pi)^{1/3}e^2, \nu = 0, C_{sy} = 217 MeV fm^3, V_0 = -1429 MeV fm, \mu = 0.5 fm$ various nuclear quantities can be calculated. First we can plot the binding energy of nuclear matter according to eq. (5). For a given A one can maximize the binding energy with respect to Z. The resulting

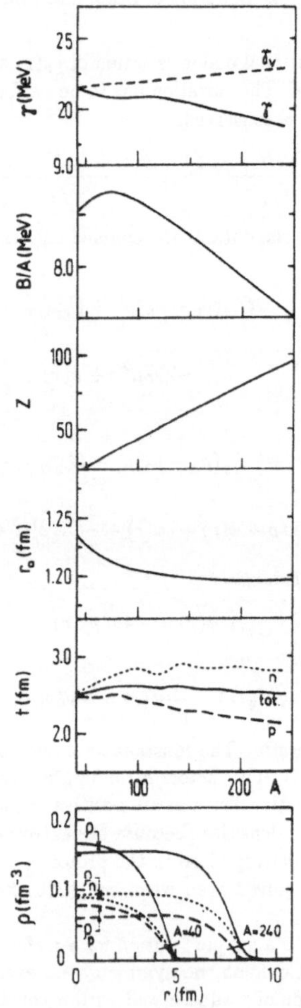

Fig. 4. The surface tension constant γ, the binding energy per nucleon, the proton number Z, the equivalent radius $R_{eq} = r_0 A^{\frac{1}{3}}$ and the surface thickness t as functions of the nucleon number A. In addition the radial density distribution for two different nuclei A = 40 and A = 240 are shown for the parameter set quoted in text.

binding energy B/A and proton number Z for the most stable isotopes are drawn as functions of A in fig. 4 for the parameter set given above. The radial density distributions for a few sample nuclei from the periodic table are included to demonstrate the quality of the description.
The constant γ of the surface energy of the mass formula (see J. M. Eisenberg and W. Greiner: Nuclear Theory 1: Nuclear Models, North Holland(1987), 3^{rd}edition) is also shown in fig. 4. It includes effects of the finite extension of the nucleus and can be defined as

$$\gamma_V = A^{-\frac{2}{3}}[B - W_0 A - E_{sym} - E_{coul}], \tag{27}$$

where E_{sym} and E_{coul} are the second to fourth contributions to $E(\rho)$ in eq. (5). In writing this formula effects arising from the asymmetry energy have been excluded. If one wants to include these, one must first find the binding energy of infinitely saturated nuclear matter under inclusion of the symmetry energy. Minimizing the following expression for infinite nuclear matter (homogeneous densities ρ_n , ρ_p)

$$\frac{E}{A} = W(\rho) + C_{sy}(\rho_p - \rho_n)^{\nu-1} = W(\rho) + C_{sy}\left(\frac{2Z}{A} - 1\right)^2 \rho^{\nu+1} \tag{28}$$

with respect to ρ one gets the binding energy of saturated nuclear matter as

$$\left(\frac{E}{A}\right)_{min} \approx W_0 + C_{sy}\left(2\frac{Z}{A} - 1\right)^2 \rho_0^{\nu+1}. \tag{29}$$

This approximation is valid as long as the asymmetry energy in infinite nuclear matter is a small corection to W_0. Even for large nuclei where 2Z - A becomes appreciable (e. g. ^{238}U) the second term in (29) is less than 2 MeV. With that, the constant of the surface energy which includes symmetry effects has the form

$$\gamma \simeq \gamma_v + \left[E_{sym} - C_{sy}\left(2\frac{Z}{A} - 1\right)^2 \rho_0^{\nu+1} A\right] A^{-\frac{2}{3}}. \tag{30}$$

This results in a slowly varying function of A as can be seen from fig. 4. The empirical mass formula of Seeger gives, for example, values for γ_V of about $\gamma_V = 19.6MeV$ (see e. g. Eisenberg-Greiner, l. c.) in good agreement with theoretical predictions. Furthermore, in fig. 4 are also shown density distributions, their equivalent radii $R_{eq} = r_0 A^{\frac{1}{3}}$, and their surface thickness t. The eqivalent radius is defined as

$$R_{eq}^2 = \frac{5}{3} A^{-1} \int r^2 \varrho d\tau. \tag{31}$$

The surface thickness t is usually defined as the distance over which the density decreases from 90 to 10% of its value $\varrho(0)$ at the origin. All curves indicate the influence of the Coulomb force for $A > 100$. The Coulomb force depresses the nuclear density in the interior region since the protons are repelled to the surface. The *specific radius* $r_0 = R_{eq}A^{-\frac{1}{3}}$ is, however, nearly independent of A, as is the surface thickness.

Nuclear Shock Waves in Relativistic Heavy-Ion Collisions

The possibility of compression of nuclear matter in high energy nucleus-nucleus collisions is one of the most exciting aspects of heavy-ion physics.Except through the internal structure of neutron stars, this is the only way to obtain access to the nuclear equation of state at densities exceeding the equilibrium value $\rho_0 \simeq 0.17 fm^{-3}$ and at high temperature T.The physics of neutron stars and supernovae will be discussed in various lectures by M. Irvine, E. Müller, W. Hillebrandt, N. Glendenning, J. Wilson and S. Kahana. Clearly, the possibility of performing heavy-ion collisions under controlled conditions and by varying a series of parameters like the beam energy and the size of the colliding nuclei should enable us to extract much more detailed information on the nuclear equation of state from such experiments than from astrophysical observations.

Nuclear Fluid Dynamics

As mentioned, nuclear compression has been discussed as early as 1968 by Scheid and Greiner in connection with the sudden nucleus-nucleus potentials and their energy dependence, which seem to

confirm the experimentally deduced heavy-ion potentials of the Yale group (see Eisenberg-Greiner, Nuclear Models(North Holland 1987, 3rd ed., chapters 16,17,18). These earlier considerations are valid as long as the relative heavy-ion velocity v_r does not exceed the velocity of first sound in nuclear matter,c_s, *i.e.* for $v_r < c_s$. Here the first sound is an isospin T = 0 compression wave while the second sound describes an isospin T = 1 wave where a local proton-neutron separation travels in nuclear matter of constant density $\rho_0 = \rho_p + \rho_n$ (see fig.5). The latter type of sound waves is important in connection with giant dipole resonances, which essentially constitute waves of second sound reflected at the nuclear surface (see Eisenberg-Greiner : Nuclear Models loc.cit.).
At higher beam energies ($E \sim 1GeV/nucleon$) the relative velocity of the two nuclei exceeds the velocity of sound. This is particularly true at ultrarelativistic heavy-ion collision energies of up to $E \approx 200GeV/nucleon$ as performed at CERN. To demonstrate the basic physics of compression effects for the case $v_r > c_s$ in head-on nucleus-nucleus collisions the following very simplified model (W.Scheid, M.Müller and W.Greiner, Phys.Rev.Lett. 13 (1974)74) can be used. It was, in fact, this specific model emerging out of the diploma thesis of Hans Müller in 1972/1973, which predicted shock waves in high energy nucleus-nucleus collisions and started the field of high energy heavy-ion physics. Assume two identical nuclei whose volume is divided into three parts (see fig.6a below), namely an ellipsoid with axes a and b sandwiched between two cut-off spheres with radius R and relative distance r. The system will clearly be rotationally symmetric around the z-axix connecting the centers of the two spheres and of the ellipsoid. The nuclear matter is assumed to be homogeneously distributed over the different volumes. Compression ($\rho > \rho_0$) should only occur in the ellipsoidal region. The four coordinates a, b, r and R define the geometry of the system. However, we take R to be constant in the collision so that we are dealing with three degrees of freedom only. This approximation is rather inessential for the results, but simplifies the theoretical considerations considerably.

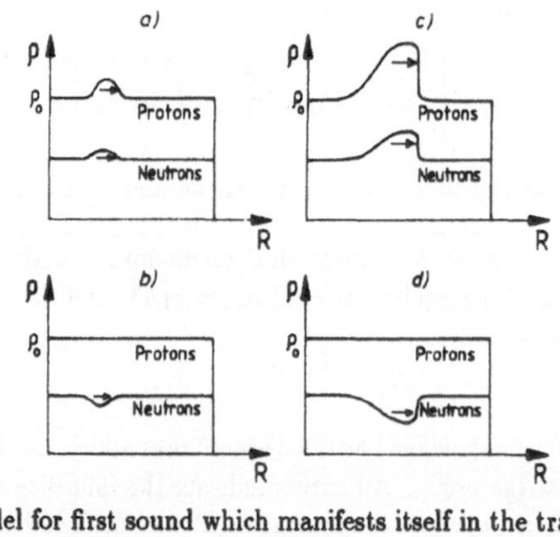

Fig.5.(a) Schematic model for first sound which manifests itself in the traveling of a T = 0 disturbance of the total density. (b) Schematic model for second sound, which characterizes the traveling of a relative proton-neutron de-admixture (T = 1 giant resonances) *without* change of total density. The arrows indicate the direction of travel of disturbance. Contrary to sound waves, where the perturbation $|\rho - \rho_0|$ is small compared to the equilibrium density ($|\rho - \rho_0| \ll \rho_0$), for shock waves the perturbation $|\rho - \rho_0|$ is comparable or larger than the equilibrium density ρ_0 in the medium. One can again distinguish T = 0 (c) and T = 1 (d) shock waves. Sound waves travel always with the sound velocity c_s, irrespective of the magnitude of the perturbation ρ, as long as it is small ($\rho \ll \rho_0$). Shock waves on the other hand travel with the shock velocity $v_s(\rho)$, which explicitily depends on the amplitude (perturbation) ρ .

The (time-dependent) density and pressure in the compressed ellipsoidal region is determined by the conservation of energy-momentum and matter during the course of the collision. Since the velocity of sound in nuclear matter is an appreciable fraction of the velocity of light (from eq.(10) with K = 210 MeV we obtain $c_s \cong 0.16c$), and we are interested in shock phenomena propagating with velocities $v_s > c_s$ a relativistic discussion of these conservation laws is appropriate. The basic relevant equation can be found in literature. The application to our present problem was worked out in (Scheid et.al. Phys. Rev. Lett. 13(1974)741, W.Scheid, J.Hofmann and W.Greiner: Nucl. Shock Waves in Relativ. Heavy Ion Coll., Proced. of 2nd High Energy Ion Summer Study

- 3675,TID - 4500 - R62 UC 34c, LBL Berkeley, Calif.(1974)(Lee Schroeder editor) Baumgardt et.al., Z.Physik A273(1975)359, Hofmann et.al. Nuovo Cimento 33(1976)343 and Phys.Rev. Lett. 36(1976)88).

The conservation of baryon number takes the form

$$\partial_\mu j^\mu = 0 \tag{32}$$

where the baryon current is given in terms of the local baryon density ρ and the local flow four-vector of the matter $u^\mu = (\gamma, \gamma\vec{v}/c)$,

$$j^\mu = \rho u^\mu \tag{33}$$

Here $\gamma = 1/\sqrt{1 - v^2/c^2}$ is the Lorentz dilatation factor. Energy-momentum conservation is expressed through the energy-momentum tensor

$$T^{\mu\nu} = (e + p)u^\mu u^\nu - pg^{\mu\nu}, \tag{34}$$

where e is the local energy density, including the nucleon mass, and p is the local pressure. The conservation law takes the form

$$\partial_\mu T^{\mu\nu} = 0, \tag{35}$$

where the metric tensor for space-time is $g^{\mu\nu} = diag(1, -1, -1, -1)$. The physical content of these equations becomes apparent when the representations (33) and (34) are inserted into the equation (32) and (35). It is convenient to introduce, in addition to the local energy and baryon densities, the analogous quantities as measured in a laboratory fixed frame, *i.e.* the frame where the target nucleus is at rest, or the nuclear center of mass in a collider experiment:

$$e_L = T^{00} = \gamma^2(e + p) - p, \tag{36}$$

$$\rho_L = j^0 = \gamma\rho, \tag{37}$$

$$M^k = T^{k0} = \gamma^2(e + p)v^k/c. \tag{38}$$

The index L indicates that the corresponding quantities are measured in the laboratory frame. The last expression denotes the momentum flow vector in the laboratory frame. (Whereas Greek indices α, ν are defined to run from 0 to 3, Latin indices k denote spacial components only and run from 1 to 3.) The Lorentz factor γ is determined by the relative velocity between the laboratory and local co-moving frame. With these expressions the conservation laws take the form of *relativistic hydrodynamic equations* (see also the lectures of J. Maruhn and D. Strottmann)

$$\partial_t e_L + \vec{\nabla}.\left(e_L\frac{\vec{v}}{c}\right) = -\vec{\nabla}.\left(p\frac{\vec{v}}{c}\right), \tag{39}$$

$$\partial_t M^k + \vec{\nabla}.\left(M^k\frac{\vec{v}}{c}\right) = -\nabla^k p, \quad k = 1, 2, 3, \tag{40}$$

$$\partial_t \rho_L + \vec{\nabla}.\left(\rho_L\frac{\vec{v}}{c}\right) = 0. \tag{41}$$

We shall refer to these equations as the *conservation laws for energy flow, momentum flow, and matter flow*, respectively. Note the source terms on the right-hand side of the eqs.(39) and (40), which tell us that pressure gradients and velocity gradients in the matter are a source of momentum and energy flow ("Newton's law"). Eqs.(39) - (41) yield five equations for the unknowns ρ, T, $\vec{v}$ and e. Hence one has to add an additional equation, which is the nuclear equation of state.

The *nuclear matter equation of state* $W(\rho)$ enters into these equations through a relationship between the local pressure and energy density and the local temperature and baryon density

$$e = e(\rho, T); \tag{42}$$

$$p = p(\rho, T). \tag{43}$$

Usually these expressions are split up into a contribution from cold, compressed nuclear matter

$$e_{comp} = \rho E_c(\rho) = \rho(W(\rho) + m_N c^2), \tag{44}$$

$$p_{comp} = \rho^2 \frac{E_c}{\partial\rho}, \tag{45}$$

and a term from thermal excitations:

$$e = e_{comp} + \rho E_{therm}(\rho, T), \tag{46}$$

$$p = p_{comp} + p_{therm}(\rho, T), \tag{47}$$

which are computed, say, by modeling the nuclear matter as a non-interacting gas of nucleons, possibly including excitation of hadronic resonances and mesons. For example, for a nonrelativistic *Fermi gas* of nucleons one has (see Eisenberg- Greiner, Nuclear Models, 3rd ed., North Holland (1987))

$$E_{therm}(\rho, T) \simeq \frac{1}{2}\left(\frac{2\pi}{3}\right)^{\frac{2}{3}} \rho^{-\frac{2}{3}} T^2. \tag{48}$$

In the general case of strongly interacting nuclear matter a simple separation like (44) and (46) into a T = 0 (compression) and a $T \neq 0$ *free* gas contribution is not theoretically justified.
In the simple model of fig. 2, where in each region all densities are assumed to be constant, the nontrivial consequences of eqs.(8)-(10) are related to the conservation of energy-momentum and matter across the discontinuous shock front separating the ellipsoidal and cut-off spherical regions. The inflow of matter from the cut-off spheres into the ellipsoidal region will cause the shock front to move outwards. The conditions of continuity required by the hydrodynamic equations are most easily expressed in the frame in which this shock front is at rest.
In this frame the fluxes into the shock front from the cutoff spheres and out of the shock front into the ellipsoid become stationary, and all time derivatives vanish. Let us denote the densities and velocity relative to the shock front in the cutoff spheres by an index 0, and the corresponding quantities in the ellipsoid by an index 1 (see fig. 3). Then we have for the component normal to the shock front

$$j_0^n = j_1^n; \tag{49a}$$

$$T_0^{nn} = T_1^{nn}; \tag{49b}$$

$$T_0^{0n} = T_1^{0n}. \tag{49c}$$

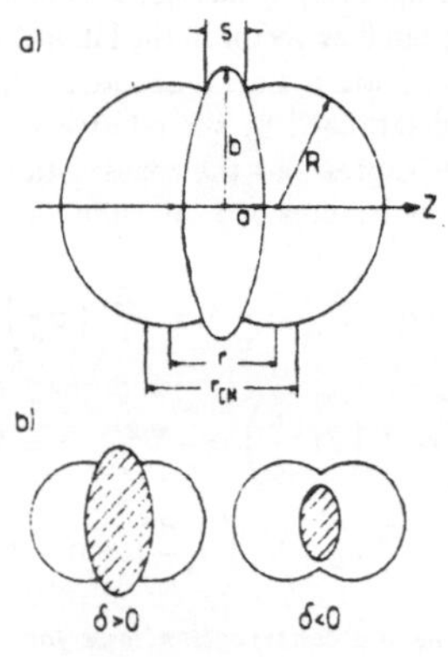

Fig.6. The basic features of the model constructed for the investigation of compression in general and shock waves in particular. The central ellipsoid serves to describe schematically the compression zone of nuclear matter.

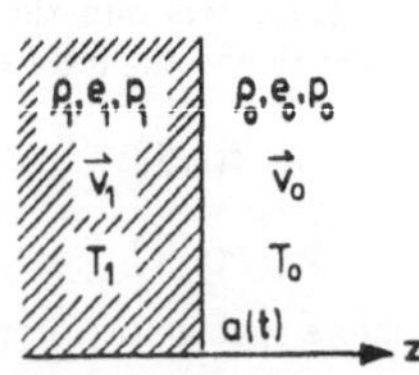

Fig.7. Illustration of the boundary situation at the endpoint of the semi-axix a of the ellipsoid. The shock front divides the unperturbed medium $(\rho_0, e_0, p_0, \vec{v}_0, T_0,)$ from the massive perturbation $(\rho_1, e_1, p_1, \vec{v}_1, T_1)$. Across the shock front the various conservation laws for matter flow , energy flow and momentum flow (see eqs. (49) and (50)) hold.

The index n denotes the normal component; at the end point of the small semi-axis of the ellipsoid in fig. (6a) the normal vector coincides with the z-axis, and n can be replaced by 3 (the normal axis n points into the 3-direction).
Dropping the index n and remembering that the following equations always apply for the normal component only, we find from (33), (34) and (49).

$$\rho_0 v_0 \gamma_0 = \rho_1 v_1 \gamma_1; \tag{50a}$$

$$(e_0 + p_0)\gamma_0^2 v_0^2 + p_0 c^2 = (e_1 + p_1)\gamma_1^2 v_1^2 + p_1 c^2; \tag{50b}$$

$$(e_0 + p_0)\gamma_0^2 v_0 = (e_1 + p_1)\gamma_1^2 v_1. \tag{50c}$$

It is straightforward to eliminate the velocities from eqs. (50) to obtain a relationship between the energy and matter densities on both sides of the shock front. This relation is known as the relativistic *Rankine-Hugoniot equation*(W. J. M. Rankine, Phil. Trans. Roy. Soc. London 160(1870)277,J. Hugoniot, Ecole Polytechn. Cahier 57(1887)1):

$$\frac{(e_1+p_1)^2}{\rho_1^2} - \frac{(e_0+p_0)^2}{\rho_0^2} = (p_1 - p_2)\left(\frac{e_1+p_1}{\rho_1^2} + \frac{e_0+p_0}{\rho_0^2}\right). \tag{51}$$

The velocities relative to the shock front are

$$\frac{v_0^2}{c^2} = \frac{(p_1-p_0)(e_1+p_0)}{(e_1-e_0)(p_1+e_0)} = \frac{p_1 W_1 \rho_1}{(W_1\rho_1 - W_0\rho_0)(W_0\rho_0 + p_1)}, \tag{52a}$$

$$\frac{v_1^2}{c^2} = \frac{(p_1-p_0)(e_0+p_1)}{(e_1-e_0)(p_0+e_1)} = \frac{p_1(W_0\rho_0 + p_1)}{(W_1\rho_1 - W_0\rho_0)W_1\rho_1}, \tag{52b}$$

$$\frac{v_F^2}{c^2} = \frac{(p_1-p_0)(e_1-e_0)}{(e_1+p_0)(e_0+p_1)} = \frac{p_1(\rho_1 W_1 - \rho_0 W_0)}{\rho_1 W_1(p_1 + \rho_0 W_0)}. \tag{52c}$$

with $p_0 = 0$ where $v_F = (v_1 - v_0)/(1 - v_1 v_0/c^2)$ is the velocity of the compressed zone relative to the cut-off spheres, and the rightmost expressions are for $T = 0$. This is frequently called ***flow velocity***. The velocity v_0 is identical with the ***shock front velocity*** v_s, *i.e.* $v_s = v_0$. Both the shock front velocity v_s and the flow velocity v_F are functions of ρ and T in general. One can depict their dependence on the density ρ in the shockfront as in fig. 8. Obviously for $\rho \to \rho_0$ the flow velocity vanishes and the shock velocity assumes a finite value, *i.e.* the velocity of sound. Both v_s and v_F depend strongly on the nuclear equation of state $e(\rho, T)$. If a density isomer (see fig.2) exists in the energy density functional, shock and flow velocity show an irregular behaviour, as indicated in fig. 8 (dotted lines). Also when one includes excitation of the hadrons involved ($\Delta - resonances, N^* - resonances, etc.$) there is an influence on the shock dynamics as indicated by dashed lines.
The relativistic Rankine-Hugoniot equations, eq.(19), can be solved analytically for the equation of state of a gas of noninteracting nucleons with $p_{therm} = \frac{2}{3}\rho E_{therm}$ (which holds for an ideal Fermi gas as well as for a classical gas). A quadratic equation in $E_{therm} \equiv E_T$ is then obtained, which yields the thermal energy (or the temperature,*e.g.* $T = \frac{2}{3}E_{therm}$ for a classical ideal gas) directly as a function of the density for different compression energies,namely

$$(1+\alpha)E_{therm}^2 + \left\{\frac{(W_0 + E_c)(2+\alpha) + \rho\frac{\partial E_c}{\partial \rho} - \alpha(\rho/\rho_0)W_0}{2(1+\alpha)}\right\} E_{therm}$$

$$+ \left\{(W_0 + E_c)^2 - W_0^2 + W_0\rho^2\frac{\partial E_c}{\partial \rho}\left(\frac{1}{\rho} - \frac{1}{\rho_0}\right) + E_c\rho\frac{\partial E_c}{\partial \rho}\right\} = 0 \tag{52d}$$

where $\alpha = \frac{2}{3}$ for a nonrelativistic ideal gas and $\alpha = \frac{1}{3}$ for an ultrarelativistic gas (*e.g.* a gas of massless quarks and gluons).
Clearly,the shock-front model is an oversimplification of the actual collision process,and a proper dynamical treatment should be based on the full hydrodynamical eqs. (39)-(41). However,the shock eqs. (51) and (52) are easy to solve (no differential equations are involved), and they give important qualitative insight on a semi-analytical level. The quality of the prediction of the shock-front model can be judged from fig. 9, where a simple one-dimensional shock-model calculation (solution of equations (51) and (52) along the z-axis, neglect of the transverse dimensions), with the nuclear equation of state (21) supplemented by the thermal properties (48) of a nonrelativistic Fermi gas,is compared with a full three-dimensional hydrodynamic solution for U-U collisions. Wee see that the maximum temperature achieved in the compressed zone is well reproduced,whereas the one-dimensional shock model somewhat overestimates compression effects because of the unrealistic nuclear density profile and the neglect of transverse flow.

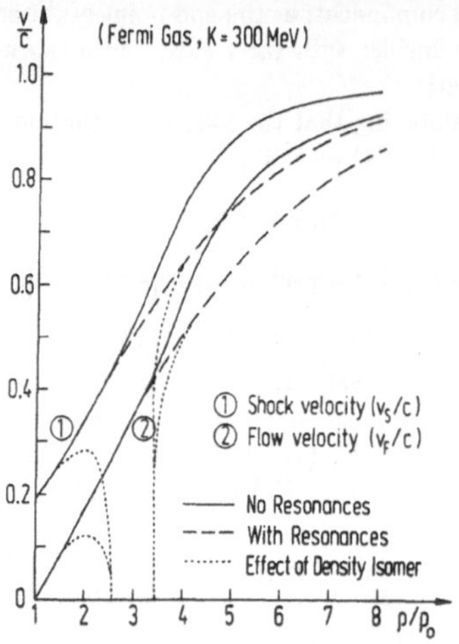

Fig.8. The dependence of the velocity of the shock front v_s and the matter flow velocity v_F on the density. Dashed lines: influence of hadronic resonances; short dashes: influence of a density isomer at $\rho = 3\rho_0$ (after Hofmann et. al. Nuovo Cim. 33(1976)343, H. Stöcker et. al., Proc. Top. Conf. on Nuclear Coll. Bled(Yugoslavia), Fizika 9 Supp. 4(1977)671 and Proc. Top. Conf. on Heavy Ion Coll., Falls Greek Falls, Oak Ridge, Tennessee(1977).).

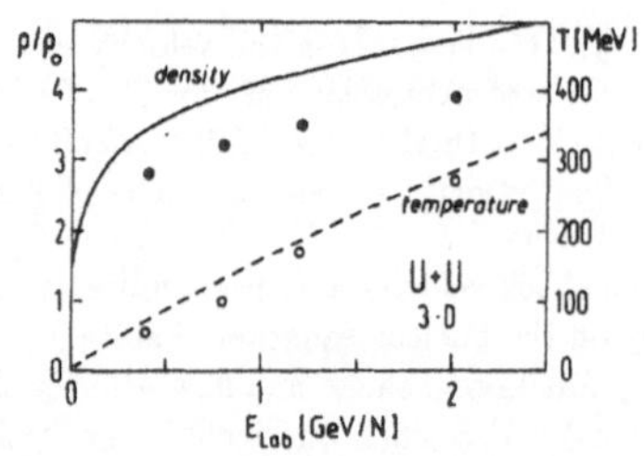

Fig.9. Test of the validity of the shock-front model: density and temperature within the shock front are shown as a function of the incident energy. The standard equation of state (21) , with K = 200 MeV supplemented by the thermal (Fermi gas) energy density (48), has been used. The points indicate the maximum density and temperature resulting from a three-dimensional fully relativistic hydrodynamical calculation for uranium on uranium (G. Gräbner, Dissertation , Inst. f. Theoret. Physik, Univ. Frankfurt a. M. (1984)). Obviously the simple shock-front model works rather well. It is, therefore, useful for obtaining essential features of nuclear hydrodynamics and easily.

Relativistic Nucleus-Nucleus Collisions

We present now some typical results of nuclear fluid dynamics applied to high-energy nucleus-nucleus collisions and confront them with experiments. Figs.10 and 11 show a calculation (Graebner loc. cit, A. Stöcker and W. Greiner, Phys. Report 137 Nr.5 and 6 (1986)278) for the Ar + Pb-collision at 770 MeV/nucleon at impact parameters b= 0 and b = 4 fm respectively. In the case of the central collision the smaller projectile is strongly compressed upon entry into the bigger target nucleus and a ***Mach shock wave*** develops. Such Mach shock waves have been predicted as a typical signature for hydrodynamical behavior (Scheid, Müller, Greiner 1974,loc. cit., Baumgardt el. al. 1975, loc. cit.). For noncentral collision this Mach shock wave travels unsymmetrically through the target nucleus, from which the projectile *"bounces off"* (fig.11). The *bounce-off effect*(H. Stöcker, J. Maruhn and W. Greiner, Phys. Rev. Lett. 44(1980)725) is even stronger for smaller projectile and larger impact parameter, as shown in figs.12 and 13. The arrows in this figure indicate the flow velocity of nuclear matter.

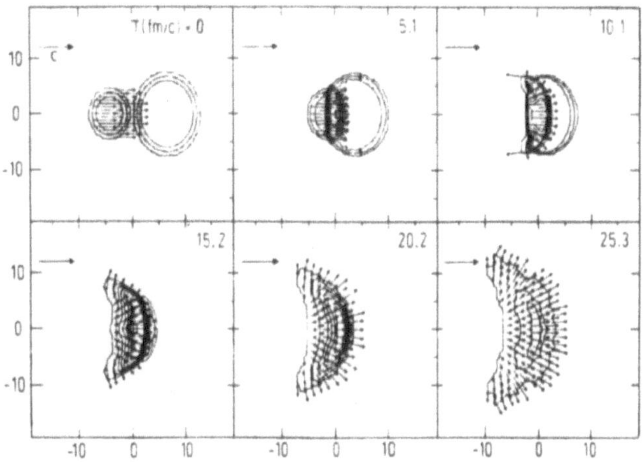

Fig.10. Collision of Ar (770 MeV/nucleon) + Pb at b = 0 fm in the Nuclear Fluid Dynamical Model

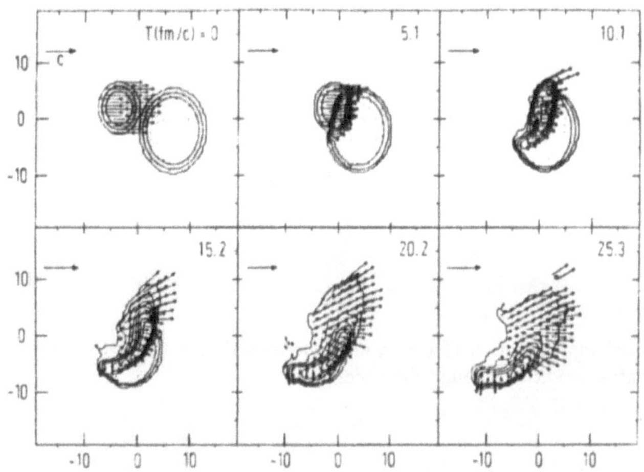

Fig.11. The same system at b = 0 in the Nuclear Fluid Dynamic Model (after G. Graebner, loc. cit.).

The hydrodynamical predictions are compared with those of nuclear cascade calculations (K. R. Gudima, A. Iwe and V.D.Toneev, J. Phys. G5(1979)229, H.Stöcker and W.Greiner, Phys. Report 137, Nr.5 and 6(1986)278) which clearly show no bounce-off effect (or only a very small one). This is due to the lack of compression in the cascade model (fig.10).
First experimental evidence for such Mach-shock emission has been reported by Baumgardt and Schopper (1975, 1978) on the basis of low-statistics particle track detector measurements: Peaks in the angular distributions of α-particles emitted from central (high multiplicities) collisions have been observed. Nuclear fluid dynamics and particularly the prediction of nuclear (Mach) shock waves and the related bounce-off effect have at first been strongly opposed because of dubious theoretical arguments and wrong early experiments. We remind with a few excerpts from the original literature in the following figs.14-17 on the extraordinary struggles during the years 1975-1980.

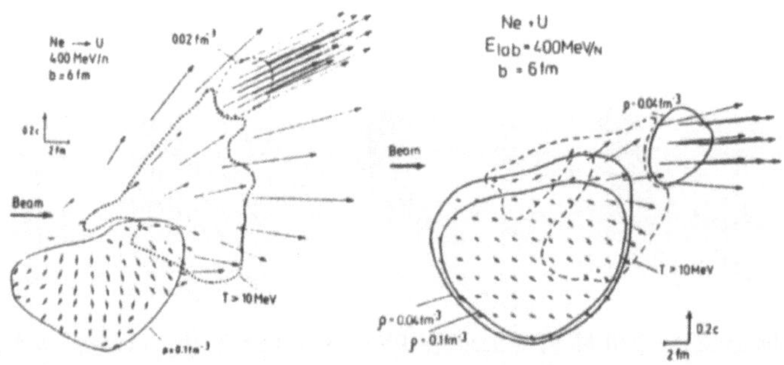

Fig.12. Final state of reaction Ne (400 MeV/nucleon) + U at b = 6 fm shown as a contour plot in the scattering plane. The experimentally observed (see the lectures by H. G. Ritter and by K. H. Kampert) strong bounce-off effect was predicted by nuclear fluid dynamics (left). It is not reproduced by the cascade model of Gudima and Toneev (right) which does not contain the repulsive nuclear interactions at high density, *i.e.* the nuclear compression energy.

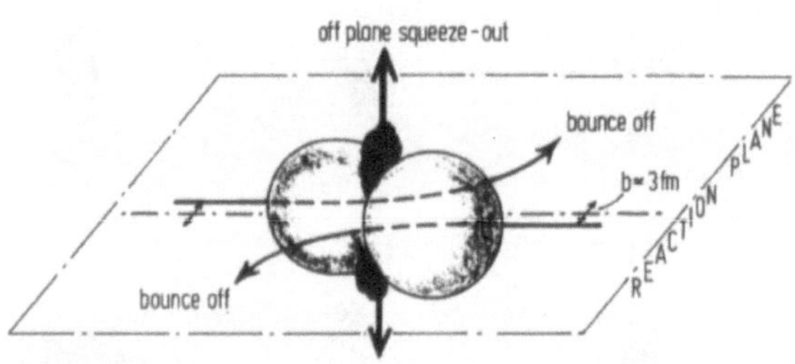

Fig.13. Pictorial representation of the in-plane bounce-off and the out-of-plane squeeze-out of participant matter predicted by fluid dynamics.

Only after the plastic ball detector of the GSI-LBL-group was installed at Berkely were the ***side-ward*** flow and ***squeeze-out*** of nuclear matter (as the ***Mach shock waves***, bounce-off effect and compressional flow of figs. 6 and 11 were now called) uniquely detected (H. G. Ritter el. al., Nucl. Phys. A477(1984)36; H. A. Gustafsson el. al. Phys. Rev. Lett. 52(1984)1590; Phys. Lett. 142B(1984)141). These pioneering experiments were performed under the leadership of H. Gutbrod, A. Poskanzer and H. G. Ritter. An impression of this success of fluid dynamics is given in fig. 18, which shows the measured flow angles for $^{98}Nb + ^{98}Nb$ at 400 MeV/nucleon for various multiplicity events. The multiplicity of fragments emitted in a single nucleus-nucleus encounter is a measure for the impact parameter in this event (Baumgardt el. al., Z. F. Physik 1975 loc. cit.). The mass dependence of the sideward flow is revealed in fig.19. Obviously the agreement between

VOLUME 34, NUMBER 11 PHYSICAL REVIEW LETTERS 17 MARCH 1975

Comment on Nuclear Shock Waves in Heavy-Ion Collisions

George F. Bertsch
Department of Physics and Cyclotron Laboratory, Michigan State University, East Lansing, Michigan 48824
(Received 30 September 1974)

Heavy-ion collisions are not likely to produce shock waves and baryon densities greater than twice the nuclear matter density, contrary to the conclusions of Scheid, Müller, and Greiner.

The hydrodynamics of high-energy heavy-ion collisions, studied recently by Scheid, Müller, and Greiner,[1] is of some interest, because high-density nuclear matter might have unusual properties. I mention the recent speculation that the compression might lead to a collapsed state of nuclear matter,[2] and also the possibility that pion condensation[3] might occur in nuclear matter at twice normal densities.[4] Scheid, Müller, and Greiner[1] and Wong and Welton[5] assume that the hydrodynamics is governed by formation of a shock wave.

The hydrodynamics of the compressed state depends crucially on the equation of state and the relative size of the mean free path of nucleons, compared to the size of the system. In this comment I wish to suggest a more realistic model than the one given by Ref. 1, which entirely changes the physical situation.

Equation of state.—Solution of the shock equations requires knowledge of the pressure as a function of density and internal energy (part of which may be heat). It is found that there is an upper limit to the maximum density achievable with a shock wave,[6] which for a gas having no internal degrees of freedom is a four-fold compression. This seems implicit in the equation following Eq. (9) of Ref. 1, but is not satisfied in the final results.

Turning to the details of the equation of state at zero temperature, the form assumed in Ref. 1 requires only 3 MeV per particle to double the density of ordinary nuclear matter. This is unreasonably low; already in a pure Fermi gas model 13 MeV per nucleon is required to double the density. A similar figure is found from the hard-core gas model[7] and from a mesonic model.[8] The realistic calculations of the nuclear equation of state by Negele and Vautherin[9] predict that a density doubling requires 16 MeV per nucleon. For an $^{16}O+^{16}O$ collision, this implies that the projectile lab energy should be of the order of 1 GeV.

Mean free path.—Numerical calculations for strong shock waves[10] show that the thickness of the transition region is typically 2.5λ, where λ is the mean free path of the particles prior to the collision. I estimate the mean free path by considering individual nucleons of the same velocity as the incident heavy ion. The mean free path is determined from the nucleon-nucleon cross section.[11] These are summarized in Table I for several energies.[12] It may be seen that the transition region will certainly encompass the whole nucleus, for such a light nucleus as ^{16}O.

TABLE I. Mean free path for ^{16}O - ^{16}O kinematics.

$E_{c.m.}(^{16}O+^{16}O)$	E_{lab} per incident nucleon (MeV)	λ (fm)
100 MeV	12.5	7
600 MeV	80	2.0
2 GeV and higher	250	2.7

The conclusion is that heavy-ion collisions with enough energy to cause the nuclei to interpenetrate will more resemble gases passing through each other than droplets splashing on each other.

[1]W. Scheid, H. Müller, and W. Greiner, Phys. Rev. Lett. 32, 741 (1974).
[2]T. D. Lee and G. C. Wick, Phys. Rev. D 9, 2291 (1974).
[3]R. Sawyer and D. Scalapino, Phys. Rev. D 7, 953 (1973).
[4]G. Bertsch and M. Johnson, unpublished.
[5]C. Y. Wong and T. A. Welton, Phys. Lett. 49B, 243 (1974).
[6]L. Landau and E. Lifshitz, *Fluid Mechanics* (Pergamon, London, 1959), p. 330.
[7]A. Bohr and B. Mottelson, *Nuclear Structure* (Benjamin, New York, 1969), p. 253.
[8]J. D. Walecka, Ann. Phys. (New York) 83, 491 (1974).
[9]J. Negele and D. Vautherin, Phys. Rev. C 5, 1472 (1972).
[10]D. Gilbarg and D. Paolucci, J. Rational Mech. and Anal. 2, 617 (1953).
[11]W. O. Lock and D. F. Measday, *Intermediate Energy Nuclear Physics* (Methuen, London, 1970), p. 270.
[12]At extremely high energies, collisions do not thermalize the motion—see for example P. L. Jain *et al.*, Phys. Rev. Lett. 33, 660 (1974).

Fig. 14. Excerpt from a paper by G. Bertsch, who - nevertheless of his arguments presented here - published a few month later a paper with Amsden, Nix and others, utilizing solely nuclear hydrodynamics.

2.N Nuclear Physics A251 (1975) 502–529; © North-Holland Publishing Co., Amsterdam

SHOCK WAVES IN COLLIDING NUCLEI

MICHAEL I. SOBEL'

Nordita, Copenhagen

PHILIP J. SIEMENS and JAKOB P. BONDORF

The Niels Bohr Institute, Copenhagen

and

H. A. BETHE''

Nordita, Copenhagen

Received 14 April 1975

(Revised 30 May 1975)

Abstract: We consider the circumstances under which nuclear matter at high densities can be produced in heavy ion collisions. We argue that lab energies of a few hundred MeV per nucleon will be suitable: the matter velocity will exceed the speed of isentropic compression waves, while the nuclear matter has sufficient stopping power to generate a shock front. From the hydrodynamic conservation laws we show that there is a maximum attainable compression ratio r_c determined by the thermal properties of the high-density matter. In an independent-fermion model, $r_c = 4$, but it can be much larger if the phase of the system changes, for example by excitation of nucleon isobars, production of π-mesons, or the scalar-field condensation conjectured by Lee and Wick. We discuss the propagation of the shock front and subsequent decompression of the dense, hot matter.

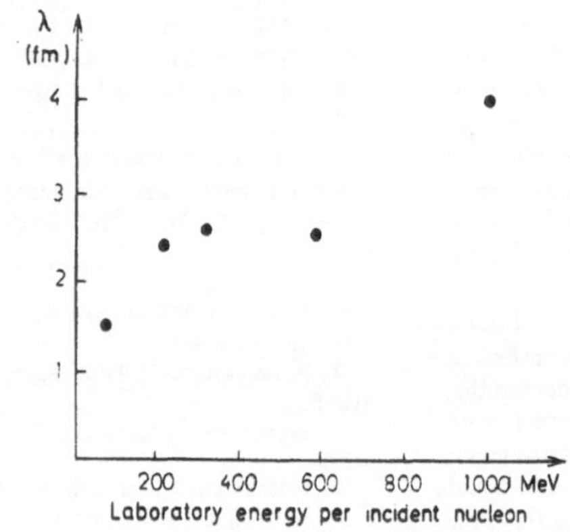

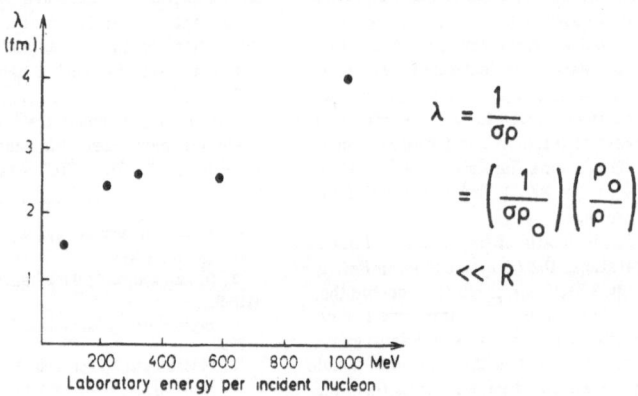

$$\lambda = \frac{1}{\sigma\rho} = \left(\frac{1}{\sigma\rho_0}\right)\left(\frac{\rho_0}{\rho}\right) << R$$

Fig. 1. Momentum decay length for various beam energies in the laboratory system. Estimates based on data in W. N. Hess, Rev. Mod. Phys. **30** (1958) 368.

V. Ruck, M. Gyulassy
Precritical
Scattering

References

1) G. Chapline, M. Johnson, E. Teller and M. Weiss, Phys. Rev. **D8** (1973) 4302; G. H. Dahlbacka, G. F. Chapline and T. A. Weaver, Nature **250** (1974) 36

2) W. Scheid, H. Müller and W. Greiner, Phys. Rev. Lett. **32** (1974) 741; C. Y. Wong and T. A. Welton, Proc. Int. Conf. on reactions between complex nuclei, vol. 1 (North-Holland, Amsterdam, 1974)

3) R. Courant and K. O. Friedrichs, Supersonic flow and shock waves (Interscience, NY, 1948)

4) J. W. Clark and C. H. Yang, Lett. Nuovo Cim. **3** (1970) 272

5) T. D. Lee and G.-C. Wick, Phys. Rev. **D9** (1974) 2291

6) G. Baym and C. Pethick, Landau Fermi liquid theory and low temperature properties of liquid ^{4}He, to be published

7) Suggestion attributed to E. Teller

8) D. Bodansky, D. D. Clayton and W. A. Fowler, Astrophys. J. Suppl. Series **16** (1968) no. 148

Fig. 15. Excerpt from a paper by Sobel et. al., who claime that nuclei might become too transparent at high energy. Nevertheless they utilize essentially the schematic model of Scheid, Müller and Greiner!

Nuclear Physics A335(1980)491-505. *North-Holland Publishing Co., Amsterdam*

HEAVY ION COLLISIONS Vancouver 1979

PHILIP J. SIEMENS

Lawrence Berkeley Laboratory, University of California,
Berkeley, California 94720, U.S.A.

ABSTRACT

We present an overview of recent attempts to understand high energy nuclear collisions.

For a while it was hoped that the passage of the participant region would excite coherent compression waves of large amplitude in the nearby spectator matter[12]. In the idealized geometry of the participant-spectator model, these waves would come off at the Mach angle

$$\Theta_M = \cos^{-1} \beta_o/v_1 \ ,$$

where v_1 is the speed of the shock front and approaches the speed of sound for small-amplitude disturbances. Unfortunately, this very interesting prediction seems to be an artifact of the oversimplified geometry, and appears neither in more sophisticated models (discussed below) nor in the current data. If the hope of creating high-density nuclear matter is to be realized, we must look to the participant matter.

References

1) J. Hüfner, Proc. 4th High Energy Heavy Ion Summer Study, Berkeley, 1978, LBL-7766, p. 135; J. Hüfner and J. Knoll, Nucl. Phys. A290, (1977) 460.
2) S. Nagamiya, Heavy Ion Collisions at Relativistic Energies, 1979, LBL-9494.
3) D.E. Greiner, P.J. Lindstrom, H.H. Heckman, B. Cork and F.S. Bieser, Phys. Rev. Lett. 35 (1975) 152.
4) A.S. Goldhaber, Phys. Lett. 53B (1974) 306.
5) H. Feshbach and K.Huang, Phys. Lett. 47B (1973) 300.
6) K. Van Bibber, D. Hendrie, D. Scott, H. Wyeman, L. Schroeder, J. Geaga, S. Chessin, R. Truehaft, J. Grossiord, J. Rasmussen and C.Y. Wong, Evidence for orbital dispersion in the fragmentation of ^{16}O at 90 and 120 MeV per nucleon,(Berkeley, 1979), Pre-print LBL-8939, (submitted to Phys. Rev. Lett.)
7) C. Gelbke, D. Scott, M. Bini, D. Hendrie, J. Laville, J. Mahoney, M. Mermaz and C. Olmer, Phys. Lett. 70B (1977) 415.
8) Y.P. Viyogi, T. Symons, P. Doll, D. Greiner, H. Heckman, D, Hendrie, P. Lindstrom, J. Mahoney, D. Scott, K. Van Bibber, G. Westfall, H. Wreman, H. Crawford, C. McParland and C. Gelbke, Fragmentation of ^{40}Ar at 213 MeV/nucleon, (1978), LBL-8267.
9) J. Hüfner, C. Sander and G. Wolschin, Phys. Lett. 73B (1978) 289.
10) J. Bendorf, G. Fai and O.B. Nielson, Physc. Rev. Lett. 41 (1978) 391.
11) T. Symons, Y. Viyogi, G. Westfall, P. Doll, D. Greiner, H. Faraggi, P. Lindstrom, D. Scott, H. Crawford and C. McParland, Observation of new neutron rich isotopes by fragmentation of 205 MeV/nucleon ^{40}Ar ions, (1978) LBL-8269.
12) H. Baumgardt, J. Schott, Y. Sakamoto, E. Schopper, H. Stocker, J. Hofmann, W. Scheid, and W. Greiner, Z. Physik A273 (1975) 359.
13) R.L. Hatch and S.E. Koonin, Phys. Lett. 81B (1978) 1; S.E. Koonin, Phys. Rev. Lett. 39 (1977) 680.
14) R. Amado and R. Woloshyn, Phys. Rev. Lett. 36 (1976) 1435; S. Frankel, Phys. Rev. Lett. 38 (1977) 1338.

Fig. 16. Excerpt from a paper by Ph. Siemens, in which he happily explains his own "more sophisticated model", according to which nuclear Mach shock waves are not possible.

Nuclear Fireball Model for Proton Inclusive Spectra from Relativistic Heavy-Ion Collisions*

G. D. Westfall, J. Gosset,† P. J. Johansen,‡ A. M. Poskanzer, and W. G. Meyer
Lawrence Berkeley Laboratory, Berkeley, California 94720

and

H. H. Gutbrod
Gesellschaft für Schwerionenforschung, Darmstadt, Germany, and Lawrence Berkeley Laboratory, Berkeley, California 94720

and

A. Sandoval
Fachbereich Physik, Universität Marburg, Marburg, Germany, and Lawrence Berkeley Laboratory, Berkeley, California 94720

and

R. Stock
Fachbereich Physik, Universität Marburg, Marburg, Germany

(Received 30 August 1976)

A simple model is proposed for the emission of nucleons with velocities intermediate between those of the target and projectile. In this model, the nucleons which are mutually swept out from the target and projectile form a hot quasiequilibrated fireball which decays as an ideal gas. The overall features of the proton-inclusive spectra from 250- and 400-MeV/nucleon ^{20}Ne ions and 400-MeV/nucleon ^{4}He ions interacting with uranium are fitted without any adjustable parameters.

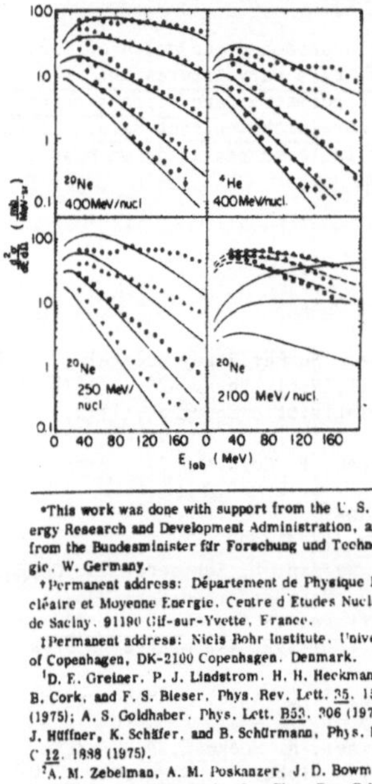

FIG. 3. Measured proton inclusive spectra from a uranium target at 30°, 60°, 90°, 120°, and 150° in the laboratory arranged in decreasing order. When the less than five sets of data are shown, the backward angles are missing. The solid lines are calculated with the fireball model. For the case with 2.1-GeV/nucleon ^{20}Ne on uranium, the data normalization is uncertain by a factor of 2. The dashed lines for this case, which have been raised by a factor of 2.5 to fit the data, are calculated assuming that there are separate, equal-temperature, projectile and target fireballs and that only 25% of their original relative longitudinal momentum in their center of mass was dissipated into heat.

*This work was done with support from the U. S. Energy Research and Development Administration, and from the Bundesminister für Forschung und Technologie, W. Germany.

†Permanent address: Département de Physique Nucléaire et Moyenne Energie, Centre d'Etudes Nucléaires de Saclay, 91190 Gif-sur-Yvette, France.

‡Permanent address: Niels Bohr Institute, University of Copenhagen, DK-2100 Copenhagen, Denmark.

[1]D. E. Greiner, P. J. Lindstrom, H. H. Heckman, B. Cork, and F. S. Bieser, Phys. Rev. Lett. 35, 152 (1975); A. S. Goldhaber, Phys. Lett. B53, 306 (1974); J. Hüfner, K. Schäfer, and B. Schürmann, Phys. Rev. C 12, 1888 (1975).

[2]A. M. Zebelman, A. M. Poskanzer, J. D. Bowman, R. G. Sextro, and V. E. Viola, Jr., Phys. Rev. C 11, 1280 (1975); A. M. Poskanzer, A. Sandoval, R. G. Sextro, A. M. Zebelman, and H. H. Gutbrod, unpublished data.

[3]H. H. Gutbrod, A. Sandoval, P. J. Johansen, A. M. Poskanzer, J. Gosset, W. G. Meyer, G. D. Westfall, and R. Stock, Phys. Rev. Lett. 37, 667 (1976).

[4]H. W. Bertini, T. A. Gabriel, and R. T. Santoro, Phys. Rev. C 9, 522 (1974); R. K. Smith, private communication.

[5]A. A. Amsden, G. F. Bertsch, F. H. Harlow, and J. R. Nix, Phys. Rev. Lett. 35, 905 (1975).

[6]J. P. Bondorf, S. Garpman, E. C. Halbert, and P. J. Siemens, to be published; A. R. Bodmer and L. Wilets, private communication.

[7]J. D. Bowman, W. J. Swiatecki, and C.-F. Tsang, Lawrence Berkeley Laboratory Report No. LBL-2908 (TID-4500-R61), 1973 (unpublished).

[8]R. Hagedorn, in *Cargèse Lectures in Physics VI*, edited by E. Schatzmann (Gordon and Breach, New York, 1973), p. 643.

[9]J. P. Alard *et al.*, Nuovo Cimento A 30, 320 (1975).

[10]G. F. Chapline, M. H. Johnson, E. Teller, and M. S. Weiss, Phys. Rev. D 8, 4302 (1973).

[11]L. D. Landau and E. M. Lifshitz, *Statistical Physics* (Addison-Wesley, Reading, Mass., 1969), p. 110.

[12]R. Hagedorn and J. Ranft, Nuovo Cimento 6, 300 (1968).

[13]To test the sensitivity to the assumption of a Maxwellian-like distribution function, a rectangular distribution function with the same first and second moments was tried, but the results bore no resemblance to the data.

Fig. 17a. Even the experimentallists, who have now become our friends, were in the 70'ies opposed to nuclear shock waves, because their inclusive data did not show anything. They did not even quote us at that time. Arthur Poskanzer, whom I invited to this meeting to talk about the history of the Bevalac - experiments, has even nowadays great difficulties to acknowledge the theoretical stimuli given to him and his group during the years 1973 up til now by the Frankfurt school which have furthered and guided the experiments (see A. Poskanzer's contribution).

Nuclear Squeeze at Lawrence Berkeley Lab

High-energy collisions at the Bevalac between niobium nuclei yield the first direct evidence that nuclear matter can be compressed

Physicists from the Laboratory for Heavy Ion Research (GSI) in Darmstadt, West Germany, and the Lawrence Berkeley Laboratory (LBL) have come up with experimental evidence for the proposition that nuclear matter can be significantly compressed. "We have made the first direct observation of compressed nuclear matter," Arthur Poskanzer of LBL told *Science*.

Even more important, the likely mechanism whereby the compression took place and its attendant experimental signature strongly suggest that physicists will eventually be able to construct a nuclear "phase diagram" spanning a wide range of temperatures and densities. If theorists are correct, lying in wait are one or more phase transitions to new states of nuclear matter, such as the so-called quark-gluon plasma or quark soup that characterized the earliest moments of the universe.

The compression itself is not surprising. Physicists know that nuclei shiver and shake. In the monopole or breathing-mode giant resonance, the entire nucleus alternately expands and contracts, implying a limited compressibility. Moreover, astrophysicists calculate that neutron stars, the superdense remnants of some kinds of burned-out stars, have a density about three times that of ordinary nuclear matter, which is close to 2.5×10^{14} grams per cubic centimeter.

However, it is one thing to conjecture about inaccessible neutron stars and another to measure properties of nuclei under similar conditions in the laboratory. Theorists have calculated that the maximum density achieved in the GSI-LBL experiment is three to four times normal, although the dense state lasted only 10^{-22} second at most.

THE GSI-LBL collaboration did its work with Berkeley's heavy ion accelerator, the Bevalac. With recent improvements, the Bevalac can create relativistic beams of nuclei as heavy as uranium that collide with stationary target nuclei. The accelerator boosts heavy nuclei up to 1 billion electron volts (GeV) per nucleon (proton or neutron).

In the course of the collision, nuclei disintegrate. In general, fragments derived from portions of the projectile and target nuclei that are farthest from the collision center respectively race away in the direction of the incident beam or sit at rest. A much hotter and more dense "nuclear fireball" comprising nucleons from both the projectile and target near the collision center also moves forward in the beam direction but at a speed intermediate between those of the projectile and target fragments.

To detect the outcome of the collisions, the physicists used a special instrument, the Plastic Ball detector, that identifies hydrogen and helium isotopes and positively charged pions emanating from the collision point and measures their energy. The Plastic Ball comprises 815 modules covering almost the entire 4π solid angle surrounding the collision point and can thereby make a detailed map of the spatial distribution of the energy flow. A companion Plastic Wall measures properties of the collision products downstream along the beam direction.

"If you want to see collective effects, you have to have lots of nucleons."

The colorful term "side-splash" describes the main effect seen by the collaboration, headed by Hans Gutbrod and Hans-Georg Ritter of GSI and Poskanzer of LBL, when the beam of niobium-93 nuclei of energy 0.4 GeV per nucleon struck niobium nuclei in a thin metal target. "Side-splash" refers to a sideways leap that is superimposed on the forward motion of the fireball. It is also an example of "collective flow." Collective means that the nucleons act in concert rather than as independent particles, even when they are not bound together.

Several years ago, theorist Walter Greiner of Frankfurt University, West Germany, and several co-workers at Frankfurt and GSI predicted that "side-splash" would be a principal signature for nuclear compression. The theorists calculated from a hydrodynamic model of the nucleus, which treats continuous nuclear density and velocity fields rather than discrete nucleons.

With regard to compression in nuclear collisions, the differences between the independent particle and hydrodynamic models are quite significant. A "gas" of nucleons, which is what the former treats, can clearly be compressed, but because the nucleons are independent (the potential energy they have does not depend on the exact positions of the neighboring nucleons), the model yields no collective effects and no "side-splash," in particular.

Greiner and his co-workers also found that the predicted collective effects [redacted]. The reason for the heavy nuclei requirement is not mysterious. "If you want to see collective effects, you have to have lots of nucleons," says Horst Stöcker, who has migrated to Michigan State University from GSI.

In the hydrodynamic model, the collision sets up a shock wave that compresses nuclear matter in the central fireball for the briefest of instants. The compressed matter then expands, giving rise to the "side-splash." The degree of "side-splash" then depends partly on the maximum compression.

Until the upgrade of the Bevalac and the construction of the Plastic Ball detector, it was not possible to test this finding in detail. Nuclear compression can and probably does occur without noticeable "side-splash," but there had been no way to measure density changes before. For example, work reported last year by another GSI-LBL collaboration under the leadership of Reinhard Stock found indirect evidence for nuclear compression in a lower than expected number of pions emanating from nuclear collisions. The energy associated with the missing pions was assumed to have gone into compressing the nuclei.

The GSI-LBL group has now tried three pairs of nuclei. Calcium-calcium collisions showed no strong "side-splash." Niobium-niobium collisions demonstrate a distinct effect. "Side-splash" should be even more pronounced in the heavier gold-gold system. Experiments with gold beams up to 1 GeV per nucleon have now been cor

MAY 1984 857

measurements and nuclear hydrodynamical theory for massive systems, $A > 40$, is astounding. The nuclear cascade model fails completely because it lacks nuclear compression energy and therefore pressure-build-up so that matter cannot be pushed aside. The observation of sideward flow in high-energy nucleus-nucleus collisions must be considered as one of the very important discoveries in nuclear physics, since it means that the *key mechanism for compressing and heating nuclear matter* has been established. Thus the experimental investigation of the nuclear equation of state can now begin. This can mainly be achieved by measuring excitation functions of observables like flow angle, pion-production rate, cluster production, *etc.* (J.Hofmann, W. Scheid and W. Greiner, Nuovo Cimento 33(1976)343, J. Hofmann el. al. Phys. Rev. Lett. 36(1976)88, H. Stöcker, W. Greiner, W. Scheid, Z. Phys. A286(1978)121, D. Hahn and H. Stöcker, MSUC1-535(1985)). The principal idea is exhibited in fig.20. The mean number of pions per nucleon $< n_\pi > /N$ produced in a nucleus-nucleus collision depends on the temperature achieved in the compression zone. One expects that the temperature rises as a function of incident energy which is indeed the case as can be seen from the lowest frame of fig.20. The temperature achieved for a certain energy depends pleted, but the analysis of the results is still under way.

Meanwhile, Greiner, Gerd Buchwald, Joachim Maruhn, and their co-workers at Frankfurt and Stocker have explicitly modeled the niobium-niobium system and found the same strong sideward peaks in the distribution of events as the experimentalists saw. At this point, the agreement is mainly qualitative.

Establishment of nuclear compression opens the way to a wide-ranging exploration of the properties of nuclear matter under conditions of varying temperature and density. By using the heaviest possible nuclei and carefully choosing the beam energy, physicists should be able to "adjust" the compression to create the temperature and density of interest. Moreover, during the expansion process, the density will "overshoot" and decrease to less than that of ordinary nuclear matter, so that low densities can be explored as well.

What will physicists find, if they carry out this program? Theorists promise them a rich variety of new states of nuclear matter. One of the most intriguing is the quark-gluon plasma. According to models of the evolution of the universe after the Big Bang, during the first microsecond of existence there were no protons, neutrons, mesons, or other heavy particles. Instead, quarks, which are the entities out of which elementary particles are constructed, and gluons, which cement the quarks together, roamed freely everywhere.

After the first microsecond, the universe had "cooled" enough that the liquid-like quark soup "froze," creating the elementary particles in the process. With a sufficiently powerful accelerator, theorists say, it is possible to reverse this process in nuclear collisions. The nucleons momentarily "melt" and generate a quark-gluon plasma. In effect, scientists would be able to study in the laboratory the kind of matter that only existed just after the Big Bang.

Although it is somewhat conjectural how high a collision energy is necessary, the Nuclear Science Advisory Committee, which counsels the Department of Energy and the National Science Foundation on facilities for nuclear physics, last fall recommended as the highest priority in the field the construction of an accelerator that would allow counter-circulating 30-GeV-per-nucleon beams of heavy nuclei to collide head on. A rough guess on the price was $250 million. The latest results from the GSI-LBL collaboration will not hurt the process of building political support for such a machine.—ARTHUR L. ROBINSON

An Impact but No Volcano

It had looked as if the arguing would go on for years. Did a huge volcanic eruption or the impact of an asteroid or comet lay down the clay layer associated with the extinction of so many species 65 million years ago? In this issue of *Science* (p. 867), Bruce Bohor and his colleagues at the U.S. Geological Survey (USGS) in Denver present "compelling evidence" that an impact did it. Experts familiar with their discovery believe that, if anything, the USGS researchers are being too modest in their claim. The quartz grains that they found in the clay layer now marking the 65-million-year-old Cretaceous-Tertiary extinctions are engraved by the apparent traces of a highly energetic impact. "It has to be an impact," says Jay Melosh of the University of Arizona. "Nothing else could do that."

For the USGS group, the search for evidence of an impact began with their studies in the Branch of Coal Resources. The mineralogy of the boundary, whether the clay began as volcanic ash or dust from an impact, should be as informative as that of the volcanic ash layers called tonsteins that they had studied in coal beds, they reasoned. Once they found their own exposure of the boundary clay near Brownie Butte in east-central Montana, the USGS workers removed the 99.99 percent of the sample that was clay, leaving mostly quartz grains about 80 micrometers in diameter. Hydrofluoric acid treatment, the last step in the removal of the clay, serendipitously left many of the grains etched by a distinctive pattern of intersecting, parallel grooves precisely oriented with respect to the crystal structure. There was the proof.

The only way known to produce such features in quartz grains is by a high-velocity impact. The shock of the impact—producing in this case pressures of over 150,000 atmospheres—disorganizes the crystalline quartz and produces amorphous glass, but only on planes having particular orientations with respect to the crystal structure. High-velocity shock experiments in the laboratory have produced these planar features and their grooves, as have nuclear explosions. Such features are also found in the debris from known impact craters on Earth, but they have never been found in volcanic ash. As Richard Grieve of Brown University explains, crustal rocks are too weak to contain pressures greater than 1000 atmospheres and, even if trapped gases could generate higher pressure during explosive release, the easily compressed gases cannot transmit that pressure to the relatively incompressible rock. "It just doesn't work," he says.

The USGS group has other evidence. X-ray diffraction analysis of the quartz grains revealed streaking of the normally sharp diffraction spots, which is typical of shocked quartz. It also detected stishovite, a form of quartz formed at pressures above 100,000 atmospheres. "We're sure it's there," says Bohor, but its low abundance has prevented the production of an x-ray diffraction pattern that is strong enough to be reproduced in a journal. They have also seen the planar shock features in quartz from Cretaceous-Tertiary boundary clays at three other sites in east-central Montana and at the now-classic European locales—two sites in Denmark, two in Italy, and one in Spain. Charles Pillmore of the USGS in Denver and his colleagues subsequently discovered similarly shocked quartz at the boundary in the Raton Basin of Colorado and New Mexico.

The unequivocal evidence of highly shocked quartz follows less compelling finds in the boundary clay. These included spherules that presumably condensed from vaporized rock and are too large to have been flung around the globe by a volcano, and osmium isotope compositions that allowed an extraterrestrial or volcanic source but not a continental source (*Science*, 11 November 1983, p. 603). Recently, Jan Smit, Frank Kyte, and John Wasson of the University of California at Los Angeles reported that they have found spheres containing geochemical markers of an extraterrestrial object and magnetite that must have crystallized from a very high-temperature liquid.

Once the shocked quartz grains settle the impact-volcano question, they may be put to further use. Since continents contain abundant quartz and ocean basins very little, these quartz grains may shed light on the question of where the impact occurred.—RICHARD A. KERR

858 SCIENCE, VOL. 224

Fig. 17b. After the nuclear flow had been discovered with the Plastic Ball, our experimental friends aknowledged the theoretical guidance we have given them (from Research News, LBL 1984).

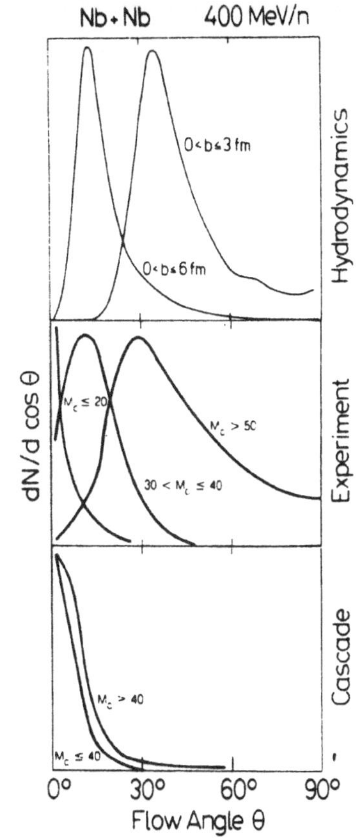

Fig.18. Distributions of flow angles $dN/cos\theta_f$ for the reaction $^{93}Nb(400AMeV)+{}^{93}Nb$. (a) Result for the hydrodynamical calculation. The finite-multiplicity distortions are taken into account. The given impact-parameter ranges correspond to the multiplicity cuts indicated at the experimental curves below (b). (b) Plastic Ball Data for various multiplicity cuts labeled on the curves. (c) Result of the cascade simulation after multiplicity selection.

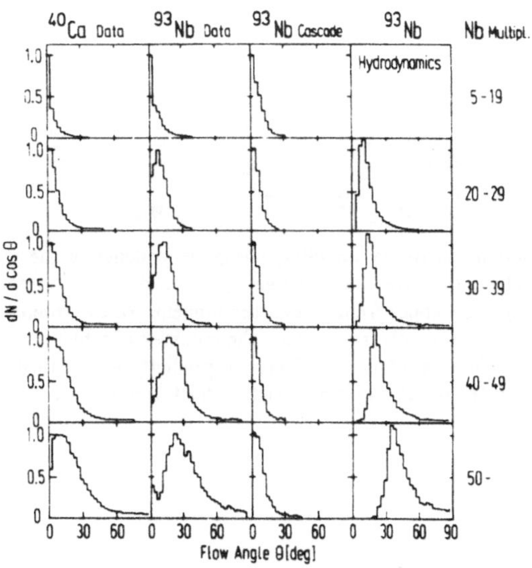

Fig.19. Systematic evidence for the nuclear side splash and hence of nuclear compression.

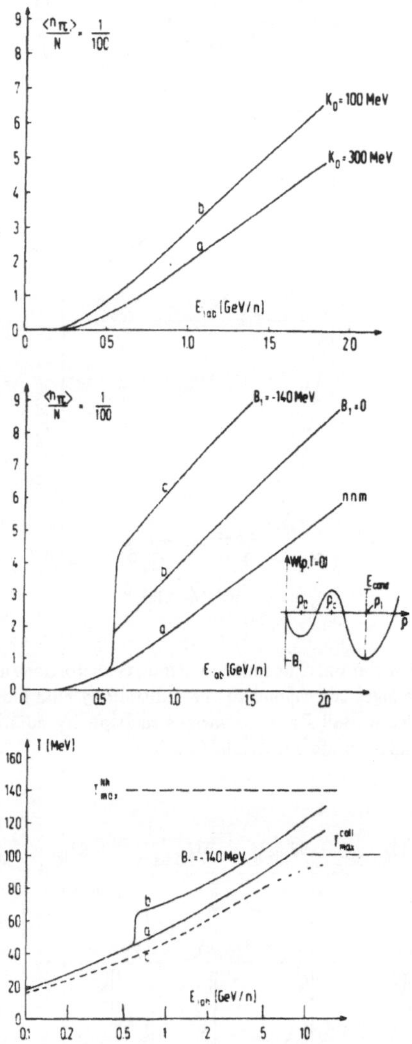

Fig.20a. Early predictions on the bombarding energy dependence of the pion multiplicities and their dependence on the nuclear compressional energy (H. Stöcker, W. Greiner and W. Scheid, Z. Phys. A286(1978)121). The upper frame shows the influence of the compression constant, using the linear $E_c(\rho)$ ansatz. The middle frame shows the influence of a hypothetical abnormal density isomeric state; threshold increase of pion production is predicted at a critical energy. Lowest frame: influence of the EOS on the temperature achieved in shock compression.

on the nuclear equation of state: For a soft equation of state (small compression constant K) the density compression and the temperature achieved will be higher than for a stiff equation of state (large compression constant K).
If in particular the equation of state were to contain a density isomer (middle frame of fig.20) one would expect sudden jumps of the temperature (and therefore of the pion-production rate) as a function of incident energy. This should be so because as soon as the density ρ achieved in the compression zone becomes larger than ρ_c (see middle frame of fig.20) a sudden release of potential energy into internal kinetic energy, *i.e.* temperature, will occur.This in turn is reflected in an increase of the number of pions produced. Sandoval et. al. (Phys. Rev. Lett. 45(1980)874) have measured the pion production rate (fig.21), but no such jumps have been found. However, information on the thermal part of the *nuclear equation of state* as well as on the *nuclear viscosity* could be obtained from these data (H. Stöcker and W. Greiner, Phys. Rep. 137 Nr.5 and 6(1986)278). Recently arguments were given, however, that the pion production rate may not be the best observable for deducing the nuclear equation of state, because the velocity dependence of the nuclear forces may also have large effects on the compression and temperature achieved (J. Aichelin, A. Rosenhauer, G. Peilert, H. Stöcker and W. Greiner, Phys. Rev. Lett. 58(1987)1926). This is not the case for the excitation function of the flow angle. Such systematic experiments are not yet available. A very beautiful review of the present status of nuclear shock waves is given by K. H. Kampert (J. Phys. G., 15, Nr. 6(1989)691; see also K. H. Kampert's lecture). An effective (mean) temperature reached in a nucleus-nucleus collision can also be deduced from the energy distribution produced for pions, protons, deuteron, ^3He, *etc.*. These spectra show a typical Boltzmann form proportional to $e^{-E/T}$, where E is the energy of the particles measured. Hence the slope of the spectra allows one to deduce a temperature T. Such T-values determined in various collisions and for various particles are exhibited in fig.22.

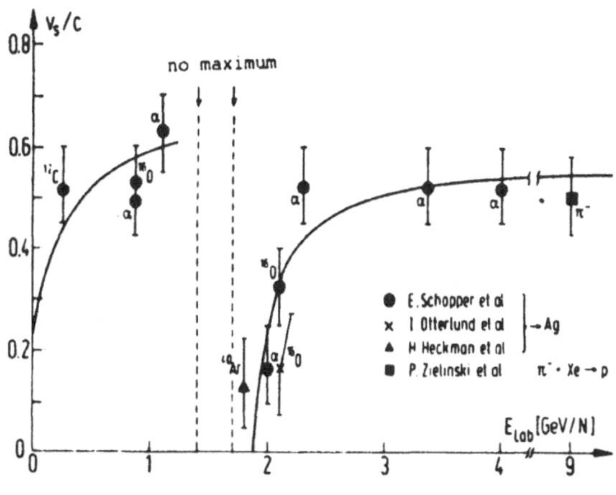

Fig.20b. A density isomer would also show up in the excitation function of the flow angle (J. Hofmann, H. Stöcker, U. Heinz, W. Scheid and W. Greiner, Phys. Rev. Lett. 36(1976)88). The only rather wide - spread excitation function of this sort existing to date is that of E. Schopper, which has the indication of a density isomer. A confirmation of this finding is urgently needed.

The theoretical curves are obtained from the shock model with various assumptions for the mass spectrum of hadrons. If, in particular, the density of excited hadron states increases strongly with excitation energy, *e.g.* exponentially,one expects that a *limiting temperature* will occur: Instead of pumping the internal energy available into kinetic hadron energy it simply will be used to create more and more excited hadron states, thus leading effectively to a leveling-off of the temperature achieved. One calls such a possible limiting temperature the *Hagedorn temperature*. Obviously, according to fig.22 the present energies available are not yet sufficient to clarify this interesting conjecture. The experiments with ultra-relativistic nucleus-nucleus collisions at CERN, where energies of more than 200 GeV/nucleon are available, should be able to contribute an answer.Theoretical models for nuclear shock waves have now become microscopic: The lectures of Aichlin, Hartnack, Peilert, Blättel tell this fascinating story. They confirm practically all the hydrodynamic predictions and go far beyond. Cluster production,squeeze out, energy distributions of fragments are some of the additional features described. There are many more.

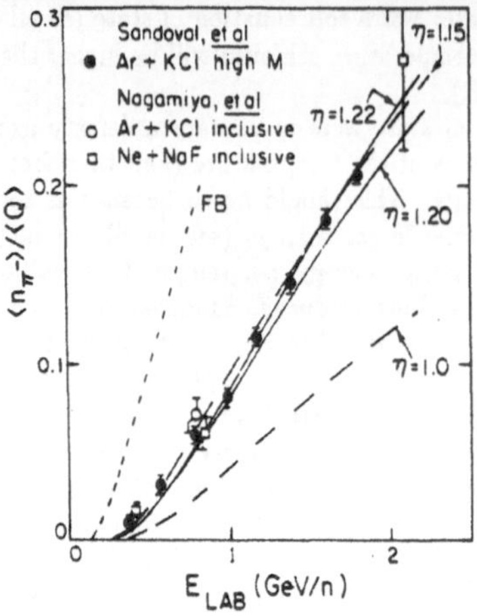

Fig.21. The observed pion multiplicities (Sandoval et. al., loc. cit.) are compared to fireball (FB) calculations (short dashed line) and non-viscous fluid dynamics with late freeze-out (long dashed line). The influence of the viscosity is shown by the lines labeled η. The fireball calculation (FB) - which neglects the compression energy $E_c(\rho)$ - employs only noninteracting Fermi- and Bose gases of nucleons, pions and deltas for the thermal part of the nuclear equation of state. All other results are obtained with the addition of a linear $E_c(\rho)$ ansatz. Obviously the pion multiplicities depend on both, the nuclear equation of state and the nuclear viscosity.

Let us close this overview with an outlook. Clearly the deduction of the nuclear equation of state is a most challenging and fundamental task for the near future. Since the key mechanism for compression has been established with the appearance of shock waves, systematic experiments of the kind described above and accompanied by subtle theoretical analysis should yield an answer. Furthermore,it is rather likely that the hydrodynamic behaviour of nuclear matter will also occur at ultrarelativistic energies,because we expect precritical scattering phenomena near phase transitions (V. Ruck, M. Gyulassy, W. Greiner, Z. Phys. A277(1976)391; M. Gyulassy and W. Greiner, Ann. Phys. 109(1977)485) and nonlinear excitation mechanism to keep the mean free path of nucleons λ small enough, $\lambda \ll 2R$, where R is the nuclear radius. A small free path is necessary for the validity of hydrodynamics. Hence very high nuclear compressions and temperatures can be reached. First experimental data taken at CERN by various groups (see the lectures of H. Gutbrod, H. Ströbele, H. R. Schmidt) indicate that indeed a large degree of stopping is obsrved. Streamer chamber pictures of ^{16}O (3200 GeV) + Pb collision are shown in H. Ströbele's lectures. The corresponding ultrarelativistic nuclear fluid dynamic calculation is illustrated in fig.23 with the $^{16}O + Pb$ central collision at 200 GeV/nucleon (G. Graebner, Dissertation, Inst. f. Theor. Phys. der J. W. Goethe Univ., Frankfurt a. Main(1984)). The normal equation of state with K = 700 MeV has been used. The high compression constant K comes from a fit to the pion excitation function measured by Harris et. al. (J. W. Harris et. al. Phys. Lett. 153B(1985)377). It must be considered as an effective value, because the pion yield is also depending on nuclear viscosity (see our discussion in conjunction with fig.21) and on momentum dependent nuclear forces. In that collision nuclear matter is compressed up to $\rho \gg 10\rho_0$ and energy densities far above $16GeV/fm^3$ are reached. The compression zone is rather small, *i.e.* of the order of $10fm^3$. It is much larger and also the time over which the strong compression appears is considerably extended (up to a factor 5) in Pb + Pb or U + U-collisions(G. Graebner, loc. cit.), for example. This is due to the greater volume of the projectile and target nucleus. Under these extreme conditions one expects the nucleons in the compression zone to be so densely packed that considerable overlap between them will appear and, consequently, the individual nucleon bags may dissolve and a *quark-gluon plasma* (QGP) may be created. This is a new phase of nuclear matter. It is of enormous interest because quarks would exist relatively free within the QGP while they are ordinarily confined to small baryon and meson bags. The information of QGP may eventually be detected by observing the exotic fragments like anti-nuclei (U. Heinz, P. R. Subramanian and W. Greiner, Z. Physik A318(1984)247 and J. Phys. G 12(1986)1232) or strange matter droplets (Carsten Greiner, P. Koch and H. Stöcker, Phys. Rev. Lett. 58(1987)1825; see Carsten Greiner's lecture). The strangeletts are a particularly appealing theoretical prediction, because they represent a qualitative (not quantitative) signal for the formation of quark matter - in fact, they would be an exotic remnant of such a hypothetical state.

The phase diagram of fig.24 illustrates qualitatively the various phases of nuclear matter. Besides the ground state of nuclear matter at $\rho = \rho_0$, T = 0 and a gas-liquid phase in its vicinity ($\rho < \rho_0, T \sim fewMeV$) a crystalline structure is expected for $\rho \geq 3\rho_0, T \sim fewMeV$. The latter is often characterized as pion condensation (A. B. Migdal, Sov. Phys. Jetp 34(1972)1181, T. D. Lee and G. C. Wick, Phys. Rev. D9(1974)2291). Clearly, the quark-gluon plasma phase is of fundamental importance.

The question is, of course, whether the QGP is a really fully deconfined phase. It is possible, for example, that the quarks might be deconfined, but the gluons might still be clustered in glue-balls. If this is the case, the QGP produced in high energy heavy ion collisions may turn out to be a "breeding place" for true glueballs. There have been many other suggestions of how to observe the QGP. J. Rafelski in particular suggested that an enhancement of particles with strangeness ($K^+/\pi^+, K^-/\pi^-$) should be taken as a signal (see the lectures by J. Rafelski and by P. Koch). Indeed such enhancement has been observed at the AGS in Brookhaven (see O. Hansen's lecture). However it should be pointed out that presently only light ion beams are available for experiments. For such beams we are in a situation similar to that in the early days of the BEVALAC. The really exciting, novel physics of collective phenomena can only be studied when massive beams become available (the AGS plus booster in ~1992, the Pb-project at CERN at ~ 1993-4).
All the present observations (including Matsui's suggestion of J/Ψ-depletion) can be described by

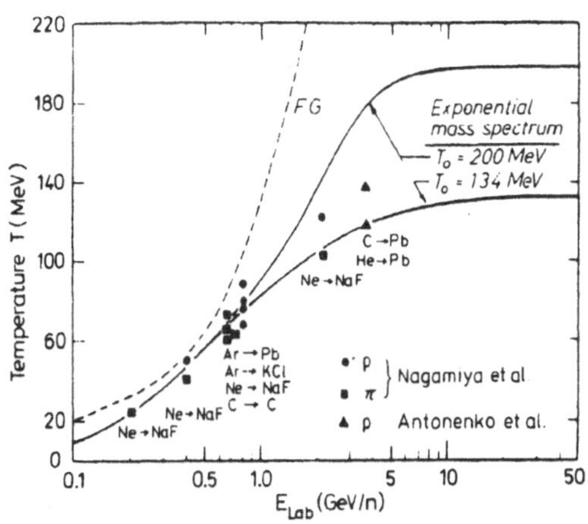

Fig.22. Bombarding energy dependence of the temperature calculated in the shock model using different equations of state: nucleon Fermi gas (dashed line) and hadron plasma with a Hagedorn mass spectrum (H. Stöcker, W. Greiner and A. A. Ogloblin, Z. Physik A303(1981)253). The data are extracted from fits to the 90° c. m. slopes of the spectra of protons (and pions) (S. Nagamiya, Nucl. Phys. A400(1983)565c.).

hadron-hadron collisions. H.Sorge's lectures on *Relativistic Quantum Molecular Dynamics* give a vivid impression on how beautiful all the presently observed features can be theoretically understood. The first selfconsistent covariant microscopic calculations (by H. Sorge, R. Matiello, A. v. Keitz and H. Stöcker, who pioneered this exciting and fastly develloping field of reasarch) for collisions of massive nuclei (Au + Au from 10 - 200 GeV/n) show indeed that there can be extreme (up to 10-20 times normal) energy and baryon number densities achieved. One has to view the newly develloped RQMD as an astounding theoretical break-through the field of high energy nucleus - nucleus dynamics.
Let us conclude: all the various phases of nuclear matter are special features of the *nuclear equation of state*. Its investigation is only possible with the compression and heating-up mechanism as furnished by nuclear shock waves. Without them an investigation of these phenomena is probably impossible.

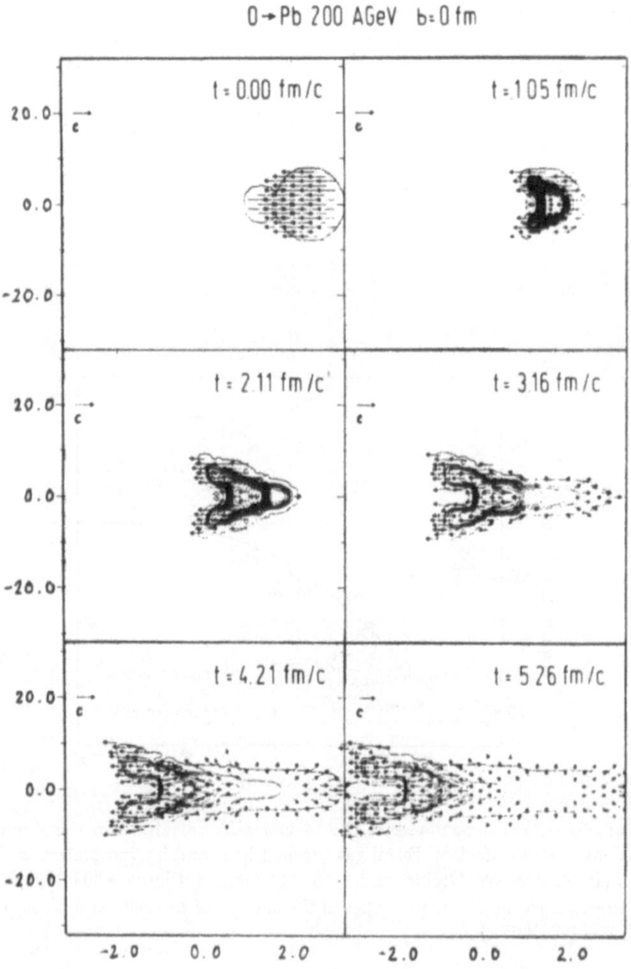

Fig.23. High energy heavy ion collision $^{16}O \rightarrow Pb$ (200 A GeV) according to nuclear hydrodynamics. The time development of the comoving densities are shown in the reaction plane. (a) Central collision with velocity arrows in the nucleon-nucleon cm frame. (b) Collision at b = 3 fm. The velocity arrows are in the nucleon-nucleon cm frame. (c) Collision at b = 3 fm. The velocity arrows are in the lab system (after T. Rentzsch, G. Graebner, J. Maruhn, H. Stöcker and W. Greiner, Mod. Phys. Lett. A2(1987)193).

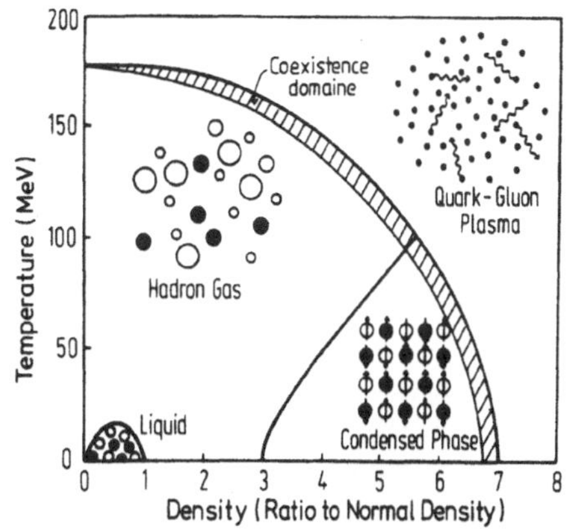

Fig.24. The phase diagram of nuclear matter shows transformations predicted by various theories and conjectures. The normal, liquid-like state of nuclear matter ocupies the lower left-hand corner of the diagram. At elevated temperature and density the liquid evaporates, forming a "hadronic resonance gas" (HRG) made up not only of nucleons but also of mesons and of nucleon resonances. With further heating and compression the hadrons themselves could break down into their constituent particles,the quarks and gluons, creating a plasma-like mixture called quark-gluon plasma(QGP). During the transition to the QGP a phase coexistence region is passed through (which is established according to the general Gibbs equilibrium conditions for coexistent phases), containing hadrons as well as quarks and gluons. At high density but comparetively low temperature nuclear matter could be frozen into a lattice structure analogous to a crystalline solid (condensed phase). Protons and neutrons with oppositely oriented spins would alternate in the lattice.

COLLECTIVE FLOW MEASURED WITH THE PLASTIC BALL

H.G. Ritter, H.H. Gutbrod, K.H. Kampert, B. Kolb, A.M. Poskanzer.
R. Schicker, H.R. Schmidt, and T. Siemiarczuk

Gesellschaft für Schwerionenforschung
D-6100 Darmstadt, West Germany
and
Nuclear Science Division, Lawrence Berkeley Laboratory
University of California, Berkeley, CA 94720, USA

INTRODUCTION

The study of the bulk properties of nuclear matter, *e.g.* the equation of state or the transport properties, is one of the main objectives of relativistic heavy ion physics. As pointed out many times during this conference, the knowledge of these properties is of fundamental interest and it is also essential for the understanding of supernova explosions and of the structure of neutron stars.

The collision of heavy ions has been described by many different approaches. Intranuclear cascade calculations are based on the assumption that a nuclear collision is equivalent to the superposition of nucleon-nucleon collisions. Hydrodynamical models on the other hand predicted the appearance of shock waves initiated by a very high energy incident particle early on [1] and other authors also considered shock waves in colliding nuclei [2, 3]. But a mechanism of shock compression in nucleus-nucleus collisions that would lead to densities 3 to 5 times higher than that of normal nuclear matter was first proposed by Scheid *et al.* [4].

A signature of the compression effects predicted by the calculations using a nontrivial equation of state is collective flow of the nuclear matter in the expansion phase [4, 5, 6]. Collective flow is the consequence of the pressure buildup in the high density zone through the short range repulsion between nucleons, *i.e.* through compressional energy. This effect leads to characteristic, azimuthally asymmetric sidewards emission of the reaction products.

Collective flow has not been observed in single particle inclusive measurements [7]. Early on a need was clearly seen for a large acceptance (4π) detector at the Bevalac. The Plastic Ball detector, which was designed to measure most of the charged particles from heavy ion reactions, is ideally suited to study the emission patterns and event shapes resulting from collective flow.

The Nuclear Equation of State, Part A
Edited by W. Greiner and H. Stöcker
Plenum Press, New York

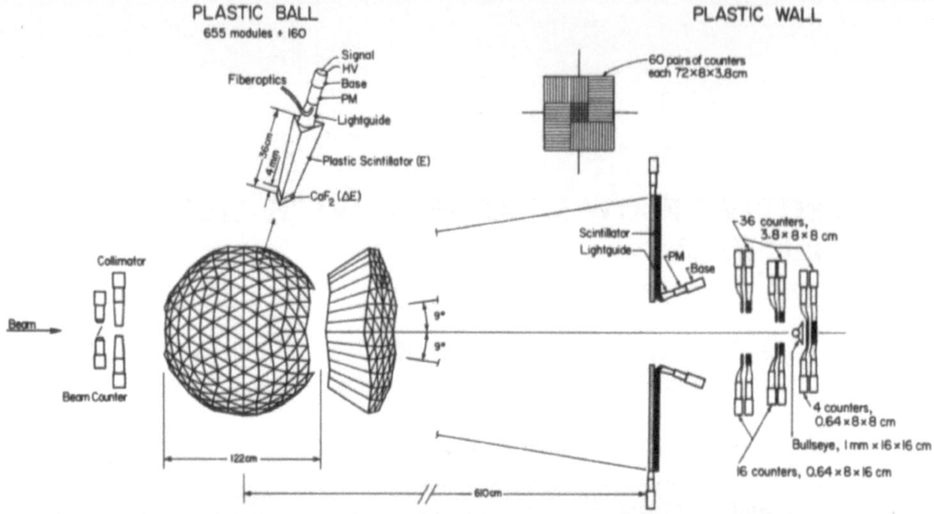

Figure 1. Schematic view of the Plastic Ball and the Plastic Wall. On the upper left is a picture of a single module and on the upper right is a view of the Wall as seen by the beam.

EXPERIMENT

For coverage of almost 4π the Plastic Ball was built completely surrounding the target except for the extreme backward angles, where the beam enters the system, and the extreme forward angles. Because of the large fragment velocities at forward angles the region from 0 to 10 deg was covered with a multielement time-of-flight system called the Plastic Wall. The complete system is shown schematically in Fig. 1 and described in detail in Ref. [8].

The Plastic Ball thus covers the region between 10 and 160 degrees, which is 96% of the total solid angle. It consists of 815 detectors where each module is a $\Delta E - E$ telescope capable of identifying the hydrogen and helium isotopes and positive pions. The ΔE measurement is performed with a 4 mm thick CaF_2 crystal and the E counter is a 36 cm long plastic scintillator. Both signals are read out by a single photomultiplier tube. Due to the different decay times of the two scintillators, ΔE and E information can be separated by gating two different ADCs at different times. Positive pions are additionally identified by measuring the delayed $\pi^+ \rightarrow \mu^+ \rightarrow e^+$ decay.

THE MULTIPLICITY PARAMETER

In measuring the proton multiplicity, N_p, we attempt to account for all participant protons, including those bound in light composites (d, t, ^{3}He, and ^{4}He). These bound protons add approximately 40% to the proton multiplicity. The proton energy threshold in the laboratory frame is approximately 15 MeV. The energy threshold for ions of charge 2 is roughly 2 or 3 times higher. The projectile spectators are largely eliminated by excluding a region in $p_\perp$-rapidity space that is identified by use of low multiplicity, peripheral events.

Since the particle multiplicity is related to the impact parameter, we classify the events according to this proton multiplicity. The average multiplicity depends on the target-projectile mass and on the bombarding energy. To allow meaningful comparisons the multiplicity bins chosen should correspond always to approximately the same range in normalized impact parameter. The best approach is to divide the multiplicity distribution into bins of constant fractions of the maximum multiplicity. The multiplicity distribution has roughly the same form for all systems and energies: a monotonic decrease with increasing multiplicity with a rather pronounced plateau before the final sharp decrease at the highest multiplicities. Therefore the maximum multiplicity (N_p^{max}) can be defined at the point where the curve drops to one half the plateau height. Table 1 contains the value of $N_p^{max}/2Z$ for all symmetric systems measured. The data accumulated with a minimum bias trigger are then divided into 5 bins, 4 equal width bins between 0 and maximum multiplicity and one bin with multiplicities larger than N_p^{max}. These multiplicity bins are labelled MUL1, MUL2, MUL3, MUL4, and MUL5 and range from peripheral collisions with few observed charges to central collisions with very high multiplicities.

Table 1. Maximum participant proton multiplicities N_p^{max} divided by the sum of the projectile and target nuclear charges for all measured symmetric systems and beam energies.

E/A (MeV/A)	150	250	400	650	800	1050
Au + Au	0.41	0.58	0.71	0.81	0.85	
Nb + Nb	0.46	0.63	0.78	0.88	0.90	0.95
Ca + Ca			0.75			0.90

STOPPING AND THERMALIZATION

Obviously the use of the concept of bulk properties of nuclear matter needs experimental justification. We must prove that the system is in global, or at least local, thermal equilibrium. This has not yet been done rigorously, but with a 4π detector it is straightforward to measure the degree of stopping that can be reached in the reaction by investigating the rapidity distribution dN/dy of the baryons which is shown in Fig. 2 for Au + Au at 250 MeV per nucleon for three multiplicity bins. In peripheral reactions (top, MUL1) most of the reaction products experience a very small momentum transfer and stay at beam rapidity ($y = 0.72$) or at target rapidity ($y = 0$). However, since target rapidity fragments are absorbed in the target and cannot be observed in the detector, the distribution is not symmetric around midrapidity ($y = 0.36$). In semi-central collisions (Fig. 2 center, MUL3) already an appreciable amount of reaction products populates midrapidity. In central collisions (Fig. 2 bottom, MUL5) we observe a distribution that is symmetric around midrapidity. This indicates that the two Au nuclei completely stop each other and form a highly excited system at midrapidity.

More detailed information can be obtained by studying how the longitudinal momentum of the projectile is transformed into transverse motion during the collision. From the measured momenta of all the particles in the center of mass system we can calculate the ratio

$$R = \frac{2}{\pi} \sum_i \left|p_\perp^i\right| / \sum_i \left|p_\parallel^i\right| . \tag{1}$$

The sums in Eq. 1 contain the perpendicular, $p_\perp$, and longitudinal, $p_\parallel$, momentum components of all particles in one event. Global stopping of the two nuclei in the center

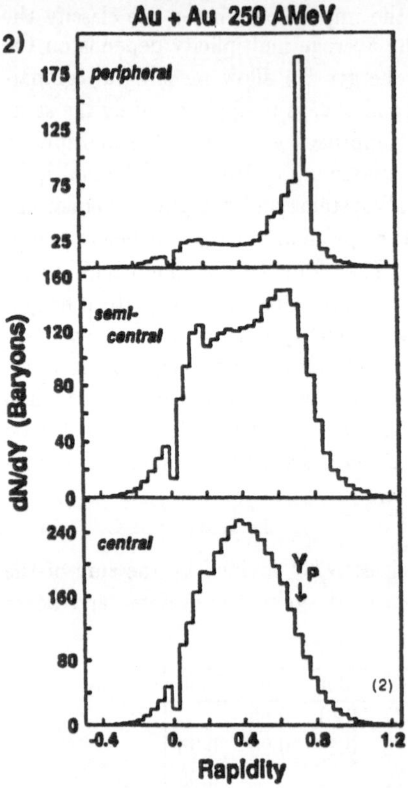

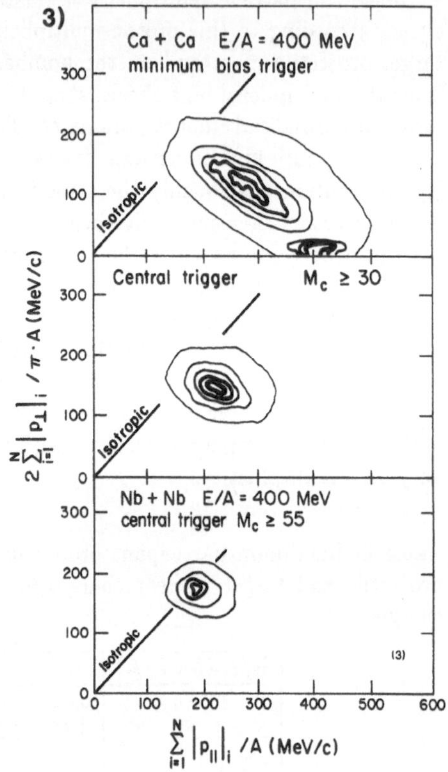

Figure 2. Baryon rapidity distributions for Au + Au collisions at 250 MeV per nucleon for three multiplicity bins. Baryons are defined as: $(1 + N/Z)n_p + 2n_d + 3(n_t + n_{^3He}) + 4(n_{^4He})$.

Figure 3. Contour plot of the average momentum components perpendicular and parallel to the beam axis for Ca + Ca (top and center) and Nb + Nb (bottom) at 400 MeV per nucleon. The diagonal line ($R = 1$) corresponds to isotropic events.

of mass system (or isotropic emission) would manifest itself by a ratio $R = 1$ [9]. Flow in the transverse direction would result in an even larger ratio. Thus, $R = 1$ is a necessary, but not sufficient condition, for thermalization. If in addition the energy distributions are of the Maxwell-Boltzmann type, the emitting system could be called thermalized.

The top part of Fig. 3 shows the yield as contour lines in the $p_\perp$-$p_{||}$ plane for minimum bias Ca + Ca events. The peak at small $p_\perp$ but large $p_{||}$ corresponds to peripheral reactions and is dominated by projectile fragments. This contribution vanishes if the trigger is changed to a central one. Figure 3 (center) shows central events with a charged particle multiplicity larger than 30. The maximum of the yield is shifted towards the diagonal line, calculated for isotropic emission, but only a few events actually reach $R = 1$, which in the limit of large multiplicity corresponds to full stopping of the two nuclei. In the lower part of Fig. 3 central events (charged particle multiplicity > 55) of Nb + Nb at 400 MeV per nucleon almost fulfill the stopping and isotropy condition ($R = 1$) on average. The multiplicity cuts applied correspond roughly to the same fraction of the total cross section for both systems. More quantitatively, R can be investigated as a function of multiplicity [10]. As expected from Fig. 3, R increases with multiplicity and reaches a value of 0.62 for Ca, whereas it comes close to one for Nb. The increase with multiplicity can be explained by the decreasing role played by the projectile spectator particles as the collisions become more central. The difference between Ca and Nb central collisions may result because either Ca nuclei may be too small to stop each other at 400 MeV per nucleon or because stopping occurs only in the central part of the nuclear volume and surface effects are less important in Nb as compared with Ca.

ENERGY FLOW

The first attempt to determine the event shapes was done by adapting the thrust [5, 11, 12, 13] and sphericity [13, 14] analyses developed in high energy physics [15, 16] to the heavy ion case. The sphericity tensor

$$F_{ij} = \sum_{\nu} p_i(\nu) p_j(\nu) w(\nu)$$

is calculated from the momenta of all measured particles for each event. It is appropriate to chose the weight factor $w(\nu)$ so that composite particles have the same weight per nucleon as the individual nucleons of the composite particle at the same velocity. Commonly, the weight $w(\nu) = 1/2m(\nu)$ as proposed in Ref. [14] (kinetic energy flow) is used. Other coalescense invariant weights such as $1/p(\nu)$ [13] have been proposed and have been used in our analysis with similar results. The sphericity tensor approximates the event shape by an ellipsoid, whose orientation in space and whose aspect ratios can be calculated by diagonalizing the tensor.

The shapes predicted by hydrodynamical and intranuclear cascade calculations are quite different. The hydrodynamical model predicts prolate shapes along the beam axis for grazing collisions. With decreasing impact parameter the flow angle increases and reaches 90 degrees (with oblate shapes) for zero impact parameter events [5, 11, 12, 14]. This behaviour is independent of projectile and target mass. Early cascade calculations, however, predicted zero flow angles at all impact parameters [14]. Later, improved cascade calculations yielded finite, but small, flow angles [17, 18, 19].

Fluctuations due to finite particle effects are a major obstacle in extracting information from a flow analysis. Danielewicz and Gyulassy [20] have shown that those distortions strongly depend on multiplicity and that the flow angle Θ, if properly weighted by the Jacobian ($\sin\Theta$), is much less severely shifted towards higher values than the aspect ratios.

In this work the energy flow tensor [14] in the center of mass system has been determined and diagonalized for each individual event. The distribution of the flow angles (angle between the major axis of the flow ellipsoid and the beam axis) [21, 22] for Ca + Ca, Nb + Nb, and Au + Au at 400 MeV per nucleon is shown on the left side of Fig. 4 as a function of multiplicity. A striking difference between the light Ca system and the heavier Nb and Au systems can be observed. For all but the highest multiplicity bins, the distribution of the flow angles for the Ca data is peaked at 0 deg. For the heavier systems, however, there is a finite deflection angle increasing with increasing multiplicity. In addition, the flow angles increase with the mass of the system. An increase with mass has been predicted qualitatively by Vlasov-Uehling-Uhlenbeck calculations [23]; however, that predicted increase is more pronounced than the one observed here.

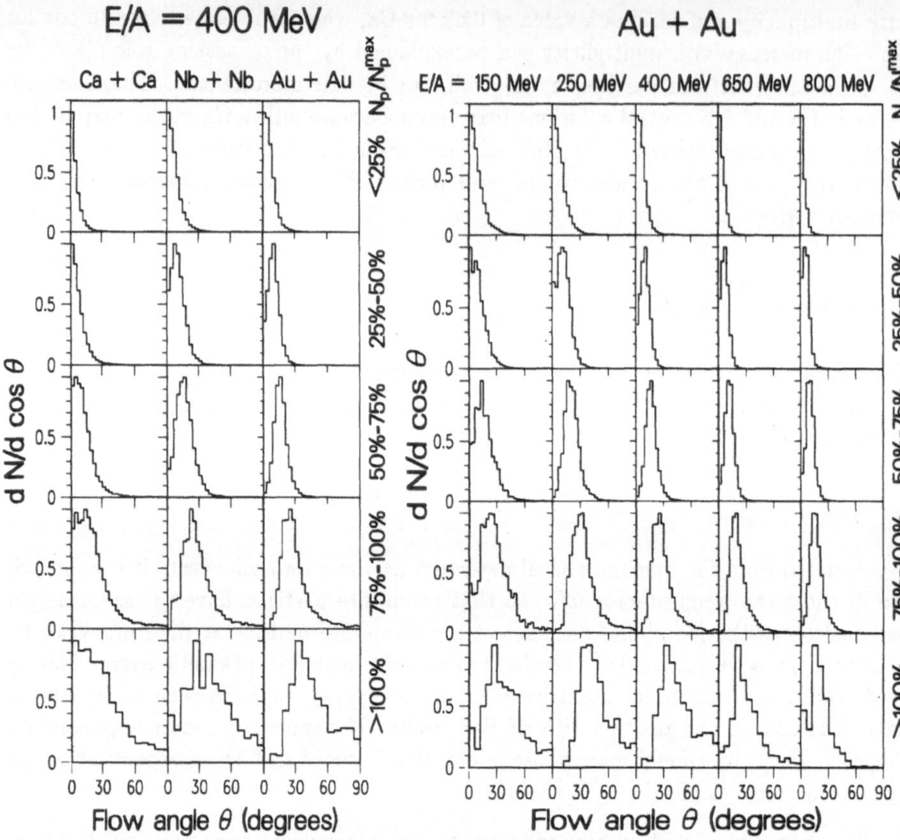

Figure 4. Distributions of the flow angles ($dN/d\cos\Theta$) in five multiplicity bins. The systems Ca + Ca, Nb + Nb, and Au + Au, all at 400 MeV per nucleon, are shown on the left, Au + Au at five energies is on the right.

Also important is the energy dependence of the flow angles. This is shown on the right side of Fig. 4 for five Au + Au energies from 150 MeV per nucleon up to 800 MeV per nucleon. The general trend observed is that the flow angles decrease with increasing energy above 250 MeV per nucleon. At the lowest energy the reaction mechanism responsible for the flow effect might lose importance in favor of other mechanisms known from low energy heavy ion reactions, such as, *e.g.*, deep inelastic scattering.

The decrease in the flow angles with increasing energy does not indicate that the flow effect gets smaller; it means, however, that the mean transverse momentum does not increase quite as fast as the longitudinal momentum. On the contrary, the mean perpendicular momentum transfer increases with energy, as will be seen from the transverse momentum analysis described in the next section.

So far, the events have been parameterized by ellipsoids, but it is of interest to study the shape in more detail. The presence of finite flow angles in the data indicates that in those events a reaction plane exists that is defined by the flow axis and the beam axis. All events can be rotated by the azimuthal angle ϕ, determined by the flow analysis, so that their individual reaction planes all fall into the x-z plane, with the z-axis being the beam axis. For those rotated events the invariant cross section in the reaction plane $d^2\sigma/dy\ d(p_x/m)$ [5, 11, 24] can be plotted, where p_x is the projection of the perpendicular momentum into the reaction plane and y is the center of mass rapidity. Figure 5 shows this plot for a selected multiplicity bin for 400 MeV per nucleon Ca + Ca and Nb + Nb data, together with filtered events from a cascade code calculation [25]. The depletion near target rapidities is due to limited experimental acceptance for low energy particles in the laboratory system. This depletion enhances the flow angles artificially but does not change the reaction plane. The cascade plot is almost symmetric around the beam axis, whereas the Ca and Nb in-plane data plots are clearly asymmetric. The highest level contour results largely from the projectile remnants and indicates a definite bounce-off effect. The multiplicity dependence of the outer contour lines seems to follow the

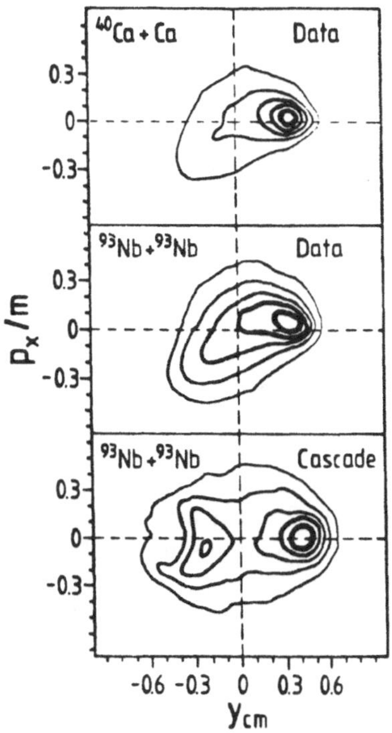

Figure 5. Contour plots (linear contours) of p_x/m as a function of the center of mass rapidity for multiplicities from 40 to 49 for Nb and 20 to 24 for Ca at 400 MeV per nucleon.

trend indicated by the flow angle distributions (Fig. 4). However, the position of the peak from the projectile remnants changes only slightly with multiplicity. Thus one can conclude that the strong sideward peaking (side-splash) seen in Fig. 4 is mainly due to the midrapidity particles. It should be noted that the bounce-off and side-splash effects are in the same plane.

The first report on the observation of finite flow angles by the Plastic Ball group [21] has been taken as proof for the existence of collective flow in relativistic heavy ion reactions [21, 24] and has stimulated, not only new experimental work and new analysis methods, but also new theoretical approaches.

TRANSVERSE MOMENTUM ANALYSIS

We have seen that the sphericity method is an extremely useful tool for establishing experimentally the existence of collective flow effects. However, reducing all the information available for each event to essentially one observable, the flow angle Θ, is a rather inclusive representation of the data. The contour plots of p_x/A versus y_{cm}, shown in Fig. 5, however, are very much influenced by experimental biases and are difficult to describe. Based on the observation that the reaction plane can also be determined from the collective transverse momentum transfer [26, 27], Danielewicz and Odyniec have proposed a better, more exclusive way to analyze the momentum contained in directed sidewards emission [27]. They propose presenting the data in terms of the mean transverse momentum per nucleon in the reaction plane $< p_x/A >$ as a function of the rapidity. By also removing autocorrelation effects this method is sensitive to the true dynamic correlations and has lead to indications for collective flow in cases where the kinetic energy flow analysis was not sensitive enough [27, 28]. Studying the momentum transfer as a function of rapidity permits one to distinguish between participant and spectator contributions and to exclude regions with large detector bias.

In the transverse momentum analysis the reaction plane is determined by the vector $\vec{Q}$ calculated for each event from the transverse momentum components $p_\perp$ of all the particles observed in the forward and backward hemispheres in the center of mass

$$\vec{Q} = \sum_i p_{\perp i}^{forw} - \sum_i p_{\perp i}^{back}.$$

Pions are not included. A similar method, using the transverse momentum unit vectors of the slow and fast particles was developed in parallel [26]. Each event can be rotated around the beam axis (z-axis) so that $\vec{Q}$ defines the x-axis of a new coordinate system. Autocorrelations are removed by calculating $\vec{Q}$ individually for each particle without including that particle. Evidently $\vec{Q}$ is only an estimate for the true reaction plane, and the projections into the estimated plane are too small by a factor $1/ < \cos\phi >$, where ϕ is the angle between the estimated and the true plane. The quantity $< \cos\phi >$ can be estimated [27] by randomly dividing the events into two subevents and averaging the cosine of one half the angle between the $\vec{Q}$ vectors of the two subevents.

Figure 6 shows the mean transverse momentum per nucleon projected into the reaction plane, $< p_x/A >$, as a function of the normalized center of mass rapidity, y/y_{proj}, for the third multiplicity bin (MUL3) of Nb + Nb collisions at a bombarding energy of 400 MeV per nucleon. The error bars reflect statistical errors only, and data points are corrected for the deviation from the true reaction plane, as described above. The curve exhibits the typical S-shape behavior demonstrating the dynamical collective momentum transfer between the forward and backward hemispheres.

It is our aim to extract quantitative information, with as little detector bias as possible, from the type of data presented in Fig. 6, thus allowing us to compare different mass systems at different beam energies with each other and with theoretical model calculations. The maximum transverse momentum transfer occurs close to the target and projectile rapidities, where there is great sensitivity to the exclusion of spectator particles and where the experimental biases are most disturbing. Therefore, the maximum value is not a good choice. However, to a good approximation, all curves are straight lines near midrapidity. If the data are plotted as a function of the normalized rapidity, the slope at midrapidity, which we call flow, has the dimensions of MeV/c per nucleon and is a measure of the amount of collective transverse momentum transfer in

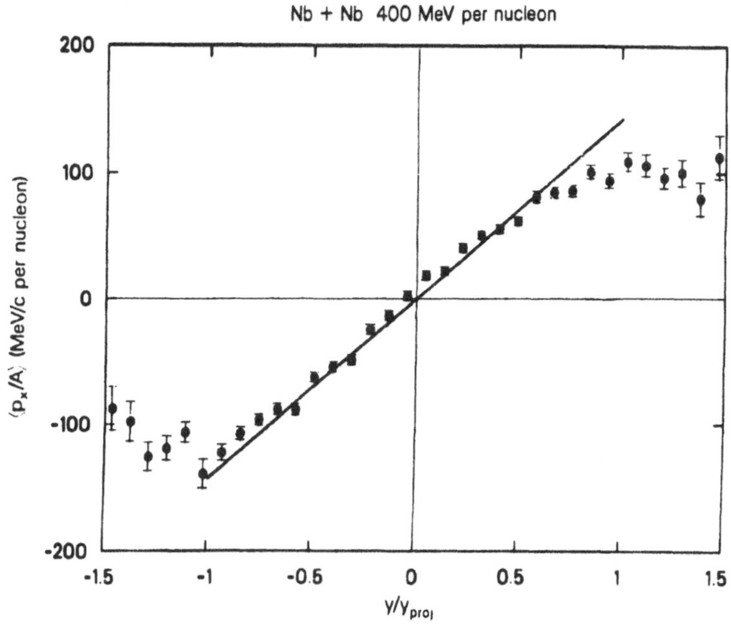

Figure 6. $< p_x/A >$ as a function of the normalized center of mass rapidity for 400 MeV per nucleon Nb + Nb in the third multiplicity bin. The slope of the solid line represents the flow obtained from fitting the data.

the reaction. Since the flow is determined at midrapidity it is a characteristic of the participants. Technically it is obtained by fitting a polynomial with first and third order terms to the S-shaped curve. The fit was done for y/y_{proj} between -1 and 1. Due to detector biases the curve is not completely symmetric about the origin; therefore a second order term has been included in the fit in cases where χ^2 can be improved considerably, as is the case for the higher energies and the heavier mass systems. The coefficient of the first order term, which is the slope of the fitted curve at $y/y_{proj} = 0$, is the flow. The straight line in Fig. 6 shows the result of this fit.

In Fig. 7 the flow, extracted from this kind of fits, is plotted as a function of the multiplicity for the three systems Ca + Ca, Nb + Nb, and Au + Au, all at a beam energy of 400 MeV per nucleon. As already seen from the distributions of the flow angle (see Fig. 4) [22], the amount of flow increases with increasing target-projectile mass. The multiplicity dependence, however, shows the flow peaking at intermediate multiplicity, whereas the mean flow angle increases monotonically with multiplicity (see Fig. 4) [21]. This is because the flow quantity goes to zero at zero impact parameter.

The multiplicity dependence shows a maximum in the directed flow between the third and fourth multiplicity bins. The mean value of the flow in these two multiplicity bins is shown in Fig. 8 as a function of the beam energy for all systems investigated. The flow increases monotonically with increasing beam energy — increasing rather rapidly up to about 400 MeV per nucleon and leveling off at the highest bombarding energies. The error bars represent statistical errors only. If in Fig. 8 the multiplicity averaged flow would be plotted instead of the maximum, then one would find a slight fall-off at beam energies above 650 MeV per nucleon, as reported in Ref. [29]. This difference is mainly due to the increasing contribution of peripheral reactions.

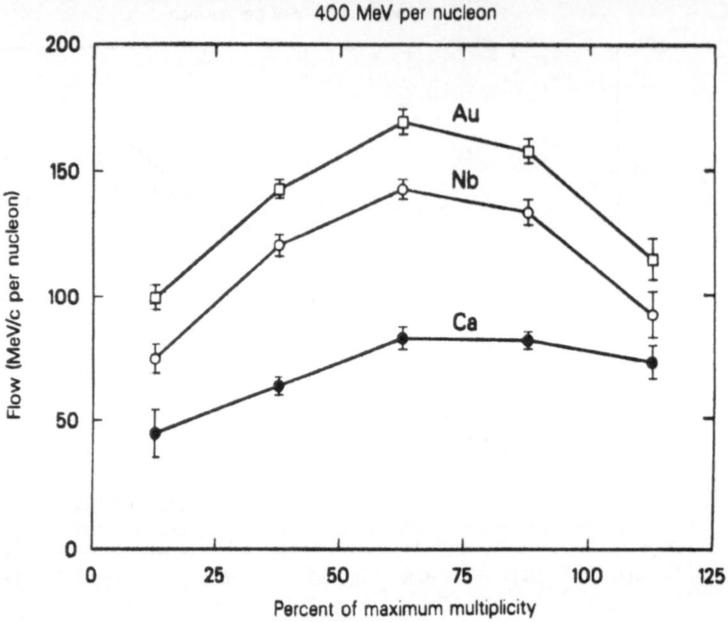

Figure 7. Flow as a function of the normalized participant proton multiplicity (N_p/N_p^{max}) for the three systems measured at a beam energy of 400 MeV per nucleon.

A different approach to analyzing the same data is to investigate multiparticle correlations globally between forward and backward hemispheres [30] in the center of mass system. This method also shows a peaking of the correlation function in semicentral collisions and a maximum value at 650 MeV per nucleon if integrated over multiplicity.

The transverse momentum transfer has been predicted by microscopic theories [31, 32] and by viscous fluid dynamics calculations [33]. Simpler models, like *e.g.* ideal fluid dynamics, cascade models, or fireball models, do not describe the data [34]. The microscopic theories [35] show a dependence of the flow on the nuclear matter equation of state. But the flow effects depend as well on the effective nucleon-nucleon cross sections which are not known. However, it might be possible to determine the effective cross sections from a systematic study of the nuclear stopping power via the dN/dy distributions [36] shown *e.g.* in Fig. 2. Once the cross sections are known, it should be possible to gain information about the equation of state from the flow effects.

TRIPLE DIFFERENTIAL CROSS SECTIONS

Over the last years we have seen a tremendous progress on the part of theoretical predictions. Many new microscopic models, all containing information about the equation of state, have been developed. The numerous contributions to this conference provide an excellent overview of this exciting development. Those models have greatly enhanced predictive power, and more quantitative comparisons between experimental results and these calculations will be needed. The comparisons will be done mainly in terms of triple differential cross sections [11]. A possibility for such a comparison is presented in Fig. 9. Here the yield, not yet the cross section, for proton emission into two differently defined cones is shown as a function of the center of mass energy of the protons for Nb + Nb

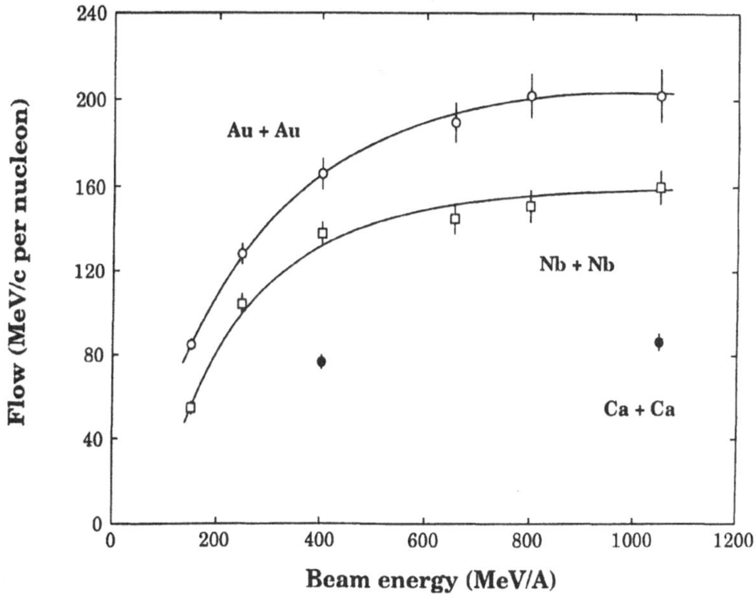

Figure 8. Flow of semicentral collisions ($50\% \leq N_p/N_p^{max} < 100\%$) as a function of beam energy.

at 400 MeV per nucleon. Only events with a flow angle between 35 deg and 55 deg have been selected, and only protons emitted into a cone around the flow axis (filled circles) and into a cone rotated by $\phi = 180$ deg (open circles) have been taken into account. This particular representation has the advantage that the protons emitted into the two cones have the same laboratory angle, thus minimizing acceptance problems. It can be clearly seen that more particles are emitted in the direction of the flow axis and that those protons on average have a higher momentum.

SUMMARY

In summary, the experimental results from the Plastic Ball detector have contributed vastly to the understanding of the reaction mechanism of nuclear collisions at several hundred MeV per nucleon. The discovery of the collective flow phenomena (bounce-off of spectator fragments, side-splash in the reaction plane, and squeeze-out out of the reaction plane described in Ref. [37]), as they were predicted by hydrodynamical models, has lead to the experimental observation of compressed nuclear matter, which is a necessary condition before one can study the equation of state in detail and search for phase transitions at higher energies.

Stimulated in part by the experimental successes, we have seen a tremendous progress on the theoretical side. There are many new microscopic models with greatly enhanced predictive power. In order to discriminate between different models and to extract more precise information about the nuclear equation of state, a comprehensive comparison with a large body of quantitative data will be necessary. It is expected that these data will mainly come from the new 4π detectors presently under construction at the new SIS/ESR accelerator at GSI [38] and at the Bevalac [39]. This will surely lead to a better understanding of the bulk properties of nuclear matter.

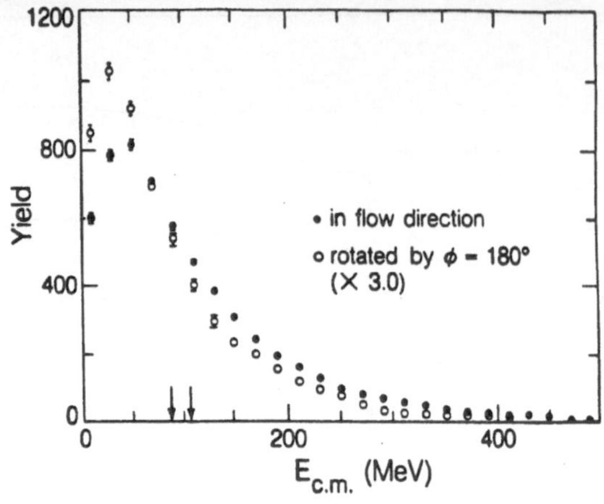

Figure 9. Energy spectra of protons emitted into a cone along the flow axis (filled circles) and into a cone rotated by $\phi = 180$ deg (open circles) for events with flow angles between 35 deg and 55 deg.

Acknowledgement

HGR would like to thank Prof. W. Greiner and Prof. H. Stöcker for the invitation to this exciting and stimulating conference. This work was supported by the Director, Office of Energy Research, Office of High Energy and Nuclear Physics, Division of Nuclear Physics of the U.S. Department of Energy under Contract DE-AC03-76SF00098.

References

[1] A.E. Glassgold, W. Heckrotte, and K.M. Watson, Ann. Phys. (NY) 6, 1 (1959).

[2] G.F. Chapline, M.H. Johnson, E. Teller, and M.S. Weiss, Phys. Rev. D8, 4302 (1973).

[3] M.I. Sobel, P.J. Siemens, J.P. Bondorf, and H.A. Bethe, Nucl. Phys. A251, 502 (1975).

[4] W. Scheid, H. Müller, W. Greiner, Phys. Rev. Lett. 32, 741 (1974).

[5] H. Stöcker, J. Hofmann, J.A. Maruhn, and W. Greiner, Progr. in Part. and Nucl. Phys. Vol.4, 133 (1980).

[6] H. Stöcker and W. Greiner, Phys. Reports 137, 277 (1986).

[7] A.M. Poskanzer, these Proceedings.

[8] A. Baden, H.H. Gutbrod, H. Löhner, M.R. Maier, A.M. Poskanzer, T. Renner, H. Riedesel, H.G. Ritter, H. Spieler, A. Warwick, F. Weik, and H. Wieman, Nucl. Instr. and Meth. 203, 189 (1982).

[9] H. Ströbele, R. Brockmann, J.W. Harris, F. Riess, A. Sandoval, R. Stock, K.L. Wolf, H.G. Pugh, L.S. Schroeder, R.E. Renfordt, K. Tittel, and M. Maier, Phys. Rev. C27, 1349 (1983).

[10] H.Å. Gustafsson, H.H. Gutbrod, B. Kolb, H. Löhner, B. Ludewigt, A.M. Poskanzer, T. Renner, H. Riedesel, H.G. Ritter, A. Warwick, and H. Wieman, Phys. Lett. B142, 141 (1984).

[11] H. Stöcker, L.P. Csernai, G. Graebner, G. Buchwald, H. Kruse, R.Y. Cusson, J.A. Maruhn, and W. Greiner, Phys. Rev. C25, 1873 (1982).

[12] J. Kapusta and D. Strottman, Phys. Lett. B106, 33 (1981).

[13] J. Cugnon, J. Knoll, C. Riedel, and Y. Yariv, Phys. Lett. B109, 167 (1982).
[14] M. Gyulassy, K.A. Frankel, and H. Stöcker, Phys. Lett. B110, 185 (1982).
[15] S. Brandt and H.D. Dahmen, Z. Physik C1, 61 (1979).
[16] S.L. Wu and G. Zoberning, Z. Physik C2, 107 (1979).
[17] J. Cugnon and D. L'Hôte, Nucl. Phys. A447, 27c (1985).
[18] J.J. Molitoris, H. Stöcker, H.Å. Gustafsson, J. Cugnon, D. L'Hôte, Phys. Rev. C33, 867 (1986).
[19] E. Braun and Z. Fraenkel, Phys. Rev. C34, 120 (1986).
[20] P. Danielewicz and M. Gyulassy, Phys. Lett. B129, 283 (1983).
[21] H.Å. Gustafsson, H.H. Gutbrod, B. Kolb, H. Löhner, B. Ludewigt, A.M. Poskanzer, T. Renner, H. Riedesel, H.G. Ritter, A. Warwick, F. Weik, and H. Wieman, Phys. Rev. Lett. 52, 1590 (1984).
[22] H.G. Ritter, K.G.R. Doss, H.Å. Gustafsson, H.H. Gutbrod, K.H. Kampert, B. Kolb, H. Löhner, B. Ludewigt, A.M. Poskanzer, A. Warwick, and H. Wieman, Nucl. Phys. A447, 3c (1985).
[23] J.J. Molitoris, D. Hahn, and H. Stöcker, Nucl. Phys. A447, 13c (1985).
[24] G. Buchwald, G. Graebner, J. Theis, J. Maruhn, W. Greiner, and H. Stöcker, Phys. Rev. Lett. 52, 1594 (1984).
[25] Y. Yariv and Z. Fraenkel, Phys. Rev. C20, 2227 (1979).
[26] H.Å. Gustafsson, H.H. Gutbrod, B. Kolb, H. Löhner, B. Ludewigt, A.M. Poskanzer, T. Renner, H. Riedesel, H.G. Ritter, T. Siemiarczuk, J. Stepaniak, A. Warwick, and H. Wieman, Z. Physik A321, 389 (1985).
[27] P. Danielewicz and G. Odyniec, Phys. Lett. B157, 146 (1985).
[28] D. Beavis, S.Y. Fung, W. Gorn, D. Keane, Y.M. Liu, R.T. Poe, G. VanDalen, and M. Vient, Phys. Rev. Lett. 54, 1652 (1985).
[29] K.G.R. Doss, H.Å. Gustafsson, H.H. Gutbrod, K.H. Kampert, B. Kolb, H. Löhner, B. Ludewigt, A.M. Poskanzer, H.G. Ritter, H.R. Schmidt, and H. Wieman, Phys. Rev. Lett. 57, 302 (1986).
[30] P. Beckmann, K.G.R. Doss, H.Å. Gustafsson, H.H. Gutbrod, K.H. Kampert, B. Kolb, H. Löhner, A.M. Poskanzer, H.G. Ritter, H.R. Schmidt, T. Siemiarczuk, and H. Wieman, Modern Phys. Lett. A2, 169 (1987).
[31] J.J. Molitoris and H. Stöcker, Phys. Lett. B162, 47 (1985).
[32] J.J. Molitoris, H. Stöcker, and B.L. Winer, Phys. Rev. C36, 220 (1987).
[33] G. Peilert, A. Rosenhauer, T. Rentsch, H. Stöcker, J. Aichelin, and W. Greiner, Proceedings of the 8th High Energy Heavy Ion Study, page 43, LBL-24580 (1988).
[34] B. Schürmann and W. Zwermann, Phys. Rev. Lett. 59, 2848 (1987).
[35] G. Peilert, H. Stöcker, W. Greiner, A. Rosenhauer, A. Bohnet, and J. Aichelin, Phys. Rev. C39, 1402 (1989).
[36] M. Berenguer, C. Hartnack, G. Peilert, A. Rosenhauer, W. Schmidt, J. Aichelin, J.A. Maruhn, W. Greiner, and H. Stöcker, Preprint UFTP-228 (1989).
[37] K.H. Kampert et al., these Proceedings.
[38] P. Kienle, these Proceedings.
[39] G. Rai et al., these Proceedings.

Relativistic Heavy Ion Collisions studied with the Streamer Chamber at the BEVALAC

H. Ströbele[1], P. Danielewicz[2]

1) Institut für Kernphysik Universität Frankfurt/Main

2) National Superconducting Cyclotron Laboratory
Michigan State University, East Lansing/USA

I. Introduction

Some of the most interesting subjects of heavy ion research at BEVALAC energies are the bulk properties of nuclear matter. In this context we refer to hot, compressed and excited rather than normal, ground state nuclear matter.

The bulk properties that can include compressibility, viscosity, particle density etc. can vary in a continuous or discontinuous manner with projectile/target mass and bombarding energy. The discontinuities can be due to the predicted pion condensate or the quark-gluon-plasma formed when critical temperature or density are reached. Further the discontinuities can be simply due to the crossing of the thresholds for particle production. The discontinuous and the continuous variation of parameters, such as associated with the hydrodynamic behaviour of nuclear matter, impose different conditions onto the experiments. To establish the appearence of the new matter properties it may suffice to select a certain class of collisions at a given energy and carry out a specific very likely inclusive measurement. The study of decompression flow, however, requires semi-exclusive measurements at different energies with narrow selection of impact parameters.

Three approaches have been used so far to measure this kind of collective effects in large scale experiments at the BEVALAC:

The Nuclear Equation of State, Part A
Edited by W. Greiner and H. Stöcker
Plenum Press, New York

The Plastic Ball[1] detects and identifies by means of several hundred plastic scintillators almost all charged particles emitted in a collision. The multi arm neutron spectrometer[2] concentrates on neutron detection with an additional multiplicity detector used for impact parameter and event plane selection. The Streamer Chamber[3], if operated in a magnetic field, allows to determine momentum and charge of all charged particles emitted in a collision. Hampered by limited particle identification and slow analysis procedures (both of which are recently being improved considerably by CCD-Camera deployment[4]) the Streamer Chamber will soon be superseded by an electronic tracking device (the HISS TPC[5]) which combines its tracking capabilities with the high data rate and particle identification abilitites of a gaseous electronic detector. Nevertheless, the Streamer Chamber is still unique in its capabilities to identify quantitatively negative charged pions and to measure particle momenta over complete kinematic phase space even at the highest BEVALAC energies.

In the next chapter we recall results from the Streamer Chamber on the pion excitation function and their use to narrow the parameters of the nuclear equation of state. In chapter 3 we present new results on flow in asymmetric collisions at high energies (1.8 GeV/nucl.)*. Finally in chapter 4 rapidity distributions of pions, protons and deuterons from central Ar + Pb collisions at 0.8 and 1.8 GeV/nucl. are presented.

II. The Pion Excitation Function

Pion production in nucleus-nucleus collisions at BEVALAC energies proceeds via excitation of baryon (N^*- or Δ-) resonances which decay into nucleons and pions. The number of produced pions increases with beam energy, because more and heavier resonances can be excited the latter having a higher probability to decay into more than one pion.

The pion excitation function gives the functional dependence of the number of produced pions, $\langle \pi \rangle$, per nucleon as function of beam- or c.m.-energy. For N-N collisions it is just the mean number of pions produced in the interaction divided by two (for the two participating nucleons). Fig. 1 shows the pion excitation function for N-N collisions (solid line) as obtained from p-p, n-p, and n-n data[6]. Also given are the results from central Ar + KCl[7] (dots) and La + La[8] (open circles) interactions. It is interesting to note that the absolute value of the difference between NN and AA collisions is relatively small in the whole range of c.m.-energies considered. The discrepancy between the two excitation functions may be attributed to "cascadelike" effects, which are multiple collisions of the nucleons, fermi motion of the nucleons, reabsorbtion of the nucleons and collective effects, which are "in-medium" corrections to the elementary cross sections, modifications of the particle momenta in the mean field, and many body interactions. The influence of the former has been simulated in cascade-type Monte Carlo calculations[9,10] the result of which are shown in Fig. 1 by the dashed line. The remaining difference between nucleus-nucleus data and such a prediction is attributed to collective effects.

At this point we would like to stress that the number of pions in the final state (i.e., the number of observed pions) is realized during the high density phase of the collisions[11]. Thus the number of pions is insensitive to the details of the expansion and decay of the interaction zone and thus a good probe of compressed and hot nuclear matter.

In order to evaluate the influence of the collective effects we change the picture of nucleus collisions from one which is characterized by a superposition of binary interactions to one which describes the interaction zone as a closed system of interacting particles in thermal and chemical equilibrium with bulk properties T (temperature), P (pressure), ρ (density) etc. The initial pion abundance in this system is approximately given by the first generation of nucleon-nucleon collisions which obviously proceed still in the binary collision scheme. Assuming that roughly half of all nucleons are subject to this kind of interaction the preequilibrium pion abundance is already within a factor of two of the final yield in the full energy range (cf. Fig. 1). Thus, the thermal collisions are only adjusting the pion yields rather than producing (all the) pions. Since only the thermal energy is involved in this process, the pion abundance is a thermometer which measures the thermal energy prevailing during the high density phase of the collision.

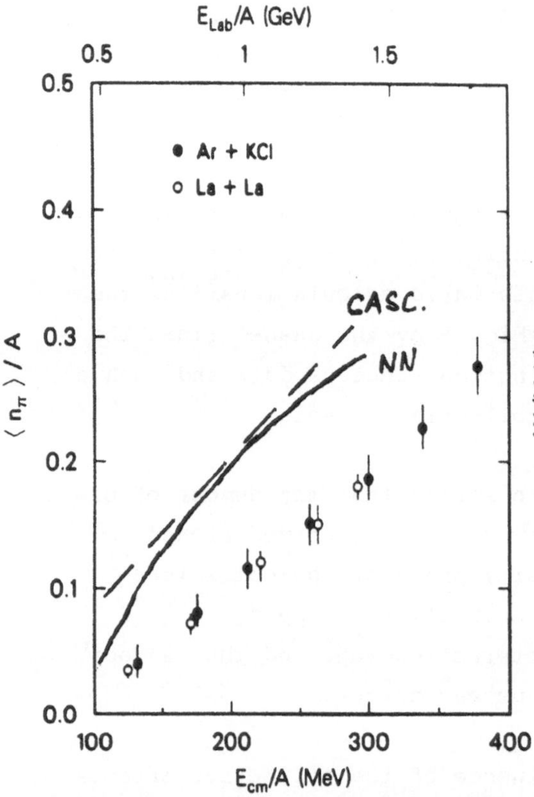

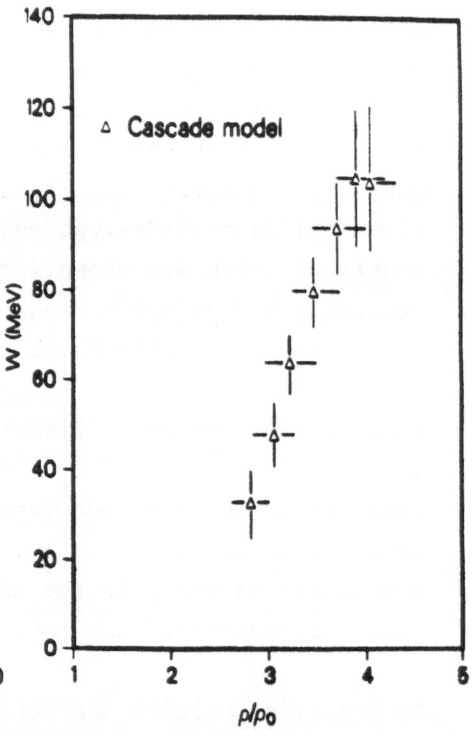

Fig. 1. Pion excitation function for Ar + KCl and La + La collisions. Also shown are the N + N → π + X results (solid line) and the prediction of the cascade model (see text.)

Fig. 2. Values of W (ρ, T=0), the ground state nuclear matter internal energy, plotted as function of the density in units of normal nuclear matter density ρ_0 (see text).

Assuming that the number of pions stays constant during expansion of a nuclear system R. Stock et al.[12,13] attempted to deduce the nuclear equation of state from the discrepancy between the data and the results of cascade calculations. The difference in c.m. energy per nucleon between the data and the calculation giving the same pion multiplicity, was identified with the energy needed to compress the matter to a specific density. The compressional energy missing in the cascade model is typically parametrized in a parabolic form.

$$E_c(\rho) = K/(18\rho_o^2)(\rho - \rho_o)^2 + Wo \qquad (1)$$

The density reached for a given bombarding energy was estimated from the cascade model calculations. A stiff equation of state, i.e. $E_c(\rho)$ rapidly increasing with density, was obtained as indicated in Fig. 2. Its qualitative finding is still compatible with the most commonly used parametrizations of the equation-of-state today.

The approach through which the compressional energy was inferred from the difference in the pion yield between experimental data and cascade-type simulations leads to an estimate of the thermal energy and the potential energy. It has been shown recently that the latter cannot be readily connected to the equation of state[13)], because in medium effects, which are density dependent too, modify the functional dependence of the potential energy with density. Sophisticated microscopoic model calculations based on the Boltzmann equation (VUU[14)], BUU[15)], QMD[16)]) with built-in potential energy have shown that the inclusion of momentum dependent interactions (potentials) reduces significantly the compression constant in (1). We conclude this section by stressing again the two most important and controversial subjects in this context: the appropriate correspondance between bulk properties of the nuclear medium and the parameters in microscopic models as well as the treatment of the compressibility at finite temperature; both questions might get promoted by further experimental results on triple differential cross sections from collisions of heavy systems at high bombarding energy.

III. Flow in the Asymmetric System Ar + Pb

The Streamer Chamber facility at the BEVALAC has been used to study Ar + Pb collisions at different energies. Data at 400 MeV/nucl.[2)] and 800 MeV/nucl.[17)] are available. Here we present preliminary results of a run at 1.8 GeV/nucl. They are derived from a sample of 926 central collision events which correspond to a cross section of ~ 0.9 barn or an impact parameter range of 0 - 5 fm. The events have been measured and analysed in the same way as the 800 MeV/nucl. data[17]. However, the higher energy allows separation of p and d over comparatively larger areas of phase space. In fact protons and deuteron are identified for rapiditites above y ~ 0.5 on the basis of ionization measurements and kinematical cuts. In case of ambiguities a decision is made on a statistical basis. As a result the deuterons are essentially bias free for $p_{lab} > 1000$ MeV/c whereas the protons can be contaminated by 20% deuterons from lower rapidities, their number being correct within a systematic error of 7%. Baryon masses below $p_{lab} = 1000$ MeV/c are assigned on a statistical basis only. In this rapidity range the deuteron to proton ratio was chosen to be $d/p = 0.5$. Positively charged pions were identified up to laboratory momenta of 350 MeV/c. The bias due to pions above this value and false identification below is negligible for protons and deuterons.

The data from 800 MeV/nucl. Ar + Pb collisions have been re-analysed in order to obtain consistency with the analysis at the higher energy. This implied a statistical assignment of protons and deuterons for laboratory momenta below 1000 MeV/c. For rapidities above y ~ 0.5 the deuterons are still bias free whereas the protons although showing the right yield are contaminated by up to 20% deuterons from lower rapidities. Protons and deuterons below $p_{lab} = 1000$ MeV/c are also identified on a statistical basis only with a d/p ratio of 0.5. This procedure causes the following bias: assuming that a particle has a certain in-plane transverse momentum p_x the magnitude and sign of which depend on mass and rapidity. If the individual particle mass assignment is false, p_x is wrong in magnitude and perhaps even in its sign; this effectively causes a reduction of the apparent $\langle p_x \rangle$. This effect will be most pronounced in the backward hemisphere ($y \leq 0.5$) of the c.m.s. and present also for protons with $y > 0.5$. In the latter case the magnitude of the bias can be estimated from Fig. 3 which shows the $\langle p_x/A \rangle$ distributions of protons and deuterons at both energies. Protons with $0.5 < y < 1.0$ are contaminated by deuterons originating from the range $0.25 < y < 0.5$. At both energies these deuterons have a net

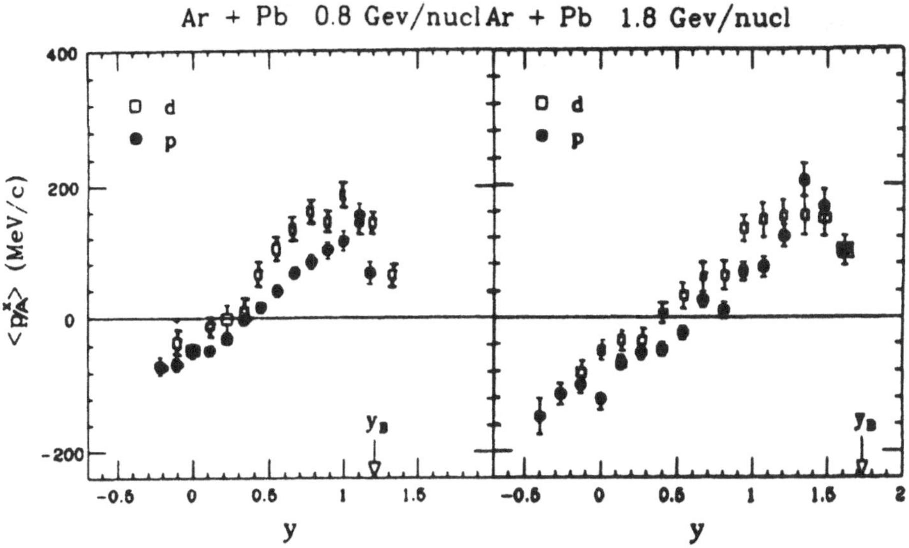

Fig. 3. Average inplane transverse momentum as function of rapidity for protons and deuterons for Ar + Pb collisions at 800 MeV/nucl. (left) and 1800 MeV/nucl. (right).

p_x near to zero; since their fraction in the proton sample ($0.5 < y < 1.0$) is 20%, they will cause a reduction of $<p_x/A>$ by 20% too. In the following we will, therefore, consider p_x/A of all protons including those bound in deuterons. Its systemaic error is estimated to be smaller than 15% based on a d/p ratio of 0.5.

So far no attempt was made to identify particles with $A > 2$. Tritons and ^3He will contaminate the deuteron sample, however, distortions in rapidity and p_x caused by both particle types are approximately the same but opposite in sign. Overall a smearing effect in y and a slight systematic reduction of $<p_x/A>$ is expected which is estimated to be of the 5% level.

^4He particles (~ 5%) are contained in the deuteron sample too. They are treated correctly because they have the same Z/A ratio as deuterons. Heavier fragments can be completey neglected above rapidities of ~ 0.4, because their abundance is below the percent level in central collision events and outside of the target fragmentation region.

The systematics of flow effects in Ar + Pb collisions at 0.8 and 1.8 GeV/nucl. are studied with the transverse momentum method[18]. In order to allow an appropriate comparison of the data at different beam energies the rapidity as measured in the laboratory frame and normalized to beam rapidity was chosen to represent the longitudinal degree of freedom in momentum space. Fig. 4 and 5 show the distribution of $\langle p_x/A \rangle$ and dp_x/dy as function of $y' = y/y_{beam}$ for the experimental

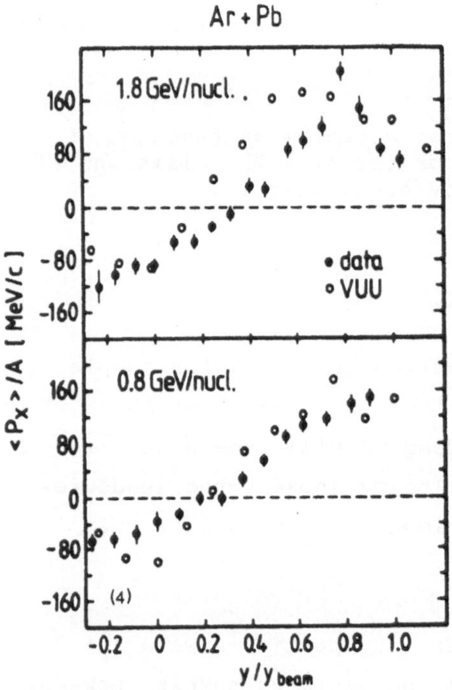

Fig. 4. Average inplane transverse momentum per nucleon of protons and deuterons as function of normalized rapidity.

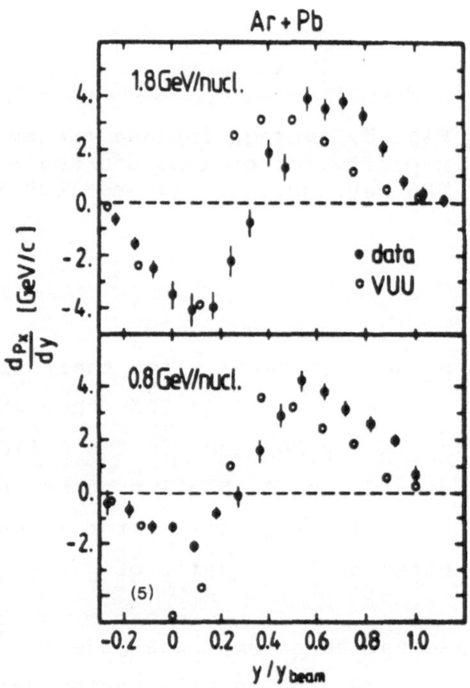

Fig. 5. Total inplane transverse momentum of protons (also those protons bound in deuterons) as function of normalized rapidity.

results (dots) and for the VUU predictions (circles)[19]. Statistical errors only are indicated. The data points below $y' = 0.3$ have been plotted for completeness only. Here the systematical errors due to missing particle identification are large. The following differences between the $\langle p_x \rangle$ distributions at the two energies stand out:

(i) The intercept y_o $(= y'(\langle p_x \rangle = 0))$ is shiftet to higher rapidities at the higher beam energy (from $y_o = 0.25$ at 0.8 to $y_o = 0.35$ at 1.8 GeV/nucl.). This indicates a weaker stopping at the higher energy.

(ii) The maximum value of $\langle p_x \rangle$ increases slightly with beam energy (from 150 MeV/c to 170 MeV/c) which means that a sort of saturation seems to occur above 800 MeV/nucl.

(iii) The position of the maximum of $\langle p_x \rangle$ is shifted to lower y' with increasing energy contrary to naive expectations. In fact even a peak structur becomes apparent at the higher energy as the result of a simultaneous shift of the intercept (see (i)) to the right and a reduced $\langle p_x \rangle$ at high rapidities.

The VUU model does not reproduce the shift of the intercept (i), but predicts correctly the moderate rise of the maximum value(ii). Its position, consistent at 800 MeV/nucl., is shifted to lower rapidities than observed at 1.8 GeV/nucl. It seems that the model overestimates the stopping power of nuclear matter at the higher energy. This may be due to relativistic effects not included in the model.

The differential distribution of the in-plane transverse momentum in protons and deuterons dp_x/dy is shown in Fig. 5. It is essentially the result of folding appropriately Fig. 4 with the rapidity distributions of protons and deuterons (see next section). The normalization is in units of y', i.e. the integral (from the intercept upwards) has to be multiplied by the bin size (0.11 at 800 MeV/nucl. and 0.1333 at 1800 MeV/nucl.) to obtain the total momentum transfer into the transverse direction to all the protons.

The observation (i) is consistently reproduced. However, the maximum value of dp_x/dy does not change significantly with energy and its position shifts to higher rapidities with increasing beam energy in contrast to (iii) above. The obvious explanation of this discrepancy is found in the differences of the rapidity distributions: at the higher energy fast particles are stopped more efficiently (see next section). This finding is qualitatively reproduced in the VUU model (open circles in Fig. 5). At 0.8 GeV/nucl. the $<p_x/A>$, which covers only free nucleons and no fragments, is well reproduced whereas at the higher energy the intercept as well as the distribution as a whole is shifted to lower rapidities. These differences become more pronounced in the dp_x/dy distributions. Here the normalization was obtained by making the integrals above y_o of the model prediction equal to the corresponding integral in the data ($\int(1.8$ GeV/nucl.) = 1.7 GeV/c und $\int(0.8$ GeV/nucl.) = 1.6 GeV/c). It is clearly seen that the model misses transverse momentum when approaching beam rapidity. This may be due to the missing fragments (deuterons) in the model predictions.

The comparison of the data and VUU (or QMD and BUU) model calculations with appropriate light fragment formation included will provide stringent constraints on several ill defined parameters in the models.

Fig. 6 summarizes this section. The in-plane transverse momentum averaged over the forward c.m. hemisphere ($y' > y_o$) $<p_x>$ is plotted as function of beam energy for Ar + KCl, Ar + Pb and Ar + BaI_2 central collisions. There is a clear

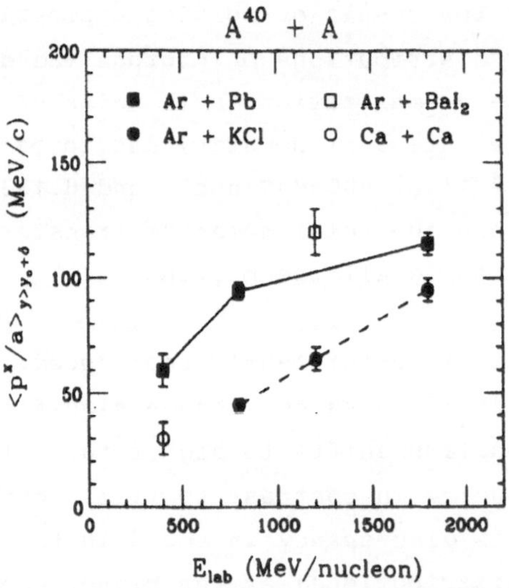

Fig. 6. Average inplane transverse momentum integrated over rapidities above the intercept (cf. Fig. 4,5) as function of beam energy.

difference between the symmetric and asymmetric projectile/ target systems. In the latter configuration a sort of saturation occurs which stimulates the question wether ~ 100 MeV/c is the limiting value for $<p_x>$ or wether it will continue to rise if symmetric systems are studied at higher energies. The Plastic Ball results[21)] are consistent with $<p_x> \lesssim 100$ MeV/c as do Dubna data on C + C collisions at 4.5 GeV/nucl.[21)].

IV. Rapidity distributions

Based on the identification procedures described in the previous section and in reference 18 the rapidity distributions of different particle species have been determined at 800 MeV/nucl. and 1800 Mev/nucl. In order to allow for a direct comparison of their shapes the rapidity has been calculated in the laboratory frame and normalized to beam rapdity ($y' = y/y$ beam). The normalization of all distributions, however, is still per unit of rapidity (per event); this means that before comparing yields at the two energies a normalization factor of 1.385 has to be applied. The dip at $y' = 0$ seen in all distributions below is an artefact of particle absorption in the target.

Positively charged pions are identified below plab = 350 MeV/c. A comparison with π^- is done by applying this same cut; the result is shown in Fig. 7. At both energies the shapes of the π^+ and π^- rapidity distributions are very similar. At 800 MeV/nucl. the yield of π^+ (Fig. 7b) is found to be 2/3 of the yield of π^- (Fig. 7a); at the higher energy this ratio is ~ 0.8 (cf. Fig. 7 c,d). In the latter case the π^+ sample shows a slight enhancement around $y' = 0.75$ as compared to π^-. Such an effect is not seen at the lower energy where the possible bias due to a significant proton background would be more likely, due to the large p/π^+ ratio. Therefore we consider this finding to be significant. It should be noted that the number of π^+ per unit rapidity above $y' = 0.6$ is roughly the same as the corresponding π^- multiplicity. This observation is in line with the projectile having a p/n ratio close to unity. At the lower beam energy the strongly reduced number of detected pions precludes such a finding.

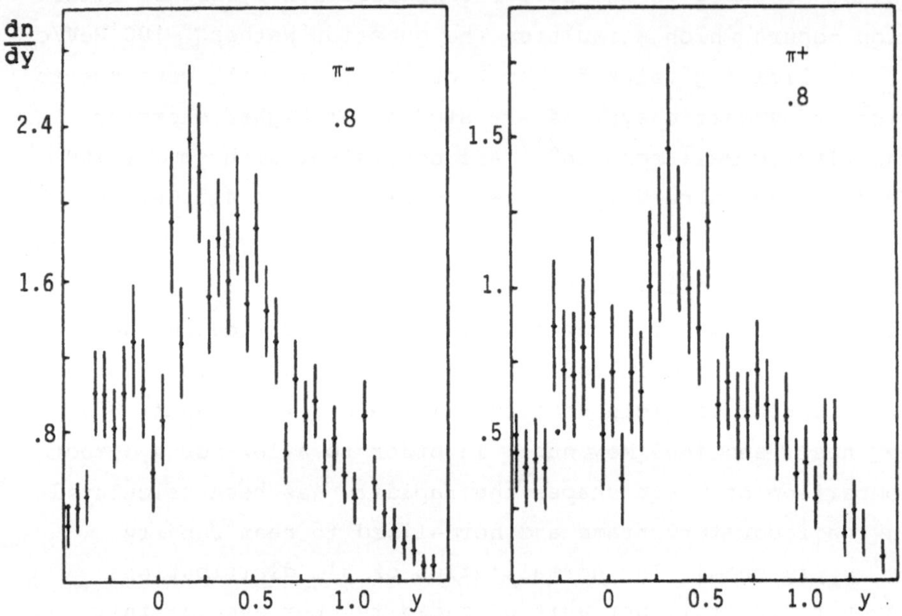

Fig. 7, a,b. Rapidity distribution for pions with p_{lab} < 350 MeV/c for Ar + Pb collisions at 800 MeV/nucl. The vertical scale is number of particles per unit rapidity per event. The abszissa is in units of normalized rapidity $y' = y/y_{beam}$.

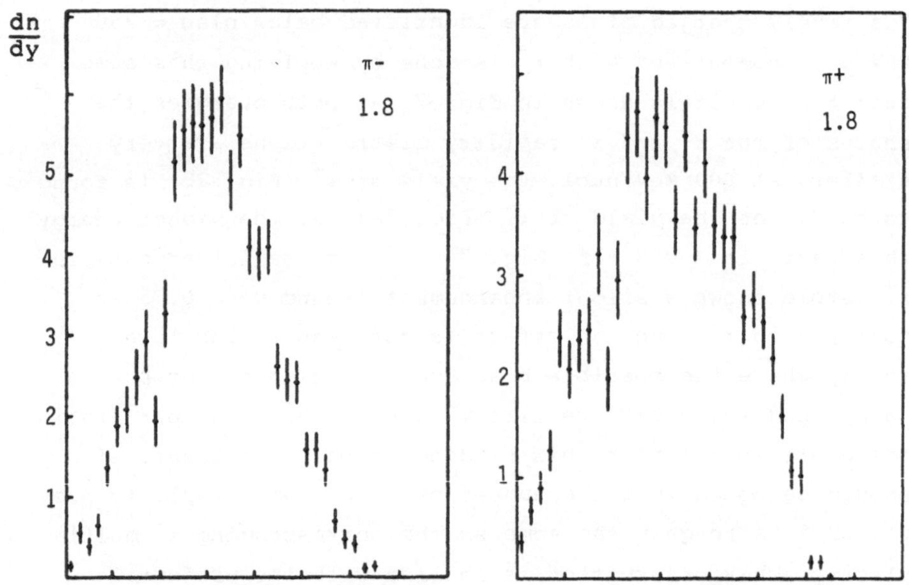

Fig. 7, c,d. Same as Fig. 7, a,b at 1.800 MeV/nucl. beam energy

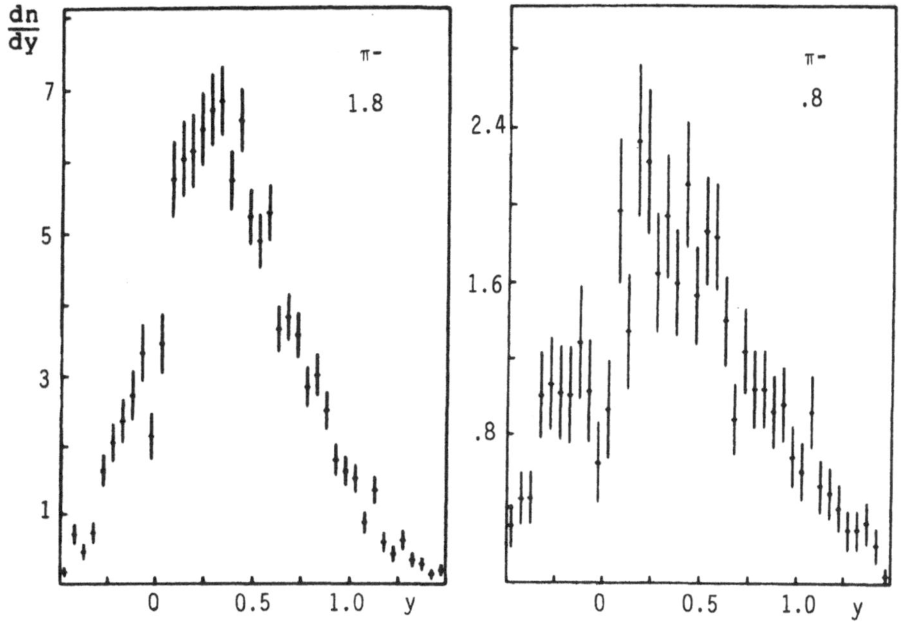

Fig. 8. Rapidity distribution of π^- at 1.8 and 0.8 GeV/nucl. For normalization see Fig. 7

In Fig. 8 the unbiased rapidity distributions of π^- at both energies are compared. The shapes are again very similar. The tails of the distributions extend well beyond projectile rapidity. This effect is not observed in the baryon spectra and can be attributed to the fact that most of the pions stem from Δ-resonance decays; the decay pions from projectile-like Δ's get boosted well beyond beam rapidity. When comparing the total pion yield from Ar + Pb collisions at 0.8 and 1.8 GeV/nucl. an increase by a factor of 3.8 is observed. This factor is slightly higher than the measured ratio of 3.5 observed for the symmetric system Ar + KCl[7)].

The baryon rapidity spectra are studied for two multiplicity selections. The total sample of events has been split by cutting at the mean multiplicity of charged particles. "HIM" ("LOM") denotes the event sample with $m_{ch} \geq \langle m_{ch} \rangle$ ($m_{ch} < \langle m_{ch} \rangle$). We first compare the proton rapidity distributions for HIM and LOM events at 0.8 GeV/nucl. (Fig. 9 a,b) and at 1.8 GeV/nucl. (Fig. 10 a,b).

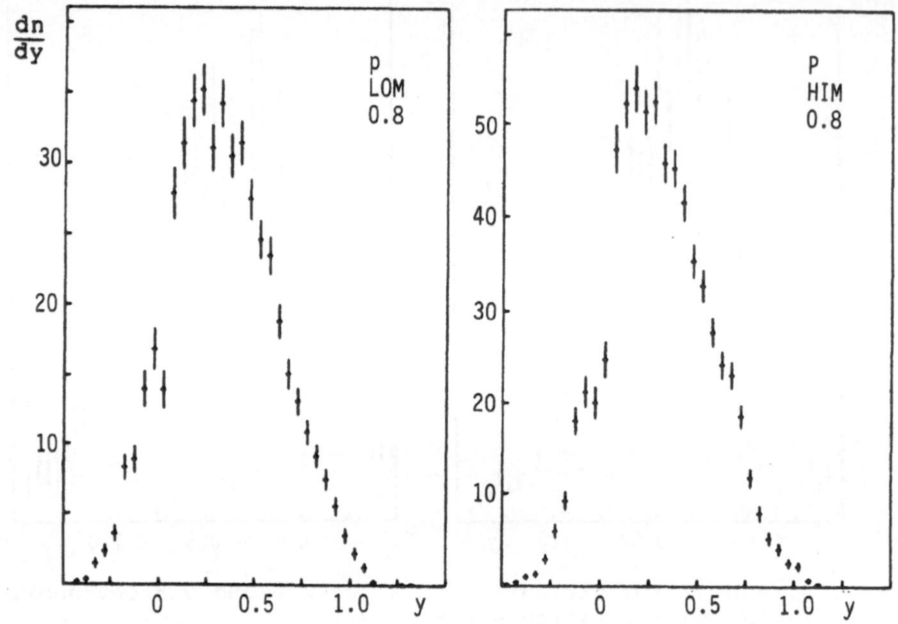

Fig. 9. Rapidity distribution of protons at 0.8 GeV/nucl. for events having low (high) multiplicity of observed charged particles LOM $\hat{=}$ m_{ch} < $\langle m_{ch} \rangle$, (HIM $\hat{=}$ m_{ch} > $\langle m_{ch} \rangle$). For normalization see Fig. 7.

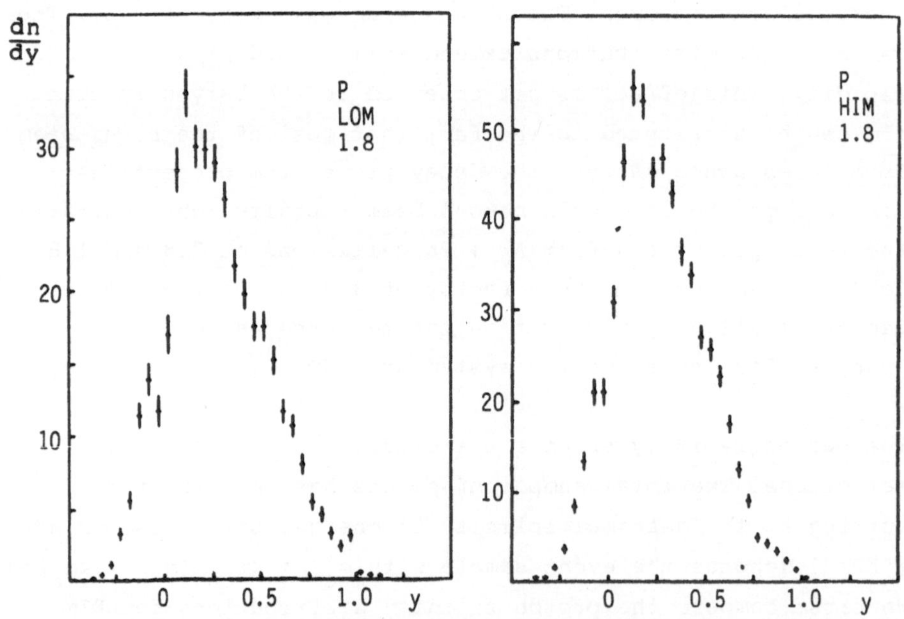

Fig. 10. Same as Fig. 9 at 1.8 GeV/nucl. beam energy

Their widths are consistently reduced by 10% from the lower to the higher energy. Otherwise the shapes are rather similar. Going into finer details we observe an excess of protons above y' = 0.4, if the LOM sample is compared to the HIM sample. This effect is strongly amplified in the deuteron spectra at 0.8 GeV/nucl. (Fig. 11 a,b) and also at 1.8 GeV/nucl. (Fig. 12 a,b). A two peak structure becomes evident which is more pronounced in the lower multiplicity sample. The dip between the two peaks coincides with the intercept at $y' = Y_o$ observed in the p_x-distributions (see the previous section).

By systematic studies of the particle identification procedures we have excluded that the dip in the deuteron rapidity distributions is due to a bias in the analysis. In fact the triple differential cross sections show that the dip is caused by reduced population of deuterons at low transverse momenta around y' = 0.3.

Turning now to the comparison of the spectra at the different energies (and taking into account the scale factor of 1.385 by which the yields at 1.8 GeV/nucl. have to be multiplied) we find that the increase in energy does not change significantly neither the shape nor the yield forward from y' = 0.3. It seems that increasing the energy causes an increase in multiplicity of baryons in the backward hemisphere of the c.m.s. only. Thus the additional longitudinal energy is rather dissipated in the target than used for acceleration of the "fireball". The yield of deuterons backward from y' = 0.3 is not a well defined observable in this experiment, since it is determined entirely by the chosen d/p ratio for laboratory momenta below 1 GeV/c. The corresponding bias in the proton spectra is much less severe. Their shapes are not affected except for the small range of rapidities above $y' \sim 0.75$; below only the yield is sensitive; making the extreme assumption that no deuterons at all are present in the corresponding laboratory momentum range would increase the proton yield by 25% only.

Analysis of rapidity distributions from symmetric systems like Ar + KCl[22)], Nb + Nb and Au + Au[23] at 400 MeV/nucl. are underway. Together with the results presented here for the asymmetric system Ar + Pb they provide for the first time results

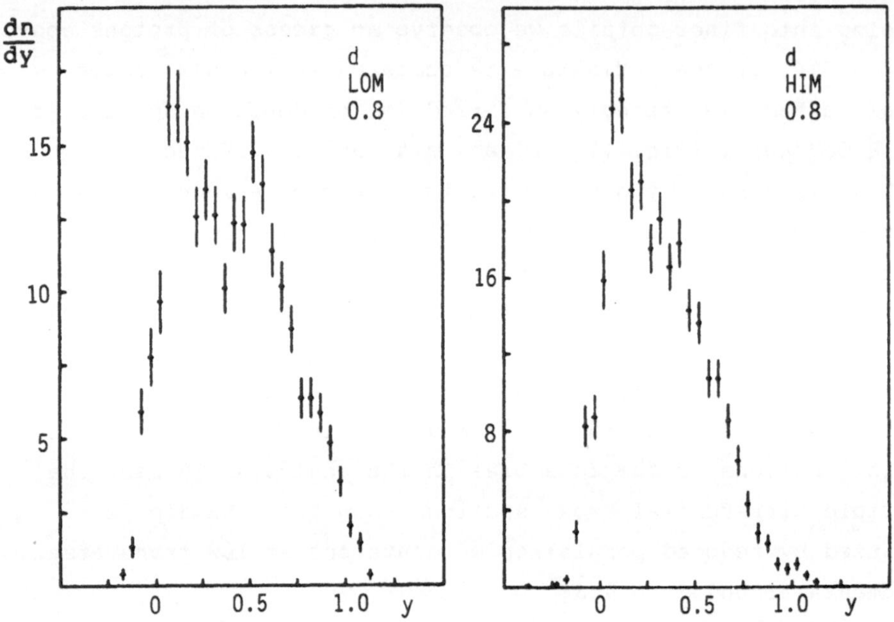

Fig. 11. Rapidity distribution of deuterons at 0.8 GeV/nucl. for events having low (high) multiplicity of observed charged particles LOM $\hat{=}$ $m_{ch} < \langle m_{ch} \rangle$, (HIM $\hat{=}$ $m_{ch} > \langle m_{ch} \rangle$). For normalization see Fig. 7.

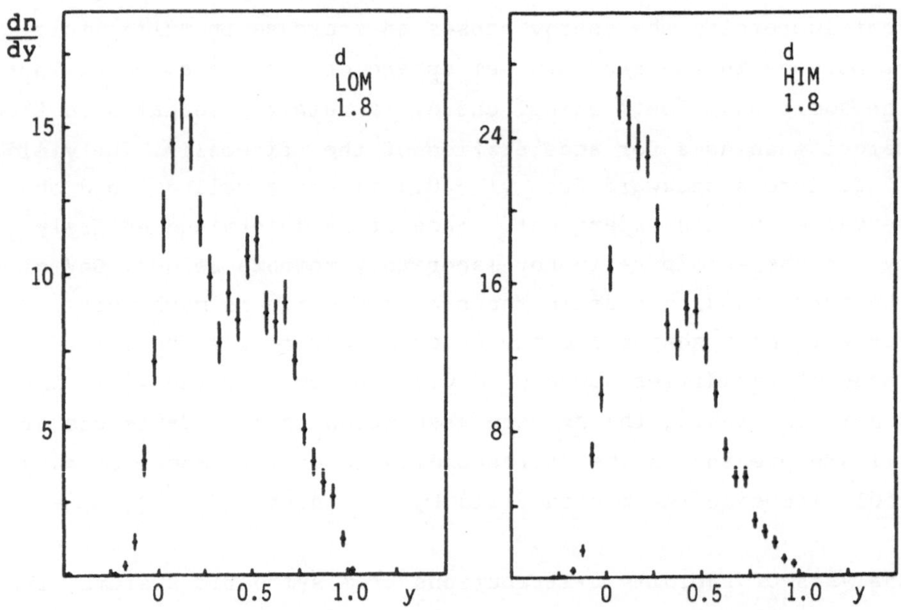

Fig. 12. Same as Fig. 11 at 1.8 GeV/nucl. beam energy.

for both the longitudinal and transverse degree of freedom of the same event samples. This will help to determine simultaneously the stopping power and compressional energy of nuclear matter under extreme conditions.

*The Streamer Chamber pictures of Ar + Pb collisions at 1.8 GeV/nucl. had originally been taken by the Riverside group for the study of pion production. Later they were used for semi-exclusive measurements in Heidelberg. In a joint effort we are currently preparing the publication of these results together with the corresponding data at 0.4 and 0.8 GeV/nucl.

References

1. A. Baden et al., NIM 203 (1982) 189
2. D. Keane et al., Proc. Winter School on Nucl. Phys., Les Houches, France, 1989, to appear in Nucl. Matter and Heavy Ion Coll., ed. M. Soyeur, Plenum N.Y.
3. U. v. Bibber, A. Sandoval, Streamer Chamber for Heavy Ions, Treatise on Heavy-Ion Science, Vol. 7, D.A. Bromley (Ed.), Plenum Press N.Y., p. 333
4. M. Tincknell et al., Optical Engeneering 26(10) (1987) 1067
5. H.G. Ritter, these proceedings
6. Particle Data Group, UCRL-20000 NN, August 1970 and JINR, P1-82-507, Dubna 1982 (GSI-tr-86-01)
7. A. Sandoval et al., Phys. Rev. Lett. 45 (1980) 874
8. J.W. Harris, Phys. Rev. Lett. 58 (1987) 463
9. J. Cugnon et al., N.P. A352 (1981) 505
10. Y. Yariv and Z. Fraenkel, Phys. Rev. C20 (1979) 2227
11. R. Stock et al., Phys. Rev. Lett. 49 (1982) 1236 and J.J. Molitoris et al., Phys. Rev. C 36 (1987) 220
12. J.W. Harris et al., Phys. Lett. 153B (1985) 377
13. G.F. Bertsch et al., Phys. Rev. C29 (1984) 673 J. Aichelin et al., Phys. Rev. Lett. 58 (1987) 1926
14. H. Kruse et al., Phys. Rev. C31 (1985) 1770
15. J. Aichelin et al., Phys. Rev. Lett. 58 (1987) 1926
16. G. Peilert et al., Phys. Rev. C39 (1989) 1402
17. P. Danielewicz et al., Phys. Rev. C38 (1988) 120
18. P. Danielewicz and G. Odyniec, Phys. Lett. 157B (1985) 146 Lawrence Berkeley Lab. Report LBL-18600, 1984
19. C. Hartnack, these proceedings
20. see e.g. H.H. Gutbrod et al., GSI-89-36
21. H. Bialkowska et al., Phys. Lett. B 173 (1986) 349
22. K.H. Kampert, Ph. D. Thesis, Univ. Münster 1986 and M. Berenguer et al., UFTP preprint 228/1989 to appear in Nucl. Matter and Heavy Ion Coll., ed. M. Soyeur, Plenum N.Y.
23. D. Keane et al., Phys. Rev. C19 and Proc. of the 8th High Energy Heavy Ion Study, LBL, Berkeley (1987)

Fragment Flow and Squeeze-Out of Hot Dense Nuclear Matter

K.H. Kampert,[a] H.H. Gutbrod,[b] J.W. Harris,[c] B.V. Jacak [d]
B. Kolb,[b] A.M. Poskanzer,[c] H.G. Ritter,[c] R. Schicker [c]
H.R. Schmidt,[b] and T. Siemiarczuk[b,†]

[a] *University of Münster, D-4400 Münster, West Germany*
[b] *Gesellschaft für Schwerionenforschung, D-6900 Darmstadt, West Germany*
[c] *Lawrence Berkeley Laboratory, University of California, Berkeley, CA 94720, USA*
[d] *Los Alamos National Laboratory, Los Alamos, NM 87545, USA*

Abstract

Experimental results are presented on the production and collective flow patterns of light nuclei ($Z = 1,2$) and intermediate mass nuclear fragments ($3 \leq Z \leq 10$) over a large solid angle. The reaction Au + Au at 200 AMeV is studied to provide information on creation and on collective sidewards flow of intermediate mass fragments into the reaction plane. Squeeze-out of hot participating baryons – observed perpendicular to the reaction plane – is in addition investigated for 400 AMeV Ca + Ca, Nb + Nb, and Au + Au collisions. Both components of the collective flow provide important information on the equation of state and viscosity of hot nuclear matter.

1 Introduction

The key mechanisms for producing hot dense nuclear matter in high-energy heavy-ion collisions were formulated first more than a decade ago in a series of papers by Chapline, Johnson, Teller, and Weiss [1], Scheid, Müller, and Greiner [2], and later also by Sobel, Siemens, Bondorf, and Bethe [3]. Several of these authors are being present again on this school. At that time only very little was known about the properties of hadronic matter at finite temperatures and densities other than the nuclear ground state density $\varrho_0 = 0.16$ fm^{-3}. Scheid *et al.* argued that high density nuclear shock waves should occur in violent head-on collisions of heavy nuclei leading to a preferential collective sidewards emission of the compressed hot matter. Particularly these suggestions set the stage for an investigation of the nuclear equation of state (EOS) by performing heavy-ion collisions at sufficiently high kinetic energies.

At laboratory projectile energies of about 1 GeV per nucleon the speed of interprenetation of the two colliding nuclei exceeds the sound velocity of ordinary nuclear matter ($v_s \cong 0.25\,c$), i.e. the penetration velocity is higher than the velocity of information propagation inside the medium. Hence, if nuclei are able stop each other,

The Nuclear Equation of State, Part A
Edited by W. Greiner and H. Stöcker
Plenum Press, New York

one encounters a typical shock scenario where the participant matter cannot escape rapidly enough from the interaction zone, resulting in a pile-up, i.e. in a travelling shock front of highly compressed nuclear matter. The goal as formulated in these early papers noted above, was to learn about the mechanism for high compression ($\varrho/\varrho_0 \geq 2$) and heating of nuclear matter, and if ultimately possible, also about the nuclear matter EOS, i.e. the response of the nuclear medium to these extreme conditions of temperature, energy- and baryon density.

Probing the nuclear matter EOS in regions of high densities and temperatures ($\varrho/\varrho_0 \simeq 2$–$5$, $T \simeq 50$–$100\,\mathrm{MeV}$) is of fundamental importance not only in nuclear physics (nuclear viscosity, heat conductivity, possible phase transitions such as liquid-vapour, pion condensation, Δ-isomers, etc.) and field theory (QCD phase transition to a Quark-Gluon-Plasma), but is also a basic prerequisite for an understanding of many astrophysical problems. For example, the dynamics of the early universe during the first fractions of a second after the 'Big Bang', as well as the dynamics of supernovae (SN) explosions, such as the recent SN1987A and the structure and stability neutron stars depend strongly on the compressibility of dense nuclear matter [4,5] over wide regions of densities, temperatures, and Z/A ratios (see also specific talks to these topics presented on this school). Heavy-ion collisions are the first opportunity to study these phenomena under controlled conditions in the laboratory. Important information would be accessible from these kind of experiments, if one could apply the knowledge about the conditions achieved for very short time spans of $t \cong 10^{-22}$ seconds, to the much larger space- and time dimensions relevant in astrophysical events.

In this contribution we shall report on two fundamental phenomena which are relevant for an understanding of the bulk properties of nuclear matter under extreme conditions; this is i) the study of fragment formation and its related question of multifragmentation, and ii) a detailed investigation of the full 3-dimensional event shape in the final state of a collision, which will manifest strong collective jet-like patterns in the particle emission, referred to in the following as 'in-plane flow' and 'squeeze-out', respectively.

2 Experiment

The experiments were performed at the Berkeley Bevalac, employing the Plastic Ball/Wall spectrometer system [6]. Some details of the detector were described in the lecture by H.G. Ritter. In order to measure intermediate mass fragments ($2 < Z \leq 10$) over a large solid angle in addition to the proton and helium isotopes in reactions of 200 AMeV Au + Au, computer-controlled high voltage supplies were implemented in one of the last experiments on the 160 Ball modules covering $\vartheta_{\mathrm{lab}} \leq 30°$. This allowed careful on-line gain matching and extension of the energy loss spectra up to neon by simultaneously maintaining mass separation of the lightest fragments. A typical *on-line* particle identification spectrum of the 160 modules (before applying any final off-line gain adjustments and scattering-out corrections) is depicted in figure 1 with an insert showing the hydrogen isotope separation. In order to be identified, fragments must traverse the 4 mm thick CaF_2 scintillator. This produces a low energy cut-off in the laboratory of about $E/A \cong 35$–$40\,\mathrm{MeV}$. Since a projectile energy of 200 AMeV in a symmetric system corresponds to a CM velocity of about 50 AMeV in the laboratory, the low energy cut-off is unimportant in the forward direction of the CM system

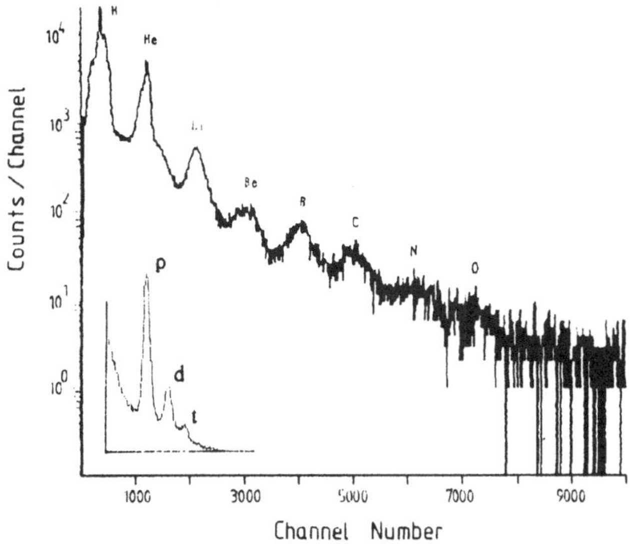

Figure 1. On-line particle identification spectrum of 200 AMeV Au + Au collisions with hydrogen isotope identification insert for 160 Ball modules in the angular region $10° \leq \vartheta_{lab} \leq 30°$.

($\vartheta_{lab} \leq 30°$). A zero degree gas proportional chamber [7,8] was employed in addition to the scintillation counters covering $\vartheta_{lab} \leq 2°$, and providing extremely high position resolution for heavy projectile remnants.

3 Fragment Formation

The study of fragment production in intermediate energy nucleus-nucleus collisions has become interesting particularly with the capability of present and future accelerators to study systems using very heavy nuclear beams. The relative abundances of light and intermediate mass fragments over protons are believed to contain information about the entropy produced in the system [9,10,11]. If the entropy stays constant during the expansion [12,13,14], composite particles contain information about the highly excited and compressed stage of the reaction, so that determination of the entropy might help to determine the EOS of hot dense nuclear matter. In fact, an analysis of composite particle ($2 \leq A \leq 4$) over proton ratios, measured as a function of the impact parameter in exclusive 4π data [15,16], has demonstrated that the pure fireball model [17], where all available kinetic CM energy is converted into heat, produces significantly too much entropy compared to data, indicating that compression effects play an important role in heavy-ion reactions.

Here, we will now discuss the event-by-event production of heavier fragments up to $Z = 10$. Such an investigation is driven, for example, by questions of how much excitation energy nuclei can support before they break apart, and how the disassembly of highly excited systems takes place. The widely differing mechanisms proposed in various models include fragment formation via purely statistical processes [18,19,20], emission from a gas of nucleons and fragments in equilibrium [21,22,23,10], breakup as a result of dynamic instabilities or partial equilibrium processes [24], or the treatment of the expansion on a microscopic level in form of TDHF methods [25], or molecular

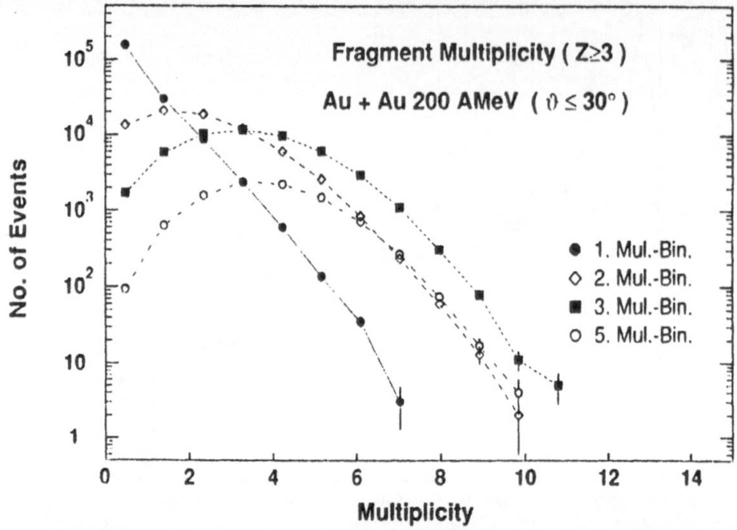

Figure 2. Fragment ($Z \geq 3$) multiplicity distributions for 200 AMeV Au + Au for four multiplicity bins of N_p, increasing from peripheral to central collisions. The numbers correspond to fragments observed in the forward part of CM-hemisphere only [30].

dynamics [26]. Another group of models, finally, assumes neither global nor local equilibrium but rather treats fragment formation by methods of percolation theory or assumes a cold breakup similar to the shattering of glass [27]. Furthermore, at intermediate bombarding energies of $E_{\rm lab} \cong 100$ AMeV, i.e. for moderate temperatures of the system, the expansion may lead to lower densities than the ground state density and possibly to a liquid-gas phase-transition [28,29]. The study of fragment yields thus offers the possibility to investigate the EOS at higher temperatures and lower densities than the ground state.

Experimentally, a total breakup of the system into a large number of nuclear fragments, known as multi-fragmentation, was first observed in proton induced reactions at bombarding energies of $E_p \geq 1$ GeV [31,32]. In contrast to sequential evaporation processes, multi-fragmentation may be characterized by nearly simultaneous breakup of a nucleus into several pieces. Until recently, heavy-ion experiments identifying IMF have only studied single particle inclusive measurements or few particle correlations [33,34,35]. Furthermore, various theoretical investigations have been able to reproduce qualitatively the fragment mass distributions, and in particular the observed power-law behaviour of the mass yield curve, $\sigma(A) \sim A^{-\tau}$, which was often interpreted as an indication of a liquid-vapour phase transition. However, as recently argued e.g. by Aichelin *et al.* [36], the inclusive mass-yield power-law is purely accidental as it basically originates from a mixing of different impact parameter dependent processes and, therefore, cannot be considered a unique candidate for revealing a possible phase-transition.

Results of IMF production as a function of impact parameter are shown in figure 2. Plotted are multiplicity distributions of fragments with $3 \leq Z \leq 10$, observed in the forward hemisphere of the CM frame of Au + Au at 200 AMeV. Events are divided into five bins (of which four are shown) of participant proton multiplicity [15], spanning

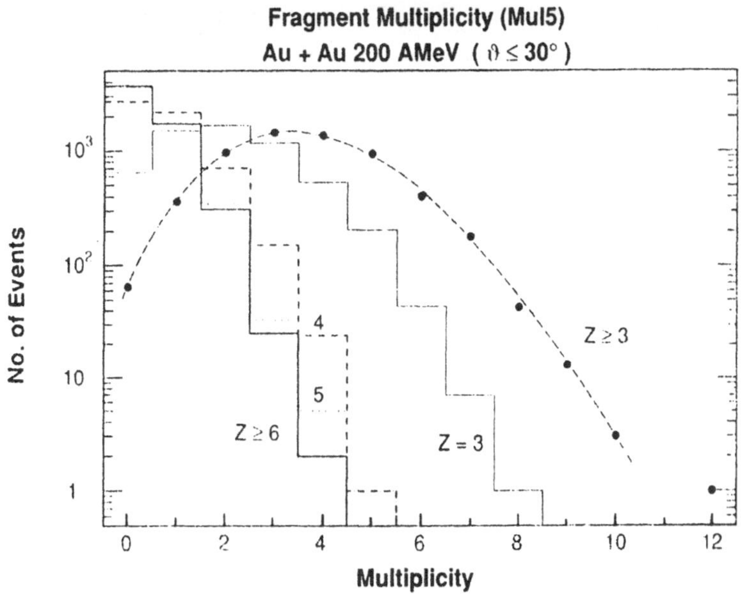

Figure 3. Decomposition of the fragment multiplicity distributions of figure 2 for central collisions of 200 AMeV Au + Au. •: $Z \geq 3$; full histogram (thin line): $Z = 3$; dashed-dotted histogram: $Z = 4$; dotted histogram: $Z = 5$; and full histogram (thick line): $Z \geq 6$.

the region from peripheral to central collisions. As can be seen from Fig. 2 most peripheral collisions (labelled 1. Mul.-Bin.) result in a low multiplicity of IMF. These fragments are observed with rapidities close to the beam rapidity, accompanied mostly by a heavy projectile remnant observed in the zero-degree detector. The data are thus consistent with expectations for fragmentation of an excited projectile spectator. Sampling more central collisions results in larger IMF multiplicities and reduces their average rapidity to values intermediate between that of projectile and target. In the most central collisions, finally, they are found to be centered with respect to y_{CM}. In such collisions practically the full projectile charge is observed in form of light and intermediate mass fragments, with no large projectile remnant remaining. Extrapolation of this measurement to 4π leads to an average of about 8 IMF per event, with a significant number of events yielding up to 20 fragments. These numbers are still slight underestimates because of the low-energy cutoff of the detector. The data clearly illustrate, that multi-fragmentation is a dominant process in central collisions of heavy nuclei at this bombarding energy.

Recently, the Quantum Molecular Dynamics approach (QMD) [37,36,38] has been employed to the same system as discussed above. The calculated charged particle multiplicities, the multiplicities of intermediate mass fragments, and their respective rapidity distributions were found to compare fairly well with the experimental data. These quantities were also reported to be quite insensitive to the EOS, but are on the other hand strongly dependent on the in-medium nucleon-nucleon cross section. The importance of this quantity which can be adjusted by means of such experimental data, will be discussed in more detail in subsequent lectures.

A decomposition of the integrated $Z \geq 3$ multiplicity distribution of Fig. 2 is shown in Fig. 3 for central events. As expected, the dominant contribution comes from lithium fragments. Note, however, that in a small fraction of events up to 4 fragments heavier than carbon ($Z \geq 6$) are observed, signifying that about a quarter of the total baryonic mass is emitted in form of such large clusters! It will be highly interesting to study those fragmentation events in further detail in order to learn about the dynamics of these reactions.

The onset of multi-fragment emission, i.e. the amount of excitation energy, E_X, in the composite system that is necessary to increase the production cross section of more than one IMF dramatically, was recently reported to be located near $E_X \approx 300\,\text{MeV}$ [39]. Furthermore, this value was found to be almost independent of the projectile-target mass system.

4 Fragment Flow

After having discussed the abundant production of IMF, let us now investigate their emission patterns, in terms of collective flow variables. Up to now collective flow effects were discussed only for the entire event by averaging over protons and composite particles with a proper weight factor (see also lectures given e.g. by H.G. Ritter, H. Ströbele, and J. Gosset). Several calculations, capable of producing nuclear fragments with $A > 1$, however, predict that a stronger flow effect should be observed for nuclear fragments than light particles emitted in the reaction [40,41,42]. This effect has been expected rather early [43] by the simple argument that heavier clusters are produced with relatively lower (undirected) thermal velocities than light particles. In order to investigate this effect experimentally, the $\langle p_x/A \rangle$ distributions were analyzed differentially as a function of the mass of the fragments. Results of such an analysis are shown in Fig. 4 for the reaction Au + Au at 200 AMeV and the multiplicity bin containing between 50% and 75% of $N_p^{\max}$ [7,44]. Fragments heavier than ^{4}He were only measured in the forward CM-hemisphere, as discussed above. The data clearly show an increasing transverse momentum flow *per nucleon* as the fragment mass increases. This rise amounts to more than 40% when comparing $Z = 1$ to $Z \geq 6$ fragments, and is particularly exciting since the azimuthally averaged transverse momenta per nucleon, $\langle p_\perp/A \rangle$, were found to decrease with increasing fragment mass.

The stronger flow of intermediate mass fragments becomes even more pronounced if the correlation is studied in position space rather than in momentum space. This is clearly visible in figure 4b. Here the same data have been analyzed, but plotted is now the fraction of the particle's transverse momentum that lies in the reaction plane, $\langle p_x/p_\perp \rangle$. If the particles were emitted exactly into the reaction plane, then this alignment function would yield plus or minus one for ideal positive or negative alignment, respectively. Note, that the azimuthal correlation with respect to the reaction plane increases now by almost a factor of three when going from hydrogen to carbon fragments. Similar results were obtained from the φ-distributions of different fragments measured relative to the reaction plane [30].

Comparing the relative increase of the alignment in momentum and position space in Fig. 4, one might interpret the increasing transverse momentum flow with increasing fragment mass as at least partly being caused by the substantially stronger *spatial* correlations of the intermediate mass fragments. This would be in line with the simple idealized interpretation of the fragment mass dependence of the collective flow, where one assumes nucleons and fragments stemming from a common thermalized source, characterized by a certain temperature, and being boosted by a common collective

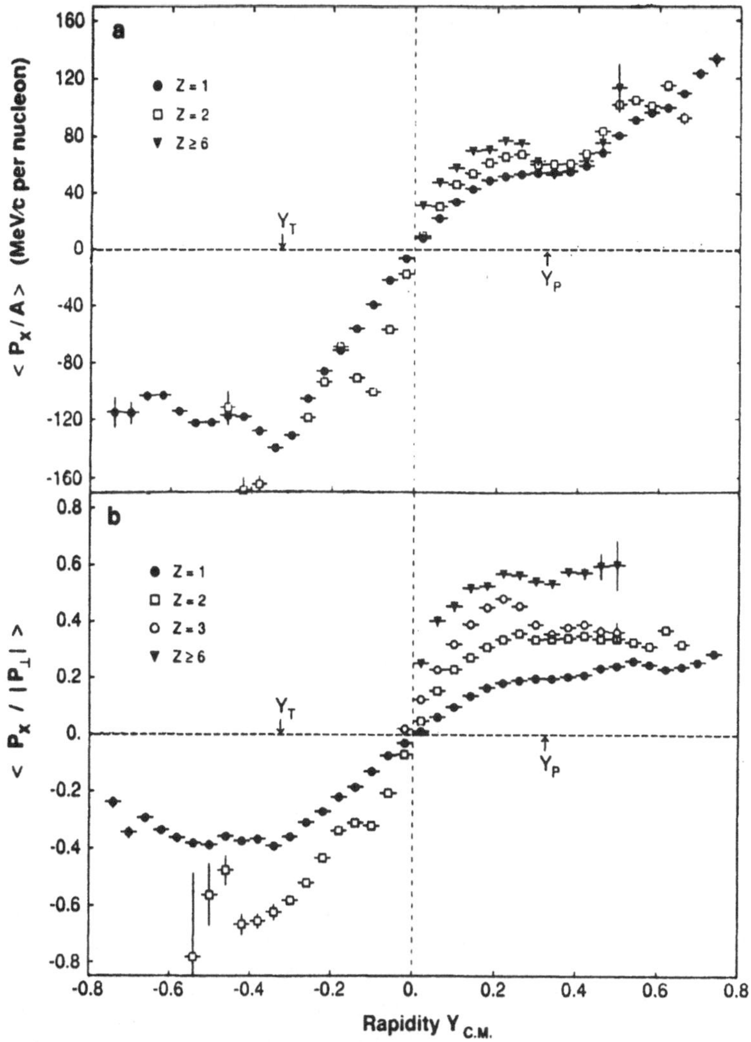

Figure 4. Mean in-plane transverse momentum per nucleon (a) and fraction of the particle's transverse momentum that is emitted into the reaction plane (b) for different fragments as a function of the CM rapidity for semi-central Au + Au collisions at 200 AMeV.

asymmetric expansion velocity caused by the inherent asymmetry in the pressure in non-zero impact parameter collisions. The thermal energy component thus defines the A-independent thermal energy per particle, whereas the flow energy, i.e. the originally built-up compressional energy, is determined by the expansion velocity and therefore should have a linear A-dependence. The flow energy therefore gains an increasingly larger fraction of the fragments energy, and the random undirected thermal motion becomes less important, as the fragment mass increases.

The dependence of the directed flow on the mass of the emitted fragment has recently been studied in the framework of the QMD model by Peilert, Aichelin, and Bohnet *et al.* [38,45,46] for Au + Au at 200 AMeV. Both, the increase of the directed momentum flow, $\langle p_x/A \rangle$, as well as the the increase in the strength of the azimuthal alignment, $\langle p_x/p_\perp \rangle$, with the mass of the fragments is fairly well reproduced by these calculations. Good agreement with the data is found when employing a stiff EOS.

Furthermore, the difference between the soft and hard EOS is most significant for heavy fragments ($Z \geq 6$), where it amounts to a doubling of p_x/A when going from the soft to the hard EOS.

5 'Squeeze-Out' of particles; another component of collective flow

The sphericity- and global transverse momentum analysis have so far been the most commonly applied methods for analyzing data in terms of collective flow variables. However, one might criticize that both methods, as well as the multi-particle particle correlation functions defined in Ref. [47], do not yet exhaust all information that is in principle available from 4π spectrometers and allow only for an investigation of collective phenomena appearing *in* the reaction plane. In particular, the successful global transverse momentum method results in $\langle p_x \rangle = 0$ at $y_{CM} = 0$ and $\langle p_y \rangle = 0 \, \forall y_{CM}$ because of symmetry reasons. On the other hand, it is anticipated that strong collective phenomena take place particularly at midrapidity, because of the large compression effects achieved in the overlap zone. Similar criticisms have been expressed also by Welke *et al.* [48], who proposed to inspect the azimuthal distributions of particles with respect to the reaction plane, as being a useful probe for the EOS. Such an analysis has been performed already by Doss *et al.* [30].

A very interesting result has very recently been reported by Hartnack *et al.* [49] (see also his conference contribution). The authors investigated the dN/dy- and $\langle p_x \rangle$ distributions for 1.05 AGeV Nb + Nb collisions in the framework of the QMD model differentially for particles which experienced from a small ($\varrho \leq 1.5\varrho_0$) and high ($\varrho \geq 2.8\varrho_0$) compression zone, respectively. The rapidity distributions exhibit the expected behaviour, i.e. particles from the low density zone are found to populate dominantly the projectile/target rapidities, and particles from the high density zone more dominantly the fireball rapidity region. However, inspecting the $\langle p_x \rangle$ distributions shows that mainly the particles from the *low* density region contribute to the experimentally observed large p_x values, i.e. to the sidewards deflection in the reaction plane, while particles from the *high* density zone result in $\langle p_x \rangle \approx 0$ over an extended region around midrapidity. In-plane transverse momentum distributions of this type therefore seem to carry only limited information on properties of highly compressed nuclear matter.

Because of this apparent lack of information we will focus now on new observables providing much more details about the collective emission pattern of emitted particles. Here will concentrate in particular on the direction *perpendicular* to the reaction plane, because this is the only coordinate where nuclear matter might escape during the whole collision time without being hindered by either the target or projectile nucleus. As a consequence, this might lead to a jet-like emission pattern – discussed very early on the basis of hydrodynamics by Scheid *et al.* [2], and being referred to as *out-of-plane squeeze-out* [50,51,14] – and allow an unique investigation of the interior of the hot dense region of the interacting system.

A first indication of such an out-of-plane peak in the particle distribution at midrapidity has recently been reported by the Diogène group for 800 AMeV Ne-induced reactions [52], and was shown also in more detail in the lecture given by J. Gosset. In order to look for such effects in data of symmetric systems, we employ a novel

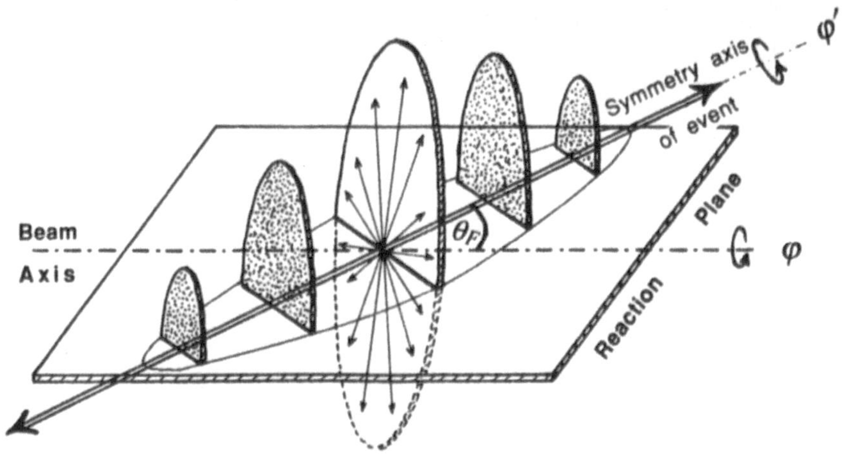

Figure 5. Schematic view of the event shape illustrating the orientation of the event in coordinate space, and indicating the difference between the azimuthal distributions φ and φ', respectively.

analysis, which combines aspects of the sphericity- and transverse momentum analysis. This method has recently been proposed and applied to 400 AMeV Au + Au data by Gutbrod *et al.* [53]. In this approach, intended to reveal new information about the 3-dimensional event shape, the azimuthal distributions of particles are analyzed in the coordinate system given by the principal axis of the kinetic energy flow tensor. A schematic view of the event shape and its orientation in coordinate space is depicted in Fig. 5 [54]. As indicated in that figure, analyzing the φ-distributions of particles around the beam-axis with respect to the reaction plane tests dominantly for the in-plane flow effects. However, if one is interested in the event shape at midrapidity, one needs to account for the well known non-zero flow angle, θ_F, i.e. one has to measure the azimuthal angle of particles around the major principal axis of the flow tensor (denoted φ' in Fig. 5). Because each event is now rotated not only by $-\varphi_F$ into the reaction plane, but also by $-\theta_F$ into the major symmetry axis of the event, the method allows for the first time to superimpose several events without effectively distorting and smearing out the characteristic individual event shapes. Thus, it evades the difficulty of the standard sphericity method to extract aspect ratios because of finite particle number distortions. Moreover, average ratios of the associated half-axes may now be inspected also as a function of rapidity.

The improvement in the data analysis resulting from the rotation of events into the direction of the flow axis – although obvious already from figure 5 – is demonstrated by figures 6 and 7. The 400 AMeV Au + Au data are selected for semi-central collisions, corresponding to the third of five multiplicity bins. A strong alignment of particles in regions of projectile and target rapidity at $\varphi = 0°$ and 180° can be seen in figure 6. This alignment corresponds to the two branches of the S-shaped curve in the conventional $\langle p_x \rangle$ analysis. However, in a narrow bin at mid-rapidity (Fig. 6b) the particles are aligned *perpendicular* to the reaction plane. The emission pattern becomes much clearer if we present the data in the coordinate system of the flow ellipsoid. This is demonstrated in figure 7; the φ' anisotropy at $p'_z \simeq 0$, as can be seen from Fig. 7b, is more pronounced as compared to Fig. 6b. Furthermore, it can be seen from the

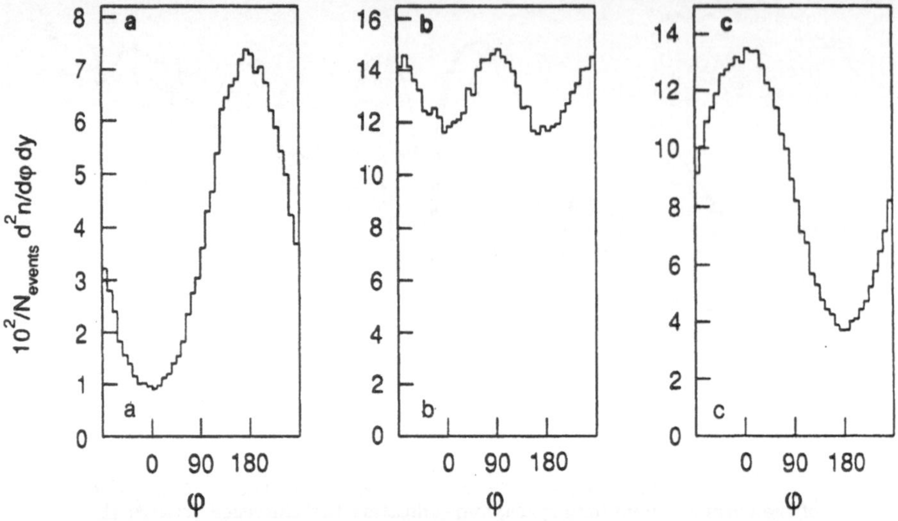

Figure 6. φ **distributions of the number of particles with respect to the reaction plane from semi-central 400 AMeV Au + Au collisions in different rapidity bins: a)** $-1.2 \leq y/y_{CM} \leq -0.8$**; b)** $-0.1 \leq y/y_{CM} \leq 0.1$**; c)** $0.8 \leq y/y_{CM} \leq 1.2$**.**

two-dimensional distribution (Fig. 7a) that the preferential out-of-plane emission of particles does not occur only at mid-rapidity, but extends over the *whole* p'_z axis. The large distortions visible at projectile and target rapidity result from in-plane bounce-off particles in the forward- and from low energy detector biases in the backward hemisphere, respectively. The isometric surface plot, shown in figure 8, illustrates the collective flow features observed along the rapidity axis in the particle emission even more impressively. The squeeze-out particles at $\varphi' = \pm 90°$ are clearly separated from the bounce-off particles at projectile rapidity and $\varphi' = 0°$ (the corresponding peak at target rapidity is absent because of detector biases for low energetic particles). Figure 7c displays the transverse momentum per nucleon $\langle p'_\perp/A \rangle$ as a function of the azimuthal angle φ' at midrapidity. Again, the same anisotropy in φ' at $p'_z \simeq 0$ is observed, revealing that not only the density of particles is enhanced in the out-of-plane direction, but that these particles are also emitted with a higher average transverse momentum per nucleon!

Results of an alternative analysis of the 90° squeeze-out are shown in figure 9, again for 400 AMeV semi-central Au + Au collisions. Plotted is now the direction of the second largest eigenvector, $\vec{e}_2$, of the standard sphericity analysis with respect to the reaction plane, which is identified as usually with the plane spanned by the beam-axis and the direction of the largest eigenvector, $\vec{e}_1$. In absence of any preferred emission perpendicular to the reaction-plane, or in other words, in case of a rotational symmetric flow ellipsoid, the angular distribution $\Delta\Phi(\vec{e}_2, \vec{e}_{\mathrm{R-plane}})$ would be uniform. The experimental observation of a strong $\mp 90°$ peaking with respect to the reaction-plane confirms again the results from above and also demonstrates the usefulness and sensitivity of the standard sphericity method in revealing more details about the event shape than had been expected originally [55].

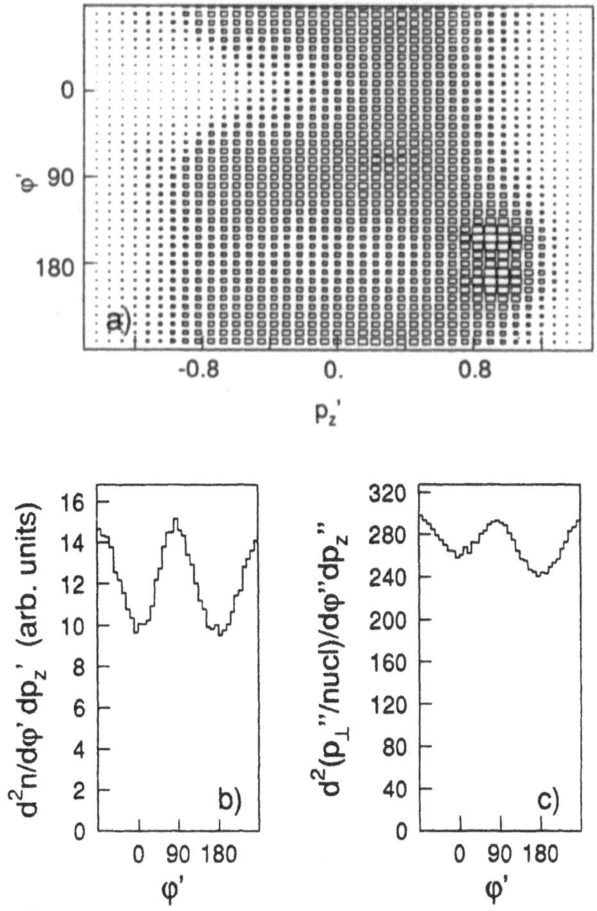

Figure 7: Same data as in Fig. 6, but presented now after the $-\theta_F$-rotation of events into the flow axis; a) Distribution of the number of particles in the φ' versus p'_z plane. The size of the squares is proportional to the number of particles; b) φ' distribution of the number of particles for $-0.1 \leq p'_z \leq 0.1$; c) φ' distribution of the average transverse momentum per nucleon for $-0.1 \leq p'_z \leq 0.1$.

5.1 Projectile-target mass and impact parameter dependence

In order to study the squeeze-out effect more quantitatively as a function of projectile-target mass, impact parameter, fragment mass, and beam energy, let us return to the φ' distributions as shown in Fig. 7b,c. The anisotropy in φ' corresponds to an elliptical distribution of the particles both in position and momentum space. Thus, fitting those distributions to a function $f(\varphi') \sim 1 + \alpha \cos(2\varphi')$, with α being the free parameter, yields the aspect ratio of the two shorter half-axes of the associated ellipsoid as extrema of f. Figure 10 (left) shows the multiplicity dependence of this aspect ratio at midrapidity for the position density (filled circles) and the average momentum per nucleon (open circles). Both ratios exhibit the same tendency; they are found to reach a maximum in semi-central collisions, resembling the impact parameter dependence of the directed in-plane flow. In the limit of impact parameter $b = 0$ (which cannot be reached in practice) symmetry requires that the ratios become equal to one. The right

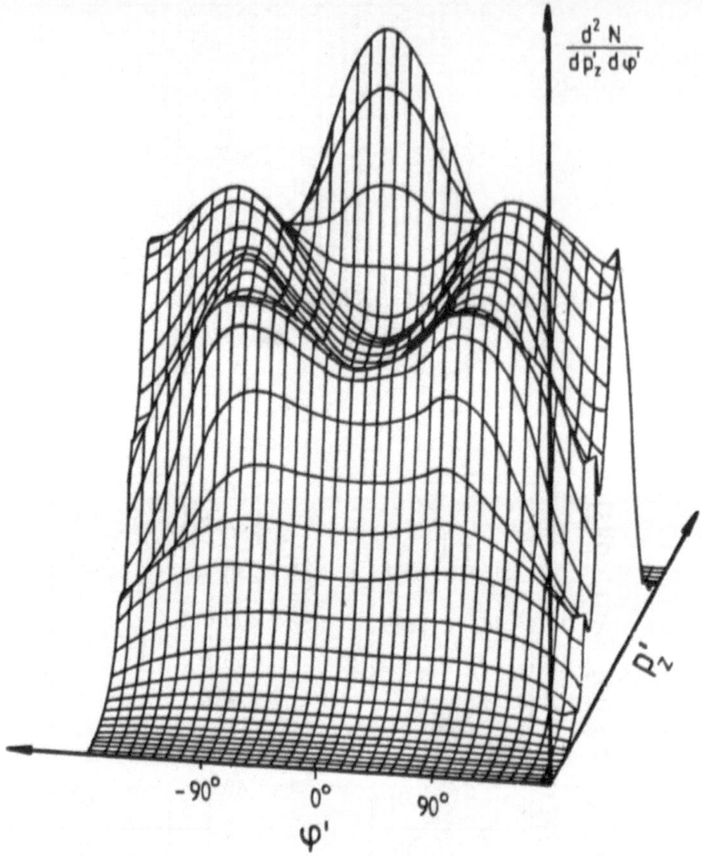

Figure 8. Isometric distribution, $d^2N \,/\, dp'_z\, d\varphi'$ of the same data as shown in figure 7.

hand part of Fig. 10 shows the mass dependence of the *energy squeeze-out* ($\sum_i E'^{,i}_{\perp}(\varphi')$) in semi-central collisions of 400 AMeV Au + Au. The aspect ratio increases here fairly proportional to A and reaches a value of about $R_{max}/R_{min} \cong 2$ for Au + Au. This means that one observes twice as much transverse energy flowing into out-of-plane direction as opposed to the in-plane direction!

Recent QMD calculations – presented on this school by C. Hartnack – are in very good agreement to the experimental data, both in absolute numbers as well as in their predicted projectile-target mass dependence, when a hard EOS is assumed. However, these calculations were carried out for a fixed impact parameter of $b = 0.25 \cdot b_{max}$. A real simulation rather needs to take into account the experimental trigger conditions in order to resemble also the experimental impact parameter distribution.

5.2 Fragment mass dependence

Since heavy fragments were found to show a stronger in-plane flow effect than light particles, it is of interest to learn also about the fragment mass dependence of the out-of-plane squeeze-out. As an example of such an analysis figure 11 displays the azimuthal distributions, $dN^2 \,/\, d\varphi'\, dp'_z$, differentially for hydrogen and helium fragments,

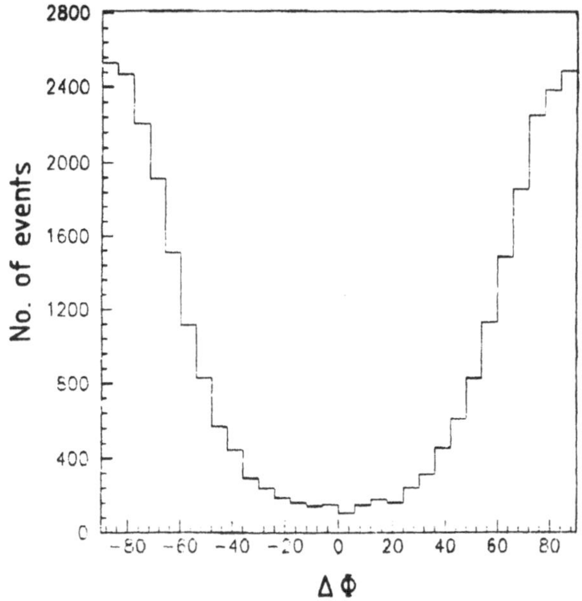

Figure 9. Azimuthal distribution of the opening angle between the second largest eigenvector of the sphericity tensor and the reaction-plane for semi-central Au + Au collisions at 400 AMeV.

together with the extracted aspect ratios, R_{max}/R_{min}, as a function of multiplicity. Again, heavier fragments are found to exhibit also the 90°-component of the collective flow much stronger than light particles; the peak-to-valley ratio in semi-central collisions increases in this case from about $R_{max}/R_{min} \cong 1.5$ for $Z = 1$, to more than 1.8 for $Z = 2$ fragments. This finding is particular interesting, as it translates into higher fragment-to-proton ratios, and consequently into lower values of entropy in the 90° out-of-plane direction as opposed to the direction into the reaction plane. Previously published results on entropy production, extracted from the φ integrated yields, thus have to be considered average numbers.

A compilation of the projectile-target mass dependence of the fragment squeeze-out is presented in figure 12. In all systems the aspect ratios, R_{max}/R_{min}, are found to be about 20 % larger for helium than they are for hydrogen fragments. This can be interpreted qualitatively with the same arguments of thermal smearing, affecting more strongly light particles because of their higher thermal velocities. Note, that the increase of R_{max}/R_{min} with increasing projectile-target mass is found to be weaker than is found for the energy squeeze-out, $d\Sigma E'_\perp/d\varphi'$ (see above). This is a consequence of the fact that the aspect ratios of *both* the $dN/d\varphi'$, as well as of the $\langle E'_\perp/\text{particle}\rangle \,/\, d\varphi'$ distributions increase like a power law with increasing projectile-target mass, so that the total energy squeeze-out, $d\Sigma E'_\perp/d\varphi'$ results in a linear rise. It would be interesting to find out by detailed comparisons with appropriate models whether the observed A^1 dependence of the energy squeeze-out is due to a basic underlying process or just an accidental result of this two contributing components.

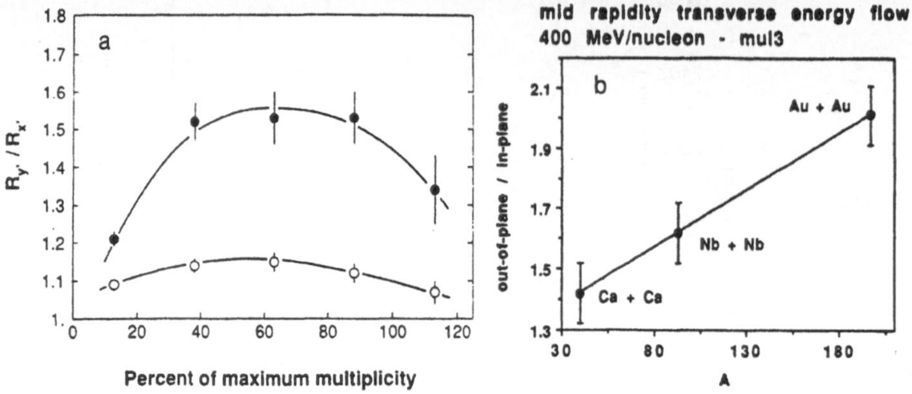

Figure 10. a) Aspect ratio of the associated half axes of an ellipsoid describing the density of particles (filled circles) and the momentum per nucleon (open circles) in a slice normal to the major symmetry axis of the event (see Fig. 5) at $p'_z \simeq 0$ as a function of multiplicity for 400 AMeV Au + Au, and b) projectile-target mass dependence of the energy squeeze-out (no. of particles × transverse energy) at 400 AMeV. The solid lines are to guide the eye.

6 Summary, Conclusions, and Outlook

Results from 4π measurements of fragment formation, multi-fragmentation, and collective flow phenomena in symmetric systems of heavy nuclei have been presented. In central collisions of Au + Au at 200 AMeV about 8–9 IMF ($Z \geq 3$) are produced on the mean, with up to 20 possible in a fraction of events. Decomposition reveals, that the dominant contribution to the IMF multiplicity results from lithium fragments, however, events are also observed with up to 8 carbon nuclei accompanied by lighter fragments. The in-plane transverse momentum distributions, characterizing the directed sidewards flow, show that both the average transverse momentum components parallel to the reaction plane are higher, and that the azimuthal alignment with the reaction plane is stronger for heavy fragments than is for light particles. The observation of a stronger flow of IMF over light particles supports early predictions of fluid dynamics and is quantitatively in agreement with recent QMD calculations.

Another very important piece of information became just available from the observation of the out-of-plané squeeze-out of particles. New viscous fluid results exhibit a strong dependence of this effect on the nuclear viscosity; for large values the directed emission is damped out, and a maximum is not visible in the $dN/d\varphi'$ distributions [56]. Thus, such data will be essential in determining this unknown property of hot nuclear matter. The same sensitivity is also found in microscopic calculations (VUU, QMD), where $dN/d\varphi'$ depends strongly on the EOS and σ_{eff}, i.e. parameters which enter directly into the coefficients of the viscosity. The anisotropy in $dN/d\varphi'$ is furthermore found to be most prominent in semi-central collisions and to increase with projectile-target mass, fragment mass, and bombarding energy. Because squeeze-out particles escape geometrically from the hot and dense reaction zone unhindered by surrounding cold target- or projectile matter, they provide a clean probe through which one can look directly at the compressed and hot fireball. It will be most interesting in the next future to see whether the successful microscopic models can consistently, i.e. using the same EOS and σ_{eff}, describe the p_x/A, $p_x/p_\perp$, dN/dy, and $dN/d\varphi'$ distributions. De-

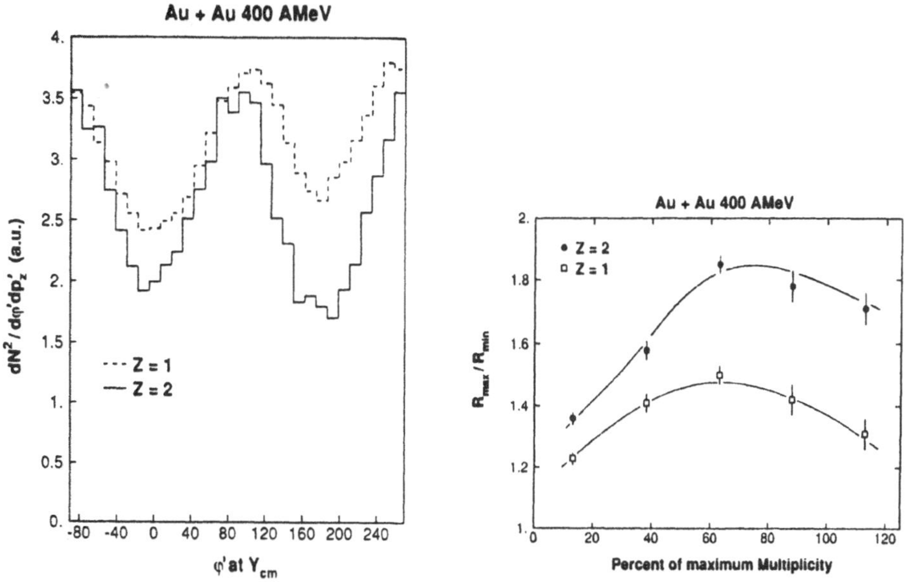

Figure 11. φ' distribution of the number of $Z = 1$ (dashed line) and $Z = 2$ (solid line) fragments observed from semi-central 400 AMeV Au + Au reactions at $-0.1 \le p_z' \le 0.1$ (left hand side), and extracted aspect ratios, $R_{\max}/R_{\min}$, as a function of multiplicity (right hand side).

tailed studies of energy spectra, π-yields, etc., might give further useful information. Simultaneous description of such a large body of exclusive experimental observables would be a large step forward in the ultimate goals of relativistic heavy-ion collisions, namely the determination of the bulk properties of nuclear matter.

There remains much work to be done, however, also on the experimental side, both in improving the quality of the data, as well as in identifying proper observables. The need for improving the accuracy of the experimental data became clear during the past one or two years, because of many unknowns in present theories, such as strong in-medium effects, which influence the results beside the EOS significantly. Fixing these parameters requires precise knowledge of very different data, as for example rapidity distributions, flow angle distributions, azimuthal asymmetries in position and momentum space, π, η, K, λ... yields, and all for a large range of bombarding energies across a range of and projectile-target masses — requirements which are hardly to be achieved by a single apparatus. Large efforts are therefore presently being made in preparing for new experiments at new facilities, most importantly the development of the HISS TPC at the Berkeley Bevalac and the SIS/ESR accelerator/storage-ring complex with it's new arsenal of experiments. Identifying proper observables for an investigation of event shapes in the *most central* collisions is another task to be addressed. Note, that the directed in-plane flow, as well as the squeeze-out are observed to be most prominent in *semi* central collisions. The highest density achieved over an extended volume, however, is expected to be achieved in the most central collisions. Questions about the possible existence of oblate events, to be observed – according to predictions by fluid dynamics – in head-on collisions of heavy nuclei, are not yet answered. This problem is also related to the question on how to trigger experimentally on very central ($b \le 1$ fm) collisions. Triggering, for example, on high multiplicity alone still samples a quite large range of impact parameters.

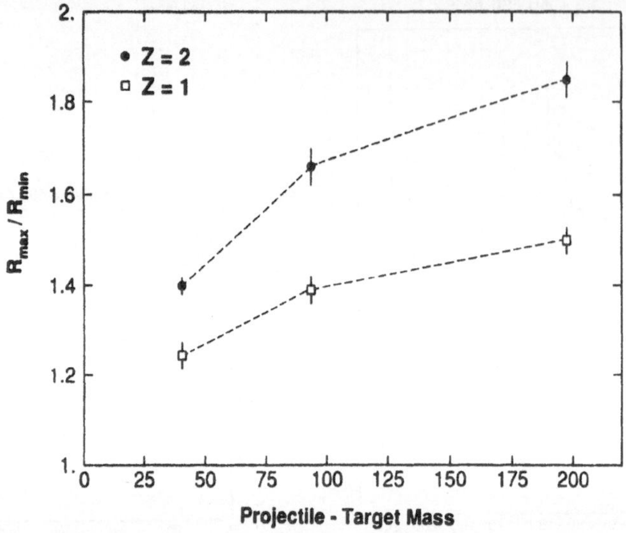

Figure 12. Extracted aspect ratios, R_{max}/R_{min} from the dN/φ' distributions at mid-rapidity and semi-central collisions of 400 AMeV Ca + Ca, Nb + Nb, and Au + Au collisions.

Nevertheless, although all obstacles on the theoretical and experimental side are not yet completely surmounted, enormous progress has been made during recent years. The most outstanding discovery of of nuclear collective flow effects, which confirmed theoretical expectations of shock-wave phenomena from nuclear fluid dynamics, must be considered the manifestation of the key mechanism for heating and compression of nuclear matter. Thus, it provided the grounds for extracting first information on transport properties and the equation of state of hot and dense hadronic matter from data of high-energy heavy-ion collisions; the result as it appears at this stage are large viscous effects and a surprisingly stiff EOS.

Acknowledgement

It is a pleasure to warmly thank the conference organizers W. Greiner and E. Oset for inviting KHK to Peñiscola and giving him the opportunity to participate on this combination of an extremely stimulating and fruitful school and conference, taking place in the pleasant and relaxed atmosphere of the Mediterranean.

† *On leave of absence from Institute for Nuclear Studies, PL-00681 Warsaw, Poland*

References

[1] G.F. Chapline, M.H. Johnson, E. Teller, and M.S. Weiss, *Phys. Rev.* **D 8** (1973) 4302.
[2] W. Scheid, H. Müller, and W. Greiner, *Phys. Rev. Lett.* **32** (1974) 741.
[3] M.I. Sobel, P.J. Siemens, J.P. Bondorf, and H.A. Bethe, *Nucl. Phys.* **A251** (1975) 502.
[4] E. Baron, J. Cooperstein, and S. Kahana, *Phys. Rev. Lett.* **55** (1985) 126.
[5] J.R. Wilson and H.A. Bethe, *Astrophys. J.* **295** (1985) 14.

[6] A. Baden, H.H. Gutbrod, H. Löhner, M. Maier, A.M. Poskanzer, T. Renner, H. Riedesel, H.G. Ritter, H. Spieler, A. Warwick, F. Weik, and H. Wieman, *Nucl. Instr. Meth.* **203** (1982) 189.

[7] Karl-Heinz Kampert, Doctoral thesis, Institut für Kernphysik, Münster, (1986).

[8] R. Albrecht, H.W. Daues, H.Å. Gustafsson, H.H. Gutbrod, K.H. Kampert, B. Kolb, H. Löhner, B. Ludewigt, A.M. Poskanzer, H.G. Ritter, H.R. Schulze, and H. Wieman, *Nucl. Instr. Meth.* **A 245** (1986) 82.

[9] P.J. Siemens and J.I. Kapusta, *Phys. Rev. Lett.* **43** (1979) 1486.

[10] H. Stöcker, G. Buchwald, G. Gräbner, P. Subramanian, J.A. Maruhn, W. Greiner, B.V. Jacak, and G.D. Westfall, *Nucl. Phys.* **A400** (1983) 63c.

[11] L.P. Csernai and J.I. Kapusta, *Phys. Rep.* (1986) 223.

[12] G. Bertsch and J. Cugnon, *Phys. Rev.* **C 24** (1981) 2514.

[13] J.I. Kapusta, *Phys. Rev.* **C 24** (1981) 2545.

[14] H. Stöcker and W. Greiner, *Phys. Rep.* **137** (1986) 277.

[15] K.G.R. Doss, H.Å. Gustafsson, H.H. Gutbrod, B. Kolb, H. Löhner, B. Ludewigt, A.M. Poskanzer, T. Renner, H. Riedesel, H.G. Ritter, A. Warwick, and H. Wieman, *Phys. Rev.* **C 32** (1985) 116.

[16] K.G.R. Doss, H.Å. Gustafsson, H.H. Gutbrod, D. Hahn, K.H. Kampert, B. Kolb, H. Löhner, A.M. Poskanzer, H.G. Ritter, H.R. Schmidt, and H. Stöcker, *Phys. Rev.* **C 37** (1988) 163.

[17] G.D. Westfall, J. Gosset, P.J. Johansen, A.M. Poskanzer, W.G. Meyer, H.H. Gutbrod, A. Sandoval, and R. Stock, *Phys. Rev. Lett.* **37** (1976) 1202.

[18] J. Randrup and S.E. Koonin, *Nucl. Phys.* **A356** (1981) 223.

[19] J.P. Bondorf, R. Donangelo, I.N. Mishustin, C.J. Pethick H. Schulz, and K. Sneppen, *Nucl. Phys.* **A443** (1985) 321.

[20] J.P. Bondorf, R. Donangelo, I.N. Mishustin, C.J. Pethick H. Schulz, and K. Sneppen, *Phys. Lett.* **150B** (1985) 57.

[21] A.Z. Mekjian, *Phys. Rev. Lett.* **38** (1977) 640.

[22] H.H. Gutbrod, A. Sandoval, P.J. Johansen, A.M. Poskanzer, J. Gosset, W.G. Meyer, G.D. Westfall, and R. Stock, *Phys. Rev. Lett.* **37** (1976) 667.

[23] L.P. Csernai, B. Lukàcs, and J. Zimànyi, *Nuovo Cimento Lett.* **27** (1980) 111.

[24] G. Bertsch and P.J. Siemens, *Phys. Lett.* **126B** (1983) 9.

[25] B.J. Strack and J. Knoll, *Z. Phys.* **A315** (1984) 249.

[26] S.M. Kiselev, *Phys. Lett.* **198B** (1987) 324.

[27] J. Aichelin, J. Hüfner, and R. Ibarra, *Phys. Rev.* **C 30** (1984) 107.

[28] P. Danielewicz, *Nucl. Phys.* **A314** (1979) 465.

[29] B. ter Haar and R. Malfliet, *Phys. Rev. Lett.* **56** (1985) 1237.

[30] K.G.R. Doss, H.Å. Gustafsson, H.H. Gutbrod, J.W. Harris, B.V. Jacak, K.H. Kampert, B. Kolb, A.M. Poskanzer, H.G. Ritter, H.R. Schmidt, L. Teitelbaum, M. Tincknell, S. Weiss, and H. Wieman, *Phys. Rev. Lett.* **59** (1987) 2720.

[31] A.M. Poskanzer, G.W. Butler, and E.K. Hyde, *Phys. Rev.* **C 3** (1971) 882.

[32] J.E. Finn, S. Agarwal, A. Bujak, J. Chuang, L.J. Gutay, A.S. Hirsch, R.W. Minich, N.T. Porile, R.P. Scharenberg, B.C. Stringfellow, and F. Turkot, *Phys. Rev. Lett.* **49** (1982) 1321.

[33] A.I. Warwick, H.H. Wieman, H.H. Gutbrod, M.R. Maier, J. Peter, H.G. Ritter, H. Stelzer, F. Weik, M. Freedman, D.J. Henderson, S.B. Kaufmann, E.P. Steinberg, and B.D. Wilkins, *Phys. Rev.* **C 27** (1983) 1083.

[34] W. Trautmann, K.D. Hildenbrand, U. Lynen, W.F.J. Müller, H.J. Rabe, H. Sann, H. Stelzer, R. Trockel, R. Wada, N. Brummund, R. Glasow, K.H. Kampert, R. Santo, E.M. Eckert, J. Pochodzalla, I. Bock, and D. Pelte, *Nucl. Phys.* **A471** (1987) 191.

[35] R. Trockel, U. Lynen, J. Pochodzalla, W. Trautmann, N. Brummund, E. Eckert, R. Glasow, K.D. Hildenbrand, K.H. Kampert, W.F.J. Müller, D. Pelte, H.J. Rabe, H. Sann, R. Santo, H. Stelzer, and R. Wada, *Phys. Rev. Lett.* **59** (1987) 2844.

[36] J. Aichelin, G. Peilert, A. Bohnet, A. Rosenhauer, H. Stöcker, and W. Greiner, *Phys. Rev.* **C 37** (1988) 2451.

[37] J. Aichelin and H. Stöcker, *Phys. Lett.* **B 176** (1986) 14.

[38] Georg Peilert, Diploma thesis, Institut für theoretische Physik, Frankfurt, (1988).

[39] R. Trockel, K.D. Hildenbrand, U. Lynen, W.F.J. Müller, H.J. Rabe, H. Sann, H. Stelzer, W. Trautmann, R. Wada, E. Eckert, P. Kreuz, A. Kühmichel, J. Pochodzalla, and D. Pelte, *Phys. Rev.* **C 39** (1988) 729.

[40] H. Stöcker, A.A. Oglobin, and W. Greiner, *Z. Phys.* **A 303** (1981) 259.

[41] L.P. Csernai, H. Stöcker, P.R. Subramanian, G. Buchwald, G. Graebner, A. Rosenhauer, J.A. Maruhn, and W. Greiner, *Phys. Rev.* **C 28** (1983) 2001.

[42] L.P. Csernai, G. Fài, and J. Randrup, *Phys. Lett.* **140B** (1984) 149.

[43] H.G. Baumgardt, J.U. Schott, Y. Sakamoto, E. Schopper, H. Stöcker, J. Hofmann, W. Scheid, and W. Greiner, *Z. Phys.* **A 273** (1975) 359.

[44] H.Å. Gustafsson, H.H. Gutbrod, K.H. Kampert, B.W. Kolb, A.M. Poskanzer, H.G. Ritter, and H.R. Schmidt, *Mod. Phys. Lett.* **A3** (1988) 1323.

[45] G. Peilert, A. Rosenhauer, H. Stöcker, W. Greiner, and J. Aichelin, *Mod. Phys. Lett.* **A3** (1988) 459.

[46] J. Aichelin, private communication, (1988).

[47] P. Beckmann, H.Å. Gustafsson, H.H. Gutbrod, K.H. Kampert, B. Kolb, H. Löhner, A.M. Poskanzer, H.G. Ritter, H.R. Schmidt, T. Siemiarczuk, and H. Wieman, *Mod. Phys. Lett.* **A2** (1987) 163–168, 169–176.

[48] G.M. Welke, M. Prakash, T.T.S. Kuo, S. DasGupta, and C. Gale, *Phys. Rev.* **C 38** (1989) 2101.

[49] Christoph Hartnack, Diploma thesis, Institut für theoretische Physik, Frankfurt, (1989).

[50] H. Stöcker, L.P. Csernai, G. Gräbner, G. Buchwald, H. Kruse, R.Y. Cusson, J.A. Maruhn, and W. Greiner, *Phys. Rev.* **C 25** (1982) 1873.

[51] G. Buchwald, G. Graebner, J. Theis, J. Maruhn, W. Greiner, H. Stöcker, K. Frankel, and M. Gyulassy, *Phys. Rev.* **C 28** (1983) 2349.

[52] Denis L'Hôte, private communication and talk at Fifth Gull Lake Nuclear Physics Conference (Gull Lake) (MI), (1988).

[53] H.H. Gutbrod, K.H. Kampert, B.W. Kolb, H.G. Ritter, A.M. Poskanzer, and H.R. Schmidt, *Phys. Lett.* **216B** (1989) 267.

[54] K.H. Kampert, *J. Phys.* **G 15** (1989) 691.

[55] P. Danielewicz and M. Gyulassy, *Phys. Lett.* **129B** (1983) 283.

[56] W. Schmidt, University of Frankfurt, private communication, 1988.

STOPPING POWER AND THE EQUATION OF STATE AT INTERMEDIATE ENERGY

D. Keane and J. Cogar
Department of Physics, Kent State University, Kent, Ohio 44242

S.Y. Chu, S.Y. Fung, and M. Vient
Department of Physics, University of California, Riverside, California 92521

Y.M. Liu and S. Wang
Department of Physics, Harbin Institute of Technology, Harbin, People's Rep. of China

If complete stopping takes place in a heavy ion collision, no memory of the incident beam direction persists among the final state nuclear fragments, and the maximum possible fraction of the initial kinetic energy is available to impart transverse momentum to the fragments and for populating other degrees of freedom, such as pion production. At intermediate energies, it is clear *a priori* that nuclear matter outside the geometric overlap region — the spectators — cannot be stopped, and so the non-trivial question is whether or not stopping occurs for the participant matter. Stopping of participants is a necessary condition for regarding the interaction region as being in global thermal equilibrium.

Although the qualitative concept of stopping in heavy ion collisions is unambiguous, there is no generally accepted quantitative definition of stopping. In the limit of the most central collisions of intermediate to heavy systems at Bevalac energies, the fact that fragment momenta are close to isotropic in the c.m. frame has been used to argue that there is complete stopping under these conditions,[1,2] and the disappearance of the peaks in rapidity spectra dN/dy at projectile and target rapidities for the most central collisions has also supported the same conclusion.[3] In the context of macroscopic models, these observations provide some justification for assuming that the nuclear matter approaches thermal equilibrium. From the perspective of microscopic transport theories like the Boltzmann-Uehling-Uhlenbeck[4] (BUU) and Vlasov-Uehling-Uhlenbeck[5] (VUU) models, and the more recent Quantum Molecular Dynamic[6,7] (QMD) approach, measurements that can constrain the effective two-body scattering cross sections in the nuclear medium, σ_{NN}^{eff}, are more relevant than tests of signatures for thermal equilibrium. Although σ_{NN} for free particles is well known, theoretical calculations of in-medium corrections[8,9] to arrive at σ_{NN}^{eff} are based on elaborate assumptions that must be experimentally justified. From this microscopic perspective, the observable that offers the best sensitivity to σ_{NN}^{eff}

The Nuclear Equation of State, Part A
Edited by W. Greiner and H. Stöcker
Plenum Press, New York

should be used to define stopping power. Complete stopping would then correspond to σ_{NN}^{eff} (and the number of collisions for participant nucleons) being large enough to generate a "thermalized" final state.

Thus, the ideal stopping observable is one that is both maximally sensitive to σ_{NN}^{eff}, and minimally sensitive to any other quantities that may need to be determined experimentally. The transport models indicate that the number of fragments at mid-rapidity (or, to include mass-asymmetric systems, the number at the rapidity where dN/dy has its participant peak) has this desirable characteristic.[10,7,11] In particular, dN/dy, like every other quantity averaged over all reaction planes, is independent of the equation of state (EOS).

The transport models indicate[12,13,7,11] that collective flow observables such as $p^x(y)$, the mean component of transverse momentum per nucleon in the event reaction plane as a function of rapidity,[14] are sensitive to both the EOS and σ_{NN}^{eff}. These findings cover various projectile and target masses, bombarding energies and impact parameters, and can be summarized by noting that a decrease in σ_{NN}^{eff} in BUU/VUU or QMD simulations by a factor of $\sim \frac{1}{\sqrt{2}}$ changes the predicted flow by an amount equivalent to changing the EOS from "stiff" to "soft"; likewise, a $\sim \sqrt{2}$ increase in σ_{NN}^{eff} is equivalent to changing the EOS from "soft" to "stiff". Thus, a minimal requirement for distinguishing between these two standard parametrizations of the EOS is that *the limits of uncertainty in* σ_{NN}^{eff} *should differ by a factor significantly smaller than* $\sqrt{2}$.

Figure 1 shows previously reported[10] dN/dy results for high multiplicity 1.2 A GeV Ar beam events in the Bevalac streamer chamber, with a mass-symmetric target KCl, and a heavier target, BaI_2 (mean mass number 130). VUU calculations that have been filtered to simulate the experimental conditions are also shown. The predicted dN/dy is very strongly dependent on impact parameter, and so it is especially important that the experimental multiplicity selection has been simulated as realistically as possible.[10] The sensitivity to σ_{NN}^{eff} is better in the case of the heavier target, and the data favor the "normal" cross sections σ_{NN}^{UU}, *i.e.*, the free particle scattering cross sections corrected for Pauli blocking of occupied final states, as per the VUU implementation[5] of the Uehling-Uhlenbeck collision term. Although we can infer that

$$\alpha = \frac{\sigma_{NN}^{eff}}{\sigma_{NN}^{UU}} \approx 1.0,$$

we still need to establish a better understanding of the various possible sources of systematic uncertainty in α. Comparisons based on data from different detectors are needed in order to probe the experimental uncertainties and the limitations of the detector filtering procedures. Likewise, comparisons using a variety of independent transport codes help to clarify theoretical uncertainties, and in order to be prepared for the possibility[9] that α is a function of the state variables, stopping power should be studied over as wide a range of conditions as possible.

Aichelin *et al.*[15] have recently compared predictions of several microscopic transport models to inclusive momentum spectra for 0.8 A GeV La + La, as reported by Hayashi *et al.*,[16] and these comparisons provide a new estimate of α as well as some insight into the systematic uncertainties discussed above. In Figure 2, inclusive p-like invariant cross sections σ_I for the experimental data at three laboratory polar angles are plotted as solid

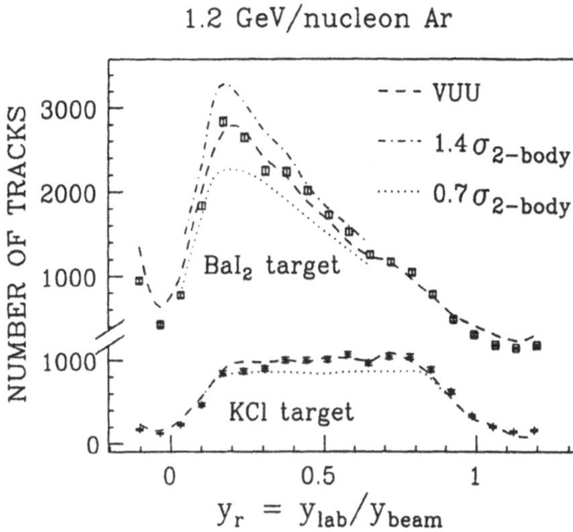

FIGURE 1. Nucleon rapidity distributions dN/dy for streamer chamber events[10] at high multiplicity, along with filtered predictions from the VUU model. The multiplicity selections correspond to 20% of the inelastic cross section in the case of the KCl target, and 35% in the case of the BaI_2 target.

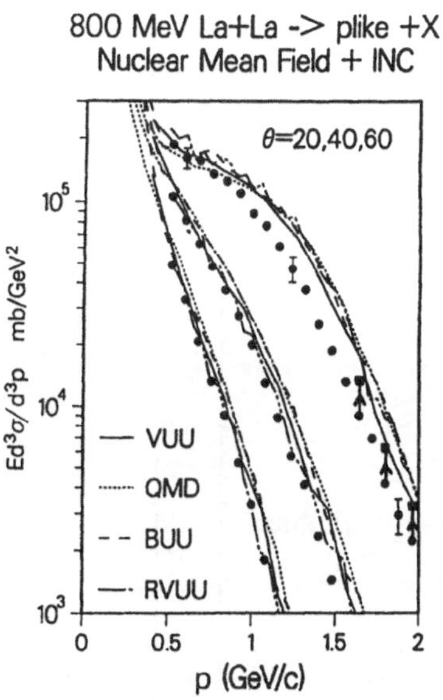

FIGURE 2. Comparison between inclusive momentum spectra from the experiment of Hayashi *et al.*[16] and four transport models,[4–7,17] as reported by Aichelin *et al.*[15]

circles. The term *p-like* signifies that the contribution from protons in all fragments (up to ^{4}He) is included, and so the comparison should be negligibly affected by the fact that final state clustering is not implemented in some of the models. The rather good agreement among the different predictions, and the discrepancy between them and the data at $\theta_{lab} = 20°$ has been interpreted as a possible failure of the models to adequately incorporate all the effects of the nuclear medium. However, the arguments outlined below favor an interpretation in terms of $\alpha \approx 1$.

Figure 3(a) shows contours of σ_I for protons from 0.8 *A* GeV La + La collisions, plotted in the plane of rapidity and transverse momentum, as reported by Hayashi *et al.*[16] The vertical set of solid circles on the left corresponds to measurements at 89° in the laboratory, and moving to the right, successive sets of points correspond to $\theta_{lab} = 75°, 60°, 50°, 40°, 30°$, and 20°. For a mass-symmetric system, the contours should be symmetric about mid-rapidity; these data satisfy this requirement within the stated systematic uncertainties. The indicated arrow for the projectile rapidity, y_p, corresponds to a beam energy between 0.77 and 0.78 *A* GeV. Assuming a nominal energy of 0.80 *A* GeV in the Bevatron, a mean energy loss of 20 to 30 *A* MeV up to the point of interaction is not inconsistent with the experimental configuration, and the overall symmetry of the contours is best when an energy loss of this magnitude is assumed. It can be seen that the 20° data span rapidities centered around the projectile rapidity, and in this region, the cross section as a function of the laboratory momentum is strongly dependent on the beam energy. In the absence of precise information about the projectile energy at the mean interaction point in the target, the most reliable prescription for testing the models is to reflect the contours at target rapidity in Figure 3 onto what the model calculations assumed to be the beam rapidity, *i.e.*, $y = 1.23$. Figure 3(b) shows the same data as Figure 3(a), with the dashed contours removed to reduce clutter. The dashed vertical line marks the beam rapidity corresponding to 0.80 *A* GeV, and the two solid squares show the intersections of the reflected $\sigma_I = 10^3$ and $\sigma_I = 10^4$ contours with this line. Without extrapolating contours, it is possible to use the experimental data near target rapidity to

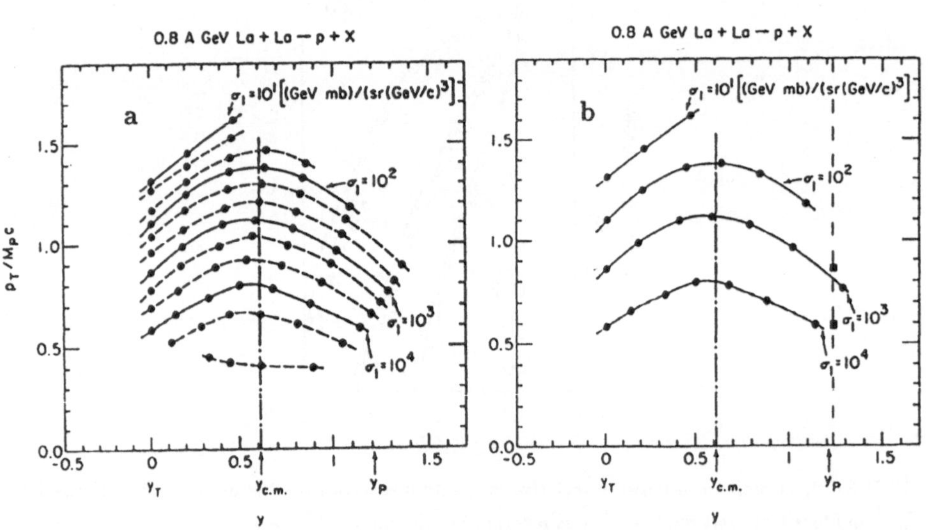

FIGURE 3. (a) Contours of invariant proton cross section, plotted in the plane of rapidity and transverse momentum, from the experiment of Hayashi *et al.*[16] (b) A subset of the contours shown in panel (a) (see text).

test the $\theta_{lab} = 20°$ predictions in Figure 2 above $p \sim 1.6$ GeV/c. The mean displacement of the reflected contours corresponds to a factor in cross section of just under 1.5. The solid squares in Figure 2 indicate that this factor is sufficient to explain most or all of the discrepancy at 20°. Although Figure 2 shows data for proton-*like* fragments and Figure 3 shows data for protons only, composite fragments make a negligible contribution to the proton-like cross section in the region being considered.

It should also be noted that a similar effect applies at $\theta_{lab} = 40°$, but in this region, the correction to the cross section is only 10 to 15%. Nevertheless, this factor serves to optimize the agreement between the models and experiment at 40°. The correction at 60° is smaller still, and is not significant.

Overall, it is doubtful that the comparisons between transport models and data at 20° provide useful information about in-medium effects in view of the large uncertainty associated with beam energy loss and the fact that $\theta_{lab} = 20°$ corresponds to a rapidity range where dN/dy and $d\sigma_I/dy$ have reduced sensitivity to σ_{NN}^{eff}. By the same token, the correction for beam energy loss is much smaller at 40° and 60°, and so the comparisons are more useful at these angles. The 40° and 60° data favor $\alpha \approx 1$ at a level of precision comparable to the variation among the models. This variation is predominantly non-statistical, and arises from differences in implementation of various dynamical features of the models. We estimate that the variation among the models is equivalent to an uncertainty $\Delta\alpha \sim \pm\ 0.1$ to 0.2 — large enough to be of some concern if these models are to be used to infer properties of the EOS.

Because the experiment of Hayashi *et al.* provides single particle inclusive data, the simulation of the event selection is much simpler than in cases where a multiplicity cut is applied. In spite of the additional complication, however, stopping data at higher multiplicities is valuable because of the importance of probing σ_{NN}^{eff} over as wide a range of conditions as possible, and particularly in the moderately central collisions where sideward flow is a maximum. The nuclear density averaged over the interaction zone for high multiplicity Ar + BaI_2 events is higher than for inclusive La + La collisions; the fact that we infer $\alpha \approx 1$ in both cases provides some indication that in-medium effects beyond σ_{NN}^{UU} are not strongly density-dependent. Of course, these empirical arguments cannot exclude the possibility that competing effects cancel each other.

In summary, we argue that not all of the effects of the nuclear medium in the context of microscopic transport models are well understood, and the need for experimental constraints on these effects provides the motivation for new, more precise measurements of stopping power in intermediate-energy heavy-ion collisions. The models indicate that the number of mid-rapidity fragments is a good indicator of relative stopping power, and measurements of stopping can be used to estimate $\alpha = \sigma_{NN}^{eff}/\sigma_{NN}^{UU}$, the effective cross section in the medium for NN collisions, in units of σ_{NN}^{UU}, the free particle cross section corrected for final-state Pauli blocking. High multiplicity data from the Bevalac streamer chamber, and inclusive data from the Bevalac B30 spectrometer are both consistent with $\alpha \approx 1$. These data probe the nucleus-nucleus collision process at significantly different mean densities, and the agreement suggests that α does not exhibit strong density-dependence. Different transport models predict momentum spectra near mid-rapidity that agree within a level corresponding to a $\sim \pm\ 10$ to 20% variation in σ_{NN}^{eff}. Such variations are large enough to be of some concern if these models are to be used to infer properties of the nuclear equation of state.

This work has been supported in part by the U.S. Department of Energy. We acknowledge computing facilities provided by the San Diego Supercomputer Center and the Ohio Supercomputer Center.

References

1. H. Ströbele, R. Brockmann, J.W. Harris, F. Riess, A. Sandoval, R. Stock, K.L. Wolf, H.G. Pugh, L.S. Schroeder, R.E. Renfordt, K. Tittel, M. Maier, Phys. Rev. C **27**, 1349 (1983).
2. H.-Å. Gustafsson, H.H. Gutbrod, B. Kolb, H. Löhner, B. Ludewigt, A.M. Poskanzer, T. Renner, H. Riedesel, H.G. Ritter, A. Warwick, and H. Wieman, Phys. Lett. **B142**, 141 (1984).
3. K.-H. Kampert, J. Phys. G (in press).
4. G. Bertsch, S. DasGupta, and H. Kruse, Phys. Rev. C **29**, 673 (1984); J. Aichelin and G. Bertsch, Phys. Rev. C **31**, 1730 (1985).
5. H. Kruse, B.V. Jacak, and H. Stöcker, Phys. Rev. Lett. **54**, 289 (1985); J.J. Molitoris and H. Stöcker, Phys. Rev. C **32**, 346 (1985); J.J. Molitoris, H. Stöcker, and B.L. Winer, Phys. Rev. C **36**, 220 (1987).
6. J. Aichelin and H. Stöcker, Phys. Lett. **B176**, 14 (1986); J. Aichelin, A. Rosenhauer, G. Peilert, H. Stöcker, and W. Greiner, Phys. Rev. Lett. **58**, 1926 (1987); G. Peilert, A. Rosenhauer, H. Stöcker, W. Greiner, and J. Aichelin, Mod. Phys. Lett. **A3**, 459 (1988); J. Aichelin, G. Peilert, A. Bohnet, A. Rosenhauer, H. Stöcker, and W. Greiner, Phys. Rev. C **37**, 2451 (1988).
7. G. Peilert, H. Stöcker, W. Greiner, A. Rosenhauer, A. Bohnet, and J. Aichelin, Phys. Rev. C **39**, 1402 (1989).
8. B. ter Haar and R. Malfliet, Phys. Lett. **172B**, 10 (1986); Phys. Rep. **149**, 207 (1987).
9. J. Cugnon, A. Lejeune, and P. Grangé, Phys. Rev. C **35**, 861 (1987).
10. D. Keane, S.Y. Chu, S.Y. Fung, Y.M. Liu, L.J. Qiao, G. VanDalen, M. Vient, S. Wang, J.J. Molitoris, and H. Stöcker, Phys. Rev. C **37**, 1447 (1988).
11. M. Berenguer, C. Hartnack, G. Peilert, A. Rosenhauer, W. Schmidt, J. Aichelin, J.A. Maruhn, W. Greiner, and H. Stöcker, Universität Frankfurt preprint UFTP 228/1989, and in *Nuclear Matter and Heavy Ion Collisions*, ed. M. Soyeur, Plenum, N.Y. (in press).
12. M. Gyulassy, K.A. Frankel, and H. Stöcker, Phys. Lett. **110B**, 185 (1982);
13. G.F. Bertsch, W.G. Lynch, and M.B. Tsang, Phys. Lett. **189B**, 384 (1987).
14. P. Danielewicz and G. Odyniec, Phys. Lett. **157B**, 146 (1985).
15. J. Aichelin, J. Cugnon, Z. Fraenkel, K. Frankel, C. Gale, M. Gyulassy, D. Keane, C.M. Ko, J. Randrup, A. Rosenhauer, H. Stöcker, G. Welke, and L.Q. Wu, Phys. Rev. Lett. **62**, 1461 (1989).
16. S. Hayashi, Y. Miake, T. Nagae, S. Nagamiya, H. Hamagaki, O. Hashimoto, Y. Shida, I. Tanihata, K. Kimura, O. Yamakawa, T. Kobayashi, and X.X. Bai, Phys. Rev. C **38**, 1229 (1988).
17. C.M. Ko, Q. Li, and R. Wang, Phys. Rev. Lett. **59**, 1084 (1987); C.M. Ko and Q. Li, Phys. Rev. C **37**, 2270 (1988); Q. Li, J.Q. Wu, and C.M. Ko, Phys. Rev. C **39**, 849 (1989).

COLLECTIVE FLOW MEASUREMENTS IN ASYMMETRIC NUCLEAR COLLISIONS

J. Gosset, M. Demoulins, D. L'Hôte, O. Valette, J.P. Alard* J. Augerat*, R. Babinet, N. Bastid*, F. Brochard** C. Cavata, P. Charmensat*, N. De Marco, P. Dupieux* H. Fanet, Z. Fodor, L. Fraysse*, P. Gorodetzky** M.C. Lemaire, B. Lucas, J. Marroncle*, G. Montarou*, M.J. Parizet*, J. Poitou, D. Qassoud*, C. Racca**, A. Rahmani* W. Schimmerling, Y. Terrien

DPhN, CEN Saclay, 91191 Gif-sur-Yvette Cedex, France
* LPC Clermont-Ferrand, BP 45, 63170 Aubière, France
**CRN, BP 20 CR, 67037 Strasbourg Cedex, France

The collective flow measurements presented in this lecture have been performed with the DIOGENE electronic 4π-detector[1] installed at the Saturne synchrotron in Saclay, which can deliver beams up to mass 40. In order to study the properties of dense (and hot) nuclear matter, it is required to measure nucleus-nucleus collisions involving a large number of nucleons That is the reason why we focused our experiment on asymmetric collisions,. with target nuclei heavier than the beam. The actual combinations of projectile, target, and energy for which we have available results are listed in table 1. It is indeed more difficult to study asymmetric collisions than symmetric ones, mainly because the center-of-mass reference frame of the participants system is not well defined. On the other hand, asymmetric collisions may follow different paths in the temperature-density phase diagram, as compared to symmetric collisions, and thus bring another independent piece of information on the nuclear matter equation of state. Our measurements are thus complementary to those of the Plastic Ball program, [ref.[2-4]] which was focused on symmetric systems. The streamer-chamber collaborations have also collected data on asymmetric systems.[5] However, with the DIOGENE electronic 4π- detector,[1] the information has much better statistical accuracy, and good separation is achieved between π^-, π^+, protons and deuterons.

Table 1
Projectile-target-energy combinations

projectile	target	energy per nucleon (MeV)
^{20}Ne	NaF, Nb, Pb	400 , 800
^{40}Ar	Ca, Nb, Pb	400
	Ca, Nb	600

The Nuclear Equation of State, Part A
Edited by W. Greiner and H. Stöcker
Plenum Press, New York

We measured triple differential cross sections of pseudoprotons (free protons as well as protons bound in light nuclei), inside the acceptance of the DIOGENE pictorial drift chamber (PDC),[1] restricted to $20° < \theta < 132°$ in polar angle and to kinetic energy larger than ~ 40 MeV. We analysed these cross sections in three different ways, which lead to various features of the collective flow : the usual flow parameter F, azimuthal angular distributions $dN/d\phi$ showing possible evidence for preferential emission transversely to the reaction plane, and finally two-dimensional Gaussian fits giving a more complete characterization of the participants collective flow, with the flow angle and two aspect ratios.

1. "MEASUREMENT" OF THE IMPACT PARAMETER VECTOR

In nucleus-nucleus collisions trivial geometrical effects provide very different initial conditions for central and peripheral collisions, especially in the case of asymmetric ones. The usual way to "measure" the impact parameter b is through the multiplicity M of light charged particles. It is quite obvious when a cylindrical clean cut between target and projectile is assumed, and it remains true in any model, that M is a monotonously decreasing function of b. In order to be able to compare our results from different projectile-target-energy combinations, we transformed all our multiplicity distributions into distributions of the (dimensionless) reduced impact parameter $\tilde{b} = b/(R_1 + R_2)$, where R_1 and R_2 are the projectile and target radii ($R = r_0 A^{1/3}$, $r_0 = 1.12$ fm), according to the simple formula :

$$\sigma(M > M_i) = \tilde{b}_i^2 \cdot \sigma_g$$

where σ_g is the total geometrical cross section, assumed to be equal to the total reaction cross section. Since the correlation between M and b has a rather large standard deviation, there is no point to make too small slices in the multiplicity distributions. We most often divided them into 4 bins of approximately equal cross sections, corresponding to 4 almost equally spaced $\tilde{b}^2$ values, between 0 and 0.5, which will be used for all the plots, as a general scale for all projectile-target combinations.

Even after impact parameter selection along those lines, if we want to measure triple differential cross sections, i.e. the full one-body momentum distribution of emitted particles, we also need to evaluate, on an event-by-event basis, the azimuth of the reaction plane, i.e. the direction of the impact parameter vector. For that purpose, we have adapted to our case the transverse momentum analysis proposed by Danielewicz and Odyniec.[6] For every event, the direction of the impact parameter vector is estimated by the azimuth of the vector $\vec{Q}_\nu$, defined as :

$$\vec{Q}_\nu = \sum_{\mu \neq \nu} Z_\mu \, (y_\mu - \langle y \rangle) \, \frac{\vec{p}_\perp^{\,\mu}}{m_\mu} ,$$

where μ and ν are indices for all charged baryons in the event, inside the acceptance of the DIOGENE PDC, with atomic number Z, mass m, rapidity y and transverse momentum $\vec{P}_\perp$. The $(y_\mu - \langle y \rangle)$ weight, where $\langle y \rangle$ is the estimated center-of-mass rapidity :

$$\langle y \rangle = (\sum_\mu Z_\mu \, y_\mu) / (\sum_\mu y_\mu) ,$$

is introduced in order to treat correctly asymmetric systems, where the participants center-of-mass rapidity is not known a priori. The reaction plane estimated in this way corresponds to the plane going through the beam axis and over which the projections of the transverse momenta of all particles in an event present the highest degree of correlation with their rapidities. The $\mu \neq \nu$ restriction of the summation is intended for removing self-correlations between the $\vec{Q}_\nu$ vector and the momentum of the particle with index ν.[6] For each particle, we obtain by projection the two components of its transverse momentum $\vec{P}_\perp^\nu$, in the estimated reaction plane ($p_{x'}^\nu = \vec{P}_\perp^\nu \cdot \vec{Q}_\nu/|\vec{Q}_\nu|$) and out of the estimated reaction plane ($p_{y'}^\nu$). Due to inaccuracies in the reconstruction of the reaction plane, these components are different from the components in and out of the true reaction plane. All the results about the measured triple differential distributions (y ; $p_{x'}$; $p_{y'}$) have to be corrected for the fluctuations $\Delta\phi$ of the azimuth of the estimated reaction plane around the azimuth of the true one. Such corrections can only be applied on average. They are based on estimates of the average values of $\cos\Delta\phi$ and of $\cos^2\Delta\phi$, made with the assumption that all correlations originate from the existence of the reaction plane. The average value of $\cos\Delta\phi$ is estimated according to reference 6. In order to analyse not only the first but also the second moments of the distributions, we need an estimate of $<\cos^2\Delta\phi>$. This is done along the lines of reference 7, using three-body correlations. Since the measured multiplicities are not very high (limited down to 6, and 40 at most), the corrections for finite number effects are rather important. Typically $<\cos\Delta\phi> = 0.4$-0.5 with a relative uncertainty smaller than 3 % and $<\cos^2\Delta\phi> = 0.55$-$0.65$ with a relative uncertainty between 5 and 15%.

2. FLOW PARAMETER F

A first simple way of analysing the triple differential cross sections obtained from the transverse momentum method, has been to look at the rapidity dependence of $<p_{x'}>$, the average value of the transverse momentum projection on the reaction plane. Around the rapidity at which $<p_{x'}>$ is equal to zero, our data show a linear dependence (Fig. 1), as previously observed by the streamer chamber[6] and the Plastic Ball[8] collaborations. In fact we use the dimensionless parameter $<p_{x'}/m>$, that we adjust linearly with a slope S and a rapidity y_0 at which $<p_{x'}>$ is equal to 0 :

$$< p_{x'}/m > (y) = S\,(y-y_0)$$

The linear fit is performed over a limited region of rapidity around y_0 (triangles in Fig. 1) where the statistics is the biggest. The slope S has to be corrected for finite number effects, and we finally get the flow parameter F :

$$F = S/<\cos\Delta\phi>$$

Our measurements of the flow parameter F are plotted in Fig. 2 as a function of $\tilde{b}^2$ for Ne + Pb at 400 and 800 MeV per nucleon, and for Ar + Ca and Ar + Pb at 400 MeV per nucleon, for all pseudoprotons and also for selected heavy fragments between deuterons and alpha particles. The values of F for pseudoprotons always lie between 0.35 and 0.45, with a slight bump between $\tilde{b}^2 = 0$ and 0.5. There is almost no change, for Ne + Pb, between the values measured at 400 and 800 MeV per nucleon. At 400 MeV per nucleon, the flow parameter F is slightly higher for the asymmetric system Ar + Pb than for the symmetric one Ar + Ca, and also slightly higher for Ar + Pb than for Ne + Pb. Finally, there is a clear trend for the flow F to increase with the fragment mass, which is in qualitative agreement with the results of the Plastic Ball.[9] Even though the particle identification with DIOGENE is not so good above deuterons, the result for deuterons is already significant.

The main conclusion at this point is that the flow parameter F shows only small variations with projectile, target, energy, or impact parameter.

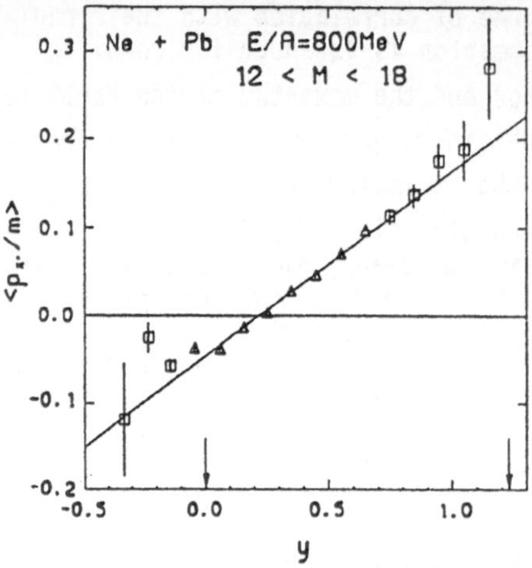

Fig. 1. Average in-plane momentum (divided by the proton mass) as a function of rapidity. The two arrows indicate the target and projectile rapidities.

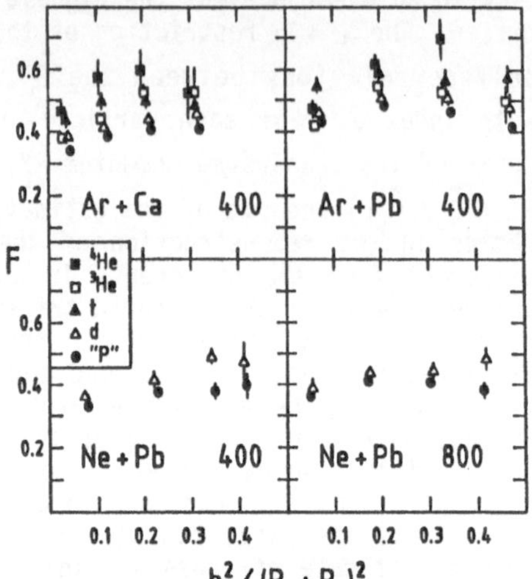

Fig. 2. Impact parameter dependence of the flow parameter.

Two reasons could explain, at least qualitatively, this behaviour. Firstly, at any value of the rapidity, we miss the central region of the p_x' distribution, where the rate is the highest. The flow parameter F that we measure is certainly affected by these cuts. In any case, it is larger than the value that would be extracted without any instrumental cut. Secondly, even without any cut, this flow parameter is not sufficient to characterize completely the collective flow of fragments emitted from nuclear collisions. In particular, if the collective flow in the reaction plane is parametrized with an ellipse shape, i.e. with a flow angle θ_F (the angle of the ellipse principal axis) and an aspect ratio r (ratio between the bigger and smaller semi-axes), the flow parameter F can be expressed as a function of θ_F and r :

$$F = tg\theta_F \ (r^2 - 1)/(r^2 + tg^2\theta_F)$$

It is clear that the flow angle is strongly correlated with F ($F = tg\theta_F$) only in the limit $r \gg 1$ and $r \gg tg\theta_F$. In most physical cases, F is a complicated mixture between θ_F and r. It would be much better to have a direct measurement of the complete emission pattern, with two parameters instead of the F parameter alone.

3. AZIMUTHAL DISTRIBUTIONS dN/dϕ

A second type of triple differential cross section analysis has been to plot, for several values of the rapidity, the distribution of the azimuth at which particles are emitted with respect to the reaction plane. The collective flow in the reaction plane should result in the peaking at 0° (resp. 180°) of these distributions for positive (resp. negative) values of the rapidity. It has been pointed out that these distributions, complementary to the flow parameter F, are strongly influenced by the nuclear equation of state.[10] At mid-rapidities, the contribution of the collective flow in the reaction plane should vanish, and the azimuthal distributions should therefore be very sensitive to out-of-plane emission of particles, and could even show a preferential emission transverse to the reaction plane (squeeze-out) as predicted by hydrodynamical calculations.[11] Our azimuthal distributions (Fig. 3 and ref. 12) indeed peak at 0° (resp. 180°) for posi-

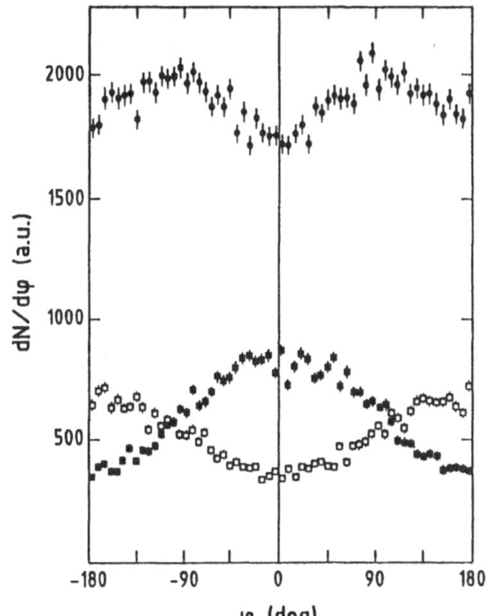

Fig. 3. Azimuthal distributions of pseudoprotons emitted at mid-rapidity ($0.15 < y < 0.35$; circles), and at larger ($y > 0.65$; filled squares) or smaller ($y < -0.05$; open squares) rapidities, for Ne + Pb collisions at 800 MeV per nucleon and $\tilde{b}^2 = 0.3$.

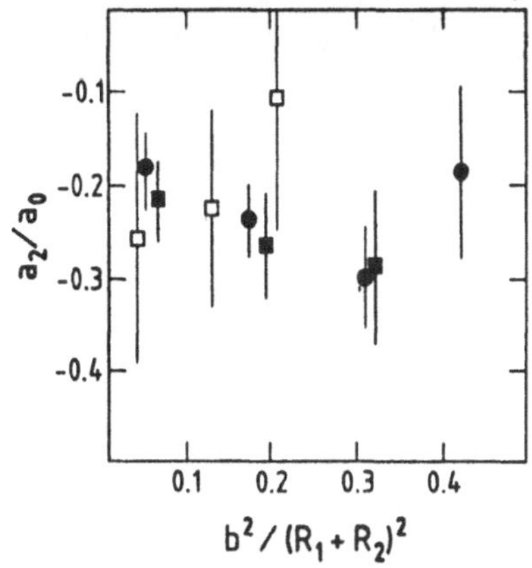

Fig. 4. Impact parameter dependence of the degree of preferential emission at 90° with respect to the reaction plane, for pseudoprotons emitted at mid-rapidity in collisions of Ne with Pb (circles), Nb (filled squares) and NaF (open squares), at 800 MeV per nucleon.

tive (resp. negative) values of the rapidity. At mid-rapidity, they are not flat at all, but rather present a wide bump at 90°, i.e. in the transverse direction with respect to the reaction plane.

A systematic study of this out-of-plane emission has been performed by fitting our azimuthal distributions to a Legendre polynomial expansion up to the second order, with parameters a_0, a_1 and a_2 depending upon the rapidity y. The a_1/a_0 ratio should mainly reflect the collective flow in the reaction plane, and the a_2/a_0 ratio the squeeze-out emission, with a negative sign corresponding to a peak at 90°. The fitting procedure is well suited to all our measurements, yielding values of χ^2 per degree of freedom close to unity. For a given system at a given impact parameter, $|a_2/a_0|$ is dependent upon rapidity, with a broad maximum around the mid-rapidity y_0 defined in the previous analysis of the flow parameter F. It means that, even though the out-of-plane emission is mostly visible at mid-rapidity, where the dominant a_1/a_0 contribution to the azimuthal distribution vanishes, it is still present at other values of the rapidity and never completely vanishes. For neon-nucleus collisions at 800 MeV per nucleon, the a_2/a_0 ratio, measured at mid-rapidity and corrected for finite number effects, is plotted in Fig. 4 as a function of $\tilde{b}^2$ for the three targets (NaF, Nb, Pb). The error bars are rather large, because of the uncertainties in the evaluation of the $\langle\cos^2\Delta\phi\rangle$ correction, which is needed since the second order Legendre polynomial contains a term in $\cos^2\phi$. Within the error bars, the Ne + Nb and Ne + Pb results present the same behaviour : $|a_2/a_0|$ slowly increases with $\tilde{b}^2$, from ~ 0.2 at $\tilde{b}^2=0$ to ~ 0.3 at $\tilde{b}^2=0.3$.

This type of analysis has also been used by the Plastic Ball group [refs.[3,13]]. It has even been pushed further. After rotation in the reaction plane by $(-\theta_F)$, they get azimuthal distributions with a vanishing a_1/a_0 coefficient, which exhibit more clearly the transverse emission with respect to the reaction plane. Due to the importance of DIOGENE cuts, this procedure cannot be applied to our data.

4. TWO-DIMENSIONAL GAUSSIAN FITS

There are several drawbacks to the previous analyses of the triple differential cross sections. Firstly they do not make full use of the three-dimensional information. In the flow parameter analysis one integrates over the transverse dimension with respect to the reaction plane. The azimuthal distributions analysed in the previous section result from integration over the transverse momentum. Even it they were sorted against both rapidity and transverse momentum, it would be difficult to summarize in a single way this complete information. Secondly the previous analyses suffer from biases due to the limitations of the detector acceptance. We have thus performed a third kind of analysis, in order to characterize more completely, and independently of the detector cuts, the three-dimensional behaviour of the nuclear collective flow.

The basic idea of this third analysis consists in fitting the $(p_z/m, p_{x'}/m)$ and $(p_z/m, p_{y'}/m)$ two-dimensional cross sections with two-dimensional Gaussian distributions.[14] The cross sections are evaluated in the center-of-mass reference frame for each multiplicity cut, and p_z is the longitudinal momentum in this frame. With this procedure there is not yet a complete use of the three-dimensional information. However, a three-dimensional Gaussian emission pattern, symmetric with respect to the reaction plane, would provide a simple summary of the complete information, and it reduces to two-dimensional Gaussian distributions for the in-plane and out-of-plane cross sections. The raw two-dimensional cross sections are strongly affected by the detector cuts at small values of $|p_{x'}|$ or $|p_{y'}|$ because, at any given value of p_z, the acceptance criterion corresponds to a minimum value of the transverse momentum, and not to a minimum value of either $|p_{x'}|$ or $|p_{y'}|$. If we apply the acceptance criterion to $|p_{x'}|$ for the in-plane cross section and to $|p_{y'}|$ for the out-of-plane cross section, we indeed lose some information, at small values of $|p_{x'}|$ or $|p_{y'}|$. However, the remaining information is completely free of any bias from the detector acceptance. The fitting procedure is thus performed on the $(p_z/m, p_{x'}/m)$ and $(p_z/m, p_{y'}/m)$ cross sections after the clean cut is applied to $|p_{x'}|$ and $|p_{y'}|$, respectively. These clean cross sections only span a limited region of phase space. However, in this limited region, at high enough transverse momentum, there should not be any contribution from the spectators of the collisions. This is another advantage when we are mostly interested in the emission pattern of the participants.

The parameters extracted from this analysis are one angle θ and two standard deviations τ_3 and τ_1 $(\tau_3 > \tau_1)$ for the in-plane cross sections, one angle θ' and two standard deviations τ_2 and τ for the out-of-plane cross sections. Since the in-plane and out-of-plane cross sections are not independent, but result from the integration of the same three-dimensional cross sections over $p_{y'}$ and $p_{x'}$ respectively, the parameters obtained from the two fits are related as follows :

$$\tau^2 = \tau_3^2 \cos^2\theta + \tau_1^2 \sin^2\theta$$

$$\theta' = 0° \text{ (if } \tau_2 < \tau) \quad \text{or} \quad 90° \text{ (if } \tau_2 > \tau)$$

The angle θ and the variances τ_3^2, τ_1^2 and τ_2^2 can also be interpreted as the parameters of a three-dimensional Gaussian distribution, closely related to the flow angle and the eigenvalues obtained from a sphericity analysis. All these parameters are finally corrected for finite number effects, using the estimates of $\langle\cos\Delta\phi\rangle$ and $\langle\cos^2\Delta\phi\rangle$, and transformed into the true flow angle θ_F and the true variances σ_3^2, σ_1^2 and σ_2^2. With a χ^2 per degree of freedom varying between 1.5 and 3, the quality of the fits is not so bad, especially if we realize that the Gaussian distribution is a very simple one. Our cross sections are not corrected for distortions due to the resolution on the momentum measurements. This could explain that the fitted source velocity, for symmetric systems, is a little smaller than the center-of-mass velocity.

Typical results for the in-plane collective flow parameters, θ_F, σ_3 and σ_1, are illustrated with ellipses in Fig. 5 for asymmetric systems and several values of the impact parameter. The semi-axes of these ellipses are equal to $\sqrt{2}$ times the standard deviations σ_3 and σ_1. The flow angle θ_F increases when the impact parameter decreases, up to values as large as about 60°. The size of the ellipses, which reflects roughly the energy available per participant nucleon in the participant center-of-mass, decreases when the impact parameter decreases, in qualitative agreement with the simple picture that the projectile nucleons encounter more and more nucleons from the heavy target when the impact parameter decreases.

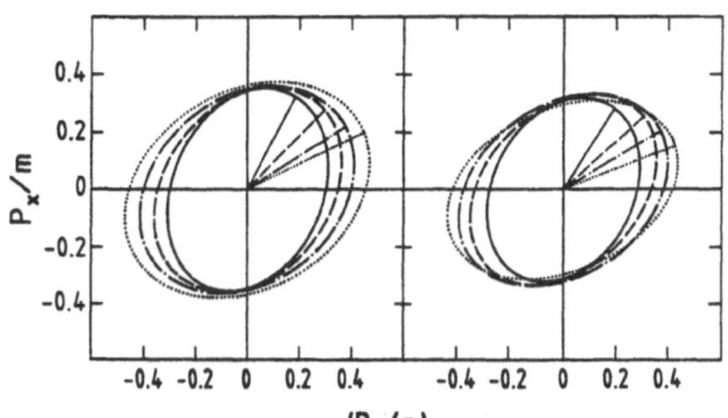

Fig. 5. Ellipses representing the two-dimensional Gaussian fits to the in-plane cross sections, for Ne + Pb collisions at 800 MeV per nucleon (left) at $\tilde{b}^2$=.05, .17, .30, .42 and for Ar + Pb collisions at 400 MeV per nucleon (right) at $\tilde{b}^2$ = .06, .20, .33, .46. Increasing $\tilde{b}^2$ values correspond successively to full, dashed, dot-dashed and dotted lines.

The shape of the emission pattern can be summarized with three quantities : the flow angle θ_F and two aspect ratios λ_{31} and λ_{32}, respectively equal to σ_3^2/σ_1^2 and σ_3^2/σ_2^2. Systematic results for these three parameters are shown in Figs. 6, 7 and 8, as a function of $\tilde{b}^2$. Except for the lightest system Ne + NaF, the flow angle θ_F is anticorrelated to $\tilde{b}^2$ (Fig. 6). For any given values of the projectile mass, incident energy and $\tilde{b}^2$, the flow angle is larger for heavier targets. The in-plane aspect ratio λ_{31} increases almost linearly with $\tilde{b}^2$ (Fig. 7), more rapidly for the argon-nucleus than for the neon-nucleus collisions, but almost independently of the tar-

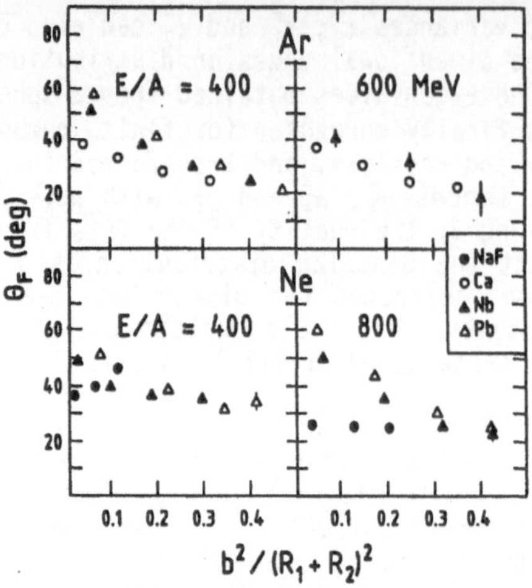

Fig. 6. Impact parameter dependence of the flow angle.

get. The out-of-plane aspect ratio λ^2_{32} also increases linearly with $\tilde{b}^2$ (Fig. 8), but less rapidly than λ^2_{31}, and with some dependence upon the target. At any value of $\tilde{b}^2$, for all projectile-target-energy combinations, the out-of-plane aspect ratio is smaller than the in-plane aspect ratio. This is in qualitative agreement with the previous observation that the mid-rapidity azimuthal distributions are peaked at 90° with respect to the reaction plane.

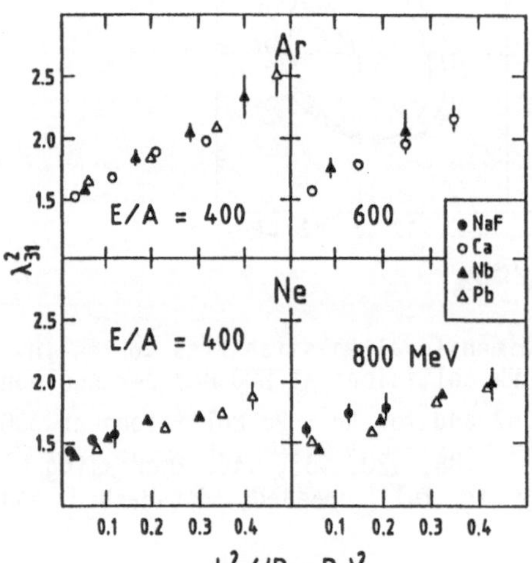

Fig. 7. Impact parameter dependence of the in-plane aspect ratio.

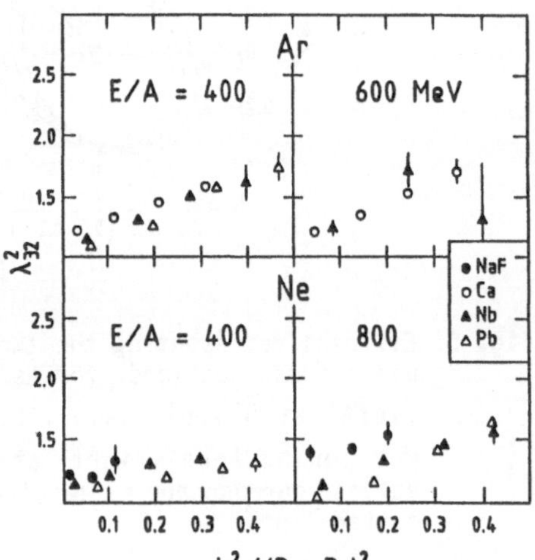

Fig. 8. Impact parameter dependence of the out-of-plane aspect ratio.

A comparison with predictions from intranuclear cascade calculations[15] is shown in Fig. 9 for the flow angle θ_F in argon-nucleus collisions. A simple filter is applied to the cascade outputs, including only the polar angle window and the kinetic energy threshold, but not including the full simulation of the DIOGENE biases, due to finite resolution of momentum measurements and inefficiencies of the track reconstruction. Anyhow there is a huge difference between the experimental results and the cascade prediction, more important for more central collisions. There is about the same difference between experimental results and cascade predictions concerning the flow parameter F. It remains to be seen whether all these results can be reproduced by more sophisticated models including the effects of the in-medium propagation of particles and the nuclear matter equation of state.

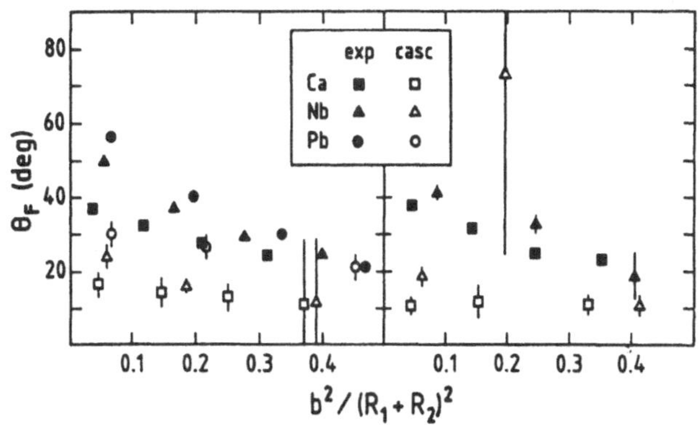

Fig. 9. Comparison between experimental results (filled symbols) and predictions of intranuclear cascade calculations (open symbols), concerning the impact parameter dependence of the flow angle, in argon-nucleus collisions at 400 (left) and 600 (right) MeV per nucleon.

When plotted in the plane $(\theta_F,\ r = 2\ \sigma_3^2/(\sigma_1^2 + \sigma_1^2))$, our data show a trend very similar to the results obtained by the streamer chamber collaboration[7] with a different method. Only the trends can be compared because the available data correspond to different energies. Our results for the three-dimensional Gaussian distributions can be used to extract the true flow parameter, which is in agreement with results from the Plastic Ball collaboration.[8] Flow angles were also reported by this collaboration[16] after an event-by-event sphericity analysis. For the rather light system Ar + Ca they did not observe any peak at finite angle in the $dN/d(\cos\theta_F)$ distributions. This could appear as a contradiction with the rather large values (20-40°) that we measured (Fig. 6). However, it should be noted that the sphericity analysis is performed on all particles detected by the Plastic Ball, thus mixing the spectators and the participants. Moreover, the results of the event-by-event sphericity analysis are more strongly influenced by finite number effects than the event-by-event transverse momentum analysis for finding the reaction plane, followed by a sphericity analysis performed on the sum of the rotated events.

5. CONCLUSION

Two-dimensional Gaussian fits, performed on both in-plane and out-of-plane cross sections in order to summarize the participant emission pattern, give results (the flow angle and two aspect ratios) that should not be influenced by the detector biases. They show some dependence upon the values of the impact parameter, projectile and target masses, and incident energy. This is in contrast with the usual flow parameter and the degree of preferential emission at 90° with respect to the reaction plane.

For argon-nucleus collisions, intranuclear cascade calculations predict flow angles, and also flow parameters, that are too small by a factor 2. There is now an urgent need for a careful and systematic comparison between all our results and the predictions of more sophisticated models that incorporate explicitly the properties, or equation of state, of dense and hot nuclear matter, together with the propagation of particles inside this dense medium. For such a comparison, especially in order to ascertain the differences due to the variation of the stiffness of the equation of state, theoretical simulation with large enough statistics are needed, which requires a large amount of computer time.

REFERENCES

1. J.P. Alard et al., Nucl. Instr. Meth. A261: 379 (1987).
2. H.G. Ritter, lecture given at this school.
3. K.H. Kampert, lecture given at this school.
4. H.H. Gutbrod, A.M. Poskanzer and H.G. Ritter, LBL-26922 (1989), submitted to Reports on Progress in Physics.
5. H. Stroebele, lecture given at this school.
6. P. Danielewicz and G. Odyniec, Phys. Lett. B157: 146 (1985).
7. P. Danielewicz et al., Phys. Rev. C38: 120 (1988).
8. K.G.R. Doss et al., Phys. Rev. Lett. 57: 302 (1986).
9. K.G.R. Doss et al., Phys. Rev. Lett. 59: 2720 (1987).
10. G.M. Welke et al., Phys. Rev. C38: 2101 (1988).
11. J.A. Maruhn, lecture given at this school.
12. D. L'Hôte, talk given at "Fifth Gull Lake Nuclear Physics Conference", Gull Lake, MI (1988).
13. H.H. Gutbrod et al., Phys. Lett. B216: 267 (1989).
14. O. Valette et al., Proc. of the XVIth International Workshop on Gross Properties of Nuclei and Nuclear Excitations, Hirschegg (1988).
15. J. Cugnon et al., Nucl. Phys. A379: 553 (1982) ;
    J. Cugnon and D. L'Hôte, Nucl. Phys. A452: 738 (1986) ;
    a new version including isospin effects, Pauli blocking and prescriptions for simulating binding energy has been used.
16. H.A. Gustafsson et al., Phys. Rev. Lett. 52: 1590 (1984) ;
    H.G. Ritter et al., Nucl. Phys. A447: 3c (1985).

4π FRAGMENT MEASUREMENTS AT MSU AND THE NUCLEAR EQUATION OF STATE

G.D. Westfall, D.A. Cebra, J. Clayton, P. Danielewizc
S. Howden, J. Karn, C.A. Ogilvie, A. Nadasen†
A. Vander Molen, W.K. Wilson, and J.S. Winfield

National Superconducting Cyclotron Laboratory and
Department of Physics and Astronomy

Michigan State University
East Lansing, Michigan 48824-1321 USA

INTRODUCTION

In the collision of two nuclei at intermediate energies (20 - 200 MeV/nucleon) a variety of phenomena can be studied including multifragmentation, the liquid-gas phase transition in nuclear matter, energy and momentum flow, the production of entropy, and the role of thermodynamics in these collisions.[1,2] These phenomena involve correlations of many particles. In such experiments it is clear that the more complete the event characterization is, the more accurate will be the understanding of the underlying phenomena. Presented here are the results from the first set of experiments using the MSU 4π Array to study nuclear collisions at intermediate energies using the NSCL K500 Superconducting Cyclotron. The systems studied are 35 MeV/nucleon $^{40}Ar+^{51}V$ and Au and 50 MeV/nucleon $^{12}C+^{12}C$ and Au. In this paper we will concentrate on the results for the symmetric systems of Ar+V and C+C. The measurements include charged particle multiplicity, total observed charge, reaction plane determination, impact parameter determination, the average transverse momentum as a function of rapidity, the average ratio of in-plane transverse momentum to the total transverse momentum, the sphericity versus the coplanarity of the each event, the production of intermediate mass fragments as a function of observed charge, and triple differential cross sections. These observations will be used to provide information about event topography selection, directed transverse momenta, multifragmentation, entropy production, and the roles of collective versus thermal energy.

The MSU 4π Array in its current configuration has 215 fast/slow phoswich scintillator detectors.[3] A schematic drawing of the Array is shown in Fig. 1. In Fig. 1 the vacuum chamber is shown completely surrounding the phoswich counters. The

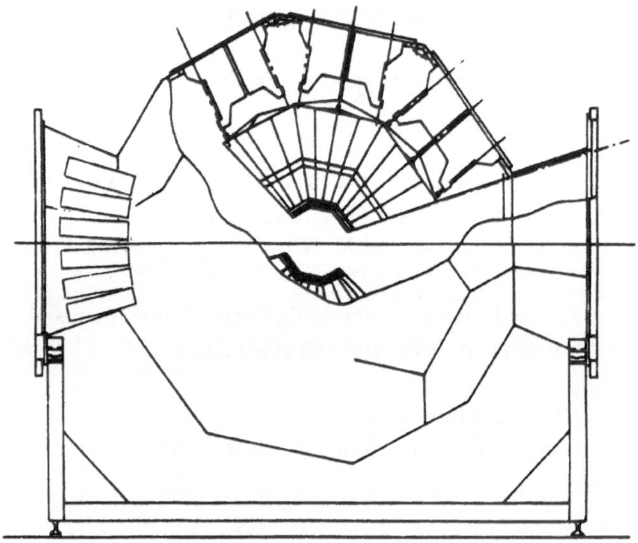

Figure 1. Schematic drawing of the MSU 4π Array showing the location of the forward array.

main ball of the Array contains thirty subarrays of phoswiches totalling 170 detectors covering angles from 20 to 160 degrees in the lab. These counters are made up of a 3 mm thick fast plastic ΔE and a 25 cm thick slow plastic E counter. The dynamic range of these counters was set to allow the detection of particles with Z up to 4. An array was placed in the forward direction consisting of 45 telescopes covering angles from 7 to 18 degrees in the lab. These counters were constructed using 1.5 mm thick fast plastic ΔE and 15 cm thick slow plastic E counters. The dynamic range of these counters was set to accept fragments with Z up to 8. The trigger for the system was based on the multiplicity in either the main ball or in the forward array. Minimum bias triggers (multiplicity ≥ 1) as well as high multiplicity triggers (multiplicity ≥ 5) based on either the main ball detectors or the forward array were used to collect the data.

Energy calibration of each telescope was accomplished off-line by matching to very complete energy calibrations for a wide range of particle types and energies combined with range-energy calculations. Thus effects due to gain shifts and changes in the energy spectra due to the event trigger timing could be corrected. The energy calibrations are accurate to within 10%. The lower energy cut-off caused by the ΔE thickness was around 17 MeV/nucleon for the main ball counters and 10 MeV/nucleon for the forward array counters. In the near future the MSU 4π Array will be completed with the addition of thirty Bragg Curve spectrometers to allow the identification of intermediate mass fragments with energies of a few MeV/nucleon. In addition thirty low pressure multi-wire proportional counters will be installed to measure fission fragments.

EVENT TOPOGRAPHY SELECTION

The first two characteristics of intermediate energy nuclear collisions to be extracted are the centrality of the collision and the reaction plane of the event. These two quantities play a vital role in the understanding of such collisions.

The impact parameter of a collision can be related to various observables including charged particle multiplicity, total observed charge, mid-rapidity charge, moments of the momentum distribution[4], the forward angle multiplicity, the observed forward angle charge, the multiplicity of intermediate mass fragments, the backward angle charge particle multiplicity, neutron multiplicity, and the fission fragment folding angle. We have simulated several of these observables using FREESCO by Fai and Randrup.[5,6] This model is capable of producing complete events including complex fragments and simulating transverse momentum using multiple moving sources. The events generated using FREESCO were passed through a filter that accounts for the charged particle particle acceptance of the MSU 4π Array including geometry, energy cutoffs, and double hits.

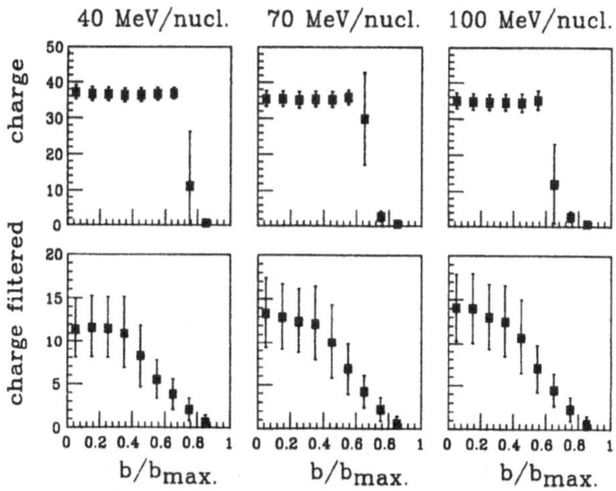

Figure 2. FREESCO simulations for Ca+Ca at 40, 70, and 100 MeV/nucleon. The mid-rapidity charge is calculated as described in the text.

In Fig. 2 a representative set of simulations is shown for three incident energies, 40, 70, and 100 MeV/nuclcon, for Ca+Ca as a function of the impact parameter, b, divided by the sum of the radii of the two nuclei, b_{max} using FREESCO. Note that the intermediate rapidity charge from the calculation is independent of the impact parameter because the choice of the mid-rapidity cuts ($0.75y_t \leq y \leq 0.75y_p$) includes most of the spectator particles except in peripheral collisions. However, when these calculations are passed thought the filter, a strong correlation between impact parameter and mid-rapidity charge, Z_{mr}, is observed. The mid-rapidity charge is better than simple charged particle multiplicity because the total charge is less sensitive to the details of the breakup of the system into fragments and thus is less model dependent. In the following work, four impact parameter bins will be presented based on the mid-rapidity charge. The four bins will be termed central, near central, near peripheral, and peripheral. The values for the various cuts in Z_{mr} versus the impact parameter are given in Table . The relative percentages of events falling into each

Table 1. The gates on mid-rapidity charge (Z_{mr}) used to select impact parameter ranges.

Reaction	Gate	Z_{mr}	Median impact
C+C 50 MeV/nucleon	central	$Z \geq 7$	0.3
"	mid-central	$5 \leq Z < 7$	0.5
"	mid-peripheral	$2 \leq Z < 5$	0.7
"	peripheral	$0 \leq Z < 1$	0.9
Ar+V 35 MeV/nucleon	central	$Z \geq 11$	0.3
"	mid-central	$7 \leq Z < 11$	0.5
"	mid-peripheral	$2 \leq Z < 7$	0.7
"	peripheral	$0 \leq Z < 1$	0.9

impact parameter bin is illustrated in Fig. 3. Note that the distributions have finite widths and that there is some overlap between the various impact parameter bins. However, each of the four bins provides a quantitative selection on impact parameter. In Fig. 3, gate 1 corresponds to the most central impact parameter and gate 4 corresponds to the most peripheral impact parameters.

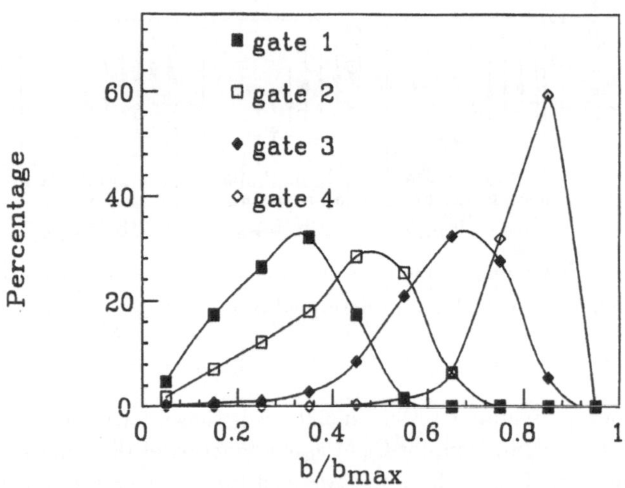

Figure 3. Percentage of events falling into four impact parameter bins from FREESCO simulations passed through the acceptance filter.

Another type of event topography selection is the determination of the reaction plane of each event. This determination is carried out assuming that if there is some collective transverse momentum in the reaction plane, then the vector $\vec{Q}$

$$\vec{Q} = \sum_{\nu=1}^{n} \omega_\nu \vec{p}_{\perp\nu} \tag{1}$$

will lie in the reaction plane. Here n is the number of particles, ω_ν is a weighting factor, and $\vec{p}_{\perp\nu}$ is the perpendicular momentum of each particle. In the original work by Danielewizc and Odyniec[7], these weighting factors were chosen to be +1 or -1 depending on whether the particle was emitted in the forward or backward direction in the center of mass system. Here the weighting factors are chosen to be the component of each particle's momentum in the beam direction, p_z, evaluated in the center-of-mass system.[8] In Fig. 4 the angle between the found reaction plane and the known reaction plane is shown for 70 MeV/nucleon Ca+Ca for a range of impact

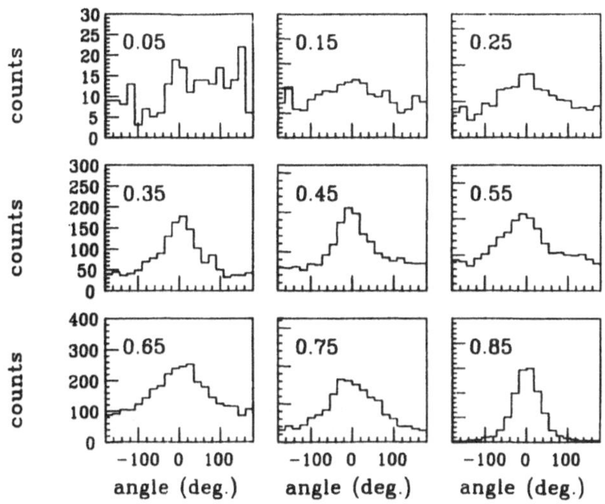

Figure 4. Difference of the azimuthal angle between the found and known reaction planes for events simulated with FREESCO for 70 MeV/nucleon Ca+Ca filtered through the acceptance of the 4π Array.

parameters. The calculations have been filtered by the acceptance of the experimental apparatus. Note that the reaction plane is poorly determined in very central collisions while the determination is very good for impact parameters greater than 0.3 times the sum of the radii of the two nuclei. The projected transverse momentum for particle i, p_{xi}, is given by

$$p_{xi} = \frac{\vec{p}_{\perp i} \cdot \vec{Q}'}{\mid \vec{Q}' \mid} \tag{2}$$

where $\vec{Q}'$ is the vector determined by all other particles except the particle being projected. This modification is necessary to remove autocorrelation effects. This vector is given by

$$\vec{Q}' = \sum_{j=1, j\neq i}^{n-1} \omega_j \vec{p}_{\perp j}. \tag{3}$$

In addition to removing autocorrelation effects, another effect was considered where the effect of momentum correlations in these light systems was minimized by adding a velocity boost[9], $\vec{v}_{boost}$, to each particle to insure that net momentum is zero in the system used to calculate the reaction plane. This boost is given by

$$\vec{v}_{boost} = \frac{\vec{p}_{\perp i}}{m_{system} - m_i} \tag{4}$$

where m_{system} is the mass of the emitting system and m_i is the mass of each particle. A further refinement for displaying these results is to plot the average of the fraction of the transverse momentum within the reaction plane to the total transverse momentum, ¡$p_x / < p_\perp >$.

In Fig. 5 the fraction of transverse momentum in the reaction plane divided by the

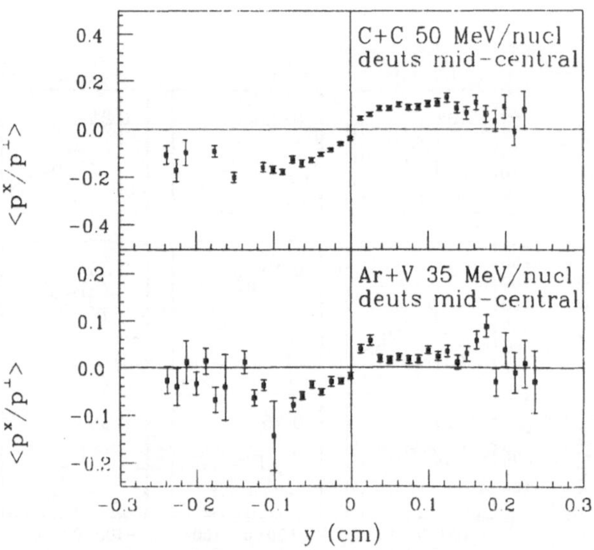

Figure 5. Tranverse momentum versus rapidity for C+C and Ar+V.

total transverse momentum is plotted as a function of rapidity for 50 MeV/nucleon C+C and 35 MeV/nucleon Ar+V. A surprising result is visible in this figure:a substantial amount of collective motion is observed for collisions in these light systems and at these relatively low energies. In order to compare these results with those at higher energies[7,10,11,12], some method should be used to remove simple beam energy effects. We have chosen to divide the observed transverse momentum per nucleon, $< p_{x/a} >$ by the center of mass momentum per nucleon of the projectile. In Fig. 6 the scaled average transverse momentum, $< p_{x/a} > / p^{cm}_{proj}$, is plotted versus the incident energy. The current results for Ar+V can be compared to the higher energy results for Ca+Ca. Note that the present value is comparable to the higher energy results. This similarity could be fortuitous. The sign of the scattering angle is predicted to be negative at energies around 35 MeV/nucleon[13,14] using VUU while at higher energies the scattering is predicted to be repulsive. Thus one expects that the average transverse momentum must go through zero and become negative with respect to the reaction plane. However, the experimental observable will always remain positive because of the method used to determine the reaction plane described above. This phenomenom would be visible in Fig. 6 by a dip towards zero average transverse momentum as a function of incident energy. Clearly this effect requires more extensive excitation functions and studies of the mass dependence.

MULTI-FRAGMENT EMISSION

In Fig. 7a the charged particle multiplicity is shown for 35 MeV/nucleon Ar+V for a minimum bias trigger using the main ball counters. This multiplicity includes fragments observed in the forward array. In Fig. 7b, the multiplicity of fragments with $Z \geq 3$ is also shown. This measurement includes fragments with $Z \leq 4$ in the main ball counters and fragments with $Z \leq 8$ in the forward array. It is clear that one is observing events with many charged particles and events with more than two complex fragments. However the question remains, how does this breakup process proceed? The final state observed experimentally could be reached in a variety of ways. For example the two nuclei could collide and form a hot system that emits particles sequentially, reequilibrating after each particle is emitted. On the other hand, the system could explode simultaneously into many fragments. The observed multiplicity distributions could be produced by either process.

However, in a multi-particle measurement such as the present experiment, one has the ability to study globally the momentum distribution of each event. López and Randrup[15] have proposed that sequential emission can be distinguished from instantaneous breakup by studying various elements of the energy/momentum tensor defined by

$$F_{ij} = \sum_n \frac{p_i^n p_j^n}{2m_n}. \tag{5}$$

Ordered eigenvectors of F, $f_1 < f_2 < f_3$ are used to define the quantities

$$q_i = \frac{f_i^2}{\sum_{j=1}^{3} f_j^2}. \tag{6}$$

In terms of these variables, sphericity and coplanarity are defined as $S = \frac{3}{2}(1-q_3)$ and $C = \frac{1}{2}\sqrt{3}(q_2 - q_1)$ respectively. López and Randrup demonstrated that by plotting the correlation between these two parameters one could differentiate between the two processes.

For sequential processes, most of the emitted fragments lie near a line corresponding to two-dimensional shapes. The prediction for events exhibiting simultaneous emission fills the entire triangular region corresponding to more spherical shapes. However, these calculations predict the emission of some particles that cannot be detected using the 4π Array in its current configuration. Therefore the calculations for each type of dynamical decay were run through the acceptance filter of the MSU 4π Array and the results are shown in Fig. 8 for the simultaneous emission of fragments and in Fig. 9 for sequential emission of fragments. There is a definite difference between the two calculations.

The experimental data for 35 MeV/nucleon Ar+V is shown in Fig. 10 for central collisions. The data look very much like the calculation describing multifragmentation. The vertical scale of the two dimensional histogram is logarithmic. Clearly values of S and C corresponding to more spherical distributions are populated.

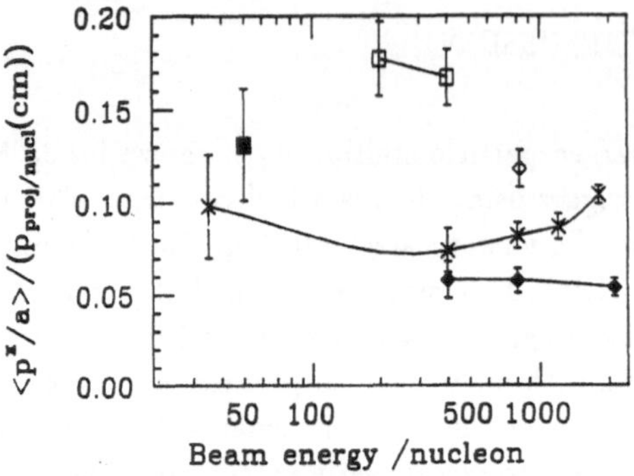

Figure 6. Scaled average transverse momentum values comparing present results for C+C and Ar+V to results at higher energies[7,10,11,12].

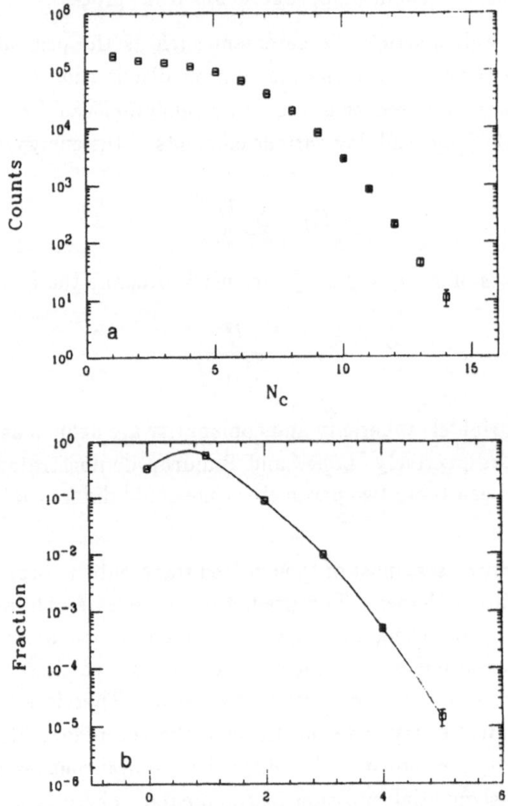

Figure 7. Charge particle multiplicity a) and complex fragment multiplicity from 35 MeV/nucleon Ar+V b).

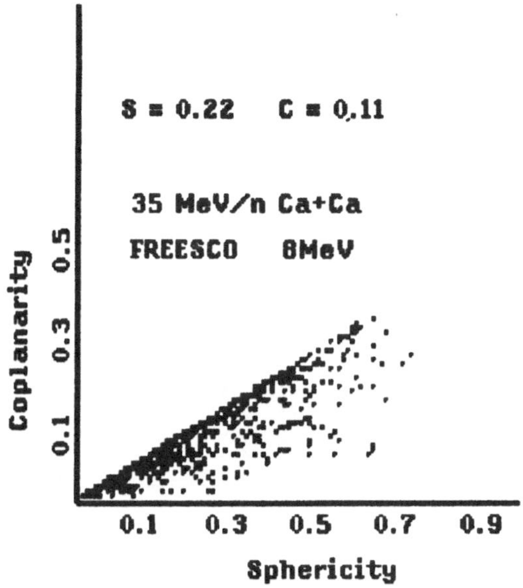

Figure 8. Theoretical predictions for simultaneous emission of fragments filtered through the acceptance of the MSU 4π Array.

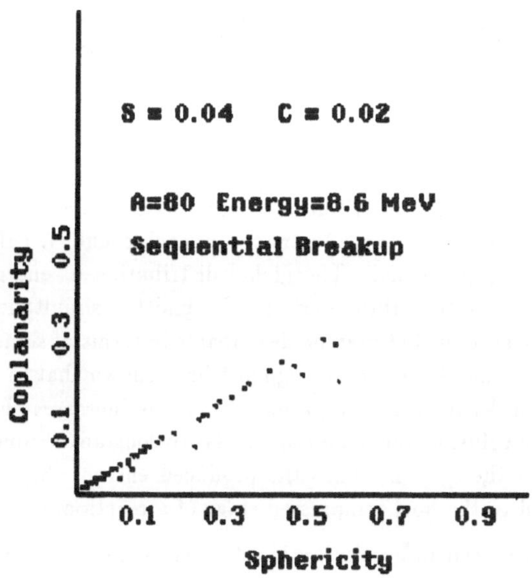

Figure 9. Theoretical predictions[15] for sequential emission of fragments filtered through the acceptance of the MSU 4π Array.

Thus we conclude that this process probably proceeds through a simultaneous breakup rather that a sequential one.

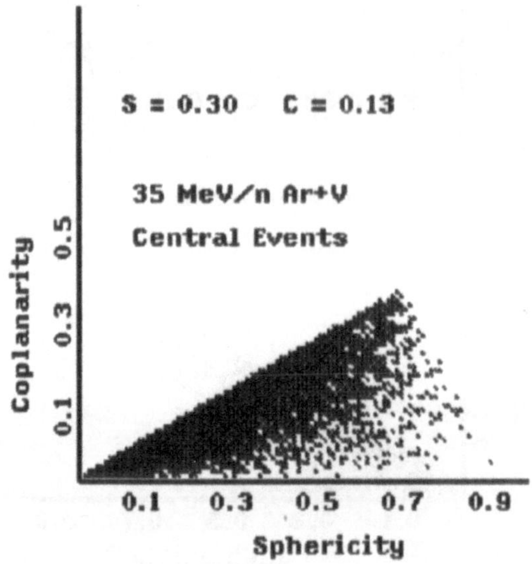

Figure 10. Measured values for sphericity versus coplanarity in central collisions of 35 MeV/nucleon Ar+V.

In contrast, the experimental results for peripheral collisions of Ar+V are shown in Fig. 11. Here most of the events occur along a line representing a rod-like shape of the momentum distribution. Presumably this shape reflects the fact that in peripheral collisions most particles are emitted from target-like or projectile-like sources. In addition the low multiplicities associated with peripheral collisions can contribute to producing the shape of the momentum distribution shown in Fig. 11.

ENTROPY PRODUCTION

As seen in the multiplicity distributions, collisions between two nuclei at intermediate energies produce many charged particles in each event. A substantial fraction of these particle are light nuclei. The global distribution of energy and momenta discussed above suggest that these particles are emitted simultaneously. This fact implies an emitting system that may be describable in terms of some kind of thermal or chemical equilibrium. Bertsch and Cugnon[19] have shown that in a collision of two Ca nuclei at 800 MeV/nucleon, the produced entropy increases sharply during the initial stages of the collision and then remains fairly constant throughout the expansion and cooling of the system. Thus the produced entropy, S, may be a quantity that can be related to the hot, compressed stage of a reaction.

The entropy produced in a collision is reflected in the resulting fragment distribution. The production of mostly light particles signals high entropy values while relatively more complex fragments argues for lower entropy values. The entropy values are obtained by comparing experimental fragment mass or charge distributions with the quantum statistical model of Hahn and Stöcker.[16] Only two of the three quantities, temperature, density, and entropy, as well as the number of neutrons and protons in the emitting system, are required to specify the quantum statistical model prediction.

In Fig. 12 the ratio of deuterons, tritons, and helium fragments to protons are plotted as a function of mid-rapidity charge for 35 MeV/nucleon Ar+V. The Plastic Ball group[17] has measured similar data at higher energies in which the ratios of composite particles to protons increase with increasing centrality. However, the present ratios are very nearly flat with mid-rapidity charge. In Fig. 13 the ratio of fragments with Z=2 to 6 to fragments with Z=1 are plotted as a function of mid-rapidity charge. Here the increase of the ratios is very strong for low values of the mid-rapidity charge and an asymtoptic value may be reached for mid-rapidity charges above 10. The values for the relative production of complex fragments is thus taken from events where the mid-rapidity charge is greater than 10. These yields then correspond to production in central collisions. A comparison of these yields to the quantum statistical model is shown in Fig. 14.

The produced entropy can be extracted from the yield of complex fragments using the quantum statistical model. The entropy is taken as that entropy which gives the best fit for different choices density or temperature. In Fig. 15 the statistical χ^2) is plotted as a function of density and entropy. In Fig. 16 the extracted entropy is given as a function of nuclear density. The extraction was done using all the observed particles, light particles only (Z=1,2), and complex fragments only (Z=3-6). Note the suppressed zero on the ordinate of Fig. Entropy. Contrary to what was observed previously in inclusive measurements[18], the extraction of entropy using only light particles gives a lower value for the entropy than using either all the fragments or the heavy fragments only for a given value of the density. The best fits in Fig. 16 are shown as solid circles. The best fit entropy is the same for all three groups of particles but the density at which the best fit occurs seems to increase with particle mass even though the differences in entropy for different densities is relatively small. This effect suggests that the heavier fragments may be freezing out at higher densities than light fragments.

TRIPLE DIFFERENTIAL CROSS SECTIONS

Another powerful tool provided by multi-particle measurements is the ability to extract cross sections for fragments emitted in-plane and out-of-plane. Collective effects not otherwise observable may manifest themselves in this detailed type of analysis. In Fig. 17 the energy spectra of helium fragments from Ar+V emitted in the reaction plane ($\phi = 0 \pm 22.5°$) versus those emitted out of the reaction plane ($\phi = 90 \pm 22.5°$) are shown. Note that the two sets of spectra have different slopes. By fitting a moving source to these spectra and to spectra for other particle types, one can study the effect of selecting the orientation with respect to the reaction plane.

Fig. 18 shows the apparent temperature from a moving source fit to triple differential spectra at a variety of angles with respect to the reaction plane. There is a distinct minimum in the extracted slope which is more pronounced with increasing mass of the observed fragment.

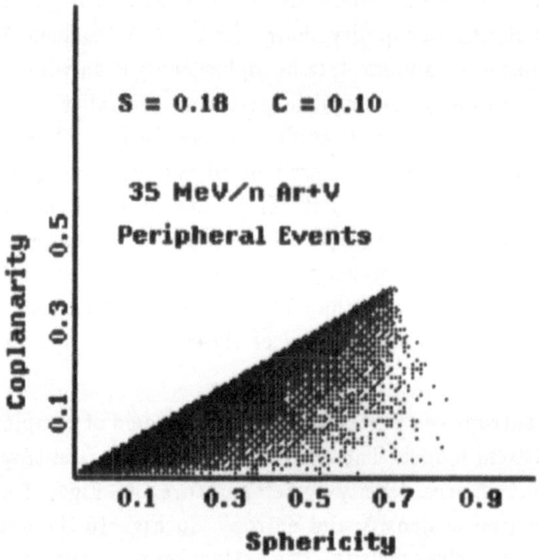

Figure 11. Measured values for sphericity versus coplanarity in peripheral collisions of 35 MeV/nucleon Ar+V.

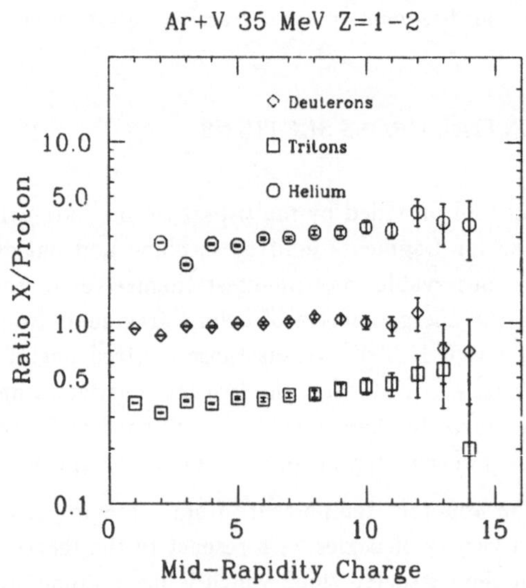

Figure 12. Ratio of light particles to protons for 35 MeV/nucleon Ar+V.

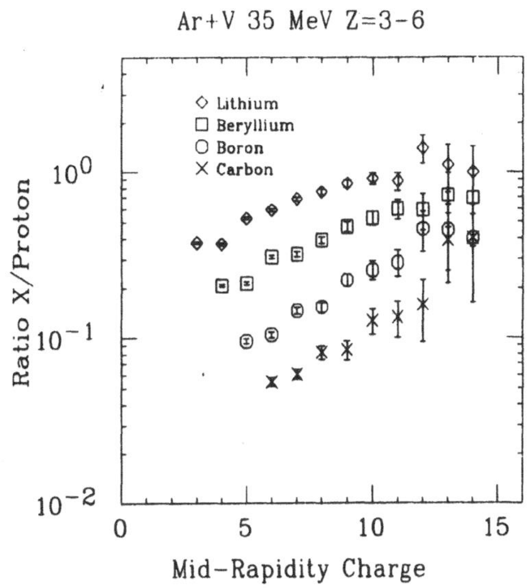

Figure 13. Ratio of complex fragments to hydrogen for Ar+V as a function of the mid-rapidity charge.

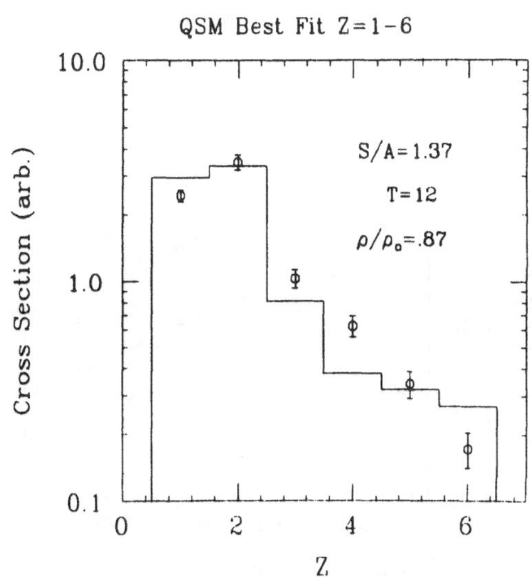

Figure 14. Relative production cross section for fragments from 35 MeV/nucleon Ar+V compared to a quantum statistical model calculation.

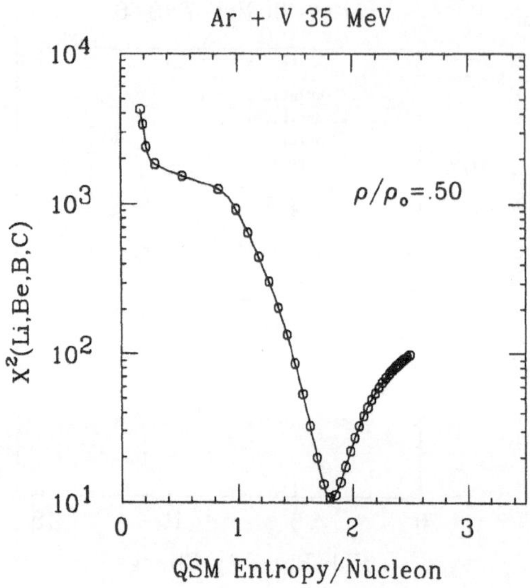

Figure 15. Goodness of fit for quantum statistical model compared to fragments from Ar+V.

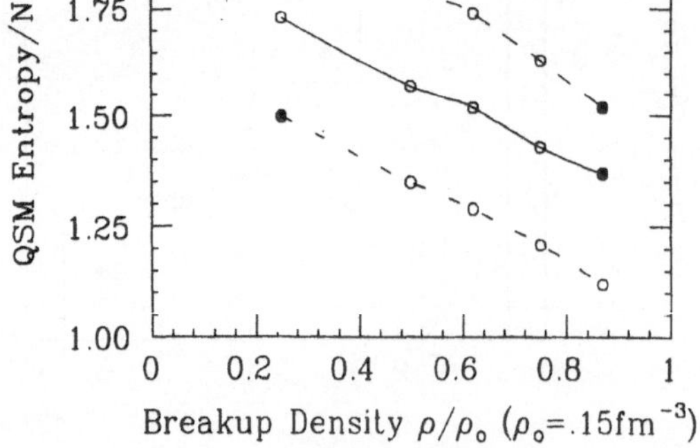

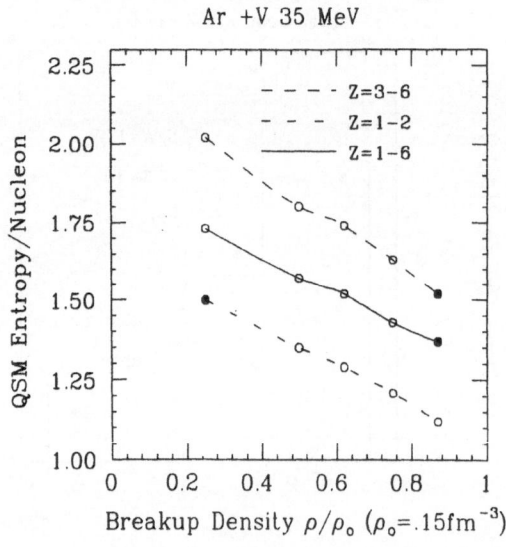

Figure 16. Extracted entropy for fragments produced in the reaction of 35 MeV/nucleon Ar+V.

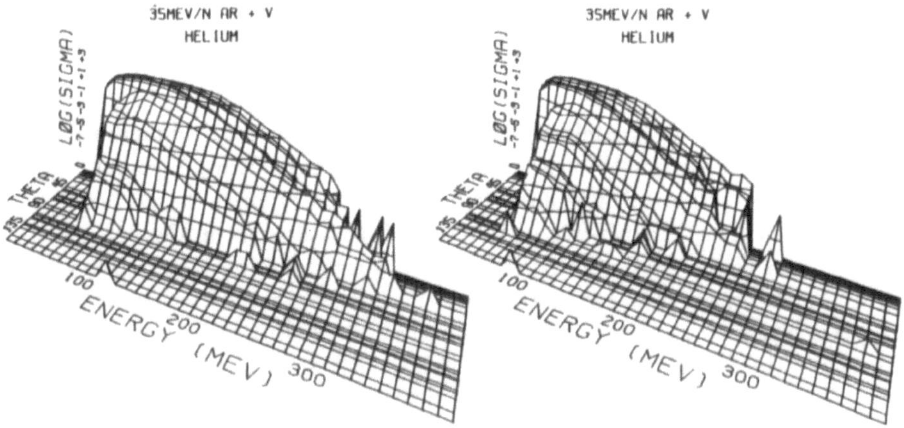

Figure 17. Triple differential cross sections for helium fragments from Ar+V in- and out-of-plane.

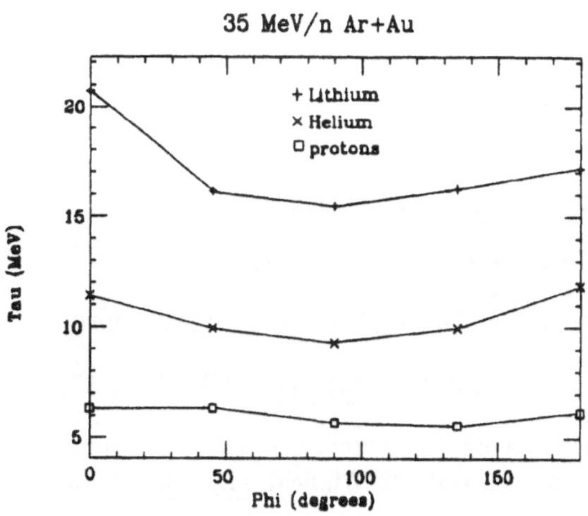

Figure 18. Slopes of triple differential spectra for helium fragments from Ar+V.

The peaking in-plane and the minimum out-of-plane suggests that there is some collective motion in the reaction plane. This type of collective phenomena may be the same as that observed in the transverse momentum analysis above or it may be due to collective rotational motion.

CONCLUSIONS

The present multi-particle measurements give a coherent picture of two nuclei colliding, forming an excited interaction zone, emitting fragments simultaneously with a statistical distribution both in energy spectra and fragment yields. Superimposed on these thermal effects is a distinct directed transverse momentum in the reaction plane. This directed transverse momentum is observed most strongly in the light system C+C. The evidence for the statistical emission comes from the fit to the fragment distributions using the quantum statistical model. The results for the sphericity versus coplanarity argue for an emission mechanism consistent with simultaneous breakup. The directed transverse momentum is detectable in plots of the average transverse momentum as a function of rapidity as well as in triple differential cross sections.

This work was supported by the National Science Foundation under Grant No. PHY-86-11210.

† Permanent address: Department of Physics, University of Michigan, Dearborn

References

[1] C.K. Gelbke, D.H. Boal Prog. Part. Nucl. Phys. 19:33 (1987)

[2] C. Ngo, Nucl. Phys. A488:233c (1988)

[3] G.D. Westfall, J.E. Yurkon, J. van der Plicht, Z.M. Koenig, B.V. Jacak, R.Fox, G.M. Crawley, M.R. Maier, B.E. Hasselquist, R.S. Tickle, and D. Horn, Nucl. Inst. Meth. A238:347 (1985)

[4] W. Bauer, Phys. Rev. Lett. 61:2534 (1988)

[5] G. Fai and J. Randrup Nucl. Phys. A404:551 (1983)

[6] G. Fai and J. Randrup, Comp. Phys. Comm. 42:385 (1986)

[7] P. Danielewizc and G. Odyniec, Phys. Lett. 157B:116 (1985)

[8] C.A. Ogilvie, D.A. Cebra, J. Clayton, S. Howden, J. Karn, A. Vander Molen, G.D. Westfall, W.K. Wilson, and J.S. Winfield, Phys. Rev. C to be published (1989)

[9] C.A. Ogilvie, D.A. Cebra, J. Clayton, P. Danielewizc, S. Howden, J. Karn, A. Nadasen, A. Vander Molen, G.D. Westfall, W.K. Wilson, and J.S. Winfield, to be published, 1989

[10] P. Danielewizc, H. Strobele, G. Odyniec, D. Bangert, R. Bock, R. Brockmann, J.W. Harris, H.C. Pugh, A. Sandoval, L.S. Schroeder, and R. Stock, Phys. Rev. C 38:120 (1988)

[11] H.Å. Gustafsson, H.H. Gutbrod, J.W. Harris, B.V. Jacak, K.H. Kampert, B. Kolb, A.M. Poskanzer, H.G. Ritter, and H.R. Schmidt, Mod. Phys. Lett. A3:1323 (1988)

[12] M. Vient, Ph.D. Thesis, University of California, Riverside (1988).

[13] Joseph J. Molitoris, Detlev Hahn, and Horst Stöcker, Nucl. Phys. A447:13c (1985)

[14] A. Bonasera, L.P. Csernai, and B. Schürmann, Nucl. Phys. A476:159 (1988)

[15] Jorge A. López and Jørgen Randrup, Nucl. Phys. A491:477 (1989)

[16] Detlev Hahn and Horst Stöcker, Nucl. Phys. A476:718 (1988)

[17] K.G.R. Doss, H.Å. Gustafsson, H.H. Gutbrod, D. Hahn, K.-H. Kampert, B.Kolb, H. Löhner, A.M. Poskanzer, H.G. Ritter, H.R. Schmidt, and H. Stöcker, Phys. Rev. C 37:163 (1988)

[18] B.V. Jacak, H. Stöcker, and G.D. Westfall, Phys. Rev. C 29:1744 (1984)

[19] G.F. Bertsch and J. Cugnon, Phys. Rev. C 24:2514 (1981)

COLLECTIVE FLOW OF CHARGED FRAGMENTS AND NEUTRONS FROM BEVALAC EXPERIMENT 848H

J. Schambach, R. Madey, W. M. Zhang, M. Elaasar, D. Keane, B. D. Anderson, A. R. Baldwin, J. W. Watson

Kent State University
Kent, OH 44242

G. D. Westfall

Michigan State University
East Lansing, MI 44823

G. Krebs, H. Wieman

Lawrence Berkeley Laboratory
Berkeley, CA 94720

Bevalac Experiment 848H was designed to probe the nuclear equation-of-state by measuring triple-differential cross sections $d\sigma/d\cos\theta\ d(\phi-\phi_R)\ dY$ for neutrons from high-multiplicity collisions of equal mass nuclei as a function of mass number and bombarding energy. In this experiment, an estimation of the azimuthal angle ϕ_R of the reaction plane was determined by measuring the transverse velocities of charged particles emitted with positive rapidities in the center-of-mass system and summing them to obtain a total transverse-velocity vector. The ϕ-distribution of these summed transverse-velocity vectors is peaked about the reaction plane. Fai *et al.* [1] showed that this transverse-velocity method, which is an adaptation of the transverse-momentum method of Danielewicz and Odyniec [2], determines the azimuthal angle ϕ_R of the reaction plane with reasonable accuracy for this experiment without the need for particle identification in a large number of detectors, such as in the Plastic Ball [3]. The purpose of this report is to present some preliminary results on the performance of the segmented time-of-flight wall for charged fragments, the observation of collective flow of charged fragments in this wall, which confirms the results from the Plastic Ball, the search of the onset of collective flow in Au - Au collisions, and the observation of collective flow of neutrons in high-multiplicity collisions. These results are based on only a small sample of all collected data. The observation [4] of collective flow is one of the most significant developments in the last five years in the field of relativistic heavy-ion physics.

Bevalac Experiment 848H was carried out for collisions of Au on Au at 650, 400, 250, 150, and 75 MeV/nucleon and collisions of Nb on Nb at 400 MeV/nucleon. The experiment was carried out in July 1988 in the HISS cave of the Bevalac experimental area. Figure 1 is a diagram of the experimental arrangement. Neutron time-of-flight (tof) spectra were

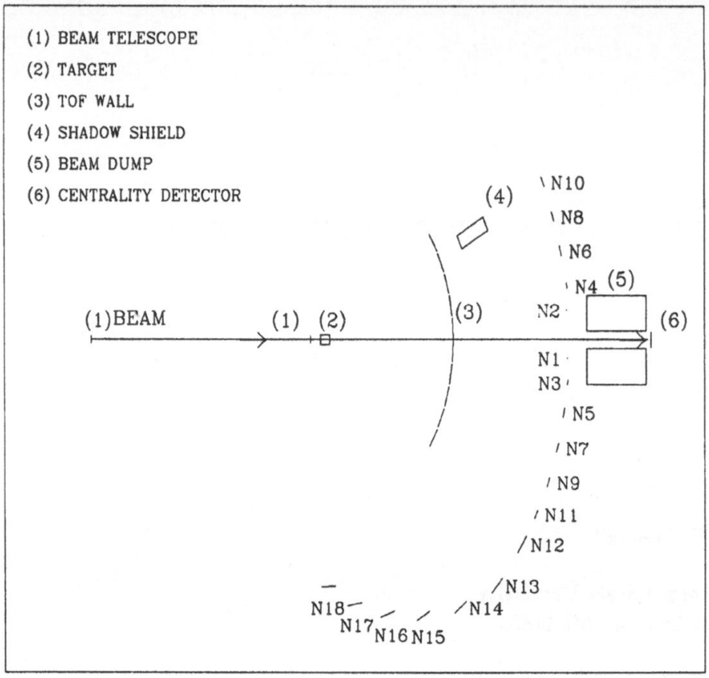

Fig. 1. Experimental arrangement of Bevalac Experiment 848H

measured at 18 angles from 3° to 90°. The tof spectra represent the time-difference between the detection of the neutron in one of the 18 mean-timed neutron detectors N1 through N18, and the detection of a projectile ion in detector S2 of the beam telescope, which consisted of two plastic scintillation counters, S1 and S2. After traversing the beam telescope and the target T, the beam was stopped in a reentrant beam dump.

An array of 184 scintillation detectors for the detection of charged particles was located 4.4 m from the target. The purpose of this segmented wall was to estimate the reaction plane for each collision, and to record the multiplicity of charged particles, which is an indication of the degree of centrality of the collision. A view of one quadrant of this "plastic wall" is shown in Fig. 2. Delta-rays produced by the beam in the air were absorbed by thin steel sheets, which covered the front side of the 24 inner detectors. Neutrons can transverse the relatively thin ($\approx 1.2\, g/cm^2$) plastic wall with an attenuation of only a few percent. From the coincident rate in a pair of neighboring detectors, the double-hit rate of charged particles was estimated to be about 25% in the inner detectors of the plastic-wall and about 15% in the outer detectors for collisions of Au on Au at 650 MeV/nucleon. The loss of information from these relatively small double-hit rates does not have a significant effect on the determination of the reaction plane.

The multiplicity of detected charged particles was measured with the target in and with the target out. Figure 3 shows the multiplicity distribution for the Au beam at a bombarding energy of 650 MeV/nucleon. The solid line in Fig. 3 is the multiplicity distribution for collisions with the Au target; the dashed line, without the target.

The event trigger consisted of a coincidence of the following three conditions: 1) A projectile ion traversed the beam telescope and was not rejected by a "beam pile-up rejection" module. 2) At least one neutron is detected in N1 through N18. 3) The multiplicity of charged particles is greater than a certain value M, where M was chosen so that the contamination from collisions of Au with air and materials in the beam telescope was less than 5% of the events with the target in (as determined from spectra such as Fig. 3). Once an event passed these trigger conditions, the tof of charged particles and neutrons and the pulse heights in the neutron detectors were recorded.

An estimation of the azimuthal angle ϕ_R of the reaction plane was determined by summing the transverse velocity vectors of all charged particles detected in a given collision with rapidities (normalized to the projectile rapidity) greater than a given normalized rapidity α_0:

$$\vec{Q} = \sum \omega_i (\vec{V}^t / V^t)_i, \qquad \omega_i = \begin{cases} 1 & \text{for } \alpha = (Y/Y_P)_{cm} \geq \alpha_0 \\ 0 & \text{otherwise} \end{cases}$$

The $\vec{Q}$ vector determines the "estimated" reaction plane through its cross product with the beam-momentum vector. The weight ω_i is set equal to unity if a particle has a normalized center-of-mass rapidity greater than α_0, where the value of α_0 was chosen to minimize the dispersion $\Delta\phi_R$ in the azimuthal angle ϕ_R. To determine the dispersion $\Delta\phi_R$, we used the method given by Danielewicz and Odyniec [2], modified for our experiment. First we determine the average of the normalized in-plane velocities of all charged particles projected on the reaction-plane estimated by the vector $\vec{Q}_i$, determined now by excluding the considered charged particle to remove autocorrelations:

$$\overline{(V^x/V^t)'_i} = \overline{(\vec{V}^t/V^t)_i \cdot \vec{Q}_i / Q_i}, \qquad \vec{Q}_i = \sum_{j \neq i} \omega_j (\vec{V}^t/V^t)_j$$

The same average, projected on the "true" reaction-plane, can be obtained from an appropriate average of the vectors $\vec{Q}$ and the sum of all weights W of all events:

$$\overline{(V^x/V^t)} = \left(\frac{\overline{Q^2 - W}}{\overline{W(W-1)}} \right)^{1/2}, \qquad W = \sum \omega_i$$

This formula is an adaptation of the equivalent equations given in [2]. For small $\Delta\phi_R$, the ratio of these two averages is equal to the cosine of $\Delta\phi_R$:

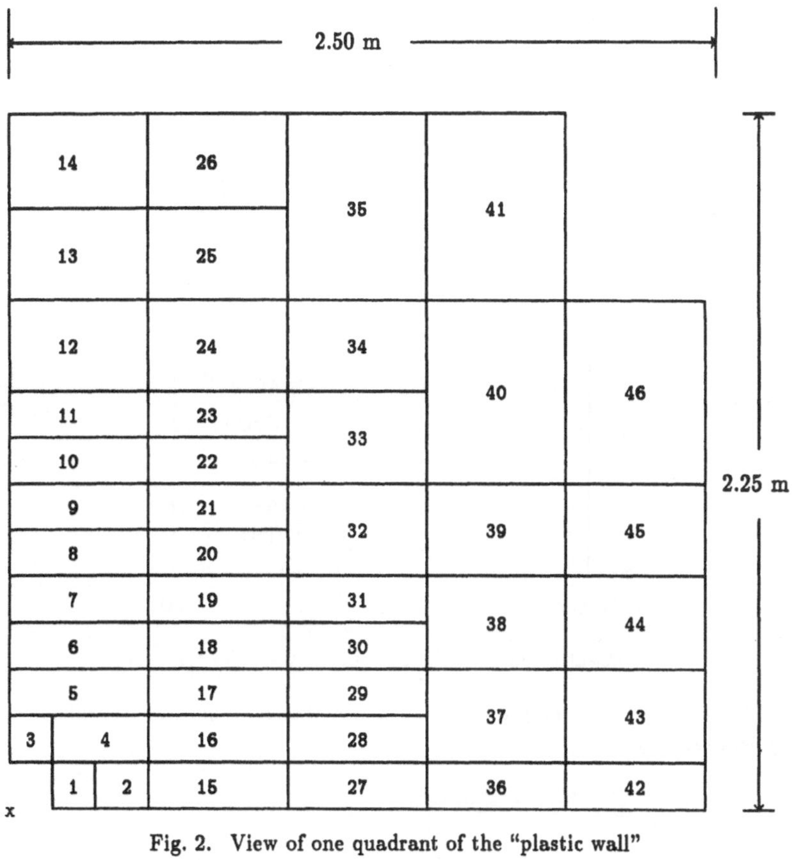

Fig. 2. View of one quadrant of the "plastic wall"

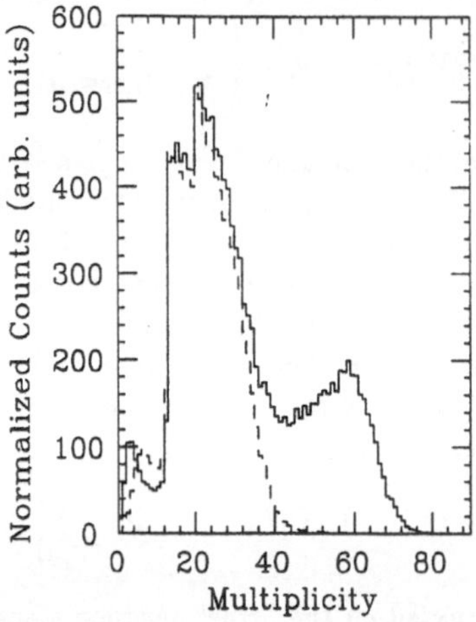

Fig. 3. Multiplicity distribution for 650 MeV/nucleon Au on Au. The solid line is with the target in; the dashed line with the target out.

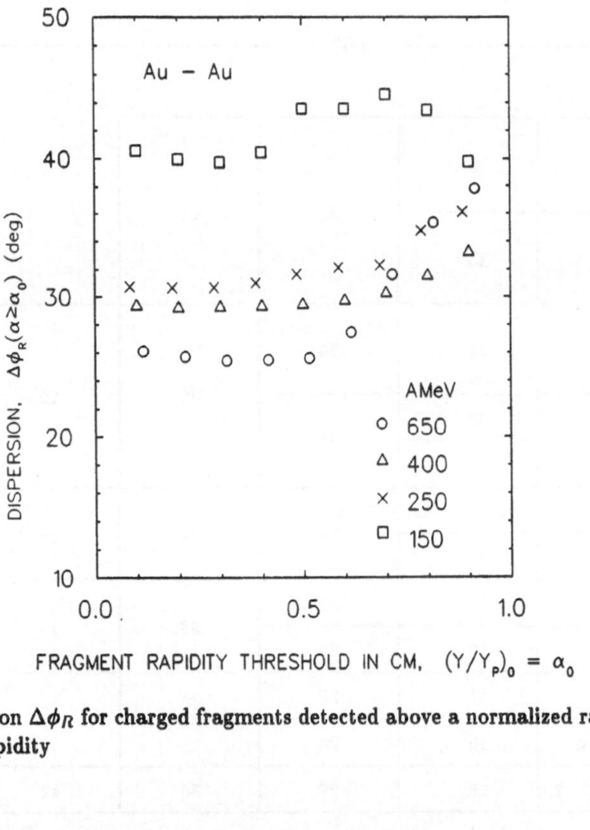

Fig. 4. Dispersion $\Delta\phi_R$ for charged fragments detected above a normalized rapidity α versus the fragment rapidity

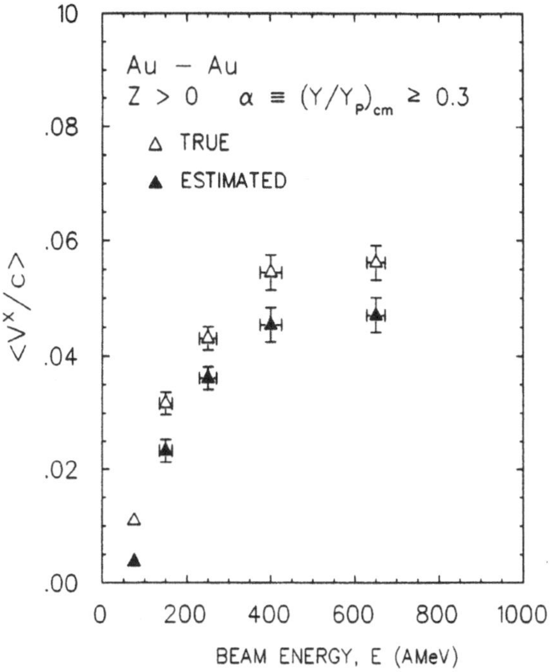

Fig. 5. Average in-plane velocities $\overline{V_x'/c}$ and $\overline{V_x/c}$ projected, respectively, on the estimated and "true" reaction-planes as a function of bombarding energies for Au - Au collisions

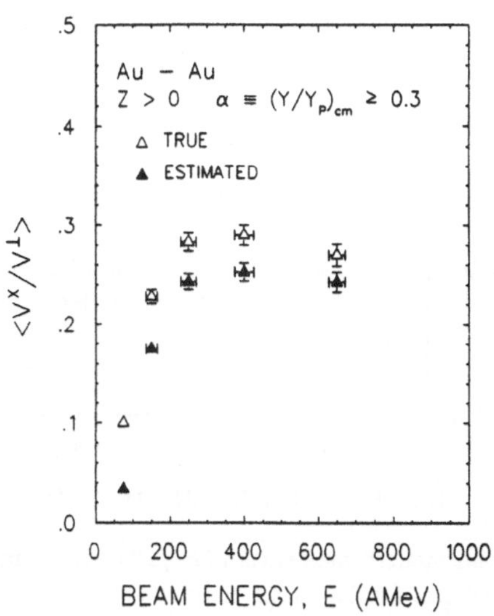

Fig. 6. Average in-plane normalized velocities $\overline{(V^x/V^t)'}$ and $\overline{(V^x/V^t)}$ projected, respectively, on the estimated and "true" reaction-planes as a function of bombarding energies for Au - Au collisions

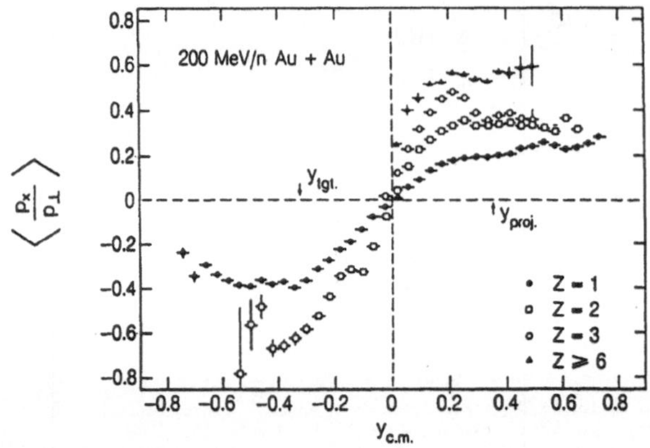

Fig. 7. Average in-plane normalized momentum versus rapidity for 200 MeV/nucleon Au - Au collisions from the Plastic Ball

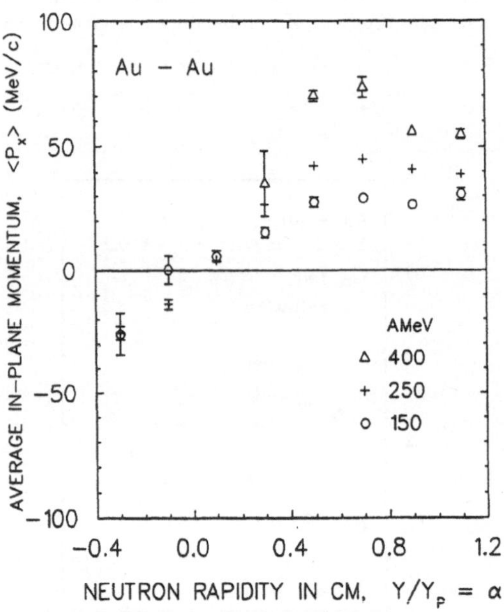

Fig. 8. Average in-plane normalized momentum $\overline{(P^x/P^t)'}$ versus normalized rapidity α as a function of bombarding energies for Au on Au collisions

$$\cos(\Delta\phi_R) = \frac{\overline{(V^x/V^t)'_i}}{\overline{(V^x/V^t)}}$$

Figure 4 shows the dispersion versus rapidity cut for the different beam energies. The minima of the dispersions $\Delta\phi_R$ occur at α_0 values between about 0.3 and 0.6. Dispersions at these minima range from about 26° for 650 MeV/nucleon Au on Au to about 40° for 150 MeV/nucleon Au on Au.

One objective of this experiment was to look for the onset of flow (i.e., the energy corresponding to minimum flow) for Au - Au collisions. Figure 5 shows the average in-plane velocities $\overline{V'_x}/c$ and $\overline{V_x}/c$ projected, respectively, on the estimated and "true" reaction-planes as a function of bombarding energies for Au - Au collisions. The average is over all charged particles with $\alpha > 0.3$. The flow indicated from the projection on the estimated reaction-plane seems to vanish below about 60 MeV/nucleon; the projections on the "true" reaction-plane reveal a lower onset energy. Work on this onset energy is still in progress and will be published in a later report [5].

Figure 6 shows the same averages normalized to the transverse velocities. These values for 150 and 250 MeV/nucleon Au - Au collisions can be compared to the normalized average in-plane momenta at the projectile rapidities for the 200 MeV/nucleon Au - Au results from the Plastic Ball group shown in Fig. 7 [6]. The values determined in 848H are 0.25 and 0.31 for 150 and 250 MeV/nucleon, respectively. The Plastic Ball values are about 0.2 for $Z = 1$ up to about 0.5 for $Z \geq 6$. Because experiment 848H doesn't distinguish between different charges, the 848H values are the result of an appropriate averaging over all charges. Figure 8 shows the average in-plane momentum, normalized to the transverse momentum, of neutrons versus the normalized neutron rapidity α in the center-of-mass system. Here, the interpolation of the plateau values at the projectile rapidity ($\alpha = 1$) of 0.20 and 0.21 for 150 and 250 MeV/nucleon Au on Au, respectively, agree with the value of 0.20 at the projectile rapidity for the protons of the 200 MeV/nucleon Au - Au system from the Plastic Ball group shown in Fig. 7. It should be noted that the influence of detector acceptances have not been addressed in these preliminary comparisons.

References

[1] G. Fai, W. M. Zhang, and M. Gyulassy *Phys. Rev.*, C36, 597 (1987).

[2] P. Danielewicz and G. Odyniec *Phys. Lett.*, 157B, 146 (1985).

[3] H. H. Gutbrod *et al. Nucl. Instrum. Methods*, 203, 189 (1982).

[4] H. Å. Gustafsson *et al. Phys. Rev. Lett.*, 52, 1590 (1984).

[5] W. M. Zhang *et al.* submitted to Phys. Rev. Lett.

[6] K. G. R. Doss *et al. Phys. Rev. Lett.*, 59, 2720 (1987).

Figure 4 shows the dispersion curves rapidly [illegible] for the different beam energies. The [illegible] of the dispersions [illegible] at 3 [illegible] between about 6.5 and 9.0 [illegible] range [illegible] for the 250 MeV [illegible] to about [illegible] for 300 MeV [illegible] A.

[illegible] of this comparison was to [illegible] on the basis of [illegible] the energy [illegible] for the collision. Figure 5 shows the average in plane [illegible] respectively [illegible] distributions [illegible] The average [illegible] 6.5 [illegible] indicated [illegible] the [illegible] plane [illegible]

Figure 6 shows the [illegible] to the [illegible] relative. These values for 150 and 250 MeV [illegible] calculations [illegible] to the formation average [illegible] for the 200 MeV [illegible] from the [illegible] group shown in Fig. 7 [illegible]. The values determined [illegible] are 4.25 and [illegible] for 150 and 250 MeV [illegible], respectively. The [illegible] values [illegible] 6.5 [illegible] between [illegible] range; [illegible] the [illegible] Figure 8 shows the [illegible] compared to the [illegible] the [illegible] of the particle [illegible] at the 150 and 250 MeV [illegible] respectively [illegible] the [illegible] of the 250 MeV [illegible] shown in Fig. 7. It [illegible] that the influence of detector acceptance [illegible] in these [illegible] comparisons.

References

1) [illegible], Phys. Rev. 130, [illegible] (1963)

2) [illegible] and [illegible], Phys. Rev. [illegible] 140 (1965)

3) [illegible], [illegible] 20, 139 (1963)

4) [illegible], Phys. Rev. Lett. 12, 1199 (1964)

5) [illegible], Phys. Rev. Lett. [illegible]

6) [illegible], Phys. Rev. [illegible] (1964)

PION AND PROTON EMISSION IN INTERMEDIATE ENERGY HEAVY ION COLLISIONS*

Walter Benenson

Cyclotron Laboratory and Department
of Physics and Astronomy
Michigan State University
East Lansing, MI 48824 USA

INTRODUCTION

The intermediate range of heavy ion bombarding energies (20 MeV<E/A<1 GeV) encompasses a large variety of reaction mechanisms from fusion at the lowest energies to almost complete explosion at the highest. Up until now, equation of state studies have been hampered by insufficient detailed information on some of the basic properties of the mechanisms. In the discussion below I will deal with a few of these problems. In particular, what can we learn about source size and shape from inclusive and semi-inclusive measurements?

PION PRODUCTION

The first data on subthreshold pion production with heavy ions are now about eleven years old. They were in fact spurious [1] and gave a cross section much bigger than could be calculated by any model which involves incoherent nucleon-nucleon collisions. This paper was published soon after a theoretical treatment by Bertsch [2] which gave a method for calculating the background incoherent process and stimulated a whole series of experiments beginning with work [3] at the Bevalac and continuing right up until now with experiments at CERN, GANIL, GSI, MSU, and ORNL. The motivation of this research has always been to observe the enhancement of pion production from a source of hot, dense nuclear matter. Only the most recent results give indications of pion production which deviate significantly from an incoherent nucleon-nucleon process.

Coulomb Effects

In the first successful experiment [3] in subthreshold pion production with heavy ions a very large difference between π^+ and π^- was observed at 0°. The effect was determined to be Coulomb and due to the charge of the projectile fragment since its was a maximum at a pion velocity

The Nuclear Equation of State, Part A
Edited by W. Greiner and H. Stöcker
Plenum Press, New York

corresponding to the beam velocity. A theory for the Coulomb effect was developed by Gyulassy and Kauffmann [4] and then used by Sullivan *et al.*[5] to determine the average projectile charge after the collision. A verification of this assumption has been obtained at HISS at the Bevalac by Hashimoto *et al.* [6]. Recently in the first pion production experiment [7] with a true heavy ion (La + La) a large Coulomb effect was observed which appears to carry some information. The ratio $R = \pi^-/\pi^+$ is peaked in this case not in the rest frame of the projectile but rather in the rest frame of the center of mass system, as can be seen in fig.1. Calculations based on ref 4 give qualitative agreement with the data, but the calculated R values are smaller than observed. A simple Coulomb shift of the data to give a reasonable R at high energies indicates a charge which is of the order of at least 60. Therefore the data seem to be saying that the pions are emitted promptly from the collision site.

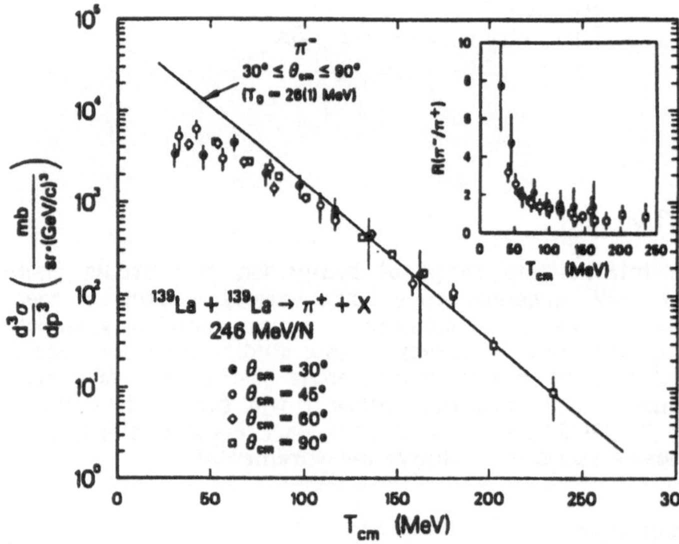

Figure 1

Comparison of π^+ and π^- spectra at various cm angles for La + La at E/A= 246 MeV. The data are from ref 7. The π^- spectra are represented by a single line.

A recent analysis using the statistical model reproduced the magnitude and pion energy dependence of R almost perfectly.[8] A treatment along the same lines for the cross section for both pion and high energy gamma production was recently published by Prakash *et al.*[9] In both case is the elementary cross section is the absorption of pions (or gammas) on a cold nucleus. In evaluating the ratio R, virtually all elements of the cross section formula drop out except for these inverse cross section values. The experimental ratio R for pion production is compared to the ratio for pion absorption on Au in fig. 2. A possible explanation for the strong difference between π^- and π^+ absorption is given by a simple geometric effect as is shown in fig. 3. The nucleus appears to be a much bigger black disk for π^- than it does for π^+ in a that depends strongly on the pion energy. A calculation based this simple model is compared to experiment in fig. 2.

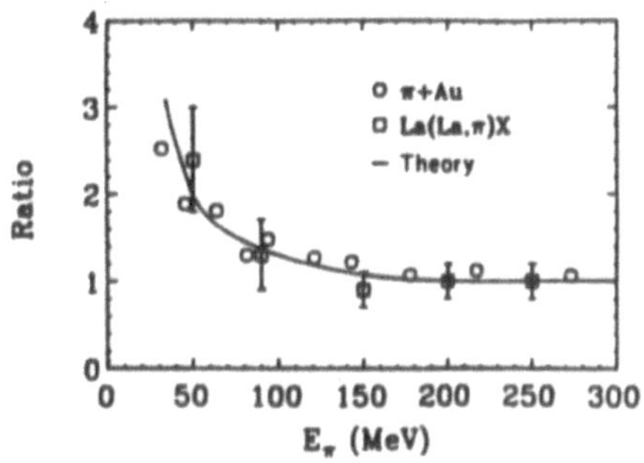

Figure 2
The ratio R= π^-/π^+ for
production and absorption.

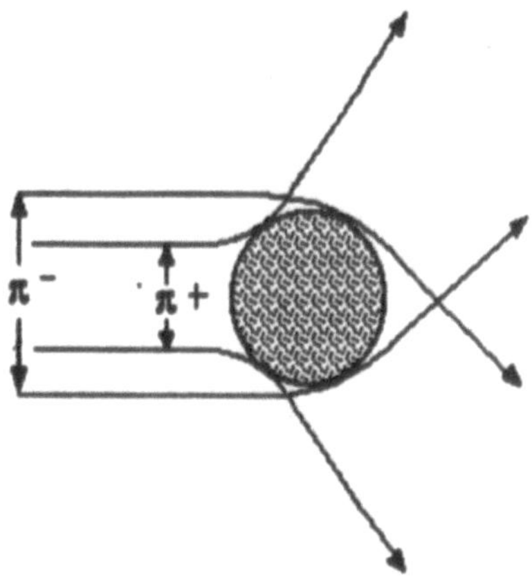

Figure 3
A model for the Coulomb
effect in pion absorption.

Cross Sections

The original idea of the pion production experiments to look for enhancements of measured yield over theories for simple mechanisms, has proven difficult to implement. This is for two reasons; first, the calculations have not settled down to a single widely believed model partly because of the difficulty of taking the Pauli exclusion principle into account. Secondly the effect of pion absorption has been shown to be non-negligible. The shape of the π^0 angular distributions are seen to be strongly dependent on target mass. This effect has been used by Grosse *et al.* [10] to determine the mean free path of pions in nuclear matter. The data also indicate that absorption effects which are usually neglected in theoretical treatments are important for pions.

There are two experimental indications that pion production does not proceed always by a simple incoherent mechanism. They are found in π^0 production near the absolute energy threshold [11] with relatively light beams and in charged pion production just below the N-N threshold with heavy beams [7]. A dramatic result is given in the La + La experiment of Miller *et al.* [7]. The total cross section for Ne + NaF is compared to La + La, and the normal scaling works perfectly above the N-N threshold as usual, but below the threshold there is a very large enhancement for the high mass system (both beam and target) as can be seen in Fig. 4.

There are still a number of interesting questions which can be answered in inclusive measurements. Besides corroborating the high mass enhancement observed by Miller *et al.* [7], there has never been a serious attempt to reconcile apparent differences between π^0 spectra and cross sections and those for π^- and π^+ In Fig. 5, you will see the latest such comparison from a paper of Norén *et al.* [12] The magnitude of the cross section seems to agree, but there is a slope difference which may point to an experimental problem in one of the techniques or may be an interesting phenomenon to be investigated further. The correct energy dependence of the total cross section and its absolute magnitude at the lowest energies have been shown by Hahn *et al.* [13] to be a sensitive test

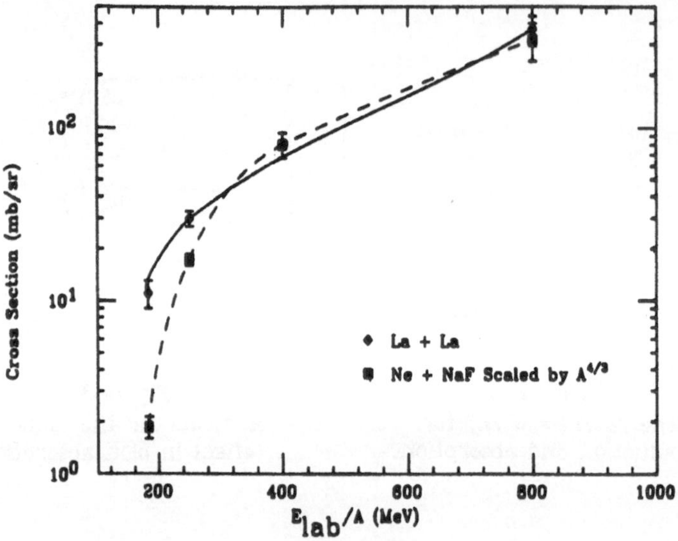

Figure 4
The π^- total cross section for symmetric A= 20 systems as compared to A=139 as a function of beam energy.

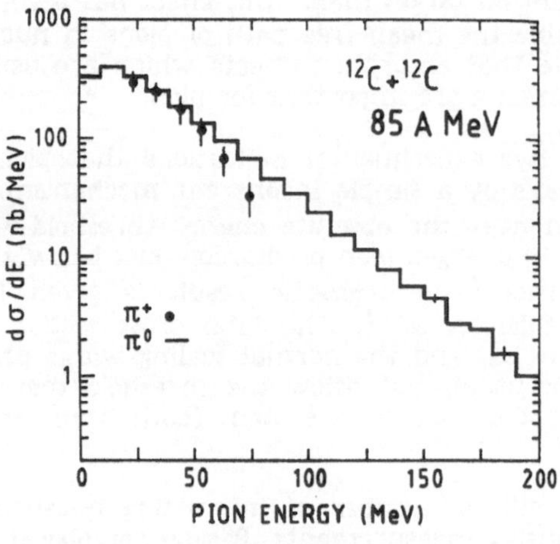

Figure 5
Comparison of π^+ and π^0 spectra for ^{12}C + ^{12}C at E/A= 85 MeV.

of the compressibility of nuclear matter, or in other words the nuclear equation of state. Another interesting topic, which receives its first attention in the paper by Norén *et al.* [12], is the dependence of $R = \pi^-/\pi^+$ on N and Z separately. This can be studied by comparing isotopes and isobars as targets. Unfortunately the detection of π^- in this experiment was by plastic scintillators and subject to large uncertainties. The result for the Sn isotopes, $R_{124}/R_{116} = 1.2 \pm 0.3$ was consistent with all models. Future work with magnetic spectrometers will be much more precise.

Coincidence Experiments

Very recently we have seen the first results of pion coincidence experiments. Besides the above mentioned projectile fragment coincidences, [6] there are three others which attempt to characterize the centrality of the pion producing event. In Miller's experiment [7] with La beams at E/A = 246 MeV the charged particle multiplicity was compared to that of protons. The protons show some evidence for peripheral production at forward angles, but the pions do not, as can be seen in Fig. 6. Very similar results were obtained by Noren *et al.*,[12] Aiello *et al.*,[14] and Erazmus *et al.* [15]. Although all of these data show centrality, they do not identify a particularly unusual or explosive nature for the pion producing collisions. Thus we are lead to the conclusion that the emission of a pion in a collision does not signify an event that is much different from one in which an 140 MeV proton (or gamma ray) is emitted.

There is still a large amount of work to be done in the field of subthreshold pion production with heavy ions. On the theoretical front, one should mention the recent BUU calculations of Bauer, [16] who is able to reproduce both high energy gamma ray and pion production results within the same framework and with realistic NN cross sections. This work tends to indicate that both pion and high energy gammas are produced in the first, or in a smaller fraction of the cases, second collision. On the experimental front, there are no charged pion data near the energy conservation threshold, and the π^0 data are sparse and have never been corroborated by independent measurements. Isotope effects in either the beam or target have never been measured. Excitation functions with reasonable fine steps have never been attempted, and there is no data at all with $A_{beam} > 139$. Therefore one can expect during the coming years to see a whole new generation of pion production experiments at MSU, GANIL, and GSI.

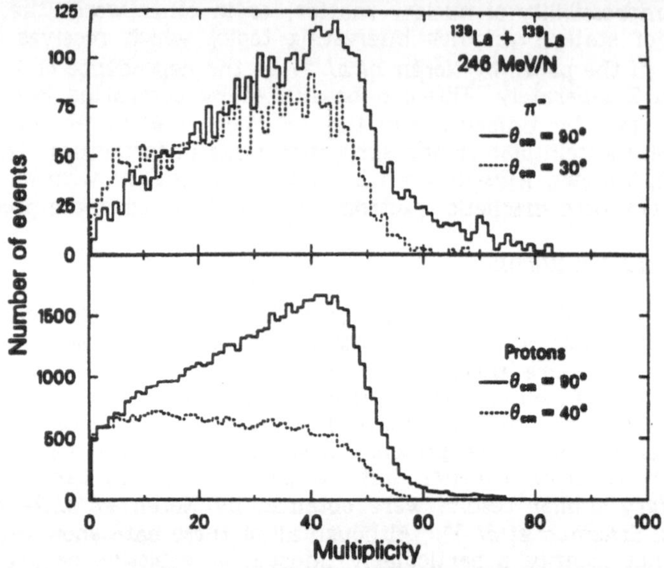

Figure 6
Charged particle multiplicity in coincidence with pions and protons for La + La at E/A = 246 MeV.

SOURCE SHAPE

Recent work on collective flow in heavy ion reactions has relied strongly on event-by-event flow tensor analysis[17]. In this treatment, multiparticle data is used to reduce very detailed information to just one or two numbers. The data includes all the observed particles and hence is a time averaged event shape since it includes preequilibrium emission as well as particles emitted at the very late stages. If one looks at the pictures of the events and considers that the typical flow angle is 30° with respect to the beam direction, then one wonders why the collective flow doesn't show up as an enhancement of the triple differential cross section at 90° in central collisions. In fact the multiplicity data in fig. 6 seem to show an enhancement of protons at 90°. This effect is discussed in more detail in a recently submitted paper by G. Claesson *et al.*.[18]

In ref 18 a 110 element hodoscope was used as a multiplicity measuring device in conjunction with a magnetic spectrometer. The data was taken during the same Bevalac run as ref 7, and therefore a description of the apparatus can be found there. The beam and beam energy were also the same as in the pion experiment of ref. 7, La + La at E/A = 246 MeV. The proton spectra do depend on multiplicity in a number of ways. For one thing, unlike the pions the slope parameter is significantly larger for central collisions than for peripherals. An interesting feature of the data is that although the unbiased data are isotropic, the central collisions (M>46) are apparently peaked at 90°. This can be seen in fig. 7, which shows an effect much greater than can be explained by statistics. This effect is not present at E/A= 800 MeV [19] in an experiment which used identical equipment to that in ref 16. The 800

MeV data also show multiplicity dependent anisotropies, but there is no 90° peaking. These data clearly show that models which give isotropic angular distributions, such as the fireball model, cannot represent the the reactions dynamics correctly. They also illustrate the dangers of using such models to extrapolate data taken over a limited dynamic range for example for the purpose of entropy estimations.

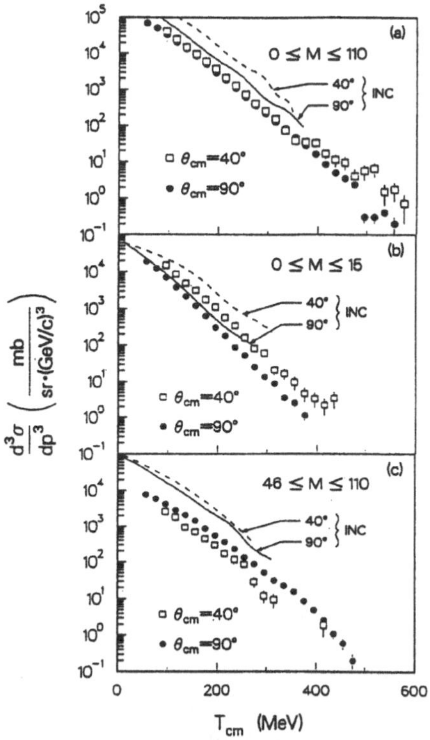

Figure 7
Cross sections for La + La -> p + X as a function of kinetic energy at Qcm = 40° and 90°. a) Inclusive data, b) M ≤15 c) M ≥ 46. The dashed and solid curves are the results from internuclear cascade calculations.

The manifestation of this 90° effect and its disappearance at higher energies can be understood in terms of theoretical models. A strong 90° enhancement in central La + La collisions is a feature of viscous hydrodynamic [20] and of quantum molecular dynamics [21] (QMD) although in the later case it is present only for proton energies above 150 MeV. The QMD model [22] is an extension of Vlasov-Uehling-Uhlenbeck (VUU) model to include isospin. It has been noted previously that VUU and fluid dynamics can be expected to converge only for heavy systems, where there is sufficient material for nuclear stopping and consequent equilibration. Note that cascade [23] and statistical models [24] do not give any 90° enhancement for central collisions, but a model [25] which is sensitive to the nucleon-nucleon (NN) cross section does reproduce the effect if the in-medium NN cross section is set to 60 mb. This is in general agreement with cascade calculations [26] in which the NN cross section was increased and the scattering made repulsive rather than stochastic. Simple experiments can be used to limit important quantities in the theoretical treatment of the reaction as was recently pointed out by Aichelin *et al.*.[27] In this paper the importance of many body effects on in-medium NN cross sections is discussed, and the need for absolute p-like cross section data is emphasized.

CONCLUSIONS

Inclusive cross section measurements over a wide dynamic range of emitted particle energies can reveal details of heavy ion reactions which may be lost in complicated multi-particle correlation experiments. However, some kind of multiplicity selection seems to be required particularly for proton measurements to eliminate peripheral reactions from the study.

This work was supported in part by the National Science Foundation.

---

*Lecture at the NATO Advanced Studies Institute, "Nuclear Equation of State", Peniscola, Spain, May 1989.

REFERENCES

[1] P. J. McNulty, *et, al.*,, Phys. Rev. Lett. 38, 1519 (1977).

[2] G. F. Bertsch, Phys. Rev. C15, 713 (1977).

[3] W. Benenson *et, al.*, Phys. Rev. Lett., 43, 683 (1979) and 44, 54 (1980).

[4] M. Gyulassy and S. K. Kauffmann, Nucl. Phys. A362, 503 (1981).

[5] J. P. Sullivan *et al.*, Phys. Rev. C25 , 1499 (1982).

[6] O. Hashimoto *et al.*, Private Communication

[7] J. Miller *et al.*, Phys. Rev. Lett. 58, 2408 (1987) and 59, 519 (1987).

[8] A. Bonasera and G. F. Bertsch, Phys. Lett. B195, 521 (1987).

[9] M. Prakash *et al.* Phys. Rev. C33, 937"(1986).

[10] E. Grosse *et al.*, Gull Lake Conference, 1987.

[11] G. R.Young *et. al.*, Phys. Rev., C33, 742 (1986).

[12] B. Norén *et al.* LUIP 8802, University of Lund and Nucl. Phys. A489, 763(1988).

[13] D. Hahn and H. Stöcker, Nucl. Phys. A452, 723 (1986).

[14] S. Aiello *et al.* , Europhys. Lett. 6, 25 (1988)

[15] B. Erazmus *et al.* , Nucl. Phys. A481.

[16] W. Bauer, MSUNSCL 672 and contribution to this conference.

[17] H.-Å. Gustafsson et al., Phys. Rev. Lett 52, 1590 (1984).

[18] G. Claesson et al.. submitted to Phys. Rev. Letters.

[19] S. Hayashi et al., Phys. Rev. C38, 1229 (1988).and Masters Thesis, U. of Tokyo (1986) unpublished.

[20] W. Schmidt, Diplomarbeit, Johann Wolfgang Goethe-Universität Frankfurt am Main, (1989) unpublished.

[21] Ch. Hartnack, Diplomarbeit, Institut für Theoretische Physik, Universität Frankfurt am Main, (1989) unpublished.
[22] J. Aichelin and H. Stöcker, Phys. Lett. B176, 14 (1986).
[23] J. Cugnon, Nucl. Phys. A387, 191c (1982)
Y. Kitazoe et al. Private communication
[24] G. Fái and J. Randrup, Nucl. Phys. A404, 551 (1983).
[25] D. H. Boal and J.N. Glosli, Phys. Rev. C38, 1870 (1988) and C38, 621 (1988).
[26] M. Gyulassy, K.A. Frankel and H. Stöcker, Phys. Rev. Lett. 110B, 185 (1983) and K. Frankel private communication.
[27] J. Aichelin et al., Phys. Rev. Lett. 62, 1461 (1989).

HARD PHOTONS AND SUBTHRESHOLD MESONS FROM NUCLEUS-NUCLEUS COLLISIONS

Eckart Grosse

GSI, Darmstadt
D-6100 Darmstadt, FRG

INTRODUCTION

The study of nucleus-nucleus collisions at an energy high enough to permit an appreciable nuclear overlap is the only experimental means to obtain information about the properties of nuclear matter at a density well above the saturation density of heavy nuclei $\rho_0 = 0.16\ fm^{-3}$. Considering the interest in reproducing ρ_0 and the nuclear matter binding energy per nucleon $\varepsilon_0 = -16$ MeV in various (non relativistic) self consistent calculations on the basis of realistic nucleon-nucleon potentials[1], experimental information about the ρ-dependence of ε, i.e. the equation of state (eos), may help to contribute to this long-standing problem of nuclear physics. Additionally, such information is extremely valuable for astrophysics, as the stability of neutron stars as well as the dynamics of a supernova of type II are strongly dependent[2] on the nuclear eos. Small density variations can be studied in the E0-giant resonance (breathing mode); the excitation energy of the resonance is related to the compressibility K_0 of nuclear matter which determines the eos near the minimum ρ_0. A recent analysis[3] of a series of new E0-energy determinations results in $K_0 = 300$ MeV, a value which is larger than the 140 MeV used predominantly[2] in astrophysical calculations.

For heavy ion reactions at about 1 GeV/u nuclear densities three times higher than that of ground state matter are predicted by recent calculations[4] on the basis of transport equations. Obviously, a large high density volume is assured only by a large spatial overlap of the two colliding nuclei, which makes the use of very heavy ions

The Nuclear Equation of State, Part A
Edited by W. Greiner and H. Stöcker
Plenum Press, New York

desirable. Of course, this high density is reached only during a collision time of a few times 10^{-23} s, and it is the key problem of the experiments to find the right observables for this high density phase of the collision. Newly created particles of high total energy probe the early collision zone since at a later stage the projectile energy is dissipated over many nucleons ("thermalized") and soft pions are produced predominantly. Because of their mass, positive kaons of all energy are such a probe, as long as the projectile energy remains sufficiently below the NN-threshold of 1.56 GeV (for $NN \rightarrow N\Lambda K^+$). During the early stage of the reaction the nuclear density may be up to $3\rho_0$ at these beam energies.

In these lectures it will be outlined how the hard photons from the early phase of nucleus-nucleus collisions - as well as high energy pions and kaons - can be detected and identified as such, which means they have to be distinguished from those emitted thermally from the remnants of the collision at later time. First the properties of photon emission will be shown to be a good model case for the application and test of recent theoretical concepts, derived for nucleus-nucleus collisions. Then the special features of pion and kaon production will be discussed and their possible use as a probe for high density nuclear matter.

HARD PHOTONS FROM NUCLEUS-NUCLEUS COLLISIONS - A MODEL CASE

Gamma rays in the range of 20 to 200 MeV ("hard photons") have become a rather widely investigated feature in the study of nucleus-nucleus collisions at beam energies of 10 - 100 MeV/u. Because of the well understood electromagnetic interaction a theoretical treatment of such radiative collision processes should be straightforward. Hard photons were first observed[5] incidentally in experiments on nucleus-nucleus collisions: In one case they showed up as uncorrelated background to the two coincident photons from π^0 decay; in other experiments[6][7] they formed a high energy tail in the γ-spectrum from the decay of giant resonances produced in heavy ion collisions (see figure 1).

The decay of nuclei highly excited in giant resonances or in the underlying quasi-continuum of closely spaced levels is described in the statistical model of compound nucleus decay; the partial γ-decay width can be calculated[7] from the γ-absorption cross section by applying the detailed balance principle

$$\Gamma_\gamma(E_\gamma, A) = (\pi\hbar c)^{-2}\sigma_\gamma(E_\gamma, A)E_\gamma^2\exp(-E_\gamma/_T)dE_\gamma$$

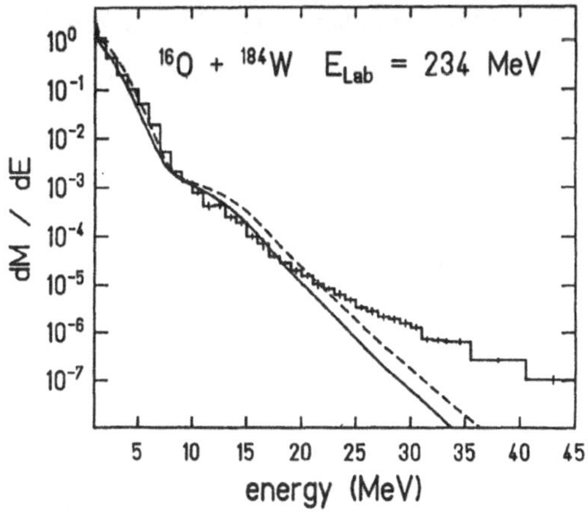

Fig. 1. Inclusive photon spectrum from 14.6 MeV/u $^{16}O + ^{184}W$. The dashed line describes the statistical decay of the compound nucleus assuming complete fusion; precompound particle emission was accounted for to obtain the drawn line (cf. ref. 6).

The exponential factor accounts for the fact, that the absorption is measured on the nucleus A in its ground state, whereas in the photon emission experiments the compound nucleus - if it is formed - has a temperature

$$T = \frac{2}{3} \cdot \frac{E_p}{4A_p}$$

The products of incomplete fusion or deep inelastic scattering processes have very similar temperatures, such that the procedure (indicated in figure 1) of extrapolating the contribution of the statistical γ-decay to higher photon energies should be rather reliable not only in the case of complete fusion. This is especially so when the steeply falling low energy section as well as the giant resonance region are well reproduced by statistical model calculations.

The surplus cross section at higher photon energies - the "hard photon" yield - shows[6] the interesting feature of an angular distribution which is strongly forward peaked in the laboratory system. A similar observation was already made for hard photons produced[8] in collisions at higher energy (see fig. 2). A Lorentz transformation aiming for symmetry around 90° leads to a source rapidity which is equal to half the beam rapidity (at these low energies the rapidity y = artanh β is not very different from the velocity β). This "half-rapidity" source of the hard photons is obviously formed from equally many nucleons out of projectile and target and can thus be identified as the initial collision zone formed in the first encounters of projectile and target nucleons.

The idea of assigning the hard photons to originate from these first collisions is supported by their systematic dependence on projectile and target mass: The "normalized" photon production cross section is proportional[9] to an exponential

$$\frac{d\sigma}{dE_\gamma d\Omega} \propto \frac{N_c\sigma_R}{E_o} \exp(-E_\gamma/E_o)$$

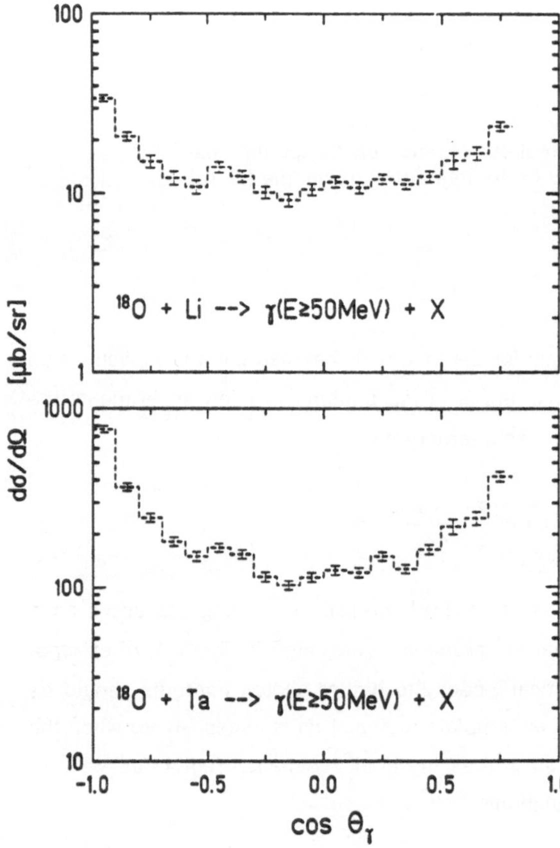

Fig. 2. Angular distributions of photons produced inclusively in $^{18}O + Li$ and $^{18}O + Ta$ at 84 MeV/u. The data have been Lorentz-transformed to a system with half the projectile rapidity; there they are symmetric around 90°.

with a slope constant E_O. All data published so far fall on the same line[9] when σ_R is the (geometrical) total reaction cross section and N_c is the number of (first chance) collisions between a photon from the projectile and a neutron from the target or vice versa. Collisions between two protons play a minor role, since the respective elementary bremsstrahlung cross section[10] is smaller by about one order of magnitude; in a multipole expansion picture dipole radiation can only be emitted from a charge asymmetric system.

The main features of the hard photon emission from colliding nuclei, i.e. the absolute yield, the scaling with projectile and target and the spectral shapes are reasonably well reproduced by calculations[10-13] based on transport equations for the nucleons within the mean field of the colliding nuclei. The nuclear mean fields not only lead to nucleon binding and Fermi motion before the collision, but also to a collisional acceleration due to the partial overlap of the two mean fields; they also induce a blocking of occupied phase space to those nucleons, which have lost part of their initial momentum in a collision, eventually accompanied by bremsstrahlung radiation. Of course, the elementary bremsstrahlung cross section sensitively enters the calculations; unfortunately there is some minor inconsistency[9] between different experiments on photon production from p+nucleus collisions. This inconsistency is

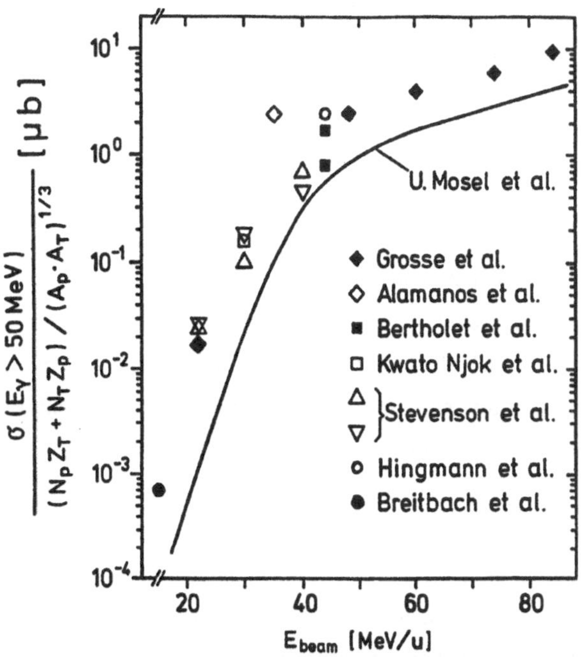

Fig. 3. Beam energy dependence of the photon production yield normalized to the number of p-n collisions. The drawn line shows a transport-equation calculation[11][12].

transferred to the theoretical description of the nucleus-nucleus data, making the comparison of absolute cross sections still somewhat marginal. But the systematics of the data is well reproduced, including the exponential fall off of the photon yield with photon energy. This exponential slope resembles a (thermal) spectrum from statistical decay, but the slope parameter E_0 is significantly larger than the temperature T expected for products of - complete or incomplete - fusion, fusion-fission or deep inelastic processes. Slopes corresponding to the temperatures T of these processes are observed in the photon spectra below 15 MeV; for the hard photon part ($E_\gamma \geq 20$ MeV) slope parameters E_0 were reported[5] to be larger than T by about 4 MeV, another clear indication of the pre-equilibrium non-thermal nature of this radiation. It should be mentioned here, that collective nuclear bremsstrahlung[14] to be emitted from the two nuclei (as a whole) during their collisional deceleration would also show up at E_γ < 15 MeV, where the photon wavelength is comparable to the dimensions of the emitting system.

The photon angular distributions are rather flat in the frame comoving with the source at half the beam rapidity; data[15] at variance to this are in contradiction to several independent recent experiments[9]. The more pronounced structure in the elementary pn → pnγ angular distribution[10] gets lost in the nucleus-nucleus collision due to Fermi motion and relativistic effects - as indicated by the transport equation calculations[12]. The overall consistency between the data and these calculations makes the photon production a sensitive test and model case for the latter and shows the reliability for their extension to the meson production processes.

PIONS FROM THE COLLISION ZONE AND THEIR RESCATTERING

In the preceding discussion about photon production the good general accord between data and calculations on the basis of transport theory was emphasized. It becomes transparent from figure 3 which displays the beam energy dependence of the photon production cross section properly normalized to the geometrical situation of a nucleus-nucleus collision. When comparing this cross section to the one for the elementary process pn → pnγ a strong enhancement at low bombarding energies is observed; this can be depicted as the result of medium effects present in the nucleus-nucleus-collisions: binding and Fermi-motion, collisional acceleration and Pauli-blocking. Medium effects in the exit channel can be neglected as the mean free path of photons in nuclear matter is huge. For pion production the situation is largely different from the photon case: there not only the coupling to the three isospin channels has to be considered but there also is the strong pion rescattering by the Coulomb and nuclear fields of the projectile and target nucleons.

A general overview[16][17] of the integrated pion production yields (fig. 4) has to cover many orders of magnitude to show the multitude of data taken so far. On that scale differences between charged and neutral pion data or data from collision systems of different mass do not become obvious. In comparison to the elementary cross section[18] there is a clear enhancement in the nucleus-nucleus collision data taken at lower beam energy; it is especially spectacular due to the much higher quality and sensitivity of these as compared to the p-p data.

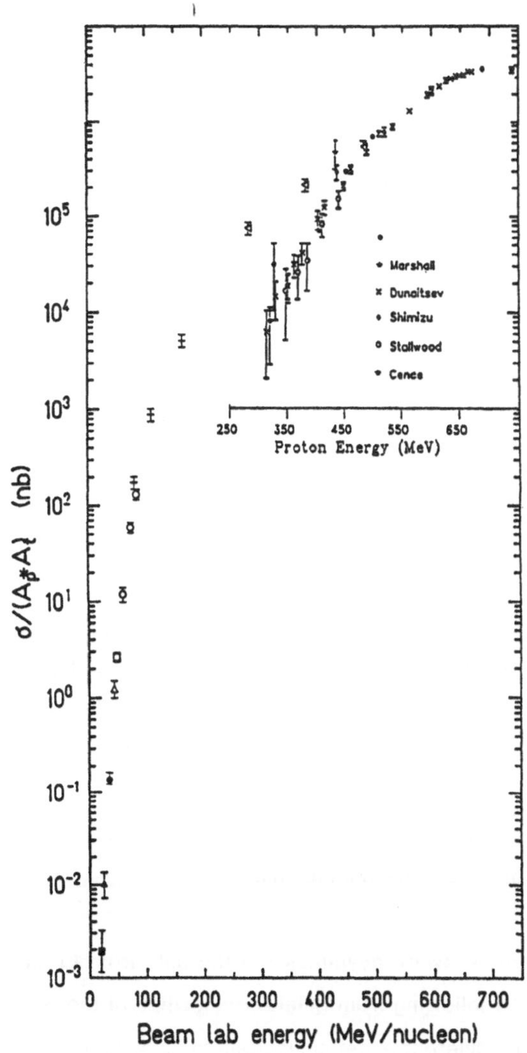

Fig. 4. Excitation function for π-production from nucleus-nucleus collisions normalized to the number of nucleon-nucleon collisions. The data[18] for the "elementary" process $pp \rightarrow pp\pi^0$ are shown for comparison.

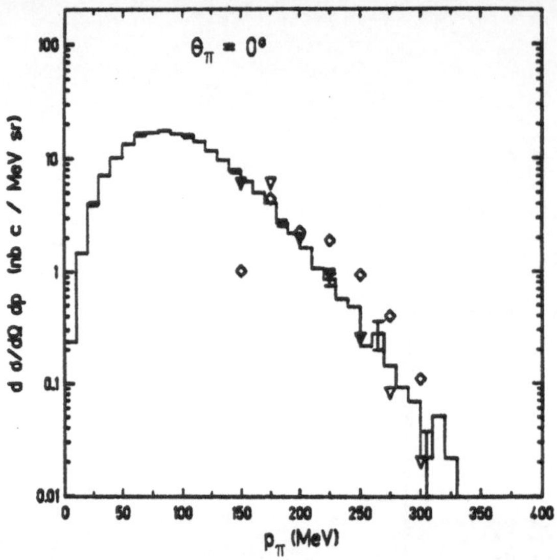

Fig. 5. Comparison of spectra of π^+, π^- and π^0 observed for $^{12}C + ^{12}C$ at 84 MeV/u.

From a comparison[17] of π^+ and π^- data to those obtained for π^0's from $^{12}C + ^{12}C$ collisions at 84 MeV/u isospin effects can clearly be seen. There are shifts in the charged pion spectra due to the finite state Coulomb interaction at high pion momenta and differences at small momenta which are possibly due to the different reabsorption of π^+ and π^-.

An inspection of the complete π^0 angular distribution for the asymmetric collision systems $^{18}O + Li$ and $^{18}O + Ta$ (cf. Fig. 6) shows such reabsorption effects very clearly: For the light Li target the cross section - Lorentz transformed to the half rapidity system - is forward peaked whereas the Ta target causes a strong decrease of the forward cross section. This can be explained straightforward by the reabsorption of the pions in the heavy nucleus after being produced on its surface, where the first chance collisions between nucleons from target and projectile take place.

In principle, the different forward-backward asymmetry in the data from Li and Ta targets could also be interpreted as following from different velocities of the pion source formed in the collisions with the different target nuclei. But an inspection of the spectral shape shows[17] that an angle independent slope parameter is only obtained after transformation into the half-rapidity frame. Additionally the similarity of the π^0 spectra to those obtained for γ's (cf. fig. 7) suggests similar production mechanisms. Similarly as for photons it has been shown[17] that in the energy range below 100 MeV/u also the pion production yields are proportional to the number of first chance nucleon-nucleon collisions, after correcttion for reabsorption.

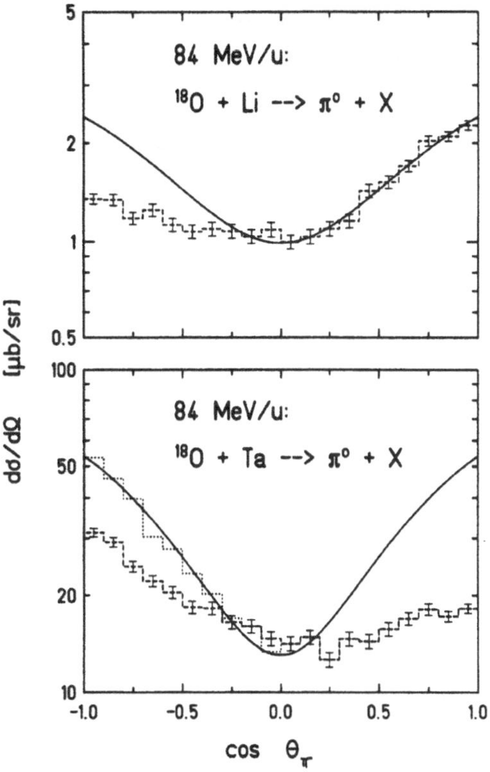

Fig. 6. π^0-angular distributions for 84 MeV/u ^{18}O projectile on the light Li (top) and the heavy Ta-target (bottom). The data are transformed into the half-rapidity system.

As can be seen from the comparison of the spectra, pions are more abundant than photons at a given total energy; taking into account, that both photon spin directions are compared to only one pion isospin and that pions are strongly reabsorbed, the primordial pion enhancement is in the order of 20-30. Regarding the coupling constants of electromagnetic and strong interaction, a larger π/γ-ratio is expected. Calculations[13)19)] have not yet been developped as far as in the photon case; the "elementary" cross sections are not well known for all isospin channels and it is not clear if s-wave production is to be added to the p-wave going through the Δ-resonance. Since also rescattering and charge exchange in the outgoing channel are not well determined the existance of more exotic phenomena like the pion-collectivity proposed recently[20)] can not yet be identified in the data.

EMISSION OF KAONS

The strong pion reabsorption observed at 84 MeV/u (see fig. 6) shows the problems arising when envisaging pions as a probe for dense nuclear matter, especially since it is not obvious if the "missing" pions are rescattered (changed in angle, energy,

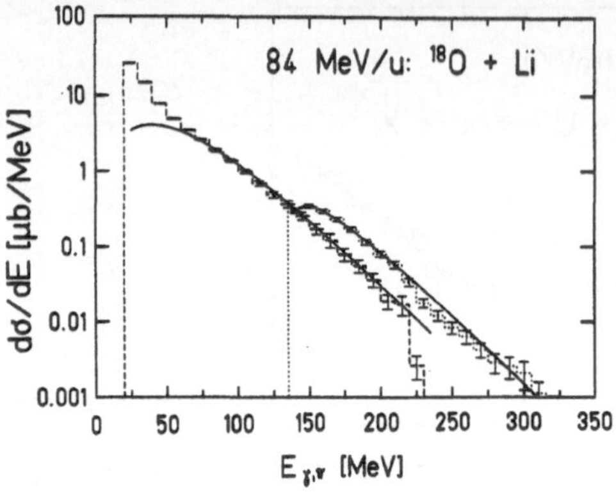

Fig. 7. Photon and π^0-spectra from $^{18}O+Li$ at 84 MeV/u. The cross sections are plotted vs. the total energy allowing a direct comparison at a given energy taken out of the relative motion. The drawn lines depict the enery dependence of phase space.

or isospin) or truly absorbed. A recent detailed analysis[21)] of proton- and pion-nucleus collisions has demonstrated the difficulties of a consistent treatment of medium effects in Δ-propagation and pion production in nuclei. In view of the small effects in the pion emission yields expected[22)] to arise from compression effects in heavy ion collisions a detailed knowledge about the nuclear eos can probably not be extracted from them alone.

Due to strangeness conservation kaons (with the exception of low momentum K^-) have a low absorption probability in nuclei, which makes them a good probe for the nuclear interior. The excitation function for K^+-production in nuclear collisions has been predicted[23)] to rise steeply up to a projectile energy of 1.5 GeV/u. This makes K^+-production especially sensitive to effects reducing the energy as available in the moment of a nucleon-nucleon encounter. The compression of the nuclear matter in the collision zone has been described[22)] as using up part of the projectile energy very early in the collision. Higher compressibility is equivalent to to a smaller energy loss of that kind and results in a higher production rate. K^+-emission could thus become a powerful probe for the nuclear eos as manifest in nucleus-nucleus collisions.

Up to now only very scarce data exist[24)], most of which have been taken above the nucleon-nucleon threshold at 1.58 GeV/u. In nuclei, the production of K^+ at low bombarding energy not only proceeds via the correlated hyperon production channel:

$$NN \rightarrow N\Lambda K \text{ or } N\Sigma K$$

but also via two step processes like

$$NN \rightarrow \pi NN$$

$$N\pi \rightarrow \Lambda K$$

or

$$NN \rightarrow \Delta N$$
$$N\Delta \rightarrow N\Lambda K$$

or

$$NN \rightarrow \eta\, NN$$
$$N\eta \rightarrow \Lambda K.$$

The rate of such two-step processes depends not only on the number of participating nucleons but also on the particle density in the collision volume and thus rises with increasing compressibility.

Apparently two different consequences of a high nuclear matter compressibility both lead to an increased kaon production as compared to expectation. The interpretation of experimental results is further complicated by the fact, that the cross section to be expected from calculations is already subject to various ambiguities: (1) The elementary production cross section near threshold is difficult to measure; it has to be known on shell and off shell. (2) Higher order processes already occur in nuclear matter at normal density. Information about points (1) and (2) can be gained from an analysis of proton-nucleus collisions; corresponding kaon yields have been measured[25] and spectra are beeing taken at present[26]. The nucleon dynamics during the heavy-ion collision is governed by the nucleon-nucleon force, which is density and momentum dependent. Model calculations are tested by the observation of nucleons from the participant and spectator regions. At lower energies the particle dynamics is further tested by a comparison of different ejectiles - pions and photons in that case, as was outlined above. (4) Of course, all cross sections have a more trivial dependence on the geometry of the nuclear collision.

Experimentally the collision volume can be controlled and defined in size and baryon number independently from the meson production yield by various methods. The number of participants can be measured by the coincident observation of nucleons (and clusters) from the participant zone or - more indirectly - the spectator region. As a means to measure the impact parameter and thus the collision volume it has been proposed[27]to observe in coincidence soft photons from collective nucleus-nucleus bremsstrahlung. Similarily, as described in the first chapter, the number of hard photons is proportional to the number of n-p-encounters during the nuclear collision. By varying projectile and target mass these geometrical quantities characterizing the collision can be altered in a rather well defined way, which should allow a test of the different methods to determine the collision zone.

Experiments at SIS are expected to start early 1990. There is great hope that the knowledge about the equation of state of nuclear matter, as it appears in heavy ion collisions, will be enlarged considerably by the upcoming experiments.

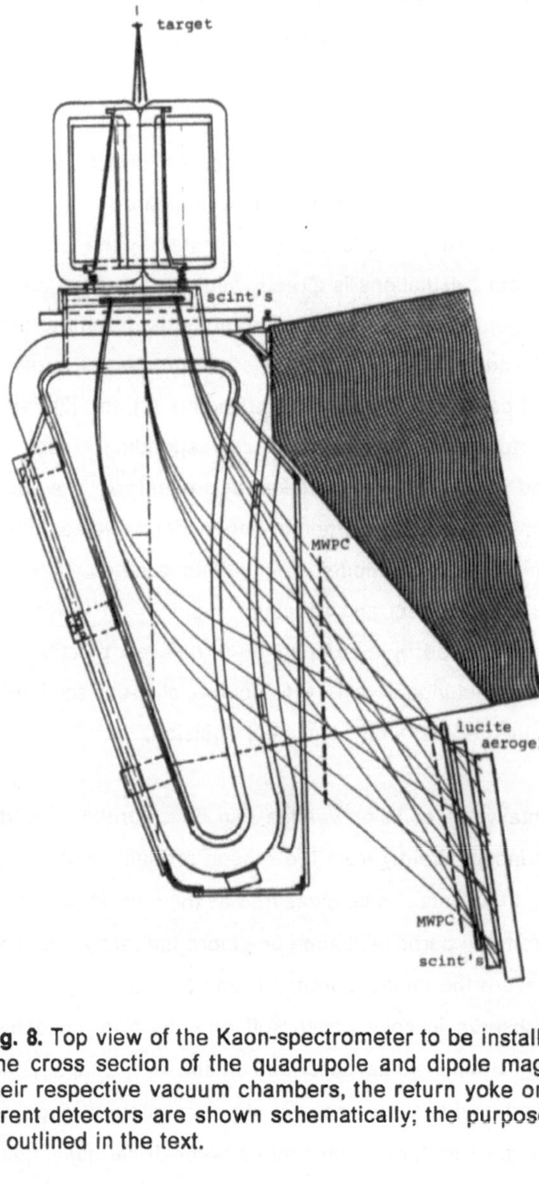

Fig. 8. Top view of the Kaon-spectrometer to be installed at SIS. The cross section of the quadrupole and dipole magnets with their respective vacuum chambers, the return yoke ond the different detectors are shown schematically; the purpose of these is outlined in the text.

The study of the emission of different probe particles from the same collision system allows a consistency check of the concepts used in the interpretation of the data. Pions will be present abundantly at energies around and above 1 GeV/u, the energy where compression effects will be studied. Very probably this will make the observation of directly produced hard photons rather difficult because of the combinatorical background from π^0's. Large arrays of scintillators with good time and energy resolution might allow to subtract this background; they also can be used to observe the production of η's. These are similar in mass to kaons but they only carry hidden strangeness. This makes them easier to produce but also more likely to be reabsorbed; e.g. by the strangeness exchange process leading to a K^+. For a complete understanding spectra for the different ejectiles are obviously needed.

At the new heavy ion synchrotron SIS a strong effort in this direction will be made by installing as first experiments a large solid angle photon detector and a magnetic spectrometer of QD type equipped with various detection systems especially suited for charged kaons and pions (later also leptons or even antiprotons). The measurement of the magnetic rigidity will be combined to a high resolution time of flight determination obtained from two granulated scintillator walls. Threshold Cherenkov detectors using aerogel radiators serve to identify pions and electrons and lucite (as well as water) is beeing used to tag all mesons (kaons and pions). To allow good statistics experiments on the rather low kaon production cross sections high countrates from background reactions have to be accepted. This is achieved by incorporating an intelligent second level trigger into the readout of the detectors and by performing the averaging needed for the position measurement in the MWPC's on line in a dedicated chain of transputers.

REFERENCES

1. C. Mahaux, Lecture Les Houches 1989
   B. Day & R. Wiringa, Phys. Rev. C 32 (1985) 1057
2. H.A. Bethe in: Unified Concepts of Many-Body Problems, p. 3,
   T.T.S. Kuo and J. Speth ed., Amsterdam 1987
3. M.M. Sharma et al., Phys. Rev. C 38 (1988) 2562 and this volume.
4. A.L. DePaoli et al., proceedings Hirschegg (1988)
5. E. Grosse et al., Europhys. Lett. 2 (1986) 9
6. G. Breitbach et al., submitted to Phys. Rev. Lett.
7. N. Hermann et al., Phys. Rev. Lett. 60 (1988) 1630
8. P. Grimm and E. Grosse, Progr. in Part. and Nucl. Phys. 15 (1985) 339
9. H. Nifenecker and J.A. Pinston, to be publ. in Reports on Progr. in Physics
10. K. Nakayama and G.F. Bertsch, Phys. Rev. C34 (1986) 2190;
    K. Nakayama, to be published in Phys. Rev. C
11. V. Metag, proceedings St. Malo 1988, Nucl. Phys. A488 (1988) 483
12. T.S. Biro et al., Nucl. Phys. A475 (1987) 579

13. W. Bauer, MSUCL-672, to be published
14. R. Heuer et al., Z. Phys. A330 (1988) 315
15. N. Alamanos et al., Phys. Lett. 173B (1986) 392
16. H. Noll et al., Phys. Rev. Lett. 52 (1984) 1284
H. Heckwolf et al., Z. Phys. A315 (1984) 243
17. E. Grosse, Nucl. Phys. A447 (1985) 611; id., Varenna lectures 1987, to be publ.
18. T. Reposeur, thesis Paris 1989; T.D.S. Stanislaus, thesis Vancouver 1987
19. M. Tohyama et al., Nucl. Phys. A437 (1985) 739
20. G.F. Bertsch et al., MSUCL-644, to be published
21. J. Cugnon and M.C. Lemaire, Nucl. Phys. A489 (1988) 781
22. R. Stock, Phys. Reports 135 (1986) 259
23. J. Aichelin and C.M. Ko, Phys. Rev. Lett. 55 (1985) 2661
B. Schürmann and W. Zwermann, Europhys. Lett. (1989)
24. J.W. Harris et al. Phys. Rev. Lett. 47 (1981) 229
S. Nagamiya, Nucl. Phys. A400 (1983) 399
S. Schnetzer et al., Phys. Rev. Lett. 49 (1982) 989
J.B. Carroll, Lecture Les Houches 1989
25. N.K. Abrosimov et al., JETP Lett 43 (1986) 270
26. E. Grosse et al. Proposal for an experiment
to be performed at SATURNE (1988).
27. J. Kapusta, Phys. Rev. C15 (1977) 1580
28. E. Grosse et al. Proposal for an experiment
to be performed at SIS (1989).
P. Senger et al., GSI-Nachrichten 88-5 (1988)
and to be published.

HARD PHOTONS FROM INTERMEDIATE ENERGY HEAVY

ION COLLISIONS

V. Metag

II. Physikalisches Institut Universität Gießen
Gießen, West-Germany

1 Introduction

High energy photons emitted in heavy ion collisions have been shown to provide direct information on the reaction dynamics. This result emerges from extensive studies performed by several groups for various target-projectile combinations at bombarding energies below 100 MeV/u. A comprehensive review of this field has recently been given in [1].

In this contribution particular emphasis is placed on the analysis of data taken in the 15 - 20 MeV/u bombarding energy range. It will be shown that hard photons predominantly originate from proton–neutron bremsstrahlung in initial nucleon–nucleon collisions of the heavy ion reaction. Measured energy spectra can directly be related to the nucleon momentum distribution in the collision zone. Photon production probabilities at a given bombarding energy are found to provide information on the geometrical overlap of the colliding heavy ions. If highly excited fragments are produced in the heavy ion reaction additional contributions to the photon spectrum arise from statistical photon emission from thermally equilibrated reaction products in the late stage of the reaction.

After discussing in detail the results for two target-projectile combinations a systematics of hard photon emission in heavy ion collisions is presented. Finally, prospects of future experiments at higher bombarding energies are outlined.

2 Competition between proton–neutron bremsstrahlung and thermal photon emission

A typical photon spectrum (fig.1), as e.g. measured for the ^{16}O + W reaction [2] is characterized by 3 features: soft photons below 10 MeV due to statistical emission from excited reaction products, a bump arising from γ-de-excitation of giant resonances, and hard photons above 25 MeV. At bombarding energies exceeding

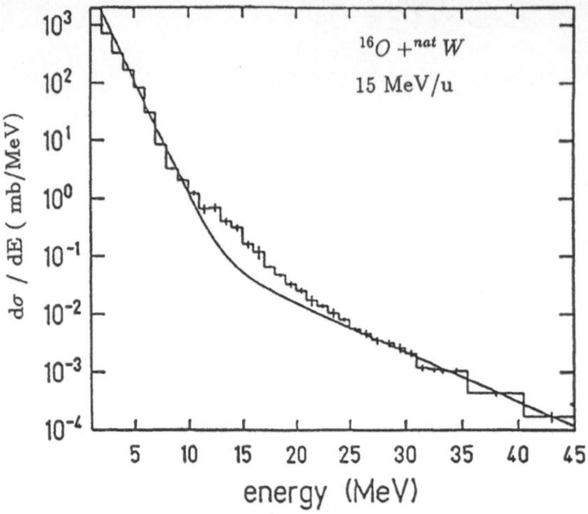

Figure 1. Inclusive photon spectrum from the reaction $^{16}O + ^{nat}W$ at 15 MeV/u. The solid curve represents a fit to the spectrum with two exponentials.

the Fermi energy (> 30 MeV/u) all experimental and theoretical investigations performed so far consistently indicate that hard photons predominantly originate from bremsstrahlung emitted in the first nucleon–nucleon collisions of a heavy ion reaction [1]

At the lower bombarding energies discussed here highly excited fragments are known to be produced. This raises the question whether in addition to nucleon-nucleon bremsstrahlung in the early phase of the heavy ion collision there are contributions to the hard photon spectrum from statistical photon emission in the late stage of the reaction after the fragments have thermally equilibrated. These photons are emitted in competition with particle evaporation. Their yield can be calculated within the statistical model once the excitation energy of the fragments is known. In mass asymmetric target-projectile combinations proton-neutron bremsstrahlung and thermal photon emission can be distinguished experimentally by measuring the photon source velocity. Photons produced in nucleon-nucleon collisions are emitted from the nucleon-nucleon center of mass system which moves in the laboratory with half the beam velocity. In contrast, thermal photon emission occurs in the frame of the fragments formed in fusion-like or deeply inelastic collisions.

The photon source velocity is determined by exploiting the Doppler-effect. The photon angular distribution is assumed to be symmetric in the source system and converts into an anisotropic angular distribution in the laboratory system with an anisotropy, which depends on the source velocity. As an example fig.2 shows the angular distribution of photons above 25 MeV for the $^{16}O+^{nat}W$ reaction at 19.5 MeV/u in comparison to the angular distributions expected for photon sources moving with the velocity of the nucleon-nucleon and nucleus-nucleus center of mass system, respectively. A fit to the data gives a source velocity consistent with half the beam velocity, suggesting an interpretation of the hard photon yield in terms of proton–neutron bremsstrahlung. This result is corroborated by similar analyses of other target projectile combinations [2-7] with $A_L/A_H < 0.5$ presented in fig.3. The amount of statistical photon emission contributing to the photon spectrum of fig.2 can be estimated by a comparison with a statistical model calculation using the code CASCADE [8]. The solid curve in fig.4 represents the fraction

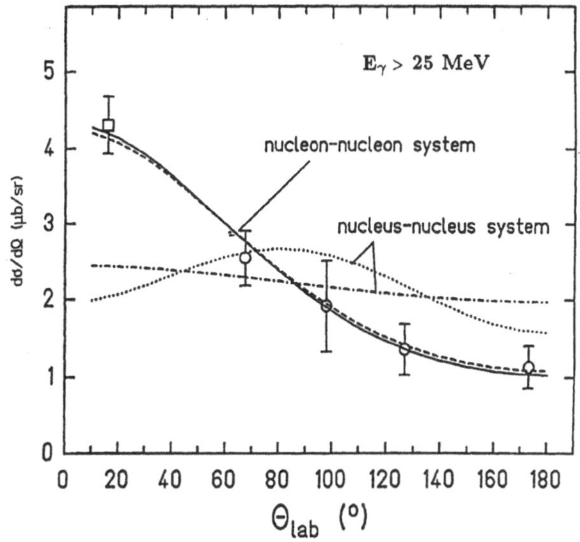

Figure 2. Measured angular distribution for photons with laboratory energies above 25 MeV in comparison to isotropic (dott–dashed) and dipole (dotted) radiation from the nucleus–nucleus system and to isotropic emission (dashed) from the nucleon–nucleon system, respectively. The error bars include both statistical and systematic uncertainties. The solid line shows the best fit to the data giving a source velocity of $\beta = 0.094 \pm 0.020$. The figure is taken from [2].

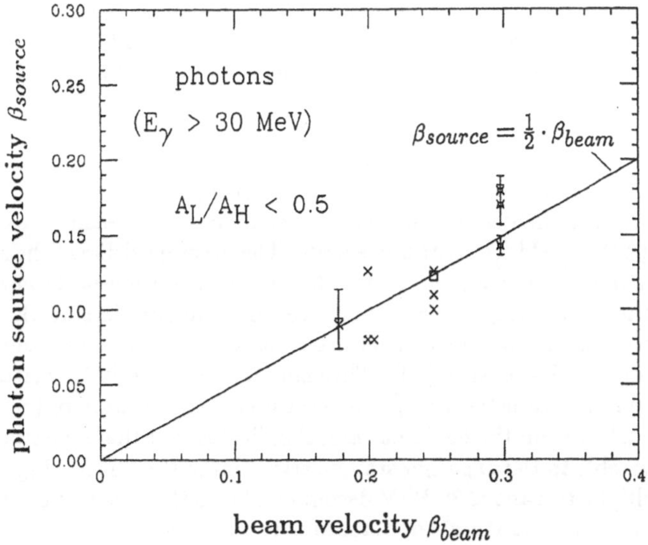

Figure 3. Photon source velocities extracted from moving source fits as a function of the beam velocity. The data are compiled from refs. [2-7]

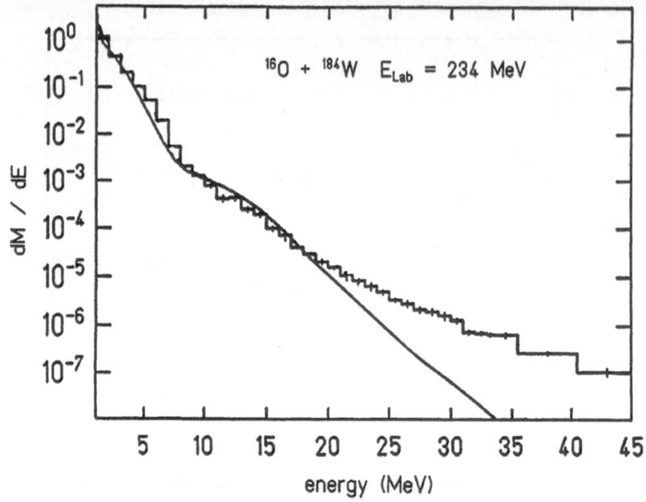

Figure 4. measured photon multiplicity (histogram) [2] compared to the calculated statistical emission multiplicity. The solid curve is the result of a CASCADE [8] calculation including pre–equilibrium and incomplete fusion effects.

of the spectum which can be accounted for by thermal photon emission. Above 30 MeV this contribution amounts to only $\approx 2\%$ of the observed hard photon yield, consistent with the conclusion drawn from the measured photon source velocity. The small thermal contribution is due to the rather moderate temperatures (T $\approx$ 2.5 MeV) of the excited reaction products.

In mass symmetric target-projectile combinations considerably higher fragment temperatures are reached for a given bombarding energy per nucleon. The importance of thermal photon emission from highly excited reaction products in the late phase of the reaction has been demonstrated by N. Herrmann et al. [9] for deeply inelastic collisions of ^{92}Mo + ^{92} Mo at 19.5 MeV/u. In this experiment an upper limit on the total excitation energy of the reaction products has been derived from a kinematically complete measurement of the outgoing fragments. In the analysis the energy removed by pre-equilibrium particle emission has to be considered. Fig.5 shows photon multiplicity spectra for different total kinetic energy losses (TKEL). The effective excitation energies and corresponding temperatures of the reaction products have been inferred by reproducing the experimental spectra below 20 MeV with CASCADE calculations, exploiting the sensitivity of the giant resonance photon yield on excitation energy. The decomposition of the spectra into proton–neutron bremsstrahlung and a thermal contribution is, however, subject to systematic uncertainties. These may arise from fluctuations in the deposition of the excitation energy and/or a possible variation of the level density parameter a with excitation energy [10]. Throughout this analysis a constant value of the level density parameter a = $\frac{A}{8}$ has been used. An estimate of the fluctuation width $\sigma^2 = aT^3$ within the Fermi gas model indicates that these fluctuations are at most comparable to the experimental uncertainties in the TKEL. Fig.6 shows the photon multiplicities above 30 MeV decomposed into thermal and bremsstrahlung contributions as a function of the total kinetic energy loss.

For TKEL > 200 MeV the thermal contributions amount to 20 – 40%. The higher thermal photon yield compared to the O+W reaction arises from the significantly higher fragment temperatures which are presented in fig.7. as a function of TKEL. The difference to the solid curve, representing complete conversion of total kinetic energy loss into excitation energy, is due to pre–equilibrium energy losses.

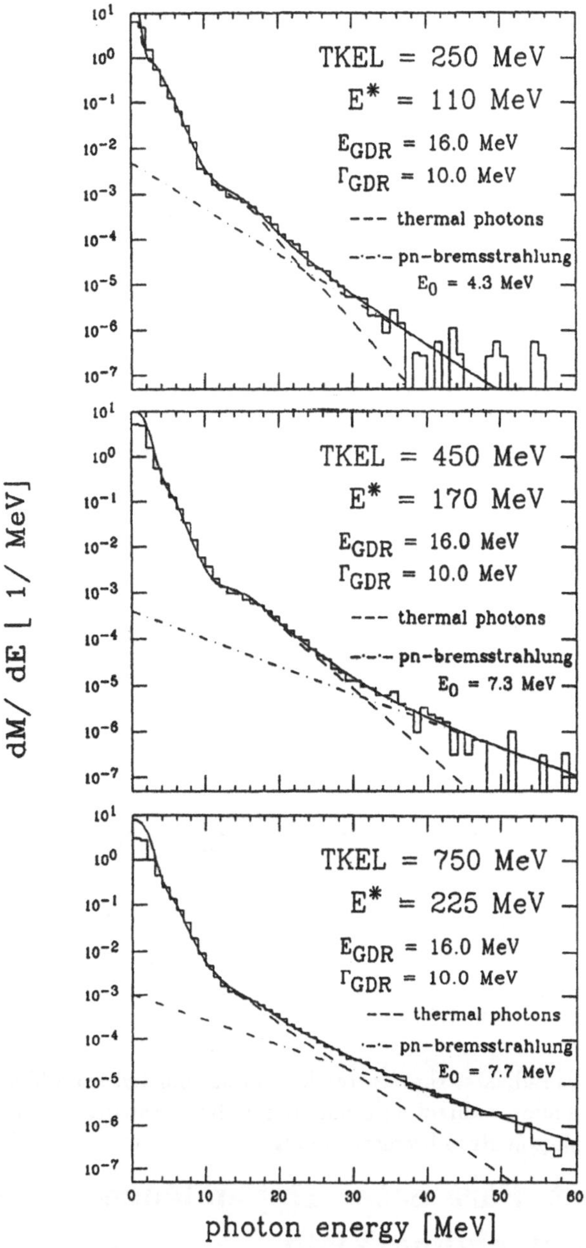

Figure 5. Photon multiplicity spectra for selected total kinetic energy losses in the system $^{92}Mo + ^{92}Mo$ at 19.5 MeV/u. The dashed curves represent statistical model calculations for the fragment excitation energies given in the figure. The dotted curves are fits to the bremsstrahlung component. The solid curves are the sum of both contributions.

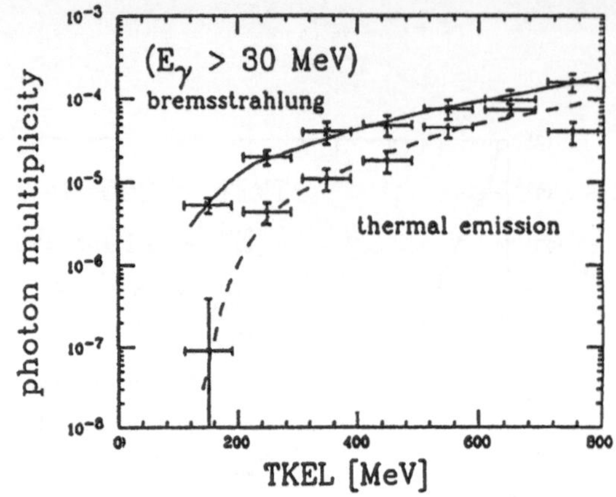

Figure 6. Multiplicity of high energy photons ($E_\gamma > 30$ MeV) decomposed into a thermal and a bremsstrahlung contribution, respectively. The multiplicities are plotted as a function of the total kinetic energy loss (TKEL) for the system ^{92}Mo + ^{92}Mo at 19.5 MeV/u. The solid and dashed curves are to guide the eye.

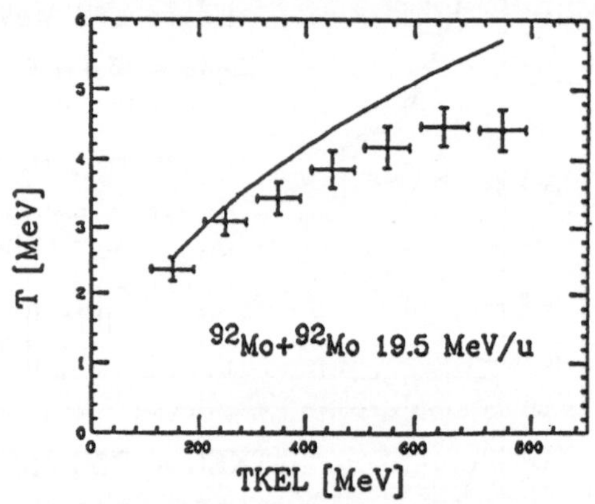

Figure 7. Fragment temperatures for different TKEL extracted by reproducing photon multiplicity spectra below 20 MeV with statistical model calculations. The solid curve corresponds to temperatures reached for complete conversion of the total kinetic loss into excitation energy.

This analysis indicates that energetic photons emitted from highly excited reaction products allow a direct determination of fragment temperatures also in the presence of proton-neutron bremsstrahlung.

3 Impact Parameter Dependence of the Hard Bremsstrahlung Yield

The bremsstrahlung part of the photon spectrum provides information on the geometrical overlap of the coiliding heavy ions. The photon multiplicity observed for different TKEL can be related to the number of proton-neutron collisions and thereby to the impact parameter.

The number of proton-neutron collisions for a given TKEL bin is determined by a comparison of the corresponding hard photon multiplicity with that for a single proton-neutron collision. Following [11] the latter is derived from the inclusive hard photon yield by

$$P_{pn}^{\gamma} = \frac{\sigma_\gamma(E_\gamma > 30MeV)}{\sigma_R < N_{pn} >_b} \quad (1)$$

Here, σ_R is the reaction cross section and $< N_{pn} >_b$ is the number of proton-neutron collisions averaged over impact parameter [12]. The number of proton-neutron collisions is directly related to the impact parameter by purely geometrical considerations [12]. Following this prescription a correlation between impact parameter and total kinetic energy loss TKEL is obtained. Other characteristic parameters of the hard photon spectrum – originally determined for different total kinetic energy losses – can thereby be plotted as a function of the impact parameter.

The inverse exponential slope parameter E_0 ($M_\gamma \sim e^{-\frac{E_\gamma}{E_0}}$) and the bremsstrahlung multiplicity are plotted in fig.8a,b as a function of the impact parameter. Photon spectra are found to become softer for more peripheral collisions (fig.8a). As pointed out in [13] this may be related to the fact that nucleons at the nuclear surface have smaller kinetic energy and thus carry smaller momenta. Consequently, nucleon-nucleon encounters in peripheral nucleus-nucleus collisions are less violent, leading to smaller slope parameters in the bremsstrahlung spectrum.

The variation in the photon multiplicity with impact parameter (fig.8b) by more than an order of magnitude opens up the possibility to use the hard photon yield as a tag for the geometrical overlap of two colliding heavy ions. Although hard photons can not be used for an on-line trigger because of the low intensity, the centrality of any heavy ion reaction can nevertheless be determined by off-line analysis of the associated hard photon multiplicity. Since bremsstrahlung is emitted in the initial phase of the collision it provides a signature independent of the subsequent evolution of the reaction. Furthermore, this information is not distorted by rescattering because of the weakness of the electromagnetic interaction. The measurement of hard photon multiplicities may thus turn out to be a new experimental approach and a valuable method for event characterization in reaction mechanism studies.

4 Systematics of hard photon emission

So far only two target-projectile combinations have been discussed in detail. A wealth of information has been gathered in a series of experiments by several groups [2-7,9,13-18]. The in-medium probability for hard photon production ($E_\gamma >$ 30 MeV) derived from inclusive cross section with eq.(1) shows a systematic increase with bombarding energy (fig.9).This trend is qualitatively reproduced in BUU-calculations by K. Niita et al.[19]. The importance of the nucleonic momentum distribution in the collision zone for the observed hard photon yield is demonstrated by a comparison with the hard photon production probability in free proton-neutron collisions which decreases rapidly below 70 MeV/u.

A rather systematic trend with bombarding energy is also found for the inverse slope parameter E_0, reflecting harder nucleon-nucleon collisions for increasing projectile velocity. Here, a comparison with BUU calculations [19] indicates an increasing discrepancy for higher bombarding energies (fig.10).

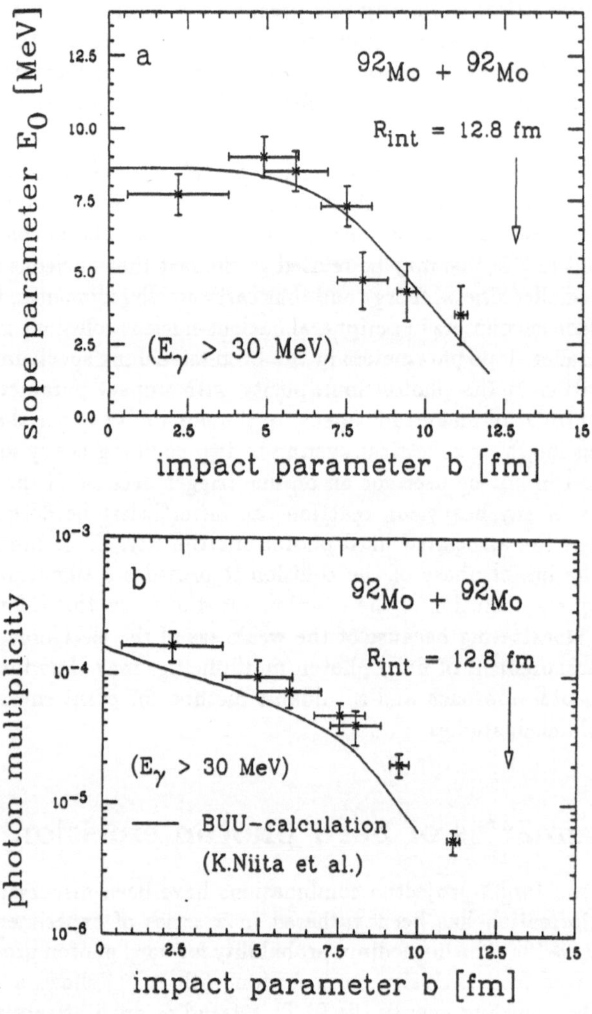

Figure 8. Impact parameter dependence of (a) the inverse slope parameter E_0 and (b) the bremsstrahlung multiplicity ($E_\gamma > 30$ MeV) for the system $^{92}Mo + ^{92}Mo$ at 19.5 MeV/u. The solid curve in (a) is to guide the eye; the curve in (b) represents the result of a BUU calculation [19].

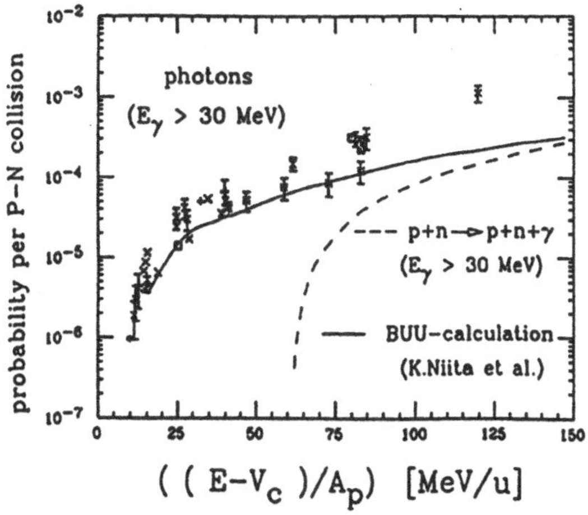

Figure 9. In–medium hard photon ($E_\gamma > 30$ MeV) production probability per proton–neutron collision as a function of the Coulomb corrected bombarding energy per nucleon. The data points are derived from inclusive cross section data [2-7,9,13-18] with eq.(1). The solid curve represents a corresponding BUU calculation by K. Niita et al.[19]. For comparison the dashed curve shows the hard photon production probability in free proton–neutron collisions.

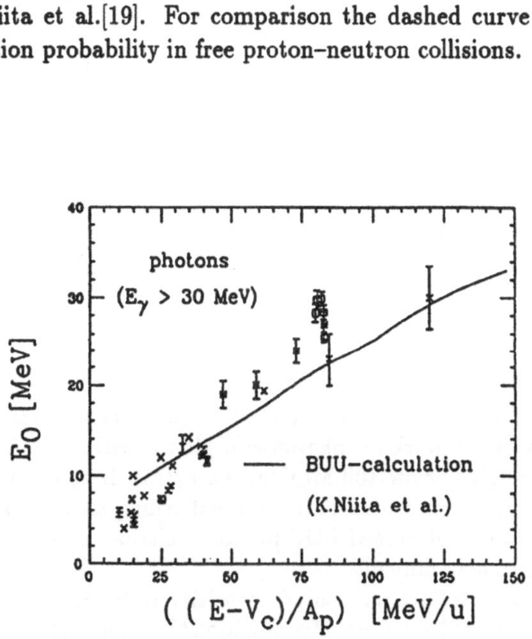

Figure 10. Inverse slope parameter of hard photon spectra as a function of the bombarding energy for different target–projectile combinations. The solid curve represents the result of a corresponding BUU calculation[19].

5 Prospects for future experiments at higher energies

Photons produced either directly or by neutral meson decay will remain a valuable probe of heavy ion collisions also at relativistic energies which will soon be available, e.g. at the heavy ion synchroton SIS (GSI Darmstadt). Fig.11 shows

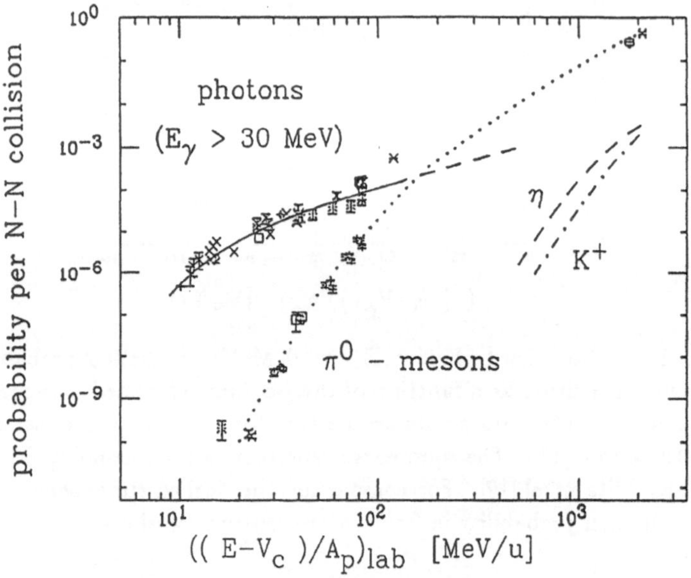

Figure 11. In–medium photon and meson production probabilities as a function of the bombarding energy. The photon data are from [2-7,9,13-18], π^0–data from [20-27]. The production probabilities for η and K^+ mesons have been calculated in BUU simulations by [19], [28].

production probabilities per nucleon-nucleon collision as a function of bombarding energy. Most of the work on photon and pion emission has so far been done at bombarding energies below 100 MeV/u. An extrapolation of the experimental data to higher energies indicates that for bombarding energies above 200 - 300 MeV pion production will exceed hard photon emission, making bremsstrahlung measurements extremely difficult.

Pion production rates are particularly sensitive to the nucleonic motion in the collision zone at energies below the π^0 production threshold in free nucleon–nucleon collisions. At energies below 280 MeV/u pion production relies on a favourable summation of Fermi velocities and the relative velocity of the colliding nuclei, leading to much more energetic nucleon–nucleon collisions than for free nucleons.

For energies of about 1 GeV/u the role of the most sensitive probe is taken over by the heavier K^+ and η-mesons ($m_{K^+}c^2 = 494$ MeV; $m_\eta c^2 = 549$ MeV). Kaons as well as η-mesons contain strange quarks and are thus expected to experience a weaker final state interaction with nucleons. In contrast to kaons, η-mesons have net strangeness zero and can thus be emitted without the associated production of a strange hadron. This lowers the production threshold in free nucleon–nucleon collisions to 1.26 GeV. Below this energy η-mesons can only be produced in heavy ion reactions, exploiting the momentum distribution of nucleons in the collision zone. Thereby, η-mesons become a sensitive probe of the reaction dynamics in the 1 GeV/u energy regime. Furthermore, BUU calculations [28] indicate that at bombarding energies of 500 - 800 MeV/u η-mesons are predominantly produced in a highly compressed themalized collision zone and are thus an ideal probe of high density ($\rho \approx 2.0$ - $2.5\ \rho_o$) nuclear matter produced in relativistic heavy ion collisions.

A modular Two Arm Photon Spectrometer (**TAPS**) [29] comprising 320 BaF_2 detectors is being set up by an international collaboration [30] for studying neutral meson production in heavy ion and photonuclear reactions. Neutral mesons will be identified by their two photon decay, requiring photon coincidences between the two arms of the spectrometer. It is hoped that this spectrometer will provide valuable information on the behaviour of nuclear matter at high densities and thereby offer a new experimental approach to study the nuclear equation of state.

I would like to thank G. Breitbach, K. Hagel, N. Herrmann, W. Kühn, and R. Novotny who carried the major load of taking and analyzing data included in this contribution. Illuminating discussions with W. Cassing, U. Mosel, H. Nifenecker, and R. Vandenbosch are gratefully acknowledged. This work was supported in part by Gesellschaft für Schwerionenforschung und Deutsches Bundesministerium für Forschung und Technologie under Contract No. 06 GI 174 I.

References

[1] H.Nifenecker and J.A.Pinston
Progress in Particle and Nuclear Physics, to be published

[2] G.Breitbach, G.Koch, S.Koch, W.Kühn, A.Ruckelshausen, V.Metag, R.Novotny, S.Rieß , D.Habs, D.Schwalm, E.Grosse, H.Ströher
submitted to Phys.Rev.Lett.

[3] J. Stevenson, K.B. Beard, W. Benenson, J. Clayton, E. Kashy, A. Lampis, D.J. Morrissey, M. Samuel, R.J. Smith, C.L. Tam, J.S. Winfield
Phys. Rev. Lett. 57 (1986) 555

[4] C.A. Gossett
Proc. XXVI Int. Winter Meeting on Nuclear Physics, Bormio Italy, 1988, edt. I. Iori

[5] M. Kwato Njock, M. Maurel, E. Monnand, H. Nifenecker, J. Pinston, F. Schussler, D. Barneoud
Phys. Lett. 175B (1986) 125

[6] R. Bertholet, M. Kwato Njock, M. Maurel, E. Monnand, H. Nifenecker, P. Perrin, J.A. Pinston, F. Schussler, D. Barneoud, C. Guet, Y. Schutz
Nucl. Phys. A 474 (1987) 541

[7] R. Hingmann, W. Kühn, V. Metag, R. Mühlhans, R. Novotny, A. Ruckelshausen, W. Cassing, H. Emling, R. Kulessa, H.J. Wollersheim, B. Haas, J.P. Vivien, A. Boullay, H. Delagrange, H. Doubre, C. Grégoire, Y. Schutz
Phys. Rev. Lett. 58 (1987) 759

[8] F.Pühlhofer *Nucl.Phys. A280 (1977) 267*
M.Harakeh, *private communication*

[9] N. Herrmann, R. Bock, H. Emling, R. Freifelder, A. Gobbi, E. Grosse, K.D. Hildenbrand, R. Kulessa, T. Matulewicz, R. Rami, R.S. Simon, H. Stelzer and J. Wessels, P.R. Maurenzig, A. Olmi, A.A. Stefanini, W. Kühn, V. Metag, R. Novotny, M. Gnirs, D. Pelte, P. Braun-Munzinger, L.G.Moretto
Phys. Rev. Lett. 60 (1988) 1630

[10] R. Wada, D. Fabris, K. Hagel, G. Nebbia, Y. Lou, M. Gonin, J.B. Natowitz, R. Billerey, B. Cheynis, A. Demeyer, D. Drain, D. Guinet, C. Pastor, L. Vagneron, K. Zaid, J. Alarja, A. Giorni, D. Heuer, C. Morand, B. Viano, C. Mazur, C. Ngô, S. Leray, R. Lucas, M. Ribrag, E. Tomasi
Phys. Rev. C 39 (1989) 497

[11] V.Metag
Nucl.Phys. A482 (1988) 159c
Nucl.Phys. A488 (1988) 483c

[12] H. Nifenecker and J.P. Bondorf
Nucl. Phys. A442 (1985) 478

[13] M.Kwato Njock, M.Maurel, E.Monnand, H.Nifenecker, P.Perrin, J.A.Pinston, F.Schussler, and Y.Schutz,
Nucl.Phys. A489 (1988) 368

[14] E.Grosse, P.Grimm, H.Heckwolf, W.F.Müller, H.Noll, A.Oskarsson, H.Stelzer, and W.Rösch
Europhys.Lett.2 (1986) 9

[15] M. Kwato Njock, M Maurel, H.Nifenecker, J.A.Pinston, F.Schussler, D.Barneoud, S.Drissi, and J.P.Vorlet
Phys.Lett.B207 (1988) 269

[16] N. Alamanos, P. Braun-Munzinger, R.F. Freifelder, P. Paul, J. Stachel, T.C. Awes, R.L. Ferguson, F.E. Obenshain, F. Plasil, G.R. Young
Phys. Lett. 173B (1986) 392

[17] J. Clayton, J.Stevenson, W. Benenson, Y. Chen, E. Kashy, A.R. Lampis, D.J. Morrissey, T.K. Murakami, B. Sherriil, C.L. Tam, and J.S. Winfield
MSU preprint 1989

[18] J.J.Gaardhøje, A.M.Bruce, J.D.Garrett, B.Herskind, M.Maurel, H.Nifenecker, J.A.Pinston, P.Perrin, C.Ristori, F.Schussler, A.Bracco, and M.Pignanelli
Phys.Rev.Lett. 59 (1987) 1409

[19] K.Niita, W.Cassing, and U.Mosel
Reprint VGI-89-03, Nucl.Phys.A, in print and private communication

[20] G.R.Young, F.E. Obenshain, F. Plasil, P. Braun-Munzinger, R. Freifelder, P. Paul, J. Stachel
Phys.Rev.C 33 (1986) 742

[21] P.Braun-Munzinger, P. Paul, L. Ricken, J. Stachel, P.H. Zhang, G.R. Young, F.E. Obenshain, E. Grosse
Phys.Rev.Lett. 52 (1984) 255

[22] H.Heckwolf, E. Grosse, H. Dabrowski, O. Klepper, C. Michel, W.F.J. Müller, H. Noll, C. Brendel, W. Rösch, J. Julien, G.S. Pappalardo, G. Bizard, J.L. Laville, A.C. Mueller, J. Péter
Z.Phys. 315 (1984) 243

[23] H. Noll, E. Grosse, P. Braun-Munzinger, H. Dabrowski, H. Heckwolf, O. Klepper, C. Michel, W.F.J. Müller, H. Stelzer
Phys.Rev.Lett. 52 (1984) 1284

[24] P.Braun-Munzinger and J.Stachel,
Ann.Rev.Nucl.Part.Sci.37 (1987) 1

[25] P. Grimm and E. Grosse,
Progr. in Part. and Nucl. Phys. 15 (1985) 339

[26] G.Roche, J.Carroll, C.C.Chang, T.Hallman, P.N.Kirk, R.Koontz, G.Krebs, L.Madansky, T.Mulera, H.G.Pugh, L.S.Schröder, J.Vicente
Nucl.Phys. A 439 (1985) 721

[27] T.Hallman, J.Carroll, W.Dejarnette, E.K.McIntyre, L.Madansky, A.Sagle, R.Semper
Nucl.Phys.A 440 (1985) 697

[28] A.DePaoli, K.Niita, W.Cassing, U.Mosel, and C.M.Ko
Phys. Lett.B 219 (1989) 194

[29] V.Metag and R.S.Simon
Technical Proposal for a Two Arm Photon Spectrometer, GSI Report 87-19

[30] **TAPS**-collaboration: GANIL, Univ. Gießen, ISN Grenoble, GSI, KVI Groningen, LMU München, Univ. Münster, DPhn Saclay, Univ. Utrecht

PRODUCTION OF LEPTON PAIRS AT THE LBL BEVALAC

G. Roche[a]

Nuclear Science Division
Lawrence Berkeley Laboratory
Berkeley, CA 94720, USA

ABSTRACT

We discuss the physics objectives of the DLS program with some emphasis on the possible use of dileptons as a probe of pion dynamics in nuclear matter. Data on p-Be reactions at 1-5 GeV and Ca-Ca at 1-2 A GeV are presented. The observation of a structure at about twice the pion mass in the e^+e^- invariant mass spectra above 2 GeV beam energy and the excitation function for the p-Be reaction suggest that pion annihilation is a significant dielectron source above 2 GeV. The dielectron mass spectrum from Ca-Ca at 1 A GeV exhibits an inverse slope larger than the one from p-Be at the same beam energy.

INTRODUCTION

I am going to present the work done by the Dilepton Spectrometer (DLS) Collaboration[b] on the production of lepton pairs at the LBL Bevalac. This study is relevant to *nuclear matter* and has nothing to do with quark matter, except perhaps that it may give some indication on the backgrounds that could be experienced in the search for the quark-gluon plasma. The energy domain is about 1 A GeV.

The DLS program deals more precisely with the production of electron pairs (dielectrons) in p-p, p-nucleus and nucleus-nucleus collisions:

$$A + B \longrightarrow e^+e^- + X \text{ (multiplicity measurement).}$$

Multiplicity information was not recorded with the first data presented herein.

I will start with a brief review of the experiments at the beginning of the DLS program and give the physics motivations with some emphasis on the aspects relevant to pion dynamics in nuclear matter. The experimental set up will be described and results obtained so far on p-Be and Ca-Ca collisions will be presented. I will end the talk with some first conclusions and a brief discussion of the possible developments of the program.

STATUS OF EXPERIMENTS AT THE BEGINNING OF THE DLS PROGRAM

At the beginning of the DLS program, there was obviously no data on nucleus-nucleus collisions. Dilepton production had been extensively studied in hadron-nucleon and hadron-nucleus collisions above 10 GeV beam energy while there was no data between 1 and 10 GeV. A low energy experiment[1] on p-p at 256 and 800 MeV found no evidence for direct[c] single electron production, at the level of 10^{-6} of the pion production rate for the 800 MeV measurement. Thus, there was a possibility for the existence of a threshold in between 1 and 10 GeV beam energy.

Fig. 1 shows a typical dimuon mass spectrum measured in high energy hadronic collisions[2]. It exhibits peaks corresponding to the various meson resonances and a continuum. The high mass region of the continuum is well interpreted in term of the Drell-Yan hard quark-antiquark annihilation process, while the low mass region, sometime referred to as "the anomalous dilepton continuum", is not well understood. Several experiments have been devoted to the study of the dilepton low mass continuum, and we list in Table 1 those performed with electron pairs at energies closer to the Bevalac domain. The KEK dielectron mass spectrum[3] is shown in Fig. 2. The estimated background due to η and ω^0 Dalitz decays cannot account for the dielectron yield. The soft parton model calculation of V. Černý *et al.*[4] or the quark-gluon plasma model of E.V. Shuryak[5] are not in very good agreement with the data points. Also shown in Fig. 2 is a fit to the data using a functional form from K. Kinoshita *et al.*[6]. We will use this fit for a comparison to the DLS data later on.

At the Bevalac, we are evidently limited to the dilepton mass region that corresponds to the anomalous dilepton continuum and the DLS results should help to clarify the situation.

PHYSICS MOTIVATIONS

p-N and p-nucleus Collisions

The DLS program aims to establish the existence of direct electron pair production in the few GeV beam energy domain and help clarify the production mechanism(s).

Nucleus-nucleus Collisions

When we submitted our first proposal five years ago, there was no theoretical study precisely relevant to the Bevalac energy range. We were relying on general arguments as follows. Dileptons should be a good probe of the primary hot stage of the fireball. They present three advantages: (i) they are a penetrating probe and do not interact much in going out of nuclear matter, (ii) their production rate is biased toward the high density phase of the collision, and (iii) their coupling to other particles is very well known so accurate calculations are in principle possible. However, they present the disadvantage of low production rates due to the smallness of the fine structure constant α (there is roughly one e^+e^- pair produced per ten thousand NN collisions). This disadvantage actually makes the experimental difficulty quite

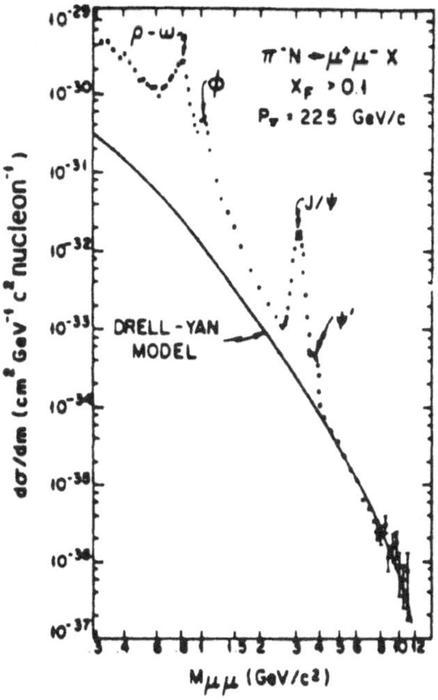

Fig. 1. Dimuon mass spectrum measured by the Chicago-Princeton group and contribution of the Drell-Yan process[2].

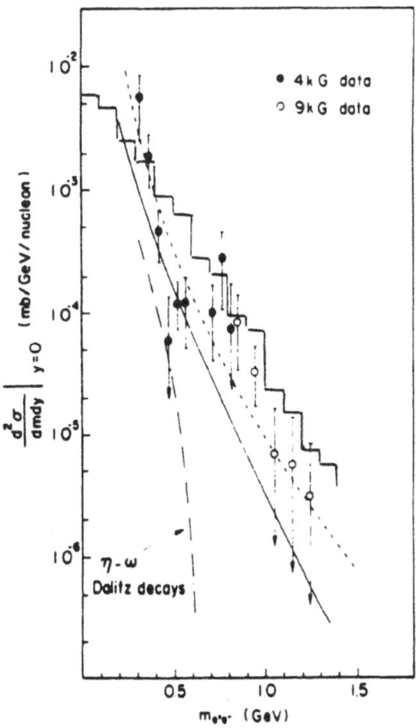

Fig. 2. The KEK dielectron mass spectrum[3], p-Be at 12.1 GeV. The dot-dashed curve is the estimated background due to η and ω^0 Dalitz decays. The histogram is the soft parton model of V. Černy *et al.*[4], the solid curve is the prediction of E.V. Shuryak[5] and the dotted curve a fit to the model of K. Kinoshita *et al.*[6].

Table 1.

Existing data at the beginning of the DLS project (electron pairs)

KEK[3]	p-Be at 12.1 GeV	144 true pairs
SLAC[7]	π^--p at 15.9 GeV	107 true pairs
BNL[8]	π^--p at 16.9 GeV	165 true pairs
LAMPF[1]	(single electron experiment)	
	p-p at 800 MeV	no signal at the level of $e/\pi \sim 10^{-6}$

serious. The use of real photons would provide the same advantages as dileptons, with in general[d] a much higher production rate ($\sim 1/\alpha$ higher), but it is then difficult to subtract the copious gamma ray yield from π^0 decay. In the case of dileptons, the combinatorial background (false pairs) can be measured directly from the like-sign pair yield, and the true pair Dalitz decay background is only important at very low masses, below about 100 MeV.

Later on, C. Gale and J. Kapusta[9] have made calculations applicable to the Bevalac energy domain and pointed out possible interesting effects relevant to the pion dispersion relation in hot-dense nuclear matter. This study is generalized by L.H. Xia *et al.*[10] who include the expansion of the fireball and consider dileptons as a probe of pion dynamics in heavy ion collisions. G. Brown[11] discusses the interest of dilepton measurements in connection with the subjects of the nuclear equation of state and pion condensation. M. Schäffer *et a.*[12] compute the pn bremsstrahlung contribution in p-Be collisions at 1 GeV. Very recently, a preprint by S. Pratt[13] seems to raise a controversy about the effect of pion dispersion on the dilepton mass cross section.

There are several theoretical talks at this meeting on the subject of dilepton production in the 1 *A* GeV energy range, so I would like to only briefly discuss the possibility of using dileptons as a probe of pion dynamics in nuclear matter.

Pion Dynamics in Nuclear Matter

Two possibly dominant processes of dilepton production in the 1 *A* GeV range are shown in Fig. 3. When dealing with pion dynamics, $\pi^+\pi^-$ annihilation is of most interest and pn bremsstrahlung is a background. The propagation of pions in nuclear matter is described by the following dispersion relationship:

$$\omega^2 = \mathbf{k}^2 + m_\pi^2 + \prod(\omega, \mathbf{k}),$$

where ω is the pion total relativistic energy and $\mathbf{k}$ its vector momentum. The effect of the nuclear medium is introduced through the term $\prod(\omega, \mathbf{k})$. It is both temperature and density dependent, the strongest dependence coming from the baryon density. There is little experimental information on the pion dispersion relation and it is mostly constructed on theoretical arguments.

A kinematical domain of special interest corresponds to lepton pairs emitted back-to-back in the center-of-mass frame of the collision. In that case, the total vector momentum of the pions is

$$\mathbf{q} = \mathbf{k}_1 + \mathbf{k}_2 = 0.$$

It results that

$$\omega_1 = \omega_2 = \omega$$

and

$$M_{e^+e^-} = 2\omega.$$

There is an almost one-to-one correspondence between a point of the dielectron mass spectrum and a point of the pion dispersion curve. In that kinematical domain, the pion dispersion effect on the dielectron mass spectrum can be expected to be the strongest and is clearly seen in Fig. 4 from L.H. Xia *et al.*[10]. The pn bremsstrahlung

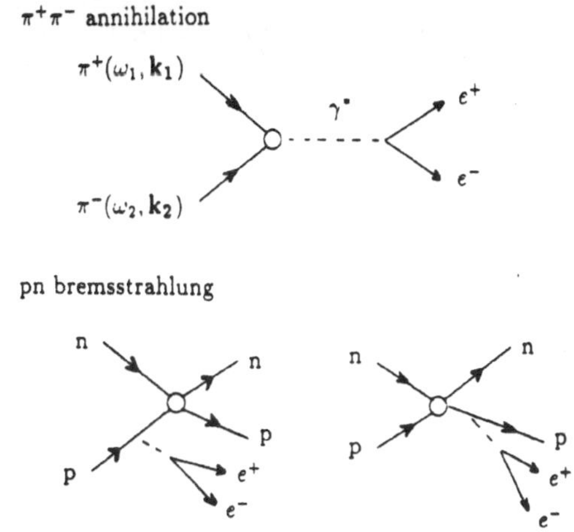

Fig. 3. Two possibly dominant processes of dilepton production in the 1 A GeV range.

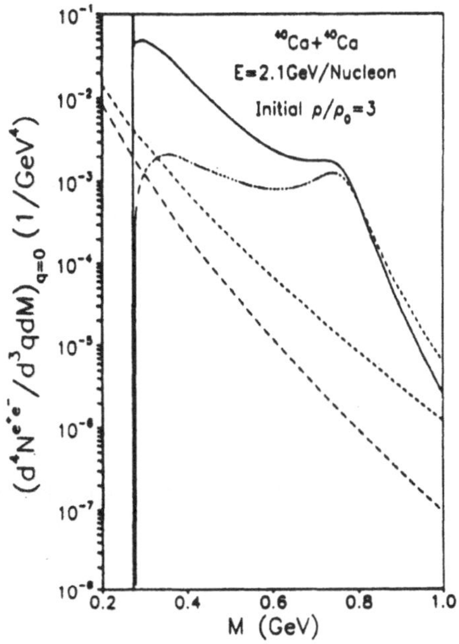

Fig. 4. The effect of the pion dispersion relation on the dielectron production rate, solid curve. The dotted curve is the $\pi^+\pi^-$ annihilation contribution without dispersion effect. The other two curves are the contribution from pn bremsstrahlung computed in the soft photon approximation with phase space included (long dashed curve) or without (dashed curve). The figure is from ref. 10.

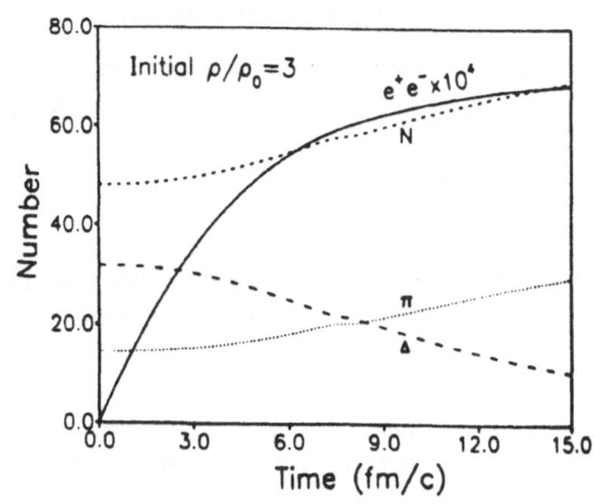

Fig. 5. Time evolution of the number of nucleons, deltas, and pions in the fireball for the same reaction as in Fig. 4. The solid curve is that for the density in momentum space of dileptons with zero total momentum. The figure is from ref. 10.

contribution is also shown in the figure. The effect of the pion dispersion amounts to an enhancement of the mass region just above the $\pi^+\pi^-$ annihilation threshold by more than an order of magnitude. Also, the same authors quantitatively establish that dileptons are created during the first stage of the nucleus-nucleus collision, within the first 6 fm/c for Ca on Ca at 2.1 A GeV (see Fig. 5).

As a conclusion to this part of the talk, I can say that ***dileptons should be a good probe of pion dynamics in nuclear matter.*** Notice that the same dispersion relation concept should actually apply to both p-nucleus and nucleus-nucleus collisions. The subject of pion condensation, which received much attention in the 1970's, is also regaining interest. Finally, the dilepton study is also useful in the more general framework of the nuclear equation of state for it should provide information (density, temperature) on the early hot-dense stage of heavy ion collisions.

EXPERIMENTAL SET UP

Design Considerations

There are some numbers that are important to better understand the experimental set up and the data. The direct electron yield as measured in high energy experiment is very low. It is usually reported as the ratio of direct electrons to pions at a given tranverse momentum p_t:

$$e/\pi \sim 10^{-4}$$

(the ratio goes up to about 10^{-3} at very low p_t's of about 100 MeV/c as measured at the CERN ISR[14]). The dielectron yield can also be expressed as the ratio of dielectrons to dipions and high energy experiments give the value

$$(e^+e^-)/(\pi^+\pi^-) \sim 10^{-5},$$

the DLS data actually providing the same value (see below). Thus, the experimental set up must fulfill the following requirements: (i) an extremely good hadron rejection power ($\gtrsim 10^5$), and (ii) a large acceptance and/or high interaction rate capability.

The main backgrounds result from π^0 decay:

$$\pi^0 \rightarrow \gamma\gamma \xrightarrow{conversion} e^+e^- \qquad BR \simeq 1$$
$$\pi^0 \rightarrow \gamma e^+e^- \ (Dalitz decay) \qquad BR \simeq 10^{-2}.$$

Difficulty can be anticipated (i) for low dielectron masses ($\lesssim$ 100 MeV) and (ii) because of a combinatorial background (false pairs).

The Dilepton Spectrometer

The DLS experimental set up is shown in Fig. 6. The target is segmented to reduce the combinatorial background from gamma ray conversion. The two large aperture dipole magnets offer each an angular acceptance of 170 msr. The electron identification with adequate hadron rejection power is provided by two gas Cherenkov counters in each arm. Tracking is achieved with drift chambers and scintillator hodoscopes provide trigger flexibility and redundant information. Details can be found in ref. 15. The multiplicity array was not yet implemented when we collected the first data presented below. The kinematical domain under investigation is approximately 0.1-1.2 GeV in invariant mass, 0.0-0.8 GeV/c in transverse momentum and 0.5-1.9 in units of laboratory rapidity (y).

The central ray of each arm is set at 40° to the beam direction, which is an adequate value for beam energies 4.9 GeV (p-Be reaction) and 2.1 A GeV (p-Be and Ca-Ca reactions). For these two beam energies, mid-rapidity electrons are emitted (in the laboratory) at 31.7° and 43.4°, respectively. However, for the lowest beam energy of 1.0 A GeV that has been used, the mid-rapidity laboratory angle of 53.9° does not match well the DLS rapidity acceptance and back-to-back pairs for instance are detected with a low efficiency.

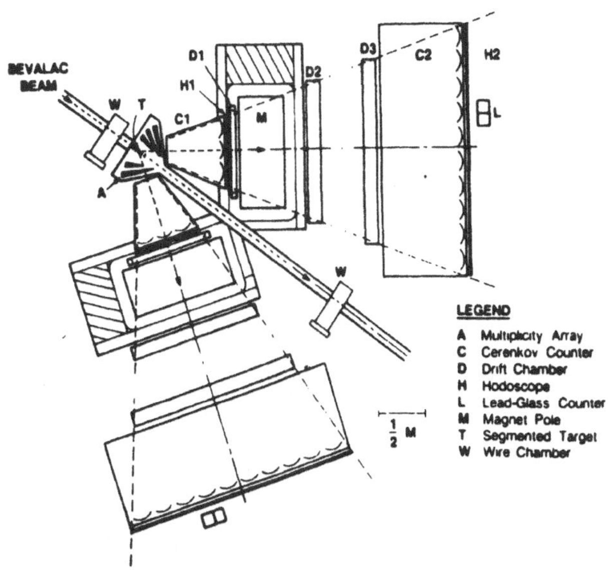

Fig. 6. The DLS experimental set up.

False Pair Subtraction

A significant combinatorial background (false pairs) originates when two uncorrelated electrons are detected, one in each DLS arm, these two electrons being mostly produced through the decay of π^0's, either directly from the Dalitz mechanism or by the conversion of their γ-ray products in the target or surrounding materials. These uncorrelated electrons are always produced in pairs and there are some cases when the two members of Dalitz or conversion pairs go through the whole system on one arm. Then we just remove the corresponding events. The presence of two electrons before one magnet can also be detected from the ADC[e] information of the front Cherenkov counters. The hodoscope and drift chamber information also helps. Therefore, a false pair mostly results when all ADC, TDC[e], time-of-flight and reconstruction conditions are fulfilled and that we miss one member of a Dalitz or conversion pair in one arm, on both sides. However, due to the charge detection symetry of the two arms (and the symetry is improved by collecting data for the four field polarity combinations in the magnets over equal periods of time), the opposite-sign false pair sample is equal to the like-sign sample and the true pair signal is simply obtained by subtraction of the like-sign pairs from the opposite-sign pairs. Evidently, the statistical accuracy on the true pairs depends on the amount of false pairs. The false pair yield increases as the square of the pion multiplicity and, depending on the origin of the true pairs, the true to false ratio can get worse when going to heavier target/projectile systems.

Notice that Dalitz pairs can be emitted with wide opening angles and detected by the system. These will not be subtracted but estimates of their contribution to the true pair signal will be given.

RESULTS ON p-Be AND Ca-Ca COLLISIONS

Table 2 gives the running conditions and Table 3 the pair statistics for the data taken so far. The interaction rates go from about 3×10^5 up to 3×10^7. The low number of reconstructed events compared to the number of recorded events is only partly due to the very simple trigger that we have been using (an eight-fold coinci-

Table 2.

DLS running conditions

(for p-Be at 1.0 GeV, the star refers to the same analysis as Ca-Ca at 1.0 *A* GeV)

Reaction	Average beam int. (proj./spill)	Acq. time (hours)	Target thick. (col. leng.)	# of recorded events	# of recons. pairs
4.9 GeV p-Be	1.2×10^8	34	0.1	1.4×10^5	933
2.1 GeV p-Be	2.3×10^8	16	0.1	7.7×10^4	715
2.0 *A* GeV Ca-Ca	2.5×10^7	30	0.01	2.8×10^5	139
1.0 GeV p-Be	3.0×10^8	26	0.1	7.3×10^3	130
					374*
1.0 *A* GeV Ca-Ca	1.0×10^8	83	0.02	2.0×10^5	1207

Table 3.

DLS pair statistics

(for p-Be at 1.0 GeV, the stars refer to the same analysis as Ca-Ca at 1.0 *A* GeV)

OS= number of opposite sign pairs, LS= number of like sign pairs,

F= number of false pairs in the OS sample (F=LS),

T= number of true pairs (T=OS−LS), $\sigma_T=\sqrt{OS+LS}$.

Reaction	OS	LS	T	T/F	T/σ_T
4.9 GeV p-Be	732	201	531±31	2.6	17.4
2.1 GeV p-Be	567	148	419±27	2.8	15.7
2.0 *A* GeV Ca-Ca	94	45	49±12	1.1	4.2
1.0 GeV p-Be	111	19	92±11	4.8	8.1
	263*	111*	152±19*	1.4*	7.9*
1.0 *A* GeV Ca-Ca	731	476	255±35	0.5	7.3

dence of the signals from the hodoscopes and Cherenkov counters). It also results from the large acceptance and the severe background conditions, and translates the difficulty of the measurements. The data is much cleaner for the lowest beam energy of 1.0 A GeV which allowed the use of looser cuts in the analysis (needless to say that the efficiency of the cuts is corrected for). The true to false ratio is about three times better for p-Be than for Ca-Ca collisions, independently of the analysis cuts and the incident energies. The existence of a dielectron signal down to 1 A GeV is established to a high level of statistical accuracy and the first goal of the DLS program is achieved.

p-Be Data

The cross section per nucleon (assuming an $A_t^{2/3}$ dependence, where A_t is the target mass) for p-Be as a function of the dielectron invariant mass is shown in Fig. 7 for the three beam energies 4.9, 2.1 and 1.0 GeV. The general shape of the 4.9 and 2.1 GeV distributions for masses above 300 MeV are similar to that seen at higher energies. For comparison to the KEK 12 GeV p-Be data[3], the fit given in Fig. 2 is plotted as a solid curve in Fig. 7(a). An enhancement in the ρ/ω region is seen in the mass spectrum from p-Be at 4.9 GeV. At 2.1 GeV, the maximum energy available in the nucleon-nucleon center-of-mass frame is 850 MeV, just barely above the ρ/ω threshold. The total Dalitz decay contributions to the dielectron cross sections (see ref. 16 for details) are shown as dashed curves in Fig. 7 for all three beam energies. The significant contributions are from π^0 and η at 4.9 and 2.1 GeV, while π^0 and $\Delta(1232)$ contribute at 1.0 GeV. At 4.9 and 2.1 GeV, the Dalitz decay background is approximately an order of magnitude smaller than the measured yield for masses above 200 MeV, in agreement with the higher energy results. For the 1.0 GeV data, the Dalitz decay contribution is less accurate due to the uncertainty in the $\Delta(1232)$ production cross section and our systematic errors.

The new observation is the structure at about 300 MeV (twice the pion mass) in the 4.9 and 2.1 GeV spectra. We have been much concerned with it, trying to answer the two questions: its statistical significance and the possibility of an experimental bias. An experimental bias could come from the acceptance due to the fact that it gets limited in y and p_t at low mass and thus more difficult to evaluate. In fact, we know that the very first mass bin (the tip of the acceptance domain) is not reliable and it is given only for a *qualitative* understanding of the mass spectra. We believe that the acceptance bias is negligible above $M = 150$ MeV and the second mass bin should already be reasonably accurate[17]. Evaluation of the statistical significance assumes the choice of a structureless model. Taking a softer functional form (such as an exponential) yields a lower statistical significance, while using a steeper functional form (such as power laws, M^{-2} or M^{-4}, reasonable for bremsstrahlung or soft parton models) yields a very high statistical significance. We finally decided not to give any number and let the readers decide. It is interesting to notice that a recently presented data[18] on very low mass dielectron production in p-p at a center-of-mass energy of 63 GeV is not inconsistent with the DLS result (see Fig. 8).

The cross section (per nucleon) for producing the low mass dielectron continuum, $200 \lesssim M \lesssim 700$ MeV, is plotted in Fig. 9 as a function of the available center-of-mass energy. The brackets around the DLS data points represent the systematic normalization errors of approximately +70/-20%. We first notice that the DLS cross

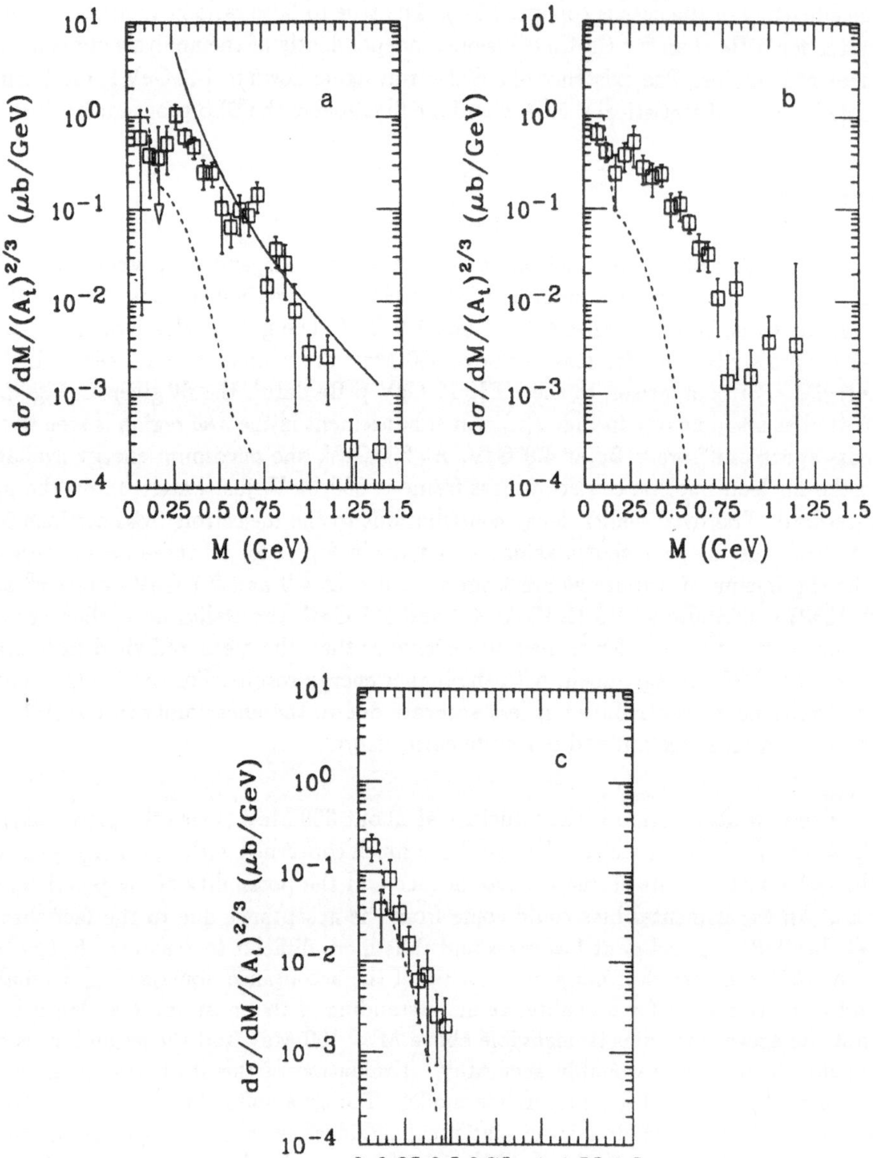

Fig. 7. The dielectron invariant mass spectra from the p-Be reaction at beam energies of (a) 4.9, (b) 2.1 and (c) 1.0 GeV. The first data point on all three spectra is qualitative. The dashed curves are the total Dalitz decay contributions. The solid curve is the fit to the KEK 12 GeV p-Be data[3] for comparison.

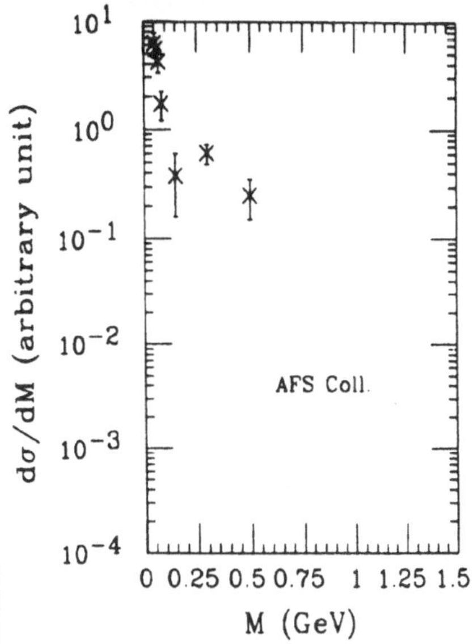

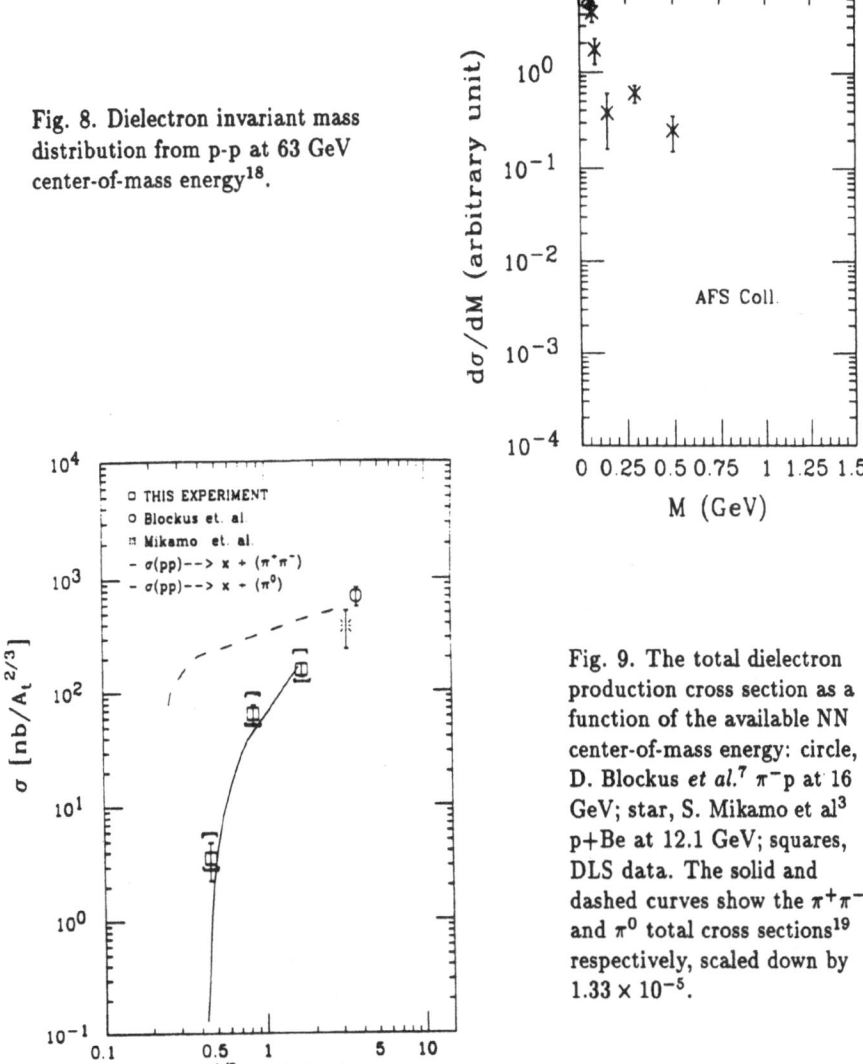

Fig. 8. Dielectron invariant mass distribution from p-p at 63 GeV center-of-mass energy[18].

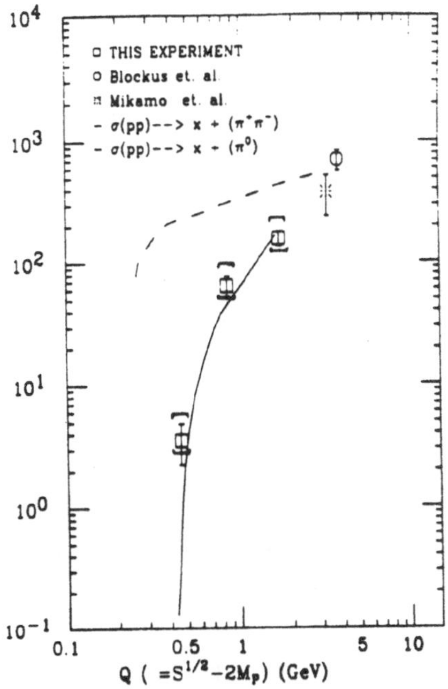

Fig. 9. The total dielectron production cross section as a function of the available NN center-of-mass energy: circle, D. Blockus *et al.*[7] π^-p at 16 GeV; star, S. Mikamo et al[3] p+Be at 12.1 GeV; squares, DLS data. The solid and dashed curves show the $\pi^+\pi^-$ and π^0 total cross sections[19] respectively, scaled down by 1.33×10^{-5}.

sections are in agreement with the higher energy results from KEK[3] and SLAC[7], and probably also with the lower energy measurement at LAMPF[1]. The e^+e^- and $\pi^+\pi^-$ cross sections are found to have similar threshold behavior while the π^0 excitation function is much flatter in the same range of available center-of-mass energy. The dielectron to dipion ratio is about 10^{-5}.

Both the mass structure at about twice the pion mass and the excitation function suggest that *pion annihilation* is a possible dominant mechanism of dilepton production for proton beam energies above 2 GeV. However, other mechanisms may also have significant contributions (e.g., Dalitz decay of the $\Delta(1232)$ resonance, bremsstrahlung).

Ca-Ca Data

Fig. 10 shows the dielectron invariant mass spectra for the reaction Ca-Ca at 2.0 and 1.0 *A* GeV. The Dalitz decay contributions shown as dashed curves in the figure have been scaled from the previous p-Be calculations and their estimated uncertainties are within ±50%. These contributions cannot account for the dielectron yield for masses above 200 MeV. The first mass bin in both spectra is again qualitative. The statistical accuracy is not good enough to draw a conclusion on the existence of a structure around twice the pion mass, even though subtraction of the Dalitz contributions would significantly change the shapes of the distributions in the low mass region. We have not performed this subtraction yet as more accurate estimates may become available (e.g., Dalitz decay of the Δ(1232) resonance[20]).

For comparison to the more accurate Ca-Ca data at 1.0 *A* GeV, a preliminary calculation by C.M. Ko[21] for the same reaction is shown in Fig. 11 (same type of a calculation as in Fig. 4). The features of the computed spectrum (the brake at twice the pion mass and the ρ enhancement) are not seen in the experimental distribution, perhaps due to the reduced statistical accuracy as indicated above. Thus we may just compare slopes. Hand-made exponential fits to the two components in Fig. 11 give inverse slopes[f] of 95 and 69 MeV for pion-pion annihilation and pn bremmsstrahlung respectively, while the exponential fit to the DLS spectrum above M = 200 MeV yields an inverse slope of 125 ± 16 MeV, see Fig. 12(a). Of course it must be reminded that the calculation is performed for back-to-back pairs and integration over p_t and y may somehow wash out the pion dispersion effect.

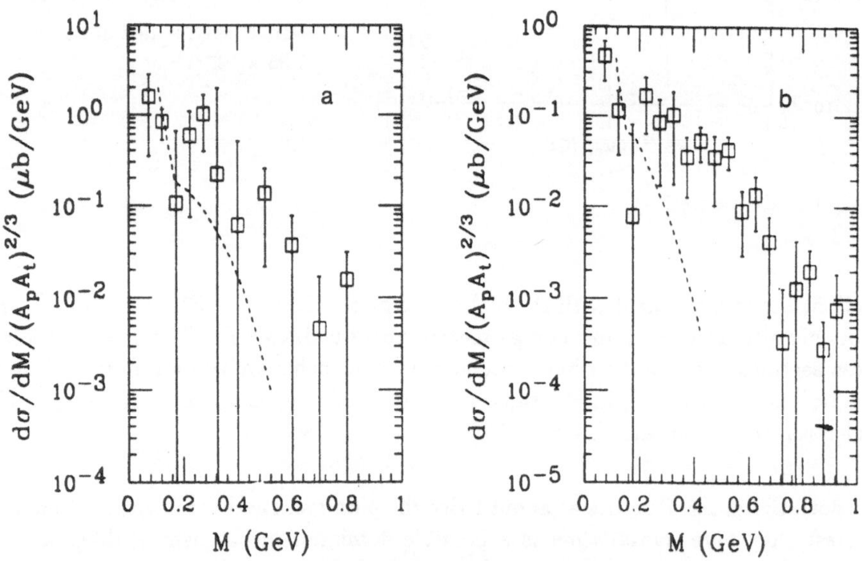

Fig. 10. The dielectron invariant mass distributions (per nucleon) from the Ca-Ca reaction at (a) 2.0 and (b) 1.0 *A* GeV. The first data point on both spectra is qualitative. The dashed curves are the Dalitz decay contributions.

Fig. 12 compares the dielectron mass spectra for both Ca-Ca and p-Be reactions at 1.0 A GeV. There is a higher yield at higher masses in the Ca-Ca spectrum compared to p-Be. The exponential fit to the p-Be data above $M = 200$ MeV yields an inverse slope of 71 ± 18 MeV, much lower than the Ca-Ca inverse slope of 125 ± 16 MeV. A first estimate indicates that this large difference in slope cannot be explained from Fermi motion.

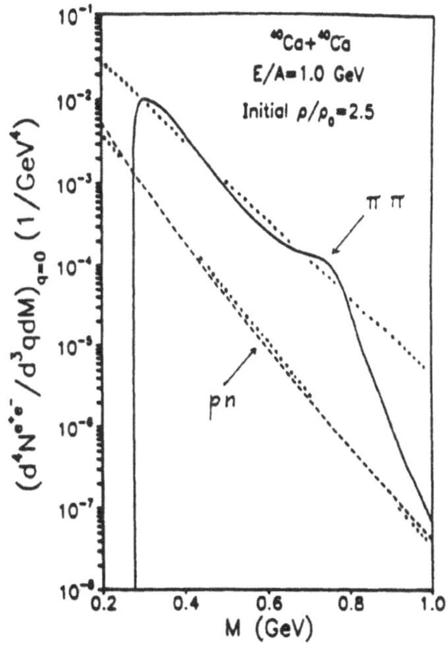

Fig. 11. The rate of back-to-back pairs computed for pion-pion annihilation and pn bremsstrahlung[21]; the dotted lines are hand-made exponential fits to the annihilation and bremsstrahlung components.

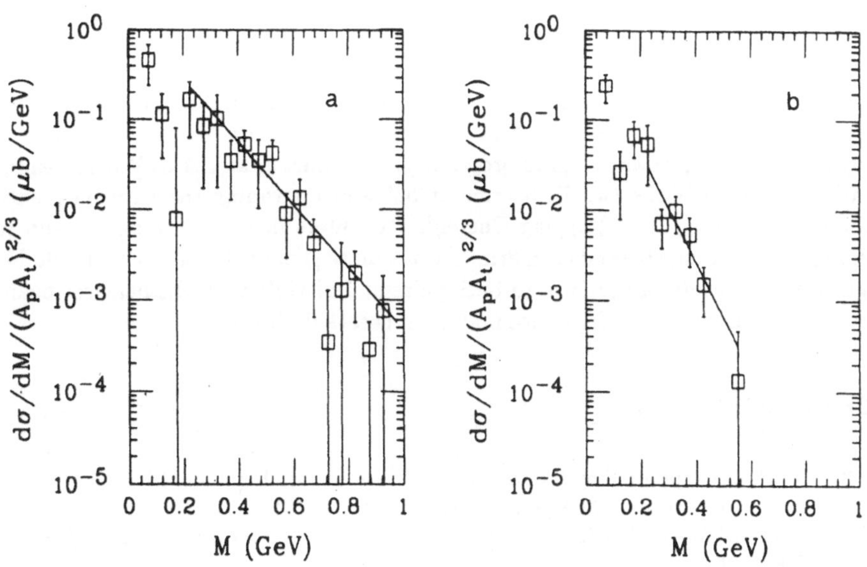

Fig. 12. Comparison of the dielectron mass distributions for both (a) Ca-Ca and (b) p-Be reactions at 1.0 A GeV. The solid lines are exponential fits to the data.

CONCLUSION

I have restricted the DLS data presentation mostly to the mass distributions. We actually measure the triple differential cross section $d\sigma/(dMdp_tdy)$ and more data is available, being published or in the process of publication, e.g., p_t distributions[15,16,22], p_t vs. M in p-Be at 4.9 GeV[17]. These should be of interest to check the model calculations. The conclusions below refer to the whole data set.

We have established the existence of a dielectron signal down to 1 A GeV incident energy.

In p-Be collisions above 2 GeV, the mass distributions ($M > 300$ MeV), p_t distributions and yields ($(e^+e^-)/(\pi^+\pi^-) \sim 10^{-5}$) are similar to those obtained at higher energies. The observation of a structure in the mass spectra at about twice the pion mass and the excitation function suggest that pion annihilation is a possible dominant production mechanism, even though other processes may have significant contributions.

Comparison of the Ca-Ca and p-Be data at 1 A GeV shows a large difference in the slopes of the mass distributions, the Ca-Ca spectrum being much flatter. The production yields are consistent with a projectile/target mass dependence as A_pA_t.

The DLS results show the *feasibility* of experiments using dileptons as a probe of nucleus-nucleus collisions in the 1 A GeV range. There is hope to obtain information on pion dynamics in nuclear matter but it needs more work, both theoretical and experimental.

Developments of the program

The multiplicity detector has just been implemented and we have collected data on Ca-Ca at 2.0 A GeV and Nb-Nb at 1.0 A GeV. In the very near future, we are going to take data on p-p and p-d reactions to gather information on the basic processes and further study the 300 MeV mass structure. On a longer term basis, the project will develop toward higher projectile/target masses at 1 A GeV beam energy. It will need setting the two DLS arms at 54° and upgrading the electron identification system, e.g., Ring Imaging Cherenkov counters in front of magnets and full calorimetric coverage behind each arm. An ultimate goal of the program should be a high statistics study of the back-to-back pairs at 1 A GeV with the heaviest beams, but the Bevalac beam performance may then create limitations.

ACKNOWLEDGEMENTS

This work was supported by the Director, Office of Energy Research, Division of Nuclear Physics of the Office of High Energy and Nuclear Physics of the U.S. Department of Energy under Contracts DE-AC03-76SF00098, DE-FG03-88ER40424, DE-FG02-88ER40413 and DE-FG05-88ER40445.

FOOTNOTES

[a]On leave from the Université de Clermont II, France.

[b]Lawrence Berkeley Laboratory: G.F. Krebs, A. Letessier-Selvon, H.S. Matis, C. Naudet, G. Roche, L. Schroeder, P.A. Seidl, A. Yegneswaran; University of California at Los Angeles: S. Beedoe, J. Carroll, J. Gordon, G. Igo; The Johns Hopkins University: T. Hallman, L. Madansky, R. Welsh; Louisiana State University: P. Kirk, Z.F. Wang; Northwestern University: D. Miller; Université de Clermont II (France): P. Force, G. Landaud.

[c]In the language of high energy physics, direct leptons are those not originating from the decay of known particles or resonances.

[d]In the case of $\pi^+\pi^-$ annihilation, dielectron and real photon production rates are of the same magnitude (see next lecture by C. Gale).

[e]The abbreviations ADC and TDC stand for analog-to-digital converter and time-to-digital converter, respectively. Signals from phototubes (hodoscopes and Cherenkov counters) are split and sent to both ADC and TDC channels which allows to record the amplitude (energy) and time information of a given signal.

[f]The inverse slope is the parameter M_0 in the exponential fit $exp(-M/M_0)$.

REFERENCES

1. A. Browman *et al.*, Phys. Rev. Lett. **37**, 246 (1976).
2. A.J.S. Smith, in *Proceedings of the Moriond Workshop on Lepton Pair Production, Les Arcs, France, 1981.*
3. S. Mikamo *et al.*, Phys. Lett. **B106**, 428 (1981).
4. V. Černý *et al.*, Phys. Rev. **D24**, 652 (1981).
5. E.V. Shuryak, Phys. Lett. **B78**, 150 (1978).
6. K. Kinoshita *et al.*, Phys. Rev. **D17**, 1834 (1978).
7. D. Blockus *et al.*, Nuc. Phys. **B201**, 205 (1982).
8. M.R. Adams *et al.*, Phys. Rev. **D27**, 1977 (1983).
9. C. Gale and J. Kapusta, Phys. Rev. **C35**, 2107 (1987).
10. L.H. Xia *et al.*, Nuc. Phys. **A485**, 721 (1988).
11. G.E. Brown, in *Proceedings of the Third International Conference on Nucleus-Nucleus Collisions, Saint-Malo, France, 1988*, and Nuclear Science Seminar, LBL, January 1989.
12. M. Schäffer *et al.*, Phys. Lett. **B221**, 1 (1989).
13. S. Pratt, University of Wisconsin preprint (1989).
14. T. Akesson *et al.*, Phys. Lett. **B153**, 419 (1985).
15. G. Roche *et al.*, Phys. Rev. Lett. **61**,1069 (1988).
16. C. Naudet *et al.*, to be published in Phys. Rev. Lett..
17. A. Letessier-Selvon *et al.*, to be published in Phys. Rev. C.
18. W.J. Willis, Nuc. Phys. **A478**, 151c (1988), see Fig. 14 of the article.
19. C. Dermer, Astrophysical Journal **307**,47 (1986); G. Alexander *et al.*, Phys. Rev. **154**, 1284 (1967).
20. C. Gale and J. Kapusta, submitted to Phys. Rev. Lett..
21. C.M. Ko, Texas A&M University, private communication.
22. G. Roche *et al.*, to be published in Phys. Lett. B.

INTERMEDIATE MASS FRAGMENT EMISSION IN THE REACTIONS Nb + Au AND Fe + Au AT 50 TO 100 MeV/A

H. C. Britt, D. J. Fields, L. F. Hansen, R. G. Lanier
D. Massoletti, M. N. Namboodiri, B. A. Remington, T. C. Sangster, G. L. Struble, and M. L. Webb
Lawrence Livermore National Laboratory
Livermore, CA 94550

M. L. Begemann-Blaich, T. Blaich, M. M. Fowler, and J. Wilhelmy
Los Alamos National Laboratory
Los Alamos, NM 87545

Y. D. Chan, A. Dacal, A. Harmon, J. Pouliot, and R. Stokstad
Lawrence Berkeley Laboratory
Berkeley, CA 94720

S. Kaufman, and F. Videbaek
Argonne National Laboratory
Argonne, IL 60439

Z. Fraenkel
Weizmann Institute of Science
76100 Rehovot, Israel

ABSTRACT

We report preliminary results from exclusive measurements on the reactions Nb + Au and Fe + Au at incident energies in the range 50 to 100 MeV/A using a detector array with large solid angle and large dynamic range. Correlations among target fragments, energetic light charged particles and projectile fragments have been measured. These results indicate that intermediate mass fragment (IMF) emission dominates the reaction cross section for all cases studied. Their cross sections, multiplicites and distributions in Z change little with bombarding energy.

I. INTRODUCTION

In experimental and theoretical studies of heavy ion collisions at intermediate energies, reactions leading to several complex fragments in the exit channel are found to become important as one goes up in energy.[1,2] Systematic studies covering a range of reaction systems and bombarding energies are essential for understanding the underlying mechanisms of such multifragment processes. Exclusive measurements leading to the determination of correlations among the observables are particularly important. Currently there is very little data of this type available especially for projectiles heavier than Ar and at energies above 50 MeV per nucleon. We report here on our recent studies of the reactions Nb + Au at 50, 75, and 100 MeV/A and Fe + Au at 50 and 100 MeV/A at the Bevalac facility at Lawrence Berkeley Laboratory. The experimental set up is summarized below and described in detail elsewhere.[3] We then present preliminary results on relative cross sections, Z distributions and event characterization for intermediate mass fragment (IMF) production and for the other major competing decay mechanisms, fission (FF) and heavy residue (HR) production.

II. EXPERIMENTAL

To measure the important correlations among the wide range of reactions products at intermediate energies, one needs a detector system with a large geometric coverage and a wide dynamic range in energy and particle type. We have designed a "logarithmic" detector system for this purpose. A detailed description of the detector system and its performance can be found in Ref. 3. A schematic diagram is shown in Fig. 1. There are three major components in the detector system: eight gas telescopes, six arrays of nine phoswich telescopes and a thirty-four element forward angle hodoscope.

A schematic of a gas module and phoswich array is shown in Fig. 1b. Each gas telescope consists of a 8 cm x 16 cm position sensitive multiwire proportional counter (MWPC), followed by a low pressure proportional counter (PC), a second MWPC (16 cm x 16 cm) located 18 cm behind the first, and a 20 cm longitudinal field Frisch grid ion chamber (IC). The timing signals from the two MWPC's give a time-of-flight of the fragment over the PC region. The PC provides dE/dx measurements with a linear response for energy losses > 200 keV. These measurements allow fragment charge and velocity identification over a large range of fragment energies, with unit charge resolution for lighter fragments. The axial field ionization chamber (IC) provides primarily dE/dx information for lighter particles (Z < 10 and total kinetic energies greater than 2-4 MeV/A) and residual energy information for slower, heavier particles. A 3 x 3 array of plastic scintillator telescopes is mounted behind each of the six forward ion chambers. Each scintillator module consists of a 1 mm thick fast plastic delta-E and a 26 cm slow plastic E-element operated in a phoswich mode. These detectors can resolve individual Z's and stop protons up to 200 MeV.

At forward angles we have a thirty four element array of fast/slow plastic phoswich telescopes. These are constructed in the same way as the large angle phoswich detectors. The energy thresholds for identification are 9 MeV for protons and 35 MeV/A for Nb, so that they can detect and identify the entire range of projectile-like fragments at the measured energies.

The eight gas detector modules are arranged about the target in a cylindrical geometry. Together, they cover approximately 10% of 4 pi between 24° and 158° in the lab and ± 12° out of the reaction plane. The large angle phoswich detectors cover about 8% of 4 pi, essentially matching the solid angle of the forward six gas modules. The forward angle hodoscope covers from 2° to 10° horizontally and from 2° to 14° vertically.

A plot of the time-of-flight vs energy loss in the PC for fragments observed in Nb + Au at E/A = 100 MeV, summed over all measured angles, is displayed in Fig. 2. In this figure, three fragment classes are apparent: intermediate mass fragments (IMF), fission fragments (FF) and heavy residues (HR). The three fragment classes are referred to in the discussion of the results below.

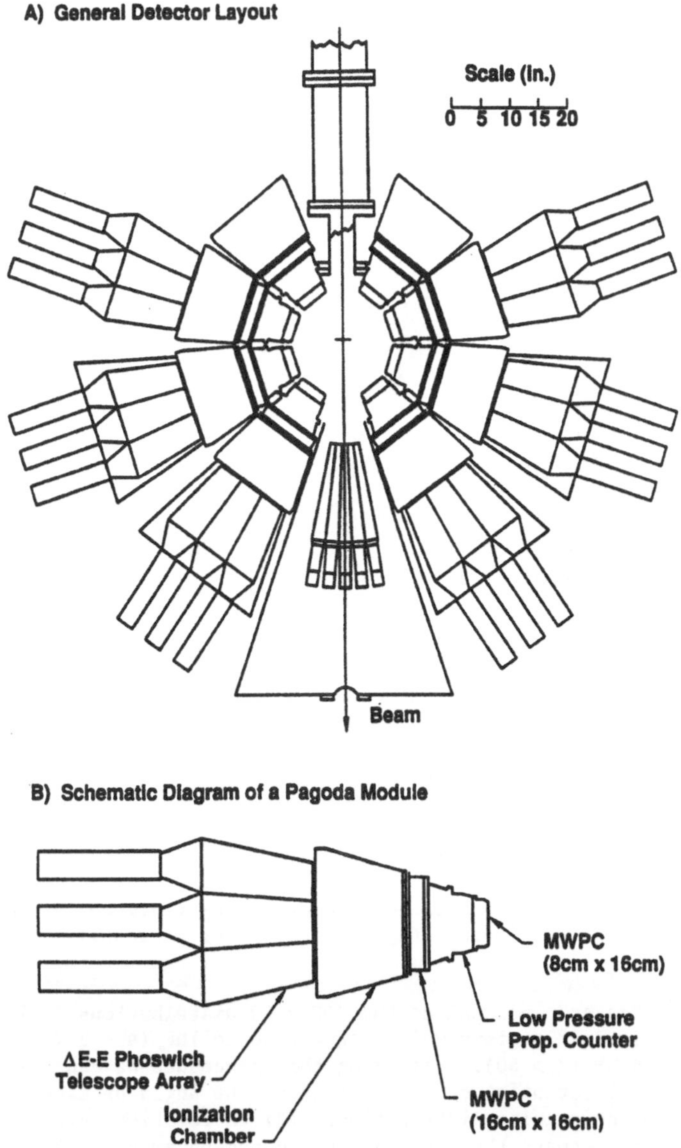

Fig. 1 A schematic diagram of the experimental setup.

III. RESULTS

The TOF x PC distribution of the type shown in Fig. 2 can be used for each gas detector module to get qualitative information on the properties of the IMF, FF and HR distributions. In this procedure two

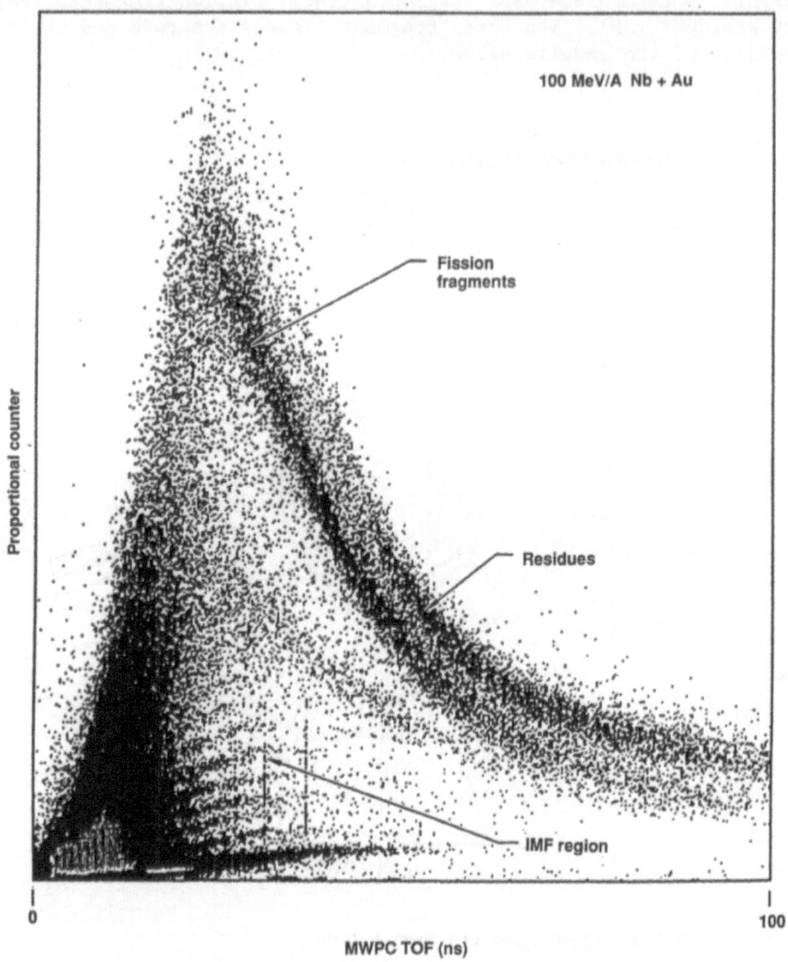

Fig. 2 The signal from the PC plotted against time of flight. The data are for Nb + Au at 100 MeV/A summed over all detectors. The major classes of emitted particles (intermediate mass fragments, fission fragments and heavy residues) are indicated.

dimensional "gates" were set on the TOF x PC distributions dividing them into three distinct regions which correspond to IMF ($4 < Z < 25$); FF ($25 < Z < 50$) and HR ($Z > 50$). Utilizing this procedure we can then look at, (1) relative differential cross sections at the angle of each gas module (i.e., a 4-point angular distribution); (2) correlations either among the gas modules or between the gas modules and the phoswich or forward hodoscope detectors, and (3) angular correlations between two particles in two different gas modules. Our preliminary analysis indicates that there are three approximately independent reaction mechanisms, each corresponding to a different range of impact parameters: fission occurring in very peripheral collisions; heavy residues, in less peripheral reactions and IMF's (with their mass distribution extending into the fission fragment mass range), in "central" collisions. Results for both Nb + Au and Fe + Au systems in the 50-100 MeV/A energy range support these conclusions.

Figure 3 shows relative angular distributions and gross two particle coincidence rates for the data from the gas modules in the case Nb + Au at 75 MeV/A. In Fig. 3 all results have the same normalization; singles are denoted by circles. The other results are the number of events of the indicated type at the indicated angle which are in coincidence with a particular type of particle (also indicated) in any of the other seven gas modules. The singles data (circles) show strong forward peaking for the IMF and HR groups contrasted to the fission coincidence results (diamonds) that are almost isotropic. From correlations with the phoswich and hodoscope detectors (more detail below) we conclude that the events in the FF gate are actually from two uncorrelated sources; about 1/2 from peripheral fission events and about 1/2 are the heavy mass tail of the IMF distribution.

Figure 4 shows a composite of the same angular distribution for the IMF gates for the five separate experiments with Fe and Nb projectiles. Table I summarizes this data in terms of the relative yields for the three types of fragments and the IMF x IMF coincidence yields summed over the data at the four angles. Most remarkable about these results is that for the Nb projectile the IMF angular distribution, the relative fraction of IMF's in our solid angle, and the IMF multiplicity as measured by the ratio of coincidences/singles are all constant with bombarding energy from 50 MeV/A to 100 MeV/A. Furthermore, the 100 MeV/A Fe data is almost identical to the Nb results and it is only at 50 MeV/A Fe that the relative IMF rate begins to drop and fission to rise. Absolute cross sections are not yet available but for a given projectile (Nb or Fe) we have determined with a simple extrapolation to 0° that the IMF cross section is constant to within 50% between 50 and 100 MeV/A. We do not yet have an absolute normalization between the Fe and Nb data.

Figure 5 shows the total mean multiplicity <M> for particles (mostly protons) detected in the 54 phoswich detectors mounted behind the gas detector modules at 36°, 72° and 108°. Relative values of <M> should give an approximate measure of the relative excitation energies associated with events satisfying the indicated singles or coincident gating conditions. These results indicate that the coincident fission events

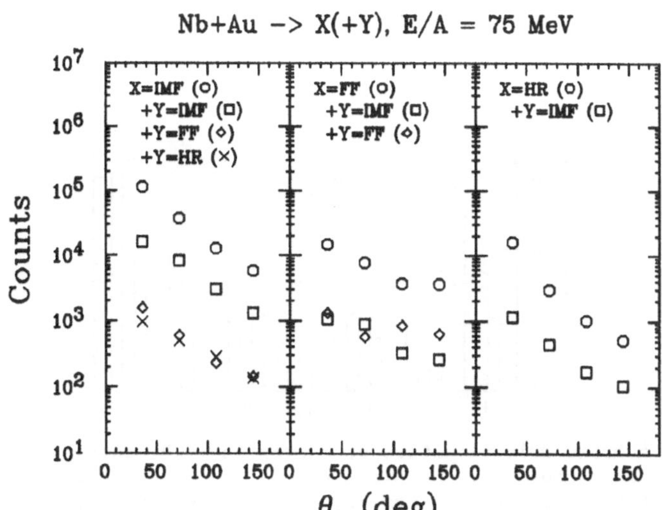

Fig. 3 Angular distributions for singles (circles) and various coincidence modes for IMF, F, HR triggers. All results have a common normalization.

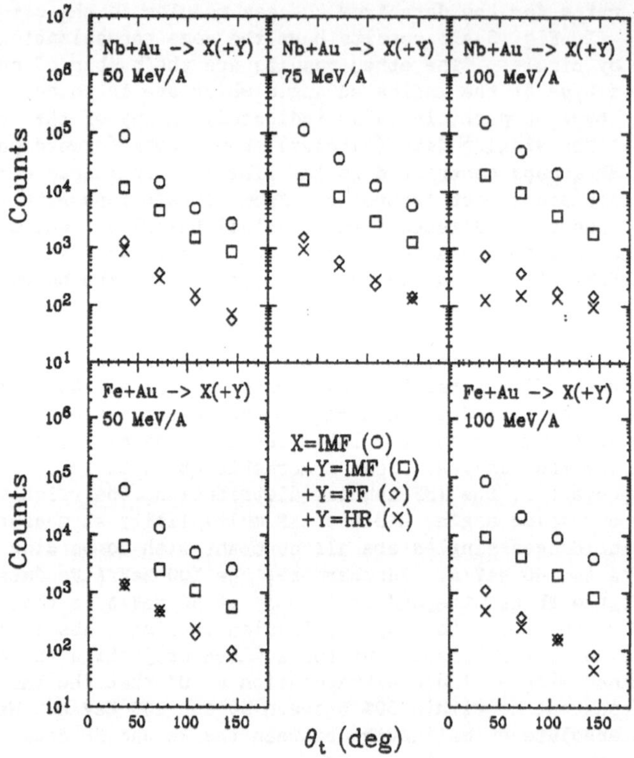

Fig. 4 Angular distributions for singles and coincident events involving IMF triggers for all the reactions studied. Normalizations among different data sets are arbitrary.

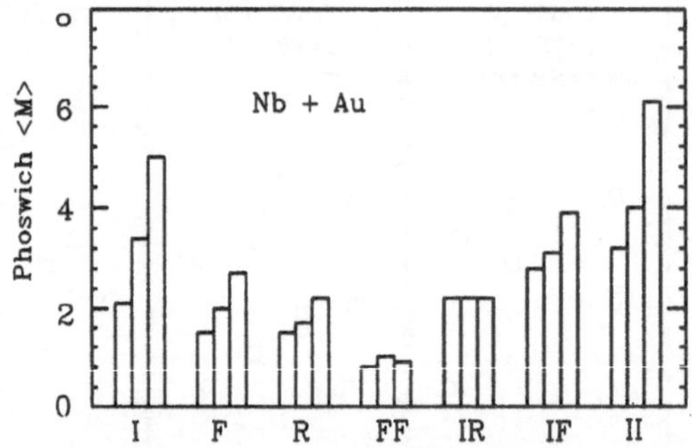

Fig. 5 Average multiplicities of energetic light charged particles for different event types for Nb + Au. For each event type, the three bars are for 50, 75, and 100 MeV/A respectively. I, F, and R refer to IMF, fission-mass fragment and heavy residue respectively. The labels FF, IR, IF, and II denote binary coincidences.

TABLE I Relative rates for various trigger conditions normalized to the total number of singles triggers and summed over all gas detector modules. I = IMF triggers; F = fission triggers; R = heavy residue triggers; I x I = two I coincidences; F x F = two fission coincidences.

Projectile	Energy MeV/A	I %	F %	R %	I x I %	$\frac{I \times I}{I}$	F x F %	$\frac{F \times F}{F}$
Fe	50	68	21	11	9	.13	4.0	.19
	100	81	13	6	11	.14	1.7	.13
Nb	50	81	11	7	14	.17	1.4	.13
	75	77	13	10	13	.17	1.4	.13
	100	83	10	7	13	.15	1.0	.10

come from the lowest excitation energies whereas coincident IMF's come from the highest. The results also indicate that the total excitation energies associated with IMF events rises approximately proportional to the beam energy whereas in the FF events it is constant.

For those events where we also observe one or more projectile fragments in the forward hodoscope an additional event characterization is obtained from the total charge of these fragments. Figure 6 shows the Z_{max} distribution gated by coincident events in the gas detectors. Z_{max} is defined as the Z of the largest projectile fragment detected in the forward hodoscope. It is seen that for F x F events there is generally a large part of the projectile remaining confirming the conclusion that these events arise from the most peripheral reactions. On the other hand I x F events have only light fragments at forward angles consistent with the conclusion that these arise from central collisions with one heavy fragment in the fission mass region.

Preliminary Z distributions for the Nb + Au results show that the IMF Z distributions become steeper with increasing angle and surprisingly do not seem to change slope grossly when the bombarding energy is increased from 50-100 MeV/A. This result is yet another indication that the IMF distributions are not very sensitive to the total amount of energy available above some threshold that may be in the vicinity of the total binding energy for a heavy system. For comparison the 50 MeV/A Fe + Au reaction has 2.3 GeV total energy available in the center of mass for a system with total binding energy of about 2 GeV.

IV. DISCUSSION

At lower energies with lighter beams previous data have shown a very rapid rise in IMF production with increasing available center of mass energy (cf. e.g. Trockel et al. Ref. 4). Excitation functions for proton

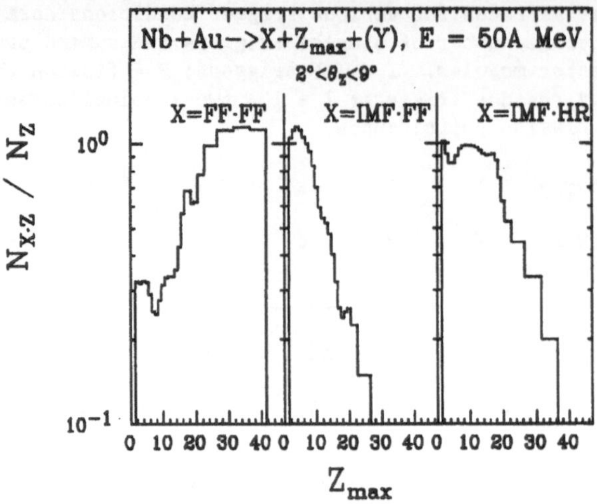

Fig. 6 The ratio between coincidence and inclusive spectra of the largest fragment detected in the hodoscope for coincidences with two fragments in the gas detectors. The ratio for events in coincidence with two fission mass fragments are shown in the left hand panel; with an IMF and a fission, in the center panel; and with an IMF and a HR, in the right hand panel.

induced reactions at the AGS and Fermilab[5] and for the production of $^{44}Sc^m$ in ^{16}O and p bombardment of uranium[6] show a relatively sharp threshold for IMF production followed by a plateau region. Comparisons of the proton and oxygen data suggest that the total center of mass energy or the total energy deposition in the target may be the most important factors in determining the onset of IMF production. The production of $^{44}Sc^m$ from both the p and ^{16}O data is consistent with a threshold of 2-3 GeV. In the case of Fe + Au at 50 MeV/A the center of mass energy is 2.3 GeV and our results show this is the only case that deviates from the higher energy plateau-like behavior. Therefore, our preliminary results look most consistent with an IMF production dependence that relates more to the total energy available than to the projectile velocity.

In our experiment we have much more detail than was available in any previous experiment and, therefore, we are able to show that not only the cross-sections for a particular IMF but also the gross properties of the angular distributions and multiplicites of IMF's seem to be showing plateau-like behavior. We plan detailed Monte Carlo calculations to see if this saturation is some artifact of our acceptance. However, a comparison of the results in Fig. 4 and Table I seem to show a plateau-like behavior. Properties of the cross sections, angular distributions and correlations for specific smaller Z ranges of the IMF fragments are currently being sorted and effects of the energy and angular acceptance of the detector modules are being estimated. Theoretical predictions for IMF emission in these systems are also being investigated with the goal to attempt to resolve whether dynamical effects in the entrance channel[2,7] or statistical coalesence in the exit channel[2] are most important in the determination of the IMF yields. Results will be presented in future publications. Preliminary results for IMF and Fission angular correlations have recently been included in conference proceedings.[8]

Work performed under the auspices of the U.S. Department of Energy by the Lawrence Livermore National Laboratory under contract number W-7405-ENG-48.

REFERENCES

1. J. W. Harris et al., Nucl. Phys. A471, 241c (1985) R. Bougault et al., Nucl. Phys. A488, 255c (1988).
2. J. Bondorf et al., Nucl. Phys. A444, 460 (1985) D. H. E. Gross et al., Phys. Rev. Lett. 56, 1544 (1986) G. Peilert et al., This Conference.
3. M. M. Fowler et al., to be published Nucl. Inst. Meth. (1989).
4. R. Trockel et al., Phys. Rev. C39, 729 (1989).
5. N. T. Porile et al., Nucl. Phys. A471, 149 (1987); Phys. Rev. C 39, 1914 (1989).
6. K. Aleklett et al., Phys. Lett. B, 197, 34 (1987).
7. J. Aichelin et al., Phys. Rev. C37, 2451 (1988).
8. D. J. Fields et al., Nucl. Phys., A495, 209C (1989); M. N. Namboodiri et al., Nucl. Phys. (to be published).

THE HISS TPC

G. Rai, F. Bieser, H.S. Matis, C. McParland, G. Odyniec, D.L. Olson
H.G. Pugh, H.G. Ritter, T.J.M. Symons, and H. Wieman

Nuclear Science Division, Lawrence Berkeley Laboratory
University of California, Berkeley, CA 94720, USA

INTRODUCTION AND PHYSICS MOTIVATION

Studies with 4π detectors at the Bevalac (the Streamer Chamber [1] and the Plastic Ball [2]) have opened a major new field of nuclear physics - the study of the dynamics of nuclear matter under extreme conditions, where baryon and meson degrees of freedom are excited. The first indications of the behavior of the nuclear matter equation of state under such conditions have been obtained. The energy regime of the Bevalac is optimally suited to such studies, yielding nuclear densities 2-4 times that of ground-state nuclear matter, with substantial excitation of low-lying hadronic states such as the Δ-resonance. However, only a modest number of such states are excited, so that microscopic analyses of the reaction dynamics can be attempted in parallel with the macroscopic analyses most directly related to questions of the equation of state. A complete understanding of the dynamics will require a consistent treatment of relativistic effects and of density-dependent and velocity-dependent forces, as well as an understanding of the time dependence of quasi-equilibrium processes. Such investigations are of fundamental interest, and are also essential for an understanding of supernova explosions and of neutron star structure. For the last several years one of the crucial questions for supernova calculations has been whether the nuclear equation of state is "soft" as predicted by most extrapolations from low energy nuclear structure physics, or "hard" as implied by observations made so far at the Bevalac.

In order to reach a unique description of the collision dynamics, experience has shown that it is necessary to make the simultaneous observation of a variety of nuclear processes and to apply a consistent theoretical formalism to fitting the data. Thus a comprehensive 4π detector system covering a large solid angle, with particle identification and momentum measurement for all the charged particles is needed. Such a detector will enable the extraction of source temperatures from the energy spectra, source pressures from collective momentum flow, source entropies from a study of the particle mix in the secondary particle spectra, and source sizes from Hanbury Brown, Twiss measurements on pairs of identical particles. Much work has already been done in this direction using the Streamer Chamber and the Plastic Ball, but the capabilities of those detectors are limited. The HISS TPC will allow this research to continue and to broaden.

The Nuclear Equation of State, Part A
Edited by W. Greiner and H. Stöcker
Plenum Press, New York

Production of strange particles represents a distinctly different probe of the collision dynamics and of the properties of the hot nuclear matter produced. The study of neutral strange particle production (Λ^0 and K^0) is a natural candidate for experiments using any 4π detector that can resolve secondary decay vertices and identify the charged particles emerging from them. Such studies have been carried out for central collisions using streamer chambers, but statistical accuracy has been limited. Furthermore, for the lower yield peripheral collisions and for the important calibrations needed from nucleon-nucleon and nucleon-nucleus collisions, almost no measurements exist. The HISS TPC will permit a comprehensive study to be made of these processes.

Multifragmentation processes are presently of great interest. A highly excited system produced in heavy ion collisions decays into a large number of fragments of various sizes. It has been pointed out that this process might be similar to a transition from the liquid to the gas phase [3]. In spite of considerable experimental and theoretical efforts this process is not yet understood. In particular, more complete experimental information is needed. It is planned to perform experiments where the HISS TPC is combined with the MUSIC chamber [4] so that the charge of the emitted fragments can be determined and a complete experiment in charge space can be performed.

In 1986 a conceptual design report for a 4π TPC detector for the Bevalac (EOS) was presented [5]. The EOS TPC was a cylinder two meters long by two meters in diameter in a solenoidal magnet. The detector was designed to study central collisions with the most energetic and heaviest beams available at the Bevalac. In that study it was pointed out that in order to cope with the high multiplicity and high particle density typical for heavy ion reactions it would be necessary to abandon the conventional wire read out of the TPC and to read out a very high number of individual pads, making the TPC a truly three-dimensional detector.

The HISS-TPC is based on this concept, but an alternative design was chosen for the magnetic field which, by using the HISS dipole and other already existing HISS facilities, greatly reduces the cost while preserving as much as possible the capabilities envisioned for the original EOS design. Such a configuration also has the advantage of being able to make use of the existing array of detectors at the HISS facility.

Two classes of experiments are considered. One class concerns flow analysis and triple differential cross sections (cross sections relative to the reaction plane). For these studies it is important to have uniform acceptance, particularly in the azimuthal (ϕ) angle about the beam axis. Large beam currents are not required, however, since cross sections are large. The other class concerns study of rare events requiring large solid angles but not necessarily completely uniform coverage.

For the first class of experiments the TPC will be operated with the beam, at an intensity of $\leq 10^3$ particles per spill, passing through the active region. This minimizes problems with non-uniform azimuthal coverage normally associated with a dipole geometry and permits excellent momentum resolution in the forward direction. The beam current must however be limited to avoid excessive distortions of the drift field due to the slowly drifting positive ions created in the primary ionization. This operation will require a very efficient gating grid to prevent leakage of the non-interacting beam tracks into the gas amplification region where they generate additional positive ions. Prototype TPC tests in the HISS magnetic field have demonstrated that a low intensity Au beam does not cause serious problems even when there is no gate to exclude the electrons. The second class of experiments, those requiring large beam intensities, will need a beam pipe inside the TPC to isolate the heavily ionizing beam particles from the active gas volume. In this mode large solid angles will still be available for studying kaons or for other low cross section measurements such as momentum distributions far out on the tails.

The specific design parameters and performance of the TPC were checked in simulations. Special attention was given to those areas where the application of TPC techniques in a high multiplicity environment are more demanding or difficult than previously encountered in existing TPC's. The pad size and layout were optimized for two-track, momentum and dE/dx (charged particle identification) resolutions, using the simulations and the measurements made with the prototype TPC.

MECHANICAL DESIGN

The HISS TPC, sketched in Figure 1, is a single rectangular box centered in the HISS dipole. The detector is configured as a drift volume enclosed with field cage panels on the sides and a single proportional wire chamber - pad plane on the bottom. The active drift volume is 150 cm long in the beam direction, 96 cm wide in the bending direction and 75 cm high in the drift direction. The detector is encased in a relatively light weight skin for gas containment and thermal isolation.

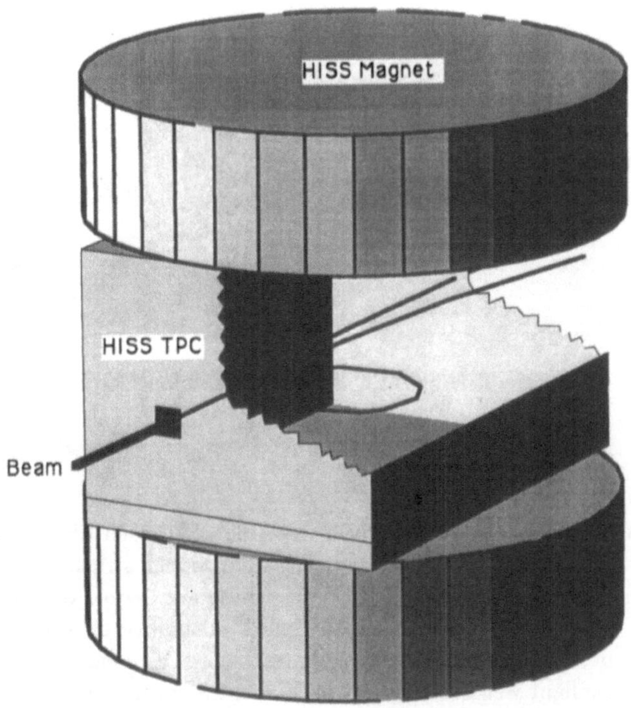

Figure 1. HISS TPC diagram. The E and B fields run vertically such that the primary ionization electrons drift down to the pad plane.

The pad plane is a single panel with an array of 1.2 cm × 0.8 cm pads covering a 96 cm by 150 cm rectangle (15,360 pads total) as depicted in Figure 2. The geometry of the wire planes over these pads will be essentially the same as used in the PEP4 and TOPAZ detectors: the first plane consists of alternating field and anode wires, the next plane is an isolation grid, and the third plane is a gating grid. The latter passes drifting electrons for the accepted events only and thereby limits positive ion build up in the drift volume. The anode wires will normally be operated at around 1170 volts to give a gas gain of 3000, but they will be divided among 16 separate power supplies to permit sections to be operated at reduced gas gain for analysis of tracks from heavier, more strongly ionizing particles.

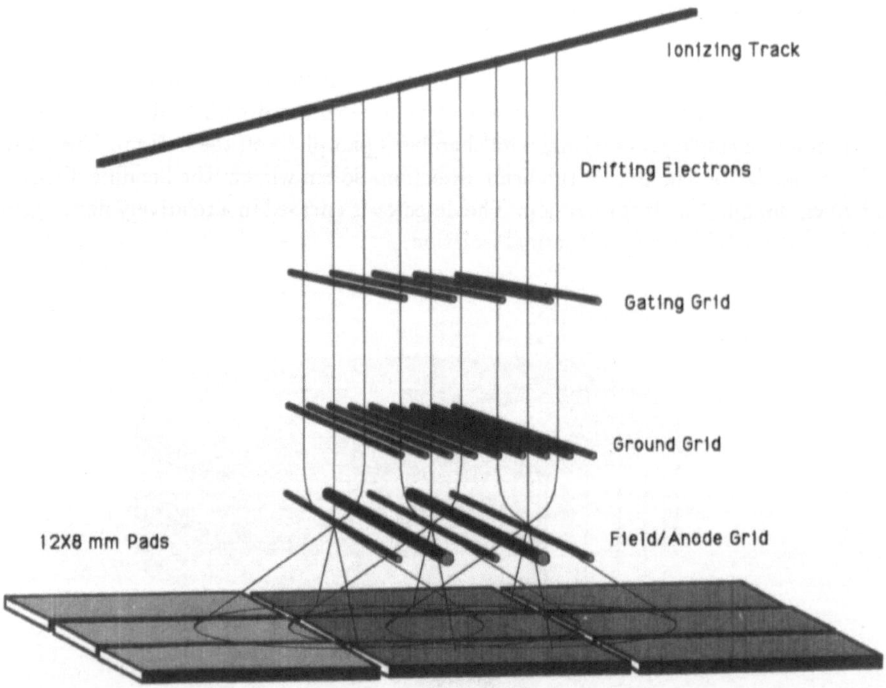

Figure 2. HISS TPC Wire and Pad plane layout.

The design of the field cage is derived from that of the ALEPH detector. It will be fabricated using kapton clad on both sides with copper strips. The copper strips alternate such that the thin exposed bands of kapton are backed on the opposite side with copper. Tests are in progress exploring the possibility of forming panels from two of these kapton/copper sheets with a Rohacell core. The cage will be constructed from four of these light weight panels set back about 5 cm from the active pad region, thus avoiding the local field distortions near the surfaces of the cage. In the low beam intensity configuration thin entrance and exit windows will be provided for the beam. In the high beam intensity configuration the field cage-box will be divided into two separate cages on either side of the beam.

The drift field for 90% Ar plus 10% CH_4 (P10) at atmospheric pressure is 130 V/cm. Thus a total bias of 9 KV will be required for the field cage.

ELECTRONICS CONFIGURATION

It is quite obvious that an electronic system for 15,360 pads cannot be constructed in the traditional way, where the preamplifier resides on the detector and the signals are transferred individually by cable off the detector for further processing. Recent progress in analog VLSI electronics encouraged us to choose a different design for the HISS TPC that would have a number of significant advantages. The most important advantage in the VLSI approach is the ability to accomplish amplification, shaping, analog storage, a high degree of multiplexing and digitization immediately on the pad plane. The resulting cabling reduction saves valuable vertical space between the pole tips. The remainder of the electronics can be contained in one or two racks, thus avoiding the need for additional housing and greatly reducing installation and maintenance problems and overall cost. Figure 3 shows a block diagram of the electronic system that will read out and process the data from 15,360 pads.

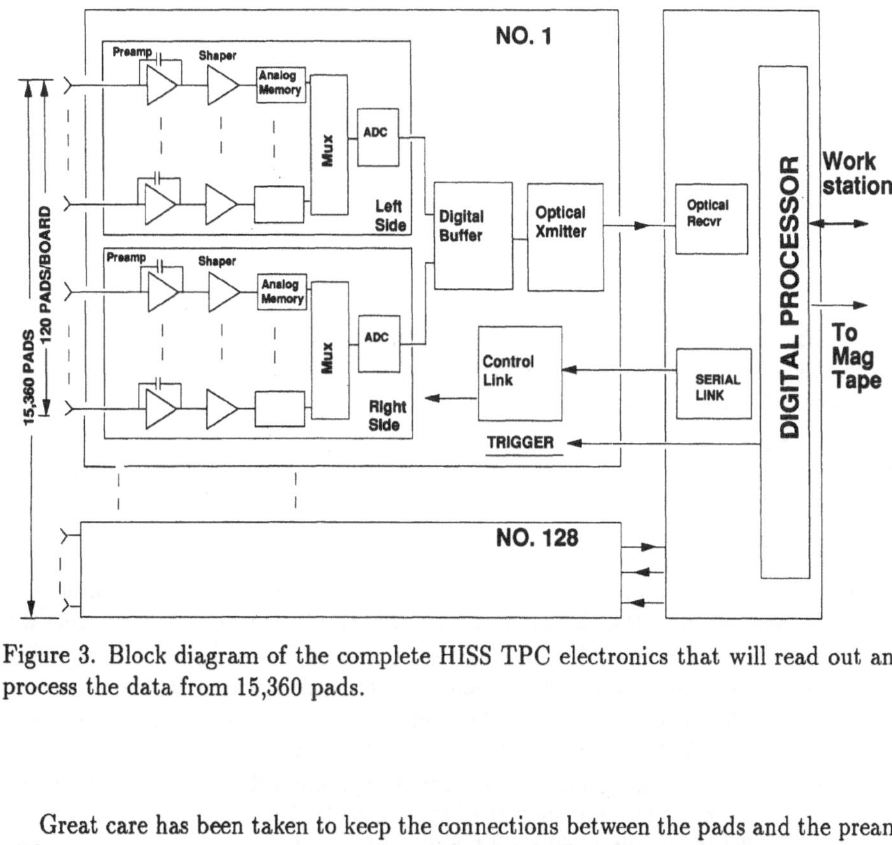

Figure 3. Block diagram of the complete HISS TPC electronics that will read out and process the data from 15,360 pads.

Great care has been taken to keep the connections between the pads and the preamplifiers as short as possible in order to obtain an optimal signal to noise ratio. This is achieved by mounting the electronic components on a "stick" that is inserted into card edge connectors (zero insertion force) directly connected to the individual pads. Each stick services two rows of pads (a total of 120 pads) across half of the width of the chamber. 15 hybrid circuits are mounted on each side. A hybrid contains an integrated 4-channel preamplifier and four discrete shaper-amplifiers. The shapers are designed to restore the baseline of the signals with a time constant of about 250 ns and to compensate for the tails generated by the slow drift of the positive ions. The hybrid occupies a height of about 10 cm between the pad plane and the pole face of the HISS magnet. In a multi-layered printed circuit board structure the output signals of the shapers are guided to the end of the stick where they are written into a 256-cell deep analog store (CCD) at a rate of 10 MHz. Thus each cell contains information about a drift space of 5 mm (100 ns). All 256 cells of the 60 CCD's on each side of the stick are multiplexed and digitized by a common ADC with a frequency of 1.25 MHz. The digital information from the two ADC's is sent off the detector via an optical link for further processing.

Two requirements drive the specification for noise and dynamic range of the pad read out electronics. Good position resolution for minimum ionizing particles is the first requirement. Position measurement is achieved by fitting a pad response function to the signals from two or more pads. This procedure places demands on both noise and digital resolution. The second requirement is a large dynamic range in dE/dx to include measurements of highly ionizing particles. To accommodate these two requirements the system has been designed with the maximum practical dynamic range and the TPC will be operated with the minimum gas gain required to achieve the desired position resolution.

The signal on a pad for a minimum ionizing track passing directly over the pad center is 11000 electrons. This number corresponds to the most probable value in the dE/dx distribution when a gas gain of 3000 is used. The expected noise on the preamp is 600 electrons rms and the final system noise is 700 electrons. This yields a most probable minimum ionizing signal/noise of 16:1 which is more than adequate to achieve 300 micron position resolution.

The dynamic range (maximum signal/noise) of the complete electronics system is 1400:1, that is the maximum signal is 90 $\times$ the most probable minimum ionizing signal. This dynamic range will allow dE/dx measurements for ions with charges as high as oxygen. By running sections of the chamber at a gas gain of 33 it will be possible to span a range in dE/dx from minimum ionizing to Au ions at 1 GeV per nucleon as demonstrated in Figure 4.

DATA ACQUISITION

Data rates are a major concern when implementing a data acquisition system for a detector of this size. Simulations with events with 200 tracks (central collisions of 1 GeV Au on Au) show that between 10-20 % of the pixels in the TPC volume will contain signals above threshold. Allowing two bytes per recorded pixel and including required addressing information, these events will be approximately 0.5-1.0 Mbytes in length. Combining the expected data rate of 10 events per spill with the expected spill rate of one spill per 5 seconds, we reach an aggregate recorded data rate of 1-2 Mbytes per second. Current large volume storage media (e.g. 8mm digital tape) will operate at sustained rates of approximately 250 Kbytes/second. Operating a number of these devices in parallel (5-10 units) will yield the desired recording speeds.

The design of a data acquisition system to achieve the above recording rates will require careful placement of processing, memory and recording elements. In order to achieve the necessary throughput we will use a VME-based system to read out, format and record events onto 8mm digital tape. With few exceptions, the system will utilize commercial components arranged in six interconnected VME crates. Four of these

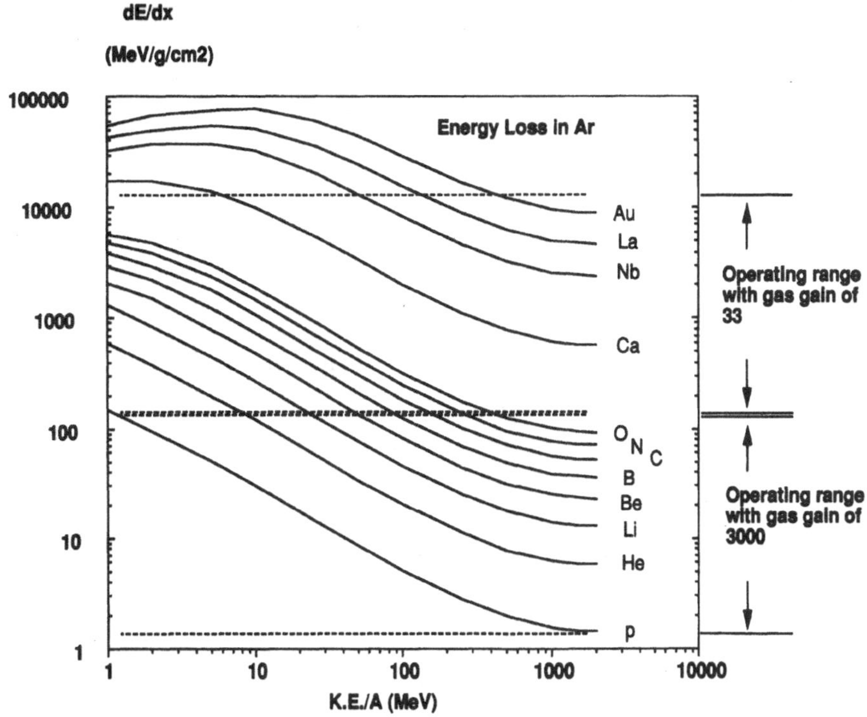

Figure 4. dE/dx versus kinetic energy per nucleon for a variety of ions. The dynamic range of the TPC electronics is shown for two different gas gains

VME crates will operate in parallel - each receiving and processing data from a different portion of the chamber. The remaining two VME systems will collect resulting data from these four systems and record it onto tape.

The progression of data from pad plane to tape is, roughly, as follows: Pad signals are amplified, shaped, stored and digitized by hybrid circuits located near the pad plane. The resulting data are then transmitted over optical fiber to off-chamber electronics located in an adjacent experimental area. Custom designed VME interfaces receive these digitized signals and store them in local memory buffers. A dedicated digital signal processor (DSP) applies pedestal and gain corrections to these signals and, after possible digital filtering and suitable data compression, moves them into dual-ported memory buffers connected to a VME bus interface. Data is read out of these buffers using a commercial VME processor and collected into one of a set of larger spill buffers. As their name implies, these large dedicated buffers will contain all the data associated with events of a single spill. Data from each of these buffers is written to an individual tape unit, thus achieving the required degree of parallelism.

Since scientific workstations currently offer substantially better computational performance than larger mini-computer systems, we plan to use workstations to perform on-line analysis. The high quality video subsystems on these systems will provide the required data display capabilities. On-line analysis programs will obtain events or, more typically, selected portions of events, via an ethernet link that interconnects all of the processors in the system. During data taking, events waiting to be taped will be available for selection and analysis by a number of workstations.

PROTOTYPE TPC

A small TPC borrowed from the PEP-4 group [6] has been modified to test the TPC design which utilizes complete pad coverage. The immediate goal was to demonstrate tracking and particle identification using only pad information. This test TPC, originally developed to study electrostatic field cage distortions, has a rectangular geometry (approximately 40 × 40 × 40 cm) with a drift length of 30 cm. Tracks drift down to the avalanche region which consists of two wire planes and a pad plane. The top wire plane, which is at ground potential, is a grid of 75 μm wires on a 2 mm pitch. This grid separates the drift and amplification region. The second plane, the avalanche plane, is located 4 mm below the grid and is composed of alternate field wires (75 μm) and sense wires (20 μm) set on a 2 mm pitch. The pad plane is located 4 mm beneath the field/sense wire plane and completes the confinement of the avalanche cells.

The pad plane consists of 256 pads arranged in two different ways. In the first set there are 8 rows of 16 pads whose size is 8 mm x 8 mm. Each row is displaced 4 mm from an adjacent pad to study a staggered pad geometry. The second group of pads consists of 12 mm x 6 mm pads.

The electronics for this test system are 256 channels of PEP-4 electronics. Each pad signal is recorded into CCD's at 10 MHz and read out into ADC's at 20 KHz. The read-out electronics can at most store 4000 words per events. As a result, in order to store data from a trigger, it is first necessary to do a threshold cut on the data in the electronics. As the pedestals from the CCD's are not constant with time, the threshold cut removes useful data. A more sophisticated cut will be made in the HISS TPC so that results from the HISS TPC should be better than the prototype.

The prototype has been studied at the Bevalac using He and Au beams. The chamber performed as expected during each of the tests. Resolution studies have shown that the measured standard deviation for a track can be as low as 0.250 mm depending on the signal to noise ratio and the track angle. Tracks were studied from 0 degrees to 90 degrees over a variety of signal to noise ratios and analysis is in progress.

Tests with a low intensity 600 MeV per nucleon Au beam have demonstrated that the chamber does not experience electrical breakdown when a highly ionizing particle passes through the active volume. This test was performed without a gating grid. From analyzing the Au tracks, we conclude that tracks which are more than a few cm away from the beam particle should be fully reconstructible.

SUMMARY

The HISS TPC will greatly expand detector capabilities at the Bevalac. It will provide the ability to completely measure most of the charged particles emitted from central collisions with the heaviest and highest energy beams at the Bevalac. Three dimensional tracking makes possible the unfolding of high multiplicity events with as many as 200 charged particles. Good tracking resolution in the HISS dipole and dE/dx information provide momenta and particle identification for most of the p, d, t, ^{3}He and ^{4}He ions emitted. A substantial fraction of the charged pions will also be measured. These capabilities can extend flow and entropy studies to full energy Au + Au collisions. They will also permit analysis of two particle correlations and make possible measurements of triple differential cross sections into the tails of the momentum distributions. In addition the system can be used to study a great variety of other interesting processes, like *e.g.* multifragmentation.

This work was supported by the Director, Office of Energy Research, Office of High Energy and Nuclear Physics, Division of Nuclear Physics of the U.S. Department of Energy under Contract DE-AC03-76SF00098.

References

[1] A. Sandoval, R. Bock, R. Brockmann, A. Dacal, J.W. Harris, M. Maier, M.E. Ortiz, H.G. Pugh, W. Rauch, R.E. Renfordt, F. Riess, L.S. Schroeder, R. Stock, H. Ströbele, and K.L. Wolf, Nucl. Phys. **A400**, 365c (1983).

[2] H.H. Gutbrod, A.M. Poskanzer, and H.G. Ritter, LBL-Preprint LBL-26922, to be published in Rep. on Prog. in Physics.

[3] R.W. Minich, S. Agarwal, A. Bujak, J. Chuang, J.E. Finn, L.J. Gutay, A.S. Hirsch, N.T. Porile, R.P. Scharenberg, B.C. Stringfellow, and F. Turkot, Phys. Lett. **B118**, 458 (1982).

[4] W.F.J. Müller, G. Bauer, H. Beeskow, F. Bieser, W. Christie, U. Lynen, H. Sann, and C. Tull, 8th High Energy Heavy Ion Study, LBL-24580 (1987).

[5] H.G. Pugh, G. Odyniec, G. Rai and P. Seidl, **LBL-22314** (1986).

[6] M. Iwasaki, R.J. Madaras, D.R. Nygren, G.T. Przybylski and R.R. Sauerwein, Time Projection Chamber Workshop, TRIUMPF, Vancouver, B.C., Canada, AIP Conference Proceedings **108**, 214 (1983).

[3] [illegible]

[4] ... Weiss, ... C. ... [illegible] ... (1991)

[5] ... Nucl. Phys. ... [illegible] (1991)

[6] ... [illegible] (1990)

THE EQUATION OF STATE AND VISCOSITY

IN THE HYDRODYNAMIC MODEL [1]

W. Schmidt, U. Katscher, B. Waldhauser
J. A. Maruhn[2], H. Stöcker, and W. Greiner

Institut für Theoretische Physik der Universität Frankfurt
Postfach 111 932, D 6000 Frankfurt am Main, West Germany

INTRODUCTION

HYDRODYNAMICS AND MICROSCOPIC MODELS

The hydrodynamic model[1,2,3,4,5,6] has been influential in the development of the theory of high-energy heavy-ion collisions since its very beginning. One of the most important insights in the field was the discovery of collective flow through a comparison between experiment, the hydrodynamic and the cascade models. Two major components of collective flow have been unambigously observed so far: Transverse momentum transfer *in the reaction plane* – the *Bounce-Off* effect [7,8], and, only recently, the off-plane (perpendicular) *Squeeze-Out*, which was also predicted by nuclear fluid dynamics [4,9], has been demonstrated in a re-evaluation of old data [10,11].Since then microscopic models have learnt how to include collective effects in such treatments as the "quantum molecular dynamics" [12,13,14], and have become very successful in quantitative predictions. A present day evaluation of the hydrodynamic model must thus motivate the continued need for such simulations in comparison to the best microscopic models.

Let us then start off with a brief comparison of assumptions and features of these model families. The following points give an overview of the contrasting ingredients:

- Validity: the hydrodynamic model is based on *instantaneous local equilibrium*, which will certainly only hold approximately for realistic situations. On the

[1]Supported by the Deutsche Forschungsgemeinschaft, D-5300 Bonn, and by the Bundesministerium für Forschung und Technologie, D-5300 Bonn

[2]Invited speaker

other hand, the microscopic models presently favored are still characterized by ***independent nucleon-nucleon collisions***, supplemented by a mean field. No provision is taken for off-shell collisions, and thus the assumption is that of a *dilute system.*

- Nuclear matter properties: in the hydrodynamic model these are directly used as input functions, encompassing both the equation of state and dissipative coefficients such as viscosity and thermoconductivity. In microscopic models the properties of nuclear matter result from the assumed collision cross sections and the mean field, with — in a rough way of speaking — the mean field determining the equation of state and the cross sections the dissipative properties. *It should be stressed, however, that e. g. the equation of state cannot be trivially computed from the mean field; such a relation requires extensive theoretical effort on its own.*

- Clusters: in the hydrodynamic model the decomposition of nuclear matter in the final state is treated in a *chemical equilibrium model*, which contains a number of simplifying assumptions of its own such as the simultaneous breakup of the total system irrespective of local conditions. While such problems could be ameliorated with more extensive computing, the breakup will probably always be a weak point of the model. The "quantum molecular dynamics" contains many-body correlations which could lead to realistic cluster formation if the clusters themselves were described well in their nuclear structure properties. At present probably only the average properties of the clusters are believable. Simpler microscopic models such as VUU do not contain cluster formation but rely on final state coalescence or similar approaches.

- Quantum effects: both types of models contain quantum effects to a limited extent: for hydrodynamics they influence the equation of state, which contains the Fermi gas kinetic energy of the nucleons for example; for microscopic models the cross sections depend on quantum properties and some of these models also simulate the Pauli principle by various mechanisms.

From these considerations it should be clear that the two classes of models are to some extent complementary and a careful comparison of both with experiment may shed light on the precise extent to which equilibration is actually attained in these reactions.

MODEL EQUATIONS

The equations of motion for the viscous, non relativistic nuclear fluid can be written as a system of five continuity equations (indices occurring twice are to be summed over)

$$\frac{\partial \rho}{\partial t} + \frac{\partial}{\partial x_i}(\rho v_i) = 0 \tag{1}$$

$$\frac{\partial(\rho v_i)}{\partial t} + \frac{\partial}{\partial x_j}(\rho v_i v_j) = \frac{\partial}{\partial x_j}\sigma_{ij} - \rho\frac{\partial \Phi}{\partial x_i} \tag{2}$$

$$\frac{\partial \varepsilon}{\partial t} + \frac{\partial}{\partial x_i}(\varepsilon v_i) = \frac{\partial}{\partial x_j}(v_i \sigma_{ij}) - \frac{\partial q_i}{\partial x_i} - \rho v_i \frac{\partial \Phi}{\partial x_i}, \tag{3}$$

where ρ, ρv_i and $\varepsilon = \rho(v^2/(2m) + W(\rho, s))$ are the local densities for baryon number, momentum and energy respectively. v_i is the local velocity, m the nucleon mass, $W(\rho, s)$ is the internal energy, which has to be provided as a function of density and specific entropy s by the equation of state, and $q_i = -\kappa \partial T/\partial x_i$ is the vector of heat transport according to Fourier's law, where κ is the coefficient of thermal conduction. Yukawa and Coulomb potential are considered by Φ.

For the time being the stress tensor will contain only a scalar pressure

$$\sigma_{ij} = -p\delta_{ij};$$

viscosity will be considered later. In the present calculations thermoconductivity was not included at all.

The equations will be closed if the pressure p is given in terms of W. If the *equation of state* is given, i. e. W is known as a function of density and entropy s, the pressure can be obtained through the standard relation

$$p(\rho, s) = \rho^2 \left. \frac{\partial W(\rho, s)}{\partial \rho} \right|_s \tag{4}$$

This is not strictly necessary, however. For the equations of motion it is sufficient to give $p(\rho, W)$, the so-called reduced equation of state, which does not allow the calculation of all thermodynamic variables. The most important problem for the field of high-energy heavy-ion collisions is of course the determination of the equation of state from the hydrodynamic flow. Some remarks concerning this *inverse problem* will be made later in connection with the one-dimensional shock model.

THE EQUATION OF STATE

To regard the function of two variables $W(\rho, s)$ as completely unknown would introduce too much uncertainty into the considerations. It has therefore become customary to assume that the thermal behaviour is essentially known and can be computed from that of a nucleon or hadron gas. This leads to the decomposition

$$W(\rho, s) = W_0(\rho) + W_{th}(\rho, s), \tag{5}$$

where $W_0(\rho)$ is the compressional part of the equation of state (often simply referred to as "the" equation of state), while $W_{th}(\rho, s)$ is the thermal part. The pressure can then also be split up as

$$p(\rho, s) = \rho^2 \frac{dW_0}{d\rho} + \rho^2 \left. \frac{\partial W_{th}}{\partial \rho} \right|_s . \tag{6}$$

The crucial assumption then is not the division into two parts, but that the thermal part is presumed to be known. Usually either a non-relativistic Fermi gas or a more comprehensive mixture of hadronic states is used [15].

For the compressional energy in most cases it is not necessary to use an unknown function and instead a parametrized form with a small number of parameters is used. A standard form is for example the quadratic equation of state

$$W_0(\rho) = \frac{K}{18\rho_0^2}(\rho - \rho_0)^2 + B_0 \tag{7}$$

or the linearized version

$$W_0(\rho) = \frac{K}{18\rho\rho_0}(\rho - \rho_0)^2 + B_0. \tag{8}$$

Both are constructed such as to reproduce a given equilibrium density ρ_0, equilibrium binding energy B_0, and incompressibility K.

It should be stressed quite strongly, however, that these assumptions have by no means beem checked in experiment. The true equation of state could have much more complicated thermal properties such as occur for example in the relativistic mean field model.

THE ONE-DIMENSIONAL SHOCK MODEL

The physical situation in a heavy ion collision is in general too complicated to allow for simple solutions. In order to get a general impression of the behavior of the solutions, though, it is exceedingly useful to have a simplified model available. In the *one-dimensional shock model*, it is assumed that the physical motion is only along one direction, viz. the beam axis, so that the physical conditions can be obtained from a solution of the one-dimensional hydrodynamic equations of motion. In this approach, of course, the deflection of matter to the sides is neglected and later results will show the shortcomings.

In the one-dimensional shock model the conditions before and behind the shock front can be linked by simple conservation equations. Assume that the thermodynamic state before the shock front is given by density ρ_0, velocity u_0 and to zero internal excitation, i. e. $p_0 = 0$, $W_0 = B_0$. Similarly the conditions in the shocked material are given by ρ_1, u_1, p_1, and W_1. The reference frame is chosen such that the shock front is at rest. Assuming a stationary shock the hydrodynamic equations of motion then reduce to conservation conditions across the shock front:

$$\begin{aligned} \rho_1 u_1 &= \rho_0 u_0 \\ p_1 + \rho_1 u_1^2 &= p_0 + \rho_0 u_0^2 \\ \rho_1(W_1 + \tfrac{1}{2}u_1^2) + p_1 &= \rho_0(W_0 + \tfrac{1}{2}u_0^2) + p_0. \end{aligned}$$

Eliminating the velocities from these equations yields the well-known *Rankine-Hugoniot relation*, which links the thermodynamic variables on both sides of the front:

$$W_1 - W_0 = \tfrac{1}{2}(p_1 + p_0)(\rho_1^{-1} - \rho_0^{-1}).$$

Solutions of this equation were used quite extensively for estimating the conditions attainable in the reactions[1,2,16]; we only summarize a few crucial findings:

- Most of the incoming energy is converted into thermal excitation. This implies that in general the final state will have $W_{th} >> W_0$. Clearly this is a problem for extracting information about the desired function W_0.

- Phase transitions lead to threshold-like increases in most observables as a function of bombarding energy. In addition the shock front may not be stable under such conditions and tend to split up into several fronts.

- Full three-dimensional simulations show a reduction of the densities and temperatures achieved by only about 10%.

Recently the shock model was also used to address the problem of measurability of the equation of state[17]. To see this, assume that one of the thermodynamic variables in the compressed state can be measured. Candidates are the density, the pressure, and the thermal excitation as given by the entropy or the temperature. In each of these cases, of course, it is a difficult matter for itself whether such a measurement is possible, how directly a measurable quantity can be linked to one of these, and whether this quantity is determined during that specific phase of the reaction corresponding to maximum density and temperature. For the present purpose let us assume that it can be done and examine whether the equation of state can then be inferred.

If the equation of state is given by a simple expression such as the quadratic or linear equation of state given above, the only unknown is the incompressibility K, and fitting it to reproduce the observable as a function of energy will yield a definite result. However, Hahn and Stöcker[18] found an ambiguity with a quadratic and a linear equation of state of very different shapes giving roughly equivalent agreement with the thermal excitations inferred from the pion yields. What is the reason for this ambiguity?

The problem is that if the equation of state $W_0(\rho)$ is not presumed to be known in its functional form, the Rankine-Hugoniot relation involves the derivative $dW_0(\rho)/d\rho$ because of the definition of the pressure. Thus in general one will end up with a differential equation for the equation of state, which has to be integrated from the only known point, the ground state of nuclear matter. Since such an integration is not practically possible (the differential equation becomes singular near the ground state and measurements are also not meaningful for very low energies), one has to assume another starting point $W_0(\bar{\rho}) = \bar{W}_0$. This was done in [17] and yielded the results in figure 1.

This dilemma is easily seen to appear whenever only one thermodynamic variable is measured. If two are known the problem is solved:

- If ρ and s are measured, $W_{th}(\rho, s)$ is known and $W_0(\rho)$ can be obtained from $W_0 = E - W_{th}$.
- If ρ and p are measured, the derivative in the Rankine-Hugoniot relation is no longer used, and it yields the entropy s, from which the equation of state follows as before.
- s and p are measured in the same way ρ can be determined from the Rankine-Hugoniot relation and the rest is again clear.
- If T or the thermal energy per nucleon are measured, the argument runs similarly to the case of the entropy; only the thermodynamic derivations become more involved.

The net result thus is that *two thermodynamic quantities in the compressed zone have to be measured* for deriving the equation of state. One should bear in mind, of course, that this is still a very simplified model assuming the exact validity of the one-dimensional shock model and also ignoring the experimental problems in measuring those variables. So its usefulness is rather in the sense of a "minimum information" condition, and in the way it showed that using a predefined functional form of the equation of state may be dangerous.

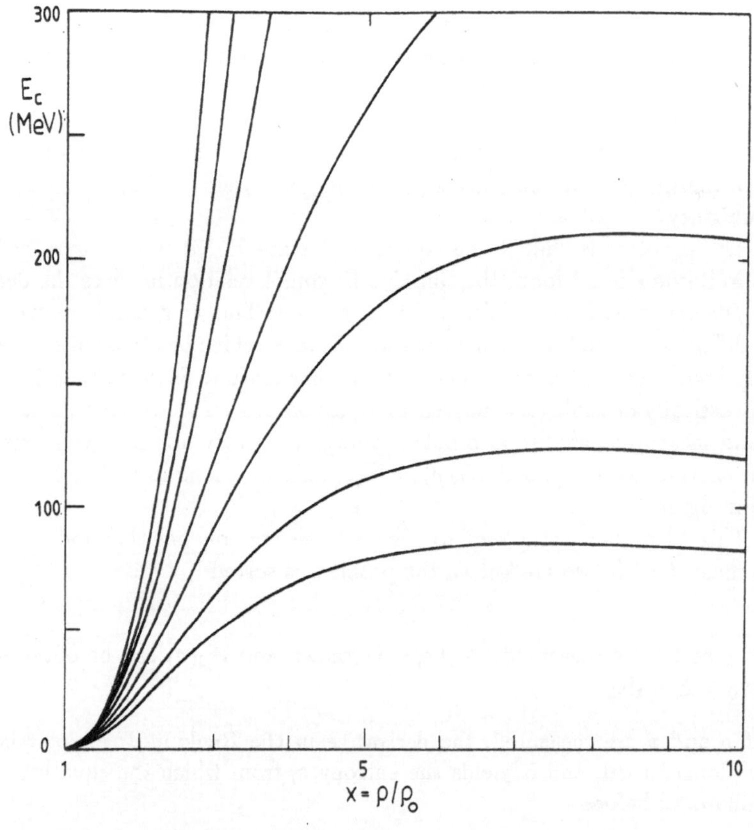

Figure 1. Equations of state reproducing the experimental pion yields of [19]. Below the region of integration the curves are defined by a fourth-order polynomial fitted to the first three points. Note that the lowest few curves have to be rejected because they bend over for high densities and that the highest one violates causality; nevertheless the remaining range of equations of state reproducing the pion yields is still considerable.

Model Formulation

Now we will present studies of the influence of viscocity on the hydrodynamic model. Using a Newtonian ansatz, the stress tensor σ_{ij} can be written as

$$\sigma_{ij} \quad = \quad -p \cdot \delta_{ij} + \eta \cdot \left(\frac{\partial v_i}{\partial x_j} + \frac{\partial v_j}{\partial x_i} - \frac{2}{3} \delta_{ij} \frac{\partial v_k}{\partial x_k} \right) + \zeta \cdot \delta_{ij} \frac{\partial v_k}{\partial x_k} \tag{9}$$

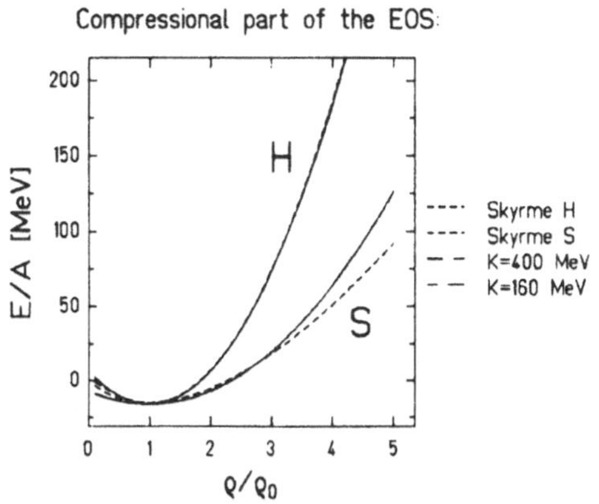

Figure 2. Sketch of the compressional energy $E_c(\rho)$, showing two different parametrizations of the equation of state as used in QMD- and nuclear fluid dynamics-calculations, both of them with a hard and a soft variant.

with η and ζ being the coefficients of shear and bulk viscosity. These are, in general, functions of density and temperature [20]. In our model, however, they are set constant.

The thermal energy of a free, nonrelativistic Fermi-gas was used in the calculations for the contribution W_{th}, and for the compressional part the quadratic form of eq. (7) was employed with $B_0 = -16$ MeV$/N$ and $\rho_0 = 0.16$ MeV/fm^3.

To allow for comparison of our results with the Quantum Molecular Dynamics (QMD) , we used two different values for the incompressibility K:

- The *soft* equation of state, where $K = 160$ MeV, and
- The *hard* equation of state, where $K = 400$ MeV.

Note that the QMD-equation of state has been derived from time dependent Hartree-Fock calculations of the ground state with Skyrme-interactions. Refer to figure 2 for a sketch of both types of equation of state.

Fragment Formation

Since the basic assumptions of nuclear fluid dynamics (i.e., local thermal equilibrium and short mean free path) are no longer justified at a late stage of the reaction, the hydrodynamic calculation is abandoned when the *maximum density* is less than ρ_0. At that time, no compressed cells are left in the nuclear fluid, so that the average density decreases to about $\rho_0/2$: The nuclear fluid is assumed to freeze out. The formation of free nucleons and clusters of nucleons is computed in chemical equilibrium, with conservation of baryon number and energy per particle. The equilibrium is established in a reduced volume $V = V_0 - \sum_i n_i V_i$, where n_i is the number of particles of sort i and V_i is their volume [21].

So far, only 6 particles are considered in the calculation, namely p, n, d, t, ^{3}He and ^{4}He. The chemical break-up calculation yields particle numbers and temperatures for each fluid cell. To calculate differential cross sections, the thermal momentum distribution for each cell has to be Lorentz-transformed to the laboratory frame, assuming that the particles in the cell are forming a free gas with isotropic expansion in the local rest frame. Particle interactions and decay of instable particles are neglected.

The resulting invariant triple differential cross section $1/p\ \partial^3\sigma/\partial E\partial\Omega$ can be used to compute particle spectra as well as various other observables. *It should be emphasized that this triple-differential cross section contains most of the information about collective flow in the final state and that analysis methods like flow only serve to reduce it to a small set of numbers.*

Observables

From the baryon density ρ, momentum density $\boldsymbol{M} = \rho\boldsymbol{v}$, energy density ε, and from the invariant cross sections $1/p\ \partial^3\sigma/\partial E\partial\Omega$ for the six particle species p, n, d, t, ^{3}He and ^{4}He one can derive certain quantities which will be called ***observables***, although it may actually be very difficult to extract them from experimental data. The observables that will be investigated in this paper are:

- The *flow angle*, Θ_F, is the angle between the beam axis and the principal axis of the weighted coalescence-invariant flow tensor.
- The *aspect ratio*, R_{13}, is the ratio between the largest and the smallest principal axis of the flow tensor F_{ij}.
- The *entropy per baryon*, S/A, can be computed from the thermodynamic relation

$$\frac{S}{A} = \frac{1}{T}\left(\frac{5}{3}\frac{E}{A} - \mu\right) \tag{10}$$

  where E is the thermal energy and μ the chemical potential. For simplicity, we used an approximation of the Fermi-integral with a difference of less than 1% compared to the exact solution.

  The experimental determination of entropy is much more difficult. For large entropies, $S/A \geq 5$, it can be approximated [22] by a simple expression, but for lower entropies, a full quantum statistical treatment including complex unstable fragments, is necessary [23].

- The distribution of longitudinal momenta parallel to the beam axis is commonly plotted as dN/dY, the baryon *rapidity distribution*.

- The *transverse momentum analysis* proposed by Danielewicz and Odyniec [24] to analyze the projection of transverse momentum transfer per particle to the reaction plane, p_x/A, which is plotted as a function of the rapidity. The slope of the s-shaped curve at CM-rapidity, dp_x/dY, is then extrapolated to projectile rapidity. This quantity is commonly (but unprecisely) denoted as *collective flow*.

- The *azimuthal angular correlation* of the fragments with respect to the reaction plane, $p_x/|p_\perp|$ as a function of rapidity. Experimental results of Kampert [30] show a strong correlation for heavy fragments, which means that those are preferredly emitted in the collective direction of motion, as had been anticipated long ago by Baumgardt *et al.* [2].

- *Cross sections* and *particle spectra*. Recent measurements of proton spectra in central collisions of La → La at 246 MeV/N have shown a strong 90°-enhancement [25], even in the double differential cross section.

- A new analysis of BEVALAC-data has shown that the baryon rapidity distribution is not azimuthally symmetric. When plotting the *angular rapidity distribution* $dN/dYd\phi$, one clearly finds a peak at Y_{CM} and $\phi = 90°$ indicating a strong off-plane squeeze out of cold nuclear matter [10], as predicted by early fluid dynamical calculations [26].

Time-Development in nuclear fluid dynamics-calculations

Figure 3 shows the time-development of some typical quantities in the reaction Au → Au at 400 MeV/N with an impact parameter of $b = 3$fm, using a hard equation of state. The dashed curve shows data for a nonviscous calculation with $\eta = 0$ and $\zeta = 0$. The solid curve was calculated with a high shear viscosity ($\eta = 60$ MeV/fm²c), but no bulk viscosity ($\zeta = 0$). The dashed-dotted curve was obtained for $\eta = 30$ and $\zeta = 30$ MeV/fm²c.

The upper picture (fig. 3a) shows the *compression*, i.e. the density of the most highly compressed cell in the numerical grid at any given time. Shortly after the collision, the compression starts to rise and reaches a maximum of about $1.6\rho_0$ for viscous and $1.9\rho_0$ for nonviscous calculations. This maximum is maintained for a certain time, as long as the shock wave runs through the nuclei. When the expansion starts, compression is decreasing rapidly. After a certain time, no compressed cells are left. This is the appropriate stage for chemical breakup. In a 400 MeV/N-reaction, it is reached after about 40 fm/c.

The lower picture (fig. 3b) shows the *entropy per baryon* in the most highly compressed cell. Note that most of the entropy is produced during the compression stage. The expansion is adiabatic, unless there is a finite bulk viscosity ζ. Therefore, if $\zeta \neq 0$, the entropy per baryon depends on the chemical break-up time.

As one can see from figure 4, the influence of the equation of state on these results is low. Here we compare calculations of Au → Au at 400 MeV/N with $\eta = 0$, $\zeta = 0$ (fig. 4a) and $\eta = 60$, $\zeta = 0$ MeV/fm²c (fig. 4b) for both the hard and the

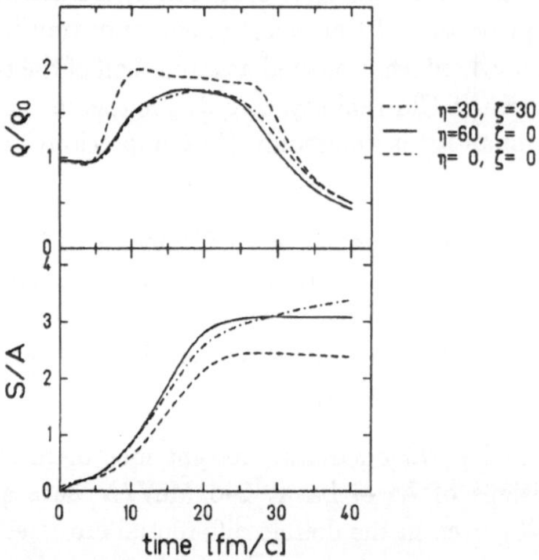

Figure 3. Compression (top) and entropy per baryon (bottom) in a Au → Au-reaction at 400 MeV/N, impact parameter $b = 3$ fm, hard equation of state:

Dotted: $\eta = 0, \zeta = 0$ MeV/fm^{2}c
Straight: $\eta = 60, \zeta = 0$ MeV/fm^{2}c
Dashed: $\eta = 30, \zeta = 30$ MeV/fm^{2}c

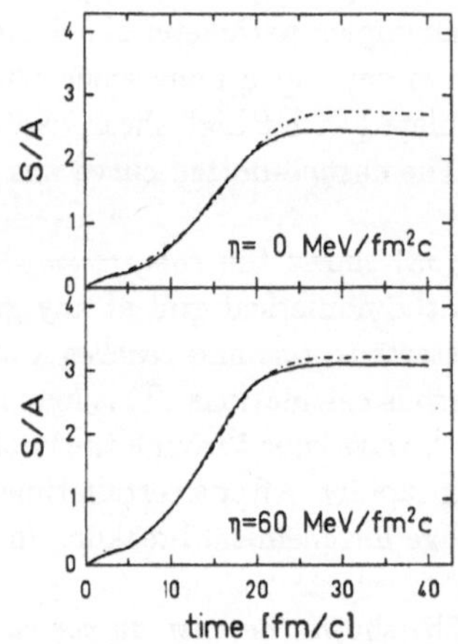

Figure 4. Influence of viscosity and equation of state on entropy per baryon in a Au → Au-reaction at 400 MeV/N, impact parameter $b = 3$ fm, for $\eta = 0$ (top) and $\eta = 60$ MeV/fm^{2}c (bottom):

Dashed: Soft equation of state
Straight: Hard equation of state

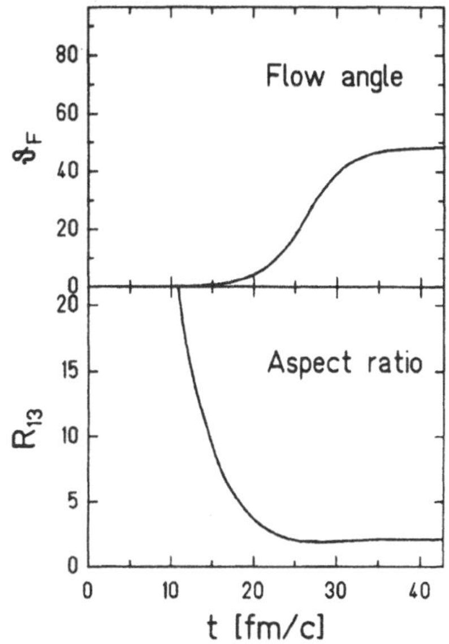

Figure 5. Time development of flow angle θ_F (top) and aspect ratio R_{13} (bottom) for Au $\rightarrow$ Au at 200 MeV/N, $b = 3$ fm, using a hard equation of state, $\eta = 60$ and $\zeta = 0$ MeV/fm^{2}c.

soft equation of state. Particularly in the case of a viscous calculation (fig. 4b), the equation of state does not show *any* influence.

From figures 3 and 4 we have seen that the entropy is almost entirely produced during the shock phase of the reaction. Figure 5 shows the time development of the flow angle θ_F (fig. 5a) and the aspect ratio R_{13} (fig. 5b) for the same reaction as in fig. 3. The kinetic flow is produced in the expansion phase, after the shock wave has run through the nuclei. Both θ_F and R_{13} finally saturate asymptotically, which gives an indication that the assumption of a fixed break-up time is justified.

Entropy

Figure 7 compares entropy data from plastic ball experiments with nuclear fluid dynamics-calculations. The excitation function of the entropy is shown for central collisions ($b = 1$ fm) of Au $\rightarrow$ Au (figure 7a) and Nb $\rightarrow$ Nb (figure 7b). Note that the difference between soft and hard equation of state is low compared to the influence of viscosity.

It can be seen that our nonrelativistic model does certainly not reproduce the data for higher energies (600 and 800 MeV/N). At low energies (200 and 400 MeV/N), however, the agreement is quite good for viscous calculations ($\eta = 60$ MeV/fm^{2}c). One is lead to the conclusion that the nuclear viscosity is in the range of $\eta \approx 60$ MeV/fm^{2}c. This is in agreement with the result of Danielewicz [20], which has been derived from the Uehling-Uhlenbeck equation, and with recent calculations of Schürmann [27].

Flow Angle and Aspect Ratio

In this section we want to study some properties of the flow tensor, which can be computed from hydrodynamical densities and momentum distributions. Neither θ_F nor R_{13} can be measured experimentally because of the fluctuations imposed

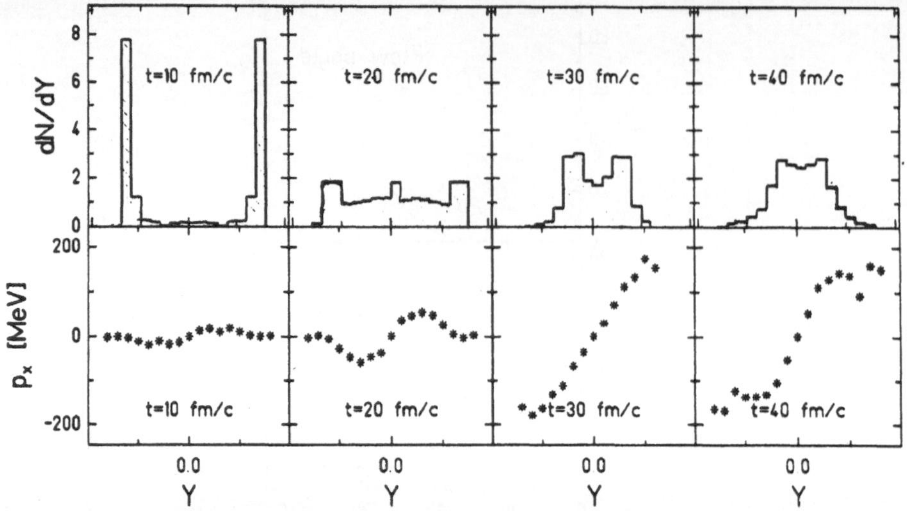

Figure 6. Time development of dN/dY (top) and p_x/A (bottom) for Au → Au at 200 MeV/N, $b = 3$ fm, using a hard equation of state, $\eta = 60$ and $\zeta = 0$ MeV/fm^{2}c.

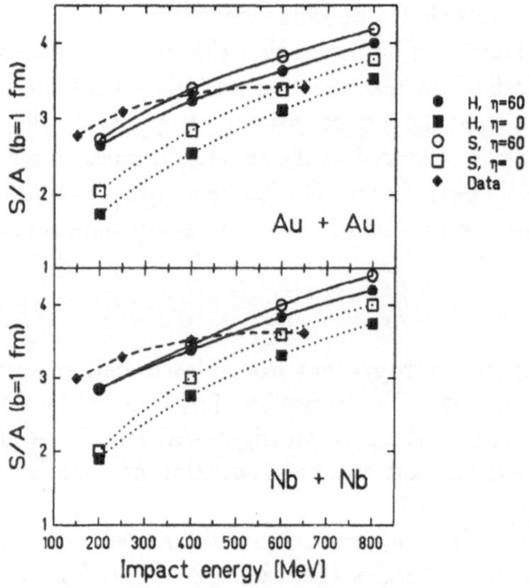

Figure 7. Excitation function of the entropy per baryon for central collisions (b = 1 fm) of Au → Au (top) and Nb → Nb (bottom). Experimental data are for the fifth multiplicity bin.

by finite multiplicities. Only the Jacobian flow angle distribution $dN/d\cos\theta$ was considered measurable so far [28]. To compare nuclear fluid dynamics-calculations with experimental data, one therefore had to compute $dN/d\cos\theta$-distributions by randomly generating numerous "events" with finite experimental multiplicities and integrate over a range of impact parameters [6]. This method was applied successfully to experimental $dN/d\cos\theta$-data measured by the Plasticball-collaboration for the system Nb → Nb at 400 MeV/N.

In this paper, we do not want to repeat this procedure and abandon the ability to compare our calculation with experiment. Instead, we focus on the pure hydrodynamical momentum distribution. In this way, flow angle and aspect ratio can be calculated exactly for each system. Hence we can display θ_F and R_{13} as a function of impact parameter, viscosity, equation of state and impact energy.

It turns out that the flow angle is almost independent of either the equation of state and viscosity [29]. It does not even vary significantly for different impact energies; however, it depends strongly on the impact parameter. This implies that the flow angle is a purely geometrical quantity.

Collective Transverse Momentum Transfer

The transverse momentum transfer, p_x/A, has been measured experimentally for the systems Ca → Ca, Nb → Nb and Au → Au at various bombarding energies. Data have been selected according to charge and multiplicity [24,30].

Again there are two different ways to calculate p_x/A in nuclear fluid dynamics:

1. It can be computed for every cell in the nuclear fluid and plotted as a function of rapidity (*macroscopic* distribution).

2. It can be calculated by integration of $1/p\ \partial^3\sigma/\partial E\partial\Omega$ (*microscopic* distribution). Different results are obtained for different particle species, which can then be compared with experiment. Note that there is little dependence on the break-up time, if the break-up condition $\rho_{max} < \rho_0 1$ is fulfilled [29] (cf. 6b).

Figure 8 provides an overview of the *macroscopic* p_x-distribution obtained for various collisions of Au → Au at 200 MeV/N. All calculations have been done with the Hard equation of state. The impact parameter b varies from left to right ($b = 1$,

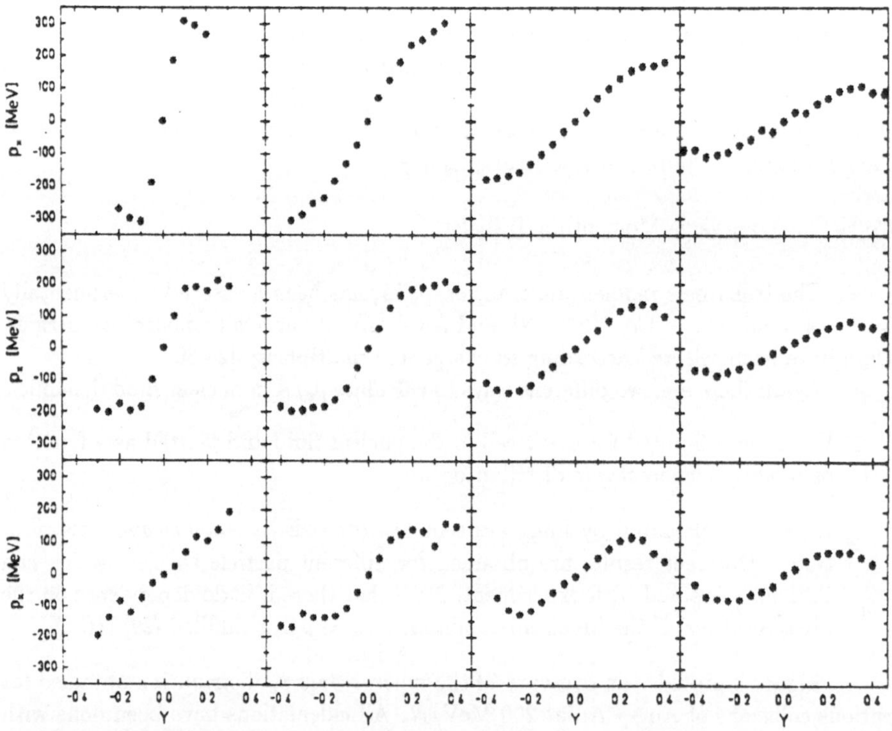

Figure 8. Macroscopic p_x-distribution for various reactions of Au → Au at 200 MeV/N, using the hard equation of state.
The impact parameter varies from left to right ($b = 1, 3, 5$ and 7 fm), shear viscosity η from top to bottom ($\eta = 0, 30$ and 60 MeV/fm^{2}c)

3, 5 and 7 fm), viscosity η from top to bottom (η = 0, 30 and 60 MeV/fm^{2}c). The influence of evaporation has been neglected. As one can clearly see, the collective flow, which is taken to be $dp_x/dY|_{Y_P}$, decreases for increasing impact parameter and viscosity. Note that the maximum flow does not occur at intermediate impact parameters, but at rather central collisions (b = 1 fm), showing a collective flow of more than 300 MeV/c^2. On the other hand, p_x vanishes *per definitionem*, if $b = 0$! Therefore, the macroscopic p_x-distribution as computed from hydrodynamical momenta and densities does not reproduce the experimentally observed b-dependence, where the maximum occurs in the fourth multiplicity bin [30], corresponding to $b \approx$ 3 fm. It also overestimates the magnitude of the flow by a factor of 2 or more compared with data.

Figure 9a demonstrates how the macroscopic hydrodynamical p_x/A is affected by the evaporation: The flow decreases drastically for protons, whereas ^{4}He still shows the strong collective flow which is predicted by plain hydrodynamics. The calculation has been done for Au → Au at 200 MeV/N and b = 3 fm, using the hard equation of state and η = 60 MeV/fm^{2}c.

To show the impact of viscosity and equation of state on p_x/A we included figure 9b, which compares the proton p_x for soft and hard equation of state with $\eta = 0$ and η = 60 MeV/fm^{2}c. It can be seen that the equation of state has no influence at all for a viscous calculation, due to the small maximum compression ($\rho_{max}/\rho_0 \approx 1.5$).

In figure 10 we compared our calculation (hard equation of state, η = 60 MeV/fm^{2}c) with experimental data for $Z = 1$ (a) and $Z = 2$ (b). There is a remarkable quantitative agreement with experimental data. The discrepancies at $Y < 0$ are due to efficiency cuts of the Plasticball-detector at target rapidity, which have been neglected in our calculation.

This figure gives another clear evidence for a fairly high viscosity of $\eta \approx$ 60 MeV/fm^{2}c. Since soft and hard equation of state give almost precisely the same p_x-distribution at this viscosity, we must conclude that the equation of state can not be extracted from flow!

Azimuthal Angular Correlation

Following the approach of Kampert [30] and Gustafsson et al. [31], we studied the dimensionless quantity $p_x/|p_\perp|$.

Figure 11a shows the strong correlation of hydrodynamic flow which is preserved by heavier fragments. Evaporation of light particles, however, leads to a much more isotropic flow.

Once again it turns out that collective flow is very sensitive to the viscosity: Figure 11b shows that the angular correlation of protons in a viscous calculation of Au → Au at 200 MeV/$N$is roughly 40% less than for the nonviscous case. Note the small influence of the equation of state in both calculations.

Angular Rapidity Distribution

Besides the collective in-plane flow discussed above there is also a completely different collective effect, namely the off-plane ($\phi = 90°$) squeeze out at $Y_{CM} = 0$, which was predicted by hydrodynamics [4,9] and recently confirmed experimentally [10,11].

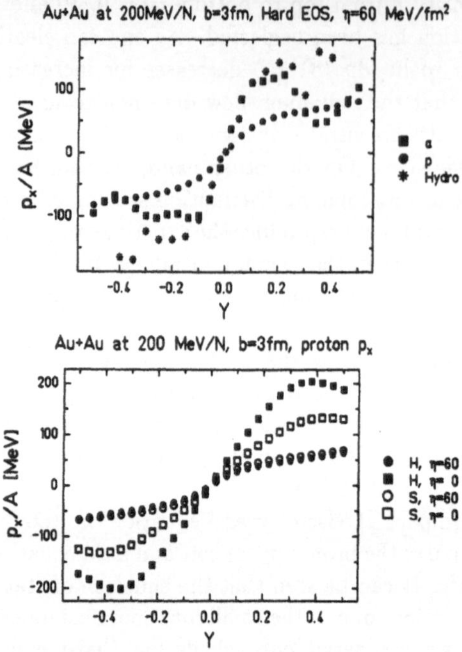

Figure 9. top: p_x-distribution in a Au → Au-reaction at 200 MeV/N, $b = 3$ fm, hard equation of state, $\eta = 60$, $\zeta = 0$ MeV/fm^{2}c,
comparing macroscopic (Hydro) and microscopic (p, α) results. bottom: microscopic p_x-distribution of protons in Au → Au at $b = 3$ fm, hard and soft equation of state, nonviscous ($\eta = 0$) and viscous ($\eta = 60$ MeV/fm^{2}c) calculation.

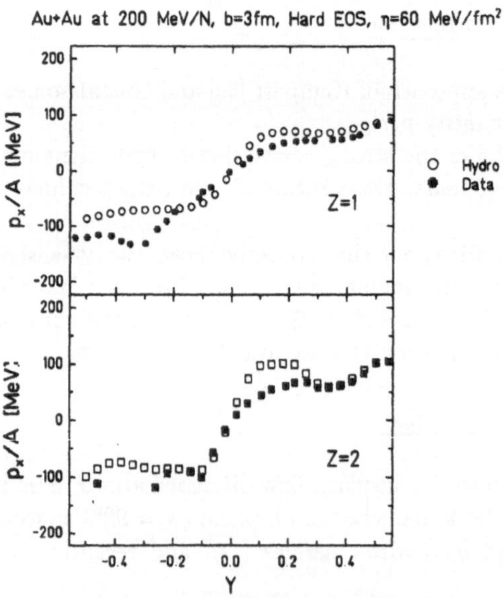

Figure 10. p_x-distribution in a Au → Au-reaction at 200 MeV/N, $b = 3$ fm, hard equation of state, $\eta = 60$, $\zeta = 0$ MeV/fm^{2}c,
comparing theory and experiment for $Z = 1$ (top) and $Z = 2$ (bottom).

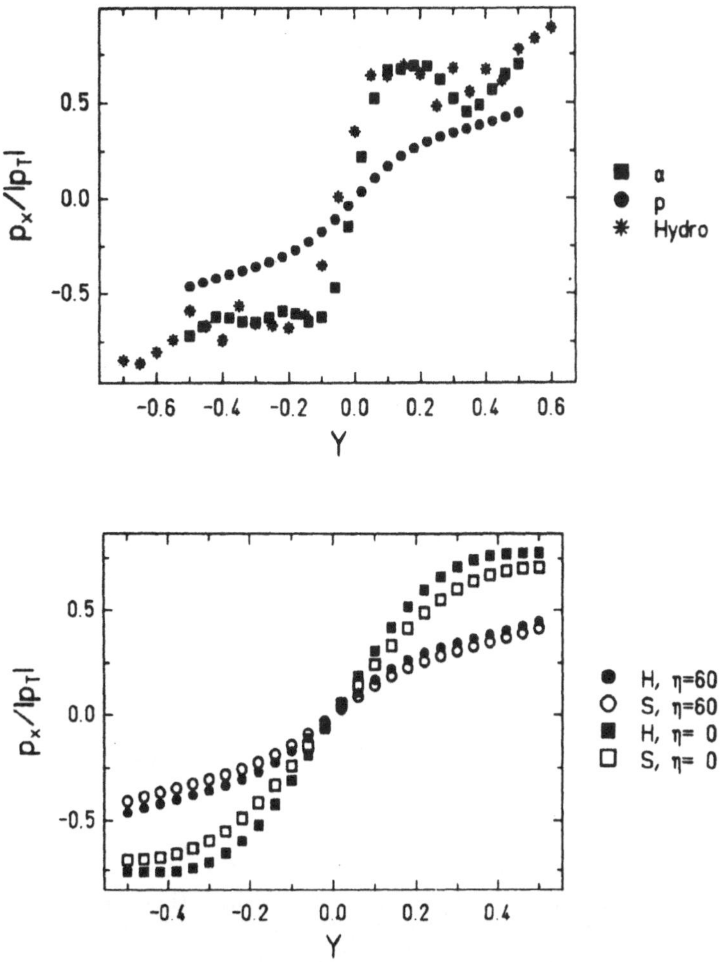

Figure 11. top: $p_x/|p_\perp|$-distribution in a Au → Au-reaction at 200 MeV/N, $b = 3$ fm, hard equation of state, $\eta = 60$, $\zeta = 0$ MeV/fm^{2}c,
comparing macroscopic (Hydro) and microscopic (p, α) results. bottom: microscopic $p_x/|p_\perp|$-distribution of protons in Au → Au at $b = 3$ fm, hard and soft equation of state, nonviscous ($\eta = 0$) and viscous ($\eta = 60$ MeV/fm^{2}c) calculation.

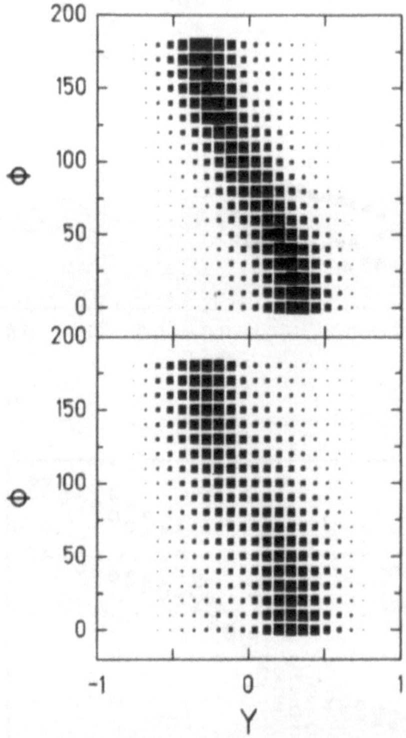

Figure 12. $dN/dYd\phi$-distribution in a Au → Au-reaction at 400 MeV/N, $b = 3$ fm, hard equation of state, $\eta = 0$ (top) and $\eta = 40$ MeV/fm^{2}c (bottom).

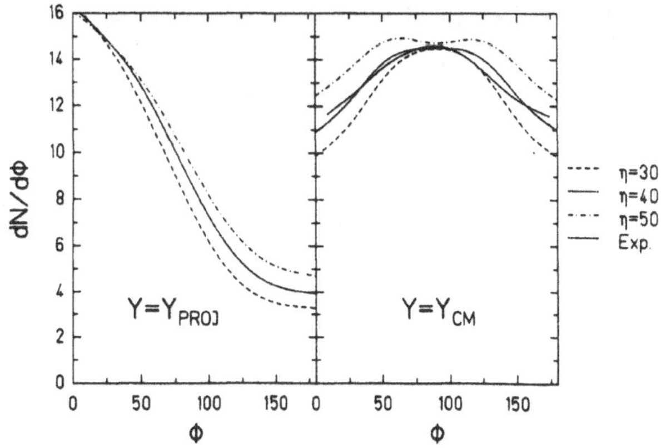

Figure 13. $dN/dYd\phi$-distribution in a Au $\rightarrow$ Au-reaction at 400 MeV/N, $b = 3$ fm, hard equation of state, $\eta = 40$ MeV/fm^{2}c, cuts at $Y = Y_{Proj}$ (left) and $Y = Y_{cm}$ (right).

Figure 12 shows a cluster plot of the angular rapidity distribution $dN/dYd\phi$ in calculations of Au $\rightarrow$ Au at 400 MeV/N for $b = 3$ fm, hard equation of state. Fig. 12a was calculated with $\eta = 0$, fig. 12b with $\eta = 40$ MeV/N. The off-plane squeeze out can be seen at $\phi = 90°, Y = 0$, i.e. in the center of each plot.

To provide for a synopsis of this effect, we have included figure 13 which displays $dN/dYd\phi|_{Y=0}$ (fig. 13a) and $dN/dYd\phi|_{Y=Y_{Proj}}$ (fig. 13b) for different viscosities ($\eta = 0, 30$ and 60 MeV/fm^{2}c). The heigth of the 90°-peak at $Y = 0$ decreases drastically as viscosity increases.

How does this effect depend on viscosity? Figure 14 shows $dN/dYd\phi|_{Y=0}$ for the soft (fig. 13a) and the hard equation of state (fig. 13b) for different viscosities η simultaneously. Experimental data from KAMPERT *et al.* have been included in the figure. It can be seen that the soft equation of state can reproduce data qualitatively, if $\eta \approx 30$ MeV/fm^{2}c. For the hard equation of state, a viscosity of $\eta \approx 50$ MeV/fm^{2}c is needed. Therefore, since $\eta = 30$ MeV/fm^{2}c can be ruled out from our entropy and transverse momentum studies, we are lead to the conclusion that the nuclear equation of state is quite hard, with $K \approx 400$ MeV being a realistic estimate. One must, however, be aware of the fact that the data in fig. 13 are not precisely scaled so far. A real quantitative comparison of our calculations with experiment is therefore not yet possible.

A further problem is the possible contribution of bulk viscosity and thermoconductivity. It is possible that the contributions of these effects add up for some observables, while they may partially cancel for others.

ULTRARELATIVISTIC HYDRODYNAMICS

The equations of motion of a relativistic fluid are quite similar to the nonrelativistic case: again there are conservation equations for density ρ, momentum density $\vec{M}$, and energy density E, with motion proceeding according to velocity $\vec{u}$,

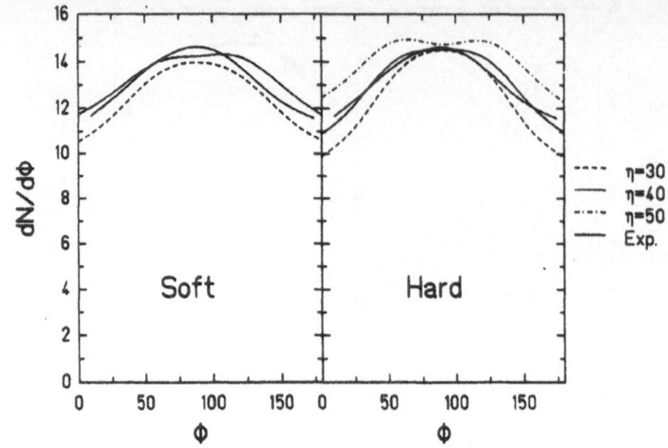

Figure 14. $dN/dY d\phi$-distribution in a Au → Au-reaction at 400 MeV/N, $b = 3$ fm, $\eta = 40$ MeV/fm^{2}c, cuts at $Y = Y_{cm}$, soft (top) and hard (bottom) equation of state, compared with data.

$$\begin{aligned} \frac{\partial \rho}{\partial t} &+ \nabla \cdot (\rho \vec{u}) = 0 \\ \frac{\partial \vec{M}}{\partial t} &+ \nabla \cdot (\vec{u}\vec{M}) = -\nabla p \\ \frac{\partial E}{\partial t} &+ \nabla \cdot (E\vec{u}) = -\nabla \cdot (p\vec{u}). \end{aligned}$$

However, there are also decided differences between the two cases:

- The equations are formulated in a fixed system of reference. The pressure, however, has to be evaluated in the local rest frame of the fluid. If we call the density in this frame n, the energy density ϵ, and if γ is the usual relativistic contraction factor,

  $$\gamma = (1 - u^2/c^2)^{-1/2}$$

  , the equations linking the two systems are

  $$\begin{aligned} \rho &= \gamma n \\ \vec{M} &= \gamma^2(\epsilon + p)\vec{u} \\ E &= \gamma^2(\epsilon + p) - p. \end{aligned}$$

  So at each time step and for each cell of the fluid these equations have to be solved for γ, n, and ϵ, so that the pressure can be computed from $p = p(n, \epsilon)$. This introduces considerable computational effort.

- Viscosity, thermoconductivity, and potentials have not yet been introduced into the calculation.

- There are considerable numeric problems. Their magnitude may be estimated by looking at the γ-factor and velocities. In the equal speed frame the γ-factor is 2.52 at 10 GeV and 10.37 at 200 GeV, while the velocity is 0.92 c and 0.995 c respectively. We thus have the situation that most of the material moves at roughly the same speed and the numerical method has to be able to keep track of minute differences.

A numerical code for this problem was set up by Graebner [32] and used by Rentzsch et al. for simulations of 200 MeV per nucleon reactions [33]. At present, however, there are still doubts concerning the accuracy of the numerical methods[34] for this extreme energy; there many cells for which the local rest frame cannot be found because the fixed-frame quantities are contradictory. Recently, therefore, Katscher and Waldhauser have reinvestigated the numerical method with a careful check in a one-dimensional case.

Fig. 15 shows a simple estimation of the O → Pb reaction (200A GeV) in a 1-d model, which is similar in physical limitations to the one-dimensional shock model but contains a full solution of the equations of motion in one direction. In

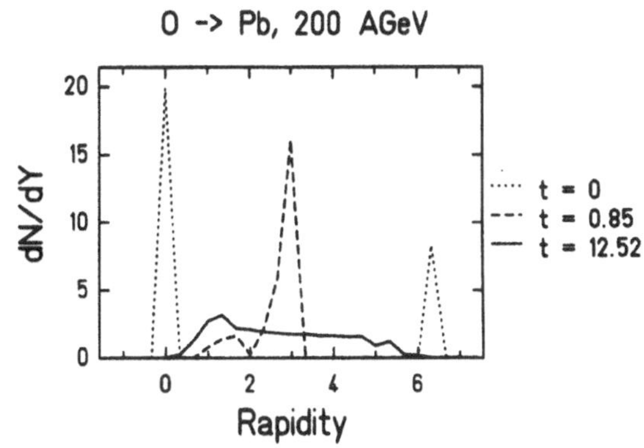

Figure 15. Rapidity distribution of the nucleons in a one-dimensional relativistic hydrodynamic simulation, showing to the evolution from separate target and projectile peaks through a "stopped" phase (t = 0.85) to the final spread in the expansion phase.

this case there should be much more acceleration of the target than in 3-d models, because of the lack of spectators, which will probably shift the maximum of the final dN/dY - distribution towards zero. To first order we can neglect effects of transverse flow, for in the compression phase (longitudinal) speed of mass is much higher than speed of sound. The equation of state for a relativistic ideal gas $p = e/3$ was used.

The solution shows a short period of extremely high density near mid-rapidity (for $t \approx$ 1fm). Afterwards during a long period of expansion up to 10 fm there is a shift of the compressed distribution in both directions. The dN/dY - distribution smears out over almost the whole initial rapidity gap, but with a clear maximum at Y = 1.3. This is in contradiction to the results of Rentzsch et al., which show most nucleons near Y = 3 (mid-rapidity) in the final state. A final settlement of these contradictory results will have to wait for the completion of the numerical studies.

CONCLUSION

We now are lead to the following conclusions from the comparison of our calculations with Plastic Ball-data:

Nuclear Hydrodynamics can quantitatively reproduce heavy ion data

We have shown that several observables, namely the entropy per baryon S/A, the p_x-distributions from transverse momentum analysis, triple differential cross sections $1/p\, \partial^3\sigma/\partial E\partial\Omega$ and rapidity distributions $dN/dY d\phi$ can be calculated in good quantitative agreement with experimental data. It is, however, necessary to treat the chemical break-up separately, using a quantumstatistical model, which takes into account the *microscopic* evaporation of fragments. Hydrodynamics *without* evaporation strongly overestimates the collective motion of light fragments, which are affected by their thermal momentum distribution. The behaviour of *heavy* fragments can be explained in terms of hydrodynamics.

The shear viscosity of nuclear matter can be determined

Both the entropy production and the kinetic flow observed in heavy ion reactions can not be explained in terms of nonviscous hydrodynamics. It turns out that the nonequilibrium properties of nuclear matter, described by the transport coefficients, play an important role in the transit to equilibrium. Calculations with a *constant* coefficient of viscosity – neglecting thermal conductivity – provide an *upper bound* of $\eta \approx 60$ MeV/fm^{2}c, which can be obtained from entropy and transverse momentum analysis. This is slightly higher than in microscopic calculations, where one gets $\eta \approx 40$–50 MeV/fm^{2}c [20,27]. Part of the entropy production may be due to the bulk viscosity ζ, which does not influence kinetic flow. Therefore, in principle it should be possible to fix the bulk viscosity from entropy data. So far, we have not been able to do so, since this task requires the study of excitation functions up to 800 MeV/N or so, which is impossible in our nonrelativistic model.

The nuclear EOS might be extracted from experiment

It was shown that the measurement of at least two independent state variables as functions of energy is necessary for a determination of the equation of state. Unfortunately, none of the observables related to the reaction plane (p_x and dN/dY-distributions) depend on the equation of state, if η has been given a realistic value. On the other hand, the *azimuthal* rapidity distribution dN/dYdϕ shows an influence of the equation of state which is of the order of 40% compared to that of viscosity.

Acknowledgments

The authors wish to express their thanks to the members of the Plastic Ball-collaborations for stimulating discussions. We also gratefully acknowledge contributions made by G. Buchwald and G. Graebner.

References

[1] W. Scheid, W. Greiner, *Z. Phys.* **226** (1969) 365
W. Scheid, H. Müller, W. Greiner, *Phys. Rev. Lett.* **32** (1974) 741

[2] H. G. Baumgardt, J. U. Schott, Y. Sakamoto, E. Schopper, H. Stöcker, J. Hofmann, W. Scheid and W. Greiner, Z. Phys. **A237** (1975) 241

[3] A.A. Amsden, A.S. Goldhaber, F.H. Harlow and J.R. Nix, Phys. Rev. **C17** (1978) 2080

[4] H. Stöcker, J.A. Maruhn and W. Greiner, Z. Phys. **A293** (1979) 173

[5] G. Buchwald, G. Graebner, J. Theis, J.A. Maruhn, W. Greiner, H. Stöcker, K. Frankel and M. Gyulassy, Phys. Rev. **C28** (1983) 2349

[6] G. Buchwald, G. Graebner, J. Theis, J.A. Maruhn, W. Greiner and H. Stöcker, Phys. Rev. Lett. **52** (1984) 1594

[7] H.-A. Gustafsson, H.H. Gutbrod, B. Kolb, H. Löhner, B. Ludewigt, A. M. Poskanzer, I. Renner, H. Riedesel, H.G. Ritter, A. Warwick and H. Wieman, Phys. Rev. Lett. **52** (1984) 1590

[8] K.G.R. Doss, H.-A. Gustafsson, H.H. Gutbrod, D. Hahn, K.-H. Kampert, B. Kolb, H. Löhner, A.M. Poskanzer, H.G. Ritter, H.R. Schmidt, H. Stöcker, *Phys. Rev.* **C37** (1988) 163

[9] H. Stöcker, L.P. Csernai, G. Graebner, G. Buchwald, H. Kruse, R.Y. Cusson, J.A. Maruhn and W. Greiner, Phys. Rev. **C25** (1982) 1873

[10] K.-H. Kampert, To be published in *J. Phys.* **G**

[11] H. Gutbrod, K.-H. Kampert, B. Kolb, A. Poskanzer, H. Ritter and H.R. Schmidt, Phys. Lett. **B216** (1989) 267

[12] A. Rosenhauer, *PhD-thesis*, Johann Wolfgang Goethe-Universität, Frankfurt am Main 1988; G. Peilert, H. Stöcker, W. Greiner and A. Rosenhauer, To be published in *Phys. Rev.* C.

[13] J. Aichelin and G. Bertsch, Phys. Rev. **C31** (1985) 1730; J. Aichelin and H. Stöcker, Phys. Lett. **176B** (1986) 14; J. Aichelin, A. Rosenhauer, G. Peilert, H. Stöcker and W. Greiner, Phys. Rev. Lett. **58** (1987) 1926

[14] G. Peilert, *Diplomarbeit*, Johann Wolfgang Goethe-Universität, Frankfurt am Main 1988; G. Peilert, H. Stöcker, W. Greiner, A. Rosenhauer, A. Bohnet and J. Aichelin, Phys. Rev. **C39** (1989) 1402

[15] J. Hofmann, W. Scheid, and W. Greiner, Nuovo Cimento **33** (1976) 343; U. Heinz, H. Stock, W. Scheid, and W. Greiner, J. Phys. **G3** (1977) 21.

[16] H. Stöcker, W. Greiner, and W. Scheid, Z. Physik **A286** (1987) 121.

[17] J. A. Maruhn and H. Stöcker, Z. Physik **A327**, 75 (1987).

[18] D. Hahn and H. Stöcker, Nucl. Phys. **A452** (1986) 723.

[19] A. Sandoval *et al.*, Phys. Rev. Lett. **45**, 874 (1980).

[20] P. Danielewicz, *Phys. Lett.* **B146** (1984) 141

[21] L.P. Csernai, H. Stöcker, P. Subramanian, G. Buchwald, G. Graebner, A. Rosenhauer, J.A. Maruhnand W. Greiner, *Phys. Rev.* **C28** (1983) 2001

[22] P. Siemens, J. Kapusta, *Phys. Rev. Lett.* **43** (1979) 1486; G. Bertsch, J. Cugnon, *Phys. Rev.* **C24** (1981) 2514; G. Bertsch, *Nucl. Phys.* **A400** (1983) 221

[23] H. Stöcker, G. Buchwald, G. Graebner, P.S. Subramanian, J.A. Maruhn, W. Greiner, B. Jacak and G.D. Westfall, *Nucl. Phys.* **A400** (1983) 63c; D. Hahn, H. Stöcker *Nucl. Phys.* **A476** (1988) 718; D. Hahn, H. Stöcker *Phys. Rev.* **C37** (1988) 1048

[24] P. Danielewicz, G. Odyniec *Phys. Lett.* **157B** (1985) 146

[25] G. Claesson, G. Krebs, J. Miller, G. Roche, L.S. Schroeder, W. Benenson, J. van der Plicht, J.S. Winfield, G. Landaud, J.-F. Gilot, H. Stöcker, submitted to *Phys. Rev. Lett.*

[26] H. Stöcker, L.P. Csernai, G. Graebner, G. Buchwald, H. Kruse, R.Y. Cusson, J.A. Maruhn and W. Greiner, *Phys. Rev.* **C25** (1982) 1873

[27] B. Schürmann, *Mod. Phys. Lett.* **A3** (1988) 1137

[28] P. Danielewicz and M. Gyulassy, Phys. Lett. **129B** (1983) 283

[29] W. Schmidt, *PhD-thesis*, Johann Wolfgang Goethe-Universität, Frankfurt am Main 1989

[30] K.-H. Kampert, *PhD-thesis*, Wilhelms-Universität zu Münster 1986

[31] H. Gustafsson, H. Gutbrod, K.-H. Kampert, B. Kolb, A. Poskanzer, H. Ritter and H.R. Schmidt,
Mod. Phys. Lett. **A**, in print.

[32] G. Graebner, PhD Thesis, University of Frankfurt 1985.

[33] T. Rentzsch, G. Graebner, J. A. Maruhn, and W. Greiner, Z. Physik **C38**, 237 (1988).

[34] J.P. Boris, D.L. Book, *J. Comp. Phys.* **11** (1973) 38

TRANSPORT THEORY, VISCOSITY

AND KAON PRODUCTION

Bernd Schürmann

Siemens AG
Corporate Research & Development
ZFE F2 INF 23
Otto-Hahn-Ring 6
D-8000 München 83
West Germany

1. INTRODUCTION

Intermediate energy heavy ion research has witnessed a rapid development in the past decade, both experimentally and theoretically. It was realized already very early that nucleus-nucleus collisions in the bombarding energy regime of several hundred MeV up to a few GeV per nucleon provide the only tool to investigate in the laboratory the behaviour of nuclear matter under extreme conditions, i.e. at high temperature and high density. This behaviour is summarized by the equation of state (EOS) which states the pressure P as a function of the density ρ and the temperature T: $P = P(\rho,T)$ (thermodynamic EOS), or, equivalently, the internal energy E as a function of ρ and T: $E = E(\rho,T)$.(caloric EOS). It has become customary to employ the latter form in intermediate energy heavy ion research.

It turned out soon that the experimental observation of the stage of high compression which lasts for only a very short time during the collision, is not so easy: one has to carefully search for observable quantities which carry the essential information of this transient stage. It took almost a decade until the experimental techniques became sufficiently sophisticated to tackle the determination of the nuclear EOS. Parallel to this progress has been the development of theoretical models for an adequate description of the experimental results.

The Nuclear Equation of State, Part A
Edited by W. Greiner and H. Stöcker
Plenum Press, New York

One approach which has accompanied the progress of the field over the years has been the near-analytical model of transport theory developed by myself and several colleagues. In section 2 I will describe this model with its successes and failures in the reproduction of experimental data from its earliest stage to its latest. This will then at the same time give the reader an impression of the history of the field or at least, of parts of it. In section 3 I will partly leave the framework of the analytic transport model to discuss the role of viscosity for the detection of the EOS. This will be done by making use of the scaling properties of the hydrodynamical equations as well as of numerical quantum molecular dynamics calculations. The information on the EOS contained in strange particle production cross sections, in particular below threshold, is discussed in section 4 by making use of the transport model again. I close with a brief epilogue in section 5.

2. TRANSPORT THEORY

Instead of entering mathematical details which can be found elsewhere[1], I rather prefer to convey the concepts and ideas underlying the model, illustrating each stage of modification and extension by comparison with experimental data.

2.1. The Basic Model

The original model was formulated together with H.J. Pirner as early as 1979,[2] with the aim to describe the measured inclusive cross sections for protons and pions obtained from collisions between two identical nuclei at 800 MeV per nucleon.[3] These data exhibited non-equilibrium features and hence could not fully be explained by the simple thermal (fireball) model. We considered a two-component system consisting of nucleons and delta resonances whose momentum distributions after a specified number of collisions were described by two coupled Fokker-Planck equations. These transport equations were constructed such that they contained the (highly non-equilibrated) single sattering process as initial condition and the thermal (fireball) distribution as asymptotic solution. Hence the model mimiced the evolution of the colliding system from the highly non-equilibrated initial up to the thermal final stage. The one-particle differential cross section was written as an incoherent sum of multiple scattering contributions,

$$d\sigma_i/d^3p = A\,\Sigma_{n=1} \int d^2b G_n(\mathbf{b})\,[M^B{}_{n,i}(\mathbf{p}) + M^T{}_{n,i}(\mathbf{p})]\ . \qquad (1)$$

Here, A denotes the nuclear mass number, and n the number of collisions. The functions $G_n(\mathbf{b})$ are known geometrical factors [4], $M^B{}_{n,i}(\mathbf{p})$and $M^T{}_{n,i}(\mathbf{p})$ are the final momentum distributions of beam and target, respectively, and the index i stands for either nucleons or pions. These distributions arise from the original solutions of the Fokker-Planck equations, as well as from contributions obtained after the decay of the delta resonances into nucleons and pions at the end of the collision.

At that stage, the model was purely heuristic. Later on, we have shown that under appropriate approximations, it is equivalent to the Boltzmann equation.[5,6] The concept behind it is the "Brownian motion" of test particles (nucleons or deltas) in a surrounding baryonic heat bath of normal nuclear matter density. An often raised criticism of this concept may be summarized by two main points: (i) the nucleons and deltas cannot be considered as Brownian particles since they have the same masses as the surrounding baryons which constitute the heat bath; (ii) it is not allowed to assume the surrounding medium to be thermalized. These objections can be met as follows. As to (i), to obtain a near-analytic treatment, we employ the eikonal approximation[7] which rests on the fact that the free elementary differential cross section in the several hundred MeV/nucleon region is strongly forward peaked and hence the deflection angle per baryon collision is small. The small deflection is the main feature of a Brownian particle, not its heavy mass. The eikonal approximation is reasonable at intermediate energy. As to (ii), it is known from statistical mechanics that the concept of minimal information on the surrounding medium often is sufficient for a quite accurate evaluation of the test particle distribution functions.

Dense nuclear matter cannot be treated with the model. Hence, a satisfactory agreement of the model calculations with experimental data signals the absence of high density phenomena in these data.

In Fig.1, the proton and pion invariant inclusive spectra calculated with the original model[2] are compared with corresponding experimental data obtained for C on C collisions at 800 MeV/nucleon.[3] It is seen that the model reproduces the proton data reasonably well, apart from a slight overestimate (Fig.1.a). This is not the case for the pion data (Fig.1.b). The slopes are much to steep and the degree of anisotropy is too large. Hence, there is not enough thermalization in the calculated pion spectra.

2.2. Inclusion of Thermal Pions and Light Fragments

As evidenced by Fig.1.b, the assumption that pions originate solely from delta decay is inappropriate. Pions can, to an appreciable amount, also be produced thermally, at a late stage of the collision. The proton primary distributions agree quite well in shape with the finally observed ones, only their magnitude is slightly overestimated. This overestimate is due to the fact that so far the formation of light fragments such as d, t, ^{3}He, and α, has not been taken into account.

I have mentioned already that the Fokker-Planck and Boltzmann equations are equivalent under certain approximations. One of these approximations is to neglect the "gentle" collisions among particles of the same class, i.e. among beam-like (B) particles on the one hand, and among target-like (T) particles on the other. It is reasonable to assume that these collisions are responsible for

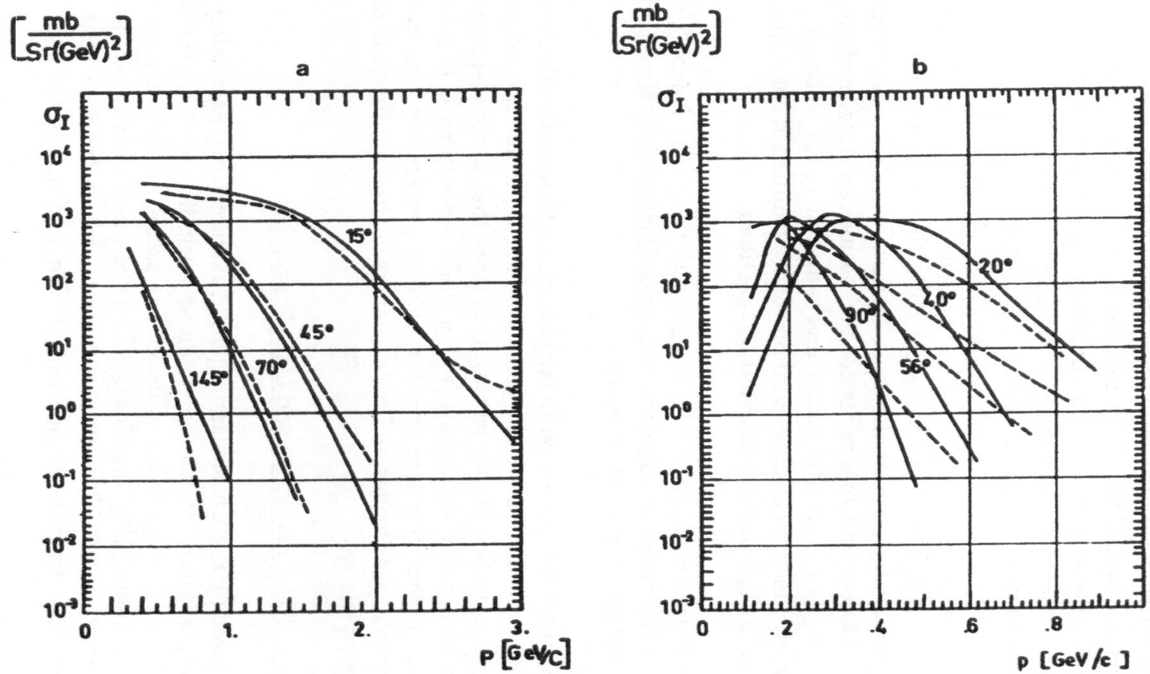

Fig.1. Inclusive differential cross section from C + C at 800 MeV/nucleon. (a) For proton production. (b) For negative pion production. From Ref.2.

composite particle production, and for pion production in the late stage. Their effect has been included in an approximate manner[6] by using methods of statistical thermodynamics along the lines of Ref.8. The underlying conception is to replace the original complex collection of strongly interacting baryons by a new ensemble of particles in local thermal and chemical equilibrium. This ensemble now includes, besides the original baryons, the products of the interactions which are either pions or bound states like deuterons, tritons, etc. The complicated dynamics is thus shifted to the density of states.

The statistical thermodynamical treatment of Ref.8 is now applied to nucleus -nucleus collisions . Incorporating light fragment and pion production as outlined above, one obtains for the final inclusive cross section of species i (nucleons, pions, deuterons...) the expression

$$d\sigma_i/d^3p^* = \Sigma_{n=1} \int d^2b \, [L_{p\to p^*}(\lambda_B^{(n)})(dN_i/d^3p)(T_B^{(n)},\mu_B^{(n)}) + L_{p\to p^*}(\lambda_T^{(n)})(dN_i/d^3p)(T_T^{(n)},\mu_T^{(n)})] \quad (2)$$

with the thermodynamical spectra dN_i/d^3p for species i depending on temperatures $T_B^{(n)}$, $T_T^{(n)}$, and chemical potentials $\mu_B^{(n)}$, $\mu_T^{(n)}$, respectively. The symbol $L_{p\to p^*}(\lambda)$ denotes the Lorentz transformation from the rest frame in which local equilibrium (with respect to collision number n) is assumed, to the moving frame (velocity $\lambda_{B,T}$) which for equal nuclei is the nucleus-nucleus c.m. frame.

In Fig.2, a comparison of the model with experimental data[3] as well as with numerical intranuclear cascade calculations[9] is made for proton inclusive production at 800 MeV/nucleon. The transport model result is fairly close to that of the numerical cascade as well as to experiment. Again, for the good agreement with the data the non-equilibrium components are decisive. Representative for light fragment production, I show in Fig.3 the deuteron inclusive cross section at 400 MeV/nucleon. The experimental data[3] are seen to be reproduced quite well by the calculations. The effect of the non-equilibrium primary nucleons is seen very clearly. At forward angle, the fireball result (dashed line) is also shown. It grossly fails with respect to the data. The shift between the collision- number-dependent rest frames and the fireball rest frame is essential in reproducing the measured points. I am not aware of any other microscopic model which yields equally good agreement with the data.

Comparing the calculated with the measured pion spectra in Fig.4, the effect of allowing for delta rescattering and pion creation in the Hagedorn stage[8] is clearly seen. In contrast to Fig.1.b, the spectra now possess the correct slopes, only the number of pions created is overestimated. This feature pertains also to intranuclear cascade models[10,11]. It has been argued that this overestimate yields information on the compressional part of the EOS [12] but other reasons for the discrepancy between the measured and calculated numbers of pions have also been discussed in the literature.

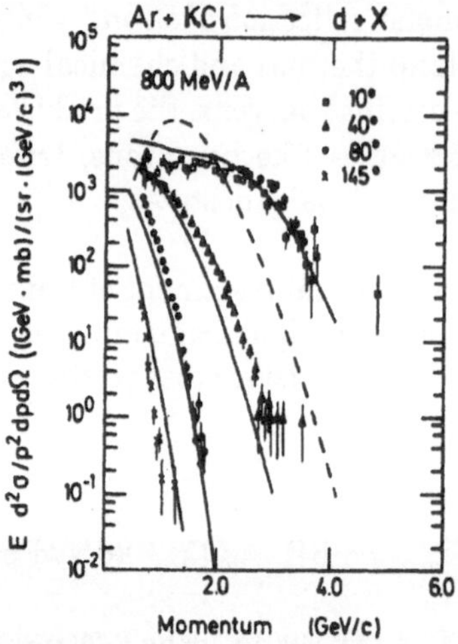

Fig.2. Proton inclusive differential cross section in the laboratory system. Transport model (full lines) in comparison with the data[3] and cascade calculations (histograms).[9] From Ref.1.

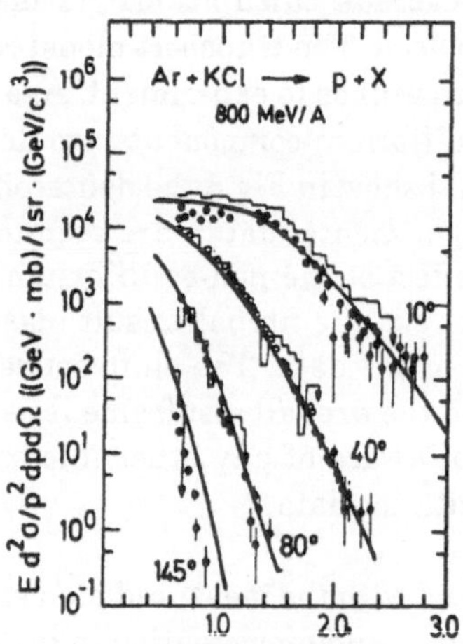

Fig.3. Deuteron inclusive differential cross section in the laboratory system.
Dashed line: fireball result. Data from Ref.3. From Ref.1.

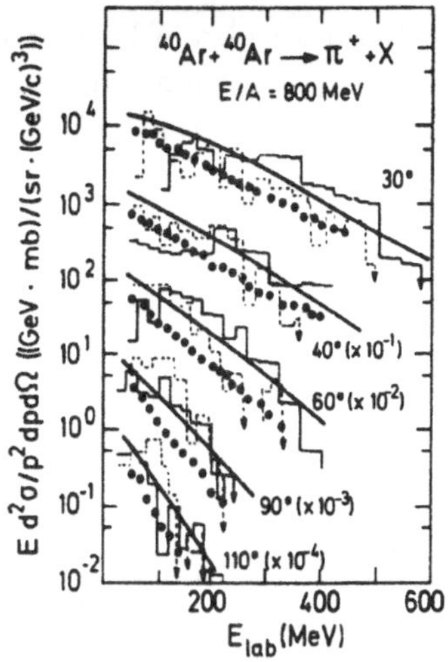

Fig.4. Positive pion inclusive differential cross section in the laboratory system. Full lines: transport model with pion rescattering. Full histograms: cascade calculation.[10] Dotted histograms: cascade calculation.[11] From Ref.1.

3. VISCOSITY

So far, the model does not acount for mean field effects. Indeed, these do not seem to be important for inclusive observables. The situation changes, however, when comparing the model calculations with exclusive experimental quantities extracted from event per event measurements like the flow angle, etc. The model as it stands, yields a flow angle of zero degrees, because of the dominance of the elementary scattering cross section which is forward peaked at the energies of interest. This is in disagreement with experimental data which yield a sizeable sideward flow.[13] This observation led to a major extension of the transport model, namely the inclusion of potential energy effects. This was achieved in collaboration with W. Zwermann.[14,15] It will be briefly alluded to in subsec.1. The transverse flow is introduced in subsec.2 and related to viscous hydrodynamics. The influence of transport coefficients, like viscosity, on the nuclear equation of state is discussed in subsec.3.

3.1. Inclusion of Mean Field Effects

Particles emitted in central high energy heavy ion reactions exhibit sideward flow. This phenomenon has been observed unambigously for the first time in event per event measurements performed by the LBL-GSI plastic ball-wall group.[13] The experimental data, analyzed in terms of the flow angle, could be explained qualitatively by the hydrodynamical model of the Frankfurt school[16] which in fact had predicted sideward flow a long time ago.[17] However, microscopic transport models, analytical as well as numerical, also reproduce the flow angle if the nuclear mean field is included in addition to binary nucleon-nucleon (NN) collisions.[14,18]

I will now briefly discuss the basic idea of how mean field effects have been included in the analytic transport model. Based on the assumption of transparency of heavy ion collisions, two classes of participant nucleons labelled "beam-like (B)" and "target-like (T)" can be distinguished . After n collisions, these nucleons on the average travel in the c.m. system in opposite directions along the z-axis with velocities $<v>_n$ and $-<v>_n$, respectively. The two classes of participant nucleons experience the influence of two correspondingly moving B- and T-nuclear mean fields (velocities v_F and $-v_F$, respectively) which separately depend on the B- and T-nuclear densities. To be specific, let me consider a test B-participant nucleon. The B- and T-fields will transfer momenta to the test participant which are opposite in direction. These momenta are different because of the different relative velocities $v_F - <v>_n$ and $v_F + <v>_n$ between the B-participant and the B- and T-fields. This leads to a generally non-zero net momentum transfer $\vec{\delta}_n$ to the B-participant. Accordingly, a net momentum transfer $-\vec{\delta}_n$ is obtained for the test T-participant.

From the momentum transfer $\vec{\delta}_n$, the flow angle can be obtained. Fig.5 displays a typical result for the flow angle calculated in this approach, employing Skyrme parametrizations for the mean field potentials with and without momentum dependence. Later on, the significance of the momentum dependence has been pointed out by others as well.

3.2. Model Comparisons with the Experimental Central Transverse Flow

An observable more sensitive than the flow angle is the average transverse momentum per nucleon $<p_x/N>$ in dependence of the c.m. parallel momentum per nucleon p_z.[19] From it the derivative of the average transverse momentum with respect to p_z/p_0 taken at $p_z = 0$ can be obtained[20] where p_0 is the incident momentum per nucleon. This quantity is called the "transverse flow". Unlike the average transverse momentum, it is not distorted by spectator particle contributions, and as such it is suited especially well for the detection of compression effects.

Based on the transverse flow, I now discuss the performance of various theoretical models with respect to experiment. An experimental study of the mass (A) and incident energy per nucleon (E_0) dependences of the transverse flow[20] leads to a discrimination between the different model discriptions.[21] The result of the comparison between theory and experiment is that two closely connected approaches consistent with the experimental data remain, as far as a comparison is possible. These are the numerical transport models "Vlasov-Uehling-Uhlenbeck(VUU)",[18] and "QuantumMolecular Dynamics (QMD)"[22] on the one hand and viscous hydrodynamics on the other, the latter being a simplified, macroscopic version of the former.

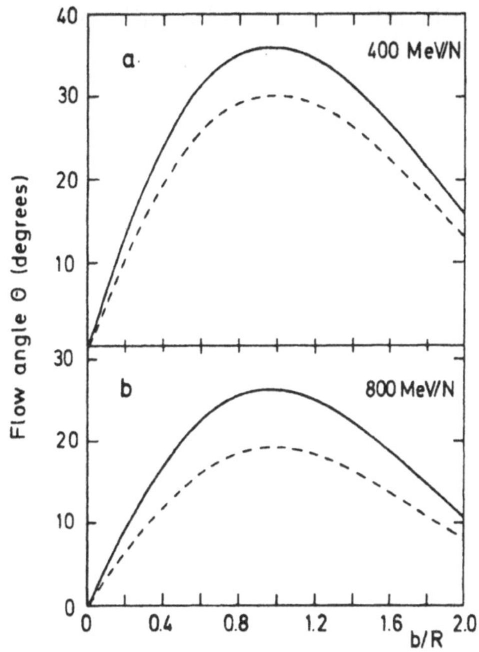

Fig.5. Flow angle Θ in dependence on the impact parameter for Nb+Nb at two different beam energies: (a) 400 MeV/nucleon, (b) 800 MeV/nucleon. Full/dashed line: with/without momentum dependence. From Ref.1.

3.3. Experimental Transverse Flow and Viscous Hydrodynamics

In Ref.20 the energy dependence of the transverse flow f has been measured for the systems Ca + Ca, Nb + Nb, and Au + Au. The unbiased data indicate a slight decrease with bombarding energy above 600 MeV/nucleon, but are also consistent with a flat behaviour. A plateau or even a slight increase is observed for high multiplicity selection.[23]

Exploiting the scaling properties of non-viscous hydrodynamics has proved successful on the inclusive level.[24] Exclusive observables such as the transverse flow are, however, more sensitive to the reaction mechanism: though non-viscous hydrodynamics exhibits the right E_0 - dependence, it yields an A-independent transverse flow, in disagreement with the data. Hence, a description based on the Euler equation is not sufficient, and viscosity has to be included at a level of at least the Navier-Stokes equation. Scaling out a characteristic length l_1, velocity u_1, and time t_1 such that the Strouhal number $S = u_1 t_1 / l_1 = 1$, the Navier-Stokes equation in principle depends on two scale-violating quantities, which are the dimensionless velocity of sound $\tilde{c}_s$ and the Reynolds number Re (cf. Ref.24). Under the assumption of a thermal EOS, $\tilde{c}_s$ is a universal constant, and the only remaining scale-violating quantity is the Reynolds number

$$Re = l_1 u_1 m \rho / \eta \qquad (3)$$

with the characteristic length $l_1 = (4\pi/3)^{1/3} r_0 (2A)^{1/3}$, the characteristic velocity $u_1 = (2E_0^{c.m.}/m)^{1/2}$, the nucleon mass m, the nuclear density ρ, and the shear viscosity η. From elementary kinetic theory one finds $\eta = m\,\rho\langle u\rangle\lambda/3$ or, with $\langle u\rangle = (3mT)^{1/2}$ and $\lambda = (\sigma\rho)^{-1}$ with the NN total cross section σ, $\eta = (mT/3)^{1/2}/\sigma$. For $\rho = \alpha\,\rho_0$ and $T = \alpha T_0$ with the normal nuclear matter density $\rho_0 = (4\pi r_0^3/3)^{-1}$ and $T_0 = 2E_0^{c.m.}/3$, one obtains

$$Re = (3/2)\,(3/\pi)^{2/3}\alpha^{1/2}A^{1/3}\sigma/r_0^2 \tag{4}$$

where $r_0 = 1.2$ fm. I have assumed the same fraction α for ρ and T because numerical transport calculations show that at any given late time, ρ/ρ_0 and T/T_0 are roughly the same. Eq. (4) exhibits two remarkable features : (i) for constant σ, $Re \propto A^{1/3}$ similar to the dimensionless experimental flow $\tilde{f}$, which suggests a close connection between $\tilde{f}$ and Re, and (ii) $Re/A^{1/3} \propto \sigma$. Even though a more realistic η depends on ρ, and also possesses a T-dependence more complicated than the one obtained from elementary kinetic theory, Eq. (4) already contains the essential physics. In actual calculations of Re an expression for η is employed, obtained from the Uehling-Uhlenbeck equation by making use of a first- order Chapman-Enskog approximation.[25] For viscous fluid dynamics to be an adequate description of central high energy heavy ion collisions, $\tilde{f}$ = const. should hold for combinations of A and E_0 for which Re = const., in other words $\tilde{f}$ should be a function of Re alone. Only low densities and temperatures will lead to such a property.[26] I choose $\alpha = \rho/\rho_0 = T/T_0 = 0.2$, a value at which the transverse flow as obtained from QMD[27] calculations saturates. This choice not only yields similar shapes of Re and $\tilde{f}$ as functions of A and E_0 (cf. Fig.6) but also leads to an approximate proportionality between $\tilde{f}$ and Re. The latter feature can be checked by comparing the numbers on the l.h.s. and r.h.s. ordinates of Fig.6. Also shown in the same figure is the result for Re obtained from the elementary formula (4) with $\alpha = 0.2$ and $\sigma = 40$ mb. It is seen to work remarkably well above $E_0^{c.m.}$ = 100 MeV/nucleon. At low energies, quantum features manifested by the Fermi statistics (which is included in the calculations of Ref.25) become important.

The analytic formula of Ref.21 interrelates the dimensionless quantities $\tilde{f}$, the flow angle θ_{fl} and the aspect ratio r:

$$\tilde{f} \approx (r-1)\sin\theta_{fl}\cos\theta_{fl}. \tag{5}$$

While the flow angle can be inferred from experimental data in a straightforward manner, the aspect ratio cannot. However, it can be determined from (5) by inserting the experimental flow angle θ_{fl}, and requiring $\tilde{f}$ to be equal to the experimental dimensionless transverse flow. This is achieved with an aspect ratio of roughly 2, for Au + Au as well as for Nb + Nb and Ca + Ca , cf. Fig.6. Since for heavy systems finite particle number distortions are small, for Au + Au and Nb + Nb, r = 2 signals significant deviations from spherical events. In fact, such a value for r is in agreement with recent viscous fluid-dynamic calculations.[28]

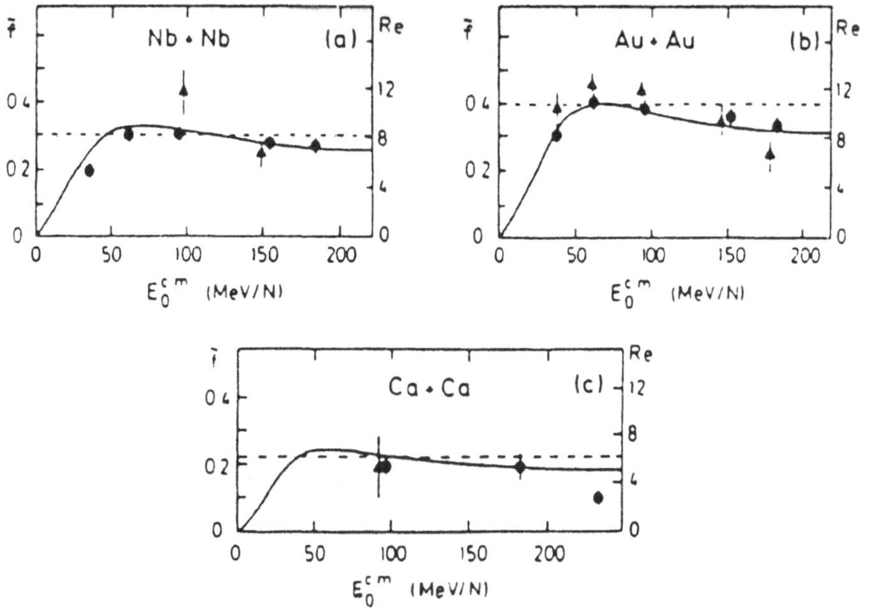

Fig.6. Dimensionless central flow $\tilde{f}$ and Reynolds number Re as functions of the incident c.m. energy / nucleon. Full lines: Re based on Ref.25. Dashed line: Re from eq.(4). Circles: data from Refs.20,23. Triangles: $\tilde{f}$ from eq.(5) with an aspect ratio r = 2. From Ref.29.

How viscous a fluid is highly excited nuclear matter? Having fixed the Reynolds number Re as a function of A and E_0 by choosing the ratio α in an appropriate manner as described, the viscosity coefficient η is also accordingly fixed. Averaging over an energy range of E_0^{Lab} between 400 and 1200 MeV/nucleon, $\eta = 30$ MeV/(fm^{2}c) is a representative lower bound value; lower bound, because in the high density - high temperature phase of the collision, the NN cross section σ presumably will be reduced by Pauli-blocking, leading to a temporarily higher η. An upper bound can be obtained from viscous hydrodynamic calculations. Non-viscous hydrodynamics yields a much too large $\langle p_x/N \rangle$ (cf. Ref.28). To reduce the calculated values to the experimental ones, a shear viscosity of at least $\eta = 60$ MeV/(fm^{2}c) is required. I consider this to be an upper bound because part of it might reflect certain non-equilibrium effects, like for instance quasi-free scattering events which in a microscopic approach would not contribute to viscosity. As one would expect from the foregoing considerations, average viscosities extracted from microscopic approaches lie in-between the lower and upper bounds. This feature is illustrated in Fig.7. The shear viscosity coefficient η has been obtained hereby, taking the density and the temperature at any given time from a QMD code[27] and inserting them in the expression for η. Large differences to values for η calculated in a way consistent with the QMD approach are not expected because the one-particle distribution function constructed from it should not differ very much from the Uehling-Uhlenbeck one. To summarize, I obtain

$$30\ \text{MeV/(fm}^2\text{c)} < \eta < 60\ \text{MeV/(fm}^2\text{c)}. \qquad (6)$$

It is remarkable that values for η and Re close to the ones obtained here, have already been guessed by Bodmer a decade ago when the field of high energy heavy ion physics was still in its infancy.[30]

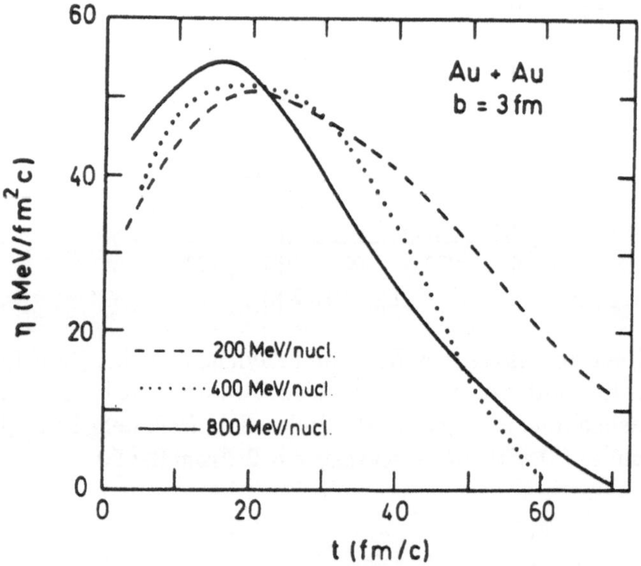

Fig.7. Shear viscosity coefficient η as a function of time t, obtained by using the Frankfurt-Heidelberg QMD-code.[27] From Ref.29.

3.3. Viscosity and The Nuclear Equation of State

The approximate validity of the similarity relation between $\tilde{f}$ and Re which is based on a thermal equation of state, indicates a relatively small effect of the compressional part of the EOS , whereas a sizeable η is mandatory. Viscous hydrodynamic calculations[28] point in the same direction: the differences between the transverse flow calculated with a hard and with a soft EOS are of the order of only 30 to 50 % whereas for $\eta = 0$ and $\eta = 60$ MeV/(fm^{2}c) they are larger than a factor of 2. Hence, the influence of the EOS on the flow is obscured by η, or more microscopically, by the NN cross section σ.

The investigation presented relies on experimental data which do not discriminate between charges of the particles detected. According to results of recent QMD calculations , the sensitivity of the transverse flow on the nature of the nuclear equation of state increases with increasing charge number [31], and a hard EOS is required to reproduce the experimental data. In any case, it appears to be apparent from the present and other complementary investigations that before the nuclear equation of state can be determined, the NN cross section in the medium has to be accurately known.

4. KAON PRODUCTION

The production of strange particles (K-mesons, Λ- and Σ-hyperons) can be described conveniently within the analytic transport model. The lowest-lying channel for strangeness production in a nucleon-nucleon collision (NN→N ΛK) has a threshold energy of 1.6 GeV in the laboratory system. Therefore, in the relevant energy regime, the production of strange particles in nucleus-nucleus collisions is either subthreshold or slightly above, and mainly the elementary strange particle production cross section in the vicinity of the threshold will contribute. In this energy range, this cross section is very small compared with the total NN scattering cross section; hence strange particle production may be treated perturbatively, i.e. its influence on the baryon distributions may be neglected.

Strange particle production in heavy ion collisions has been one of the main research topics of Winfried Zwermann and myself over the last few years. In the limited space available, I can here only discuss some of the results obtained. I restrict myself to *total* production cross sections, and furthermore only to kaon production (for a review on strange particle production in general and antikaons in particular, cf. Refs. 32,33).

In subsec.1, I address the description of kaon inclusive production cross sections within the transport approach. The elementary production cross section which plays a key role, is discussed in subsec.2. Finally, in subsec.3, I present the model's predictions for the K- excitation function as well as some speculations on compression effects.

4.1. The Kaon Inclusive Cross Section

Within the multiple collision approach described in section 2, the integrated inclusive cross section for kaon production in a reaction of two nuclei with mass number a is given by the expression[34]

$$\sigma_K(A,E_0) = \sum_i \sum_j N_i(A)\, N_j(A) \sum_{m=1}^{i} \sum_{n=1}^{j} \sigma^K{}_{mn}(E_0). \tag{7}$$

The quantities $N_i(A)$ are the numbers of nucleons which leave the interaction zone after i binary collisions. They are related to the weight factors of eq.(1) by

$$N_i(A) = \int d^2b\, G_i(\mathbf{b})/\sigma \tag{8}$$

with the NN cross section σ. The weights $N_i(A)$ obey the sum rule

$$\sum_i i\, N_i(A) = A\,. \tag{9}$$

The partial cross sections $\sigma^K_{mn}(E_n)$ are the elementary yields for kaons, folded with the pre-interaction momentum distributions of the two nucleons producing this particle.

Eq.(7) describes the inclusive kaon production rate perturbatively, i.e. it is assumed that the nucleon momentum distributions are unaffected by the creation of the kaon. This restricts the range of validity of (7) to cases where the elementary partial cross section for kaon production is small as compared to the total NN scattering cross section. This condition is satisfied for the energy range of concern here. The most relevant beam energy range for subthreshold kaon production presumably is between roughly 300 and 1000 MeV/nucleon because of its suitability for detecting strong compression effects by the use of kaons.[35,36]

4.2. The Elementary Kaon Production Cross Section

A crucial input for the computation of the inclusive cross section for kaon production is the corresponding elementary cross section $\sigma^K_{elem}(S)$ where S is the invariant energy. I am mainly concerned here with subthreshold kaon production. Hence, the cross sections in the vicinity of the kaon threshold for the elementary process are most crucial.

The simplest procedure is to employ three-particle phase space for kaon production. Near threshold this leads to

$$\sigma_{elem}(S) \propto (p^K_{max})^4 , \tag{10}$$

where p^K_{max} denotes the maximum available momentum of the kaon.

Fig.8 displays how the expression (10), with an appropriately chosen constant, compares with the measured elementary cross sections.

4.3. The Kaon Excitation Function and Compression Effects

Other parametrizations of the elementary kaon production cross section fit the few experimental points equally well, in particular the one of Ref.38 which is almost linear in p^K_{max}. These uncertainties affect the kaon excitation function σ_K obtained from heavy ion collisions. A corresponding calculation is shown in Fig.9.a. The parametrization of Ref.38 leads to the dotted, and eq.(10) to the full curve in this figure. Both curves fit, within the experimental error, the single measured point at 2.1 GeV/nucleon.[39] At low energies, the differences become as large as a factor of about 2.5. Because of the approximate A^2-dependence of σ_K obtained from (7) and (9), in the transport model these differences are of the same size for heavier colliding nuclei.

For heavy systems, compression effects should, however, lead to a measured number of kaons substantially smaller than the calculated one, taken at the same

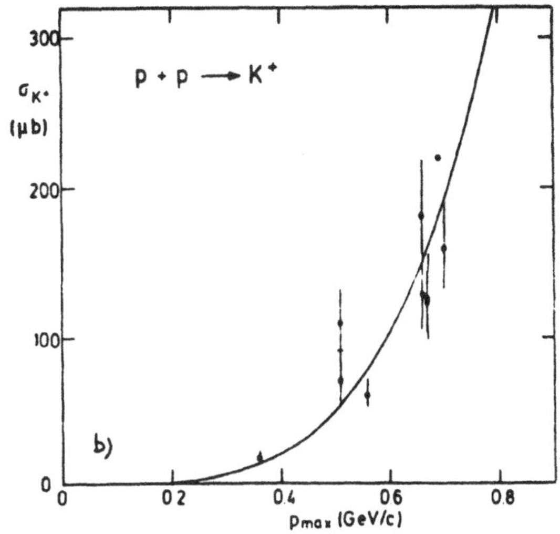

Fig.8. Elementary cross section for kaon production in proton-proton collisions: phase space parametrization (10) compared with the experimental data.[37] From Ref.36.

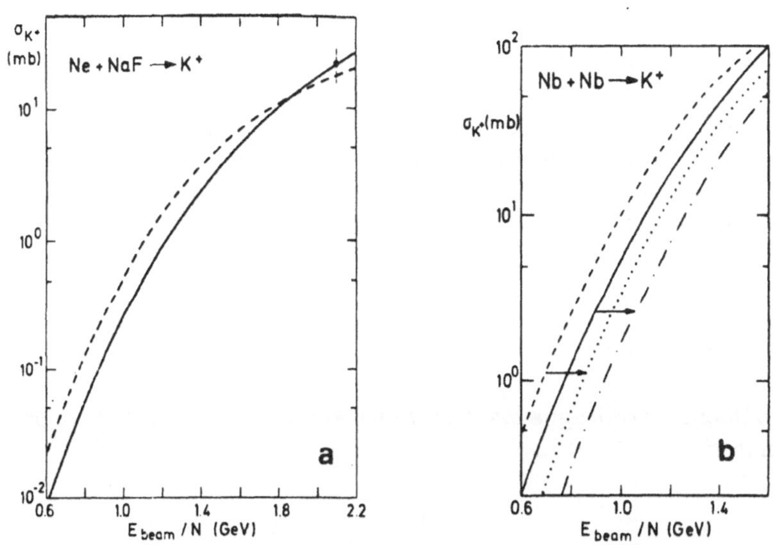

Fig.9. The kaon inclusive yield in dependence on the beam energy per nucleon.(a) For the reaction Ne + NaF, obtained with the parametrizations of Ref.38 and of eq.(10) for the elementary production cross section (dashed and full lines, respectively). (b) For the reaction Nb + Nb. Dashed and full lines: same as in(a); dotted and dashed-dotted lines: excitation functions shifted by 40 MeV/nucleon c.m. energy, indicated by the arrows.

energy. Despite the uncertainties in the calculations, these effects still should be visible. This is indicated in Fig.9.b for a Nb on Nb collision. Using 40 MeV as a rough estimate for the c.m. compressional energy[12], at a beam energy of 800 MeV/nucleon we predict a difference of a factor of 4 which exceeds the uncertainties introduced by the elementary cross section.

To summarize, experimental subthreshold kaon yields may provide a useful complementary tool to probe the compressional part of the nuclear matter equation of state. Precise measurements of the elementary kaon production cross section near the lowest threshold would be of great help.

5. EPILOGUE

In this contribution, I have summarized those aspects of my research work which are closely connected to an analytic transport theory for intermediate energy heavy ion collisions. It is gratifying to see that this approach , whose origins go back to the early days of heavy ion research, still is a useful tool to date. I just mention our predictions for subthreshold kaon production which are used by experimentalists as guide lines for the order of magnitudes to be expected from measurements to be performed.

However, witnessing failures of the model is just as, if not more. gratifying. Its inability to explain the systematics of the experimental central collective flow demonstrates that intermediate energy heavy ion collisions are more than a mere superposition of elementary baryon baryon encounters.

I am confident that the experiments to be performed with the SIS machine at GSI Darmstadt in the near future will continue the success achieved with the Bevalac at Berkeley and bring us a great deal closer to the determination of the equation of state of hot and dense nuclear matter.

ACKNOWLEDGMENT

I wish to thank my colleague Manfred Weick at Siemens for his help in preparing the manuscript.

REFERENCES

1. B. Schürmann, W. Zwermann and R. Malfliet, Phys. Reports 147 (1987) 1.
2. H.J. Pirner and B. Schürmann, Nucl. Phys. A316 (1979) 461.
3. S. Nagamiya et al., Phys. Rev. C24 (1981) 971.
4. R. Glauber and G. Matthiae, Nucl Phys. B21 (1970) 135.
5. B. Schürmann and Don Pil Min, Nucl. Phys. A370 (1981) 496.
6. R. Malfliet and B. Schürmann, Phys. Rev C31 (1985) 1275.

7. R.J. Glauber, "Lectures in Theoretical Physics", Vol.1, Interscience, New York (1959).
8. R. Hagedorn and J. Ranft, Nuovo Cimento Suppl. 6 (1968) 169.
9. J. Cugnon, Phys. Rev. C22 (1980) 1885.
10. J. Cugnon, T. Mizutani and J. Vandermeulen, Nucl. Phys. A352 (1981) 488.
11. Y. Yariv and Z. Fraenkel, Phys. Rev. C20 (1979) 2227.
12. R. Stock et al., Phys. Rev. Lett. 49 (1982) 1236.
13. H.A. Gustafsson et al., Phys. Rev. Lett. 52 (1984) 1590.
14. B. Schürmann and W. Zwermann, Phys. Lett. 158B (1985) 366.
15. W. Zwermann and B. Schürmann, Phys. Rev. C34 (1986) 1318.
16. G. Buchwald et al., Phys. Rev. Lett. 52 (1984) 1594.
17. W. Scheid, H. Müller and W. Greiner, Phys. Rev. Lett. 32 (1974) 741.
18. H. Kruse, B.V. Jacak and H. Stöcker, Phys. Rev. Lett. 54 (1985) 289.
19. P. Danielewicz and G. Odyniec, Phys. Lett. 157B (1985) 289.
20. K.G.R. Doss et al., Phys. Rev. Lett. 57 (1986) 302.
21. B. Schürmann and W. Zwermann, Phys. Rev. Lett. 59 (1987) 2848.
22. J. Aichelin and H. Stöcker, Phys. Lett. 176B (1986) 14.
23. K.H. Kampert, Ph. D. thesis, Univ. Münster (1986).
24. N.L. Balazs et al., Nucl. Phys. A424 (1984) 605.
25. P. Danielewicz, Phys. Lett. 146B (1984) 168.
26. A. Bonasera, L.P. Csernai and B. Schürmann, Nucl. Phys. A476 (1988) 159.
27. A. Rosenhauer, Ph. D. thesis, Univ. Frankfurt (1988).
28. W. Schmidt, Ph. D. thesis, Univ. Frankfurt (1989).
29. B. Schürmann, Mod. Phys. Lett. A3 (1988) 1137.
30. A.R. Bodmer, Proc. Top. Con. on Nuclear Collisions, Fall Creek Falls, Tennessee, Conf-770602 (1977) 309.
31. G. Peilert et al., Phys. Rev. C 39 (1989)1402.
32. B. Schürmann and W. Zwermann, Proc. 7th High Energy Heavy Ion Study, GSI-report 85-10 (1985) 275.
33. B. Schürmann and W. Zwermann, Z. Phys. A330 (1988) 233.
34. W. Zwermann and B. Schürmann, Nucl. Phys. A423 (1984) 525.
35. J. Aichelin and C.M. Ko, Phys. Rev. Lett. 55 (19859 2661.
36. B. Schürmann and W. Zwermann, Phys. Lett. 183B (1987) 31.
37. V. Flaminio et al., CERN-HERA-report 84-01 (1984).
38. J. Randrup and C.M. Ko, Nucl. Phys. A343 (1980) 519; A411 (1983) 537.
39. S. Schnetzer et al., Phys. Rev. Lett. 49 (1982) 989.

Study of Observables from Heavy Ion Data using VUU and QMD Calculations

C. Hartnack, H. Stöcker and W. Greiner

Institut für Theoretische Physik, Johann Wolfgang Goethe–Universität
D-6000 Frankfurt am Main, Germany

Abstract

We give a short overview over some microscopic models including long and short range interactions. The idea of extracting the EOS is explained and its influence on physical observables is discussed. The collective sidewards flow of nuclear matter gives an indication to a repulsive high density region in the center of the reaction. This behaviour was already predicted by the hydrodynamical model. A second prediction was the stopping of the nuclei at central collisions. Here the momentum transfer via nucleon-nucleon collisions plays an important role. A new component of the collective flow is the squeeze out of highly energetic particles perpendicular to the reaction plane. From this observable one may hope to gain a direct view into the hot high density region. All three predictions, transverse flow, nuclear stopping and 90 degree out of plane squeeze are confirmed by experiment. The influence of the nuclear EOS on these observables and other influences like the nucleon-nucleon cross section are discussed. We conclude that the considered microscopic models are able to explain experimental data.

1. Microscopic models for the description of RHIC's

The approach to heavy ion collisions can be done at a macroscopic level as well as in a microscopic one. In a macroscopic model nearly all degrees of freedom are integrated, only the essential ones (e.g. energy, momentum, baryon number) are propagated. In a microscopic model all single particles of a heavy ion collision are propagated and the districution function is represented by the superposition of all single particle distribution functions, which are (like in VUU[1]) δ functions or (like in QMD[2]) Gaussians.
The interactions between the particles are split up into a long range potential part and binary collisions for the short range part. Especially for this part quantum mechanical features like angular distributions of the collisions and the inclusion of

The Nuclear Equation of State, Part A
Edited by W. Greiner and H. Stöcker
Plenum Press, New York

the Pauli principle for the final states are very important.
The first simple descriptions were intranuclear cascades (INC-models) [3]taking only the short range collision term into account. These models are applicable at energies higher than 1 GeV. At lower energies those models have problems in the description of collective effects.
The models described below take care of the three main ingredients into microscopic approaches:

1. the (long range) potential
2. the (short range) collision term
3. the inclusion of Pauli's principle

All three ingredients are found in the Vlasov-Uehling-Uhlenbeck[1] equation, a transport equation for the time development of the single particle distribution function f in phase space:

$$\begin{aligned}\frac{\partial f}{\partial t} &+ \vec{v}\cdot\vec{\nabla}_r f - \vec{\nabla}_r U\cdot\vec{\nabla}_p f = -\int \frac{d^3p_2\, d^3p_1'\, d^3p_2'}{(2\pi)^6}\sigma v_{12} \\ &\cdot \left[f f_2(1-f_1')(1-f_2') - f_1' f_2'(1-f)(1-f_2)\right]\delta^3(p+p_2-p_1'-p_2') \quad .\end{aligned}$$

The VUU equation contains on the l.h.s. a mean field potential which is normally described by a Skyrme interaction $U = \alpha'\varrho + \beta'\varrho^\gamma$. The r.h.s. is the collision term with the Uehling-Uhlenbeck factors $(1 - f_1')$ etc. describing the Pauli blocking for the final states of the collision.
In the VUU equation there are two input parameters which cause changes on the observables of the simulation:

1. For the potential U on the (left) Vlasov part we can use different compression constants for the equation of state and introduce additional potentials (e.g. momentum dependent interactions-MDI).
2. The cross section σ in the (right) Uehling-Uhlenbeck part is assumed to be the free nucleon-nucleon cross section. There are suggestions to reduce its value to take into account in medium-effects [9].

One problem of the VUU approach is that fragment formation and multifragmentation cannot be calculated in a single particle theory. This can only be done in N-body theories with many body correlations like the quantum molecular dynamics model (QMD)[2], which combines the NN-correlation of classical molecular dynamics [4] with quantum features in the representation of its single particle Wigner-density and in the collision term.

Here every nucleon is represented by its Wigner densitiy in phase space corresponding to a boosted Gaussian wave packet.

$$f_i(\vec{r},\vec{p},t) = \frac{1}{\pi\hbar^3}\exp\left\{-(\vec{r}-\vec{r}_{i0}(t))^2\frac{1}{2L}\right\}\exp\left\{-(\vec{p}-\vec{p}_{i0}(t))^2\frac{2L}{\hbar^2}\right\}$$

The centre of gravity of each nucleon is propagated by classical equations of motion with relativistic kinematics. The particels interact via a Skyrme-type, a Yukawa- and a Coulomb- interaction. Also momentum dependent interactions can be used. For the collision term stochastic scattering is done if the distance d of two incorporated nucleons fulfills $d < \sqrt{\sigma/\pi}$. Elastic and inelastic channels of the collisions are available and the Pauli blocking is taken into consideration by regarding the

phase space densities of the final chanels before a collision is allowed.
A disadvantage of QMD is that isospin is not explicitely included as it is done in VUU. A recently developed version of QMD, the so-called IQMD, has successfully done this inclusion. Thus, isospin is taken into account in the collision term (remember the difference of σ_{pp} and σ_{pn} at low energies and the different isospin channels e.g. for the delta decay) as well as in the potential. Charged particles can interact via Coulomb interaction and an asymmetry potential corresponding to the Bethe-Weizsäcker-formula is introduced by $V_{asym} = 25\text{MeV}\ \text{sgn} I_3(\varrho_p - \varrho_n)/\varrho_0$ where I_3 denotes the isospin projection of the particle. Free pions are moving under influence of the Coulomb-forces.
Here are the differences of VUU, QMD and IQMD in one scheme:

	VUU	QMD	IQMD
isospin	yes	no	yes
theory	one body	N body	N body
particles	pointlike	Gaussians	Gaussians
events	parallel	single	single
fragments	no (?)	yes (A)	yes (Z, A)
potential	mean field	NN-correlations	NN-correlations
Coulomb	no	mean	explicit
Yukawa	no	yes	yes
MDI	no	yes	yes
asymmetry pot.	no	no	yes
σ_{NN}	pp,pn	mean	pp,pn
free pions	yes	no	yes
forces on free π	none	no π's	Coulomb

2. What do we expect from the nuclear equation of state?

The idea of the equation of state is the split up of the total energy of nuclear matter into a compressional part, that depends only on the density, and a thermal part with additional dependence on temperature . [5]. For the description of the compressional part commonly the Skyrme ansatz is used

$$U = \alpha \left(\frac{\varrho}{\varrho_0}\right) + \beta \left(\frac{\varrho}{\varrho_0}\right)^\gamma \quad .$$

For the determination of α, β, γ we use the knowledge of the properties of nuclear matter at $\varrho = \varrho_0$, i.e. the ground state energy, the vanishing derivative with respect to the energy and the compression modulus. For our calculations we normally use two parameter sets.

1. a repulsive potential with high compressibility (380 MeV), a so-called hard EOS with $\alpha = -124MeV, \beta = 70.5MeV, \gamma = 2$.

2. a less stiff potential with a lower compressibility (200 MeV), the soft EOS with $\alpha = -356MeV, \beta = 303MeV, \gamma = \frac{7}{6}$.

Both parametrizations yield a different contribution of the compressional part to the total energy at a given density. Nevertheless, one should keep in mind that in calculations with different equations of states different densities are reached. Figure 1 shows on the l.h.s. the multiplicity distribution of the maximal density a nucleon has reached during the whole collision for the reaction Nb + Nb. We see

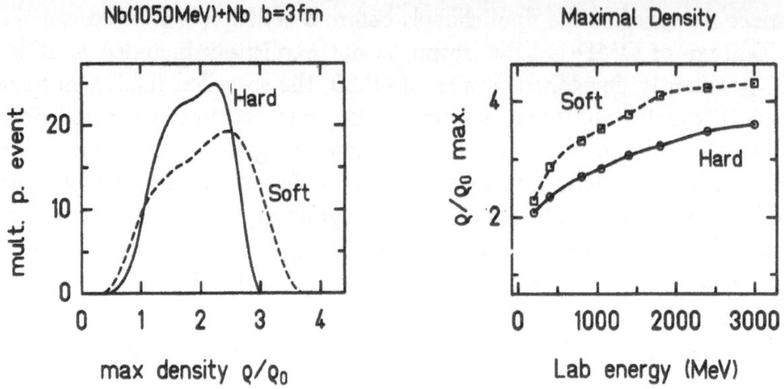

Figure 1. Multiplicity distribution of the maximal density of the nucleons (l.h.s.) and the maximal density of the collision as a function of time (r.h.s.), both for a hard and soft EOS

that with a soft EOS higher densities are reached than with a hard EOS since the repulsion of the compressed matter is smaller. We also see that not all particles are reaching maximal densities. On the r.h.s. the maximal density reached during the whole reaction is plotted as a function of the incident energy for several systems for a hard and a soft EOS.

The values of ϱ_{max} for a hard and a soft EOS at 400 and 800 MeV are taken and marked on the l.h.s. of Figure 2 showing the compressional energy for a hard and a soft EOS. Now the difference in compressional energy between hard and soft EOS becomes very small. The repulsion of the potential on the r.h.s. of Figure 2 indeed shows still a visible difference between both equations of state. Therefore we assume observables related to the repulsion (e.g. collective flow and squeeze out) to be sensitive to the nuclear EOS.

Now it is interesting to know whether the significant flow originates from particles of this high compression region or more from spectator matter. The l.h.s. of Figure 3 shows the rapidity distribution of the nucleons of the reactions Nb (1050 MeV) + Nb selected according to their maximal density. Particles from the high compression region (full line) are stopped to CM-rapidity, whereas particles from the low compression regions (dashed) still carry the rapidities of projectile and target. Particles from the intermediate region show intermediate behaviour. From the right column of Figure 3 we extract that the transverse flow p_X is highest for particles from low density regiions (dashed) whereas it is smaller for particles from the intermediate region (dash-dotted) and nearly vanishes for the particles from the high density region (full line). These particles show a higher tendency to isotropy. The transverse momentum seems more to be more sensitive the density gradient of the boundary of the high compression region than to the compression region itself [15].

This corresponds to the behaviour of fragments. Large fragments that steem from the spectator matter also are at rapidities near projectile and target rapidity and show the highest flow in the reaction. This has been found as well by experimental data as by QMD-calculations. [8,10,11]

3. Collective flow

As already stated the average momentum transfer $p_X(Y)$ is regarded to be a signal of the repulsion of the compressed nuclear matter. By construction the analysis of

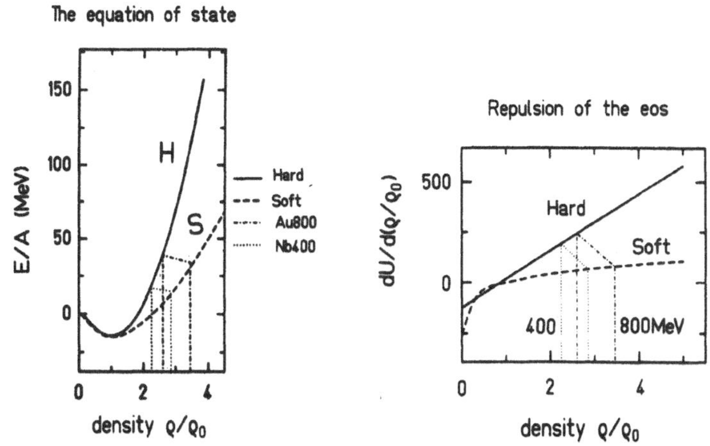

Figure 2. The compressional energy (left) and the repulsion of the potential (right) as a function of density for a hard (full line) and a soft (dashed) EOS.

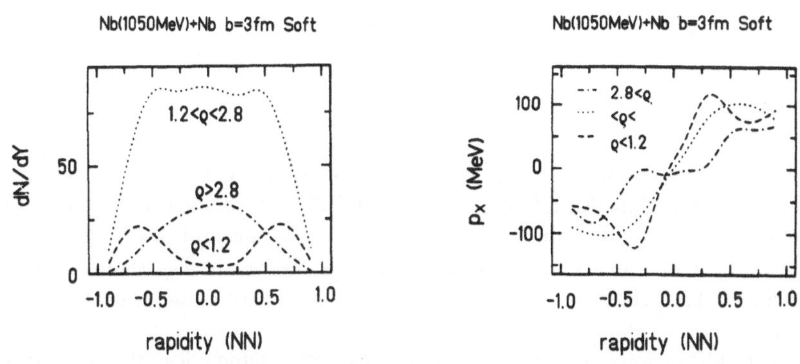

Figure 3. Rapidity distribution (left) and transverse flow $p_X(Y)$ (right) selected according to particles, that reached high (dash-dotted), medium (dotted) and low (dashed) maximal density during the reaction.

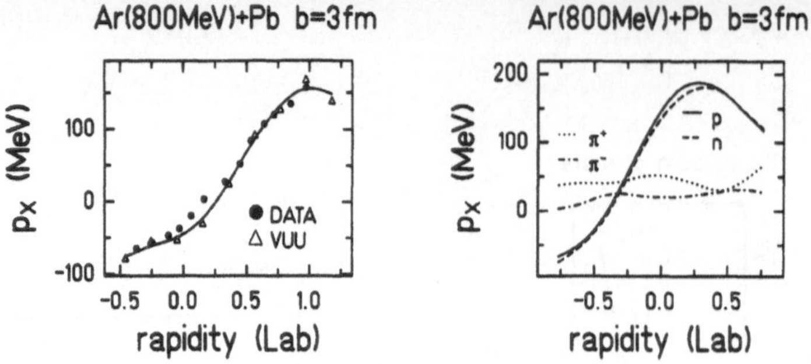

Figure 4. Transverse flow for the system Ar(800MeV)+Pb. On the left column a comparison between Streamerchamber data (bullets) and VUU calculations (triangles, full line). On the right part a IQMD calculation with distinction between protons and neutrons and between the charged pions.

transverse momentum fails at most central collisions because of symmetry. For a symmetric system the mean transverse flow at the target rapidity is up to a minus sign the same than at projectile rapidity. For asymmetric systems the transverse flow at the rapidity of the (light) projectile is much higher than the absolute value at the rapidity of the (heavy) target due to momentum conservation. On the l.h.s. of Figure 4 we see the transverse flow as a function of rapidity for the system argon on lead at 800 MeV. Note the good agreement of data taken by the streamer chamber collaboration (bullets) and VUU calculations (triangles) using a hard EOS. On the r.h.s. we see a IQMD-calculation with an analysis of the importance of the isospin. Protons (full line) and neutrons (dashed) show a very similar behaviour so that it is no mistake to carry out the analysis with all nucleons instead of protons only. Hence calculations of QMD without isospin remain trustful under this view. Further we see a slightly different behaviour of negative (dash-dotted) and positive (dotted) pions. In asymmetric systems pions show an overall poistive p_X; this has also been found for negative pions by the measurements of the streamer chamber group. Measurements of the Diogéne group confirmed the higher positive p_X of positive pions in comparison to negative pions for another asymmetric system, Ne (800 MeV) on Pb - in qualitative agreement to IQMD calculations. Now let us return to symmetric systems. The l.h.s. of Figure 5 shows VUU calculations for different systems at 800 MeV incident energy using a hard EOS. Heavier systems show a larger transvers momentum than smaller ones. For 100 MeV incident energy the flow becomes negative for small systems like neon and calcium. The transition from positive to negative flow changes by the mass number.

For further calculations we introduce the quantity p_X^{dir} , which describes the mean transverse momentum per particle at the positive part of the $p_X(Y)$ distribution.

$$p_X^{dir} = < p_X \operatorname{sgn}(Y - Y_{CM}) >$$

Figure 6 shows on the l.h.s. the relation between p_X^{dir} and $p_X(Y)$ taken at projectile rapidity for the reaction Nb (400 MeV) + Nb. Throughout the whole range of the impact parameter p_X^{dir} is smaller but the differences decrease for larger impact parameters. This corresponds to the broader rapidity distribution in peripheral collisions. For symmetry reasons both values vanish to zero for

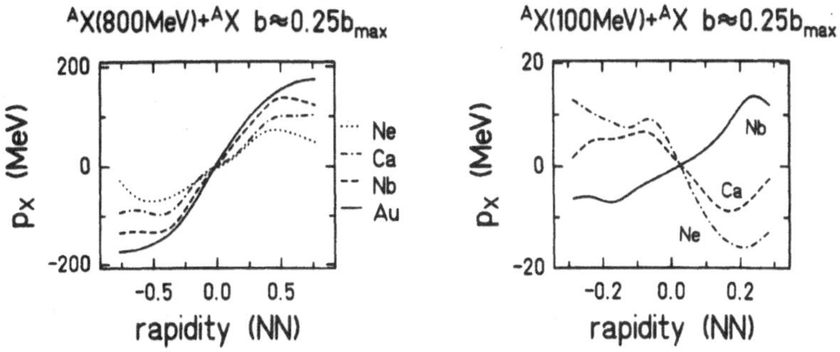

Figure 5. Transverse flow $p_X(Y)$ as a function of rapidity for different systems at 800 MeV (left column) and 100 MeV (right column) incident energy.

$b = 0$, for large impact parameters they also decrease to zero and get maximal for semicentral collisions at $b \approx 0.2 - 0.3b_{max}$. The r.h.s. shows the impact parameter dependence of p_X^{dir} for a light systems at lower energies of 140 MeV (dashed) and 100 MeV (full line). For very small and very large impact parameters p_X^{dir} also vanishes to zero but now it has as well positive values for small impact parameters and negative values for large ones. For these low energies the transition from positive to negative p_X^{dir} happens at smaller impact parameters, also the values themselves are becoming smaller.

For lower energies the repulsive Coulomb forces play a more important role. In Figure 7 we see on the l.h.s. the IQMD calculation of the dependence of p_X^{dir} on the impact parameter b for calculations with (full line) and without (dashed) Coulomb interaction. The repulsive Coulomb interaction shifts the transition to higher energies and enlarges p_X^{dir} especially for larger impact parameters.
The r.h.s. of Figure 7 shows the energy dependence of p_X^{dir} for a small ($b = 2fm$, full) and a rather large ($b = 5fm$, dashed) impact parameter for the reaction Ca + Ca using a hard EOS and Coulomb interactions. The onset of the positive flow happens at different energies. Thus, it is not clear to define a unique transition point for the onset of flow.

Figure 8 shows on the l.h.s. the energy dependence of p_X^{dir} scaled with the incident momentum per nucleon in the CM-frame p_P for Nb + Nb for fixed impact parameter $b = 2$ calculated with VUU. We see that there is a plateau at about 250-1000 MeV where the scaled momentum stays constant and decreases for higher and lower energies. The hard EOS (full line) yields higher p_X^{dir} than the soft one (dashed). Also the onset of flow happens at slightly different energies. But the difference is small compared to the influence of the impact parameter. On the r.h.s. we see the energy dependence of the flow angle Θ_{flow}. It stays roughly constant at about 250-1000 MeV in correspondence to the scaled p_X^{dir} . The ratio of longitudinal to transverse momentum is constant. At the energies of the onset of flow the angle has a minimum and increases rapidly at smaller energies. This corresponds to the positive definition of the flow angle.

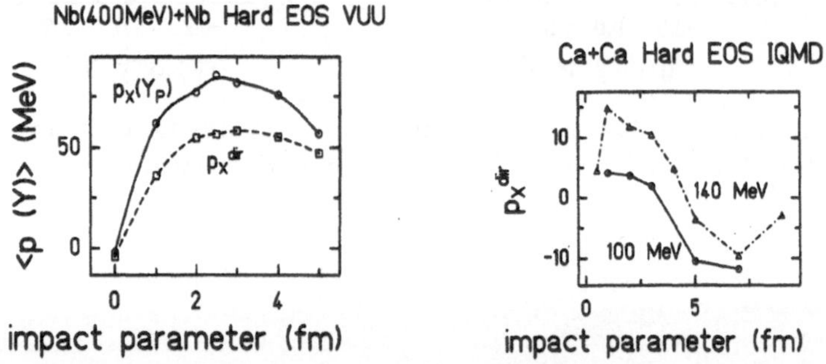

Figure 6. Dependence of p_X^{dir} on the impact parameter. On the left column $p_X(Y_{Proj.})$ (full line) and p_X^{dir} (dashed) for the system Nb+Nb at 400 MeV, the right column shows p_X^{dir} for the system Ca+Ca at 100 MeV (full line) and 140 MeV (dashed).

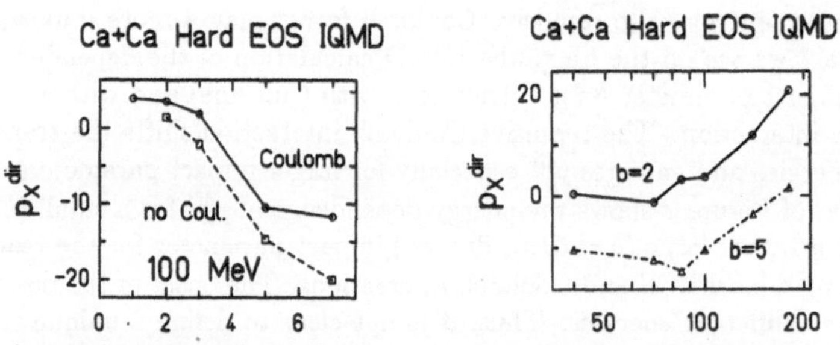

Figure 7. p_X^{dir} as a function of the impact parameter for Ca+Ca at 100 MeV for a calculation with (full line) and without (dashed) Coulomb interactions (left). The energy dependence of p_X^{dir} for different impact parameters is shown on the right side.

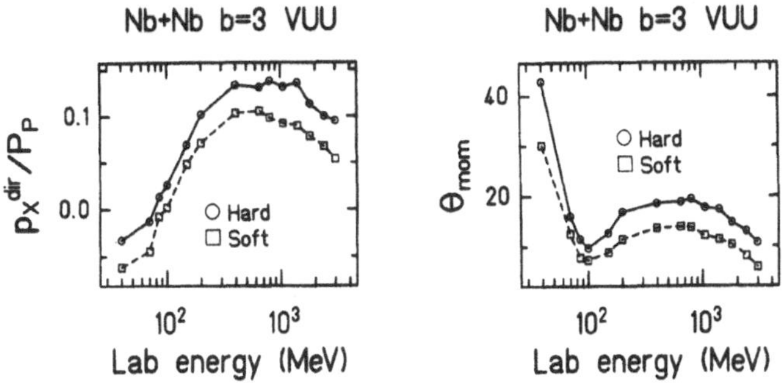

Figure 8. Energy dependence of p_X^{dir} scaled with p_P (left column) and of the flow angle Θ_{flow} (right column) for a hard (full line) and a soft (dashed) EOS.

4. Nuclear stopping and rapidity distributions

In the discussion of the flow one parameter has not been mentioned, namely the influence of the cross section. The cross section used in the simulation is the free nucleon-nucleon cross section with an additional Pauli blocking mechanism. A reduction of the cross section decreases the collective flow, an enlarged cross section enlarges the flow. Thus, a calculation with a hard EOS plus reduced cross section can yield similar results than a soft EOS with full cross section as it can be seen on the l.h.s. of Figure 9 , where the reaction of Nb (400 MeV) + Nb at $b = 3fm$ is plotted for a hard (full line) and a soft (dashed) EOS both with full cross section and for a hard EOS with a cross section reduced by a factor of 0.7. The r.h.s. shows the corresponding rapidity distributions. A hard and a soft EOS with full cross section yield nearly similar rapidity distributions, whereas for the calculation with reduced cross section we see a slight dip at mid rapidity. The change of the used cross section causes a change in the rapidity distribution.

A reduced cross section enlarges the mean free path $\lambda = \frac{1}{\sigma \cdot \varrho}$ and thus leads to less stopping while an enlarged cross section increases the stopping. The l.h.s. of Figure 10 illustrates this effect. An enlarged cross section (dashed line) causes a smaller width and higher maximal value of the rapidity distribution than the normal free cross section (full line). A reduced cross section yields a broader distribution with a clip at CM-rapidity. This effect could be used to extract the cross section by means of the rapidity distribution. Comparison of VUU-calculations and experimental data performed by D. Keane in 1987 gave indications for the normal unreduced cross section. There are additional influences which have to be taken into account. The r.h.s. of Figure 10 shows the rapidity distribution for the same system (Nb (400 MeV) + Nb) with a hard EOS for different impact parameters. A very central collision (full line) yields very much stopping, while the rapidity distribution gets two peaks near projectile and target rapidity for peripheral collisions. Experimental data of the Plastic Ball collaboration show a similar

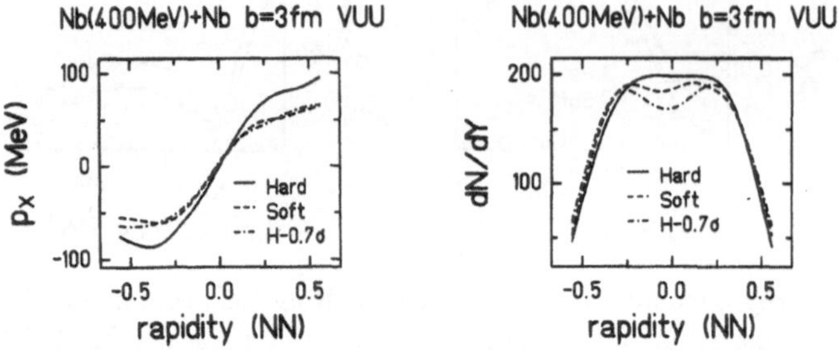

Figure 9. Transverse flow (left) and rapidity distribution (right) for the system Nb(400MeV)+Nb calculated with a hard (full line) and a soft (dashed) EOS and full cross section and with a hard EOS with a reduced cross section.

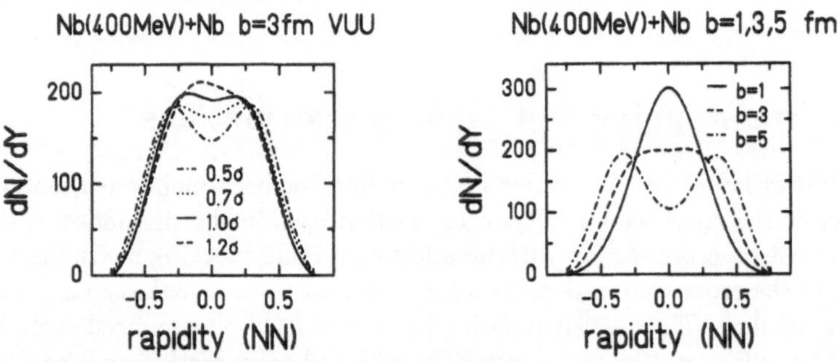

Figure 10. Dependence of the rapidity distribution on the cross section (left) and on the impact parameter (right).

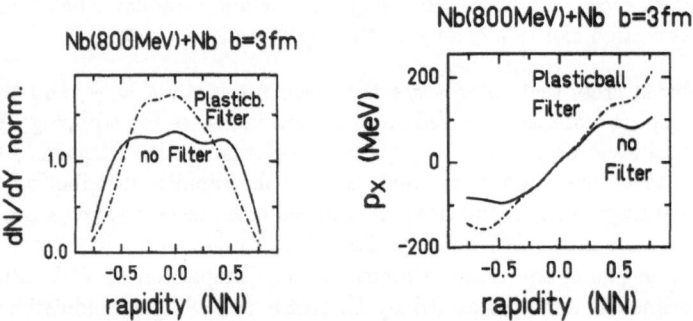

Figure 11. The effect of the detector efficiencies on the rapidity distribution (left column) and the transverse flow (right column)

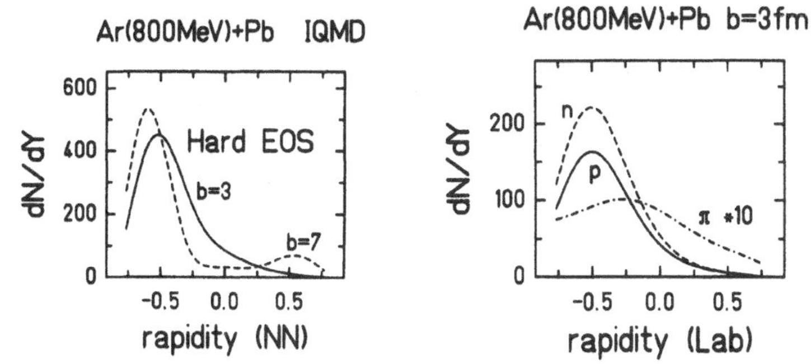

Figure 12. Rapidity distribution for an asymmetric system. The left column shows the difference between central and peripheral collisions, the right column the rapidity distribution of protons, neutrons and pions for a central collision.

behaviour concerning stopping at central collisions and less stopping in more peripheral collisions. For a distinguished quantitative discussion one has to keep in mind that the experiment selects data according to multiplicity. A relation beween multiplicity and impact parameter can only be obtained within some limits, thus comparison between the data of a multiplicity bin with calculations using a fixed impact parameter seem to be doubtful for observables depending sensitively on the impact parameter.

Further on, the experimental efficiency cuts influence the observables. Figure 11 shows l.h.s. the rapidity distribution and r.h.s. the transverse flow p_X for calculations without efficency cuts and with a simple Plastic Ball filter. The experimental efficency cuts simulate a stronger stopping and higher transverse momenta.

Now let us turn back to the analysis of IQMD-simulations without experimental filter. For asymmetric systems we see again on the l.h.s. of Figure 12 the stopping for central collisions and less stopping for peripheral collisions. In the right column we can see that protons and neutrons have similar rapidity distributions. For the pions the rapidity distribution is more shifted to midrapidity. The pions are produced in the first high energetic collisions, thus the shift towards midrapidity ($Y = 0$ in the NN system) can be understood.

Finally, we return to symmetric systems. Figure 13 shows on the l.h.s. the normalized rapidity distribution versus $Y/Y_{proj.}$. We see that the stopping is nearly independent on the energy in the range about 200 MeV to 800 MeV. The mass dependence of the stopping is shown on the r.h.s. of Figure 13 . Large systems like gold (full line) show more stopping than small ones like calcium (dash-dotted) which corresponds to the different ratio of the mean free path λ to the size of the nucleus. Similar results are obtained from the experiments of the Plastic Ball group. A comparison can be taken from the contribution of G. Peilert to this proceeding.

5. 90 degree squeeze out

Now let us focus on the behaviour of the particles stopped at CM-rapidity. Theoretical predictions for a squeeze out of high energetic particles in central collisions are based on the idea of the formation of shock waves in nuclear matter formed in

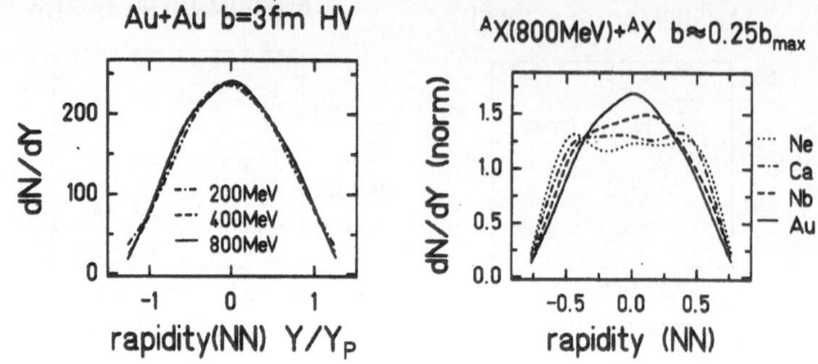

Figure 13. Dependence of the rapidity distribution on the energy (left) and on the mass number of the system (right).

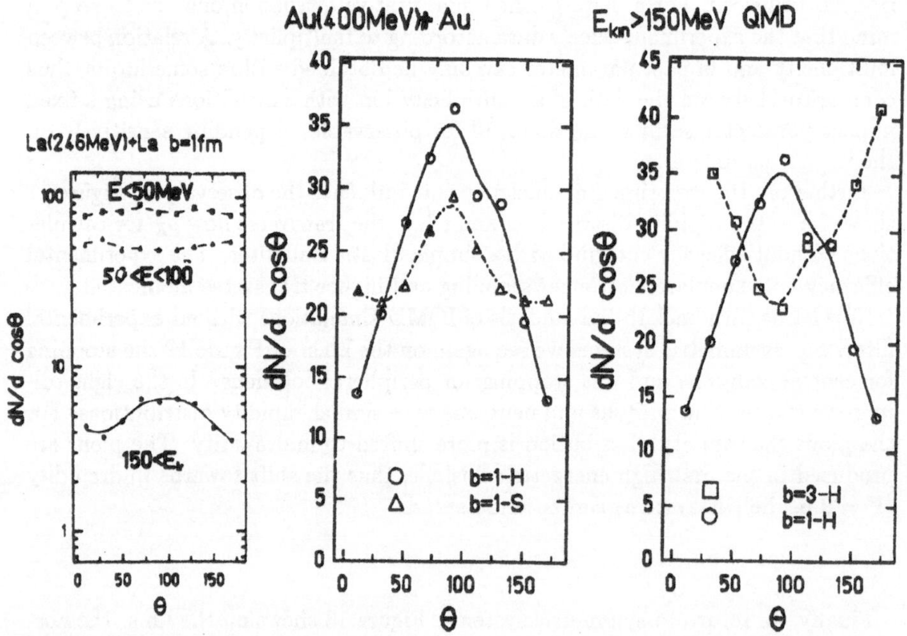

Figure 14. Theta-distribution for the system La(246MeV)+La selected to the kinetic energy of the particles (left column) and for high energetic particles for the system Au(400MeV) on Au. In the middle column for very central collisions with a hard (full line) and a soft (dashed) EOS, right column calculation with a hard EOs for $b = 1$fm (full line) and $b = 3$fm (dashed).

heavy ion collisions. In hydrodynamical simulations [6] nuclear matter is pressed out perpendicular to the beam axis (for central collisions) and - for semicentral collisions - perpendicular to the reaction plane [12].

First we regard the θ-distribution of the particle. In the following θ denotes the lateral CM angle for each particle, it should not be misinterpreted with the distribution of the flow angle Θ_{flow} obtained from the event-by-event analysis. Figure 14 shows on the left part the lateral distribution for the system La(246 MeV)+La for a very central collision $b = 1fm$. The particles are selected according to their kinetic energy in the CM-frame. Low energetic particles ($E < 50MeV$ dashed line) and particles of medium kinetic energy ($50 < E < 150MeV$ dash-dotted) show no significance. High energetic particles ($E > 150MeV$ full line) show a 70-80% enhancement at 90 degree lateral angle. This means that high energetic particles are pressed out perpendicular to the beam axis. Recent experiments with a strong multiplicity selection taken by Claesson et al. yielded analogous results.[13]
For the qualitative discussion let us consider the reaction Au(400MeV)+Au. The middle part of Figure 14 shows the lateral distribution for very central collisions ($b = 1fm$) using a hard (full line) and a soft (dashed line) EOS. Again we find a clear peak for the high energetic particles. For a hard EOS the peak is stronger corresponding to the higher repulsion.The right part shows the dependence on the impact parameter for a calculation with a hard EOS. For a very central collision ($b = 1fm$ full line) there is a clear maximum at 90 degree which changes into a clear minimum for semicentral collisions ($b = 3fm$ dashed line) [11,15]. This effect is extremely sensitive to the impact parameter and therefore it is not clear whether a quantitative comparison with data, e.g. those from the Plastic Ball, can be done using fixed impact parameters.

Now consider semicentral collisions ($b = 3fm$) and look on the azimuthal distributions at CM-rapidity. The azimuthal angle φ is the inclination angle between the $\vec{p}_T$ vector and the X-axis. 0 degree denotes the X-axis, 180 degree the negative X-axis and 90 resp. 270 (-90) degree the out of plane (Y) direction.

The left part of Figure 15 shows the azimuthal distribution for a VUU-calculation of Au(400MeV)+Au at $b = 3fm$ in the unrotated (normal) system (full line) and in a system rotated by the flow angle Θ_{flow}, i.e. in the eigensystem of the momentum ellipsoid (dashed line). The maximum at 90 degree increases in the rotated system, the particle flow out of the reaction plane is clearer in the rotated system. In the middle part of Figure 15 we see that also the average tansverse momentum per particle (full line) has also a maximum at 90 degree. If we combine this with the multiplicity distribution, i.e. if we calculate the total transverse momentum per azimuthal bin (dashed line), this signal becomes very clear. For the total transverse energy it becomes even larger.

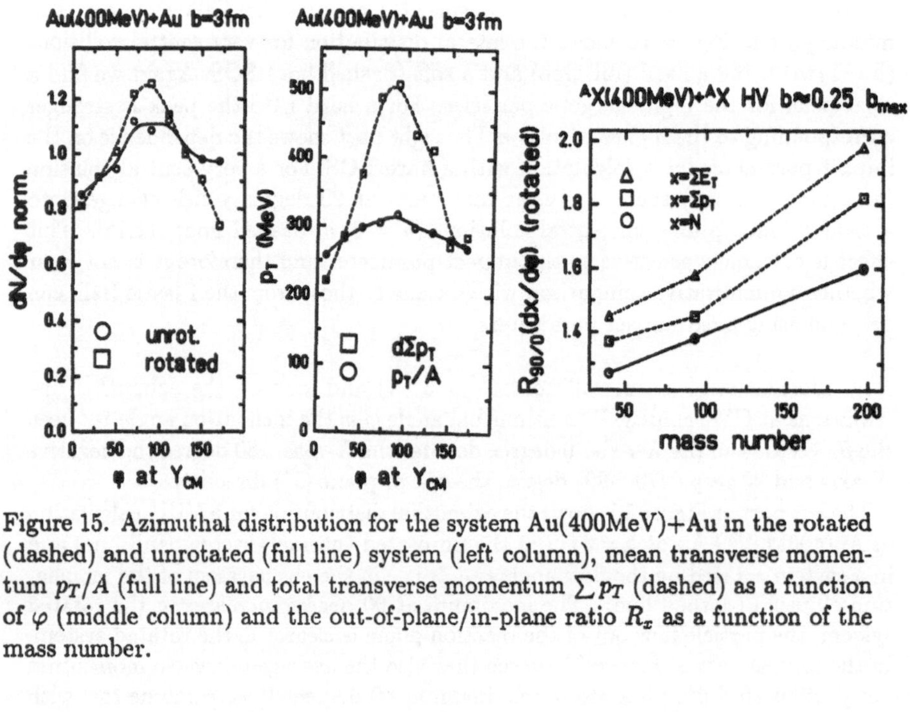

Figure 15. Azimuthal distribution for the system Au(400MeV)+Au in the rotated (dashed) and unrotated (full line) system (left column), mean transverse momentum p_T/A (full line) and total transverse momentum $\sum p_T$ (dashed) as a function of φ (middle column) and the out-of-plane/in-plane ratio R_x as a function of the mass number.

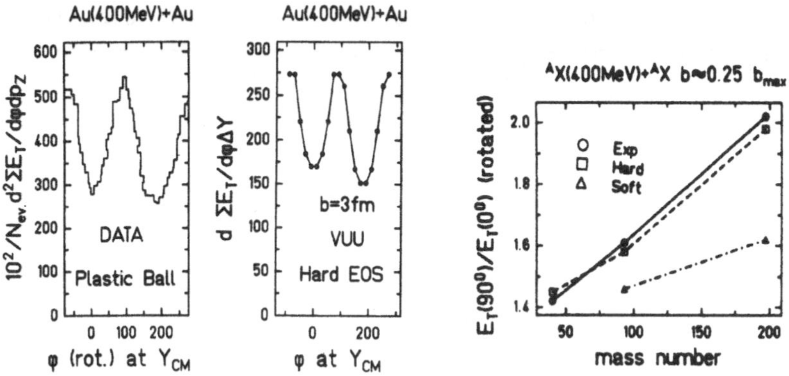

Figure 16. Azimuthal distribution of the total transverse energy $\sum E_T$. The left column shows experimental data of the Plastic Ball group, the middle column shows a VUU calculation using a hard EOS. In the right column the mass dependence of R_E for calculations with a hard (dashed line) and a soft (dash-dotted) EOS is compared to experimental Plasic Ball data (full line).

The peak can now be described by the ratio

$$R_x = \left. \frac{\frac{dx}{d\varphi}(\varphi = 90^0, 270^0)}{\frac{dx}{d\varphi}(\varphi = 0^0, 180^0)} \right|_{Y=Y(CM)} \qquad x = N, P = \sum p_T, E = \sum E_T$$

The right part of Figure 15 describes the dependence of the ratios of the multiplicity N (dashed line), the total transverse momentum (dash-dotted) and the total transverse energy (full line) of the mass number the rotated system. We see a nearly linear dependence.

For Figure 16 we used a simple Plastic Ball filter for the comparison. Left the azimuthal distribution of the total transverse energy measured by the Plasticball group (Mul 3) [14] is plotted. In the middle we see a VUU-calculation for $b = 3fm$ using a hard EOS. Besides the different scaling, the data and the calculation yield similar results. In the right column we see the mass dependence of the ratio R_E of the total transverse energy E_T measured by the Plasticball group (full line) in comparison with VUU-data using hard and soft EOS,respectively. The hard EOS seem appropriate to reproduce the data than the soft EOS. First preliminary calculations using MDI show enhanced ratios of R_E, but nevertheless the soft EOS with MDI seems not to reach these high values.

Finally, a prediction for 800 MeV using VUU is given in Figure 17 [11,15]. For 800 MeV we expect as well a visible squeeze out of plane with a stronger signal from a hard EOS (full line) than from a soft one (left). The dependence on the cross section is not so large but still visible (mid). An enlarged cross section (dash-dotted) enlarges the squeeze, a reduced cross section (dashed line) reduces it in comparison to a normal free cross section (full line). A comparison (right) between VUU (full line) and IQMD (dashed line) shows similar results. The azimuthal distribution does not yet solve the problem of extracting the nuclear EOS, but a correct description is an additional constraint for a simulation of heavy ion collisions.

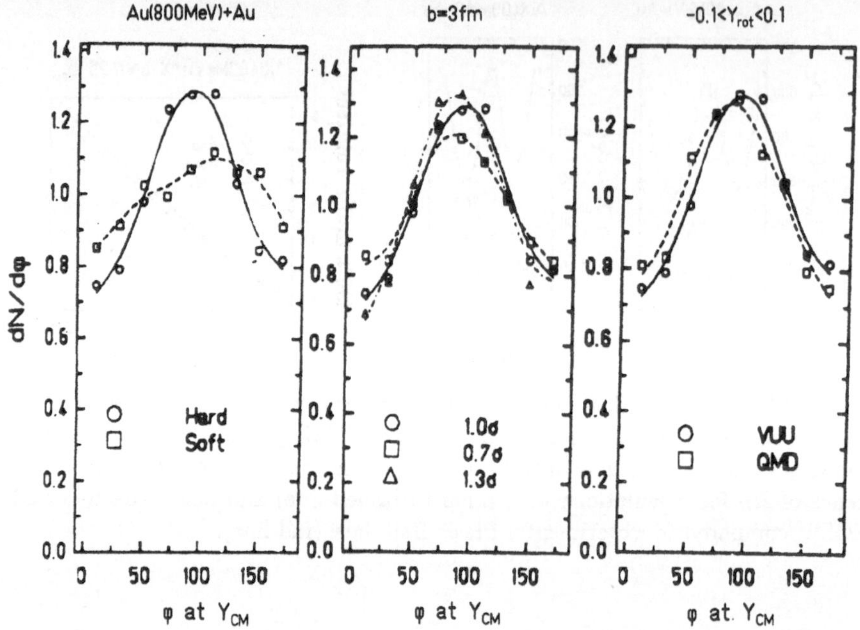

Figure 17. Azimuthal distributions for the reaction Au(800MeV)+Au. On the left side a comparison between hard (full line) and soft (dashed) EOS, in the middle comparison of different cross sections and on the right side a coparison between VUU and QMD both using a hard EOS.

6. Conclusion

The presented simulation models VUU and (I)QMD can fairly well explain most of the observables. For the flow we find a dependence on the EOS and the cross section. The stopping depends on the cross section only. The squeeze out depends as well on the EOS as and on the cross section. Protons and neutrons show similar behaviour. In all cases we find a general dependences on the impact parameter. Therefore, a further step is to perform inclusive calculations with random impact parameter, filter them with the experimental efficencies and sort them according to multiplicity biases. There is still a lot of work to do till we may be able to pin down the nuclear EOS explicitly.

References

[1] H. Kruse et al., Phys.Rev. C31 (1985) 1770;
G.F. Bertsch, H. Kruse and S.Das Gupta, Phys. Rev. C29 (1984) 673;
J. Aichelin and H. Stöcker, Phys. Lett. 163 (1985) 59;
C. Gale, G.F. Bertsch and S.Das Gupta, Phys. Rev. C35 (1987) 1666;
C. Gregoire, B. Remaud, F. Sebille and L. Vinet, Nucl. Phys. A465 (1987) 317.

[2] J. Aichelin and H. Stöcker, Phys. Lett. B176(1986) 14;
A. Rosenhauer et al., J. Physique C4 (1986) 395;
J.Aichelin et al., Phys. Rev. Lett. 58 (1987) 1926;
G. Peilert et al., Mod. Phys. Lett. A3 (1988) 459. and Phys. Rev C39 (1989) 1402;
J.Aichelin et al., Phys. Rev. C37 (1988) 2451.

[3] Y. Yariv and Z.Fränkel, Phys. Rev. C20 (1979) 2227 and Phys. Rev. C24 (1981) 488;
J. Cugnon, Phys. Rev C22 (1980) 1885.

[4] A.R. Bodmer and C.N. Panos, Phys. Rev. C15 (1977) 1342, and Phys. Rev C22 (1980) 1023;
S.M. Kiselev and Y.E. Prokovskil, Sov. J. Nucl. Phys. 38(1) (1983) 46;
J.J. Molitoris et al., Phys. Rev. Lett 53 (1984) 899.

[5] H. Stöcker, J.A. Maruhn and W. Greiner, Phys. Rev. Lett 44 (1980) 725;
H. Stöcker et al., Phys. Rev. Lett 47 (1981) 1807;
H. Stöcker et al., Phys. Rev. C25 (1982) 1873;
G. Buchwald et al., Phys. Rev. Lett 52 (1984) 1594.
H. Stöcker and W. Greiner, Phys. Rep. 137 (1986).

[6] W. Scheid, H. Müller and W. Greiner, Phys. Rev. Lett. 32 (1974) 741;
C.Y. Wong and T. Welton, Phys. Lett B49 (1974) 243;
A.A. Amsden, G.F. Bertsch, F.H.Harlow and J.R. Nix, Phys. Lett. 35 (1975) 905;
H.G. Baumgardt et al., Z. Phys. A273 (1975) 359.

[7] J. Aichelin et al., Phys. Lett B224 (1989) 34.

[8] K.H.Kampert, Thesis, University of Münster (1986), J. Phys. G. (1989)

[9] B.ter Haar, R. Malfliet and W. Botermans, Phys. Rep. 149 (1987) 207 and Phys. Lett. 172B (1986) 10.

[10] G. Peilert diploma thesis University Frankfurt (1988).

[11] M. Berenguer et al., Proceedings of the winter school on nuclear physics in Les Houches, France 1989.

[12] W. Schmidt, thesis, University of Frankfurt (1989).

[13] C. Claesson et al. subm. to Phys. Rev. Lett.

[14] H. Gutbrod et al. Phys. Lett. 216B (1989) 267.

[15] Ch. Hartnack diploma thesis University Frankfurt (1989).

MEDIUM EFFECTS AND NUCLEAR TRANSPORT THEORY

Joseph Cugnon

University of Liège
Institut de Physique au Sart Tilman, B.5, B-4000 LIEGE 1
(Belgium)

1. INTRODUCTION

It is commonly believed nowadays that an acceptable nuclear transport theory for heavy ion collisions in a very broad energy range (from a few MeV/u to several GeV/u) should include a correct description of : (1) mean field effects ; (2) two-body collisions ; (3) particle production. A complete theory embodying these features does not exist yet. However, considerable progress has been made in the recent years. On the theoretical side, it was recognized that the Landau-Vlassov equation phenomenologically introduced by Landau for Fermi fluids, had many attributes of a good nuclear theory (for a review, see Ref. 1), despite the fact that it appears as a truncation[2] of the BBGKY hierarchy of the density matrices[3], which is the exact theory in a nonrelativistic approach with potential forces. On the practical side, a variety of models (BUU, VUU, QMD, quasi-particle methods,...) aiming at the solution of the Landau-Vlassov equation, give more and more satisfactory descriptions of the experimental data (for a review, see Ref. 4). The appealing feature of the Landau-Vlassov theory is the presence in the mean field term of physical quantities directly related to the equation of state. Therefore it is a priori a suitable tool to investigate the nuclear matter equation of state at several times normal nuclear matter density. However, the equation of state is not the only ingredient of the Landau-Vlassov equation : collision cross-sections are also needed. At the beginning free space cross-sections were used, despite the fact that for a long time corrections were introduced in studies of p-nucleus collisions.[5] In fact, baryons (and hadrons) do not interact in the nuclear medium as in free space. The simplest way of accounting for that consists in introducing effective cross-sections, which incorporate medium effects. By this term,

The Nuclear Equation of State, Part A
Edited by W. Greiner and H. Stöcker
Plenum Press, New York

we here mean genuine medium effects which change the interaction, and which do not merely originate from Pauli blocking of final states, something which is trivially (if not satisfactorily) included now in all numerical simulations. Our purpose is to discuss these medium effects and to study the sensitivity of the results upon them. We will be interested in medium effects occurring in elastic scattering as well as in production processes. We will say a few words on medium effects influencing other quantities than the cross-sections.

2. THE TRANSPORT THEORY

As we said in the introduction, a suitable formulation of the nuclear transport theory has the structure of the Landau-Vlassov equation

$$\{\frac{\partial}{\partial t} + \frac{1}{m^*}\vec{p}.\vec{\nabla} - (\vec{\nabla}U).\vec{\nabla}_p\}f_1(\vec{r},\vec{p},t)$$

$$= \int \frac{d^3p_2}{(2\pi)^3}\frac{d^3p_3}{(2\pi)^3}\frac{d^3p_4}{(2\pi)^3}[W(\vec{p}_3\vec{p}_4 \to \vec{p}\vec{p}_2)\tilde{f}_3\tilde{f}_4(1-\tilde{f})(1-\tilde{f}_2)$$

$$-W(\vec{p}\vec{p}_2 \to \vec{p}_3\vec{p}_4)\tilde{f}\tilde{f}_2(1-\tilde{f}_3)(1-\tilde{f}_4)]\delta(\vec{p}+\vec{p}_2-\vec{p}_3-\vec{p}_4)\delta(e(\vec{p})+e(\vec{p}_2)-e(\vec{p}_3)-e(\vec{p}_4)) \quad , \tag{2.1}$$

where f_1 is the Wigner representation of the one-body density matrix, U is the average field, m* is the effective mass and e(p) is the single particle density. We used the notation

$$\tilde{f}_i = f_1(\vec{r},\vec{p}_i,t) \quad . \tag{2.2}$$

The structure of this equation implies that some parts of physics are somewhere neglected : relativity, some aspects of the quantum motion, retardation effects, three-body forces, two-body correlations,.. Nevertheless, the remaining physics is sufficient for the description of many aspects of heavy ion collisions (except of course those which are related to many-body distribution functions), as testified by the success of the actual calculations.

At the beginning, the transition probability W was taken as in free space, i.e. related to the experimental cross-sections :

$$W(\vec{p}_i\vec{p}_j \to \vec{p}_k\vec{p}_\ell) = v_{ij}\frac{d\sigma_{ij \to k\ell}}{d\Omega} \quad , \tag{2.3}$$

where v_{ij} is the relative velocity in the entrance channel. Although this approach was successful, at least at high energy, it was proposed to improve the situation by replacing free cross-sections by medium cross-sections, as calculated in local Brueckner approximation.[6-8]

This procedure was justified later by Botermans and Malfliet[7], who derived eq. (2.1) from BBGKY including properly short range two-body correlations by summing particle-particle ladder diagrams, establishing so the link between the Landau-Vlassov transport theory and the usual (perturbative) theory for static properties of nuclear matter.

The medium corrections to particle production is a much harder problem for which one has only embryonic theories[10,11], although some aspects are very much documented for hadron propagation in equilibrium nuclear matter.[11,12] We will discuss some of these aspects but mainly concentrate on the sensitivity of the results on medium corrections to some ingredients entering the description of particle production.

3. MEDIUM EFFECTS IN ELASTIC SCATTERING

If one adopts as starting point the Brueckner approach in the local density approximation, one should replace the free cross-sections by

$$\frac{d\sigma}{d\Omega}(\vec{k},\vec{k}_2) = \frac{m^2}{4\pi^2}\,|\langle\vec{k}_1\vec{k}_2|G(\rho,T)|\vec{k}_3\vec{k}_4\rangle|^2 \quad , \qquad (3.1)$$

where $\vec{k}_3$ points in the angle $d\Omega$ and corresponds to elastic scattering and where G is the Brueckner matrix at the local density ρ and temperature T. Of course, this approximation is better and better for smaller and smaller deviations from local equilibrium. In general, one is interested to have an expression which more or less averages over the struck particle (k_2) :

$$\frac{d\sigma}{d\Omega}(\vec{k}) = \frac{m^2}{4\pi^2}\int\frac{d^3k_2}{(2\pi)^3}\,n(\vec{k}_2)|\langle\vec{k}_1\vec{k}_2|G(\rho,T)|\vec{k}_3\vec{k}_4\rangle|^2 \quad , \qquad (3.2)$$

where $n(\vec{k}_2)$ is the equilibrium occupation number at density ρ and temperature T. Such quantities have been calculated by several authors[13,14]. These quantities are still hard to handle since they depend upon four variables (k,Θ,ρ,T). In Ref. 15, a simplification was proposed, which averages the correction factor over the final phase space available. One so introduces the factor

$$\alpha_m = \frac{\bar{\sigma}_{med}}{\sigma_{free}} = \frac{\int\frac{d^3k_2}{(2\pi)^3}\frac{d^3k_3}{(2\pi)^3}\frac{d^3k_4}{(2\pi)^3}\,|\langle\vec{k}\vec{k}_2|G|\vec{k}_3\vec{k}_4\rangle|^2 n(k_2)(1-n(k_3))(1-n(k_4))}{\int\frac{d^3k_2}{(2\pi)^3}\frac{d^3k_3}{(2\pi)^3}\frac{d^3k_4}{(2\pi)^3}\,|\langle\vec{k}\vec{k}_2|T|\vec{k}_3\vec{k}_4\rangle|^2 n(k_2)(1-n(k_3))(1-n(k_4))} \qquad (3.3)$$

where T is the free transition matrix and where we skipped in the summations the energy-momentum conservation delta functions of eq. (2.1). The results are shown in Fig. 1 in the case of the Paris potential, for several densities and two temperatures. At low density and low

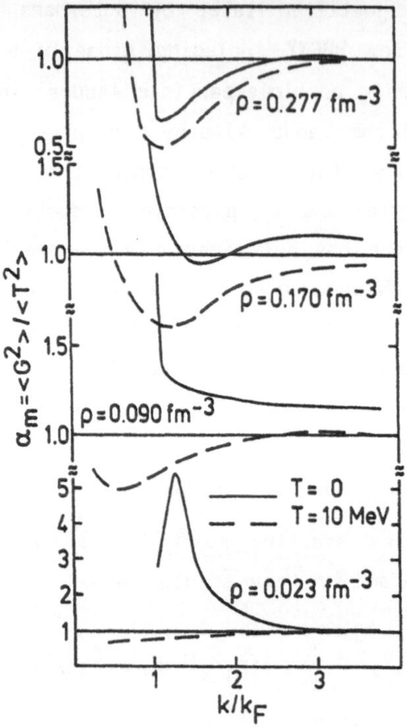

Figure 1. Schematic representation of medium correction (eq. (3.3)) for elastic scattering suffered by a nucleon of momentum k travelling inside uniform nuclear matter of density ρ and temperature T. The quantity α_m is slightly different from the quantity α' defined in Ref. 15, where constant effective mass approximation was used. See text for detail.

temperature, and for k just above the Fermi momentum k_F, the correction α_m is quite large. This is due to a strong medium reduction of the attractive part of the potential (corresponding to the contribution to the T-matrix). The repulsive part is not so affected at low density because of its shorter range. The large compensation between repulsive and attractive effects is therefore destroyed. At large density, both contributions are equally renormalized (when going from T- to the G-matrix) and then α_m is smaller than unity at intermediate k. Of course the free space value should be recovered at large k.

The medium effects have strong influence on the equilibration times, on the viscosity and on flow properties. We just illustrate this point in Fig. 2, where we show the flow angle (calculated in an intranuclear cascade calculation) in a typical case, for free space and medium corrected cross-sections. A similar conclusion was obtained in a BUU study by Bertsch et al.[16] They observe a stronger dependence of the so-called transverse momentum[17] upon the value of the nucleon-nucleon cross-section than upon the equation of state. Recently[18], a similar claim was made concerning small angle inclusive proton cross-section in heavy systems.

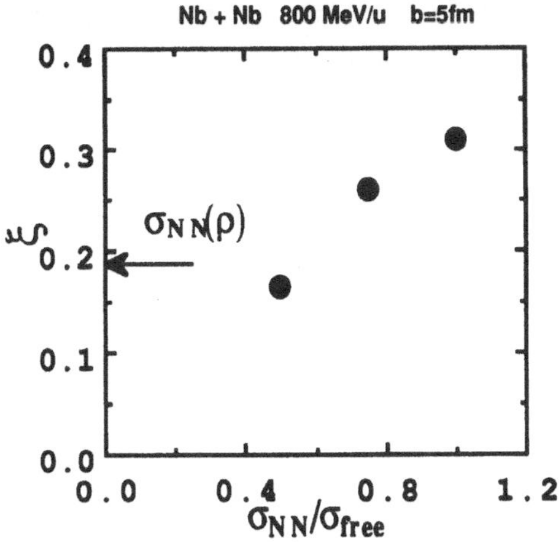

Figure 2. Dependence of the quantity ξ, roughly equal to the flow angle, upon a variation of the NN cross-sections, as given by an intranuclear cascade model. The arrow indicates the value obtained if one takes account of density dependence of the medium correction as calculated in Ref. 24.

Medium effects do not only change the transition probability. Because particles are feeling a mean field the phase space is changed. In other words, the single particle energies e(p) appearing in r.h.s. of eq. (2.1) are different from $p^2/2m$, what they reduce to in free space processes. In the local density approximation, one can write $e(p) = p^2/2m + U_B(p)$, where U_B is the mean field evaluated in the Brueckner approximation at density ρ and temperature T. If $U_B(p)$ is a quadratic function for all values of p, $e(p) = p^2/2mm^*$, m* being the effective mass. In that case the last delta function appearing in the r.h.s. of eq. (2.1) may be replaced as

$$\delta(e(p)+e(p_2)-e(p_3)-e(p_4)) \rightarrow m^*\delta(\frac{p^2}{2m}+\frac{p_2^2}{2m}-\frac{p_3^2}{2m}-\frac{p_4^2}{2m}) \quad . \tag{3.4}$$

Therefore, one can make, as usual, simulations with conservation of kinetic energies provided the cross-section is multiplied by the effective mass. At low momenta $U_B(p)$ is quadratic, giving in that range $m^* \approx 0.8$. However, $U_B(p)$ is not quadratic for the momentum range experienced in heavy ion collisions at ~ 100 MeV/u or more. In simulations, the distortion of phase space is usually not implemented. In Ref. 15, medium corrections on the transition probability and on the phase space are put together in a parameter α similar to (2.3) :

$$\alpha = \frac{\int \frac{d^3k_2}{(2\pi)^3}\frac{d^3k_3}{(2\pi)^3}\frac{d^3k_4}{(2\pi)^3}\,|\langle\vec{k}\vec{k}_2|G|\vec{k}_3\vec{k}_4\rangle|^2 n(k_2)(1-n(k_3))(1-n(k_4))\delta^3(\vec{k})\delta(e(k))}{\int \frac{d^3k_2}{(2\pi)^3}\frac{d^3k_3}{(2\pi)^3}\frac{d^3k_4}{(2\pi)^3}\,|\langle\vec{k}\vec{k}_2|T|\vec{k}_3\vec{k}_4\rangle|^2 n(k_2)(1-n(k_3))(1-n(k_4))\delta^3(\vec{k})\delta(\frac{k^2}{2m})}$$

where we used shorthand notation for the energy-momentum conservation δ functions. Roughly speaking $\alpha \approx 0.8\,\alpha_m$ at low k and tends to α_m at large k. See Ref. 15 for more detail and for suitable parametrization of α and α_m as functions of ρ and T.

4. MEDIUM EFFECTS IN PION PRODUCTION

4.1. Introduction

The importance of the pion yield has been a subject of great controversy since R. Stock et al.[19] proposed that it could be a good thermometer, and through an analysis not really free of hypotheses[20], serve as a mean to determine the compression energy as a function of density. The method has been criticized, but not from the point of view of the medium distortions on pion production. Furthermore, pion production and pion absorption are generally pictured, in numerical studies, as due to the following processes

$$NN \rightleftarrows N\Delta \quad , \quad \Delta \rightleftarrows \pi N \quad , \tag{4.1}$$

for which free space data are rather well known. Medium corrections are not so well known, except for pion propagation (and absorption) in cold normal nuclear matter, which has been under intensive study for several years.[11,21] However, this is of limited help, since the propagation of pion implies repeated iteration of the second mechanism in (4.1) and if one sticks with this description, one can have trouble with double counting. This is also true for the so-called three nucleon absorption[22,23], which has been presented as an important source of absorption and which is overlapping with iteration of processes (4.1) in the above scheme. Furthermore, the influence of the medium effect on the Δ-production mechanism has not been studied very much. Therefore, it is very hard to estimate with reliability the medium corrections to pion yield. But, the sensitivity of the pion yield on these corrections can be evaluated since their order of magnitude are more or less known. This is the philosophy of the recent work of Ref. 24, that we summarize below. This work also studied in particular the proton-nucleus case which can be a good test ground, since there compression effects are negligibly small.

4.2. The proton-nucleus system

In the model description for pion production/absorption described above, the dynamical input related to pion production are the inelastic cross-sections $NN \rightleftarrows N\Delta$ (linked by detailed balance), the angular distribution, the mass spectrum for produced Δ's and the Δ lifetime. One should also add the pion mean field. Below, we examine the influence

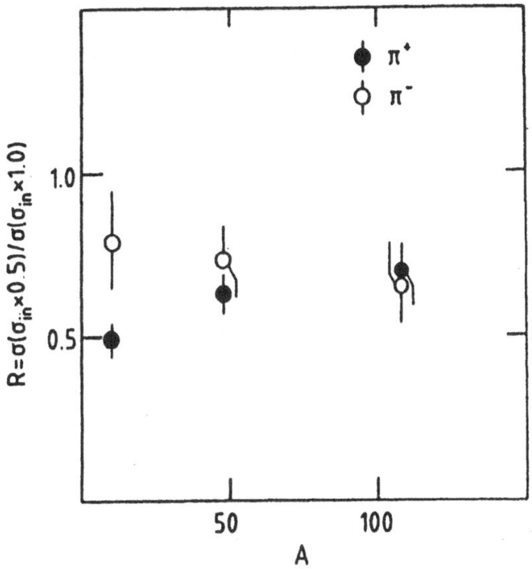

Figure 3. Ratio between pion yield calculated with half inelastic nucleon-nucleon cross-sections and the one calculated with the full cross-section in interactions between 730 MeV protons with several targets. The results are given by the intranuclear cascade calculation from Ref. 24.

of some of them. In Fig. 3, we show the influence of the inelastic cross-section on the pion yield in the interaction of 730 MeV protons with nuclei. A reduction of a factor 2 on the inelastic cross-section directly reflects on a decrease of a factor 2 on the π^+ yield for the ^{12}C target case. For heavier masses, the reduction diminishes to about 30 %. The different behaviour of π^+ and π^- can be understood in terms of multiple scattering. Obviously, π^+'s can be produced by a single pp interaction whereas π^- production necessarily requires further interaction of a proton. The influence of Δ-mass is illustrated by Fig. 4, where it has been artificially raised by 50 MeV. This leads to an overall reduction of the pion yield.

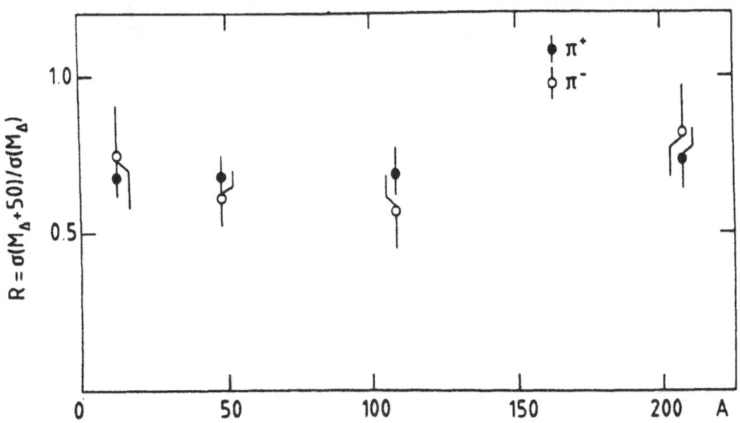

Figure 4. Same as Fig. 3 for a modification of the Δ-mass.

It is shown in Ref. 24 that a good agreement can be reached only if one modifies the pion absorption separately from the production mechanism, a feature which is hardly acceptable at first sight. Roughly speaking, an amplification of the $N\Delta \rightarrow NN$ cross-section by a factor 2-3 and a lengthening of the Δ-lifetime by a factor 2-3 is sufficient to achieve a good description of the pion yield. This may be understood as due to the fact that in the scheme above a finite time is required for a pion to be absorbed after it has first interacted. A modification of the time sequence may be a better alternative to the possibility envisaged above. The usual scheme is quite able to describe the trend of the excitation functions (see Fig. 5), which once again may point to a need for a better description of pion absorption.

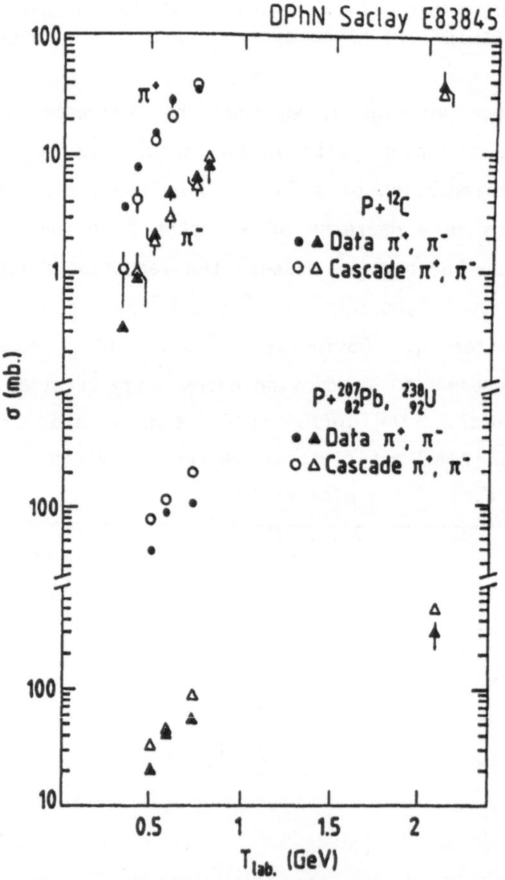

Figure 5. Comparison between cascade calculation (Ref. 24) and experimental data for the excitation function of pion production. From Ref. 24.

4.3. The heavy ion case

As we said above, the average pion multiplicity might be related to the equation of state. In particular, emphasis was put on rather complete data for negative pion multiplicity in Ar + KCl collisions. It is thus interesting to know the sensitivity of this quantity upon medium correction of the dynamical input. Some of the results of Ref. 24 are contained in Fig. 6. One can first see that the more or less realistic treatment of binding energy (left part) already removes half of the discrepancy observed for the original cascade calculations. We recall that this discrepancy was presented as an evidence for a stiff equation of state.[19,27,28] Fig. 6 also shows that a modification of the Δ-mass by 30 MeV or a reduction of the NN ⇄ NΔ cross-sections are sufficient to bring the numerical results close to the experimental data. In Ref. 24, other modifications are considered. The conclusion is that the pion yield is much more sensitive in the heavy ion case than in the proton-nucleus case, and therefore medium corrections of the expected order of magnitude are largely sufficient to achieve good agreement. The larger sensitivity in the heavy ion case is easy to understand, as in this case the final pion yield results from a large number of pion creations and destructions. If a modification does not change by the same factor (the raising of the Δ-mass is a typical one), the creation and the destruction rates, it will modify the final pion yield more importantly in the heavy ion case.

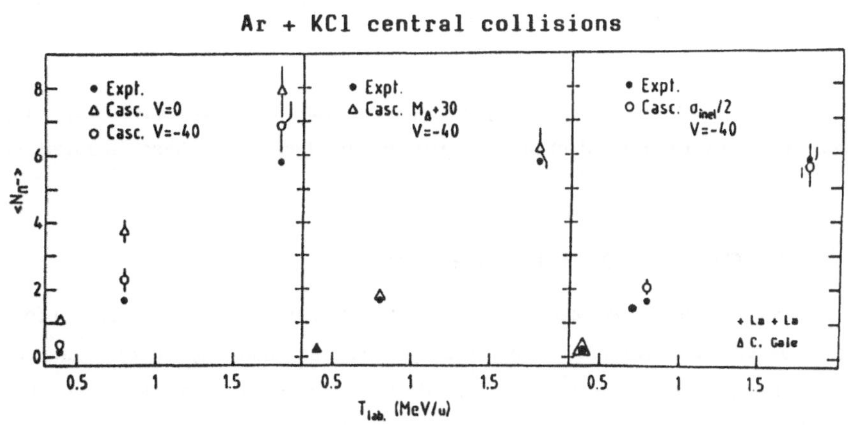

Figure 6. Comparison between cascade calculations of Ref. 24 and experimental data (Ref. 19) for π^- production in central Ar + KCl collisions. The influence of some medium effects is displayed. See text for detail. In the right part of the figure, the cross indicates the La + La data of Ref. 25 (scaled by the ratio of the masses of the two systems) and the triangle gives the results of a similar calculation by C. Gale.[26]

4.4. Discussion

The work of Ref. 24 studies the sensitivity to any modification without presuming of the direction of this modification. However, at least for some of the input data, the direction and the size of the medium modification is known, to some extent. The medium $NN \to N\Delta$ cross-section has been calculated in a detailed Brueckner-Dirac approach in Ref. 14. The Pauli blocking of the final states is included automatically in the calculation, which is then mixed with genuine medium corrections. It seems however that the latter only slightly reduce ($\sim$ 20 %) the cross-section for T=0 and density ρ up to $2\rho_0$. Calculations of the shift of the Δ-mass are rather contradictory.[14,29-31] The work of ter Haar and Malfliet[14], which is the most trustworthy, indicates that altogether the Δ-mass is shifted downward by $\sim$ 15 MeV at ρ_0. However, in usual numerical treatments (see Ref. 24), the Δ is supposed to move in the same mean field as the nucleons. Therefore, in such a scheme, the results of Ref. 14 are equivalent to an upward shift of $\sim$ 25 MeV. This shift is however rather strongly density-dependent. Ref. 14 also indicates a strong reduction of the Δ-width in nuclear matter, much stronger than in conventional nonrelativistic approaches.[11,32]

Let us finally mention that the medium corrections on the pion propagators could be more important than those we discuss here. In particular Bertsch et al.[33] suggested that medium corrections on the pion dispersion relation could increase the pion production by a factor 2 at $\rho = 2\rho_0$. This is rather disturbing since, as we indicated above, the current approaches generally overpredict the pion yield. The authors of Ref. 33 also found a singularity in the clothed pion propagator opening the possibility of pion Cerenkov radiation and of stronger stopping power.

As for elastic scattering, the distortion of phase space may play some role in pion production. This aspect has never been studied so far.

5. MEDIUM EFFECTS IN OTHER PARTICLE PRODUCTION

In the previous two sections, we discussed medium corrections to two-body collisions. In other words, it was implicitly assumed that the scattering wavefunction of the two colliding particles has the time and the place to become asymptotic before another collision takes place. Nevertheless, the presence of the surrounding medium may importantly distort the collision process, because the two particles are feeling the mean field created by their neighbours (and the blocking of the phase space in the intermediate case in the case of fermions).

In some production processes, the medium can also act differently. Take for instance the production of a low pion, say ~ 20 MeV kinetic energy. Its de Broglie wave length is 2.7 fm, i.e. longer than NN interdistance in normal nuclear matter. This may have several consequences : (1) the source of pion production may not only be the correlated NN scattering system, but involve the (appropriate) current of several nucleons. This could explain why the subthreshold pion yield in heavy ion collisions are so large compared to the collision contribution[34] ; (2) there may be a screening of the pion source due to the surrounding medium. This would deduce the source strength, but more importantly limit the long wave length (or equivalently the low frequency) part of the spectrum. The latter would then appear flatter than the free space spectrum. This might very well explain why the pion spectrum seems to show a large and constant temperature (slope parameter) for a wide range of incident energy in heavy ion collisions (see Fig. 7) ; (3) if production is partly due to meson exchange currents (MEC) and not simply to baryon current, this contribution may be considerably changed inside matter.

These considerations may apply more clearly to photon production, a topic which has been under intensive study these last years (for a review, see Ref. 36). Point (1) above corresponds to gamma emission by the cooperative part of the electromagnetic current carried by several nucleons. In its extreme form, this phenomenon is the elusive coherent bremsstrahlung process.[37] Point (2) has been nicely investigated by Knoll and collaborators.[38,39] They showed that photon production rate (in a thermalized system) should fulfill some sum rule, which

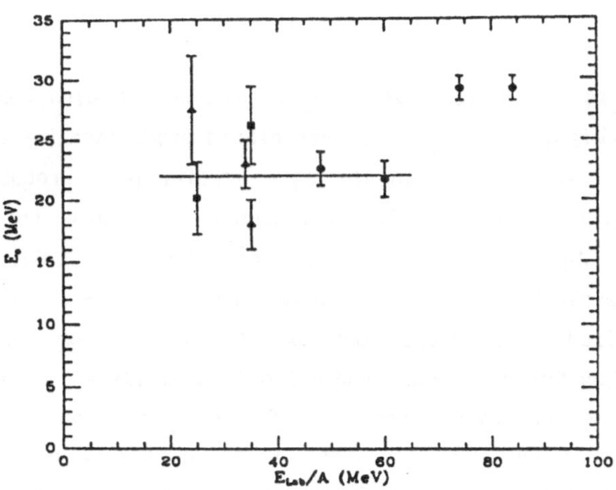

Figure 7. Beam energy dependence of the slope parameter E_0 of pion kinetic energy spectrum for several heavy ion systems. Adapted from Ref. 35.

can be considered as a generalisation of the ordinary dipole sum rule. They also showed that kinetic models using free NN bremsstrahlung cross-section are bound to violate the sum rule for low energy photons ($\lesssim$ 35 MeV in their example). They relate this violation to the diverging bremsstrahlung spectrum at low frequency. In this case, the medium effect is very clear. The bremsstrahlung does not lead to electromagnetic excitations of the vacuum, but to those of the matter, or in this case, to the even more restrictive spectrum of the finite nuclear system. The argument above has been obtained in the soft photon limit, which automatically favours long wave lengths. However, the problem seems to persist if this limit is abandonned. It is not known yet how to correct for this medium effect in collision models, but the question is certainly worth to be investigated. Point (3) above is not very well documented for the photon case, because the importance of MEC in photon production is unclear.[40,41] New measurements of $p+n \rightarrow p+n+\gamma$ reaction[42] are in progress, which will help to settle this question.

These kinds of medium effects should be investigated more carefully. In our opinion, it would be instructive to conduct this investigation in parallel for various particle production. Besides pions and gamma's, we think to dileptons (basically related to virtual photons), which are just starting to be studied experimentally[43,44], and to heavier mesons.

The case of K-meson production could perhaps be different from the other ones. The reason is that kaons appear in associated production, whose simplest mechanisms are

$$NN \rightarrow N\Lambda K \quad , \quad NN \rightarrow N\Sigma K \quad , \qquad (5.1)$$

it is to say reactions with a high threshold energy, thus occurring in violent collisions only. Therefore, it is expected that associated production is highly localized. This property is also consistent with the underlying quark mechanism. Associated production basically results from an $s\bar{s}$ excitation. In the same spirit, pion production results from a mixture of $u\bar{u}$ and $d\bar{d}$ excitations. According to modern views of hadronic physics, the pion can be considered as a collective (rather extended) excitation of the vacuum, whereas $s\bar{s}$ excitations are much more localized. Therefore, one expects less medium correction for associated production. Kaon production (especially K^+) is thus presumably more sensitive to the equation of state than pion production.[45-47] However, before any conclusion can be drawn, medium corrections, even if smaller than those described in sections 3 and 4 should be evaluated. Furthermore, other uncertainties are still to be removed. First, the $NN \rightarrow N\Lambda K$ cross-section is badly known close to threshold as illustrated

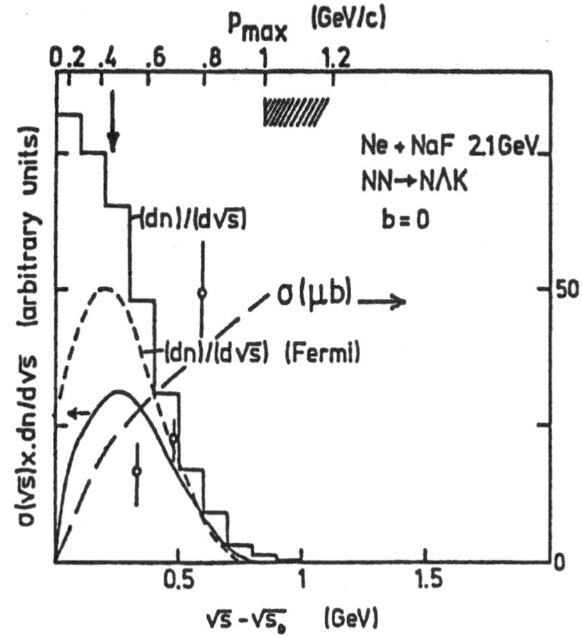

Figure 8. The histogram gives the frequency $dn/d\sqrt{s}$ of baryon-baryon collisions with c.m. energy $\sqrt{s}$ ($\sqrt{s}_0$ is the threshold for NN → NΛK) in the central Ne + NaF collisions at 2.1 GeV/u. The dots indicate the only available measurements (actually pp → pΛK^+) in this energy range. The long dashed is the customarily used parametrisation of the cross-section. The full line is the reaction probability distribution, i.e. the frequency distribution multiplied by the cross-section. The dotted line would correspond to the same quantity (arbitrarily normalized) if only first collisions contributed. The shaded area indicates the border of the kinematical region accessible due to Fermi motion and the vertical arrow indicates the c.m. energy of the pp system with the same kinematics as the heavy ion system. Finally, p_{max} is the maximum K^+ momentum for the corresponding c.m. energy. Adapted from Ref. 46.

by Fig. 8. The latter applies to 2.1 GeV/u (symmetric) heavy ion collision. For subthreshold production in heavy ion collisions, i.e. in the GeV/u range, the most important value of the c.m. energy, i.e. the one which corresponds to the maximum of probability for the above mentioned reaction, is even smaller. Second, it was recently shown that the three-body phase space is strongly distorted[48] close to the threshold because of the ΛN rescattering. One thus expects some further distortion inside matter.

6. CONCLUSION

We have reviewed the question of medium corrections in the dynamics of heavy ion collisions. We started from the Landau-Vlassov equation, which seems to provide a satisfactory framework for the description of nuclear transport in the few MeV/u to few GeV/u energy range. Medium effects appear as mainly renormalizing collision cross-sections. However they may also modify the available phase space. They can also lead to an effective mass different from the bare mass. We did not consider this last aspect here. We discussed carefully the medium effects on elastic scattering, which are the best known by far, though in the limit of local density approximation. We underlined the strong sensitivity of flow properties upon medium effects, apparently stronger than the one related to the equation of state. We discuss the influence of medium effects on the pion yield, which also shows a great sensitivity. We qualitatively discussed the current ideas concerning medium effects in other production mechanisms. Our final conclusion is that a great effort is needed to understand and evaluate reliably medium effects (not only in the collision term) before being capable of extracting nuclear matter equation of state. As La Bruyère[49] told us a long time ago : "Tout le mal vient de ce que nous ne sommes pas seuls" or "The bad thing comes from the fact that we are not alone". But we think also that medium effects should be studied for themselves since they involve an interesting physics.

REFERENCES

1. C. Grégoire, Proc. of "Ecole de Physique des Houches - Nuclear Matter and Heavy Ion Collisions", Plenum Press, to be published.
2. L.P. Kadanoff and G. Baym, "Quantum Statistical Mechanics", W.A. Benjamin, New York (1962).
3. R. Balescu, "Equilibrium and Nonequilibrium Statistical Mechanics", J. Wiley, New York (1975), chapter 3.
4. M. Berenguer et al., Proc. of "Ecole de Physique des Houches - Nuclear Matter and Heavy Ion Collisions", Plenum Press, to be published.
5. F.A. Brieva and J.R. Rook, Nucl.Phys. A297:206 (1978).
6. J. Cugnon, Proc. of "Société française de Physique, Congrès National de Nice", SFP Editions, (1985), p. 323.
7. A. Lejeune, P. Grangé, M. Martzolff and J. Cugnon, Contribution to the International Conf. on Heavy Ion Nuclear Collisions in the Fermi Energy Domain, Caen, (1985).
8. H.S. Köhler and B.S. Nilsson, Journal de Physique Colloque C2, suppl. nº 6 (1987), p. 225.
9. W. Botermans and R. Malfliet, Phys.Lett. 171B:22 (1986).
10. Ph. Siemens, Proc. of "Ecole de Physique des Houches - Nuclear Matter and Heavy Ion Collisions", Plenum Press, to be published.
11. E. Oset, H. Toki and W. Weise, Phys.Rep. 83:287 (1982).
12. T.E.O. Ericson and W. Weise, Pions and Nuclei, Clarendon Press, Oxford (1988).
13. J. Dąbrowski and W. Piechocki, Acta Phys. Polonica B16:1095 (1985).
14. B. ter Haar and R. Malfliet, Phys.Rev. C36:1611 (1987).
15. J. Cugnon, A. Lejeune and P. Grangé, Phys.Rev. C35:861 (1987).
16. G. Bertsch et al., Phys.Lett. 189B:384 (1987).
17. P. Danielewicz and G. Odyniec, Phys.Lett. 157B:146 (1985).
18. J. Aichelin et al., Phys.Rev.Lett. 62:1461 (1989).
19. R. Stock et al., Phys.Rev.Lett. 49:1236 (1982).
20. J.W. Harris et al., Phys.Lett. 153B:377 (1985).
21. L.L. Salcedo, E. Oset, M.J. Vicente-Vacos and C. Garcia-Recio, Nucl. Phys. A484:557 (1988).
22. G.E. Brown, Nucl.Phys. A488:689c (1988).

23. E. Oset, L.L. Salcedo and D. Strottman, Phys.Lett. 165B:13 (1985).
24. J. Cugnon and M.-C. Lemaire, Nucl.Phys. A489:781 (1988).
25. J.W. Harris et al., Phys.Rev.Lett. 58:1611 (1987).
26. C. Gale, Phys.Rev. C36:2152 (1987).
27. R. Stock, Phys.Rep. 135:259 (1986).
28. J.J. Molitoris and H. Stöcker, Phys.Rev. C32:346 (1983).
29. K. Dreissigacker et al., Nucl.Phys. A375:334 (1982).
30. R. Cenni and G. Dillon, Nucl.Phys. A392:438 (1983).
31. R. Cenni and G. Dillon, Nucl.Phys. A422:527 (1984).
32. M. Hirata, F. Lenz and Y. Yuzaki, Ann.Phys. (N.Y.) 108:116 (1977).
33. G. Bertsch, G.E. Brown, V. Koch and B. Li, Nucl.Phys. A490:745 (1988).
34. C. Guet and M. Prakash, Nucl.Phys. A428:119c (1984).
35. P. Braun-Munzinger and J. Stachel, Ann.Rev.Nucl.Sci. 37:97 (1987).
36. V. Metag, Nucl.Phys. A488:483c (1988).
37. D. Vasak, H. Stöcker, B. Müller and W. Greiner, Phys.Lett. 93B:243 (1980).
38. J. Knoll and C. Guet, preprint GSI-88-25.
39. M. Durand and J. Knoll, to be published.
40. D. Neuhauser and S.E. Koonin, Nucl.Phys. A462:163 (1987).
41. W. Bauer et al., Phys.Rev. C34:2127 (1986).
42. J.A. Pinston et al., to be published.
43. G. Roche et al., Phys.Rev.Lett. 61:1069 (1988).
44. C. Naudet et al., Phys.Rev.Lett. 62:2652 (1989).
45. J. Aichelin and C.M. Ko, Phys.Rev.Lett. 55:2661 (1985).
46. J. Cugnon, Proc. of the JES4 Meeting, ed. by CEA Saclay, (1986).
47. J. Aichelin et al., Phys.Rev.Lett. 58:1926 (1987).
48. R. Frascaria et al., Few-Body Systems, sup. 2 (1987), p. 425.
49. J. de La Bruyère, "Caractères".

MANY BODY MODES OF EXCITATION IN HEAVY ION COLLISIONS

E. Oset [a], M. Vicente-Vacas [b]

[a] Depto. de Física Téorica and IFIC
University of Valencia, Burjasot, Spain
[b] Inst. of Theoretical Physics
University of Regensburg, W.-Germany

ABSTRACT

The one step excitation of two particle holes in heavy ion collisions is shown to be an important channel in the reaction. Pion production has to compete against this new channel, not present in the NN free reaction, and is reduced sensibly. The effective NN cross section is increased leading to a more effective stopping of the nucleons.

COUPLED CHANNEL EXCITATION: ONE EXAMPLE -Λ DECAY IN NUCLEI-

Let us study how a Λ particle decays inside a slab of nuclear matter. The Λ will adquire a selfenergy in the medium, Σ_Λ, and the modulus squared of the Λ wave function will be

$$| \psi_A |^2 \;\alpha\; | e^{-i\Sigma_\Delta t} ... |^2 \;\alpha\; e^{2\, Im\, \Sigma_\Delta t} \equiv e^{-\Gamma t} \tag{1}$$

showing the exponential decay law, with

$$\Gamma = -2\, Im\, \Sigma_\Lambda \tag{2}$$

We must then evaluate the Λ selfenergy in the nuclear medium and particularly its imaginary part. Since we know that Λ decays into πN, the Λ selfenergy diagram accounting for this channel is given in fig. 1a. We should recall that the imaginary part of the selfenergy comes when in the integrations over the momenta of the internal states all the intermediate states cut by a straight line are placed on shell. This is depicted by the dotted line in fig. 1a. These are the states into which the Λ decays,

The Nuclear Equation of State, Part A
Edited by W. Greiner and H. Stöcker
Plenum Press, New York

i.e., we would be considering the decay $\Lambda \to \pi N$.

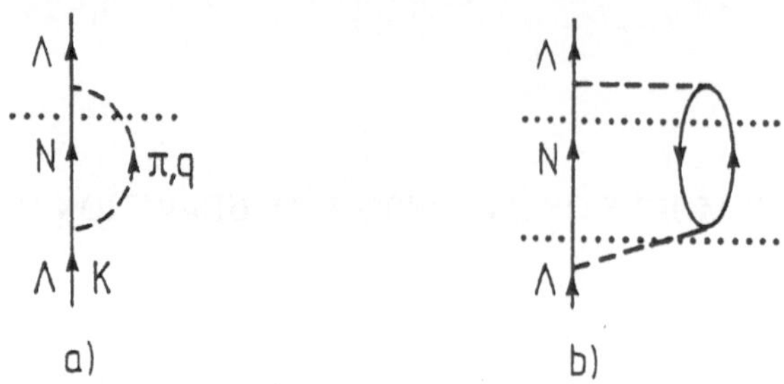

fig. 1

In order to illustrate that we note that

$$\Sigma^{(\pi)}(k) \; \alpha \int \frac{d^4q}{(2\pi)^4} G(k-q) D(q) \tag{3}$$

and by means of a Wick rotation one can see that

$$Im\Sigma^{(\pi)}(k) \; \alpha \int \frac{d^3q}{(2\pi)^4} ImG(k-q) ImD(q) \theta(k^0 - q^0) \theta(q^0) \tag{4}$$

with

$$G(k-q) = \frac{1}{k^0 - q^0 - E(\vec{k} - \vec{q}) + i\epsilon}; \; ImG(k-q) = -i\pi\delta(k^0 - q^0 - E(\vec{k} - \vec{q}))$$

$$D(q) = \frac{1}{q^{o2} - \vec{q}^{\,2} - \mu^2 + i\epsilon}; \; ImD(q) = -i\pi\delta(q^{o2} - \vec{q}^{\,2} - \mu^2) \tag{5}$$

and $E(\vec{k} - \vec{q})$ the on shell energy of the nucleon of momentum $\vec{k} - \vec{q}$. One can then realize that the contribution to $Im\Sigma^{(\pi)}$ comes from situations where the energy and momenta of the nucleons and pions correspond to their on shell values (arguments of the δ functions equal zero).

So far the effect of the medium is not yet accounted for. This would be done when we renormalize the nucleon and the pion propagators. In fig. 1b we show one of the diagrams coming from the renormalization of the pion propagator, where the pion excites a ph in the medium. Now we also know that the nucleon line should have a momentum above the nucleon Fermi momentum in the Λ decay. When placing intermediate particles on shell in fig. 1b we have two possibilities, as depicted there by the dotted lines. The upper cut places a N and 1ph on shell, the lower cut places again the π and the N on shell. Hence this lower cut contribution together with the one in fig. 1a would give a contribution to the $\Lambda \to \pi N$ decay channel in the medium. The upper cut however introduces a new channel not open in free Λ decay. This channel is the one $\Lambda N \to NN$, (also a weak interaction decay since it involves change of strangeness).

A detailed study of these processes has been carried out in refs.[1,2] and one finds out that the $\Lambda \to \pi N$ decay (mesonic decay) is highly suppressed in nuclei (totally suppressed in nuclear matter at normal density for Λ at rest) and that the $\Lambda N \rightarrow NN$ reaction (non mesonic decay) dominates the Λ decay process. The nuclear medium produces several effects: 1) the Pauli blocking limits severely the possibilities of Λ mesonic decay, 2) the pion renormalization in the medium acts two fold, on the one hand for a pion in a virtual state it can give rise to a ph physical excitation and this gives rise to the Λ non mesonic decay channel. On the other hand, through the lower cut in fig. 1b, it leads to important corrections in the mesonic channel making the Pauli blocking less effective, since the attraction of the nuclear medium on the pion makes this one carry less energy in the Λ decay, thus allowing the nucleon to carry more energy and have more chances to be on top of the Fermi energy.

In summary this study shows us three things: 1) The free reaction channels can be highly renormalized in a nuclear medium. 2) Now reaction channels can appear and 3) there is a coupling between these channels which is made clear when a systematic many body expansion in terms of Feynman diagrams is done (for instance, the presence of the $\Lambda \rightarrow \pi N$ correction from the lower cut of fig. 1b appears necessarily tied to the existence of the $\Lambda N \rightarrow NN$ channel of the upper cut in the same diagram).

NUCLEON SELFENERGY IN A NUCLEAR MEDIUM

Let us follow the analogy and construct the N selfenergy from the diagrams equivalent to fig. 1, depicted now in fig. 2.

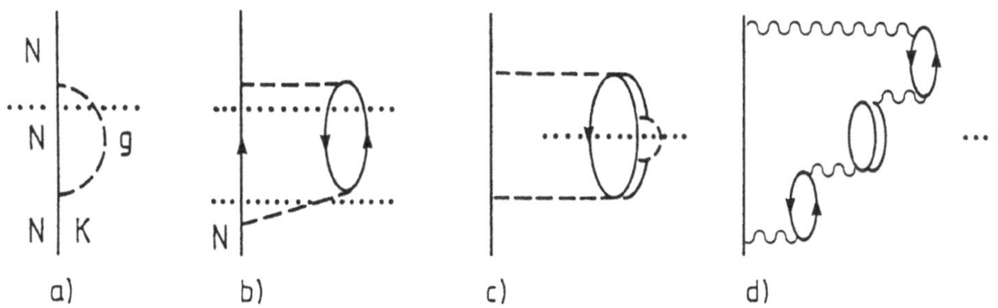

fig. 2

The analogy is clear but there are also differences. The first one is that $N \rightarrow N\pi$ in free space cannot occur for reasons of energy and momentum conservation, hence there is no imaginary part associated with the cut in fig. 2a nor with the lower cut in fig. 2b. However, there is an imaginary part associated with the upper cut of fig. 2b corresponding to $N \rightarrow N\, ph$ or equivalently $NN \rightarrow NN$, the nucleon does not decay but the meaning of $Im\ \Sigma$ is here that the nucleus is excited from its ground state, or equivalently that the incoming nucleon is removed from the elastic flux. Now we said that new channels can appear in the nucleus. One of them is the $N \rightarrow \pi N$ if the relationship of momentum to energy of a pion in the nuclear medium is appreciably altered, a thing that happens at the nuclear densities reached in heavy ion reactions. In order to envisage this, one has to look at the whole diagramatic series in fig. 2 which indicates that the pion propagator would be dressed by picking up a pion selfenergy

$$D_0(q) \rightarrow D(q) = \frac{1}{q^{o2} - \vec{q}^{\,2} - \mu^2 - \Pi(q^0, \vec{q}\,)} \tag{6}$$

with Π, the pion selfenergy, due to ph or Δ h excitation. Assuming $\Pi(q^0, \vec{q}\,)$ real for the moment, a renormalized pion could be produced if we have

$$k^0 = E(\vec{k} - \vec{q}) + \tilde{\omega}(\vec{q}) \tag{7}$$

where $\tilde{\omega}(\vec{q})$ is the energy in the medium for a pion of momentum $\vec{q}$, satisfying

$$\tilde{\omega}(\vec{q})^2 - \vec{q}^{\,2} - \mu^2 - \Pi(\tilde{\omega}(\vec{q}), \vec{q}) = 0 \tag{8}$$

or equivalently giving rise to poles of the renormalized pion propagator.

We have seen that dressing up the pion with its selfenergy is necessary to eventually produce such pionic modes. However, Π is not real, since both real and virtual pions have an appreciable probability for been absorbed in the nucleus or undergoing quasielastic scattering, which removes them from the elastic flux, as in the case of the nucleons. A realistic picture should incorporate these pion interaction channels. The model of fig. 2 for the pion selfenergy already incorporates some imaginary part for Π. Indeed, in fig. 2b the pion couples to a ph excitation. For certain off shell situations of the pion the ph excitation can be placed on shell and this is why the upper cut in fig. 2b gives a contribution to $Im\ \Sigma$, corresponding to the $NN \rightarrow NN$ scattering as we indicated. On the other hand the Δh excitation in fig. 2c can also be placed on shell, its physical meaning being that a pion excites a ph component together with a pion, or equivalently that the pion undergoes a quasielastic scattering. Correspondingly, the cut in fig. 2c for the nucleon selfenergy would account for $NN \rightarrow NN\pi$, via Δ excitation, a leading mechanism in pion production. However fig. 2c only considers the mechanism of Δ excitation in the target while we should also have Δ excitation in the projectile. This is depicted in fig. 3.

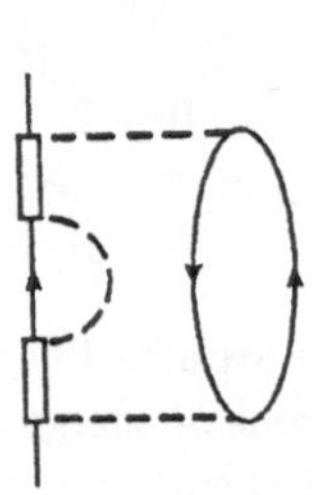

fig. 3

So far the sources of $Im\ \Sigma$ involve two initial nucleons the incoming one and one from the nuclear medium. At this point we should recall the low density theorem which states that

$$\Gamma = -2\, Im\, \Sigma = \sigma\, v_{rel}\, \rho \tag{9}$$

where σ is the total NN cross section, including NN scattering and pion production. Our calculations allow us not only to evaluate σ, by taking the limit $\rho \rightarrow 0$ in $Im\ \Sigma$, but also to separate its contribution from the NN scattering and from π production since they correspond to the contribution of different cuts in the Feynman diagrams discussed. This will always be a check of consistency of our models. Eq. (9) is used by us to define a modified cross section in the medium in the case of finite ρ.

Coming back to the pion selfenergy, Π, we have seen that $Im\,\Pi$ in fig. 2 comes from the coupling of the pions to the ph continuum or the channel $\pi \rightarrow \pi$ ph, representing the quasielastic channel in the case of real pions. However, there is a reaction channel still missing: pion absorption. Indeed, we know that in pion nuclear reactions in medium and heavy nuclei pion absorption is the dominant channel[3]. Pion absorption appears from the coupling of the pion to two particle two hole or three particle three hole components[4], with p-wave, Δ driven pion absorption being the most important part at the energies and momenta involved in the process that we discuss. Such mechanisms would lead to the series of diagrams depicted in fig. 4 for the nucleon selfenergy. However, at this point we introduce a new element of realism in the picture and recall that the Δ is excited by the spin-isospin effective interaction, of which π exchange is only one of the ingredients.

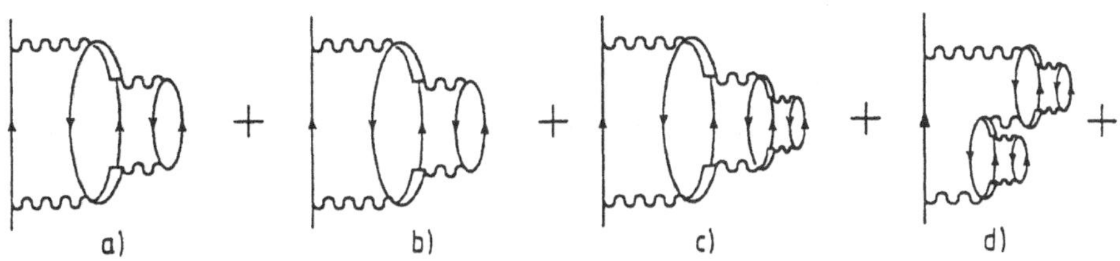

fig. 4

Hence we use such interaction, instead of just π exchange, in our study. This is depicted by the wavy line in fig. 4. If we look now at the cuts in the figure we observe new excitation channels. Fig. 4a contains the $N \rightarrow N'2p2h$ excitation, a channel with no parallel in NN collisions because it involves three incoming nucleons. Fig. 4b contains the $N \rightarrow N'\,2p2h\ \pi$ excitation, again with no parallel in the NN interaction. Fig. 4c contains the excitation $N \rightarrow N'3p3h$, also with no analogy in the free NN collision. Analogous channels can be obtained by exciting the Δ in the target and coupling the Δ to 2p2h components etc., as depicted in fig. 5.

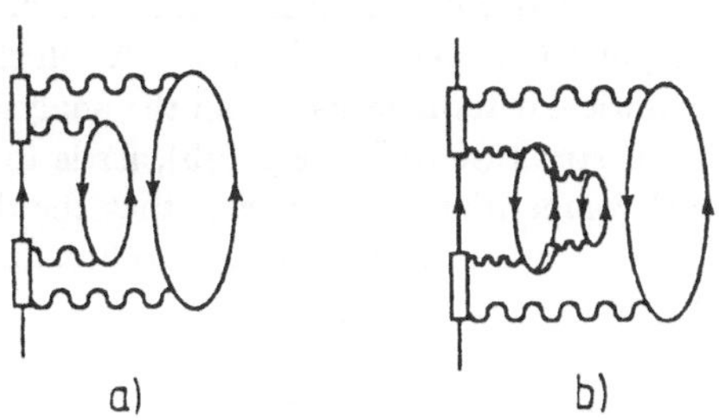

fig. 5

The spin-isospin interaction used by us is

$$V(q) = [V_\ell(q)\hat{q}_i\hat{q}_j + V_t(q)(\delta_{ij} - \hat{q}_i\hat{q}_j)]\sigma_i\sigma_j\vec{\tau}\vec{\tau}$$

$$V_\ell(q) = \frac{f^2}{\mu^2}(-q^2 D_o(q)F_\pi^2(q) + g'(q)) \tag{10}$$

$$V_t(q) = \frac{f^2}{\mu^2}(-q^2 D_\rho(q)F_\rho^2(q)C_\rho + g'(q))$$

which is separated into a longitudinal, V_ℓ, and a transverse part V_t. $D_0(q)$ is the pion propagator, $D_\rho(q)$ the ρ meson propagator, C_ρ the ratio squared of ρ and π coupling constants, $F_{\pi,\rho}(q)$ the πNN and ρNN form factors and g' the Landau-Migdal parameter. We take

$$F(q) = \frac{\Lambda^2 - m^2}{\Lambda^2 - q^2} \;;\; g'(q) = \frac{g'}{1 - q^2/\Lambda} \tag{11}$$

When substituting a ph by a Δh excitation we must substitute $\sigma_i \rightarrow S_i$, $\tau_i \rightarrow Ti$, $f \rightarrow f^*$, with S_i, T_i the spin, isospin transition operators and f^* the $\pi N\Delta$ coupling constant.

Although the sum of the series of diagrams shown might look complicated, for the case of no Δ excitation in the projectile the sum is done easily and one obtains[5]

$$\begin{aligned} Im\ \Sigma(k) =& 3(\frac{f}{\mu})^2 Im \int \frac{d^3q}{(2\pi)^3}[1 - n(\vec{k} - \vec{q})]\theta(\omega)\times \\ & [\frac{V_\ell(\omega,\vec{q})}{1 + V_\ell(\omega,\vec{q})U(\omega,\vec{q})} + \frac{2V_t(\omega,\vec{q})}{1 + V_t(\omega,\vec{q})U(\omega,\vec{q})}]_{\omega=\epsilon(\vec{k})-\epsilon(\vec{k}-\vec{q})} \end{aligned} \tag{12}$$

and similar expressions for the terms with Δ excitation in the projectile, with $U = U_N + U_\Delta$, the Lindhard function for ph excitation, U_N, and Δh excitation, U_Δ. In our calculation we have

$$U_\Delta(q) \propto \frac{1}{\sqrt{s} - M_\Delta + i\frac{\Gamma(q)}{2} + i\frac{\Gamma^{(abs)}(q)}{2}} \tag{13}$$

where Γ is the $\Delta \rightarrow \pi N$ width corrected by Pauli blocking and $\Gamma^{(abs)}$ accounts for the coupling of the Δ to the 2p1h channel (2p2h in the pion nuclear interaction). We thus keep the 2p1h components of the model for the Δ selfenergy of ref. 6, which is in agreement with empirical determinations from the analysis of elastic π nucleus scattering[7]. The resonant structure of U_Δ, eq. (13), leads to a peculiar feature of these reactions: channel competition. This means that the different new reaction channels, rather than accumulate, compete with each other. Indeed, in the Δ driven pieces $Im\ \Sigma$ will be proportional to $Im\ U_\Delta$ given by

$$Im\, U_\Delta\ \alpha\ \frac{\Gamma + \Gamma^{abs}}{(\sqrt{s} - M_\Delta)^2 + (\Gamma + \Gamma^{abs})^2} \tag{14}$$

and the term proportional to Γ will account for π production while the one proportional to Γ^{abs} will account for 2p2h excitation. Now, the inclusion of the coupling to these ph excitations reduces the pion production rate. Indeed, at $\sqrt{s} \sim M_\Delta$, in the absence of Γ^{abs} the rate for pion production is proportional to $1/\Gamma$, while in the presence of Γ^{abs} it is proportional to $\Gamma/(\Gamma+\Gamma^{abs})^2$, which can be sensibly smaller given the importance[6] of Γ^{abs}. This is a reminder of what is observed in photonuclear reactions around resonance where not only the (γ,π) cross section per nucleon, but also the total photonuclear cross section per nucleon is decreased with respect to the free value for the $\gamma N \rightarrow \pi N$ reaction[8]. The widening of the Δ width has this consequence around the resonance peak but the energy distribution of the cross section is accordingly widened and the integrated cross section does not change much for the sum of the two channels; however, the appearance of the 2p2h excitation is then done at the expenses of the pion production channel.

RESULTS

In ref. 5 the calculations have been done for two sets of parameters: $I(g'_{NN} = 0.55, g'_{N\Delta} = g'_{\Delta\Delta} = 0.35, \Lambda_a = 1.0GeV, \Lambda_\rho = 1.1GeV, \lambda \rightarrow \infty)$ and $II(g'_{NN} = g'_\Delta = g'_{\Delta\Delta} = 0.7; \Lambda_\pi = 1.2GeV, \Lambda_\rho = \lambda = 1.GeV)$ which are inside the range of reasonable values but on opposite extremes and lead approximately to good free cross sections via eq. (9).

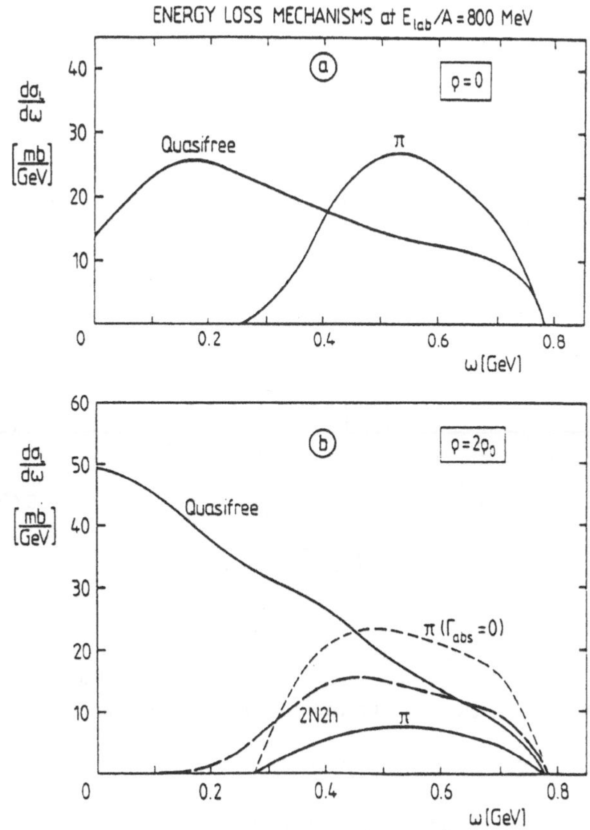

fig. 6

We show in fig. 6 $d\sigma_\ell/d\omega$ for $E_{lab}/A = 800 MeV$, with σ_ℓ the part of σ coming from the longitudinal part (see, eq. 12), since there is no interference between the longitudinal and transverse parts. In fig. 6a we show the contribution from the $NN \to NN$ (quasifree) and π production channels. In fig. 6b we show the results at $\rho = 2\rho_0$. The quasifree cross section is modified, the 2p2h excitation channel becomes quite important and, as a consequence of the discussion in the former section, the pion production cross section is sensibly reduced. One can observe there that the reason is the competition with the 2p2h excitation channel, because if Γ^{abs} is set to zero one has a rate of π production similar the free case. The reduction of the pion production rate is a feature to be welcomed in view of the persistent theoretical overcounting of pions produced in heavy ion collisions[9]. In figure 7 we plot σ_ℓ, σ_T, σ as a function of the density at E_{lab} = 800 MeV/A. We observe that while the transverse part of the cross section does not change much with the density, σ_ℓ increases with the density. This is a reminder of the role played by the attraction of the medium on the pion and in the longitudinal channel of the reaction as a consequence. The total cross section, or equivalently, the rate of nucleon loss from the elastic channel, or rate of nuclear excitation, increases as a function of the density and this would be translated in heavy ion reactions into a more efficient stopping of the nucleons.

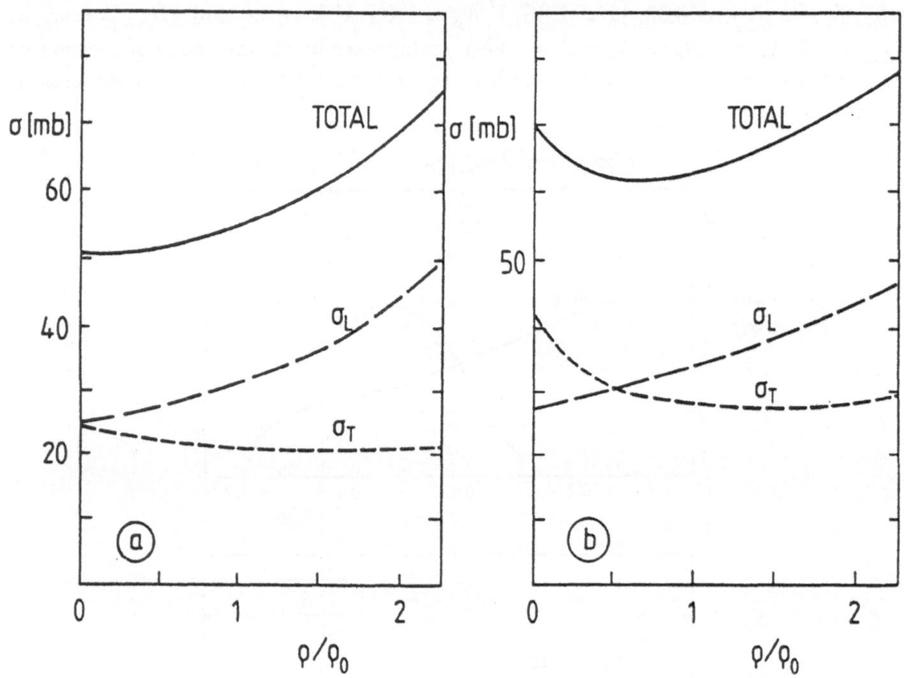

fig. 7

CONCLUSIONS

The main conclusions that we can draw from this paper are the following:

1) The many body excitation channels are very important at $\rho \geq 2\rho_0$.

2) These channels compete strongly with each other and have to be considered simultaneously. At densities $\rho \simeq 2\rho_o$ the presence of the 2p2h excitation channel reduces the pion production rate by about a factor three.

3) We find an increased cross section, or rather an increased rate for nuclear excitation, which leads to a more efficient stopping of the nucleons.

4) The enhancement in σ is found in the longitudinal channel, indicating that the nuclear attraction on the pions makes easier the production of pionic modes, which later on decay in the nucleon excitation channels.

This talk reports on parts of ref. 5 complementing some aspects of it. We are grateful to G.E. Brown and W. Weise with whom these matters have been discussed and worked out.

REFERENCES

1. Oset and L.L. Salcedo, Nucl. Phys. A 443 (1985) 704

2. E. Oset, L.L. Salcedo and R. Brockmann, Phys. Reports, to be published

3. D. Ashery and J.P. Schiffer, Ann. Rev. Nucl. Part. Scie. 36 (1986) 207

4. E. Oset, Y. Futami and H. Toki, Nucl. Phys. A 448 (1986) 597

5. G.E. Brown, E. Oset, M.J. Vicente-Vacas and W. Weise, submitted to Nucl. Phys. A

6. E. Oset and L.L. Salcedo, Nucl. Phys. A 468 (1987) 631

7. Y. Horikawa, M. Thies and F. Lenz, Nucl. Phys. 345 (1980) 386

8. J. H. Koch, E. J. Moniz and N. Ohtsuka, Ann. Phys. 154 (1984) 99

9. J. Cugnon and M.C. Lemaire, Nucl. Phys. A 489 (1988) 781

How to extract the nuclear equation of state from heavy ion data

G. Peilert[a], M. Berenguer[a], A. Rosenhauer[b], J. Aichelin[c], H. Stöcker[a] and W. Greiner[a]

[a]) Institut für Theoretische Physik, Johann Wolfgang Goethe–Universität, D–6000 Frankfurt am Main, Germany

[b]) on leave to the Physics Department, University of Bergen, N-5007 Bergen Norway

[c]) Institut für Theoretische Physik, Universität Heidelberg and Max Planck Institut für Kernphysik, D-6900 Heidelberg,Germany

Abstract: The Quantum Molecular Dynamic method is used to study the influence of the equation of state (EOS), in–medium n–n cross sections and momentum dependent interactions (MDI) on the main observables of heavy ion collisions.
The procedure how to extract the equation of state from these observables is presented. Therefore we show, that the in–medium n–n cross sections can be measured by comparing the rapidity distributions of singles and fragments with the results of microscopic calculations for light systems ($A < 40$).
The transverse flow distribution $p_x(Y)$, in contrast to the flow angle distributions $dN/d\cos\theta_F$, is shown to depend strongly on the equation of state but also on the n–n cross sections and the MDI. We show, that the MDI have only minor influences at lower bombarding energies ($E \leq 200 MeV/nucl.$). Higher bombarding energies do not lead to an drastic increase of the central density and therefore the EOS could be fixed at least with a finite uncertainty by comparing the excitation function of the transverse flow with all available microscopic calculations.

In the last two decades a lot of experiments have been done with heavy ion accelerators in the energy regime from several MeV/nucl. up to 2 GeV/nucl. [1,2]. One of the basic questions which could be answered from such experiments is the question about the nuclear equation of state at high ($\varrho > \varrho_0$) densities. The present and also further experiments in this energy regime lead in very central collisions of symmetric collision partners to maximum compressions in the order of $\varrho \approx 1-3\varrho_0$. Therefore we can hope that these experiments will provide us with information about the behaviour of nuclear matter at these densities.
In order to deduce nuclear matter properties from such experiments one has to compare the experimental data with the results of different theoretical models. Parallel to the experimental progress in this field also a lot of theoretical approaches have been developed to describe heavy ion collisions adequately. The most important approaches include the Vlasov–Uehling Uhlenbeck model [3], the

The Nuclear Equation of State, Part A
Edited by W. Greiner and H. Stöcker
Plenum Press, New York

Cascade Model [4], Hydrodynamic calculations [5,6,7] and Molecular Dynamic Calculations in the classical form [8] or with the implementation of some quantum features [9].

In the present work we want to outline the actual status of the simulation of heavy ion collisions in the Bevalac energy regime with the Quantum Molecular Dynamic model (QMD) [9]. In this context we want to show how far we can compare our results with experiments in order to learn something about the equation of state (EOS).

The QMD model has been developed from the former VUU approaches and the classical Molecular Dynamic calculations in order to describe also multifragmentation processes in heavy ion collisions. This is not possible in the VUU framework, because the test particle method smears out all correlations and fluctuations which are necessary to describe the fragmentation process. However, the fragments are very important in this energy regime and one finds experimentally several intermediate mass fragments (up to 20) and a lot of light fragments $(p, d, t, ^3He, ^4He, ..)$ in each collision [11]. Also it has been proposed from fluid dynamic calculations [6] that the fragments should, due to their lower thermal excitation, show all the flow observables which are most sensitive to the EOS, most clearly.
In the QMD approach the nucleons are not represented by pointlike testparticles but by a density distribution of Gaussian form:

$$\rho_i(\vec{r},t) = \frac{1}{(2\pi L)^{3/2}} e^{-\frac{(\vec{r}_i-\vec{r}_{i0}(t))^2}{2L^2}} \tag{1}$$

The propagation of the centroids of the gaussians $(\vec{r}_{i0}(t))$ takes place via two and threebody local Skryme interactions plus a long-range Coulomb potential (the width of the gaussians keeps constant in this process).

$$V(\vec{r}_i,\vec{r}_j,\vec{r}_k) = t_0\delta(\vec{r}_i-\vec{r}_j) + t_3\delta(\vec{r}_i-\vec{r}_j)\cdot\delta(\vec{r}_i-\vec{r}_k) + \frac{e^2}{|\vec{r}_i-\vec{r}_j|} \tag{2}$$

With the finite width distribution (eq. (1)) this local Skyrme interaction is equivalent to a finite range interaction between pointlike particles.
One disadvantage of this procedure can be found in the determination of the range parameter L in eq. (1). This parameter has no physical relevance in nuclear matter calculations but it causes problems for finite nuclei where the density in the center of the nuclei and also the surface thickness depend on L. But this dependence gives us also the opportunity to adjust L on the properties of finite nuclei. For the present calculations we use L = 2.1 fm.

The microscopic two and three particle interactions in eq. (2) can be identified with a density dependent potential determining the EOS in infinite nuclear matter.

$$U_{local} = \alpha\frac{\varrho}{\varrho_0} + \beta\frac{\varrho}{\varrho_0}^{\gamma} \tag{3}$$

where the third parameter γ can be used to adjust several compressibilities with the compressibility constant

$$K = 9\varrho\left.\frac{\partial\frac{E}{A}}{\partial\varrho}\right|_{\varrho=\varrho_0} \tag{4}$$

Another important but up to now not exactly known property of the n–n interaction is its momentum dependence, which leads to an additional repulsion in heavy ion collisions [9,13]. For the computation of those momentum dependent interactions (MDI) we parametrized the real part of the experimental p-nucleus optical potential in the following way [9]

$$V_{MDI} = t_4 \ln^2(t_5(\vec{p}_i - \vec{p}_j)^2 + 1)\delta(\vec{r}_i - \vec{r}_j) \tag{5}$$

with the parameters $t_4 = 1.57\,MeV$ and $t_5 = 500 \cdot 1/GeV^2$.

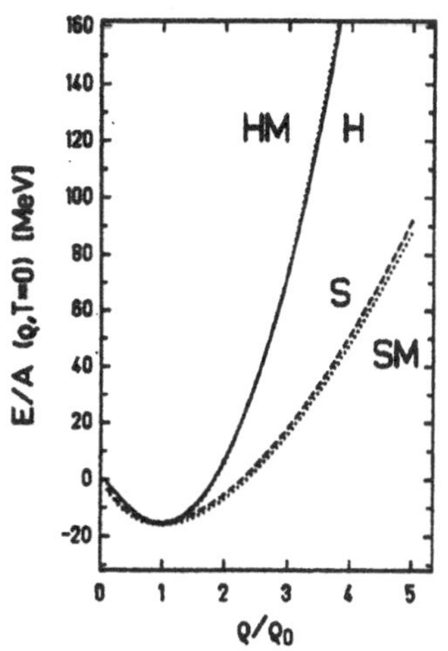

Fig.1. The equation of state in nuclear matter for a hard (H) and a soft (S) equation of state. The dashed curves show the same interaction with additional momentum dependent interactions

The EOSs which result from these interactions in nuclear matter are shown in Fig. 1. For the compression constant K we choose K = 380 MeV (H) and K = 200 MeV (S). The inclusion of the MDI (HM,SM) leads, for nuclear matter at rest, to no major change in the EOS.

So far the QMD model is equivalent to the solution of the Vlasov equation (see ref [10]). In addition to the propagation due to the n–n forces two nucleons can scatter if their centroids come closer than $r = \sqrt{\sigma/\pi}$. The scattering angle for a single n–n scattering is chosen randomly in a way that the integral over the scattering angle agrees with the experimental angular and energy distribution parametrized by Cugnon et al. [4].

In order to obey the Pauli principle in the collision the n–n collision is blocked in the same way as in the VUU model.

Up to now it is a still open question whether the free n–n cross section can be used in the high density regime of a heavy ion collision. Recent calculations in a Dirac–Brueckner theory indicate an additional reduction of the in-medium scattering cross section of about 30% due to the Pauli blocking of intermediate scattering states [14].

A second uncertainty in the present cross sections lies in the isospin dependence of the n–n scattering. The parametrization of Cugnon, which is used in this model and also in all the VUU/BUU models, is a parametrization of the free proton–proton cross section which is also about 30% lower than the neutron–proton scattering cross setion. A detailed investigation of the influence of the cross sections on the observables is in work [15].

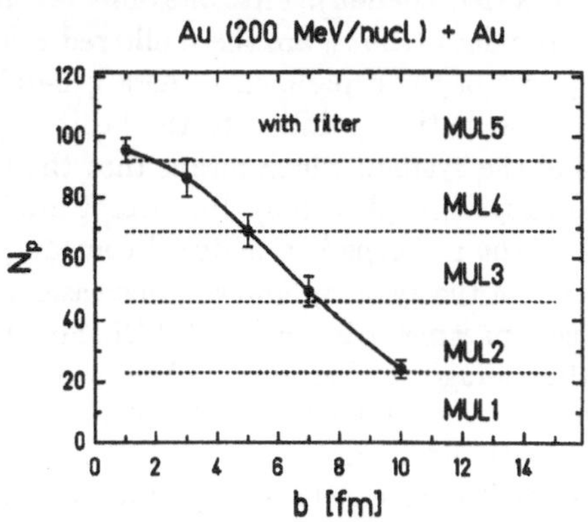

Fig.2. Impact parameter dependence of the participant proton multiplicity for the same reaction as in Fig. 1. The horizontal dashed lines show the cuts which have been used in the experiment [11]

The procedure of pinning down the EOS from heavy ion data is to compare the data with all available models which use the same input (EOS, n–n cross sections, MDI etc.). One crucial point in this procedure is to find observables which depend in a unique way on the different input variables. In the present work we use the QMD approach to heavy ion collisions and study the dependence of the main observables of heavy ion reactions on the EOS, the MDI and the cross sections.
In order to compare the calculations with the data one first has to find a way to relate the "multiplicity trigger" of the experiment with the different impact parameter ranges of the calculations. One of the observables which have been used to trigger on the centrality of the reaction is the participant proton multiplicity N_p.

Fig. 2 shows the dependence of N_p (i.e. the free protons plus the protons bound in light fragments up to mass 4) on the impact parameter b for the reaction Au (200 MeV/nucl.) + Au. In order to compare with the Plastic ball data a Plastic Ball filter (see refs. [9]) has been applied to the theoretical results. The dashed lines show the cuts which have been used in the experiment to distinguish between five different multiplicities which were labeled with MUL1 – MUL5. From this relation one can see that for not central collisions ($b \geq 3fm$) there is a linear depence of N_p on b, but the curve flattens out for the most central collisions. This indicates a general problem both for the experiment and for the theories because almost all variables which are sensitive to the high compression stage of the reaction can be seen best for the most central collisions were the highest densities are achieved. But as one can see here it is impossible to trigger on the most central collisions because they lead to the same multiplicities as the more peripheral ones.
If one wants to measure the n–n cross section (or the mean free path of the nucleons) one has to be careful with the system under investigation. The general purpose is to look for the stopping power of the reaction which can be measured via the the rapidity distributions dN/dY where the rapidity Y is given by $Y = \frac{1}{2}\ln\frac{E+p_z}{E-p_z}$.

These distributions provide us with information about how much of the incoming longitudinal energy has been transformed into other forms of energy, e.s. thermal and transversal energy.

Fig. 3 shows the calculated dN/dY distributions (right column) for the reactions Au + Au, Nb + Nb and Ca + Ca at 400 MeV/nucl. bombarding energy together with the Plastic Ball data (left column). First one observes that the QMD results agree quite good with the data (shown are the unfiltered calculations in order to compare the efficiancy loss of the experiment). As a general behaviour one sees that the width of the distribution (relative to the beam rapidity) gets broader with decreasing mass of the system, which means that the complete stopping is only seen for the massive systems (A > 100). However it has been shown [17] that the mean free path λ of the nucleons is a material constant, i.e. independent of the system. The volume of the system, however, decreases with decreasing mass and the crucial smallness parameter is not λ but λ/R. For the heavy system this number is small and the average number of 5 to 10 n–n collisions per nucleon leads to a complete stopping. But in an equilibrated system where every nucleon has scattered several times with other nucleons we cannot learn much about the n–n cross sections. For the smaller systems (A < 40) however, λ/R is of the order of 1 and therefore we observe much less equilibration and the behaviour of the system is very sensitive on the n–n cross section.

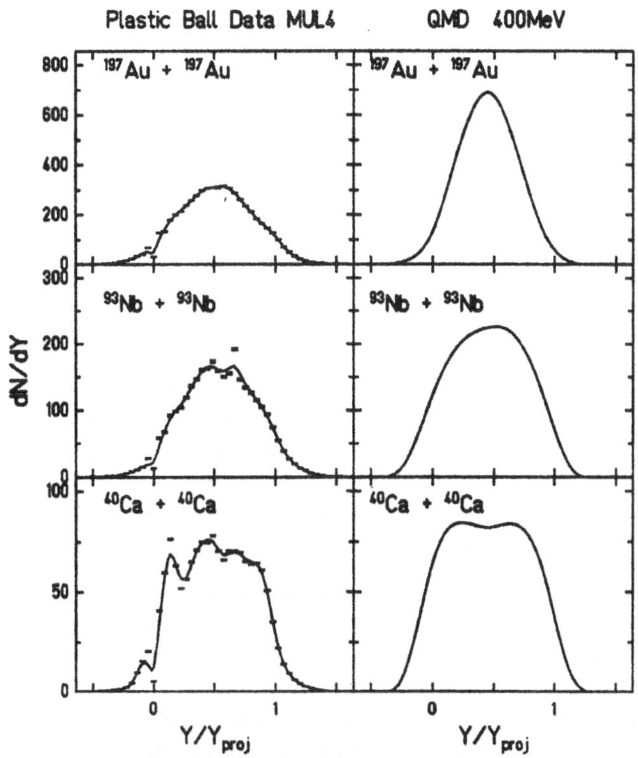

Fig.3. Rapidity distributions of the reactions Au + Au, Nb + Nb and Ca + Ca at 400 Mev bombarding energy and the same scaled impact parameter $\tilde{b} = b/(R_{Pr} + R_{Ta}) = 0.25$. The left side shows the experimental results [12] while the right side presents the QMD calculations.

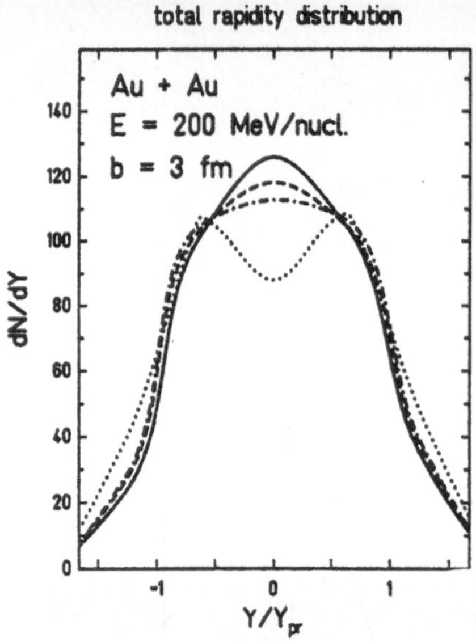

Fig.4. dN/dY distributions for the reaction Au(200 MeV/nucl., b = 3 fm) + Au for the interactions H (full line), S (dashed line), SM (dashed dotted line) and SIM (dotted line).

Fig. 4 shows now for a fixed impact parameter the dependece of the dN/dY distribution on the interaction used. Here one observes that the rapidity distribution does not depend on the EOS used (compare the full and the dashed line for the hard and the soft EOS) and is only weakly influenced by the MDI (dashed dotted line). Recent investigations have also shown that the rapidity distributions do not sensitively depend on L.
The only variable which has been found to have drastic influences on the rapidity spectra is the n–n scattering cross section. The 30% reduction of the cross section (dotted line) together with the soft EOS and the MDI leads to an drastic increase of the transperancy of the reaction. Now the rapidity spectra can clearly be decomposed into two gaussians with their centroids near by the beam rapidities.

Therefore the correct way to extract information about the EOS out of heavy ion data has to start with a deduction of the in–medium cross sections from the dN/dY distributions over a broad range of bombarding energies and for all available symmetric and asymmetric collision partners. The most promising sytems for measuring the mean free path of the nuclei are therefore central collisions of light nuclei up to the mass of Calcium.

Up to now we have seen only observables which do not depend on the underlying EOS at all. If we want to yield information about the EOS we have to use observables which are directly related to the high densitiy stage of the reaction because only there we are far away from the groundstate density. But on the other hand also the temperature in this zone is very high so that all the nucleons which come directly from this reaction zone are thermally smeared out and it is a very difficult procedure to disentangle the effects of the EOS, i.e. the compression energy per particle at temperature 0, from thermal effects. Therefore the best observables are those ones which don't come directly out ouf the high density zone but newertheless are sensitive on the EOS.

There is up to now only one observable which shows a dependence on the EOS and this is the collective sidewards flow of the matter out of the high density zone. This flow can be measured via the sphericity analysis in form of the flow angles Θ_F or the directed transverse flow $p_x(y)$ [2] and has been predicted from hydrodynamical calculations [5,6]. Recently all the flow effects have experimentally been observed [1]. Here we want to show how these observables can be reproduced with the QMD model and how they depend on the EOS and also on different other variables.

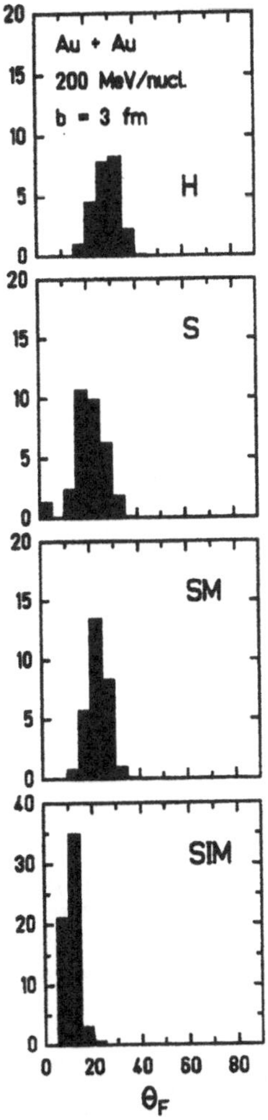

Fig.5. $dN/d\cos\Theta_f$ distributions for the reaction Au(200 MeV/nucl., b = 3 fm) + Au for the different interactions.

First we show in Fig. 5 the $dN/d\cos\Theta_F$ distributions for the reaction Au(200 MeV/nucl., b = 3fm) + Au for the different EOS's and cross sections. Unfortunately we observe that this variable, although it shows clearly a peak at $\Theta_F \approx 30°$ is also rather insensitive to the EOS alone.

The soft EOS leads only to slightly lower flow angles than the hard EOS. Also the addition of the MDI to the soft EOS (SM) leads to almost no change in the distribution. Only the reduction of the cross section (SIM) results in a a decrease of the average flow angle. Therefore we switch now over to the directed flow p_x/A. In the upper part of Fig. 6 we show for the same interactions as in Fig. 5 the corresponding p_x/A–Y distributions for particles with mass 1 and 2 together with the data. If one now compares the flow for the pure soft with the pure hard EOS one clearly observes an increase of p_x/A of 40% when going from the soft to the hard EOS. The inclusion of the MDI leads almost to no increase of the flow but this influence becomes more important at higher bombarding energies. An additional reduction of the cross section leads to a totally decrease of the transverse flow.

However, this figure shows the transverse flow of the nucleons and light fragments. However, it has been proposed that the heavy fragments should show the flow effects even stronger than the lighter particles because they are subjected to less thermal motion [6].

Therefore we show in the lower part of Fig. 6 the p_x/A–Y distributions splitted up into the three mass bins A = 1,2 , A = 3,4 and A = 12–20 and one can observe clearly that the absolute value of the transverse momenta per nucleon increases with increasing mass of the fragments. This effect has also been found experimentally [11]. Therefore the flow of the complex fragments is exactly the measure of the EOS that we are looking for, because these fragments don't come directly out of the reaction zone but are most sensitive on the EOS. This can be understand if one mentions that this fragments come from the outer and therefore colder parts of the sytem and are pushed away from the high pressure in the "fireball" zone.

In conclusion we have shown that the present QMD model can describe the dynamics of heavy ion collisions and also the formation of complex fragments. If one wants to extract the EOS from heavy ion data we have up to now only the collective sidewards flow as an observable which is sensitive to the EOS, but unfortunately this variable is also very sensitive on all other input variables of the

Au (200 MeV/nucl., b = 3 fm) + Au

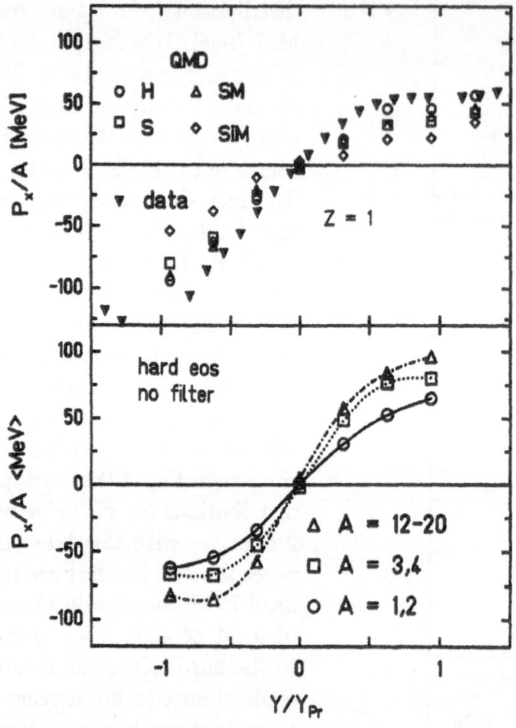

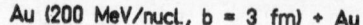

Fig.6. p_x/A distributions for reaction Au(200 MeV/nucl., b = 3 fm) + Au. The upper part shows the px–Y distributions for the charge 1 fragments for the different interactions together with the data. This curves have been filtered according to the Plastic Ball efficiencies. The lower curve however shows the (unfiltered) mass dependece of the fragment flow for the same reaction.

microscopic models which are for example the n–n scattering cross sections and the momentum dependence of the n–n interaction.

But newertheless this flow observables are very sensitive on the EOS and especially the flow of the complex fragments can be used to extract the EOS from the data. The best way to do so is to pin down all the different input variables from different observables. For exampple the in–medium n–n scattering cross section can be deduced from the rapidity spectra. This should be done with all available models (VUU, QMD, viscous Hydrodynamics etc.) and should give us at least a finite range for the EOS which is compatible with all the present experiments and models.

References

[1] H.A. Gustafsson et al., Phys. Rev. Lett. 52 (1984) 1590;
R.E. Renfordt et al., Phys. Rev. Lett. 53 (1984) 763;
D. Beauvis et al., Phys. Rev. C27 (1983) 2443;
H.G. Ritter et al., Nucl. Phys. A447 (1985) 3c;
K.G.R. Doss et al., Phys. Rev. Lett. 57 (1986) 302.

[2] P. Danielewicz and G. Odyniec, Phys. Lett 157B (1985) 146.

[3] H. Kruse et al., Phys.Rev. C31 (1985) 1770;
G.F. Bertsch, H. Kruse and S.Das Gupta, Phys. Rev. C29 (1984) 673;
J. Aichelin and H. Stöcker, Phys. Lett. 163 (1985) 59;
C. Gale, G.F. Bertsch and S.Das Gupta, Phys. Rev. C35 (1987) 1666;
C. Gregoire, B. Remaud, F. Sebille and L. Vinet, Nucl. Phys. A465 (1987) 317.

[4] Y. Yariv and Z.Fränkel, Phys. Rev. C20 (1979) 2227 and Phys. Rev. C24 (1981) 488;
J. Cugnon, Phys. Rev C22 (1980) 1885.

[5] W. Scheid, H. Müller and W. Greiner, Phys. Rev. Lett. 32 (1974) 741;
C.Y. Wong and T. Welton, Phys. Lett B49 (1974) 243;
A.A. Amsden, G.F. Bertsch, F.H.Harlow and J.R. Nix, Phys. Lett. 35 (1975) 905;
H.G. Baumgardt et al., Z. Phys. A273 (1975) 359.

[6] H. Stöcker, J.A. Maruhn and W. Greiner, Phys. Rev. Lett 44 (1980) 725;
H. Stöcker et al., Phys. Rev. Lett 47 (1981) 1807;
H. Stöcker et al., Phys. Rev. C25 (1982) 1873;
G. Buchwald et al., Phys. Rev. Lett 52 (1984) 1594.
H. Stöcker and W. Greiner, Phys. Rep. 137 (1986).

[7] W. Schmidt, thesis, University of Frankfurt (1989).

[8] A.R. Bodmer and C.N. Panos, Phys. Rev. C15 (1977) 1342, and Phys. Rev C22 (1980) 1023;
S.M. Kiselev and Y.E. Prokovskil, Sov. J. Nucl. Phys. 38(1) (1983) 46;
J.J. Molitoris et al., Phys. Rev. Lett 53 (1984) 899.

[9] J. Aichelin and H. Stöcker, Phys. Lett. B176(1986) 14;
A. Rosenhauer et al., J. Physique C4 (1986) 395;
J.Aichelin et al., Phys. Rev. Lett. 58 (1987) 1926;
G. Peilert et al., Mod. Phys. Lett. A3 (1988) 459 and Phys. Rev C39 (1989) 1402;
J.Aichelin et al., Phys. Rev. C37 (1988) 2451.

[10] J. Aichelin et al., Phys. Lett B224 (1989) 34.

[11] K.H.Kampert, Thesis, University of Münster (1986), and J. Phys. G 15 (1989) 691.

[12] R. Schmidt, private communication.

[13] T.L. Ainsworth et al., Nucl. Phys. A464 (1987) 740.

[14] B.ter Haar, R. Malfliet and W. Botermans, Phys. Rep. 149 (1987) 207 and Phys. Lett. 172B (1986) 10.

[15] Ch. Hartnack diploma thesis University Frankfurt (1989).

[16] J.E. Finn et al., Phys. Rev. Lett. 49 (1982) 1321;
A.D. Panagiotou et al., Phys. Rev. Lett. 52 (1984) 496;
L. Csernai and J. Kapusta, Phys. Rep. 131 (1986) 223.

[17] M. Berenguer et al., Proceedings of the winter school on nuclear physics in Les Houches, France 1989.

TRANSVERSE MOMENTUM GENERATION AND DILEPTON EMISSION IN HIGH ENERGY HEAVY ION COLLISIONS

Charles Gale*

Physics Dept., McGill University, 3600 University St.
Montréal, Qué., Canada H3A-2T8

Abstract. We investigate the generation of transverse momentum in high energy heavy ion reactions and its relation to the nuclear equation of state. We also study the emission of electron-positron pairs from hot nuclear matter and from nucleon-nucleon collisions. We argue that the study of dilepton spectra will yield valuable information on nuclear dynamics in general and on interacting pions in particular.

1. INTRODUCTION

The interest in the study of high energy heavy ion collisions stems from the fact that little is known about the behavior of nuclear matter in regions of temperature and density far removed from equilibrium. However, the amount of information one can obtain from such reactions may be considerably obscured by the transient nature of the colliding system. To extract the nuclear equation of state has been a major motivation of the experimental program. It is fair to say that this goal has not been completely achieved although substantial progress has ben made in theory as well as in experiment. Also, the strongly interacting many-body system has several other interesting facets. In particular, the dynamical properties of π's and Δ's in hot and dense nuclear media remain unclear.

Of the many observables suggested to obtain limits on the nuclear compressibility, measurements of transverse momenta[1,2] have raised much optimism. Other related observables are flow angles[3], azimuthal distributions about the reaction plane[4] and inclusive particle spectra for heavy colliding partners[5]. More recently, it has been suggested[6,7] that the spectrum of emitted dilepton pairs could be used to probe some up to now inaccessible features of the collision dynamics and especially to reveal the dispersion relation for pions in interaction. We shall discuss in turn these two classes of observables: strong and electromagnetic. First, we address the issue of the magnitudes of generated transverse momenta as studied via the simulations of high energy heavy ion collisions in the Boltzmann-Uehling-Uhlenbeck (BUU) approach[8,9]. Partic-

*Lecture given at the NATO Advanced Study Institute on the Nuclear Equation of State, Peñiscola Spain, 21 May – 3 June 1989.

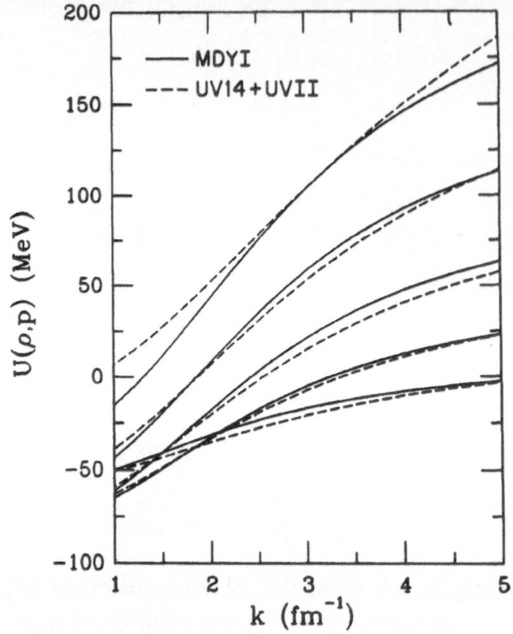

Figure 1. A comparison of the single particle potential from MDYI with the microscopic calculation of Wiringa (Ref. 10). The abscissa show wave numbers. Starting from the bottom at right, the curves are for $\rho = 0.1, 0.2, 0.3, 0.4$ and $0.5\ fm^{-3}$.

ular attention will be paid to the parametrization of the nuclear mean field. Next we will discuss the relevance of dilepton measurements in high energy heavy ion collisions. We will finally discuss the emission of lepton pairs from nucleon-nucleon reactions in a simple model and then in a more complete covariant framework.

2. THE BUU MODEL

The BUU approach is a numerical simulation procedure which describes the complete time evolution of the one-body phase space density for nucleons. The equation describing this is the Boltzmann equation with a Uehling-Uhlenbeck collision integral:

$$\frac{\partial f}{\partial t} + \vec{v}\cdot\frac{\partial f}{\partial \vec{r}} - \frac{\partial U}{\partial \vec{r}}\cdot\frac{\partial f}{\partial \vec{p}} = -\frac{1}{(2\pi)^6}\int d^3p_2 d^3p_{2'}\, d\Omega\, \frac{d\sigma_{NN}}{d\Omega}\, v_{rel} \quad (2.1)$$
$$\{[f f_2(1-f_{1'})(1-f_{2'}) - f_{1'}f_{2'}(1-f)(1-f_2)]\,(2\pi)^3\,\delta(\vec{p}+\vec{p}_2-\vec{p}_{1'}-\vec{p}_{2'})\}\ .$$

The factors $(1-f)(1-f_2)$ and $(1-f_{1'})(1-f_{2'})$ incorporate the effects of final state Pauli blocking. An energy density is defined, from which one calculates an equation of state (EOS). The single particle potential is calculated by taking a functional derivative with respect to the occupation function: $U(\rho,\vec{p}) = (\delta V(\rho)/\delta f)|_{\vec{p}}$. We introduce a realistic parametrization of $V(\rho)$, which accounts for the well known fact that the real part of the optical model potential is momentum-dependent. We use

$$V(\rho) = \frac{a}{2}\frac{\rho^2}{\rho_0} + \frac{b}{\sigma+1}\frac{\rho^{\sigma+1}}{\rho_0^{\sigma}} + \frac{c}{\rho_0}\int\int d^3p\, d^3p'\, \frac{f(\vec{r},\vec{p})f(\vec{r},\vec{p}')}{1+\left(\frac{\vec{p}-\vec{p}'}{\Lambda}\right)^2}, \quad (2.2)$$

which leads to a potential

$$U(\rho,\vec{p}) = a\left(\frac{\rho}{\rho_0}\right) + b\left(\frac{\rho}{\rho_0}\right)^{\sigma} + 2\,\frac{c}{\rho_0}\int d^3p'\,\frac{f(\vec{r},\vec{p}')}{1+\left(\frac{\vec{p}-\vec{p}'}{\Lambda}\right)^2}\,. \tag{2.3}$$

The five constants in Eq. (2.3) are found by requiring that E/A = -16 MeV, ρ_0 = 0.16 fm^{-3}, K = 215 MeV, $U(\rho_0, p=0) = -75MeV$, $U(\rho_0, \frac{p^2}{2m} = 300\text{ MeV}) = 0$. Their values are then a= - 110.4 MeV, b = 140.9 MeV, c = -64.95 MeV, σ = 1.24 and Λ = 1.58 $p_F^{(0)}$. With these parameters, the potential saturates to an asymptotic 30.5 MeV and the effective mass is $\frac{m^*}{m}$ = 0.67 at the Fermi surface. As the form of this interaction can be identified with the exchange term of a Yukawa force in the local approximation, we refer to it as the momentum dependent Yukawa interaction (MDYI). We claim that MDYI is a realistic interaction and it is therefore pertinent to compare it with more microscopic calculations. Wiringa[10] has calculated the single particle potential in nuclear matter using realistic Hamiltonians. We show the comparison of MDYI with the UV14 + UVII interaction in Fig. 1. The agreement is quite good over a wide range of momenta for all the values of density shown: 0.1, 0.2, 0.3, 0.4 and 0.5 fm^3.

3. THE EQUATION OF STATE AND TRANSVERSE MOMENTUM

We can obtain the equation of state associated with MDYI for different values of temperature by writing down the grand potential

$$\Omega = E - TS - \mu N, \tag{3.1}$$

knowing that $P = -(\frac{\partial\Omega}{\partial V})_{\mu,T}$. Note that in the above, V is now the volume. One obtains the results shown on Fig. 2 after going through a simple self-consistent procedure[11]. Also shown are the results associated with momentum-independent hard (HBKD) and soft (SBKD) interactions used in previous BUU calculations[8,9]. We also display the pressure calculated from another momentum dependent interaction, GBD[12]. Note that SBKD, GBD and MDYI have nearly identical compressibility coefficients and are quite close to each other; HBKD stands apart even for T = 80 MeV. It is however known that HBKD and momentum-dependent interactions can generate about the same transverse momentum[4,12,13,11] and that SBKD produces less. We then conclude that two EOS's (HBKD and MDYI) which are rather different over a wide range of temperatures can produce similar transverse momenta, while two EOS's (SBKD and MDYI) which are similar produce different transverse momenta.

In an effort to clarify this, we study the growth of transverse momentum as a function of time for La + La at 800 MeV/n, for a single impact parameter. On Fig. 3 we compare the contributions of MDYI, HBKD and SBKD. One thus realizes that a sizeable amount of transverse momentum is generated quite early in the reaction, when equilibrium can not possibly exist. At such early times the very concept of EOS is clearly ill-defined and one is then led to question the relevance of $\langle p_x/a\rangle$ in this discussion. Furthermore, turning off the relaxation mechanism provided by the two-body collisions, one sees that the mean field propagation alone can generate large transverse momenta (Fig. 4). This is understood in terms of the asymptotically positive optical potential and of the tendency for momentum-dependent forces to preserve initial state correlations. A similar "coherence" effect is observed when displaying the Wigner function for low energy head–on collisions between slabs with a momentum dependent interaction of a quadratic (Skyrme) form[14]. Therefore in the Vlasov equation, MDYI

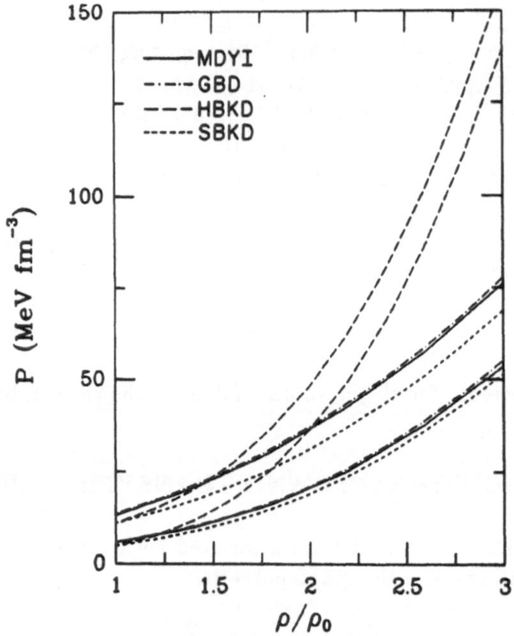

Figure 2. Equilibrium pressure versus density for different interactions introduced in the text. For each curve, the lower (upper) curve corresponds to T = 40(80) MeV.

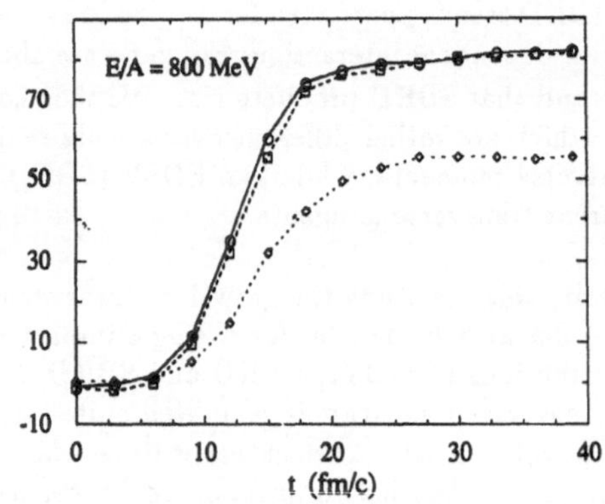

Figure 3. Average transverse momentum *vs.* time from BUU calculations of La + La at b = 2.67 fm. Results from MDYI (solid), HBKD (long dashes) and SBKD (short dashes) are shown.

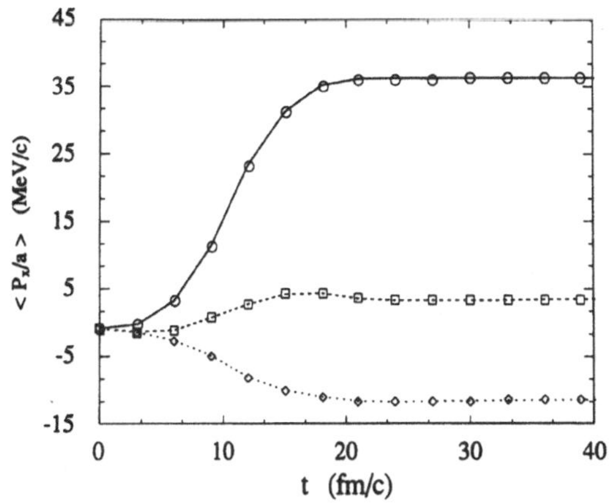

Figure 4. Same caption as the previous figure, but considering transverse momentum generation in the Vlasov mode only.

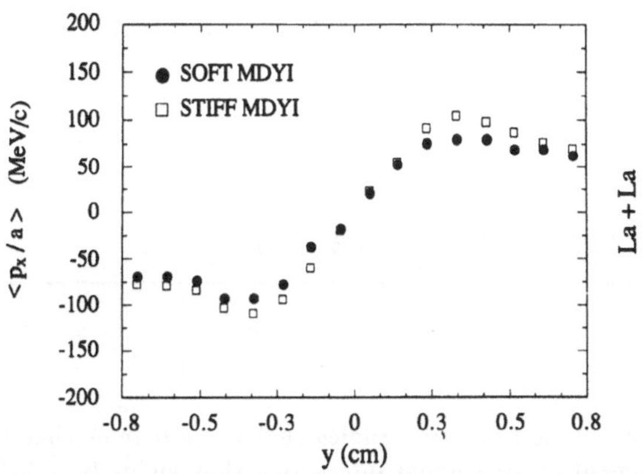

Figure 5. We compare the transverse momenta per nucleon obtained with a soft (K = 215 MeV) and a stiff (K = 380 MeV) equation of state. The range of impact parameter integration is adjusted to the experimental trigger cross sections (see text).

produces more transverse momentum than HBKD. Two-body collisions tend to equilibrate the participants and in complete equilibrium HBKD would have relaxed to a stiffer distribution than MDYI. The combination of these two mechanisms then leads to similar results. In this energy regime the interplay between mean field and two-body scattering can be quite complex. It follows from the above discussion that there are important non-equilibrium features involved in the production of transverse momentum in high energy heavy ion collisions. A direct corollary of this is that the EOS is at best only probed *indirectly* by the study of $\langle p_x/a \rangle$ vs. y distributions. To substantiate this, we verify on Fig. 5 that using a compressibility coefficient of K = 380 MeV induce only small changes in our impact parameter integrated distribution. It is interesting to note that results obtained in a relativistic approach can lead to similar conclusions[15]. We next turn to the comparison of BUU results with the MDYI interaction with streamer chamber data for near central collisions of Ar + KCl, La + La and Ar + Pb. Impact parameter integrated results are displayed on Fig. 6 and on Table 1. As in the calculations shown on the previous figure, the range of integration is determined by requiring a simple geometrical overlap model to match the experimental trigger cross sections[11,2]. We see that the agreement with data in the transverse momentum plots is quite remarkable. This is true even for the large values associated with the Pb target. The flow angle and integrated $\langle p_x \rangle$ fits are slightly less spectacular. Note however that these single numbers show more sensitivity to details of the impact parameter–trigger cross section mapping.

TABLE 1. We compare the values obtained with our BUU approach with the experimental data of Ref. 2, for symmetric systems. We estimate the numerical uncertainties to be ≈ ± 2 MeV/c for the transverse momentum results and ≈ ± 0.5 deg. for the flow angles. We use $b_{max.}$ = 3.8 fm(Ar + KCl) and 8.5 fm(La + La).

System	$(\theta_f)^{\text{This work}}$	$(\theta_f)^{\text{Exp.}}$	$\langle p_x \rangle_{y>0.15}^{\text{This work}}$	$\langle p_x \rangle_{y>0.15}^{\text{Exp.}}$
	(deg.)		(MeV/c)	
Ar + KCl	12	9.6 ± 0.8	61	50 ± 4
La + La	11	16.5 ± 1.7	76	72 ± 6

We can then conclude from our studies that the streamer chamber data are consistent with a momentum dependent interaction that yields K = 215 MeV. However, because of the demonstrated small sensitivity to the EOS, it is difficult to give quantitative support to a stronger statement.

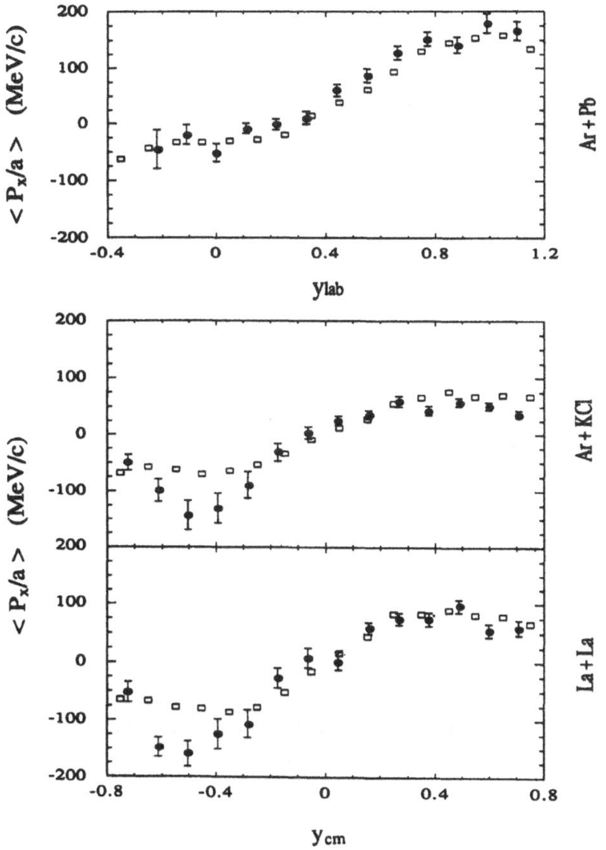

Figure 6. Transverse momentum per nucleon as a function of rapidity in reactions of 800 Mev/n. Results of BUU simulations with the MDYI interaction (open squares) are compared with the data of Ref. 2 (filled circles).

4. DILEPTON RADIATION FROM HOT NUCLEAR MATTER

In the search for penetrating probes of heavy ion collisions, measurements of photon spectra have generated considerable activity on the experimental as well as on the theoretical side[16]. More recently, lepton pairs have been put forward as a revealing signal for the relativistic strongly interacting many body system[6,7,17]. In general, electromagnetic signals offer some obvious advantages over the strongly interacting probes discussed earlier:

(a) They suffer little final state interaction, *i.e.* they travel relatively unscathed from the interaction zone. If the dileptons are produced mainly in incoherent processes at the single particle level, they carry valuable information about the space-time region where those interactions were most frequent: the high temperature and density phase.

(b) Their coupling to other particles is fairly well known.

The main drawback is of course the extremely low counting rates, owing mainly to the size of α, the fine structure constant. Note that the rate of dilepton emission is of order α^2, whereas photon emission is of order α. The advantages of dileptons over photons will become clear when we consider annihilation processes.

In our first works[6,7], we concentrated on the evaluation of thermal rates for the processes $np \to npe^+e^-$ and *particle antiparticle* $\to e^+e^-$. Generally, in the GeV/n regime, the channel $\pi^+\pi^- \to e^+e^-$ is dominant for $M_{e^+e^-} > 400$ MeV. When considering low dilepton invariant masses the bremsstrahlung contribution has to be included. We also emphasized the importance of looking at the spectrum of back to back e^+e^- pairs in the nuclear rest frame. With this configuration the connection between the mass spectrum and the pion dispersion relation is direct. One can show that, neglecting the imaginary part of the pion self energy, the rate for emission of back to back electron positron pairs from pion annihilation is in fact[6]

$$\left.\frac{dR^{e^+e^-}_{\pi^+\pi^-}}{dMd^3p}\right|_{\vec{p}=0} = \frac{\alpha^2}{3(2\pi)^4}\frac{|F_\pi(M)|^2}{(e^{\omega/T}-1)^2}\sum_k \frac{k^4}{\omega^4}\left|\frac{d\omega}{dk}\right|^{-1}, \tag{4.1}$$

where $\omega(k)$ is the energy of a pion with in-medium momentum k and F_π is the pion electromagnetic form factor. The sum is over those values of k for which $2\omega(k) = M$, the dilepton invariant mass. From the inverse proportionality to the group velocity in the medium, it follows that if the group velocity vanishes at non-zero k_0 then there will be a peak in the dilepton spectrum at $M_0 = 2\omega(k_0)$. Because of the strong coupling to the delta-hole excitation in the nuclear medium, the pion dispersion relation is very likely flattened out and may in fact develop a broad minimum[18]. Following ref. 6 we plot the thermal rate for producing e^+e^- pairs with invariant mass M and zero total momentum on Fig. 7, for a temperature of 100 MeV and for several values of the baryon density. Notice the correlation between the position of the dilepton peak and the density value. The threshold also moves *below* twice the free pion mass. Therefore the lepton pairs open a window on the interacting pions, which are *not* directly observed. Note that similar arguments also hold for the reaction $\pi^+\pi^- \to \gamma\gamma$. The thermal rate for this process is calculated and shown on Fig. 8. This involves calculating photon emission from direct, crossed and seagull diagrams, as opposed to the case with dileptons in the final state, where only one diagram contributes. However, for the case of photon

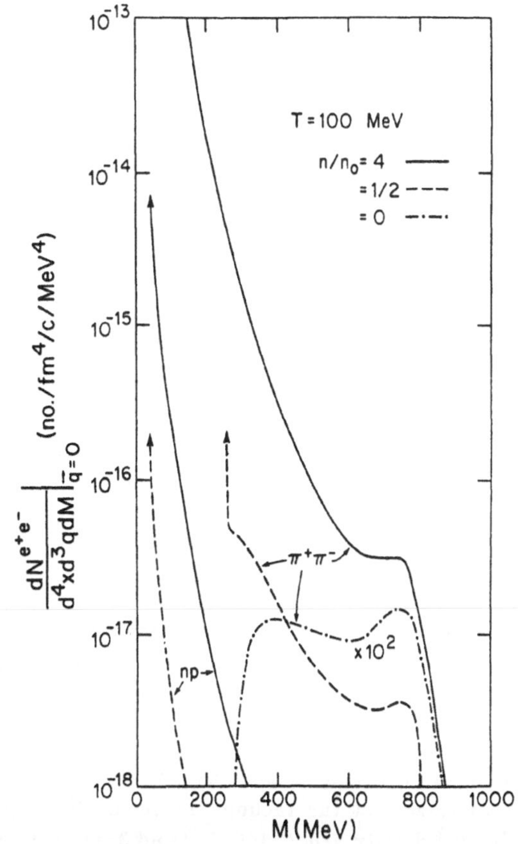

Figure 7. Comparison of thermal rates for producing e^+e^- pairs of invariant mass M and zero total momentum coming from np collisions and from $\pi^+\pi^-$ annihilation.

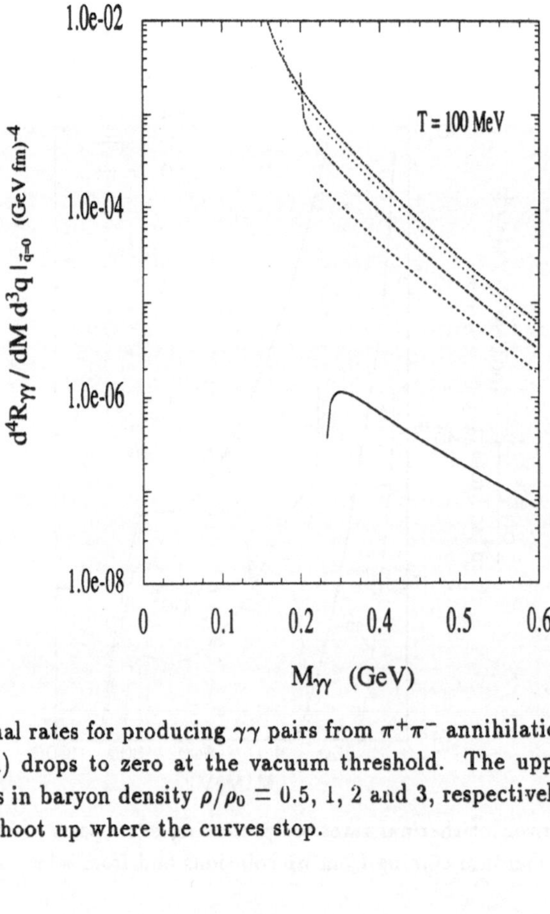

Figure 8. Thermal rates for producing $\gamma\gamma$ pairs from $\pi^+\pi^-$ annihilation. The rate from free pions (solid) drops to zero at the vacuum threshold. The upper curves are for interacting pions in baryon density $\rho/\rho_0 = 0.5$, 1, 2 and 3, respectively from bottom to top. The rates shoot up where the curves stop.

emission the background problems are expected to be severe[19]. The pionic excitations discussed above can have drastic implications on the heavy ion dynamics.

5. DILEPTON RADIATION FROM NUCLEON–NUCLEON COLLISIONS

In the nuclear many-body environment there will be a profusion of processes involving nucleons, pions and deltas for heavy ion bombarding energies of the order of 1 GeV/n. As far as the net contribution to the measured dilepton yield is concerned, there will be components associated with charged particle bremsstrahlung, particle–antiparticle annihilation and particle decay. The signature of single particle decays (*e.g.* π^0's) should be characteristic and identifiable. In general, a detailed comparison with experiment will ultimately require a complete enumeration of all such sources and their inclusion in a time dependent approach such as the BUU model described earlier. A calculation of dilepton production from elementary particle-particle interactions thus appears desirable for several reasons: First and more importantly, one has to understand the process at a fundamental level before the extrapolations to many-body systems can be made. Secondly, these basic cross sections will be used as inputs in the simulations programs in order to evaluate which portion of the measured signal can be accounted for in terms of a superposition of incoherent processes. Because of its small coupling to the dynamical system, the dilepton generation can be calculated perturbatively.

We have proposed such a program some time ago[6] and we will now turn to the evaluation of the elementary processes. With this in mind we consider the existing data on dilepton production in $p + Be$ collisions, recently acquired by the DLS collaboration at the Bevalac[20]. Be being a small system, we do not expect many-body effects to be important and furthermore we shall neglect rescattering contributions. We first focus upon the 4.9 GeV data[21]. At this energy, we consider the relevant channels to be the bremsstrahlung contribution and the radiative decay of the delta into a nucleon and a virtual photon which in turn internally converts into an electron-positron pair. The hadronic bremsstrahlung piece can be calculated in the soft photon approximation[22]. In this limit the radiated photon has $\omega \to 0$, radiation from the external legs of a diagram dominate the bremsstrahlung amplitude and only the classical characteristics of the electromagnetic current are important. The proton–proton rate is negligible compared to the proton–neutron one[†]. One then analytically continues from a soft photon to a soft lepton pair. Although this approximation is extremely convenient we will see later on, when we consider a more detailed approach, that it is strictly valid only at threshold.

It is well known that the delta has a radiative decay channel $\Delta \to N\gamma$. The measured branching ratio for this process is 0.6 %. We then calculate the decay into an electron-positron pair[23], using a Vector Dominance Model prediction for the delta structure functions. Adding this to the bremsstrahlung contribution and correcting for the DLS acceptance we bin our results in regions of p_T, the pair transverse momentum. The low p_T bin is shown on Fig. 9. The overall agreement is quite good, keeping in mind that our calculation has no free parameters. At higher transverse momenta, the fit is less spectacular[23] and we attribute this to a failure of the soft photon approximation in the hadronic bremsstrahlung sector.

In order to relax this approximation, we have turned to a covariant evaluation of the

---

[†]In fact in this limit the pp rate is identically zero.

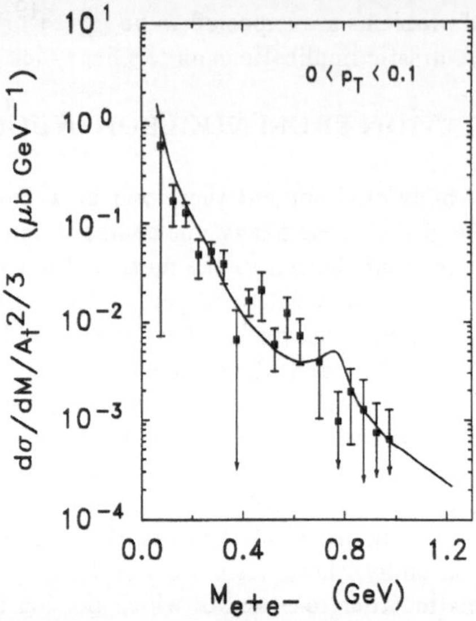

Figure 9. **The acceptance corrected dilepton invariant mass spectrum from the reactions $NN \rightarrow NNe^+e^-$ (bremsstrahlung) and $NN \rightarrow \Delta X \rightarrow e^+e^-X'$ for $0 < p_T < 0.1$ GeV/c. The filled squares are the DLS data.**

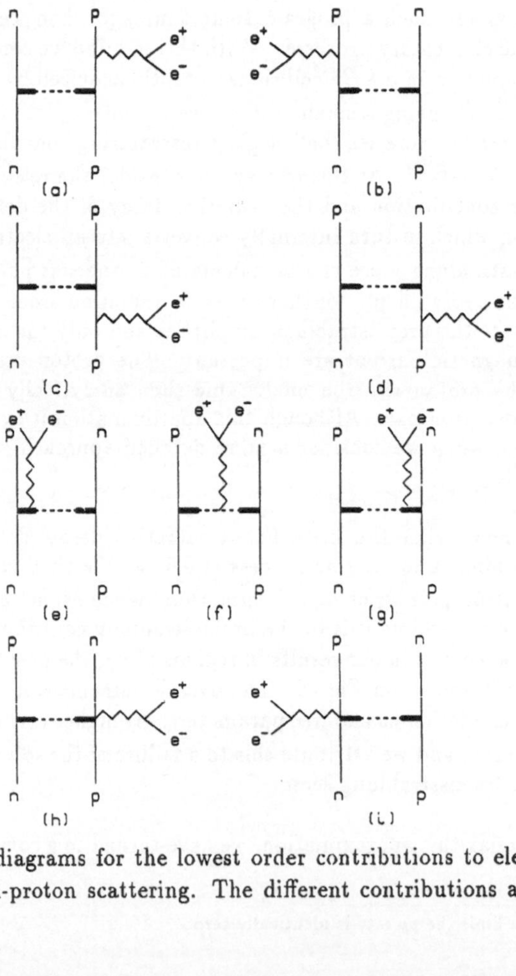

Figure 10. Feynman diagrams for the lowest order contributions to electron-positron production in neutron-proton scattering. The different contributions are enumerated in the text.

e^+e^- emission one pion exchange Feynman diagrams[24]. For the interaction Lagrangian we use a pseudovector coupling of the form

$$\mathcal{L}_{int.} = g_{pv}\bar{\psi}\gamma_5\gamma^\mu\vec{\tau}\psi \cdot \partial_\mu\vec{\pi} \quad . \tag{5.1}$$

The lowest order contributions to the differential cross sections are shown in Figure 10. There are direct and exchange diagrams contributing to the radiation from external lines ("shake–off" diagrams, (a–d)). Minimal substitution in the free pion Lagrangian also gives a coupling of the photon with the exchanged pion (f). To maintain a covariant theory, one must also allow a minimal substitution in the pion–nucleon interaction term. This allows the photon to couple directly to the πNN vertex ((h) and (i)). The two extra diagrams, (e) and (g) follow from the use of strong interaction form factors and from the requirements of gauge invariance.

We now turn to a comparison with the 2.1 GeV DLS data. In our first such calculation, we do not include contributions from the delta. Work to include it in our covariant formalism is in progress. For the aforementioned energy of 2.1 Gev the kinematical limit is near the ρ mass. Therefore, we do not include electromagnetic form factors for the γNN and $\gamma\pi\pi$ vertices as their effects are suppressed. At higher energies one must include such electromagnetic form factors with time-like arguments. First, we make a comparison among (1) the four "shake–off" diagrams, which should be dominant for soft dileptons, (2) the soft virtual photon approximation mentioned and used earlier[†] and (3) the full set of diagrams shown in Fig. 10. We plot $d\sigma/dMdyd^2p_T$ *vs.* $M_{e^+e^-}$ for e^+e^- pairs which are emitted back to back ($p_T = 0$) at y = 0 (Fig. 11). At low invariant masses, all three converge as expected. At higher masses, the exact treatment of the radiation from the external legs yields more pairs than the soft photon approximation and the complete calculation even more so. Folding in the DLS acceptance, we compare with experimental data on Fig. 12. We integrated over all measured y and over the p_T range 0 – 100 MeV/c. The bending over at small invariant mass is a result of experimental cut–offs. The data goes up as $M_{e^+e^-}$ goes down and this is consistent with what one would expect from Dalitz decay. We have not attempted to take this into account. We see that the complete calculation produces too many pairs. However, there are improvements of the calculation we see could lead to better agreement. First, the exchange of heavier mesons should be considered. This has now been done and results will be presented elsewhere[25]. Second, electromagnetic form factors should be inserted, including one for the neutrons as they can interact through the Pauli term. Given the small cross section for e^+e^- emission, it is comforting to see that a simple OPE can come within an order of magnitude of representing the data. This gives hope that further measurements and calculations for heavy nucleus collisions will probe the properties of hot and dense nuclear matter and reveal features that were up to now inaccessible.

6. CONCLUSIONS

Within the framework of the numerical BUU model, a momentum dependent mean field parametrization was introduced that matches quite well the properties of more sophisticated microscopic nuclear Hamiltonians and also agrees with the optical potential data. BUU calculations with this relatively soft interaction (K = 215 MeV) reproduce the transverse momentum and flow angle data from the Streamer Chamber. However, an important consequence of our studies is that the transverse momentum analysis at intermediate energies is not a *direct* probe of the nuclear equation of state, as previously thought.

[†]Note that we correct the differential three-body phase space so that energy conservation is guaranteed and the cross section vanishes at the kinematical limit.

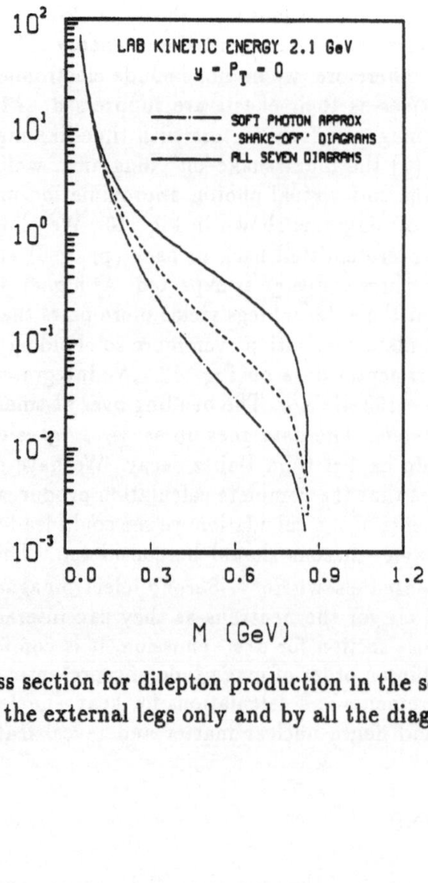

Figure 11. Differential cross section for dilepton production in the soft photon approximation, by radiation from the external legs only and by all the diagrams shown in Fig. 10.

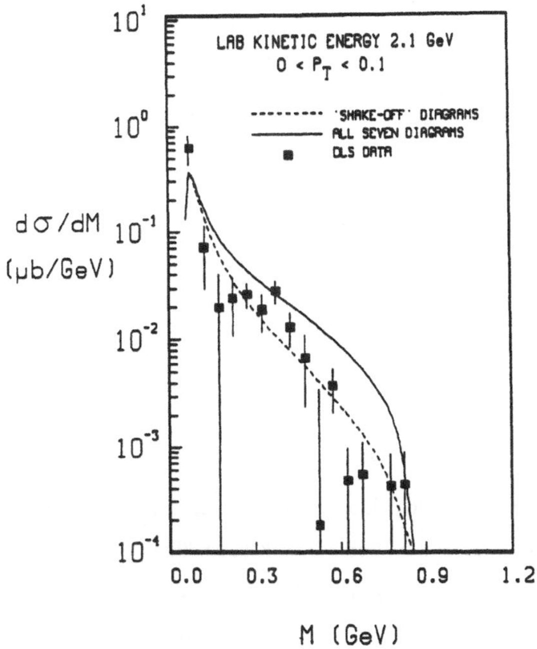

Figure 12. Acceptance corrected differential cross section *vs.* invariant mass for the "shake–off" diagrams and for all the Feynman diagrams. The DLS data are the filled squares.

We have also discussed how the production of electron–positron pairs can carry information on the nuclear many-body environment in general and on the properties of pions in particular. We have stressed that the spectrum of back to back dileptons can in principle be related to the dispersion relation of interacting pions, which is not known. One would also be able to study possible in-medium modifications of the ρ parameters. We have described how a systematic program of such investigations would combine calculations and measurements of elementary dilepton production cross sections. The extrapolation to the nucleus–nucleus cases would then be handled by simulation approaches such as the model described above. With respect to this last point, note that non-local effects in the nuclear interactions could then be important as they can affect the instantaneous scattering rates which in turn determine the dilepton yields. It is not at all clear that the momentum dependence of the mean field can be neglected in any limit and we have seen earlier how this plays an important role in the generation of net transverse momentum. Finally, both for fundamental and pragmatic reasons we have performed calculations of nucleon–nucleon processes first in a simple approach and then we insisted that a more exact theory be covariant and gauge invariant. More theoretical work along these lines clearly needs to be done. In addition to the improvements outlined earlier, one needs calculations of the processes $\pi N \to Ne^+e^-$ and the (off-shell) deltas will also have to be incorporated. Work on this and related topics is in progress.

ACKNOWLEDGEMENTS

It is a pleasure to acknowledge that much of the work discussed above has been done in collaboration with S. Das Gupta, K. Haglin, J. Kapusta, S. J. Lee, M. Prakash and G. Welke.

References

[1] P. Danielewicz and G. Odyniec, *Phys. Lett.*, **157B** (1985)146.

[2] P. Danielewicz et al, *Phys. Rev.*, **C38** (1988)120.

[3] H. A. Gustafsson et al, *Phys. Rev. Lett.*, **52** (1984)1590.

[4] G. Welke, M. Prakash, T. T. S. Kuo, S. Das Gupta, and C. Gale, *Phys. Rev.*, **C38** (1988)2101.

[5] S Hayashi et al., *Phys. Rev.*, **C37** (1988)1229.

[6] C. Gale and J. Kapusta, *Phys. Rev.*, **C35** (1987)2107.

[7] C. Gale and J. Kapusta, *Phys. Rev.*, **C38** (1988)2659.

[8] G. F. Bertsch, H. Kruse, and S. Das Gupta, *Phys. Rev.*, **C29** (1984)673.

[9] H. Kruse, B. V. Jacak, and H. Stöcker, *Phys. Rev. Lett.*, **54** (1985)289.

[10] R. B. Wiringa, *Phys. Rev.*, **C38** (1988)2967.

[11] C.Gale, G. Welke, M. Prakash, S. J. Lee, and S. Das Gupta, Preprint, Chalk River Nuclear Labs., 1989.

[12] C. Gale, G. F. Bertsch, and S. Das Gupta, *Phys. Rev.*, **C35** (1987)1666.

[13] J. Aichelin, A. Rosenhauer, G. Peilert, H. Stöcker, and W. Greiner, *Phys. Rev. Lett.*, **58** (1987)1926.

[14] H. S. Köhler, private communication.

[15] C. M. Ko and Q. Li, In *Proceedings of the 8th High Energy Heavy Ion Study*, page 256, Lawrence Berkeley Laboratory Report No. LBL-24580, 1988; eds. J. W. Harris and G. J. Wozniak.

[16] Proceedings of the International Workshop on Nuclear Dynamics at Medium and High Energies, (Bad Honnef, W. Germany, October 1988, *Nucl. Phys.*, **A495** (1989).

[17] L. H. Xia, C. M. Ko, L. Xiong, and J. Q. Wu, *Nucl. Phys.*, **A485** (1988)721.

[18] T. Ericson and W. Weise, *Pions and Nuclei*, Oxford University Press, Oxford, 1988.

[19] V. W. Metag, these proceedings.

[20] G. Roche, these proceedings and references therein.

[21] G. Roche et al, *Phys. Rev. Lett.*, **61** (1988)1069.

[22] J. D. Jackson, *Classical Electrodynamics*, John Wiley and Sons, 1975.

[23] Charles Gale and Joseph Kapusta, Preprint, Chalk River Nuclear Labs., 1989.

[24] K. Haglin, J. Kapusta, and C.Gale, *Phys. Lett. B*, in press.

[25] K. Haglin, to be published.

[15] G. M. [illegible] in *Proceedings of the* [illegible] *Summer Study*, [illegible] page 288, Lawrence Berkeley Laboratory Report No. LBL-[illegible] (1977); [illegible] editors [illegible] and [illegible] Winick.

[16] *Proceedings of the International Workshop on* [illegible] *and* [illegible] *High Energies* [illegible] Germany, [illegible]

[17] [illegible] *Phys. Rev.* [illegible]

[18] [illegible]

[19] [illegible]

[20] [illegible]

[21] [illegible]

[22] [illegible]

[23] [illegible]

[24] [illegible] *Rev.* [illegible] in press.

[25] [illegible] be published.

RELATIVISTIC VLASOV-UEHLING-UHLENBECK MODEL FOR HIGH-ENERGY HEAVY-ION COLLISIONS *

C. M. Ko, Q. Li, J. Q. Wu, and L. H. Xia

Cyclotron Institute and Center for Theoretical Physics
Texas A&M University
College Station, Texas 77843

1. INTRODUCTION

One of the main motivations for carrying out research in heavy-ion collisions is to create nuclear matter at various densities and excitation energies in order to map out the nuclear phase diagram. The normal nuclei have a density of $\rho_0 = 0.16\ \mathrm{fm}^{-3}$ and is at zero temperature. To extend beyond this requires the compression and deposition of energy in the nuclear matter. Experiments carried out so far indicate that heavy-ion collisions indeed offer such a possibility. Depending on the incident energy per nucleon in the collision, differernt regions of the nuclear phase diagram can be probed.

For heavy-ion collisions at low and intermediate energies, i.e. below about 100 MeV per nucleon, a hot nucleus of temperature up to many MeV is formed. From the decay of this hot compound nucleus, its properties can be learnt. In particular, the observation of multifragments in the final state of the reaction offers the possibility of investigating the interesting phenomenon of the nuclear gas-liquid phase transition. For collisions at these energies, nucleons move with velocities comparable to the Fermi velocity and the collision dynamics is thus determined by both the nuclear mean field and the nucleon-nucleon collisions. A very successful model for describing the collision has been the one based on the Vlasov-Uehling-Uhlenbeck (VUU) equation[1]. In this model, the mean field is derived from the Skyrme interaction and the nucleon-nucleon collision is given by its cross section in the free space, corrected by the Pauli principle. This model has been extensively used in studying heavy-ion collisions at this energy regime[2-6].

For heavy-ion collisions at high energies, i.e. between about 100 MeV/nucleon to about a few GeV/nucleon, the projectile is still stopped by the target, resulting in the formation of a highly compressed and excited nuclear matter in the initial stage of the collision. The property of such a high-density nuclear matter is relevant to the study of supernova explosion and neutron star. At these energies, nucleons move with a velocity not negligible with respect to the velocity of light and it is thus important to take into account the relativistic effects. To include the relativistic kinematics is trivial and has already been taken into account in the normal VUU model. Other relativistic effects include the explicit mesonic-exchange nature of the nucleon-nucleon interaction, the small component of the nucleon wave function, and the existence of negative-energy states. Also at very high density and/or temperature, the restoration of the chiral symmetry is expected to play an important role in high-energy heavy-ion

* Work supported in part by NSF Grant No. PHY-8608149 and the Robert A. Welch Foundation Grant No. A-1110

collisions. These effects can be included via the quantum hadrodynamics (QHD)[7] in which the nuclear matter is treated as a system of interacting baryons and mesons.

For heavy-ion collisions at ultrarelativistic energies, i.e. above a few 10 GeV/nucleon, a deconfined phase of quark-gluon matter is expected to form. Such a phase of matter existed during the early evolution of the universe. The dynamics of the quark-gluon matter is described by the quantum chromodynamics. Due to the color confinement, the quark-gluon matter transforms into the hadronic matter when it expands and cools. It is thus a great challenge to heavy-ion physicists to find the signatures for the formation of the quark-gluon matter in the initial stage of heavy-ion collisions.

In this talk, the derivation of a relativistic VUU equation from the QHD and some results from its application to high-energy heavy-ion collisions are reported. It is based on our recent work published in refs.8-11). This talk is organized as follows. In Sec. 2, the quantum hadrodynamics is briefly reviewed. Sec. 3 decribes how the relativistic Vlasov equation is obtained from QHD. The extension of QHD to include the scalar meson self-interaction is described in Sec.4. The collision term is introduced in Sec. 5. The application of the relativistic VUU model to heavy-ion collisions is given in Sec. 6. In Sec. 7, the effect of meson collectivity on the nucleon-nucleon cross section in dense nuclear matter is discussed. Finally, conclusions are given in Sec. 8.

2. QUANTUM HADRODYNAMICS

Quantum hadrodynamics is a phenomenological field-theoretical model for the nuclear system. In principle, all mesons that can be exchanged between two nucleons should be included in constructing the Lagrangian of this model. The parameters in this model, such as the masses of the mesons and their coupling constants to the nucleon, are determined by fitting to the nucleon-nucleon scattering data[12]. This involves thus solving the Bethe-Salpeter equation. For two nucleons in the nuclear matter the solution of the relativistic Brueckner-Bethe-Goldstone equation is required. This study is being pursued by Malfliet *et al.*[13] and Faessler *et al.*[14]. Our approach is more modest in that we shall include only a few mesons and treat the Lagrangian as an effective one, i.e. we shall only consider up to second order in the coupling contants. The four mesons in our Lagrangian are the isoscalar scalar and vector mesons, denoted by σ and ω, respectively, and the isovector pseudoscalar and vector mesons, denoted by π and ρ, respectively. With the nucleon field given by ψ, our effective Lagrangian has the following form,

$$\begin{aligned}L(x) =&\bar{\psi}(\gamma^\mu i\partial_\mu - m)\psi + \frac{1}{2}(\partial_\mu\sigma\partial^\mu\sigma - m_\sigma{}^2\sigma^2) - \frac{1}{4}F^{\mu\nu}F_{\mu\nu} + \frac{1}{2}m_\omega^2\omega^\mu\omega_\mu \\ &+ \frac{1}{2}(\partial_\mu\pi\cdot\partial^\mu\pi - m_\pi{}^2\pi\cdot\pi) - \frac{1}{4}\mathbf{G}^{\mu\nu}\cdot\mathbf{G}_{\mu\nu} + \frac{1}{2}m_\rho^2\rho^\mu\cdot\rho_\mu \\ &+ g_\sigma\bar{\psi}\psi\sigma - g_\omega\bar{\psi}\gamma_\mu\psi\omega^\mu - (g_\pi/2m)\bar{\psi}\gamma^5\gamma^\mu\tau\psi\cdot\partial_\mu\pi - g_\rho\bar{\psi}\gamma_\mu(\tau/2)\psi\cdot\rho^\mu, \quad (1)\end{aligned}$$

where $F_{\mu\nu} = \partial_\mu\omega_\nu - \partial_\nu\omega_\mu$ and $\mathbf{G}_{\mu\nu} = \partial_\mu\rho_\nu - \partial_\nu\rho_\mu$. The mass of nucleon and its isospin are denoted by m and τ, respectively. The masses of the mesons are given by m_σ, m_ω, m_π, and m_ρ, and their coupling constants to the nucleon are denoted by g_σ, g_ω, g_π, and g_ρ. All the meson masses except the σ meson are taken to be their measured masses, i.e. $m_\omega = 783$ MeV, $m_\pi = 138$ MeV, and $m_\rho = 770$ MeV. For the σ meson mass and the coupling coustant g_π, we use the value adopted from fitting the nucleon-nucleon scattering data[12], i.e. $m_\sigma = 550$ MeV and $g_\pi = 13.45$. The other coupling constants g_σ, g_ω, and g_ρ are determined by Walecka[15] to be $g_\sigma = 9.58$, $g_\omega = 11.68$, and $g_\rho = 6.06$ from the nuclear and neutron matter properties in the mean-field approximation to QHD.

The important quantities for our discussion are the Green's functions for the particles. They are defined, respectively, by

$$\begin{aligned}iG(x_1, x_{1'}) =& <|\, T[\psi(x_1)\bar{\psi}(x_{1'})]\,|> \\ =& \,\theta(t_1 - t_{1'})iG^>(x_1, x_{1'}) + \theta(t_{1'} - t_1)iG^<(x_1, x_{1'}), \quad (2a)\end{aligned}$$

for the nucleon and

$$iD_a(x_1,x_{1'}) = <| T[\phi_a(x_1)\phi_a^+(x_{1'})] |>, \tag{2b}$$

for the mesons, where ϕ_a denotes any one of the four mesons. In the above, $< \cdots >$ denotes the expectation value in the nuclear many-body state and T is the time-ordering operator defined on a time contour as discussed in refs.16,17). The nucleon Green's function satisfies the following equation

$$(i\gamma_\mu \partial_{x_1}^\mu - m)G(x_1,x_{1'}) = \delta(x_1 - x_{1'}) + \int d^4u \Sigma(x_1,u)G(u,x_{1'}), \tag{3}$$

where Σ is the nucleon self-energy and has a perturbative expansion. Diagrammatically, it is shown in Fig.1 up to the second order in the coupling constants. The dotted line in Fig.1 denotes any one of the four mesons.

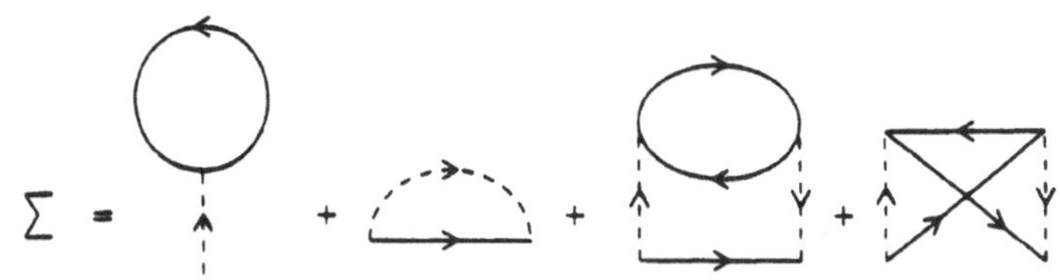

Fig. 1. First and second order nucleon self-energies. The solid line denotes the nucleon while the dashed line represents the mesons.

3. RELATIVISTIC VLASOV EQUATION

The first order nucleon self-energy is normally written as[18]

$$\Sigma_{HF}(x) = \Sigma_s(x) + i\gamma_\mu \Sigma_v^\mu(x). \tag{4}$$

It can be absorbed into the effective mass and momentum of the nucleon, i.e.

$$m^* = m + \Sigma_s, \qquad p_\mu^* = p_\mu + \Sigma_{v\mu}. \tag{5}$$

In this limit, the equation satisfied by the nucleon Green's function can thus be written as

$$\{i\gamma_\mu[\partial_{x_1}^\mu - \Sigma_v^\mu(x_1)] - m(x_1)\}G^<(x_1,x_{1'}) = 0. \tag{6}$$

Let us introduce the Fourier transform of the nucleon Green's function

$$G^<(x,p) = \int \frac{d^4y}{(2\pi)^4} e^{ipy} G^<(x+y/2, x-y/2). \tag{7}$$

In the semiclassical approximation, it is related to the seven-dimensional on-shell nucleon's phase-space distribution function $f(x,\mathbf{p}^*)$, i.e.

$$Tr[G^<(x,p)] = i16\pi\delta(p^{*2} - m^2)mf(x,\mathbf{p}^*). \tag{8}$$

From eq.(6) a relativistic Vlasov equation follows[19] and has the form

$$\{[\partial_x^\mu - (\partial_x^\mu \Sigma_v^\nu - \partial_x^\nu \Sigma_v^\mu)\partial_\nu^{p^*}]p_\mu^* + m(\partial_x^\mu m)\partial_\mu^{p^*}\}f(x,\mathbf{p}^*) = 0, \tag{9}$$

This equation can be more conveniently written as[10]

$$\frac{\partial}{\partial t}f + \mathbf{v}\cdot\nabla_x f - \nabla_x U\cdot\nabla_p f = 0, \tag{10}$$

where

$$\mathbf{v} = \mathbf{p}^*/E^*, \qquad \nabla_x U = \nabla_x(E^* + (g_\omega/m_\omega)^2\rho_N), \tag{11}$$

with $E^* = [\mathbf{p}^{*2} + m^{*2}]^{1/2}$ and ρ_N being the nuclear matter baryon density. Neglecting the exchange contribution, we can write

$$\Sigma_s = -g_\sigma < \sigma > = -(g_\sigma/m_\sigma)^2\rho_s, \qquad \Sigma_{v\mu} = -g_\omega < \omega_\mu > = -(g_\omega/m_\omega)^2\rho_\mu, \tag{12}$$

where ρ_s and ρ_μ are the nucleon scalar and current densities, respectively. In terms of the phase-space distribution function, the scalar and current densities in the local-density approximation can be expressed, respectively, as

$$\rho_s = \int d^3\mathbf{p}^* f(x,\mathbf{p}^*)m^*/E^*, \tag{13a}$$

and

$$\rho_\mu = \int d^3\mathbf{p}^* f(x,\mathbf{p}^*)p^*_\mu/E^*. \tag{13b}$$

To solve the Vlasov equation, we use the method of pseudoparticles, which was first introduced to heavy-ion collisions by Wong[20]. In this method, each nucleon is replaced by a collection of test particles and the one-body phase-space distribution function is given by the distribution of these test particles in the phase space. To solve the Vlasov equation is then equivalent to the solution of the classical equations of motion, eq.(11), for all these test particles. The time component of the current density is just the baryon density and is simply determined by the density of the nucleons in the coordinate space. For the scalar density and the spatial components of the current density, it is nontrivial to evaluate them exactly due to their nonlinear forms. We shall therefore calculate them approximately by substituting $\mathbf{p}^*$ and $\mathbf{p}^{*2}$ with their mean values at x and multiplying the resulting expressions by the local baryon density[8].

Results obtained with this equation for high-energy heavy-ion collisions reproduce very well those from the time-dependent Dirac approach based on the same Lagrangian[21]. In particular, both lead to an appreciable transverse momentum distribution.

4. SCALAR MESON SELF-INTERACTION

The original Walecka mean-field model gives a nuclear compressibility of 540 MeV which is much larger than that determined from the energy of the giant monopole resonances in nuclei. It also leads to a nucleon effective mass of 0.56 m at the saturation density which is smaller than the value extracted from the optical-model analysis of nucleon-nucleus scattering[22]. Including the exchange term, the vacuum and particle-hole polarizations reduces the nuclear compressibility to 360 MeV and increases the nucleon effective mass to 0.77 m[23]. To obtain similar results without carrying out such a complicated calculation, we shall include the self-interaction of the scalar meson, i.e.

$$U(\sigma) = b\sigma^3/3 + c\sigma^4/4. \tag{14}$$

The two parameters b and c make it possible to obtain different values of the nuclear compressibility and the nucleon effective mass as shown in ref.24). In this case, the relation between the value of the scalar meson and the nuclear scalar density becomes

$$m_s{}^2 < \sigma > + b < \sigma >^2 + c < \sigma >^3 \approx g_s\rho_s. \tag{15}$$

Indeed, the results of ref.23) can be obtained with the parameters, $C_\sigma = (g_\sigma/m_\sigma)m = 12.60$, $C_\omega = (g_\omega/m_\omega)m = 9.36$, $B = b/(g_\sigma{}^3 m) = 1.5 \cdot 10^{-3}$, and $C = c/g_\sigma{}^4 = 2.88 \cdot 10^{-2}$. Including the scalar meson self-interaction reproduces not only the nuclear compressibility and the nucleon effective mass, but also the density dependence of the nuclear binding energy and the nucleon effective mass.

5. COLLISION TERM

If we include the second-order Born term in the nucleon self-energy, then its imaginary part can be shown to lead to the collision integral on the right hand side of the relativistic Vlasov equation[11],

$$I_c = \int \frac{d\mathbf{p}_2^*}{(2\pi)^3} \int \frac{d\mathbf{p}_3^*}{(2\pi)^3} \int \frac{d\mathbf{p}_4^*}{(2\pi)^3} | M |^2 (2\pi)^4 \delta^3(\mathbf{p}^* + \mathbf{p}_2^* - \mathbf{p}_3^* - \mathbf{p}_4^*)$$
$$\cdot \delta(E^* + E_2^* - E_3^* - E_4^*)\{f(x,\mathbf{p}_3^*)f(x,\mathbf{p}_4^*)[1 - f(\mathbf{p}^*)][1 - f(x,\mathbf{p}_2^*)]$$
$$- f(x,\mathbf{p}^*)f(x,\mathbf{p}_2^*)[1 - f(x,\mathbf{p}_3^*)] \cdot [1 - f(x,\mathbf{p}_4^*)]\}, \qquad (16)$$

where $| M |^2$ is the spin and isospin average of the square of the nucleon-nucleon invariant scattering amplitude M in the Born approximation,

$$M = i \sum_{a=\sigma,\omega,\pi,\rho} [\bar{U}_x^{s_3}(p_3^*)\Gamma_a U_x^s(p^*) D_a(x, p^* - p_3^*) \bar{U}_x^{s_4}(p_4^*)\Gamma_a U_x^{s_2}(p_2^*)$$
$$- \bar{U}_x^{s_4}(p_4^*)\Gamma_a U_x^s(p^*) D_a(x, p^* - p_4^*) \bar{U}_x^{s_3}(p_3^*)\Gamma_a U_x^{s_2}(p_2^*)], \qquad (17)$$

and is related to the nucleon-nucleon cross section. In the above, $U_x^s(p^*)$ is the spinor wave function for spin state s and is normalized according to $Tr \sum_s U_x^s(p^*)\bar{U}_x^s(p^*) = 4m/E^*$; $D_a(x, p^* - p_3^*)$ is the Fourier transform of the meson propagator; and $\Gamma_a = ig_\sigma, -ig_\omega\gamma_\mu, -i(g_\pi/2m)\gamma^5\gamma^\mu\partial_\mu, -ig_\rho\gamma_\mu/2$ for σ, ω, π, ρ, respectively. The collision term contains both the Pauli-blocking factors and the nucleon-nucleon cross section and has the familiar form as that used in the normal VUU model.

As in the normal VUU model, we shall first use in the collision term the isospin-averaged nucleon-nucleon cross section in the free space. At high energies, it can be parametrized by[25]

$$\frac{d\sigma}{d\Omega} = ae^{2bp^2(cos\theta - 1)}, \qquad (18)$$

where a and b are energy-dependent parameters and are given in ref.25). The empirical nucleon-nucleon cross section can be reproduced by our effective meson model if we introduce the from factors in the meson-nucleon vertices. We use the following form

$$F_a(q^2) = \frac{\Lambda_a^2}{\Lambda_a^2 - q^2}, \qquad (19)$$

where q^2 is the four momentum transfer, a labels the mesons, and Λ_a's are the cutoff parameters which we take as free parameters. Since all the form factors have values of one at $q^2 = 0$, the mean-field results are not affected. We have calculated the nucleon-nucleon cross section in the Born approximation with the four effective mesons of coupling constants $g_\sigma = 6.885$, $g_\omega = 7.546$, $g_\pi = 13.45$, and $g_\rho = 8.217$. These coupling constants are different from those of the original Walecka model and are obtained, as shown in the next section, from the inclusion of the scalar meson self-interaction such that the nuclear matter compressibility is 380 MeV and the nucleon effective mass is $0.83m$. The nucleon-nucleon differential cross section at 1 GeV obtained with the cutoff parameters $\Lambda_\sigma = 1.098$ GeV, $\Lambda_\omega = 0.592$ GeV, and $\Lambda_\pi = 0.576$ GeV and $\Lambda_\rho = 0.792$ GeV is shown in Fig.2 by the solid curve. The empirical cross section of eq.(18) is given by the dotted curve. We see that the calculated cross section agrees very well with the empirical one.

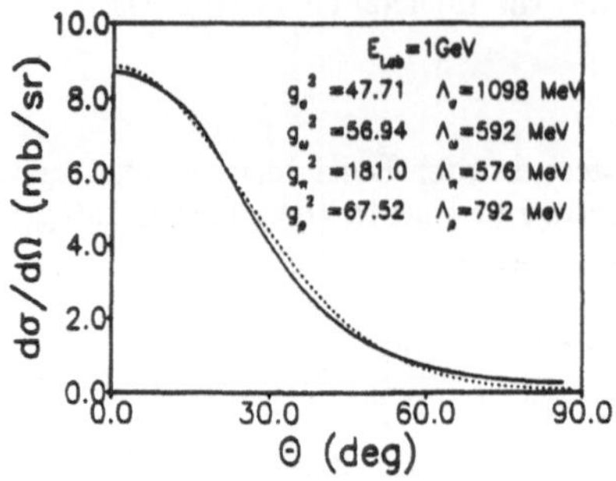

Fig. 2. Nucleon-nucleon elastic cross section in free space. The experimental cross section is given by the dotted curve while the Born approximation gives the solid curve.

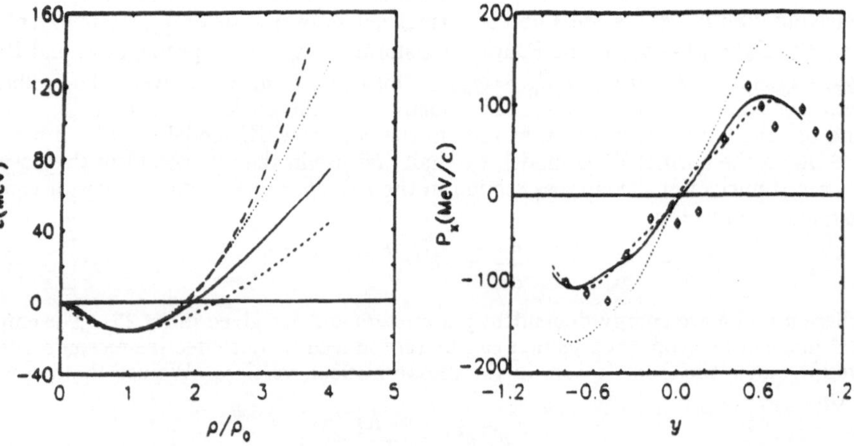

Fig. 3. Left panel: the binding energy per nucleon ϵ as a function of the nuclear matter density. Right panel: the transverse momentum projected onto the reaction plane and averaged over all perpendicular momentum P_x, as a function of the rapidity y for the reaction ${}^{40}Ca+{}^{40}Ca$ at 1.8 GeV/nucleon. The three cases are: solid curve (m^*=0.83 m, K=380 MeV), dashed curve (0.83 m, 200 MeV), and dotted curve (0.7 m, 380 MeV). The long-dashed curve in the left panel corresponds to a stiff equation of state used in the normal VUU model[2] while the open diamonds in the right panel are the experimental data from ref.26).

6. APPLICATION TO HEAVY-ION COLLISIONS

Using the relativistic VUU model, we have studied the reaction $^{40}Ca + ^{40}Ca$ at an incident energy of 1.8 GeV/nucleon and at an impact parameter of 2 fm corresponding to a central collision[9]. The nucleon transverse momentum distribution after the collision has been measured[26]. An earlier study of this reaction with the normal VUU model has led to the conclusion that the nuclear equation of state is rather stiff with a compressibility of K=380 MeV[2]. This equation of state, expressed as the binding energy per nucleon as a function of the nuclear density, is shown in the left panel of Fig.3 by the long-dashed curve. We have calculated the transverse momentum distribution for this reaction with the same compressibility but with an effective nucleon mass of 0.83 m. This equation of state is obtained with the following values for the parameters, $C_\sigma = 11.78$, $C_\omega = 9.04$, $B = -2.59 \cdot 10^{-2}$, and $C = 0.169$. It is given by the solid curve in the left panel of Fig.3 and is seen to be much softer than the previous one at high densities. The transverse momentum distribution as a function of the rapidity calculated from the relativistic VUU model with these parameters is shown in the right panel of Fig.3 by the solid curve. It agrees reasonably with the experimental data as shown by the open diamonds. When a momemtum-dependent potential is included in the normal VUU model, a similar result can be obtained for the transverse momentum distribution with a compressibility of about 200 MeV[27,28]. We have also carried out calculations with this value of the compressibility and an effective nucleon mass 0.83 m. This equation of state is shown by the dashed curve in the left panel of Fig.3 and is obtained with the parameters, $C_\sigma = 14.5$, $C_\omega = 9.04$, $B = 1.73 \cdot 10^{-2}$, and $C = -1.20 \cdot 10^{-3}$. The resulting transverse momentum distribution is shown in the right panel of Fig.3 by the dashed curve and is similar to the previous one obtained with a compressibility of 380 MeV. It seems therefore that in the relativistic model the transverse momentum distribution in high-energy heavy-ion collisions is not sensitive to the value of the compressibility at the normal nuclear matter density.

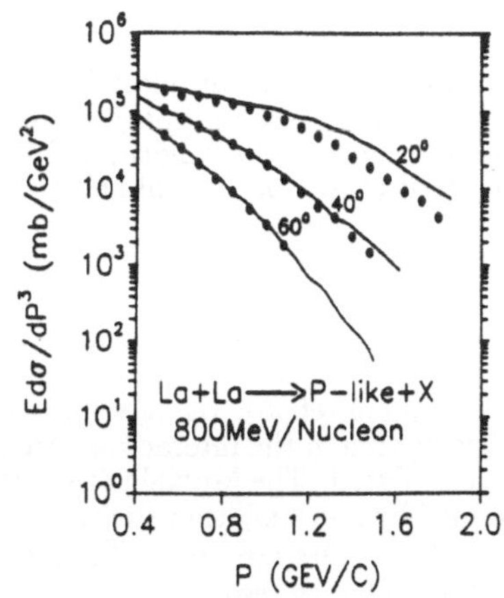

Fig. 4. Inclusive p-like cross section for La+La collision at 800 MeV per nucleon. The experimental data denoted by the solid circles are from ref.30). The calculated cross section using the relativistic VUU model is given by the solid lines.

On the other hand, the transverse momentum distribution changes drastically if a different value of the nucleon effective mass is used as shown by the dotted curve in the right panel of Fig.3 which corresponds to an effective mass of 0.7 m and a compressibility of 380 MeV. The corresponding equation of state is given by the dotted curve in the left panel of Fig.3 and is obtained by using the parameters, $C_\sigma = 16.05$, $C_\omega = 13.0$, $B = -1.69 \cdot 10^{-4}$, and $C = 5.18 \cdot 10^{-3}$. The reason for this is because a smaller effective mass implies a larger value of the vector meson coupling constant which leads thus to a stronger sideways pressure on the particles. Similar studies with the relativistic VUU model have also been carried out in ref.29).

We have also calculated the inclusive p-like cross sections for La+La collision at 800 MeV/nucleon with the parameters corresponding to K=380 MeV and $m^* = 0.83m$. The results are shown by the solid lines in Fig.4 for the three angles of 20^0, 40^0, and 60^0. Compared with the experimental data of Hayashi *et al.*[30] denoted by the solid circles, we find that the theoretical results fail to reproduce the low cross section yields at 20^0. This is the case also for all other transport models[31]. The common feature of all models is the assumption that the nucleon-nucleon cross section can be taken from the free space data. However, many-body effects can modify the in-medium cross section. If the present data are free from systematic errors, then a better understanding of nuclear transport at high densities is called for.

7. MESON COLLECTIVITY

Because of the strong interactions of the mesons with nucleons, we expect that their propagators in the nuclear matter will be very different from those in the free space. In the nuclear matter, the meson propagator satisfies the Dyson equation

$$D_a(q) = D_{0a}(q) + D_{0a}(q)\Pi_{ab}(q)D_b(q), \tag{20}$$

where $D_0(q)$ is the free meson propagator and $\Pi_{ab}(q)$ is the self-energy of the meson in nuclear matter. In terms of the particle-hole polarization of the nuclear matter, it is given by

$$\Pi_{ab}(q) = i\int \frac{d^4k}{(2\pi)^4} Tr[\Gamma_a G_0(k)\Gamma_b G_0(k+q)]. \tag{21}$$

Using these dressed meson propagators in the Born amplitude of eq.(17), we can determine the density dependence of the nucleon-nucleon cross section in the nuclear matter. Such studies have already been explored in Ref.32) for the kaon production cross section.

The meson self-energies can be carried out relativistically as in ref.33). As shown in ref.34), singularities appear in the meson propagators even at the normal nuclear matter density because of the occurence of longitudinal modes of excitation in the nuclear matter. These singularities are unphysical and will be removed if the short-range correlation and the exchange interaction between the nucleons are taken into account. We include these effects via the introduction of a zero-range Migdal-type interaction. In terms of the phenomenological Migdal parameters f and g', the meson self-energies can be written as $\tilde{\Pi} = \Pi/(1-f\Pi)$ and $\tilde{\Pi} = \Pi/(1-g'\Pi/q^2)$, for the isoscalar and the isovector mesons, respectively. To estimate the value for the Migdal parameters, we follow the method of ref.35). Using a short-range correlation function with a range of 0.7 fm, the reduction in the interaction strength of two nucleons from the meson exchange can be calculated. The Migdal parameters due to the short-range correlation are then determined by requiring that the same reduction is obtained with the zero-range interaction. The results are about 0.018 fm^2 and 0.38, respectively, for f and g'. As in ref.35), the exchange term can be expressed in terms of the direct term when averaging over the spin and isospin and leads to further contribution to the Migdal parameters. The exchange contribution depends on the nuclear density. At the normal nuclear matter density, it increases the value of f to 0.054 fm^2 and that of g' to 0.48. Our value of g' has a similar magnitude as that used in ref.36).

We have calculated the meson propagators in the nuclear matter with the particle-hole polarization renormalized by the Migdal-type interaction and have found that they are indeed free of singularities. The ratio of the total nucleon-nucleon cross section calculated with these meson propagators to that in the free space is shown in Fig.5 as a function of the nuclear matter density. It is seen that the nucelon-nucleon cross section in the medium increases with the nuclear matter density. This is similar to the conclusion obtained by Brown *et al.*[36] in a model with pion and rho meson only.

To see the effect of the enhanced nucleon-nucleon cross section in the nuclear matter on the dynamics of heavy-ion collisions, we have carried out a calculation for the collision between two Ca nuclei at 1 GeV/nucleon and at an impact parameter of 2 fm using the relativistic transport equation. The flow angle, which is given by the angle between the major axis of the momentum tensor of all particles with respect to the beam axis, is about 9^0 and is larger than the 7^0 obtained with the free nucleon-nucleon cross section. This effect is expected to be more appreciable for the collision between heavier systems.

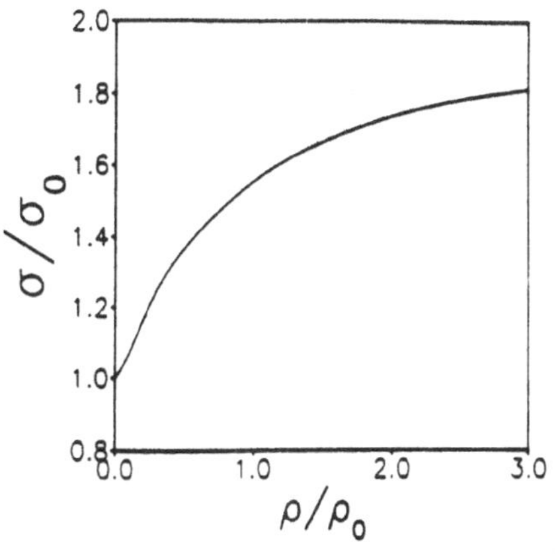

Fig. 5. The ratio of the nucleon-nucleon elastic cross sectin in the nuclear matter to that in the free space as a function of the ratio of the nuclear matter density ρ to its normal value ρ_0.

8. CONCLUSIONS

To describe high-energy heavy-ion collisions, the quantum hadrodynamics is used to include the relativistic effects. In the semiclassical and local approximations, a relativistic Vlasov equation can be derived from QHD by keeping only the Hartree term of the nucleon self-energy. To obtain various values of nuclear compressibility and the nucleon effective mass in the nuclear matter, we have included the scalar meson self-interaction. In the mean-field approximation, it reproduces the more complicated QHD calculations with the exchange term, and the vacuum and particle-hole polarizations. Including the second-order term of the nucleon self-energy in the nuclear matter, a collision integral can also be derived and has the familiar form as that in the normal VUU equation. It contains both the Pauli-blocking factors and the nucleon-nucleon cross section with the latter given by the Born diagram involving meson exchanges. We have found that the nucleon-nucleon cross section in the free space can be very well reproduced with these mesons in the Born approximation.

The relativistic VUU model has been used to study both the transverse momentum distribution and the p-like inclusive cross section in heavy-ion collisions. From comparing with the available experimental data on transverse momentum distribution, we have seen that a relatively soft equation of state is required. In the relativistic model, the magnitude of the transverse momentum from heavy-ion collisions is more sensitive to the value of the nucleon effective mass at saturation density than the value of the compressibility at this density. The failure of the model to reproduce the low yields at small angles for the p-like inclusive cross section calls for a better understanding of the nuclear transport at high densities.

In the relativistic VUU model, the density dependence of the nucleon-nucleon cross section can be incorporated by dressing the meson propagators through the particle-hole polarization of the nuclear matter. Preliminary studies indicate that the in-medium nucleon-nucleon cross section increases with the nuclear matter density. Such an enhanced cross section leads to a larger collective flow angle in heavy-ion collisions than that obtained with the free nucleon-nucleon cross section.

The relativistic VUU model will be very useful in studying heavy-ion collisions. However, many problems remain to be solved in the future. In deriving the collision integral, we have only take into account the imaginary part of the second-order nucleon self-energy. It is important to consider also its real part as it contributes to the nuclear mean field. The nuclear mean field obtained form the original Walecka model does not have the correct energy dependence and improvement may be achieved by including the second-order correction. The inclusion of the short-range correlation in the evaluation of the meson self-energies requires a more consistent relativistic treatment. Also the mesonic degrees of freedom should be included explicitly. In particular, a relativistic kinetic equation is needed for pions as they are copiously produced in high-energy heavy-ion collisions. For the same reason, both the Δ and N^* resonances require careful studies as well. Finally, the effect of vacuum polarization on both the nuclear mean field and the meson propagations have to be addressed. Only then, we shall be able to extract important information about the properties of hot and dense hadronic matter formed in high-energy heavy-ion collisions.

REFERENCES

1. G. F. Bertsch, H. Kruse, and S. Das Gupta, Phys.Rev. C29, 673 (1984).
2. J. J. Molitoris and H. Stöcker, Phys. Rev. C32, 346 (1985).
3. J. Aichelin and G. F. Bertsch, Phys. Rev. C31, 1730 (1985).
4. W. Bauer *et al.*, Phys. Rev. C34, 2127 (1986).
5. C. M. Ko and J. Aichelin, Phys. Rev. Lett. 55, 2661 (1985); Phys. Rev. C35, 1976 (1987).
6. H. Kruse, B. V. Jacak, and H. Stöcker, Phys. Rev. Lett. 52, 289 (1985).
7. B. D. Serot and J. D. Walecka, in Advances in Nuclear Physics, edited by J. W. Negale and E. Vogt (Plenum, New York, 1986), Vol. 16.
8. C. M. Ko, Q. Li, and R. Wang, Phys. Rev. Lett. 59, 1084 (1987).
9. C. M. Ko and Q. Li, Phys. Rev. C37, 2270 (1988).
10. Q. Li and C. M. Ko, Mod. Phys. Lett. A3, 465 (1988).
11. Q. Li, J. Q. Wu, and C. M. Ko, Phys. Rev. C, Phys. Rev. C39, 849 (1989).
12. K. Holinde, Phys. Rep. 68, 122 (1981).
13. B. ter Haar and R. Malfliet, Phys. Rep. 149, 207 (1987).
14. A. Faessler, Nucl. Phys. A495, 103c (1989).
15. J. D. Walecka, Ann. Phys. 83, 491 (1974).
16. P. Danielewicz, Ann, Phys. 152, 239 (1984).
17. L. P. Kadanoff and G. Baym, "Quantum Statistical Mechanics", Benjamin, New York, 1962.
18. S. A. Chin, Ann. Phys. 108, 301 (1977).
19. H. Elze *et al.*, Mod. Phys. Lett. A2, 451 (1987).
20. C. Y. Wong, Phys. Rev. C25, 1461 (1982).
21. R. Y. Cusson *et al.*, phys. Rev. Lett. 55, 2786 (1985).
22. C. H. Johnson, D. J. Horen, and C. Mahaux, Phys. Rec. C36, 2252 (1987).
23. X. Ji, Phys. Lett. B208, 19 (1988).
24. J. Boguta and H. Stöcker, Phys. Lett. B120, 289 (1983).
25. J. Cugnon, D. Kinet, and J. Vandermeulen, Nucl. Phys. A352, 505 (1981); A379, 553 (1982).
26. H. Ströbele *et al.*, Phys. Rev. C27, 1349 (1983).
27. J. Aichelin *et al.*, Phys. Rev. Lett. 58, 1926 (1987).
28. C. Gale, G. Bertsch, and S. Das Gupta, Phys. Rev. C35, 1666 (1987).
29. B. Blättel, V. Koch, W. Cassing, and U. Mosel, Phys. Rev. C38, 1767 (1988).
30. S. Hayashi *et al.*, Phys. Rev. C38, 1229 (1988).
31. J. Aichelin *et al.*, Phys. Rev. Lett. 62, 1461 (1989).
32. J. Q. Wu and C. M. Ko, Nucl. Phys. A, in press.
33. H. Krusawa and T. Suzuki, Nucl. Phys. A445, 685 (1985).
34. B. L. Friman and P. A. Henning, Phys. Lett. B206, 579 (1988).
35. A. Arima *et al.*, Phys. Lett. B122, 126 (1983).
36. G. E. Brown, E. Oset, M. Vicente Vacas, and W. Weise, Stony Brook preprint (1989).

The Relativistic BUU Approach - Analysis of Retardation Effects and Thermal Properties *

Bernhard Blättel, Volker Koch, Andreas Lang, Klaus Weber
Wolfgang Cassing and Ulrich Mosel

Institut für Theoretische Physik, Universität Giessen
6300 Giessen, West Germany

1 Introduction

The study of the nuclear equation of state (EOS) is one of the major goals in high energy heavy-ion collisions. However, it has turned out that the extraction of reliable information about the EOS is a nontrival task. Theoretical investigations, like simulations of the Boltzmann-Uehling-Uhlenbeck (BUU) equation have to include the momentum dependence of the mean-field potential [1] and should also study the sensitivity of the observables on medium modifications of the nucleon-nucleon cross section [2]. Therefore, at bombarding energies in the order of 1 GeV/nucleon a relativistic treatment, which consistently includes these features, is necessary. In this spirit, a relativistic transport approach has been proposed and investigated [3,4,5,6]. Together with a relativistic generalization of the BUU transport equation one obtains equations of motion for the meson-fields which mediate the interaction. In the first part of this contribution we present results which include the full solution of these field-equations and compare to different approximations. In particular we address the question of possible retardation effects in the case of fast moving nucleon sources. In the second part we investigate to which extend thermodynamic approaches are suited for the description of heavy-ion collisions at the considered energies and masses. We therefore calculate the pressures reached during the simulation in our microscopic approach and extract the size and the temperature of the equilibrated regions.

2 Analysis of retardation effects

The relativistic BUU approach is based on the transport equation [5]

$$\Big(\Pi_\mu \partial^\mu_x + \big(g_v \Pi_\nu F^{\mu\nu} + m^*(\partial^\mu_x m^*) \big) \partial^\Pi_\mu \Big) f(x,\Pi) =$$
$$\int \frac{d^3\Pi_1}{\Pi_1^0} \frac{d^3\Pi'}{\Pi'^0} \frac{d^3\Pi'_1}{\Pi_1'^0} (\Pi + \Pi_1)^2 \frac{d\sigma}{d\Omega} \delta^{(4)}(\Pi + \Pi_1 - \Pi' - \Pi'_1)$$
$$\Big(f(x,\Pi')\, f(x,\Pi'_1)\, (1 - \frac{(2\pi)^3}{4} f(x,\Pi))\, (1 - \frac{(2\pi)^3}{4} f(x,\Pi_1))$$

*Supported by BMFT and GSI Darmstadt

$$- f(x,\Pi)\, f(x,\Pi_1)\, (1 - \frac{(2\pi)^3}{4} f(x,\Pi'))\, (1 - \frac{(2\pi)^3}{4} f(x,\Pi_1')) \Big) \qquad (1)$$

which propagates the nucleonic phase space distribution function $f(x,\Pi)$ with the additional mass shell constraint $(\Pi^2 - m^{*2})f(x,\Pi) = 0$. Here $\Pi_\mu = p_\mu - g_v\omega_\mu$ is the kinetic momentum, $m^* = M - g_s\Phi$ the effective mass and $F^{\mu\nu} = \partial^\mu\omega^\nu - \partial^\nu\omega^\mu$ the field strength tensor. The Vlasov part can be derived as the classical limit of a mean-field theory with a vector field $\omega_\mu(x)$ and a scalar field $\Phi(x)$ mediating the forces between the nucleons.

As equation (1) shows, this relativistic approach automatically leads to a momentum dependence of the mean-field potential. This momentum dependence can be investigated experimentally via nucleon-nucleus scattering and one observes a repulsive potential which for lower bombarding energies (< 300 MeV) increases with energy and saturates for higher energies. In our relativistic model the optical potential increases linearly with energy. The slope of this increase depends on the inverse of the fermi-liquid effective mass which is close to the relativistic effective mass $m^* = M - g_s\Phi$ in nuclear matter [7,8]. Thus, especially for a low effective mass (e.g. $m^* = 0.56$ M in the linear Walecka model without σ-selfinteractions) the repulsive force may be overestimated [9]. However, for the only qualitative comparisons we are interested in at this stage this is no severe shortcoming.

In addition to the transport equation (1) for $f(x,\Pi)$ the dynamics of the meson mean-fields is determined by the following equations

$$\begin{aligned} \frac{\partial^2}{\partial t^2}\Phi - \vec{\nabla}^2\Phi + m_s^2\Phi + B\Phi^2 + C\Phi^3 &= g_s\rho_s \\ \frac{\partial^2}{\partial t^2}\omega_\mu - \vec{\nabla}^2\omega_\mu + m_v^2\omega_\mu &= g_v j_\mu \end{aligned} \qquad (2)$$

with the nucleon current

$$j_\mu(x) = \int d^3\Pi \, \frac{\Pi_\mu}{\Pi_0} f(x,\Pi), \qquad (3)$$

where $j_0(x)$ is normalized to the total nucleon number, and the scalar density

$$\rho_s(x) = \int d^3\Pi \, \frac{m^*}{\Pi_0} f(x,\Pi). \qquad (4)$$

Results from time-dependent quantum mechanical mean-field calculations [10] have led to the speculation that these equations may predict new phenomena in heavy-ion collisions, such as meson radiation, an increased stopping as well as an enhanced transverse momentum. We have therefore compared simulations of the Vlasov-equation (eq. 1 without the collision term) including the full solution (FS) of eq.(2) with simulations using different approximations. In the "quasistatic" approximation (QSA) the time derivative of the meson-field is approximated in the cm-frame by the value for a field moving in z-direction with a constant velocity v

$$\begin{aligned} v^2\frac{\partial^2}{\partial z^2}\Phi - \vec{\nabla}^2\Phi + m_s^2\Phi + B\Phi^2 + C\Phi^3 &= g_s\rho_s \\ v^2\frac{\partial^2}{\partial z^2}\omega_\mu - \vec{\nabla}^2\omega_\mu + m_v^2\omega_\mu &= g_v j_\mu \end{aligned} \qquad (5)$$

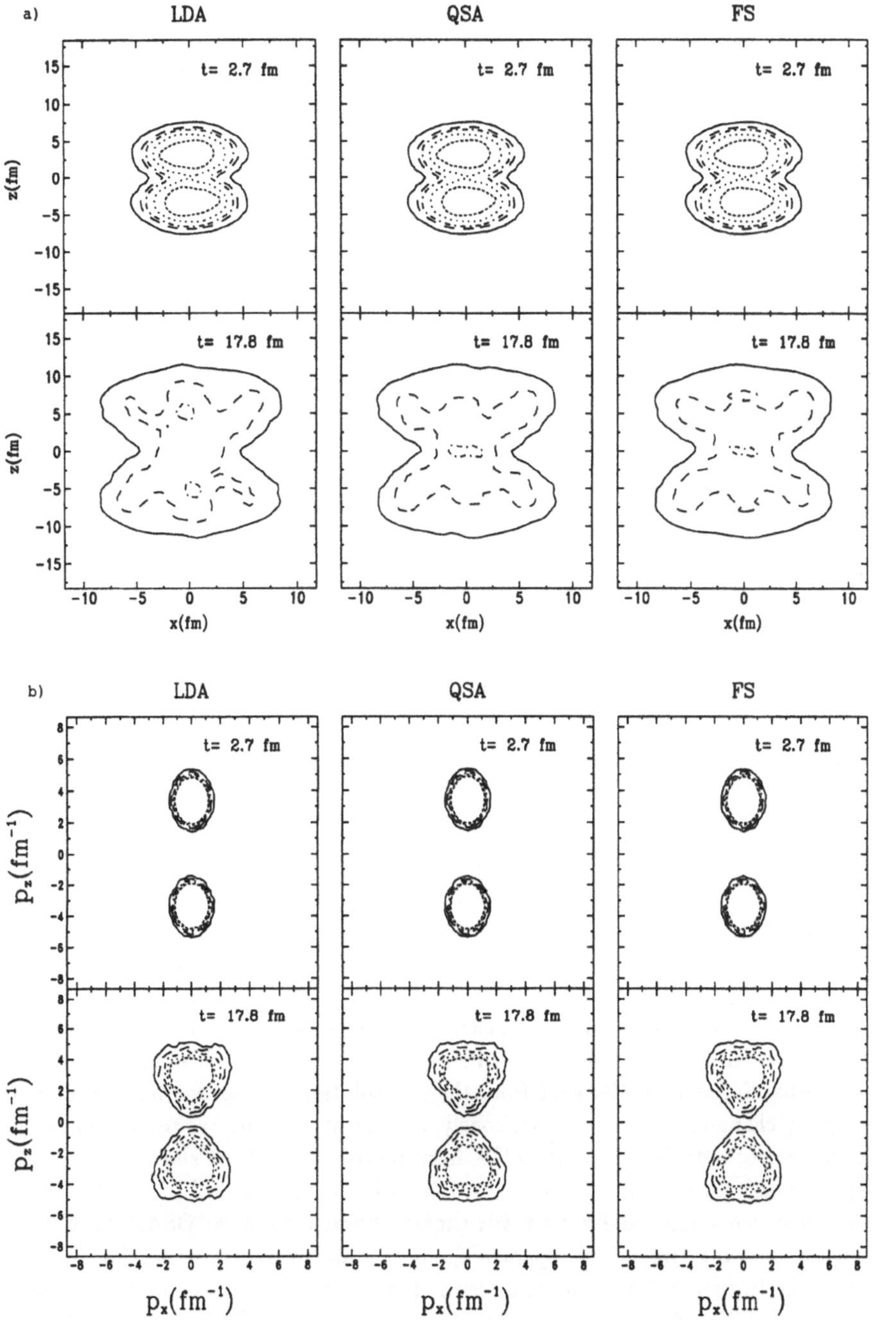

Figure 1. $^{40}Ca + ^{40}Ca$ at 1 GeVA and impact parameter $b = 0$ fm for the pure Vlasov part; left: local density approx., middle: quasistatic approx., right: full solution (a) nucleon density, contour lines: 0.003, 0.03, 0.06, 0.13, 0.2 fm^{-3} (b) momentum density integrated over p_y, contour lines: 0.01, 0.05, 0.1, 0.2, 0.3 fm^2

This limit takes the finite range of the meson-fields into account and is expected to be reasonable as long as the colliding nuclei are not significantly stopped in the collision. Furthermore, the local density approximation (LDA)

$$\begin{aligned} m_s^2\Phi + B\Phi^2 + C\Phi^3 &= g_s\rho_s \\ m_v^2\omega_\mu &= g_v j_\mu \end{aligned} \tag{6}$$

neglects all derivatives of the meson-fields and should be valid for smoothly varying field distributions.

In fig.1 we compare the initial and final nucleon densities as well as momentum densities for a collision of two ^{40}Ca nuclei at 1 GeVA in the Lab-system and zero impact parameter for the Vlasov case. For the calculations we used the parameterset QHD I [11]: $g_s = 9.57, m_s = 2.79$ fm^{-1}, $g_v = 11.67, m_v = 3.97$ fm^{-1}, $B = 0, C = 0$ leading to an effective mass $m^* = 0.56M$ and a compressibility of $K = 540$ MeV in saturated nuclear matter.

We observe that at $t = 17.5$ fm/c the nuclei have almost passed through each other. Although the final nucleon densities show some differences in the interior, we find that the final momentum distributions are very similar. The triangle-shaped momentum distributions can be understood as a result of the Lorentztype-force mediated by the vector-meson field. Due to this force, the particles with a high longitudinal momentum obtain a higher transverse momentum than particles with small initial momenta.

In fig.2 we show the corresponding ω_0-distributions for the same reaction. We observe some differences for the three cases, especially for the full solution. However, also for the FS the field is rather localized at its source, i.e. the nucleon density, and only for the lowest contour we see small deviations which are within the numerical uncertainty.

The transverse momentum distribution for the same reaction at an impact parameter $b = 2$fm is presented in fig.3. We again do not observe significant differences between the full solution and the two approximations. The same has been obtained for the longitudinal momentum distribution (dN/dp_z).

We have further applied our model to the much higher bombarding energy of 14 GeVA for $^{40}Ca + ^{40}Ca$. Again, the momentum distributions $(p_x/A, dN/dp_z)$ look very similar for the full solution(FS) and the quasi-static approximation (QSA) whereas we see a slightly increased stopping in the local density approximation (LDA) (see fig.4). Although we don't find a significant effect of the full solution on the observables considered so far one may still expect differences in the meson-fields. Especially, at the high bombarding energy the field from the full solution should not be able to follow the rapidly changing source instantaneously, i.e. one should observe retardation effects. One thus expects deviations of the QSA from the FS. In fig.5 we show the scalar field along the z-axis for various time steps. At the initial stage of the collision, when the two nuclei begin to overlap, the FS is already slightly different from the QSA. The difference increases when the two nuclei separate again. Finally, a σ-field oscillation builds up between the two separated nuclei as a result of meson-field retardation. However, if one follows the collision for a longer time this oscillation is damped down. The similarity of the dN/dp_z distributions (fig.4) shows that the oscillations do not contain sufficient energy to give rise to an enhanced stopping. Furthermore fig.5 shows small oscillations at the tails of the field which may be a hint for radiation and have to be analyzed further. Also heavier systems will be studied to look for possible field-theoretical instabilities

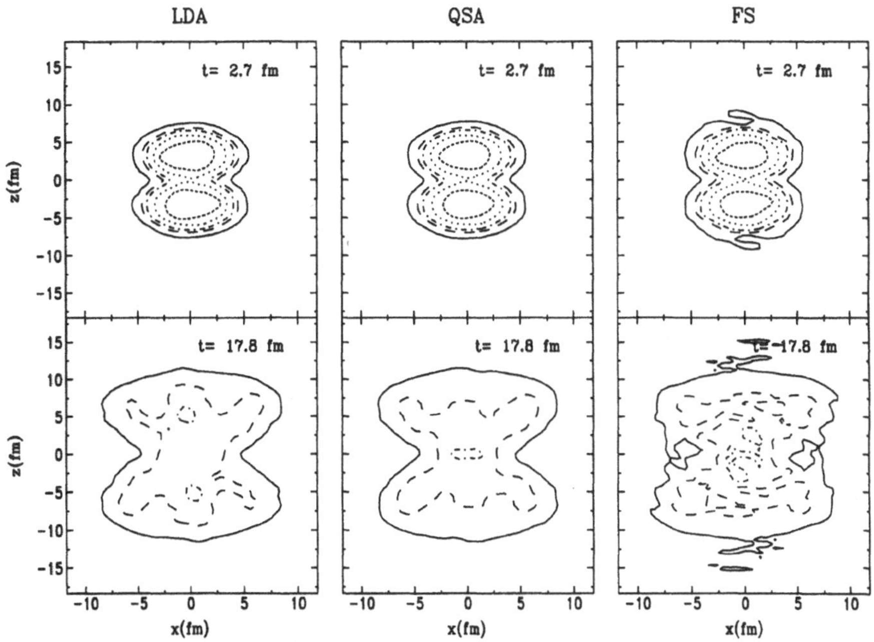

Figure 2. Initial and final ω_0-field for the same reaction as fig.1. The contour lines are the same as for the density in fig.1a but multiplied with $g_v/m_v^2 = 0.74$ fm^2.

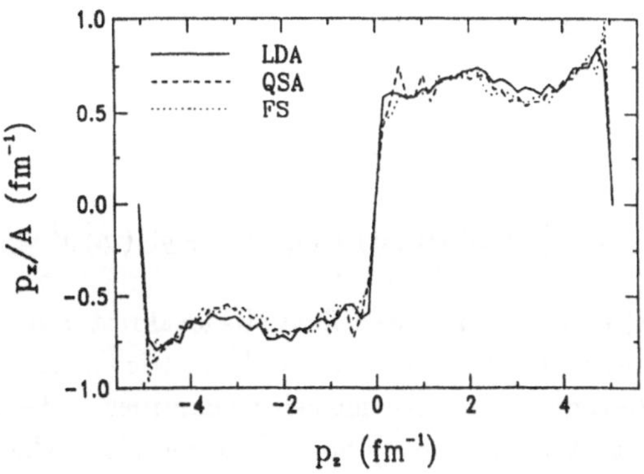

Figure 3. Transverse momentum per nucleon in the reaction plane for $^{40}Ca + ^{40}Ca$ at 1 GeVA and impact parameter $b = 2$fm obtained within phase-space simulations of the relativistic Vlasov equation (l.h.s. of (1)).

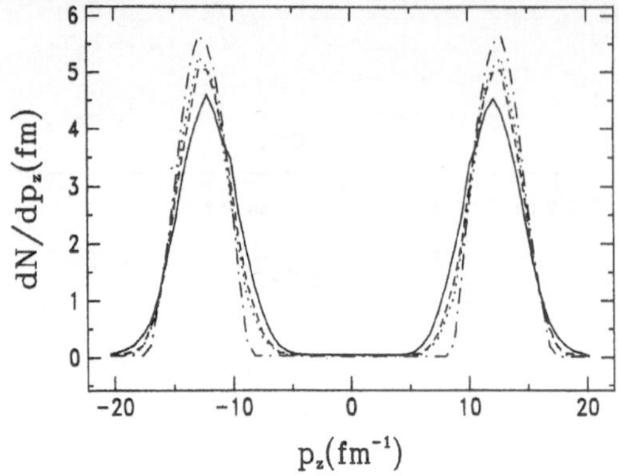

Figure 4. Number of nucleons per longitudinal momentum (dN/dp_z) for $^{40}Ca + ^{40}Ca$ at 14 GeVA and impact parameter $b = 0$fm obtained with the Vlasov equation; solid line: LDA, dashed line: QSA, dotted line: FS; dashed-dotted line: initial distribution

[12]. The most interesting question to be investigated in the future is however, how much of the retardation and radiation effects will show up if the collision-term and thus large stopping is included in the calculation.

3 Thermodynamical analysis

We now focus on the time evolution of thermodynamic quantities such as temperature and pressure during the collision. For the cross section appearing in (1) we use the free values as parameterised by Cugnon [13].

In order to get some local information, we evaluate the energy-momentum tensor $\Theta^{\mu\nu}$ at each gridpoint for a gridsize of $\Delta x = 1.0$fm. For the given Lagrangian (without nonlinear self-interaction) it reads in the local density approximation:

$$\Theta^{\mu\nu} = \int \frac{d^3\Pi}{\Pi^0} \Pi^\mu \Pi^\nu f(x,\Pi) + m_v^2 \omega^\mu \omega^\nu + g^{\mu\nu}(m_s^2 \phi^2 - m_v^2 \omega^\rho \omega_\rho)$$

As motivated by [14] we transform each cell into its local rest frame. This is done by diagonalizing the energy-momentum tensor using the metric tensor $g_{\mu\nu} = (1, -1, -1, -1)$. From this diagonalization we get the following quantities: The velocity of the cell, the energy density and three (different) pressure eigenvalues which we will denote by P_i, $(i = 1, 2, 3)$ in the following.

In fig. 6 a) we show the average pressure $\langle P \rangle = \frac{1}{3} \sum_i p_i$ for a central collision of $^{93}Nb + ^{93}Nb$ at 1 GeVA in the laboratory frame. At $t = 9$fm/c one can see a high pressure zone developing in the early stage of the reaction. The maximum pressure reached at this time is about 150 MeV fm^{-3}.

We further obtain the temperature from the locally known pressure and density by inverting the function $P(\rho, T)$ for equilibrated nuclear matter [11]. It is thus important to consider only spacial cells with an equilibrated momentum distribution. We,

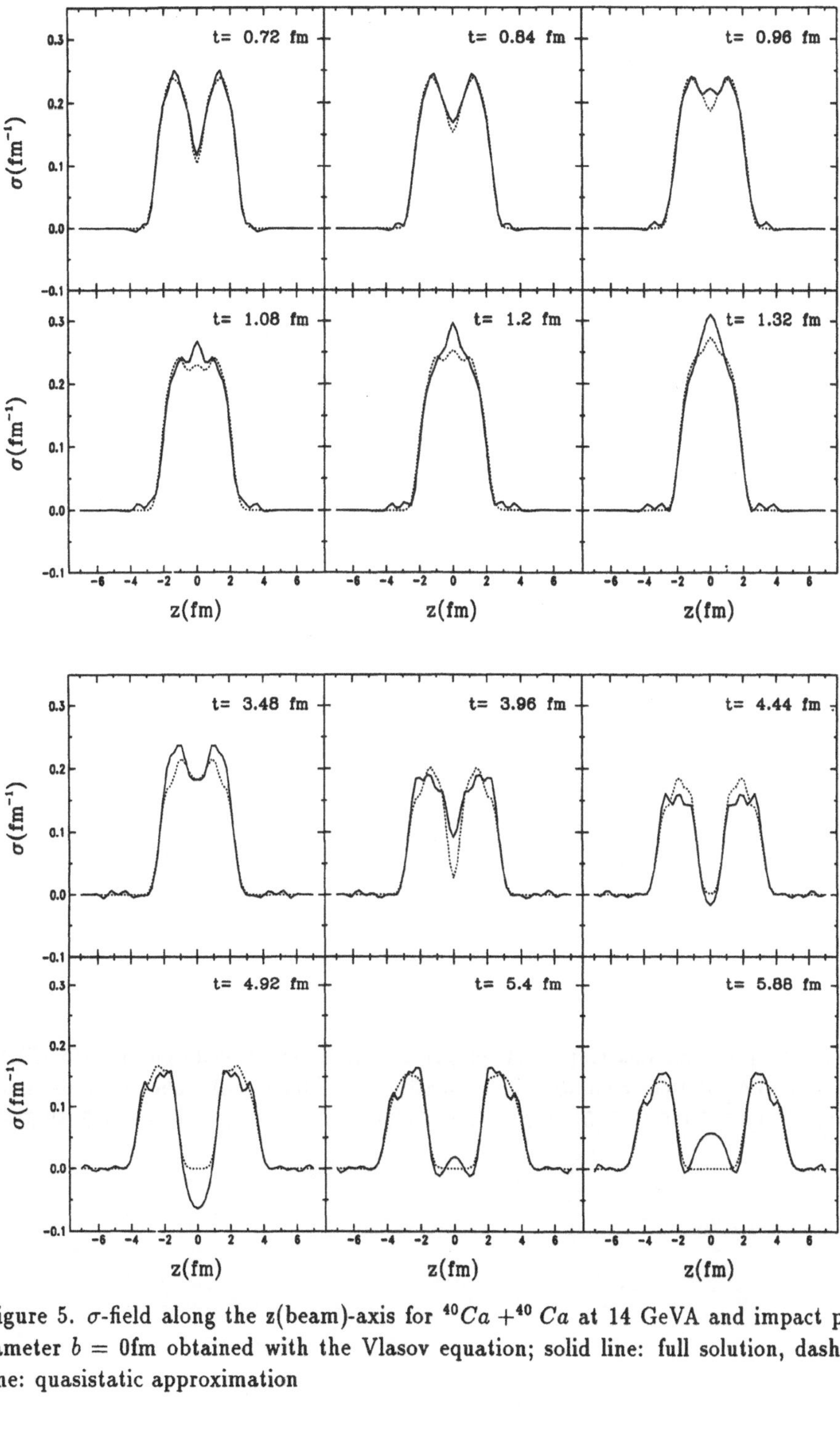

Figure 5. σ-field along the z(beam)-axis for $^{40}Ca + ^{40}Ca$ at 14 GeVA and impact parameter $b = 0$fm obtained with the Vlasov equation; solid line: full solution, dashed line: quasistatic approximation

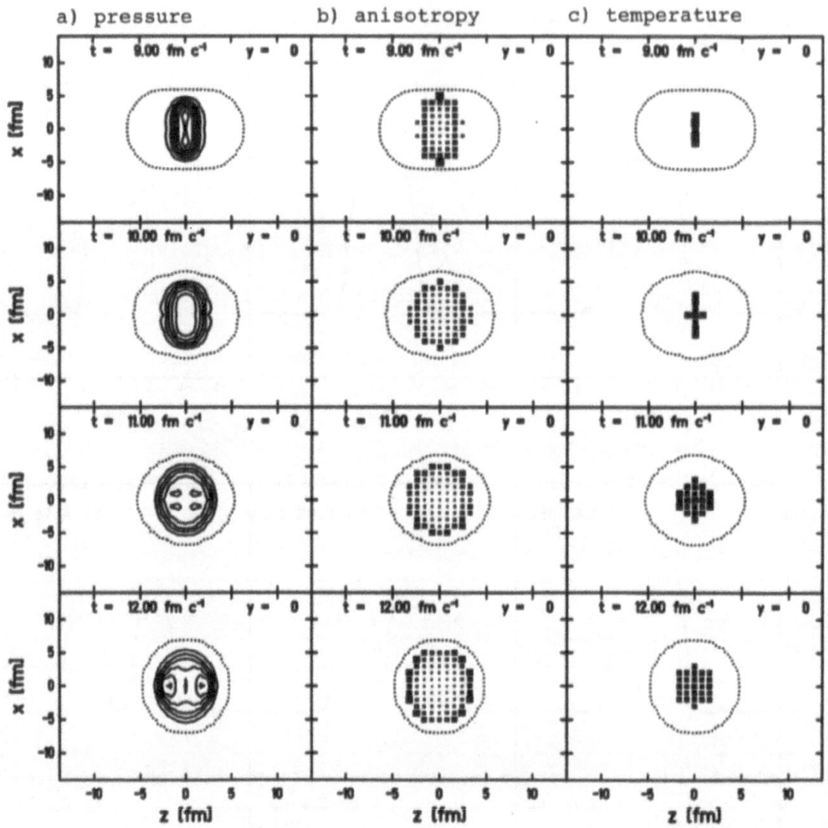

Figure 6. pressure, anisotropy and temperature in a central collision of $^{93}Nb + ^{93}Nb$ at 1 GeVA; dotted contour line: density 0.01 fm^{-3}; a) solid contour lines: 25,50,75,100,125,150 MeV fm^{-3}, b) range of cluster 0..1, c) range of cluster 0..80 MeV

therefore, define the anisotropy

$$A = \frac{1}{3}\sum_i \frac{|\langle P\rangle - P_i|}{\langle P\rangle} \tag{7}$$

For the case of $p_1 = 1, p_2 = p_3 = 0$ which represents the maximal anisotropy we obtain $A = \frac{4}{3}$. Fig. 6b) shows the local anisotropy and fig 6c) the local temperature T. For the evaluation of T(x), we choose only those cells of $A < 0.15$.

As can be extracted from figures 6b) and 6c), indeed a quite substantial volume of approximately equilbrated baryonic matter is produced in the central collision of $^{93}Nb +^{93} Nb$ at $t < 10$fm/c. It reaches about 120fm^3 at $t = 12$ fm/c, however, with an average temperature of 45 MeV, only. This value is small as compared to slope parameters of pion spectra at this energy [15] and thus may raise doubts about a thermal production mechanism and may favour first chance collisions as the dominant production channel. Furthermore, the maximum temperature reached is about 70 MeV at $t = 9$fm/c in a volume of a few fm^3 and drops by 25 MeV within 3 fm/c. We note that for more pheripheral collisions or smaller systems like $^{16}O+^{16}O$ or $^{40}Ca+^{40}Ca$ no sizeable volumes of equilibrated baryonic matter have been found [16].

4 Summary

In this contribution we have investigated the effects of meson retardation in the framework of a relativistic Vlasov equation based on a relativistic mean-field theory proposed by [11]. The numerical analysis shows that an explicit propagation of the mesons via eq.(2) leads to negligeable modifications of the time-evolution of the baryonic phase-space distribution up to at least several GeV per nucleon bombarding energy for heavy-ion collisions. How much this result is altered by the inclusion of the collision-term will be studied in the future.

We have, furthermore, investigated the thermalization in heavy- ion reactions at 1 GeVA on the basis of the relativistic BUU equation (1). While for central collisions of $^{93}Nb+^{93} Nb$ we find quite substantial regions of equilibrated baryonic matter with pressures up to 150 MeVfm^{-3} and temperatures around 50 MeV, more peripheral reactions or collisions of lighter nuclei show only negligeable signs for hot and compressed nuclear matter. These findings raise severe doubts for applying thermodynamical models to these type of reactions.

Since the nonequilibrium phase of the reaction is dominantly influenced by the momentum-dependent mean-field and might also be affected from in-medium corrections to the nucleon-nucleon cross section future work has to be performed to determine these corrections properly.

References

[1] C. Gale, G. Bertsch, and S. Das Gupta, Phys.Rev.C35 1966 (1987); J. Aichelin, A. Rosenhauer, G.Peilert, H. Stöcker, and W. Greiner, Phys. Rev. Lett. 58 1926 (1987)

[2] G.F. Bertsch, G.E. Brown, V. Koch, and B.-A. Li, Nucl.Phys. A490 745 (1988)

[3] H.-Th. Elze, M. Gyulassy, D. Vasak, H. Heinz, H. Stöcker, and W. Greiner, Mod.Phys.Lett. 2 451 (1987)

[4] C.M. Ko, and Q.Li, Phys.Rev. C37 2270 (1988)

[5] B. Blättel, V. Koch, W. Cassing, and U. Mosel, Phys.Rev. C38 1767 (1988)

[6] H.Feldmeier, M.Schönhofen, and M.Cubero, Nucl.Phys. A495 337c (1989)

[7] T. Matsui, Nucl.Phys. A370 365 (1981)

[8] V. Koch, U. Mosel, T. Reitz, Chr. Jung, and K. Niita, Phys. Lett. 206B 395 (1988)

[9] C.J.Horowitz, Proc. of the International Workshop on Gross Properties of Nuclei and Nuclear Excitations XVI, Hirschegg(Austria) ed. by H. Feldmeier, p.10, (1988)

[10] R.Y. Cusson, P.-G. Reinhard, J. J. Molitoris, H. Stöcker, and W. Greiner, Phys. Rev. Lett. 55 2786 (1985)
J.J Bai, R.Y. Cusson, J. Wu, P.-G. Reinhard, H. Stöcker, and W. Greiner, Z.Phys. A326 269 (1987)

[11] B.S. Serot, and J.D. Walecka, Adv.Nucl.Phys.16 1 (1986) J.W. Negele and E. Vogt, eds. (Plenum Press)

[12] Y.B.Ivanov, Nucl.Phys. A474 693 (1987); P.A. Henning, and B.L. Friman, Nucl.Phys. A490 689 (1988)

[13] J.Cugnon, T.Mizutani, and J.Vandermeulen, Nucl.Phys. A352 505 (1981)

[14] R.U.Sexl, and H.K.Urbantke, "Relativität Gruppen Teilchen", Springer Verlag, Wien, New York (1976)

[15] J.Harris et al., Phys.Lett. 153B 377 (1985), Phys.Rev.Lett. 58 463 (1987)

[16] A.Lang et al., to be published

HIGH ENERGY PHOTONS AND PIONS IN INTERMEDIATE ENERGY HEAVY ION REACTIONS

Wolfgang Bauer

National Superconducting Cyclotron Laboratory and
Department of Physics and Astronomy
Michigan State University
East Lansing, Michigan 48824-1321
USA

INTRODUCTION

Particles produced during the high compression phase of heavy ion collisions can serve as a probe of the nuclear equation of state. This is why the emission of particles like high energy ($E_\gamma > 20$ MeV) photons, pi-mesons, kaons, di-lepton pairs, anti-protons, ... has attracted so much attention during the last few years. In this chapter attention shall be focussed on the production of pions and high energy photons.

The production of pions in heavy ion collisions at beam energies far below the nucleon-nucleon threshold was discovered by Benenson et al.[1] High energy gamma rays from heavy ion collisions were first observed by Beard and co-workers[2] and by Grosse and collaborators[3]. Since then, a large number of experimental and theoretical studies have been performed in order to understand these phenomena.[4]

A first attempt to explain the production of pions and high energy photons was put forward by the Frankfurt group. Vasak et al.[5] parametrized the heavy ion reaction in terms of a deacceleration time τ, assumed a collective production mechanism, and obtained reasonable fits to experimental data for pion energy spectra. Shyam and Knoll[6] proposed another cooperative model in considering the contributions of many-nucleon clusters to pion production. However, using a TDHF-based model and the collective production assumption, our group found that it was not possible to reproduce the experimental high energy photon cross sections in light heavy ion systems.[7] The theoretical calculations underpredicted the production cross section for photons with $E_\gamma > 40$ MeV by an order of magnitude.

Alternatively, the production of high energy photons and pions can be explained as a superposition of the contributions of individual nucleon-nucleon collisions. Here subthreshold production is essentially obtained due to the Fermi motion of the individual nucleons inside the nuclei.[8] However, simple estimates with Fermi distributions also underpredicted the pion yields,[8] and an attempt to explain the high energy

The Nuclear Equation of State, Part A
Edited by W. Greiner and H. Stöcker
Plenum Press, New York

gamma ray production in heavy ion reactions on the basis of individual neutron-proton collisions[9] had to use an unrealistically high number (≈ 20) of nucleon-nucleon collisions per nucleon to match the existing experimental data.

Using the Boltzmann master equation, Blann[10] was able to reproduce the experimental values of the total pion production cross section at intermediate beam energies. Our ability to calculate particle production cross sections from heavy ion reactions was dramatically improved by the introduction of numerical solution techniques (BUU/VUU/LV) for nuclear transport theories on a one- and two-body level by Bertsch et al.[11] Aichelin[12] carried out the first dynamical calculation of pion production in intermediate energy collisions using the BUU theory, and Cassing[13] also was able to reproduce the total pion production cross section. All of these calculations were based on the assumption that subthreshold pions are produced in individual nucleon-nucleon collisions. Recently, we[14] have been able to also reproduce the experimental energy spectra for neutral pions produced in collisions of carbon nuclei at beam energies between 60 and 84 MeV/nucleon.

For high energy photons, the first calculations based on the nuclear BUU transport theory were done by our group.[15] We used the assumption that high energy gamma rays are produced in individual collisions between protons and neutrons during the course of the heavy ion reaction; and we were able to reproduce energy spectra and angular distributions, as well as total cross sections[15,16] for beam energies below 60 MeV/nucleon. Later, Aichelin and Ko[17] and Heuer et al.[18] made similar calculations and also came to the conclusion that the production of high energy photons in intermediate energy heavy ion collisions is dominated by the contribution from bremsstrahlung of individual proton-neutron collisions.

In the following, I will briefly introduce the numerical techniques used to perform a dynamical simulation of intermediate energy heavy ion reactions and to calculate the production of gamma rays and pions during the course of them. A presentation of some new and selected old results will follow. In the summary section, I will attempt to draw conclusions on what we have learned by studying the emission of pions and high photons which are produced in intermediate energy heavy ion reactions.

DETAILS OF THE CALCULATIONS

The particle production calculations presented here are based on a numerical solution of the Boltzmann-Uehling-Uhlenbeck (BUU) equation for Wigner's phase space distribution function $f(\vec{r},\vec{p},t)$,

$$\frac{\partial}{\partial t}f(\vec{r},\vec{p},t) + \frac{\vec{p}}{m}\vec{\nabla}_r f(\vec{r},\vec{p},t) - \vec{\nabla}_r U\,\vec{\nabla}_p f(\vec{r},\vec{p},t) = \left[\frac{\partial f(\vec{r},\vec{p},t)}{\partial t}\right]_{\text{coll}} . \qquad (1)$$

The left-hand side set equal to 0 is known as the Vlasov equation, the semiclassical equivalent to the time-dependent Hartree-Fock theory. The right-hand side was first introduced by Nordheim.[19] and is the modified Boltzmann-collision term

$$\left[\frac{\partial f(\vec{r},\vec{p},t)}{\partial t}\right]_{\text{coll}} = \frac{g}{2\pi^3 m^2}\int d^3q_{1'}\, d^3q_2\, d^3q_{2'} \tag{2}$$
$$\cdot\delta(E+E_2-E_{1'}-E_{2'})\delta^3(\vec{p}+\vec{q}_2-\vec{q}_{1'}-\vec{q}_{2'})\frac{d\sigma}{d\Omega}$$
$$\cdot\Big\{\overline{f}(\vec{r},\vec{q}_{1'},t)\,\overline{f}(\vec{r},\vec{q}_{2'},t)\left(1-\overline{f}(\vec{r},\vec{p},t)\right)\left(1-\overline{f}(\vec{r},\vec{q}_2,t)\right)$$
$$-\overline{f}(\vec{r},\vec{p},t)\,\overline{f}(\vec{r},\vec{q}_2,t)\left(1-\overline{f}(\vec{r},\vec{q}_{1'},t)\right)\left(1-\overline{f}(\vec{r},\vec{q}_{2'},t)\right)\Big\}.$$

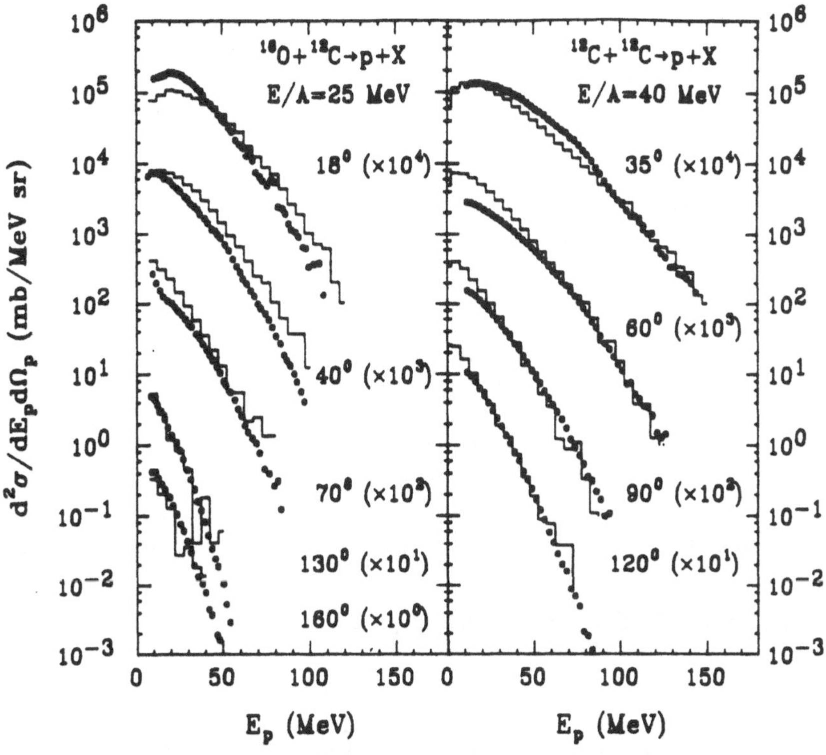

Figure 1. Comparison of experimental (circles) to theoretically calculated (histograms) proton energy spectra.[22] The experimental data are taken (left) from the work of Chitwood et al.[23] and (right) Fox et al.[24]

This equation is solved in the semiclassical test particle approximation, originally proposed by Wong[20] for the solution of the left-hand side and completed by Bertsch et al.[11] who properly took the effects of nucleonic cascading and the influence of the Pauli principle on final state momenta of scattering nucleons into account. The differential nucleon-nucleon scattering cross section, $d\sigma/d\Omega$, which appears in equation 2 is parametrized using experimental cross sections as input and following the prescription of Cugnon et al.[21]

The mean field potential U is parametrized as a density functional and is given by

$$U(\rho(\vec{r},t)) = -218\,\mathrm{MeV}\,\frac{\rho(\vec{r},t)}{\rho_0} + 164\,\mathrm{MeV}\left(\frac{\rho(\vec{r},t)}{\rho_0}\right)^{4/3}. \qquad (3)$$

This mean field potential is fitted to an infinite nuclear matter binding energy of −16 MeV, saturation at $\rho(\vec{r},t) = \rho_0$, and a value of $\kappa = 235$ MeV for the nuclear compressibility. More details of the numerical simulation can be found in references 11 and 22.

It is important to notice that the BUU theory correctly reproduces the momentum distributions of nucleons in heavy ion collisions. Experimental angle and energy dependence of proton cross sections from intermediate energy heavy ion collisions are properly reproduced within a factor of two by the calculations. As one representative example, the comparison between theory and experiment is shown for the reactions $^{16}O + ^{12}C \rightarrow p + X$ at a beam energy E/A = 25 MeV and $^{12}C + ^{12}C \rightarrow p + X$ at 40 MeV. Other comparisons to nucleon emission cross sections can be found in recent review articles by Stöcker and Greiner[25] and by Bertsch and DasGupta.[26]

From the phase space distribution function $f(\vec{r},\vec{p},t)$, the production cross section for a given particle ν can be calculated from summing the probability of producing this particle in each individual nucleon-nucleon collision over all collisions during the course of the heavy ion reaction and integrating over impact parameter:

$$\frac{d^2\sigma}{dE_\nu d\Omega_\nu} = \int d^2b \sum_{\mathrm{coll(b)}} \int \frac{d\Omega^{if}}{4\pi}\mathcal{L}\left[\frac{d^2P_\nu(\vec{p}_1,\vec{p}_2)}{dE'_\nu d\Omega'_\nu}\right](1-f(\vec{r},\vec{p}_3,t))(1-f(\vec{r},\vec{p}_4,t)). \quad (4)$$

Here $\mathcal{L}$ indicates a Lorentz-transformation from the nucleon-nucleon center of mass frame to the frame of calculation. Ω^{if} is the pair of relative angles between the initial momenta $\vec{p}_1$ and $\vec{p}_2$ of the two scattering nucleons and their final momenta $\vec{p}_3$ and $\vec{p}_4$ after emission of the particle ν, which is not determined by energy and momentum conservation.

$d^2P_\nu(\vec{p}_1,\vec{p}_2)/dE'_\nu d\Omega'_\nu$ is the elementary production probability for particle ν in a nucleon-nucleon collision. For the case of pions, ample nucleon-nucleon data are available, and the experimental production probabilities can be used. However, at least at present, no reliable data for the reaction $p + n \rightarrow p + n + \gamma$ exist. In earlier work,[15] we therefore used a theoretical classical approximation for the elementary cross section which was given by Jackson[27]

$$\frac{d^2P_\gamma(\vec{p}_1,\vec{p}_2)}{dE_\gamma d\Omega_\gamma} = \frac{\alpha\,R^2}{\sigma_{nn}\,12\pi\,E_\gamma}(2\beta_f + 3\sin^2\theta_\gamma\beta_i), \qquad (5)$$

where α is the fine structure constant, σ_{nn} is the (energy dependent) total nucleon-nucleon cross section, and β_i and β_f are initial and final velocity of the proton in the neutron-proton c.m. system.

Recently, we have also started to use elementary production probabilities which are based on a covariant meson-exchange approximation for the pn scattering amplitude[28] and G matrix calculations using the Bonn one-boson-exchange potential.[29] Here, the goal is to obtain better results for higher beam energies and to incorporate contributions from, for example, meson-exchange currents. These contributions can result in differences of more than a factor of 2 in the elementary production probability.[29]

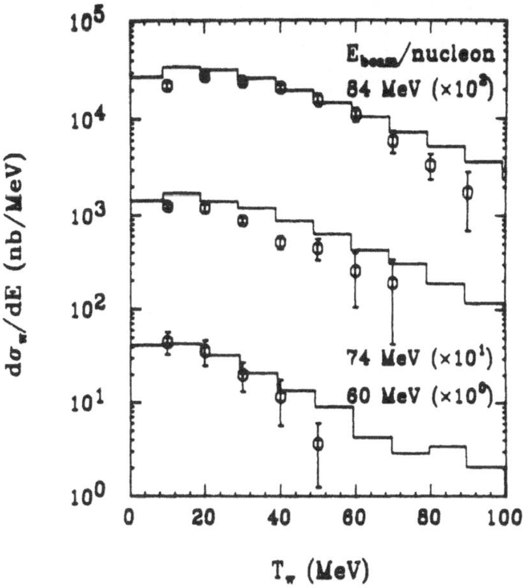

Figure 2. Energy spectra of neutral pions from collisions of ^{12}C + ^{12}C at beam energies per nucleon of 60, 74, and 84 MeV. The histograms represent the data and are taken from Ref. 30. The plot symbols with statistical error bars are the result of our calculations.

SOME RESULTS

In figure 2, we display the results of our calculation[14] (plot symbols with statistical error bars due to the Monte Carlo integration procedure applied) for the energy spectra of neutral pions in heavy ion collisions at various beam energies. The data of Noll et al.[30] are represented by histograms. To obtain this good agreement, it is important to properly take into account the role of the Δ resonance. In the calculation we followed the model of Mandelstam.[31] The mass distribution of the Δ resonance is obtained using the parametrization of Kitazoe et al.[32] Calculations which use only directly produced pions result in much steeper energy spectra. Following Aichelin's approach,[12] we do not attempt to calculate the effects of reabsorption of Δ's and π's in a dynamical way, but rather use a mean pion absorption length of $\lambda_0 = 3$ fm, independent of pion energy. This results in an overall reduction of the cross section by about a factor of 2 for the case of C + C.

Figure 2 shows that it is possible to reproduce experimental pion energy spectra in our approach. Even though our spectra are slightly steeper than the experimental ones and terminate at some maximum energy due to the semiclassical approximation for the nucleonic momentum distribution, the agreement between theory and experiment is quite good. We can therefore state that subthreshold pion production can be explained on the basis of a superposition of the contributions of individual nucleon-nucleon collisions.

A similar agreement between experiment and theory can be reached for the production of high energy Bremsstrahlung photons in heavy ion collisions. In figure 3, again the assumption of production in individual nucleon-nucleon collisions is used. In part (b), we show as one representative example the agreement between the results of the calculation and experimental data[33] for the reaction $^{14}N + ^{12}C \rightarrow \gamma + X$ at

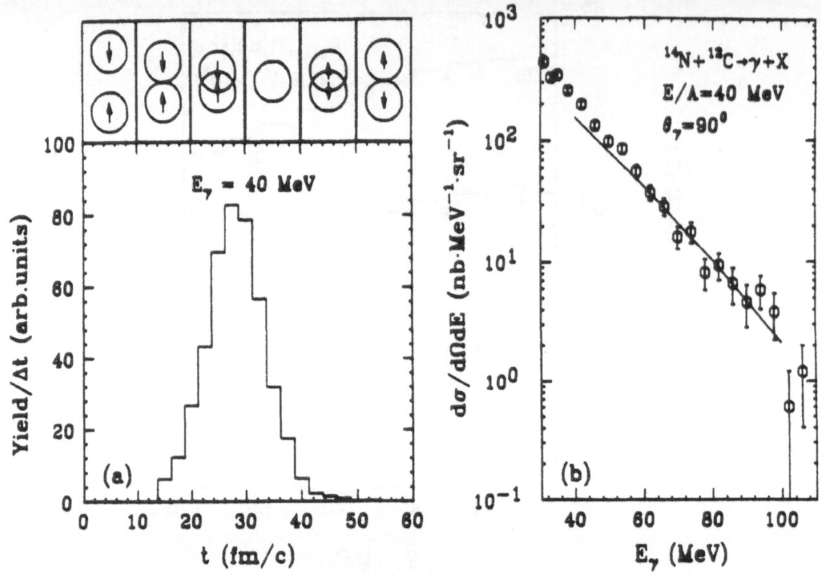

Figure 3. Production of high energy photons in the reaction $^{14}N + ^{12}C$ at a beam energy E/A = 40 MeV. (a) Theoretical calculation for the temporal dependence of the photon emission probability (histogram). Sketched above is the time evolution of the collision. (b) Comparison of the result of the theoretical calculation (line) with experimental data[33] (plot symbols) for the high energy photon spectrum at an emission angle of 90^0.

E/A = 40 MeV. Comparison to other sets of data at beam energies between 15 and 44 MeV/nucleon results in similarly good agreement.[16]

High energy photons emitted in heavy ion collisions contain important information on the early high compression phase of the heavy ion collision. This is shown in figure 3 (a). For the same reaction, we display the theoretical calculation for the differential production probability for a photon with E_γ = 40 MeV as a function of time (histogram). Sketched above is the time evolution of the heavy ion collision. Target and projectile are represented by the circles and their velocities are indicated by the arrows. It can be observed that most of the high energy photons are produced in the early stage of the collision, between the time when the nuclei first touch (t = 15 fm/c) and the time when they have achieved maximum overlap (t = 35 fm/c). Since high energy gamma rays also experience very little final state interaction on their way out of the nucleus, they are ideally qualified as probes for the high compression stage of the heavy ion reaction.

The reason for the limitation of photon production to the early part of the reaction is twofold. Firstly, on the path to partial thermalization, high momentum components in the nucleus-nucleus c.m. frame rapidly become depopulated as soon as nucleons of the different nuclei start to scatter.[34] The second reason is a collective dynamical effect that temporarily creates empty phase space for final state nucleons to scatter into.[15]

This is illustrated in figure 4. Displayed here is the time evolution of the phase space distribution function $f(\vec{r}, \vec{p}, t)$ as a function of longitudinal coordinate and momentum for a collision of two ^{12}C nuclei at an energy E/A = 40 MeV. The time

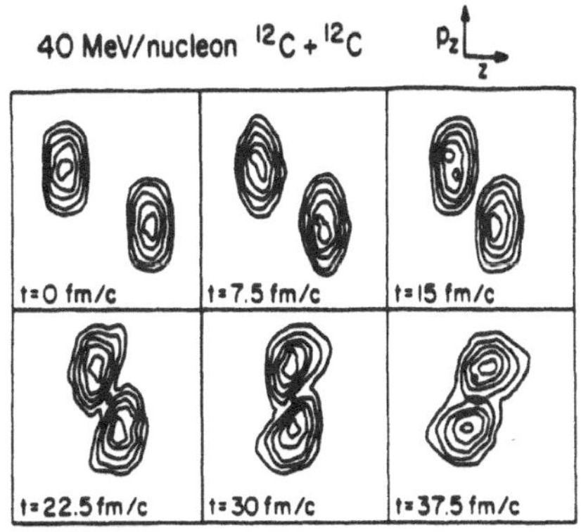

Figure 4. Time evolution of the phase space density $f(\vec{r},\vec{p},t)$, integrated over transversal coordinates and momenta, as a function of longitudinal coordinate (horizontal axis) and longitudinal momentum (vertical axis) in a collision of two ^{12}C nuclei at 40 MeV/nucleon. The total area displayed is 20 fm × 1 GeV/c. The contour lines correspond to cuts at values of 0.1, 0.3, 0.5, 0.7 and 0.9 for f.

evolution was calculated using equation 1 with the collision term set to 0 (Vlasov limit). For two noninteracting Fermi gases confined to potential wells, the initial (t=0) distributions would simply move horizontally towards each other and have a very large overlap at t = 30 fm/c. However, it can be seen from figure 4 that this is not at all the case. Since the right hand-side of equation is of the Liouville type, the two nuclei behave like incompressible fluids in phase space. The result is that a partially empty region in phase space is formed around $p = 0$. This reduces the Pauli blocking factors in equation 4 for the final state momenta of the nucleons which produce high energy photons in their collisions. However, a calculation with the full collision term shows that elastic nucleon-nucleon collisions rapidly close this central hole in the phase space distribution, thus inhibiting the production of high energy photons in later stages of the heavy ion reaction.

Recently, the question of the influence of the Δ resonance on high energy gamma ray spectra has received some attention. Prakash et al.[35] have suggested applying the statistical model to photon emission, relating photon emission rates by detailed balance to the measured photoabsorption cross section of nuclei. They find that the Δ enhances the gamma production cross section in heavy ion collisions with beam energies as low as 44 MeV/nucleon is enhanced by a factor of ≈ 10 for photon energies of $E_\gamma \approx 200$ MeV.

Using the BUU equation, we have calculated[36] the production cross section of photons due to the process $N + N \rightarrow N + \Delta$; $\Delta \rightarrow \gamma + N$. As mentioned above, we treat the production of Δ's in nucleon-nucleon collisions using the experimental pion production cross sections, following the ansatz of Mandelstam.[31] For the decay $\Delta \rightarrow \gamma + N$ we use the experimentally measured branching ratio of 0.6 %, which we assume to be independent of the energy of the Δ.

In figure 5, we display the resulting contribution of the decay of the Δ resonance in the collision of two ^{12}C nuclei at an energy of E/A = 200 MeV to the energy spectrum of gamma rays (histogram). It is compared to the contribution to the photon spectrum resulting from the decay $\pi^0 \rightarrow 2\gamma$ (dashed line). Above an energy of 200 MeV, both contributions are of the same magnitude.

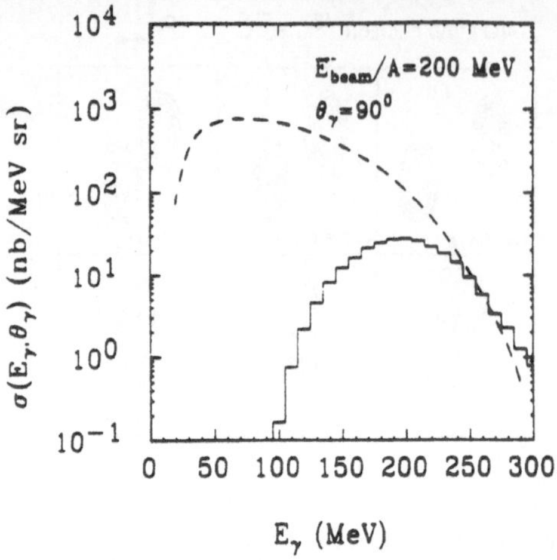

Figure 5. Comparison of the contributions from the decay of the π^0 (dashed line) and the one-photon decay of the Δ resonance (histogram) to the photon spectrum at a beam energy per nucleon of 200 MeV. The calculations were performed in the laboratory system for the system ^{12}C + ^{12}C.

It is known[37] that for beam energies greater than approximately 150 MeV/nucleon the contribution of the two-photon decay of the π^0 to the high energy gamma ray spectrum dominates over the bremsstrahlung contribution. In this case, it should therefore be possible to eliminate the contribution of the π^0 decay from the photon spectrum by subtraction or anticoincidence methods; and the photons from the electromagnetic decay of the Δ resonance should indeed be visible as a bump on top of an otherwise exponentially falling photon spectrum. However, in our calculation the photons from the Δ decay were hidden under the Bremsstrahlung background for a beam energy of 75 MeV/nucleon. Beam energies which are higher than tho. e presently used are therefore neccesary to study this predicted effect.

CONCLUSIONS

In heavy ion reactions at intermediate beam energies, high energy photons and subthreshold pions are mainly produced in incoherent nucleon-nucleon collisions. Using experimental nucleon-nucleon pion production cross sections, it is possible to reproduce the production cross sections and energy spectra of pions in heavy ion reactions. Similarly, it is possible to explain the yield of high energy gamma rays from the superposition of individual proton-neutron collisions. Here, there is some disagreement between theory and experiment for beam energies greater than 60 MeV/nucleon. We speculate that this is due to the fact that meson exchange currents, which are not taken care of in the classical approximation to the elementary photon production probability, gain a greater importance. Work in this direction is in progress.

High energy photons are a good probe for the early stage of the heavy ion reaction in which high compressions are achieved. We have shown that gamma rays are emitted in the early stage of the heavy, ion collision. Their small final state interaction thus makes high energy gamma rays valuable probes of the compression phase. A better description of the elementary production probability and a refinement of the numerical techniques used should lead to a better understanding of gamma ray production and therefore of the early phase of heavy ion reactions. We expect to find the total photon production cross section to be sensitive to the nuclear compressibility in heavy systems.

The contribution of the decay of the Δ resonance should be visible in the high energy photon spectrum at beam energies above 200 MeV/nucleon. This prediction can be tested using the new SIS-18 facility at GSI. One possible effect to be studied could be a change in the width of the Δ mass peak. This could come about due to a change in lifetime of the Δ due to the effect of the Pauli exclusion principle on the nucleon in the final state.

Valuable discussions with G.F. Bertsch, W. Benenson, T.S. Biro, W. Cassing, J. Clayton, V. Metag, U. Mosel, K. Niita, J. Stevenson, and M. Tohyama during various stages of this work are gratefully acknowledged.

REFERENCES

[1] W. Benenson et al., Phys. Rev. Lett. 43:683 (1979).
[2] K. Beard et al., Phys. Rev. C 32:1111 (1985).
[3] E. Grosse et al., Europhys. Lett. 2:9 (1986).
[4] For further references to experimental papers and new experimental results please see the contributions by W. Benenson, E. Grosse, V. Metag, and J. Stevenson.
[5] D. Vasak et al., Nucl. Phys. A428:291c (1984); D. Vasak et al., J. Phys. G 11:1309 (1985).
[6] R. Shyam and J. Knoll, Phys. Lett. 136B:221 (1984); R. Shyam and J. Knoll, Nucl. Phys. A426:606 (1984).
[7] W. Bauer et al., Nucl. Phys. A456:159 (1986).
[8] G.F. Bertsch, Phys. Rev. C 15:713 (1977).
[9] H. Nifenecker and J.P. Bondorf, Nucl. Phys. A442:478 (1985).
[10] M. Blann, Phys. Rev. Lett. 54:2215 (1985). M. Blann, Phys. Rev. C 32:1231 (1985).
[11] G.F. Bertsch, et al. Phys. Rev. C 29:675 (1984).
[12] J. Aichelin, Phys. Lett. 164B:261 (1985).
[13] W. Cassing, Z. Phys. A 329:487 (1988).
[14] W. Bauer, Preprint MSUCL-672, to be published in Phys. Rev. C.
[15] W. Bauer et al., Phys. Rev. C 34:2127 (1986).
[16] T.S. Biro et al., Nucl. Phys. A471:604 (1987); K. Niita et al., in: "Proceedings of the 8th High Energy Heavy Ion Study", Lawrence Berkeley Laboratory Report (1988).
[17] J. Aichelin and C.M. Ko, Phys. Rev. C 35:1976 (1987).
[18] R. Heuer et al., Z. Phys. A 330:315 (1988).
[19] L.W. Nordheim, Proc. Roy. Soc. A 119:689 (1928).
[20] C.Y. Wong, Phys. Rev. C 25:1460 (1982).
[21] J. Cugnon et al., Nucl. Phys. A352:505 (1981).
[22] W. Bauer, Nucl. Phys. A471:604 (1987).
[23] C.B. Chitwood et al., Phys. Rev. C 34:858 (1986).

[24] D. Fox et al., Phys. Rev. C 33:1540 (1986).
[25] H. Stöcker and W. Greiner, Phys. Rep. 137:277 (1986).
[26] G.F. Bertsch and S. DasGupta, Phys. Rep. 160:189 (1988).
[27] J.D. Jackson, "Classical Electrodynamics", Wiley, New York (1962).
[28] M. Schäfer et al., Preprint UGI-89-1.
[29] K. Nakayama, Phys. Rev. C 39:1475 (1989).
[30] H. Noll et al., Phys. Rev. Lett. 52:1284 (1984).
[31] S. Mandelstam, Proc. Roy. Soc. A 244:491 (1958).
[32] Y. Kitazoe et al., Phys. Lett. 166B:35 (1986).
[33] J. Stevenson et al., Phys. Rev. Lett. 57:555 (1986).
[34] W. Bauer, Phys. Rev. Lett. 61:2534 (1988).
[35] M. Prakash et al., Phys. Rev. C 37:1959 (1988).
[36] W. Bauer and G.F. Bertsch, Preprint MSUCL-677.
[37] V. Metag, Nucl. Phys. A488:483c (1988).

PHOTON AND PION PRODUCTION IN HEAVY ION COLLISIONS

Marshall Blann

Physics Department, E-Division
Lawrence Livermore National Laboratory
Livermore, CA 94550

INTRODUCTION

Why has so much effort been expended in the measurement and interpretation of the phenomena of photon and γ-ray yields in heavy ion collisions? Several motivations may be given:

1. These data may give information on the early reaction time scale during the most violent part of the reaction.

2. They may provide a probe of the nuclear equation of state (EOS).

3. Their interpretation via reaction models may suggest interesting new mechanisms for these reactions, and in any case provide a better understanding of mass and energy transport.

Of these (3) may be the strongest motivation, in that much more straightforward measurements of emitted nucleons probably will give better answers to (1) and (2) than the more convoluted interpretation of the production of secondary reaction products such as photons and mesons.

The next question is how these experiments have been interpreted. Many different approaches have been taken. These include a set applying equilibrium phase space arguments to a space or subspace; among these the thermal, statistical, and fireball models.[1-5] Approaches have also been suggested in which collective nucleus-nucleus,[6-8] or nucleon-cluster[5] interactions have been treated with the pions or photons produced in the deceleration following these collective interactions.

Perhaps the largest total effort has been based on the more familiar microscopic description that the emissions result from nucleon-nucleon collisions during the evolution of the nucleus-nucleus interaction; these models have been treated most frequently as a sum of incoherent processes,[9-19] with the range of validity of this simplification having been investigated recently by Heuer et al.[20] This straightforward physics has or may be treated in a wide range of nuclear transport calculations each with its own simplifying approximations. In the next section we describe different formulations for treating the nucleon-nucleon transport physics. These will all be semi-classical treatments; however considerable work has been done considering the relationship between quantal and semi-classical formulations.[21-23] In the following section we discuss additional input specific to calculation of pion and photon yields, and present comparisons

The Nuclear Equation of State, Part A
Edited by W. Greiner and H. Stöcker
Plenum Press, New York

between calculated and experimental results, mostly for high energy photons. Conclusions and suggestions for future work are presented in the last section.

DIFFERENT FORMULATIONS FOR HI REACTION/TRANSPORT CALCULATIONS

Event by Event Calculations

Intranuclear cascade (INC). Follow the trajectories of nucleons in 3-dimensions, versus the time increment after the nuclei begin to interpenetrate.[24] Treat all nucleons in the overlap region until all have left the region. Assume the nucleons are bound within a central nuclear potential well. When there is a NN collision, use probability of N-N-$\gamma(\pi)$ vs NN elastic collisions to estimate the $\gamma(\pi)$ production cross sections and spectra.[25] Extreme approximations are made regarding the effect on nuclear densities following nucleon rearrangement so that it is difficult to follow the time dependent nuclear density evolution in this approach. Cluster formation may be predicted.[15] Because it is an event by event calculation, a very large number of events may have to be followed in order to generate adequate statistics for rare events. Nucleons move in straight line trajectories except for surface refraction and reflection processes. Only excited nucleons are followed in the cascades.

Quantum molecular dynamics model. The nucleons are bound in a potential which is explicitly calculated from the nucleon-nucleon force based on the total number of target/ projectile nucleons. The reaction is followed in 3-dimensions readjusting (x, y, z, p_x, p_y, p_z) of each nucleon in small time steps. During each time step the long range interaction of each nucleon with every other nucleon is followed; the nuclear density also changes, and the nucleon-nucleon interaction changes according to density and EOS. The nucleons move in curved trajectories due to the long range N-N force. When nucleons come within a fixed distance $\sigma_{NN}(E)$ of one another, a 'hard' collision takes place (as in the INC). Then pions or photons may be produced with a probability based on elementary N-N-$\gamma(\pi)$ cross sections.[20] The fate of each nucleon is followed as in the INC, so many events must be followed to generate satisfactory statistics for rare events.[26-28] As in INC, cluster formation may be estimated; unlike INC, the EOS influences the reaction dynamics in a quite natural way. The photon emission process has been followed in the QCD approach with phase relationships maintained in order to test the range of validity of the more common incoherent amplitude summation.[20] Nucleon momenta are given a Gaussian width during the selection process following a collision in order to mimic quantal effects of the Heisenberg principle.

Nucleon exchange transport model. Vandenbosch and Randrup have used the one body dissipation model of Randrup to treat nucleon and photon emission in HI collisions.[17] In this approach one nucleon at a time is allowed to transfer between the two nuclei as they approach and begin to overlap along some trajectory. The momentum of the nucleon is selected in Monte-Carlo fashion from a distribution based on the Fermi distribution with energy transfer based on the one body dissipation formula.[29] Once selected the fate of the nucleon is followed much as in the INC model, until it is emitted or undergoes a two body elastic or inelastic (NNγ) interaction. Then another nucleon is allowed to transfer, etc. The momentum of the two interacting heavy ions is decremented following the transfer of each nucleon.

Continuous (Semi-Continuous) Nucleon Distributions

The Boltzmann-Uehling-Uhlenbeck model. This approach[13,14,16,21,31-33] is in some ways similar to the QMD treatment except that the nucleons are bound in a mean field determined by local density and the EOS rather than one generated explicitly via the long range N-N interaction, and the nucleons are each divided into an arbitrary number of 'test particles' (typically 50-100) which have a spread in position and momenta about the mean nucleon value to approximate requirements of the Heisenberg principle. It is, as in QMD and INC, necessary to run a sufficiently large number of events to generate satisfactory statistics for calculated results. The evolution is, as in QMD, followed in small time steps with nucleon trajectories readjusted in each time step with a density dependent central potential. Thus the EOS is included in this calculation. The INC, QMD and BUU approaches each follow (x, y, z, p_x, p_y, p_z) of the nucleons (or test particles), and so predict angular distributions. In each case a coalescence criterion may be applied for treating cluster emission.[34] All approaches include some estimate of the influence of the Pauli exclusion factor.

Boltzmann master equation (BME). The BME approach[35-37] simplifies the transport calculation in several ways with respect to the other models described; this leads to much greater computational speed and flexibility, while retaining many of the main elements of physics. Its success depends on the assumption that the spatial evolution is of secondary importance to the energy relaxation history. This approach uses a continuous nucleon (probability) distribution which avoids the costly event mode method of computation; entire distributions may be followed as probability flux on a time dependent basis. One price paid for this simplification is that clusters cannot be followed in a coalescence approach, whereas other physical ideas may easily be tested. Angular distributions have not generally been calculated within this model, although there is no reason this cannot be done.[38]

The main simplifications of the model are (1) following evolution of the excited nucleons in an energy space only, with collision and emission rates calculated from energy and isospin dependent nucleon-nucleon and nucleon-nucleus phase space, and (2) the assumption that the result of coupling the entrance channel projectile energy with the Fermi energy may be given by few-quasiparticle distribution functions based on the assumption that every energy conserving partition occurs with equal a-priori probability.[36] This result is consistent with earlier precompound model analyses of α and ^{3}He induced reactions.[39] It has an advantage over the other models described in that the exciton distribution function used allows a probability distribution which goes smoothly to the full energy available, consistent with experimental results, whereas some of the other semi-classical approaches use a classical sharp-cutoff of this distribution due to the coupling of Fermi and beam momenta. The QMD and BUU approaches offer some relief from this classical approximation in use of some nucleon momentum width about the mean value. The BME has also been used by Scobel with a Fermi sphere coupling distribution rather than using the exciton distribution function.[40] The detailed formulation of the BME model has been presented elsewhere.[35-37,9-11]

MODEL APPLICATION TO EXPERIMENTAL RESULTS

Test of Nucleon Energy Distributions

Before using transport codes to calculate secondary processes such as N-N-π or N-N-γ, we need to know that the correct energy distributions for nucleons prior to N-N collisions are going into the calculation. This may be checked by comparisons with (HI, xn) spectra, which result when nucleons are emitted after the nucleus-nucleus interaction, and before a N-N interaction (higher emission energy regime) or after one or more N-N collisions (medium emission energy regime). Confirmation of the validity of the initial distributions is shown in Figs. 1-3, where the BME may be seen to give excellent agreement with a very broad range of experimental results, using a single standard parameter set.[41-45]

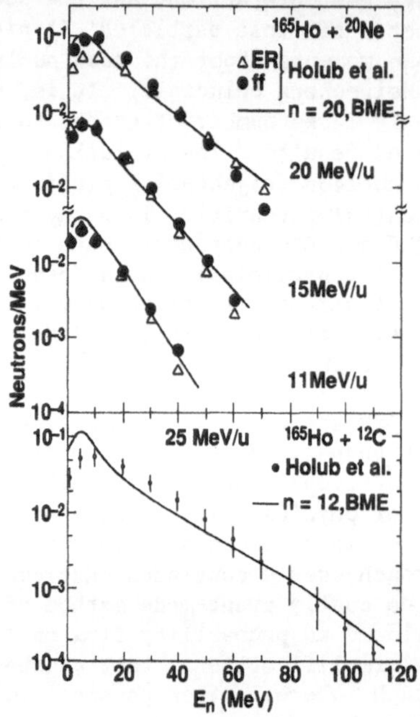

Figure 1

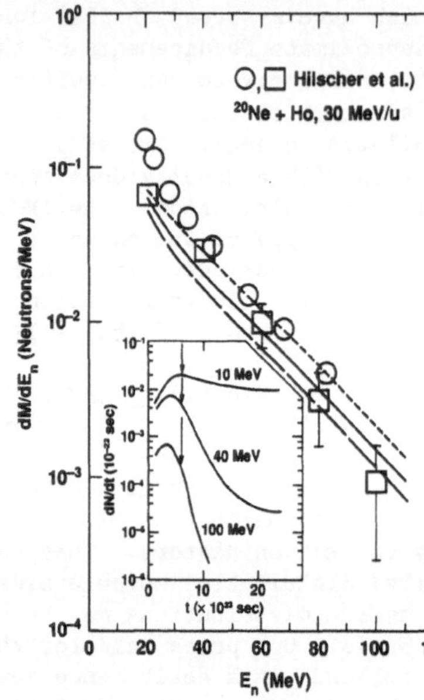

Figure 2

Fig. 1 Calculated and experimentally deduced spectra for the $^{165}Ho(^{20}Ne,n)$ and the $^{165}Ho(^{12}C,n)$ systems. Experimental points from (41) and (42) result from an integration of a moving source fit to experimental yields for the fast component only. Experimental yields for ^{20}Ne projectiles were gated on evaporation residues (ER) as represented by open triangles, and on fission fragments (FF) shown by closed circles. Results for ^{12}C were gated on ER. Calculated results are shown for the BME with $n = A_p$ in the exciton distribution function, where we assume total excitation is shared by n excitons with equal a-priori probability.

Fig. 2 Experimental and calculated neutron energy spectra from 30 MeV per nucleon ^{20}Ne on Ho. The squares represent the preequilibrium yields deduced by Hilscher (43) by fitting an assumed isotropically emitting moving source to the high energy data. The circles represent the total differential data of Hilscher et al., integrated directly. The solid line is the BME result; the short and long dashed lines correspond to increasing and decreasing the nucleon mean free path by 50%. The insert shows the calculated time dependence of the emission of 10, 40 and 100 MeV neutrons. The arrows represent the time at which fusion is assumed complete in the calculation.

There are several additional points to note in Figs. 1 and 2. The insert to Fig. 2 shows that the high energy neutrons are emitted in less than 10^{-22} sec; if N-N collisions occur before emission, the emission energies are decreased and lifetimes increase. Thus these spectra represent the nucleus-nucleus stopping process through the time/energy dependence of the emitted nucleon spectra. The second point is the observation that the nucleon spectra go smoothly to energies which exceed the semiclassical limit of Fermi plus beam momentum coupling.

Subbarrier Pion Emission

Experimental N-N-π cross sections may be used in the nucleon transport codes described to estimate the pion yields in heavy ion reactions. This

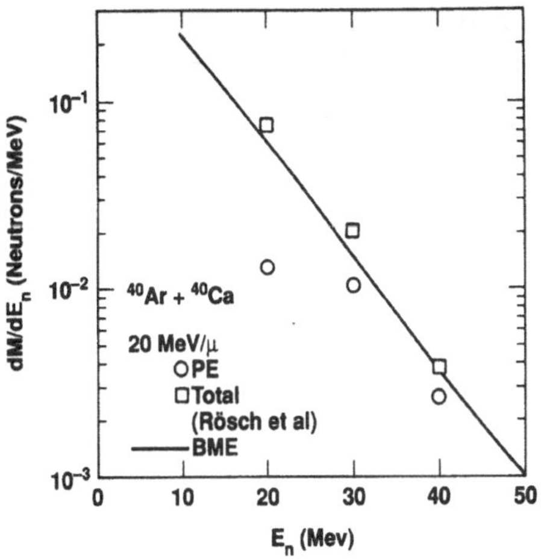

Fig. 3 Calculated and experimental $^{40}Ca(^{40}Ar,n)$ spectra for 20 MeV/u incident energy, gated on evaporation residues. Data are from Rösch (44). The solid line is the result of the BME calculation assuming 40 excitons.

was done successfully with the BME[9]; results are summarized in Table II of Ref. 9. Other transport calculations have similarly found good agreement with measured pion yields.[12,14,19] For pions, there is an ambiguity due to the reabsorption prior to escape from the nuclei. Within this uncertainty the several nucleon transport calculations give good agreement with data, confirming that one possible interpretation of the pion emission is as being due to N-N collisions with nucleons boosted in energy by coupling of beam and Fermi momenta.[46]

Photons do not have the strong reabsorption expected for pions, and so many experimental groups chose to study photon emission. In the following subsection we consider the requirements to enter this capability into the transport codes.

High Energy γ Emission in Heavy Ion Collisions

For the nucleon transport codes described in the previous section, photon production spectra may be calculated if the differential (or double differential) neutron-proton-bremsstrahlung cross sections are known. The evaluation of this basic input to the calculation has an interesting history, and somewhat divides the results obtained thus far. We note that neutron-neutron or proton-proton quadrupole bremsstrahlung are expected to be one to two orders of magnitude smaller than n-p electric dipole bremsstrahlung, and are therefore generally ignored.[47,48]

A semi-classical description of the p-n-γ differential cross section is given in several approximations in the textbook of Jackson; one form is[49]

$$\frac{d^2N}{dE_\gamma d\Omega_\gamma} = \frac{1}{E_\gamma}\frac{\alpha^2}{(2\pi)^2}\sum_{k=1}^{2}\left|\frac{\hat{\epsilon}_k \cdot \beta_i}{1 - \hat{q} \cdot \beta_i} - \frac{\hat{\epsilon}_k \cdot \beta_f}{1 - \hat{q} \cdot \beta_f}\right|^2 \times P_{fac}(1 + X) \quad . \qquad (1)$$

Here, $\alpha = \frac{1}{137}$ is the fine structure constant; $\hat{q}$, $\hat{\epsilon}_1$, and $\hat{\epsilon}_2$ are the unit vectors denoting the γ-ray direction of propagation and two directions of polarization, and $\beta_{i,f}$ denote the initial and final velocities of the proton. $P_{fac} = (\gamma_f\beta_f)/(\gamma_i\beta_i)$ is a quantal correction for the reduced final-state phase space in nucleon-nucleon vs. nucleon-heavy nucleus scattering, where $\gamma_{i,f}$ represents the relativistic contraction factor.[50,51] The (1 + X) factor represents the uncertainty due to neglecting meson exchange effects in the radiation formula.[47,48] A relativistic quantal calculation including meson exchange effects was given in early works by Brown[47] and by Brown and Franklin,[48] but the results were not cast in a form easily used to replace Eq. (1). These authors concluded that neglect of meson exchange in Eq. (1) resulted in an underestimate of the differential cross section by roughly a factor of two. Neuheuser and Koonin[52] integrated the equations of Brown and Franklin[48] non-relativistically to calculate the differential γ-ray spectra, and gave a prescription for scaling the semiclassical result of Eq. (1) to approximate the quantal result. We will refer to their results shortly. Other authors have also recently reconsidered the problem with inclusion of meson exchange.[53-54]

The only set of reasonably extensive experimental data with which to make comparisons were, until recently, the nucleus (p, γ) measurements due to Edgington and Rose using 140 MeV protons.[55] Deuterium was one of the targets used, which except for the internal momentum of the neutron in deuterium, gives the p-n-bremsstrahlung spectrum. Therefore most groups doing the heavy ion-γ-ray transport calculation began by comparing their results with the data of (52); e.g. in Fig. 4 we show results of Remington et al.,[10] using Eq. (1), and Eq. (1) multiplied by two and corrected for the internal momentum of the deuteron. The results of Neuheuser and Koonin (N-K) are also shown.[52] Based on these comparisons Eq. (1) multiplied by two was used by Remington et al.[10]; the N-K results[52] were also used in many of their calculations. In Fig. 5 we show a similar result from Ref. 12 using a semiclassical radiation formula.[49]

More recently nucleus (p, γ) measurements have been made by Kwato Njock et al., for 72 and 168 MeV protons on several targets.[56,57] When these are scaled and compared with the results of Edgington and Rose,[55] they suggest that for photon energies in excess of 50 MeV, the results of (55) are low by a factor of 3-4 versus the newer data (Fig. 6). This in turn suggests that the results of N-K (see Fig. 4) may be more realistic than the scaled results of Eq. (1) which have been used frequently in transport code analyses. In Fig. 4, we show the $^{12}C(p,\gamma)$ measurement of (57) for 168 MeV protons, scaled in magnitude by 140/168, but not scaled in energy, versus the N-K and semi-classically calculated spectra at 140 MeV. That the N-K results are in better agreement with the 168 MeV experiment is obvious. The calculations of Nakayama, with inclusion of meson exchange effects, should also be in better agreement with the newer p-nucleus measurements,[53-54] based on improved results in applications to nucleus-nucleus gamma ray yields.

In Figs. 4 and 7, we show comparisons between calculated and experimental high energy γ ray data from several sources,[58,60] calculated with the BME and with the BUU model in Fig. 5. The data of Stevenson et. al.[58,60] have been replotted with a recent detector efficiency recalibration.[60] Using the newer data of Stevenson et. al., it may be seen in Figs. 4, 5 and 7 that the semi-classical radiation expression underestimates the experimental γ-ray spectra, and this underestimation becomes worse as the incident heavy ion energy increases. The N-K quantal calculation for p-n-γ, on the contrary, may be seen in Figs. 4 and 7, to give a satisfactory agreement for all energies for which it has been tried, as well as being in better agreement with the newer experimental p-nucleus measurements of Kwato-Njock et al.

These results, as well as similar results by other groups using formulations briefly summarized in Section II, confirms that incoherent N-N-bremsstrahlung processes provide one viable explanation of the high energy

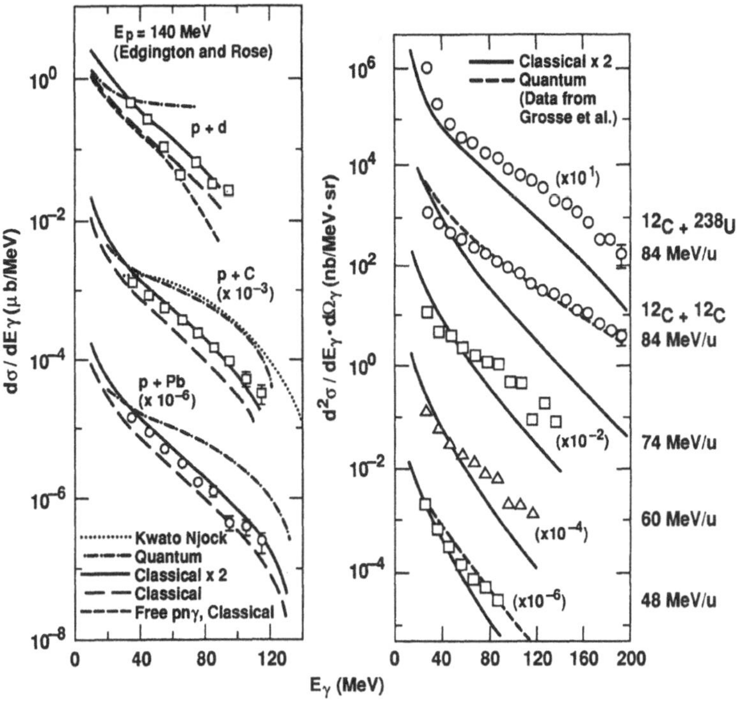

Fig. 4 (Left): Comparisons of calculations with high energy γ-ray data of Edgington and Rose (55) for a 140 MeV proton beam. At the top, the p+d data are used as a standard with which to "normalize" the pnγ bremsstrahlung equation. The short-dashed line corresponds to a semi-classical bremsstrahlung formula for free neutron-proton scattering. The long-dashed line is the same, only with the momentum distribution of the target neutron in deuterium taken into account. The solid line corresponds to the deuteron calculation multiplied by two to crudely account for the effect of meson exchange. The dotdashed line corresponds to folding the quantum bremsstrahlung result of Neuheuser and Koonin (52) into the deuteron calculation. The lower two spectra show γ-ray data for p+^{12}C and p+Pb. The curves represent calculations with the master equation using the semiclassical bremsstrahlung cross sections (dashed lines), the semi-classical cross sections multiplied by two for meson exchange (solid lines), and the quantum bremsstrahlung cross sections (dot-dashed lines). For C+p, the dotted curve is the experimental result of Kwato Njock et al. (57) for 168 MeV protons, scaled in magnitude by 140/168, but not scaled for energy.
(Right): γ-ray data (61) for ^{12}C + ^{238}U at 84 MeV/nucleon (top, ^{12}C + ^{12}C at 84, 74, 60 and 48 MeV/nucleon. The solid lines represent the master equation calculation for a sharp-cutoff initial exciton distribution, while the dot-dashed line is for a continuous exciton distribution. The dashed lines represent master equation calculations using the quantum bremsstrahlung cross sections.

γ-rays observed in HI reactions. The success of the BME in reproducing these data suggests that nuclear compressibility properties are not yet an issue. Data at higher energies may yet provide results for which the EOS is an essential ingredient in interpretation. This should be checked by model calculations. Meanwhile, Heuer et al.,[20] are confirming the regime in which the problem may be modeled as a semiclassical sum of incoherent N-N-γ processes via the QMD approach.

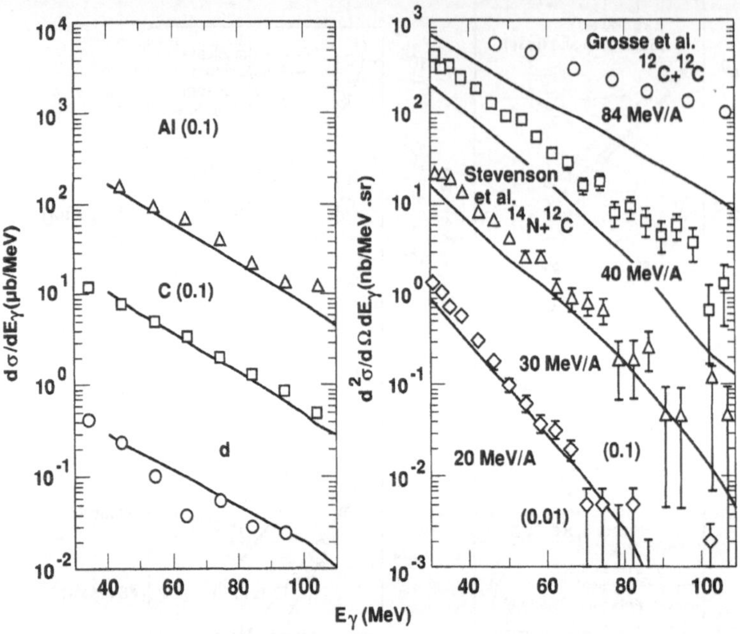

Fig. 5 (Left) Comparison of p, nucleus bremsstrahlung spectra calculated using the semi-classical formula as reported in (13) (solid lines) compared with the data of Edgington and Rose (55). Right hand side shows spectra from the $^{14}N + ^{12}C$ (58,60) and $^{12}C + ^{12}C$ reactions (59) compared with BUU calculations using the same semi-classical radiation formula (13).

CONCLUSIONS

Several transport codes have been described which interpret pion or photon production as an incoherent result of nucleon-nucleon scattering, with nucleons having a distribution of energies resulting from Fermi plus beam momentum coupling. All of these provide a description of the time dependence of the nucleus-nucleus interaction resulting from nucleon-nucleon scattering in a potential due to the nuclear matter. All give a generally satisfactory reproduction of experimental pion and photon yields and spectra. Analyses have been performed for the production of mesons other than pions, but these results are not discussed in this work.[61,62]

Some of the codes discussed are limited by the semi-classical Fermi plus beam velocity coupling limit. This point should be addressed and understood for those particular phenomena which are sensitive to the approximation (e.g., pion production near the thermodynamic threshold). A problem for all nucleon transport codes when used to calculate photon production is a lack of a sufficient body of data (e.g., d(p, γ)) to test and if necessary modify theoretical models which provide the basic n-p-γ input to the transport codes. Recent results of Kwato Njock et al.,[56,57] suggest that values frequently used are significantly in error. Better model calculations await a more comprehensive determination of the basic p(n, γ) differential cross sections over a sufficient range of incident nucleon energies. These data will be essential if γ-rays are to be used as a tool for answering questions of significance regarding the EOS based on results of heavy ion collisions. The general overall agreement of different calculations with and without the EOS included suggests that in the incident energy regime up to 84 MeV/A the EOS does not have a great influence on results.[63,64] Once the basic p(n,γ) process is understood in a reasonably quantitative fashion, transport calculations may be performed to see if

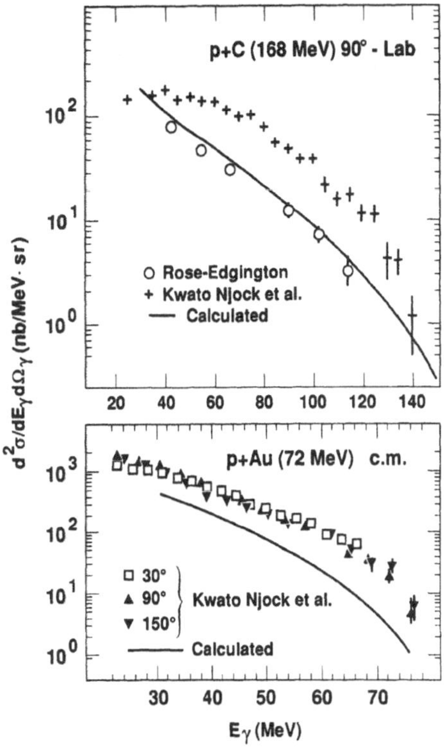

Figure 6

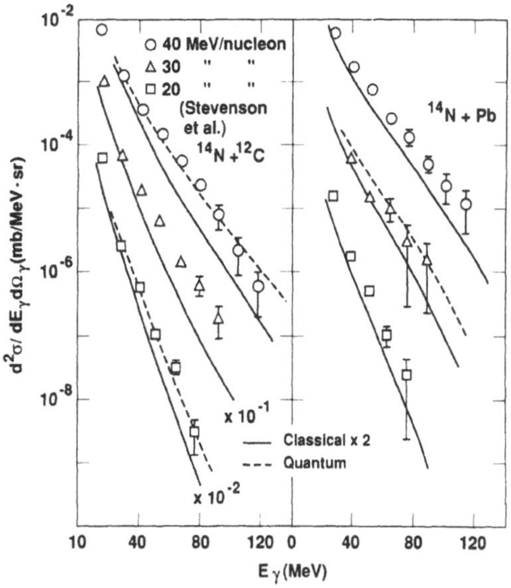

Figure 7

Fig. 6 The solid curves are results of a semi-classical radiation formula from (49) compared (upper) with 168 MeV p + ^{12}C gamma ray spectra from (57). The results of Rose and Edgington scaled from 140 MeV by multiplication of ordinate and abscissa by 168/140 are shown for comparison. In the lower section data for p + Au with 72 MeV protons from (56) are shown versus predictions of the same semi-classical radiation formula. All results are taken from (56, 57).

Fig. 7 The γ-ray data of Stevenson (58,60) are shown for $^{14}N+^{12}C$ (left side) and for ^{14}N+Pb (right side) at 20, 30 and 40 MeV/nucleon. The solid lines correspond to the BME calculation using the semi-classical bremsstrahlung cross sections with X=1 in Eq. 1 while the dashed lines correspond to quantum bremsstrahlung cross sections of Ref. (52). The results of Stevenson et al. have been plotted to include the recent detector calibration results (60).

the γ-ray yields in nucleus-nucleus collisions should be (and at what level) sensitive to the EOS. Making extensive experimental measurements of nucleus-nucleus results when there are nearly no nucleon-nucleon results on which to base an interpretation brings to mind the idiomatic expression "don't put the cart before the horse."

Since completion of this work a very comprehensive review paper on the same topic by Nifenecker and Pinston has been received; the author wishes to call this outstanding review to the attention of interested parties.[65]

During the course of this work the author has greatly benefited from discussions with W. Greiner, R. Heuer, H. Stöcker, G. F. Bertsch and H. Nifenecker.

Work performed under the auspices of the U.S. Department of Energy by the Lawrence Livermore National Laboratory under contract number W-7405-ENG-48.

REFERENCES

1. J. Aichelin, Phys. Rev. Lett. 52, 2340 (1984).
2. J. Aichelin and G. Bertsch, Phys. Lett. B138, 350 (1984).
3. H. Nifenecker and J. P. Bondorf, Nucl. Phys. A442 (1985) 478.
4. D. Hahn and H. Stocker, Nucl. Phys. A452, 723 (1986).
5. R. Shyam and J. Knoll, Nucl. Phys. A448 (1986) 322.
6. D. Vasak, B. Müller and W. Greiner, Physica Scripta 22, 25 (1980).
7. D. Vasak, B. Müller and W. Greiner, J. Phys. G11, 1309 (1985).
8. T. Stahl, M. Uhlig, B. Müller, W. Greiner, D. Vasak, Z. Phys. A327, 311 (1987).
9. M. Blann, Phys. Rev. Lett. 20, 2215 (1985); Phys. Rev. C32, 1231 (1985).
10. B. A. Remington, M. Blann and G. F. Bertsch, Phys. Rev. Lett. 57, 2909 (1986); Phys. Rev. C35, 1720 (1987).
11. B. A. Remington and M. Blann, Phys. Rev. C36, 1387 (1987).
12. W. Bauer, G. F. Bertsch, W. Cassing and U. Mosel, Phys. Rev. C34, 2127 (1986).
13. T. S. Biro et al., Nucl. Phys. A475, 579 (1987).
14. J. Aichelin, Phys. Lett. 164B, 261 (1985).
15. Che Ming Ko, G. Bertsch and J. Aichelin, Phys. Rev. C31, 2324 (1985).
16. Che Ming Ko and J. Aichelin, Phys. Rev. C35, 1976 (1987).
17. J. Randrup and R. Vandenbosch, Nucl. Phys. A 474 (1987) 219; Nucl. Phys. A 490, 418 (1988).
18. W. Cassing, T. Biro, U. Mosel, M. Tohyama and W. Bauer, Phys. Lett. B181, 217 (1986).
19. W. Cassing, Z. Phys. A329, 487 (1988).
20. R. Heuer, B. Müller, H. Stocker and W. Greiner, Z. Phys. A330, 315 (1988).
21. G. Bertsch, "Nonrelativistic Theory of Heavy Ion Collisions," School in Heavy Ion Physics, Erice, Sicily (1984).
22. W. Cassing, Z. Phys. A327, 87 (1987); A326, 21 (1987); A327, 447 (1987); Z. Phys. A329, 471 (1988); A329, 487 (1988).
23. W. Cassing, K. Niita and S. J. Wang, Z. Phys. A331, 439 (1988).
24. E. Braun and Z. Fraenkel, Phys. Rev. C34, 120 (1986); Z. Fraenkel, Nucl. Phys. A428, 373c (1984); Z. Fraenkel, Nucl. Phys. A374, 475c (1982).
25. Z. Fraenkel, G. Mamane and M. R. Clover, in "Fundamental Problems in Heavy Ion Collisions," Singapore, World Scientific: 365 (1984).
26. J. Aichelin and H. Stöcker, Max Planck Inst. Report MPIH-1986-V6 (Quantum Molecular Dynamics - A Novel Approach to N-Body Correlations in Heavy Ion Collisions).
27. J. Aichelin, G. Peilert, A. Bohnet, A. Rosenhauer, H. Stöcker and W. Greiner, Phys. Rev. C (1988).
28. G. Peilert, H. Stöcker, W. Greiner and A. Rosenhauer, UFTP preprint 214/1988.
29. J. Randrup, Nucl. Phys. A 327, 490 (1979); A 383, 468 (1983); T. Dossing and J. Randrup, Nucl. Phys. A 433, 215 (1985).
30. G. F. Bertsch, H. Kruse and S. Das Gupta, Phys. Rev. C29, 673 (1984).
31. J. Aichelin and G. Bertsch, Phys. Rev. C31, 1730 (1985).
32. H. Kruse, B. V. Jacak, G. D. Westfall and H. Stöcker, Phys. Rev. C31, 1770 (1985).
33. J. J. Molitoris and H. Stöcker, Phys. Rev. C32, 346 (1985).
34. W. Bauer, S. Das Gupta and G. F. Bertsch, Phys. Rev. Lett. 58, 867 (1988).
35. G. D. Harp, J. M. Miller and B. J. Berne, Phys. Rev. 165, 1166 (1968); G. D. Harp and J. M. Miller, Phys. Rev. C3, 1847 (1971).
36. M. Blann, A. Mignerey and W. Scobel, Nukleonika 21, 335 (1976); M. Blann, Phys. Rev. C23, 205 (1981).
37. M. Blann, Phys. Rev. C31, 1245 (1985).
38. B. A. Remington and M. Blann, Phys. Rev. C 36, 1387 (1987).

39. M. Blann, Ann. Rev. Nucl. Sci. 25, 123 (1975) and references therein.
40. W. Scobel, E. Mordhurst and M. Strecker, "Frontiers of Heavy Ion Physics," World Scientific, Singapore (1987) p. 267 (ed. N. Cindro, W. Greiner and R. Caplar).
41. E. Holub et al., Phys. Rev. C28, 252 (1983).
42. E. Holub et al., Phys. Rev. C33, 143 (1986).
43. D. Hilscher et al., Phys. Rev. C36, 208 (1987).
44. W. Rösch et al., Phys. Lett. B197, 19 (1987).
45. M. Blann and B. A. Remington, Proceedings of the 20th Summer School on Nuclear Physics, Mikolajki, Poland (1988) to be published.
46. G. F. Bertsch, Phys. Rev. C 15, 713 (1977).
47. V. R. Brown, Phys. Rev. 177, 1498 (1969).
48. V. R. Brown and J. Franklin, Phys. Rev. C8, 1706 (1973).
49. J. D. Jackson, "Classical Electrodynamics," (Wiley, New York) 1975, p. 703.
50. Che Ming Ko, G. Bertsch and J. Aichelin, Phys. Rev. C31, 2324 (1985).
51. K. Nakayama and G. Bertsch, Phys. Rev. C34, 2190 (1986).
52. D. Neuheuser and S. E. Koonin, Nucl. Phys. A462, 163 (1987).
53. K. Nakayama and G. F. Bertsch, "Potential Model Calculations of Energetic Photon Production From Heavy Ion Collisions," in press, Phys. Rev. C.
54. K. Nakayama, "High Energy Photons in Neutron-Proton and Proton-Nucleus Collisions," in press, Phys. Rev. C.
55. J. A. Edgington and B. Rose, Nucl. Phys. 89, 523 (1966).
56. M. Kwato Njock et al., Phys. Lett. B207, 269 (1988).
57. H. Nifenecker, M. Kwato Njock and J. A. Pinston, "High Energy Gamma Ray Production in Heavy Ion Reactions," Proceedings 20th Summer School in Nuclear Physics, Mikolajki, Poland, September 1988, to be published.
58. J. Stevenson et al., Phys. Rev. Lett. 57, 555 (1986).
59. E. Grosse, Int'l Workshop on Gross Properties of Nuclei (1985), Hirschegg; E. Grosse et al., Europhys. Lett. 2, 9 (1986).
60. J. Stevenson, private communication (1987).
61. A. L. De Paoli, K. Niita, W. Cassing and U. Mosel, Phys. Lett. B (in press).
62. J. Aichelin and Che Ming Ko, Phys. Lett. B (in press).
63. M. Cubero, M. Schönhofen, H. Feldmeier and W. Norenberg, Phys. Lett. B201, 11 (1988).
64. B. M. Waldhauser, J. A. Maruhn, H. Stöcker and W. Greiner, Z. Für Physik A328, 19 (1987).
65. H. Nifenecker and J. A. Pinston, "High Energy Gamma Ray Production in Nuclear Reactions," in press, Progress in Particle and Nuclear Physics (1989).

38. M. Block, Ann. Rev. Nucl. Sci. 17, 227 (1967) and references therein.
40. W. Selove, L. Hutchinson and G. Shenkman, [illegible] Physics, [illegible] (1968) [illegible] 264, 268 [illegible]
[illegible] and [illegible]
41. [illegible] Phys. [illegible] 722 (1967).
42. [illegible] Phys. Rev. [illegible] 144 (1966).
43. [illegible] 299 (1967).
44. [illegible] 819, 33 (1967).
45. [illegible] Proceedings of the [illegible]
[illegible]
[illegible]
47. [illegible]
[illegible]
[illegible]
[illegible]
[illegible]
[illegible]
[illegible]
[illegible] "[illegible] Collisions," to press, Phys. Rev.
53. [illegible] "High Energy [illegible] and [illegible] Nucleon Collisions," [illegible] Phys. Rev.
54. [illegible] Phys. [illegible] 569 [illegible]
55. [illegible] 253, 253 [illegible]
57. [illegible] "High [illegible]
[illegible] Production in [illegible] Conference on [illegible] Nuclear [illegible] September 1965, [illegible]
[illegible]
[illegible] et al., Phys. Rev. Lett. [illegible]
58. [illegible] (1967).
[illegible] et al., [illegible] Lett. [illegible]
59. [illegible], private communication (1970).
61. [illegible] and [illegible], to be published.
62. [illegible] Phys. Lett. [illegible]
63. [illegible] Phys. [illegible]
[illegible] (1970).
64. [illegible] and [illegible]
[illegible] (1968).
65. [illegible] "High Energy [illegible] Production [illegible] Nucleon [illegible]" in [illegible] Progress in Particle and Nuclear Physics [illegible]

DILEPTON PRODUCTION IN HEAVY ION COLLISIONS†

G. Batko*, T. S. Biró, W. Cassing, U. Mosel, M. Schäfer, G. Wolf

Institut für Theoretische Physik, Universität Giessen
D–6300 Giessen, West Germany

Nucleus–nucleus collisions provide probes of dense and hot nuclear matter far from the groundstate. One of the major aims is to obtain information about the nuclear equation of state and to extract signals which can directly be related to the high compression zone of the system. Hadronic probes often suffer from strong final state interactions, such that the primary information carried by these particles might be washed out due to rescattering in the expansion phase. Alternatively, electromagnetic probes are expected to provide a more unique signal since they leave the hadronic environment after their production without sizable disturbances. In fact, the observation of hard photons from intermediate–energy heavy–ion collisions has led to widespread activities, both on the experimental[1,2] and theoretical side[3,4]. Since the use of real photons is restricted to bombarding energies below about 300 MeV/u due to the experimental difficulty to distinguish the direct photons from the decay photons of the copiously produced π^0's [2], it is natural to use at higher energies virtual photons as they manifest themselves in the production of dileptons. After many years of difficult experimental developments, the Berkeley group has recently obtained first results on dilepton production at BEVALAC energies[5]. It is therefore a natural challenge to see how well these yields can be understood theoretically on the same basis as the hard real photons.

Apart from dilepton production in proton–neutron bremsstrahlung, π^0 decay and π–nucleon scattering, e^+e^- or $\mu^+\mu^-$ pairs are also created in $\pi^+\pi^-$ annihilation, which dominates in the high density compression phase. As has been stressed by

†Work supported by BMFT and GSI Darmstadt.

*Supported by the German Academic Exchange Service (DAAD) and by the CONICET of Argentina.

Gale and Kapusta[6], the mass distribution of back-to-back e^+e^- pairs is expected to provide for the first time experimental information on the pion dispersion relation $\omega_\pi(\mathbf{k})$ in dense and hot nuclear matter.

While preliminary estimates for dilepton production were performed in refs. [6,7], we have recently published first results of a calculation of the elementary $p + n \rightarrow p + n + e^+ + e^-$ cross section based on a covariant meson exchange description of pn scattering[8]. In this paper we apply this cross section in a dynamical simulation of heavy-ion collisions at 1 GeV/u and compare with the available experimental data to see how well pn bremsstrahlung dileptons can explain the observed yield. We also present here for the first time dynamical calculations of the $\pi^+\pi^-$ annihilation radiation to assess its importance on the same basis.

For the calculation of the np scattering amplitude at energies between 100 and 400 MeV laboratory energy we use a parametrization of the T-Matrix in terms of OBE amplitudes by Horowitz and coworkers[9] that provides a good fit to the scattering data in this energy range. At higher energies we use a fit to the experimental np scattering cross section in terms of σ, ω, ρ and π mesons, that we have performed ourselves, and that gives a good description both of the absolute value as well as of the scattering angular distribution[8]. The radiation field is then introduced by minimal coupling, so that the dominant Feynman-diagrams which contribute to the process

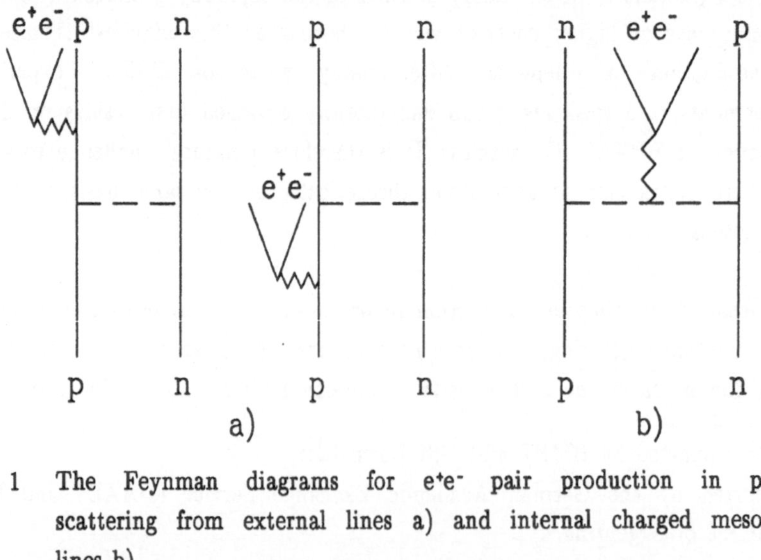

Fig. 1 The Feynman diagrams for e^+e^- pair production in pn scattering from external lines a) and internal charged meson lines b).

$n + p \rightarrow n + p + e^+ + e^-$

can be evaluated in a covariant way. The relevant diagrams are shown in Fig. 1; they indicate radiation from the external lines (a) as well as from internal charged meson exchange (b); although we have not gauged the cut–off factors, a check has shown that this yields only corrections of the order of 6 – 8 %. We restrict our calculations to pn bremsstrahlung since pp bremsstrahlung is lower by about an order of magnitude due to the destructive interference of the radiation from the two outgoing proton lines[10]. The differential cross section then is given by

$$\frac{d\sigma^{e^+e^-}}{dM^2_{e^+e^-}\, d\Omega_N\, d\Omega_L\, d\Omega_P\, dP} = \frac{1}{(E_+ + E_-)\, E_p\, |p_p|\, (2\pi)^8} \times |T|^2 \times$$

$$\frac{2\, m_N^4\, m_L^2\, Q_N^2\, Q_L^2\, P^2}{|Q_L\, (E_+ + E_-) + P\cos\beta\, (E_- - E_+)|\; |Q_N\, (E'_p + E'_n) + P\cos\alpha\, (E'_p - E'_n)|} \qquad [1]$$

where $\vec{Q}_L = \vec{p}_- - \vec{P}$, $\vec{Q}_N = \vec{p}'_p - \vec{P}$ and $\vec{P}$ is half of the total lepton momentum; the transition matrix element T is obtained by summing all diagrams of order e^2g^2 (such as the ones illustrated in Fig. 1) over all exchanged mesons. The term $|T|^2$ already includes a sum over final spin and average over initial spin degrees of freedom of all particles. α and β are the angles between $\vec{P}$ and $\vec{Q}_N$ and $\vec{P}$ and $\vec{Q}_L$, respectively.

The kinematical situation is illustrated in Fig. 2. In order to obtain the differential cross section for dilepton production $d\sigma/dM^2$ with

$$M^2 = (p_+ + p_-)^2 \qquad [2]$$

we integrate the differential cross section over the residual degrees of freedom by Monte Carlo techniques. Due to energy and momentum conservation they amount to the angles shown in Fig. 2 and half of the c.m. momentum of the e^+e^- pair, $\vec{P}$ (cf. Fig. 2). The possible range for $\vec{P}$ is restricted as a function of invariant mass M; note, however, that events with large invariant masses M can have small momenta $\vec{P}$ and vice versa.

In Fig. 3 we show the results calculated from the graphs in Fig.1 for two energies, 200 MeV (left solid curve) and 1000 MeV (right solid curve), together with an approximate expression (dashed lines) used by Gale et al. in ref. 6. In this comparison we use for $\sigma_{np}^{elastic}$ the values obtained from the OBE T–Matrix approximation and not the parametrization employed by Gale et al. Whereas this approximate expression[6] yields a reasonable description of the exact expression at 200 MeV, it differs quite significantly from it at 1000 MeV. Here the approximate cross section is too low at very low invariant masses ($M \simeq 20 - 40$ MeV) and by up to an order of magnitude too high at larger M. In ref. 8 we investigated the

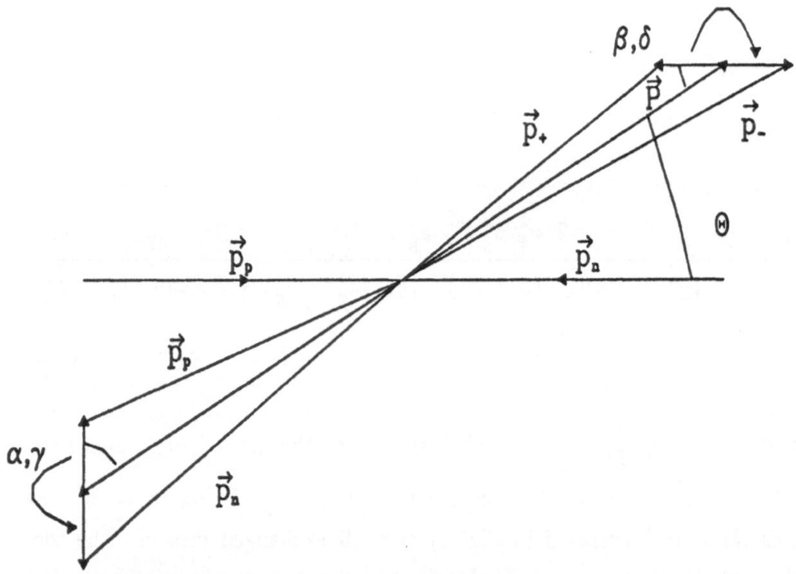

Fig. 2 Illustration of the kinematic situation for e^+e^- pair production in pn collisions. $\vec{P}$ is half of the total dilepton momentum $\vec{p}_+ + \vec{p}_-$. θ is the angle between the beam axis, given by the initial proton momentum $\vec{p}_p$, and the momentum $\vec{P}$. α and β are the angles between $\vec{P}$ and the relative momentum between both nucleons and leptons, respectively. γ and δ are the additional rotational degrees of freedom around the $\vec{P}$ direction.

reasons for this disagreement and attributed it mainly to the long–wavelength approximation used in ref. 6 which is not reliable at any invariant mass. The long–wavelength approximation involves two ingredients; one is the treatment of the T–matrix and the other that of the final state phase space. For the latter the long–wavelength approximation leads clearly to an overestimate of the available phase space since the nucleons retain their initial energy. Gale and Kapusta have recently corrected for this deficiency by calculating the phase space factor for the

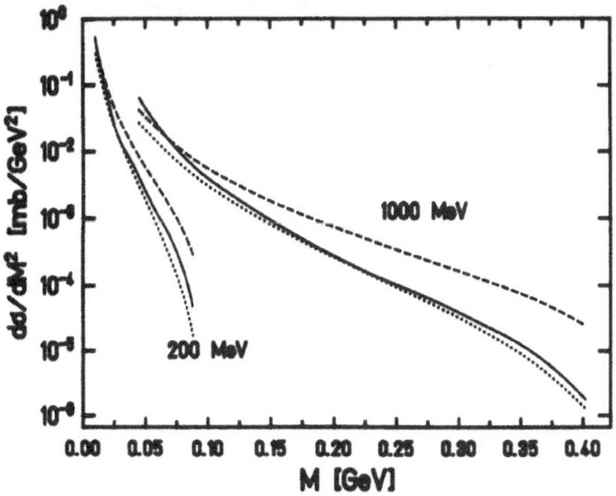

Fig. 3 Calculated differential dilepton cross section for two energies, 200 MeV (left) and 1000 MeV (right). The full lines corresponds to the cross section of eq. (1), the dashed ones to those obtained in the long–wavelength approximation of ref. 6, and the dotted ones to the results of ref. 11.

correct final energies[11] whereas the long–wavelength approximation is retained for the matrix element. It is seen from Fig. 3 that this approximation (dotted lines) gives excellent agreement with the full calculation.

We now turn to a comparison of our calculations with experimental data. Since nucleon–nucleon data are to our knowledge not available, we perform this comparison with data for the reaction p + Be at 1 GeV, obtained very recently at the BEVALAC[5], by assuming, as in the case of hard photons[3], that the observed yield is due to an incoherent superposition of the dilepton yield from independent collisions of the proton with the target neutrons; in order to take the Fermi motion of the target nucleons into account we have performed an average over neutron momenta in a Fermi–sphere of radius $k_F = 1.1\ fm^{-1}$. In Fig. 4 we show the result of this calculation, together with the experimental data for p + Be at 1 GeV from ref. 5, divided the number of neutrons in ^{9}Be; to make this comparison possible we have employed the experimental acceptance filter[13]. As can be seen in Fig. 4 we find a quite remarkable agreement with these data except for the strikingly low experimental point at 125 MeV. This indicates that in p + Be at this 'low' energy primarily e^+e^- – pairs from pn bremsstrahlung are observed. Haglin, Kapusta and Gale reach the same conclusion for the data at 2.1 GeV in a recent, quite similar calculation that, however, takes only pion exchange into account[14].

The calculation just described neglects any distortions of the target momentum distribution during the course of the collision. We have, therefore, in a next step used this microscopic cross section in a BUU simulation of the reaction. The dileptons are created here perturbatively, i. e. without taking the energy of the emitted dileptons out of the nuclear system. The final-state phase space blocking is, however, calculated correctly for the inelastic scattering. This is thus the same method as the one used in calculations of hard photon production[3] and finds its justification in the very small probability for dilepton production at these energies. The results of this BUU calculation are shown in Fig. 4; again very good

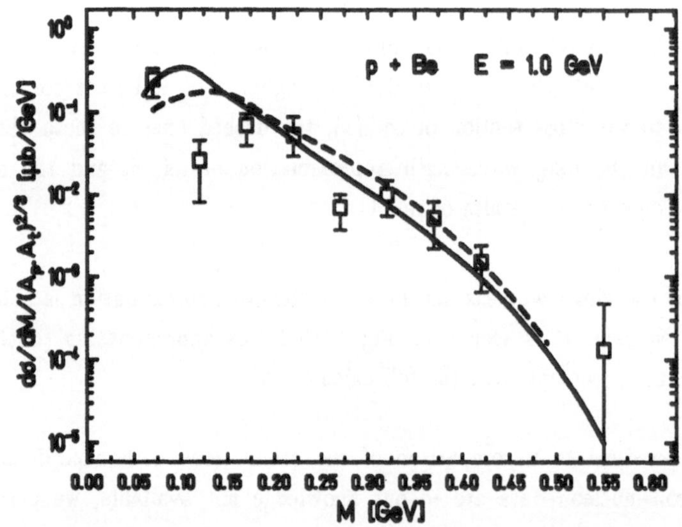

Fig. 4 Comparison of our microscopic calculation for the differential e^+e^- yield with the available experimental data from ref. 5 for p + Be at 1 GeV. The dashed line corresponds to the differential dilepton yield from pn bremsstrahlung, which has been calculated within the first collision model, averaged over the neutron momenta in the Fermi sphere of Be and corrected by the experimental acceptance filter, while the solid line corresponds to the calculated dilepton spectra obtained in a BUU calculation as described in the text.

agreement with the data is reached over the whole range of invariant masses. Note that also in this result, as well as in the ones to be discussed below, the effects of the experimental filter are taken into account.

Since many-body effects and distortions of the phase-space distributions are expected to be very small in such a light system as p + Be, we have then performed the same calculations for the heavier system $^{40}Ca + ^{40}Ca$, also at 1.05 GeV/u, for which there are also data available[5]. The result of this calculation is shown in Fig. 5 by the dashed line. Whereas the observed spectrum is very well reproduced at the small invariant masses, there is a clear discrepancy seen at the higher M (around 300 MeV and above) where the calculated cross section underestimates the experimental one.

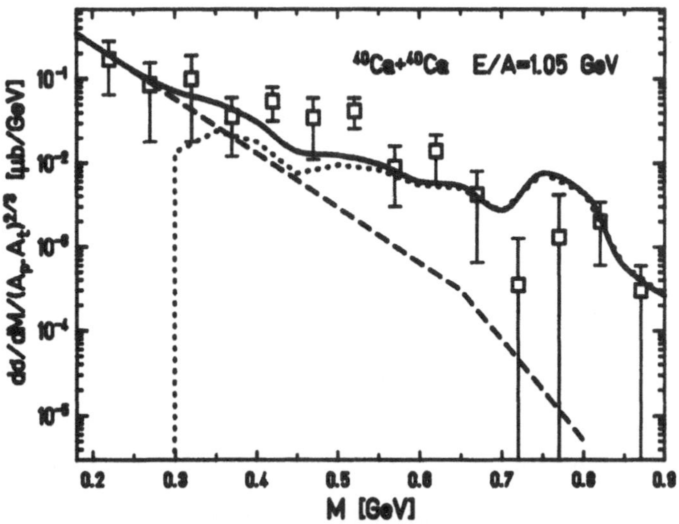

Fig. 5 Results from a BUU calculation of the dilepton yield for $^{40}Ca + ^{40}Ca$ at 1.05 GeV/u. The dashed curve shows the contribution from pn bremsstrahlung only, the dotted curve gives the contribution from pion annihilation, and the solid curve is the incoherent sum of both.

It is suggestive that this underestimate of the experimental cross section appears only at energies higher than the invariant mass of a $\pi^+\pi^-$ pair. The missing dilepton yield could therefore be just that coming from pion annihilation. The total number of pions produced in the heavier system is much larger than in the small system p + Be ($N_\pi \cong 0.7$).

In order to check this hypothesis we have performed BUU calculations in which the pions are produced nonperturbatively and explicitly propagated[15] with their free dispersion relation; pion reabsorption and rescattering at nucleons and Δ's as well as Δ decay are all taken into account. The total pion production cross sections and spectra obtained in this way are in good agreement with experiment[16]. At present we do not distinguish between the different charge states of the pion; we thus assume that on average all three charge states are produced with equal probability. This hypothesis is in reasonable agreement with experimental results on pion production at these energies. The pion annihilation cross section is then assumed to be given by

$$\sigma(M) = \frac{4\pi}{3} \alpha^2 \frac{1}{M^2} (1 - 4 m_\pi^2/M^2)^{1/2} |F_\pi(M)|^2 \qquad [3]$$

with

$$|F_\pi(M)|^2 = \frac{m_\rho^4}{(M^2 - m_\rho'^2)^2 + m_\rho^2\Gamma_\rho^2} \qquad [4]$$

$$m_\rho = 775 \text{ MeV}, \quad m_\rho' = 761 \text{ MeV}, \quad \Gamma_\rho = 118 \text{ MeV}$$

as used by Gale and Kapusta[6]; here $F_\pi(M)$ is the formfactor for coupling to the virtual photons through the ρ meson (vector meson dominance).

Since the pions are followed in their dynamical evolution we can determine their probability to collide by converting this cross section in the usual way into a distance of closest approach, inside which annihilation takes place. Dividing the dilepton production cross section from such a process by the pion–pion scattering cross section yields the dilepton production probability; by summing over all $\pi^+\pi^-$ collisions and integrating over impact parameter of the heavy ions, the cross section for producing a dilepton pair from pion annihilation in a heavy–ion collision is calculated.

The resulting cross section for this process is shown by the dotted line in Fig. 5. It is seen that this process of pion annihilation indeed accounts quantitatively for the missing cross section in the bremsstrahlungs process at higher M. We stress that this result is obtained with the free dispersion relation for the pions.

A closer analysis of this pion annihilation contribution shows a markedly different impact parameter dependence than the bremsstrahlung component: the

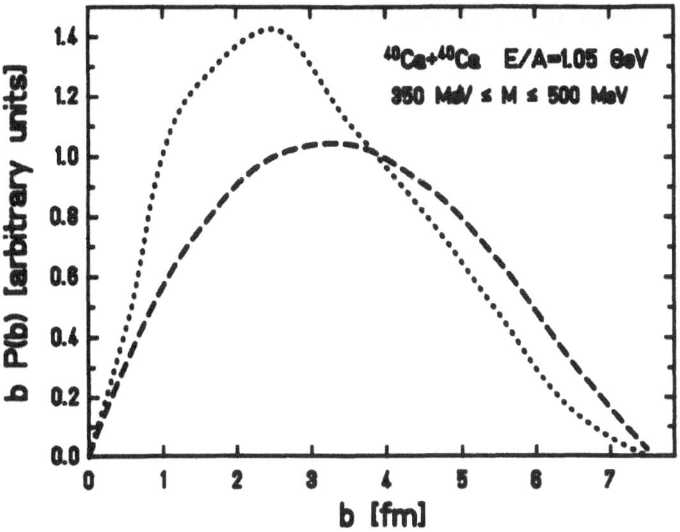

Fig. 6 Comparison of the dilepton yield multiplied by the impact parameter b for invariant masses 350 MeV $\lesssim$ M $\lesssim$ 500 MeV in collisions of ^{40}Ca + ^{40}Ca at 1.05 GeV/u, as a function of b. The dotted curve shows the contribution from $\pi^+\pi^-$ annihilation while the dashed curve correspond to the process $p + n \rightarrow p + n + e^+ + e^-$.

$\pi^+\pi^-$ annihilation dominates at smaller b (cf. Fig. 6). This suggests to enhance this component over the pn bremsstrahlung by using a centrality trigger in such studies; this would have the additional advantage of triggering on the high-density events where possible dispersion relation effects are expected to become strongest.

In summary, we have calculated the differential cross section for e^+e^- – pair production from pn bremsstrahlung in a covariant way based on a realistic approximation of the pn scattering amplitude via boson exchange amplitudes. We find that the previously used long-wavelength approximation gives qualitatively correct spectra but overestimates the exact cross section at the higher beam energies and invariant masses, whereas a phase-space corrected version of this approximation gives much better results.

The calculations based on a first collision model and on a BUU simulation of the process describe the experimental dilepton spectra for p + Be at 1 GeV/u very well but yield too little cross section for the higher invariant masses in the heavier system ^{40}Ca + ^{40}Ca at the same energy. We have shown in a BUU calculation, which takes pion production and their annihilation into account, that this missing yield can be explained by the dileptons produced in pion annihilation.

Nearly quantitative agreement with the data is obtained, even though the pions are propagated as free particles.

References

1. R. Bertholet, M. K. Kwato, M. Maurel, E. Monnand, H. Nifenecker, P. Perrin, J. A. Pinston, F. Schussler, D. Barneoud, C. Guet, Y. Schutz, Nucl. Phys. A474 (1987) 541, and refs. therein
2. V. Metag, Nucl. Phys A482 (1988) 159c, and refs. therein
3. T. S. Biró, K. Niita, A. L. de Paoli, W. Bauer, W. Cassing, U. Mosel, Nucl. Phys. A475 (1987) 579, and refs. therein
4. G. F. Bertsch and S. Das Gupta, Phys. Rep. 160 (1988) 190, and refs. therein
5. J. Carroll, Nucl. Phys. A495 (1989) 409c
   G. Roche, G. F. Krebs, E. Lallier, A. Letessier–Selvon, H. S. Matis, C. Naudet, L. Schroeder, P. A. Seidl, A. Yegneswaran, J. Bystricky, J. Carroll, J. Gordon, G. Igo, T. Hallman, L. Madansky, R. Welsh, P. Kirk, Z. F. Wang, D. Miller, G. Laudaud, Nucl. Phys. A488 (1988) 477c,
   C. Naudet et al., LBL preprint LBL–26461 (Dec. 1988)
6. C. Gale and J. Kapusta, Phys. Rev. C35 (1987) 2107
7. L. H. Xia, C. M. Ko, L. Xiong, J. Q. Wu, Nucl. Phys. A485 (1988) 721
8. M. Schäfer, T. Biró, W. Cassing, U. Mosel, Phys. Lett. 221B (1989) 1
9. C. F. Horowitz, Phys. Rev. C31 (1985) 1340
   D. P. Murdock and C. F. Horowitz, Phys. Rev. C35 (1987) 1442
10. R. Baier, H. Kühnelt, P. Urban, Nucl. Phys. B11 (1969) 675
    P. F. M. Koehler, K. W. Rothe, E. H. Thorndike, Phys. Rev. Lett. 18 (1967) 933
11. C. Gale and J. Kapusta, Nucl. Phys. A495 (1989) 423c
12. M. Schäfer, T. Biró, W. Cassing, U.Mosel, University of Giessen preprint UGI–89–1, to be published
13. G. Roche, private communication
14. K. Haglin, J. Kapusta, C. Gale, University of Minnesota preprint, April 89
15. G. Wolf et al., to be published
16. J. Harris et al., Phys. Rev. Lett. 58 (1987) 463; Phys. Lett. 153B (1985) 377

On the Derivation of the Quantum Molecular Dynamics Approach *

J. AICHELIN

Institut für Theoretische Physik der Universität Heidelberg

Abstract. We investigate the formal structure of a quantal n -body theory. Employing a product ansatz and the impuls approximation we obtain a time evolution equation for the n - body system which contains only the measured free nucleon nucleon T matrix. For a limiting case the equation can be solved analytically. We discuss the different features of the result. Finally we investigate how the time evolution may change if we take the fermionic nature of the nucleons seriously.

1 Introduction

The Quantum Molecular Dynamics (QMD) [1] approach has been very successful for describing the dynamics of heavy ion reactions in the energy regime between 25 AMeV and 1AGeV. Almost all measured observables like the double differential cross sections for charged particles, the squeeze out and the flow of matter and the distribution of fragments have been reproduced quantitatively in this approach.

In this approach the n - body nucleon density is a direct product of n Gaussian wave packets. They interact in two ways: via a potential interaction of a Skyrme type and via collisions. Conceptually the Quantum Molecular Dynamics was an ad hoc ansatz, not considered as a solution or a rigorous approximation to the quantal n-body equation. It incorporates some quantum features, like the Fermi momentum of the nucleons and the Pauli suppression of the nucleon nucleon scattering in already occupied phase space regions. Other features, however, like the antisymmetrization due to the fermionic nature of the nucleons, are not considered.

A straight forward derivation of the QMD approach is not at hand yet (and due to the ambiguities even in the static n - body problem at high nuclear densities not to expected for the near future) . Here we will give account on the recent results in solving some of the most eminent questions:

a) Classical molecular dynamics yields the Hamilton equations: $\mathbf{p}_i = \sum_j \nabla_i V_{ij}$ and $\mathbf{r}_i = \mathbf{p}_i/m$, if we assume the two body potentials being momentum independent. Is it

* This work has been funded in part by the German Federal Minister for Research and Technology (BMFT) under the contract number 06 HD 776 and by the Gesellschaft für Schwerionenforschung (GSI)

correct that in the quantal counterpart the NN interactions acts in two ways, as a potential and as a scattering?

b) The QMD approach uses free scattering cross sections. Under which circumstances is this approximation valid?

c) Nucleons are Fermions. How does this modify the time evolution equations obtained for distinguishable particles?

2 Wigner densities

In order to compare classical and quantal theories it is most convenient to work in the Wigner density representation [2]. The Wigner density $f(\mathbf{R},\mathbf{P},t)$ of a density matrix $|\psi(t)><\psi(t)|$ is defined as the Fourier transform of the density matrix:

$$\begin{aligned} f(\mathbf{R},\mathbf{P},t) &= \int \frac{d^3p}{(2\pi)^3} e^{i\mathbf{R}\mathbf{p}} < \mathbf{P}+\mathbf{p}/2|\psi(t)>< \psi(t)|\mathbf{P}-\mathbf{p}/2 > \\ &= \int \frac{d^3r}{(2\pi)^3} e^{-i\mathbf{P}\mathbf{r}} < \mathbf{R}+\mathbf{r}/2|\psi(t)>< \psi(t)|\mathbf{R}-\mathbf{r}/2 > \end{aligned} \quad (1)$$

where $<\mathbf{R}|\psi(t)>$ is the wave function in coordinate space representation.

With this normalization we obtain immediately

$$\int d^3R f(\mathbf{R},\mathbf{P},t) =< \mathbf{P}|\psi >< \psi|\mathbf{P} >= |\psi(\mathbf{P},t)|^2 \quad (2)$$

and

$$\int d^3P f(\mathbf{R},\mathbf{P},t) =< \mathbf{R}|\psi >< \psi|\mathbf{R} >= |\psi(\mathbf{R},t)|^2 \quad (3)$$

This allows to interpret the Wigner densities as probability distributions.

The time evolution of the Wigner density is given by Wigner transformation of the von Neumann equation

$$i\hbar\frac{\partial}{\partial t}\rho = [H,\rho]. \quad (4)$$

Applying eq. (1) and (2) we obtain:

$$\begin{aligned} \frac{\partial}{\partial t} f(\mathbf{R},\mathbf{P},t) &= \int \frac{d^3p}{(2\pi)^3} e^{i\mathbf{R}\mathbf{p}} \left(\frac{(\mathbf{P}+\mathbf{p}/2)^2}{2m} - \frac{(\mathbf{P}-\mathbf{p}/2)^2}{2m}\right) \\ & < \mathbf{P}+\mathbf{p}/2|\psi(t)>< \psi(t)|\mathbf{P}-\mathbf{p}/2 > + \int \frac{d^3r}{(2\pi)^3} e^{-i\mathbf{P}\mathbf{r}} \\ & (V(\mathbf{R}+\mathbf{r}/2) - V(\mathbf{R}-\mathbf{r}/2)) < \mathbf{R}+\mathbf{r}/2|\psi(t)>< \psi(t)|\mathbf{R}-\mathbf{r}/2 > \\ &= -i\frac{\mathbf{P}}{m}\nabla_R f(\mathbf{P},\mathbf{R},t) \\ & + \int \frac{d^3r d^3P'}{(2\pi)^3} e^{-i(\mathbf{P}-\mathbf{P}')\mathbf{r}} (V(\mathbf{R}+\mathbf{r}/2) - V(\mathbf{R}-\mathbf{r}/2)) f(\mathbf{P}',\mathbf{R},t) \end{aligned} \quad (5)$$

Formally the time evolution equation of the Wigner density can be written as:

$$\left(\frac{\partial}{\partial t} + \frac{\mathbf{P}}{m}\nabla_R\right) f(\mathbf{P},\mathbf{R},t) = \int d^3P' K_1(\mathbf{P}-\mathbf{P}',\mathbf{R}) f(\mathbf{P}',\mathbf{R},t) \quad (6)$$

The function $K_1(\mathbf{P}-\mathbf{P}',\mathbf{R})$ can be cast into different forms. For pointing out the semiclassical limit, the form

$$K_1(\mathbf{P}-\mathbf{P}',\mathbf{R}) = \frac{2}{\hbar}\sin(\frac{\hbar\nabla_R\nabla_P}{2})V(\mathbf{R})\delta(\mathbf{P}-\mathbf{P}') \tag{7}$$

is most convenient. Here we have restored the $\hbar$. We see that K_1 can be viewed as a series with the expansion coefficient $\nabla_R\nabla_P\hbar$. Hence the Schrödinger equation is equivalent to the classsical Vlasov equation

$$(\frac{\partial}{\partial t} + \frac{\mathbf{P}}{m}\nabla_R)f(\mathbf{P},\mathbf{R},t) = (\nabla_R V(\mathbf{R}))\nabla_P f(\mathbf{P},\mathbf{R},t) \tag{8}$$

provided $\nabla_R\nabla_P\hbar$ is small compared to one, i.e if the potential and the momentum distribution are smooth.

For later use we display a different form of eq.(6)

$$(\frac{\partial}{\partial t} + \frac{\mathbf{P}}{m}\mathbf{p})f(\mathbf{P},\mathbf{p},t) = \int \frac{d^3P'}{(2\pi)^3}V(\mathbf{p})(\delta(\mathbf{P}'-\mathbf{P}+\mathbf{p}/2) - \delta(\mathbf{P}'-\mathbf{P}-\mathbf{p}/2))f(\mathbf{P}',\mathbf{p},t). \tag{9}$$

So far we have gained nothing. We have only rewritten the Schrödinger equation into an equivalent phase space equation which contains exactly the identical information.

3 Scattering on a bound particle

3.1 Scattering into an free state

We start with the first non trivial case, the scattering of a incoming nucleon on a bound particle. Later we will generalized the result.

A incoming particle (1) scatters with particle (2) which builds a bound state together with particle (3) . We assume that the interaction between particle (2) and (3) can be described by a potential, either because particle (3) is infinitely heavy or by constructing an effective potential which yields the same wave function of particle (2) as the full solution of the bound state Schrödinger equation. This allows to reduce the three body to a soluble two body problem. If the potential between (1) and (2) depends on the relative coordinates only the Schrödinger equation reads as follows:

$$\left(\frac{-\hbar^2\nabla_1^2}{2m_1} - \frac{-\hbar^2\nabla_2^2}{2m_2} + U(\mathbf{r}_2) + V(\mathbf{r}_1-\mathbf{r}_2)\right)\Psi_a(\mathbf{r}_1,\mathbf{r}_2) = E_a\Psi_a(\mathbf{r}_1,\mathbf{r}_2) \tag{10}$$

The wave function of the struck target particle ϕ_n satisfies the equation:

$$(\frac{-\hbar^2\nabla^2}{2m_2} + U(\mathbf{r}_2))\phi_n(\mathbf{r}_2) = W_n\phi_n(\mathbf{r}_2) \tag{11}$$

The solution of the first equation (10) for the proper boundary condition is:

$$\Psi_a = \Phi_a + \frac{1}{E_a - K - U + i\epsilon}V\Psi_a \tag{12}$$

where Φ_a ist the product of the wave functions of the incident particle and the target particle which satisfies

$$(\frac{-\hbar^2\nabla_1^2}{2m_1} - \frac{-\hbar^2\nabla_2^2}{2m_2} + U(\mathbf{r}_2))\Phi_a(\mathbf{r}_1,\mathbf{r}_2) = E_a\Phi_a(\mathbf{r}_1,\mathbf{r}_2) \tag{13}$$

Similarily the solution of equation (11) has the form

$$\phi_n = \chi_n + \frac{1}{W_n - K + i\epsilon}U\phi_n \tag{14}$$

with χ_n being the solution of $K\chi_n = W_n\chi_n$

Chew [3] has suggested to replace the T matrix of the exact solution

$$T = V + \frac{1}{E_a - K - U + i\epsilon} T \tag{15}$$

by

$$t = V + \frac{1}{E_a - K + i\epsilon} t. \tag{16}$$

the T - matrix calculated from eq.(14) . This approximation is called impuls approximation and assumes that only the first term of the expansion

$$\frac{1}{A+B} = \frac{1}{A} + \frac{1}{A} B \frac{1}{A} \cdot\cdot \tag{17}$$

is important. In this approximation the struck nucleon is considered as free during the collision, the binding potential determines the momentum distribution of the struck particles but for the scattering process the free Transition matrix is employed.

This approximation should be good as long as the collision time is small compared with the characteristic time of of a nucleus. This condition is certainly fulfilled in collisions of protons with heavy ions at energies above 100 MeV/N. The impuls approximation is especially useful in the Wigner density formalism as we will see below.

We go back to the Schrödinger equation (5) :

$$\begin{aligned}(\frac{i\partial}{\partial t} + \frac{i\mathbf{P}}{m}\nabla_R) f(\mathbf{P},\mathbf{R},t) &= \int \tfrac{d^3r}{(2\pi)^3} e^{-i\mathbf{P}\mathbf{r}} (< \mathbf{R}+\mathbf{r}/2|V|\psi(t) >< \psi(t)|\mathbf{R}-\mathbf{r}/2 > \\ &\quad - < \mathbf{R}+\mathbf{r}/2|\psi(t) >< \psi(t)|V|\mathbf{R}-\mathbf{r}/2 >)\end{aligned} \tag{18}$$

and implement our results. We make now use of the identities

$$< \psi(t)|\mathbf{R}-\mathbf{r}/2 > = < \phi(t)|\mathbf{R}-\mathbf{r}/2 > + < \psi(t)|VG^-|\mathbf{R}-\mathbf{r}/2 > \tag{19}$$

and

$$T|\phi(t) > = V|\psi(t) > \tag{20}$$

and integrate over the center of mass momentum assuming

$$< \mathbf{p}'_1, \mathbf{p}'_2|V|\mathbf{p}_1, \mathbf{p}_2 > = \delta(\mathbf{p}'_1 + \mathbf{p}'_2 - \mathbf{p}_1 - \mathbf{p}_2) < \mathbf{p}'_1 - \mathbf{p}'_2|V|\mathbf{p}_1 - \mathbf{p}_2 > \tag{21}$$

Finally we obtain the seminal equation:

$$\begin{aligned}(\frac{\partial}{\partial t} + \frac{\mathbf{P}}{m}\nabla_R) f(\mathbf{R},\mathbf{P},t) &= \tfrac{1}{(2\pi)^3} \int d^3p d^3Q d^3q e^{i\mathbf{R}\mathbf{p}} f_0(Q,q,t) \\ &\quad (< \mathbf{P}+\mathbf{p}/2|T|\mathbf{Q}+\mathbf{q}/2 > \delta(\mathbf{P}-\mathbf{p}/2-\mathbf{Q}+\mathbf{q}/2) \\ &\quad - < \mathbf{P}-\mathbf{p}/2|T^\dagger|\mathbf{Q}-\mathbf{q}/2 > \delta(\mathbf{P}+\mathbf{p}/2-\mathbf{Q}-\mathbf{q}/2)) \\ &\quad + < \mathbf{P}+\mathbf{p}/2|T|\mathbf{Q}+\mathbf{q}/2 >< \mathbf{P}-\mathbf{p}/2|T^\dagger|\mathbf{Q}-\mathbf{q}/2 > \\ &\quad (\tfrac{2\mu}{(\mathbf{P}-\mathbf{p}/2)^2-(\mathbf{Q}-\mathbf{q}/2)^2-i\epsilon} - \tfrac{2\mu}{(\mathbf{P}+\mathbf{p}/2)^2-(\mathbf{Q}+\mathbf{q}/2)^2+i\epsilon})\end{aligned} \tag{22}$$

$\mathbf{P}, \mathbf{Q}, \mathbf{p}, \mathbf{q}$ and $\mathbf{R}$ are the relative coordinates between the scattering partners. $f_0(Q,q,t)$ is the Fouriertransform of $f_0(Q,R,t)$, which can be obtained by integrating the Wigner density of the asymptotic wave function $\Phi_a = exp(-p_1 r_1) \cdot \phi_n(r_2)$ (comp. eq. (11)) over the center of mass coordinates. In deriving this equation we have made an approximation to the progagator which is discussed by Goldhaber [4].

Inspecting eq. (22) we see the big advantage of the impuls approximation: The only term which contains information about the fact that the struck particle is bound is $f_0(Q,q,t)$. The scattering process itself is described by the measurable real and imaginary parts of the free nucleon nucleon T matrix.

The equation, however, is far from being transparent. In order to investigate it in detail, we calculate the different terms for an explicit form of the Wigner density. We choose a Gaussian form which easily allows to perform the limit to states of sharp momentum.

$$f_0(Q,R,t) = (\frac{1}{4\pi L})^{3/2}(\frac{4L}{\pi})^{3/2} e^{-(\mathbf{R}-\mathbf{R}_0-\mathbf{Q}(t-t_0)/m)^2/4L} e^{-(\mathbf{Q}-\mathbf{Q}_0)^2 4L} \tag{23}$$

The measured nucleon nucleon scattering amplitude is also of a Gaussian form:

$$f(p,p') \propto < p|T|p' >= (A+iB)e^{-a(p-p')^2} \tag{24}$$

With a Gaussian wave packet the calculation cannot be performed analytically. Therefore we make the approximation $\mathbf{Q}t/m \approx \mathbf{Q}_0 t/m$. Whereas the width of a Gaussian wave packet increases as a function of time with $4L(t) = 4L(0) \cdot \sqrt{1 + (\frac{\hbar t/m}{4L(0)})^2}$ this replacement leaves the width constant $L(t) = L(0)$. This approximation is therefore valid, if during the time τ the nucleons need to pass the potential, the width does not change considerably. For energies larger than 100 MeV this requires $4L > .4 fm^2$.

We have still the freedom to choose the t_0. It is convenient to assume $t = t_0$ the particles are closest. This allows to interpret $\mathbf{R}_0$ as the impact parameter with respect to the centers of coordinate space distributions of the particles.

With this approximation we can perform the integrations analytically and finally we obtain(after additional integrations over t and R)

$$f(P,t=\infty) = f(P,t=-\infty) + I^{lin}(P) + I^{quad}(P) \tag{25}$$

with

$$\begin{aligned} I^{lin}(P) = \; & \tfrac{16\mu}{P}\sqrt{\pi(8L+4a)}(\tfrac{L}{2L+a})^{3/2} e^{-(\mathbf{P}-\mathbf{P}^0)^2 4L} e^{-\frac{R_0^2}{8L+4a}} \\ & [A \sin \tfrac{(\mathbf{P}-\mathbf{P}_0)\mathbf{R}_0 2L}{2L+a} + B \cos \tfrac{(\mathbf{P}-\mathbf{P}_0)\mathbf{R}_0 2L}{2L+a}] \end{aligned} \tag{26}$$

$$\begin{aligned} I^{quad}(P) = \; & +\tfrac{2\mu^2}{P}(A^2+B^2)\sqrt{(8L+4a)}(\tfrac{4L\pi}{2L+a/2})^{3/2} e^{-P^2(4L+4a)} e^{-P_0^2 4L} e^{-\frac{R_0^2}{8L+4a}} \\ & \int d^3m m\delta(m-P) e^{2\mathbf{m}(2a\mathbf{P}+4L\mathbf{P}_0)} \cos(\tfrac{2a\mathbf{P}-4L\mathbf{P}_0}{2a+4L} - \mathbf{m})\mathbf{R}_0 \end{aligned} \tag{27}$$

3.2 Scattering of two free particles

The equations (25-27) become much more transparent if we perform the limit $4L \to \infty$, i.e. if we assume sharp momentum eigenstates of the particles outside the scattering region. We recall that

$$(A^2+B^2)e^{-(\mathbf{m}-\mathbf{P})^2 2a} = TT^* = \frac{1}{(2\pi)^4\mu^2}\frac{d\sigma}{d\Omega} \tag{28}$$

and

$$B = -\frac{P}{2(2\pi)^3\mu}\sigma_{tot}. \tag{29}$$

Performing the limit we notice, that in the linear expression the sin term does not contribute and the cos term yields:

$$I^{lin}(P) = -\frac{1}{8\pi L}\delta(\mathbf{P}-\mathbf{P}_0)\sigma_{tot} \tag{30}$$

The quadratic part gives:

$$I^{quad}(P) = \frac{1}{8\pi L P^2}\delta(P-P_0)\frac{d\sigma}{d\Omega} \tag{31}$$

Hence as the final result we have ($f(P,t=-\infty)=\delta(\mathbf{P}-\mathbf{P}_0)$)

$$f(P,t=\infty)=\frac{1}{8\pi LP^2}\delta(P-P_0)\frac{d\sigma}{d\Omega}+(1-\frac{\sigma_{tot}}{8\pi L})\delta(\mathbf{P}-\mathbf{P}_0) \tag{32}$$

Before we discuss the physics we have to clarify the meaning of $\frac{1}{8\pi L}$. In our approximation the particles move with the velocity $\mathbf{v}=\mathbf{P}_0/m$. The incoming current is therefore

$$\mathbf{j}=\frac{1}{8\pi L}\frac{\mathbf{P}_0}{\mu}e^{(\mathbf{R}-\mathbf{R}_0-\mathbf{P}_0t/\mu)^2/8L} \tag{33}$$

Hence the time integrated flux in beam direction is

$$\int \mathbf{j}\hat{P}_0 dt == \frac{1}{8\pi L}e^{-(\mathbf{R}_\perp-\mathbf{R}_{0\perp})^2/8L} \tag{34}$$

where $\perp$ means perpedicular to $\mathbf{P}_0$. Hence in the limit $L\to\infty$ the factor $\frac{1}{8\pi L}$ is nothing but the time integrated current.

3.3 Discussion

This derivation teaches us several important issues:

1) In the limit of momentum eigenstates at $t\to\infty$,i.e if we describe the scattering of two free particles, the equation (25) allows an very easy interpretation: With a ratio $\frac{\sigma_{tot}}{8\pi L}$, i.e. the ratio of the cross section to the normalization area perpedicular to the beam direction, the particles are scattered out of the initial state. The scattering conserves energy. Therefore the new relative momentum has the same absolute magnitude but a different direction. The probability for scattering into a given direction is determined by $\frac{d\sigma}{d\Omega}$. The overall particles number is conserved.

2) For a finite width of the momentum distribution the process is more complicated:
a) The depopulation of the initial momentum space distribution is governed by the imaginary part of the T -matrix and varies with the momentum of the particles. It is more pronounced for small momenta where $cos\approx 1$ as compared to large momenta. The cos term is an effective form factor.
b) The *sin* term which is proportional to the real part of the transition matrix does not vanish but acts like a potential as can be seen by comparison with eq.(9) . In lowest order in $\hbar$ (compare eq. (7)) the real part of the T - matrix acts as a potential which moves the particles on their classical trajectories. In can be shown that this term conincides with the standard expression of the optical potential

$$V(r_1)=(2\pi)^3T(0)\rho(r_1) \tag{35}$$

if the recoil of the scattered particle is neglected.
c) Both, the linear and the quadratic part, have an exponential cutoff in R_0, which is the average impact parameter of the distribution. Thus particles being further apart than the sum of twice the width of the distribution and the range of the interaction do not interact with each other.
d) Formally the equation has the structure of the Boltzmann equation(although it is not the equation for the 1 body phase space distribution).
e) This equation yields informations on the phase space coordinates of the scattering partners. We see a decrease of the interaction with increasing impact parameter. Hence the equation describes the scattering in phase space and -because phase space is familiar to us - is therefore quite transparent.

Thus the scattering of bound particles is described by an equation which contains both, a potential (ReT) and the scattering cross section (ImT and $|T|^2$). Both parts have to be calculated from the same underlying transition matrix. This observation is in clear contradistinction to the classical molecular dynamics where only potentials are acting (which can, if the average distance between the scattering partner is large compared to the potential range alternatively but not simulataneously expressed as scattering).

The use of the measured free nucleon nucleon scattering cross section is legitimate as long as the impuls approximation remains valid. Calculations of the corrections in the eikonal approximations shows the validity of this approach down to beam energies of 100 MeV/N.

3.4 Scattering into a bound state

To describe the situation where the originally bound nucleon is not free finally but in an excited state of the target nucleus we start from the time evolution equation:

$$
\begin{aligned}
&(\tfrac{\partial}{\partial t} + \tfrac{\mathbf{P}_1}{m_1}\nabla_1 + \tfrac{\mathbf{P}_2}{m_2}\nabla_2) f(\mathbf{R}_1, \mathbf{P}_1, \mathbf{R}_2, \mathbf{P}_2, t) \\
&= \tfrac{1}{(2\pi)^6} \int d^3p_1 d^3p_2 e^{i\mathbf{R}_1\mathbf{p}_1 + i\mathbf{R}_2\mathbf{p}_2} < \mathbf{P}_1 + \mathbf{p}_1/2, \mathbf{P}_2 + \mathbf{p}_2/2 | V\rho^{(2)} - \rho^{(2)}V | \mathbf{P}_1 - \mathbf{p}_1/2, \mathbf{P}_2 - \mathbf{p}_2/2 > \\
&\equiv [V, \rho^{(2)}]_W
\end{aligned} \tag{36}
$$

$\rho^{(2)}$ is the density matrix of the scattering partners. The eigenstates of the bound particle form complete basis of the Hilbert space.

$$
\sum_n |\phi_n >< \phi_n| = 1 \tag{37}
$$

Making use of this fact we obtain after integration over the coordinates of particle 2

$$
(\frac{\partial}{\partial t} + \frac{\mathbf{P}_1}{m_1}\nabla_1) f(\mathbf{R}_1, \mathbf{P}_1,) = \sum_n \int d^3R_2 d^3P_2 (|\phi_n >< \phi_n \cdot [V, \rho^{(2)}])_W \tag{38}
$$

Under the integral, the Wigner density of the product of two operators is just the product of two Wigner densities [5]

$$
\int drdp(AB)_W = \int drdpA_W B_W \tag{39}
$$

Finally we obtain:

$$
(\frac{\partial}{\partial t} + \frac{\mathbf{P}_1}{m_1}) f(\mathbf{R}_1, \mathbf{P}_1,) = \sum_n \int d^3R_2 d^3P_2 g_n(\mathbf{P}_2, \mathbf{R}_2) \cdot [V, \rho^{(2)}]_W \tag{40}
$$

where g_n is the Wigner transform of $|\phi_n >< \phi_n|$. Thus in order to obtain the time evolution equation for a projectile which excites a target nucleus into the state $|\phi_n >$ we have to multiply the unprojected time evolution equation with the Wigner density of the desired state followed by a integration over the coordinates target particle.

4 Extensions to the n body system

In the impuls approximation the extension of the 3 body system to the n body system is straight forward [3]. The total transition matrix is in lowest order just the sum of the individual transition matrices:

$$
T = t_1 + t_2 + ... + t_n \tag{41}
$$

It can be shown that there are two kinds of higher order corrections: a) the higher order terms of eq.(17) and b) corrections due to multiple scattering of the form $t_1 G^0 t_2$. In the quantum molecular dynamics approach we neglect the corrections a) and keep the corrections b) . We define the total T matrix as

$$T = \sum_i t_i + \sum_{i,j} t_i G^0 t_j.... \tag{42}$$

We assume, however, that we can neglect any interference effects. This is correct if the average distance between scattering centers are large compared to the range of the scattering. Double scattering is then just proportional the product of single scattering times a geometrical factor. The equation solved in the QMD approch has the same form as eq.(22) however the two body T matrix is replaced by eq.(42) and instead of the two body distribution we have to employ a n body Wigner density. The involved approximations are discussed in [4]. The major approximation is now the assumption that the $f_0^N(\mathbf{R}_i, \mathbf{Q}_i, t)$ is given by the product of n Gaussians of the form:

$$f_0(Q, R, t) = (\frac{1}{2\pi L})^{3/2} (\frac{2L}{\pi})^{3/2} e^{-(\mathbf{R}-\mathbf{R}_0-\mathbf{Q}_0 t/m)^2/2L} e^{-(\mathbf{Q}-\mathbf{Q}_0)^2 2L} \tag{43}$$

where the inital values for $\mathbf{Q}_0$ and $\mathbf{R}_0$ are chosen to reproduce the measured momentum and density distribution of the nuclei. Instead of ReT we use a zero range interacting whose parameters are adjusted to give the right binding energy in nuclear matter. The equation is solved with the test particle method. In this approach the kinematical factors for double, triple and higher order scattering can be obtained in a very natural way by a Monte Carlo procedure. Details can be found in [1] .

5 Fermions

Recently it has been conjectured by Feldmaier [6]that the dynamics of Fermions cannot been described by Hamiltons equations. This claim was based on the application of a variational principle orginally suggested by Kerman and Koonin [7]. This variational principle is applicable to a system whose wave function ψ depends on time only via parameters. Feldmaier applied this methods to the Gaussian Wigner density in momentum and coordiante space which are employed in the QMD approach. For these systems a Langrange density can be defined

$$L(\mathbf{r}_0(t), \mathbf{p}_0(t), \dot{\mathbf{r}_0}(t), \dot{\mathbf{p}_0}(t)) = \int d^3r < \psi | i\hbar \frac{d}{dt} - H | \psi > \tag{44}$$

where ψ is the coherent state wave function. The time evolution equation of the parameters can be obtained by requiring L to be stationary with respect to variations of $|\psi>$ and $<\psi|$ between fixed end point t_1 and t_2:

$$(\frac{d}{dt}\frac{\partial}{\partial \dot{p}_0} - \frac{\partial}{p_0}) L = 0. \tag{45}$$

and

$$(\frac{d}{dt}\frac{\partial}{\partial \dot{r}_0} - \frac{\partial}{r_0}) L = 0. \tag{46}$$

A lengthy but straight forward calculation yields

$$L = \frac{-\mathbf{r}_0 \dot{\mathbf{p}}_0 + \dot{\mathbf{r}}_0 \mathbf{p}_0 Y - H}{1 - Y} \tag{47}$$

with

$$Y = e^{-p_0^2 \cdot 4L - r_0^2/4L} \tag{48}$$

and

$$H = \frac{3}{8L} + \frac{p_0^2 + r_0^2/(4L)^2 \cdot Y}{1 - Y} \tag{49}$$

Feldmaier [6] proved now that there exist no Hamilton function $\tilde{\mathrm{H}}$ which satisfies the Hamilton equations, i.e. where $\dot{p_0} = \frac{-\partial \hat{H}}{\partial r_0}$ and $\dot{r_0} = \frac{\partial \hat{H}}{\partial p_0}$.

One can, however take a different approach to the fermionic motion. Starting from the time evolution equation of the Wigner density for two free particles

$$\left(\frac{\partial}{\partial t} + \frac{\mathbf{P}_1}{m_1}\nabla_1 + \frac{\mathbf{P}_2}{m_2}\nabla_2\right) f(\mathbf{R}_1, \mathbf{P}_1, \mathbf{R}_2, \mathbf{P}_2, t) = 0 \tag{50}$$

we obtain immediately the solutions:

$$f(\mathbf{r}_1, \mathbf{r}_2, \mathbf{p}_1, \mathbf{p}_2, t) = \frac{1}{\pi^6} e^{-(\mathbf{r}_1 - \mathbf{r}_{10} - \mathbf{p}_1 t/\mu)^2/4L - (\mathbf{r}_2 - \mathbf{r}_{20} - \mathbf{p}_2 t/\mu)^2/4L - (\mathbf{p}_1 - \mathbf{p}_{10})^2 \cdot 4L - (\mathbf{p}_2 - \mathbf{p}_{20})^2 \cdot 4L} \tag{51}$$

The Gaussians in coordinate and momentum space are the Wigner densities of coherent states:

$$\psi(\mathbf{r}_1, \alpha = \mathbf{p}_{10}, \mathbf{r}_{10}) = e^{-(\mathbf{r}_1 - \mathbf{r}_{10} - \mathbf{p}_{10} t/\mu)^2/(4L + it/\mu) + i\mathbf{r}_1 \mathbf{p}_{10}} \tag{52}$$

For simplification we neglect the time dependence of the width. This can always be done be choosing L large enough that the width does not change during the passing of the two fermions (compare page 5) . The antisymmetrization of the wave function is straight forward. After separating the center of mass motion and after transforming back to the Wigner densities as a function of the relative coordinates we obtain

$$\begin{aligned} f(\mathbf{r}, \mathbf{p}, \mathbf{r}_0, \mathbf{p}_0, t) \;=\; & C(e^{-(\mathbf{r} - \mathbf{r}_0 - \mathbf{p}_0 t/\mu)^2/4L - (\mathbf{p} - \mathbf{p}_0)^2 \cdot 4L} \\ & + e^{-(\mathbf{r} + \mathbf{r}_0 + \mathbf{p}_0 t/\mu)^2/4L - (\mathbf{p} + \mathbf{p}_0)^2 \cdot 4L} \\ & + 2 \cdot e^{-\mathbf{r}^2/4L - \mathbf{p}^2 \cdot 4L} cos \mathbf{r} \mathbf{p}_0 * cos \mathbf{p}(\mathbf{r}_0 + \mathbf{p}_0 t/\mu) \end{aligned} \tag{53}$$

The time evolution of the Wigner density in coordinate and momentum space is obtained by integration over the complementary variable and displayed for the one dimensional case in figure 1. There we have chosen as initial condition $r_0 = 2(fm)$ and $p_0 = .1(1/fm)$.

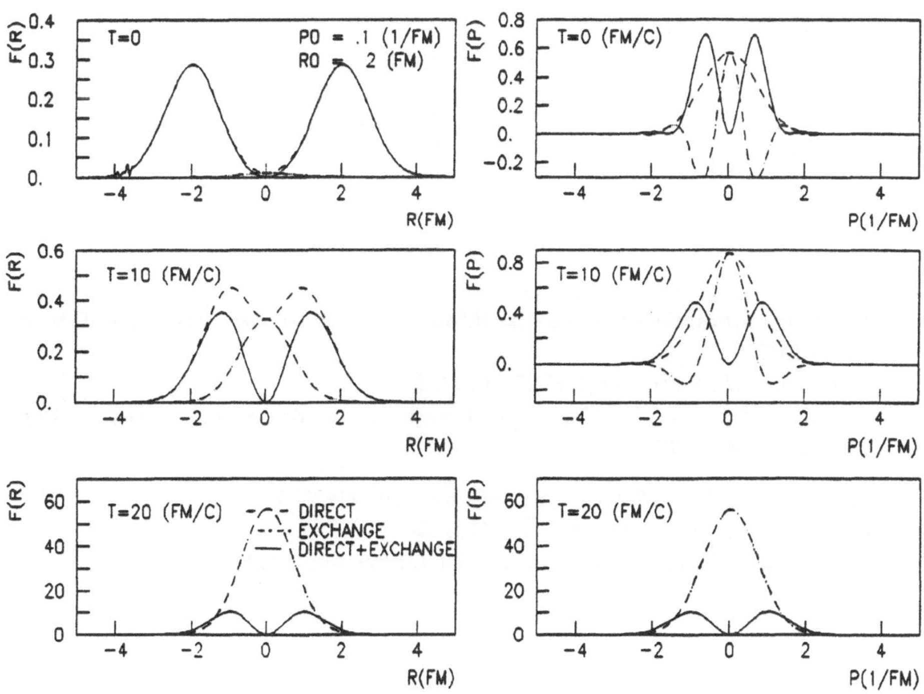

Fig1. time evolution of the Wigner density of two fermions in momentum and coordinate space. The figure displays the distribution of relative momentum and relative distance at three different times

We display in this figure the direct term, the exchange term and the sum of both. For large distances (t = 0) we see two separate Gaussian in coosdinate space and the exchange term has little influence there. In momentum space, however, the exchange term acts much stronger and generates two Gaussians out of the one peak made by the direct term. At 10 fm/c the direct term overlaps strongly now also in coordinate space. The exchange term also separates the peaks in coordinate space. The width of both peaks has decreased and the width in momentum space has to compensate for this in order to fulfill the uncertainty relation. Finally, when the direct terms merge into one Gaussians, both the direct and the exchange term increase in magnitude tremendously. The peaks genaerated by the exchange term are still quite separated, and the width of the peaks in momentum space has increased once more.

We see already from equation (52) that the parameters r_0, p_0 in the exact solution obey Hamiltons equations with $H = p_0^2/2\mu$. Therefore, the non Hamiltonian dynamics of the solution of eq.(49) is an artifact of the ansatz and not physical. It merely states that if one restricts the time dependence of your wave functions to less parameters than needed for the exact solution of the Schrödinger equation than these parameters do not have to evolve according to Hamiltons equations even if the parameters of the exact solution follow Hamiltons equations. The conjecture, that an aditional time dependence of the width yields again Hamiltons equations for p_0 and r_0 remains to be investigated.

Certainly more investigations are necessary to understand the dependence of the dynamics of the parameters on the choice of the parameters. If one has understood this relation the variational principle may turn out as a powerful tool to describe the dynamics of Fermions in a interacting system, a subject which is not well investigated up to now.

6 Conclusions

We have presented the derivation the the time evolution equation for scattering on bound states. We have found that this equations contains potential and scattering terms which are closely related with each other, in contrast to the classical case. We discussed the further approximations which are needed to derive the QMD approach. Finally, we have shown that the usefulness of variational principles for the determination of the time evolution of a fermionic has still to be investigated.

References

[1] J. Aichelin and H. Stöcker,Phys. Lett. **B176** (1986) 14.
J.Aichelin, A.Rosenhauer, G.Peilert, H.Stöcker and W.Greiner, Phys. Rev. Lett. **58** (1987) 1926.
J.Aichelin et al., Phys. Rev. **C37** (1988) 2451.
G. Peilert, H. Stöcker W. Greiner, A. Rosenhauer, A. Bohnet and J. Aichelin Phys. Rev. **C39** (1989) 1402

[2] E.A. Remler and A.P. Sathe, Annals of Physics **91** (1975) 295.
E.A. Remler Annals of Physics **119** (1978) 326.
P. Carruthers and M. Zachariasen, Rev. Mod. Phys. **55** (1983) 245.

[3] G. F. Chew, Phys. Rev. 80 (1950) 196
G. F. Chew and M. Goldberger Phys. Rev. 87 (1952) 778

[4] M. Goldhaber and K. Watson Collision theory Wiley and Sons New York (1964) p.692

[5] K. Imre, E. Özizmir, M. Rosenbaum, and P.F. Zweifel J. Mathematical Physics 8 (1967) 1097

[6] H. Feldmaier private communication and article in this proceedings

[7] A. K. Kerman and S. E. Koonin, Annals of Physics 100 (1976) 332

FERMIONIC MOLECULAR DYNAMICS

H. Feldmeier

GSI, Gesellschaft für Schwerionenforschung mbH
Postfach 110552, D-6100 Darmstadt, West-Germany

A new type of molecular dynamics is proposed to solve approximately the many-bo problem of interacting identical fermions with spin $\frac{1}{2}$. The interacting system is rep resented by an antisymmetrized many-body wave function consisting of single-particle states which are localized in phase space. The equations of motion for the parameters characterizing the many-body state (e.g. positions, momenta and spin of the particles) are derived from a quantum variational principle. The proposed Fermionic Molecular Dynamics (FMD) model is illustrated with help of two examples.

1 INTRODUCTION

Early in heavy ion research it has been proposed to describe the collision of two nuclei by molecular dynamics in which classical nucleons interact via two-body forces [1, 2]. These ideas have been pursued since then [3, 4, 5, 6]. The major problem with these models which are appropiate for distinguishable classical particles is the fact that nucleons in nuclei and in nuclear collisions are close enough in phase space that the Pauli principle is important. The most common approach to improve on that is to mock the influence of the Pauli correlations by an additional momentum-dependent potential which becomes strongly repulsive if the fermions get too close in phase space [7, 8, 9]. Other approaches add fluctuating forces (collision term) which randomly change the momenta of pairs of nulceons [10, 11, 12, 13]. Here the Pauli exclusion principle is explicitly taken into account by not allowing scattering into occupied phase spase. The motion in between collisions is, however, again that of distinguishable particles.

One knows from thermodynamics that even for macroscopic objects, where all other quantum effects are absent, it may make a big difference in macroscopic quantities like the specific heat whether one has Boltzmann, Bose-Einstein or Fermi-Dirac statistics. Not only for that reason it seems quite important to include the correct statistics also in molecular dynamics where other quantum effects are neglected.

In this contribution I would like to present for the first time a non-relativistic model called Fermionic Molecular Dynamics (FMD) which combines from the very beginning Fermi-Dirac statistics with a semi-quantal trajectory picture. In the limit where the sytem becomes dilute in phase space (low density and/or high temperature) it turns into classical molecular dynamics. The FMD model is not fully quantum

The Nuclear Equation of State, Part A
Edited by W. Greiner and H. Stöcker
Plenum Press, New York

mechanical as it does not describe wave-mechanical interference effects. The Fermi-Dirac statistics is however included on the many-body level.

In section 2 the model is defined in general terms for identical fermions (not only for nucleons). For identical bosons an analogue model may be defined by using symmetrized wave functions. Section 3 demonstrates the ideas of the model in two examples. The first is a single fermion moving on a trajectory in an external electromagnetic field. There it is shown how the dynamics in the spin degree of freedom can be described in a classical language although it is in accordance with quantum mechanics in the two-component spinor space. The second example considers the scattering of two fermions with equal spin. Section 4 explores further the many body system.

2 THE FERMIONIC MOLECULAR DYNAMICS MODEL

The first step in approximating the exact solution of the Schrödinger equation is to parametrize the wave function in terms of a finite number of parameters. The idea behind this is that the subspace of the infinite dimensional Hilbert space which is expected to house the exact solution is represented well by the set of parametrized wave functions. For the case of molecular dynamics each molecule (here each nucleon) is described by a wave packet $|q(t)\rangle$ which is localized in phase space. For convenience one may take a gaussian form

$$\langle\vec{x}|q(t)\rangle = \exp\left\{-\sum_{i,j=1}^{3}(x_i - r_i(t))A_{ij}(t)(x_j - r_j(t)) + i\vec{p}(t)\vec{x} + \eta(t)\right\} |\chi(t), \phi(t)\rangle\,. \tag{2.1}$$

The time dependent parameters $q(t) = \{\vec{r}(t), \vec{p}(t), A_{ij}(t), \eta(t), \chi(t), \phi(t)\}$ define the wave packet. Its mean position is $\vec{r}(t)$, its mean momentum is $\vec{p}(t)$, and the complex symmetric matrix $A_{ij}(t)$ describes the spreading. The real part of A_{ij} has to be positive. The complex parameter $\eta(t)$ takes care of the normalization and contains a time dependent phase. This phase is in our application of no consequence for the dynamics. Therefore it is dropped in the following. The spin of the fermion is represented by a two-component spinor $|\chi(t), \phi(t)\rangle$.

$$|\chi(t), \phi(t)\rangle = \begin{pmatrix} \cos\frac{\chi(t)}{2} \\ \sin\frac{\chi(t)}{2}e^{i\phi(t)} \end{pmatrix}\,. \tag{2.2}$$

For N fermions the ansatz for the parameterized many body wave function $|\{q(t)\}\rangle$ is a normalized Slater-determinant formed from N different single-particle states $|q_k(t)\rangle$ as given in eq. (2.1)

$$|\, q_1, q_2, \ldots, q_N\,\rangle \equiv |\{q\}\rangle = \frac{1}{\langle\widehat{\{q\}}|\widehat{\{q\}}\rangle^{\frac{1}{2}}}|\widehat{\{q\}}\rangle \tag{2.3}$$

where

$$|\widehat{\{q\}}\rangle = \sum_{all\ P} sgn(P)\ |q_{P(1)}\rangle \otimes |q_{P(2)}\rangle \otimes \cdots \otimes |q_{P(N)}\rangle\,. \tag{2.4}$$

As the single-particle states are in general not orthogonal the many body state in eq. (2.4) has to be normalized. Nevertheless it is completely antisymmetric with respect to particle exchange and thus a proper wave function for identical fermions.

The next step is to apply the quantum mechanical variational principle [14, 15]

$$0 = \delta \int_{t_1}^{t_2} dt \, \langle \{q(t)\}| \, i\frac{d}{dt} - \underset{\sim}{H} \, |\{q(t)\}\rangle \quad , \tag{2.5}$$

where $\underset{\sim}{H}$ denotes the hamiltonian and the variation is with respect to all $q_k(t)$ with $\delta q_k(t_1) = \delta q_k(t_2) = 0$. The equations of motion for the parameters $q_k(t)$ are obtained from the quantum variational principle (2.5). If $|\{q(t)\}\rangle$ spans the whole Hilbert space then the parameters $\{q(t)\}$ which result from their equations of motion provide the exact solution. If for certain initial states the exact solution of the Schrödinger equation can be written in terms of $|\{q(t)\}\rangle$ then the resulting equations of motion for $\{q(t)\}$ are also exact. The equations of motion for $\{q(t)\}$ will provide meaningful values if at all times the exact solution can be approximated well by $|\{q(t)\}\rangle$.

Let us define a "Lagrange function" by

$$\mathcal{L}(\{q(t)\}, \{\dot{q}(t)\}) := \langle \{q(t)\}| \, i\frac{d}{dt} - \underset{\sim}{H} \, |\{q(t)\}\rangle \quad . \tag{2.6}$$

This is not a usual Lagrange function. It contains the "velocities" $\dot{q}(t) \equiv dq(t)/dt$ only linearly and depends on coordinates as well as on momenta and on spin variables. The equations of motion resulting from the variational principle are given by the Euler-Lagrange equations

$$\frac{d}{dt}\frac{\partial \mathcal{L}}{\partial \dot{q}_{k,j}} = \frac{\partial \mathcal{L}}{\partial q_{k,j}} \tag{2.7}$$

where $k = 1, 2, ..., N$ enumerates the particles and j denotes the different parameters for each single particle state. These equations may be non-linear and not resemble at all a Schrödinger equation. They can be regarded as semi-classical equations of motion which for special cases may turn into classical equations. Please note that the Hamiltonian may contain spin-dependent parts. If it were only for this reason the equations of motion will not be those of classical particles on trajectories.

The above equations define the model uniquely. The equations of motion (2.7) for $q_{k,j}(t)$, must be worked out, solved, and the results must be interpreted for each application.

Before discussing the model further in general terms let us be more specific and present two examples. In the following section I shall look at a single electron (or proton) in an electromagnetic field and after that I shall discuss the two-body problem.

3 APPLICATIONS

In order to simplify the equations let us restrict the freedom in the possible shapes of the coordinate space part in eq. (2.1) by assuming $A_{ij}(t) = \delta_{ij}/a(t)$ so that the wave packet is spherical at all times. Let us for the sake of simplicity go even further and assume the width parameter a to be real and time-independent. With that we forbid the wave packet to spread or shrink as it evolves in time.

In the following two examples we shall work with single particle states

$$|q(t)\rangle = |\vec{r}(t), \vec{p}(t), \chi(t), \phi(t)\rangle = |\vec{r}(t), \vec{p}(t)\rangle \otimes |\chi(t), \phi(t)\rangle \; , \tag{3.1}$$

where the coordinate space part $|\vec{r}(t), \vec{p}(t)\rangle$ is given by

$$\langle \vec{x}|\vec{r}(t), \vec{p}(t)\rangle = \frac{1}{(\pi a/2)^{\frac{3}{4}}} \exp\left\{-\frac{(\vec{x}-\vec{r}(t))^2}{a} + i\vec{p}(t)\vec{x}\right\} . \quad (3.2)$$

A fermion described by this wave function has a mean position $\vec{r}(t)$, a mean momentum $\vec{p}(t)$ and its spin has the polar angles $\chi(t)$ and $\phi(t)$. While these parameters are time dependent the width a is supposed to stay constant in time. In cases where only the mean positions and momenta of the molecules are relevant this approximation should be acceptable. If it turns out that a time independent a constrains the dynamics too much the next step is to introduce for each wave packet a complex $a_k(t)$.

3.1 A CHARGED FERMION IN AN ELECTROMAGNETIC FIELD

To get a feeling how the proposed model for fermionic motion works let us consider the time evolution of a single charged fermion in an external time-independent electromagnetic field. The non-relativistic hamiltonian is

$$\underset{\sim}{H} = \frac{1}{2m}\left(\underset{\sim}{\vec{k}} - \frac{e}{c}\vec{A}(\underset{\sim}{\vec{x}})\right)^2 + \frac{e}{c}A^0(\underset{\sim}{\vec{x}}) - \frac{1}{2}\mu_B \underset{\sim}{\vec{\sigma}}\vec{B}\left(\underset{\sim}{\vec{x}}, \underset{\sim}{\vec{k}}\right) \quad (3.3)$$

where e is the charge, m the mass, and $\mu_B = \hbar e/(2mc)$ is the magnetic moment of the fermion. $\underset{\sim}{\vec{k}}$ and $\underset{\sim}{\vec{x}}$ denote the momentum and coordinate space operator, respectively. The charge of the particle interacts with the external vector potential $(A^0, \vec{A})$ while the interaction of the spin with the electromagnetic field is obtained from a Foldy-Wouthuysen transformation [16] as

$$\frac{1}{2}\mu_B \underset{\sim}{\vec{\sigma}}\vec{B}\left(\underset{\sim}{\vec{x}}, \underset{\sim}{\vec{k}}\right) = \frac{1}{2}\mu_B \underset{\sim}{\vec{\sigma}}\left(g_l \vec{\nabla} A^0(\underset{\sim}{\vec{x}}) \times \underset{\sim}{\vec{k}} + g_S \vec{\nabla} \times \vec{A}(\underset{\sim}{\vec{x}})\right) . \quad (3.4)$$

The term with the electric field $\vec{\nabla} A^0$ is responsible for the spin-orbit interaction ($g_l = 1$). The other term is the energy of the spin in the magnetic field $\vec{\nabla} \times \vec{A}$ ($g_S \approx 2$ for electrons). $\underset{\sim}{\vec{\sigma}}$ are the Pauli spin matrices which act on the spin part $|\chi, \phi\rangle$ of the wave function (3.1). In order to apply the variational principle (2.5) we have to calculate first the time derivative part

$$\langle \vec{r}, \vec{p}, \chi, \phi|\, i\frac{d}{dt}\, |\vec{r}, \vec{p}, \chi, \phi\rangle = -\vec{p}\dot{\vec{r}} - \dot{\phi}\left(\sin\frac{\chi}{2}\right)^2 \quad (3.5)$$

and then the hamilton function

$$\mathcal{H}(\vec{r}, \vec{p}, \chi, \phi) := \langle \vec{r}, \vec{p}, \chi, \phi|\, \underset{\sim}{H}\, |\vec{r}, \vec{p}, \chi, \phi\rangle. \quad (3.6)$$

For simplicity and in the spirit of a semi-classical approximation we assume that the width a of the wave packet is small compared to typical variations in $A^\alpha(\vec{x})$. This implies that we use only the first term of the Taylor expansion

$$\left\langle A^\alpha(\underset{\sim}{\vec{x}})\right\rangle = A^\alpha(\vec{r}) + \frac{1}{2}\Delta A^\alpha(\vec{r})\frac{a}{4} + \cdots . \quad (3.7)$$

With that the hamilton function becomes

$$\begin{aligned}\mathcal{H}(\vec{r}, \vec{p}, \chi, \phi) &= \frac{1}{2m}\left(\vec{p} - \frac{e}{c}\vec{A}(\vec{r})\right)^2 + \frac{e}{c}A^0(\vec{r}) + \frac{3}{2}\frac{1}{ma} \\ &- \frac{1}{2}\mu_B\left[\sin\chi\left(\cos\phi B_1(\vec{r}, \vec{p}) + \sin\phi B_2(\vec{r}, \vec{p})\right) + \cos\chi B_3(\vec{r}, \vec{p})\right] \quad (3.8)\end{aligned}$$

where $(B_1, B_2, B_3) = \vec{B}$ denote the three components of the magnetic field as seen by the spin of the moving fermion. From the Lagrange function

$$\mathcal{L}(\vec{r}, \dot{\vec{r}}, \vec{p}, \dot{\vec{p}}, \chi, \dot{\chi}, \phi, \dot{\phi}) = -\dot{\vec{p}}\vec{r} - \dot{\phi}\left(\sin\frac{\chi}{2}\right)^2 - \mathcal{H}(\vec{r}, \vec{p}, \chi, \phi) \tag{3.9}$$

we get the following Euler-Lagrange equations

$$0 = \frac{d}{dt}\frac{\partial \mathcal{L}}{\partial \dot{\vec{p}}} - \frac{\partial \mathcal{L}}{\partial \vec{p}} = -\dot{\vec{r}} + \frac{\partial \mathcal{H}}{\partial \vec{p}} \quad \text{or} \quad \dot{\vec{r}} = \frac{\partial \mathcal{H}}{\partial \vec{p}}(\vec{r}, \vec{p}, \chi, \phi) \tag{3.10}$$

$$0 = \frac{d}{dt}\frac{\partial \mathcal{L}}{\partial \dot{\vec{r}}} - \frac{\partial \mathcal{L}}{\partial \vec{r}} = \dot{\vec{p}} + \frac{\partial \mathcal{H}}{\partial \vec{r}} \quad \text{or} \quad \dot{\vec{p}} = -\frac{\partial \mathcal{H}}{\partial \vec{r}}(\vec{r}, \vec{p}, \chi, \phi) \tag{3.11}$$

$$0 = \frac{d}{dt}\frac{\partial \mathcal{L}}{\partial \dot{\chi}} - \frac{\partial \mathcal{L}}{\partial \chi} = \dot{\phi}\frac{1}{2}\sin\chi + \frac{\partial \mathcal{H}}{\partial \chi} \quad \text{or} \quad \dot{\phi} = -\frac{1}{\frac{1}{2}\sin\chi}\frac{\partial \mathcal{H}}{\partial \chi}(\vec{r}, \vec{p}, \chi, \phi) \tag{3.12}$$

$$0 = \frac{d}{dt}\frac{\partial \mathcal{L}}{\partial \dot{\phi}} - \frac{\partial \mathcal{L}}{\partial \phi} = -\dot{\chi}\frac{1}{2}\sin\chi + \frac{\partial \mathcal{H}}{\partial \phi} \quad \text{or} \quad \dot{\chi} = \frac{1}{\frac{1}{2}\sin\chi}\frac{\partial \mathcal{H}}{\partial \phi}(\vec{r}, \vec{p}, \chi, \phi) \tag{3.13}$$

The first two equations (3.10) and (3.11) are the usual Hamilton's equation of motion for position and momentum. We see, however, that inclusion of the spin variables χ and ϕ gives an addition to the electrostatic and the Lorentz force, namely, the interaction of the spin with the electromagnetic field. It is also interesting to see that the equations (3.11) and (3.12) for the spin degrees of freedom can be cast into Hamilton's equations if one chooses as conjugate variables

$$\phi \text{ and } s := \frac{\hbar}{2}\cos\chi . \tag{3.14}$$

With the projection of the spin on the 3-axis, s, and the azimuthal angle, ϕ, the equations of motion (3.12) and (3.13) transform to

$$\dot{\phi} = \frac{\partial \mathcal{H}}{\partial s}(\vec{r}, \vec{p}, s, \phi) \tag{3.15}$$

and

$$\dot{s} = -\frac{\partial \mathcal{H}}{\partial \phi}(\vec{r}, \vec{p}, s, \phi) . \tag{3.16}$$

Unlike a classical angular momentum, s is not a vector and its range is limited between $-\hbar/2$ and $+\hbar/2$. Furthermore, there is no kinetic energy of the classical type $s^2/(2M)$. By replacing $\sin\chi$ and $\cos\chi$ with s in the hamilton function (3.8) there is no $\hbar$ anymore so that also in this respect it looks like a classical system, but it is not. Application of Ehrenfest's theorem would not lead to these equations. In so called $\hbar$ expansions the spin degree of freedom is regarded to be first order in $\hbar$ and usually dropped. The equations of motion (3.10) - (3.13) describe the Thomas precession of the spin and the motion in an inhomogenous magnetic field in a straightforward way.

3.2 SCATTERING OF TWO IDENTICAL FERMIONS

As another example I would like to apply the Fermionic Molecular Dynamics (FMD) model to the scattering of two identical fermions (with equal spins). The two-body wave function according to eq. (2.3) is given by

$$|q_1, q_2\rangle = \frac{1}{\sqrt{2N}}\{|q_1\rangle \otimes |q_2\rangle - |q_2\rangle \otimes |q_1\rangle\} \tag{3.17}$$

with $q_k = (\vec{r}_k, \vec{p}_k, \chi_k, \phi_k)$, $k = 1, 2$ and the single particle states defined by eqs. (3.1) and (3.2). To make the system more transparent let us consider the case of equal spins, i.e. $\chi_1 = \chi_2$ and $\phi_1 = \phi_2$ and transform to center of mass and relative coordinates

$$\vec{r} = \vec{r}_2 - \vec{r}_1, \quad \vec{R} = \frac{1}{2}(\vec{r}_1 + \vec{r}_2), \quad \vec{p} = \frac{1}{2}(\vec{p}_1 - \vec{p}_2), \quad \vec{P} = \vec{p}_1 + \vec{p}_2 \tag{3.18}$$

and

$$\vec{x} = \vec{x}_1 - \vec{x}_2, \; \vec{X} = \frac{1}{2}(\vec{x}_1 + \vec{x}_2), \; \vec{k} = \frac{1}{2}(\vec{k}_1 - \vec{k}_2), \; \vec{K} = \vec{k}_1 + \vec{k}_2 \, . \tag{3.19}$$

With that the two-body wave function in coordinate representation takes the form

$$\langle \vec{x}, \vec{X} | q_1, q_2 \rangle = \Phi(\vec{X}) \cdot \Psi(\vec{x}) \cdot |\chi, \phi\rangle \otimes |\chi, \phi\rangle \tag{3.20}$$

where the center-of-mass part $\Phi(\vec{X})$ is given by

$$\Phi(\vec{X}) = \frac{1}{(\pi a/4)^{\frac{3}{4}}} \exp\left\{ -\frac{\left(\vec{X} - \vec{R}\right)^2}{a/2} + i\vec{P}\vec{X} \right\} \tag{3.21}$$

and the part for the relative coordinate by

$$\Psi(\vec{x}) = \frac{1}{\sqrt{2N}} \frac{1}{(\pi a)^{3/4}} \left[\exp\left\{ -\frac{(\vec{x} - \vec{r})^2}{2a} + i\vec{p}\vec{x} \right\} - \exp\left\{ -\frac{(\vec{x} + \vec{r})^2}{2a} - i\vec{p}\vec{x} \right\} \right] . \tag{3.22}$$

The normalisation factor N is given by

$$N = 1 - exp\left\{ -\vec{r}^2/a - \vec{p}^2 a \right\} \, |\langle \chi_1, \phi_1 | \chi_2, \phi_2 \rangle|^2 . \tag{3.23}$$

As we do not want to treat in this example the two spins as independent variables let us assume a simple hamiltonian which does not act on spin:

$$\underset{\sim}{H}_{tot} = \frac{1}{4m} \underset{\sim}{\vec{K}}^2 + \underset{\sim}{H} = \frac{1}{4m} \underset{\sim}{\vec{K}}^2 + \frac{1}{m} \underset{\sim}{\vec{k}}^2 + V(\underset{\sim}{\vec{x}}) \tag{3.24}$$

It consists of the center-of-mass kinetic energy, the relative kinetic energy and a two-body interactions $V(\underset{\sim}{\vec{x}})$. Now we are ready to calculate the Lagrange function defined in eq. (2.6) which we split in three parts

$$\mathcal{L} = \mathcal{L}_{\mathrm{cm}} + \mathcal{L}_{\mathrm{rel}} + \mathcal{L}_{\mathrm{spin}} \, . \tag{3.25}$$

The center-of-mass part is

$$\mathcal{L}_{\mathrm{cm}} = -\dot{\vec{P}}\vec{R} - \frac{\vec{P}^2}{4m} - \frac{3}{2}\frac{1}{ma} \, . \tag{3.26}$$

The part for the relative motion is not as simple because it contains exchange effects which originate from the antisymmetrized wave function (3.22).

$$\mathcal{L}_{\mathrm{rel}} = \frac{-\dot{\vec{p}}\vec{r} + \dot{\vec{r}}\vec{p}\, e^{-\xi}}{1 - e^{-\xi}} - \mathcal{H}(\vec{r}, \vec{p}) \tag{3.27}$$

The hamilton function $\mathcal{H}(\vec{r}, \vec{p})$ for the relative motion is the expectation value of the corresponding hamiltonian $\underset{\sim}{H}$

$$\mathcal{H}(\vec{r}, \vec{p}) \equiv \langle \Psi | \underset{\sim}{H} | \Psi \rangle = \frac{1}{m} \frac{\vec{p}^2 + \vec{r}^2/a^2 \, e^{-\xi}}{1 - e^{-\xi}} + \frac{3}{2}\frac{1}{ma} + \langle \Psi | V(\underset{\sim}{\vec{x}}) | \Psi \rangle \, . \tag{3.28}$$

Here and in the following we use the abbreviation

$$\xi = \vec{r}^{\,2}/a + \vec{p}^{\,2}a \tag{3.29}$$

which is a measure of the distance in phase space. The expectation value $\langle\Psi|V(\underset{\sim}{\vec{x}})|\Psi\rangle$ depends not only on $\vec{r}$ but due to the exchange part also on $\vec{p}$. For narrow wave packets ($\Delta V(\vec{r})a \ll V(\vec{r})$) it can be approximated by

$$\langle\Psi|V(\underset{\sim}{\vec{x}})|\Psi\rangle \approx \frac{V(\vec{r}) - V(ia\vec{p})\, e^{-\xi}}{1 - e^{-\xi}} . \tag{3.30}$$

For broad wave packets it has to be calculated explicitly.

Recall that narrow wave packets are <u>not</u> an assumption in the FMD model. This is in contrast to other approaches [1, 2, 3, 4, 7] which use classical equations of motion. In the case of nuclei the finite width of the wave packets in momentum space is essential and will provide the Fermi motion, therefore $a \approx 1/k_F$.

The spin part of the Lagrangian is for this simplified case

$$\mathcal{L}_{\rm spin} = -2\dot{\phi}(\sin\frac{\chi}{2})^2 \tag{3.31}$$

The Euler-Lagrange equations (2.7) for $\vec{P}$ and $\vec{R}$ are

$$0 = \frac{d}{dt}\frac{\partial\mathcal{L}}{\partial\dot{\vec{P}}} - \frac{\partial\mathcal{L}}{\partial\vec{P}} = -\dot{\vec{R}} + \frac{\vec{P}}{2m} \quad \text{or} \quad \dot{\vec{R}} = \frac{\vec{P}}{2m} \tag{3.32}$$

$$0 = \frac{d}{dt}\frac{\partial\mathcal{L}}{\partial\dot{\vec{R}}} - \frac{\partial\mathcal{L}}{\partial\vec{R}} = -\dot{\vec{P}} \quad \text{or} \quad \dot{\vec{P}} = 0. \tag{3.33}$$

This is just free motion, as it should be. The Euler-Lagrange equations for the spin variables χ and ϕ are

$$0 = \frac{d}{dt}\frac{\partial\mathcal{L}}{\partial\dot{\chi}} - \frac{\partial\mathcal{L}}{\partial\chi} = -\dot{\phi}\sin\chi \quad \text{or} \quad \dot{\phi} = 0 \tag{3.34}$$

$$0 = \frac{d}{dt}\frac{\partial\mathcal{L}}{\partial\dot{\phi}} - \frac{\partial\mathcal{L}}{\partial\phi} = -\dot{\chi}\sin\chi \quad \text{or} \quad \dot{\chi} = 0 . \tag{3.35}$$

which means that the two equal spins do not change their direction. The equations of motion for the relative coordinate turn out to be quite interesting, namely

$$\begin{aligned} 0 &= \frac{d}{dt}\frac{\partial\mathcal{L}}{\partial\dot{\vec{p}}} - \frac{\partial\mathcal{L}}{\partial\vec{p}} \quad \text{or} \\ \dot{\vec{r}} &+ \frac{2e^{-\xi}}{1-e^{-\xi}}\left(\dot{\vec{r}} - \frac{\vec{r}(\vec{r}\dot{\vec{r}})/a + \vec{p}(\vec{p}\dot{\vec{r}})a}{1-e^{-\xi}}\right) - \frac{2e^{-\xi}a}{(1-e^{-\xi})^2}\left(\vec{r}(\vec{p}\dot{\vec{p}}) - \vec{p}(\vec{r}\dot{\vec{p}})\right) \\ &= \frac{\partial}{\partial\vec{p}}\mathcal{H}(\vec{r},\vec{p}) \end{aligned} \tag{3.36}$$

and

$$\begin{aligned} 0 &= \frac{d}{dt}\frac{\partial\mathcal{L}}{\partial\dot{\vec{r}}} - \frac{\partial\mathcal{L}}{\partial\vec{r}} \quad \text{or} \\ \dot{\vec{p}} &+ \frac{2e^{-\xi}}{1-e^{-\xi}}\left(\dot{\vec{p}} - \frac{\vec{r}(\vec{r}\dot{\vec{p}})/a + \vec{p}(\vec{p}\dot{\vec{p}})a}{1-e^{-\xi}}\right) + \frac{2e^{-\xi}}{(1-e^{-\xi})^2 a}\left(\vec{r}(\vec{p}\dot{\vec{r}}) - \vec{p}(\vec{r}\dot{\vec{r}})\right) \\ &= -\frac{\partial}{\partial\vec{r}}\mathcal{H}(\vec{r},\vec{p}) . \end{aligned} \tag{3.37}$$

The first point to note is that eqs. (3.36) and (3.37) reduce to Hamilton's equations for $\xi \gg 1$ or in other words when the two fermions are far from each other in phase space they obey classical equations of motion. When they get close ($\xi < 1$) the new terms grow like $1/\xi$ and $1/\xi^2$ and become the leading terms.

To bring the coupled eqs. (3.36) and (3.37) into a useful form one has to solve for $\dot{\vec{r}}$ and $\dot{\vec{p}}$. The result is

$$\dot{\vec{r}} = \alpha_1(\xi)\frac{\partial \mathcal{H}}{\partial \vec{p}} + \alpha_2(\xi)\left\{a\left(\vec{p}\,\frac{\partial \mathcal{H}}{\partial \vec{p}} + \vec{r}\,\frac{\partial \mathcal{H}}{\partial \vec{r}}\right)\vec{p} + \left(\frac{\vec{r}}{a}\frac{\partial \mathcal{H}}{\partial \vec{p}} - a\vec{p}\,\frac{\partial \mathcal{H}}{\partial \vec{r}}\right)\vec{r}\right\} \tag{3.38}$$

and

$$\dot{\vec{p}} = -\alpha_1(\xi)\frac{\partial \mathcal{H}}{\partial \vec{r}} + \alpha_2(\xi)\left\{\left(\frac{\vec{r}}{a}\frac{\partial \mathcal{H}}{\partial \vec{p}} - a\vec{p}\,\frac{\partial \mathcal{H}}{\partial \vec{r}}\right)\vec{p} - \left(\vec{r}\,\frac{\partial \mathcal{H}}{\partial \vec{r}} + \vec{p}\,\frac{\partial \mathcal{H}}{\partial \vec{p}}\right)\frac{\vec{r}}{a}\right\}, \tag{3.39}$$

where $\alpha_1(\xi)$ and $\alpha_2(\xi)$ are functions of $\xi = \vec{r}^{\,2}/a + \vec{p}^{\,2}a$ given by

$$\alpha_1(\xi) = \frac{1 - e^{-\xi}}{1 + e^{-\xi}} \tag{3.40}$$

$$\alpha_2(\xi) = \frac{2e^{-\xi}(1 - e^{-\xi})}{(1 + e^{-\xi})^2(1 - e^{-\xi}) - 2\xi e^{-\xi}(1 + e^{-\xi})}. \tag{3.41}$$

If the two fermions are far from each other in phase space, which implies that they could be far in coordinate space with small relative momentum or close by with large relative momentum (in any case $\xi \gg 1$), the equations of motion (3.38) and (3.39) turn into Hamilton's equation as $\alpha_1(\xi \gg 1) = 1$ and $\alpha_2(\xi \gg 1) = 0$. In that limit the identical fermions behave like classical distinguishable particles although their wave function is of course still antisymmetrized. When they get close in phase space ($\xi < 1$) $\alpha_1(\xi) \to \xi/2$ and $\alpha_2(\xi) \to 3/\xi^2$ which means that the Hamilton-like parts in (3.38) and (3.39) vanish like $\xi/2$ but the remaining parts increase like $3/\xi^2$.

It is interesting to note that for $\xi \lesssim 2$ the equations of motion which result from the parameterization (3.22) cannot be cast into Hamilton's form where $\vec{r}$ and $\vec{p}$ are canonical variables. To prove this statement let us suppose that a hamilton function $\mathcal{H}_{\text{Pauli}}$ exists such that

$$\dot{r}_i = \frac{\partial \mathcal{H}_{\text{Pauli}}}{\partial p_i} \quad \text{and} \quad \dot{p}_i = -\frac{\partial \mathcal{H}_{\text{Pauli}}}{\partial r_i}\ , i = 1, 2, 3\,, \tag{3.42}$$

where i denotes the three spatial directions. Let us now disprove the existence of $\mathcal{H}_{\text{Pauli}}$ by calculating the mixed derivatives of the equations of motion (3.38) and (3.39). It is easy to verify that

$$\frac{\partial \dot{r}_i}{\partial r_k} \neq -\frac{\partial \dot{p}_k}{\partial p_i}\ . \tag{3.43}$$

The mixed derivatives should be the negative of each other if eqs. (3.42) where true. One cannot find a hamilton function $\mathcal{H}_{\text{Pauli}}(\vec{r},\vec{p})$ such that eqs. (3.38) and (3.39) could be rewritten as Hamilton's equations of motion. This disproves the existence of a hamiltonian $\mathcal{H}_{\text{Pauli}}(\vec{r},\vec{p})$ which would describe the fermionic dynamics derived from the ansatz (3.22). In how far the above statements still hold if we allow for a time dependence of the parameter a remains to be seen.

Equations (3.42) represent the ansatz often made in literature to incorporate the effects of the Pauli principle [7, 8, 9]. Different forms of momentum dependent potentials have been added to the classical hamiltonian. Here we see that this method differs in several aspects from fermionic dynamics. Firstly, as discussed above the dynamical behaviour of $\vec{r}$ and $\vec{p}$ needs not be of Hamilton's type. This will be discussed further in the following section on free scattering. Secondly, the finite width of the wave packet has not been taken into account in ref. [7, 8, 9]. so that the term $3/(2ma)$ in eq. (3.28) is missing. This constant is not important for the equations of motion but it is a large contribution to the kinetic energy, namely, Fermi motion. Also in the expectation value of the potential energy the zero point energies are not negligible. And finally, one should not replace $\langle V(\underset{\sim}{\vec{x}})\rangle$ simply by $V(\vec{r})$ because even for narrow wave packets there is an exchange term which is not small for $\xi < 1$. Since in Fermionic Molecular Dynamics (FMD) a quantum variational principle based on wave functions is involved these quantum effects are included.

3.2.1 Free motion of two identical fermions

For convenience let us switch in this section to dimensionless variables by substituting

$$\hat{\vec{p}} = \sqrt{a}\vec{p} \quad \text{and} \quad \hat{\vec{r}} = \vec{r}/\sqrt{a} \tag{3.44}$$

and let us omit the "hats" again. If there is no interaction the equations of free motion are

$$\dot{\vec{r}} = \frac{2}{ma}\left\{ \frac{1}{2}\left(\alpha_1(\xi) + 1 - \alpha_2(\xi)(\vec{r}^{\,2} - \vec{p}^{\,2})\right)\vec{p} \; + \; \alpha_2(\xi)(\vec{r}\vec{p})\vec{r} \right\} \tag{3.45}$$

$$\dot{\vec{p}} = \frac{2}{ma}\left\{ \frac{1}{2}\left(\alpha_1(\xi) - 1 + \alpha_2(\xi)(\vec{r}^{\,2} - \vec{p}^{\,2})\right)\vec{r} \; + \; \alpha_2(\xi)(\vec{r}\vec{p})\vec{p} \right\} \tag{3.46}$$

For $\xi = \vec{r}^{\,2} + \vec{p}^{\,2} \gg 1$ we retrieve the free motion in Hamilton's form

$$\dot{\vec{r}} = \frac{2}{ma}\vec{p} \quad \text{and} \quad \dot{\vec{p}} = 0 \tag{3.47}$$

For $\xi \ll 1$ we find a completely different set:

$$\dot{\vec{r}} = \frac{3}{ma}\left(-\frac{\vec{r}^{\,2} - \vec{p}^{\,2}}{(\vec{r}^{\,2} + \vec{p}^{\,2})^2}\vec{p} + \frac{2(\vec{r}\vec{p})}{(\vec{r}^{\,2} + \vec{p}^{\,2})^2}\vec{r} \right) \tag{3.48}$$

$$\dot{\vec{p}} = \frac{3}{ma}\left(\frac{\vec{r}^{\,2} - \vec{p}^{\,2}}{(\vec{r}^{\,2} + \vec{p}^{\,2})^2}\vec{r} + \frac{2(\vec{r}\vec{p})}{(\vec{r}^{\,2} + \vec{p}^{\,2})^2}\vec{p} \right) \tag{3.49}$$

From eqs. (3.48) and (3.49) we see that the "velocities" $\dot{\vec{r}}$ and $\dot{\vec{p}}$ become very large in the vicinity of $\vec{r} \approx 0$ and $\vec{p} \approx 0$. This is completely different if we use Hamilton's equations (3.42) with $\mathcal{H}_{\text{Pauli}} = \langle \underset{\sim}{H} \rangle$ also for $\xi \ll 1$ as has been proposed for example in [9, 13]. In this case we get

$$\mathcal{H}_{\text{Pauli}} = \frac{1}{ma}\left(\vec{p}^{\,2} + \frac{\xi}{e^{\xi} - 1}\right) \rightarrow \frac{1}{ma}\left(\frac{1}{2}(\vec{p}^{\,2} - \vec{r}^{\,2}) + 1\right)$$

and herewith Hamilton's equation of motion are

$$\dot{\vec{r}} = \frac{\partial \mathcal{H}_{\text{Pauli}}}{d\vec{p}} = \frac{1}{ma}\vec{p}$$

$$\dot{\vec{p}} = -\frac{\partial \mathcal{H}_{\text{Pauli}}}{\partial \vec{r}} = \frac{1}{ma}\vec{r}$$

We see that for $\vec{r} \approx 0$ and $\vec{p} \approx 0$ also the velocities are very small at complete variance with the FMD result given in eqs. (3.48) and (3.49).

Further investigations of eqs. (3.48) and (3.49) are on the way. I should like to check how strongly these equation for free motion scatter the particles (in the ideal case they should not) and what the time delay is compared to the exact solution of the Schrödinger equation (ideally there should be none). If these tests fail one may think about a complex time-dependent width $a(t)$ in the ansatz (3.2) for the wave function. Then the equations of motion might become more complicated but for free scattering we would obtain the exact solution of the Schrödinger equation.

3.2.2 Conservation laws

Since the equations of motion (3.38) and (3.39) are not of Hamilton's form one may wonder if the conservation laws are fulfilled. Considering the symmetries of the Lagrange function (3.25) one constructs with Noether's theorem the following conserved quantities:

Energy ($\mathcal{L}$ does not explicitly depend on time)

$$\frac{d}{dt}\mathcal{H}(\vec{r},\vec{p}) = \frac{d}{dt}\left\langle \underset{\sim}{H} \right\rangle = 0 \tag{3.50}$$

Total momentum ($\mathcal{L}$ is translational invariant)

$$\frac{d}{dt}\vec{P} = \frac{d}{dt}\left\langle \underset{\sim}{\vec{K}} \right\rangle = 0 \tag{3.51}$$

Center-of-mass angular momentum ($\mathcal{L}_{cm}$ is rotational invariant)

$$\frac{d}{dt}(\vec{R} \times \vec{P}) = \frac{d}{dt}\left\langle \underset{\sim}{\vec{X}} \times \underset{\sim}{\vec{K}} \right\rangle = 0 \tag{3.52}$$

Relative angular momentum ($\mathcal{L}_{rel}$ is rotational invariant)

$$\frac{d}{dt}\left(\frac{\vec{r} \times \vec{p}}{\alpha_1(\xi)}\right) = \frac{d}{dt}\left\langle \underset{\sim}{\vec{x}} \times \underset{\sim}{\vec{k}} \right\rangle = 0 \tag{3.53}$$

All conservation laws can easily be proven by using the equations of motion. The proof for the relative angular momentum is somewhat more involved due to the $\alpha_1(\xi)$ in the denominator ($\vec{r} \times \vec{p}$ is not conserved).

4 THE MANY BODY SYSTEM

In nuclei one has two kinds of fermions, protons and neutrons, therefore the many-body wave function splits into two antisymmetric parts

$$|q\rangle = |q_{\text{protons}}\rangle \otimes |q_{\text{neutrons}}\rangle \tag{4.1}$$

The hamiltonian

$$\underset{\sim}{H} = \underset{\sim}{T} + \underset{\sim}{V} \tag{4.2}$$

consists of the kinetic energy $\underset{\sim}{T}$ and a two-body interaction $\underset{\sim}{V}$ which may be spin and isospin dependent. These degrees of freedom are present in the many-body wave function (4.1).

4.1 THE GROUND STATE OF NUCLEI

In the Fermionic Molecular Dynamics model, as usual in quantum mechanics, the ground state is the parametrized state which has the lowest energy, thus

$$\frac{\partial}{\partial q_{k,j}} \langle\{q\}| \underset{\sim}{H} |\{q\}\rangle = 0 \quad \text{for all} \quad q_{k,j} \, . \tag{4.3}$$

In the ground state all derivates of $\mathcal{H} \equiv \langle \underset{\sim}{H} \rangle$ are zero. Looking at the equations of motion (2.7) we see that the ground state is stationary, in particular $\dot{\vec{r}}_k = 0, \dot{\vec{p}}_k = 0$. Nevertheless the Fermi motion is present. It is the zero point motion in each of the gaussian wave packets.

Here, the FMD model differs conceptually from all other molecular dynamics models for heavy ion collisions. There, one usually gives the distinguishable classical particles a random motion to mock the Fermi motion. One then has to fight against particle evaporation from a state which is considered to be the ground state. Even if one succeeds with some tricks the model is still only for Boltzmann particles obeying Boltzmann statistics and not Fermi-Dirac statistics. Only if the chemical potential is largely negative the two ensembles become equal. For intermediate energy heavy-ion collisions this is never the case. The FMD model obeys the Fermi-Dirac statistics since it is defined with antisymmetrized states.

The ground state in the FMD model is expected [9] to consist of single particle states which are densely packed in phase space. This gives us also an idea about the choice of the width parameter a. In momentum space the spread of the wave packet is $2/\sqrt{a}$ which of course should be smaller than $k_F \approx 1.4 fm^{-1}$. On the other hand $\sqrt{a}/2$ should not be larger than the surface thickness of nuclei ($\approx 1 fm$). From these two limits we see that $\sqrt{a} \approx 1.4 fm$. For an optimal choice of a one may minimize the energy $\langle \underset{\sim}{H} \rangle$ for a representative set of nuclei with respect to a to obtain its numerical value.

In the case where $a_k(t)$ is a dynamical variable in each single-particle state $|q_k\rangle$ the energy has to be minimized also with respect to the real and the imaginary part of all a_k. This is of course the better approach because then the model is free of parameters.

4.2 THE LAGRANGE FUNCTION

The single particle states are in general not orthogonal. Therefore the inverse of the overlap matrix, $\mathcal{O}_{kl}$, appears when calculating the expectation values.

$$\left(\mathcal{O}^{-1}\right)_{kl} := \langle q_k | q_l \rangle \tag{4.4}$$

The lagrange function (2.6) contains the time derivative part

$$\langle\{q\}|\, i\frac{d}{dt}\, |\{q\}\rangle = \frac{i}{2}\sum_{k,l}\left(\langle q_k|\,\frac{d}{dt}q_l\rangle - \langle\frac{d}{dt}q_k|q_l\rangle\right) \mathcal{O}_{lk}\,, \tag{4.5}$$

the kinetic energy

$$\langle\{q\}|\, \underset{\sim}{T}\, |\{q\}\rangle = \sum_{k,l} \langle q_k|\underset{\sim}{T}|q_l\rangle\, \mathcal{O}_{lk}\,, \tag{4.6}$$

and the two body interaction

$$\langle\{q\}|\, \underset{\sim}{V}\, |\{q\}\rangle = \frac{1}{2}\sum_{k,l,m,n} \langle q_k, q_l|\, \underset{\sim}{V}\, |q_m, q_n\rangle\, (\mathcal{O}_{mk}\mathcal{O}_{nl} - \mathcal{O}_{nk}\mathcal{O}_{ml})\,, \tag{4.7}$$

where $|q_k, q_l\rangle = |q_k\rangle \otimes |q_l\rangle$. The overlaps $\langle q_k|q_l\rangle$ and the matrix elements $\langle q_k|\frac{\partial}{\partial t}|q_l\rangle$ and $\langle q_k|\underset{\sim}{T}|q_l\rangle$ can be written down analytically for the gaussian wave packets (2.1). This helps in evaluating the equations of motion (2.7).

I should like to acknowledge stimulating and helpful discussions with L. Wilets and P. Manakos.

References

[1] L. Wilets, A.D. MacKellar and G.A. Rinker, Jr., Proc. IV Int. Workshop on Gross Properties of Nuclei and Nuclear Excitations (Hirschegg, Austria, 1976), ISSN 0720-8715, p. 111

[2] A.R. Bodmer and C.N. Panos, Phys. Rev. **C15** (1977) 1342, and A.R. Bodmer, C.N. Panos and A.D. MacKellar, Phys. Rev. **C22** (1980) 1025

[3] J.J. Molitoris, J.B. Hoffer, H. Kruse, and H. Stöcker, Phys. Rev. Lett. **53** (1984) 899

[4] T.J. Schlagel and V.R. Phandharipande, Phys. Rev. **C36** (1987) 162

[5] E. Betak, preprint DUBNA E-86-701

[6] R. Schmidt, B. Kämpfer, H. Feldmeier, and O. Knospe, submitted to Phys. Lett. B, preprint GSI-89-45 (1989)

[7] L. Wilets, E.M. Henley, M. Kraft, and A.D. MacKellar, Nucl. Phys. **A282** (1977) 341, and L. Wilets, Y. Yariv and R. Chestnut, Nucl. Phys. **A301** (1978) 359

[8] C. Dorso, S. Duarte, and J. Randrup, Phys. Lett. **B188** (1987) 287, and C. Dorso and J. Randrup, Phys. Lett. **B215** (1988) 611, and preprint LBL-27075, to be published in Phys. Lett. B

[9] D.H. Boal and J.N. Glosli, Phys. Rev. **C38** (1988) 1870

[10] J. Aichelin, A. Rosenhauer, G. Peilert, H. Stöcker, and W. Greiner, Phys. Rev. Lett. **58** (1987) 1926, and Phys. Rev. **C37** (1988) 2451

[11] G. Peilert, H. Stöcker, W. Greiner, A. Rosenhauer, A. Bohnet, and J. Aichelin, Phys. Rev. **C39** (1989) 1402

[12] G.E. Beauvais, D.H. Boal, and J.C.K. Wong, Phys. Rev. **C35** (1987) 545

[13] D.H. Boal and J.N. Glosli, Phys. Rev. **C38** (1988) 2621

[14] A.K. Kerman and S.E. Koonin, Annals of Phys. **100** (1976) 332

[15] E. Caurier, B.Grammaticos, and T. Sami, Phys. Lett. **B109** (1982) 150, and S. Drozdz, J. Okolowicz, and M. Ploszajczak, Phys. Lett. **B109** (1982) 145

[16] J.D. Bjorken and S.D. Drell, Relativistische Quantenmechanik, B.I. Wissenschaftsverlag, Mannheim 1966

Mean Field and Collisions in Hot Nuclei

H. S. Köhler[1]

Nuclear Science Division, Lawrence Berkeley Laboratory
University of California, Berkeley, CA 94720

Collisions between heavy nuclei produce nuclear matter of high density and excitation. Brueckner methods are used to calculate the momentum and temperature dependent mean field for nucleons propagating through nuclear matter during these collisions. The mean field is complex and the imaginary part is related to the "two-body" collisions, while the real part relates to "one-body" collisions. A potential model for the N-N interactions is avoided by calculating the Reaction matrix directly from the T-matrix (i.e. N-N phase-shifts) using a version of Brueckner theory previously published by the author. Results are presented for nuclear matter at normal and twice normal density and for temperatures up to 50 MeV.

1 Introduction

The primary purpose of Heavy-Ion (H.I.) collisions is to explore the properties of hot nuclear matter; both static and dynamic. It is of particular interest to analyse these collisions in order to learn about nucleonic degrees of freedom. But before proceeding to introduce some model of these into a theory of hot nuclei it seems appropriate to investigate the consequence of incorporating important many-body effects in a "nucleons only" theory. This has not been done satisfactorily. A first step is to find the "effective" force V_{eff} in hot nuclear matter from the known "free" N-N interaction. Most "microscopic" calculations have been done with forces that are fitted to zero temperature properties like binding and compressibility and depend on local density only but not on other variables of the medium. This might be adequate for low energies. It is however well known that the effective force in nuclear matter is momentum dependent (non-local). Collisions at higher energy result in strong deformations in momentum space. A force that depends on density only, gives an

[1]Permanent address: Physics Department, University of Arizona, Tucson, AZ 85721

The Nuclear Equation of State, Part A
Edited by W. Greiner and H. Stöcker
Plenum Press, New York

energy-functional that is independent of such deformations, while a momentum-dependent force leads to a dependence on this deformation. It follows that it may be necessary to incorporate the momentum dependence of the force especially in a theory of high energy H.I. collisions. Furthermore,the strength of the force depends not just on density but also on excitaion; it is "temperature"-dependent. This investigation is a contribution to our understanding of the effective forces in hot nuclei both as regards the mean field and the two-body dissipative collisions.

Let us go into some more detail as regards the arguments that were given above.

The momentum distribution in the interior of a ground state nucleus is essentially isotropic and with a sharp Fermi-surface as in a Thomas-Fermi approximation. In the initial stage of a high energy collision between two nuclei the momentum distribution at some point where the two nuclei overlap in coordinate space, is strongly deformed. It is roughly that of two Fermi-spheres separated by the relative momentum of the colliding ions. Our earlier calculations [1], incorporating two-body collisions by the relaxation-time method, show this explicitly to be the case for the distribution averaged over coordinate space. These calculations also show that the mean field distorts this averaged distribution only slightly during the course of a collision, while two-body collisions thermalizes it.

In a model of H.I. collisions where the effective two-body interaction V_{eff} is local i.e. independent of relative N-N momentum the resultant mean field will also be local and independent of the distribution in momentum space; other than of the zero moment of this distribution, i.e. the local density. Most calculations like VUU [2], BUU [3] or TDHFRX and VRX [4] have used this model. In a calculation of the properties of nuclear matter from realistic nuclear forces the momentum dependence is very important for obtaining the correct saturation density and in calculating the compressibility. In calculations of collisions between heavy ions made by Stöcker et al [5] using a local N-N interaction V_{eff}, it was found necessary to use quite a large compressibility in order to obtain the experimentally observed perpendicular flow.

One can however argue that the increase in compressibility is necessitated by neglecting the (known) momentum dependence. This can be understood as follows. By deforming or heating a zero-temperature Fermi-distribution (while keeping the density constant) the repulsive part of the energy due to the momentum dependence of the force will increase. Collisions between the ions result in such deformations in momentum space. If the energy is assumed to be independent of the deformation this increased repulsion has to be compensated for by increasing the compressibility, making use of the fact that the density also increases in the interaction region.

The calculations by Welke et al [6] (using a momentum dependence resembling the Yukawa force in momentum-space) bear this out, although the simple argument given above may not be correct. It may not just be a question of the energy-functional. Rather it appears that the dynamics especially as regards the perpendicular flow is different in the two separate cases [7].

Increasing the stiffness of the equation of state by incresing the density dependence of V_{eff} results in a larger repulsion in the mean field in the region of overlap between the ions. Nucleons hitting this repulsion in a non-central collision will be reflected out sideways. If the force, and consequently the mean field, is momentum dependent a different mechanism described as a 'coherence' in phase-space seems to enter[7]. A similar effect was observed when displaying the Wigner functions for low energy (head-on) collisions between slabs with a momentum-dependent force of quadratic (Skyrme) form[8]. A density dependent force tended to break up the distribution in phase-space much more than the momentum-dependent although the compressibility was the same for both.

Whatever the mechanism is, the calculations do indicate that the momentum dependence is important[6] and it is one purpose of this paper to calculate this macroscopically.

We have stressed that in the case of a momentum dependent V_{eff}, the mean field will change if the distribution of nucleons in momentum-space is deformed. There is however another effect of deformation to consider. In a microscopic theory V_{eff} will itself depend on the distribution; not just on density but also on the deformation under constant density. In Brueckner's formulation of the many-body problem this is explicit in that the Reaction-matrix, which in our context here is the same as the effective interaction V_{eff}, is a functional of the Pauli-operator which expresses a dependence on which states are occupied. There is also a dispersion-effect in that the nucleons propagate through a mean field while interacting. All of these effects are included in the Reaction-matrix equation defined below in Sect. 2. They are all important when calculating the Reaction-matrix at zero temperature. There is no reason to beleive they should be less important at higher temperatures.

Although the Reaction-matrix is formally defined for any distribution of occupied states, calculations are in practice restricted to some simple cases. The simplest is the zero-temperature Fermi-distribution for which of course many calculations have been done. Non-zero temperature Brueckner calculations have also been done [9,10] and in relativistic Brueckner theory by Malfliet [11]. While a heated Fermi-sphere would be the appropriate representation for the final stages of the collision, the distribution in the initial stage of a H.I. collision may best be represented by two spheres separated by the relative momentum of the two ions. Calculations for this system has also been done as a step towards calculating an optical model potential for ion collisions [12,13].

In this paper we shall present results of calculations of the mean field potential in hot nuclear matter using a method that was presented in an earlier publication [14]. We have used this method for the two sphere problem as well, but shall reserve those results for later presentation. For completeness a short description of the method is given in sect. 2 and the results are shown in sect. 3. Sect. 4 contains a summary and conclusions.

2 The Effective Interaction in Nuclei

The calculations in this paper are based upon the Brueckner many body theory of the effective interaction V_{eff}; in this theory referred to as the Reaction matrix K. A difference from the conventional formalism is that we shall calculate the K-matrix directly from the scattering T-matrix. The first order approximation to K will be T rather than a N-N potential as in conventional approaches. In order to calculate K from T one needs the off-diagonal matrix-elements of the related reactance-matrix $\mathcal{K}$ (see below) that are not accessible from N-N scattering. To overcome this problem we simply assume $\mathcal{K}$ to be separable in momentum-space. It is shown below that it is then rather easy to solve the equations for diagonal elements of K. For nuclear matter calculations we only need these diagonal elements of the effective interaction. The assumption of separability may seem ad hoc. We think it is justified by the agreement with calculations from N-N potentials and by the comparative ease with which calculations can be done for rather complicated distributions.

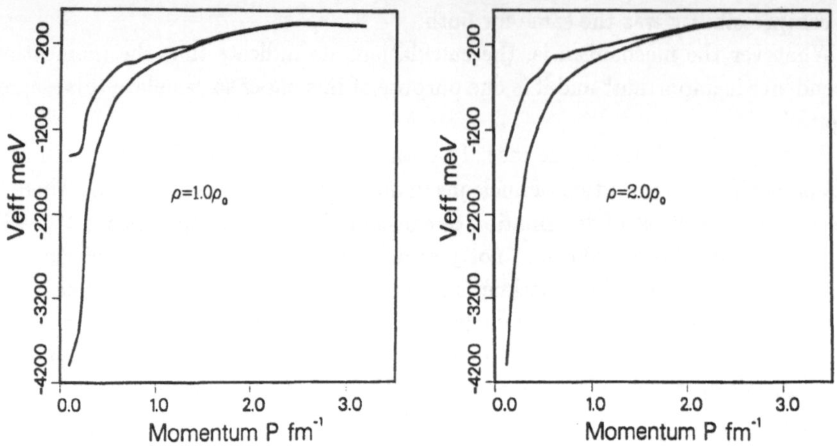

Figure 1. The effective interaction at normal (left curves) and double density (right curves) as a function of relative momentum P. Lower curves are from T- and upper is from K-matrix. See text.

Our method should not be confused with the method of using a separable potential. We are in fact not using a potential model (at least not explicitly) and we consider this an advantage in itself.

Our approach using the T-matrix as a first approximation is also made more sensible if we note that in the low density limit, and in the limit of large relative momentum, $K \Rightarrow T$. This is shown in fig. 1 at normal (left) and double (right) density. The lower curve in each figure is the T-matrix approximation (in some works referred to as the phase-shift approximation) and the more repulsive curve is the K-matrix effective interaction.

We now describe some details of calculating K following the procedure outlined above. It is defined by

$$K = v + v\frac{Q}{e + i\eta}K. \tag{1}$$

Here Q is the Pauli-operator and the energy-denominator contains the (bubble-)interactions with all other nucleons. The traditional procedure is to calculate K from this equation assuming the interaction v, the 'free' N-N interaction to be known. Instead we proceed by first defining a reactance-matrix $\mathcal{K}$ by

$$\mathcal{K} = v + v\frac{P}{e_0}\mathcal{K} \tag{2}$$

where P denotes that the principal value is to be taken when e_0, which is kinetic energy only, has a pole. This reactance-matrix has the useful property that it is directly related to the phase-shifts. In an angular momentum decomposition one has

$$\mathcal{K} = tan\delta_l(k)/k \tag{3}$$

for free scattering states while for bound states [15]

$$\mathcal{K} = \delta_l(k)/k \tag{4}$$

with $\delta_l(k)$ being the phase-shifts. The relation between K and $\mathcal{K}$ is

$$K = \mathcal{K} + \mathcal{K}\Big(\frac{Q}{e+i\eta} - \frac{P}{e_0}\Big)K \tag{5}$$

The first approximation to this equation is obtained by putting $Q = 1$ and $e = e_0$ in (eq.5) to give

$$K_1 = T$$

which is usually referred to as the impulse approximation. This implies using free cross-sections.

In order to solve eq.5 for K we first introduce another matrix K_p by

$$K_p = v + v\frac{QP}{e}K_p \tag{6}$$

related to K by ($\delta(e)$) is here *delta*-function of e)

$$K = K_p(1 - iK_pQ\delta(e))^{-1} \tag{7}$$

and to $\mathcal{K}$ by

$$K_p = \mathcal{K} + \mathcal{K}\Big(\frac{QP}{e} - \frac{P}{e_0}\Big)K_p. \tag{8}$$

This equation is readily solvable in momentum space for the diagonal elements K_p if we assume $\mathcal{K}$ to be separable. In fact we find $K_p = D(k)\mathcal{K}$ where D is a function of relative momentum k only and is related to the integral

$$I(k) = \int_0^\infty |\mathcal{K}|\Big(\frac{QP}{e} - \frac{P}{e_0}\Big)d\mathbf{k}'. \tag{9}$$

by $D(k) = 1/(1 - I(k))$. The Reaction-matrix K is then obtained from Eq. (7). It is in general complex. A non-zero imaginary part is obtained whenever we have a pole, that results in energy-conserving transitions. To first approximation Eq. (7) is solved by

$$K = K_p + i|K_p|^2Q\delta(e). \tag{10}$$

If the absorption is small i.e. if the imaginary part of K is small compared to the real part of K, this approximation which has often been used is adequate, but we

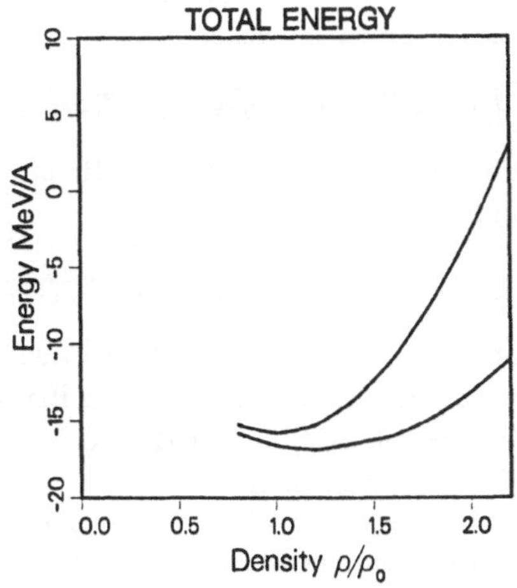

Figure 2. Upper curve is the empirical binding energy curve saturating at ρ_0. The lower is from our Brueckner-calculation.

shall find later that for large nucleon momenta with large absorption, corrections can be quite large. It is also to be noted that the total energy now is complex, except in case of a zero-temperature distribution.

The method descibed above has been tested against calculations of the Reaction matrix from the Reid HC potential model and was found quite satisfactory[14]. The ^1S states were reproduced essentially exactly. For some states ($^1D_2$,$^1P_1$,$^3P_0$ and $^3P_1$) the T-matrix approximation is actually already very good and was therefore used. The coupled states ($^3S -^3 D$) require a special treatment. For these states the contribution from the coupling was simply added to the right hand side of Eq. (9) and this procedure was found quite adequate.

The integration in Eq. (9) was cut off at $20 fm^{-1}$. The phase-shifts used in the calculations presented below are those of Arndt et al [16] which are for labenergies below 500 MeV only. For higher energies a straight line extrapolation was used with a slope that was considered a parameter. In a potential model this procedure corresponds to the choice of short-range repulsion. With a reasonable value of this parameter a binding energy of about 16 MeV/A is obtained at normal saturation-density.

The phase-shifts, the stipulation of a separable matrix and this parameter constitutes our 'potential' model although the potential itself is not obtainable.

The binding-energy that we obtain is shown by the lower curve in fig. 2. The upper curve is the empirical saturation curve. Just as in other non-relativistic Brueckner and HNC calculations saturation is not obtained in this model at the empirical point. Following the ideas in ref.[17] it is assumed that the difference between the calculated saturation curve and the "experimental" is due to 3-body and/or higher order terms. We remedy the situation by adding a density-dependent term E_3 to the Nuclear Hamiltonian. This Hamiltonian will contribute to the mean field. It is repulsive.

3 Calculations

Following the procedure just outlined in the previous section we assume that nuclear matter saturates at -15.8 MeV/A at a saturation density of 0.166 fm^{-3} and with a compressibility of 235 MeV. We achieve this by adding to the microscopic calculation of the energy per particle a function E_3 defined by

$$E_3 = 0.46 - 5.79(\rho/\rho_0) + 5.33(\rho/\rho_0)^2 \tag{11}$$

where ρ_0 is the saturation density. This part of the Hamiltonian will contribute a quantity V_3 to the mean field with

$$V_3 = 0.46 - 11.58(\rho/\rho_0) + 15.99(\rho/\rho_0)^2 \tag{12}$$

including rearrangement terms. It is to be noted that V_3 does not have to be added to the energy-denominator e in the calculation of the Brueckner Reaction-matrix because it is independent of momentum and therefore cancels when taking the difference between particle- and hole-energies.

The Reaction-matrix is calculated with on-energy shell insertions in both hole and particle lines, i.e. with a continuous spectrum. The Pauli operator for $T > 0$ is calculated in the angle averaged approximation [18].

The mean field at normal and double density is shown in fig. 3 at T =0 and 50 MeV. In addition to the energy V_3 the third oder rearrangement energy V_{Rh} has been added. It is well approximated by[19]

$$V_{Rh} = -\rho \mathcal{I}_w V_h \tag{13}$$

where V_h is the first order K-matrix energy. The "wound-integral" $\mathcal{I}_w$ is a model-dependent factor and is given the values 1.2 and 1.0 at normal and double density respectively.

The results (at T=0) shown in fig. 3 have been compared with the results of Wiringa[20] who used HNC methods and various N-N interactions to calculate single particle potentials at zero temperature . The agreement is satisfactory although his potential at double density (interpolating between his 0.3 and 0.4 fm^{-3} results) is about 25 MeV more repulsive than ours at the highest momenta. Part of the difference can be explained by the fact mentioned above after Eq. (10), that we are including propagations in an absorptive medium, even when calculating the real part of the mean field. We have estimated this contribution by calculating K from Eq. (10) instead of from Eq. (7). This increases the repulsion by about 10 MeV at double density and high momenta where the absorption is large. The propagation through an absorptive potential has therefore a non-negligble effect on the real potential , but this effect is not included in ref.[20]. Another point is that Wiringa "normalised" his curves to go through some common point, so that a direct comparison is not really relevant.

In addition to being momentum-dependent the mean field is also temperature-dependent. This is of course at least partly a consequence of the momentum dependence of the two-body interaction V_{eff}. To investigate this further we compare with the predictions of the parametrization used by Welke et al[6] who use a temperature-independent phenomenological interaction. Fig. 4 shows the mean field calculated with their interaction at normal and double density and indicated temperatures. Comparing with fig.3 we see that the overall agreement is good but there is a noticeable difference in temperature dependence especially at the higher density. This is understood as follows. We may consider the ref.[6] effective interaction as the first order contribution in the separation method of Moszkowsi and Scott[21]. This is (essentially) independent of the medium. The second order contribution is of the form $v(Q/e)v$ and depends on temperature (and density) through the Pauli Q-operator which blocks scattering into occupied states. At zero temperature the blocking

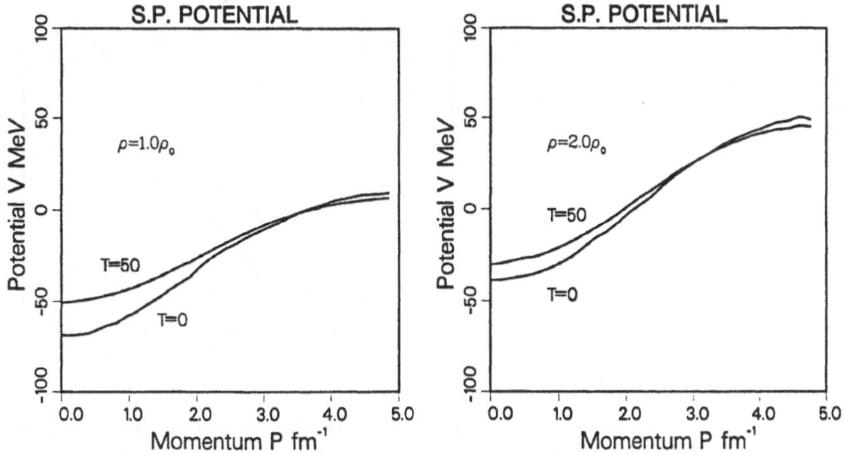

Figure 3. Mean field at normal(left) and double (right) density at temperatures T=0 and 50 MeV as a function of momentum P.

increases with density. As temperature is increased blocking is decreased and the interaction becomes more attractive. Referring to fig 1, the effective interaction will approach the T-matrix interaction as temperature increases. On the basis of our results one could consider including the temperature dependence by a temperature dependent factor applied to the ref.[6] interaction.

The imaginary part of the mean field relates to dissipative two-body collisions[22]. The left parts of fig. 5 and fig. 6 show the results at normal and double density respectively. The right parts of the same figures show the absorption calculated from

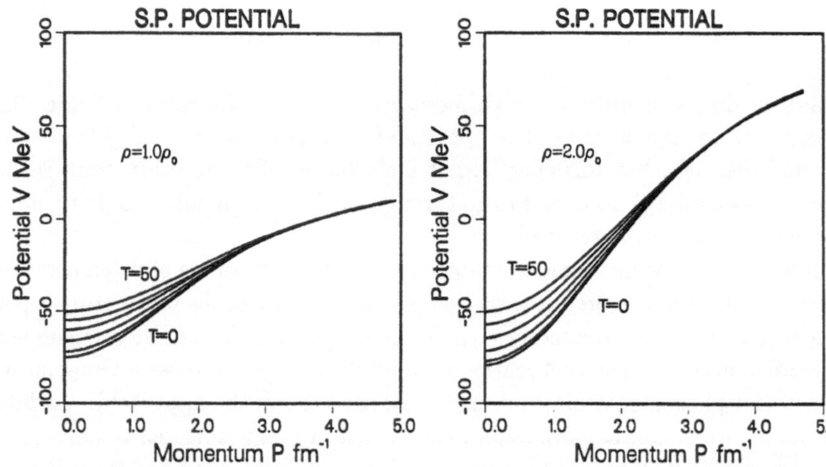

Figure 4. Mean field at normal(left) and double (right) density calculated from the interaction in Welke et al[6]. Temperatures are at 10 MeV interval.

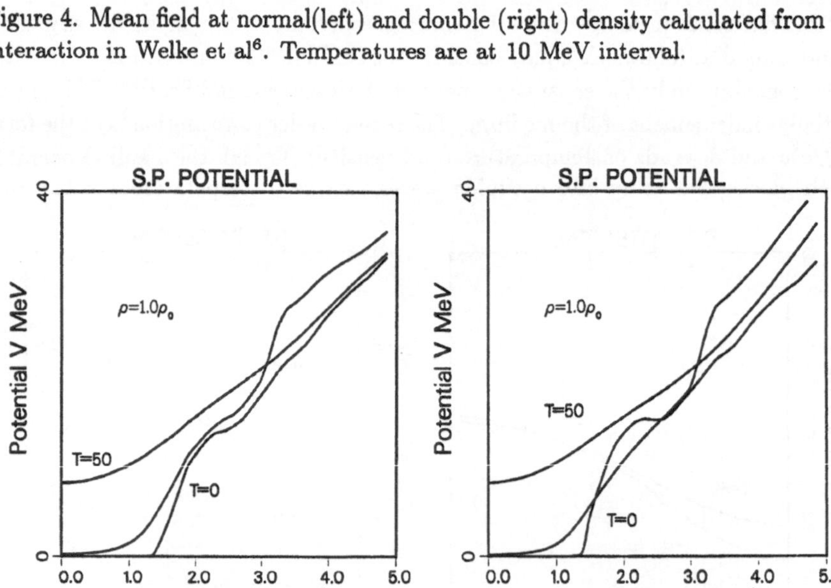

Figure 5. Left curves show the absorptive part of the mean field at normal density as a function of momentum P. Right curves in T-matrix (free cross-section) approximation. Temperatures are 0,10 and 50 MeV.

Eq. (10) and with K_p replaced by T. This is equivalent to using the "free" cross-section in the collision term. We find that it is in fact a quite acceptable approximation, especially at the higher temperatures that we are concerned with in H.I. collisions. This approximation has also been used in optical model calculations[13]. It is of course also used in practically all calculations of the Uehling-Uhlenbeck term. There is however a very important note to make here. The mean field enters through the delta function in Eq. (10). In all our calculations above, this is calculated with the selfconsistent mean field. To a first approximation one may consider using the effective mass approximation and the absorption is then proportional to m^*. The result of such an approximation is shown in fig 7. The effective mass is here calculated from [13] to get $m^* = 0.71$.

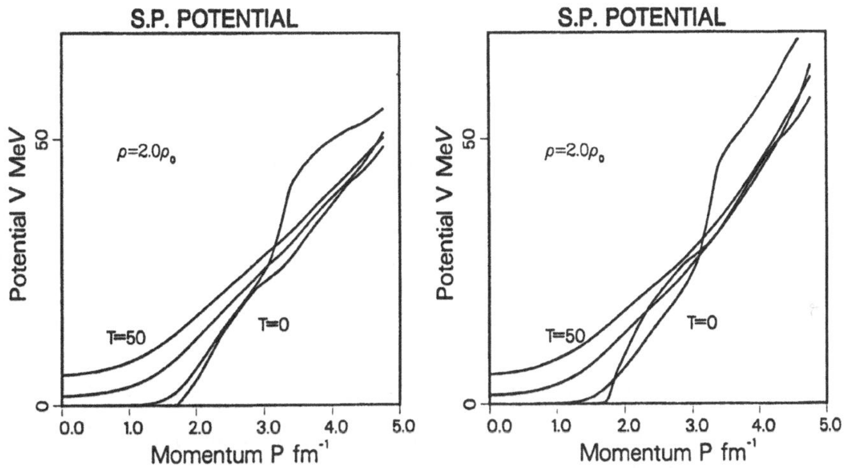

Figure 6. Same as fig. 5 but at double density and temperatures 0,10,30 and 50 meV.

4 Summary and Conclusions

The mean field in nuclear matter has been calculated as a function of momentum and temperature. At zero temperature the momentum dependence of the mean field is well approximated by deriving it from an effective interaction such as used in ref. [6]. It is suggested that the interaction should be allowed to increase in strength with temperature. Our results justify the use of free cross-sections in the collision term for hot nuclear matter. We like however to stress the more important issue of choosing the correct propagator (e in Eq. (10)) in the collision operator. The simplest choice is an effective mass m^*, as in fig. 7, but the problem is then to find the correct value of m^*. Another matter is that this propagator is complex[23].

The results presented here are for a thermally equilibrated system while the initial un-equilibrated state is really the more interesting because this is when im-

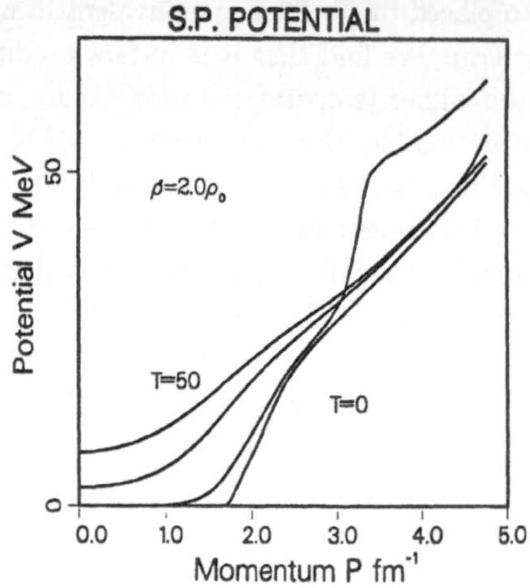

Figure 7. Same as fig. 6 but in the effective mass approximation.

portant dynamics takes place. We have however also made calculations by the same method for this case, namely for the system of two Fermi spheres. The general conclusions appear to be the same however. The main difference is really only in complexity; in the case of two spheres the mean field is not isotropic in momentum-space and the results are more complicated to display.

The results presented here are obtained by a somewhat unconventional method. We beleive however that the essential ingredients of the many-body effects are included. Density and temperature-dependence is included by the Pauli and selfconsistency effects. We do of course not imply that this method supersedes the more conventional potential-model approaches. It has an advantage because of its relative simplicity which allows us to estimate effects otherwise very hard to investigate.

I wish to express my thanks to LBL and its Nuclear Theory group with special thanks to Jørgen Randrup for their hospitality. This work has evolved from discussions with Janusz Dabrowski in Tucson and in Warsaw. I wish to thank him. This work was supported in part by the National Science Foundation, under grant PHY86-04602; and in part by the Director, Office of High Energy Research, Office of High Energy Nuclear Physics, Division of High Energy Physics, of the U.S. Department of Energy under Contract No. DE-AC03-76SF00098.

References

[1] H.S. Köhler and B.S. Nilsson, Nucl. Phys. **A477** (1988) 318; H.S. Köhler, Nucl. Phys. (1989) to be published.

[2] G.F. Bertsch, H. Kruse, and S. Das Gupta, Phys. Rev. C **29** (1984) 673.

[3] W. Bauer, Nucl. Phys. **A471** (1987) 604.

[4] H.S. Köhler and W. Bauer, LBL preprint submitted to Phys.Rev

[5] H. Stöcker and W. Greiner, Phys. Rep. **137** (1986) 279.

[6] G.M. Welke,M.Prakash,T.T.S. Kuo and S. Das Gupta, Phys. Rev. C **38** (1988) 2101.

[7] S. Das Gupta, private communication.

[8] H.S. Köhler, unpublished.

[9] J.R. Buchler and S.A. Coon, Astrophys. J. **212** (1977) 807.

[10] P. Grangé, J. Cugnon and A. Lejeune, Nucl. Phys. **A373** (1987) 365.

[11] R. Malfliet, Progress in Particle and Nuclear Physics **21** (1988) 207.

[12] F. Beck,K.-H. Mueller and H.S. Köhler, Phys. Rev. Lett. **40** (1978) 837; T. Izumoto,S. Krewald and A. Faessler,Nucl. Phys. **A357** (1981) 471;

[13] J. Dabrowski and H.S. Köhler,Nucl. Phys. **A489** (1989) 303.

[14] H.S. Köhler, Nucl. Phys. **A415** (1984) 37.

[15] B.S. DeWitt, Phys. Rev. **103** (1956) 1558

[16] R.A. Arndt,R.H. Hackman and L.D. Roper,Phys. Rev. **C15** (1977) 1002; R.A. Arndt,L.D. Roper,R.A. Bryan, R.B. Clark,B.J. Verwest and P. Signell,preprint VPISA-2(82),October 1982.

[17] I.E. Lagaris and V.R. Pandharipande, Nucl. Phys. **A359** (1981) 349.

[18] J. Dabrowski and W. Piechocki, Acta Phys. Polon. **B16** (1985) 1095.

[19] H.S. Köhler, Nucl. Phys. **128** (1969) 273.

[20] R.B. Wiringa, Phys. Rev. **C38** (1988) 2967.

[21] S.A. Moszkowki and B.L. Scott, Ann. Phys. **11** (1960) 65.

[22] H.S. Köhler and B.S. Nilsson, Nucl. Phys. **A417** (1985) 541.

[23] P. Danielewicz, Ann. of Phys. **152** (1984) 239 and 305.

NUCLEAR MULTIFRAGMENTATION AND THE EQUATION OF STATE

Jørgen Randrup

Nuclear Science Division, Lawrence Berkeley Laboratory
University of California, Berkeley, California 94720

INTRODUCTION

This lecture reviews the essential elements in the theoretical description of nuclear multifragment systems within the framework of statistical mechanics and summarizes recent progress on formulating a transition-state treatment of nuclear multifragmentation.

Nuclear multifragment systems are relevant for our understanding of the equation of state of matter at densities near the nuclear saturation value ($\approx 3 \cdot 10^{14} g/cm^3$). Such densities are thought to occur in certain astrophysical environments and the associated equation of state plays a key role in stellar collapse. At subsaturation densities, nuclear matter tends to cluster and it therefore appears economical to develop a description in terms of interacting nuclear fragments, possibly embedded in a nucleon vapor. We shall here review a description that has been developed in order to allow a statistical simulation of such environments and the extraction of associated thermodynamic quantities, such as the equation of state.[1] This description also provides the formal tools for addressing statistical nuclear dissassembly.

Nuclear collisions offer unique opportunities for probing the nuclear equation of state in the laboratory. A major difficulty inherent in such studies is the fact that any particular collision process explores an entire region of the nuclear phase diagram, thus obscuring the interpretation of the observable outcome. This feature ties the measurement of the equation of state to our understanding of the collision dynamics.

Central collisions of heavy nuclei at intermediate energies ($\approx$ 100 MeV per nucleon) typically results in the production of many complex fragments. Although the mechanisms leading to such multifragmentation events are not yet well understood, the possibility exists that a transient highgly excited system is formed, expands to subsaturation densities, and subsequently breaks up into receding nuclear fragments. Experimental studies of nuclear multifragmentation may then yield information about the thermodynamic properties of the environment in which the fragments were formed, and hence teach us about the nuclear equation of state.

The second part of this lecture is a summary of current efforts aimed at formulating a statistical theory for multifragment breakup of an idealized nuclear source.[2] The treatment adapts the transition-state approximation for the calculation of the rate at which the source breaks into specified prefragments, which still interact significantly and consequently experience non-trivial further evolution.

The Nuclear Equation of State, Part A
Edited by W. Greiner and H. Stöcker
Plenum Press, New York

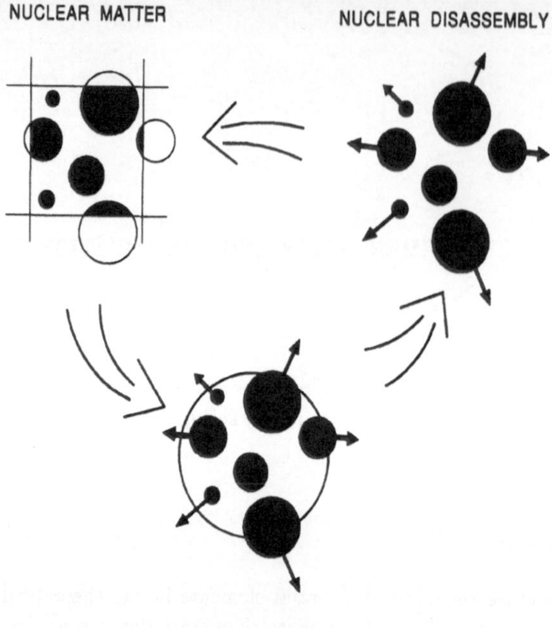

Fig. 1. Illustration of the relationship between the statistical properties of hot and dilute nuclear matter and fragment production in nuclear collisions. A major motivation for studying nuclear multifragmentation processes is to learn about the statistical properties of nuclear matter far from ordinary conditions. Conversely, knowledge of those properties can be used to predict fragment production in nuclear collisions. To do that one calculates the fragment distribution within a specified freeze-out volume Ω, assuming that these fragments cease to interact and propagate freely away from the source zone.

MULTIFRAGMENT STATES

Our presentation adheres to the formalism developed by Koonin et al. for the microcanonical simulation of nuclear multifragmentation.[1] The first part of this lecture covers results from that work.

It is convenient to define a fragmentation F as a set of elementary multifragment states that all have the same fragment masses, positions, momenta, and internal excitations. Thus a given fragmentation F contains the following specific information,

$$F \; : \; \{A_n, \; \epsilon_n, \; \mathbf{r}_n, \; \mathbf{p}_n, \; n = 1, ..., N_F\} \; . \tag{1}$$

Here the number of fragments is denoted by N_F, they are labeled by the index $n, n = 1, ..., N_F$, and they have the mass numbers A_n. Furthermore, their internal excitation energies are ϵ_n, and their positions and momenta are $\mathbf{r}_n$ and $\mathbf{p}_n$, respectively. Although we do not distinguish between neutrons and protons in the present discussion, it would be formally simple to incorporate the isospin degree of freedom by simply interpreting the nucleon number A_n as a two-dimensional quantity $\mathbf{A}_n$ whose components are the respective neutron and proton numbers, N_n and Z_n. It should also be noted that in the present exposition we do not incorporate the nucleon vapor in which the fragments may be immersed.

The fragmentations associated with a given mass partition can be mapped into a $3N$-dimensional hyperspace, in which the hyperposition is $(\mathbf{r}_1, ..., \mathbf{r}_N)$, and the metric is designed so as to compensate for the fact that the fragments have different masses. This provides a very convenient tool for formal developments on multifragmentation problems.[2]

An "interval" ΔF in the space of fragmentations is generated by the tolerances $\Delta\mathbf{r}_n$, $\Delta\mathbf{p}_n$, and $\Delta\epsilon_n$ for position, momentum, and excitation, respectively. The number of elementary multifragment states in the interval ΔF is then

$$\Delta\nu_F = \prod_{n=1}^{N_F} \left[\rho_n(\epsilon_n)\, \Delta\epsilon_n\, \frac{\Delta\mathbf{r}_n\, \Delta\mathbf{p}_n}{h^3}\right] . \tag{2}$$

Here h is Planck's constant and $\rho_n(\epsilon_n)$ is the density of states in the fragment n at the excitation energy ϵ_n. When the temperature in the system exceeds several MeV, the contribution to $\rho_n(\epsilon_n)$ from unbound levels become important. Moreover, when the specified mean density is not small in comparison with the saturation value, the fragments can not be considered as isolated and so the internal level density is modified. Indeed, a consistent treatment must incorporate also the nucleon vapor. These problems are both complicated and quantitatively important and in need of further consideration. Some recent developments were reported in [3].

The summation over all possible multifragment states can be expressed in terms of a sum over fragmentations, with each fragmentation weighted according to the measure (2):

$$\sum_F (\cdot) = \sum_{N_F=1}^{\infty} \frac{1}{N_F!} \prod_{n=1}^{N_F} \left[\sum_{A_n=A_{\min}}^{A_{\max}} \int \rho_n(\epsilon_n) d\epsilon_n \int \frac{d\mathbf{r}_n d\mathbf{p}_n}{h^3}\right] (\cdot) . \tag{3}$$

The factor $1/N_F!$ appears because multifragment states that differ only by a permutation of the fragment labels n are physically identical.

MICROCANONICAL DENSITY OF STATES

The fundamental microcanonical assumption is that all multifragment states consistent with overall conservation laws are equally probable. In the present discussion we shall, for simplicity, consider only conservation of the total nucleon number A, the total energy E, and the total momentum $\mathbf{P}$. Conservation of the center-of-mass position can also be incorporated, and will be in the transition-state treatment to be summarized later. The incorporation of angular-momentum conservation is also possible.

The properties of the system can then be expressed in terms of the density of states, $\rho(\Omega, A, E, \mathbf{P})$, which is obtained by performing the above summation using the appropriate weight of a given fragmentation,

$$\rho_{\text{microcan}}(\Omega, A, E, \mathbf{P}) = \sum_F \delta(A_F - A)\, \delta(E_F - E)\, \delta(\mathbf{P}_F - \mathbf{P}) . \tag{4}$$

Here $A_F = \sum_n A_n$ is the total number of nucleons in F, E_F is its total energy, and $\mathbf{P}_F = \sum_n \mathbf{p}_n$ is the total momentum. Since the dependence on $\mathbf{P}$ is physically uninteresting (see problem 1), it suffices to consider the system in its CM frame, where the total momentum vanishes.

In order to develop a well-defined description, it is necessary to impose constraints on the fragment positions. This is often done by requiring their centers to be confined within a specified volume Ω. (However, it may be preferable to specify the overall rms extension in stead, as in the theory of multifragmentation to be summarized later.) When one is simulating a translationally invariant system, as is of relevance for the general discussion of matter at subsaturation densities and may approximate conditions in certain astrophysical sites, it is convenient to let the volume be of cubic shape and impose periodic boundary conditions. It should be noted that such systems still exhibit finite-size effects.

In the present exposition, the total energy of a given fragmentation F is taken to be of the form

$$E_F = \sum_{n=1}^{N}[\frac{p_n^2}{2mA_n} + E_n^0 + \epsilon_n] + V . \tag{5}$$

Here $p_n^2/2mA_n$ is the kinetic energy of a fragments, E_n^0 is its ground-state energy, and ϵ_n is the internal excitation energy of the fragment. The last term represents the potential energy associated with the particular positioning of the fragments. For well-separated fragments it is given in terms of pairwise interactions, but for more compact configurations V is a complicated quantity and in need of further considerations.

Because E_F depends quadratically on the fragment momenta, the constrained integrals over $\mathbf{p}_n$ in (4) can be expressed analytically, (see, for example, [2]),

$$\prod_{n=1}^{N}\left[\int d\mathbf{p}_n\right] \delta\Big(\sum_{n=1}^{N}\frac{p_n^2}{2m_n} - K\Big)\, \delta\Big(\sum_{n=1}^{N}\mathbf{p}_n\Big) = \frac{2\pi m_0^{-\frac{3}{2}}}{\Gamma(\frac{3}{2}(N-1))}\prod_{n=1}^{N}[m_n]^{\frac{3}{2}}\,(2\pi K)^{\frac{3}{2}N-\frac{5}{2}} , \tag{6}$$

where $m_0 = \sum_n m_n$. Moments of the fragment momentum distribution can also be evaluated by elementary means (see problem 6).

The above analytical result may be employed to eliminate the explicit appearance of the momentum variables. The density of states (4) can then be written in the form

$$\rho(\Omega, A, E) = \sum_{N=1}^{\infty}\prod_{n=1}^{N}\left[\sum_{A_n=A_{\min}}^{A_{\max}}\int_0^{\infty} d\epsilon_n \int \frac{d\mathbf{r}_n}{\Omega}\right] W \equiv \sum_C W(C) , \tag{7}$$

where $C : \{A_n, \epsilon_n, \mathbf{r}_n\}$ denotes a multifragment configuration. Any such configuration has a statistical weight given by

$$W(C) = \frac{1}{N!}\frac{(2\pi m_0)^{-\frac{3}{2}}}{\Gamma(\frac{3}{2}N-\frac{3}{2})}\prod_{n=1}^{N}\left[\Omega\left(\frac{mA_n}{2\pi\hbar^2}\right)^{\frac{3}{2}}\rho_n(\epsilon_n)\right]\delta(\sum_{n=1}^{N}A_n - A)\,K^{\frac{3}{2}N-\frac{5}{2}} , \tag{8}$$

where $K = E - E_{1\ldots N}^0 - \epsilon - V$ is the energy available for the motion of the N fragments, with $E_{1\ldots N}^0 \equiv \sum_n E_n^0$ and $\epsilon \equiv \sum_n \epsilon_n$. Note that in (7) the integration over fragment positions has been changed into an average by dividing by the volume Ω; this change, which is made merely for dimensional convenience, introduces an explicit factor of Ω^N into the statistical weight (8). Also note that in (8) we have, for notational simplicity, suppressed the truncation function $\theta(-K)$, which ensures that the statistical weight vanish for configurations with negative values of the available kinetic energy K.

Observables can be evaluated as averages over a representative sample of configurations, $\{C\}$. For example, the mean value of the fragment multiplicity N is given by $\bar{N} \equiv \langle N \rangle = \sum_C W(C) N_C / \sum_C W(C)$. Suitable samples of multifragment configurations can be generated by adaptation of the Metropolis method.[4]

In order to obtain expressions for the thermodynamic quantities characterizing the microcanonical ensemble, we recall that the entropy associated with a microcanonical ensemble is equal to the logarithm of the number of elementary states in the ensemble, $S = \ln \nu$. Since the number of elementary states associated with a given configuration C is proportional to its statistical weight, $W(C)$, we have that $S = \ln \nu = \ln \sum_C W(C)$ plus a constant proportional to the tolerance in the specified total energy and momentum of the system. If the characterizing quantities, or "external parameters", Ω, A, E, are varied the statistical weights change and the induced change in the entropy is $dS = \sum_C dW(C) / \sum_C W(C) = \langle d \ln W \rangle$. The various thermodynamic quantities can then be obtained by invoking the basic relation $dE = \tau dS - P d\Omega + \mu dA$.

Using this basic thermodynamic relation, one can readily obtain an expression for the ensemble average of the inverse temperature,

$$\bar{\beta} = (dS/dE)_{\Omega,A} = \langle (d \ln W/dE)_{\Omega,A} \rangle = \langle (\frac{3}{2}N - \frac{5}{2})/K \rangle , \tag{9}$$

where the expression (8) for the statistical weight has been utilized. There is no contribution from the derivative of the θ-function, since the weight vanishes smoothly at the boundary. Thus, an inverse temperature can be defined for a given configuration C as $\beta_C = (\frac{3}{2}N_C - \frac{5}{2})/K_C$. The mean inverse temperature for the microcanonical ensemble, $\langle \beta_C \rangle$, is then equal to the above expression. In the microcanonical ensemble, where the energy is fixed, the temperature β_C is configuration-dependent and therefore exhibits fluctuations. The associated variance is given by $\sigma_\beta^2 = \langle \beta_C^2 \rangle - \langle \beta_C \rangle^2$, where $\langle \beta_C^2 \rangle = \langle ((\frac{3}{2}N - \frac{5}{2})/K)^2 \rangle$. This quantity contributes to the specific heat (see problem 7).

We turn now to the calculation of the pressure which can be obtained as $P = \tau(dS/d\Omega)_{E,A}$. Thus the pressure depends on the response of the system to changes in the confining volume Ω. This property is most conveniently determined in terms of an overall scaling of the fragment positions, $\mathbf{r}_n \to \lambda \mathbf{r}_n$. Then we have

$$\begin{aligned} P\Omega &= \frac{\tau}{3}\left(\frac{dS}{d\lambda}\right)_{E,A} = \frac{\tau}{3}\sum_C \frac{dW(C)}{d\lambda} / \sum_C W(C) = \frac{\tau}{3} \langle \frac{d \ln \sum_C W}{d\lambda} \rangle \\ &= \frac{\tau}{3} \langle 3N + (\frac{3}{2}N - \frac{5}{2})\frac{d \ln K}{d\lambda} \rangle = \tau \bar{N} - \frac{\tau}{3} \langle \beta_C \frac{dV}{d\lambda} \rangle \\ &= \bar{N}\tau - \frac{\tau}{3} \langle \beta_C \sum_{n<n'} r_{nn'} \frac{dV_{nn'}}{dr_{nn'}} \rangle \approx \bar{N}\tau + \langle \frac{1}{3} \sum_{n<n'} r_{nn'} F_{nn'} \rangle , \end{aligned} \tag{10}$$

where we have used that $\Omega \sim \lambda^3$. As was the case in the calculation of $\bar{\beta}$, there is no contribution from the kinematical boundary defined by the truncation function $\theta(-K)$, because of the weight $W(C)$ vanishes at the boundary. The first term is the usual ideal-gas contribution while the second is the correction due to the interfragment forces $F_{nn'} = -dV_{nn'}/dr_{nn'}$ and is a simple generalization of the standard form.

The formulation reviewed in the preceding makes it possible to study the statistical mechanics of nuclear multifragment systems. In particular, using periodic boundary conditions, the calculation of the pressure for given density and energy (or temperature) yields the equation of state.

It is possible to approximate the conditions prevailing in an idealized nuclear fireball by employing a spherical volume Ω. By specifying a value of Ω that corresponds to the effective "freeze-out" conditions, it is thus possible to mimic the disassembly process. In this approach, the statistical weights calculated for the freeze-out conditions are taken to represent the relative yield of final fragments, possibly corrected for subsequent decay processes.

Statistical models for nuclear multifragmentation have been developed along these lines by a number of groups. Most relevant to the present discussion are the works of Koonin et al.[1], Gross et al.[5], and Bondorf et al.[6], which all formulate phase-space simulation models for nuclear multifragment systems. There are two inherent problems with this type of approach. One is that the freeze-out volume Ω is not given within the model, but must be specified separately, or fitted to data. The other is that the final yield is identified with the statistical weights at freeze-out, rather than by some suitable rate. In the following we outline how a formally well-founded treatment of nuclear multifragmentation can be developed by a generalization of the transition-state approximation adapted by Bohr and Wheeler to ordinary binary fission.[7] A more complete account of this work is given in [2].

We consider an idealized "source", corresponding to a very excited compound nucleus, and wish to study its statistical breakup into multifragment channels. We shall concentrate on the breakup into a specified mass partition $A_1, ..., A_N$ and derive a transition-state expression for the corresponding partial width $\Gamma_{A_1 \cdots A_N}(E)$. The total width $\Gamma_A^N(E)$ for breakup into any N prefragments can then be obtained by performing a summation over the various contributing mass partitions,

$$\Gamma_A^N(E) = \frac{1}{N!} \prod_{n=1}^{N} \left[\sum_{A_n} \right] \delta(\sum_{n=1}^{N} A_n - A) \, \Gamma_{A_1 \cdots A_N}(E) \,, \tag{11}$$

and the total breakup width is $\Gamma_A^{\text{total}}(E) = \sum_n \Gamma_A^N(E)$. As already emphasized, the disassembling system is expected to experience significant further development subsequent to the breakup transition into interacting prefragments and this feature must be taken into account before observable quantities can be calculated.

Disassembly variables

As already noted, in order for the density of states (7) to yield a finite result, the fragment positions must be somehow confined. This is ordinarily accomplished by requiring the fragment positions to be within a specified volume Ω. While such a scenario is appropriate for studies of infinite matter, which can be approximated by periodic boundary conditions, the nature of the confining agency is less obvious for an isolated finite system, such as may be formed in a nuclear collision. In our present treatment, we shall replace the somewhat artificial volume Ω by a suitable generalized fission (or disassembly) coordinate whose function is to constrain the overall spatial extension of the multifragment system so that the position integrals remain convergent. The corresponding density of states is well-defined and can be considered as a function of the disassembly variable. In this manner the breakup problem can be reduced to a one-dimensional form and is then amenable to a transition-state treatment in analogy with ordinary binary fission.

For this purpose we define, for a given fragmentation F, the disassembly coordinate q and its conjugate momentum p as follows,

$$q_F^2 = \frac{1}{m_0}\sum_{n=1}^{N} m_n r_n^2 \,, \qquad p_F = \frac{1}{q_F}\sum_{n=1}^{N} \mathbf{p}_n \cdot \mathbf{r}_n \,. \tag{12}$$

These variables represent the radial position and momentum in the multifragment hyperspace mentioned earlier. It then follows that q and p are conjugate variables, as is elementary to verify by evaluating their Poisson bracket, $\{q, p\} = 1$. Furthermore, the associated inertial mass is given by $m_0 = \sum_n m_n$, since the kinetic energy in the disassembly degree of freedom is $k = \frac{1}{2}p\dot{q} = p^2/2m_0$. The quantity q is simply related to the rms radius of the mass distribution of the total system and provides a general and convenient measure of the overall linear dimension of the multifragment system. Its conjugate momentum p is a simple measure of the outwards directed momentum of the fragments (the "radial flow").

There are $3N$ degrees of freedom associated with the position of the N fragments. Three of these are associated with an overall translation. Furthermore, the variables q and p introduced above are associated with an overall rescaling of the fragment positions. The remaining $3N$-4 positional degrees of freedom can be regarded as generalized angular variables. Three of these degrees of freedom can be chosen as the usual Euler angles (ϕ, θ, ψ) describing the overall spatial orientation of the system, with the rest describing the relative locations of the fragments (the "shape"), for given values of the CM position $\mathbf{R}_F$, the size q, and the orientation (as defined by means of the overall inertial tensor, for example).

Transition current

In order to formulate a transition-state approximation to the disassembly problem, one may consider the outwards probability current, i.e. the number of elementary multifragment states that pass by a given value of q per unit time. This quantity is given by

$$\begin{aligned} \nu_{A_1 \cdots A_N}(E) &= h^3 \frac{(2\pi m_0)^{-2}}{\Gamma(\frac{3}{2}N-2)} \prod_{n=1}^{N} \left[(\frac{m_n}{2\pi\hbar^2})^{\frac{3}{2}} \int d\mathbf{r}_n \right] \\ &\times \int \frac{dp}{h}\frac{p}{m_0} \int d\epsilon\, \rho_{1 \cdots N}(\epsilon)\, \kappa^{\frac{3}{2}N-3}\, \delta(\mathbf{R}_F)\, \delta(q_F - q) \end{aligned} \tag{13}$$

The flux factor p/m_0 in the p-integration can be thought of as arising from an integration over values of q extending from 0 to p/m_0, the distance covered by q per unit time. After division by h, the p-integral then yields the number of elementary states that pass the specified value of q per unit time. Since $(p/m_0)dp = dk$, where k is the kinetic energy associated with the outwards flow, the integrations over k and ϵ may be interchanged, so the former one can be performed analytically.

In the further developments, integrals of the form $\mathcal{I}_n = \int d\epsilon \rho(\epsilon)(\epsilon_0 - \epsilon)^n$ enter. These can be evaluated by making a stationary-phase approximation, resulting in the formula $\mathcal{I}_n \approx \Gamma(n+1)\rho(\epsilon_0)\bar{\tau}^{n+1}$. The logarithmic expansion should be made around the most probable internal excitation energy $\bar{\epsilon} = \epsilon_0 - n\bar{\tau} = a\bar{\tau}^2$, and

$$\bar{\tau} = \tau_0 \left([1 + (\frac{n\tau_0}{2\epsilon_0})^2]^{\frac{1}{2}} - \frac{n\tau_0}{2\epsilon_0} \right) \,. \tag{14}$$

It is advantageous to express the result in terms of an average over the constrained fragment positions. Invoking the formula for the surface of a hypersphere, we then find

$$\nu_{A_1\cdots A_N}(E)$$

$$\doteq \frac{1}{\hbar}\frac{\sqrt{4\pi}}{\Gamma(\frac{3}{2}N-\frac{3}{2})}\,\frac{1}{\Gamma(\frac{3}{2}N-1)}\,\langle\, (\frac{m_0 q^2_{1\cdots N}}{2\hbar^2})^{\frac{3}{2}N-2}\int_0^{\epsilon_{1\cdots N}} d\epsilon\ \rho_{1\cdots N}(\epsilon)\ [\epsilon_{1\cdots N}-\epsilon]^{\frac{3}{2}N-2}\,\rangle'$$

$$\approx \frac{1}{\hbar}\,\frac{\sqrt{4\pi}}{\Gamma(\frac{3}{2}N-\frac{3}{2})}\,\langle\, (\frac{m_0 q^2_{1\cdots N}\bar{\tau}}{2\hbar^2})^{\frac{3}{2}N-2}\ \rho_{1\cdots N}(\epsilon_{1\cdots N})\ \bar{\tau}\,\rangle'\ . \qquad (15)$$

The second relation is based on the stationary-phase approximation described above. The mean temperature $\bar{\tau}$ of the particular configuration can in practice be replaced by the local maximum temperature $\tau_{1\cdots N}$. The prime indicates that the average is over multifragment posiitons $\{\mathbf{r}_n\}$ that have been constrained to have trheir center-of-mass position at the origin and the specified rms extension q.

Breakup width

For given values of the constrained positions $\{\mathbf{r}_n\}$, the integrand in the flux (13) has a minimum at some value $q_{1\cdots N}$. This key feature is easy to understand since the potential energy has a maximum as the system is stretched from a compact configuration towards

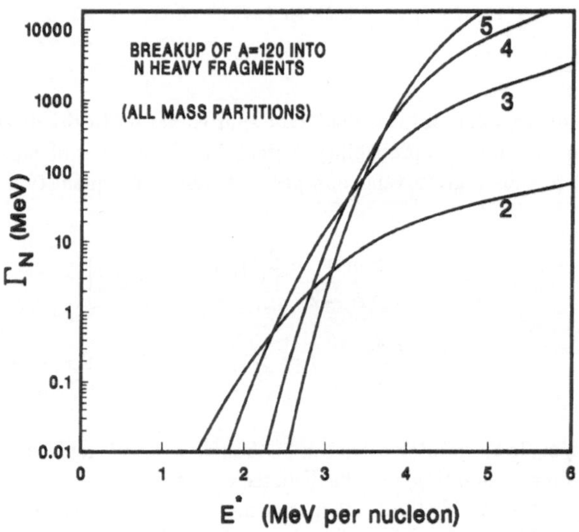

Fig. 2. This figure shows calculated results for the partial width Γ_N for breakup of a specified source into N prefragments with $A > 10$ (from [2]). It is seen that binary breakup dominates at low excitations and that the slope increases with the multiplicity. This is a consequence of the fact that the statistical weight depends on the energy as a power proportional to n. The ternary events become more preponderant than binary events at a certain threshold energy, and above this energy the various higher multiplicites dominate in rapid succession, thus producing in effect a rather sudden transition from predominantly binary decays to breakup into many fragments. It is expected that significant post-transition fusion will occur, so that the final states will have lower values of the multiplicity.

separated fragments. This barrier top is a generalization of the conditional saddle point for asymmetric binary fission. The minimum in the integrand will be shifted slightly inwards relative to the barrier top because the geometrical factor q^{3N-4} biases the statistical weights toward larger sizes. As in the treatment of binary fission, it is natural identify the value $q = q_{1\cdots N}$ with the local "bottle neck" in the evolution towards breakup. Accordingly, the total rate at which the system makes an irreversible transition towards disassembly is approximated by the above current (15), with the proviso that the local value of q be chosen as that for which the integrand has a minimum, i.e. the transition value $q_{1\cdots N}$.

Invoking the usual statistical assumption, the breakup rate of the system (into the specified mass partition) is given by the magnitude of the transition current, $\nu_{A_1\cdots A_N}(E)$, divided by the total compound level density, $\rho(A, E)$, which represents the total number of elementary states in the decaying compound system. (Both of these refer to states with a total energy within an infinitesimal interval dE around the specified value E.) We then obtain the following relation for the partial width for breakup into specified prefragments,

$$\Gamma_{A_1\cdots A_N}(E) = h\,\frac{\nu_{A_1\cdots A_N}(E)}{\rho(A,E)} \tag{16}$$

$$\approx \frac{1}{\rho(A,E)}\,\frac{\sqrt{4\pi}}{\Gamma(\frac{3}{2}N-\frac{3}{2})}\,\langle\,(\frac{m_0 q_{1\cdots N}^2 \bar{\tau}}{2\hbar^2})^{\frac{3}{2}N-2}\,\rho_{1\cdots N}(\epsilon_{1\cdots N})\,\bar{\tau}\,\rangle' .$$

We remind of the fact that the average should be taken over the <u>reduced</u> fragment positions describing configurations constrained to have a fixed center-of-mass position and a specified (but arbitrary) overall rms extension.

The formula (16) has an intuitive interpretation. It expresses the partial disassembly width Γ as the outwards transition current relative to the total number of states, as in the ordinary transition-state method. The transition current is obtained by adding the contributions from all possible reduced positions of the fragments, corresponding to an integration over the generalized orientation in hyperspace. For each such generalized orientation, the local transition flux is a product of a macroscopic and a microscopic factor. The macroscopic factor $N_{\rm macro} \sim \sqrt{4\pi}(m_0 q^2 \tau/2\hbar^2)^{\frac{3}{2}N-2}/\Gamma(\frac{3}{2}(N-1))$ is the effective number of states associated with the macroscopic degrees of freedom, i.e. those associated with the overall motion of the individual prefragments, while the second factor $\nu_{\rm micro} \sim \rho\tau$ is the outwards probability current for each such macroscopic state.

At low excitation, channels with only two fragments dominate and the formula (16) for the decay width reduces to a form rather similar to the standard Bohr-Wheeler expression, but with an extra factor arising from the orbital motion of the binary complex. The dominant multiplicity increases with excitation and at high excitation the treatment acquires considerable formal similarity with standard statistical multifragmentation models, although certain notable differences are present. An important advantage of the treatment is that it automatically provides the constraint on the fragment positions so that a finite result is obtained; in this regard it is a significant advance relative to current statistical models in which the freeze-out volume must be prescribed separately. It should also be added that conservation of angular momentum can readily be incorporated into the formulation. Its main effect is to reduce the available energy by the amount tied up in overall rotation.

PROBLEMS

1: Show that the dependence of the density of states on the total momentum $\mathbf{P}$ is given by $\rho(\Omega, A, E, \mathbf{P}) = \rho(\Omega, A, E - P^2/2m_0, \mathbf{0})$, with $m_0 = \sum_n m_n$.

2: Show by elementary means that

$$\prod_{n=1}^{N}\left[\int d\mathbf{p}_n\right]\delta\Big(\sum_{n=1}^{N}\frac{p_n^2}{2m_n}-K\Big)=\frac{2\pi}{\Gamma(\frac{3}{2}N)}\prod_{n=1}^{N}[m_n]^{\frac{3}{2}}\ (2\pi K)^{\frac{3}{2}N-1}\ .$$

Hint: First use dimensional arguments to determine the dependence on K, then multiply by $\exp(-K/\tau)$ and integrate over K.

3: Show that the microcanoncial statistical weight is given by the following expression when momentum conservation is ignored,

$$W(C)=\frac{1}{N!}\,\frac{1}{\Gamma(\frac{3}{2}N)}\prod_{n=1}^{N}\left[\Omega\left(\frac{mA_n}{2\pi\hbar^2}\right)^{\frac{3}{2}}\rho_n(\epsilon_n)\right]\delta\Big(\sum_{n=1}^{N}A_n-A\Big)\ K^{\frac{3}{2}N-1}\ .$$

Hint: Use the formula quoted in problem 2.

4a: Show that the canonical partition function $Z_{\rm can}=\sum_F\exp(-E_F/\tau)$ can be written in the form (7) and that the weights are given by

$$W_{\rm can}(C)=\frac{1}{N!}\prod_{n=1}^{N}\left[\Omega\left(\frac{mA_n\tau}{2\pi\hbar^2}\right)^{\frac{3}{2}}\rho_n(\epsilon_n)\ {\rm e}^{-(E_n^0-\epsilon_n)/\tau}\right]\delta\left(\sum_{n=1}^{N}A_n-A\right){\rm e}^{-V/\tau}\ .$$

4b: Show that the form (7) also encompasses the grand-canonical partition function $Z_{\rm grand}=\sum_F\exp(-(E_F-A_F\mu)/\tau)$, with the weights given by

$$W_{\rm grand}(C)=\frac{1}{N!}\prod_{n=1}^{N}\left[\Omega\left(\frac{mA_n\tau}{2\pi\hbar^2}\right)^{\frac{3}{2}}\rho_n(\epsilon_n)\ {\rm e}^{-(E_n^0-\epsilon_n+\mu A_n)/\tau}\right]{\rm e}^{-V/\tau}\ .$$

In these expressions, V is the total interfragment potential energy of the system, as given by the last term in (5). When these forces are neglected, the grand canonical statistical weight factorizes and the treatment simplifies significantly.[9]

5: Show how the canonical description emerges from the microcanonical one for configurations that are not too close to the kinematic boundary.
Hint: Consider two neighboring configurations with large multiplicities and employ Stirling's formula to evaluate the power of K in the statistical weights.

6: The mean kinetic energy of any one of the fragments is $<p_n^2/2m_n>=<K/N>$. Show that the corresponding covariance is $\sigma_{K_1K_2}=-<(K/N)^2/(\frac{3}{2}N+1)>$ when momentum conservation is ignored.

7a: Ignoring momentum conservation, show that the *heat capacity* is given by

$$C_V\equiv\left(\frac{dE}{d\tau}\right)_{\Omega,A}=\left[1-<\frac{(\frac{3}{2}N-1)(\frac{3}{2}N-2)}{K^2}\bar{\beta}^{-2}>\right]^{-1}\ .$$

Hint: Use that $(dE/d\tau)_{\Omega,A}=-\bar{\beta}^2/(d\beta/dE)_{\Omega,A}$ together with the result (9).

7b: Show that when the temperature fluctuations can be neglected (so that $\beta_C=(\frac{3}{2}N-1)/K$ can be replaced by $\bar{\beta}$) then $C_V\approx<1/(\frac{3}{2}N-1)>^{-1}\approx<\frac{3}{2}N-1>$, as is expected for an ideal fragment gas.

8: Verify the following detailed derivation of the expression for the pressure.
For a given volume Ω, the statistical weight of a given configuration C is given by the expression (8) and the sum over all the configurations is given by (7). In order to calculate the pressure, we need to determine how the logarithm of the sum of weights changes when the volume is changed. Consider therefore a slightly different volume $\Omega' = \lambda^3\Omega$. It is possible to bring the configurations of the new volume into a one-to-one relationship with those associated with the old volume by a simple scale transformation, i.e. each configuration C in Ω is mapped onto the configuration C' in Ω' by the coordinate scaling $\mathbf{r}_n \to \mathbf{r}'_n = \lambda\mathbf{r}_n$. The sum over the configurations in Ω' can then be organized in a form identical to the sum over the original configurations in Ω by using that $d\mathbf{r}'_n = \lambda^3 d\mathbf{r}_n$. This introduces a factor λ^{3N} in the sum over configurations. In addition, the statistical weight $W(C')$ of a given scaled configuration C' is modified relative to $W(C)$ by the fact that the corresponding kinetic energy K' is changed as a result of the induced changes in the interfragment potential energies $V'_{nn'} = V(\lambda r_{nn'})$. We therefore have

$$\sum_{C'} W(C') - \sum_{C} W(C) = \sum_N \frac{1}{N!}\frac{1}{\Gamma(\frac{3}{2}N)} \prod_{n=1}^{N} \left[\sum_{A_n} \int_0^\infty d\epsilon_n \rho_n(\epsilon_n) \int \frac{d\mathbf{r}_n}{\Omega}\Omega \left(\frac{mA_n}{2\pi\hbar^2}\right)^{\frac{3}{2}}\right]$$
$$\times\, \delta(\sum_{n=1}^{N} A_n - A)\, [\lambda^3 (K')^{\frac{3}{2}N-\frac{5}{2}} - K^{\frac{3}{2}N-\frac{5}{2}}]\,,$$

where the difference in the curly brackets can be approximated by a differential,

$$\lambda^3 (K')^{\frac{3}{2}N-1} - K^{\frac{3}{2}N-1} \approx \left[\frac{d}{d\lambda}\lambda^{3N} K^{\frac{3}{2}N-1}\right]_{\lambda=1} \Delta\lambda$$
$$= [\frac{1}{3N} + (\frac{3}{2}N - \frac{5}{2})\frac{d\ln K}{d\lambda}]\,\Delta\lambda = \frac{d\ln W(C)}{d\lambda}\,\Delta\lambda\,.$$

The second-to-last relation then leads directly to the quoted result (10), even though the last relation holds only when the volume Ω appears explicitly in the statistical weight (8), as is our chosen convention.

9: Devise a method for picking random fragment positions subject to conservation of center of mass and rms extension, as is required for the calculation of the average occurring in the expression for the transition current (15).

CONCLUDING REMARKS

The transition-state treatment of multifragmentation provides a well-founded means for calculating the partial widths for transition of the given source into a number of interacting prefragments with specified masses. The post-transition evolution of the prefragments is expected to have a significant effect and is presently being studied.[8]

A special advantage of the developed treatment is that it bridges the gap between the realm of ordinary fission characteristic of low-energy nuclear reactions and that of high-energy collisions where the systems disassemble into many simple and complex fragments. The model is therefore well-suited for exploring the expected transition to multifragmentation, a central theme in current experiments on nuclear collisons at intermediate energies.

As has been the case with transition-state models of binary fission, the quantitative utility of the formulation depends on the refinement of the calculation of the various physical ingredients. Most important are the potential-energy landscape V and the internal level density ρ. The reliable calculation of these quantities presents a considerable challenge.

Finally, it should be emphasized that the role of statistical models in dynamical processes, such as nuclear reactions, is to provide well-defined reference calculations against which both dynamical theories and experimental data can be compared and discussed. Indeed, the most fascinating aspect of heavy-ion collisions as a physics tool is the opportunities provided for studying dynamical phenomena in unusual environments.

This work was supported by the Director, Office of High Energy and Nuclear Physics of the Department of High Energy and Nuclear Physics of the Department of Energy under contract DE-AC03-76SF00098.

References

[1] S.E. Koonin and J. Randrup, Nucl. Phys. **A474** (1987) 173;
J. Randrup and S.E. Koonin, Nucl. Phys. **A471** (1987) 355c

[2] J.A. López and J. Randrup, LBL-26164 (1989), submitted to Nucl. Phys. A

[3] G. Fai and J. Randrup, Nucl. Phys. **A487** (1988) 397

[4] N. Metropolis, A.W. Rosenblut, M.N. Rosenblut, A.H. Teller, and E. Teller, J. Chem. Phys. **21** (1953) 1087

[5] D.H.E. Gross, Zhang Xiao-ze, and Xu Shu-yan, Phys. Rev. Lett. **56** (1986) 1544; A.Y. Abul-Magd, D.H.E. Gross, Xu Shu-yan, and Zheng Yu-ming, Z. Phys. **A325** (1986) 373

[6] J.P. Bondorf, R. Donangelo, I.N. Mishustin, C.J. Pethick, and K. Sneppen, Phys. Lett. **150B** (1985) 57; J.P. Bondorf, R. Donangelo, I.N. Mishustin, C.J. Pethick, H. Schulz, and K. Sneppen, Nucl. Phys. **A443** (1985) 321; J.P. Bondorf, R. Donangelo, I.N. Mishustin, and H. Schulz, Nucl. Phys. **A444** (1985) 460; J.P. Bondorf, R. Donangelo, H. Schulz, and K. Sneppen, Phys. Lett. **162B** (1985) 30

[7] N. Bohr and J.A. Wheeler, Phys. Rev. 56 (1939) 426

[8] J.A. López and J. Randrup, LBL-26810 (1989), work in progress

[9] S.E. Koonin and J. Randrup, Nucl. Phys. **A356** (1981) 223

MEDIUM ENERGY COLLISIONS AND MULTIFRAGMENTATION

A. Adorno, M. Colonna, M. Di Toro, and G. Russo

Dipartimento di Fisica and I.N.F.N., Catania
57, Corso Italia
95129 Catania, Italy

Abstract

Complex fragment production in medium energy heavy ion collisions is analysed on the basis of sequential binary decays from primary equilibrated sources. Transitional reaction mechanisms for dissipative reactions are developped in order to relate entrance channel properties to the de-excitation stage and therefore to make possible the calculation of absolute values for yields, angular distributions and velocity spectra. Due to the high energy and angular momentum dissipation, a strong competition between asymmetric fission and evaporation decay is revealed. Exclusive data for the reactions $^{84}Kr+^{27}Al$ at 34.4 MeV/A and $^{86}Kr+^{197}Au$ at 43 MeV/A are reproduced without any need for direct multifragmentation.

1. INTRODUCTION

In medium energy heavy ion collisions the enormous growth of final channels, and in particular the large production of complex fragments have induced several groups in experiment as well as in theory to look for a clear evidence of a new reaction mechanism, a direct multifragmentation of the nuclear system, expected on the basis of various physical pictures, from dynamical instabilities to liquid - gas phase transitions[1,2].

However it is important to see how far one can go in explaining the data following already existing reaction mechanisms, from deep - inelastic to partecipant - spectator processes, suitably modified in this transitional region, and considering successive binary decays of highly excited primary fragments. This could be also very useful in order to focus more clearly where to look carefully for a signature of a sudden multifragment explosion. The main point is that if we form primary nuclear systems at high excitation energy and high spin, a sequential decay through some asymmetric fission with emission of massive fragments becomes to be in competition with the well known light particle evaporation[3]. However in order to perform a quantitative comparison of the results obtained through such "more conventional" mechanisms, it is important to relate the production yield and the structure of primary fragments to entrance channel properties (ion masses, beam energy, impact parameters). In this contribution we use a simple phase space model, based on the competition between one-body and collisional dynamics [4], to link the de-excitation stage to the initial con-

The Nuclear Equation of State, Part A
Edited by W. Greiner and H. Stöcker
Plenum Press, New York

ditions, which will eventually allow to get absolute values of various differential cross sections for emitted intermediate mass fragments (IMF).

In section 2 we present the model to describe the heavy ion reaction mechanism in the transitional energy region, from 10 to 200 MeV/A beam energy. In section 3 we extend the nuclear evaporation theory to allow a statistical emission of IMF's. Particular attention is paid to the analysis of fission barriers, with temperature and angular momentum, and to the kinematics of sequential emissions. Finally some results are shown in section 4, compared with recent exclusive measurements. Conclusions are drawn in section 5.

2. A TRANSITION REACTION MECHANISM FOR DISSIPATIVE COLLISIONS AT MEDIUM ENERGIES

As already stressed in the introduction, it is extremely important to relate entrance channel properties to the de-excitation stage, i.e. to the characteristics of the primary products of the reaction. In a dynamics ruled by the competition between mean field an NN collisions, microscopic treatments require large numerical efforts with some fundamental problems of difficult solution: consistency between mean field and collision term, medium effects, accurate description of phase space distributions[5,8]. In recent years a macroscopic phase space model has been introduced, based on safe physical grounds, able to reproduce a large volume of data and to make reliable predictions, with no free parameters[4,6]. A two stage mechanism is considered. In the first stage we have a one-body dissipation process (incoherent nucleon exchange through a neck) quite well described within the window formula, suitably corrected due to the partial overlap of momentum distributions. We can evaluate energy and angular momentum losses, orbiting and prompt particle emissions. Coherent one-body dissipation is also included in the approaching phase through the excitation of high frequency collective motions, giant resonances, expected to play and important role just on the basis of time matching conditions[7]. If enough energy is still available we can enter a second stage, an abrasion mechanism with formation of a highly excited intermediate source, the participant zone, whose properties (mass, energetics) are substantially affected by a still active Pauli blocking and one-body corrections. Now three primary bodies projectile - like (PLF), target - like (TLF) and "fireball" are formed in the exit channel with well defined properties: excitation and kinetic energies, spins, masses, deflection angles. All possible correlations can be worked out, and sequential statistical emission of particles can be evaluated.

In this way we can smootly describe the transition from fusion (incomplete) to deep-inelastic and to participant-spectator mechanisms with increasing beam energy (at a fixed impact parameter) or with increasing impact parameter (at a fixed beam energy). Fig.1 shows a typical example of these predicted transitions for $^{86}Kr+^{197}Au$ collisions. In Fig.2 a comparison is made with fully microscopic Landau-Vlasov calculations[8] with an improved collision simulation prescription[9,10], for the reaction $^{84}Kr+$ $+^{27}Al$ at 34.4 MeV/A. The density profile at t=100 fm/c for a collision initiated at impact parameter b=1fm clearly shows a dinuclear rotating system typical of deep inelastic collisions, in agreement with the predictions of our phase space model.

Depending on the initial conditions we have several ways to produce primary nuclear systems at high excitation energy and high spin. The reaction model, based on the concept of impact parameter, gives the absolute yields of the various primary fragments and univocally fixes the kinematics.

Assuming an equilibrium reached before prompt nucleon emission we can try to analyse the predictions of a statistical de-excitation stage.

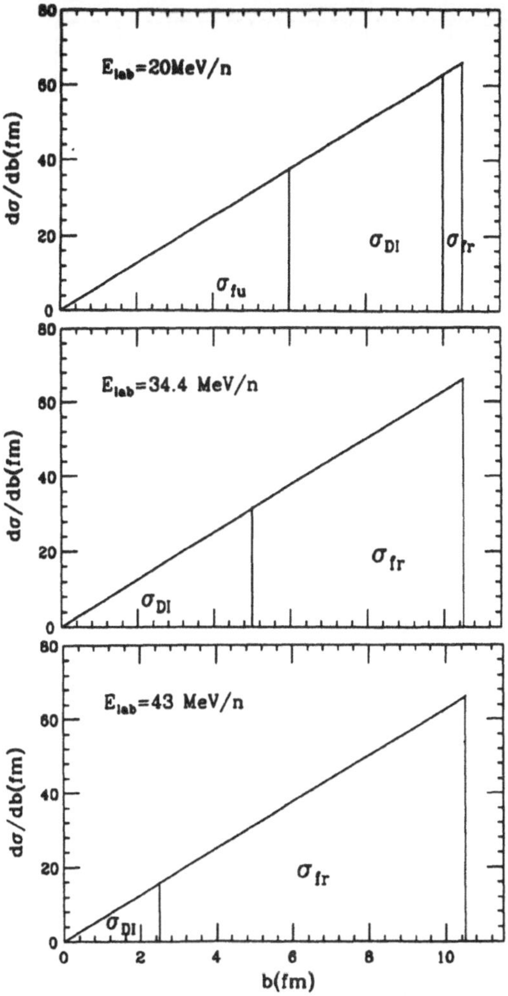

$^{86}Kr+^{197}Au$

Fig.1

3. IMF EMISSION FROM EQUILIBRATED SOURCES

From the experimental standpoint the production of particles with mass intermediate between fission fragments and alpha particles presents a cross section rapidly increasing with excitation energy and angular momentum. The apparent distinction between particle evaporation and fission, stressed in the formalism commonly used in the calculation of the decay widths, is really rather artificial. A unified statistical decay process allows to describe the generally asymmetric binary decay of a excited system.

The following expression provides the differential decay width[11]

$$\Gamma \, d\alpha d\varepsilon = \frac{1}{2\pi\rho(E)h} \left(2\pi m_{\alpha} T^{\frac{1}{2}}_{\alpha} \right) \rho(E-B_{f}-\varepsilon) d\varepsilon d\alpha \qquad (1)$$

where α is the asimmetry parameter, ε is the kinetic energy of the fission-like mode, B_f is the fission barrier depending on the angular momentum, E is the excitation energy, T_α is the saddle point temperature.

The fission barriers are calculated in the two tangent spheroids approximation:

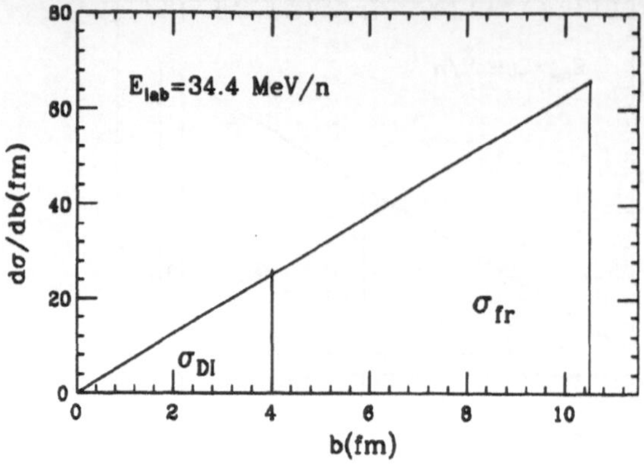

$^{84}Kr+^{27}Al$

b = 1.00 FERMI

t = 100.00 FERMI/C

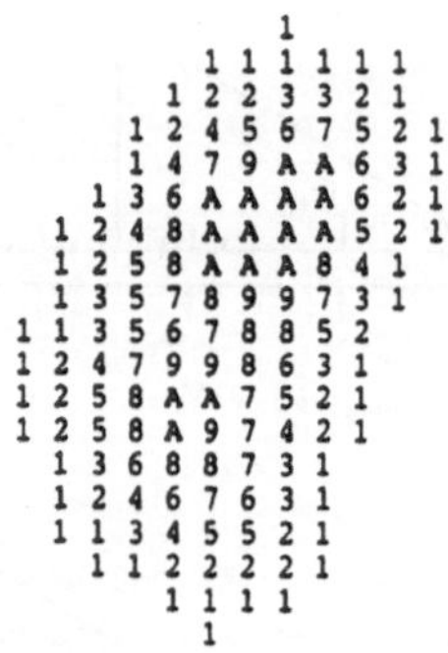

Fig.2

$$B_f = 17.8\,(A_1^{\frac{2}{3}}+A_2^{\frac{2}{3}})(1+\frac{2}{5}\eta^2) - 17.8A^{\frac{2}{3}}+0.71(\frac{Z_1^2}{A_1^{\frac{1}{3}}}+\frac{Z_2^2}{A_2^{\frac{1}{3}}})(1-\frac{\eta^2}{5}) + 1.44\,\frac{Z_1Z_2}{r_{12}} - 0.71\,\frac{Z^2}{A^{\frac{1}{3}}} + \frac{\hbar^2J(J+1)}{2I_{\perp}} - \frac{\hbar^2J(J+1)}{2I_o} \qquad (2)$$

where η is the deformation parameter, A_1A_2, Z_1,Z_2 are respectively the mass and atomic numbers of the two fission fragments, A and Z are the mass and atomic numbers of the primary fragment and I_o is its inertia moment at the ground state. This represents a good description of fission barriers for fissility parameter $x\leq 0.4$, i.e. for medium mass number ($A\leq 120$). E.g. we get a Businaro Gallone point at $x\sim 0.47$ to compare with the observed $x_{BG}\sim 0.39$. We should keep in mind this when we will analyze the emission for heavy primary fragments.

The differential probability to observe a fission event characterized by a particular α-value at a angle θ and a given ε is[12]:

$$\frac{d^2P}{d\Omega d\varepsilon} = \frac{\frac{1}{\sigma_f^2}T_\alpha^{\frac{1}{2}}\exp(\frac{E-B_f-\varepsilon}{T_\alpha})}{\Sigma_\alpha\frac{\hbar^2}{I_{\perp\alpha}}T_\alpha^{\frac{1}{2}}(\exp\frac{E-B_{f\alpha}}{T_\alpha}-1)}W(\theta) \qquad (3)$$

with $\sigma_f^2 = \frac{I_\perp T_\alpha}{\hbar^2}$

$$W(\theta) = \frac{\frac{1}{4\pi}\frac{2}{\sqrt{\pi}}x \exp(-x^2\frac{\sin^2\theta}{2})J_o(ix^2\frac{\sin^2\theta}{2})}{\mathrm{erf}(x)} \tag{4}$$

$$x^2 = \frac{J(J+1)}{2K_o^2} \qquad K_o^2 = \frac{I_{eff}T}{\hbar^2} \qquad I_{eff} = \frac{I_\parallel I_\perp}{I_\perp - I_\parallel}$$

J_o is the modified Bessel function of zero order $I_\parallel$ and $I_\perp$ are the principal inertia moments at the saddle point. The expression (4) has two interesting limits: as x^2 tends to infinity one obtains

$$\lim_{x^2\to\infty} W(\theta) = \frac{1}{2\pi^2\sin\theta}$$

on the other hand, as x^2 tends to zero one obtains

$$\lim_{x^2\to o} W(\theta) = \frac{1}{4\pi}$$

In the nucleon emission case ($I_\perp \simeq I_\parallel$) x^2 tends to zero recovering the isotropy of the angular distribution.

The total kinetic energy of the two fission fragments can be expressed as:

$$E_k = E_o + \frac{\mu r^2}{I_\perp} E_R + \varepsilon \qquad \text{with} \quad 0 \leq \varepsilon \leq E - B_f$$

where E_o is the coulomb interaction energy at the saddle point, μ the reduced mass of the system, r the distance between the centroids of the two fragments and E_R the total rotational energy.

In the emitter reference frame the velocity of a given fission fragment A_1 results

$$v_1 = \sqrt{\frac{2}{M_1}\frac{(A-A_1)}{A}E_k}$$

Therefore it exists a finite range of allowed values for v_1 which also limits velocity range in the laboratory system, partioularly for heavy emitted fragments. From simple velocity combinations this implies a clear decrease of heavy emitted fragments with increasing laboratory angles.

Finally we remark that for the systems analyzed here, at medium energy, we always get relatively low temperatures for the emitting sources, below 6 MeV. This justifies the neglect of particle emission, with relative excitation energy losses, before the equilibrium is reached.

4. RESULTS

We have analyzed IMF production data in two experiments on reactions induced by Kr beam at different energies.

In the $^{84}Kr + ^{27}Al$ reaction at 34.4 MeV/A, IMF spectra were taken at various laboratory angles for different mass bins[13]. Clear kinematics effects showing binary emission from an equilibrated source were inducing the interpretation as incomplete fusion-fission events. At variance, in our reaction mechanism study, we do not have any evidence of fusion reactions. Deep inelastic processes are dominating the more dissipative reactions, up to impact parameters b≃4 fm (see fig.2), while for

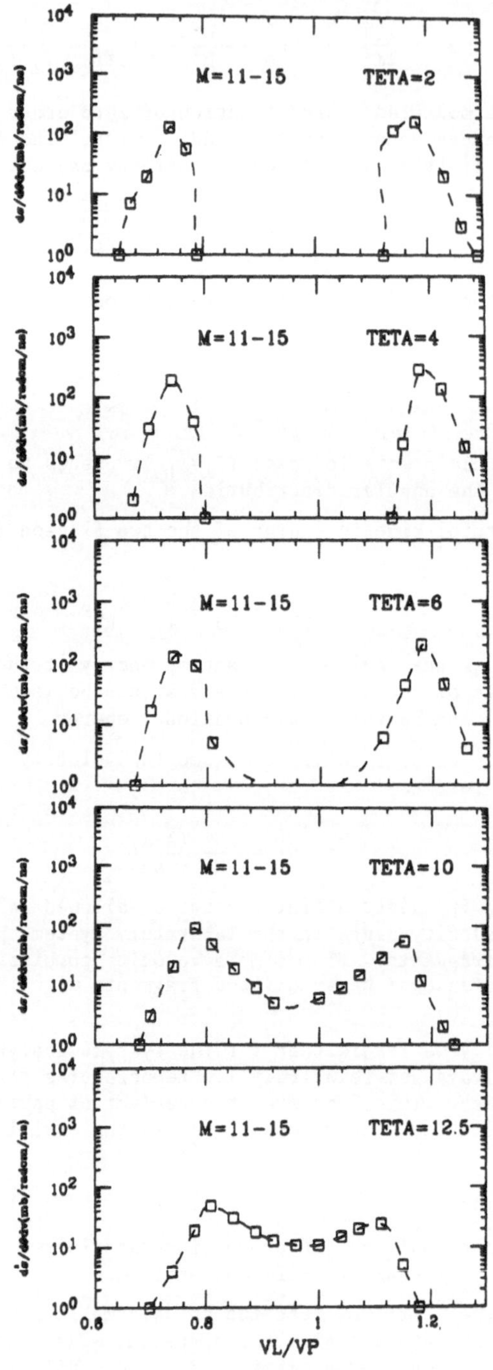
M=11-15
TETA=2
M=11-15
TETA=4
M=11-15
TETA=6
M=11-15
TETA=10
M=11-15
TETA=12.5
VL/VP

Fig.3

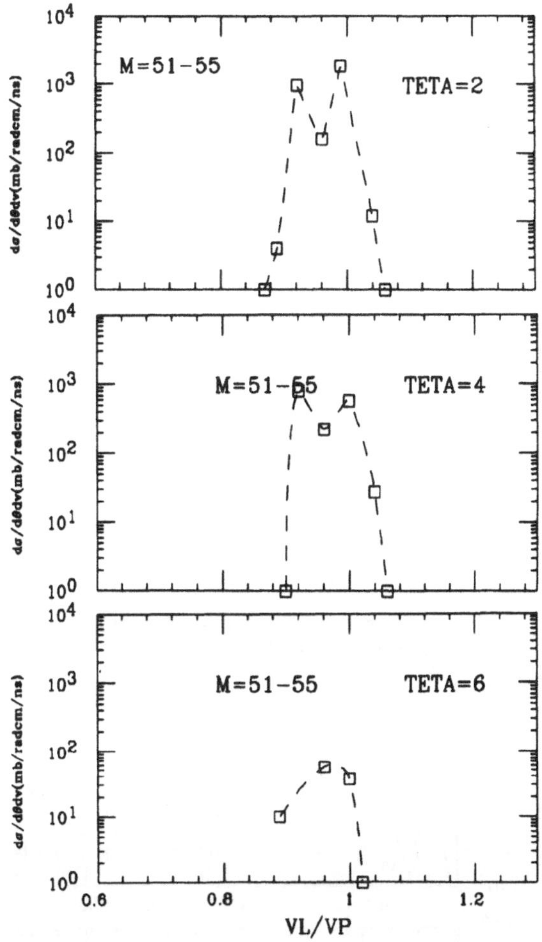

dσ/dθdv(mb/radcm/ns)
M=51-55
TETA=2
TETA=4
TETA=6
10^0
10^1
10^2
10^3
10^4
0.6
0.8
1
1.2
VL/VP

Fig.4

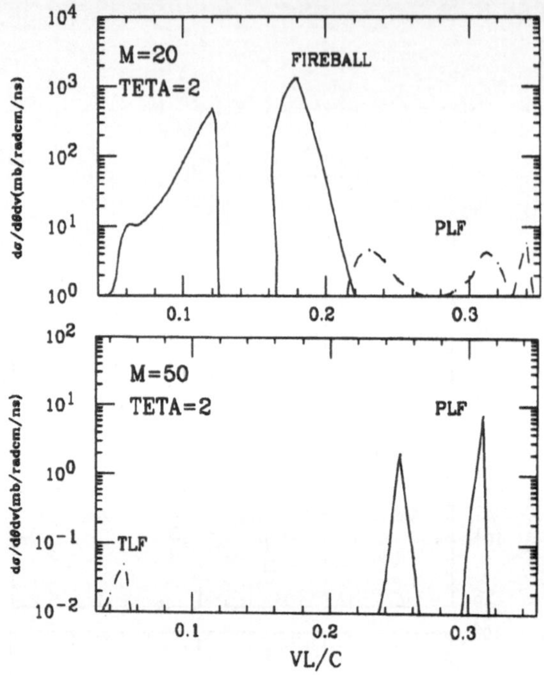
M=20
TETA=2
FIREBALL
PLF
dσ/dθdv(mb/radcm/ns)
M=50
TETA=2
PLF
TLF
VL/C

Fig.5

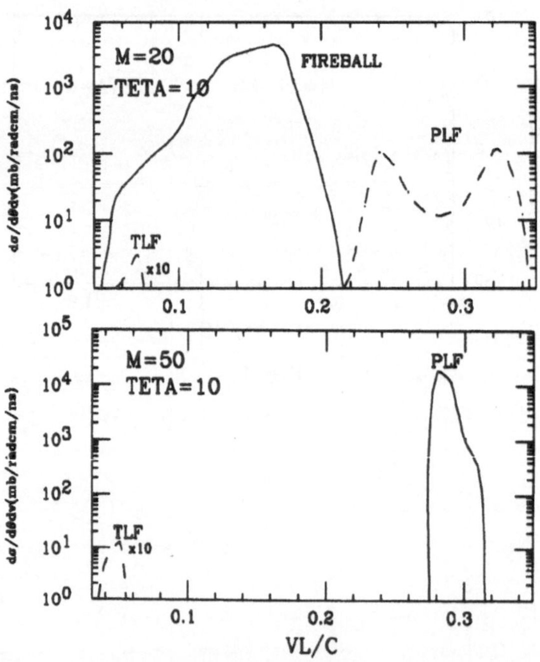
M=20
TETA=10
FIREBALL
PLF
TLF
x10
dσ/dθdv(mb/radcm/ns)
M=50
TETA=10
PLF
TLF
x10
VL/C

Fig.6

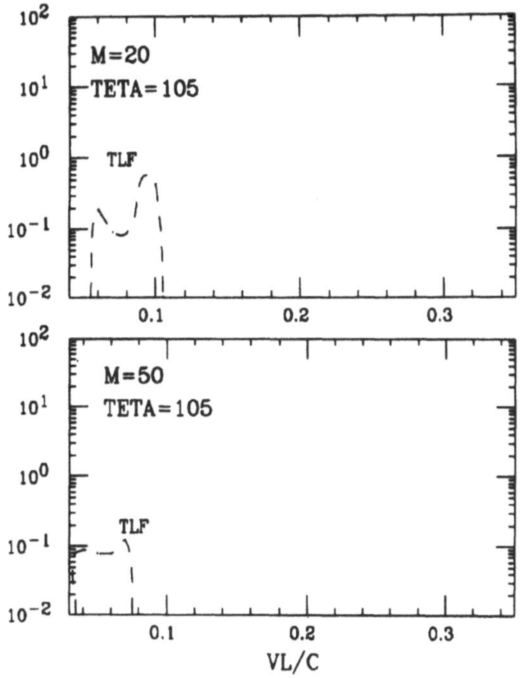

Fig.7

more peripheral collisions we have some evidence of a formation of a small participant zone. Landau-Vlasov calculations are confirming our model results. The IMF spectra, for different mass bins, shown in fig.s 3,4, which are in very good agreement with the data of ref.13, are obtained from projectile-like fragment de-excitations in the deep inelastic part of the reaction cross-section, where a large amount of energy and angular momentum is dissipated. We show absolute values for the cross sections with no normalization parameters.

For the reaction $^{86}Kr+^{197}Au$ at 43.0 MeV/A IMF's were detected in an almost 4π geometry[14]. Angle integrated velocity spectra show three distinct regions, around the projectile velocity, around the recoiling target one and, for lighter mass fragments, around some intermediate source velocity. For this projectile-target system, as already discussed, our model predicts a strong dominance of a fragmentation mechanism at this energy, with formation of a relatively large participant zone. From the subsequent de-excitation of primary fragment we get the spectra at $\theta_L=2°$, 10° and 105°, for IMF masses 20 and 50, shown in fig.s 5,6 and 7. At forward angles we have mainly PLF contributions, and for IMF mass=20, a very important fireball emission also down to relatively low velocities. At backward angles the TLF emission dominates: we get some reduced production yield for M=50 from our fission barriers which are not correct for fissility parameters $x \geq 0.4$. The strong contribution from "fireball fission" to forward emitted IMF's has been recently remarked in the same experiment from the Strasbourg group[5] through an analysis of velocity and angle correlations between two intermediate mass fragment emitted between 3° and 25°. These events, as expected, are related to the most violent collisions.

5. CONCLUSIONS

We have shown that it is possible to account for several features of complex fragment production at intermediate energies (absolute values of yields, angular distributions and velocity spectra) without considering direct multifragmentation processes. We have developped a model which describes transitions from incomplete fusion to deep-inelastic and participant-spectator mechanisms, allowing us to evaluate all the characteristic of primary fragments. The subsequent de-excitation stage is analyzed as statistical emission through a sequential symmetric fission decays. High excitation energies and angular momenta available for the sources allow the competition between complex fragment emission and usual evaporation. One should probably go to higher bombarding energies to see clear signatures of dynamical instabilities. Moreover, due to the dominance of participant-spectator mechanisms such analysis shoud be perfomed for the different sources.

REFERENCES

1. J.P.Bondorf Nucl.Phys.A 488(1988) 31c
2. E.Suraud, Ch. Grégoire and B. Tamain "Birth, life and death of hot nuclei" Progress of Nuclear and Particle Science 1989, in press
3. L.G.Moretto and G.J.Wozniak, Nucl.Phys. A488(1988) 337c
4. A Bonasera,M.Di Toro and Ch. Grégoire , Nucl.Phys. A463(1987)653 and A483(1988) 738
5. G.F.Bertsch and S.Das Gupta, Phys.Rep. 160(1988)189
6. A.Adorno,A.Bonasera,M.Di Toro,Ch.Gregoire and F.Gulminelli, Nucl.Phys. A488(1988) 415c and Bormio 1988, p.491
7. M.Di Toro,G.Lanzanò and A.Pagano, Phys. Rev.C37(1988) 1485
8. G.Grégoire,B.Rémaud,F.Sebille,L.Vinet and Y.Raffray, Nucl.Phys. A465 (1987) 315
9. A Bonasera,G.F.Burgio,F.Gulminelli and H.H.Wolter "Phase space model of hard phton production in heavy ion collisions" Munich LMU preprint 1989
10. A.Bonasera,G.F.Burgio and M.Di Toro, Phys.Lett. B221(1989)233
11. L.G.Moretto, Nucl.Phys. A247(1975) 211
12. R.Bass "Nuclear Reactions with Heavy Ions" Springer Verlag, Berlin 1980
13. B.Faure,F.Auger,J.P.Wieleczko, W.Mitting,A.Cunsolo,A.Foti,E.Plagnol, J.Québert,J.P.Pascaud,Phys.Rev. C35(1987)190 and "XXVI Winter Meeting on Nuclear Physics" Bormio 1988 p.133
14. R.Bougault et al. Nucl.Phys. A488(1988) 255c and Proc. "Symposium on Nuclear Dynamics and Nuclear Dysassembly" Dallas 1989
15. G.Rudolf and L.Stuttge private communication and Colloque GANIL,1989.

INTRODUCTORY REMARKS TO THE LECTURES OF ERWIN SCHOPPER AND ARTHUR M. POSKANZER

André Gallmann

Ladies and gentlemen, welcome to the afternoon session !

Early shock wave experiments . . . . History of central collisions . . . . : It all sounds like a session on the history of nuclear physics. This promises to be very stimulating, because this subject is an important and integral part of the whole field, and is interesting for us and for the future generation of scientists.

Concerning the first of the next two talks, we have the great honor to have with us Professor Erwin SCHOPPER who has been, for many years, the director of the "Institut für Kernphysik der Johann Wolfgang Goethe Universität in Frankfurt am Main." He has been involved in nuclear and cosmic ray physics since the early days of these sciences. He is a living witness to those days. He has touched the history of nuclear physics with his hands. Therefore it will be a great benefit for us to listen to his talk which will, I am sure, include personal anecdotes of these early days of nuclear science.

Compressibility of nuclear matter . . . . , shock waves . . . . , Walter GREINER and his collaborators predicted them 16 years ago. But, as Walter himself recalled at the opening session of this Conference, other physicists did not believe in the compressing and heating of nuclear matter. Instead they thought that nuclear matter was more or less transparent : that the collision of one heavy ion with another could be compared to one swarm of bees flying through another swarm of bees. In such a case, as always, if two theories are in contradiction, one needs an experimentalist "to turn the page." I will come back to this point shortly.

The Nuclear Equation of State, Part A
Edited by W. Greiner and H. Stöcker
Plenum Press, New York

Today we all know who was right : nuclear matter is indeed compressible. Hans GUTBROD, Arthur POSKANZER and their collaborators, and other physicists have proven the existence of shock waves in nuclear matter without possible doubt. The experiments they did needed heavy equipment, a lot of manpower and a significant amount of money. This was the price that had to be paid for the good statistics which led to a clear answer.

Let's now come back to 1973 : Walter GREINER had predicted the compressibility of nuclear matter, and needed an experimental proof of his idea. He addressed himself to our first speaker : why could not one look for this phenomenon in heavy ion collisions with AgCl and/or emulsion detectors, using heavy ion beams or cosmic rays as a source ? Cosmic rays . . . . , AgCl and emulsion detectors . . . . These are fields in which our next speaker has been a pioneer and a widely recognized expert. With non-sophisticated apparatus, very few people and little money, Erwin SCHOPPER, in 1975, found indications for shock waves if a very small nucleus interpenetrates a silver nucleus. The statistics of his experiments were poor, but the starting signal for an extensive study of this field had been given. Then as a consequence, there came many new results, and . . . . a conference at this nice spot of the Spanish Mediterranean coast with, finally, sunshine.

It's time now to listen to the talk of our colleague and friend, talk entitled *"EARLY SHOCK WAVE EXPERIMENTS WITH AgCl AND EMULSION DETECTORS."*

Professor Erwin SCHOPPER !

Our next speaker will be Dr. Arthur POSKANZER, senior physicist at the Lawrence Berkeley Laboratory, where he heads a big group. He has been a worldwide leader and colleague in heavy ion physics for a quarter of a century. He was one of the first to initiate physics at the Bevalac, namely high energy heavy ion physics. I think one can undoubtedly say that experimental heavy ion physics was initiated at Berkeley.

This happened together with a group from GSI, namely Professor Hans GUTBROD, Professor Reinhardt STOCK, Dr. Hans Georg RITTER and Dr. Karl Heinz KAMPERT and others. Professor Rudolf BOCK from GSI played a substantial role in initiating this collaboration which started in the mid 70's and is still very active today - even though they moved to CERN.

Art POSKANZER was well known in our community before he went to Berkeley. At BNL he was active in high energy proton spallation where he used clever methods to separate exotic isotopes. He also initiated fragmentation studies in the mid nineteen-sixties. Arthur has always been at the forefront of the areas of research he has chosen to pursue.

We are all eager to listen to his talk entitled *"HISTORY OF CENTRAL COLLISIONS AT THE BEVALAC."*

Dr. Arthur POSKANZER !

Early History of Shock Waves in Heavy Ion Collisions (The Frankfurt Group)

Erwin Schopper

University of Frankfurt am Main

1. Prologue

1.1 Concerning heavy ions

When I was invited by the Chairman of this summer school to speak on the early history of shock waves in heavy ion collisions, I was tempted to go a little further back into the past, back to the time when we first encountered energetic heavy ions: Let us return to the middle thirties; Erich Regener, my teacher and a famous cosmic ray physicist at Stuttgart, was investigating the Cosmic Radiation – called Ultrastrahlung – with ionization chambers and Geiger counters in the stratosphere with balloon–borne instruments, on ground and down to the depths' of the Lake of Konstanz.

One of the outstanding questions was the occurrence and the role of heavy particles: protons (and nuclei) in the primary component. The most direct experiment would have been to record their tracks in the upper atmosphere; the only trackrecording instrument existing at that time, however, was the cloud chamber, which already had been used successfully in mountain altitudes. Regeners idea was to built up a "permanent" cloud chamber. With the technical means available in those days, however, it would have been a very difficult problem to operate it in a balloonflight, even in the hands of an ingenious experimentalist like Regener.

Hence I suggested to use thick layers of photographic emulsion, as a permanently working servicing–free solid state analogon to the cloud chamber. Encouraged by the successfull work with thick emulsion layers of Myssowsky and Tshishow in Leningrad /1,4/ and by Blau and Wambachers work in Vienna /2/. I asked Prof. John Eggert, Head of the Zentrallaboratorium of the Agfa Photographic Plant for his help. In 1936 we succeeded in finding an emulsion of good sensitivity for nuclear particles – the highest available at that time, recording protons up to 35 MeV /3/ – (afterwards available as Agfa K–plate).

With these nuclear plates we observed in the stratosphere /5,6/ high energetic nuclear disintegrations , emitting energetic heavy fragments ($Z > 2$), we observed nuclear spallations and also single tracks hinting at charges $Z > 2$. One of these disintegrations, shown in Fig 1, recorded in 1937 in a balloonflight at about 26 km of altitude, shows the surprising break–off of the hit nucleus into heavy fragments, besides the emission of protons (track b) and of α–particles. Even the emission of fast α–particles was a surprise. We are in the era of the nuclear drop–model, with nuclear binding energies of ~ 8 MeV; the emission of aggregates of nucleons seemed rather improbable. A paper by Heisenberg of 1937 /7/ dealing with high energetic proton–nucleus interactions assumes a knock–on mechanism for fast emitted nucleons, and a smooth evaporation of single nucleons from the residual nucleus, without taking into consideration complex fragments.

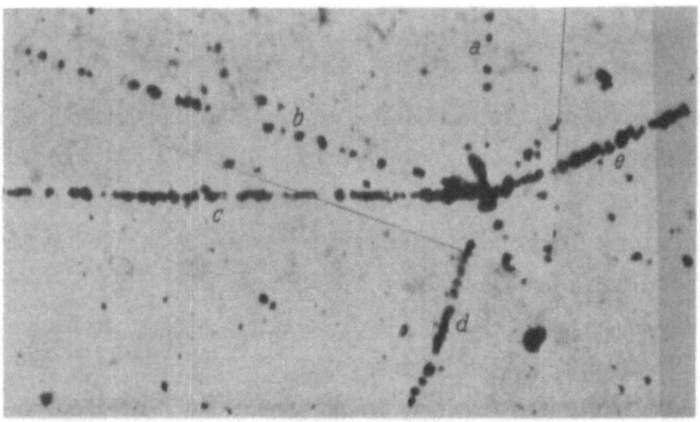

Figure 1. Vertex of a nuclear interaction emitting fast heavy fragments, recorded in a balloon flight (Schopper 1937).

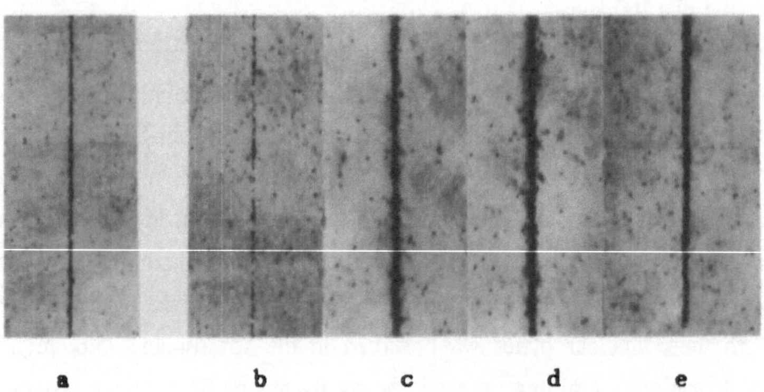

Figure 2. Micrographs of tracks of heavy ions in AgCl detectors at various energies:
a) O^{16} 2 GeV/A ; b) C^{12} 2 GeV/A ; c) C^{12} 90 MeV/A;
d) C^{12} 25 MeV/A ; e) stopping

It was our first encounter with energetic heavy ions. Coming from and produced by the Cosmic Radiation in the upper atmosphere discovered as part of the primary component in 1948 by Freier et al /8/, they remained the only source for a couple of years, until they could be produced in the laboratory by accelerators. The Cosmic Radiation has been an important tool for high energy physics and particle physics: many data concerning nucleon – nucleus and nucleus – nucleus interactions up to very high energies have been obtained with cosmic ray projectiles /13/ unfortunately a mixture of masses, charges and energies. Let me remind the situation at the beginning of my own experiments: the only "laboratory projectiles" available for the calibration of the detectors were protons of 11 MeV from the Heidelberg Cyclotron and α–particles from the radioactive decay of Ra and Th (max. α–energy 8.78 MeV) What an embarrasing offert of particles today!

1.2 Concerning the detectors

The nuclear emulsion, improved after the war by the Laboratories of Ilford Ltd (England) and Kodak (USA) became one of the most important instruments in cosmic ray physics, high energy and elementary particle physics in the fifties and sixties. The π–Meson has been discovered by Powell in 1948, inaugurating the grand era of elementary particle physics.

Nuclear emulsion consists of small silverhalide crystals – grains – embedded in gelatin. Its simultanous use as target and detector which is often practised, for instance in our cosmic ray experiments when we observed the disintegration of one of the nuclei of the emulsion (Fig 1), suffers from its complex nuclidic composition: of heavy nuclei as Ag and Br with $A > 100$, and light nuclei of the gelatin with $A < 16$. Another disadvantage is the wet processing required for revealing the tracks.

Eggert and myself had tried, in 1938, to avoid the gelatin and to record particle tracks in macroscopic crystals of AgBr, without satisfying result at that time. When in 1961 Childs and Slifkin in Raleigh reported promising results of recording particle tracks in macroscopic pieces of crystals of silver–chloride, I took up again, together with my colleagues Haase and Granzer in Frankfurt, research on silverhalide crystal detectors. We succeeded in growing crystallin sheets of Cd – doped AgCl, usually as foils of about 150 microns thickness – adapted to microscopic scanning – with rather unique properties as track detectors /9/.

As they are less known as nuclear emulsion and as they were playing a role with the early shock wave experiments, I want to say some words about them: They are "instant" detectors. The nuclear tracks are revealed to microscopic visibility by simple exposure to blue light (~ 400 nm). The undevelopped (latent) tracks are unstable in time and fading, unless they are stabilized by irradiation with yellow light (~ 560 nm), during or soon after the passage of the particles. Recording of tracks thus can be switched "on or off" by will, id est for a desired period of time for instance. Their sensitivity corresponds to that of K2 – Ilford emulsion (Fig 3), recording tracks of protons up to 40 MeV, He–ions up to 200 MeV/A and all particles of $Z \geq 3$ up to relativistic energies. Fig 2 shows some micrographs of tracks in AgCl–detectors. They are quite similar to those in nuclear emulsion.

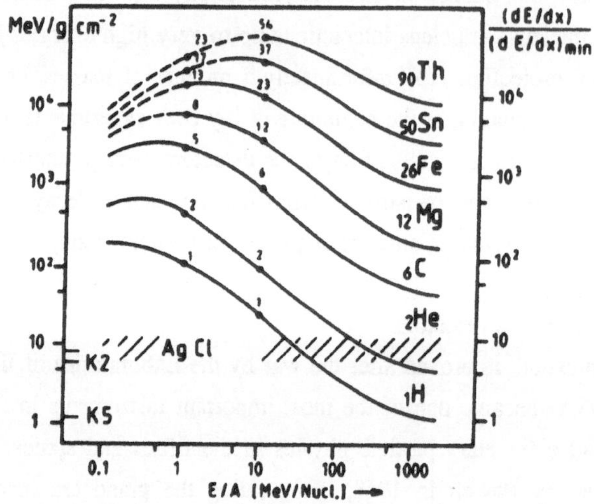

Figure 3. Sensivity thresholds of AgCl detector; K2 and K5 emulsion.

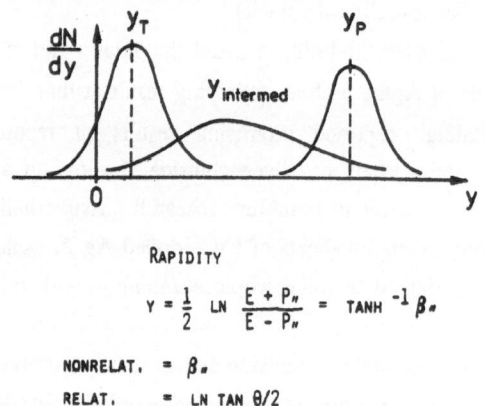

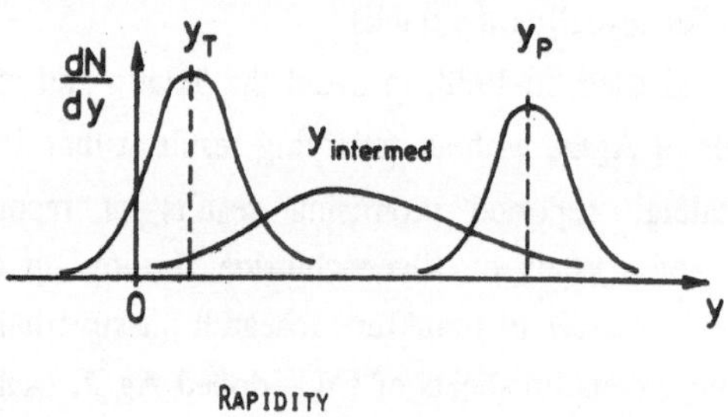

$$Y = \frac{1}{2} \ln \frac{E + P_{\parallel}}{E - P_{\parallel}} = \tanh^{-1} \beta_{\parallel}$$

NONRELAT. $= \beta_{\parallel}$

RELAT. $= \ln \tan \theta/2$

Figure 4. Rapidity distribution of reaction products in nuclear collisions (schematic).

For the evaluation of the tracks in nuclear emulsion and AgCl–detectors we have built up at Frankfurt a semi–automatic, computer controlled microscope combined with a videoelectronic shape recognizing system /10/. A "geometric" programm evaluates geometric parameters of particles, including angular distributions or the extrapolation of vertices of nuclear interactions from emitted fragments; a "densitrometric" programm, evaluating lateral density profiles of tracks, determines the ionization (LET) and the parameters Z, β, A.

With compact tracks we are measuring in narrow consecutive steps along the track the width of their lateral density profiles, their mean (MTW) and their variance; the latter, influenced by the δ–ray penumbra, discriminates Z and β at the same Z^2/β^2 with a charge resolution ΔZ of 1 – 2 units from a piece of the track of only 100 microns length, which is automatically followed. In the case of grainy tracks we obtain from the determination of the gap–length distribution – also obtained from the lateral density profiles – a charge resolution ΔZ of a fraction of a unit, if β is known elsewhere.

These properties and the performance of the AgCl–detectors, described above made them a useful instrument for experiments in Space Biology, for the investigation of the influence of H(igh) ZE–particles of the cosmic radiation on biologic objects under spaceflight conditions, such as microgravity for instance. We were participating in Space flight experiments since 1972 on Apollo 17 until the D1–Mission in 1986, and we are just preparing a similar experiment on the Sowjetic Biokosmos 9 – satellite, starting in september of this year.

The exposure of our detectors to fast heavy ions at the LBL–Bevalac–accelerator for their calibration for heavy ion experiments and for "ground" experiments in comparison to the space–flights directly lead us to the observation of heavy ion collisions in the detectors, and to the topical problem of the occurrence of shock waves in these interactions.

2. Shock waves

The acting persons:

theory:	W. Greiner	experiment:	E. Schopper
	W. Scheid		H. G. Baumgardt
	J. A. Maruhn		
	H. Stöcker		

It was Walter Greiner who induced me to search for collective mechanisms in those heavy ion collisions, in particular for signatures of shock waves, which by him, W. Scheid and H. Müller had been predicted in 1974 /11/, as a consequence of the compression of the nuclear matter during the impact of two fast nuclei, and derived from a hydrodynamical model. He has summarized the conceptual background and the precendent ideas in his introductory speech to this conference. Hence I would like to pass on to the concept of the early experiments and to their results and conclusions, mainly those obtained by the Frankfurt group. We are now in 1974.

2.1 The state of knowledge

As I already mentioned many investigations on nucleon–nucleus and also nucleus–nucleus collisions had been made with cosmic ray projectiles in the GeV/A region, meantime enlarged and completed by data from investigations with proton and pion beams from the big accelerators. Before entering experiments, we had to sift the literature for hints on collective phenomena or shock waves.

Proton – nucleus – collisions. The early paper by Glassgold, Heckrotte and Watson /12/ predicting shock waves in p–A collisions, seemed not to have been followed by successfull experimental observations. Nevertheless, before considering the mechanism of the interpenetration of two fast nuclei, which expectedly is more complex, it was useful to throw a glance onto the more simple situation of p–A collisions.

We shall classify the particles involved in p – A and A – A interactions, following the tradition of the early investigators, according to their ionization (and the apparition of their prongs in emulsion; (Fig 1 and 3) with the following notation:

$I \leq 1.4\, I_{min}$	$\beta \geq 0.7$	s shower particles
$I > 1{,}4\, I_{min}$	$\beta < 0.7$	h heavy particles $M \geq M_p$
	$0.3 < \beta < 0.7$	g (grey) group
	$\beta < 0.3$	b (black) group

In proton–nucleus collisons up to projectile energies of 400 GeV, non collective mechanisms were observed to be dominant at least for the production of baryonic secondaries /13/ which can be traced to the recoil nucleons of the "primary" collisions and their subsequent low–energy evaporation. As a dominant picture of the mechanism of p–A interactions the cascade evaporation model had still kept its actuality:

It begins with the interaction of the incident proton with the nucleons of the target met on its path, generating for the main part pions, and strange particles and isobars like $p\bar{p}$ and excited states; in the rapidity picture they have projectile rapidity or little less (Fig 4). The models of the reaction tube and the energy flux cascade, the diffractive excitation models /14/ the fireball – model /15/ are variants of this primary interaction /16/. A hydrodynamical model had been proposed by Belenkij and Landau /17/.

Particles dropping out of the interaction tube are scattered within the surrounding matter of the target nucleus, producing the so called g–group of knock–on nucleons followed by an internuclear cascade of nucleon–nucleon impacts of intermediate rapidity. (participants)

In a final step nucleons and complex fragments are evaporating isotropically from the residual target, "the spectator", forming the b–group (black tracks; $\beta < 0.3$). The residual nucleus – excited up to the boiling temperature of the nucleus of ≤ 16 MeV – is moving with a small forward–momentum corresponding to $\beta \sim 0.02$, or a target rapidity near zero.

This fireball–cascade–evaporation model has a long persistent history and has proved its validity and usefullness for the understanding of the general features of high–energy p–A interactions over a wide range of projectile energies; with some modifications and its flexibility it allows to systematize experimental data, such as the multiplicities and the

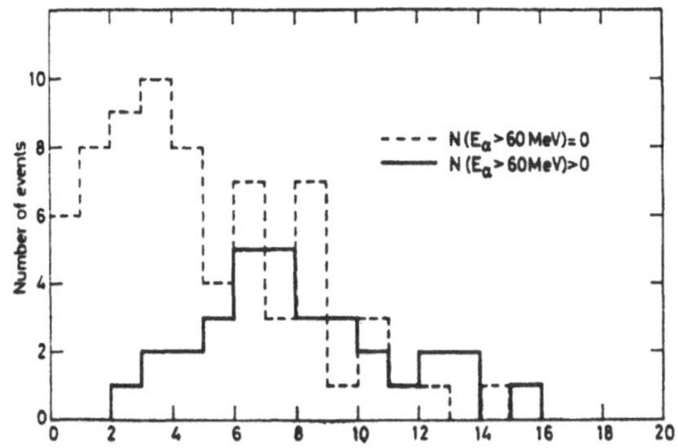

Figure 5. Nh - distribution of different types of nucleus-nucleus collisions with and without emission of He - fragments.

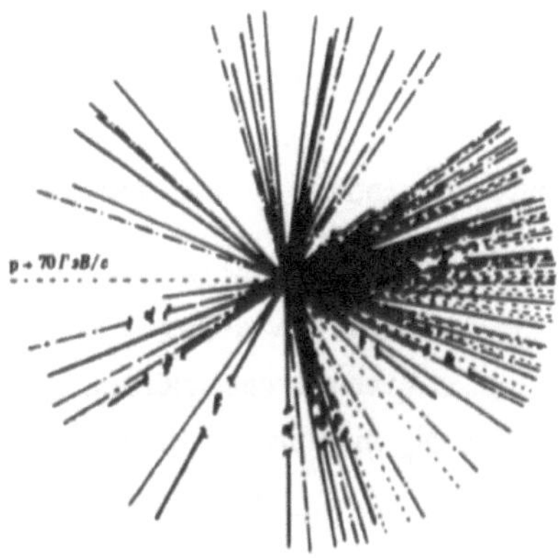

Рис. 6. Звезда с $N_h = 66$, $N_\alpha = 22$.

Figure 6. Total disintegration of a Pb nucleus in emulsion in a p - Pb interaction (courtesy of K.D. Tolstov).

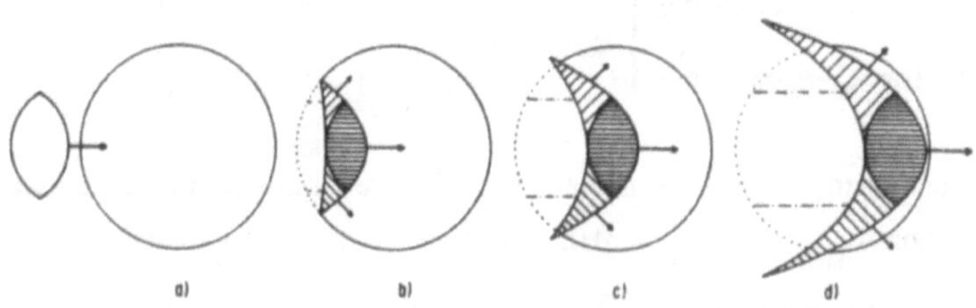

Figure 7. W.Greiner's vision of a shock wave.

angular distributions of the s, g and b group, for different impact parameters, peripheral and central collisions. It is not surprising therefore, that physicists tried to find agreement with the cascade model, before – in the case of deviating observations – introducing new mechanisms. The cascade model, however, contradictory to collectivity, did not leave space for hydrodynamics.

It is beyond the scope of my talk to go into more details of the cascade model. It was our aim to find or not indications for collective behaviour; I shall mention in the following, paradigmatically, some facts which – I thougt – could show us the way. /18,19/

Apart from earlier investigations with Cosmic Ray particles, accelerator experiments of several collaboration groups with protons up to 400 GeV – energy on emulsion had yielded valuable recent data (ref. see /18/). Some salient observations possibly relevant for our search for collective phenomena have been the following:

He–fragments: In p–Em (= emulsion) collisions the distribution of the N_h–multiplicity was shifted to higher multiplicities in events combined with the emission of He–fragments of intermediate energy above the evaporation spectrum (g–group), compared to those events without these He–fragments.

The same but still more pronounced effect had been observed formerly with heavy projectiles from the Cosmic Radiation by the Lund–group (Fig. 5, /20/). I shall come back to this point again.

The occurence of this He–group, preferentially in large stars (of high N_h–multiplicity) and missing in small ones, could not be reproduced by a cascade mechanism.

The sticking proton: Among the central p – A collisions (we are considering p–Ag and –Br collisions) a small percentage of events is characterized by a complete disintegration of the target nucleus, with a high N_h multiplicity, filling up the range of intermediate rapidity without a distinct b–group. In these "violent" collisions the projectile is transferring its total energy onto the target; it is sticking.

The frequency of these sticking–events has been determined by Tolstov /21/ and by Badawy et al /22/

as 3% for p–A collisions, practically constant in the energy range $6.3 < E \leq 400$ GeV

$$\left.\begin{array}{l}8\% \text{ for He–A collisions} \\ 16\% \text{ for C–A collisions}\end{array}\right\} \text{at } 3.85 \text{ A GeV}$$

of all collisions with Ag and Br nuclei. The observed ratio of the cross–section of a sticking collision $\sigma_{stick}/\sigma_{tot\ inel} = 0.03$ for these p–A collisions is smaller than the calculated geometric ratio for central–collisions. We supposed that with these sticking collisions we were dealing with a (rare) stochastic transmutation of the impinging proton (and/or its reaction partners) into a state of strongly increased interaction cross–section, firing the nucleus. The collision is then not necessarily connected to a very small impact–parameter, since the length of the interaction path of the projectile proton in the target nucleus which determines the probability of the transmutation, varies only smoothly within a certain range of small impact–parameters. The sticking proton could help to initiate hydrodynamics. Fig 6 shows a proton–induced violent interaction with a lead nucleus embedded in nuclear emulsion, recorded by K. D. Tolstov. It was Tolstov, who was repeatedly calling our attention to this process. /23/.

Nucleus–Nucleus collisions: The decisive question was: Are A_1–A_2 collisions dominantly behaving like simple superpositions of A_1 nucleon – nucleon collisions?

Or can we expect following the predictions of Greiner et al. that the interpenetrating projectile will generate sufficiently high energy density and excited hadronic states in the reaction zone, comparable to the fireball–state of proton–projectiles, leading to compression– and shock wave phenomena. Walter Greiner's view is shown in Fig. 7.

In contrast to p–A interactions we have a priori two disintegrating systems, the projectile nucleus and the target nucleus, covering different rapidity–ranges. We were excluding from our considerations peripheral collisions which either are leading to a pure projectile fragmentation or to a mixed event from the overlap–area. We were focussing our interest to non–peripheral events, defined as collisions with full overlap of the projectile and target area. They were selected by a required large number of N_h–secondaries.

The first look into the literature was slightly disillusionating our expectations of very new phenomena. Early investigations of heavy ion collisions from the Cosmic Radiation (mostly C, N, O) in the energy range of 100 A MeV up to 15 A GeV referred in a paper by Andersson et al /24/ were hinting at only slight differences of the behaviour of the g– and b–group, compared to p–A collisions, id est the A–A collisions could be described grosso modo with the same models as p–A collisions, dominantly by non–collective mechanisms. A negative signal with respect to expected collectivity was the N_h multiplicity, which was increasing in A_1–A_2 collisions less than A_1 times N_h of p–A_2 collisions, caused by shadowing effects.

A hopeful new aspect was the already mentioned observation of He–nuclei of intermediate energy, emitted in large stars in heavy ion interactions; first observed by Fowler et al in 1957, they had been observed and investigated ty the Lund group /25,26/ (Fig 5 and Fig 8).

What are now the signatures for shock waves we could observe with He, C and O–projectiles and Ag – targets? We should find them in the projectile system as well as in the target–system. We expected the preferential angular emission of target fragments as a consequence of the propagating Mach–shock, as candidates the He–fragments discussed above.

The projectile would propagate the Head–shock and show effects of its re–expansion and de – exitation at the end of its passage through the target, if both systems remain "separated". We assumed this case as dominant compared to the sticking events, which should occur in O–Ag collisions in about 10 % of all events /21/ at the Berkeley projectile energies /27/.

As the sensitivity of our AgCl–detectors did not allow us to record all projectile fragments (Fig 3) we concentrated our interest onto the target.

With a O projectile (on Ag–target) of E = o.87 A GeV we expected /18, 19/ for instance

$$V_{Hs} \sim V_{proj} = 0.86\,c \quad \text{and} \quad V_{Ms} = 0.53\,c$$

$$\Theta_{Mach} = \text{arc cos } V_{MS} / V_{HS} = 50^{o}$$

The energy of the mach shock ejecticles with $\beta \leq 0.53$, decreasing behind the shockfront, corresponds to a spectrum above the evaporation energies (Fig 13), id est $E_p > 30$ MeV, $E_{He} > 30$ A MeV.

This, however, was the energy–spectrum of the He–group observed in large stars. We therefore regarded them as candidates for shock–wave signatures.

2.2 Experiments

Angular distribution of target fragments: The preceding estimate showed us, that we had to seek the Mach shock ejectiles within an energy range corresponding to $\beta \leq 0.53$, and among them the mentioned He–component.

The procedure follows from Fig 13; it contains the schematic energy spectra of the b–and g–groups of the H and He–fragments.

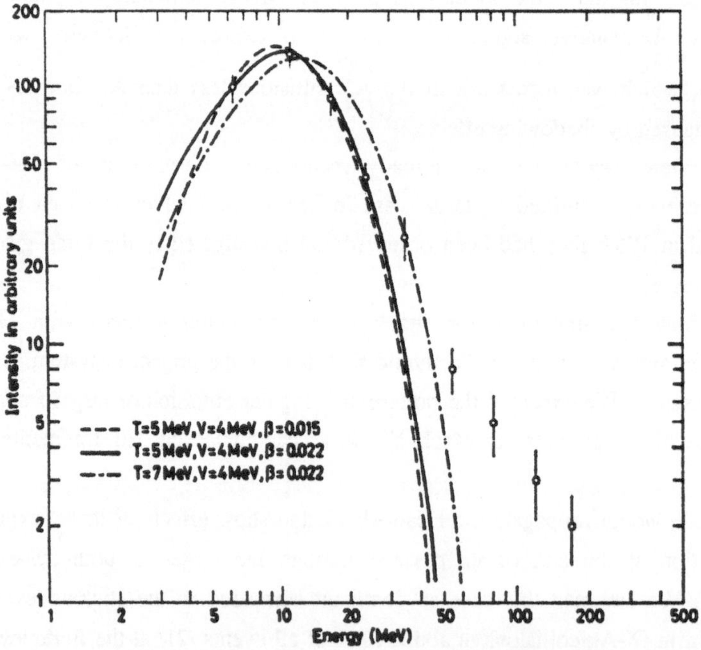

Figure 8. Energy spectra of He - fragments /25/

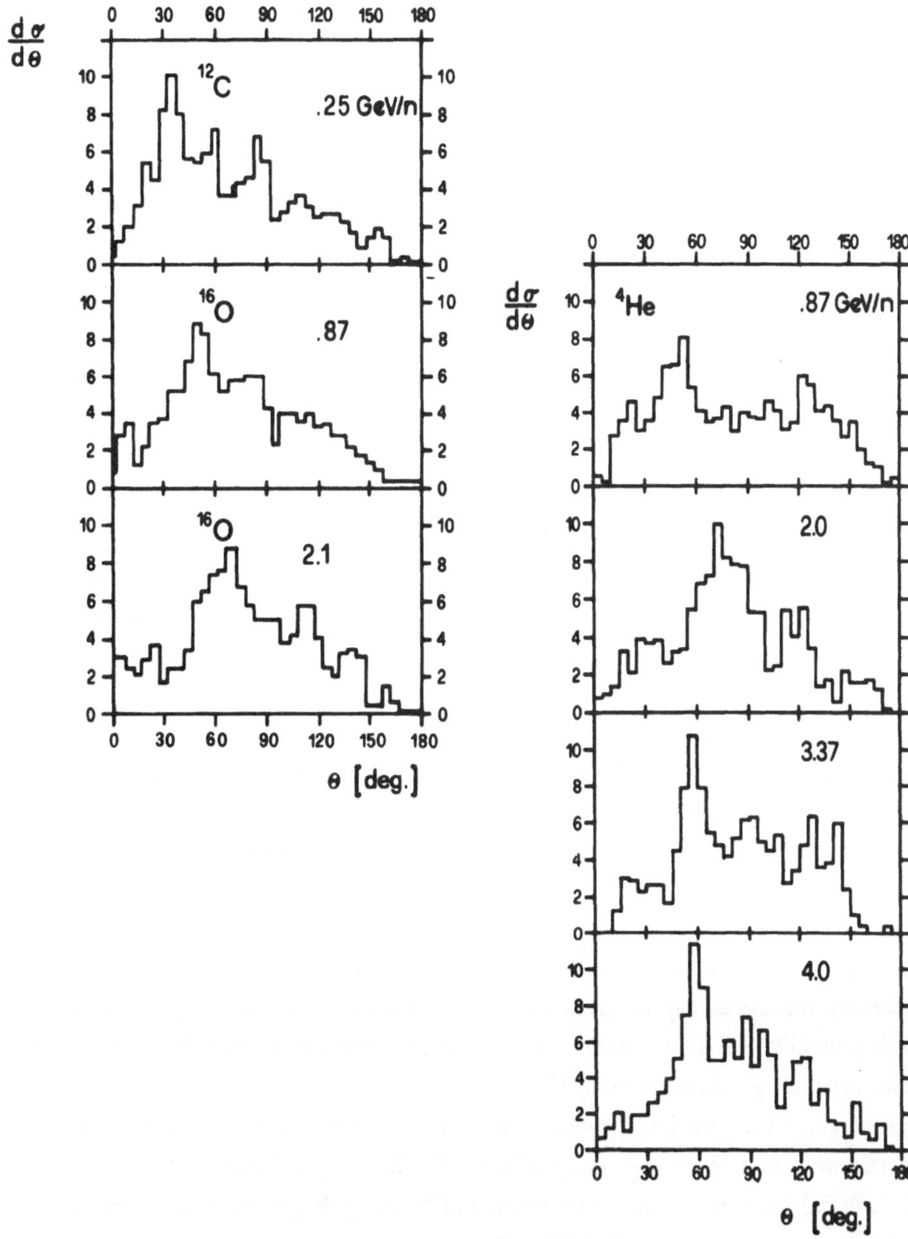

Figure 9. Angular distribution of Nh-fragments at various energies of He^4, C^{12}, O^{16} projectiles.

Our detectors were recording the b + g–group of the He component up to energies of 200A MeV, corresponding to $\beta \sim 0.5$, whereas they were only grasping the b–group of the H–fragments, dropping the g–group. (Fig 3 and 13)

Hence we had simply to evaluate all visible tracks, event by event in 4π geometry. Taking into account the occurrence of the g–group of He–fragments in events of high multiplicity, we selected events with N_h multiplicities > 12; they correspond together with the dropped g–group of H–fragments to N_h > 25, which in nuclear emulsion defines central collisions with Ag–targets.

We obtained angular distributions with pronounced peaks; their position was depending on the projectile energy, and practically independent from the mass of the projectile (Fig 9) up to the available energy of 4A GeV. In order to excerpt the distribution of the He–ejectiles we had to subtract the evaporation fragments of the H and He–component (Fig 10). This was done by fitting the evaporation–component at the backward angles of the distribution, where the influence of the Mach shock was expected to dissappear /18, 19, 29/:

We could further show, that the angular distribution of the He–component was inconsistent with a knock–on mechanism of independent particles.

The observed peaks and their position dependent on the projectile energy could be described by the hydrodynamical concept of W. Greiner and his group in Frankfurt. We therefore interpreted them as signatures of compression and shock–waves in these nucleus–nucleus collisions /18/.

The presentation of our first results on the Symposium on Physics with Relativistic Heavy Ions at LBL Berkely in March 1975 was a challenge to the physicists convened to discuss the facilities and chances of observing shock waves in Heavy Ion collisions and caused an increasing and vivid theoretical and experimental activity. Confirming and contradictory results have been published in the following period. Theoretical models like fireball, firestreak, cascades, coalescence models or hydrodynamics besides microscopical concepts have been applied and proposed. It is impossible, of course, to discuss or even to mention the many interesting papers appearing in this early period. A paper by Goldhaber and Heckmann /30/ reviews data and interpretations up to the end of 1977. Now, in 1989, we know about the great success of hydrodynamics.

Let me come back for just a moment to 1976: During a visit to Dubna we established contact with the High energy group of Prof. Baldin /23/. W. Toneev showed us Monte Carlo calculations of the angular distribution of b– and g–fragments from C, resp O – Ag collisions at various energies /28/. (Fig 11)

They are reflecting the situation of Fig 10: the difference at the position of our peaks between our experimental data and the calculations based on a purely cascading process. Soon after the presentation of our data at Berkeley (1975) Poskanzer et al /31/ published angular distributions from inclusive measurments at the reaction O^{16}–Ag $\longrightarrow$ He^3 at 1.05 A GeV. (Fig 12)

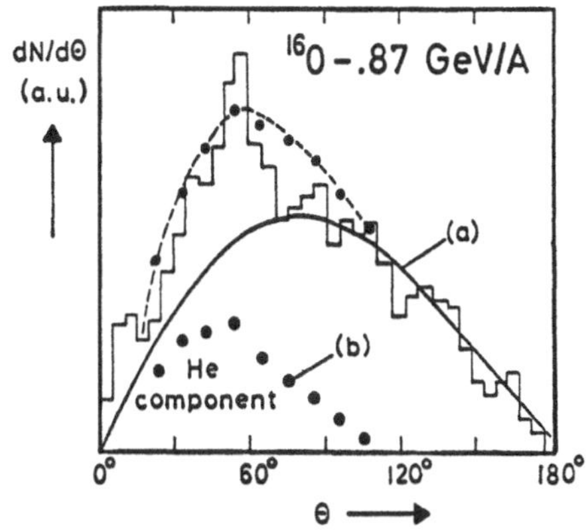

Figure 10. Angular distribution,fitted evaporation component and residual peak.

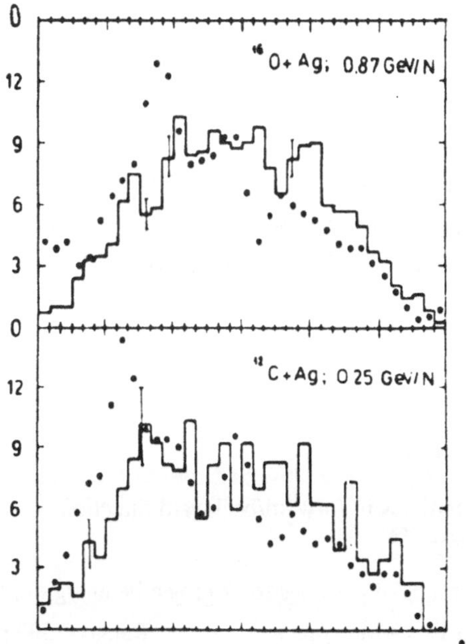

Figure 11. Angular distribution of h -particles; points - experimental data, histogram - calcutation /28/.

It turned out, that the smooth peaks observed depended on the energy window of the He^3–fragments.

It is an inherent property of the shockwave that pressure and velocity behind its front are decreasing, bending it and smearing it out as can be seen at the angular distribution of the He component in Fig 10, derived from an averaged histogram. It was, I think, commonly understood, that careful and comprehensive measurments of all available parameters and observables regarding target and projectile were indispensable for the understanding of the complex mechanism of nucleus nucleus interactions.

So far I think, would it be worthwhile, to repeat experiments in the energy range between 100 A MeV and 4 A GeV in small energy steps and with projectiles of various – not only large – masses.

The sticking probability of the projectile as a function of its energy and mass also is an outstanding problem with respect to the onset of collectivity.

This type of investigations requires instrumentation and manpower, as we can see at the number of about 400 scientists and technicians working at CERN in the high energy heavy ion project.

My own experimental group at Frankfurt was applying its small (man)power to a less extensive problem. We were interested in the sources of the He–component.

Walter Greiners group was continuing – steadfast and successfully – hydrodynamics. The bounce–off, observed in 1980, removed the ban from the shock waves.

The sources of the He–component

The target source: We investigated C^{12}–Em collisions at 0.87 and 2.1 A GeV. We were interested in the temperature and the velocity $\beta\|$ of the source of the He–component. Assuming a Maxwell Botzmann–distribution of the form

$$N(E)\,dE \sim \sqrt{E}\ \exp\left(\frac{-E}{T}\right) dE$$

in the source frame, we find for the Lab–System:

$$dN/d\Theta \sim \sin\Theta\ (F/B)\cos\Theta$$

$$\text{with } F/B \sim \exp 4\ \beta\|/\beta_0\ \sqrt{\pi}$$

F/B = particles in forward/backward direction

$$\text{or } dN/d\cos\Theta \sim (F/B)^{\cos\Theta}$$

In Fig 13 we plotted 3 groups of particles for 2 projectile energies /32/

f.	H + He	$0 > E < 9$ MeV/nucl	Range < 500 μm
2.	H	$0 \leq E < 30$ MeV	Range $500 < R < 4000$ μm
3.	He	$0 \leq E < 30$ MeV/nucl	Range $500 < R < 4000$ μm

groups 1 + 2 are representing the evaporation spectra; they correspond to the same ratio of $T/\beta\|$, whereas group 3 deviates from the expected straightline, id est: either T or $\beta\|$ are different from that of the evaporation groups at both projectile energies. Assuming the same β for all groups we find $T_{1,2} \sim 10$ MeV, $T_3 \sim 40$ MeV for the He group 3.

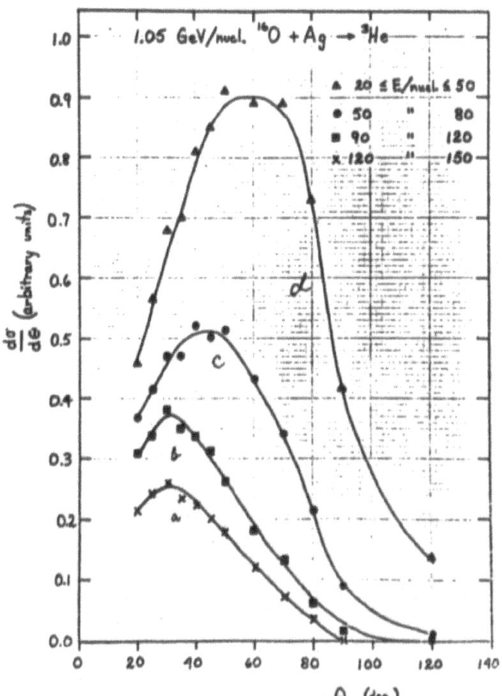

Figure 12. Angular distribution (Poskanzer et al).

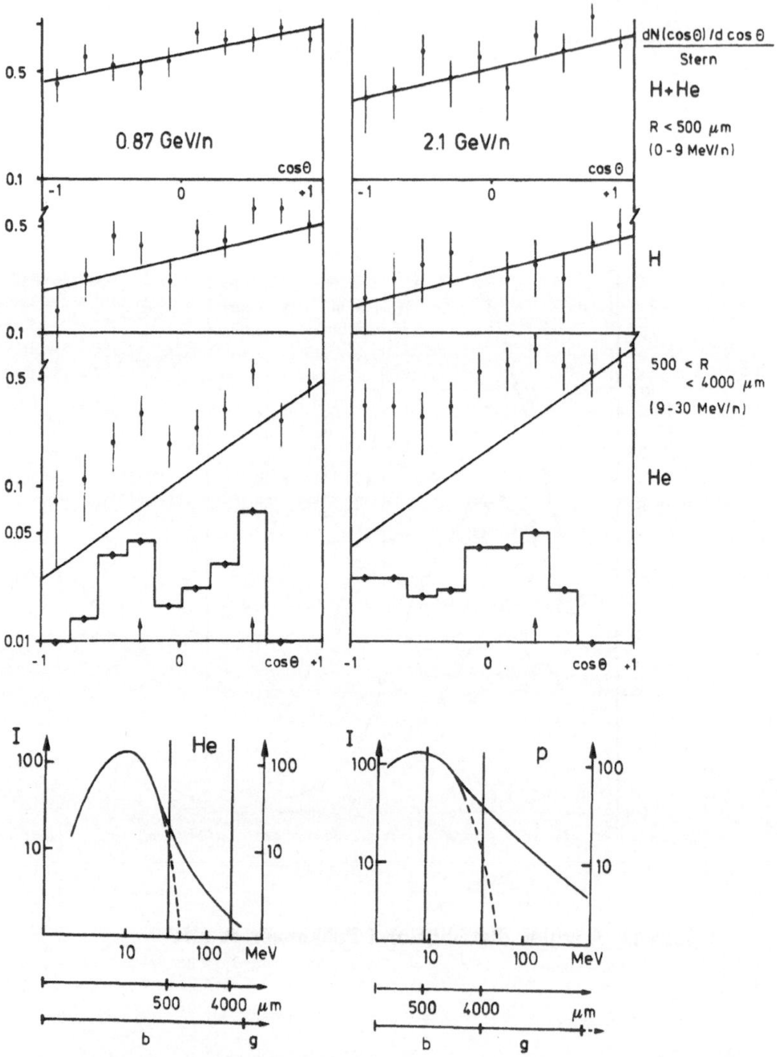

Figure 13. Angular distribution dN / d(cos ϑ) for a Maxwellian spectrum ⧫ experimental data (upper); energy spectra of target fragments, schematic with window cuts (below).

The projectile source of He–fragments: In order to get rid of the knowledge of $\beta||$ we investigated the He–fragments emitted from the projectile of Fe–Em collisions at 1.8 A GeV./33/ To avoid any superseding longitudinal motion, the analysis was restricted to the perpendicular momenta P_T

$$P_T = A\, m_0\, \beta\, \gamma \sin \Theta$$

Assuming the momentum distribution as Gaussian, having the width σ, their temperature can be defined by

$$\sigma^2 = A\, m_0\, T$$

with an integral distribution

$$\ln F\,(>P_T) = 1/2\sigma^2 p_T^2 + \text{const.}$$

The experimentally found distribution was deviating from the expected straightline belonging to one single temperature of the He source. The detailed analysis /32, 33/ revealed that the events were falling either into one or the other of two distinct classes belonging to assoziated temperatures of

$$T_1 = 10 \text{ MeV } (\sim 80\ \% \text{ of events})$$

$$T_2 = 40 \text{ MeV } (\sim 20\ \% \text{ of events})$$

The events are either hot or cold; mixed events or intermediate temperatures do not occur. The cold events have the same N_h–multiplicity distribution as p–A collisions, the hot ones belong to higher multiplicities. This phenomenon, too, requires further attention /34/.

Conclusions:

In order to find out experimental evidence for compression phenomena and shock waves in hith energy nucleus–nucleus collisions, we had investigated collisions of light projectiles He, C, O in the energy range of 0.2 to 3.8 A GeV with Ag nuclei of AgCl and nuclear emulsion track detectors, which were used simultanously for recording the collisions and as targets.

The experiments were based on two particular phenomena observed at high energy p–A and A–A interactions and reported in the literature:

a) the occurrence of a group of He fragments of intermediate energy in events of high N_h multiplicity. They were supposed to be candidates for signatures of collective mechanisms and shock waves. The investigation of this group allowed a very simple experiment with the AgCl–detectors.

b) The phenomenon of the sticking proton in violent p–A collisions with a total disintegration of the target. Its contribution to A–A collisions with our light projectiles in the considered energy range was estimated as ≤ 10 % of all non–peripheral events. It increases with increasing projectile and target size.

The analysis of the mentioned He–group exhibited peaks in the angular distributions, compatible with the predictions of the hydrodynamical model; they were interpreted as signatures of shock waves.

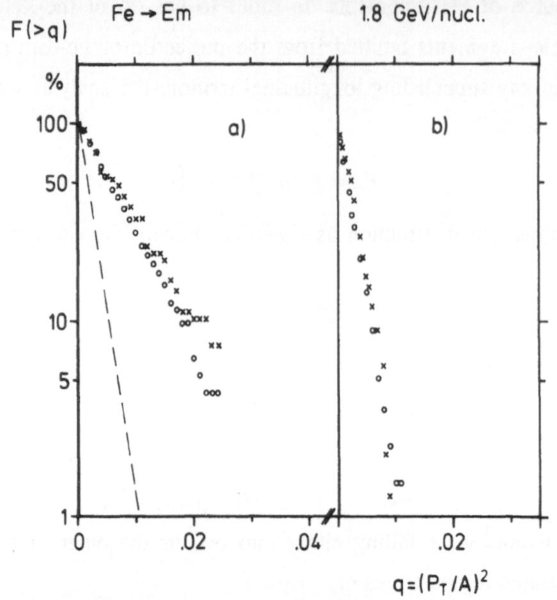

Figure 14. Normalised integral frequency distributions of square transverse momenta q (laboratory frame; croses) and ρ (' bounce' frame; circles) a) for 'hot' events and b) for 'cool' events.

The range of information from these first observations was limited by

a) the presumption of a special mechanism

b) poor statistics.

c) biasing the energy window of the fragments

These investigations should be repeated in small energy steps of the projectile, and with projectiles of different sizes, in order to determine the contribution of the sticking mechanism.

We were also looking for characteristics of the emitting sources, with respect to the onset of collective, hydrodynamical behaviour. We observed large transverse momenta of He–fragments in a distinct class of events.

As outstanding problems released from our experiments I would like to repeat the following:

the sticking proton

the two distinct temperatures of target or projectile–fragments with respect to a possible transmutation of the interaction mechanism.

I would like to express my great satisfaction about the big progress achieved by the scientists working in a field, to which we could contribute in its beginning. I am happy to see the great success of hydrodynamics in the description of high energy nucleus–nucleus collisions based on the equation of state of nuclear matter.

Acknowledgments
I would like to thank all colleagues who have contributed to our investigations, H. G. Baumgardt and, in particular E. Friedländer for critical and stimulating cooperation and K. D. Tolstov; last but not least Walter Greiner for our cooperation and friendship lasting over more than two decades.
The work described has been sponsored by the German Ministery of Research and Technology.

REFERENCES

/1/ Myssowsky, L.and Tshishov, P., 1927, Z. Phys. 44, 408
/2/ Blau, M., Wambacher, H., Mit. Rad. Inst. Wien (1932), 299, (1934), 339; 1937, Nature 140, 585
/3/ Schopper, E., 1938, Z. Wiss Photographie 1, 21, 1939, Veröff. Agfa VI 170, 1939
/4/ Filippov, A., Idanov, A., Gurewich, I., 1939, Journal of Physics 1, 51,
/5/ Schopper, E., 1937, Naturwiss 25, 557
/6/ Schopper, E., Schopper, E. M., 1939, Phys. Z. 40, 22
/7/ Heisenberg, W., 1937, Naturwiss. 46, 749
/8/ Freier, P., Lofgren, E. J., Ney, E. P., Oppenheimer, F., Bradt, H. L., Peters, B., 1948, Phys. Rev. 74, 213
/9/ Schopper, E.,Wendnagel, Th., Baican, B., 1986, Nuclear Tracks 12, 71
Baican, B., Schopper, E., Schott, J. U., 1986, Nuclear Tracks 12, 519
Bücker, H. et al, 1984, Science 225, 226
/10/ Baumgardt, H. G., Amend, W., Staudte, R., Schopper, E., 1986, Nuclear Tracks, 12, 265
Baumgardt, H. G., Baican, B., Schopper, E., 1986, Adv. Space Res 6, 83
/11/ Scheid, W.,Müller, H., Greiner, W., 1974, Phys. Rev. Lett. 32, 741
Proc. Symp. on Physics with Relativistic Heavy Ions, LBL Berkeley, July, 1974
/12/ Glassgold, A. E., Heckrotte, W., Watson, K. M., 1959, Annals of Physics, 6, 1
/13/ Cohen, I. et al, 1974, Lett. Nuovo Cimento 9, 249
Friedländer, E. M., Friedmann, A. A., 1967, Nuovo Cimento A 52, 912
/14/ Gottfried, K., 1973, 5th Int. Conf. on High Energy Physics, Uppsala CLNS – 260,
Friedländer, E. M., 1974, Letter al Nuovo Cimento 9, 349
Fishbane, P. M., Trefil, I. S., 1973, Phys. Rev. Lett. 31, 734
/15/ The picture of the fireball is rather old and described e. g. in the article by
Sitte, K., 1961, Encyclopedia of Physics Vol XLVI/1 Berlin, Springer,
Hayakawa, S., Fujimoto, Y., ibid Vol XLVI/2, 115
/16/ Summary Reports see with:
Hayakawa, S., 1969, Cosmic Ray Physics, Wiley–Interscience, New York 1969,
Feinberg, E. L., Physics Reports 5 C, 5, 1972
/17/ Belenkij, S. Z., Landau, L. D., 1956, Nuovo Cimento, Suppl 3, 15
/18/ Baumgardt, H. G., Schott, J. K., Sakamoto, Y., Schopper, E., Stöcker, H.,

Hofmann, J., Scheid, W., Greiner, W., 1975, Z. Phys A 273, 359

Baumgardt, H. G.; Schopper, E., Schott, J. K., Kocherov, N. P., Voronov, A. V., 1976, Proc Int. Workshop IV, Hirschegg EAD–Conf 76 – 056 000; 105

/19/ Schopper, E., Baumgardt, H. G., Obst, E., 1977, Proc Topical Conf on Heavy Ion Collisions, Fall Creek Falls (Tennessee, USA), 398

Schopper, E., Baumgardt, H. G., 1978, Proc. Symp on Relat. Heavy Ion Research GSI Darmstadt, GSI–P–5, 104

Baumgardt, H. G., Schopper, E., 1979, I. Phys 6, 5, L231

/20/ Kullberg, R.,Otterlund, I., 1973, Z. Phys 259, 245

/21/ Tolstov, K. D., 1965 Dubna, JINR, R1, 2016; Tolstov, K. D., 1973 Dubna, JINR, R1, 6897

/22/ Badawy, O. E., Hussein, A. M., Metwalli, N., 1976, Z. Phys. A 279, 407

/23/ We are indebted to Prof. Bogoljubov and Prof. Baldin, JINR Dubna, for arranging with us the Dubna–Frankfurt cooperation, which gave us the oppurtunity to cooperate with Prof. Tolstov and Prof. Toneev at Dubna and to expose detectors at the Dubna synchrophasotron accelerator.

/24/ Andersson, B., Otterlund, I., Kristiansson, K., 1966, Arkiv Fysek 31, 527

/25/ Otterlund, I., Resmann, R., 1969, Arkiv Fysik 39, 265

/26/ Kullberg, R., Otterlund, I., 1973, Z. Physik 259, 245

/27/ We are indebted to our colleagues at the LBL Berkeley for having given us the opportunity of exposures with C and O–projectiles at a variety of energies.

/28/ Gudima, K. K., Toneev, W. D., JINR–Report, R2, 10431, Dubna

/29/ Baumgardt, H. G., Diploma Thesis, 1977, Frankfurt am Main

/30/ Goldhaber, S., Heckman, H. H., 1978, LBL 6570

/31/ Poskanzer, A. M., Sextro, R. G., Zebelmann, A. M., Gutbrod, H. H., Sandoval, A. Stock, R., 1975, Phys Rev Letters 35, 1701

/32/ Stöcker, H. et al; Baumgardt, H. G., Schopper, E., 1980 Proc Int. Conf. on Extreme States in Nucl. Systems Dresden, Vol 2, 33

Baumgardt, H. G., 1983, PhD–Thesis, Frankfurt am Main

/33/ Baumgardt, H. G., Friedländer, E. M., Schopper, E.,

1981, I. Phys G 7, L175

1984, 1. Int. Workshop on Local Equilibrium in Strong Interactions LESIPI, Bad Honnef, 258

/34/ Kallies, H., 1987 PhD–Thesis, Marburg

A HISTORY OF CENTRAL COLLISIONS AT THE BEVALAC

Arthur M. Poskanzer

Lawrence Berkeley Laboratory
Berkeley, CA 94720

You have heard a great deal about Plastic Ball results at this conference. There were talks on the first morning by Hans-Georg Ritter and Karl-Heinz Kampert on the Plastic Ball at Berkeley, there will be a talk next week by Rudi Schmidt on the Plastic Ball at CERN, and many other speakers have mentioned Plastic Ball results. The young students may think that when the new field of relativistic heavy ion physics opened up, an ideal detector was designed and built, data immediately analyzed, and results produced. The theme of my talk is to show that this is incorrect. The experiments proceeded in logical stages, one building upon the other, increasing in complexity and sophistication. The analysis techniques and the theory developed along with the experiments. If the more senior people in the audience easily remember this history of the development of the relativistic heavy ion field, they may spend their time during this talk thinking about what is happening now and what will happen in the future in the ultrarelativistic heavy ion field, where I believe history is repeating itself.

EARLY HISTORY

My story (which only concerns experiments with which I have been associated) starts in the late 60's at the Bevatron, where I was studying high energy proton-nucleus collisions such as 5.5 GeV protons on uranium. With small silicon dE/dx-E telescopes, we were measuring energy spectra at various angles to the beam of what were then called the fragmentation products produced from high deposition energy reactions. These reactions we would now call central proton-nucleus collisions. The so-called fragmentation products, helium up through carbon or nitrogen, we would now call intermediate mass fragments. Earl Hyde and I studied these reactions systematically (see Fig. 1), measuring the energy spectra and angular distributions of all the isotopes.[1] There were several anomalies shown which are still not completely understood. These included the apparent low Coulomb barriers, the high "temperatures", and the higher "temperatures" for the neutron deficient isotopes. In addition, the angular distributions were forward peaked in the moving system deduced from the shifts in the energy spectra. In 1974 deuteron and alpha particle beams arrived at 2.1 GeV/nucleon. Although the yields of these fragmentation products were higher, the energy spectra displayed apparent temperatures that were only very slightly higher.[2] The small differences produced by the alpha beams compared to proton beams were disappointing.

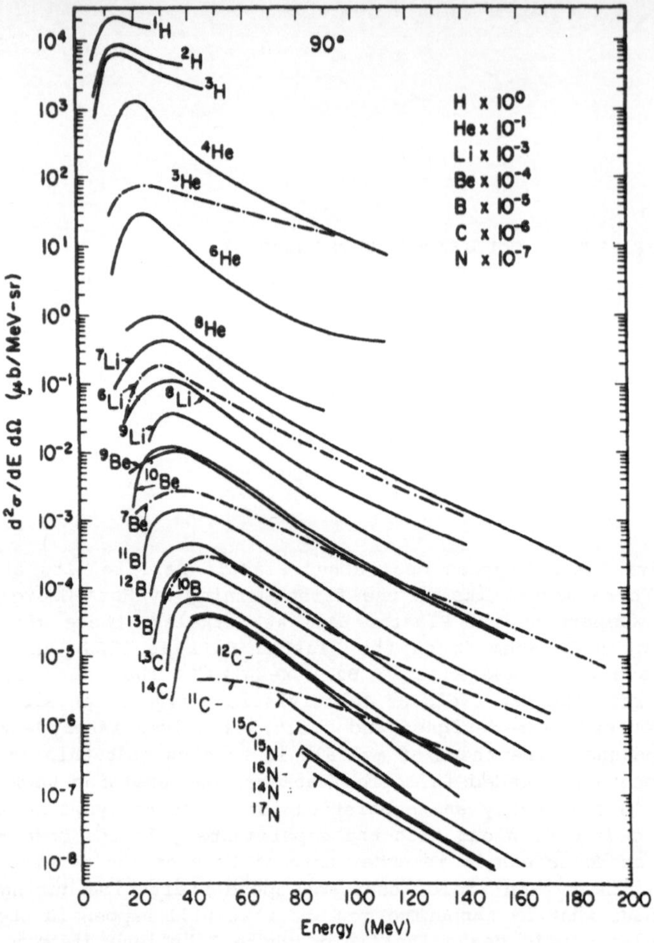

Fig. 1. Laboratory energy spectra at 90° to the beam for 5.5 GeV protons on an uranium target. The most neutron deficient isotopes are shown as dot-dash lines.

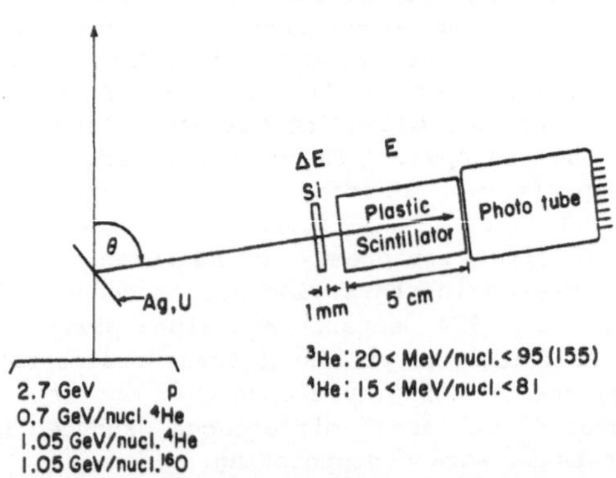

Fig. 2. The first silicon-plastic telescope.

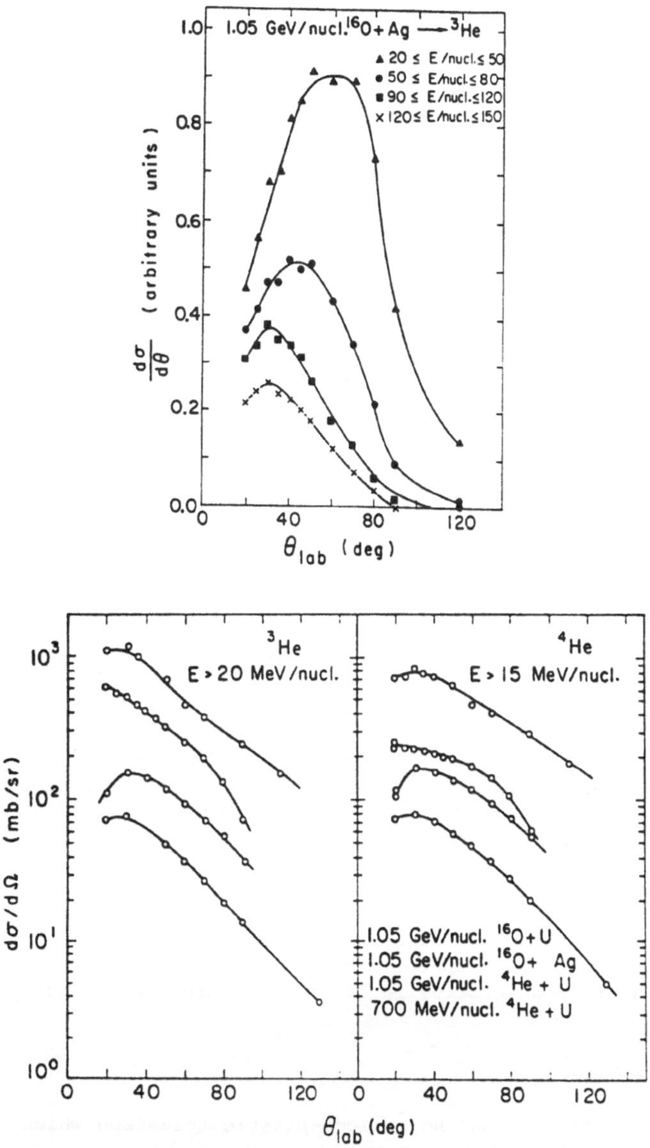

Fig. 3. Laboratory angular distributions plotted as $d\sigma/d\theta$ in the top figure, and as $d\sigma/d\Omega$ in the bottom figure. The top curve in the top figure is reploted as the second from the top curve in the lower left figure.

THE GSI-LBL COLLABORATION

In July 1974 Reinhard Stock and Rudolf Bock came to visit me to talk about starting a collaboration. We spent three hours in my office calculating what would be the best experiment to do with the upcoming heavier ion beams. Because of the expected low beam intensities, time of flight techniques with small solid angle were ruled out and we decided on a large area dE/dx-E telescope made of an array of silicon and germanium detectors. The plan was to continue the study of the fragmentation products with the heavier beams. We decided to go ahead with this GSI-LBL collaboration and in the fall of 1974 Hans Gutbrod and Andres Sandoval arrived.

In 1974 the transfer line was completed connecting the SuperHILAC to the Bevatron, creating the Bevalac which could produce 2 GeV/nucleon heavy ions, an energy more than 100 times higher than previously available. Initially the heaviest beam was oxygen, but the mass increased later to neon. In this new energy regime it was hard to imagine what to expect. Because this field of relativistic heavy ion physics was completely new to all of us, group meetings were very exciting, as everybody contributed on an equal basis. Nobody had any experience and the young people, as you will see, participated equally with the more senior people.

SHOCK WAVES?

While the large area silicon-germanium telescope was being planned and built, Prof. Schopper reported his shock wave peaks in AgCl track detectors irradiated by 0.9 GeV/nucleon oxygen ions.[3] In his angular distributions he observed a broad peak with small narrower peaks superimposed. We decided to proceed quickly to verify this. We took a 5 cm thick piece of plastic scintillator, slapped it on a phototube, put a 1 mm thick Li drift silicon detector in front of it, and placed it in my existing scattering chamber (see Fig. 2). This silicon-plastic telescope was designed to look for ^{3}He and ^{4}He fragments. We used a 1 GeV/nucleon oxygen beam to irradiate a silver target in order to approximate the conditions of the silver chloride track detectors. Our results[4] in 1975 when plotted (see Fig. 3 top) as $d\sigma/d\theta$ showed a nice broad peak at about 60 degrees in the laboratory, similar to the data of Prof. Schopper. However, plotting the data (see Fig. 3 bottom) as $d\sigma/d\Omega$, the way we thought it should be plotted, produced curves which were exponentially decreasing smooth lines with increasing laboratory angle. No peaks were observed – nothing that we could call a shock wave. It must be remembered that Prof. Schopper's small peaks were obtained by subtracting an evaporation component and using a method of three channel smoothing which is known to make statistical fluctuations appear more significant. In fact the narrow peaks have never been verified in later data which have clearly shown the collective flow effect.

ROUGH UNDERSTANDING

In 1976 we built a second silicon-plastic telescope which could measure protons as well as helium fragments. It consisted of a 2 mm thick Li drifted silicon detector and a 10 cm thick plastic scintillator. The old silicon-plastic telescope was used as a monitor detector. In addition a thin rounded dome was put on the scattering chamber and 15 plastic scintillators were placed outside the chamber as an associated multiplicity tag array (see Figs. 4 & 5). We quickly learned that a multiplicity array which covers only part of the solid angle is very sensitive to fluctuations in the number of particles which go into the array, so that one could not select on high multiplicity events. However, having for the first time a telescope which could measure protons in addition to the

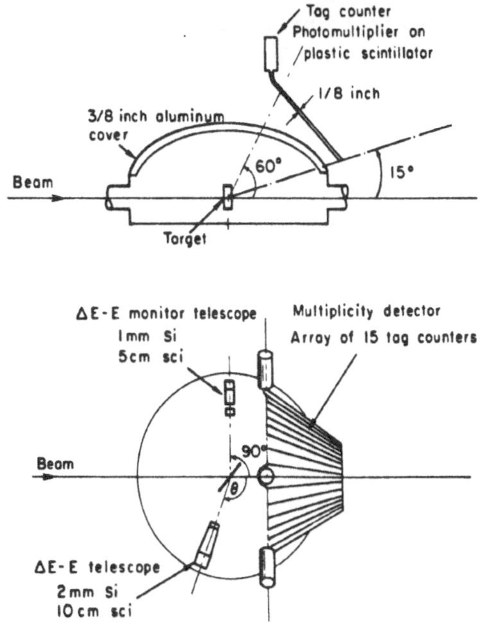

Fig. 4. The scattering chamber containing the second silicon-plastic telescope with the 15 counter tag array.

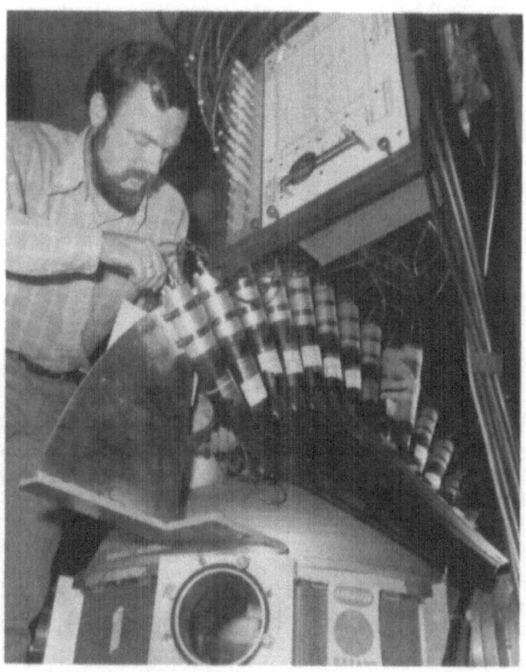

Fig. 5. Hans Gutbrod with the 15 counter tag array and scattering chamber.

heavier fragments, opened up to us a whole new world of physics. A young theoretician working with us, Jannik Johansen, discovered in 1976 that the spectra of the deuterons, tritons, ^{3}He, and ^{4}He particles could be explained in terms of the proton spectra raised to the Ath power (see Fig. 6). This was the first use of the Coalescence Model in relativistic heavy ion physics[5], and changed the whole emphasis of the work away from the fragmentation products to the protons. Only a few months later a young post-doc, Gary Westfall, combined the spectator-participant model of Wladek Swiatecki with a thermal breakup of the participant nucleons and found he could qualitatively describe the proton data (see Fig. 7). This was the birth of the Nuclear Fireball Model[6] (see Fig. 8), and suddenly we were in the exciting position that a rough description of all the data was at hand. Of course this was a tremendous stimulus to the theoreticians to try to do better.

By now the large area silicon-germanium telescope was ready (see Fig. 9), but we were now no longer interested in the intermediate mass fragments! However, we made systematic measurements of these fragments in 1976 and published them together with a summary of all our other work in a big paper in 1977. This paper, called "Central Collisions of Relativistic Heavy Ions"[7], is strongly recommended for the student. The Bevalac beams had only gotten as heavy as ^{20}Ne at that time, but the paper describes the use of rapidity plots, methods to analyze associated multiplicity, and a complete summary of the Coalescence and Fireball models. Of the authors of that paper, Jean Gosset and Gary Westfall have gone on to build their own 4π detectors, Hans Gutbrod, Reinhard Stock, Andres Sandoval and myself are in the CERN program, and Bill Meyer has gone to Bell Labs.

Since the main interest now was no longer the heavier fragments but was concentrated in finding out where the total mass from the reaction was going, we decided to concentrate on the hydrogen and helium isotopes. A new thick silicon-germanium telescope was built in 1977, consisting of a 5 mm lithium drifted silicon detector followed by seven cm of germanium (see Fig. 10). This was placed in a thin spherical scattering chamber designed by Hans Gutbrod, on the outside of which were 80 scintillation counters covering the whole forward hemisphere (see Figs. 11 & 12), because we had learned that in order to select high multiplicity events you have to detect almost all the charged particles. An exhaustive set of data was published for protons[8] and pions[9,10] with beams as heavy as ^{40}Ar, but it was soon discovered that the many different theoretical models then available all agreed with the proton data within a factor of two. Clearly, single particle inclusive measurements with associated multiplicity were not sufficient to distinguish between the models.

THE PLASTIC BALL

Thoughts then turned to measuring all the particles in each event. However, in 1979 the GSI-LBL collaboration split up, with Reinhard Stock and Andres Sandoval starting a group to work at the Bevalac Streamer Chamber. Just before, Reinhard had discouraged us from building a 4π liquid argon detector because of the difficulties involved. Then, one morning, Hans arrived at work and said he knew how to build a 4π detector. The key concept was putting two scintillators on one phototube. Later on we learned that Denys Wilkinson had published this idea 27 years earlier and called it phoswich. Hans-Georg Ritter was brought in to help build this detector. Thus, Hans, Hans-Georg, and myself formed the leadership of this other GSI-LBL collaboration, now called the Plastic Ball group.

While the Ball was being designed the outer part of the Plastic Wall was built at LBL. It was first used in 1979 as a forward angle array in conjunction with the existing thick silicon-germanium telescope in its

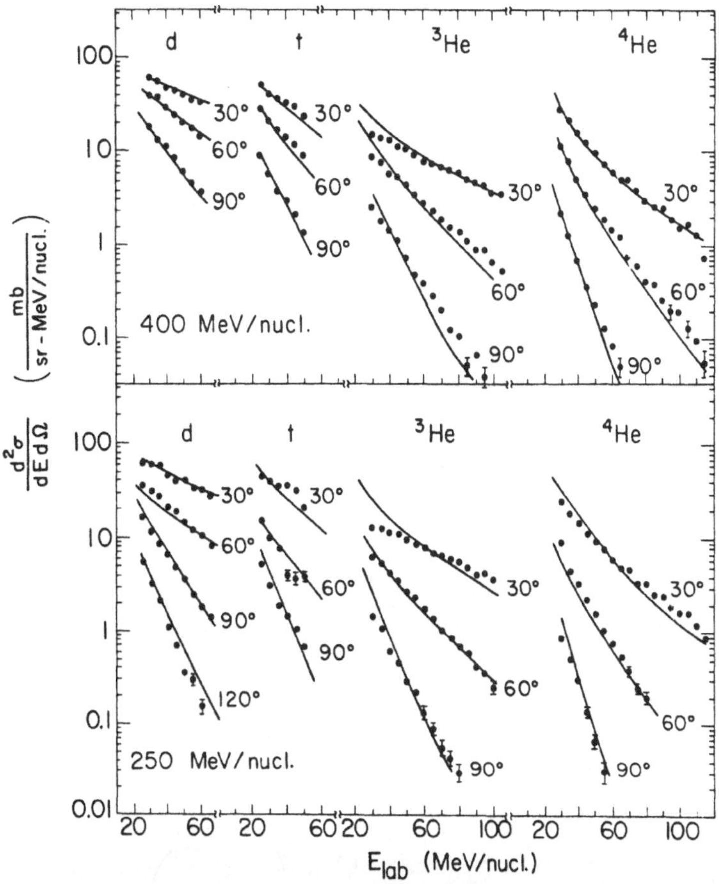

Fig. 6. Double differential cross sections for d, t, ^{3}He, and ^{4}He from ^{20}Ne + U at two bombarding energies. The solid curves from the Coalescence Model are calculated from the proton data raised to the Ath power.

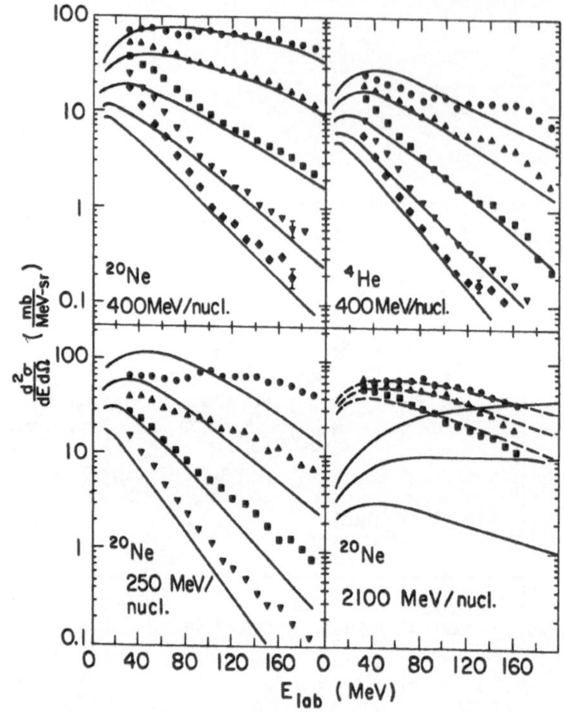

Fig. 7. Double differential cross sections for protons from ^{4}He and ^{20}Ne on uranium targets. The curves from top down as far as they go are for the laboratory angles of 30°, 60°, 90°, 120°, and 150°. The solid curves are calculated by the Fireball Model. The dashed curves on the lower right are fits from a two source Fireball Model.

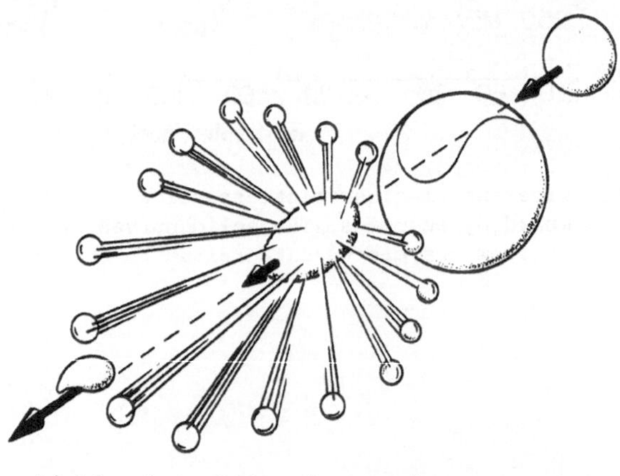

Fig. 8. An artist's view of the Fireball Model showing the spectator-participant geometry.

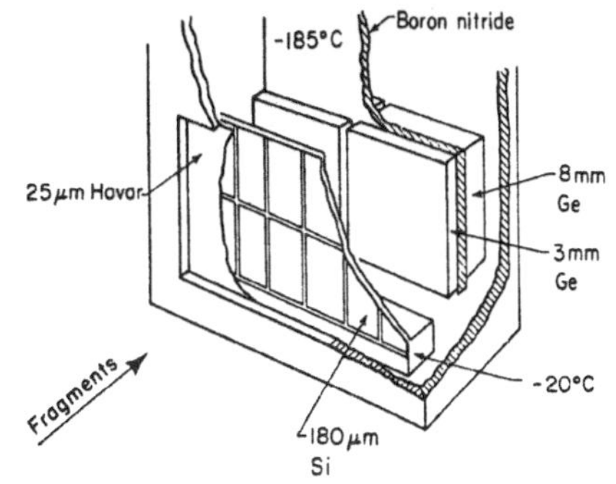

Fig. 9. The large area (20 cm^2) silicon-germanium telescope.

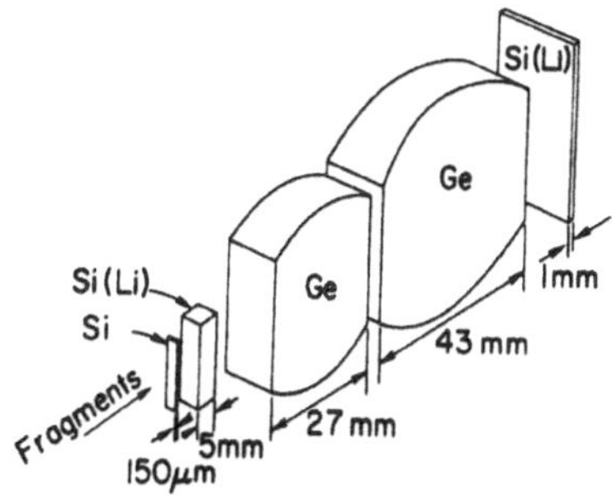

Fig. 10. The thick silicon-germanium telescope.

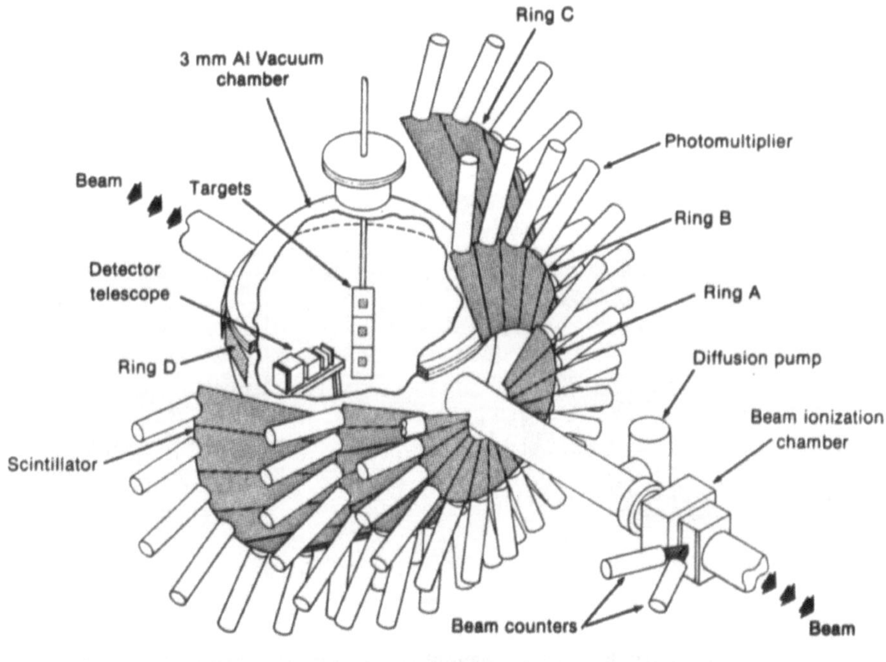

Fig. 11. The spherical scattering chamber containing the thick silicon-germanium telescope with the 80 counter associated multiplicity array.

Fig. 12. Art Poskanzer (left) and Reinhard Stock (right) next to the spherical scattering chamber.

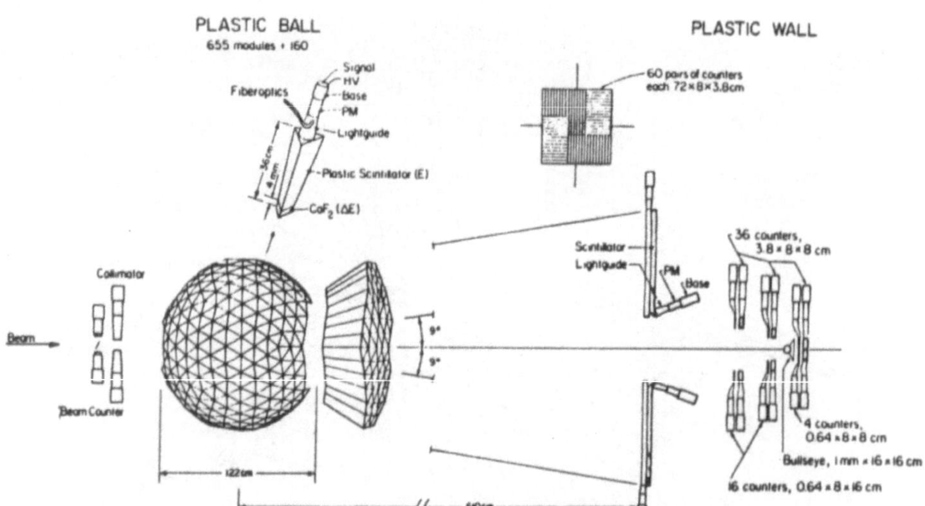

Fig. 13. The Plastic Ball/Wall layout. The modules from 9° to 30° are called the Mall. On the top are a view of a single module (left) and a view of the Wall as seen by the beam (right).

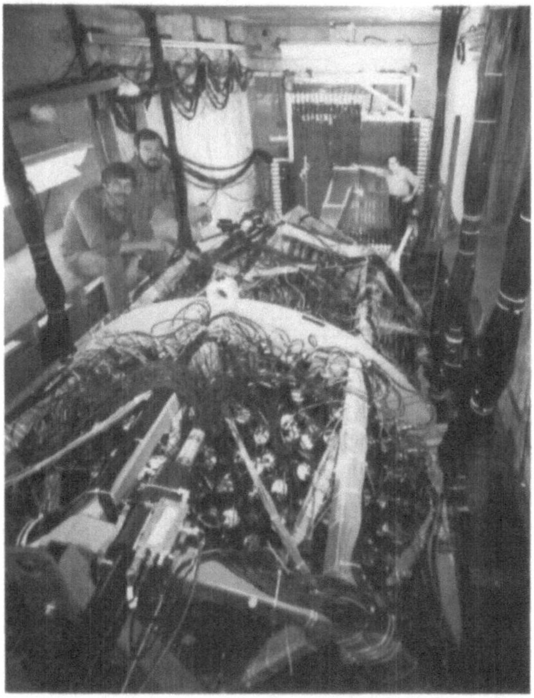

Fig. 14. The Plastic Ball/Wall. Hans-Georg Ritter and Hans Gutbrod are above the Ball, and Art Poskanzer is in front of the Wall.

Fig. 15. The Plastic Ball group in 1983. From left to right front row: Hans-Georg Ritter, Mohan Doss, Howard Wieman, and Burkhard Kolb. Middle row: Tim Renner, Tony Warwick, and Karl-Heinz Kampert. Back row: Herbert Löhner, Hans-Åke Gustafsson, Bernhard Ludewigt, Hans Gutbrod, and Art Poskanzer.

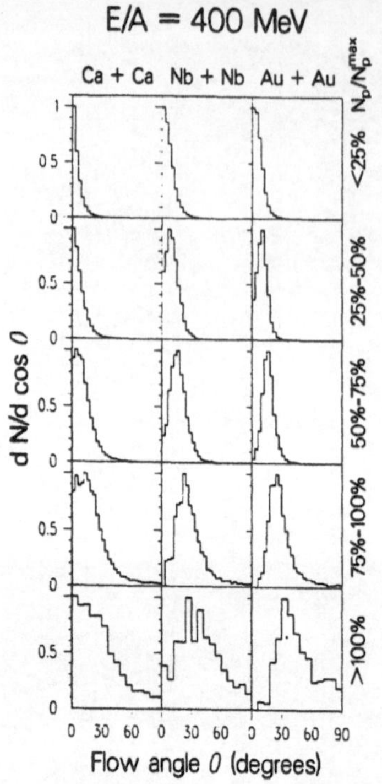

Fig. 16. Flow angle distributions for three equal mass target-projectile combinations and five normalized multiplicity bins at 400 MeV/nucleon.

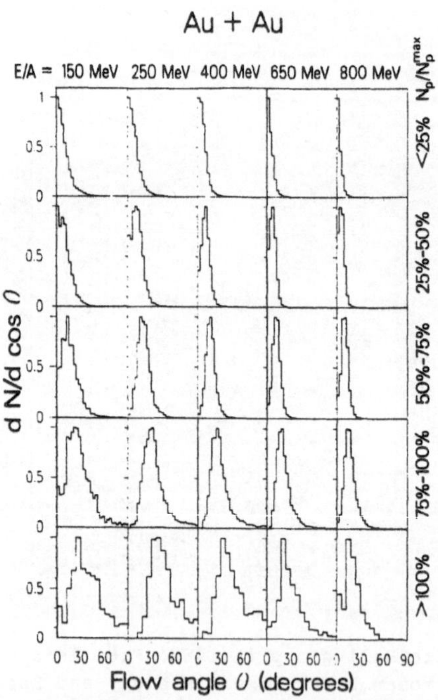

Fig. 17. Flow angle distributions for five beam energies and five normalized multiplicity bins for Au + Au.

spherical scattering chamber. The Ball took one year to design and to obtain the scintillators from Japan, and one year for the actual construction[11]. In 1981 it was ready to take beam with an augmented Plastic Wall (see Figs. 13 & 14). In 1982 the new Bevatron vacuum tank was installed and the beams delivered began to increase in mass. Thus the Plastic Ball proved to be the right detector in the right place at the right time.

Of the Plastic Ball group in 1983 (see Fig. 15)[12], Hans Gutbrod, Hans-Georg Ritter, Karl-Heinz Kampert, Hans-Åke Gustafsson, Burkhard Kolb, and myself are still working together in WA80 at CERN. Howard Wieman is at Berkeley developing a new 4π detector, a TPC, in conjunction with Hans-Georg Ritter. Tim Renner, Bernhard Ludewigt, and Tony Warwick are still at Berkeley working in other divisions of the Laboratory. Mohan Doss is at the University of Saskatchewan.

COLLECTIVE FLOW

We now faced the new challenge of how to analyze the 4π data. We tried looking at a few events in the center of mass but that was not very instructive. Hans-Georg Ritter, using a global analysis method called sphericity, was able to show in 1984 from our data with projectiles as heavy as niobium that, in fact, there were finite flow angles away from 0° (see Figs. 16 & 17). If you looked at the medium high multiplicity events in the niobium plus niobium collisions, you could see the matter flowing off to one side. This was the discovery of the collective flow of nuclear matter.[13] The effect was not readily visible in the calcium plus calcium data. A year later when we had gold plus gold data, the effect was very evident.

This collective flow effect was studied systematically as a function of multiplicity, target-projectile mass, and beam energy.[14,15] Our premise in doing this was that if one is ever going to learn about the equation of state of nuclear matter at high density, it is going to be through collective properties of high energy nuclear collisions such as this collective flow.

In 1984 we had the tenth anniversary of a very productive GSI-LBL collaboration. To commemorate this, we issued a souvenir photo which contained one picture of a Plastic Ball event and one picture of a streamer chamber event, representing the two GSI-LBL collaborations.

Improvements were made to the Plastic Ball in 1985 so that the Mall, which was approximately the forward hemisphere in the center of mass for 200 MeV/nucleon equal mass collisions, could detect intermediate mass fragments all the way from protons up through fluorine nuclei. Barbara Jacak and John Harris joined us and we saw that the heavy fragments have more collective flow than the lighter ones[16]. There is still more information in this data which has not been analyzed yet, which I think will lead to a better understanding of multifragmentation.

SUMMARY OF THE PLASTIC BALL

Collective flow was not the only result from the Plastic Ball. A summary of all the Plastic Ball physics would include the following: the average momentum in the reaction plane, which is an indication of the pressure built up in the reaction; rapidity distributions, dN/dy, which are an indication of thermalization or stopping; ratios of clusters, like deuterons to protons, from which we obtained information about entropy; even two-particle correlations, which led to information about the size of the source; and finally the intermediate mass fragments, which are expected to lead to information about multifragmentation. We have actually

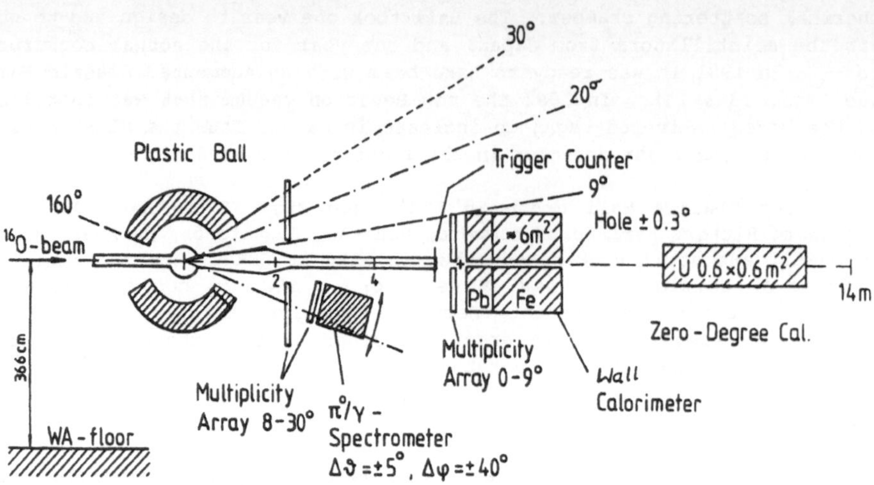

Fig. 18. Layout of the WA80 experiment showing the Plastic Ball.

just finished writing a review paper called "Plastic Ball Experiments."[17] The Plastic Ball filter, describing the acceptance of the Plastic Ball, is available as a Fortran program on EARN/Bitnet from POSK@LBL. We believe that a systematic comparison of filtered theoretical simulations with the Plastic Ball data as a function of multiplicity, target-projectile mass, and beam energy, including not only collective flow, but also other observables such as dN/dy and cluster production, will eventually lead to a quantitative understanding of the nuclear equation of state.

As a slight digression, I want to talk about the terminology used at this conference. The experimentalists have talked about participant side splash and collective flow while the theoreticians, for the same phenomenon, have been using words like shock waves, bounce off, and squeeze out. Experimentalists should be allowed to name the effects that they discover. Shock waves, a theoretical concept, is probably very important in understanding collective flow, but collective flow describes experimentally what is observed. For the more quantitative description, the slope at mid-rapidity on a p_x vs. y plot, we have used the words flow or flow parameter. In retrospect, I think we should have used the words directed flow. The recent observation that there is more flow perpendicular to the reaction plane than in the reaction plane we have called squeeze out, in agreement with the theoretical prediction, which was called off-plane squeeze-out. The word bounce off, which we use for the spectator recoil, so far has had no word assigned to it by the theoreticians.

CERN

In 1986 the Plastic Ball was put in a big box and shipped off to CERN to start a new life as a target rapidity detector for heavy ion reactions at energies up to 200 GeV/nucleon (see Fig. 18) in experiment WA80 at the SPS at CERN.[18] Again, this is a factor of 100 increase in beam energy, opening up a whole new exciting field where we have to learn how to do the experiments and analyze the data, with theoretical developments going on in parallel.

The other GSI-LBL collaboration, the steamer chamber group, also went to CERN to do an experiment called NA35. This year there is hope that these two GSI-LBL collaborations will rejoin for the expected program with lead beams at the CERN SPS. This will bring back together these very fruitful and friendly collaborations, continuing the long history of GSI-LBL cooperation.

ACKNOWLEDGEMENTS

I would like to thank all my colleagues who made this exciting history possible, and the theoreticians who gave this work meaning. Also, I would like to thank Walter Greiner for inviting me to present this history talk, and then encouraging me to write it up.

REFERENCES

1. A.M. Poskanzer, G.W. Butler, and E.K. Hyde, Phys. Rev. C3, 882 (1971).
2. A.M. Zebelman, A.M. Poskanzer, J.D. Bowman, R.G. Sextro, and V.E. Viola, Phys. Rev. C11, 1280 (1975).
3. H.G. Baumgardt, J.U. Schott, Y. Sakamoto, E. Schopper, H. Stöcker, J. Hoffmann, W. Scheid, and W. Greiner, Z. Physik A273, 359 (1975).
4. A.M. Poskanzer, R.G. Sextro, A.M. Zebelman, H.H. Gutbrod, A. Sandoval, and R. Stock, Phys. Rev. Lett. 35, 1701 (1975).
5. H.H. Gutbrod, A. Sandoval, P.J. Johansen, A.M. Poskanzer, J. Gosset, W.G. Meyer, G.D. Westfall, and R. Stock, Phys. Rev. Lett. 37, 667 (1976).
6. G.D. Westfall, J. Gosset, P.J. Johansen, A.M. Poskanzer, W.G. Meyer, H.H. Gutbrod, A. Sandoval, and R. Stock, Phys. Rev. Lett. 37, 1202 (1976).
7. J. Gosset, H.H. Gutbrod, W.G. Meyer, A.M. Poskanzer, A. Sandoval, R. Stock, and G.D. Westfall, Phys. Rev. C16, 629 (1977).
8. A. Sandoval, H.H. Gutbrod, W.G. Meyer, R. Stock, Ch. Lukner, A.M. Poskanzer, J. Gosset, J.-C. Jourdain, C.H. King, G. King, Nguyen Van Sen, G.D. Westfall, and K.L. Wolf, Phys. Rev. C21, 1321 (1980).
9. K.L. Wolf, H.H. Gutbrod, W.G. Meyer, A.M. Poskanzer, A. Sandoval, R. Stock, J. Gosset, C.H. King, G. King, Nguyen Van Sen, and G.D. Westfall, Phys. Rev. Lett. 42, 1448 (1979).
10. K.L. Wolf, H.H. Gutbrod, W.G. Meyer, A.M. Poskanzer, A. Sandoval, R. Stock, J. Gosset, J.-C. Jourdain, C.H. King, G. King, Nguyen Van Sen, and G.D. Westfall, Phys. Rev. C26, 2572 (1982).
11. A. Baden, H.H. Gutbrod, H. Löhner, M.R. Maier, A.M. Poskanzer, T. Renner, H. Riedesel, H.G. Ritter, H. Spieler, A. Warwick, F. Weik, and H. Wieman, Nucl. Instr. and Meth. 203, 189 (1982).
12. Other members of the Plastic Ball group have included Rudolf Albrecht, Andrew Baden, Peter Beckmann, Göran Claesson, John Harris, Barbara Jacak, Per Kristiansson, Francois Lefebvres, Herbert Löhner, Michael Maier, Hubertus Riedesel, Rainer Schicker, Hans Rudolf Schmidt, Teodor Siemiarczuk, Helmuth Spieler, Joana Stepaniak, Larry Teitelbaum, Mark Tincknell, Friedeman Weik, Steve Weiss, and Wojciech Wislicki.
13. H.-Å. Gustafsson, H.H. Gutbrod, B. Kolb, H. Löhner, B. Ludewigt, A.M. Poskanzer, T. Renner, H. Riedesel, H.G. Ritter, A. Warwick, F. Weik, and H. Wieman, Phys. Rev. Lett. 52, 1590 (1984).
14. K.G.R. Doss, H.-Å. Gustafsson, H.H. Gutbrod, K.H. Kampert, B. Kolb, H. Löhner, B. Ludewigt, A.M. Poskanzer, H.G. Ritter, H.R. Schmidt, and H. Wieman, Phys. Rev. Lett. 57, 302 (1986).
15. H.-Å. Gustafsson, H.H. Gutbrod, J.W. Harris, B.V. Jacak, K.-H. Kampert, B. Kolb, A.M. Poskanzer, H.G. Ritter, and H.R. Schmidt, Modern Physics Lett. 3, 1323 (1988).

16. K.G.R. Doss, H.-Å. Gustafsson, H. Gutbrod, J.W. Harris, B.V. Jacak, K.-H. Kampert, B. Kolb, A.M. Poskanzer, H.G. Ritter, H.R. Schmidt, L. Teitelbaum, M. Tincknell, S. Weiss, and H. Wieman, Phys. Rev. Lett. 59, 2720 (1987).
17. H.H. Gutbrod, A.M. Poskanzer, and H.G. Ritter, to be published in Reports in Progress in Physics (1989).
18. Albrecht et al., CERN Report CERN/SPSC/85-39 (1985).

The SIS/ESR–Project at GSI - Present and Future

P. Kienle
Gesellschaft für Schwerionenforschung mbH
D-6100 Darmstadt, West Germany

This is a report on the status of the construction of the SIS/ESR– accelerator complex and the experimental equipment at GSI as of May 1989.

1 Introduction,Overview

In 1985, GSI received from the BMFT of the FRG the approval and funds to extend its heavy ion acceleration facilities to relativistic energies. Fig. 1 gives an overview of the heavy ion accelerator complex under construction [1]. It consists of an upgraded UNILAC used as an injector into a medium energy (1–2 GeV/u) heavy ion synchrotron SIS 18 [2,3] which is connected with a storage cooler ring ESR [4] of half the circumference of SIS 18.

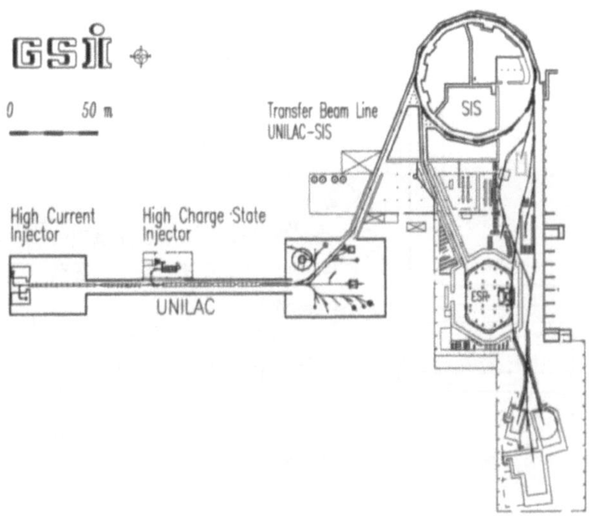

Fig.1. Layout of the upgraded UNILAC, SIS, ESR. The fragment separator (FRS) and the new experimental area are also indicated.

The combination of these two rings will allow to produce completely stripped heavy ion beams up to U^{92+} with the highest possible phase space densities, achieved by various beam cooling techniques. In addition SIS/ESR will provide beams of radioactive nuclei in the energy range from several MeV/u up to 1–2 GeV/u, again cooled to the highest possible phase space densities. The beams of the ESR may be used either circulating with high currents or extracted with a great variety of time structures and intensities. They may also be reinjected into SIS for further acceleration or deceleration. There will be a large experimental area with several experiments set up on beams from both SIS and ESR. Further experimental areas are located directly behind SIS, between SIS and ESR, and around the ESR. In future one can think of injecting the high phase space density beams of completely stripped ions into superconducting collider rings with small apertures, modest size and cost to achieve very high cm–energies ($\geq$ 20 GeV/u).

2 Accelerators

2.1 UNILAC Upgrade

Very recently we changed our injector concept into the UNILAC in such a way that we can run a truly independent low energy program with a free choice of ion species and energy parallel to a low duty–factor high current injection cycle into SIS 18. The SIS injection is based on recently developed high intensity ion

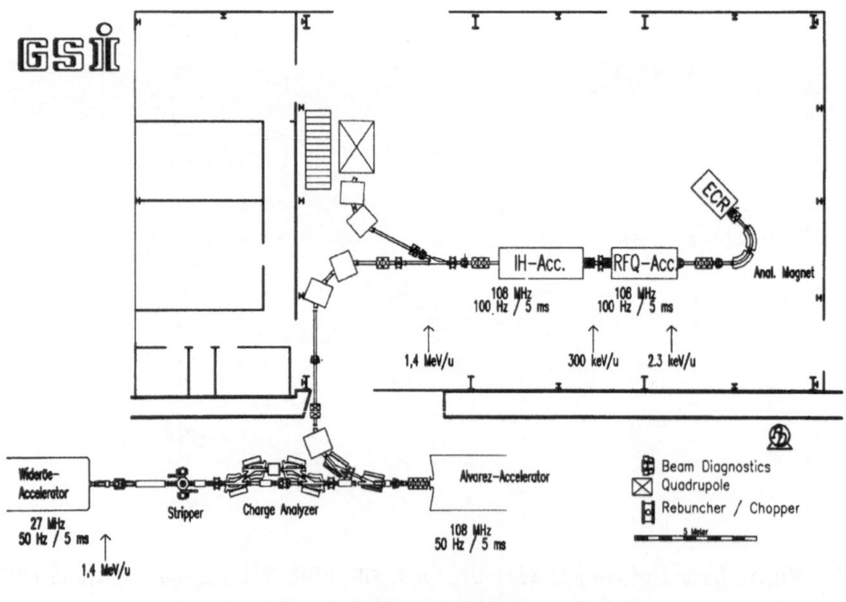

Fig.2. Layout of the new high charge injector.

sources [5] for low charge states (U^{2+}) which will be accelerated by 27 MHz RFQ structures up to 130 keV/u and after stripping injected straight into the second Wideröe tank. This high current injector will be operated with a duty-factor of 1%, which is sufficient for synchrotron injection. It can provide 100–1000 times more injection current than the present UNILAC.

For the low energy UNILAC-program we construct an independent injector (Fig.2) [6], which consists of a 14.5 GHz ECR-source, a RFQ linac for energies up to 300 keV/u, followed by an interdigital line structure up to the injection energies of the Alvarez section (1.4 MeV/u). These structures will be operated at 108 MHz and a duty-factor of 50%. The U^{28+}-current is specified as 5μA, which is more than an order of magnitude larger than the one presently available. It also has an improved microstructure delivering a pulse each 9 ns, which is favourable for coincidence experiments with fast detectors. In addition it can be debunched, thus achieving dc beams, if desired.

2.2 Heavy-Ion Synchrotron SIS

Since April 1985, a heavy-ion synchrotron (SIS 18) is under construction. It has a bending power of 18 Tm and a circumfence of 216.72 m.

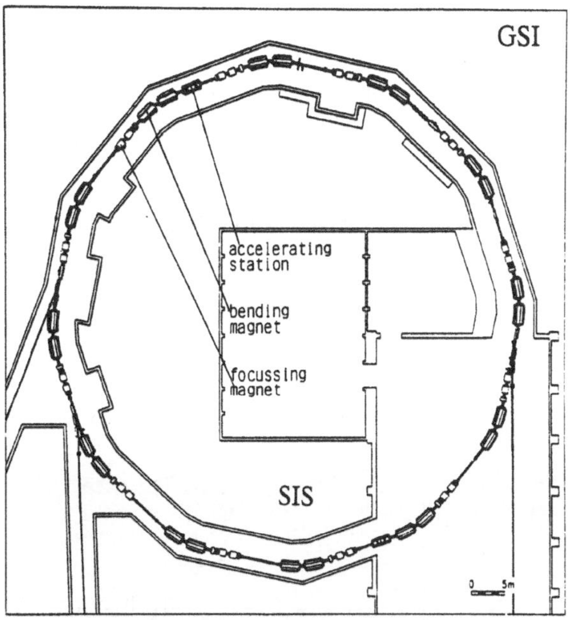

U = 216,72 m (B·ρ)m = 18 Tm

Fig.3. Layout of the synchrotron for heavy ions (SIS 18).

The heavy ion beam accelerated in the UNILAC up to 11.6 MeV/u, and stripped to an adequate high charge state for the desired energy and intensity, is injected into SIS 18 during 10 to 30 turns and then accelerated with a repetition rate between 3 Hz (up to 1.2 T) and 1 Hz (up to 1.8 T) to maximum energies, depending on the charge states of the ions as shown in fig.4.

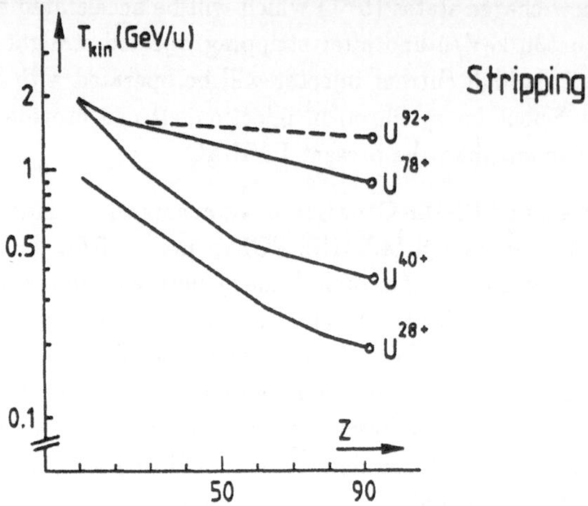

Fig.4. Maximum achievable energies at SIS 18 as a function of nuclear charge. The energies are given for a gas or a foil-stripper at an energy of 1.4 MeV/u, resulting in a relatively low degree of ionization. If a second stripper at 11.4 MeV/u is added or if completely ionized particles from the experimental storage ring ESR are reinjected into the synchrotron, higher energies can be achieved.

For uranium ions at a charge state of q=78, after stripping at 11.4 MeV/u with a foil target, 1 GeV/u is achieved as maximum energy. The maximum beam intensities from SIS 18 are shown in fig.5 for Ne– and U–ions of various charge states, depending on the stripping procedure, as a function of their specific energies. The decrease of intensities towards higher energies is caused by a small decrease of the synchrotron repetition rate. The drop for 1 GeV/u Ne and 500 MeV/u U is due to a change of the repetition rate from 3 to 1 Hz.

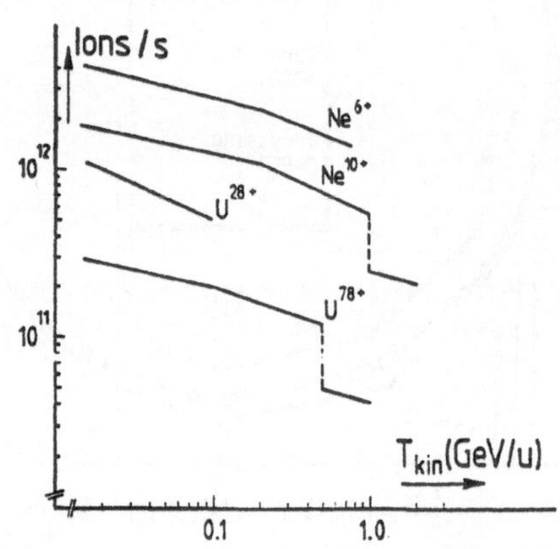

Fig.5. Beam currents for various charge states of Ne– and U–ions, obtained by different stripping procedures as a function of the energy. The intensity drops for Ne^{10+} and U^{78+} are due to a decrease of the repetition rate from 3 Hz to 1 Hz.

Between SIS and ESR the beam may be stripped once more to the highest desired charge state. The ESR with a bending power of Bρ = 10 Tm allows to store ions up to U^{92+} with the following maximum energies: Ne^{10+} (834 MeV/u), Ar^{18+} (709 MeV/u), Kr^{36+} (656 MeV/u), Xe^{54+} (609 MeV/u), and U^{92+} (556

MeV/u). The uranium ions can be fully stripped at this energy in a Cu-target of 100 mg/cm^2 thickness with an efficiency of 60% [7]. The stripping yield increases strongly with decreasing nuclear charge, thus one expects a yield of 70% for Pb^{82+}-ions (574 MeV/u), and already 100% for Xe^{54+} (609 MeV/u). Alternatively one can install a reaction target for projectile fragmentation. The favourable kinematic focusing of the products around the beam direction and velocity allows an effective mass separation in the projectile fragment separator (FRS; see sect.(3.1) below) between SIS and ESR, followed by accumulation of radioactive beams with the ESR, which accepts beams with $\Delta p/p = \pm\ 0.5\%$ and transverse emittances of 20 πmm mrad.

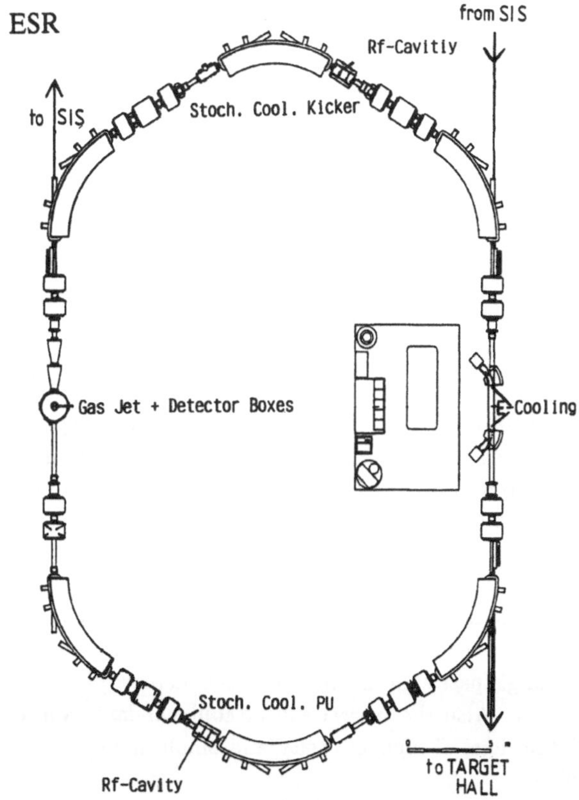

Fig.6. Layout of the experimental storage ring (ESR) for heavy ions.

The whole SIS-ring has been installed in the tunnel and connected to power supplies and cooling circuits. The transfer line between UNILAC and SIS was taken in operation in fall of 1988. On May 18,1989 a $20\mu A$ beam of $^{132}Xe^{21+}$, injected at 11.6 MeV/u was accelerated for the first time to energies above 100 MeV/u. The experimental program will start in December 1989.

2.3 The Storage and Cooler Ring ESR

The ESR (Fig.6) with a circumfence of 108.36 m has two 9.5 m long straight experimental sections, in one of which an electron cooling device will be installed. The other four straight sections will be used for the installation of rf cavities, and slow and fast extraction elements. The rf cavities are used for acceleration, deceleration, and especially also for bunching the beam together with the electron cooling for reduction of the occupied longitudinal phase space volume. With the fast extraction system of the ESR one can transfer a highly ionized and cooled beam back to SIS 18 for further acceleration. The optics of the ring allows three modes of operation: one with a moderate dispersion along the ring specially suited for the accumulation of beams with large momentum spread ($\Delta p/p$ = 1%) and emittance (E_{hv} =20 πmm mrad), one with zero dispersion in the straight sections, which allows multi-charge operation ($U^{89+} - U^{92+}$), and one with large dispersion to accomodate two beams of slightly different momenta, which then may be brought to merge with a well defined angle of about 100 mrad [8]. This can be used to study collisions of *two* highly ionized beams at fixed target equivalent energies of up to 7.2 MeV/u and an energy definition better than 10%.

The most important facilities of the ESR are various cooling devices which can be applied complementary. For secondary beams with low phase space density stochastic pre-cooling may be used. For cooling to very high phase space density, electron cooling of completely stripped heavy ions is foreseen in an interaction zone of 2 m length. A "cool" electron beam of 5–10 A is focussed within an area of 5 cm diameter collinearly along the ion beam at the corresponding average velocity. For cooling of ion beams between 30 MeV/u and 560 MeV/u, electron energies in the range of 16.5 keV and 310 keV are required. With an electron beam current density of up to 1 A/cm^2 and ion beams of initially $\Delta p/p$ = 0.1 % and 4 πmm mrad cooling times of 30 ms for U^{92+} at 500 MeV/u are expected. Heavy ion beams with emittances as small 0.1 πmm mrad and momentum spreads of less than 10^{-5} may be produced. Space charge effects limit the number of ions to be cooled in a circulating beam [9].

While the cooled beam circulates in the ring, it may be used in the second straight section for the study of collision processes with internal targets, which may be atomic or electron beams (unpolarized or even polarized), gas jets or fibres. For all experiments which need thin targets a large gain in luminosity may be achieved compared with a single pass experiment due to the increase of the circulating beam current ($\approx 2 \times 10^6$). Also the interaction of collinear laser and electron beams with the circulating ions of high intensity and small momentum spread may be favourably studied.

The ESR is constructed parallel to SIS. The magnets, power supplies, UHV systems, and the electron cooling device are under construction. In summer 1988 the installation of the ESR magnets has begun with the goal to start with commissioning and cooling experiments after SIS has started to be used for experiments. Already the initial experimental program at the ESR will be quite diversified. It will start with electron cooling of heavy ion beams to find out the limits of phase space densities which can be achieved. The ultimate aim will be to reach a plasma parameter $\Gamma = (Z^2e^2/a)/kT$ (with Z being the charge of the ion, a the Wigner–Seitz radius, and kT the temperature) as large as possible in order to come close to a critical value ($\Gamma_c \approx 170$) at which a transition to a crystalline state should occur. Unfortunately the lattice of the ESR will introduce shears due to the bend-

ing of the beam in the dipole magnets so high that a complete order may not be achieved.
Another goal is the cooling of radioactive beams which are delivered from the fragment separator. These beams which have low phase space densities, will be pre-cooled by stochastic cooling and then brought to still lower temperature by electron cooling.
By performing Schottky-scans of the revolution frequency of the radioactive cooled beams, mass determinations with an accuracy of 10^{-5} to 10^{-6} should be possible. For very short lived activities mass measurements by time of flight techniques in an achromatic operation mode of the ESR are proposed.
With cooled radioactive beams various nuclear reaction experiments are proposed. One class being precision reaction spectroscopy experiments at energies up to 200 MeV/u, using inverse reaction kinematics. Proton, deuteron or $^{3,4}He$-targets may be bombarded with cooled radioactive beams. By measurement of the emission angle and the energy of the light reaction products, reaction spectroscopy with a resolution of about 50 keV may be carried out on radioactive nuclei.
With completely ionized nuclei a new decay mode, namely β-decay to bound states, may become observable in the ESR. Such processes may occur in highly ionized plasmas of stars and thus may be used as thermometers. Electron capture is the inverse process, therefore by studying bound state β-decay one may be able to measure interesting weak coupling matrix elements.

3 Experimental Facilities

3.1 Projectile Fragment Separator (FRS)

SIS will provide high-intensity beams of relativistic heavy ions. Secondary beams of radioactive isotopes can be produced by projectile fragmentation and electromagnetic dissociation. The projectile fragments are emitted in forward direction with velocities close to those of the projectiles, so that they can be separated in flight and injected into the storage ring ESR. The separator for projectile fragments (FRS, see fig. 7) is planned to separate isotopes up to uranium according to A and Z at specific energies in the range between 0.1 and 1.0 GeV/u. The principle of the isotopic separation is based on a combination of magnetic analysis ($B\rho$) and energy loss (ΔE) of the fragments in matter.

The keys for the separation are an achromatic magnetic system, characterized by a high resolving power independent of the velocity spread of the fragments ($\Delta v/v = 1\%$) and a profiled degrader at the dispersive focal plane, providing the separation in Z and A. The separator, located in a beam transfer line from SIS to ESR (see fig.1) can deliver separated beams of projectile fragments either to the experimental area or to the ESR.
The FRS consists of 4 magnetic dipole stages with focussing quadrupoles. The production target is positioned at the entrance of the FRS. Radioactive beams with intensities up to $5 \times 10^8/s$ can be produced using target thicknesses of $\approx 1g/cm^2$. The primary beam has to be separated from the fragments at the focal plane of the first dipole section. Because all fragments have similar velocities the first separation selects only those with a certain A/Z-ratio i.e. all fragments with the same magnetic rigidity are focused on the degrader. The energy loss of the fragments in the degrader provides the additional selection needed for the separation of a

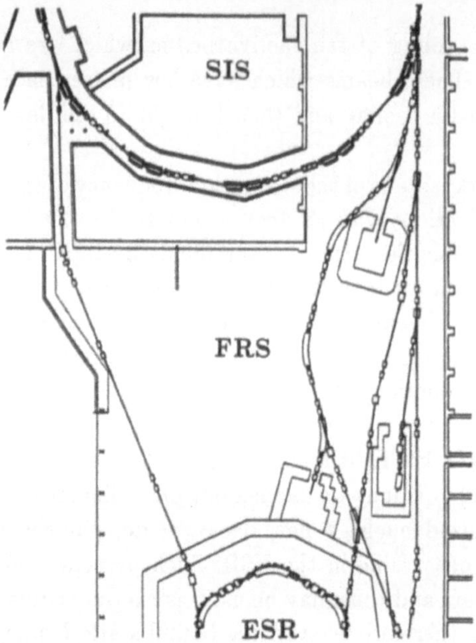

Fig.7. Layout of the fragment separator (FRS).

nucleid. The shape of the degrader is chosen to preserve the velocity achromatism of the system. The separation quality at the final focus is strongly dependent on the ion–optical resolution of the separator, the thickness and perfection of the degrader and the energy of the fragments. Separation in A and Z of all fragments up to mass 240 seems possible with low background.

The physics program at the FRS is expected to become very diversified. First we will focus on the study of the fragmentation process especially with heavy masses including Coulomb–dissociation using the relativisticly enhanced field of fast ions. Besides the measurement of A and Z distributions the study of the momentum and energy transfer on the fragment should shed light on the reaction dynamics. With separated isotopes detailed nuclear structure studies of heavy neutron rich nuclei should be posssible. Then of course high energy radioactive beams may be produced and used for reaction studies, in particular also in combination with the ESR, in which they may be cooled and decelerated.
A very different class of reactions which may be favourably investigated with the FRS are fusion reactions at high energies using inverse kinematics, like for example $^{12}C(p;\gamma,\pi^0)^{13}N$. The heavy fusion products emitted in a small forward cone may be identified and completely momentum analyzed with 100% detection efficiency. Thus, very rare processes can be investigated. Another interesting field is connected with the proposed study of Δ–production in quasielastic collisions, for which the FRS may be used as a high resolution spectrometer. Very rare processes like subthreshold production of K^- and antiprotons can advantageously be studied at the FRS.

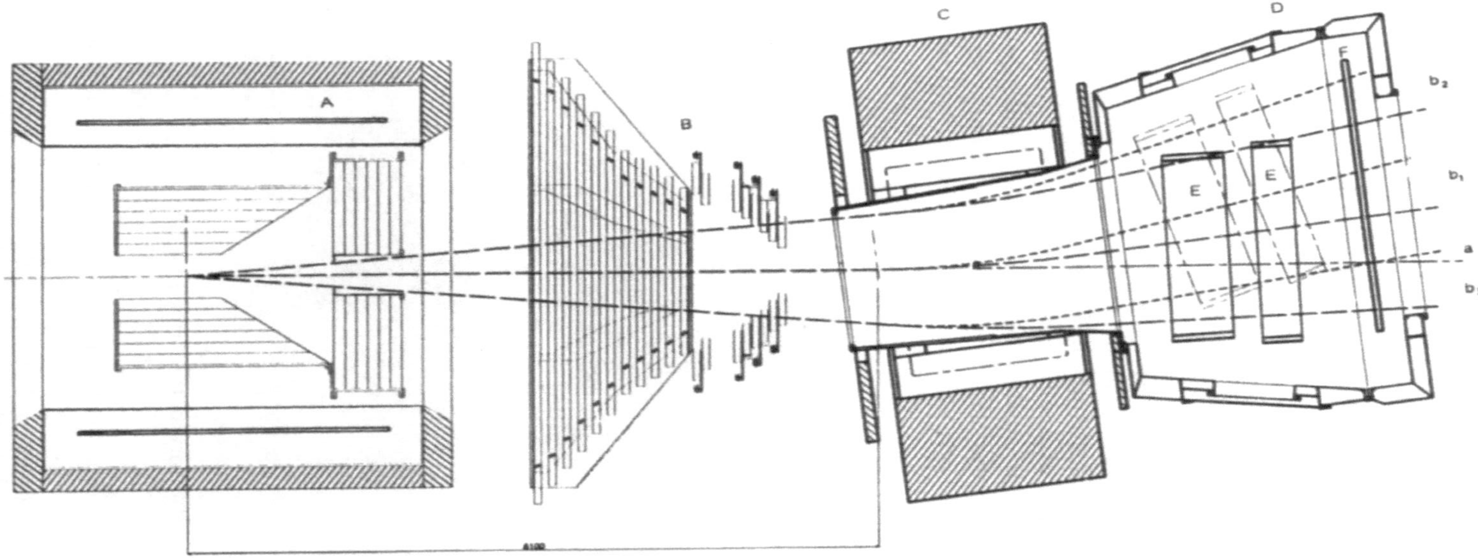

Cross sectional view of the combined setup. The symbols denote **A**: *the central drift chamber;* **B**: *the forward plastic wall;* **C**: *return yoke of the* **ALADiN** *magnet;* **D**: *detector vacuum chamber;* **E**: **TP-MUSIC** *detector;* **F**: *the* **TOF** - *wall;* **a** *indicates the symmetry axis of the setup;*

Fig.8. Layout of the complete 4π–detector including ALADIN.

3.2 4π–Detector

For the study of central collisions an advanced 4π–detector for charged particles including a forward–spectrometer (ALADIN, see sect.(3.3) below) a large BaF_2-detector array (TAPS, see sect.(3.5) below) for high energy photon spectroscopy and a large area neutron detector (LAND, see sect.(3.6) below)is under construction. This device is designed to measure the complete momentum flow ($d^3\sigma/d\vec{p}$) for all charged particles originating from a hard collision, which will allow to analyze in substantial detail the collective nuclear matter flow first observed in exclusive experiments by Gustafson et al. [10].
A schematic lay–out of the complete detector system is shown in fig.8. The target is placed in a large solenoid (2.4 m diameter and 3.34 m length) with superconducting coils, which produce a uniform magnetic field ($\Delta B/B \leq 2.5\%$) up to 0.6 T. The target is surrounded by a central drift chamber and a barrel of plastic scintillators for time of flight measurements. There is additional space for Cherenkov–detectors, which may be necessary for K^+–identification. In forward direction between 7° and 30° a drift chamber which measures the transversal momentum components of the particles and a "spaghetti–type" time of flight detector is placed. It consists of 512 rods of plastic scintillators with cross sections of $20 \times 24mm^2$ with fast photomultipliers on both sides. They are arranged in an octogon shaped roof–like structure as indicated in fig.9.

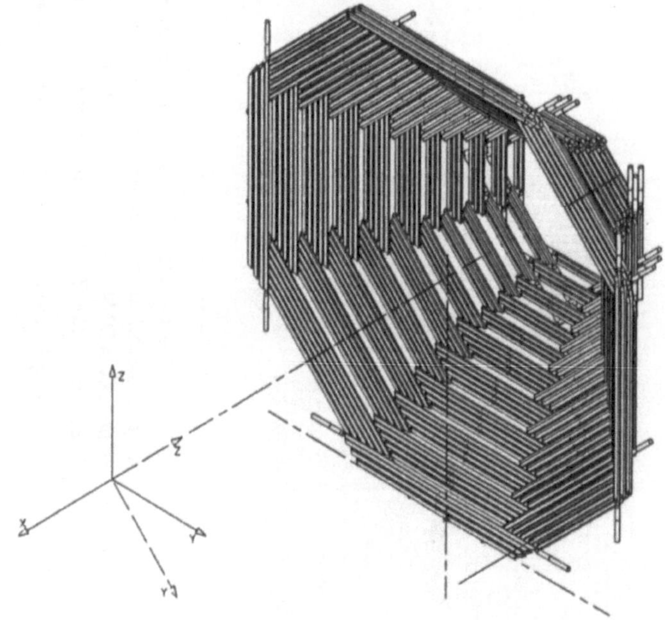

Fig.9. Perspective view of the forward plastic wall.

For a better Z–identification of larger fragments emitted in forward direction, ionisation chambers also are included.
Thus, the 4π–detector as a whole allows a momentum, velocity and energy loss determination of all charged particles with high granularity. All particles, including K^+–mesons can be identified and their momentum be determined on an event–

by–event basis. To complete such an analysis a forward–spectrometer is added in the forward direction which covers the angular range between 0° and 5°.

3.3 Forward–Spectrometer (ALADIN)

The Forward-Spectrometer will be built from an already existing magnet which has been obtained from CERN as a temporary loan (MNP21). It has an aperture of $1.5 \times 0.5\ m^2$ and a bending power of 2.3 Tm (see fig.8). This spectrometer, capable of detecting and identifying nuclear fragments up to the largest masses and momenta to be expected at SIS will complement the 4π–facility in the forward direction (±2° vertically,±5° horizontally).
As a tracking and Z–identification device a multiple sampling position sensitive ionisation chamber will be used. Time of flight is measured with a wall of 2×100 vertical plastic scintillator rods of 110 cm length.The momentum resolution of the forward spectrometer should be better than 1%, the ionisation chamber gives a complete Z resolution up to the heaviest atomic number and from the long time of flight (10 m) a good velocity resolution is expected.
The forward spectrometer will be used to study the mechanisms of multifragmentation. A special application will be the investigation of exclusive multifragmentation using inverse reaction kinematics.

3.4 Two Arm Photon Spectrometer (TAPS)

The production of γ–rays, π^0- and η^0–mesons will be studied with a Two Arm Photon Spectrometer (TAPS) consisting of four to six arrays of 64 BaF_2–crystals, each being 12 radiation lengths deep, which are arranged in 2 or 3 tower structures. High energy γ–spectroscopy may be a useful probe to investigate the temperature and possibly the energy density of the hot nuclear matter in an unambiguous way. It may also be possible to study directly the production and decay of barionic resonances. At higher bombarding energies the combinatorial background of many γ–rays from π^0–dacay may prevent single photon spectroscopy or will make it very difficult. At these energies it may be possible to study the subthreshold production of η_0 at high compression.

3.5 Large Area Neutron Detector (LAND)

A large area neutron detector (LAND) will be installed at the SIS- facility. It consists of a structure of subsequent layers of converter material (Fe, thickness 5 mm) and active plastic material (BC408, 5 mm thick) with a total thickness of 1 m. The total area of 2×2 m^2 is subdivided into 200 separate cells (paddles) of 200×10×10 cm^3. Subsequent layers of the paddles are arranged perpendicular to each other. Each detector is read out by two 2" photomultipliers at the small front faces of the paddle. With a time of flight path of 16 m, LAND may be used as a high resolution neutron spectrometer.
The LAND will be used to study extremely peripheral collisions where multiple excitation of the Giant Dipole Resonance may populate exotic high–lying nuclear states, which will decay mainly by neutron evaporation. In these experiments the neutron detector will be placed at 0° and operated together with ALADIN. For the investigation of central collisions it may also be operated at various angles in conjunction with the 4π–detector.

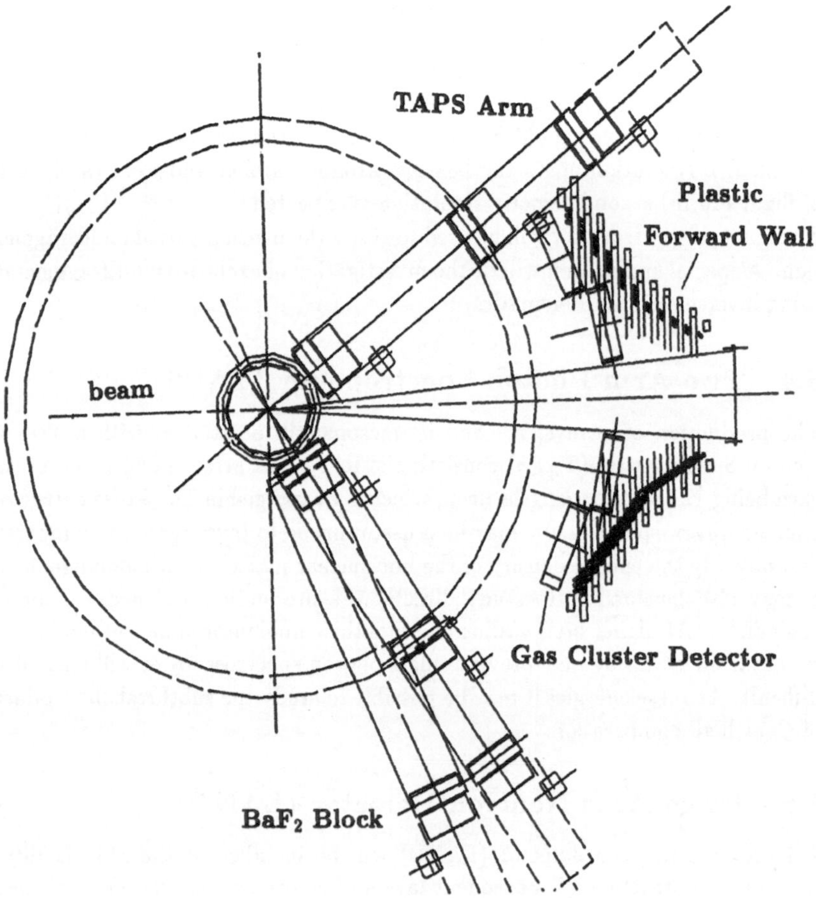

Fig.10. Schematic picture of TAPS in combination with the forward plastic wall.

3.6 K–Spectrometer

A double focussing QD magnetic spectrometer (fig.10) is being built to be installed in a separate cave in the experimental hall (see fig.1).

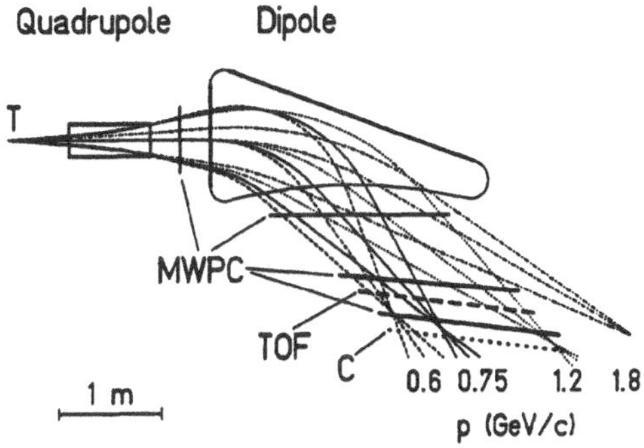

Fig.11. Schematic picture of the K–spectrometer.

Its primary purpose is to study in detail meson production in energetic collisions between nuclei. The compact design is especially matched to the requirements for kaon detection with short flight path ($\approx 5m$), large solid angle (20 - 35 msr), wide momentum acceptance ($\pm 30\%$), maximum momentum at 1.2 GeV/c (1.8 GeV/c at 10 msr), and modest momentum resolution ($\approx$ 1% without and $\approx 10^{-3}$ with raytraycing). A focal plane length of about 1.5 m allows the efficient use of the detectors necessary for particle identification and raytraycing, involving wire–chambers, time–of–flight scintillators, aerogel and water Cherenkov detectors and segmented calorimeters for particle decay. While the primary purpose for the construction of the spectrometer is the measurement of kaons, it can as well be considered a general purpose magnetic spectrometer for other hadrons and for leptons. Its large solid angle also allows the study of two–particle (e.g. $\pi - \pi$) correlations. The subthreshold production of K^+–mesons is considered as the best probe for compressed excited nuclear matter.

Acknowledgement

The help of Dr.H.Ströher in preparing this manuscript is gratefully acknowledged.

References

[1] Die Ausbaupläne der GSI, March 1984

[2] SIS–Ein Beschleuniger für schwere Ionen hoher Energie, GSI–Bericht 82-2

[3] K.Blasche, D.Böhne, B.Franzke, H.Prange, 1985 Particle Acc. Conf., Vancouver, IEEE Trans. NS 32 (1985)

[4] B.Franzke et al., Zwischenbericht zur Planung des Experimentier Speicherrings (ESR) der GSI, GSI–SIS–INT/84-5, August 1984, and Information about ESR, GSI–ESR–TN/87-02

[5] R.Keller et al., Proc. Int. Ion Engineering Congress, Kyoto (1983)

[6] N.Angert, internal GSI–report

[7] H.Gould et al., Phys. Rev. Lett. 52 (1984) 180

[8] B.Franzke, Ch.Schmelzer, GSI–Scientific Report 1984, p.341

[9] I.Hofmann, GSI–Scientific Report 1985, p.387

[10] H.A.Gustafson et al., Phys. Rev. Lett. 52 (1984) 1590

HIGH ENERGY HEAVY ION COLLISIONS AND THE RHIC PROJECT AT BROOKHAVEN*

T.W. Ludlam

Brookhaven National Laboratory
Upton, New York 11973

I. INTRODUCTION: Expanding Horizons

High energy and nuclear physics have undergone extraordinary changes during the past fifteen years. The existence of quarks as the elementary constituents of hadronic matter has been confirmed, and the field of nuclear physics has embraced a wide-ranging program of research to study the structure of nuclear matter in terms of the quark degrees of freedom. As a result the characteristic energy of nuclear collision experiments is no longer confined to the few MeV range of traditional nuclear structure experiments, exploring a system of neutrons and protons, but extends into the many-GeV range which once was the exclusive domain of particle physics. This has led to a new generation of large accelerator facilities for nuclear physics. One example of these is the CEBAF 4 GeV electron beam facility now under construction in New Port News, Virginia.

The most spectacular of the new high energy approaches to nuclear physics is in the field of relativistic heavy ion collisions, for which it is anticipated that construction of the Relativistic Heavy Ion Collider (RHIC) facility will begin soon. Here the goal is to subject large volumes of nuclear matter to such extreme conditions of temperature and pressure that a new form of matter is produced in which the recognizable components are not the familiar neutrons and protons, but are quarks.

In this state of matter the densities are so high that the constituents interact through nuclear forces. Once this state is achieved it will allow us to study the thermodynamics of strongly interacting particles in macroscopic systems (i.e., systems whose size is large compared to that of a single hadron). Heavy ion collisions offer the only means for producing matter of such densities in terrestrial experiments.

The thermodynamic conditions required to bring about this <u>phase transition</u> from ordinary nuclear matter to a plasma of quarks and gluons can be estimated from the observed properties of quarks in high energy scattering experiments, and detailed theoretical studies can be carried

---

* Work performed under the auspices of the U.S. Department of Energy.

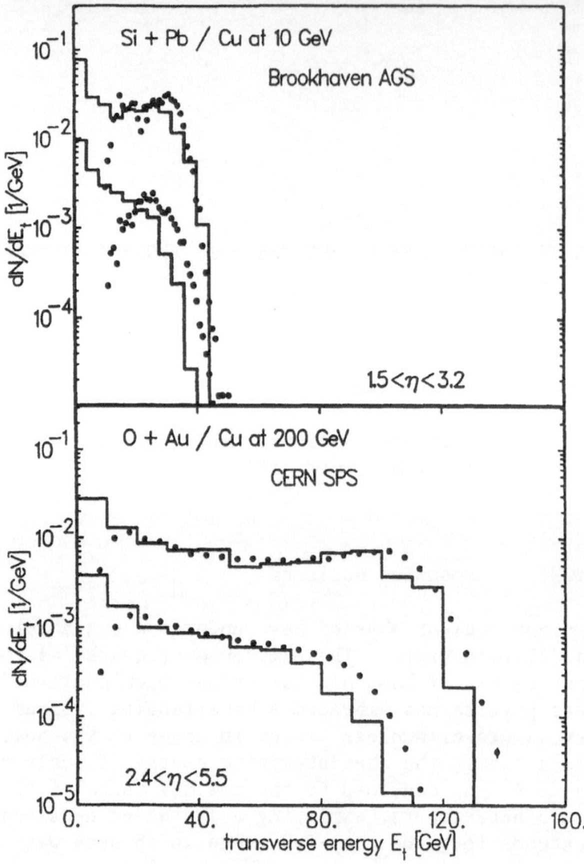

Fig. 1 Transverse energy spectra, for the indicated pseudo-rapidity (η) intervals, as measured with silicon beam at 10 GeV/u at the AGS (experiment E814, Ref. 2) and with oxygen beam at 200 GeV/u at the SPS (experiment WA80, Ref. 3). In each case the results are shown for two different target nuclei. The solid curves are the result of the VENUS Monte Carlo Model (Ref. 4).

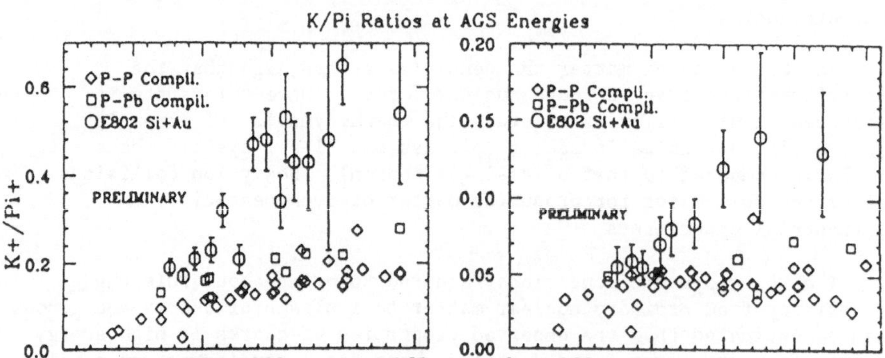

Fig. 2 A compilation of K/π ratios as a function of transverse momentum in p-p and p-A collisions at AGS energies compared with Si + Au data obtained in AGS experiment E802 (Ref. 5).

out via quantum chromodynamics. The interesting temperatures are of the order of 100-200 MeV, and the corresponding densities about 10 times that of ordinary nuclear matter. These conditions are characteristic of the expanding universe a few microseconds after the Big Bang. The relevant densities might be attained in the cores of neutron stars. The possibility of exploring, in the laboratory, nuclear matter under such extreme conditions has sparked the recent widespread interest in colliding heavy nuclei at very high energies.

During the past three years the first steps have been taken to carry out this kind of research. Ion beams have been accelerated in the Brookhaven AGS (14.5 GeV/u) and the CERN SPS (200 GeV/u). These beams, impinging on fixed targets, provide c.m. collision energies which are about an order of magnitude below the values which are ultimately desired. To date, the accelerated ion beams have been limited to nuclei of relatively low mass (A<32). Nonetheless, important new data have resulted:[1] These experiments have established that states of compressed nuclear matter can be created and studied under laboratory conditions in which extreme values of energy density are achieved. The general characteristics of the observed events bear out the theoretical expectations for large thermal energy deposition in violent nuclear collisions, and they point the way for extending further the range of thermodynamic conditions over which to study new forms of dense hadronic matter. The most exciting results are the several indications that, indeed, when these extreme conditions are reached, there are new physical phenomena to be explored.

Measurements made so far are of two important types. Global measurements of average properties such as multiplicity and transverse energy serve to characterize events according to such properties as the entropy and the initial energy density produced in the collision. An example of this kind of data is illustrated by the transverse energy spectra shown in Fig. 1. The transverse energy is a measure of the extent to which the kinetic energy of the beam goes into compressing and heating the colliding nuclei. At the very high energies of the RHIC collider the transverse energy should be directly related to the energy density achieved in the collision. At RHIC the predicted energy densities are in excess of 20 times that of ordinary nuclear matter. The fact that color string models, such as the VENUS calculations shown in Fig. 1, give good agreement with the AGS and SPS data is an indication that the essential nuclear dynamics of these collisions are indeed in agreement with the assumptions which underlie the extrapolations to collider energies.

A second type of measurement involves specific observables which may signal the onset of new phenomena. One example is the suppression of heavy vector mesons such as J/ψ, one of the most widely discussed results of the CERN heavy ion program. At the AGS a surprisingly high abundance of strange particle production has been observed in the heavy ion collisions, as illustrated in Fig. 2. Both of these trends were predicted as signatures of new physics, and since their observation, the interpretation of each is a subject of active debate. These and other measurements of this type are laying the groundwork for future experiments with heavier ions and, ultimately, much higher energies.

II. The RHIC Project

The planning for future facilities in both the U.S. and Europe call for a continued pushing back of the frontiers of this new field. Within a few years it will be possible to accelerate the heaviest nuclei (e.g., gold, lead, uranium) at both the SPS at CERN and the AGS at Brookhaven. The next step is the realization of RHIC in which high energy colliding beams will be available in a dedicated facility for heavy ion research.

The Relativistic Heavy Ion Collider at Brookhaven will extend the present heavy ion capabilities of the AGS into an energy domain not available at any other laboratory within the foreseeable future. The Brookhaven site map in Fig. 3 shows the accelerators and connecting beam tunnels involved in the heavy ion program, i.e., the Tandem Van de Graaff, the Heavy Ion Transfer Line, the AGS, the Booster Synchrotron presently under construction, and the existing ring tunnel for the proposed collider. Operation of the AGS for heavy ion experiments started in October 1986 with the delivery of O^{8+} beams. Subsequently, the mass range was extended with the AGS delivering typically 2×10^{8} Si^{14+} ions/pulse at an energy of 14.5 GeV/u. Completion of the AGS Booster Synchrotron in 1991 will extend the mass range to the heaviest ions, typically ^{179}Au, with ^{238}U a definite possibility.

The collider[6] consists of two rings of superconducting magnets for accelerating and storing circulating beams of ions. The facility will have the flexibility of using the full range of ion species from protons to gold, and will cover the entire range of collision energies from those available at the AGS up to the top collider energy (100 GeV/AMU for gold beams). Average luminosities (particle fluxes) of 10^{26} - 10^{27} $cm^{-2}sec^{-1}$ are predicted for gold-on-gold collisions at full energy, depending on the details of the experimental layout. Proton energies up to 250 GeV in each beam will be available at luminosities of about 10^{31} $cm^{-2}sec^{-1}$. The capability for collisions between different species will also be provided. The RHIC ring lattice has six crossing regions where the beams collide and experiments can be performed. Four of these experimental areas will be utilized initially, with the remaining two available for future expansion of the research program. RHIC will use the existing AGS and Tandem Van de Graaff accelerator complex at Brookhaven as an injector. The new accelerator will be built in the existing CBA tunnel (3.8 km circumference), and will utilize the experimental halls, support building and liquid helium refrigerator from the partially completed CBA project. The RHIC project is proposed to begin construction in 1991, with the first collider experiments beginning in 1996. A plot of the design luminosity vs. energy for various ion species, with corresponding beam lifetime is shown in Fig. 4.

III. Experiments and Detectors at a Heavy on Collider

Of the six crossing regions built into the RHIC rings, those at the 2, 4, 6 and 8 o'clock positions (see Fig. 1) have completed experimental halls, including support buildings and (except in the 4 o'clock "open area") crane coverage. The RHIC plan calls for mounting experiments initially in these four areas, leaving the remaining two unfinished until some later time.

The recording of events at this collider, will require the extension of techniques for particle detection beyond the present ranges of application in elementary particle and nuclear physics. With beam energies up to 100+100 GeV/nucleon and ion masses up to ≥200, the total energy in each collision can reach up to 40 TeV in the center-of-mass: a range far beyond that of any present accelerator or any existing detector system. Unlike the proposed SSC collider, which will accelerate elementary particles to such an energy and produce hundreds of very high energy particles in the final state, the most interesting events at the RHIC collider are expected to produce tens of thousands of final-state particles in each collision, with proportionately less energy carried away by each particle. Thus, while the basic detector technology will have much in common with the detector systems developed for colliding beams of elementary particles, the design of detector systems for the heavy ion collider must address a different set of problems.

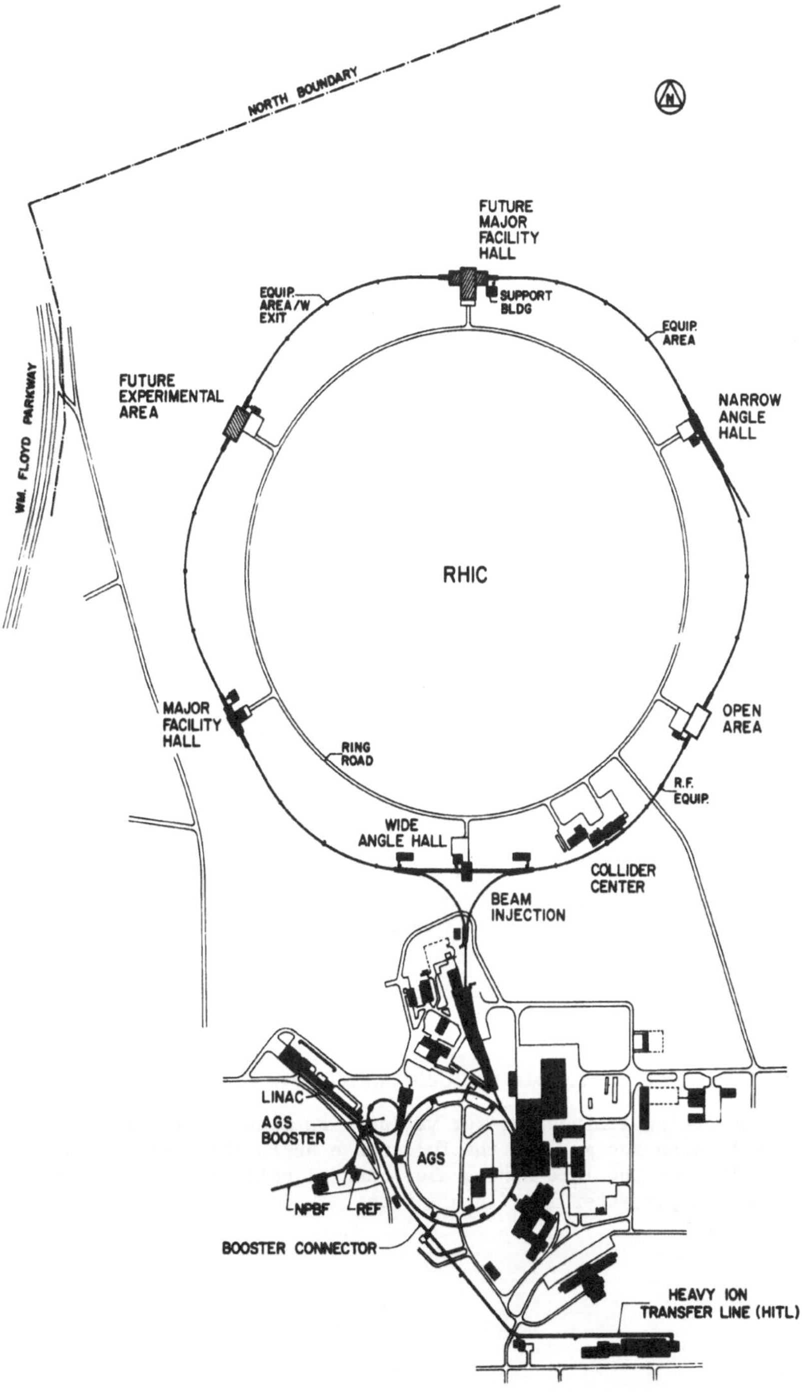

Fig. 3 Site map of present and proposed accelerators at Brookhaven. The Tandem Van de Graaff and the AGS with its linac injector are existing machines. TheBoosterSynchrotron for pre-injector to the AGS is currently under construction. The RHIC colliding beams accelerator to the north of the AGS complex is a proposed construction project.

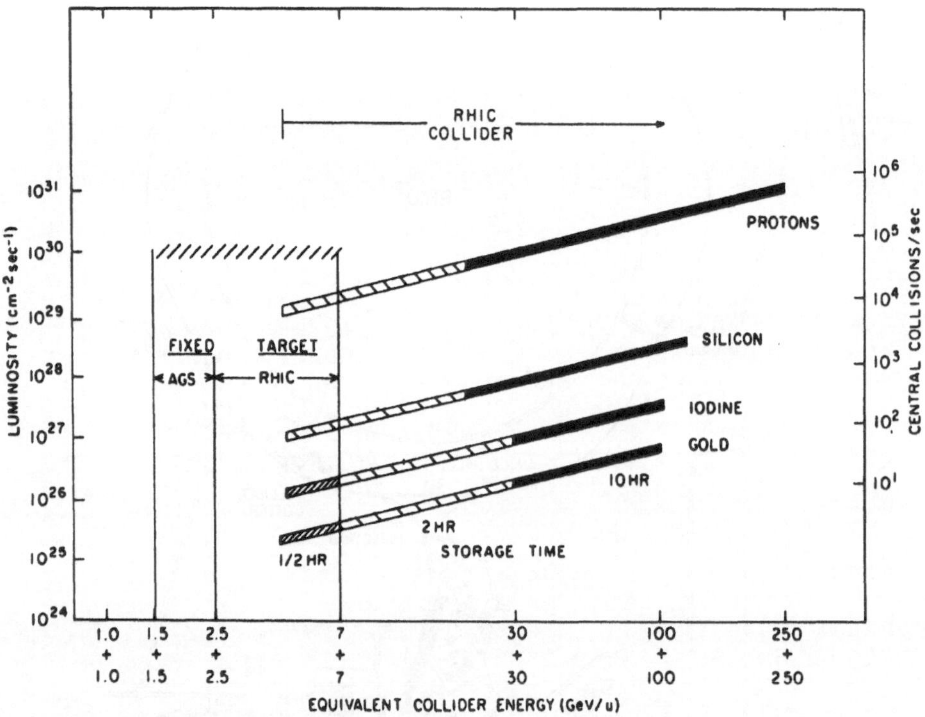

Fig. 4 The design luminosity, for various ion masses, as a function of collision energy over the full range accessible with AGS and RHIC. On the left-hand scale, central collisions correspond to impact parameter less than 1 fermi.

In addition to extraordinarily high particle multiplicities, experiments in a heavy ion collider must deal with the fact that the signals which characterize equilibrium phenomena in a hot plasma of quarks and gluons are expected to be carried by particles whose transverse momenta and thermal energies are relatively small, and similar to that of the background radiation. This is in contrast to the detection of hard-scattering phenomena in elementary particle colliders, where sensitive triggers can be made blind to the vast majority of background particles by selecting large transverse momenta. Detector systems to deal with high energy nucleus-nucleus collisions will require new approaches to the technologies of tracking, calorimetry, particle identification, fast trigger decisions and on-line data processing.

In April 1985 a workshop involving about 100 nuclear and high-energy physicists provided preliminary designs and cost estimates for a first-round suite of detectors for RHIC. The second RHIC workshop was held in May, 1987 at Lawrence Berkeley Laboratory. This week-long meeting culminated a year of effort carried out by individual working groups holding meetings at BNL, CERN and elsewhere. The proceedings from these workshops[7,8] are available, and provide a detailed discussion of physics goals and conceptual designs for detector systems.

At the May, 1987 workshop a few of the designs for large detector systems were examined in detail to study their technical feasibility, their impact on the design and modes of operation of the collider and the range of capabilities for physics research represented by each. Out of these studies came the realization that the measurement of lepton pairs, (electron pairs and muon pairs) radiating from heavy ion collisions at RHIC cannot be adequately measured if the luminous interaction length of the crossing beams is of the order of a meter or more, as it was in the original RHIC design. Since these lepton pair measurements are of fundamental importance to the research program, the design of the RF system of the machine has been revised to provide shorter beam bunches and ensure a useable interactions length over the full beam lifetime. These workshop studies also showed that the most useful improvements to the machine performance - from the point of view of extending the physics reach - would be those which increased the luminosity of the machine, rather than extending the energy range. Subsequent accelerator physics studies have explored possible future upgrades of the machine which could increase the luminosity from all ion species by about a factor of ten beyond the current design values.[9]

With several well-specified concepts now in hand for large detectors at RHIC, attention in the user community has turned to the need for R&D on detector techniques. Past workshops have pointed out a number of specific areas where detector technology must be advanced beyond the present state of the art in order to realize the needed capability for experiments at RHIC. In July, 1988 a workshop was held at BNL to examine the needs and priorities for detector R&D for RHIC.[10]

Among the areas which have been identified as most urgently requiring study before the design of large scale detector systems for a heavy ion collider can be completed are:

- Tracking and particle identification techniques in a very high multiplicity environment
- Calorimeter response to high densities of soft particles
- High density readout electronics
- Methods of muon and electron identification suitable for studies at transverse momenta and pair masses at the < 1 GeV scale.

This year, Brookhaven National Laboratory, with the cooperation of the U.S. Department of Energy, has initiated a program of detector R&D which is providing support for a large number of user groups from universities and National Laboratories to begin a technical effort which will grow with increased funding as the project reaches the construction stage.

References

1. For summaries of these results see G. Baym, P. Braun-Munzinger and S. Nagamiya (eds.), Proc. Int. Conf. on Ultra-Relativistic Nucleus-Nucleus Collisions, Lenox, MA, Sept. 1988, Nucl. Phys. A498.
2. P. Braun-Munzinger et al., Z. Phys. C 38, 45 (1988).
3. S.P. Sorensen et al., Z. Phys. C 38, 3 (1988).
4. K. Werner, Phys. Lett. B208, 510 (1988).
5. P. Vincent et al., Nucl. Phys. A498, 67C (1989).
6. Conceptual Design of the Relativistic Heavy Ion Collider, BNL Report 2195, (1989).
7. P. Haustein and C. Woody, eds., Proc. of the Workshop on Experiments and Detectors for a Relativistic Heavy Ion Collider, Brookhaven National Laboratory Report BNL 51921 (1985).
8. H.G. Ritter and A. Shor, eds., Proc. of the Second Workshop on Experiments and Detectors for RHIC, LBL-24604 (1988).
9. F. Khiari et al., eds., "Workshop on the RHIC Performance", BNL 41604 (1988).
10. B. Shivakumar and P. Vincent, eds., Proc. of the Third Workshop on Relativistic Heavy Ions, BNL 52185 (1988).

ADAPTATION OF THE THEORY OF SUPERCONDUCTIVITY TO THE BEHAVIOR OF OXIDES

Edward Teller

Lawrence Livermore National Laboratory
Livermore, CA 94550

ABSTRACT

An adaptation of the conventional theory to high-temperature superconductors is proposed. Excitation of electrons from below the Fermi surface to above the Fermi surface (according to Bardeen, Cooper and Schrieffer) is replaced by excitation from a filled energy band into an empty one. The energy bands are constructed from two-dimensional Bloch functions in neighboring layers of the oxide lattices. Strong coupling with lattice displacements is due to the removal of the topmost electrons from the O^{--} ions in the perovskite planes. The main methods of the BCS theory are retained. The formation and observability of a superlattice is discussed.

DISCUSSION

High-temperature superconductivity is found in a wide variety of oxides of an apparently complicated lattice structure. But comparison of the various examples makes the situation appear simpler, and the accumulating evidence is remarkable and suggestive.

The positive ions form, in good approximation, a body-centered cubic lattice that is distorted into a layered structure. The layers have the chemical composition KO_x, where K is the cation and x is 0, 1, or 2. For x = 2, the composition is CuO_2 forming a perovskite layer, which is shown in Figure 1. Such a layer is present in the new high-temperature superconductors. So are layers of composition KO, where the oxygen is in the center of the squares formed by the cations. In that case, an oxygen has four rather than two neighbors in the layer. If oxygen-free layers are absent, the critical temperature is below 40° K. If all types of layers (x = 2, 1, and 0) are present, then critical temperatures may be reached that remain near 100° K with remarkable consistency. Any x = 0 layers are always bracketed by two x = 2 (perovskite) layers.

In the new materials, most electrons may be accommodated in closed bands, forming an ionic lattice. The incomplete d-shell on the copper ions is an exception. There are nine rather than ten d-electrons per Cu ion in the perovskite plane, described in first approximation by the Hubbard model. That gives rise to antiferromagnetic behavior above the

The Nuclear Equation of State, Part A
Edited by W. Greiner and H. Stöcker
Plenum Press, New York

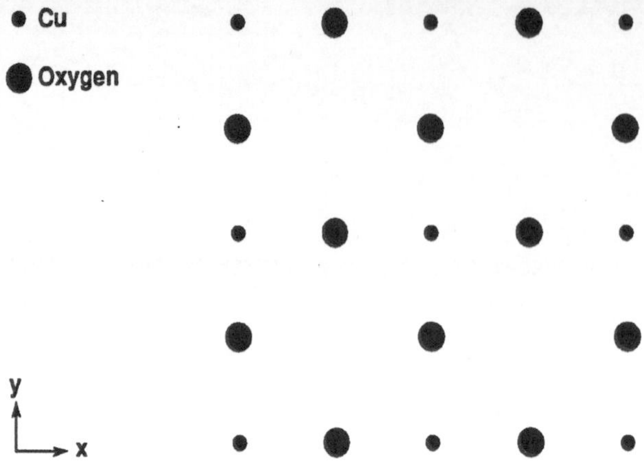

Figure 1. The Cu O_2 perovskite plane.

critical temperature (T_c). In many attempted theories of superconductivity, the electronic configuration of the perovskite layer is used to explain superconductivity. There can be little question that the perovskite layers are involved in superconductivity and that changes in the electron configuration in those layers have an effect on the d-electrons. I believe, however, that high-temperature superconductivity is more likely to be explained by an interaction of electrons in the perovskite layer with electrons in other layers and that the changes in paramagnetism above T_c are not essential for the explanation.

The first step in the Bardeen, Cooper, Schrieffer (BCS) theory of superconductivity is to emit and reabsorb a phonon while lifting two electrons from below the Fermi surface to above the Fermi surface. I would like to suggest that the phonon should be a two-dimensional compression-expansion wave in the perovskite plane. Its propagation in a direction perpendicular to the planes is slow.

The model I propose will consider the perovskite layer coupled with one additional electronic state in neighboring layers. The discussion is carried out as a two-dimensional problem in the perovskite plane. The emission or absorption of the phonon is coupled with throwing an electron from the top of the O^{--} band in the perovskite layer into a low state of an appropriate band in a neighboring layer, for instance, in a BaO layer or a Ca layer.

The topmost Bloch-state in the O^{--} band has a node between each pair of neighboring oxygens, as shown in Figure 2 by the dotted lines. Those nodes cross on each Cu ion of the perovskite plane. But considered as planes perpendicular to the paper, they also cross on every cation in the lattice, as indicated by the intersections in the center of the Cu squares in Figure 2. As wave functions around the cations, the two nodes will correspond to an angular momentum $2\hbar$ (with appropriate distortions due to the crystal lattice) around the z-axis. Near the cation, the wave function can be written approximately as $x^2 - y^2$, where the coordinates are indicated in Figure 2. That would correspond to a d-state in the free ion. An alternative is to consider an f-state, which near the cation behaves as $z(x^2 - y^2)$.

It is remarkable that in all cations within the crystals in which oxide superconductivity has been observed, unoccupied d- or f-states of

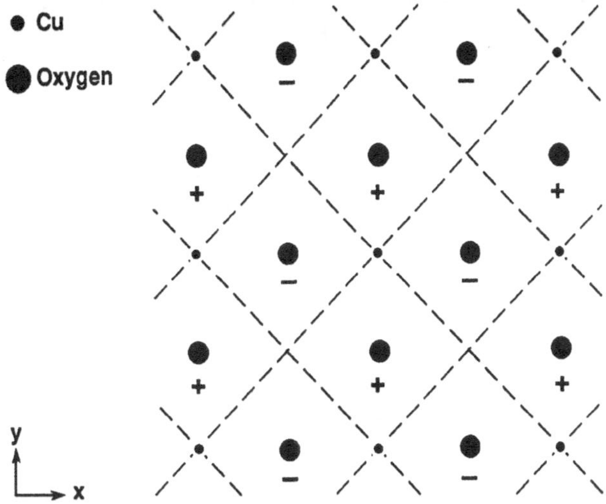

Figure 2. The perovskite plane with the dotted lines showing nodes of the electrons of highest energy in the O^{--} band.
The + and – signs refer to the amplitudes of the wave functions of the electrons. The nodes also cross in the center of the Cu squares. Cations are located above and below these crossing points.

a relatively low energy are available. Those include alkali earth ions, yttrium, lanthanum and the lanthanides, bismuth, thallium, and, of course, copper. In the following discussion, I will consider just one KO_x-layer, with x = 1 or 0, which is supposed to be next to the perovskite layer. In reality, all layers with x = 0 or x = 1 must participate, and in all of them, d- or f-states are involved. A full discussion of superconductivity must describe the participation of those several layers.

In the BCS theory, electrons and holes participate, the electrons having a slight excess of energy above the Fermi surface, while the holes are just below the Fermi surface. In the model I propose, the electrons are at the bottom of an unoccupied energy band of d- or f-electrons on the cations. The holes are at the top of the oxygen band in the perovskite layer. Specifically, I propose holes resulting from the missing p_z electrons with a node in the perovskite plane.

A further needed modification may be the admixture of a wave function of a 2s-electron on the O^{--} ion, which will cause the extension of the wave function of the hole in the positive or the negative z-direction to produce a stronger overlap with wave functions in the appropriate layer of cations. It should be noted that in order to involve copper d-electrons in the perovskite layer itself, holes on the oxygen of the p_x or p_y type are needed. Such holes are apt to exist, and their coupling with the tenth d-electron of the perovskite copper shell will affect both the paramagnetism and the superconductivity. I assume, however, that this coupling is not an essential or sufficient reason for the superconductivity.

In Figure 3, the energies of individual Bloch states in the perovskite plane are plotted in two parabolas as a function of the quasi-momentum k. Even in a simplified model, k should be two-dimensional, but we are plotting only one component, for instance, k_x. One of the parabolas with lesser curvature opens toward high energies. That represents the energy states at the bottom of the band of d- or f-

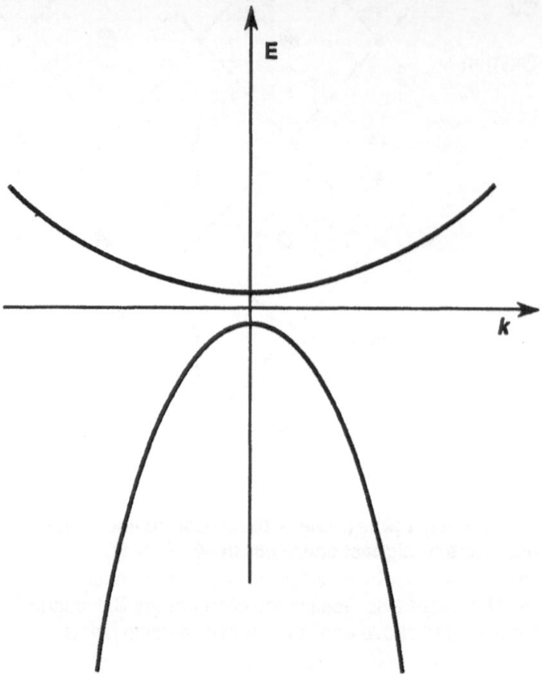

Figure 3. Energy (E) as a function of quasi-momentum (k) in the O^{--} band (lower curve) and in the d– or f–band (upper curve).

electrons on cations. The other parabola represents the top of the band of the electrons on the O^{--} ions. It is assumed that the crowding of the O^{--} ions in the perovskite layers gives rise to a low effective mass (i.e., sharp curvature) and also to an energy maximum that approaches the energy minimum of an appropriate d- or f-band of a plane of cations (as shown in the upper parabola).

The assumptions inherent in Figure 3 make the oxide an insulator or, if there is a small gap between the filled and empty bands, an intrinsic semiconductor. (Conductivity decreases exponentially as one approaches absolute zero.) In reality, the superconducting oxides are poor conductors above the superconducting transition temperature T_c with the resistance roughly proportional to the absolute temperature. It is therefore clear that free electron charges exist above T_c with their motion impeded in proportion to the energy of the lattice vibrations. The easiest way to remove the discrepancy of those facts with my assumption would be to assume that the maximum of the lower (O^{--}) parabola lies above the minimum of the upper (d- or f-state) parabola so that the two intersect.

The difference between intersecting and non-intersecting curves appears smaller if one takes into account the interaction of the states represented by the parabolas, as will be done in connection with Figure 4. I have chosen to discuss the case with the gap, partly because it makes the argument a little simpler, and partly because I want to emphasize that superconductivity might occur even if the substance, when above T_c is an insulator or an intrinsic semiconductor.

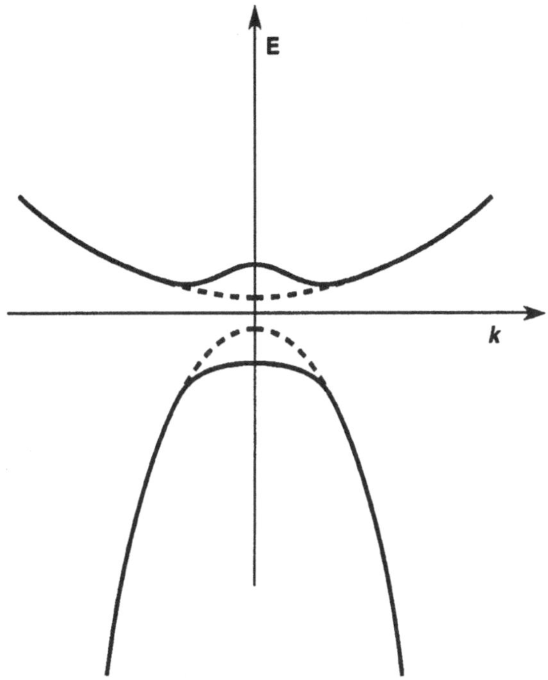

Figure 4. Change of electron energies due to interaction of wave functions in neighboring layers.

Another simple but interesting possibility remains. In the presence of doping, as in a $YBa_2Cu_3O_{7-\delta}$ compound, where δ is small but not equal to 0, the electrons present due to the missing oxygens should make the assumption of a small gap acceptable. It would be very interesting to see whether, in the case of good crystals or well-ordered thin films, the conductivity above T_c and (perhaps) the maximum value of the superconducting current would decrease with decreasing δ while the value of T_c would remain unchanged.

We return to Figure 3, where the two parabolas do not overlap. If the composition is stoichometric, one may assume that all states belonging to the lower parabola representing O^{--} electrons are filled, and all states in the top parabola representing d- or f-electrons are empty. That would result in an insulator. Any changing of the k-values in the occupied parabola will amount merely to a permutation of electrons and will have no effect. It is from this model that I want to start.

As a next step, I shall assume that the p-wave functions from the lower parabola, and the d- or f-functions from the upper parabola will overlap and therefore interact. The overlap will be strongest near k = 0. With increasing k-values, the nodes for the p-functions (see Figure 2) will be displaced and will no longer coincide with the nodes of the d- and f-wave functions. The overlap and interaction will be strongest near k = 0 and diminish at higher k-values.

Furthermore, for increasing values of |k|, the energy difference between the upper and lower parabolas will increase, and, therefore, their interaction will diminish. Finally a contraction of the perovskite plane will raise the energy of the lower parabola, which represents the O^{--} wave functions. Indeed, the O^{--} ions in the perovskite plane are overcrowded, and the van der Waals repulsion depends sensitively on compression. The topmost electrons with the most nodes will respond most sensitively.

The appropriately perturbed parabolas are shown in Figure 4. The original parabolas reappear as dotted lines that extend the simple parabolic behavior of the energy dependence on k into the perturbed region. One will observe that the upper curve has developed a small maximum near k = 0 due to the influence of the lower curve, while the lower curve has obtained a flat top for similar reasons. It may be noted that if one assumes that the curves in Figure 3 intersect, the two curves will avoid each other because of the interaction. The result is a sharper maximum near k = 0 for the upper curve and probably also a minimum for k = 0 in the lower curve. The ultimate change in Figure 4 may not be particularly pronounced. Thus, Figure 4 does not depend in a sensitive way on the relative positions of the curves in Figure 3. One may conclude that superconductivity and even the value of T_c may not depend strongly on the relative positions of the energy bands in neighboring layers.

In the model to be introduced, the properties of electrons near the minima of the upper curve play an important role. If one assumes for the sake of simplicity that the interaction between the two wave functions does not diminish with k but stays constant, and if one also assumes that the effective masses do not depend on k, then one obtains a simple statement on the mixture of the wave functions at k_{min}, that is, the k-value at which the minima of the upper curve occur. At those points, the probability of the electron in the lower curve being in the d- or f-state (that is, the absolute square of its wave function) will be $m_\ell/(m_\ell + m_u)$, where m_ℓ and m_u are the effective masses for the lower and upper parabolas. Similarly, the probability for the electron being found in the p-state of oxygen will be $m_u/(m_\ell + m_u)$. For k = 0, the analogous probabilities are

$$1/2\left[1 - \frac{C}{(C^2 + M^2)^{1/2}}\right] \quad \text{and} \quad 1/2\left[1 + \frac{C}{(C^2 + M^2)^{1/2}}\right]$$

where C is one-half the energy difference between the two parabolas at k = 0, and M is the matrix element of the interaction. The difference between those two probabilities is always less at k = 0 than at k_{min}, e.g., considering the lower curve, the electron is less likely to be found on the oxygen at k = 0 than at k_{min}. That means that $(m_u - m_\ell)/(m_u\, m_\ell)^{1/2} > 2C/M$. That relation also happens to be the condition required for the upper curve to have a maximum.

The next step is to transfer an electron from the neighborhood of the maximum of the lower curve to the neighborhood of the minima of the upper curve. That costs some energy. The replacement of the O^{--} electrons in the perovskite plane by d- or f-electrons in neighboring planes relieves the van der Waals repulsion, the more so if the perovskite plane is contracted. The additional van der Waals energy due to the contraction is what compensates the energy needed to excite the electrons. The distortion responsible for that compensation is a contraction of the lattice in some regions and an expansion in others. Then the electrons should be expelled from the contracted regions of the perovskite plane and introduced in the expanded regions of the neighboring planes.

What I am describing here are the consequences of the process initiated by electron lattice interactions, as in the BCS theory. That can be carried to a higher approximation by the use of the variational principle. In applying that principle, one should be considering the creation of holes and electrons only near k = 0. If one forms wave packets from wave functions of different k-values, one can localize electrons or holes to a limited extent.

For $|k| = 0.2$, one may localize the packet to approximately five lattice distances. Over a region of approximately five cells, one might contract the lattice. One may then replace a linear combination of electron wave functions in the lower curve of Figure 4 by a wave packet of electrons in the upper curve of Figure 4 with somewhat higher k-values. At the same time, the lattice is contracted around one point but expanded in a ring around the contracted region. In the ring, electronic states from the upper curve predominate, which still contain wave amplitudes (of smaller amplitudes) of the O^{--} (repulsive) wave functions.

For the purpose of illustration, one may construct a wave packet in the lower curve from a Gaussian around k = 0. In the upper curve, one may use the difference of two Gaussian distributions in k-space with a maximum contribution near k_{min}. One may set the amplitudes equal at k = 0 so that, for the difference, the resultant amplitude vanishes. In configuration space there will be a minimum at r = 0, but the value of the wave function will not vanish. The packet of holes (electrons missing from the lower curve) will be concentrated at the geometric center and will have a greater contribution from O^{--} wavefunctions, so near the center, the van der Waals repulsion will be diminished.

All that may be repeated around every point of a super-lattice in a two-dimensional, hexagonal arrangement with a spacing of several lattice distances. In the actual crystal, columns along the z-axis (which is perpendicular to the planes) with slightly increased ion density and a slight deficiency of negative charge will be produced. The columns will be surrounded by annular regions of decreased ion density with a slight excess of negative charge.

The arrangement described may still have a higher energy than the one without any lattice distortions as long as only single electrons are considered together with their interaction with the distorted lattice. However, one should place two holes with opposite spins and symmetrized spatial wave functions in each region of excess density together with the pair of electrons in the expanded regions.

A pair of electrons will, of course, require twice the energy in order to be lifted from the lower curve into the upper one. That will be connected with twice the lattice displacement and produce four times the negative interaction with the lattice as compared to the interaction energy of a single electron.

In the k-space, that means that we replace wave packets around k = 0 from the lower curve with wave packets of appropriate linear combinations of wave functions from the upper and lower curves. Those wave packets then can be fitted together with appropriate phase-relations within the perovskite planes and also along the z-axis. Decreased occupation of the top of the O^{--} band concentrated along the axes of the super-lattice gives the needed energy release when coupled with the contraction of the perovskite planes around the locations of those axes. Overlap of the d- or f-electron wave functions from neighboring regions of the super-lattice will provide an appropriate correlation energy that produces minima for even spacing in the super-lattice.

Along the z-axis, the correlations are apt to be weaker, but each electron pair displaced from the perovskite plane may be found in a phase-related f- or d-wave function in the other planes. Because there are more planes with x = 1 or 0 than perovskite planes, one can limit the k_z-values to the lower portions of the band of those wave functions.

There is a significant quantitative difference between high-temperature superconductivity of the oxides and the low-temperature superconductivity explained by Bardeen, Cooper, and Schrieffer. In the BCS theory, electrons are lifted from just below the Fermi surface to just above it. At the same time, phonons of all energies were emitted and absorbed. Because those phonons have frequencies comparable to the Debye limit, the phonon energies were much larger than the excitation energies of the electrons. Therefore, it makes sense to describe the contributing electron states alone in calculating the behavior of the lattice and to eliminate the virtual states describing the vibrations from calculation.

In my model, the relevant vibrations have one-fifth to one-tenth the Debye energy. At the same time, the electron excitations are apt to be higher. The two could be comparable in energy. However, the excitation energies of the electrons can hardly predominate because in order to establish stability, the negative energy of the accompanying lattice distortions must overcompensate the electron excitation. At the same time, the energy of lattice distortion appears formally as the energy of phonons. Indeed, it is possible that we approach the limit where the phonons and lattice distortions can be treated by a classical model.

The result is that in a variational approach, one may start from lattice distortions with slightly denser regions along the z-axis, which may be arranged in a hexagonal super-lattice of 30 or 100 Å-spacing in the x-y plane. One can then determine by variational calculation what kind of superposition of holes and electrons will minimize the energy. That state will then approximate the lowest energy of the crystal.

At the temperature of absolute zero, states forming appropriate superpositions of the perovskite O^{--} bands and the d- or f-bands in the other planes will be occupied. With each excitation, the equilibrium distortion of the lattice will be lower, and the energy contribution that had forced the electrons into the appropriate linear combination between the two curves will be decreased. The situation is precisely analogous to the behavior of ferromagnets, where the average magnetic fields and the magnetization are the cause of each other and decrease together as the curie point is approached from the low-temperature side.

As has been indicated, d- or f-wave functions must be available in each of the planes if the phase-relations in the z-direction are not to be interrupted. (The perovskite planes may be an exception.) If, in the x-y plane, strong interactions are established, even a weak coupling along the z-direction will suffice to produce three-dimensional, long-range order. But the coupling in the z-direction may have little influence on the value of the transition temperature T_c.

What has been described so far is the construction of a non-current-carrying state of the superconductor. In such a state, there occurs for each value k of an electron or a hole, a corresponding value of -k. Therefore, the current vanishes for reasons of symmetry. In the BCS theory, that compensation actually occurs in the individual Cooper pairs. In the present model, the same situation may hold, although it is possible that the compensation of the k-values occurs only on the average.

To obtain a current-carrying state, one has to assume that with the excitation of each k-value, there is a corresponding excitation of $-k+\epsilon$, either individually or on the average. That procedure could be used in the variational calculation as a condition imposed on the trial functions. In that way, one may obtain current-carrying states analogous to the one that does not carry a current.

In such an arrangement, the super-lattice will reappear in such a way that all the dense regions, and indeed the whole super-lattice, will participate in the same common motion. In that way, one may explain sustained currents in the x-y plane. One may actually look at the situation by considering the motion of the super-lattice as the primary cause of currents. If the electrons and the holes are swept along to a different degree, a current will ensue. In principle, there is the possibility of finding a phase-transition of the type discussed, but no current-carrying state, if there should be a reason why electrons and holes move with the same velocity. It seems very much more likely that states where the average value of k does not vanish will carry a current. For the stability of the current, it is necessary that the energies of the current-carrying state should differ only very slightly from those of the non-current-carrying-state.

If one associates one pair of holes and one electron pair with each dense region in every perovskite plane, one will have a few times 10^{19} electrons per cm^3, moving with a velocity, probably less than the velocity of sound, having values up to a few km/sec. That can give currents up to a few million amp/cm^2.

The relevant lattice distortions are composed of phonons of quanta near .01 eV; at the transition point, kT_c is of same order of magnitude, and so is the coupling with the lattice. One is approaching the conditions where the lattice vibrations (though, of course, not the behavior of electrons) may be described in a classical fashion.

Strong interactions between electrons and lattice displacements raise the danger of a Peierls instability. That occurs when the distortion of the lattice becomes great enough that, upon displacement by a lattice period, the relative lattice positions lie outside the zero point vibration.

If lattice vibrations with wavelength λ equal to several lattice periods play the decisive role, the conditions required for the Peierls instability may not be fulfilled. Indeed, if the vibrations are excited by approximately one quantum, then lack of overlap will commence if we displace by $\lambda\!\!\!^-$ and not at the shorter displacement by one lattice cell.

My model is obviously related to the description of charge density waves in the crystal. The latter phenomenon may be considered in one, two or three dimensions. I shall attempt to clarify the relation between these models.

If a pair of electrons and holes is distributed over N lattice cells, the lattice distances will be disturbed by $\sim N^{-1}$. The lowering of the energy per cell (taking into account the coupling of the wave-functions and the lattice distortion but neglecting the Coulomb energy) will be proportional to N^{-2} and the total stabilizing energy summed over N cells will be $\sim N^{-1}$.

In one dimension, i.e., over a string of cells, this will amount to $- D^{-1}$ where D is the linear dimension in question. At the same time, the kinetic energy needed for the localization is D^{-2}.

In two dimensions, the analogous quantities will be:

$E_{stabilization} \sim - D^{-2}$, $E_{kinetic} \sim D^{-2}$.

In three dimension, we have

$E_{stabilization} \sim - D^{-3}$, $E_{kinetic} \sim D^{-2}$.

The consequence is that in three dimensions, any tendency of stabilization by a localized lattice distortion will tend to go to the smallest possible D-value, that is to a single lattice cell. The result is Peierls instability. In one dimension, charge density waves may occur at high D-values extending over many coordinated cells. Such cases are well known.

In two dimensions depending on the comparison of the size of $E_{stabilization}$ and $E_{kinetic}$ one may seem to have an opportunity for both large and small D-values. But we have mentioned that for very small D-values, that is large k-values, the matrix element which is responsible for the coupling will decrease. Therefore, we escape the danger of the Peierls instability. On the other hand, for very large D-values, we get into the region of small k-values for which the energy dependences of k tends to be flat as seen in Figures 3 and 4. Therefore, there is no reason for the development of very long density waves.

The question remains whether an approximation in which a density wave is produced is compatible with the phenomena of superconductivity that is with a sharply-defined, second-order phase transition, and the existence of stable current carrying states. While this paper contains qualitative arguments for the compatibility of these ideas, further investigation is needed.

It is interesting to compare the picture described above with the general behavior of low-temperature superconductivity as observed for elements and as described by the BCS theory. Here it is often assumed that electrons are moved arbitrarily by substantial k-values near the Fermi surface. There is, however, in all probability, a limit to this arbitrary movement. One does not observe superconductivity for monovalent elements, like the alkalis, or like copper, silver, or gold. Superconductivity may depend in those cases on the specific interactions that are connected with the position of the Fermi surface in the Brillouin zones. If, for instance, the Fermi surface is partly in a Brillouin zone connected with s-wave functions and partly in a zone connected with p-wave functions, the relevant electron excitations should occur by lifting s-type electrons into p-type wave functions or vice versa.

That results in the superpositions of s- and p-type wave functions and the extension of the wave function toward a neighboring element; at the same time, neighboring atoms move toward each other. All of that is apt to result in virtual formations of bonding orbits occupied by a pair of electrons. Since transient bonds will not be found in a particularly ordered fashion, no formation of a super-lattice is expected at T_c. Indeed none is found.

By contrast in the model I am proposing, wave packets are playing a role, where k-values cover a narrow range. The formation of an ordered system of dense regions extending over a few lattice distances in the perovskite planes becomes possible. In the current-carrying case, those regions will move with velocities that may approach, in extreme cases, the velocity of sound. All that is due to the circumstance that strong coupling between electrons and lattice is limited to small k-values. On the other hand, the energy changes due to the van der Waals repulsive force in the perovskite layer are strong enough so that even one missing electron pair, in a two-dimensional region of a few dozen lattice cells, will produce an appreciable effect.

If the model is valid in the simple form I have described, the super-lattice of contracted and expanded regions should be detectable by x-ray diffraction. The expected changes of densities should be less than one percent, and the intensity of corresponding maxima in the diffraction pattern should be less than 10^{-4} of the Bragg peaks and should be near these peaks. In a normal x-ray diffraction picture, those satellite peaks may be hard to detect and should indeed become very weak near the critical temperature. There is a better chance to get evidence for the super-lattice, if x-rays of longer wavelengths are used. At one keV, the normal Bragg peaks will no longer be present. On the other hand, the diffraction maxima from the super-lattice will be widely spaced. They should still remain quite weak compared to specular reflection.

It will be very interesting to see by variational calculations how greatly the free energy differs if one assumes different values for the spacing in the super-lattice. As one approaches T_c from the high-temperature side, limited regions of the super-lattice may be formed, and there may be fluctuations in the corresponding lattice distances. The result could well be that at T_c, a range of super-lattice distances will be found instead of just the optimal value. As a result, the diffraction pattern might still be there, but the peaks would be broadened. Because the peaks are weak in any case, it will be difficult to find the diffraction structures.

What has been attempted here is a crude description of how a trial wave function for a variational calculation may be selected. That falls far short of an explanation of high-temperature superconductivity. I am also trying to point out a strange relation between the theories of low-temperature and high-temperature superconductivity. The mathematical structure of the BCS theory need hardly change, but the physical content does. At low temperatures, the approximation of independently moving electrons suffices as the field of applications of the mathematics of superconductivity. For high-temperature superconductivity, restrictions on the changes in the values of the quasi-momentum k make it possible to describe the phase-transition to a much greater extent in terms of changes that occur in real space. In that way, it may become possible to give a special example of the BCS theory that can be visualized more easily.

I would like to express my thanks for most helpful discussions to: Valdimir Kresin, Stephen Libby and Richard More. Work performed under the auspices of the U.S. Department of Energy by the Lawrence Livermore National Laboratory under contract number W-7405-ENG-48.

HIGH PRESSURE EQUATION OF STATE IN VARIOUS AREAS OF PHYSICS - OVERVIEW

Shalom Eliezer

Plasma Physics Department
SOREQ NRC
Yavne 70600, ISRAEL

ABSTRACT

The equation of state (EOS) is the relation between the pressure, the temperature and the density (specific volume) of a physical system and is related both to fundamental physics and applied science. The study of equations of state under extreme conditions is an interdisciplinary subject with important applications to material science, astrophysics, geophysics, nuclear physics, plasma physics etc. EOS describes nature over all possible values of pressure, density and temperature where local thermodynamic equilibrium can sustain. Various domain of the EOS are given by analyzing a temperature-pressure diagram for various states of matter.

In this paper three subjects are discussed: (a) The Thomas-Fermi-Dirac (TFD) EOS, (b) EOS problems in inertial confinement fusion (ICF), (c) The virial theorem and EOS. The TFD equation of state plays a major role in the physics of high pressure. The use of EOS in inertial confinement fusion is a typical example of the importance of EOS in the study of compressing small pellets with liquid deuterium (+tritium) by powerful lasers or ion beams. Finally, by using the virial theorem the EOS for ideal gasses is derived. Moreover, we expose the relation between the virial theorem and the EOS for a gravitational system and for the TFD system.

1. Introduction

The equation of state (EOS) is the relation between the pressure, the temperature and the density (specific volume) of a physical system and is related both to fundamental physics and applied sciences. Important branches of physics were developed or originated from the equation of state while in return more and more complex formulations of the equations of state were due to the developments of modern physics. The ideal gas law was the first quantitative treatment of chemical kinetics and the beginning of thermodynamics and statistics. Furthermore, as mechanics was extended to take account of relativity and quantization, thermodynamics and statistics was developed to describe states of matter in extreme density and temperature domains.

The study of equations of state under extreme conditions is an interdisciplinary subject with important applications [(1)] to material science, astrophysics, geophysics, nuclear physics, plasma physics and in applied sciences such as fission, hypervelocity impact, weapons development, etc. In geophysics EOS knowledge helps to understand the structure of our planet (and other planets). In astrophysics the EOS

The Nuclear Equation of State, Part A
Edited by W. Greiner and H. Stöcker
Plenum Press, New York

studies are an integral part of the description of stellar evolution like white dwarfs, neutron stars and black holes. EOS is used in nuclear physics to investigate the dynamics of high energy heavy ion collisions. In general, any system which is described by a "fluid model" requires the knowledge of the EOS in order to be able to solve the equations of motion.[1]

The physics of high pressure is studied experimentally in the laboratory by using static and dynamic techniques. In static experiments the sample is squeezed between pistons or anvils. The maximum static pressures obtained so far are below 2 Mbar ($=2\cdot 10^6$ Atmospheres) and the temperatures are up to a few hundred degrees Celsius. The conditions in these static experiments are limited by the strength of the construction materials. In the dynamic experiments shock waves are created. Since the passage time of the shock is short in comparison to the disassembly time of the (shocked) sample, one can do shock wave research for any pressure that can be supplied by a driver. Chemical explosives have been used to create shock waves up to about 10 Mbar (in metals) with accompanying temperatures of the order of 10^4°K. The high power lasers developed for inertial confinement fusion have achieved up to 10^2 Mbar pressures and temperatures of the order of 10^7°K. Using underground nuclear explosions pressures up to 10^3 Mbar have been achieved. Other shock wave generators are also used (such as two stage light gas gun, exploding foils and rail guns, magnetic compression) in the study of the equations of state.

The modern age of systematic investigations of high pressure physics was started by P.W. Bridgman[2] (from 1906 to 1954) who received the Nobel Prize for (high pressure) physics in 1946. During and immediately after World War II the high pressure research was mainly associated with weapons laboratories, for the most part, in the United States[3] and the USSR.[4] The increase in interest in the last two decades in high pressure physics is associated with the development of modern computers and the emergence of powerful sources of concentrated energy (e.g. high explosives, lasers, ion beams, etc.). The high pressure physics is the physics of high energy densities described theoretically by the equations of state.[1-11]

The EOS describes nature over many orders of magnitude; actually over all possible values of pressure, density and temperature where local thermodynamic equilibrium can sustain. It is therefore difficult to describe the interparticle interaction of a many body system for any value of coupling and dimensions by a simple model. Therefore, one has to introduce simplified methods whose range of applicability is limited (Ideal gas EOS, Thomas-Fermi EOS, etc.). For this purpose it is convenient to describe various domains in the calculations of equations of state. See the Pressure-Temperature diagram (Fig. 1). Region 1 represents the normal state of matter where, in general, the equations of state are controlled by ordinary chemical forces. This region includes phase transitions such as liquid-solid transitions, critical phenomena (such as ferromagnetism), phase transitions due to changes of symmetry in solids etc. The origin of figure 1 includes the phenomena of superfluidity and superconductivity.

We increase the temperature corresponding to region 1 without increasing the pressure and we enter region 2. At about $kT\approx 1\text{eV}(T\approx 10^4\,^\circ\text{K})$, the solid matter is dissociated into atoms which are described to a good approximation by the equation of state for an ideal gas.

$$PV = Nk_BT\ , \tag{1.1}$$

$$E = \frac{3}{2}Nk_BT\ , \tag{1.2}$$

where P, V, T, N and E represent respectively the pressure, volume, temperature, total number of particles and total energy of the gas and k_B ($\approx 1.38\times 10^{-16}$erg/°K) represents the Boltzmann constant.

Further increase of the temperature (see region 3) leads to electron dissociation (i.e., to ionization) and one may use the Saha equation to describe the equation of state,[1]

$$\frac{nx^2}{1-x} = 2\frac{g_2}{g_1}\left[\frac{2\pi m}{h^2}\right]^{3/2}\frac{(k_BT)^3}{2}\exp(-I/kT) \tag{1.3}$$

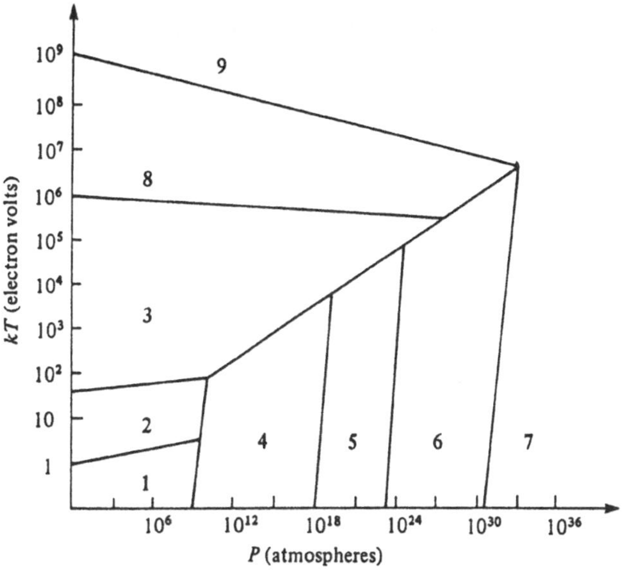

FIGURE 1. Scheme of a temperature-pressure diagram for various states of matter.

where g_1 and g_2 are the degeneracies of the two states under consideration, I is the ionization energy, x is the degree of ionization, n is the electron density, h is Planck's constant and m is the electron mass. Taking for example x=0.9 (i.e. 90% ionization) and using the ideal EOS (eq.(1.1) with $n=N/V$), one gets from Saha equation (1.3),

$$P \approx 10^{38}(k_B T)^{5/2}\exp(-I/k_B T) \ [\text{dyne/cm}^2] . \tag{1.4}$$

For a pressure of one atmosphere ($\approx 10^6 dyne/\text{cm}^2$) one obtains from eq.(1.4),

$$\frac{0.4I}{k_B T} \approx 32 + \frac{5}{2}\log(k_B T) . \tag{1.5}$$

In order to ionize most of the electrons in different elements we choose $I \approx 500$eV as an order of magnitude for the ionization energies, so that eq. (1.5) yields $k_B T \approx 40$eV as a transition point from region 2 to region 3 in Fig. 1. At temperatures larger than 40eV (note: this is an order of magnitude calculation only) one has a plasma with about 90% ionization. The electrons in the plasma (region 3) are a non-degenerate gas satisfying the ideal EOS to a good approximation.

Before discussing region 4 we would like to mention that a gas consisting of free electrons is said to be non-degenerate when,

$$k_B T \gg \varepsilon_{F_0} , \tag{1.6}$$

where ε_{F_0} is known as the Fermi energy of the electron gas at T=0°K and is given by

$$\varepsilon_{F_0} = \frac{h^2}{8\pi^2 m}(3\pi^2 n)^{2/3} = 5.84\times 10^{-27} n^{2/3} \ [erg] \tag{1.7}$$

where n (measured in cm^{-3}) represents the number of electrons per unit of volume. When (1.6) is satisfied, the electron gas behaves as a classical gas and the pressure and total energy will be given by (1.1) and (1.2) respectively. On the other hand, when $k_B T \ll \varepsilon_{F_0}$, quantum effects dominate and the gas is said to be highly degenerate. At T=0°K, the gas is said to be completely degenerate. If we assume the electron gas to be completely degenerate then elementary application of Fermi-Dirac statistics gives

$$\varepsilon = \frac{3}{5}\varepsilon_{F_0} , \ P = \frac{2}{5} n \varepsilon_{F_0} , \tag{1.8}$$

where ε represents the average energy of the electron. Assuming, $n \approx 10^{26}\text{cm}^{-3}$, we get

$$\varepsilon_{F_0} \approx 7.9\times10^2\text{eV}\,,\ \varepsilon \approx 4{\cdot}7\times10^2\text{eV}\,,\ P \approx 5\times10^{10}\text{atm}\,.$$

Thus the average energy of an electron is much greater than the binding energy of the electrons in atoms and the electrons behave as a degenerate gas. Region 4 corresponds to $P>10^9$ atm and $(k_BT/\varepsilon_{F_0}) \ll 1$ and the equation of state is satisfactorily described by the Thomas-Fermi model. However, the transition regions (between 1 and 4, 2 and 4) are most difficult to describe satisfactorily and, in general, corrections to the Thomas-Fermi model are necessary. In this immediate region which can be as broad as $10^5 < P < 10^{12}$ atm, one tries to describe the electrons by a Thomas-Fermi model modified by quantum corrections.[12] In this region the nuclei can be considered as a classical gas or can be described by the Gruneisen model.[1]

The boundary between the non-degenerate gas (region 3) and the degenerate electron gas (region 4) can be estimated by equating the energy associated with the ideal gas (1.2) and the energy associated with the degenerate Fermi gas (1.8)

$$\frac{3}{2}k_BT = \frac{3}{5}\varepsilon_{F_0}$$

$$n \approx \frac{1}{3\pi^2}\left(\frac{20\pi^2 m}{h^2}\right)^{2/3}(k_BT)^{3/2} \approx 9\times10^{39}(k_BT)^{\frac{3}{2}}\ [\text{cm}^{-3}]\,, \tag{1.9}$$

where (k_BT) is measured in ergs. Thus,

$$P \approx nk_BT \approx 9\times10^{39}(k_BT)^{5/2}[\text{dynes/cm}^2] \tag{1.10}$$

For $k_BT \approx 10^2$eV we get $P \approx 3\times10^9$ atm. For $k_BT \approx 10^4$eV, $P \approx 3\times10^{14}$ atmospheres. These estimates indicate the order of the pressures and temperatures at the boundary of regions 3 and 4.

The Thomas-Fermi EOS is given by,

$$P = \frac{h^2}{20\pi^2 m}(3\pi^2)^{\frac{2}{3}}\bar{n}^{\frac{5}{3}}\left[1 - \frac{1}{2\pi a_0}\left[\frac{4Z^2}{\bar{n}}\right]^{\frac{1}{3}} + \cdots\right], \tag{1.11}$$

where $a_0\ (=h^2/4\pi^2 me^2)$ is the Bohr radius, Z is total number of electrons in an atom and $\bar{n} = \dfrac{Z}{V}$ represents the average number of electrons per unit volume; V is the volume associated with each atom. Obviously, at high pressures, $\bar{n}$ becomes very large and (1.11) goes over to (1.8) implying the applicability of the simple Fermi-Dirac model. However, when the pressures become very high (region 5), the electrons become relativistic and we should therefore use the proper relativistic theory for a degenerate electron gas which has been extensively used in many astrophysics problems.[13] We should mention that in many stars, the temperature T may be as high as 10 million degrees ($k_BT \approx 10^3$eV); however, the matter may be so compressed that the Fermi energy ε_F may be 10^6eV. Thus the condition $k_BT \ll \varepsilon_F$ will be satisfied and we may apply the theory corresponding to the completely degenerate electron gas (i.e., assume $T=0$). In region 4, for $k_BT \ll \varepsilon_F$, one can approximate (1.11) by

$$P \approx 3\times10^{39}\varepsilon^{5/2}[\text{dyne/cm}^2]\,. \tag{1.12}$$

This relation is not appropriate for energies $\varepsilon \gtrsim mc^2$, where m is the electron mass, however the boundary between regions 4 and 5 may be estimated by substituting $\varepsilon = mc^2$ in eq. (1.12). This implies a region 5 defined by $P>10^{18}$ atmospheres.

For a relativistic gas the EOS scaling laws change. This can be seen on an elementary level from the uncertainity principle and the special relativistic relations: $\Delta x{\cdot}\Delta p \sim h$, where $\Delta x \sim x \sim V^{1/3}$, ($V$ = volume) implying $\Delta p \sim p \sim V^{-1/3}$. The relation between energy (E) and momentum (p): $E=(p^2c^2+m^2c^4)^{1/2}$ for the nonrelativistic limit (NR) yields $E \approx mc^2+p^2/2m+\cdots$ while for the extreme relativistic (ER) limit gives $E \approx pc+m^2c^3/2p+\cdots$. These formulae yield

$$\text{NR: } E \sim V^{-2/3}\,,\ P = \frac{\partial E}{\partial V} \sim V^{-5/3}\,, \tag{1.13}$$

$$\text{ER:}\quad \begin{cases} P\sim V^{-4/3}\ ,\ P\approx\dfrac{ch}{24\pi^3}\left[3\pi^2 n\right]^{4/3} \\ E\sim V^{-1/3}\ ,\ E\approx\dfrac{3hc}{8\pi}(3\pi^2 n)^{1/3} \end{cases} \tag{1.14}$$

In region 6, the electrons are degenerate and relativistic while the protons and neutrons of the matter are degenerate and in region 7 the pressures are so high that the following reaction becomes possible:

$$e^-+p\rightarrow n+\nu \tag{1.15}$$

and we have a neutron star. Substituting in eq.(1.14) for E the mass of a neutron ($\approx10^9$)eV, one gets a pressure of $P\approx10^{31}$ atm, which was chosen as the boundary between region 6 and 7.

We have seen that going along the pressure axis, from low pressure to high pressure, the energy of the elementary particles of matter is increasing even for zero temperature. Those increase in energy is inducing relativistic effects. Moreover, because at high pressure the density is higher, the distance between two identical particles is becoming smaller up to the point where quantum effects are dominant. So by increasing the pressure the quantum effects become more significant. On the other hand, by increasing the temperature (at a constant pressure) the distance between identical particles is increasing leading to the limit of ideal gasses. However, the temperature increase leads to relativistic effects.

In region 8, $\varepsilon\approx k_B T\approx 2\text{mc}^2$, one gets pair production of electrons and positrons; the neutron disintegrates ($n\rightarrow p+e^-+\bar{\nu}_e$) and the protons ($p$) satisfy the ideal EOS. The boundary of region 9 is defined by $\varepsilon\approx k_B T\gtrsim 2\text{MC}^2$, where M is the nuclear mass. In this region proton- antiprotons are created in equilibrium with radiation. In this regime the ideal relativistic EOS describes a system of elementary particles. The question about the existence of a maximum temperature is related to this regime of physics.[14]

Due to practical space limitations I shall discuss only the following subjects:(a) The Thomas-Fermi-Dirac (TFD) EOS, (b) EOS problems in Inertial Confinement Fusion (ICF) and (c) The virial theorem and EOS. I believe that the TFD-EOS is important to the physics of high pressure (in astrophysics, in geophysics, shock wave studies, etc.). The use of EOS in ICF is a typical example of how a new and growing field of research has to cope with the knowledge of proper equations of state. Last, but not least, is the demonstration of how useful a general theorem can be in dealing with EOS. The virial theorem is a good example how one can relate different uses and perspectives of EOS.

2. The Thomas-Fermi (TF) and the Thomas-Fermi-Dirac (TFD) Models.

The Thomas-Fermi model is essentially a statistical model for the atomic electrons put forward by Thomas[15] (1927) and Fermi[16] (1928). Originally, the model was introduced to study a many electron atom system, however, since then, it has found important applications[1,6,7] in molecular theory, solid state theory and in determining the contribution from the electrons to equations of state of matter at high pressures (P>10^7 atmospheres). The last application is of considerable interest in the inertial confinement fusion problem. The advantages of the Thomas- Fermi model over other models are, its simplicity, clarity and validity over a wide range of densities and temperatures.

The Thomas Fermi model of the atom is based mainly on the following two assumptions; (i) the electrons are considered as a degenerate gas placed in a self-consistent electrostatic field described by the electrostatic potential V(r) which varies little over a de-Broglie wavelength of the electron and (ii) the field varies slowly enough so that we can consider a volume element $\vec{dr}$ which contains a large number of particles and at the same time the field can be assumed to be approximately constant in this volume $\vec{dr}$.

The Thomas-Fermi model describes the electronic systems (in an atom, in a molecule, in a perfect or defect solid, in a compressed gas or liquid, etc.) in terms of the electron density $n(\vec{r})$, $\vec{r}$ denoting the position in space. In general, this electron density is an observable (e.g. $n(r)$ can be measured by X-ray scattering). In quantum mechanics the density is obtained from solving the Schroedinger equation for the electronic wave function $\psi_i(r)$,

$$n(r)=\sum_i \psi_i^*(r)\psi_i(r) \tag{2.1}$$

where the wave functions are normalized and the sum is over the occupied levels. To calculate n(r) in general by using this procedure is a rather very complex task. Therefore, the Thomas-Fermi model is attractive and very useful because it calculates the density n(r) directly from the knowledge of the potential V(r), by avoiding Schroedinger equation and without using eq.(2.1)

First, we briefly review the Thomas-Fermi model at zero temperature. This can be described as an electron fluid model of an atom system. The energy in this model is given by [1]

$$E=E_K+U_{eN}+U_{ee}\equiv E_K+\int V(r)d^3r \tag{2.2}$$

where E_K is the kinetic energy ($\propto \int n^{5/3}d^3r$), U_{eN} and U_{ee} are the Coulomb interactions of electrons-nucleus and electron-electron accordingly. This energy is minimized under the constraint that the electron number is constant (=Z)

$$Z=\int n(r)d^3r \tag{2.3}$$

Introducing the Lagrange multiplier μ, one varies

$$\delta(E-\mu Z)=0 \tag{2.4}$$

to obtain the relation between the potential $V(r)$ defined in (3) and the electron density $n(r)$:

$$n(r)=\begin{cases}\dfrac{8\pi(2m)^{3/2}}{3h^3}\{\mu-V(r)\}^{3/2} & \text{for } \mu\geq V(r)\\[2ex] 0 & \text{for } \mu<V(r)\end{cases} \tag{2.5}$$

where m is the electron mass, h is the Planck constant and μ has the physical meaning of the chemical potential ($\mu=\frac{\partial E}{\partial Z}$). Using Poisson equation (note that μ is not a function of r).

$$\nabla^2(\mu-V(r))=4\pi e^2 n(r), \tag{2.6}$$

and the relation between $n(r)$ and $V(r)$ given by equation (2.5), one derives the Thomas-Fermi equation:

$$\frac{d^2y}{dx^2}=\frac{y^{3/2}}{x^{1/2}} \tag{2.7}$$

where x and y are defined in a spherical symmetry system by

$$r=\left[\frac{1}{4}\left(\frac{9\pi^2}{2}\right)^{1/3}a_0\right]Z^{-1/3}x\equiv Z^{-1/3}a_0x \tag{2.8}$$

$$\mu-V(r)=\frac{Ze^2}{r}y(x) \tag{2.9}$$

where a_0 is Bohr radius and Z is the atomic charge number. For an isolated atom the boundary conditions for the Thomas-Fermi equation (2.7) are given by (note that μ=0 in this case)

$$y=1 \quad \text{at } x=0 \tag{2.10}$$

$$y=0 \quad \text{at } x\to\infty \tag{2.11}$$

while for a gas of atoms the boundary conditions are

$$y(0)=1 \tag{2.12}$$

$$\left[\frac{x}{y(x)}\cdot\frac{dy}{dx}\right]_{x=x_0}=1 \tag{2.13}$$

where x_0 is related to r_0 by equation (2.8), and r_0 defines the atom boundary in the gas, $4\pi r_0^3/3 = 1/n_a, n_a$ being the nucleus density. The boundary condition given by eq. (2.13) is equivalent to the charge neutrality of an atom

$$Z = \int_0^{r_0} n(r) 4\pi r^2 dr \tag{2.14}$$

For a non-zero temperature the same basic formalism as described above is used. In this case, for the spherical symmetric case one defines a function

$$\psi(x) = \left\{ \frac{\mu + eV(r)}{k_B T} \right\} \frac{r}{r_0} \tag{2.15}$$

$$x = \frac{r}{r_0} \tag{2.16}$$

$$4\pi r_0^3/3 = \frac{N}{V} = \text{number of atoms/cm}^3 \tag{2.17}$$

so that the Thomas-Fermi equation is given by

$$\frac{d^2\psi}{dx^2} = ax\, I_{½} \left[\frac{\psi(x)}{x} \right] \tag{2.18}$$

where $a = (r_0/c)^2$, $c \cong 1.6 \cdot 10^{-9} cm/[k_B T/keV]^{¼}$ and I_n is the Fermi-Dirac function

$$I_n(x) = \int_0^{\infty} \frac{y^n dy}{\exp(y-x)+1} \tag{2.19}$$

The boundary conditions for eq. (2.18) are:

$$\psi(0) = \frac{Ze^2}{k_B T r_0} \tag{2.20}$$

$$\frac{d\psi(1)}{dx} = \psi(1) \tag{2.21}$$

where eq. (2.21) is the "cell" neutrality condition.

The Thomas-Fermi equation is very attractive and convenient due to the scaling laws available in this model. In particular, the atomic volume V scales as Z^{-1}, the temperature T scales as $Z^{4/3}$, the energy/atom ε scales as $Z^{7/3}$ and the pressure P scales as $Z^{10/3}$, the entropy S scales as Z^{-1} and the chemical potential μ scales as $Z^{-4/3}$. For example, the following relations can be applied in the Thomas-Fermi model:

$$\varepsilon Z^{-7/3} = F_1 \left[VZ \,,\, TZ^{-4/3} \right] \tag{2.22}$$

$$PZ^{-10/3} = F_2 \left[VZ \,,\, TZ^{-4/3} \right] \tag{2.23}$$

After solving the Thomas-Fermi model for one element (e.g. hydrogen gas), then the energy, the pressure and all the other thermodynamic functions (S, μ) are known for all the elements in the Mendeleyev's table.

Thomas-Fermi (TF) model was generalized by Dirac in order to take into account the effect of the antisymmetrization of the (electronic) wave functions of the identical particles. While the TF kinetic energy per volume behaves as $n^{5/3}$($\varepsilon_k \propto \nabla^2/\text{volume} \propto 1/r^5$ and $r \propto n^{-1/3}$), the Dirac's "exchange" energy contribution per volume behaves as $n^{4/3}$ ($\varepsilon_{ex} \propto \frac{1}{r}/\text{volume} \propto 1/r^4$). More specifically, in eq. (2.2) one has to add the term

$$E_{ex} = -\int C_{ex} n^{4/3} d^3r \qquad (2.24)$$

where $C_{ex} = \frac{3}{4}e^2(3/\pi)^{1/3}$. The disadvantage of TFD equation is the loss of scaling laws of TF equation and thus it is necessary to solve the TFD model separately for every element.

In the Thomas-Fermi-Dirac (TFD) model, the electron density $n(r)$ at a distance r from the nucleus is conveniently calculated by minimizing the free energy density as suggested within the formulation of Kohn and Sham.(17) In particular, a three step procedure was developed(18) for conveniently solving the TFD model. For fixed boundary conditions of volume and temperature and a given chemical potential μ, one integrates from the surface to the center of the atom by performing the following three steps:

(a) Compute the potential energy $V(r)$ at a distance r of the nucleus by means of the integral representation of the TFD equation

$$-eV(r) = \mu + \frac{4\pi e^2}{r}\int_r^{r_0} n(r')(r'-r)r'dr', \qquad (2.25)$$

($V(r)$ contains also the Dirac exchange term described above).

(b) Use the calculated $V(r)$ from step (a) in order to find ξ in the following energy equation(18)

$$\xi = -eV(r) + \frac{e^2}{\pi}\left[\frac{8\pi^2 mk_BT}{h^2}\right]^{1/2} I'_{1/2}(\xi/kT) \qquad (2.26)$$

where $I'_{1/2}$ is the derivative of the Fermi-Dirac function (eq. (2.19) of the order ½).

(c) Use the calculated ξ to find the local electron density $n(r)$.

$$n(r) = \frac{1}{2\pi^2}\left[\frac{8\pi^2 mk_BT}{h^2}\right]^{3/2} I_{1/2}(\xi/k_BT). \qquad (2.27)$$

This value of $n(r)$ is used in step (a) in order to follow the above procedure up to the center of the atom (r=0). The chemical potential μ is changed systematically in the above scheme in such a way that the neutrality condition is satisfied (eq. (2.14)).

The pressure P in the TFD model is conveniently written in the following analytic form:

$$P = \frac{k_BT}{3\pi^2}\left\{(8\pi mk_B\frac{T}{h^2}\right]^{3/2} I_{3/2}(\xi_0/kT) \qquad (2.28)$$

$$-\frac{e^2}{2\pi^3}\left[\frac{8\pi^2 mk_BT}{h^2}\right]^2 \left\{I_{1/2}(\frac{\xi_0}{k_BT})\cdot I'_{1/2}(\frac{\xi_0}{k_BT}) - X(\frac{\xi_0}{k_BT})\right\}$$

$$X(x) \equiv \int_{-\infty}^{x}\left[I'_{1/2}(y)\right]^2 dy \qquad (2.29)$$

where ξ_0 is the energy parameter at the atom's surface (see eq. (2.26)) and X is the Dirac's energy exchange contribution. Similarly, the TFD energy is given by

$$E = E_k + E_V + E_{ex} \qquad (2.30)$$

where the kinetic term is

$$E_k = k_BT\int \frac{I_{3/2}(\eta)}{I_{\frac{1}{2}}(\eta)} n(r)d^3r, \qquad (2.31)$$

the potential term

$$E_V = \frac{e}{2}\int\left[V(r) + \frac{\mu}{e} - \frac{Z}{r}\right] n(r)d^3r \qquad (2.32)$$

and the exchange interaction term is

$$E_{ex} = -\frac{e^2}{\pi}\left[\frac{8\pi m\, k_B T}{h^2}\right]^{1/2}\int\frac{X(\eta)\, n(r) d^3 r}{I_{1/2}(\eta)} \tag{2.33}$$

where $\eta = \xi / k_B T$.

The TFD model was further improved by taking into account the density gradient corrections[1]

$$E_{\nabla} = \frac{h^2}{288\pi^2 m}\int\left[\frac{\nabla n}{n}\right]^2 n d^3 r. \tag{2.34}$$

In the above models the motion of the ions was not taken into account. This effect was considered by some authors[12] by using the so-called quantum statistical model where the ion-electron potential V_{ei} and the ion-ion potential V_{ii} are given by

$$V_{ei} = -\sum_j \int \frac{Ze^2 n(r) d^3 r}{|\vec{r} - \vec{R}_j|} \tag{2.35}$$

$$V_{ii} = \frac{1}{2}\sum_{\substack{i,j \\ i \neq j}} \frac{Z^2 e^2}{|\vec{R}_i - \vec{R}_j|} \tag{2.36}$$

where $\vec{R}_i$ is the vector radius of ion i.

3. Equation of State Problems in Inertial Confinement Fusion

The main idea of inertial confinement fusion is the aim of achieving very high compression using laboratory facilities, up to ρ/ρ_o=10,000! This concept, can be easily understood by using a "realistic" equation of state. Using for example the Thomas-Fermi model for hydrogen isotopes; a D-T mixture with initial (liquid) density of ρ_0=0.2cm^3, one needs an energy of 3.0 keV per atom to increase the temperature of the fuel to 1 keV ($\approx 10^7$ °K) without changing its density. However, for an extra energy of 0.2 keV/atom (i.e. an extra energy of about 7 percent!) at 1 keV temperature one gets a compression of ρ/ρ_0=20! Using elementary knowledge of nuclear reaction cross sections it is evident that it is necessary to use the driver's energy in order to compress the material as much as possible instead of heating it by "brute force". The energy gain is proportional to $(\rho/\rho_0)^2$. Although the idea seems to be self-evident from the equation of state data, the way to put it into practice is still unresolved.

The most effective compression is isentropic and this might be achieved approximately if the ablation pressure could be increased according to an ideal time profile. Besides the time shaping of the driver pulse, structured targets have to be used to vary the time evolution of the pressure on the compressed material an to avoid preheating of the pellet core in order to achieve the desired isentropic compression. These considerations have shown that high compression can be achieves by (a) shaping in time the input energy of the driver, and (b) "clever" pellet design. For example, Livermore designers[19] suggested for a Nd:YAG laser driver with maximum irradiance of 10^{14} W/cm^2, "double shell" pellet.

The properties of matter at high density and high temperature are important to explain and to calculate the compression process. In particular the equation of state data for different materials are necessary to calculate the shock wave propagation into the pellet. For inertial confinement fusion a knowledge of the properties of matter is needed[1] for temperatures up to 100 keV and densities up to 10^4 times solid density.

The corresponding pressures are enormous. For example, the pressure of the degenerate electrons of hydrogen with 10^4 times liquid density (i.e. an electron density of about n $\approx 5\times10^{26}$ cm^{-3}) is about 10^{12} atmospheres ($\approx 10^6$ Mbars). This estimate is obtained using the expression for Fermi degenerate electron pressure at zero temperature. For comparison, the thermal pressure of the non-degenerate ions P=n kT with $n = 5{\cdot}10^{26}$cm^{-3} and a temperature of $k_B T$ = 10 keV is about 5×10^{12} atmospheres. Thus, it can be summarized that for inertial confinement purposes, one needs the data of equations of state for many materials in the domain of

$$0<T<100\ \text{keV}; 10^{-4}<\frac{\rho}{\rho_0}<10^4\ ; 0<P<10^{13}\ \text{atm} \tag{3.1}$$

where ρ_0 is the initial liquid or solid density of the material under consideration.

The physics of inertial confinement fusion is based on the hydrodynamic equations of one or more fluids[20]. In order to solve these equations a knowledge of the equation of state and transport coefficients (such as thermal conductivity, electrical conductivity, radiation opacities, etc.) is necessary.

For example, in a simple model, the electron energy equation is

$$\left[\frac{\partial E_e}{\partial T_e}\right]_V \frac{dT_e}{dt}+\left[\frac{\partial E_e}{\partial V}\right]_T\left[-\frac{1}{\rho^2}\right]\frac{d\rho}{dt}+P_e\frac{dV}{dt}=\phi_e\left[\frac{\text{erg}}{\text{sec cm}^2 g}\right] \tag{3.2}$$

where the quantities with subscript e refer to the electrons, E is the energy, e.g., $E=E(T,V)$, the specific volume V is related to the "fluid" density $\rho=1/V$, P_e is the electron pressure and ϕ_e is the energy source term. A similar equation to (3.2) is written for the ions. The source term ϕ includes: (a) energy absorption from the laser in the domain of laser absorption, (b) thermal conduction of energy, (c) electron-ion energy exchange, (d) radiation losses, (e) thermonuclear energy absorption. These energy sources depend not only on equation of state data but also on ionization average $\bar{Z}$ and its higher moment $\overline{Z^2}$. For example, the inverse bressmatrahlung laser absorption contribution to (a), the thermal conductivity coefficient describing (b) and the electron-ion collision frequency for (c) are functions of $\bar{Z}$ and $\overline{Z^2}$. In particular, the thermal conductivity and the inverse bremsstrahlung absorption coefficients are functions of $\overline{Z^2}/\bar{Z}$ while the electron-ion classical collisions depends on $\overline{Z^2}$. In general, in order to know $\bar{Z}$ and $\overline{Z^2}$ for elements with any charge number Z, one has to solve the "rate equations" and to find the different distribution densities of the ions with one, two, three, etc. degree of ionization, and finally to average over these quantities in order to derive $\bar{Z}$ and $\overline{Z^2}$. In order to calculate $\bar{Z}$ and $\overline{Z^2}$ in a compact and effective way, we suggest[18] that the "free electrons" in TF or TFD models are defined as those having a positive energy exceeding the value of the chemical potential on the atom boundaries. This definition is expressed by the formula

$$\bar{Z}=\frac{1}{2\pi^2}\left[\frac{8\pi^2 m k_B T_e}{h^2}\right]^{3/2}\int F_{1/2}\left[\frac{\xi}{k_BT}\,,\,\left|\frac{\xi-\mu}{k_BT}\right|\right]d^3r \tag{3.3}$$

where $F_{1/2}(\alpha,\beta)$ is the incomplete Fermi-Dirac function of order ½ defined by

$$F_i(\alpha,\beta)=\int_\beta^\infty\frac{y^i dy}{\exp(y-\alpha)+1} \tag{3.4}$$

and ξ was defined by eq. (2.26). Results from eq. (3.3) have been recently compared[18] with those obtained from the Saha model and good agreement was obtained. The advantage of the TFD model over the Saha equation in calculating $\bar{Z}$ is that the TFD equations can be used also at very high compressions for which the Saha model is inappropriate.

Using eqs. (2.14), (2.27) and (3.3), $\bar{Z}$ may be cast in the following form:

$$\bar{Z}=Z\frac{\int F_{1/2}(\alpha,\beta)/I_{1/2}(\alpha)n(r)d^3r}{\int n(r)d^3r} \tag{3.5}$$

where $\alpha=\xi/k_BT$, $\beta=|\xi-\mu|/k_BT$ and $F_{1/2}(\alpha,0)=I_{1/2}(\alpha)$. Equation (3.5) can be written symbolically

$$\bar{Z}=Z\langle F_{1/2}(\alpha,\beta)/I_{1/2}(\alpha)\rangle \tag{3.6}$$

This equation has a simple interpretation, namely the atom is divided into an infinite number of spherical layers, each of thickness dr. These layers are characterized by an electron density $n(r)$. The average fractional ionization number in the layer, is $F_{1/2}(\alpha,\beta)/I_{1/2}(\alpha)$. Averaging these ratios over the ionic volume yields the average fractional degree of ionization of the atom, which multiplied by Z results in $\bar{Z}$.

It is plausible to generalize the above model in order to calculate higher moments,

$$\overline{Z^k} = C_k \langle F_{k/2}(\alpha,\beta)/I_{k/2}(\alpha) \rangle \tag{3.7}$$

In the limit of a fully ionized atom, $\beta = 0$ and therefore in this case $\overline{Z^k} = Z^k = C_k$. Since C_k is a constant one always has $C_k = Z^k$. Therefore, eq. (3.7) implies

$$\overline{Z^2} = Z^2 \langle F_1(\alpha,\beta)/I_1(\alpha) \rangle \tag{3.8}$$

Our method may be applicable for the computations of various atomic processes at extreme conditions of pressure and temperature. In particular, the electron thermal conductivity, bremsstrahlung emission, photoionization cross section and other quantities which depend on the moments of the charge state distribution can be estimated. Moreover, the first moment gives the average charge distribution, the second moment is connected to its width and the third moment is related to the asymmetry of the distribution. By calculating the first few moments one may characterize the whole charge distribution with reasonable accuracy. This distribution may be used to derive various plasma parameters, total photoabsorption cross section and plasma opacities.

Hydrodynamic codes which perform simulations of inertial confinement fusion require the use of equation of state models which describe the thermodynamic functions of both electrons and ions.[1] The main values of interest are the pressure and internal energy, as well as their respective temperature and volume derivatives as a function of material density and temperature. The electronic equation of state can be taken, to be a good approximation, from the corrected Thomas- Fermi-Dirac model. The ion contribution to the equation of state can be described by the Debye-Gruneisen equation of state, with the appropriate density variations of the Debye temperature and the Gruneisen coefficient. At sufficiently high temperature and low densities the ion contribution can be described by the ideal gas equation of state. To join these two limiting cases a semi-empirical interpolation method can be used.[20] The extent to which the inclusion of different models into the computer codes will influence the inertial confinement fusion results is a subject of examination and research. In general, since the heating mechanisms, the energy transport and the effects occurring in the corona are not very dependent on the ion parameters, one would expect not much dependence of these phenomena on the ion equation of state model. However, for processes which may be more sensitive to the ion parameter, such as those taking place in the compressed solid, e.g. shock wave phenomena, one may expect changes in the calculated results due to use of different models

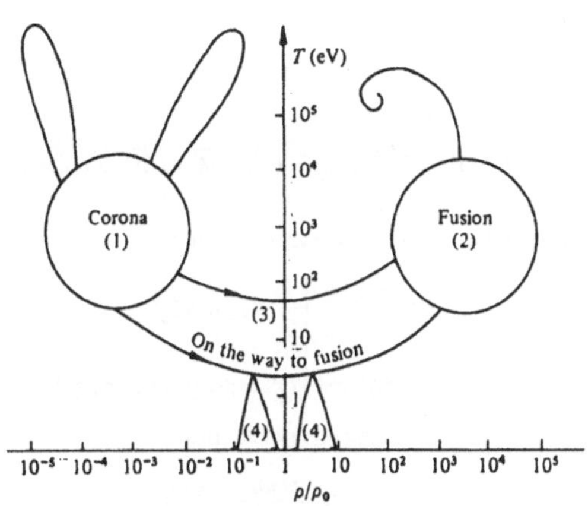

FIGURE 2: The difficulties for high gain pellet fusion due to the equation of state (EOS). (1) *The corona*: 2 or 3 temperature EOS; Ideal gasses; Weak Coulomb interaction; Non-Local-Thermal-Equilibrium. (2) *Fusion*: Strong Coulomb interaction; T-F or T-F-D model. (3) *On the Way to Fusion*: Most difficult. (4) *Easy Experiments regions.*

in computing the ion contribution to the equation of state. The equation of state determines also the velocity of shock waves and therefore the time scale of the whole implosion occurring in inertial confinement fusion phenomena.

For a summary and a conclusion see figure 2 with a rather negative aspect of the central ignition scheme.

4. The Virial Theorem

This theorem was first given by Poincare.[21] Consider a system of N particles. Let (x_i, y_i, z_i) represent the (instantaneous) Cartesian coordinates of the ith particle (i=,1,2,3,...,N) and let X_i, Y_i, and Z_i represent the Cartesian components of the force (F_i) acting on it. The equations of motion for the ith particle are

$$m_i\frac{d^2x_i}{dt^2}=X_i;\; m_i\frac{d^2y_i}{dt^2}=Y_i;\; m_i\frac{d^2z_i}{dt^2}=Z_i \tag{4.1}$$

These equations imply

$$\frac{1}{2}\frac{d^2I}{dt^2}=2K+\sum_i\vec{F}_i\cdot\vec{r}_i \tag{4.2}$$

where

$$K=\sum_i\frac{1}{2}m_i\left[\left(\frac{dx_i}{dt}\right)^2+\left(\frac{dy_i}{dt}\right)^2+\left(\frac{dz_i}{dt}\right)^2\right] \tag{4.3}$$

represents the total kinetic energy of the system and

$$I=\sum_i m_i(x_i^2+y_i^2+z_i^2)=\sum_i m_i r_i^2 \tag{4.4}$$

represents the moment of inertia of the system about the origin of the coordinate system. For a system in a steady state we may set d^2I/dt^2 equal to zero so that eq.(4.2) takes the form (note: for quasi steady state conditions, e.g. a star with turbulent motion, one may take a time average of (4.2) and set the left hand side equal to zero)

$$2K+W=0 \tag{4.5}$$

where

$$W=\sum_i\vec{F}_i\cdot\vec{r}_i \tag{4.6}$$

is known as the *virial* of the system. Equation (4.6) is usually referred to as the *virial theorem.*

Let us use eqs.(4.5) and (4.6) for an ideal gas. In this case, $d\vec{F}=-PdA\vec{n}$, where $\vec{n}$ is a unit vector perpendicular to the area A and P is the pressure. Therefore,

$$2K=-\sum_i\vec{F}_i\cdot\vec{r}_i=P\oint\vec{n}\cdot\vec{r}\,dA=P\oint\vec{r}\cdot d\vec{A}=P\int\vec{\nabla}\cdot\vec{r}\,dV=3PV\,,$$

(where the surface integral was taken over a closed area) implying

$$PV=\frac{2}{3}K=\frac{2}{3}\left[\frac{3}{2}Nk_BT\right]=Nk_BT$$

which is the ideal equation of state. Thus the ideal EOS is derived by using the virial theorem.

If we assume that gravitational forces are acting on particles then

$$\vec{F}_i=-\sum_j G\frac{m_im_j}{|\vec{r}_i-\vec{r}_j|^3}(\vec{r}_i-\vec{r}_j) \tag{4.7}$$

where the summation is over all possible values of j with the constraint $j\neq i$. Thus

$$W=\sum_i\vec{F}_i\cdot\vec{r}_i=-G\sum_i\sum_j\frac{m_im_j}{|\vec{r}_i-\vec{r}_j|^3}\vec{r}_i\cdot(\vec{r}_i-\vec{r}_j) \tag{4.8}$$

In the above summation we can interchange the indices i and j without changing W. This implies

$$W=-\frac{G}{2}\sum_i\sum_j\frac{m_i m_j}{|\vec{r}_i-\vec{r}_j|^3}(\vec{r}_i-\vec{r}_j)\cdot(\vec{r}_i-\vec{r}_j)=-\frac{G}{2}\sum_i\sum_j\frac{m_i m_j}{|\vec{r}_i-\vec{r}_j|} \tag{4.9}$$

The above expression is simply Ω, where Ω represents the total gravitational potential energy of the system. Thus the virial theorem is given by $W=\Omega$ and (4.5) becomes (In general, for a force varying as r^{ν} one gets $W=-(\nu+1)\Omega$; for the Coulomb force $\nu=-2$)

$$2K+\Omega=0 \tag{4.10}$$

(For highly relativistic particles one gets $K+\Omega=0$). Now if we consider a small mass dm of a perfect gas (consisting of dN particles) then the kinetic energy associated with it would be given by[1]

$$dK=\frac{3}{2}k_BT\,dN=\frac{3}{2}RT\frac{dm}{\mu} \tag{4.11}$$

where μ represents the mean molecular weight. Since $c_p-c_V=R$, we may write

$$dK=\frac{3}{2}(\gamma-1)\frac{c_VT}{\mu}dm \tag{4.12}$$

where $\gamma=c_p/c_V$ represents the ratio of specific heats. Now, c_VT/μ represent the internal energy per unit mass; thus

$$dK=\frac{3}{2}(\gamma-1)dU \tag{4.13}$$

For the whole system we have

$$K=\frac{3}{2}(\gamma-1)U \tag{4.14}$$

It may be noted that for an ideal monatomic gas, $\gamma=\frac{5}{3}$ and $K=U$ (obviously!).

Substituting (4.14) in (4.10) we have

$$U=-\frac{1}{3(\gamma-1)}\Omega \tag{4.15}$$

The total energy of the system would therefore be given by

$$E=U+\Omega=\frac{\gamma-\frac{4}{3}}{\gamma-1}\Omega \tag{4.16}$$

The above equation shows that for $\gamma>\frac{4}{3}$, E is negative and therefore the star is stable (Ω is a negative quantity).

It is interesting to note that since $\Omega\approx-GM^2/R$, as the star contracts Ω becomes more negative and the change in total energy would be

$$\Delta E=\frac{\gamma-\frac{4}{3}}{\gamma-1}\Delta\Omega \tag{4.17}$$

Since $\Delta\Omega$ is negative, ΔE would also be negative and the amount $-\Delta E$ is radiated away. Further, (4.15) tells us that

$$\Delta U=-\frac{1}{3(\gamma-1)}\Delta\Omega \tag{4.18}$$

is positive, which is responsible for the increase in temperature of the star. Thus, as the star contracts, its internal energy increases and it gets hotter and at the same time it radiates energy! Physically, this is due to the fact that the loss in potential energy is *greater* than the increase in its internal energy and the difference is radiated away. For example, for $\gamma=\frac{5}{3}$

$$\Delta U = -\frac{1}{2}\Delta\Omega \tag{4.19}$$

i.e., only half of the decrease in the potential energy is used up in increasing its internal energy.

We end this section by using the virial theorem in the Thomas-Fermi model. In this case the virial W is defined by

$$W = \langle \sum_i \vec{r}_i \cdot \vec{F}_i \rangle = W_{ee} + W_{en} + W_{eb} , \tag{4.20}$$

where $\vec{F}_i$ denotes the force acting on the ith electron whose position is given by $\vec{r}_i$ and the angular brackets denote a space (or time) average. The force acting on an electron may be due to electron-electron interaction (W_{ee}), electron-nucleus interaction (W_{en}) and the electron interaction with its atomic boundary (W_{eb}). Since the force applied by the boundary is directed towards the center of the sphere and occurs at the atomic boundary where $r=r_0$ we have,

$$-W_{eb} = -\langle \sum_i \vec{r}_i \cdot \vec{F}_i^{eb} \rangle = r_0 \langle \sum_i F_i^{eb} \rangle = r_0 \left[4\pi r_0^2 P \right] ,$$

where in the last step we have used the fact that the time average of the boundary force is just the pressure P times the area of the boundary, $=4\pi r_0^2$. Since $V=(4\pi/3)r_0^3$ we have

$$-W_{eb} = 3PV . \tag{4.21}$$

We next consider the electron nuclear interaction for which,

$$\vec{F}_i^{en} = -\frac{Ze^2}{r_i^3}\vec{r}_i ,$$

the negative sign showing that the force is directed towards the origin. Thus,

$$-W_{en} = \langle \sum_i \frac{Ze^2}{r_i} \rangle = -E_{pot}^{en} , \tag{4.22}$$

where E_{pot}^{en} is the potential energy arising out of the electron-nuclear interactions. Finally, we consider the force due to electron-electron interactions which will be given by,

$$\vec{F}_i^{ee} = e^2 \sum_j \frac{\vec{r}_i - \vec{r}_j}{|\vec{r}_i - \vec{r}_j|^3} ,$$

where the summation is over all possible values of j with $j \neq i$. Thus,

$$-W_{ee} = -e^2 \langle \sum_i \sum_j \frac{\vec{r}_i \cdot (\vec{r}_i - \vec{r}_j)}{|\vec{r}_i - \vec{r}_j|^3} \rangle = -\frac{e^2}{2} \langle \sum_i \sum_j \frac{(\vec{r}_i - \vec{r}_j)\cdot(\vec{r}_i - \vec{r}_j)}{|\vec{r}_i - \vec{r}_j|^3} \rangle$$

$$= -\frac{e^2}{2} \sum_i \sum_j \frac{1}{|\vec{r}_i - \vec{r}_j|} = -E_{pot}^{ee} , \tag{4.23}$$

where E_{pot}^{ee} represents the potential energy due to electron-electron interaction. Substituting the expressions for W_{eb}, W_{en} and W_{ee} in (4.20) and using (4.5) we get

$$2K + E_{pot}^{en} + E_{pot}^{ee} = 3PV . \tag{4.24}$$

Furthermore, the virial theorem can be applied equally well to calculations with exchange effects; this follows from the fact that the exchange energy is proportional to e^2/r. Thus (4.24) takes the following form for the Thomas-Fermi-Dirac model,

$$2K + E_{pot}^{en} + E_{pot}^{ee} + E_{ex} = 3PV . \tag{4.25}$$

Denoting the total potential energy by Ω_{eff}

$$E_{pot}^{en}+E_{pot}^{ee}+E_{ex}=\Omega_{eff} \tag{4.26}$$

one gets for the TFD model (or TF model)

$$2K+\Omega_{eff}=3PV\,. \tag{4.27}$$

Equation (4.27) (for the TF or TFD model) is similar to eq. (4.10) (for the gravitational case) if the following analogy is made

$$\Omega=\Omega_{eff}-3PV\,. \tag{4.28}$$

The virial theorem is an example of how the same mathematics and the same concepts can be introduced to explain the common properties of matter.

References

(1) S. Eliezer, A. Ghatak and H. Hora, *An Introduction to Equations of State: Theory and Applications* (Cambridge University Press, Cambridge) (1986).

(2) P.W. Bridgman, *The Physics of High Pressure* (G. Bell and Sons, London) (1958).

(3) M.H. Rice, R.G. McQueen and J.M. Walsh, *Solid St. Phys.* **6**, 1 (1958).

(4) L.V. Altshuler, *Sov. Phys.-Usp* **8**, 52 (1965).

(5) Y.B. Zeldovich and Y.P. Raizer, *Physics of Shock Waves and High Temperature Hydrodynamic Phenomena*, vols I and II (Academic Press, New York) (1967).

(6) N.H. March, *Adv. Phys.* **6**, 1(1057).

(7) S.G. Brush, *Progress in High Temperature Physics and Chemistry* (Ed., C.A. Rouse) **1** 1(1967).

(8) A.V. Bushman and V.E. Fortov, *Sov. Phys. -Usp.* **26**, 465 (1983).

(9) B.K. Godwal, S.K. Sikka and R. Chidambara, *Phys. Rep.* **102**, 121 (1983).

(10) M. Ross, *Rep. Prog. Phys.* **48**, 1 (1985).

(11) A.V. Bushman and V.E. Fortov, *Sov. Tech. Rev. B. Therm. Phys.* **1**, 219 (1987).

(12) R.M. More, Lawrence Livermore Laboratory Report UCRL 84991 (Parts i and ii) (1981).

(13) S. Candrasekhar, *Introduction to the Study of Stellar Structure* (University of Chicago, Chicago) (1939).

(14) R. Hagedorn, Thermodynamics of Strong Interactions, CERN 71-12 (1971).

(15) L.H. Thomas, *Proc. Camb. Phil. Soc.* **23**, 542 (1927).

(16) E. Fermi, *Zeits f. Phys.* **48**, 73(1928).

(17) W. Khon and L.J. Sham, *Phys. Rev.* **A140**, 1133(1965).

(18) H. Szichman, S. Eliezer and D. Salzman, *J. Quant. Spectrosc. Radiat. Transfer* **38**, 281 (1987).

(19) H.G. Alstrom, *Phys. of Laser Fusion*, vol ii (Lawrence, Livermore) p. 331 (1982).

(20) S. Eliezer, *Plasma Astrophysics*, Eds. T.D. Gugenne and J.J. Hunt (European Space Agency Publication, Noordwijk) (SP-285 vol i). p. 353 (1988).

(21) J. Jeans, *Astronomy and Cosmogony*, (Cambridge University Press)(1928).

SCALE TRANSFORMATIONS, THE ENERGY-MOMENTUM TENSOR

AND THE EQUATION OF STATE

P. Carruthers

Department of Physics
University of Arizona
Tucson, Arizona 85721

ABSTRACT

The Equation of State (EOS) relates diagonal elements of the energy-momentum tensor $\theta_{\mu\nu}$. The first moment of the energy-momentum tensor generates scale transformations. The virial theorem, a consequence of the behavior of the energy density under scale transformations, allows one to eliminate the kinetic energy in terms of the potential terms. The trace theorem for the energy-momentum tensor expresses ϵ-3p in terms of ensemble averages of scale-breaking operators, allowing a new approach to the EOS.

I. THE VIRIAL THEOREM IN QUANTUM THEORY

Consider a single-particle nonrelativistic bound state $\psi_n(x)$ with potential $V(r) = Ar^{-d}$ ($A > 0$ for $d < 0$) and $A < 0$ for $d > 0$). The virial operator

$$D \equiv \mathbf{x} \cdot \mathbf{p} \rightarrow -i\, r \frac{\partial}{\partial r} \tag{1.1}$$

is not conserved (here $T = p^2/2m$ is the usual kinetic energy);

$$i[D,H] = -2T - dV \quad , \tag{1.2}$$

but its average in the eigenstate does vanish, since both D and H are Hermitian (a consequence of the absence of surface terms);

$$\begin{aligned} &\langle \psi_n | [D,H] | \psi_n \rangle = 0 \\ &\langle T \rangle_{nn} = -\frac{d}{2} \langle V \rangle_{nn} \quad . \end{aligned} \tag{1.3}$$

This simplest example of the "virial theorem" allows us to re-express the kinetic energy contribution to E_n in terms of the average potential:

$$\begin{aligned} E_n &= \langle T \rangle_{nn} + \langle V \rangle_{nn} \\ &= (1 - \frac{d}{2}) \langle V \rangle_{nn} \quad . \end{aligned} \tag{1.4}$$

The Nuclear Equation of State, Part A
Edited by W. Greiner and H. Stöcker
Plenum Press, New York

Historically the most famous example is the $1/r$ potential (gravity or Coulomb potential) for which $\langle T \rangle = -\frac{1}{2} \langle V \rangle$, $\langle E \rangle = \frac{1}{2} \langle V \rangle$.

Note that D measures the dimension of the components of H (see Eq. (1.2)). The coefficients 2 and d are just the dimensions of ∇^2 and r^{-d} in the usual sense. (Here and later we use "mass dimension" instead of "length dimension.") Thus D has a simple geometric meaning. This result, generalized to many-particle and field-theoretic systems, allows one to simplify the computation of the equation of state by eliminating the kinetic energy in favor of potentials or polynomials of fields whose coefficients are dimension-dependent.

In the one-particle problem just reviewed, the dimensions of the bound system are fixed by the dynamics. As long as the system has a $(\text{radius})^3$ much smaller than the confining volume Ω, it is not appropriate to ask about pressure, much less temperature.

Next consider a degenerate Fermi gas of N neutral non-interacting spin-$\frac{1}{2}$ particles confined to a volume Ω. As is well-known, when Ω is decreased, the average kinetic energy increases, so that one speaks of the "Fermi pressure." Computing the volume dependence of the energy gives for $-\partial E/\partial \Omega$ the zero-temperature equation of state

$$p = A \rho^{5/3} \quad , \tag{1.5}$$

where $\rho = N/\Omega$ is the density.

This simple case provides a nice example of subtleties connected with surface effects. Naive application of (1.2) would indicate that the kinetic energy also vanishes. Close investigation shows that the assumed Hermiticity used in (1.3) is equivalent to the vanishing of a surface term. This is automatically satisfied for bound states but not for the Fermi gas. The failure of formal Hermiticity is in fact a very common occurrence. The most familiar example[1] is the orbital angular-momentum component $L_z = -i\hbar\partial/\partial\phi$ satisfying $[\phi, L_z] = i\hbar$. Taking the diagonal matrix element in the angular-momentum eigenstate $|m\rangle$, Hermiticity of L_z would give $\langle m|[\phi, L_z]|m\rangle = 0$, generating a paradox. The resolution can be expressed as follows: L_z is Hermitian only in the space of periodic functions. The azimuth ϕ is continuous, not periodic, for the form of the assumed commutator. A calculation designed to verify the Hermiticity of L_z will generate a non-vanishing surface term which resolves the "paradox."

Now consider a system of N identical particles interacting through two-particle potentials $v(r_{ij})$. The Hamiltonian is

$$H = \sum_i \frac{p_i^2}{2m} + \sum_{i<j} v_{ij} \quad , \tag{1.6}$$

where a,b run from 1 to N, and the N-particle virial operator is

$$D = \sum_i x_i \cdot p_i \quad . \tag{1.7}$$

D is not conserved (i.e., the system is not invariant under a scale transformation), as shown by

$$i[D,H] = -2T + \left[\sum_i x_i \cdot \nabla_i \right] \sum_{k<\ell} v_{k\ell} \quad . \tag{1.8}$$

In order to simplify the potential energy, we use the identity

$$\sum_i \mathbf{x}_i \cdot \nabla_i = \frac{1}{N-1} \sum_{i<j} d_{ij}$$

$$d_{ij} = \mathbf{x}_i \cdot \nabla_i + \mathbf{x}_j \cdot \nabla_j \quad . \tag{1.9}$$

We can regard d_{ij} as the dilation operator for the pair ij. The contributing terms to $\Sigma d_{ij}\ \Sigma v_{k\ell}$ have either $(i,j) = (k,\ell)$ or one of i,j equal to one of k,ℓ. Noting that $(d_{ij} + d_{j\ell})v_{ij}$ equals $d_{ij}\ v_{ij}$, we easily find

$$\frac{1}{N-1} \sum_{i<j} d_{ij} \sum_{k<\ell} v_{k\ell} = \sum_{i<j} d_{ij}\ v_{ij}$$

$$= \sum_{i<j} r_{ij}\ \partial v/\partial r_{ij} \quad . \tag{1.10}$$

Taking an ensemble average (e.g., with density matrix $\rho \propto \exp(-\beta H)$), we can write (1.8) as

$$2\langle T\rangle = \sum_{i<j} \mathbf{r}_{ij} \cdot \frac{\partial v(\mathbf{r}_{ij})}{\partial \mathbf{r}_{ij}} + \langle\frac{dD}{dt}\rangle \quad . \tag{1.11}$$

If v is a simple power, as assumed in Eq. (1.2), we obtain

$$2\langle T\rangle = -d\langle V\rangle + \langle\frac{dD}{dt}\rangle \tag{1.12}$$

for the N-particle system. If the interactions are attractive, it is possible to have a bound N-particle system, in which case $\langle T\rangle = -d\langle V\rangle/2$ results. For a thermalized system, the temperature must also be sufficiently low so that continuum states (representing breakup) are not important, if the term $\langle dD/dt\rangle$ is to be dropped.

The energy of the system is

$$\langle H\rangle = \langle T\rangle + \langle V\rangle \quad ;$$

$$E = \langle\sum_{i<j} (1 + \mathbf{r}_{ij} \cdot \frac{\partial}{\partial \mathbf{r}_{ij}})\ v(\mathbf{r}_{ij})\rangle + \frac{1}{2} \langle\frac{dD}{dt}\rangle \quad . \tag{1.13}$$

As for the one-particle problem, (1.13) is not an equation of state. However, if we enclose the system in a volume Ω, and adjust Ω so that interactions with the container influence the wave functions $\Psi_n(\Omega)$, (1.13) contains all the ingredients of the equation of state. It is necessary to include the interaction of the particles with the "walls" of the container by a term like $\Sigma_i V(\mathbf{r}_i)$ (for a static wall). As shown by the example of the Fermi gas, careful evaluation of the term $\langle\dot{D}\rangle$ is necessary to obtain a correct $E(\Omega)$ and corresponding pressure $p = -\partial E/\partial\Omega$.

II. SCALING AND DILATIONS

In Section I we saw that the virial operator $\mathbf{x} \cdot \mathbf{p}$ measures the dimension of operators which are powers of x or p. A component of x(p) counts as dimension -1(1) and a product $x^m p^n$ has dimension n-m. Such objects exhibit pure scaling behavior (in contrast to a function like exp(-x), for example). We now establish the necessary machinery[2] to describe scale transformations of functions and field operators.

As an example, consider the parabola $f(x) = x^2$. The value of the function f depends on the arbitrary convention about the scale, or ruler, used to label points on the x axis. Now suppose that we examine the same function using a ruler with spacings one-half the former. The same point now has (on the new ruler) the value $x' = 2x$ and the function $f'(x') = (x')^2 = 4x^2$. Although this gedanken measurement may seem like a hard way to express the idea that the dimension of x^2 is -2, it leads naturally to the definition of a scaled function.

Suppose that the function $f(x)$ (tacitly implying a ruler defining the convention for the values of x) is observed using a new ruler whose values describing the same fixed point in space are $x' = \lambda x$. The scaled function observed in the primed system is defined by

$$f'(x') = \lambda^{-d} f(x) \quad . \tag{2.1}$$

In order to compare the transformed function f' to f at the same value of their arguments, we write

$$f'(x) = \lambda^{-d} f(\lambda^{-1} x) \quad . \tag{2.2}$$

Consider now the infinitesimal form of the transformation $f'(x) = U^{-1}(\lambda) f(x) U(\lambda)$, where $U(\lambda) = \exp(iD\ln\lambda)$. Setting $\lambda = 1 + \epsilon$, we find for $\epsilon \to 0$

$$\delta f = f'(x) - f(x) = i\epsilon[f(x), D] \quad ;$$

$$\delta f = -\epsilon(d + x \cdot \partial) f \quad ;$$

$$[f(x), D] = i[d + x \cdot \partial] f \quad . \tag{2.3}$$

Hence the local change in the form of the function induced by the scale transformation is composed of two parts: d comes from the multiplicative change of scale and $x \cdot \partial$ from the change in argument on the right-hand side of (2.2). By the symbol x we mean a vector, and $x \cdot \partial$ is the scalar $x \cdot \partial / \partial x$ with whatever metric is appropriate.

The particular structure (2.2) is the most natural in field theory, since unitary transformations of the form $\phi(x) \to \phi'(x) = U^{-1} \phi(x) U$ are most natural. (For conserved generators, U is independent of x.)

As is well-known,[2,3] the Poincaré group generators are constructed from various moments of the energy-momentum tensor $\theta_{\mu\nu}$. (θ_{oo} is the energy density, $\theta^{oi} = \theta^{io}$ is the momentum density, θ^{ij}, i,j = 1,2,3 stresses. Our metric is diag $g_{\mu\nu} = (1,-1,-1,-1)$ for $\mu,\nu = 0,1,2,3$.) The geometric scale transformation operator, although not conserved in general, is also a moment:[3,4]

$$D = \int d^3x \, x^\mu \, \theta_{\mu o}$$

$$= tH + \int d^3x \, x^i \, \theta_{io} \quad . \tag{2.4}$$

The final term in (2.4) is just the virial for a momentum *density*, and could be guessed by generalizing Eq. (1.7). The explicit time dependence of D is given by the trace θ^μ_μ

$$\frac{dD}{dt} = \int d^3x \left[\theta_{oo} - x^m \frac{\partial\theta_{mo}}{\partial t}\right]$$

$$= \int d^3x \left[\theta_{oo} - x^n \frac{\partial}{\partial x^m} \theta^{mn}\right]$$

$$= \int d^3x \left[\theta_{oo} - \sum_n \theta_{nn}\right] \quad . \tag{2.5}$$

using successively the conservation law $\partial_\mu \theta^{\mu\nu} = 0$ and integrating by parts. Hence we have

$$\frac{dD}{dt} = \int d^3x \; \theta^\mu_\mu(x) \quad , \tag{2.6}$$

showing that the non-conservation of the dilation operator depends directly on the trace of the energy-momentum tensor. The field operators ϕ transform exactly as (2.3);

$$[\phi(x),D] = i(d + x \cdot \mu)\phi(x) \quad . \tag{2.7}$$

In general, D is not conserved, so that D(t) is at the same time as the variable $x = (t,\mathbf{x})$.

We remark that the form (2.4) depends on choosing $\theta_{\mu\nu}$ to be the "improved energy-momentum tensor," which has particularly advantageous renormalization properties. We give simple examples of canonical (i.e., prior to renormalization) dimension. Since $\int d^4x \, i\bar{\psi}\partial\!\!\!/\psi$ and $\int d^4x \, \phi^4$ are dimensionless contributors to Lagrangians, we expect (correctly) that the Dirac field ψ has dimension $^3/_2$ and the spinless Bose field ϕ has dimension 1. ϕ^2, ϕ^3, and $\phi\bar{\psi}\psi$ have, respectively, dimensions 2, 3, and 4. Note that a mass term $m\bar{\psi}\psi$ (or $m^2\phi^2$) has scaling dimension 3 (or 2), even though the "engineering" dimension is 4. When the theory is renormalized,[4] the dimensions shift from their canonical values and are called anomalous. In this paper we shall only use canonical dimensions.

III. CONNECTION BETWEEN THE EQUATION OF STATE AND THE ENERGY-MOMENTUM TENSOR

An equation of state, say $p = p(\epsilon)$ with p the pressure and ϵ the energy density, relates components of the expectation values of the energy-momentum tensor. If ρ is the density matrix (most simply $\propto e^{-\beta H}$), the observed energy-momentum tensor $T_{\mu\nu}$ is

$$T_{\mu\nu}(x) \equiv \mathrm{Tr}(\rho \; \theta_{\mu\nu}(x)) \equiv \langle\theta_{\mu\nu}(x)\rangle \quad . \tag{3.1}$$

The diagonal components are

$$\begin{aligned} \epsilon &= T_{oo} = \langle\theta_{oo}(x)\rangle \\ p_{xx} &= T_{xx} = \langle\theta_{xx}(x)\rangle \\ p_{yy} &= T_{yy} = \langle\theta_{yy}(x)\rangle \\ p_{zz} &= T_{zz} = \langle\theta_{zz}(x)\rangle \quad . \end{aligned} \tag{3.2}$$

Most current work assumes isotropy ($p_{xx} = p_{yy} = p_{zz}$), although this approximation is probably wrong in applications to heavy ion collision geometries.

In order to compare (3.2) with the notations of relativistic fluid mechanics, we imagine the expressions (3.2) to be evaluated in the local rest frame. The four-velocities themselves are determined[5] from $T^{\mu\nu}$. The classical tensor for a perfect relativistic fluid is

$$T^{c\ell}_{\mu\nu} = (\epsilon + p)\, u_\mu u_\nu - p g_{\mu\nu} \quad , \tag{3.3}$$

where u_μ is the four-velocity ($u^\mu u_\mu = 1$) and ϵ,p are Lorentz scalar densities, i.e., by definition $\epsilon'(x') = \epsilon(x)$, $p'(x' = p(x)$ for Lorentz transformations $x' = \Lambda x$ to the primed frame. The expression

$$\epsilon \equiv u^\mu T^{c\ell}_{\mu\nu} u^\nu \tag{3.4}$$

exhibits both the scalar character of ϵ and $\epsilon = T^{c\ell}_{0}$ in the local rest frame. Similarly the pressure field is given by projecting $T_{\mu\nu}$ on directions transverse to u_μ:

$$\begin{aligned} p_{\mu\nu} &\equiv \Delta_{\mu\rho} T^{\rho}_{c\ell} \Delta_{\sigma\nu} \\ \Delta_{\mu\nu} &= g_{\mu\nu} - u_\mu u_\nu \\ u^\mu \Delta_{\mu\nu} &= 0 \\ \Delta_{\mu\lambda}\Delta^{\lambda}_{\nu} &= \Delta_{\mu\nu} \quad . \end{aligned} \tag{3.5}$$

In the local rest frame $\Delta_{\mu\nu} = g_{ij} = -\delta_{ij}$ (i,j = 1,2,3 only) and zero otherwise. Hence only p_{ii} terms survive. The pressure scalar is then

$$p = -\frac{1}{3} p^\mu_\mu \quad . \tag{3.6}$$

The trace of Eq. (3.3) is

$$T^\mu_\mu = \epsilon - 3p \quad . \tag{3.7}$$

The "perfect" EOS,

$$p = \frac{1}{3}\epsilon \quad , \tag{3.8}$$

easily proven for free relativistic particles (masses → 0), is seen to be equivalent to the vanishing of the trace. We shall extend this result to assert that scale invariance implies the perfect EOS (3.8) even in the presence of interactions. Unfortunately, we shall find that genuine scale invariance is hard to come by because of trace anomalies.

Evaluating (3.1) in the local rest frame (assuming this concept makes sense) gives

$$T^\mu_\mu = \epsilon - 3p \quad . \tag{3.9}$$

If we can evaluate this equation as a function of temperature, for example, the corrections to (3.8) are all contained in T^μ_μ. Notice that formal scale invariance $\theta^\mu_\mu = \theta_{00} - \Sigma_i \theta_{ii} = 0$ is an operator identity version of (3.8).

Here we restrict our attention to LTE (local thermodynamic equilibrium), although attempts have been made[5] to obtain closure without an LTE EOS. We especially stress that conceptually, *a perfect EOS (Eq. (3.8)) does not imply the absence (or weakness) of interactions.* The simplest example is the scalar field with mass m and ϕ^4 interaction. The Lagrangian density is $\frac{1}{2}(\partial\phi)^2 - \frac{1}{2} m^2\phi^2 + \lambda\phi^4$. The trace theorem (see Eq. (3.12) below) gives

$$\langle\theta_{oo}\rangle - \sum_i \langle\theta_{ii}\rangle = m^2\langle\phi^2\rangle \quad . \tag{3.10}$$

Note that this expression does not depend on the (scale invariant) coupling $\lambda\phi^4$. We can let $m^2 \to 0$, obtaining the perfect EOS (3.8) with $\lambda \neq 0$.

In order to generalize and to put the foregoing in context, we next explain the "trace theorem," which expresses the trace in terms of scale-breaking operators (i.e., those with dimension not equal to 4). We refer to Refs. 2 and 4 for the nontrivial details. The procedure is as follows: decompose $\theta_{\mu\nu}$ into $\bar{\theta}_{\mu\nu}$ plus other terms. $\bar{\theta}_{\mu\nu}$ is obtained from $\theta_{\mu\nu}$ by discarding terms with $d \neq 4$ and using the equations of motion to eliminate time dervatives. $\bar{\theta}_{\mu\nu}$ is not a true tensor and the dimensional analysis is done with the frame-dependent operator D(t). The trace theorem states[6] that if the energy density has the form

$$\theta_{oo} = \bar{\theta}_{oo} + \sum_d w_d \quad , \tag{3.11}$$

where w_d are scalars of dimension d, then the trace is

$$\theta^\mu_\mu = \sum_d (4-d)\, w_d + \text{anomalies} \quad . \tag{3.12}$$

This formula (perhaps senza anomalies) shares with the virial theorem of Section I the elimination of kinetic energy in terms of the interaction terms. The anomaly term is non-canonical (coming from fermion loops) and must be added for QCD.

Consider some examples of (3.12). In QED, the mass term $w_3 = m\bar{\psi}\psi$ and $\theta^\mu_\mu = (4-3)w_3 = m\bar{\psi}\psi$. For a scalar field theory with $\alpha = \frac{1}{2}(\partial\phi)^2 - \frac{1}{2} m^2\phi^2 + g\phi^3 + \lambda\phi^4$, we find $w_2 = \frac{1}{2} m^2\phi^2$, $w_3 = -g\phi^3$ and $\theta^\mu_\mu = m^2\phi^2 - g\phi^3$. In the case of interest here, QCD, we have[7]

$$\theta^\mu_\mu = \sum_q m_q\bar{q}q - \frac{b\alpha}{8\pi} F^a_{\mu\nu} F^{a\mu\nu} \quad . \tag{3.13}$$

The anomaly term[7] has $\alpha = g^2/4\pi$, $b = \frac{11}{3} N_c - \frac{2}{3} N_f$; $F^a_{\mu\nu}$ is the usual field tensor and $F^2 = (\mathbf{E}^2 - \mathbf{B}^2)$. N_c, N_f are the number of quark colors and flavors. Equation (3.13) shows that even when the Lagrangian is scale invariant, the trace does not vanish. Of course, a length has crept into the theory via the renormalization process.

Suppose that we apply (3.13) to an equilibrium QCD plasma. The equation of state is

$$\epsilon - 3p = \sum_q \langle\bar{q}q\rangle m_q - k\langle F^a_{\mu\nu} F^{a\mu\nu}\rangle \quad , \tag{3.14}$$

where $k = b\alpha/8\pi$. First consider the mass term (as in QED where there is only $m\bar{\psi}\psi$ on the right-hand side). If we evaluate $\langle\bar{\psi}\psi\rangle$ in lowest order, we easily find

$$\epsilon - 3p = \frac{m^2}{\pi^2} \int_m^\infty \frac{d\sqrt{x^2-m^2}}{\exp(\beta(x-\mu)+1)} \quad , \tag{3.15}$$

where $\beta = 1/kT$ and μ is the chemical potential. Equation (3.15) is easily derived from the free-particle thermodynamic expressions for ϵ and p.

For u and d quarks, the right-hand side of (3.15) is negligibly small when $(m_q\beta)^2 = (m_q/T)^2 << 1$. We do not know the value of $\langle F^2\rangle$, however. Clearly (3.14) can be evaluated in perturbation theory, the right-hand side being expressed as a function of temperature.

In Ref. 8 we argued that the anomaly term might be small, thereby justifying a nearly ideal EOS and hydrodynamic sound velocity $1/\sqrt{3}$. Presently we are not confident that this is correct. Comparison with recent lattice gauge theory calculations is under way.

The equation of state allows closure of the equations of motion for relativistic hydrodynamics. Since LTE cannot be expected for all space-time domains of relativistic heavy ion collisions, it is worth exploring[5] the relation of $T_{\mu\nu}$ to the phase-space distributions of relativistic field theory.[9-10] Since much of the dynamical content of the hydrodynamic equations is local energy-momentum conservation, much of the behavior should survive the loss of LTE. Thus far, the problem has been how to obtain closure of generalized kinetic-hydrodynamic equations. Luckily there is a lot of interesting theoretical work to be done on these problems.

ACKNOWLEDGMENTS

This work has been supported in part by the High Energy Phsyics Division and the Nuclear Physics Division of the United States Department of Energy.

REFERENCES

1. P. Carruthers and M. M. Nieto, Rev. Mod. Phys. 40:411 (1968).
2. P. Carruthers, Phys. Repts. 1C:1 (1971).
3. G. Mack and A. Salam, Ann. Phys. 53:174 (1969).
4. C. Callan, Jr., S. Coleman, and R. Jackiw, Ann. Phys. (NY) 59:42 (1970).
5. P. Carruthers and F. Zachariasen, in: AIP Conf. Proceedings No. 123 "Intersections Between Paricle and Nuclear Physics," R. E. Mischke, ed., American Institute of Physics, New York (1984).
6. M. Gell-Mann, in: "Proc. of the 3rd Hawaii Topical Conference on Particle Physics," Western Periodicals Co., N. Hollywood (1970).
7. J. C. Collins, A. Duncan, and S. D. Joglekar, Phys. Rev. D 16:438 (1977).
8. P. Carruthers, Phys. Rev. Lett. 50:1179 (1983).
9. P. Carruthers and F. Zachariasen, Phys. Rev. D 13:950 (1976).
10. P. Carruthers and F. Zachariasen, Rev. Mod. Phys. 55:245 (1983).

DENSITY FUNCTIONAL THEORY AT FINITE TEMPERATURES

Reiner M. Dreizler

Institut für Theoretische Physik
Universität Frankfurt/Main
West Germany

Standard density functional theory at T = 0 deals (mainly) with the discussion of groundstate properties of quantum many particle systems. It has been applied to atoms, molecules, solids and nuclei[1]. The simpler versions of this theory (extensions of the Thomas-Fermi model) yield quite acceptable insight at reasonable labour, while the more sophisticated versions (inclusion of correlation contributions within the Kohn-Sham scheme) produce high quality results.

The extension of the density functional approach for the discussion of systems at T > 0 then follows quite naturally and it has, indeed, been used to establish equations of state as well as to discuss other aspects of thermal quantum systems[2]. In order to approach the topic in general terms, we shall consider a nucleus (or any other many particle system) in thermal equilibrium with an unspecified reservoir of heat. With the usual assumption that the coupling between the system and the bath is weak, we characterise the system as a canonical ensemble, for which the statistical density operator is determined by the (or a) nuclear Hamiltonian

$$\hat{\rho}_c = \frac{1}{Z_c} \quad e^{-\beta\hat{H}} \quad , \beta = \frac{1}{k_B T}$$

and the free energy

$$F = E - TS$$

is the minimal thermodynamic potential. Explicitly this signifies that the mean value of the free energy

$$F = \mathrm{tr}\ (\hat{\rho}_c \hat{H}) + \frac{1}{\beta} \ \mathrm{tr}\ (\hat{\rho}_c \ln \hat{\rho}_c)$$

evaluated with the canonical density operator is lower than the value one obtains, if one evaluates this quantity with any other statistical density operator $\hat{\rho}$ with tr $\hat{\rho} = 1$.

The Nuclear Equation of State, Part A
Edited by W. Greiner and H. Stöcker
Plenum Press, New York

Temperature dependent density functional theory claims (see below for details) that one can represent the free energy in this situation as a unique functional of the equilibrium density

$$F = F\,\{T,N,\ldots,\,[n_o(\underline{r},\,T)]\}.$$

It also depends explicitly on the macroscopic parameters of the system. In view of the minimal property of the free energy one can determine the equilibrium density via solution of the variational principle

$$\frac{\delta\, F[n]}{\delta n(r,T)} = 0 \qquad \rightarrow n_o(\underline{r},T).$$

After one has evaluated the free energy with the solution of the variational principle

$$F(T,N,\ldots) \equiv F\,\{T,N,\ldots[n_o(\underline{r},T)]\}\ ,$$

one may calculate any thermodynamical variable of interest, as e.g. the chemical potential

$$\mu = \left(\frac{\partial F}{\partial N}\right),$$

and in this fashion establish the equations of state of the system, or one may calculate any thermodynamical response function as e.g. the heat capacities

$$C_x = -T\left(\frac{\partial^2 F}{\partial T^2}\right)_x .$$

A more detailed statement is required concerning volume and pressure. For a homogeneous system (e.g. nuclear matter in an unspecified reservior of heat) one may consider the free energy density or the free energy per particle

$$f = \frac{F}{V} \qquad \tilde{f} = \frac{F}{N}.$$

The standard thermodynamic relation for the pressure then gives in the thermodynamic limit

$$P = -\left(\frac{\partial F}{\partial V}\right)_{T,N} = n_o^2\,\frac{\partial \tilde{f}}{\partial n_o} = N_o\,\frac{\partial f}{\partial n_o} - f$$

The constant internal pressure is given by the equilibrium density squared times the derivative of the free energy per particle with respect to this density. The bulk compressibility (a quantity of constant controversy in nuclear physics) is given by the second derivative of the free energy density with respect to the equilibrium density

$$\frac{1}{\chi_\infty} = -V\left(\frac{\partial P}{\partial V}\right)_{T,N} = n_o^2\,\frac{\partial^2 f}{\partial n_o^2}$$

The situation is different for an inhomogeneous system with a diffuse boundary. If one wishes to speak about volume and pressure in this case, one must enclose the system within a specified volume and impose suitable boundary conditions for the solution of the energy eigenvalue problem (e.g. periodic boundary conditions) within this volume. The free energy then depends on this volume and the standard relation

$$P = - \left(\frac{\partial F}{\partial V} \right)_{T,N}$$

determines the external pressure, which is necessary to keep the system within the specified volume V. The compressibility determined in this case describes the response of the system to a change of the boundary and is in general not identical with the bulk value.

An obvious extension of this scenario is the grand canonical ensemble. The system is in equilibrium with a reservoir at a given temperature and a given chemical potential. One has fluctuations in energy as well as particle number. The grand potential

$$\Omega = < \hat{H} > - T < \hat{S} > - \mu < \hat{N} >$$

is minimised by the grand canonical density operator in this case. Applications include a nucleus or nuclei in a gas of nucleons, leptons etc., atoms or ions in a gas of electrons or in another plasma and doped semiconductors.

Before we turn to the discussion of density functional theory proper, a caveat seems necessary. We only discussed statistical density operators, which are solutions of the stationary Liouville equation

$$[\hat{H}, \hat{\rho}] = 0 ,$$

and thus deal with equilibrium thermodynamics. On the basis of the resulting equations of state we may discuss phase transitions in single or many component systems, as e.g. the phase transitions between a nucleon gas and a nucleon liquid or in the quark-gluon plasma. The discussion of nonequilibrium processes, as for instance nucleus-nucleus collisions, should be based on (statistical) density operators that are solutions of the full Liouville equation

$$\frac{\partial \hat{\rho}(t)}{\partial t} = \frac{1}{i\hbar} [\hat{H}, \hat{\rho}(t)] .$$

As this is a tougher proposition, compromises have to be accepted, but the use of equilibrium equations of state for nonequilibrium processes has to be regarded with due caution. As far as density functional theory is concerned, one would have to formulate a time (and temperature)dependent variant[3], which is not yet available on the practical level.

The discussion of any variant of density functional theory involves three levels. The first is the level of existence theorems. Here one sets the stage, but does not address the practical aspects of the problem. Practical aspects are treated on the second level: The derivation of suitable functionals. Obviously this is the point, where one has to face the difficulties of the many body problem at hand. Quite summarily one may state that any functional available so far is only an approximation to the real world. Level three then deals with the actual calculation of system properties on the basis of the functionals obtained in step two.

The basic existence theorem of density functional theory at T = 0 is the theorem of Hohenberg and Kohn[4]. This theorem has been extended to the temperature dependent case by Mermin[5]. The Mermin-Kohn-Hohenberg theorem can be stated as follows.

Consider a grand canonical ensemble at a given temperature T and chemical potential μ, which is characterised by the Hamiltonian

$$\hat{H} = \hat{K} + \hat{V} + \hat{W}$$

(kinetic energy, external potential and two body interactions). Then the grand potential

$$\Omega = \mathrm{tr}\, \hat{\rho}_G \left(\hat{H} + \frac{1}{\beta} \ln \hat{\rho}_G - \mu\hat{N}\right)$$

is a unique functional of the temperature dependent equilibrium density of the system

$$\Omega[n_o(\underline{r},T,\mu)] = K[n_o(\underline{r},T,\mu)]$$
$$+ W[n_o(\underline{r},T,\mu)] - T\, S[n_o(\underline{r},T,\mu)]$$
$$+ \int d^3r (v(\underline{r}) - \mu)\, n_o(\underline{r},T,\mu).$$

The equilibrium density is defined as

$$n_o(\underline{r},T,\mu) = \mathrm{tr}\, \hat{\rho}_G\, \hat{\psi}^+(\underline{r})\, \hat{\psi}(\underline{r}).$$

For the proof of this theorem, which follows the arguments of the T = 0 case rather closely, I refer to the original literature.

Important for applications is the corollary: As a consequence of the minimal property of the grand potential it follows that

$$\Omega[n_o] \leq \Omega[n]$$

for any density n obtained with an arbitrary density operator

$$n(\underline{r},T,\mu) = \mathrm{tr}(\hat{\rho}(T,\mu)\hat{\psi}^+(\underline{r})\hat{\psi}(\underline{r})).$$

This statement is the basis for the variational principle

$$\frac{\delta\Omega[n]}{\delta n(\underline{r},T,\mu)} = 0 \rightarrow n_o(\underline{r},T,\mu),$$

provided suitable conditions for the existence of the functional derivative are satisfied. Such conditions have been discussed at some length[6] for the case T = 0, but less so for T>0 .

The direct variational principle illustrates the appeal of density functional theory. Given a suitable functional the discussion of the statistical many particle problem is reduced to the minimal number of degrees of freedom possible. In particular one can emphasise the fact that the solution of the variational problem at hand is independent of the mean number of particles involved. The catch is, however, to find sufficiently accurate functionals.

An alternative to the direct variational approach is the Kohn-Sham scheme[7], which can also be established rigorously. It essentially states that the equilibrium density (and hence the equilibrium grand potential) can also be obtained via solution of a set of selfconsistent, local single particle equations. A possible formulation is:
Consider the expansion

$$n(\underline{r},T,\mu) = \Sigma_l\, f_l(T,\mu)\phi_l^*(\underline{r},T,\mu)\, \phi_l(\underline{r},T,\mu)$$

and the partitioning of the grand potential

$$\Omega[n] = K_s[n] - T\, S_s[n]$$
$$+ \int d^3r (v(\underline{r}) - \mu)\, n(\underline{r},T,\mu)$$

$$+ \frac{1}{2} \iint d^3r\, d^3r'\, w(\underline{r},\underline{r}')\, n(\underline{r},T,\mu)\, n(\underline{r}',T,\mu) + \Omega_{xc}[n] \,.$$

The first two terms represent the kinetic energy and the entropy of an equivalent noninteracting system with the same density as the interacting system. The exchange correlation contribution to the grand potential is defined as

$$\Omega_{xc}[n] = K[n] - K_s[n]$$
$$+ W[n] - W_H[n] - T(S[n] - S_s[n]) \,.$$

The reason for this particular partitioning is obvious. Both K_s and S_s can also be represented in terms of the orbitals ϕ_1 and the weights f_1^s. All serious many body effects are contained in Ω_{xc}.
The variational principle for the grand potential then leads to the statement, that the orbitals determined by the local, selfconsistent single particle equations

$$\{ -\frac{\hbar^2}{2m} \Delta + v_{eff}(\underline{r},T,\mu) - \mu \}\phi_1(\underline{r},T,\mu) = \varepsilon_1(T,\mu)\phi_1(\underline{r},T,\mu)$$

with

$$v_{eff}(\underline{r},T,\mu) = v(\underline{r}) + \int d^3r' w(\underline{r},\underline{r}')\, n(\underline{r}',T,\mu) + \frac{\delta\Omega_{xc}[n]}{\delta n(\underline{r},T,\mu)}$$

$$= v(\underline{r}) + v_H(\underline{v},T,\mu) + v_{xc}(\underline{r},T,\mu)$$

and (for the case of fermions) the Fermi function

$$f_1(T,\mu) = [\, 1 + \exp\{\beta(\varepsilon_1 - \mu)\}]^{-1}$$

yield the exact equilibrium density.

Some comments on this scheme are in order.
(i) As stated, all serious many body effects are contained in Ω_{xc}. If this quantity is neglected entirely one has a temperature dependent Hartree-approximation. The simplest approximation for Ω_{xc}, the local approximation for exchange, leads to a temperature dependent Hartree-Fock-Slater scheme. Correlations effects can however and have been included in some approximation. The popular Skyrme-Hartree-Fock approach[8)] is really a Kohn-Sham approach, with an educated and constantly corrected guess for Ω_{xc}.

(ii) Neither the orbitals nor the orbital energies are (as a matter of principle) physical observables. They only serve as a vehicle for the construction of the equilibrium density. This warning is usually ignored, as one has found (in particular for the T = 0 case) that the orbital energies do give consistent removal energies and affinities. The temperature dependent Kohn-Sham scheme has been used to calculate transition energy shifts of atoms enbedded in hot plasmas (as a tool for plasma diagnostics).

(iii) For the application of the direct variational approach one usually retains the partitioning of Ω indicated above. In this case the density functional representations of K_s and S_s (not required in the Kohn-Sham scheme) are needed. On the other hand the Kohn-Sham scheme is more involved than the direct variational approach. In the temperature dependent Kohn-Sham scheme <u>all</u> single particle solutions (discrete and continuous) have to be calculated in order to generate the equilibrium density.

For the derivation of explicit functionals there exist, as in the T = 0 case, essentially two methods.

(a) If one is only interested in a density functional representation of K_s, S_s and Ω_x, one may use gradient expansion techniques based on the Dirac or the Bloch form of the one particle equilibrium density matrix. As the Dirac and Bloch forms are simply related by Laplace transformation one obtains the same results with either starting point.

(b) If one wishes to address in addition correlation contributions, one uses gradient expansion techniques based on expansion about the limit of a homogeneous system. The lowest order contribution of this expansion, the socalled local density approximation, is obtained via

$$\Omega^{[0]}\ [\ n = \text{const}\ ]\ \underset{n \to n(\underline{r})}{\to}\ \Omega^{[0]}\ [n(\underline{r})]\ .$$

That is one calculates the grand potential of the homogeneous system (nuclear matter, electron gas) in either perturbation theory, the random phase approximation, cluster expansion, hypernetted chain or Monte Carlo methods. One then replaces the constant density of the homogeneous system by the density of the corresponding inhomogeneous system under consideration The lowest order gradient corrections can be calculated by subjecting the homogeneous system to an external perturbation in linear response. One demonstrates quite generally that all gradient corrections in second order can be generated from the irreducible polarisation insertion of the (temperature dependent) homogeneous system

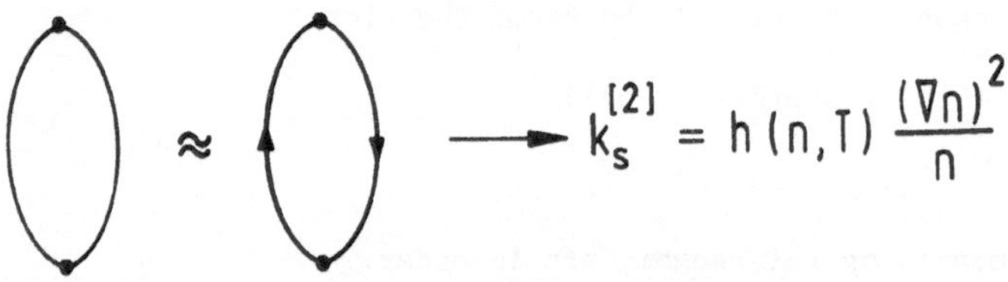

The polarisation insertion is not known exactly. However, any approximation for Π will yield a corresponding gradient correction. For instance the temperature dependent Lindhard function

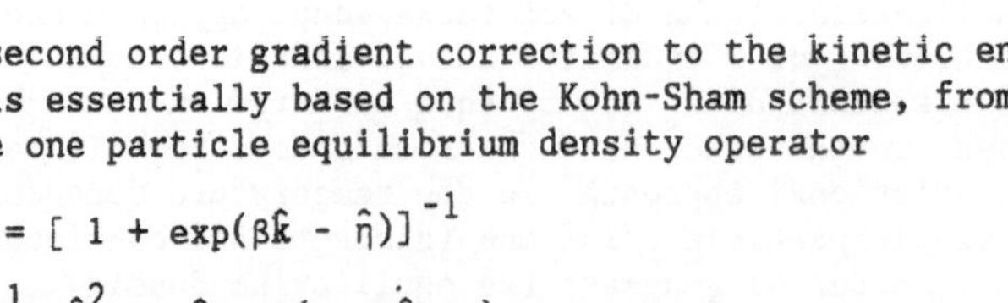

yields the second order gradient correction to the kinetic energy K_s. Method (a) is essentially based on the Kohn-Sham scheme, from which one extracts the one particle equilibrium density operator

$$\hat{n}(T,\mu) = [\ 1 + \exp(\beta\hat{k} - \hat{\eta})]^{-1}$$

$$\hat{k} = \frac{1}{2m}\,\hat{p}^2\ ,\quad \hat{\eta} = \beta(\mu - \hat{v}_{eff})\ .$$

The noninteracting kinetic energy and entropy are then represented as

$$\hat{k}_s(T,\mu) = \frac{1}{2}(\hat{k}\hat{n} + \hat{n}\hat{k})$$

$$\hat{S}_s(T,\mu) = - k_B \{\hat{n} \ln \hat{n} + (1-\hat{n}) \ln(1-\hat{n})\}.$$

In order to evaluate the spatial representation of these operators, one uses plane wave expansion, e.g.

$$n(\underline{r},T,\mu) = \int d^3k < \underline{r}|\hat{n}(T,\mu)|\underline{k}><\underline{k}|\underline{r} > .$$

The technical problem that arises, is the evaluation of the action of a function of two noncommuting operators on the eigenstate of one of the operators

$$\hat{n}(T,\mu)|k> = f(\hat{k} - \hat{\eta})|k > .$$

This is handled by the expansion

$$= \Sigma_{r=0}^{\infty} f^{(r)}(\frac{\hbar^2 k^2}{2m} - \hat{\eta}) \hat{O}_r|\underline{k} >$$

involving derivatives of the function f and operators $\hat{O}_r$, which are determined by multiple commutators of $\hat{k}$ and $\hat{\eta}$, that is gradients of the effective potential. The final results that one obtains at this stage, are

$$n(\underline{r},T,\mu) = n(\eta(\underline{r}), \partial_i\eta(\underline{r}), \partial_{ij}\eta(\underline{r}),..)$$

and similar expressions for the kinetic energy and the entropy densities, explicitly to second order

$$n(\underline{r},T,\mu) = \frac{1}{2\pi^2} (\frac{2m}{\hbar^2})^{3/2} \{ \beta^{-3/2} J_{1/2}(\eta)$$

$$+ \frac{1}{24} \frac{\hbar^2}{2m} [\frac{3}{4} \beta^{3/2} J_{-5/2}(\eta)(\underline{\nabla} v_{eff})^2$$

$$+ \beta^{1/2} J_{-3/2}(\eta) \Delta v_{eff}]\}$$

$$k_s(r,T,\mu) = \frac{1}{2\pi^2} (\frac{2m}{\hbar^2})^{5/2} \{ \beta^{-5/2} J_{3/2}(\eta)$$

$$- \frac{1}{8} \frac{\hbar^2}{2m} [\frac{3}{8} \beta^{1/2} J_{-3/2}(\eta)(\underline{\nabla} v_{eff})^2$$

$$+ \frac{5}{6} \beta^{-1/2} J_{-1/2}(\eta) \Delta v_{eff}] \}$$

$$s_s(\underline{r},T,\mu) = - \eta\, n(\underline{r},T,\mu) + \frac{1}{2\pi^2} (\frac{2m}{\hbar^2})^{3/2}$$

$$\{ 5/3\, \beta^{-3/2} J_{3/2}(\eta) + \frac{1}{24} \frac{\hbar^2}{2m} [1/4\, \beta^{3/2}$$

$$J_{-3/2}(\eta) (\underline{\nabla} v_{eff})^2 - \beta^{1/2} J_{-1/2}(\eta) \Delta v_{eff}]\}.$$

The functions J_α are standard Fermi-integrals defined by

$$J_\alpha(\eta) = \int_0^\infty \frac{x^\alpha}{[1+\exp(x-\eta)]} \, dx \quad \text{for } \alpha > -1$$

$$= \frac{1}{(\alpha+1)} \frac{d}{d\eta} J_{\alpha+1}(\eta) \quad \text{for } \alpha < -1 \; .$$

These results do not represent the final answer one is looking for. In order to obtain the kinetic energy and entropy densities as a function of $n(\underline{r},T,\mu)$ one needs to express the quantity η (containing the unknown effective Kohn-Sham potential) in terms of the density in a consistent manner and insert the inversion into the expressions for the energy and entropy densities. The result of this procedure can be stated as follows. Define the quantity $\bar{\eta}(n,T)$, which is a unique solution of the equation

$$n = A(T)\, J_{1/2}(\eta) \; , \quad A(T) = \frac{1}{2\pi^2} \left(\frac{2m}{\hbar^2} \right)^{3/2} \beta^{-3/2},$$

then one finds to second order

$$k_s(n) = \frac{2m}{\hbar^2} A(T)\, J_{3/2}(\bar{\eta}) + \gamma(\bar{\eta}) \frac{(\nabla n)^2}{n} + \frac{1}{6} \Delta n$$

$$s_s(n) = \frac{5}{3} A(T)\, J_{3/2}(\bar{\eta}) - \bar{\eta} n - \nu(\bar{\eta}) \frac{\hbar^2}{2m} \beta \frac{(\nabla n)^2}{n}$$

with

$$\gamma(\bar{\eta}) = \xi(\bar{\eta}) - \nu(\bar{\eta})$$

$$\xi(\bar{\eta}) = \frac{1}{12} J_{1/2}(\bar{\eta}) \frac{d}{d\bar{\eta}} \left(J_{-1/2}(\bar{\eta})^{-1} \right)$$

$$\nu(\bar{\eta}) = \frac{\beta}{2} \frac{\partial}{\partial \beta} \xi(\bar{\eta}) \Big|_n$$

The first terms in these results correspond to the temperature dependent Thomas-Fermi model[9]. Gradient correction terms are necessary for the description of surface structures. Various limits (high and low temperature) can be extracted. The results were first obtained by Perrot[10] on the basis of the electron gas approach (method (b)) and subsequently by Brack[11] on the basis of the expansion of the Bloch density matrix. Results to fourth order given by Bartel, Brack and Durand[12] and by Polischuk[13].

Potential contributions to the grand potential beyond the Hartree term have been calculated by Gupta and Rajagopal in the local density approximation for Coulomb exchange[14] and for Coulomb correlation contributions in the random phase approximation[15] on the basis of temperature dependent perturbation theory of the homogeneous electron gas. The diagrams to be evaluated are

Ω_x = 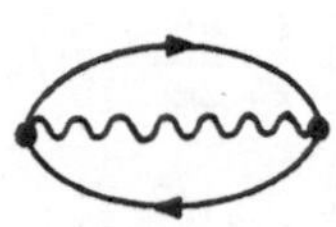

where  represents the Coulomb interaction and $\longrightarrow$ the free temperature dependent Green's function, and

Ω_c = 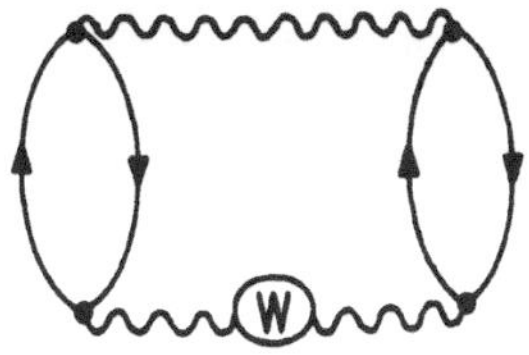

with the temperature dependent screened Coulomb interaction in the ring approximation

For the exchange energy density one obtains

$$\omega_x(n,T,\mu) = [\, \frac{3}{4} (\frac{3}{\pi})^{1/3} e^2 \, n(\underline{r},T,\mu) \,]^{4/3} \quad I(\bar{T}),$$

where the first factor corresponds to the T = 0 result (with $n(\underline{r})$ replaced by $n(\underline{r},T,\mu)$), while the second factor indicates an interesting scaling behaviour

$$\bar{T} = a \frac{T}{n^{2/3}} , \qquad a = \frac{2mk_B}{\hbar^2} \quad (\frac{1}{3\pi^2})^{2/3} .$$

The function $I(\bar{T})$ is given by a more involved integral over Fermi distributions

$$I(\bar{T}) = 2 \iint dx_1 \, dx_2 \, \ln \left| \frac{x_1 - x_2}{x_1 + x_2} \right|$$

$$[(\, 1+ \exp [\frac{x_1^2}{\bar{T}} - \beta n]) \, (\, 1 \, x \, \exp[\frac{x_2^2}{\bar{T}} - \beta n])]^{-1} .$$

Its variation with the argument $\bar{T}$ has been tabulated (see Fig. 1) and the limiting values are known analytically.

$$I(\bar{T}) \underset{\bar{T} \to 0}{\to} 1 + \frac{\pi 2}{6} \bar{T}^2 \ln \bar{T} + .., I(\bar{T}) \underset{\bar{T} \to \infty}{\to} \frac{4}{9} \frac{1}{\bar{T}} + ...$$

If the result exhibited in Fig. 1 is an indication of the general trend, one ought to view the transfer of the T = 0 Skyrme interactions to the temperature dependent situation via the simple replacement

$$n(\underline{r}) \to n(\underline{r},T,\mu)$$

with obvious caution.
For the correlation contribution only numerical results for $\omega_c(n,T)$ can be given. (The same statement applies for T = 0). It is instructive to plot the explicit temperature (or $\bar{T}$) dependence of ω_c for fixed density values and compare it to the corresponding curves for the exchange contribution (Fig. 2). One finds that (in contrast to the wellknown situation for the low temperature limit) correlation contributions dominate over exchange in the high temperature regime. Neglect of correlation effects would thus lead to substantial errors for this range of the n-T domain.

One should emphasise that the functionals presented above constitute the bare minimum for a reasonable discussion of temperature dependent many particle systems. Temperature dependent functionals available are so far less sophisticated then T = 0 functionals. Nonetheless a few remarks should be added on applications. Concerning the application of the direct variational principle in nuclear physics, there is the statement that the (second order) extended Thomas-Fermi model is able to reproduce the more elaborate temperature dependent Skryme-Hartree-Fock results with reasonable accuracy[16]. For Coulomb systems one should mention in particular the discussion of atomic properties of atoms in electron plasmas at a given density n_o and temperature T[17] with the Kohn-Sham scheme. (see Fig. 3). On the Hartree-level one establishes quite generally[18] that the Hartree term induces asymptotically either Thomas-Fermi (T→ 0)or Debye(T→∞) screening

$$v(\underline{r}) + v_H(\underline{r}, [n]) \underset{\substack{r \to \infty \\ T \to 0}}{\to} v(\underline{r}) \exp\left[-r\, n^{1/3}\, \xi_{TF}\right]$$

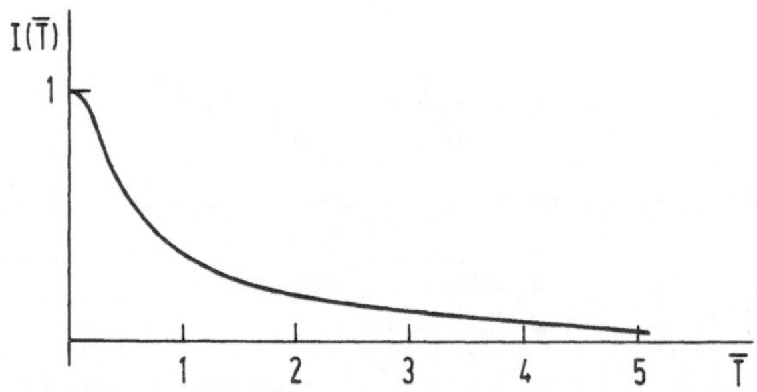

Fig. 1. The variation of the function $I(\bar{T})$ with $\bar{T}$

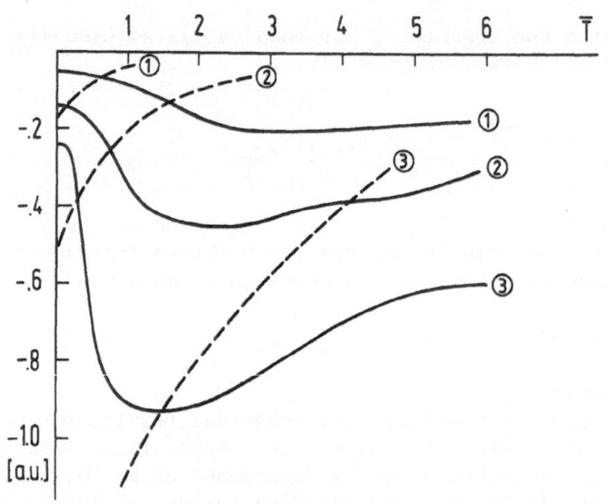

Fig. 2. The variation of ω_c(solid line) and ω_x(dashed line) with T for the density values $n = 10^{22}\ cm^{-3}$(1), $10^{24}\ cm^{-3}$(2), $10^{26}\ cm^{-3}$(3).

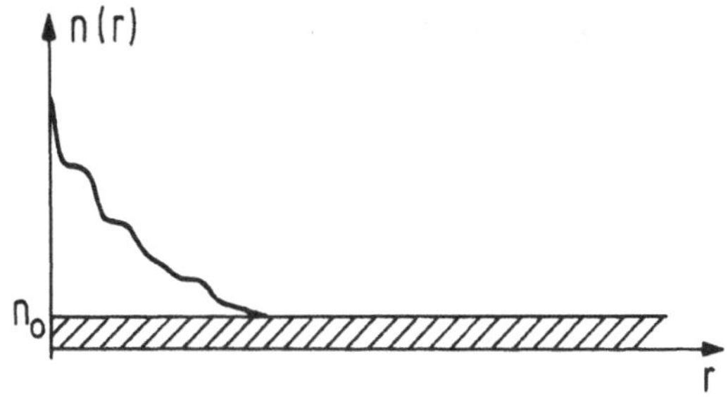

Fig. 3. Indication of the variation of the density with r for an atom in a thermal electron plasma.

$$\underset{\substack{r\to\infty\\ T\to\infty}}{\rightarrow} \quad v(\underline{r}) \exp\{-r(\frac{n}{T})^{1/2}\,\xi_D\}$$

and thus interpolates correctly between these limits. Results on the Hartree-Fock-Slater level

$$v_{eff} = v + v_H + v_x$$

are available, but one knows from the T = 0 case that selfinteraction corrections are required for better accuracy. Inclusion of v_c^{RPA} (or variants[19]), is possible, but again appeal to the T = 0 case tells us that a more extensive assessment of the correlation contribution is required for high quality results. To at least offer one statement on the trends observed for atoms enbedded in an electron plasma, we note[17] that the single particle orbits become more strongly bound for increasing temperature and fixed plasma density. The number of electrons in bound orbitals decreases.

Besides the need to investigate the functionals available more closely and to establish their merits and shortcomings, one can envisage several extensions. Neither the investigation of nonequilibrium situations nor the treatment of relativistic (field theoretical[20]) systems (of interest e.g. in astrophysics) on the basis of the temperature dependent density functional approach has been attempted so far.

REFERENCES

1. See e.g.
   a.) S. Lundqvist and N.H. March, eds. "Theory of the Inhomogeneous Electron Gas", Plenum Press, New York, 1983
   b.) R.M. Dreizler and J. da Providencia, eds." Density Functional Methods in Physics", Nato ASI Series 123, Plenum Press, New York, 1985.
2. For a more detailed review see:
   U. Gupta and A.K. Rajagopal, Phys. Rep. 87:259(1982).
3. Some indications are given in:
   R.M. Dreizler and H. Kohl, in: Lecture Notes in Physics, 256
   J. Broeckhove, L. Lathouwers and P. van Leuven eds.
   Springer Verlag, Berlin , 1986, p. 51
4. P. Hohenberg and W. Kohn, Phys. Rev. 136:B864(1964).
5. N.D. Mermin, Phys. Rev. 137:A1441(1965).
6. See e.g. E.H. Lieb in 1b, p. 31.
7. W. Kohn and L.J. Sham, Phys. Rev. 14o:A1133(1965).
8. See e.g. P. Bonche and D. Vautherin, Nucl. Phys. A372:496(1981).
9. R. Feynman, N. Metropolis and E. Teller, Phys. Rev. 75:1561(1949) and R.E. Marshak and H. Bethe, Astrophys. J. 91:239(1940).

10. F. Perrot, Phys. Rev. A26:1035(1979).
11. M. Brack, Phys. Rev. Lett. 53:119(1984); erratum: ibid, 54:851(1985).
12. J. Bartel, M. Brack and M. Durand, Nucl. Phys. A445:263(1985).
13. A.Y. Polischuk, Sol. St.Comm. 61:193(1987).
14. U. Gupta and A.K. Rajagopal, Phys. Rev. A21:2064(1980).
15. U. Gupta and A.K. Rajagopal, Phys. Rev. A22:2792(1980).
16. M. Brack, C. Guet and H.B. Hakansson, Phys. Rep. 123:275(1985).
17. F. Perrot, Phys. Rev. A25:489(1982) and Phys. Rev. A26:1035(1982).
18. U. Gupta and A.K. Rajagopal, J. Phys. B12:L703(1979).
19. M.W.C. Dharma-Wardana and R. Taylor, Can.J.Phys. C14:629(1981).
2o. For the case T = 0 see:
E. Engel and R.M. Dreizler, Phys. Rev. A35:3607(1987) and R.M.Dreizler, E. Engel and P. Malzacher, in: "Physics of Strong Fields", Nato ASI Series 153, W. Greiner, ed. Plenum Press, New York, 1987, p.565.

THE COLLECTIVE SPIN OF NUCLEAR SURFACE VIBRATIONS*

Martin Greiner and Werner Scheid

Institut für Theoretische Physik der Justus-Liebig-Universität, Giessen, F.R. Germany

ABSTRACT

The free Schrödinger equation for collective multipole degrees of freedom is linearised so that energy and momentum operators appear only in first order. The wave functions solving the linearised equation carry a collective spin depending on multipolarity. For nuclear quadrupole surface degrees of freedom a scalar coupled collective potential and an additional spin-dependent potential are introduced and some aspects of even-odd nuclei are described.

1. INTRODUCTION

It is well known that the Schrödinger equation does not incorporate spin degrees of freedom. On the other hand, the Dirac equation describes particles with a spin 1/2 from a more fundamental point of view. Because the Dirac equation is a relativistic wave equation, it has been a wide spread opinion in textbooks that the appearance of the spin degree of freedom is due to the theory of relativity. But this seems not to be true! Lévy-Leblond[1)] showed that the free three-dimensional Schrödinger equation can be linearised in the same way as Dirac[2)] constructed his equation from the free Klein-Gordon equation, where energy and momentum finally appear only in first order, and that this procedure already yields the spin-1/2 degree of freedom. Therefore, the appearance of a spin degree of freedom is not due to the theory of relativity, but due to the linearisation of the equation of motion.

In nuclear physics Schrödinger equations for multipole degrees of freedom have a wide range of application for the description of collective phenomena, like the collective surface or density vibrations. In this paper we linearise the corresponding free collective Schrödinger equations; that means we construct equivalent equations with energy and momentum appearing only in first order. The new equations contain spin degrees of freedom, which we denote as collective spins[3)].

In Sect.2 we sketch the linearisation procedure for a free Schrödinger equation of an arbitrary multipole degree of freedom and deduce the

* Work supported by BMFT (06 GI 709) and GSI (Darmstadt)

collective spin from the linearised equation. For nuclear quadrupole surface degrees of freedom a scalar coupled collective potential is introduced into the linearised Schrödinder equation in Sect.3 and first applications are presented in order to describe collective aspects of even-odd nuclei. Additional spin-dependent potentials are discussed in Sect.4.

2. LINEARISATION OF THE FREE $(2\lambda+1)$-DIMENSIONAL SCHRÖDINGER EQUATION

Schrödinger equations describing $2\lambda+1$ multipole degrees of freedom are used for example for the treatment of nuclear collective surface vibrations. In this case the vibrations of spherical nuclei are described by collective surface coordinates $\alpha_{\lambda\mu}$ defined by the expansion of the nuclear surface as follows[4]

$$R(\theta,\phi) = R_o[1+ \sum_{\lambda,\mu} (-1)^{\mu} \alpha_{\lambda-\mu} Y_{\lambda\mu}(\theta,\phi)]. \tag{1}$$

The Hamiltonian for surface vibrations of multipolarity λ is rotationally invariant and has the following structure in lowest order in these coordinates

$$H_{\lambda} = \frac{1}{2B_{\lambda}} \sum_{\mu} (-1)^{\mu} \pi_{\lambda\mu}\pi_{\lambda-\mu} + \frac{C_{\lambda}}{2} \sum_{\mu}(-1)^{\mu}\alpha_{\lambda\mu}\alpha_{\lambda-\mu}. \tag{2}$$

The quantities $\pi_{\lambda\mu}$ are the canonically conjugate momenta. Since the coordinates $\alpha_{\lambda\mu}$ are complex, the introduction of real coordinates $x_i^{(\lambda)}$ and the corresponding momenta $p_i^{(\lambda)}$ fulfilling the usual commutation relations $[p_i^{(\lambda)},x_j^{(\lambda)}] = -i\hbar\delta_{ij}$ yields a familiar expression for the Hamiltonian(2):

$$H_{\lambda} = \frac{1}{2B_{\lambda}} \sum_{i=1}^{2\lambda+1} p_i^{(\lambda)2} + \frac{C_{\lambda}}{2} \sum_{i=1}^{2\lambda+1} x_i^{(\lambda)2}. \tag{3}$$

In a completely analogous manner as the three-dimensional Schrödinger equation is linearised[1] or the Dirac equation is derived from the Klein-Gordon equation[2], we linearise the $(2\lambda+1)$-dimensional free Schrödinger equation

$$(i\hbar\frac{\partial}{\partial t} - \frac{1}{2B_{\lambda}} \sum_{i=1}^{2\lambda+1} p_i^{(\lambda)2})\psi = 0. \tag{4}$$

The linearised free Schrödinger equation must have the following structure:

$$\theta_{\lambda}\psi:=[P^{(\lambda)}(i\hbar\frac{\partial}{\partial t})+ \sum_{i=1}^{2\lambda+1} Q_i^{(\lambda)}p_i^{(\lambda)}+R^{(\lambda)}]\psi = 0. \tag{5}$$

In this equation the time derivative and momenta appear only in first order. We assume that the wave function ψ solves the free Schrödinger equation (4) and the linearised free Schrödinger equation (5). This is achieved by the definition of a second linear operator

$$\theta_{\lambda}':=P^{(\lambda)'}(i\hbar\frac{\partial}{\partial t})+ \sum_{i=1}^{2\lambda+1} Q_i^{(\lambda)'}p_i^{(\lambda)}+R^{(\lambda)'} \tag{6}$$

with the property

$$\theta_{\lambda}'\theta_{\lambda} = 2B_{\lambda}(i\hbar\frac{\partial}{\partial t}-\frac{1}{2B_{\lambda}} \sum_{i=1}^{2\lambda+1} p_i^{(\lambda)2}). \tag{7}$$

A comparison of the terms on the left and right hand side of Eq.(7) determines the unknown operators $P^{(\lambda)}$, $Q_i^{(\lambda)}$ and $R^{(\lambda)}$, so that the linearised

(2λ+1)-dimensional free Schrödinger equation (5) finally reads

$$\begin{pmatrix} I & 0 \\ 0 & 0 \end{pmatrix} (i\hbar\frac{\partial}{\partial t})\psi = H_L^{(\lambda)}\psi, \quad \text{where } \psi = \begin{pmatrix}\phi \\ \chi\end{pmatrix}, \tag{8}$$

$$H_L^{(\lambda)} = -\begin{pmatrix} 0 & 0 \\ 0 & I \end{pmatrix} 2B_\lambda - \sum_{i=1}^{2\lambda+1} \begin{pmatrix} 0 & i\gamma_i^{(\lambda-1)} \\ -i\gamma_i^{(\lambda-1)} & 0 \end{pmatrix} p_i^{(\lambda)}. \tag{9}$$

The matrices $\gamma_i^{(\lambda)}$ have to fulfill the Clifford algebra

$$\gamma_i^{(\lambda)}\gamma_j^{(\lambda)} + \gamma_j^{(\lambda)}\gamma_i^{(\lambda)} = 2\delta_{ij}I, \quad i,j=1, \ldots, 2\lambda+2. \tag{10}$$

Both functions ϕ and χ satisfy the free Schrödinger equation (4). It is observed that the linearised (2λ+1)-dimensional free Schrödinger equation can be regarded as the "nonrelativistic limit" of a (2λ+1)-dimensional free Dirac equation, i.e. $E_{Dirac}=2B_\lambda+E$ with $E<<2B_\lambda$. This gives an anschaulich argument for Eq.(10).

The linearised free Schrödinger equation (8) contains spin degrees of freedom which we denote as collective spin degrees of freedom. This notation is justified since Schrödinger equations for multipole degrees of freedom generally describe collective degrees of freedom, e.g. the nuclear surface vibrations. In order to deduce the collective spin we investigate the commutator between the angular momentum operator and the linearised free Hamiltonian.

The Cartesian components of the angular momentum operator[4], expressed within the new coordinates $x_i^{(\lambda)}$ and momenta $p_i^{(\lambda)}$, are specific linear combinations of elements of the angular momentum tensor

$$L_{mn}^{(\lambda)} = x_m^{(\lambda)}p_n^{(\lambda)} - x_n^{(\lambda)}p_m^{(\lambda)}. \tag{11}$$

For example, for quadrupole degrees of freedom the angular momentum operators are

$$L_x^{(2)} = L_{41}^{(2)} + L_{52}^{(2)} + \sqrt{3}L_{43}^{(2)}, L_y^{(2)} = L_{12}^{(2)} + L_{54}^{(2)} + \sqrt{3}L_{23}^{(2)}, L_z^{(2)} = 2L_{51}^{(2)} + L_{42}^{(2)}.$$

Each element of the angular momentum tensor $L_{mn}^{(\lambda)}$ is of course Hermitean. The following commutator algebra between two elements is valid:

$$[L_{ij}^{(\lambda)}, L_{mn}^{(\lambda)}] = -i\hbar(\delta_{jm}L_{in}^{(\lambda)} + \delta_{jn}L_{mi}^{(\lambda)} + \delta_{im}L_{nj}^{(\lambda)} + \delta_{in}L_{jm}^{(\lambda)}). \tag{12}$$

The commutator between the free linearised Hamiltonian $H_L^{(\lambda)}$ of (9) and $L_{mn}^{(\lambda)}$ yields

$$[H_L^{(\lambda)}, L_{mn}^{(\lambda)}] = i\hbar\left[\begin{pmatrix} 0 & i\gamma_m^{(\lambda-1)} \\ -i\gamma_m^{(\lambda-1)} & 0 \end{pmatrix} p_n^{(\lambda)} - \begin{pmatrix} 0 & i\gamma_n^{(\lambda-1)} \\ -i\gamma_n^{(\lambda-1)} & 0 \end{pmatrix} p_m^{(\lambda)}\right]. \tag{13}$$

Introducing a Hermitean spin tensor

$$S_{mn}^{(\lambda)} = -\frac{1}{4}i\hbar[\gamma_m^{(\lambda)}, \gamma_n^{(\lambda)}], \tag{14}$$

we find

$$[H_L^{(\lambda)}, L_{mn}^{(\lambda)} + S_{mn}^{(\lambda)}] = 0. \tag{15}$$

In addition the commutator between two elements of the spin tensor yields the same algebra as in the case of the orbital angular momentum tensor (see Eq.(12)):

$$[S_{ij}^{(\lambda)},S_{mn}^{(\lambda)}] = -i\hbar(\delta_{jm}S_{in}^{(\lambda)}+\delta_{jn}S_{mi}^{(\lambda)}+\delta_{im}S_{nj}^{(\lambda)}+\delta_{in}S_{jm}^{(\lambda)}). \tag{16}$$

As a consequence one can immediately define the Cartesian components of the spin operator analogously to the angular momentum operator. For quadrupole degrees of freedom the spin operators are then

$$S_x^{(2)}=S_{41}^{(2)}+S_{52}^{(2)}+\sqrt{3}S_{43}^{(2)},S_y^{(2)}=S_{12}^{(2)}+S_{54}^{(2)}+\sqrt{3}S_{23}^{(2)},S_z^{(2)}=2S_{51}^{(2)}+S_{42}^{(2)}. \tag{17}$$

The spins as functions of the multipolarity λ of the coordinates can be calculated by diagonalising the corresponding irreducible spin operators. The results are shown for $\lambda\leq 4$ in Table 1.

λ = 1	2	3	4
s = 1/2	3/2	3	5
		0	2

Table 1

The collective spin s for different multipolarities λ.

3. SCALAR COUPLING OF A COLLECTIVE POTENTIAL

In the preceding part it has been shown that the linearised free $(2\lambda+1)$-dimensional Schrödinger equation incorporates spin degrees of freedom. In the following we consider some possible applications of the concept of a collective spin. For simplification we only treat nuclear quadrupole surface degrees of freedom. One may hope to describe some collective aspects of even-odd nuclei if a collective potential is introduced into the linearised Schrödinger equation via a scalar coupling; this collective potential should be the same as the one used for the neighbouring even-even nuclei.

For the scalar coupling of a collective potential $V(x_i^{(2)})$ we replace the linearised free Hamiltonian $H_L^{(\lambda=2)}$ (see Eq.(9)) by

$$H_{L,V}^{(\lambda=2)} = H_L^{(\lambda=2)} + A^{(\lambda=2)}\cdot V. \tag{18}$$

The potential is assumed to be Hermitean, parity and time reversal invariant and a SO(3)-scalar. If V only depends on the coordinate $\beta^{(2)}=(\sum_{i=1}^{5} x_i^{(2)2})^{1/2}$, it is in addition also a SO(5)-scalar. The coupling matrix $A^{(\lambda=2)}$ is determined by the requirements that the Hamiltonian $H_{L,V}^{(\lambda=2)}$ is Hermitean, parity and time reversal invariant and a SO(3)- or SO(5)-scalar, respectively. Then the coupling matrix $A^{(\lambda=2)}$ results to be

$$A^{(\lambda=2)} = \begin{pmatrix} \alpha_1 I & \alpha_2 I \\ \alpha_2 I & \alpha_3 I \end{pmatrix}, \tag{19}$$

in which the α_i are real coupling constants. For the case that the Hamiltonian $H_{L,V}^{(2)}$ is a SO(5)-scalar, a further requirement can be established, namely that $H_{L,V}^{(2)}$ should be invariant under a SO(5) parity transformation. This leads to $\alpha_2=0$ in the coupling matrix $A^{(\lambda=2)}$ of Eq.(19). Then the linearised five-dimensional Schrödinger equation with a SO(5)-scalar coupled collective potential finally reads with $\psi=\binom{\phi}{\chi}\exp(-i/\hbar\, Et)$

$$(E-\alpha_1 V(\beta^{(2)}))\phi = -i\sum_{i=1}^{5}\gamma_i^{(1)}p_i^{(2)}\chi,$$

$$(2B_2-\alpha_3 V(\beta^{(2)}))\chi = i\sum_{i=1}^{5}\gamma_i^{(1)}p_i^{(2)}\phi. \tag{20}$$

Before solving Eq.(20) exactly we have first treated it approximatively in order to show the inherent interactions. Let $\alpha_1=1$, $\alpha_3=\alpha$ and

$$\alpha_3 V(\beta^{(2)}) << 2B_2. \tag{21}$$

Substituting the second equation of (20) into the first one and expanding up to first order terms in α, we derive the following Schrödinger equation for the wave function ϕ $(\beta^{(2)}=\beta)$:

$$E\phi = [\frac{1}{2B_2}\sum_{i=1}^{5} p_i^{(2)2} + V(\beta) + \frac{\alpha}{4B_2^2}\frac{1}{\beta}\frac{\partial V(\beta)}{\partial\beta}\sum_{i,j=1}^{5} S_{ij}^{(1)}L_{ij}^{(2)}$$

$$-\frac{i\hbar\alpha}{4B_2^2}\frac{1}{\beta}\frac{\partial V(\beta)}{\partial\beta}\sum_{i=1}^{5} x_i^{(2)}p_i^{(2)} + \frac{\alpha}{4B_2^2}V(\beta)\sum_{i=1}^{5} p_i^{(2)^2}]\phi. \tag{22}$$

The two last terms of Eq.(22) are spin-independent and give rise to a level shift. The third term represents a collective SO(5) spin-orbit coupling and gives rise to a splitting of the degenerate levels.

For the exact treatment of Eq.(20), the following ansatz for the wave function is made:

$$\phi = g(\beta)[(N^{+0}_{+1},0) \times (1/2,1/2)]^{(N+1/2,1/2)}_{\tau_1,\tau_2,d}$$

$$\chi = -f(\beta)[(N^{+1}_{+0},0) \times (1/2,1/2)]^{(N+1/2,1/2)}_{\tau_1,\tau_2,d} \tag{23}$$

$$\kappa := \mp(N+2). \tag{24}$$

(N,0) are SO(5) spherical harmonics and (1/2,1/2) is the lowest irreducible spinor multiplet of SO(5) which corresponds to the spin 3/2. Inserting the ansatz (23) into Eq.(20) yields coupled "radial" equations which are solved numerically.

$$\frac{\partial g(\beta)}{\partial\beta} + (2+\kappa)\frac{g(\beta)}{\beta} - \frac{1}{\hbar}(2B_2 - \alpha V(\beta))f(\beta) = 0,$$

$$\frac{\partial f(\beta)}{\partial\beta} + (2-\kappa)\frac{f(\beta)}{\beta} + \frac{1}{\hbar}(E - V(\beta))g(\beta) = 0. \tag{25}$$

It should be underlined once again that the potential of the even-odd nucleus is fixed by those of the two neighbouring even-even nuclei. For the even-odd Ir nuclei, which are the most promising candidates, the neighbouring even-even nuclei indicate that the corresponding collective potential is nearly γ-soft. Such a potential is schematically sketched in Fig.1.

The energy eigenvalues resulting from the solution of Eq.(25) are shown in Fig.2 as functions of the coupling constant α. Compared to the experimental energy values of the lowest excited states, the splitting of the first group of excited levels is by far too small. Even an additional small

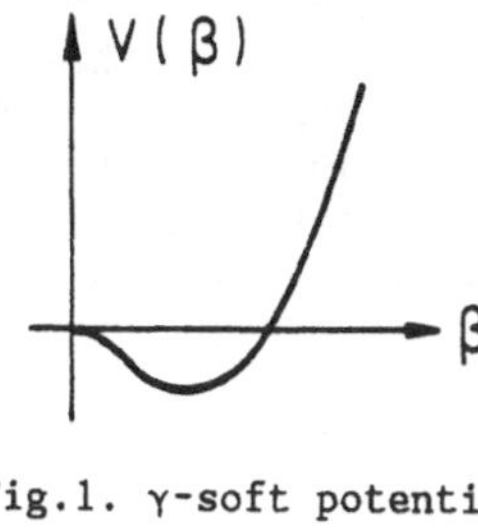

Fig.1. γ-soft potential

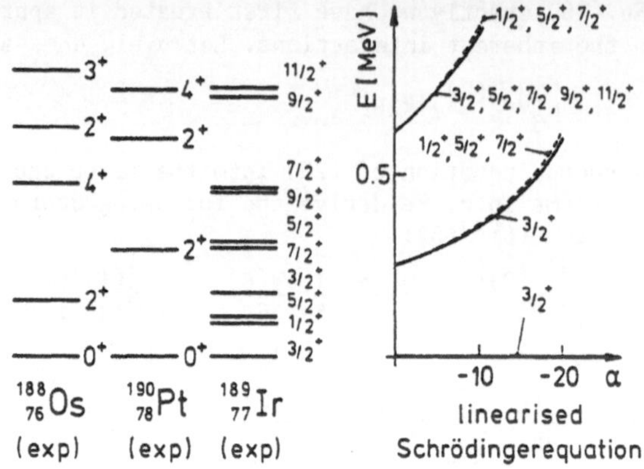

Fig.2. Experimental enery spectra of $^{189}_{77}$Ir and the two neighbouring e-e nuclei $^{188}_{76}$Os and $^{190}_{78}$Pt[5] and the theoretical energy spectrum obtained from the linearised Schrödinger-equation (25) as a function of the coupling constant α.

γ-dependent perturbation of the potential does not change this situation. Therefore, one has to state that the simple scalar coupling of a collective potential as given in Eq.(18) can not describe the collective aspects of even-odd nuclei.

4. CONCLUDING REMARKS

Up to now we have only considered coordinate-dependent potentials. On the other hand collective spin-dependent potentials represent a much more general class of potentials which could obviously simulate the interaction between the valence nucleon and the core of the even-odd nucleus. Already a simple version of the collective spin-dependent potentials, i.e. a $\vec{L}\cdot\vec{S}$-coupling term, can explain the splitting of the lowest excited levels of some of the even-odd Ir nuclei. A quantitative treatment has to show whether this new concept of collective spin-dependent potentials is also able to describe the higher excited energy levels and the electromagnetic transition probabilities correctly.

Referring to some of the algebraic models, like the IBFM[6], used to describe low energy collective aspects of even-odd nuclei, we observe that spin-dependent interactions are already inherent in these models. Hence it is not surprising that collective spin-dependent potentials are indeed needed in our approach.

In the IBFM a chosen spin value is used in order to couple the fermionic degrees of freedom to the bosonic degrees of freedom. In contrast to this method the linearisation of the Schrödinger equation as presented here yields definite values of the collective spin, and thus the spin value has not to be chosen. In this context we mention a similar linearisation scheme[7] as presented in this contribution which has been applied to the IBM and leads to some versions of the IBFM in a straightforward manner. As a consequence we conclude that the linearisation scheme gives a natural explanation of models which are based on what has become known as supersymmetry in nuclear structure physics.

REFERENCES

1) J.M. Lévy-Leblond, Math.Phys. 6: 286 (1967)
2) P.A.M. Dirac, Proc.Roy.Soc. A117: 610 (1928)
3) M. Greiner, W. Scheid and R. Herrmann, Mod.Phys.Lett. A3: 859 (1988) and J.Phys.A: Math.Gen. 21: 3227 (1988)
4) J.M. Eisenberg and W. Greiner, Nuclear Theory: Nuclear Models, Vol.1 (third edition), Amsterdam: North-Holland (1987)
5) C.M. Lederer and V.S. Shirley, Table of Isotopes (seventh edition), John Wiley & Sons, Inc., New York (1978)
6) F. Iachello, S. Kuyucak, Ann.Phys. (N.Y.) 136: 19 (1981)
7) H. Wu, M. Greiner and D.H. Feng, Phys.Rev. C 39: 1059 (1989)

Review of the Current Status of Cold Fusion *

D. Harley, M. Gajda and J. Rafelski

Department of Physics, University of Arizona, Tucson, AZ 85721

Abstract

KEYWORDS: / cold fusion / review / nuclear fusion / deuterium /

We review the current status of cold fusion experiment and theory. Recent experiments connected with neutron production, heat and fusion products are discussed. Conventional nuclear physics and it's application to the problem of cold fusion is reviewed.

1 Overview

The two months of intensive activity following the reporting of cold fusion neutrons [1] and possible excess heat [2] has resulted in a proliferation of ideas and experiments designed to verify the phenomenon and understand the underlying mechanism. Following the first major workshop devoted to Cold Fusion, held in Santa Fe, New Mexico on May 23-25, the situation remains confused, although it is possible to draw a few preliminary conclusions. The observation of 2.5 MeV neutrons by Jones *et al* [1] has been confirmed by experiments performed at Los Alamos [3] and Grand Sasso [4]. The Los Alamos experiment in particular also observe neutron 'bursts', as suggested by an earlier report from Frascati [5]. There were also two experiments reporting negative results at the level of sensitivity originally reached by Jones *et al* . However, both experiments (Yale and Bougy), and more recently the work reported from Harwell, differed from the prescription given by Jones in certain key details for most of their runs. This appears to be due to the desire to observe a nuclear signature related to heat production reported in ref. [2].

A great investment has been made in an effort to detect the presence of reaction products resulting from fusion reactions. A significant level of tritium was, and continues to be, reported from the Texas A&M university neutron experiment led by Wolf [6], at levels far greater ($10^5 - 10^8$) than the concurrently observed number of neutrons. No successful searches for ^{3}He or ^{4}He products have so far been reported. We are not aware of any extensive searches for products heavier than Helium.

Fusion rates of heavy *molecules* have been known to be exceedingly small. However, the new idea pursued in the experiments of Jones *et al* has been the possibility that the key quantity η governing the smallness of the nuclear reactions will be substantial altered when hydrogen is implanted in metallic lattice. Several metals were identified as particularly suitable, based on the known absorption and mobility of hydrogen. η is in particular sensitive to the integrated strength of the (Coulomb) repulsion in the classically forbidden domain:

*Presented at the NATO Summer Institute on Nuclear Equations of State, Peniscola, Spain, May 28 1989

$$\eta = \frac{1}{\pi} \int_{forbidden} dR \sqrt{2\mu(E_{eff} - V_{eff})}$$

In a metallic environment, V_{eff} is governed by the presence of conduction electrons and lattice ions, whilst E_{eff} can be influenced by nonequilibrium processes. Since the reduced mass μ is most favorable for the *pd* reaction, this reaction was initially regarded as more promising than the *dd* reaction. In either *pd* or *dd* fusion reactions, the final ash of the reaction is typically ^{3}He, while γ's and neutrons are produced at a specific energy, and can easily penetrate out of the experimental cell. A significant branch of the *dd* reaction produces tritium (t), which is easily detectible by the emission of an e^- during its decay into ^{3}He. In normal low-level fusion activity, one would expect a random fusion signal ('singles') in the form of a γ (*pd* reaction) or n (*dd* reaction). Tritium was much less likely to be produced for detection, without a dangerously high level of neutron activity, while neutron 'bursts' were not foreseen unless a chain-reaction of fusions were to develop.

Three key observations, vis. a) random 'singles' neutron production, b) neutron bursts of approximately 50-250 neutrons with duration of less than 50 μs, and c) probable anomalous tritium production far greater than the neutron level, leaves the theoretical explanation of the cold fusion phenomenon more confused than ever. It may well be that there are several different cold fusion mechanisms in operation, each with a different signature and sensitivity to the conditions under which the experiment is performed.

With regard to the observation of the anomalous heat reported by Fleischmann and Pons [2], it has been shown that the concurrent γ-radiation measurements they performed were incorrectly interpreted [7]. There is therefore no direct experimental evidence that the heat is attributable to nuclear reactions and the application of the term "cold fusion" in reference to the observation of excess heat production, without nuclear reaction products, is inappropriate. Two further heat experiments have been performed at Stanford [8] and Texas A&M universities [9], and both seem to confirm the heat effect. The Stanford group lead by Huggins claims to have reached the point where the amount of heat output exceeds that of the *entire* power input, so that uncertainty regarding the degree of recombination of the evolved gases cannot be the source of an erroneous measurement of excess heat. The Texas A&M experiments reported by Appleby have been performed using microcalorimetry in which power outputs of the level of less than 1 mW can be measured, much smaller than levels actually observed.

We continue in the next section to review in more detail the partially verified experimental observations that have actually emerged from the laboratories since March 23, concentrating particularly on observation of neutrons and fusion products. We then discuss some of the theoretical limits on cold fusion, and show that an approach based on conventional two-body nuclear and molecular physics is unlikely to explain this phenomenon. As this paper is also intended to an account of the events surrounding cold fusion in the past few months, we have also appended a few personal comments on the (brief!) history of cold fusion, and a perspective on a subject that has become confused by too much premature and misinformed publicity.

2 Experimental Status

The cold experiments reported should be divided into those addressing the first observation of neutrons by Jones *et al* , and those addressing the excess heat production announced by Fleischmann and Pons. In keeping with the distinction between these two types of cold fusion (see Appendix B for an elaboration of this) we have attempted to divide the experiments into two categories and we discuss them separately. We also include a separate section on fusion products.

2.1 Cold Fusion Neutrons

The observation of neutrons appears to be firmly established experimental 'fact'. The origin of the 2.5 MeV neutrons is generally presumed to be the reaction:

$$d + d \rightarrow {}^3He(0.82\ MeV) + n(2.45\ MeV) \quad (1)$$

Jones *et al* first reported a 2.5 MeV neutron signal with a confidence of 4.5 σ and an overall rate of a few times 10^{-2} counts/sec. If the process is a volume phenomenon, this corresponds to a fusion rate of $10^{-23} - 10^{-24}\ s^{-1}$ per deuteron; if however only the surface participates, the rate would have to be much higher. An indirect observation of neutrons was also reported by Fleischmann and Pons, in which they measured a 2 MeV gamma spectrum, presumed to be the 2.224 MeV peak resulting from the capture of neutrons on protons in the tank of water surrounding the cell. Serious doubts have since been cast upon Fleischmann and Pons' nuclear measurements [7], and they are now generally believed to be either incorrect or perhaps unrelated to electrolytic processes, as suggested by Bailey [13]. It is worth noting here the great differences between the Jones *et al* and FP experiments. The former used high surface to volume ratio cathodes with a complex "mother earth soup" electrolyte, modeled after the salt abundances in the earth's crust. Fleischmann and Pons used 'solid' Pd rods with a LiOD electrolyte in pure heavy water (D_2O). Jones *et al* obtained their highest neutron production with titanium cathodes, whereas Fleischmann and Pons used only palladium. Jones used gold anodes, while Fleischmann and Pons used platinum anodes. The Jones cells were open to the atmosphere, whereas the FP cells were sealed (although the evolved gasses were allowed to escape).

Since these two original cold fusion experiments, a number of others have been performed. One of the first confirmations was reported from Frascati by Scaramuzzi *et al* [5], in which the dynamical nature of the fusion process was exploited; a sample of spongy titanium weighing approximately 500 g was pressurized at 50 atm in a gas of deuterium. The sample was cooled from room temperature to liquid nitrogen temperature (77 °K), and then was allowed to heat back up to room temperature. A neutron count rate of the order of 100/min was observed, vanishing once equilibrium at room temperature had been attained. An intriguing feature of these observations was that the neutrons appeared to come in intense bursts of less than 50 μs in duration, sufficiently intense to saturate the neutron counter and result in a spectrum with 'quantized' counts. This experiment clearly demonstrated that the reaction resulting in neutrons does not depend upon some process specifically connected with electrolysis, and suggests the presence of some dynamical component in the reaction. Further experiments along similar lines, only this time heating the sample from room temperature, have also resulted in neutron bursts [1].

Neutron bursts has also been observed at Los Alamos by Menlove *et al* [3]. These experiments used the Frascati pressure/temperature variation methods with Ti chips and sponge, with pressures varying from 20 atm to 50 atm during charging. Again, no neutrons were observed while the sample was charged and then cooled to liquid nitrogen temperatures, but when the sample was heated neutron bursts were observed with a time spread of less than 50 μs. The bursts could be correlated with a sample temperature of approximately −30 °C; as the deuterated titanium was heated through this temperature, the bursts were observed. In addition to the bursts, random emissions were also observed for at least twelve hours after the sample had been heated up to room temperature. These random emissions fluctuated between values of 0.05 − 0.2 neutrons s^{-1} , the best sample giving a significance level of 11σ. Significantly, it was found that neutron emission ceased after the sample had been run through the temperature cycle several times.

An experiment more in keeping with the original Jones electrolytic experiment is an experiment performed in the Grand Sasso tunnel by Bertin *et al* [4]. The electrolyte used was identical to that of Jones *et al*, and the cathode consisted of fused titanium pellets. One hour after beginning the electrolysis, the neutron signal was observed to increase significantly, go through a maximum, and then drop back to the background level after about three

[1] We have recently been informed that these experiments cannot be reliably reproduced.

hours. This was observed in three separate experiments. The overall counting rate was estimated to be 875 ± 183 neutrons per hour which, taking into account the differences in the experimental set-ups, compares to that estimated by Jones. The significance of this particular experiment is that it is one which follows very closely the electrolytic prescription of Jones, the only notable differences between the experiments being the neutron detection schemes and the low cosmic particle background in the Grand Sasso tunnel.

Despite these successes, several careful experiments have been performed in which no neutrons were observed. One particularly notable negative experiment is by Gai *et al* [14]. Gai *et al* obtained an upper limit on the neutron emission of 2×10^{-25} at 98% confidence, one order of magnitude below that suggested by Jones. Three possible reasons for this failure to observe neutrons are: i) firstly, that only four runs of the Jones type were performed. By now it has been established that for some unknown reason, not all runs will produce neutrons; ii) Secondly, the electrode used in the Jones-type experiment was a Ti parallelepiped of dimensions $3 \times 3 \times 450$ mm, giving an effective working area of only 2.8 cm^2, whereas Jones obtained his best results with fused titanium pellets with a very much larger surface area; iii) Lastly, the final count rates were computed over time periods of 7-29 hours, and since the indications are that most of the neutron production occurs within the first few hours, the final count rates may have been suppressed by as much as a factor of five.

In conclusion to this discussion of neutron production, it appears that:

i) A background dd fusion rate of the order of $10^{-24}s^{-1}$ has been confirmed by at least three different experiments;

ii) In addition, when a deuterated metal sample is cycled trough liquid nitrogen to room temperature, neutron bursts of the order of 100 neutrons within less than 50 μs have been observed by at least two different experiments; and

iii) The fusion rate is erratic and possibly related to some dynamical process occurring within the lattice, such as a flow of deuterium within the metal or a phase change occurring between different phases of the deuteride.

2.2 Excess Heat

The connection between heat production and nuclear reactions is a lot less certain than the observation of cold fusion neutrons. In addition, heat production of nuclear origin, occurring without significant amounts of radiation, is considerably more difficult to understand. For these reasons, we shall pay only cursory attention to the heat phenomenon, and discuss it only because it is that aspect of cold fusion that has captured the wold's imagination.

The first experiment reporting the observation of heat was performed by Fleischmann and Pons [2], although interestingly there are reports of the observation of heat and Helium products dating as far back as 1926. [2] Like the Jones *et al* experiment, the Fleischmann and Pons experiment is now too well known for it to be necessary to discuss in any great detail here, but we will briefly summarize the essential points to allow comparison with other experiments that have since been performed. Firstly, they used sheet, rod and cube samples of Palladium immersed in a LiOD + D_2O electrolyte contaminated by a 0.5% content of H_2O. The D/Pd ratio achieved is not specified in their preliminary paper, but is believed to be close to 1. The currents densities on the electrodes varied from 1.6 mA/cm^2 to 512 mA/cm^2, which was found to strongly influence the generation of heat. The best

[2] In 1926, F. Panneth and K. Peters claimed to have observed the production of Helium in hydrated palladium [10]. After a series of meticulous experiments, Panneth withdrew this claim of Helium production, concluding that his Helium was ultimately due to contamination from the atmosphere but noting in the German version of his paper that one series of his experiments still lacked an explanation. In 1927, J. Tandberg applied for a Swedish patent for "A method to produce helium and useful reaction energy", in which he used light water on palladium [15]. The patent application was rejected, the reason presumably being that at the time the nuclear physics was largely unknown and hence the process appeared to violate conservation of energy.

results obtained was a heat production in the electrode of 10 W/cc, maintained for a period of 120 hours. This gives a total energy output of 4 MJ/cc, or more than 100 eV per lattice site, from which the authors conclude that chemical processes cannot be responsible for the heat production.

Many attempts have since been made to reproduce this experiment, mostly without success. There are however at least two experiments that claim similar results; the first by Huggens *et al* [8] of Stanford University, and the second by Appleby *et al* [9] from Texas A&M University. Since the Stanford experiment was historically the first to confirm Fleischmann and Pons, we shall discuss this first.

The experiment performed by Huggins *et al* [8] is similar to the work of Fleischmann and Pons experiment, and a great deal of emphasis has been placed by the Stanford group on the method of preparing the electrodes. The palladium electrodes are prepared by melting and re-melting the samples approximately 10 times within an arc furnace, in an atmosphere of argon. The electrodes are then placed within a LiOD electrolyte, carefully insulating the contents of the cell from the air in order to avoid contamination by protium; Huggins claims that his heat effect vanishes when even small amounts of H_2O is present. The electrolyte is shaken during the experiment, and the heat output is measured using standard calorimetry. What is observed is that the heavy water jars behave similarly to the light water control cells for a period of several hours. After this period, the heavy water cell heat production begins to climb above the endothermically suppressed temperature to *above* the temperature one would expect had no gasses been evolved. The highest figure quoted was a total heat excess of 12% above electrical input, although the indications were that the excess continued to climb with time. The significance of this report is that the total heat out is greater than the electrical energy in, so that even if all the evolved gasses had recombined there would still be an excess of heat.

The second, well publicized experiment reporting heat [9] has been performed in quite a different manner. Palladium electrodes in the form of a wire or sphere were placed, with about 7.5-8.0 ml of electrolyte, within a microcalorimeter capable of measuring heat generation of less than 1 mW. The electrodes were initially charged in a LiOD electrolyte at low current densities of about $60mA/cm^2$, for up to 40 hours. The current was then increased to $600ma/cm^2$, upon which the heat excess was observed to rise to as much as 40 mW, a magnitude well within the resolution of the instrument. When the electrolyte was replaced with LiOH, no effect was observed. An interesting study was made of the influence of the type of electrolyte; when Li was replaced by Na, the heat excess fell to less than 10% of the Li value. When isotopically pure ^7Li was used, the heat excess was found to be depressed by about 10%. Experiments are now under way to perform the same experiment with ^6Li. A significant difference between the ^6Li and ^7Li measurements would add an interesting new twist to the cold fusion puzzle [3].

2.3 Cold Fusion Products

The only fusion product in macroscopic quantities that have been reported thus far is tritium, by Wolf *et al* [6], from Texas A&M University. Energetic charged particles have also been observed by Cecil *et al* [11], of the Colorado School of Mines. Both these results are still preliminary at the time of writing, and we have chosen to emphasize these observations above others as they are very different but complimentary.

Wolf *et al* performed an analysis for tritium on electrolytic cells that had been run for several days in a search for neutrons. Tritium was detected in nine out of ten cells by observing tritium beta decay within the cells' electrolyte, at a level of between 10^4 to 10^6 decays/min/ml. The maximum level of tritium reported presently is at a level of 10^{16} tritons per cell, the latest findings indicating that the tritium content is higher in those cells in which the evolved gasses have been recombined. This may hint at the possibility

[3] We have since heard preliminary reports that this has not materialized.

that the tritium is produced at the surface of the cathode, rather than within its bulk. The quantity of tritium observed is far in excess of the number of neutrons observed, by a factor of 10^5-10^8. The tritium content of the cells was verified at Los Alamos, the measurements agreeing well within an order of magnitude. However, the *initial* tritium content of the electrodes and electrolytes was not known, so the possibility of contamination cannot be ruled out as yet although contamination at the observed levels is extremely unlikely.

The experiment by Cecil *et al* [11] was performed in an altogether different manner. A $1\ cm^2 \times 1.50\mu m$ Pd target, coated with a $0.05\mu m$ layer of silver, was bombarded with 95 keV deuterium ions in a high vacuum until the d(d,p)t reaction began to saturate. This occurred after the target had been bombarded with about 3.5 Coulombs of deuterium, taking 36 hours with a beam current of 30 μA. The charged particles emitted from the target were measured using a silicon barrier detector. Once the target was saturated and the external beam was turned off, a particle group at 5 MeV and of strength 30 ± 7 counts/90 hours was observed when an electrical current was passed through the sample. This signal was *not* seen when no current was passed through the sample, or in control experiments in which the sample was not deuterated. Attempts to identify the particles by inserting thin foils and observing the resulting line shift are presently in progress. The charged particles have tentatively been identified as protons, not tritons.

In conclusion, the evidence for nuclear products (apart from neutrons) is still very uncertain and preliminary. We feel that the Cecil *et al* [11] experiment is particularly noteworthy, as it offers the opportunity to examine cold fusion on a microscopic level and actually identify the fusion reactions as (if) they occur, under conditions that can be carefully controlled. In order to be consistent with the tritium production observed by Wolf *et al* , however, it seems that tritium production must arise from a different mechanism than that resulting in the charged particles of Cecil *et al* .

3 Status of Cold Fusion Theory

3.1 Introduction

One of the critical elements in understanding the "background" rate of fusions in a hydride is the understanding of the interaction potential between the hydrogen nuclei in metals such as palladium or titanium. Following the first experimental reports, it was quickly realized that solid state effects can greatly influence the interaction potential and hence the fusion rate [16,17]. One particularly important mechanism at low energies is electron screening, in which the high electron densities lead to a substantial softening of the Coulomb barrier between the fusion nuclei in the region of half an electronic Bohr radius, or about 0.25 Å. Due to many other subtle (and perhaps as yet un-thought of) collective interactions, the long-range coulomb potential may be further modified. One of the conclusions arising from these quantitative studies is that even a very substantial modification of the Coulomb barrier will result in fusion rates that are many orders of magnitude smaller than the Jones rate. This conclusion has also been reached by other authors arguing on more general grounds [18].

A second effect that has been given much attention very recently is the confinement of two hydrogen nuclei within a lattice site. The primary motivation for this interest are the recent experimental indications that a stoichiometric ratio of greater than 1 is required [19] in order to obtain the heat production as reported by Fleischmann and Pons. Some preliminary molecular dynamics calculations for $PdD_{1.1}$ have already been reported [20], but it is easy to obtain an order of magnitude estimate of the effect of confinement simply by placing two hydrogen nuclei in a box and by squeezing. This exercise again leads to the conclusion that the hydrogen fusion rates that can be realistically achieved are far too small to explain even the Jones rate. This has been discussed in detail elsewhere, so we give confinement only a passing mention here.

The final item completing the list of "conventional" ingredients is energy or temperature. Two specific mechanisms have been discussed in this context. The first is the formation of thermalized "hot-spots" in the metal, formed for example during the collapse cavities when stress within the material is suddenly relieved, or during the fracture of the material which could result in the acceleration of deuterons to high energies. The second is the acceleration of deuterons across keV potentials caused by charge separation as a material fractures. These particular mechanisms could conceivably explain the neutron bursts that have been reported, if the required peak rates can be achieved. In addition, such high-temperature phenomena would have distinct energy-dependent signatures in terms of the relative fusion rates between different hydrogen isotopes. However, after an analysis of the fusion rates to be expected from high energy deuterons within a deuterated metallic lattice, we again come to the conclusion that the rates are too small.

In the next section, we shall outline the conventional processes by which fusion reactions amongst hydrogen isotopes is believed to occur. We then examine what sort of effective interaction one might expect to find between two hydrogen nuclei embedded within a metallic lattice, and to what degree the usual coulomb barrier existing between the nuclei may be softened. Lastly, we take a more detailed look at 'hot' cold fusion, in which deuterons are accelerated to high energies or temperature.

3.2 Fusion rates of Hydrogen isotopes

3.2.1 Standard fusion rates and tunnelling

The nuclear reaction rate in a static system consisting of two reacting nuclei described by a relative wavefunction $\Psi(R)$ is simply proportional to the probability amplitude of the nuclei being close enough to fuse. Generally, when the amplitude is small the wavefunction does not vary significantly within the nuclear region and we may express the fusion rate as:

$$\lambda = K_0 |\Psi(0)|^2 \tag{2}$$

$|\Psi(0)|^2$ is the probability amplitude that the two hydrogen nuclei come together, and K_0 is the fusion constant when the fusing nuclei are in a relative S-wave. Fusion from states of higher angular momentum are strongly suppressed due to the centrifugal barrier, and is important only under very special circumstances. For the d(d,n) ^{3}He reaction, generally considered the best candidate reaction for neutron production, it is found that $K_0 = 1.5 \times 10^{-16}$ cm^{-3}s^{-1}. This implies that in order to achieve the fusion rate of $\lambda = 10^{-24}$s^{-1} per deuteron pair reported by Jones, we require a probability amplitude of $|\Psi(0)|^2 \sim 10^{-8}$ cm^{-3}. One should be cautioned not to take the magnitudes of these extremely small fusion rates too seriously, since the wave function on which they are based is of the magnitude 10^{-22} in natural units. At this level many 'quantum leak' effects may have to be incorporated.

The fusion constant K_0 in eq. (2) is obtained from scattering experiments at $E \geq 10$ keV, in which the fusion cross section can be measured. In this context, it is customary in literature to refer to the so called astrophysical function $S(E)$, which plays a very similar role as K_0 and which is also a slowly varying function of energy and is related to the fusion cross section (in free space, for the Coulomb potential) by:

$$\sigma(E) = \frac{S(E)}{E} e^{-2\pi\eta_0} \tag{3}$$

where $\eta_0 = \alpha/v$ is the Sommerfeld parameter, $\alpha \simeq 1/137$ being the fine-structure constant. (All units used from here on and above are with $c = \hbar = k_B = 1$.) The link between equations (2) and (3) is straight forward in the regime in which the WKB approximation is valid. In this semi- classical approximation, the amplitude of the wavefunction at the origin is given by:

$$|\Psi(0)|^2 \simeq \frac{1}{\Omega} 2\pi\eta e^{-2\pi\eta} \tag{4}$$

where Ω is the volume occupied by the two nuclei, and

$$\eta = \frac{1}{\pi}\int_{r_{min}}^{r_{max}} \sqrt{2\mu(V(r) - E)}dr \tag{5}$$

μ is the reduced mass of the reacting bodies, $V(r)$ the barrier potential and E the relative energy. For the Coulomb potential, η reduces to the usual Sommerfeld parameter. The range of the integral is in the classically forbidden region of motion, in which the root within the integrand is real. For the small rates considered here the fusion rate is relatively insensitive to the nuclear channel radius r_{min}, of order fm, and we have set $r_{min} = 0$ for simplicity. For larger rates the finite size could become significant, as for example in the muonic molecular systems [21].

By comparing the equations (2) and (3) in the context of a coulomb scattering experiment in which the fusion rate is given by the usual expression $\lambda = \sigma\rho v$, we obtain the correspondence between K_0, the fusion constant at low energies, and the astrophysical S-factor obtained from scattering experiments:

$$K_0 = \lim_{E\to 0} \frac{S(E)}{\pi\alpha\mu} \tag{6}$$

The values of K_0 vary by several orders of magnitude for the different hydrogen isotopes, depending upon whether the fusion rate is mediated by the strong, electromagnetic or weak interaction. The important physical parameters governing the fusion rate are gathered together in table 1. (Note that the astrophysical S-factors are extracted by extrapolating the S-factor measured at keV energies or by performing an R-matrix fit, also at keV energies. The fusion rate at eV energies is not directly measureable in scattering experiments due to the minute fusion cross sections.)

Table 1

Reaction	μ (MeV)	$S(0)$ (MeVb)	K_0 (cm^3s^{-1})
$p+d \to {}^3He + \gamma$	625.411	2.5×10^{-7}	5.2×10^{-22}
$p+t \to {}^4He + \gamma$	703.336	2.6×10^{-6}	4.8×10^{-21}
$d+d \to {}^3He + n$	937.807	5.36×10^{-2}	7.48×10^{-17}
$d+d \to {}^3H + p$	937.807	5.58×10^{-2}	7.80×10^{-17}
$d+d \to {}^4He + \gamma$	937.807	2.2×10^{-10}	3.1×10^{-25}
$d+t \to {}^4He + n$	1124.65	1.16×10^{1}	1.35×10^{-14}

The reactions of primary interest are those involving deuterons. Inserting the values from table 1 into equations (2) and (5), we find that for the different reactions the two effects of reduced mass and the fusion rate constant vie with one another as the parameter dominating the fusion rate. For low energies, the root of the reduced mass factors out in eq. (5) and it is the reduced mass that dominates the fusion rate. For higher energies, the barrier is reduced to a level of insignificance and it is the K-factor which dominates. In between, there is a cross-over in which the different fusion rates correspond. For the bare pd, pt, dd and dt reactions this occurs in the vicinity of 220 eV (CM energy), in which the different effects, each influencing the fusion rate by many orders of magnitude, conspire to cancel and bring the fusion rates to within an order of magnitude of one another. In figure 1 we present the fusion cross section as computed from eq. (3), which illustrates this rather remarkable point. Although 220 eV is much above the energies one might believe to be present in the environment of the palladium, it is important to note that in thermalized systems the primary contribution for the fusion rate comes from the high energy tail of the energy distribution function. We shall return to this point in a following section.

3.2.2 The effective Coulomb Interaction

The environment of the lattice considerably influences the long-range Coulomb force surrounding charged nuclei. In Palladium, the immobile Pd nuclei are largely stripped of their valence electrons and are ions, with a charge of four. A large fraction of the hydrogen atoms in the lattice are also stripped of their valence electron and are therefore able to move freely as nuclear, rather than atomic, entities. The delocalized electrons are highly mobile, and participate by modifying the long-range interaction between the positive ions. Even when the ions possess energies of several 100 eV, their motion is essentially adiabatic with respect

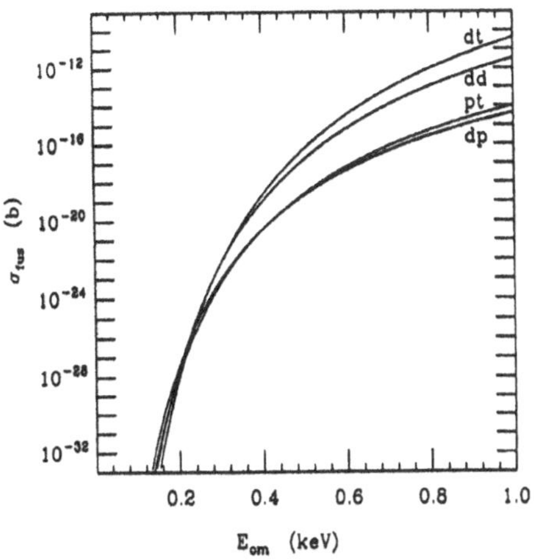

Figure 1. The fusion cross section in barn for various hydrogen-hydrogen reactions,as a function of the CM energy in keV.

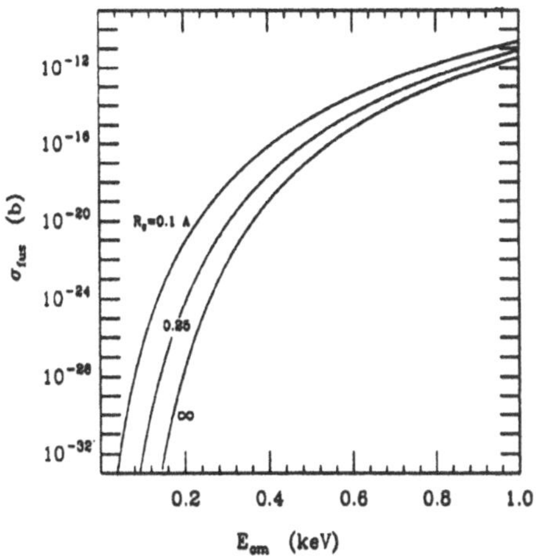

Figure 2. The screened fusion cross-section for the neutron reaction $d + d \rightarrow {}^3$H+n for different screening lengths in Å.

to that of the electrons and the electrons are able to respond to the changing configuration of the nuclei. The consequence of most interest to us here is that the coulomb potential between two hydrogen ions is modified to a Yukawa-like potential with an exponentially falling repulsion at intermediate distances, of the form [22]:

$$V(R) = -\alpha \frac{e^{-R/R_s}}{R} \tag{7}$$

R_s is the screening length, and introduces a new length scale into the usual Coulomb potential. For distances much smaller than R_s the potential returns to the usual Coulomb potential. For distances much greater than this, the exponential behavior is replaced by a $\sin(2k_f/R_s)/R^3$ behavior, where k_f is the Fermi momentum. The potential at this range is so small that it is not of much relevance to us here.

The actual value of R_s depends upon the density of electrons at the particular location within the material. If R_s simply derives from the density of states of the conduction electron gas, one finds that R_s can be as small as 0.14 Å for palladium, taking the density of states from the heat capacity of electrons [23]. However, a more realistic value appears to be closer to 0.25 Å. Actually the screening length is not uniform throughout the material, but varies according to the presence of local crystal defects, grain boundaries etc. For this reason, we shall treat the screening length as a parameter in this analysis, and take values ranging from the vacuum value of $R_s = \infty$, to a very optimistic $R_s = 0.1$ Å. In figure 2 we present the fusion cross section for the reaction $d + d \rightarrow {}^3$H+n, as computed from equation (3) for the screened potential (7). Figure 2 illustrates the impact screening has on the cross section. At low energies, below 100 eV, the impact is of several orders of magnitude and increases considerably towards the eV range (not shown). At energies above 100 eV, however, the screening becomes increasingly irrelevant. This is a point worth noting when computing the fusion rates within a thermalized system, as the largest contribution to the fusion rate comes from the high energy tail of the energy distribution function. As for the low energy regime, where screening *does* have a significant impact on the fusion rate, it has been found that the fusion rates are simply too small to explain the Jones rate, unless screening lengths for deuterium nuclei moving between ions in the metallic lattice can be reduced to less than 0.04 Å [17]. Even 0.1 Å is beyond the borders of credibility, and for this reason, we turn next to the 'high energy' domain of cold fusion.

3.2.3 'Hot' Cold Fusion?

Cold fusion neutron production requires a mechanism that is able to produce of the order of 0.01 neutrons s^{-1} cm^{-3} of deuterated metal. When considering the number of deuterons within the sample, of the order of $10^{22}cm^{-3}$, it is clear that only a few deuterons need actually be responsible for the production of such a small number of neutrons, and suggests that perhaps some mechanism exists that accelerates a small number of deuterons to higher energies which then have a substantially enhanced fusion rate. The simplest model is that of a thermalized hot-spot consisting entirely of deuterium, heated perhaps by the collapse of a cavity within the lattice. The exact temperature of the hot-spots depends upon the specific model chosen, but we can estimate the order of magnitude by considering the energy densities within a stressed palladium crystal. The stiffness constants of metals are typically of the order of $C = 10^{11}\mathrm{N/m^2}$, and the energy density of a stressed sample is $E/V = \frac{1}{2}C\sigma$ where $\sigma = \Delta L/L$ is the strain. Hydrides are known to distort by as much as 10% during the infusion of the hydrogen, so a strain of $\sigma = 0.1$ is reasonable. This results in energy densities of the order of $10^{11}\mathrm{J/m^3}$, or about 1 eV/$\mathrm{Å}^3$, as being available for compressing and heating pockets of deuterons within the crystal. It is not possible to estimate the final temperature and lifetime of the resulting plasma without specifying the initial density, geometry and volume of the deuterium pocket, details which are beyond the scope of this paper, but a simple and realistic model places an upper limit on the attainable temperature of $T = 10$ eV.

Within a thermalized plasma, the cross section in eq. (3) must be folded with the thermal distribution in energies, which is simply the Maxwell-Boltzmann distribution:

$$f_{B_i}(E;T) = N(T,m_i)\, e^{-E/T} \tag{8}$$

where $N(T,m_i) = (m_i/2\pi T)^{\frac{3}{2}}$, and m_i is the mass of the thermalized particle of species i at temperature T. The distribution function satisfies the normalization condition

$$\int_0^\infty dE f_{B_i}(E) = 1 \tag{9}$$

The net fusion rate between two species of nuclei i and j, of density ρ_i and ρ_j, is then:

$$\lambda_{fus}(T) = \rho_i\rho_j \int_0^\infty d^3v_1 d^3v_2 f_{B_i}(v_1) f_{B_j}(v_2)|v_1 - v_2|\sigma_{fus}(|v_1 - v_2|) \tag{10}$$

The velocity of the particles v is related to their energy E by the usual $v = \sqrt{2Em}$. Once the center of mass motion has been removed, eq. (2) becomes:

$$\lambda_{fus} = \rho_i\rho_j N(T,\mu_{i,j})\, \frac{1}{2}(2\mu_{i,j})^2 \times \int_0^\infty dEE\, e^{-E/T}\, \sigma_{fus}(E) \tag{11}$$

where $\mu_{i,j} = (1/m_i + 1/m_j)^{-1}$ is the reduced mass. We see from eq. (11) that the fusion rate in such a system is thus a product of a steeply decreasing probability of occupying a state of high energy, here $E\, e^{-E/T}$, and a steeply increasing fusion cross-section $\sigma_{fus}(E)$. Generally, the product peaks at an energy very much above the temperature, depending on the penetration barrier determining the fusion cross section. Almost the entire contribution to the fusion rate therefore comes from a narrow region of energy far above the mean energy of particles in the system; this is the well-known Gamow-Teller peak. This fact implies that the relatively long-range modifications to the barrier potential, such as the screening discussed above, are relatively ineffective in increasing the fusion rates of thermalized systems

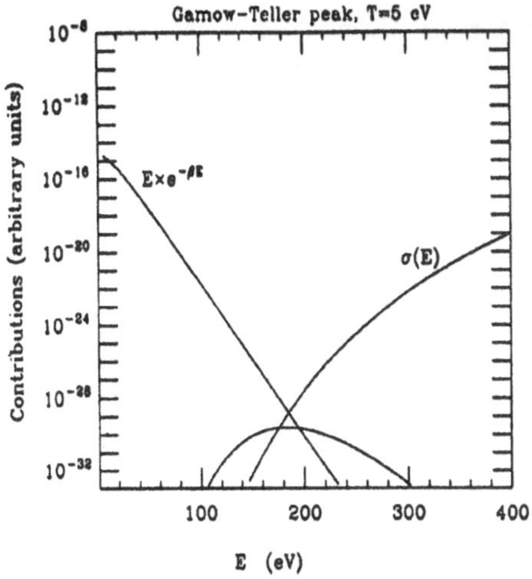

Figure 3. The Boltzmann two-particle distribution, *dd* fusion cross section and their product for a temperature of 5 eV.

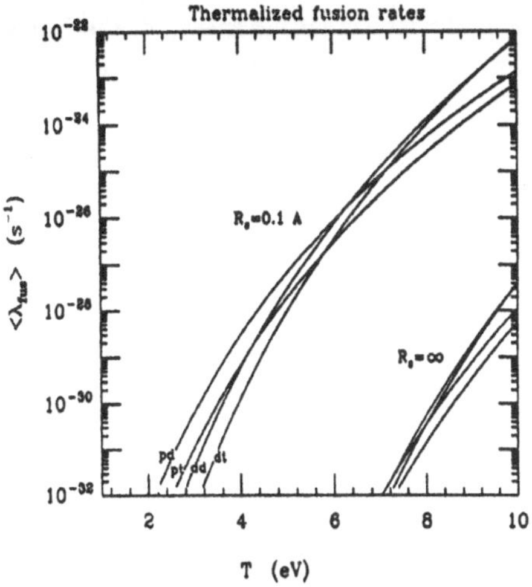

Figure 4. The fusion rate per particle in a plasma containing a 50:50 mixture of hydrogen isotopes (dt,dd,pt,dp), at an overall density of 4×10^{22}/cc, for no screening and an extreme screening of 0.1 Å

of even very moderate energies of the order of eV. In figure 3 we illustrate this by plotting the Boltzmann distribution, the cross section and their product for a temperature of 5 eV. Although the temperature is only 5 eV, the Gamow peak is found to lie at about 190 eV.

In figure 4 we plot the fusion rate per particle in a plasma for a 50:50 mixture of hydrogen isotopes, at an overall density of 4×10^{22}/cc, for no screening and an extreme screening of 0.1 Å. For the extreme screening case $r_s = 0.1$ Å we can indeed reach 10^{-24} fusions per particle pair when the temperature is between 8 and 10 eV. However, remembering that only a small fraction of the deuterons within an electrode can possibly be in such an extreme state, it is clear that high temperatures are not going to come even close to explaining the neutron flux reported by Jones and others.

This result does not yet exclude high-energy fusion as being responsible for cold fusion. As noted above, only a small fraction of the deuterons within the Gamow-Teller peak actually contribute to the fusion rate. It is not impossible that a mechanism exists that produces only high energy deuterons with energies in the keV region. There are several experiments that have been performed in which materials have been fractured or adhesive bonds broken in which the emission of x-rays in the keV region has been reported [24]. In addition, experiments have been performed in which LiD or D_2O have been fractured, and neutron emission has occurred, although the statistical significance of the neutron emission is only of the 2σ level and therefore cannot be taken as strong evidence for nuclear reactions [25]. Given that there is a source of high energy deuterons it is possible to compute the fraction of deuterons likely to yield neutrons, when incident upon a deuterated metal. This situation is somewhat different to that of the thermalized region of deuterons, in that the fusion occurs within the lattice itself and the high-energy deuteron rapidly looses energy as it passes through the lattice. Most of the energy is lost to electron ionization at energies below 1 keV, but for higher energies recoil against the lattice nuclei and deuteron nuclei within the lattice becomes increasingly important. The rate of energy loss is generally expressed in terms of the stopping powers S_i of each particle species i making up the target. When the ratio of fusing paricles to incident particles N_{fus}/N_{inc} is small, the stopping powers are approximately related to the incident particle's energy loss rate by

$$\frac{dE}{dt} \simeq -\sum_i v\rho_i S_i(E), \tag{12}$$

E is the instantaneous energy of the insident particle slowing down in the material, and $v = \sqrt{2mE}$ is its velocity. ρ_i is the number density of each species of particle in the target. The stopping power is typically of the order of $10^6 - 10^7$ keV-barn and varies from material to material, generally increasing dramatically from below 100 eV, but then levelling off and becoming almost flat up to energies of 10 keV [26].

When the energy loss in (12) is taken into account, the fraction of particles of energy E that will actually fuse when incident upon a metal deuteride AD_x of stoichiometric ratio x is:

$$\frac{N_{fus}}{N_{inc}} = x \int_0^{E_{inc}} dE \, \frac{\sigma_{fus}(E)}{S_{AD_x}(E)} \tag{13}$$

with

$$S_{AD_x}(E) \simeq S_A(E) + xS_D(E) \tag{14}$$

being the stopping power of the deuteride for the incident deuterium. E_{inc} is the energy of the incident particles. The ratio in eq. (13) is thus an expression of the efficiency of the target in allowing the incident deuterons to collide with and fuse with deuterons in the target, before the incident deuterons are slowed to a stop. This ratio is presented in figure 5. For energies in the region of 5-10 keV, the fusion ratio is of the order of $10^{-14} - 10^{-11}$. Thus, in

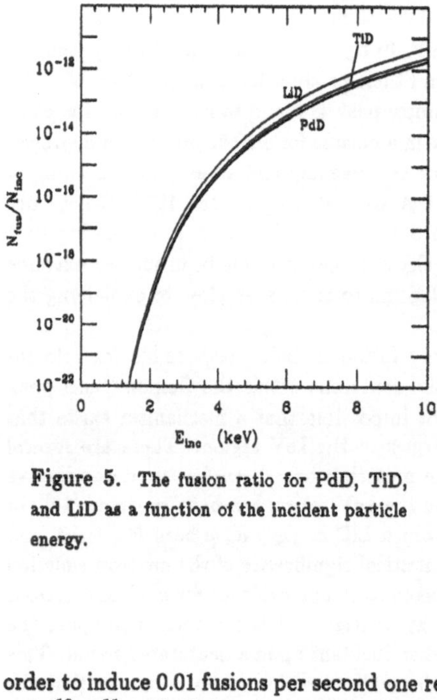

Figure 5. The fusion ratio for PdD, TiD, and LiD as a function of the incident particle energy.

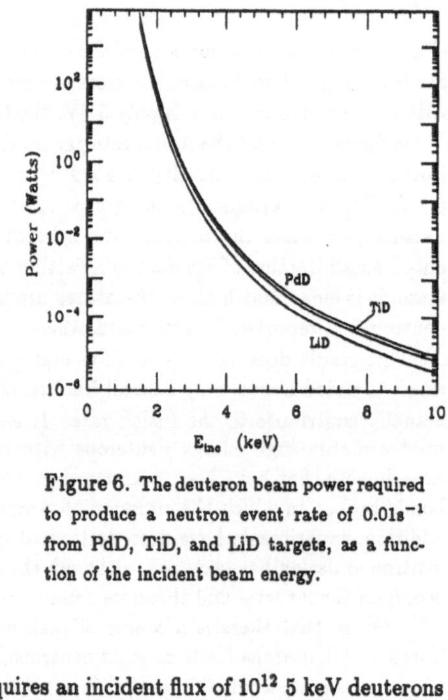

Figure 6. The deuteron beam power required to produce a neutron event rate of $0.01s^{-1}$ from PdD, TiD, and LiD targets, as a function of the incident beam energy.

order to induce 0.01 fusions per second one requires an incident flux of 10^{12} 5 keV deuterons or 10^{10}-10^{11} 10 keV deuterons per second. Numbers of this magnitude are comparable to the estimated number of keV electrons per square cm produced when an adhesive bond is broken as a material is torn from a metallic surface [27], although the great majority of these electrons will be considerably below 5 keV and some efficient mechanism would have to be found to transfer the kinetic energy from electrons to the heavier deuterons.

The number of required deuterons at a given energy may be re-expressed in terms of the power required to induce a neutron event rate of $0.01s^{-1}$; the power required is simply:

$$P = \frac{0.01\,s^{-1}}{(N_{fus}/N_{inc})} \times E_{inc} \tag{15}$$

This quantity is plotted in figure 6. The fusion rate increases exponentially with incident energy, and consequently the power required to produce the desired neutron event rate decreases dramatically with increasing energy. The deuteron beam power at 2.8 keV is about 1 W, and decreases to about 10^{-5}-10^{-6} W for a beam energy of 10 keV. As the electrical power being passed trough a typical electrolytic cell is of the order of 1 W, we can conclude that if the electrical current in the cell is indeed driving the fusion, we are faced with an almost insurmountable problem of finding a mechanism to accelerate deuterons to keV energies.

We conclude that 'hot' cold fusion is unlikely to be responsible for the neutrons that have been observed by Jones and others. Thermalized hot spots are unable to reach the required temperature for the required periods of time, and a monoenergetic mechanism requires such high energies for the particles that it is difficult to imagine a mechanism capable of accelerating against the stopping power of the crystal. All of the above discussions is of course another 13 orders of magnitude short of explaining the observations of heat.

4 Closing Remarks

The most significant problem facing cold fusion is the inability of some competent experimentalists to observe it. This could indicate that the important parameters have not yet been identified, and that some crucial element governing the effect differs from laboratory to laboratory. This uncertainty has had a very negative impact on many experimentalist and theorists, many of whom have chosen to discount those few experiments supporting cold fusion as a matter of personal preference. Nevertheless, there *is* a definite problem facing us; how to accommodate the—albeit few—experiments that *do* see the 'cold fusion effect', either confirming and explaining cold fusion theoretically, or discovering the omissions in the experimental procedure. In consideration the experimental evidence that has already been accumulated, it appears that attaining either objective will be equally difficult.

Acknowledgements: We wish to thank H. Rafelski for invaluable help in obtaining some of the data presented here. This work was supported by the Advanced Energy Projects Division of the US Department of Energy.

Appendices

A A Brief History of Cold Fusion

'Cold Fusion' is a term that originally applied to the field of Muon Catalyzed Fusion, which experienced a revival in the United States in 1981. Speculation that fusion might be occurring in ordinary electronic systems, *without* the benefit of a heavy muon to squeeze hydrogen isotopes together an make them fuse, occurred to Jones and Rafelski, as an extension of their work in muon catalyzed fusion and speculation arising from anomalous ^{3}He levels associated with geological events (noted by Palmer, co-worker of Jones and Rafelski). On the other hand, Fleischmann and Pons were lead to study the possibility of cold fusion in connection with their own work in electrochemistry and apparently anomalous heat observations in experiments with deuterated Palladium that had already been noted as long as 60 years ago [10].

These two collaborations, coincidentally both in Utah, worked independently of each other for several years up to 1989, only becoming aware of each other during November 1988. An informal agreement was reached that each would publish their results in the journal *Nature* towards the end of March 1989. Unfortunately events did not proceed as planned and Fleischmann-Pons phenomenon burst upon the world in a blase of publicity on 23 March 1989, when the University of Utah decided to hold a press conference and announce Fleischmann and Pons' observations prior to their work undergoing the usual peer review. On the same date, Jones and his colleagues submitted their paper to *Nature*. There has since been much (justified) criticism of the decision to hold a press conference, rather than have Fleischmann and Pons submit to the usual process of peer review before the public proclamation of a scientific discovery of perhaps the greatest importance.

Since that date, the scientific community has been struggling with a flood of contradictory findings, hotly disputed 'facts' about cold fusion, and in addition, a complete lack of any theory that can draw together the two historically different paths by which Cold Fusion first came to be studied; the path leading from the domain of nuclear physics, and the second from the domain of chemistry.

B The Cold Fusion Confusion

The first important point which should be recognized is that there appears to be at least two different *kinds* of cold fusion. This seems to be the only way to reconcile the various different observations that have been reported. They are:

i) Fusion resulting in neutrons, both random emission and 'bursts' [1,3,4,5]

ii) Excess heat *without penetrating radiation*, attributed to nuclear fusion. [2,8,9] [4]

Both of the above have been referred to as 'cold fusion', but the existence of one kind does not in any way prove the existence of the other. These two types of experiments lie in very different physical domains, and the distinction has not always been made clear. One might add a second list containing a few of the preliminary findings that have recently appeared:

i) Fusion resulting in tritium, but apparently without the conventionally expected number of neutrons or heat [6]

ii) Heat production without Helium products (!) [9]

iii) Fusion resulting in high energy protons [11]

Each of these last types of fusion seem to have some support, but are based on preliminary results that have not yet been reproduced by independent laboratories.

There are several other positive reports of cold fusion, and in addition, a great number of experiments that have tried but failed to see any signature of cold fusion at all. This apparent lack of reproducibility in cold fusion experiments, even within a single experimental group that does report positive results, is one of cold fusion's most confounding and perplexing features and has given rise to much contentious debate as to whether cold fusion even exists as an experimentally verifiable physical phenomenon. However, even though the situation is not clear, the existence of *some* level of repeatability should exclude remarks that have been made that cold fusion is a 'pathological' science. The opinion of the authors is that we have simply not yet identified the important parameters governing the process of cold fusion. Nevertheless, repeatability is a very important issue and hopefully the era of hurried, 'jury-rigged' cold fusion experiments is passed and we can hope to see some careful and systematic studies of sufficient detail to be built upon by independent researchers; indeed, a few experiments of this kind have already been performed.

It appears that at least two 'miracles' are required in order to explain the experiments. The first is a *molecular* or *solid state* miracle, required in order to explain the enormous fusion rates. In a standard hydrogen electro- molecule, the fusion rate is of the order $10^{-65} - 10^{-75}$ s^{-1}, depending on the isotopic content of the molecule. By comparison, one requires a fusion rate of $10^{-24} s^{-1}$ and approximately 10^{-11} s^{-1} for the observed neutron and heat production, respectively. Following the first reports of cold fusion, it was pointed

[4] The works cited above are those that are judged by us to be of most interest, and for which the experimental details are available to us in the form of a publication, preprint or have been presented at the May 23-25 1989 meeting in Santa Fe, New Mexico. There are however a great number of experiments that have been performed however, most of them yielding negative results.

out by several authors [12] that the fusion rate within a molecular system is critically sensitive to the physical parameters defining the system, and that factors such as confinement radius, effective electronic mass, molecular excitation energy etc. can increase the fusion reaction rate by as much as twenty to thirty orders of magnitude. Nevertheless, even the fusion rate of $10^{-24}s^{-1}$ per deuteron pair reported by Jones appears to be unattainable by conventional equilibrium physical processes, and the heat production that has been observed and attributed to fusion remains utterly inexplicable.

Heat or tritium production, without penetrating radiation, requires a *nuclear physics* miracle in which energies of order MeV are converted somehow to heat, without the penetrating radiation (n or γ) normally associated with nuclear processes. In the particular case of dd fusion, the biggest problem is explaining the enormous asymmetry between the isospin-symmetric reactions

$$\begin{aligned} d+d &\rightarrow {}^3\text{He}\ (0.82\text{MeV}) + n\ (2.45\ \text{MeV}),\ \text{and} \\ &\rightarrow {}^3\text{H}\ (1.01\text{MeV}) + p\ (3.02\ \text{MeV}), \end{aligned}$$

Which would be needed if the tritium reported were to originate from the *dd* fusions. Unless the first reaction is largely suppressed, the expected neutron flux from a 1 W reactor operating on this these reactions would be lethal in just a few hours.

References

[1] Jones, S.E., Palmer, E.P., Czirr, J.B., Decker, D.L., Jensen, G.L., Thorne, J.M., Taylor, S.F. and Rafelski, J. "Observation of cold nuclear fusion in condensed matter" *Nature* **338**, 737-740 (1989).

[2] Fleischmann, M., Pons, S. and M. Hawkins "Electrochemically induced nuclear fusion of deuterium", *J. Electroanal. Chem.* **261**, 301-308 (1989) and erratum.

[3] Menlove, H.O., Fowler, M.M., Garcia, E., Mayer, A., Miller, M.C., Ryan, R.R., and Jones, S.E. "Measurement of neutron emission from Ti and Pd in pressurized D_2 gas and electrolysis cells", Los Alamos preprint, submitted to Nature June 1989

[4] Bertin A., Bruschi M., Capponi, M., De Castro, S., Marconi, U., Moroni, C., Piccinini, M., Semprini-Cesari, N., Trombini, A., Vitale, A., Zoccoli, A.; Jones, S.E., Czirr, J.B., Jensen, G.L. and Palmer, E.P. "Experimental evidence of cold nuclear fusion in a measurement under the Grand Sasso massif", to appear in *Il Nuovo Cimento*

[5] De Ninno, A., Frattolillo, A., Lollobattista, G., Martinis, L., Martone, M., Mori, L., Podda, S., Scaramuzzi, F. "Neutron Emission from a Titanium-Deuterium System"

[6] Wolf, K.L., Packham, N., Shoemaker, J., Cheng, F., Lawson, D. "Neutron Emission and the Tritium Content Associated with Deuterium Loaded Palladium and Tritium Metals", presented at May 23-25 1989 cold fusion meeting in Santa Fe, New Mexico.

[7] Petrasso, R.D., Chen, X., Wenzel, K.W., Parker, R.R., Li, C.K. and Fiore, C. *Nature* **339**,183-185 (1989); and **339**,667-669 (1989)

[8] Huggins, R.A., Maly-Schreiber, M., Crouch-Baker, S., Gür, T.M., *et al*, presented at Workshop on Cold Fusion, May 23, 1989, Santa Fe, New Mexico.

[9] Srinivasan, S., Kim, Y.J., Murphy, O.J., Martin, C.R., Appleby, A.J. "Evidence for Excess Heat Generation Rates during Electrolysis of D_2O in LiOD Using a Palladium Cathode - A Microcalorimetric Study", presented at May 23-25 1989 cold fusion meeting in Santa Fe, New Mexico.

[10] Panneth, F., Peters, K. "Über die Verwandlung von Wasserstoff in Helium", *Die Naturwissenschaften* **14**, 936 (1926);
Later withdrawn in *Nature* **119**, 706 (1927)

[11] Cecil, F.E., Ferg, D., Furtak, T.E., Mader, C., McNiel, J.A. and Williamson, D.L. "Study of energetic charged particles from thin deuterated palladium foils subject to high current densities", presented at May 23-25 1989 cold fusion meeting in Santa Fe, New Mexico.

[12] Several parametric studies of the fusion rates in hydrogen isotopes have been performed since the announcement of cold fusion. Among them are:
Rafelski, J., Gajda, M., Harley, D. and Jones, S.E. "Limits on Cold Fusion in Condensed Matter: a Parametric Study" Arizona Preprint AZPH-TH/89-19.2 and "Theoretical Limits on Cold Fusion in Condensed Matter" Arizona Preprint AZPH-TH/89-19.1;
Szalewicz, K., Morgan III, J.D., Monkhorst, H.J. "Fusion Rates for Hydrogen Isotopic Molecules of Relevance for Cold Fusion", Preprint
Koonin, S.E. and Nauenberg, M. "Cold fusion in hydrogen molecules", Submitted to *Nature* April 7,1989
An earlier study of the fusion rate in dd molecules as a function of temperature is:
Van Siclen and C., Jones, S.E. "Piezonuclear fusion in isotopic hydrogen molecules", *J. Phys. G: Nucl. Phys.* **12**, 213-221 (1986)

[13] Bailey, D.C. "Gammas from Cold Nuclear Fusion", submitted to *Phys. Rev. C*

[14] Gai, M., Rugari, S.L., France, R.H., Lund, B.J., Zhao, Z., Davenport, A.J., Isaacs, H.S., Lynn, K.G. "Upper limits on emission rates of neutrons and gamma-rays from 'cold fusion' in deuterated metals", Yale-3074-1025, *Nature* **340** 29-34 (1989)

[15] Tandberg, J., Swedish patent application for "A method to produce helium and useful reaction energy" (1927).

[16] A number of groups had recognised in early April the important role of the damping of the long range of the Coulomb potential. Among the papers we have received are:
Tajima, T, Iyetomi, H. and Ichimaru, S. "Influence of Attractive Interaction between Deuterons in Pd on Nuclear Fusion" (April 14, 1989);
Feng, Shechao "Enhancement of Cold Fusion Rate by Electron Polarisation in Palladium Deuterium Solid" (ca mid April 1989);
Horowitz, Ch.J. "Cold Nuclear Fusion in Metallic Hydrogen and Normal Metals"(ca mid April, 1989);
Henis, Z., Elieser, S. and Ziegler, A. "Cold Nuclear Fusion Rates in Condensed Matter: a Phenomenological Analysis" (ca mid April, 1989);
Dharma-wardana, M.W.C. and Aers, G.C. "Theoretical estimates of the enhancement of cold fusion of deuterium in deuterated Palladium systems" (April 24, 1989);
Christensen, O.B., Ditlevsen, P.D., Jacobsen, K.W., Stolze, P., Nielsen, O.H. and Norskov, J.K. "H-H interactions in Pd" (April 27, 1989);
Baldo,M. and Pucci, R. "Plasmons Enhance Fusion" (May 1989);
Turko, L "Easier Way to pass the Coulomb Barrier" (May 1989);

[17] Burrows, A. "A Speculation on Cold Fusion in Metal "Hydrides" ", to appear in *Phys. Rev. B*

[18] Leggett, A.J., Baym G., "A rigorous upper bound on barrier penetration probabilities in many-body systems; application to "cold fusion" ", Submitted to *Phys. Rev. Lett.* May 3, 1989

[19] The necessity for a stoichiometric ratio of the order of one is not an experimental observation, but a speculation that was extensively discussed at the May 23-25 1989 cold fusion meeting in Santa Fe, New Mexico, as being a possible parameter distinguishing the successful from the unsuccessful experiments.

[20] Richards, P., "Molecular Dynamics Simulation of $PdD_{1.1}$: How close can Deuterons get?", presented at Workshop on Cold Fusion, May 23, 1989, Santa Fe, New Mexico.

[21] Harley, D., Müller, B., Rafelski, J. "MuCF with Z¿1 nuclei", Aip Conf. Proc. *Muon-Catalyzed Fusion* (1988) **181**,239

[22] There are numerous text-book references to the phenomenon of electron screening. Two of them are:
Ziman, J.M. "Introduction to solid-state physics"
March, N.H., Young, W.H. and Sampanthar, S. "The Many-body Problem in Quantum Mechanics", Section 5.2 (Cambridge University Press)

[23] Kondo, J. "Cold fusion in metals", Submitted to *J. Phys. Soc. Japan*

[24] Klyuev, V.A., Lipson, A.G., Toporov, Yu.P., Aliev, A.D. and Chalykh, A.E., "Characteristics of X-ray radiation observed in the fracture of solids and the breaking of adhesional bonds in a vacuum", *Kolloidnyi Zhurnal* **49**(5) 1001

[25] Klyuev, V.A., Lipson, A.G., Toporov, Yu.P., Deryagin, B.V., Lushchikov, A.V. and Shabalin, E.P. "High energy processes accompanying the fracture of solids", *Sov. Tech. Phys. Lett.* **12**(11) 861
Deryagin, B.V., Klyuev, V.A., Lipson, A.G. and Toporov Yu.P. "Possibility of nuclear reactions during the fracture of solids" *Kolloidnyi Zhurnal* **48**(1) 12

[26] Ziegler, J.F., Biersack, J.P. and Littmark, U. "The stopping and ranges of ions in solids", Pergamon Press (1985)

[27] Klyuev, V.A., Lipson, A.G., Toporov, Yu.P., Aliev, A.D., Chalykh, A.E. and Deryagin, B.V., *Dokl. Akad. Nauk SSSR* **279**(2) 415 (1984)

COULOMB ASSISTED COLD FUSION

Michael Danos

Center for Radiation Research
National Institute of Standards and Technology
Gaithersburg, MD 20899

INTRODUCTION

By now several experiments have been reported which suggest that light-element fusion occurs in solids having the capability of carrying hydrogen. The present lecture contains a formulation for the computation of the fusion rates, and an estimate of the order of magnitude of the rates which could be expected in semi-heavy water (HDO) and in hydrides.

Consider that a hydrogen, denoted by m_2 (say, a proton), is contained in a trap, next to a lattice nucleus, M, denoted by m_1. If a second hydrogen, m_3 (say, a deuteron), drifts past this site, as it reaches a position such that m_1, m_2, and m_3 are on a line with the Coulomb repulsion can impart a momentum to m_2 "propelling" it towards m_3 and at the same time transferring the recoil $\vec{q}$ and the energy E to m_1, keeping it on the mass shell. Hence m_2 now is off the mass shell. As m_2 reaches m_3, fusion takes place yielding the nucleus m_4 (here ^{3}He), and all particles are restored to be on the mass shell. The momentum space Feynman graph is shown in Fig. 1. Writing ($\hbar = c = 1$)

$$N(t) \sim \psi_0(t) \frac{1}{{t_1}^2 - {m_1}^2} \frac{1}{{t_2}^2 - {m_2}^2} \tag{1}$$

$$M \sim e^2 F(q) f(q) \psi(p) \frac{1}{q^2} \frac{1}{p^2 - {m_2}^2} \tag{2}$$

the matrix element of the graph is

$$R \simeq M \int dt\, N(t) \quad . \tag{3}$$

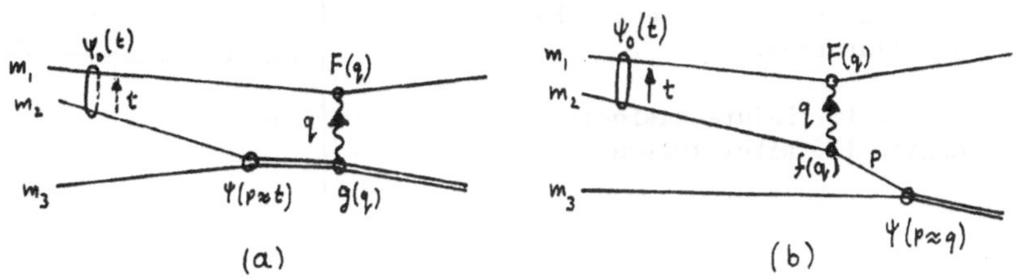

Fig. 1. Feynman graphs for Coulomb assisted fusion. m_1: lattice nucleus; m_2,m_3: fusing nuclei.

The Nuclear Equation of State, Part A
Edited by W. Greiner and H. Stöcker
Plenum Press, New York

In (1) $\psi_0(t)$ is the trapping wave functions; very little is known about it except that the momenta $|t|$ are "small," appropriate to lattice physics, i.e., of the order d^{-1} where d is the lattice spacing. F(q) is the Coulomb vertex function of m_1, i.e., of the nucleus M, f(q) that of m_2 (the proton), and $\psi(p)$ is the nuclear fusion vertex function. Of these quantities F(q) is known from electron scattering; f(q) and $\psi(q)$, being half off-the-mass-shell, must be estimated from on-the-mass-shell experiments.

Even though here a time-independent treatment most probably is not practicable, it nevertheless is useful to take a look at the qualitative features of such a description as an aid in achieving an understanding of the process. Thus, being a stationary state description, the system would have standing waves in all channels. In other words, the time-independent treatment is a superposition of the reaction and its time-reverse. In order to give an illustration of the process, and to approach the description of the graphs of Fig. 1, we replace the full-fledged three-body Coulomb problem by a two-stage description, in the first stage dropping the Coulomb interaction between m_1 and the other particles, and regarding its effect in the second stage. Before fusion we have a three-body system, after the fusion a two-body system. This final state system is kinematically fully determined. Thus, taking A_1 to be ~ 100, we have the following kinematic conditions for the recoil to the lattice nucleus: for the p+d → ^3He reaction: $p_1' \approx 1\ fm^{-1}$, $E_1' \approx 0.16$ MeV; for d+d → ^4He: $p_1' \approx 2\ fm^{-1}$, $E_1' \approx 0.95$ MeV. Thus $t/p_1' \sim 10^{-5}$ and can be neglected. The energetics of the reaction is schematically illustrated in Fig. 2, where the (9-dimensional) configuration space is symbolically split into region I containing the 3-body system and region II containing the 2-body system. (Region I excludes the space $r_{dp} \lesssim a \simeq 5$ fm where nuclear forces become important). Three of the infinite number of configurations are illustrated. Figure 2a shows the distribution of energies corresponding to the initial state of Fig. 1. Figure 2b shows the energy distribution of the configuration corresponding to the intermediate state of Fig. 1b, i.e., the state of the system after exchange of the "Coulomb photon" q, but before fusion. Thus particle 1, the Pd nucleus, has its final energy and momentum while 2 carries the momentum of the final state of the fused (^3He) system. Particle 3 is still in its initial state. Figure 2c shows another of the possible configurations, where particles 1 and 3 have undergone a further energy-momentum exchange. This state thus requires the exchange of 2 "Coulomb photons" for it to be reached from the "initial" configuration, Fig. 1a.

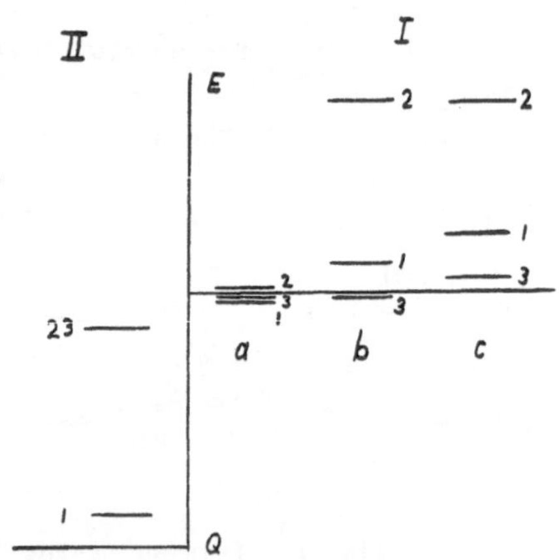

Fig. 2. Energies for different configurations: a. initial state; b. final state; c. some other configuration.

Region I: before fusion;
Region II: after fusion.

We now look at the qualitative behavior of the relative wave function of the m_2-m_3 subsystem. Consider first the configuration Fig. 2a. There the relative wave function is essentially exponentially damped owing to the Coulomb barrier (Fig. 4a). This wave function usually is computed by the WKB method. However, in contrast to the WKB wave function, at $r_{23} \sim a$ the solution requires the presence of the irregular Coulomb solution in order to allow the fulfilling of the matching conditions at the nuclear surface $r_{23} = a$. This admixture can for certain reconance conditions change the barrier penetrability by many orders of magnitude from the WKB value.

On the other hand, the wave function which corresponds to the configuration of Fig. 2b, being above the Coulomb barrier, remains oscillatory down to $r_{23} \approx a$. A very small admixture resulting from the action of V_{12}, V_{13}, will thus "circumvent" the influence of the Coulomb barrier. This is precisely the content of the graph Fig. 1b. In the time-independent Schrödinger picture the admixture of this configuration thus will be proportional to

$$A \sim \frac{\langle a|(V_{12} + V_{13})|b\rangle\langle b|(V_{12} + V_{13})|a\rangle}{\varepsilon_b - E}$$

which may be a small, but not "exponentially" small, number.

In order to estimate the fusion rate we must address the characteristics of the reaction. We observe that the wavelengths of the recoiling particles m_1 and m_2 are $\sim q^{-1} \sim 1$ fm, while the size, d, of the system is $\sim$ Å. Hence the motion of the particles is properly described by geometric (ray) optics rather than by wave optics. (Formally, this is related to the angular parts of the Fourier transformation.) This is the basis of the observation that m_1, m_2, and m_3 must be aligned. Only then can the Coulomb force, which has the direction given by $\vec{r}_2 - \vec{r}_1$, impart a momentum $\vec{q}$ (which thus also has the direction $\vec{r}_2 - \vec{r}_1$) which can carry m_2 towards m_3 resulting in fusion; m_3 must lie in the "shadow" of m_2. For uncorrelated positions this leads to a reduction factor

$$\eta_0 \sim (qd)^{-2} \approx 10^{-8}$$

Owing to the presence of correlations the actual reduction factor for fusion in such crystals thus can be very much smaller than η_0. For fusion to occur, therefore a special class of crystal imperfections is needed which would provide traps for hydrogen such that fusion alignment might occur. Simply using geometrical reasoning one finds that η in some cases could be as large as 10^{-4}. On the other hand, for example in HDO the alignment will be suppressed even further; here η may be as small as 10^{-14}.

Turning now to (3) we note that if the propagators in (1) were absent, the integral in (3) would yield $\psi(r_{12} = 0)$, i.e., it would essentially vanish. This, of course, is not the case. However one sees that the actual result depends rather sensitively on the (unknown) details of $\psi_0(t)$ in a rather involved manner since the propogators in position space are integral operators.

To estimate the reaction rates note that the central mechanism of Fig. is the (off-the-mass-shell) Coulomb scattering of m_2 on the much heavier m_1. The other parts of the graph yield factors which decrease the reaction rate differing amounts. Dropping numerical factors like P we estimate for the ca of pd fusion in a palladium lattice: $(F(q))^2 \sim 10^{-2}$, $(f(q))^2 \sim 1$, $[\psi(p)/(p^2-m^2)] \approx 10^{-2}$, $[\int dt\ N()t)]^2 \sim 10^{-5} \times 10^{\pm 3}$, and $\rho v \sim 10^{-18}\ fm^{-3}$. Wit the above estimate for the angular factor 10^{-4} we obtain

$$\text{Rate(pd, lattice)} \approx 10^{-12} \times 10^{\pm 3}\ (\text{trap sec})^{-1} \quad .$$

For dd fusion we estimate

$$(F(q))^2 \sim 10^{-4},\ (f(q))^2 \sim 10^{-3},\ \text{and}\ [\psi(p)/(p^2-m^2)]^2 \sim 10^{-4}.$$

Thus

$$\text{Rate(dd, lattice)} \sim 10^{-17} \times 10^{\pm 3}\ (\text{trap sec})^{-1} \quad .$$

To conclude, we give an estimate for heavy water. The main difference resides in the above mentioned angular suppression factor. Hence the HDO fusion rate is estimated to be around $10^{-24} \times 10^{\pm 3}$ (molecule sec)$^{-1}$, while i D_2O the rate may be a factor 10^5 smaller.

COLD CONFUSION

George Chapline

Lawrence Livermore National Laboratory
P.O. Box 808
Livermore, CA 94550

INTRODUCTION

On March 23 two chemists, Martin Fleischmann and Stanley Pons startled the world with a press conference at the University of Utah where they announced that they had achieved nuclear fusion at room temperatures. As evidence they cited the production of "excess" amounts of heat in an electrochemical apparatus and observation of neutron production. While the production of heat in a chemical apparatus is not in itself unusual the observation of neutrons is certainly extraordinary. As it turned out, though, careful measurements of the neutron production in electrochemical apparatus similar to that used by Fleischmann and Pons carried out at dozens of other laboratories has shown that the neutron production fails by many orders of magnitude to support the assertion by Fleischmann and Pons that their discovery represents a new and cheap source of fusion power. In particular, independent measurements of the neutron production rate suggest that the actual rate of fusion energy production probably does not exceed 1 trillionth of a watt.

Actually Fleischmann and Pons acknowledged in even their first press conference that there was a serious discrepancy between the observed number of neutrons and their claim that they were producing useful amounts of fusion energy. They explained this discrepancy by suggesting that the excess heat they were observing was due to a new kind of fusion reaction. However, to date Fleischmann and Pons have presented absolutely no evidence that any of the heat being produced is due to fusion reactions.

In fact, apart from the entertainment value of various remarks by Fleischmann and Pons it does not appear that their work has produced much of value. It is clear that they, and apparently also the group at Brigham Young University led by Steve Jones, deluded themselves into believing they had discovered a way of producing fusion reactions by compressing deuterium inside the electrodes of an electrochemical cell.

Historically the possibility of producing fusion reactions by compressing cold hydrogen has been of interest to astrophysicists for some time. In 1959 Alistar Cameron named such fusion reactions pycnonuclear reactions.[1] The rate of such reactions depends sensitively on electron screening; however, by 1969 the effect of electron screening was well enough understood,[2] so that one could make fairly accurate estimates of how the fusion rate depends on compression. The results of these calculations show that even if one could produce pressures as high as 10 million atmospheres, i.e., on the order of the central pressure in Jupiter, the rate of pycnonuclear fusion would still be much too low to explain the "fusion" neutrons reported by Steve Jones and his collaborators.[3] This theoretical remark is bolstered by the observation that the heat radiated by Jupiter can be entirely explained as the heat generated by gravitational contraction[4] - leaving no room for fusion energy production at the level reported by Jones.

As for the possibility of using pycnonuclear fusion to produce energy here on earth one can prove the following:

Theorem: It is impossible to produce energy by compressing small amounts of deuterium.

The point is that to get a useful fusion rate one would first of all have to produce a much higher pressure than could be contained by any real material. One could imagine transiently producing very high densities in deuterium, e.g., using explosives to produce a spherical implosion, but then the question arises whether the fusion energy gain exceeds the energy used to compress the deuterium. Using reasonable estimates of the energies involved one finds that the compressional energy always exceeds the fusion energy (see Appendix).

10^{23} ATMOSPHERES

Of course, if one had some magical way of producing extremely dense deuterium in a laboratory apparatus then cold fusion might be possible. In their appearance before the U.S. House of Representatives Committee on Science and Technology Fleischmann and Pons claimed that they were producing an effective pressure in their electrochemical apparatus of 10^{23} atmospheres! They based their claim on the well known formula for the free energy per particle of an ideal gas

$$\mu = kT \ln(\lambda^3 p/kT) \tag{1}$$

where λ is the thermal DeBroglie wavelength and p is the pressure. For the left side of this equation Fleischmann and Pons apparently took μ equal to 0.8 eV, the chemical potential of D^+ ions in the palladium electrode, even though the physical origin of this free energy is obscure. With T equal to 300° Eq. (1) yields something like 10^{23} atm (I don't get exactly 10^{23} atm). Needless to say, this is an absurd conclusion. In his presentation Professor Pons did not explain why an activity of unknown origin should be identified with the entropy of an ideal gas. In any case hydrogen does not behave like an ideal gas at high pressures, and it is easy to check that a free energy per atom of 0.8 eV corresponds to a pressure on the order of 10^3 atmospheres - a much more reasonable number.

The next question is whether a few thousand atmospheres would be sufficient to give an observable pycnonuclear fusion rate.

In pure deuterium at a pressure of a few thousand atmospheres, the deuterium atoms are still bound in molecules, so the smallest separation between deuterons is essentially the same as it is in isolated molecules; i.e., about 0.7 Å. The fusion rate in an isolated deuterium molecule has recently been calculated by Steven Koonin and Michael Nauenberg,[5] using the by now very accurately known wavefunctions for a hydrogen molecule. Their result is 10^{-64} per second. To visualize the magnitude of this result image a mass of molecular deuterium equal to the mass of the sun. The calculated rate corresponds to one cold fusion event per year!

Of course, the hope for palladium mediated cold fusion is that the deuterons in saturated palladium deuteride are somehow closer together than they are in a deuterium molecule. let's think about that for a moment. If solid hydrogen is compressed with a pressure greater than about 3 million atmospheres the molecules of hydrogen in the solid dissolve[6] and hydrogen (or deuterium) exists in a metallic form where the electrons are no longer localized in molecules, but instead form a conduction band. In first approximation the electrons form a uniform background in which the bare ions move. It is amusing that something similar apparently happens in saturated palladium hydride. In pure palladium the palladium atoms have a $(4d)^8(5s)^2$ configuration, where the 5s electrons form a conduction band. However, in saturated palladium hydride, i.e., Pd H_x where $x \geq 0.6$, the 5s electrons recombine with the palladium ions to form neutral atoms with a $(4d)^{10}$ configuration.[7] The electrons in the s-wave conduction band in the saturated hydride are contributed by the hydrogen atoms. Thus saturated palladium hydride can be thought of as "metallic hydrogen" which is stablized by the palladium atoms. The important point for us is that in first approximation the probability of cold fusion in saturated palladium deuteride will be no different than that in metallic deuterium, which as noted in the introduction is a problem that has long been studied in astrophysics. Even without referring to the astrophysical literature though it is easy to see that the pycnonuclear fusion rate in palladium deuteride will be negligible.

The lattice spacing in palladium deuteride is about 2 Å. Therefore even for a stochiometric composition Pd D_2 the average separation at the deuterons is $2^{-1/3} \cdot 2$ Å, which is more than twice the spacing in a deuterium molecule. Because of the larger deuteron separation and <u>smaller</u> electron screening in the palladium hydride the cold fusion rate in palladium deuteride should be much smaller than it is molecular deuterium, where according to the calculation of Koonin and Nauenberg it is already much too small to be detectable.

It is perhaps worth noting[6] that in the center of Jupiter the interhydrogen separation is about the same as in a deuterium molecule. As mentioned in the introduction we have direct evidence in the case of Jupiter that the fusion rate is smaller than that claimed to have been observed by Steven Jones. Of course, one must take into account that only a small fraction of the hydrogen in Jupiter is deuterium. However, one of the interesting features of cold pycnonuclear fusion is that proton-deuteron fusion should be much faster than deuteron-deuteron fusion.[5]

CAN ONE LOSE WEIGHT BY EXERCISE?

Even if one accepts that the neutrons being reported by various laboratories are not due to pycnonuclear reactions what about Fleischmann's and Pons' claims that some of the heat being produced in their apparatus cannot be accounted for by chemical reactions? A notable statement in this connection is the claim by Professor Fleischmann that (as I recall goes something like) "it is inconceivable that one could store 4 MJ cm^{-3} in the electrodes of the apparatus by chemical means." Indeed, if it were literally true that 4 MJ cm^{-3} were being stored in the electrodes of their apparatus, then that would be difficult to explain. On the other hand, an energy like 4 MJ (10 watts for 120 hours) is not totally beyond the realm of chemistry. For example, the calories in the food one eats during a single day normally exceeds 4 MJ. Although it may be rather arrogant for a physicist to assert that two professors of chemistry don't know how to do chemistry, the assertion of Professor Fleischmann that 0.8 MJ per day cannot be accounted for does invite scepticism.

I'm sure that many if not most of you have had some experience with attempting to lose weight by exercise. Vigorous exercise for an hour or more will dissipate at least a megajoule. In addition, one may vary one's diet. For example, instead of a knockwurst in the morning one might eat an orange. Unfortunately, too often the result of such an experiment is that one doesn't lose weight. One may even gain weight. I suppose that under these circumstances Professor Fleischmann would conclude that when orange juice is in your guts nuclear fusion reactions are occurring.

What this illustrates, of course, is that the law of conservation of energy can be experimentally elusive.

Incidently, in 1832 a German chemist, Johan Dobereiner, discovered that palladium will spontaneously catalyze the oxidation of hydrogen. This discovery led to an invention, the Feuerzeug, which has been commercially marketed in Germany as a cigarette lighter. The working fluid in a Feuerzeug is called Columbian Spirits. I haven't been able to ascertain the chemical composition of Columbian Spirits; however, cold fusion enthusiasts may want to delve into this as this liquid may work better than Canadian heavy water.

CONCLUSION

Despite the attention Fleischmann and Pons have drawn to their work they have produced no credible evidence that a portion of the heat being produced in their apparatus is due to nuclear fusion reactions. This is all the more remarkable because they could have easily produced credible evidence if nuclear fusion reactions were really occurring.

For example, they could have submitted the electrodes of their apparatus for independent analysis of their helium content. Their failure to do so has given new meaning to the phase "reputable chemist."

What about the possibility that under some circumstances palladium or titanium saturated with deuterium will emit neutrons? If these neutrons are real, their origin may or may not be interesting. It should be kept in mind though that a few neutrons don't represent an immediate and cheap source of electrical power. The most likely resting place for the discovery of neutrons coming from an electrochemical apparatus is in a footnote in a future history of scientific hoaxes.

Work performed under the auspices of the U.S. Department of Energy by the Lawrence Livermore National Laboratory under contract number W-7405-ENG-48.

APPENDIX

Assuming no losses the energy required to compress deuterium to high densities is 0.2 $\rho^{2/3}$ MJ/gm, where ρ is the density in units of gm cm^{-3}. The theoretical fusion energy yield for deuterium is about 10^5 MJ/gm. However, in practice the yield will be substantially less because of depletion and hydrodynamic expansion of the compressed deuterium. A more practical number would be about 10^4 MJ/gm. Thus the maximum density allowed is about 10^7 gm cm^{-3}.

Using the formula [2]

$$\text{Rate} = 9 \cdot 10^{45} \rho \lambda^{7/4} \exp(-2.638\ \lambda^{-1/2})\ cm^{-3}\ s^{-1} \qquad (A)$$

where $\lambda = 7.7 \cdot 10^{-5}\ \rho^{1/3}$, one finds that at a density of 10^7 gm cm^{-3} the cold fusion rate is 10^{40} cm^{-3} s^{-1}. This means that at a density of 10^7 gm cm^{-3} the deuterium will be consumed in $3 \cdot 10^{-10}$ sec. Sounds good? Unfortunately though the speed of sound in deuterium at a density of 10^7 gm cm^{-3} is about 200 times that in normal metals, so that the compressed sample of deuterium would have to be at least 1 mm in radius. In other words in order to achieve fusion energy breakeven we would have to compress more than 40 kg of deuterium to a density 10^8 times normal solid density. Needless to say this is not very practical, and examination of formula (A) reveals that as the density is lowered the amount of deuterium required for fusion breakeven increases very rapidly.

REFERENCES

1. A. G. W. Cameron, Ap. J. 130, 916 (1959).
2. E. E. Salpeter and H. Van Horn, Ap. J. 155, 183 (1969).
3. S. E. Jones, et al., Nature 338, 737 (1989).
4. A. Grossman, et al., Icarus 42, 358 (1980).
5. S. Koonin and M. Nauenberg, ITP preprint (Santa Barbara, 1989).
6. G. Chapline, Phys. Rev. B6, 2067 (1972).
7. M. S. Daw and M. I. Baskes, Phys. Rev. B29, 6443 (1984).

THE NUCLEAR MATTER SATURATION PROBLEM

J M Irvine

Department of Physics
The University
Manchester, M13 9PL, UK

INTRODUCTION

The general systematics of nuclear binding energies are well parameterised by the semi empirical mass formula. The volume term in this formula predicts that, in absence of coulomb interactions, an infinite, homogeneous fluid of equal numbers of neutrons and protons will be self bound with a binding energy per nucleon of $E_0 \simeq 16\pm0.5$ MeV. Measurements of the central densities of heavy nuclei suggest that the infinite nuclear matter fluid will saturate at a density $\rho_0 \simeq 0.17\pm0.01$ nucleons per fm^3 corresponding to a Fermi wave number $k_F \simeq 1.37\pm0.03\ fm^{-1}$. While it is difficult to unambiguously separate bulk and surface effects, the observation of 'breathing mode' excitations at 12-25 MeV excitation energy in nuclei suggests a bulk compressibility modulus for the nuclear matter fluid

$$K = 9\rho^2 \left.\frac{\partial^2 E}{\partial \rho^2}\right\} \simeq 250 \pm 50 \text{ MeV} \qquad (1)$$

Recently it has been suggested that the dynamics of supernova collapse imply a much smaller value of the compressibility with $K \lesssim 150$ MeV.

Attempts by nuclear theorists to calculate these basic saturation parameters and resolve the compressibility question will be reviewed in this short communication.

BRUECKNER AND THE COESTER LINE

The simplest model for nuclear matter is a fluid of non relativistic neutrons and protons interacting through a two-body potential. The hamiltonian is then

$$H = \sum_i - \frac{\hbar^2}{2m} \nabla_i^2 - \sum_{i>j} V_{ij} \tag{2}$$

The two-body potential must fit the nucleon-nucleon scattering phase shifts and the bound state deuteron properties. There are a number of such 'phase equivalent' potentials on the market. They are either completely phenomenological in nature or guided to a greater or lesser extent by meson exchange models. In general these potentials have a long range one-pion-exchange tail, an intermediate range at which they are attractive, and which is frequently attributed to an enhanced two-pion-exchange, and a short range repulsive core, which may be attributed to vector ρ- and ω- meson exchange. There is clear evidence for the non central nature of the interaction and the observed quadrupole moment of the deuteron would appear to require a tensor force, although there is continuing debate as to the strength of the tensor term.

Whatever the details of the interaction, the strength of the repulsive core alone precludes a direct application of the Hartree-Fock prescription for handling many fermion systems. In the Hartree-Fock prescription a trial wave function in the form of a Slater determinant

$$\Phi = (A!)^{-\frac{1}{2}} \det\phi_i(\underline{x}_j) \tag{3}$$

is chosen and the expectation value of the hamiltonian (2) is optimised with respect to the single-particle basis ϕ_i. This leads to the equations

$$-\frac{\hbar^2}{2m} \nabla_1^2 \phi_i(\underline{x}_1) - \sum_j \int \phi_j^*(\underline{x}_2)\, V(\underline{x}_1,\underline{x}_2)\, [\phi_i(\underline{x}_1)\phi_j(\underline{x}_2) - \phi_i(\underline{x}_2)\phi_j(\underline{x}_1)] d\underline{x}_2$$

$$= \varepsilon_i\, \phi_i\, (\underline{x}_1) \tag{4}$$

for the single-particle basis, and for the ground-state energy expectation value

$$E_0 = \frac{1}{2} \sum_i \int \phi_i^*(\underline{x}) \left[-\frac{\hbar^2}{2m} \nabla^2 - \varepsilon_i \right] \phi_i(\underline{x})\, d\underline{x} \tag{5}$$

The trial wave function (3) describes a system of particles whose motion is correlated with that of its neighbours only through a mean field. In the nuclear matter problem the strong short-range forces induce two-body correlations which cannot be ignored.

Keith Brueckner attempted to account for those correlations by allowing pairs of particles to undergo multiple scatterings and summing the so called 'ladder diagrams', thus generating a reaction matrix.

$$\langle ij|G|ij\rangle = \langle ij|V|ij\rangle + \sum_{ab}' \frac{|\langle ij|V|ab\rangle|^2}{E_{abij}} \qquad (6)$$

The prime on the summation in eqtn (6) denotes that the intermediate states ϕ_a, ϕ_b must honour the Pauli principle and be unoccupied in the uncorrelated ground state Φ of eqtn (3). The energy denominators of eqtn (6) can be written

$$E_{abij} = \varepsilon_a + \varepsilon_b - \varepsilon_i - \varepsilon_j \qquad (7)$$

The basis of single-particle states occupied in Φ is now given by

$$-\frac{\hbar^2}{2m}\nabla_1^2\phi_i(\underline{x}_1) + \sum_j \int \phi_j^*(\underline{x}_2)\, G(\underline{x}_1,\underline{x}_2)\, [\phi_i(\underline{x}_1)\phi_j(\underline{x}_2) - \phi_i(\underline{x}_2)\phi_j(\underline{x}_1)]d\underline{x}_2$$

$$= \varepsilon_i\, \phi_i\, (\underline{x}_1) \qquad (8)$$

and the lowest order expression for the ground-state energy expectation value is

$$E_0^{(1)} = \tfrac{1}{2} \sum_i \int \phi_i^*\, (\underline{x})\, [\, -\frac{\hbar^2}{2m}\nabla^2 + \varepsilon_i]\, \phi_i(\underline{x})d\underline{x} \qquad (9)$$

Because of the structural similarity of eqtns (8) and (9) to eqtns (4) and (5), this approach is called the Brueckner-Hartree-Fock prescription (BHF). It must be remembered, however, that the true Hartree-Fock prescription is based upon the variational principle while the BHF prescription simply represents the lowest order term in a partially summed perturbation theory expansion. The discerning reader will have noted that I have not fully defined the BHF approach because I have not defined the basis of single-particle states unpopulated in Φ which are required in the summation of eqtn (6). The only safe choice is to use a system of plane waves orthogonalised to the states occupied in Φ and to calculate the single particle energies ε_a and ε_b as expectation values of the kinetic energy operator. A justification of this approach may be that the interaction V is so strong that it scatters particles so far out of the Fermi sea that their potential energy is negligible compared with their kinetic energy.

Going beyond lowest order perturbation theory it is conventional to re-arrange the series so that it is no longer in order of ascending powers of the reaction matrix but rather in ascending order of the number of vacancies in the Fermi sea - the so called 'hole-line' expansion. For a central, repulsive hard-core potential of radius r_c the hole-line expansion can be shown to converge with powers of the small parameter

$$\kappa = ({}^{r}c/r_0)^3 \tag{10}$$

where the density of the nuclear fluid is

$$\rho = [{}^4/_3\, \pi r_o{}^3]^{-1} \tag{11}$$

At the nuclear saturation density ρ_o we have ${}^{r}c/r_o \simeq 1/3$ and convergence of the hole-line expansion would seem assured.

The results of BHF calculations for a wide range of phase equivalent potentials are summarised in Figure 1.

The results for a wide range of phase equivalent potentials scatter about the line AB, known as the Coester line, and most certainly do not account for the semi empirical mass formula data represented by the black triangle. It is possible to choose a nucleon-nucleon potential which will saturate at the correct density but this fails to give sufficient binding energy. It is possible to choose a potential which will provide sufficient binding energy, but then it will saturate at too high a density.

A closer examination of the results of BHF calculations shows that it is the tensor force which plays an important role in producing saturation. Thus, it is no surprise to learn that those bare nucleon-nucleon interactions which have a strong tensor force saturate first and yield results at the upper end of the Coester line while the results with interactions containing weak tensor forces cluster towards the lower end B.

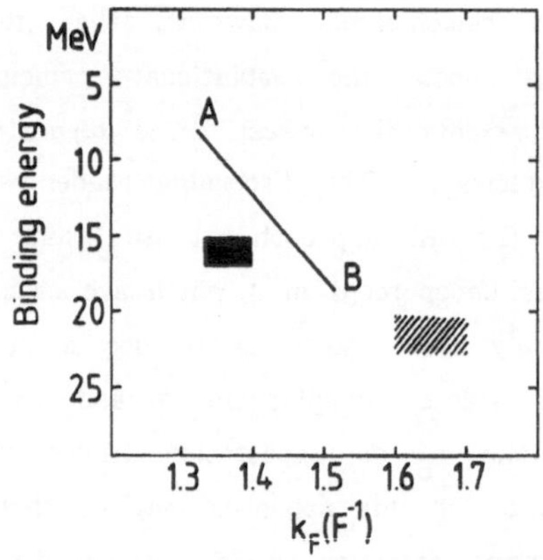

Fig 1 The saturation of nuclear matter. The black rectangle represents the predictions of the semi empirical mass formula. The line AB is known as the Coester line.

JASTROW CORRELATIONS: BRUECKNER JUSTIFIED?

Given the complexity of a fully self-consistent BHF calculation, particularly if there is need to go beyond lowest order as might be required at supernuclear densities, eg, in order to describe the core of a neutron star, it is attractive to explore alternative calculational procedures.

One possibility would be to stay with the variational principle and improve the choice of trial wave functions. Instead of the Slater determinant (3) consider

$$\Psi = F\Phi \tag{12}$$

where F is a symmetric product of two-body correlation functions

$$F = \prod_{i>j} f_{ij} \tag{13}$$

designed to accommodate the effect of particularly strong components of our nucleon-nucleon interaction. In the Jastrow approximation the f_{ij} are central functions of the separation between the pair of particles which reduce the two-body wave function at short distances, where the repulsive core forces the particles apart, and which heal to unity at large distances where the interactions are extremely weak.

The full wave function (12) is extremely difficult to work with and it is usual to make some form of cluster expansion of F based upon the relatively low density of the nuclear matter problem as measured by the small parameter κ (10). The problem then is that, if the cluster expansion is terminated at any finite order and the correlation functions f_{ij} freely varied, the energy functional

$$E([F]) = \langle\Psi|H|\Psi\rangle_{Approx} \tag{14}$$

will not saturate, the so called Emery problem. The notation $\langle\ \rangle_{Approx}$ means that the expectation value in some limited cluster expansion approximation. This problem can be traced to the fact that, in searching for a lower bound to the energy in the variational calculation, one can increase the radius of the wound in the correlated two-body wave function to completely eliminate the repulsive core and at the same time reduce the rate of healing of the correlation function so that the kinetic energy is negligible. Then we can adjust the long range part of the correlation function to give unbounded weight to the attractive tail of the interaction. Of course, such long range correlations are inconsistent with a termination of the cluster expansion.

What is required in the variational calculations is a constraint on the correlation functions f_{ij} which will ensure the convergence of the cluster expansion. Then for a constraint

$$g\ ([f]) = 0 \tag{15}$$

the variational functional is

$$\delta \left\{ \langle\Psi|H|\Psi\rangle_{Approx} - \lambda g\ ([f]) \right\} = 0 \tag{16}$$

with λ an undetermined Lagrange multiplier. A natural choice for the constraint is that it should retain the convergence properties of the hole-line expansion. This can be achieved by requiring that on average the wound volume around any given nucleon contain at most one further nucleon, thus making the restriction to two-body clusters very plausible. Such calculations are generally called lowest order constrained variational calculations (LOCV).

When LOCV calculations are performed for a wide range of phase equivalent nucleon-nucleon interactions the results agree to an impressive extent with BHF calculations using the same potentials. The Coester line is reproduced and once more the saturation predictions disagree with the semi-empirical mass formula.

We note that both LOCV and BHF calculations not only fail to reproduce nuclear matter saturation but that they predict very high compressibilities, typically $K \simeq 400$ MeV which is inconsisent with the observation of breathing mode excitations at 12-15 MeV in nuclei.

TENSOR CORRELATIONS: THE MYSTERY, AND THE BINDING-ENERGY, DEEPENS

The full variational principle states that the results of some restricted variational calculation will be an upper bound to the true ground state energy and that the calculated ground state should approach the hamiltonian eigenstate as the restrictions on the scope of the variational calculation are removed. Within the class of LOCV calculations the simplest relaxation would be to free the form of the two-body correlation functions f_{ij}.

The form of our two-body potentials is

$$V_{ij} = \sum_{\lambda} V^{\lambda}{}_{ij} \tag{17}$$

where in different reaction channels λ we have the freedom to include central, tensor, spin-orbit and quadratic spin-orbit forms as appropriate. It seems natural to accord the correlation functions the same degrees of freedom and write

$$f_{ij} = \sum_{\lambda} f_{ij}^{\lambda} \tag{18}$$

with

$$f_{ij}^{\lambda} = f_c^{\lambda}(rij) + f_T^{\lambda}(rij)\, S_{ij} + f_{LS}^{\lambda}(r_{ij})\, \underline{L}.\underline{S} + f_Q^{\lambda}(r_{ij})\, (\underline{L}.\underline{S})^2 \tag{19}$$

For all singlet states there are no tensor or spin-orbit terms and thus λ is simply the partial wave L for the triplet states with J=L again λ is simply the partial wave but triplet states are coupled by the tensor force, eg $^3S_1$-$^3D_1$, $^3P_2$-$^3F_2$ etc, and in these cases λ refers to the two orthogonal reaction channels associated with the pairs of partial waves.

Since the new correlation functions are designed not simply to take into account the wound in the two-body wave functions produced by the repulsive core of the two-body interaction but also any other possible strong correlations induced by other elements of the force, the constraint used in the Jastrow cluster expansion no longer seems appropriate.

The single-particle energies ε_i in the original Hartree-Fock approach (4) are themselves Lagrange multipliers chosen to assure the orthonormality of the single particle functions ϕ_i and hence the normalisations of the trial many-body function Φ (3). In this we shall use as our constraint (15) the normalisation of two-body correlated wave functions.

$$\Psi_{ij} = f_{ij}\,(2!)^{-\frac{1}{2}} \det \phi_i\,(\underline{x}j). \tag{20}$$

Later we shall check the convergence of our cluster expansion and show that our termination in lowest order is valid. One consequence of our change in constraint is that because Ψ_{ij} contains a wound at short distances, then, in order to be normalised, f_{ij} must overshoot unity in a region where the interaction is attractive and hence we shall get a slightly increased binding energy.

It has been known for many years that in perturbation theory the tensor force converges relatively slowly. This was recognised by Kuo and Brown in their analysis of the effective central $^3S_1$ interaction in nuclei where they wrote

$$\langle ^3S_1 | V^{eff} | ^3S_1 \rangle = \langle ^3S_1 | V_c | ^3S_1 \rangle + \sum_n \frac{|\langle ^3S_1 | V_T\, S_{12} | n^3D_1 \rangle|^2}{\Delta E_n} \tag{21}$$

Using an approximate closure argument $\langle ^3S_1 | S_{12}^2 | ^3S_1 \rangle = 8$ they obtained the result

$$V^{eff}_{^3S_1} \simeq V_c(^3S_1) + 8V_T^2/E_{eff} \tag{22}$$

Kuo and Brown reasoned that E_{eff} was 200-300 MeV and that this second order tensor contribution yielded a considerable increase in the attractiveness of the $^3S_1$ effective interaction. By putting tensor components directly into our trial wavefunction we include these effects and many more. It transpires that it is the tensor components in the correlation functions (19) which are the really significant new feature and we shall not refer further to the spin-orbit components.

When the new LOCV calculations are carried out for the same range of phase equivalent potentials the results cluster in the cross hatched box in figure 1. Now the results for various phase equivalent potentials are very similar, instead of being spread along the Coester line they are all grouped together. This is not surprising, we would expect the effect of tensor correlations to be largest for those interactions with strong tensor forces, ie those which were at the end A of the Coester line, and least for those with the weakest tensor forces. At the same time the predicted compressibility is greatly reduced $K \leq$ 150 MeV.

Recently perturbation calculations based on the hole-line expansion have been extended to fourth and fifth order confirming the results of these generalised LOCV calculations. The bad news is that the results of the cross hatched area do not overlap the empirical black rectangle.

Perhaps the most alarming aspect of the results is that the variational calculations predict binding which is greater than the empirical value. This phenomena has happened at least once before in the history of physics. As variational calculations for atomic structure calculations were refined there came a time when the Schrödinger equation for the helium atom yielded a ground state energy below that which was experimentally observed. This violates the upper bound condition of the variational principle. The solution to this apparent paradox is that the model being used must incompletely describe the physics being investigated. In the case of the helium atom the breakdown was traced to the inadequacy of the non relativistic Schrödinger equation. In the next section we shall explore the missing element in the nuclear matter problem.

ISOBARS TO THE RESCUE

Let us review our model which consists of a fluid of non relativistic nucleons interacting through pair potentials. The potentials necessarily contain tensor forces and these we know can be generated by the exchange of pions. So at root our model is a soup of nucleons and pions. The π-nucleon scattering cross section is dominated by a broad resonance at ~ 300 MeV. This is variously referred to as the 3-3 resonance, the Δ particle or the $N^*(1232)$

isobar. Thus, it would seem natural to generalise our trial wave function to allow for the possibility that our nucleons may become excited into N* isobar states. This economically achieved by extending our correlation function (19)

$$f_{ij}^{\lambda} \rightarrow f_{ij}^{\lambda} = f_c^{\lambda}(r_{ij}) + f_T^{\lambda}(r_{ij})S_{ij} - f_*^{\lambda}(r_{ij})\, S_{ij}^{II} \qquad (23)$$

where S_{ij}^{II} is the generalised tensor operator connecting the two-component nucleon spinner to the four component spin 3/2 isobar. Such a term arises first in the coupling of the $^1S_0$(NN) channel to the $^5D_0$(NN*) channel. The analogy with the tensor correlations is now complete and is exhibited graphically in figure 2. Naturally our hamiltonian must now be extended to include the isobar channels and this is done using the N-N* potentials of Green, Sanio and Niskenen based on the π-N production of N* as analysed by Arenhovl.

The contribution of figure 2b to the N-N $^1S_0$ scattering amplitude is subsumed into the phenomenological potential fits and is responsible for a significant part of the attraction in this channel. Note that one-boson-exchange models of the N-N interaction which try to mimic the two-pion-exchange with a single fictitious σ-meson exchange miss the richness of this phenomena and thus dangerously hide real physics. When two nucleons interact in the nucleus the amplitude in figure 2b must be modified, the range of momenta allowed to the intermediate nucleon being restricted by the Pauli principle to lie outside the Fermi sea. Thus, we must subtract the contribution of this amplitude for all intermediate nucleon momenta $k<k_F$. This subtraction turns this attractive amplitude into a repulsion. Such an effect had been investigated many years earlier by Green and co workers who found it to be a small effect or order 1-2 MeV at the empirical saturation density. Starting from point A on the Coester line this further increased the discrepancy between BHF calculations and the semi empirical mass formula. Now we see that the repulsion increases rapidly as k_F increases with the results summarised in figure 3.

At last we have a curve that saturates at the empirical binding energy and density. An added bonus is that the curvature of the saturation curve now yields a compressibility modulus K $\simeq$ 200-300 MeV. Further studies of asymmetric fluids predict a symmetry term for the semi empirical mass formula of 24 MeV.

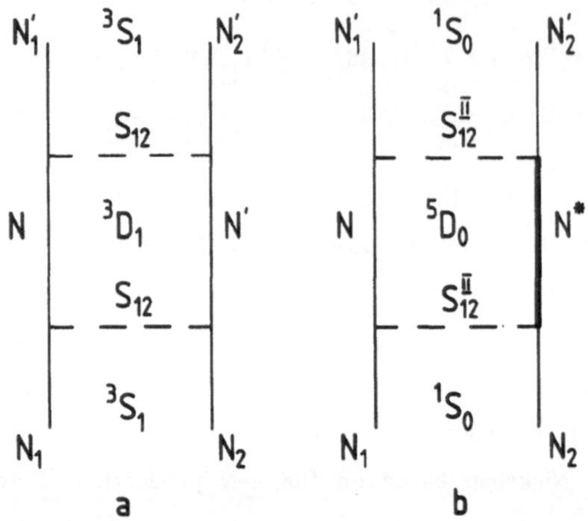

Fig 2 a) The intermediate D-state admixture in the two nucleon $^3S_1$ channel induced by a tensor force, eg, one-pion-exchange.

b) The intermediate D-state N-N* (represented by the heavy line) induced by a one-pion-exchange interaction in the ^1S_o nucleon-nucleon channel.

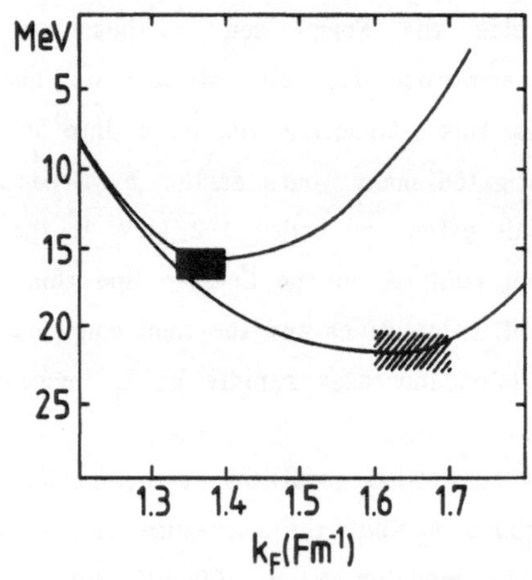

Fig 3 The lower curve represents the results of a typical LOCV calculation including tensor correlations but excluding isobar effects. The upper curve results when isobars are included.

We have calculated the three-body cluster contributions to the saturation curve and have found them to be negligible ($\lesssim$1 MeV) at densities up to $\rho=3\rho_0$ after which they grow rapidly. Thus, at higher densities LOCV calculations are expected to fail and higher terms in the cluster expansion must be included.

There are still many nuclear structure calculations carried out using as an effective interaction the BHF reaction matrix. Our analysis of the nuclear matter saturation problem suggests that this will underestimate the attraction in the $^3S_1$ channel and overestimate the attraction in the $^1S_0$ channel. While this latter effect is not great for $\rho\lesssim\rho_0$ it is density dependent. It is my belief that authors who have found a need for effective repulsive three-body forces are simply rectifying their omission of the dependence of the isobar effect and we would stress that such a density dependence is hidden in one-boson exchange potentials incorporating a scalar σ-meson.

SN 1987 A

A few comments with regard to the size of the compressibility modulus and supernova dynamics are in order. Most computer simulations of supernova collapse assume spherical symmetry and ignore rotations. It is assumed that the pulsar/neutron star created in the ashes of the explosion has a large period >> 1ms and may subsequently be spun up by accreting material. The stability of a rapidly rotating neutron star is governed by the balance between centrifugal forces, gravity and the nuclear equation of state. The larger the compressibility modulus the faster the pulsar can spin before it breaks up.

On February 24 1987 the supernova SN 1987 A was observed in the Large Megallanic Cloud. On January 18 1989 an optical pulsar was detected at the centre of the remnant of SN 1987 A. Two things were unexpected in the pulsar observation. First, the period was very short only half a millisecond and second, the pulses were modulated by an eight hour period which could be interpreted as the pulsar having a binary companion 'Jupiter'.

The short period of the pulsar and the existence of a companion star argue against a spherical collapse in which angular momentum can be ignored. So that the dynamics of supernovae collapse might be much more complicated than originally envisaged. The stability of a 0.5 ms pulsar requires a stiff equation of state with K $\simeq$ 300 MeV. Finally, the pulsar now appears to have vanished. This could be because of simple obscuration by the supernova remnant or it could be because the binary stars have coallesced and degenerated into a black-hole.

Having a scheme which predicts reasonable values of the volume and symmetry energies together with the compressibility and saturation density for bulk nuclear matter, it is tempting to apply it to a realistic description of matter at, or above, nuclear densities. The only scenario for producing bulk cold matter at, or above, nuclear densities is in the interior of neutron stars.

Neutron star material is supposed in beta equilibrium with respect to

$$n \longleftrightarrow p + e^- + \bar{\nu}_e \tag{24}$$

It is assumed that neutrinos can easily escape the star and that no finite density of neutrinos accummulates within the stellar interior. The Fermi energy (chemical potential) of the relativistic, cold, highly degenerate electrons is

$$\mu_e = (3n_e/\text{Fm}^3)^{1/3}\,193 \quad \text{MeV} \tag{25}$$

which exceeds the neutron-proton mass difference well before the nuclear saturation density is reached.

Because the matter is electrically neutral the number density of electrons must equal the number density of protons. Hence, because it is too expensive in energy terms, to maintain electrons and hence protons in the system at nuclear densities we develop a large neutron excess - the traditional reason for naming pulsars 'neutron' stars. However, as the density climbs and μ_e exceeds 105 MeV a new channel for maintaining charge neutrality opens up

$$n \longleftrightarrow p + \mu^- + \bar{\nu}_\mu \tag{26}$$

This slows the rate of neutronization with increasing density. Again, we assume that the muon neutrinos freely stream out from the star. However, when the chemical potential exceeds the negative pion rest mass 139.6 MeV (strictly rest mass less symmetry energy ~ 115 MeV) yet another channel opens up with

$$n \longleftrightarrow p + \pi^- \tag{27}$$

Furthermore, since the pions, unlike the electrons and muons, are bosons, there is no increasing Fermi energy to inhibit them as the density increases still further. Hence, this becomes the favoured mode and pions condense into the lowest state. What is the nature of this state?

Pions interact with nucleons principally through a p-wave interaction

$$V_{\pi N} \sim - g_{\pi N}\, \underline{\sigma}_N \cdot \underline{k}_\pi \tag{28}$$

Hence, the energy of a pion in a nucleon background would be

$$E(k_\pi) \frac{\hbar^2 k_\pi^{\,2}}{2m} + \Sigma g_{\pi N}\, \underline{\sigma}_N \cdot \underline{k}_\pi \tag{29}$$

We note that if the nuclear spins are randomly orientated the interaction term sums to zero. If there is an ordering of the spins so that $\Sigma g_{\pi N} \underline{\sigma}_N = \underline{S}$ then the condensate momentum is given by

$$\nabla E(\underline{k}_\pi) = 0 \tag{30}$$

or

$$\underline{k}_\pi = \frac{m_\pi}{\hbar^2}\, \underline{S} \tag{31}$$

To see the origin of this spin ordering we return to the role of the tensor correlations required to describe the nuclear saturation. The tensor operator is

$$S_{12} = 3\underline{\sigma}_1 \cdot \underline{r}\ \underline{\sigma}_2 \cdot \underline{r} - \underline{\sigma}_1 \cdot \underline{\sigma}_2 \tag{32}$$

which is exactly the dipole-dipole interaction which produces spin correlations in a paramagnetic material. Thus, the law of corresponding states predicts for nucleon matter a nuclear phase transition anologous to a ferro magnetic transition in magnetic materials. One twist to the story is that the tensor force arises from pion exchange. Isospin symmetry and the isovector nature of pions requires that the pion exchange potential must contain in addition to the spin dependence (32) the isoscalor factor $\underline{\tau}_1 \cdot \underline{\tau}_2$ with the result that the ordered nuclear phase has the structure given in figure 4. The low energy excitations of the spin-½ ferromagnet are spin-1 waves. The nuclear analogue are spin-0, isovector spin waves, or pions.

Access to the pion opens the door to the restoration of neutron-proton symmetry schematically represented in figure 5. We can translate this into the equation of state in figure 6 going from neutron rich matter at densities $\lessapprox \rho_0$ to the pion condensed phase at densities $\leq 2\rho_0$. The region between ρ_0 and $2\rho_0$ is phase unstable. Thus, a neutron star with a central density in this range would suffer condensation of the core as indicated by figure 7. A uniform magnetisation $\sim \mu_N \rho_0$ over a sphere of radius R_c would generate a magnetic field at the surface of a star of radius R of magnitude

$$B \simeq \mu_o \mu_N \rho_o \left(\frac{Rc}{R}\right)^3$$

$$\simeq 10^{12} \left(\frac{Rc}{R}\right)^3 \ T$$

and it is interesting to note that surface magnetic field of order 10^{10} T are associated with strong magnetic pulsars and that $(Rc/R)^3$ is $\sim 10^{-2}$.

TENSOR CORRELATIONS AND DEEP INELASTIC LEPTON SCATTERING

The use of deep inelastic scattering establish the existence of three valence spin-½ quarks in nuclei is now well known. Similarly, the fact that deep inelastic scattering of nuclear targets shows that the structure function per nucleon in the nucleus deviates from the structure function of the nucleon (the EMC effect) is well tabulated. In figure 8 we present a summary of the 'nuclear' explanation of the effect. An alternative explanation of the nuclear binding dip can be given in terms of QCD renormalisation (colour delocalisation).

The Bjorken scaling variable

$$x = AQ^2/2M_A\nu$$

where $-Q^2$ is the momentum transfer and ν is the energy transfer, is defined for a nuclear target on the range $0 \leq x \leq A$. Hence for a nucleon $0 \leq x \leq 1$ and a nucleon structure function cannot exhibit cluster correlation effects. Thus, if we wish to seek evidence of cluster correlation effects in a nucleus of mass A we might look at the cluster function in the region $x > 1$.

In figure 9 we compare the theoretically calculated F_2 structure function for carbon compared with experimental data. Curve A using only central Jastrow correlation functions and curve B includes the tensor correlations. A word of warning is, however, in order, both the data and the calculations should be considered as preliminary.

POSTSCRIPT

We have tried to present a coherent account of the cold matter equation of state at nuclear and slightly higher than nuclear densities. We have not commented on cosmology or relativistic heavy ion collisions - in both cases one should deal with hot (or at least warm) equations of state. Already a temperature of ~5 MeV would be sufficient to destroy the pion condensed phase.

Our language has been that of old fashioned nuclear physics in which the nucleus is composed of nucleons and mesons and the resonances that can be constructed from them. We believe that the underlying strong interaction is QCD with its structure of quarks and gluons. Support for this view comes from

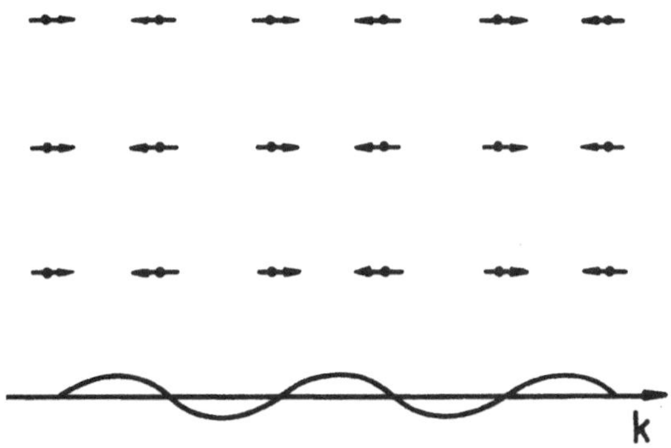

Fig 4 The baryonic layered spin lattice associated with a pion condensate of momentum k.

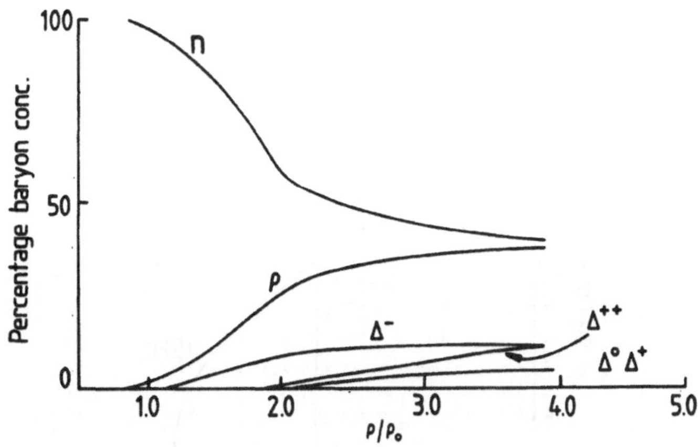

Fig 5 Abundances of baryons predicted in an interacting hadronic fluid allowing for pion condensation.

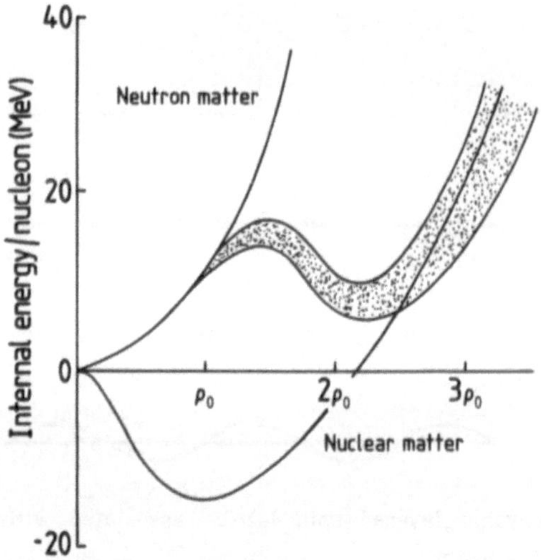

Fig 6 Equation of state for nucleon matter allowing for a pion condensate. The two solid curves represent the neutron matter and nuclear matter equations of state. The shaded region represents the uncertainty allowed in the equation of state of pion condensed matter.

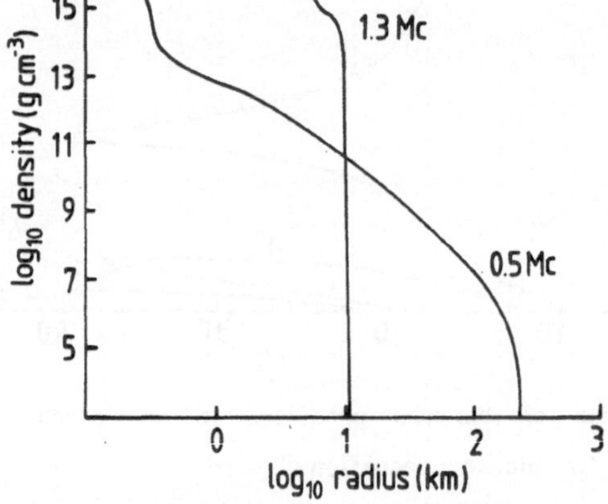

Fig 7 Pion condensation produces a density discontinuity in the core of a neutron star.

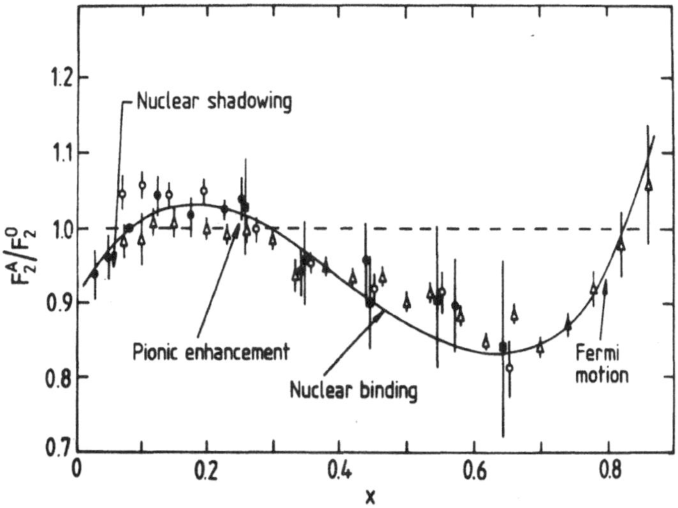

Fig 8 The nuclear explanation of the general features of the EMC ratio $R^A(x)$ in the range $0 \le x \le 1$.

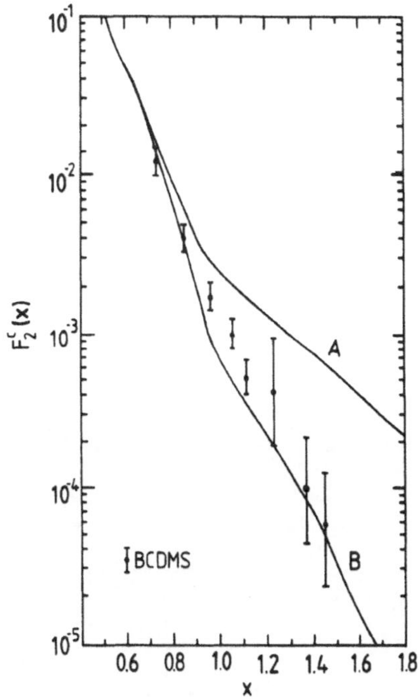

Fig 9 A comparison of our calculated structure function for ^{12}C with the deep inelastic muon scattering data; A including only central correlations; B including tensor correlations.

deep inelastic nucleon scattering. While the strict confinement of quarks in free hadrons is extremely probably, already at nuclear densities, we might expect to see some partial deconfinement in nuclei. To date there is no clear evidence for this and as we suggest in figures 8 and 9 a conventional 'nuclear' explanation may suffice to describe the data. Even if a 'nuclear' explanation of saturation densities is adequate, what about the higher densities in neutron star cores?

For a solar mass star the Schwarzschild radius is ~ 3km and the neutron star radius is ~ 10km. The Schwarrzschild radius is directly proportional to the mass of the star while the neutron star radius is a decreasing function of mass, the details of which depend on the equation of state. However, it would appear unlikely that the central density of stable neutron stars will ever greatly exceed $3\rho_0$. This means that the mean separation between nucleons is always greater than $(3)^{-1/3}\, 2r_0 \simeq 1.5$ fm. With a nucleon bag radius of 0.8 fm and a softening of the equation of state at each phase transition it is probable that a cold quark-gluon plasma is never produced in a neutron star core. However, the study of cold nuclear equations of state may be a guide to the behaviour of low energy QCD.

VARIATIONAL THEORY OF NUCLEAR AND NEUTRON MATTER

V. R. Pandharipande

Department of Physics, University of Illinois at Urbana-Champaign
1110 West Green Street, Urbana, IL 61801 USA

and

R. B. Wiringa

Physics Division, Argonne National Laboratory
Argonne, IL 60439-4843 USA

I. INTRODUCTION

In the many-body theory of nuclei, nuclear and neutron-star matter it is assumed that the nucleons are the main degrees of freedom, and their dynamics is given by the Hamiltonian:

$$H = \sum_i - \frac{\hbar^2}{2m} \nabla_i^2 + \sum_{i<j} v_{ij} + \sum_{i<j<k} V_{ijk} + \ldots.. \qquad (1.1)$$

Here v_{ij} and V_{ijk} are two- and three-nucleon interactions that contain the effects of all mesonic and quark degrees of freedom. In principle we can also have many-body interactions (more than three) between the nucleons, and we will study such a model. This approach to the theory of nuclei and nuclear matter was pioneered in the fifties by Brueckner and Bethe, and many researchers have contributed to its development over the past forty years.

In this theory the properties of nuclei and nucleon matter can be studied from the solutions of the many-body Schrödinger equation:

$$H\ \Psi_n\ (1,2....A) = E_n\ \Psi_n\ (1,2....A), \qquad (1.2)$$

where Ψ_n is a wave function of the positions, spins and isospins of all the A nucleons in the system. The A=2, two-body problem can be solved by simple numerical integration of the radial Schrödinger equation.[1] The three-body problem can also be solved exactly for the ground state[2] and low-energy continuum states[3] with the Faddeev

The Nuclear Equation of State, Part A
Edited by W. Greiner and H. Stöcker
Plenum Press, New York

equations. The four-body ground state can in principle be calculated with the Faddeev-Yakubovsky equations.[4] The A=4 to ∞ problems can be treated approximately by perturbative methods based on the Brueckner-Bethe theory.[5,6]

In these lectures we will discuss attempts to solve the A=3 to ∞ nuclear many-body problems with the variational method.[7,8] We choose the form of a variational wave function $\Psi_V(1,2...A)$ to describe the ground state. The Ψ_V and the ground-state energy E_V are obtained by minimizing

$$E_V = \langle\Psi_V|H|\psi_V\rangle/\langle\psi_V|\psi_V\rangle \tag{1.3}$$

with respect to variations in Ψ_V. If the form of the variational wave function is chosen properly we can expect $\Psi_V \approx \Psi_o$ and $E_V \approx E_o$ where Ψ_o and E_o are the exact ground-state wave function and energy. In general $E_V \geq E_o$ in variational calculations.

Some of the advantages of the variational method are as follows. (i) With the Green's function Monte Carlo method[9,10,11] it is possible to start from the variational Ψ_V and obtain the exact Ψ_o and E_o; so far however, this has only been done for the A≤4 nuclei[11] due to the complexities of the nuclear forces. (ii) It is simpler to treat both many-body forces and high density matter with the variational method. (iii) It is possible to treat excited states either directly[13] or by using orthogonal correlated states[14,15] based on Ψ_V. (iv) It is simple to estimate expectation values of various observables[11,15] with the Ψ_V. An overview of the variational method is given in section II.

Only the very long-range properties of the interactions v_{ij} and V_{ijk} can be calculated from the known pion-nucleon coupling. Thus both v_{ij} and V_{ijk} have to be determined from experimental observations. Since the two-body problem can be solved exactly the available N-N scattering data can be used to develop realistic models of v_{ij}. The ground state energies of ^3H, ^3He and ^4He and the binding energy and equilibrium density of nuclear matter have been used to construct models of V_{ijk}. However, since the many-body calculations are not exact, their errors influence the parameters of V_{ijk}. The recent models of nuclear forces are discussed in section III, and some of the results obtained for nuclear and neutron matter are given in sections IV-VI.

II. THE VARIATIONAL METHOD

It is most important to choose the form of the variational wave function Ψ_V correctly. For example, nuclei and nuclear matter are not even bound if Ψ_V is chosen as a Slater determinant Φ, as in Hartree-Fock theory,

$$\Phi = \mathcal{A} \prod_i \phi_i \,(x_j), \tag{2.1}$$

or as a Jastrow wave function:

$$\Psi_J = \left(\prod_{i<j} f^c\,(r_{ij}) \right) \Phi, \tag{2.2}$$

with realistic forces. In the above equations x_j denotes position, spin and isospin of particle j, $\phi_i(x_j)$ are single-particle wave functions, $\mathcal{A}$ denotes antisymmetrization and $f^c(r_{ij})$ are central spatial correlation functions.

The tensor force, dominated by the one-pion exchange, provides a crucial part of the attraction with which nuclei are bound. The wave functions (2.1) and (2.2) do not contain tensor and other spin-isospin correlations, and hence are not suitable for variational calculations of nuclei. A reasonable Ψ_V for nuclear systems has the form:

$$\Psi_V = \left(S \prod_{i<j} F_{ij} \right) \left[S \prod_{i<j<k} (1+u_{ijk}) \right] \Phi, \tag{2.3}$$

where the pair correlation operator F_{ij} is given by:

$$\begin{aligned} F_{ij} &= f^c(r_{ij}) + f^\tau(r_{ij})\, \tau_i\cdot\tau_j + f^\sigma(r_{ij})\, \sigma_i\cdot\sigma_j + f^{\sigma\tau}(r_{ij})\sigma_i\cdot\sigma_j\, \tau_i\cdot\tau_j \\ &\quad + f^t(r_{ij})S_{ij} + f^{t\tau}(r_{ij})S_{ij}\tau_i\cdot\tau_j + f^b(r_{ij})\; L\cdot S + f^{b\tau}\,(r_{ij})\; L\cdot S\; \tau_i\cdot\tau_j \\ &\equiv \sum_{p=1,8} f^p(r_{ij})\; O^p_{ij}. \end{aligned} \tag{2.4}$$

The N-N interaction has strong components associated with the above eight operators. Not much of work has been done on three-body correlations u_{ijk} in nuclear matter. In light nuclei these three-body correlations can reduce the energy by several percent.[8,16] They are generated by both the two-body forces[17] and three-body forces.[16] The correlation operators F_{ij} and u_{ijk} do not commute with each other, and hence their products have to by symmetrized.

In principle the single particle functions $\phi_i(x)$, the correlation functions $f^{p=1,8}(r)$ and u_{ijk} should be obtained by minimizing the ground state energy. It is difficult to treat this variational problem exactly. The calculations are simpler for light nuclei (^3H, ^3He and ^4He), where $\phi_i(x)$ can be taken as the four spin isospin states $\chi^{\sigma\tau}$ = $n\uparrow, n\downarrow, p\uparrow$ and $p\downarrow$ with no spatial dependence, and for nuclear matter, where $\phi_i(x)$ are plane waves exp $(i\vec{k}\cdot\vec{r})\chi^{\sigma\tau}$ with $k \le k_F$ the Fermi momentum. For simplicity and brevity we will consider only these systems, discuss only the parametrization of F_{ij} in nuclear matter and furthermore ignore the effects of Fermi motion on F_{ij}; they are given in ref. 18.

One can generally assume that F_{ij} is given by:

$$-\frac{\hbar^2}{m} \nabla^2_{ij} F_{ij} + v_{ij} F_{ij} = \omega_{ij} F_{ij}. \tag{2.5}$$

This equation serves to define ω_{ij} which can be considered as the change in v_{ij} due to many-body effects. The ω_{ij} is responsible for obtaining the correct long-range behavior of F in nuclear matter; $f^c(r \to \infty) = 1$ and $f^{p \neq c}(r \to \infty) = 0$, and in many ways it is more convenient to vary F_{ij} by varying ω_{ij}. In a commonly used form the parameters of F are denoted by d_p, α_p and β_p; d_p is the range at which f^p attains its asymptotic value. The ω_{ij} is taken as:

$$\omega_{ij} = \sum_p \left[(1-\alpha_p)\, v^p(r_{ij}) + \lambda^p \right] O^p_{ij}, \tag{2.6}$$

and λ^p are adjusted to obtain the prescribed d_p's. The number of parameters can be greatly reduced by assuming that

$$d_t = d_{t\tau}, \tag{2.7}$$

$$d_c = d_\sigma = d_\tau = d_{\sigma\tau} = d_b = d_{b\tau}, \tag{2.8}$$

$$\alpha_c = 1, \text{ and } \alpha_{p \neq c} = \alpha. \tag{2.9}$$

These approximations are found to be reasonable in several calculations.[19]

Let the F obtained from equations (2.5-9) by minimizing the energy with respect to d_t, d_c and α be denoted by $\tilde{F}$. The F used in calculations including three-body forces is defined as:

$$F_{ij} = \tilde{f}^c(r_{ij}) + \sum_{p=2,8} \beta_p \tilde{f}^p(r_{ij})\, O^p_{ij} \tag{2.10}$$

with the "reasonable" approximations:

$$\beta_t = \beta_{t\tau}, \tag{2.11}$$

$$\beta_\sigma = \beta_\tau = \beta_{\sigma\tau} = \beta_b = \beta_{b\tau}. \tag{2.12}$$

The β_p parameters are meant to take into account the effect of V_{ijk} on F.[19] When V_{ijk} is zero all β_p's are close to unity, and it is sufficient to vary only d_c, d_t, and α. The two-pion exchange part of V_{ijk} increases the magnitude of tensor and spin-correlations significantly, thus $\beta_{t\tau}$ and $\beta_{\sigma\tau}$ obtain equilibrium values > 1.

The expectation value of H has to be calculated and minimized in variational calculations. In the light nuclei <H> is calculated exactly by the Monte Carlo method.[20] Unfortunately the computational time required to calculate <H> with the currently available Monte Carlo methods increases approximately as $2^A A!/N!Z!$, where A, N and Z are respectively the number of nucleons, neutrons and protons in the system. Thus this exact method is practical only for nuclei having $A \lesssim 8$, with presently available computers. The expectation value $\langle H \rangle$ in nuclear matter is calculated with chain summation techniques.[7,19] Even though this method is not exact its accuracy is believed to be very good[21]; for example the error in the calculation of E_v for the ground state of nuclear matter at equilibrium density is expected to be < 1 MeV per nucleon.

Let us denote the ground state of noninteracting Fermi gas by $|0]$. The normalized variational ground state $|0)$ can then be written as:

$$|0), = G|0]/[0|G^\dagger G|0]^{1/2}, \tag{2.13}$$

where G represents the product of correlation operators in eq. (2.3). The G can be used to transform all noninteracting states $|I]$ into correlated states $|I)$. For example a correlated two-particle two-hole state $|p_1p_2h_1h_2)$ is obtained as

$$|p_1p_2h_1h_2) \equiv G|p_1p_2h_1h_2]/[p_1p_2h_1h_2|G^\dagger G|p_1p_2h_1h_2]^{1/2}, \tag{2.14}$$

where $|p_1p_2h_1h_2]$ is a Fermi-gas state with particles p_1 and p_2 and holes h_1 and h_2. These correlated states have reasonable energies, but they are not orthogonal to each other. They have been used in nonorthogonal basis perturbation theory[13,22] to calculate various quantities such as corrections to E_v, the real and imaginary parts of the optical potential of a nucleon in nuclear matter, properties of matter at finite temperatures[23], etc.

Recently[14] new methods have been developed to orthogonalize the correlated states while preserving their energies. Let $|I\rangle$ be the orthogonal correlated state that corresponds to the correlated state $|I)$ or Fermi-gas state $|I]$. The orthogonalization preserves energies:

$$\langle I|H|I\rangle = (I|H|I), \tag{2.15}$$

which are determined variationally. A perturbation theory can be easily developed using the orthogonal states $|I\rangle$.

$$H = H_o + H_I, \tag{2.16}$$

$$\langle I|H_o|J\rangle = \langle I|H|J\rangle\, \delta_{IJ}, \tag{2.17}$$

$$\langle I|H_I|J\rangle = \langle I|H|J\rangle\, (1-\delta_{IJ}). \tag{2.18}$$

It is expected to have good convergence when the correlation operator G is realistic so that |I> are close to the eigenstates of H and <I|H|J> is small when I≠J. This orthogonal correlated functions theory has been used to study the response of nuclear matter,[24] spectral functions,[25] etc., and it can also be used to study the optical potentials and many other properties.

III. NUCLEAR FORCES

Nuclear forces are the only input in the many-body theory; however, they are not yet well understood. Nucleons are finite objects made up of quarks and gluons, and they interact in a fairly complex manner. Realistic models of two-nucleon-interaction potential v_{ij} are obtained in several towns like Argonne[26], Bonn[27], Nijmegen[28], Paris[29], and Urbana[30], by fitting the deuteron properties and two-nucleon scattering data at $E_{lab} \lesssim 400$ MeV. All these models have a one-pion exchange long-range part and more or less phenomenological intermediate and short range parts. The Nijmegen potential gives the best χ^2 fit to the data, while the Argonne model gives the best description[31] of the deuteron form factors $A(q^2)$ and $B(q^2)$.

The Urbana and Argonne models are simpler to use in many-body calculations. The Argonne potential is a more recent version of the Urbana; the tensor force in the Argonne model is much stronger than that in the Urbana at small r (fig. 1). In other respects these two models are similar. The tensor forces in Argonne and Paris models are similar (fig. 1); the D-states account for 5.2, 5.8 and 6.1% of the Urbana, Paris and Argonne deuteron wave functions. The properties of nuclear matter depend to some extent on the strength of the tensor force. For example, neutron matter has a transition to a phase in which π^0-condensation appears likely, when the Argonne model is used, but not if the Urbana model is used.[19]

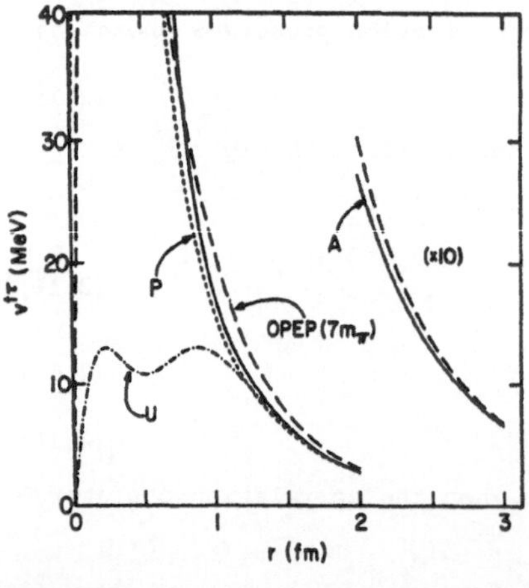

Fig. 1. The $v^{t\tau}(r)$ in Urbana, Paris and Argonne models is compared with the one-pion-exchange potential with a cutoff $\Lambda = 7m_\pi$.

The binding energy of the three-body nucleus ^3H can be very accurately calculated for each model from 34-channel Faddeev calculations.[2,32] The experimental energy is -8.48 MeV, while the Paris, Argonne and Nijmegen potentials give -7.47, -7.67 and -7.62 MeV respectively. The coordinate-space Bonn potential gives a better triton energy[32,33] of -8.29 MeV, but this result may not be very significant since the improvement in the energy probably comes from the unrealistic energy dependence of the coordinate space Bonn $^3S_1$-$^3D_1$ mixing phase[34] ε_1.

The energies of nuclei having four or more nucleons can not as yet be calculated accurately, i.e. with <1% error. Quite recently Carlson[11,35] has made significant advances in exact calculations of ^4He binding energy with the Green's function Monte Carlo method. Exact calculations have been carried out with the Reid-v_8 interaction[35] for which the exact energy (-24.5 MeV) is ~6% below the variational energy of -23 MeV. Currently available variational calculations of ^3H also have ~6% error; more recently Wiringa[16] has used better parameterization of the F_{ij} and u_{ijk} to reduce the error to ~3%. Available variational calculations[8,35] give energies of -23 ± .2 MeV with the Urbana, Argonne and Nijmegen potentials for the α particle.

Brueckner theory[36] and the variational method[36,19] have been used to study nuclear matter (NM) with the Paris, Argonne and Urbana interactions. Day[36] obtained almost identical NM energies with the Argonne and the Paris models by summing two-, three- and some four-hole line terms in Brueckner theory. His results are compared with those obtained for the Argonne and Urbana potentials by the variational method[19] in fig. 2. The Brueckner theory energies are ~10% below the variational energies.

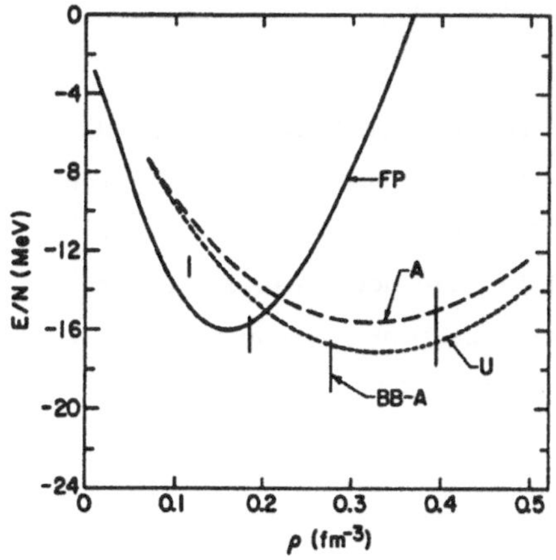

Fig. 2. The nuclear matter energy obtained with the Urbana and Argonne models in variational calculations. The vertical lines BB-A show results obtained with the Brueckner-Bethe theory and the Argonne model by Day. The curve FP has correct empirical saturation point.

The results obtained with the Argonne, Nijmegen, Paris and Urbana models can be summarized as follows. All these potentials underbind ^3H and ^4He, they give reasonable value for the binding energy of nuclear matter, but the calculated equilibrium density ($\sim 0.3 fm^{-3}$) has about twice the empirical value of $\sim 0.16 fm^{-3}$. Most importantly the differences between the results of different potential models are smaller than those between experiment and theory. It thus appears likely that the simple model of nucleons interacting with only two-body forces is not realistic for many-nucleon systems.

It is well known that there are three-nucleon forces[37] V_{ijk}. Particularly the two-pion exchange $V_{ijk}^{2\pi}$ is theoretically well established[38], and it can provide the additional attraction necessary to obtain the correct binding energy of ^3H and ^4He.[39] However, if one takes only the attractive $V_{ijk}^{2\pi}$, NM gets overbound and also has too large an equilibrium density. Hence it is necessary to have other terms in V_{ijk} to obtain reasonable NM. Our phenomenological approach[8] has been to consider a model of V_{ijk}:

$$V_{ijk} = V_{ijk}^{2\pi} + V_{ijk}^{R} \tag{3.1}$$

in which V_{ijk}^{R} is a spin-isospin independent repulsive interaction of reasonable range, and to adjust the strengths of $V_{ijk}^{2\pi}$ and V_{ijk}^{R} to obtain the experimental binding energies of ^3H, ^4He and reasonable nuclear matter properties in variational calculations. The obvious problems with this approach are that the variational calculations are not exact, and the assumed form of V_{ijk}^{R}, which should contain all the components of V_{ijk} other than $V_{ijk}^{2\pi}$ may be simplistic. Better calculational methods are being developed, and attempts to calculate the binding energies of more nuclei like ^6Li, ^8He, ^{16}O etc. are underway to address these problems.

The latest model of the type (3.1) is numbered VII; it gives the experimental ^3H and ^4He binding energies[8], when used with either Argonne or Urbana interactions, and NM results[19] shown in fig. 3. Arguments suggesting that the parameters of model VII are not unrealistic are given in ref. 40. Unfortunately neither Urbana or Argonne $v_{ij} + V_{ijk}$ (VII) models give the empirical NM energy (-16 MeV) or density (ρ_o = 0.16 fm^{-3}), but they come quite close.

One can develop even more phenomenological models that are constructed to explain the equilibrium properties of NM. In a model developed by Lagaris and Pandharipande (LP) the Urbana model of v_{ij} is expressed as a sum of pion exchange v^{π}_{ij}, two-pion exchange v^{I}_{ij} and short range v^{S}_{ij} parts:

$$v_{ij} = v^{\pi}_{ij} + v^{I}_{ij} + v^{S}_{ij}. \tag{3.2}$$

The V^{R}_{ijk} essentially makes the attractive v^{I}_{ij} weaker as the density increases. LP assume that the energy of NM (per nucleon) is given by:

$$E(\rho) = E[v_{ij}(\rho)] + \gamma_2\rho^2\exp(-\gamma_3\rho) \;(3-2\;(\frac{N-Z}{A})^2), \tag{3.3}$$

where $E[v_{ij}(\rho)]$ is the energy obtained with the density dependent interaction

$$v_{ij}(\rho) = v^{\pi}_{ij} + v^{I}_{ij}\exp(-\gamma_1\rho) + v^{S}_{ij}, \tag{3.4}$$

The term linear in ρ in $v_{ij}(\rho)$ contains the effects of V^{R}_{ijk}, while the γ_2-term in eq. (3.3) represents the contributions of $V^{2\pi}_{ijk}$ and all other three-body interactions. The ρ^2 and higher terms in $v_{ij}(\rho)$ represent four and more body interactions.[18] The parameters γ_1, γ_2 and γ_3 are varied to obtain E_o = -16 MeV, ρ_o = 0.16 fm^{-3} and compressibility K=240 MeV for NM. This model has been used by Friedman and Pandharipande[41] (FP) to study properties of hot and cold nuclear and neutron matter. Their results are also shown in figs. 2 and 3.

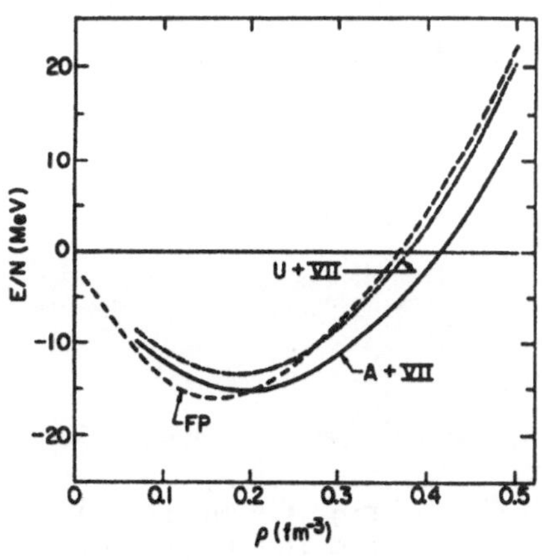

Fig. 3. The nuclear matter energy obtained with the three models.

Unfortunately the LP model cannot be used to study the light nuclei. In this respect the models with explicit three-body forces are superior. Secondly the compressibility of nuclear matter is not well known. The values of K extracted from observed breathing mode energies of nuclei range from ~220 MeV[42] to 300 MeV[43], and it has been suggested that there may be large uncertainties in these extractions.[44]

IV. NEUTRON STAR STRUCTURE

The structure of neutron stars is determined by the equation of state $P(\varepsilon)$ of electrically neutral matter at zero temperature. The pressure P and energy density ε of matter is related to the energy $E(\rho)$ per nucleon in matter at nucleon density ρ:

$$\varepsilon(\rho) = \rho(E(\rho) + mc^2), \tag{4.1}$$

$$P(\rho) = \rho^2\, \partial E(\rho)/\partial\rho, \tag{4.2}$$

and the $E(\rho)$ is calculated from chosen nuclear forces. There have been many calculations of neutron star structure in the past thirty years. We will discuss here the most recent results obtained by Wiringa, Fiks and Fabrocini[19] with the Argonne and Urbana v_{ij} and model VII of V_{ijk}, and with the density-dependent interaction model of LP. These models are denoted by A+VII, U+VII and FP respectively. The $\varepsilon(\rho)$, $P(\rho)$ and the sound velocity

$$s(\rho) = (\partial P(\varepsilon)/\partial\varepsilon)^{1/2}|_{\varepsilon=\varepsilon(\rho)} \tag{4.3}$$

are shown in fig. 4. Both the U+VII and A+VII models give similar results; in these models $s(\rho)$ exceeds c at $\rho \gtrsim 0.8$ fm^{-3} primarily due to the contribution of V_{ijk}. The FP model gives a softer equation of state, and its $s(\rho)$ does not exceed c at least up to ρ=2fm^{-3}. The main difference in the U+VIII and FP models is that the latter include many-body (more than three) forces which help keep $s(\rho) \leq c$.

The A+VII equation of state has a structure at ρ~0.2 fm^{-3} that is most visible in the $s(\rho)$. In the density region ρ=0.2 to 0.25 fm^{-3} the

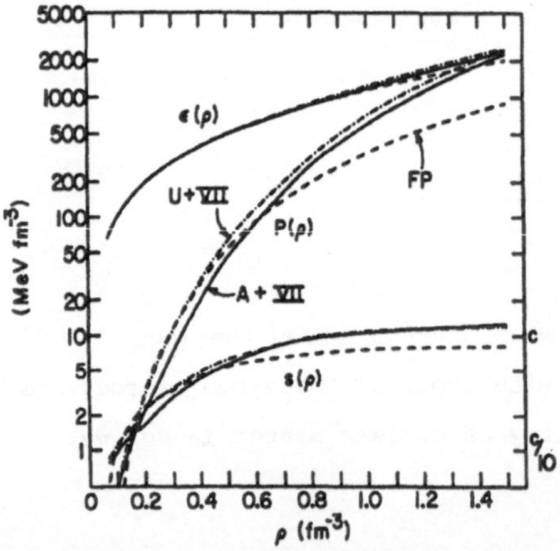

Fig. 4. The $\varepsilon(\rho)$, $P(\rho)$ and $S(\rho)$ for the three models. The scale on the right side is for $S(\rho)$.

tensor and spin correlations change very rapidly[19] in this model of neutron matter, and the contribution of pion exchange interactions suddenly increases as expected in a phase transition to a state with π^0 condensation. This transition does not occur in nuclear matter with the Argonne or Urbana models with or without three-nucleon interactions. It also does not occur in neutron matter with the Urbana model with or without V_{ijk} or with the Argonne model without V_{ijk}. Thus, in the models studied it occurs for only the A+VII. It is very sensitive to the details of the forcs, and does not have a large effect on the static properties of neutron stars. It can however have dynamical effects, such as enhanced cooling rates, that can be observed.

The properties of neutron stars calculated with these models are compared with some of the available observational data in fig 5. For comparison the results obtained with only the Urbana v_{ij} are also shown by curves labeled U. Masses of seven binary x-ray pulsars, believed to be neutron stars, have been determined[45]; they are all near 1.4 $M_\odot$ as expected from evolutionary considerations. The largest minimum mass is 1.55 $M_\odot$ for 4U0900-40 ($M = 1.85^{+0.35}_{-0.3}$ $M_\odot$) as indicated in fig. 5. The mass and radius obtained for the source MXB 1636-536, by assuming that it is a neutron star whose x-ray bursts are due to thermonuclear flashes[46] in the accreted surface matter, are also shown in fig. 5 by a box. The surface red-shift inferred from the gamma-ray bursts of many stars[47] suggests that they are neutron stars with $M \sim 1.4 M_\odot$ and $R \sim 10.5$ km; consistent with the values assigned to MXB 1636-536. It appears that the neutron stars obtained from models including the three-nucleon interaction V_{ijk} are in better agreement with the observational data than those obtained from models with only two-nucleon interactions.

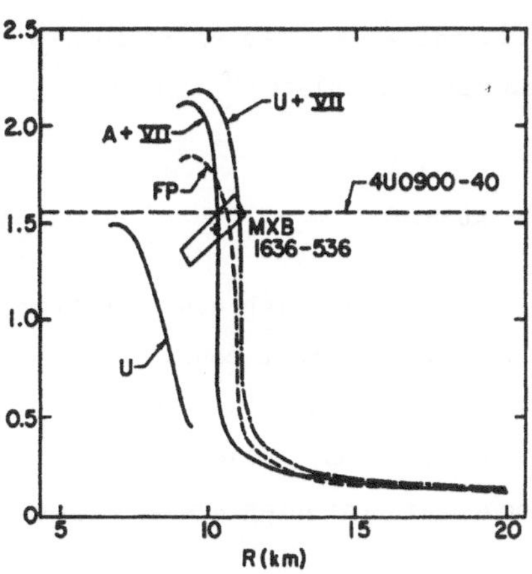

Fig. 5. The mass-radius relation for neutron stars.

The properties of the neutron star that may have formed in supernova 1987A have been studied by many authors. For example analysis[48,49] of the neutrinos observed during this supernova indicates that the binding energy of the star could be 2-5 10^{53} ergs which is consistent with 3×10^{53} ergs predicted for a 1.4 $M_\odot$ star by these models. The reported observation of a half millisecond period in SN1987A has generated renewed interest in the maximum rotational velocity of neutron stars. Recent calculations[50,51] indicate that if the half millisecond period is indeed the period of rotation of the neutron star in SN1987A, and if our present understanding of the equations of state is even approximately correct, the mass of the neutron star in SN1987A is probably greater than the Chandrashekhar mass limit, it is stable due to its large angular momentum, and it will collapse into a black hole on slowing down.

V. MOTION OF NUCLEONS IN NUCLEAR MATTER

The motion of nucleons in nuclear matter is generally described with an optical potential whose real and imaginary parts $U(\rho,e)$ and $W_o(\rho,e)$ give the momentum and mean free path of nucleons of energy e in nuclear matter at density ρ:

$$k^2(\rho,e) = \frac{2m}{\hbar^2}\,[e - U(\rho,e)], \tag{5.1}$$

$$\lambda(\rho,e) = \hbar^2\, k(\rho,e)/[2mW_o(\rho,e)]. \tag{5.2}$$

The optical potential is used to study nucleon-nucleus scattering[52], and heavy-ion collisions.[53]

In the variational method the state of a nucleon of momentum p moving in nuclear nuclear matter is described by the wave-function:

$$|p\rangle = |p) = G\, a^\dagger_p\, |0]/[0|a_p\, G^\dagger G\, a^\dagger_p|0]^{1/2}, \tag{5.3}$$

following the notation of equations 2.13 and 2.14. The variational approximation $U_v(\rho,e)$ to the $U(\rho,e)$ has been calculated[54,55] from $\langle p|H|p\rangle$,

$$e(p) = \frac{\hbar^2}{2m}p^2 + U_v(\rho,e) = \langle p|H|p\rangle - \langle 0|H|0\rangle. \tag{5.4}$$

The motion of the nucleon is damped by collisions which generate two-particle-one-hole states $|p'p''h\rangle$, and the second-order contributions of $\langle p|H|p'p''h\rangle$ give the imaginary part $W(\rho,e)$ of the single-particle energy as well as a correction ΔU to $U(\rho,e)$. Thus, optical potentials correct up-to second order in correlated basis theories are given by:

$$U_2(\rho,e) = U_v(\rho,e) + \Delta U(\rho,e), \tag{5.5}$$

$$W_o(\rho,e) = m^*(\rho,e)\; W(\rho,e)/m \tag{5.6}$$

$$\frac{m^*(\rho,e)}{m} = 1 - \frac{dU(\rho,e)}{de}. \tag{5.7}$$

The effective mass factor reduces the $W_o(\rho,e)$ significantly at low energies where $m^* \sim 0.7m$, and it is necessary to obtain the empirical mean-free paths in nuclear matter.[56,57] The $U(\rho,e)$ and $W_o(\rho,e)$ obtained with the FP model[22] are compared with the empirical data at $\rho=\rho_o$ in figures 6 and 7. The U_v $(\rho > \rho_o,e)$ obtained with A+VII, U+VII and DDI models are given in ref. 55.

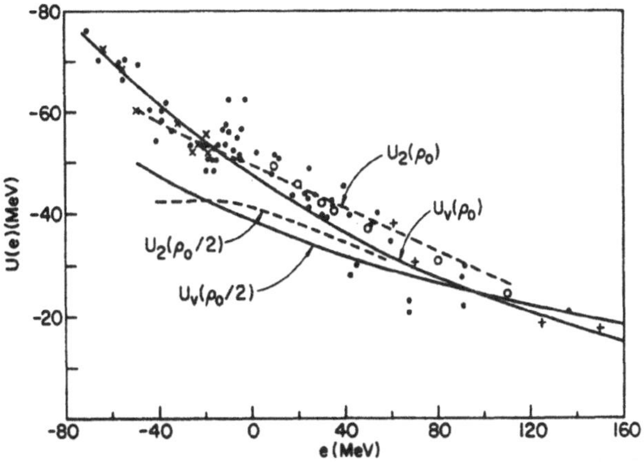

Fig. 6. The real part of the optical potential calculated with the FP model. Full lines give results of variational calculations and dashed lines include second order correlated basis corrections. The sources of empirical data at ρ_o are given in Ref. 54.

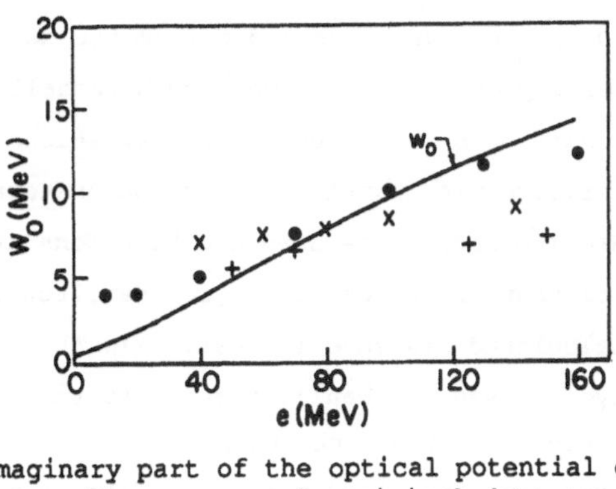

Fig. 7. The immaginary part of the optical potential calculated with the FP model at ρ_o. The sources of empirical data are given in ref. 56.

The two nucleon scattering cross section $d\sigma(\rho,\theta,\phi)$ in matter at density ρ is important in theoretical simulations of heavy-ion collisions.[53] It is given by:

$$\left(\frac{\hbar\vec{p}_1}{m^*(p_1)} - \frac{\hbar\vec{p}_2}{m^*(p_2)}\right) \quad d\sigma(\rho,\theta,\phi)$$

$$= \frac{2\pi}{\hbar} |< p_1 p_2 | H | p_1' \ p_2' >|^2 \ \rho_D, \tag{5.8}$$

where the factor multiplying $d\sigma$ is the relative velocity, $|p_1p_2>$ and $|p_1' \ p_2'>$ are correlated orthogonal many-body states, and ρ_D is the density of final states $|p_1' \ p_2'>$. At low energies $d\sigma(\rho_o,\theta,\phi)$ is approximately half the corresponding cross section for free nucleons. Much of the difference is due to effective masses; ρ_D in matter is smaller than that in free space by a factor of m^*/m, and so, if the interaction matrix elements are similar in vacuum and matter,

$$d\sigma(\rho,\theta,\phi) \approx (m^*/m)^2 \ d\sigma_{free} \tag{5.9}$$

The effective mass $m^*(\rho,e)$ depends upon both the density and energy. At low energies and equilibrium density $m^* \sim 0.7m$ and hence $d\sigma$ is reduced by almost a factor of two from its free space value. At higher densities m^* is even smaller and hence the difference could be larger. At larger energies $m^* \sim m$ and hence differences will be smaller.

VI. THE EQUATION OF STATE

The free-energy

$$F(\rho,T) = U(\rho,T) - TS(\rho,T) \tag{6.1}$$

has been calculated variationally[41] with the DDI model of LP at selected values of ρ,T for nuclear and neutron matter. In principle all thermodynamic quantities like pressure $P(\rho,T)$, entropy $S(\rho,T)$ internal energy $U(\rho,T)$ etc. can be obtained from the derivatives of $F(\rho,T)$. A very useful procedure, developed by Ravenhall[58], is to use a Skyrme type Hamiltonian density fitted to the variational $F(\rho,T)$ to calculate these thermodynamic quantities. The parameters of the Ravenhall-Skyrme Hamiltonian, whose low- and high-density behavior is quite different from that of the common Skyrme Hamiltonians, are given in ref. 58. The calculated pressure isotherms $P(\rho,T)$ are shown in fig. 8; they have the typical van-der-Waal's shape. At temperatures above the critical temperature T_c=17 MeV the $P(\rho,T)$ increases monotonically with ρ and matter is stable at all densities. Below T_c there exists a density range $\rho_g(T)<\rho<\rho_\ell(T)$ in which matter separates into gas at density $\rho_g(T)$ and liquid at $\rho_\ell(T)$. The densities $\rho_\ell(T)$ and $\rho_g(T)$ are found by standard Maxwell construction, and their loci form the liquid-gas coexistence curve shown in fig. 9. Matter is unstable against small isothermal density fluctuations in the region where $\partial P/\partial\rho|_T$ is negative. This region is bounded at a given T by isothermal spinodal densities $\rho_g'(T)$ and $\rho_\ell'(T)$ at which $\partial P/\partial\rho|_T = 0$, and their loci is shown in fig. 9 as the isothermal spinodal curve. Matter at ρ,T in the region between the isothermal spinodal and coexistence curves is metastable, i.e. it is stable against small fluctuations, but will phase separate via nucleation in the presence of large fluctuations. This is the region in which supercooled liquid or gas can exist, and cloud and bubble chambers operate.

Figure 9 also shows the contours for equal energy $U(\rho,T)$ and entropy $S(\rho,T)$. These contours are very useful in estimating the expansion and fragmentation of hot matter formed in heavy-ion collisions.[59] The curve labeled P=0 is the loci of points $P(\rho,T) = 0$. It has a maximum temperature $T_m \approx 12$ MeV, and one can have liquids cooling by evaporation in vacuum only at $T<T_m$. Matter at ρ,T inside the adiabatic spinodal curve in fig. 9 is unstable against small adiabatic fluctuations.[58]

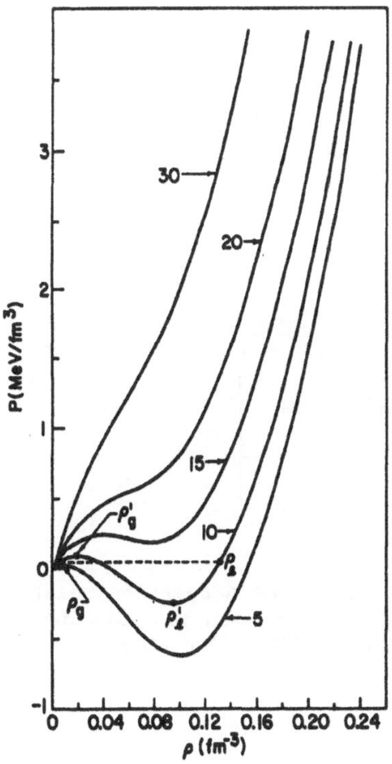

Fig. 8. The $P(\rho)$ isotherms of the FP model of nuclear matter. THe coexistance and spinodal points at T=10 MeV are denoted by ρ_ℓ, ρ_g, ρ_ℓ' and ρ_g' respectively.

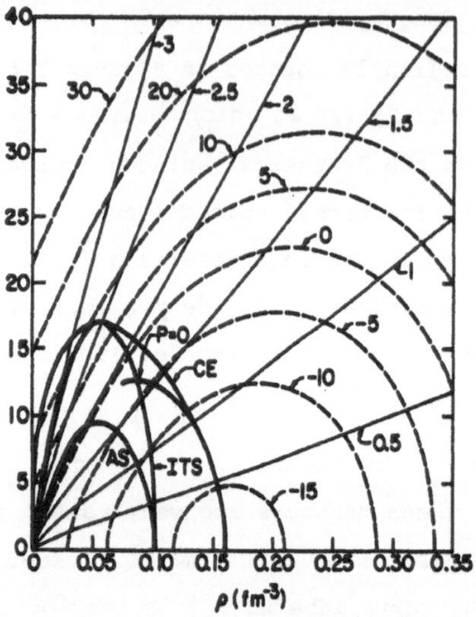

Fig. 9. The phase diagram of nuclear matter showing coexistance (CE), P=0, isothermal (ITS) and adiabatic (AS) spinodal curves. The thin full lines are isentropes labeled with entropy, while the dashed lines are contours of equal internal energy labeled with $U(\rho,T)$ per nucleon.

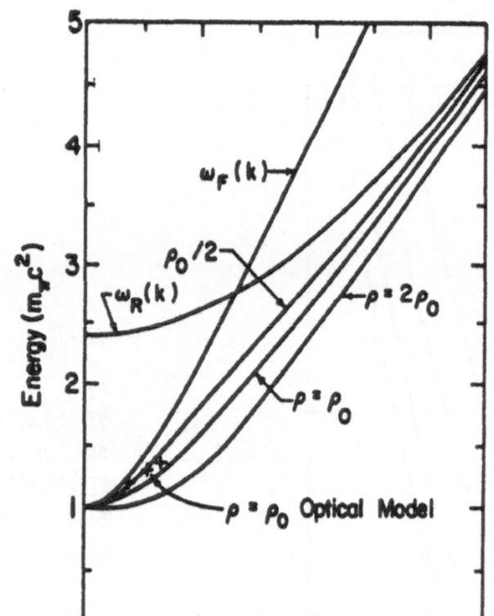

Fig. 10. The curves $\omega_F(k)$ and $\omega_R(k)$ give the dispersion relations for free pions and $N \rightarrow \Delta$ excitations. The curves labeled with density show pisobar dispersion relations, and the + signs show empirical data at $\rho = \rho_0$ from ref. 62.

At high temperatures the non-nucleonic degrees of freedom are excited and contribute to the properties of matter. The two excitations that appear to be most relevant in the T~50 MeV region are pion production and $N \rightarrow \Delta$ (nucleon-hole Δ-particle) excitation. The spectrum of noninteracting pions $\omega_F(k)$ and that of $N \rightarrow \Delta$ excitation $\omega_R(k)$ (neglecting the momentum of nucleon hole and Δ-decay width) are shown in fig. 10. At small k the pion excitation has lower energy, while at large k the $N \rightarrow \Delta$ excitation has lower energy. The two modes are coupled with the strong $\pi N\Delta$ coupling and they naturally mix. The coupled low-energy excitation, recently named pisobar[44], is essentially a pion at small k, a $N \rightarrow \Delta$ excitation at large k, and a mixture of the two at intermediate values of k. The calculation of the spectrum $\omega(k,\rho)$ of this pisobar excitation is discussed by Migdal[60] and the results obtained in ref. 61 are shown in fig. 10. The $\omega(k,\rho_o)$ at small k can be estimated from the pion-nucleus optical potential, and the values obtained from that of ref. 62 are also shown in fig. 10.

The effects of pisobar excitations on the equation of state have been estimated[61] by assuming that hot hadronic matter can be treated as a mixture of hot nuclear matter and pisobar gas. The typical results obtained are shown in fig. 11, where E_π is the energy of pisobars per nucleon, and $R_{\pi N}$ is the ratio of pisobars to nucleons. The E_N^* (S_n) and E^* (S) are the nuclear and total excitation energies (entropies) per nucleon ($E^* = E_N^* + E_\pi$ for example). The pisobars are created at T ~ 30 MeV and dominate the properties of hot matter at T~100 MeV.

The approximations in these calculations are probably reasonable at lower temperature where $R_{\pi N}$ is small. The calculation neglects the interaction between pisobars, which would be important when $R_{\pi N}$ ~ 1. At $T \gtrsim 100$ MeV heavier mesons and higher energy baryon resonances may get excited before matter turns into a quark-gluon plasma. At zero baryon density the transition to quark-gluon plasma is expected to occur at T ~ 155 MeV.[63] We may expect this transition to occur at $T \lesssim 155$ MeV at densities ~ ρ_o. At some high density ρ_{max} matter will become a deconfined quark liquid even at T=0. We do not know what ρ_{max} is, but the many-body theory discussed in these lectures will certainly break down as ρ approaches ρ_{max}.

This work was supported by the US National Science Foundation under Grant PHY84-15064, and by the US Department of Energy (Nuclear Physics Division) under contract No. W-31-109-ENG-38.

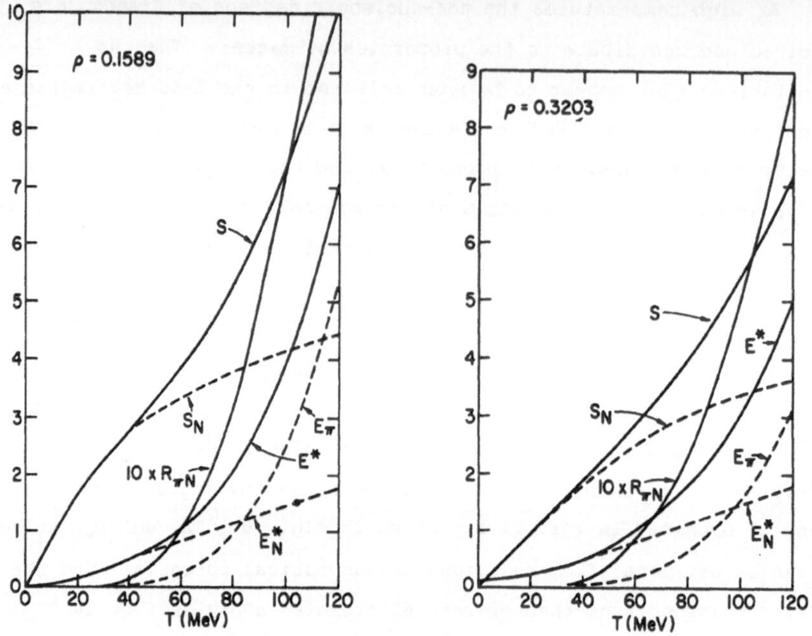

Fig. 11. the properties of very hot matter at ρ_o and $2\rho_o$. The energies are shown in units of 100 MeV per nucleon.

REFERENCES

1. Reid, R. V., Ann. of Phys. 50, 411 (1968).
2. Chen, C. R., et al., Phys. Rev. C31, 2266 (1985).
3. Witala, H., Cornelius, T., and Glöckle,W., Few Body System. 3, 123 (1988).
4. Tjon, J. A., Phys. Rev. Lett. 40, 1239 (1978).
5. Day, B. D., Phys. Rev. C24, 1203 (1981).
6. Kümmel, H., Lührmann, K. H., and Zabolitzky, J. G., Phys. Reports C36, 1 (1978).
7. Pandharipande, V. R., and Wiringa, R. B., Rev. Mod. Phys. 51, 831 (1979).
8. Schiavilla, R., Pandharipande, V. R., and Wiringa, R. B., Nucl. Phys. A449, 219 (1986).
9. Schmidt,K. E., and Kalos, M. H., Applications of the Monte Carlo Method in Statistical Physics, Ed. K. Binder, Springer (1984).
10. Reynolds, P. J., et al., J. Chem. Phys. 77 L(1982) 5593.
11. Carlson, J., Phys. Rev. C36, 2026 (1987).
12. Carlson, J., Pandharipande, V. R., and Wiringa, R. B., Nucl. Phys. A424, 47 (1984).
13. Feenberg, E., Theory of Quantum Liquids, Academic Press, 1969.
14. Fantoni S., and Pandharipande, V. R., Phys. Rev. C37, 1697 (1988).
15. Schiavilla, R., Fabrocini, A., and Pandharipande, V. R., Nucl. Phys. A473, 290 (1987).
16. Wiringa, R. B., private communication (1989).
17. Pandharipande, V. R., Phys. Rev. B18, 218 (1978).
18. Lagaris, I. E., and Pandharipande, V. R., Nucl. Phys. A59, 349 (1981).
19. Wiringa, R. B., Fiks, V., and Fabrocini, A., Phys. Rev. C38, 1010 (1988).

20. Lomnitz-Adler, J., Pandharipande, V. R., and Smith, R. A., Nucl. Phys. A361, 399 (1981).
21. Wiringa, R. B., Nucl. Phys. A338, 57 (1980).
22. Fantoni, S., Friman, B. L., and Pandharipande, V. R., Nucl. Phys. A399, 57 (1983).
23. Schmidt, K. E., and Pandharipande, V. R., Phys. Lett. 87B, 11 (1979).
24. Fantoni,S., and Pandharipande, V. R., Nucl. Phys. A473, 234 (1987).
25. Benhar, O., Fabrocini, A., and Fantoni, S., preprint (1989).
26. Wiringa, R. B., Smith, R. A., and Ainsworth, T. L., Phys. Rev. C19, 1207 (1984).
27. Machleidt, R., Holinde, K., and Elster, Ch., Phys. Repts. 149, 1 (1987).
28. M. M., Nagels, T. A., Rijken and J. J., DeSwart, Phys. Rev. D17, 768 (1978).
29. M., Lacombe, et al., Phys. Rev. C21, 861 (1980).
30. I. E., Lagaris and V. R., Pandharipande, Nucl. Phys. A359, 331 (1981).
31. P. L., Chung, et al., Phys. Rev. C37, 2000 (1988).
32. J. L., Friar, B. F., Gibson and G. L., Payne, Phys. Rev. C37, 2869 (1988).
33. T., Sasakawa, Nucl. Phys. A463, 327C (1987).
34. S., Ishikawa and T., Sasakawa, Phys. Rev. C36, 2037 (1987).
35. J., Carlson, Phys. Rev. C38, 1879 (1988).
36. B. D., Day and R. B , Wiringa, Phys. Rev. C32, 1057 (1985).
37. B. F., Gibson and B. H. J., McKellar, Few Body SYst. 3, 143 L(1988).
38. Fujita, J., and H., Miyazawa, Prog. Theo. Phys. 17, 360 (1957).
39. Lect. Notes in Physics, Vol. 260 Ed. B. L., Berman and B. F., Gibson (1986).
40. Pandharipande, V. R., Lect. Notes in Physice 260, 59 (1986).
41. Friedman, B. A., and Pandharipande, V. R., Nucl. Phys. A361, 502 (1981).
42. Blaizot, J. P., Phys. Repts. 64, 171 (1980).
43. Sharma, M. M., et al., Phys. Rev. 38C, 2562 (1988).
44. Brown, G. E., Nucl. Phys. A488, 689C (1988).
45. Joss,P. C., and Rappaport, S. A., Ann. Rev. Astron. and Astrophys. 22, 537 (1984).
46. Fujimoto, M. Y., and R. E., Taam, Astrophys. J. 305, 246, (1986).
47. E. P., Liang, Astrophys. J. 304, 682 (1986).
48. Lamb, D. Q., Melia,F., and Loredo, T. J., in Supernova SN1987A in the Large Magellanic Cloud, Ed. M. Katatos and A. G. Michalitsianos (1988).
49. Burrows,A., and Latimer, J., Astrophys. J. 318, L63 (1987).
50. Sato, K., and Suzuki, H., preprint (1989).
51. Friedman, J. L., Ipser, J. R., and Parker, L., preprint (1989).
52. Interaction Between Medium Energy Nucleons and Nuclei, AIP Conf. Proc. Ed. H. O., Meyer, (1983).
53. Bertsch, G. F., and DasGupta, S., Phys. Rept. 160, 189 (1988).
54. Friedman, B., and Pandharipande, V. R., Phys. lett. B100, 205 (1981).
55. Wiringa, R. B., Phys. Rev. C38, 2967 (1988).
56. Fantoni, S., Friman, B. L., and Pandharipande, V. R., Phys. Lett B104, 89 (1981).
57. Negele, J. W., and Yazaki, K., Phys. Rev. Lett. 47, 71 (1981).
58. Pandharipande, V. R., and Ravenhall, D. G., Proc. of Les Houches Winter School (1989).
59. Heiselberg, H., Pethick, C. J., and Ravenhall, D. G., Phys. Rev. Lett. 61, 818 (1988).

60. Migdal, A. B., Rev. Mod. Phys. 50, 107 (1978).
61. Friedman, B., Pandharipande, V. R., and Usmani, Q. N., Nucl. Phys. A372, 483 (1981).
62. Stricker, K., McManus, H., and Carr, J. A., Phys. Rev. C19, 929 (1979).
63. Gottlieb, S., et al., Phys. Rev. Lett. 59, 1513 (1987); Grady, M., et al., Phys. Lett. B200, 1148 (1988).

EXPLORING MANY-BODY THEORIES IN LIGHT NUCLEI

R.F. Bishop and M.F. Flynn
Department of Mathematics, UMIST
P.O.Box 88,Manchester M60 1QD England
and
M.C. Boscá, E. Buendía and R. Guardiola
Departamento de Física Moderna, Universidad de Granada
18071 Granada, Spain

Abstract

A brief review of several many-body theories is presented with applications to the ^4He nucleus in a simplified model which only considers Wigner type nucleon-nucleon interactions. Exact results are obtained with the Diffusion Monte Carlo method, and the results from the Jastrow variational method as well as linearized version of coupled cluster theory are compared with them.

1. Introduction

The Quantum Many-Body Problem presents us with a simply-stated but otherwise quite formidable task, namely the determination of the ground state (and eventually also the excitations) of a system of many mutually interacting particles, in the limit as this number of particles grows to infinity at a constant density. The actual solution of the many-body Schrödinger equation is moreover complicated by the strong coupling character of the usual two-body interactions, which thereby prevents the use of naive perturbative methods. The time-honored Brueckner-Bethe-Goldstone (BBG) theory was invented as a special perturbation expansion technique to solve this problem. In addition, the formalism has a deep physical content.

The optimistic belief at the beginning of the seventies [1] was that BBG theory gave the appropriate tools to solve the many-body problem, at least in nuclear matter, but soon some variational calculations of Jastrow type showed the deficiencies of lowest-order BBG theory. This was called the *crisis in nuclear matter theory* (see [2] for a concise but precise historical review) and during the last decade several new many-body theories were developed and brought into use. Here we will refer to three of them, namely the Green Function Monte Carlo method [3], the hypernetted chain HNC/FHNC massive summation scheme for the

The Nuclear Equation of State, Part A
Edited by W. Greiner and H. Stöcker
Plenum Press, New York

Jastrow variational description [4] and the exp(S) method or Coupled Cluster theory [5]. A resolution of the crisis was illustrated during the first conference on Recent Progress in Many-Body Theories held in Trieste in 1978 [6].

Unfortunately, the plettora of technicalities required for the application of these many-body theories to infinite systems, obscure their underlying simplicity. Here we will attempt a brief description of these theories through a simpler yet still very challenging problem, with a basic didactical objective. We thus concentrate on the study of the ^4He nucleus, and all complicated questions regarding discrete degrees of freedom (spin and isospin) will be absent by limiting ourselves to Wigner-type nucleon-nucleon interactions. Because of the absence of spin/isospin degrees of freedom in the hamiltonian the four nucleons may be considered as distinguishable, or, in other words, we will deal effectively with a system of four bosons. Thus another of the technical complications will be absent, namely antisymmetrization.

Given that our objective is to show the usefulness of many body theories in a wide range of physical situations, we will nevertheless consider four kinds of nuclear interactions, ranging from mild two-body potentials to realistic interactions with a strongly repulsive core at short distances. We describe them below in order of increasing complexity.

The simplest interaction considered was introduced by Kalos [7] in a pioneering work on Green Function Monte Carlo method. The aim of that work was the determination of the coupling constant of a fixed-shape two-body potential which would produce the experimental binding energy of 28.3 MeV of ^4He, i.e. an inverse many-body problem. The interaction is purely attractive and has gaussian shape. Next we consider the well known Brink Boeker BB1 potential [8], which is an *effective* interaction with parameters adjusted so as to obtain the proper saturation conditions in ^4He, ^{16}O and nuclear matter when using *uncorrelated* wave functions. This interaction has been widely used in spectroscopic calculations, and it will certainly overbind ^4He when used as a bare microscopic interaction, as considered here. Finally we have considered two *realistic* interactions known as S3 [9] and MTV [10]. These two interactions are only defined in the $\ell = 0$ channel, so that the word *realistic* should be considered somewhat sceptically in so far as the interactions do not contain spin-orbit nor tensor components. We will use these interactions as local potentials acting in *all* partial waves. The S3 and MTV interactions are of particular interest because of the large amount of work devoted to them in the study of three- and four-nucleon systems.

As said above, we have only considered the Wigner part of the above interactions, so that we will actually obtain upper bounds to the ground state (g.s.) energy corresponding to the full interactions. The interactions will be referred to respectively by the letters K, B, S and M, or sometimes, in the case of the latter three, by the more usual acronyms BB1, S3 and MTV. Potentials K, B and S are a combination of gaussians, and this fact is particularly useful because all of the required matrix elements can be computed by means of semi-numerical algorithms, which both considerably reduce the calculational time and avoid unpleasant cumulative numerical errors. This will permit us to carry out calculations in a very large configuration space. The MTV potential, on the other hand, is a combination of yukawians.

The shape of these four interactions is shown in Figure 1, which also illustrates their various degrees of difficulty to handle. This is basically related to the importance of the short range repulsion.

2. The diffusion Monte Carlo method

Stochastic methods can be used to integrate the many body Schrödinger equation exactly in the simplified case of boson systems. These methods are known as Green Function Monte Carlo [3] and Diffusion Monte Carlo [11,12,13]. We will consider only the latter and will refer to it by the acronym DFMC.

Three are three basic ingredients for DFMC theory. First is the consideration of the time-dependent Schrödinger equation in terms of imaginary time. Starting from an initial state $\Psi(0)$ its imaginary-time evolution is governed by a linear combination of exponentially increasing or decreasing functions, instead of the normal oscillatory evolution of amplitudes in real time. Moreover, if the hamiltonian is shifted by an amount $\mathcal{E}$ equal to the g.s. energy E_{gs}, all the amplitudes will drop to 0 at $t \to \infty$ with the exception of the g.s. component, provided that $\Psi(0)$ has a nonzero projection on the exact ground state. In other words, the norm of $\Psi(t)$ will stabilize when $\mathcal{E} = E_{gs}$ at $t \to \infty$. This is a way of determining E_{gs} known as the *grow estimator.*

The imaginary-time equation is solved with the help of the time-dependent Green function, in such a form that the equation

$$\Psi(\mathcal{R}, t_0 + t) = \int d\mathcal{R}' G(\mathcal{R}, \mathcal{R}', t)\Psi(\mathcal{R}', t_0) \tag{1}$$

gives the wave function after a time step t. In this and subsequent equations $\mathcal{R}$ will represent the set of all coordinates of the many body system. The time-dependent and energy-shifted Green function is given by the matrix element

$$G(\mathcal{R}, \mathcal{R}', t) = \langle \mathcal{R} | \exp[-(H - \mathcal{E})t] | \mathcal{R}' \rangle. \tag{2}$$

Certainly, the determination of the Green function is a very formidable problem, which is actually more difficult than determining only the ground state. It is for this reason that one introduces the second basic ingredient of DFMC theory. One may approximate $G(\mathcal{R}, \mathcal{R}', t)$ for a sufficiently small value of $t = \tau$ by the form (see [14] for a detailed derivation)

$$G(\mathcal{R}, \mathcal{R}', \tau) = \frac{\exp\{-[\mathcal{R} - \mathcal{R}']^2/4D\tau\}}{(4\pi D\tau)^{3A/2}} \exp[(\mathcal{E} - (V(\mathcal{R}) + V(\mathcal{R}'))/2)\tau] \tag{3}$$

In this equation $D = \hbar^2/2m$ is called the diffusion constant, A is the number of particles and $V(\mathcal{R}) = \sum_{i<j} V(r_{ij})$ is the full interaction potential.

Given that we want the $t \to \infty$ limit, eq. (1) is applied repeatedly to an initial $t = 0$ state using the approximate form eq. (3) up to a sufficiently large time. In parallel, $\mathcal{E}$ is adjusted to stabilize the norm, so that we obtain in this manner the ground-state energy.

We arrive now to the third basic ingredient of DFMC theory, which is the way in which the wave function is represented. This hinges on the special property of the positivity of both the (bosonic) ground-state wave function and the small time Green function. Because of this one can interpret $\Psi(t)$ as a probability distribution function and, in turn, represent $\Psi(t)$ by a set of $3A$ dimensional random vectors. In normal Monte Carlo practice one knows the functional form of the distribution function and the task is to get random numbers corresponding to this distribution. Here one works in the inverse way: the distribution function is not known but it is represented by a set of random points.

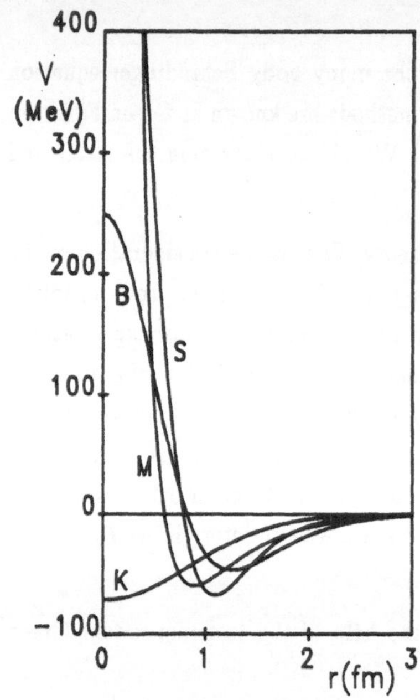

Figure 1. The radial dependence of the two body interactions

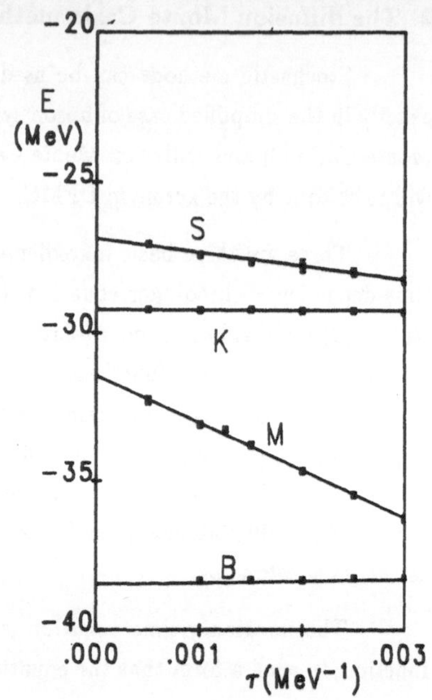

Figure 2. The variation of the sampled energy as a function of the time step

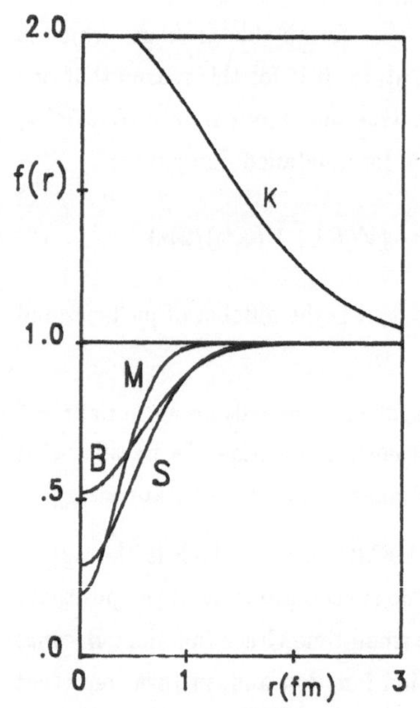

Figure 3. Radial dependence of the correlation factors corresponding to the four interactions

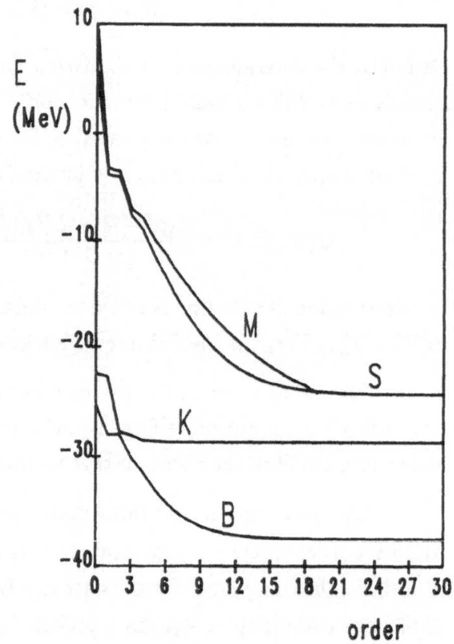

Figure 4. The convergence of linear CC calculations as a function of the number of basis states

With this representation of $\Psi(t)$ in mind, the action of eq. (1) is to obtain a new set of random vectors corresponding to $t+\tau$ from the previous set at time t, and the rules of this evolution are contained in eq. (3). The first term of eq. (3) is a normalized gaussian and corresponds to an isotropic diffusion centered at $\mathcal{R}'$. The second term is an unnormalized exponential and corresponds to a source or sink of points. By adjusting $\mathcal{E}$ to the g.s. energy, the average population remains stable, in correspondence with the stability of the norm of the wave function.

There are still three technical points to be mentioned. First, the results must be extrapolated to $\tau \to 0$, and several runs corresponding to different values of τ must be carried out. Second, one has to distinguish the first set of time steps, which serves to stabilize $\mathcal{E}$ from another subsequent set of steps to control the statistical error of the result. Finally we should mention a very important improvement of the previous description known as *importance sampling*, which incorporates in the algorithm all information corresponding to an assumed good trial function, and has the effect of accelerating the convergence and lowering the statistical uncertainty of the result [12,13].

Our results are presented in Figure 2 for the four interactions and several values of the time step τ. In all cases we have found a linear dependence with the time step, and the intercept of these straight lines with the vertical axis gives the exact (within statistical errors) value of the ground state energy of ^4He for the mentioned interactions. We have used the Jastrow function to be described in the next section as the importance sampling function, and a total of 900.000 points was used to sample each point of Figure 2.

The numerical values of the energy are shown in the last row of Table 1. Our result for the MTV potential is compatible with the value of $-31.3 \pm 0.2\,MeV$ obtained by Kalos and Zabolitzky [15]. Our result for the Kalos potential is $1\,MeV$ lower than the value of $-28.3\,MeV$ obtained in the old paper of Kalos [3]. On the other hand the statistical error in Ref. [3] is surely quite high, and our value is certainly in full agreement with the best variational estimates.

3. Jastrow variational method

The so-called Jastrow trial function was introduced as early as 1940 by Bijl [16] and afterwards by Dingle [17] and Jastrow [18]. One may view this theory as a kit to taylor the variational wave function in such a way that physical requirements related to the nature of the system or to the properties of the two body interaction are manually incorporated into the trial function.

Consider our simple problem. Since we deal with a system of finite size, we may take a so a shell model-like wave function which localizes all particles around the center of mass. Second, we have in some cases interactions which are strongly repulsive at short distances so that the wave function should be very small or even null when $r_{ij} \to 0$ for any pair (ij) of particles. Moreover one would not like to spoil the independent particle motion when one of the particles moves far away from the rest. To fulfill these last two characteristics we may put for each pair of particles a Jastrow correlation factor $f(r_{ij})$ which is small when $r_{ij} \to 0$

Table 1. The ground-state energy of ^4He computed with various many body techniques

METHOD	KALOS	BB1	S3	MTV
HOWF	-23.15	-28.16	-5.89	-6.40
HOWF+J	-29.11	-36.44	-24.29	-29.48
EXP+J	-27.3±0.2	-37.6±0.2	-26.0±0.2	-30.2± 0.3
CC2-L	-28.74	-37.80	-25.31	-24.17
DFMC	-29.25±0.05	-38.5±0.1	-26.9±0.2	-31.5±0.2

and goes to a constant, e.g. 1, when $r_{ij} \to \infty$. Thus, the Jastrow variational function for our problem is

$$\Psi_J = \prod_{i<j} f(r_{ij}) \prod_i \phi(r_i). \tag{4}$$

In more complex systems one may also introduce such other properties as statistics, angular momentum or isospin coupling, spin- and isospin-dependent correlations, triplet correlations and so on. All of these properties are incorporated into the trial wave function in the same way as above.

In eq. (4) there appear single-particle wave functions referred to the origin of coordinates, so that Ψ_J is not translationally invariant. The obvious way of restoring translational invariance is to consider that all r_i appearing in eq. (4) are actually distances from the center of mass of the system. However, this will create serious technical problems when computing the expectation value of the hamiltonian. Another way of dealing with this question is to consider single particle wave functions which are eigenstates of the harmonic oscillator potential (HOWF). Then Ψ_J factorizes into two terms, one of which depends only on the center of mass coordinate and the other being translationally invariant. Then, the true energy of the system will be the difference of the energy computed with the function Ψ_J and the kinetic energy of the center of mass. This is the procedure that we will follow. Note that the factorization property comes from the shell model part of the wave function, and this property will be preserved in so far as such additional factors , as the Jastrow factor $f(r_{ij})$, depend only on relative distances.

The most straightforward calculation with Ψ_J consists in assuming functional forms for f and ϕ which depend on several parameters and then searching for the values of these parameters which give a minimum for the expectation value of the hamiltonian. We have assumed the very simple forms

$$\begin{aligned} \phi(r) &= \exp(-\tfrac{1}{2}\alpha^2 r^2) \\ f(r) &= 1 + a\exp[-(r/b)^2] \end{aligned} \tag{5}$$

which have three free parameters: α, which controls the size of the nucleus, and a and b which correspond to the depth and range of the correlation. The value of the depth a is mainly determined by the nature of the interaction alone. However, α and b are strongly correlated so that one cannot search for the position of the minimum by considering independent variations of these parameters.

The energies obtained by means of this simple variational function are shown again in Table 1. The row labelled HOWF corresponds to the expectation value in the absence of correlations, and the row labelled HOWF+J incorporates the two body correlations in the way above described. Note that the values of the harmonic oscillator parameter α for these two rows are not the same. It is very satisfying to realize that even with so simple an ansatz for the wave function one obtains almost the entire binding energy of the system, being only $2\,MeV$ above the exact DFMC result except for the Kalos potential, in which case our results are almost exact. Part of this energy difference is certainly related to the very sharp shape of the nuclear surface produced by the harmonic oscillator central field. If in eq. (5) we replace $\phi(r) \to 1$ and at the same time multiply $f(r)$ by $\exp(-r/c)$ we gain 1 or $1.5\,MeV$, with the exception of K interaction which prefers a well defined surface. The values corresponding to this calculation are shown in the row labelled EXP+J in Table 1. Note that all integrals have been computed with the Monte Carlo method, and the errors appearing in Table 1 are the statistical errors of the computation.

The gain in energy due to Jastrow correlations is impressive, ranging from $6\,MeV$ for the Kalos potential up to more than $20\,MeV$ for the S3 and MTV potentials. The flexibility of this simple form of the correlation is also noticeable, being able to adapt itself to this wide range of interactions. Figure 3 shows the shape of the correlation of Eq. (5) for the four potentials considered, showing a clear relationship between the interaction and f.

We have thus seen how simple it is to obtain up to 90% or even 95% of the binding energy. A natural question arises: How far may we go with a Jastrow ansatz using only two-body correlations?. To answer this question one has to compute the energy with a wave function like eq. (4) but allowing the maximum freedom in the variational determination of f. In other words, one has to carry out an Euler-Lagrange calculation with respect to f. Calculations of this kind for the MTV potential have been carried out at Urbana [19] by solving an Euler-Lagrange equation for en energy functional obtained at second order of the cluster expansion with the result of $-31.19 \pm 0.05\,MeV$, and at Pisa [20] using the exact functional, to obtain $-31.35\,MeV$. This small gain of $1\,MeV$ is very costly to obtain. Comparing these numbers with the DFMC results one may conclude that the Jastrow form describes almost completely the physical system.

4. Coupled cluster theory

Exp (S) or Coupled Cluster (CC) theory was first proposed by Coester and Kümmel [21]. Its most important feature is the incorporation from the very outset of Goldstone's linked cluster theorem. A full review of the theory may be found in [5]. The exact wave function is written in the following form. Assume a filled Fermi sea, and a series of one-body S_1, two-body S_2, three-body S_3, ... operators which correspond to coherent 1p-1h, 2p-2h, 3p-3h ... excitations, respectively. Then the ansatz for the ground state wave function is

$$\Psi_{CC} = \exp[S_1 + S_2 + S_3 + \ldots]\Psi_0, \tag{6}$$

with Ψ_0 corresponding to the Fermi sea. Expanding in powers the above exponential one may understand the physical content of the theory. For example, to produce 3p-3h excitations one has three possibilities, represented by the operators S_1^3, S_1S_2 and S_3. We then realize that S_3 will only contain true linked 3p-3h excitations, the unlinked ones being contained in the other two operator products.

In the form given by eq. (6) the theory is not translationally invariant (in finite systems), because the exponential factor spoils the basic factorizability property of the harmonic oscillator Ψ_0 wave function. For a system of bosons we have found [22] a simple modification of eq. (6) which préserves the translational invariance. Up to two body clusters it reads

$$\Psi_{CC2} = N[\exp \mathcal{S}_2]\Psi_0 \tag{7}$$

where N represents the normal ordering, and $\mathcal{S}_2$ is an admixture of the previous S_1 and S_2 given by

$$\mathcal{S}_2 = \sum_{n=1}^{\infty} S_2^n \Omega_2^n \tag{8}$$

with

$$\Omega_2^n = \sum_{n_i \ell_i n_j} \langle n0, 00, 0 | n_i \ell_i, n_j \ell_i, 0 \rangle [a^\dagger_{n_i \ell_i} \times a^\dagger_{n_j \ell_i}]^0 a_0 a_0 \tag{9}$$

where the coefficients S_2^n are numerical amplitudes to be determined, $a^\dagger_{n\ell m}$ creates a single-particle harmonic oscillator state with quantum numbers $(n\ell m)$, the subscript $0 \equiv (000)$, and the sum over n_i, ℓ_i and n_j has no other restrictions than those implied by the Brody-Moshinsky bracket. In particular, $(n_i\ell_i)$ may be equal to (00) and eq. (8) contains both 1p-1h and 2p-2h excitations. The particular combination shown will guarantee translation invariance. Note that Ω_2^n corresponds globally to a excitation energy of $2n\hbar\omega$.

Perhaps it is more interesting to look at the coordinate representation of the action of $\mathcal{S}_2$ on the vacuum state. Using standard Fock algebra one may obtain the correspondence,

$$\Omega_2 \Psi_0 = Const \times \sum_{i<j} L_n^{(1/2)}(\alpha^2 r_{ij}^2/2)\Psi_0 \tag{10}$$

where $L_n^{(1/2)}$ is the associated Laguerre polynomial corresponding to an $(n, \ell = 0)$ relative motion of a pair of nucleons. Quadratic terms like $\Omega_2^n \Omega_2^m$ correspond to one pair (ij) in relative state $(n0)$ and another pair (kl) in relative state $(m0)$ with both particles k and l different from i and j.

Restricting ourselves to linear terms only from the exponential in eq. (7) , we may rewrite the simplified CC translationally invariant wave function in the general coordinate form

$$\Psi_{CC2-L} = [1 + \sum_{i<j} g(r_{ij})]\Psi_0 \tag{11}$$

to bring out the similarity with the Jastrow ansatz. One can then work with eq. (8) using standard shell model and Fock algebra machinery, or with equation (11) in the same manner as in the case of Jastrow form. We have preferred the first form, to maintain ourselves as close to the original ansatz as possible, but for the case of gaussian shaped interactions a special algorithm based on the properties of the generating function of Laguerre polynomials was used, allowing us to extend the calculation up to incredibly high values of $n \simeq 30$, or equivalently up to $60\hbar\omega$ excitation quanta. Solving the linear CC equations in the usual form [5] is equivalent to diagonalize a finite matrix, so that the linear CC method gives actually a variational upper bound to the binding energy. The variation of the results with the number of excitations n is shown in Figure 4, and the convergence is strongly tied to the complexity of the interaction, being very rapid for the Kalos potential and quite slow for the S3 interaction. The results for the MTV potential were obtained by using the standard shell-model machinery and we could not go further than $n = 19$. The energy at this value is not yet stabilized. This simply means that we have used a poor basis in which to solve CC equations, which would perhaps have been better solved in coordinate representation [23].

The numerical results of the binding energy are shown again in Table 1, in the row labelled CC2-L (L for linear). Note that the value for MTV is still far from convergence. It was a nice surprise to find that the CC2-L results, corresponding to an ansatz which is much simpler than the Jastrow variational method, produced results of comparable quality. Unfortunately, the only case with which we may compare corresponds to the MTV potential, where we have not attained convergence. For this interaction Zabolitzky [24] obtained the value of $-31.24\,MeV$, very close to the exact DFMC result, by using the full CC theory. There is also a configuration interaction calculation [25] for MTV potential, which considers all possible p-h excitations up to $10\hbar\omega$ (a total of 2765 basis states) with the result of $-18.31\,MeV$, which is very far from both our n=19 result (20 basis states) and, obviously, from the fully converged value. We could say that CC theory also gives a rather clever selection of the important basis states.

5. Summary

This work was planned with a basic didactical purpose. The hope is that after reading it, the reader will be convinced of the basic simplicity of the theories discussed here, as well as of their high quality, even with the drastic simplifications and simple forms used for the calculations. Nevertheless, we should not conceal the fact that even with all these simplifications the calculations still involve a large amount of algebraic as well as computational effort. Certainly, the many-body problem is not a simple question.

Acknowledgements

This work was developed under an *Acciones Integradas* program between Spain and the United Kingdom. The authors acknowledge the financial support from the corresponding joint committee. M.C.B., E.B. and R.G. are also supported by the *Comisión Interministerial de Ciencia y Tecnología, Spain* under contract 969/87. R.F.B. and M.F.F. also acknowledge the support of a research grant from SERC of Great Britain.

References

[1] H.A. Bethe, *Ann. Rev. Nucl. Sci* 21 (1971) 93.

[2] J.W. Clark, *Prog. Part. Nucl. Phys.* 2 (1979) 89.

[3] D.M. Ceperley and M.H. Kalos, *Quantum Many Body Problems* in *Monte Carlo Methods is Statistical Physics*, K. Binder editor (Springer Verlag, New York 1979).

[4] S. Rosati and S. Fantoni, *Correlations in Infinite Systems* in *The Many Body Problem: Jastrow Correlations versus Brueckner Theory*, R. Guardiola and J. Ros editors (Springer Verlag, New York 1981), p. 1.

[5] H. Kümmel, K.H. Lührmann and J.G. Zabolitzky, *Phys. Rep.* 36C (1978) 1.

[6] Proceedings of the *First International Conference on Condensed Matter Theories*, C. Ciofi degli Atti, A. Kallio and S. Rosati editors. *Nucl. Phys.* A328 (1979).

[7] M.H. Kalos, *Phys. Rev.* 128 (1962) 1791.

[8] D.M. Brink and E. Boeker, *Nucl. Phys.* A91 (1967) 1.

[9] I.R. Afnan and Y.C. Tang, *Phys. Rev.* 175 (1968) 1337.

[10] R.A. Malfliet and J.A. Tjon *Nucl. Phys.* A127 (1969) 161.

[11] J.B. Anderson, *J. Chem. Phys.* 63 (1975) 1499, 65 (1976) 4121 and 73 (1980) 3879.

[12] P.J. Reynolds, D.M. Ceperley, B.J. Alder and W.A. Lester Jr, *J. Chem. Phys.* 77 (1982) 5593.

[13] R. Guardiola, *Monte Carlo Techniques in the Many Body Problem*, in *First International Course on Condensed Matter*, D. Prosperi, S. Rosati and G. Violini editors. (World Scientific, Singapore, 1988).

[14] J. Vrbik and S.M. Rohtstein, *J. Comput. Phys.* 63 (1986) 130.

[15] M.H. Kalos and J.G. Zabolitzky, *Nucl. Phys.* A356 (1981) 114.

[16] A. Bijl, *Physica* 7 (1940) 869.

[17] R.B. Dingle *Phil. Mag.* 40 (1949) 573.

[18] R. Jastrow, *Phys. Rev.* 98 (1955) 1479.

[19] J. Carlson and V.R. Pandharipande, *Nucl. Phys.* A371 (1981) 301.

[20] L. Bracci, S. Rosati and M. Viviani, in *Proc. Secondo Convegno su Problemi di Fisica Nucleare Teorica*, (ETS Editrice, Pisa 1988), p. 34.

[21] F. Coester and H. Kümmel, *Nucl. Phys.* 17 (1960) 477.

[22] R.F. Bishop, M.F. Flynn, M.C. Boscá, E. Buendía and R. Guardiola, in preparation.

[23] J.G. Zabolitzky, *Nucl. Phys.* A228 (1974) 272.

[24] J.G. Zabolitzky, *Phys. Lett.* 100B (1981) 5.

[25] R. Ceuleneer and P. Vandepeutte, *Phys. Rev.* C 31 (1985) 1528.

SPECTRAL FUNCTIONS AND THE MOMENTUM DISTRIBUTION OF NUCLEAR MATTER*

A. Ramos, A. Polls

Departament d'Estructura i Constituents de la Matèria Universitat de Barcelona, 08028-Barcelona, Spain

W. H. Dickhoff

Department of Physics, Washington University, St. Louis MO63130, USA

A self-consistent Green's Function approach is used to study the influence of short-range correlations beyond the mean field picture of nuclei. The ladder equation, including both particle-particle and hole-hole propagation, is solved in nuclear matter for a semi-realistic interaction derived from the Reid soft core potential. The nucleon spectral functions are calculated from the momentum and energy dependent self-energy. An important fraction of the single particle strength is moved at very high energy due to the strong short-range repulsion in the interaction. The momentum distribution is calculated from the hole spectral function and an average depletion of normally fully occupied states of about 13% is found at normal density.

1. Introduction

A microscopic study of the nucleon spectral functions is of special interest to establish the extent to which a single particle (sp) description of the nucleus, implied by the mean field models, is valid. The hole spectral function $S_h(k,\omega)$ is the probability of removing a particle with momentum k from the target system of A particles leaving the resulting $(A-1)$-system with an energy $E^{A-1} = E_o^A - \omega$, where E_o^A is the ground state energy of the target. Analogously, the particle spectral function $S_p(k,\omega)$ is the probability of adding a particle with momentum k and leaving the resulting $(A+1)$-

* This research was supported in part by the Condensed Matter Theory Program of the Division of Materials Research of the US National Science Foundation under Grant No. DMR-8519077 (at Washington University) which also provided computer time for the calculations which were performed on the Cray X-MP of the Pittsburgh Supercomputing Center and in part by NATO under Grant No. RG 85/0684 and in part by CAYCIT Grant No. PB85-0072-C02-00 (Spain).

system with an energy $E^{A+1} = \omega + E_o^A$. The Lehmann representation for the *sp* propagator gives

$$S_h(k,\omega) = \sum_n |< \psi_n^{A-1} \mid a_k \mid \psi_o^A >|^2 \, \delta\left(\omega - (E_o^A - E_n^{A-1})\right)$$
$$S_p(k,\omega) = \sum_m |< \psi_m^{A+1} \mid a_k^\dagger \mid \psi_o^A >|^2 \, \delta\left(\omega - (E_m^{A+1} - E_o^A)\right), \quad (1)$$

where $\mid \psi_o^A >$ is the ground state of the target, and $\mid \psi_n^{A-1} >$ and $\mid \psi_m^{A+1} >$ are excited states of the $A-1$ and $A+1$ systems with energies E_n^{A-1} and E_m^{A+1} respectively.

In a system of independent particles moving in an average field each nucleon of momentum k has a well defined energy $\epsilon(k)$ and, therefore, the corresponding spectral function reduces to a delta function $\delta(\omega - \epsilon(k))$. In the presence of correlations this strength is distributed over several energies as a result of the interaction which couple the *sp* motion to other more complicated states. The shape of the spectral function is then broadened and its strength quenched with respect to the delta behavior characteristic of a system of independent particles. Experimental information on the spectral functions in finite nuclei can be obtained from inelastic scattering experiments, such as exclusive electron scattering, pick-up, knock-out and stripping reactions.[1] It is in general found that the total strength corresponding to a *sp* orbital is distributed over several states, except for few levels near the Fermi energy in double closed shell nuclei plus or minus a nucleon. These sharp spectral functions do not carry, however, 100% of the *sp* strength.

In this contribution special attention is paid to the influence of short-range correlations in the breakdown of the independent particle picture of nuclei. Since short-range correlations are expected to have similar effects in finite and infinite systems this study will be made for nuclear matter around normal density. An interesting quantity that can be derived from the hole spectral function is the occupation probability, or momentum distribution in nuclear matter

$$n(k) = \int_{-\infty}^{\epsilon_F} d\omega S_h(k,\omega), \quad (2)$$

where the Fermi energy ϵ_F is the energy difference $E_o^A - E_o^{A-1}$.

From the experimental point of view, recent $(e, e'p)$ scattering data[2] , analyzed together with relative charge density distributions,[3] indicate an 80±10% occupation of the $3s_{1/2}$ proton shell in ^{208}Pb. Quenching of magnetic form factors at high momentum transfer in the lead region can be interpreted in a similar picture of partially occupied shell model states.[4] This is in global agreement with an analysis based on combining the effect of short-range correlations deduced from nuclear matter results and long-range RPA correlations in ^{208}Pb.[5] Similar results are obtained in ref.6. Direct comparison of $(e, e'p)$ cross sections with shell model momentum distributions suggests even substantially lower occupation numbers.[7] At present the precise occupation numbers are not certain but these results demonstrate the breakdown of the mean field picture of nucleons moving in the average attractive field of all the other nucleons.

Our study in nuclear matter will be based on the Green's function formalism which is summarized in section 2. Results on the hole and particle spectral functions are presented and discussed in section 3, together with the resulting momentum distribution. Finally, a summary and conclusions are given in section 4.

2. Self-consistent Green's Function formalism in the ladder approximation

The Lehmann representation for the *sp* propagator provides direct information on the spectral functions. In a correlated many-body system the *sp* propagator is determined by the interaction with the other particles by means of the self-energy as implied by the well known Dyson equation. In turn one can relate the self-energy to the propagation of two particles in the medium. This scheme can be continued leading to a hierarchy of coupled equations which express the n-body propagator in terms of $n+1$-body propagators. In the case of a two-body interaction between the particles, the *sp* propagator, and consequently the self-energy, is completely determined by the two-particle propagator,[8] which can be written in terms of a vertex function Γ representing the interaction in the medium.

This exact formulation contains the concept of non-linearity or self-consistency explicitly since the equation for the effective interacction involves fully dressed propagators. Any approximation scheme that fulfills this self-consistency principle will be referred to as belonging to a self-consistent Green's function ($SCGF$) treatment of the many-body problem. We have developed a method[9] which embeds in a Green's function formalism the original suggestion of Brueckner[10] of summing the so-called ladder diagrams in the perturbative expansion of the effective interaction to treat the influence of short-range correlations. The ladder equation includes, as a natural consequence of the Green's function treatment, the hole-hole (hh) propagation on the same footing as the particle-particle (pp) propagation characteristic of Brueckner theory. The calculation of the resulting self-energy requires then the numerical use of dispersion relations over the imaginary part of the effective interaction.[11,12] In principle, a complete self-consistent calculation implies the use of fully dressed *sp* propagators in the solution of the ladder equation for the effective interaction. At present, self-consistency has been established only for the quasi-particle energy, i.e., at the level of the real on-shell part of the self-energy

$$\epsilon(k) = \frac{k^2}{2m} + Re\ \Sigma(k, \epsilon(k)). \tag{3}$$

When this averaged self-consistency is achieved, the complete energy dependence of the self-energy can be studied. By combining the Lehmann representation, which expresses the *sp* propagator in terms of the spectral functions, with the Dyson equation, which relates the propagator to the proper self-energy, one obtains the following equations

$$S_h(k,\omega) = \frac{1}{\pi}\frac{Im\ \Sigma(k,\omega)}{\left(\omega - \frac{k^2}{2m} - Re\ \Sigma(k,\omega)\right)^2 + (Im\ \Sigma(k,\omega))^2} \qquad \omega < \epsilon_F \tag{4}$$

$$S_p(k,\omega) = -\frac{1}{\pi}\frac{Im\ \Sigma(k,\omega)}{\left(\omega - \frac{k^2}{2m} - Re\ \Sigma(k,\omega)\right)^2 + (Im\ \Sigma(k,\omega))^2} \qquad \omega > \epsilon_F, \tag{5}$$

which show that the spectral functions are completely determined by the full energy dependence of the self-energy. We note that an imaginary contribution to the self-energy, and consequently a non-vanishing hole spectral function S_h, can only be obtained for values of ω below ϵ_F if hh propagation, appearing naturally in the Green's function version of the ladder equation, is considered.

3. Results

In this section we discuss results for spectral functions obtained at a density corresponding to $k_F = 1.4 \text{ fm}^{-1}$. Calculations have been performed for the $v_2^{l=0}$ interaction, which is the central part in the $^3S_1 - ^3D_1$ channel of Reid soft core potential,[13] considered only for S waves. Although this interaction is not fully realistic, it should be emphasized that, on the one hand, it avoids the strong pairing instability observed in the deuteron channel[14] and, on the other hand, the short-range correlations it induces due to the repulsive core are expected to survive in a complete calculation for the full Reid soft core interaction.

Results for the hole spectral function are shown in fig.1 for momenta k =0.828, 1.258, 2.220 and 4.376 fm^{-1} as a function of the excitation energy $\omega - \epsilon_F$. A strong peak located at the quasi-particle energy (see eq.(3)) is observed for the two momenta smaller than k_F, being a reflexion of the delta behavior characteristic of a mean field description. A smooth distribution is observed for momenta larger than k_F indicating the ocurrence of knock-out processes from states above the Fermi sea, which would be empty in an uncorrelated system.

In fact, the hole spectral function gives information on the coupling of a single hole state to more complicated states of 2h-1p, 3h-2p etc. type. As such it does not give direct information on the influence of short- range correlations. This information is contained in the particle spectral function which shows how the *sp* state couples to high lying states of 2p-1h, 3p-2h etc. nature. This coupling is possible since the nature of the short-range repulsion implies that the interaction couples low momentum states to high momentum states which have such high energies.

The particle spectral functions are shown in fig.2 for the same momenta mentioned above. These functions are practically identical for energies above 500 MeV in the case of the two momenta below k_F considered here. This fact, together with the enormous energy range over which *sp* strength is redistributed, naturally reflects the effect of the high momentum correlations implied by the strong short-range repulsion of the two-body interaction. The difference between figs. 2a and 2b is mainly concentrated just above the Fermi energy ϵ_F, which shows that the momentum which is closer to k_F can mix with the low energy excitations with greater ease. The result of this is a larger depletion for momenta the closer they come to the Fermi momentum k_F.

These results are particularly important for the discussion of quenching phenomena in nuclei. They show that short-range correlations really remove *sp* strength to high energy. Similar effects are expected in finite nuclei. Therefore, the missing strength for the particle-hole (ph) excitations of the system at low energy can not be completely recovered from correlations induced by ph interactions since part of the *sp* strength is simply at too high energy.[15,16]

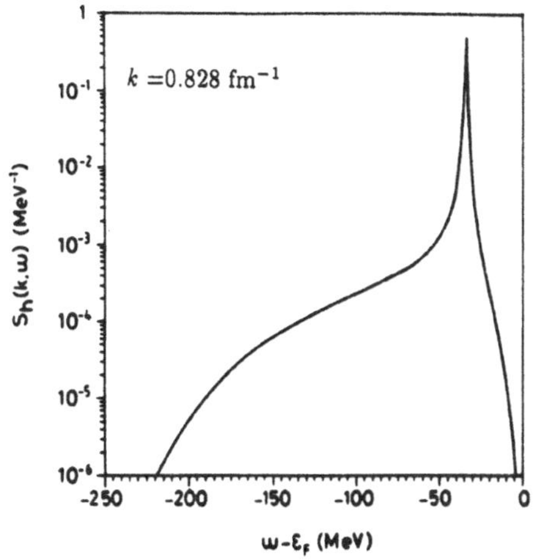

(a)

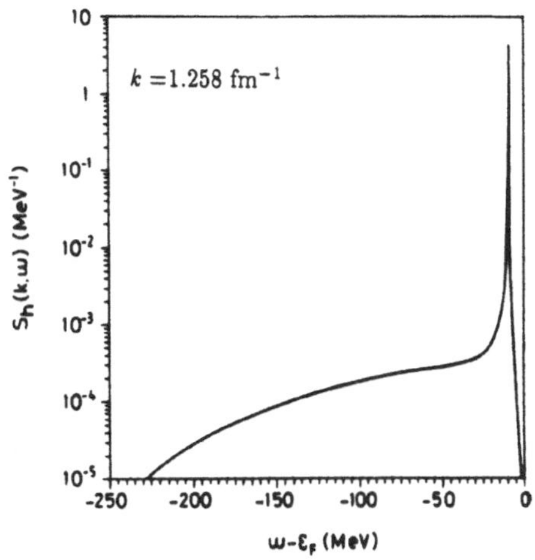

(b)

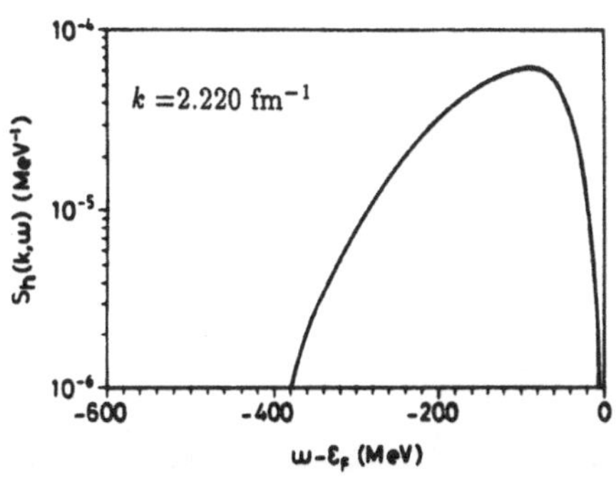

(c)

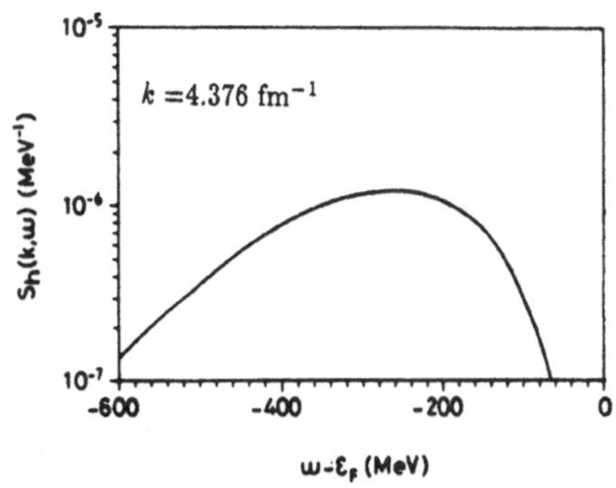

(d)

Figure 1. Hole spectral functions at $k_F = 1.4$ fm^{-1} for k =0.828, 1.258, 2.220 and 4.376 fm^{-1}

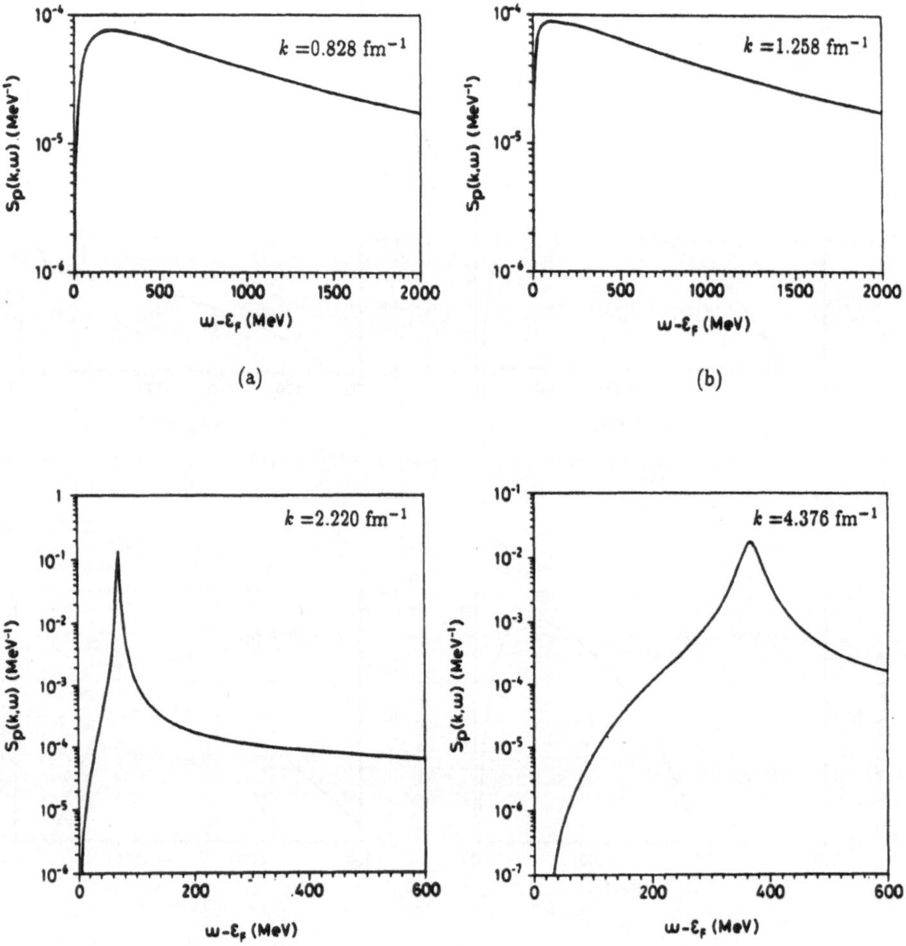

Figure 2. Particle spectral functions at $k_F = 1.4$ fm^{-1} for k =0.828, 1.258, 2.220 and 4.376 fm^{-1}

The broadening of the spectral function increases for momenta further away from the Fermi momentum as is evident from the comparison of figs. 1a and 1b. To obtain a more quantitative measure of the broadening of the *sp* strength it is useful to study the quasi-particle properties associated with the spectral functions. By expanding the self-energy $\Sigma(k,\omega)$ around the quasi-particle energy (see eq.(3)), one obtains the following quasi-particle contribution to the spectral function

$$S_{QP}(k,\omega) = \frac{1}{\pi}\frac{z^2(k) \mid W(k) \mid}{(\omega-\epsilon(k))^2+(z(k)W(k))^2} \tag{6}$$

with $W(k) = Im\ \Sigma(k,\epsilon(k))$, $z(k) = \left\{1-\left(\frac{\partial Re\ \Sigma(k,\omega)}{\partial\omega}\right)_{\omega=\epsilon(k)}\right\}^{-1}$ and $\epsilon(k)$ given by eq.(3).

The on-shell imaginary part of the self-energy W is related to the width of the quasi-particle peak. The $v_2^{l=0}$ interaction is not very efficient in admixing 2h-1p states and therefore produces widths for small momenta which are of the order of 1 MeV much smaller than second order G matrix results for the Paris potential.[17] This emphasizes the importance of the details of the interaction to describe the mixing of *sp* degrees of freedom with low-lying excitations within a few hundred MeV of the Fermi energy. Tensor components are very important in this respect since these can redistribute the momentum distribution around the Fermi momentum substantially[18] implying that they are able to mix in low-lying as well as intermediate energy excitations with the *sp* states. The results in ref.17 confirm this and lead to widths which are as large as 40 MeV. Similar qualitative features for the hole spectral functions have been recently obtained for a variational treatment of nuclear matter improved by perturbative corrections based on correlated basis functions.[19] The larger widths obtained in ref.19 can be partly ascribed to the tensor correlations that are missing in the present study. It is therefore mainly the influence of the short-range correlations on the spectral functions which are addressed in this work.

The quantity $z(k)$, called the quasi-particle strength, gives a measure of the total strength below the quasi-particle peak. Whereas the quasi-particle spectral function provides an excellent approximation to the peak of the spectral function for momenta not too far from the Fermi momentum, it should be emphasized that it can never replace the full spectral function. To illustrate this point, one obtains for example $z(k=0.9k_F) = 0.83$ (see fig.1b) which represents essentially strength below ϵ_F due to the small width W for this momentum. Integration of the hole spectral function, $S_h(k,\omega)$, over all possible excited states of the $(A-1)$-system gives the occupation probability

$$n(k=0.9k_F) = \int_{-\infty}^{\epsilon_F} d\omega S_h(k=0.9k_F,\omega) = 0.86, \tag{7}$$

which indicates that a 3% of the strength is not located below the quasi-particle peak and is found mainly at lower energies. The remaining 14% of strength is moved beyond the Fermi energy and belongs to the particle spectral function shown in fig.2b. This has been numerically checked by integrating the particle spectral function $S_p(k,\omega)$ over all excited states of the $(A+1)$-system.

The momentum distribution derived from eq.(2) is shown in fig.3 for the $v_2^{l=0}$ interaction at a Fermi momentum $k_F = 1.4$ fm^{-1}. The analytical properties of the self-energy for normal Fermi systems imply a discontinuity[20] $z(k_F) = n(k_F^-) - n(k_F^+)$ in the momentum distribution. The value $z(k_F)$=0.83 is compatible with those for realistic potentials which give a discontinuity of the order of 0.7 at normal density.[22,18,21] The smaller value obtained for realistic interactions can be ascribed to the tensor correlations which modify substantially the momentum distribution around the Fermi momentum k_F.[18] We observe a decrease of the gap as the density of the system increases ($z(k_F)$=0.75 and 0.66 at k_F=1.6 and 1.8 fm^{-1} respectively) as a consequence of the larger amount of interparticle correlations present in a denser system. No evidence for an exponential decrease of the momentum distribution as a function of k is observed in the present calculation.

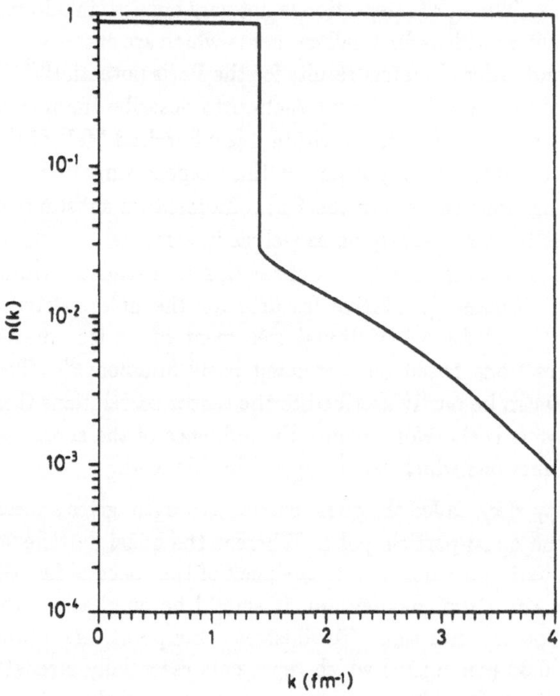

Figure 3. Momentum distribution at $k_F = 1.4$ fm^{-1}.

As discussed before most of the depletion effect studied in this work for the $v_2^{l=0}$ interaction is due to short-range correlations. As a reasonable measure of these correlations one can consider the 13% depletion found at $k_F = 1.4$ fm^{-1} for $k = 0$. This number is more or less the same for all momenta below k_F and should therefore provide a reasonable comparison with occupation numbers obtained in finite nuclei. On top of this effect of short-range correlations one still has to consider the effect of the tensor force as well as the influence of long-range correlations in a particular nucleus due to surface excitations or giant resonances taking care to avoid double counting.

4. Summary and Conclusions

The influence of short-range correlations on the nucleon spectral functions has been studied in the framework of a Self-Consistent Green's Function formalism. The ladder equation for the effective interaction has been solved treating *hh* propagation on the same footing as *pp* propagation. This allows a complete study of the k- and ω- dependence of the self-energy and, consequently, of the particle and hole spectral functions.

We have found that the short-range repulsion of the NN interaction redistributes the strength over a wide range of energies. For the $v_2^{l=0}$ interaction, representing the central part of the $^3S_1 - ^3D_1$ component of the Reid potential, about a 10% of the *sp* strength is moved to very high energies at normal density. A better estimate of the widths of the spectral functions, and consequently of the validity of a single particle motion picture, can be made by studying the quasi-particle properties associated to the spectral functions. For momenta close to k_F we have found $z(k) = 0.83$ to be compared to $n(k) = 0.86$. These results show that 3% of the strength is not located below the quasiparticle peak but moved at lower values of ω as a consequence of the coupling of the *sp* states to 2h-1p, 3h-2p,... configurations. The additional 14% of strength belongs to the particle spectral function. This also reflects the existence of correlations since it indicates that it is possible to add a nucleon into a state that would be fully occupied in a mean field description.

The integral over ω of $S_h(k,\omega)$ yields the momentum distribution $n(k)$ in nuclear matter. Around normal density, we have obtained an average depletion of 13% which is mostly due to the short-range correlations effects. Although the effect of the tensor force and the influence of long-range correlations are expected to produce a further reduction, we can conclude from the present study that a substantial portion of the total depletion is due to short-range correlations.

REFERENCES

1. S. Frullani and J. Mougey, *J. Adv. Nucl. Phys.* **14** (1984) 1
2. E. N. M. Quint et al., *Phys. Rev. Lett.* **58** (1987) 1088
3. J. M. Cavedon et al., *Phys. Rev. Lett.* **49** (1982) 978
4. B. Frois and C. N. Papanicolas, *Ann. Rev. Nucl. Part. Sci.* **37** (1987) 133
5. V. R. Pandharipande, C. N. Papanicolas and J. Wambach, *Phys. Rev. Lett.* **53** (1984) 1133
6. M. Jaminon, C. Mahaux and H. Ngo, *Nucl. Phys.* **A440** (1985) 228
7. J. W. A. den Herder et al., *Phys. Lett.* **184B** (1987) 11
8. A. A. Abrikosov, L. P. Gorkov and I. E. Dzyaloshinski, *Methods of quantum field theory in statistical physics* (Dover, New York, 1975)
9. A.Ramos, A.Polls and W.H.Dickhoff, preprint
10. K. A. Brueckner, C. A. Levinson and H. M. Mahmoud, *Phys. Rev.* **95** (1954) 217
11. A. Ramos, A. Polls and W. H. Dickhoff, *Condensed Matter Theories Vol.3*, eds. J. S. Arponen, R. F. Bishop and M. Manninen (Plenum, New York, 1988)
12. A. Ramos, W. H. Dickhoff and A. Polls, *Phys. Lett.* **219B** (1989) 15
13. R. V. Reid, *Ann. Phys.* **50** (1968) 411

14. W. H. Dickhoff, *Condensed Matter Theories Vol.4* (Plenum, New York, 1989) in press
15. M. G. E. Brand, K. Allaart and W. H. Dickhoff, *Phys. Lett.* **214B** (1988) 483
16. W. H. Dickhoff, M. G. E. Brand, K. Allaart, A. Ramos and A. Polls, *Momentum distributions*, R. N. Silver and P. E. Sokol, eds. (Plenum, New York, 1989), in press
17. P. Grangé, J. Cugnon and A. Lejeune, *Nucl. Phys.* **A473** (1987) 365
18. S. Fantoni and V. R. Pandharipande, *Nucl. Phys.* **A427** (1984) 473
19. O. Benhar, A. Fabrocini and S. Fantoni, Preprint INFN-ISS 88/4 (1988)
20. A. B. Migdal, *Soviet Phys. JETP* **5** (1957) 333
21. M. F. Flynn, J. W. Clark, R. M. Panoff, O. Bohigas and S. Stringari, *Nucl. Phys.* **A427** (1984) 253
22. J. Jeukenne, A. Lejeune and C. Mahaux, *Phys. Reports* **25C** (1976) 83

MESON AND QUARK DEGREES OF FREEDOM

IN NUCLEAR MATTER AND FINITE NUCLEI

Qi-Ren Zhang

Center of Theoretical Physics CCAST (World Laboratory)
Department of Technical Physics
Peking University, Beijing, China

INTRODUCTION

Since the middle of fifties, people had tried to build a fundamental nuclear theory on the basis of nonrelativistic quantum many-body theory for nucleons interacting each other by two-body nuclear forces. But in the late seventies, it turned out that this is impossible[1,2]. Both relativity and meson degrees of freedom should be considered. Because of the quark substructure of nucleon, logically the quark degrees of freedom in nuclei should also be considered. This last point is strengthened by the discovery of EMC effect.

In this lecture, I would report some works on meson and quark degrees of freedom in nuclear matter and finite nuclei, done in Peking University and in Frankfurt University.

MESON DEGREES OF FREEDOM

Pi, scalar, and vector meson degrees of freedom in nuclear matter have been considered since early seventies[3-7]. Among those, the scalar meson condensation played an important role. By condensation we mean a nonzero expectation value of the corresponding field.

Scalar Meson Condensation

The simplest Lagrangian density of a nucleon-scalar meson system is

$$L = -\bar{\Psi}\left[\gamma_\mu \partial_\mu + (m - g\Phi)\right]\Psi - \frac{1}{2}\left(\partial_\mu \Phi \partial_\mu \Phi + m_\sigma^2 \Phi^2\right). \tag{1}$$

We see the scalar field Φ may directly cancel the nucleon mass m, therefore may develop a nuclear binding. From (1) we also see that the Φ dependence of nucleon energy is of the first order, while the scalar field energy is, at least, of the second order of Φ. It means there must be a nonzero Φ value for extremizing the energy of the nuclear state. In other words, if there is a scalar meson field interacting with nucleons in the sence of (1), there will always be a scalar meson condensation in nuclear matter and in finite nuclei.

The Nuclear Equation of State, Part A
Edited by W. Greiner and H. Stöcker
Plenum Press, New York

Assuming nuclear binding is dominated by the scalar meson condensation, one can calculate nuclear properties. To achieve the right nuclear saturation, one needs a repulsive core of nucleon or a nonzero vector meson field.

Nuclear Matter. In the nature units of $\hbar = c = m = 1$, energy per-nucleon in a zero-temperature infinite nucleon-scalar meson system is

$$\varepsilon = \chi + \frac{3p^2}{10\chi} + \frac{4}{9\pi\alpha}(1-\chi)^2 r_0^3, \tag{2}$$

in which

$$\chi = 1 - \frac{g\Phi}{m} \tag{3}$$

is the ratio between the effective nucleon mass in nuclear matter and the free nucleon mass m,

$$p = (9\pi)^{\frac{1}{3}} / 2r_0 \tag{4}$$

is the Fermi momentum of nucleon in nuclear matter,

$$v = \frac{4}{3}\pi r_0^3 \tag{5}$$

is the average volume occupied by a nucleon, and

$$\alpha = \frac{2}{3}\left(\frac{gm_\sigma}{\pi m}\right)^2 \tag{6}$$

In (2), the sum of first two terms is the average energy of a nucleon in nuclear matter, and the last term is the scalar field energy per-nucleon. If there is a hard repulsive core of radius r_c in a nucleon, Fermi momentum (4) should be replaced by

$$p = \frac{(9\pi)^{\frac{1}{3}}}{2(r_0 - 0.8r_c)}, \tag{7}$$

according to the Van der Waals approximation[3,8]. Taking the volume energy 15.677MeV from the empirical formula of nuclear binding energy[9], we obtain the 'empirical' value of $\varepsilon = 0.9833$. The corresponding 'empirical' value of r_0 is 1.2049fm for a ground state nuclear matter. Substituting these values and (7) into (2), we obtain

$$\chi + \frac{0.3(9\pi)^{2/3}}{4\chi(r_0 - 0.8r_c)^2} + 4(1-\chi)^2 r_0^3/(9\pi\alpha) = 0.9833\,. \tag{8}$$

The stationary condition for the ground state nuclear gives

$$\frac{\partial\varepsilon}{\partial\chi} = 1 - \frac{0.3(9\pi)^{2/3}}{4\chi^2(r_0 - 0.8r_c)^2} - 8(1-\chi)r_0^3/(9\pi\alpha) = 0\,, \tag{9}$$

$$\frac{\partial\varepsilon}{\partial r_0}=\frac{-0.6(9\pi)^{2/3}}{4\chi(r_0-0.8r_c)^3}+4(1-\chi)^2r_0^2/(3\pi\alpha)=0. \tag{10}$$

Solving equations (8)-(10), we obtain $\chi = 0.85$ for ground state nuclear matter,

$$r_c = 0.57fm\,, \qquad \text{and} \qquad \alpha = 8.82\,. \tag{11}$$

From (9) we may solve χ as a function of r_0. Substituting this result, (7) and (11) into (2), we may get the nuclear equation of state in this model.

If the repulsive core is replaced by a vector meson field, (1) becomes

$$L=-\bar{\Psi}[\gamma_\mu\partial_\mu+(m-g\Phi)]\Psi-\frac{1}{2}(\partial_\mu\Phi\partial_\mu\Phi+m_\sigma^2\Phi^2)-\frac{1}{4}F_{\mu\nu}F_{\mu\nu}-\frac{1}{2}m_v^2V_\mu V_\mu+ig_v\bar{\Psi}\gamma_\mu\Psi V_\mu\,, \tag{12}$$

with $F_{\mu\nu} = \partial_\mu V_\nu - \partial_\nu V_\mu$. The energy per-nucleon in nuclear matter now is

$$\varepsilon = \chi + 0.3p^2/\chi + (1-\chi)^2/(2\alpha p^3) + \gamma p^3/2. \tag{13}$$

The last term is the contribution of the vector field, in which

$$\gamma = \frac{2}{3}\left(\frac{g_v m_v}{\pi m}\right)^2 . \tag{14}$$

Fermi momentum is expressed in (4). The correspondence of (8)-(10) is

$$\chi + 0.3p^2/\chi + (1-\chi)^2/(2\alpha p^3) + \gamma p^3/2 = 0.9833, \tag{15}$$

$$\frac{\partial\varepsilon}{\partial\chi} = 1 - 0.3p^2/\chi^2 - (1-\chi)/\alpha p^3 = 0, \tag{16}$$

$$\frac{\partial\varepsilon}{\partial p} = 0.6p/\chi - 3(1-\chi)^2/(2\alpha p^4) + 3\gamma p^2/2 = 0. \tag{17}$$

Solving these equations, we obtain $\chi = 0.56$ for ground state nuclear matter,

$$\alpha = 25.3, \qquad \text{and} \qquad \gamma = 19.4. \tag{18}$$

From (16) one may solve χ as a function of r_0 . Substituting it with (4) and (18) into (13), we may get the nuclear equation of state in this model.

Finite Nuclei.[10,11] We may test the nuclear equation of state by comparing its prediction on properties of finite nuclei with experiments. Assume (2) and (7) or (13) and (4) valid locally. It means we would use the Thomas-Fermi method. In the model with a hard repulsive core, we have for a spherical nucleus

$$A = \alpha\beta\int_0^{\xi_0}\frac{p^3}{(1+bp)^3}\xi^2 d\xi, \tag{19}$$

$$Z = \alpha\beta \int_0^{\xi_0} \frac{1-\iota}{2} \frac{p^3}{(1+bp)^3} \xi^2 d\xi, \tag{20}$$

$$E_1 = \alpha\beta \int_0^{\xi_0} \{\chi + 0.15[(1+\iota)^{5/3} + (1-\iota)^{5/3}]p^2/\chi\} \frac{p^3}{(1+bp)^3} \xi^2 d\xi\,, \tag{21}$$

$$E_2 = \frac{\beta}{2} \int_0^{\infty} [\left(\frac{d\chi}{d\xi}\right)^2 + (1-\chi)^2]\xi^2 d\xi\,, \tag{22}$$

$$b = 1.6r_c/(9\pi)^{1/3}\,, \qquad \beta = 4\pi m/(g^2 m_\sigma). \tag{23}$$

A, Z, E_1, and E_2 are mass number, charge number, nucleon energy and scalar field energy of the nucleus respectively. ι is the local neutron excess. ξ is the distance between the point under consideration and the center of the nucleus measured by the compton wave length of the scalar meson. ξ_0 is the nucleus radius in the same unit. Minimizing the nucleus energy $E = E_1 + E_2$ under fixed A and Z, we obtain

$$\frac{d^2\chi}{d\xi^2} + \frac{2d\chi}{\xi d\xi} - \chi = \frac{\alpha p^3}{(1+bp)^3}\{1 - 0.15[(1+\iota)^{5/3} + (1-\iota)^{5/3}]\frac{p^2}{\chi^2}\} - 1 \tag{24}$$

from $\delta E/\delta\chi = 0$,

$$[(1+\iota)^{5/3} + (1-\iota)^{5/3}](2bp^3 + 5p^2) + 20\chi(\chi - \chi_0 + \lambda\frac{1-\iota}{2}) = 0 \tag{25}$$

from $\delta E/\delta p = 0$, and

$$[(1+\iota)^{2/3} - (1-\iota)^{2/3}]p^2 - 2\lambda\chi = 0 \tag{26}$$

from $\delta E/\delta\iota = 0$. χ_0 and λ are Lagrange multipliers.

The correspondence of (19)-(26) in the model with a vector meson field instead of the repulsive core is

$$A = \alpha\beta \int_0^{\xi_0} p^3 \xi^2 d\xi, \tag{27}$$

$$Z = \alpha\beta \int_0^{\xi_0} \frac{1-\iota}{2} p^3 \xi^2 d\xi, \tag{28}$$

$$E_1 = \alpha\beta \int_0^{\xi_0} \{\chi + 0.15[(1+\iota)^{5/3} + (1-\iota)^{5/3}]p^2/\chi\} p^3 \xi^2 d\xi\,, \tag{29}$$

$$E_2 = \frac{\beta}{2} \int_0^{\infty} [\left(\frac{d\chi}{d\xi}\right)^2 + (1-\chi)^2]\xi^2 d\xi, \tag{30}$$

$$E_3 = \alpha\beta \int_0^{\xi_0} p^3 u \xi^2 d\xi - \frac{\alpha\beta}{2\gamma\eta} \int_0^{\infty} [\left(\frac{du}{d\xi}\right)^2 + \eta u^2]\xi^2 d\xi, \qquad \eta = \left(\frac{m_v}{m_\sigma}\right)^2, \tag{31}$$

$$E = E_1 + E_2 + E_3 \,, \tag{32}$$

$$\frac{d^2\chi}{d\xi^2} + \frac{2d\chi}{\xi d\xi} - \chi = \alpha p^3\{1 - 0.15[(1+\iota)^{5/3} + (1-\iota)^{5/3}]\frac{p^2}{\chi^2}\} - 1 \,, \tag{33}$$

$$\frac{d^2u}{d\xi^2} + \frac{2du}{\xi d\xi} - \eta u = -\gamma\eta p^3, \tag{34}$$

$$[(1+\iota)^{5/3} + (! - \iota)^{5/3}]p^2 + 4\chi(\chi - \chi_0 + \lambda\frac{1-\iota}{2} + u) = 0, \tag{35}$$

$$[(1+\iota)^{2/3} - (1-\iota)^{2/3}]p^2 - 2\lambda\chi = 0. \tag{36}$$

A Simple Relation Between Nuclear Surface Energy And The Mass Of Scalar Meson[10]. For symmetric nuclei, λ and ι equal zero, (26) and (36) become identities. In this case, a regular solution $\chi(\xi)$ of (24) and (25) satisfying boundary condition $\chi(\infty) = 1$ can be found for a given ξ_0 and a characteristic value of χ_0. Parameters (11) should be used. From (25) and $\chi(\xi)$, we may find $p(\xi)$.

Using equation (24), we may, after a partial integration, write (22) in the form

$$E_1 = \alpha\beta\int_0^{\xi_0} \frac{p^3}{(1+bp)^3}\{1 - 0.15[(1+\iota)^{5/3} + (1-\iota)^{5/3}]\frac{p^2}{\chi^2}\}\frac{1-\chi}{2}\xi^2 d\xi. \tag{37}$$

Substituting the solved $\chi(\xi)$ and $p(\xi)$ into (19), (21), and (37), putting $\iota = 0$, we may calculate A and $E = E_1 + E_2$ for the symmetric nucleus up to a factor

$$\alpha\beta = \frac{8m^3}{3\pi m_\sigma^3}. \tag{38}$$

The binding energy per-nucleon

$$B(MeV)/A = 939(1 - E/A) \tag{39}$$

is independent of this yet unknown factor, and is a function of

$$J = \int_0^{\xi_0} \frac{p^3}{(1+bp)^3}\xi^2 d\xi. \tag{40}$$

According to the phenomenological liquid drop model for nuclei,

$$B/A = E_v - E_s A^{-1/3} = E_v - E_s(\alpha\beta)^{-1/3}J^{-1/3} \tag{41}$$

should be a linear function of $J^{-1/3}$. This is indeed the case in our numerical results. From (41) and (38) we further see

$$E_s m_\sigma = (8/3\pi)^{1/3} m\sigma, \tag{42}$$

$-\sigma$ is the slope of the straight line $(J^{-1/3}, B/A)$. Substituting the calculated value $\sigma = 9.86$ and the empirical value $E_s = 18.56 MeV$ into (42), and using (6) and (11), we obtain

$$m_\sigma = 472 MeV\ , \qquad\qquad g^2/4\pi = 2.63. \tag{43}$$

This is quite reasonable. I would emphasize that the calculated scalar meson mass in (43) is just in the range required by the OBEP theory of nuclear force. It shows a kind of selfconsistency in the theory of nuclear force and nuclear structure on the basis of nucleon-meson field theory. This point is confirmed by the latter Hatree-Fock calculations for finite nuclei[12].

Nuclear Binding Energy Formula[11]. For a given set of parameters $\xi_0 \geq \xi_1$, there is a regular solution $\chi(\xi)$, $p(\xi)$, and $\iota(\xi)$ of equations (24)-(26), satisfying $\chi(\infty) = 1$, $p(\xi \geq \xi_0) = 0$, and $\iota(\xi_1 \leq \xi < \xi_0) = 1$. It means there is a neutron skin for nonsymmetric nuclei having $\xi_1 < \xi_0$. Substituting the solution into (19)-(21) and (37), we calculated A, Z, and $E = E_1 + E_2$. The result has been well fitted by a nuclear binding energy formula

$$\begin{aligned} B(MeV) =& [15.677 - 18.56A^{-1/3} - 5.346A^{-2/3} + 9.327A^{-1} + 2.128A^{-4/3} \\ & - (32.634 - 63.318A^{-1/3} + 49.35A^{-2/3})I^2 \\ & + (10.915 - 5.915A^{-1/3})I^4 - 1.874I^6]A - E_c, \end{aligned} \tag{44}$$

$$\begin{aligned} E_c(MeV) =& 0.71775[1 + 0.00847A^{-1/3} - 1.1264A^{-2/3} + 1.440A^{-1} \\ & - 0.873A^{-4/3} - (0.00646 - 1.0313A^{-1/3} + 2.061A^{-2/3} \\ & - 1.694A^{-1})I - (0.2282 - 1.0093A^{-1/3} + 1.730A^{-2/3})I^2 \\ & - (0.0546 - 0.834A^{-1/3})I^3 + 0.4I^4]Z^2/A^{1/3}\ , \end{aligned} \tag{45}$$

in which $I = (A - 2Z)/A$ is the neutron excess of the nucleus, the coulomb energy E_c was calculated by perturbation from the solved charge distribution. The deviation of our 'derived' formula (44)-(45) from the 'empirical' formula[9] is less than 4% for all thousands existing nuclei. Since there are only three free parameters in our model, namely the scalar meson mass m_σ, the scalar meson-nucleon coupling constant g , and the hard core radius r_c, and since their values are quite reasonable, this result is not bad. Further more, the calculated binding energies are systematically smaller than the empirical ones. It means, a part of deviation may be removed by the minimization of coulomb energy.

Pi-Condensation[13–15]

Unlike scalar mesons, pseudoscalar mesons may be condensed only under some special conditions, such as high density, high temperature and so on. To find out the critical condition for pi-condensation is interesting.

Exact Solution Of Dirac Equation For A Nucleon In Plane Pi-Field. The Dirac equation of a nucleon in the pi-field $\vec{\Phi}$ is

$$(\gamma_\mu \partial_\mu + m - ig\gamma_5 \vec{\tau} \cdot \vec{\Phi})\Psi = 0 \qquad \text{(pseudoscalar coupling)}, \tag{46}$$

or

$$(\gamma_\mu \partial_\mu + m - if\gamma_5 \gamma_\mu \partial_\mu \vec{\Phi} \cdot \vec{\tau}/m_\pi)\Psi = 0 \qquad \text{(pseudovector coupling)}\ . \tag{47}$$

g and f are corresponding coupling constants. Substituting plane wave

$$\vec{\Phi}(x) = \phi \begin{pmatrix} \cos k_\mu x_\mu \\ \sin k_\mu x_\mu \\ 0 \end{pmatrix}, \qquad \phi = constant \tag{48}$$

into (46), and using the transformation

$$\Psi = \exp(-i\tau_3 k_\mu x_\mu/2)\psi, \tag{49}$$

we obtain

$$[\gamma_\mu(\partial_\mu - i\tau_3 k_\mu/2) + m - ig\gamma_5\tau_1\phi]\psi = 0. \tag{50}$$

This equation is already x_μ independent, therefore its solution may be written in the form

$$\psi = \exp(ip_\mu x_\mu)SU, \tag{51}$$

in which U is a constant spinor, and

$$S = [(A + p_\mu k_\mu)/2A]^{1/2} - g\phi\tau_1\gamma_5\gamma_\mu k_\mu/[2A(A + p_\mu k_\mu)]^{1/2}, \tag{52}$$

$$A = [(p_\mu k_\mu)^2 + g^2\phi^2 k_\mu k_\mu]^{1/2}. \tag{53}$$

Substituting (51) with (52) and (53) into (50), we have

$$(i\gamma_\mu Q_\mu + m)U = 0, \tag{54}$$

$$Q_\mu = p_\mu + [\frac{g^2\phi^2}{(A + P_\nu k_\nu)} - \tau_3/2]k_\mu. \tag{55}$$

(54) has the form of the Dirac equation for a free particle, therefore has been exactly solved:

$$Q_\mu Q_\mu = -m^2 \qquad or \qquad (Q_\mu) = (\vec{Q}, i\sqrt{Q^2 + m^2}), \tag{56}$$

$$U = (i\gamma_\mu Q_\mu - m)U_0, \tag{57}$$

U_0 is an arbitrary bispinor in Minkowski space-time and a spinor in isospin space, independent of space-time coordinates x_μ.

Substituting (48) into (47), using the transformation

$$\Psi = \exp(-i\tau_3 k_\mu x_\mu/2)S_1\psi \tag{58}$$

with

$$S_1 = [\frac{1+B}{2B}]^{1/2} - \frac{if}{m_\pi}\phi[\frac{2}{B(1+B)}]^{1/2}\tau_1\gamma_5 , \tag{59}$$

$$B = (1 + 4f^2\phi^2/m_\pi^2)^{1/2}, \tag{60}$$

we obtain

$$[\gamma_\mu(\partial_\mu - i\tau_3 k'_\mu/2) + m' - ig'\gamma_5\tau_1\phi]\psi = 0, \tag{61}$$

$$m' = m/B, \qquad g' = 2m'f/m_\pi, \qquad k'_\mu = Bk_\mu. \tag{62}$$

(61) is almost identical with (50) except that m, g, and k_μ are replaced by m', g', and k'_μ respectively. Therefore the solution (51)-(57) for (50) is also applicable to (61), if this same replacement has been made.

Lorentz Covariant Mechanism For Pi-Condensation. We see that the nucleon in a pi- field looks like a free Dirac particle. This quasiparticle has two different quasimomenta, namely, (p_μ) and (Q_μ). The former is responsible for the state numeration , while the latter is responsible for the energy calculation. The appearance of two different quasimomenta distorts the Fermi sea of the quasiparticle system, therefore raises its energy. On the other hand, the pseudovector pi-nucleon coupling lowers the effective mass of the quasiparticle, therefore lowers its energy. This is the mechanism for the pi-condensation. The competition between these two tendencies determines the critical condition of pi-condensation.

It is well known that a pure pseudovector coupling is unrenormalizable. However, there are renormalizable models, such as σ models, in which the pi-nucleon interaction may be reduced to an effective pseudovector coupling, sothat the above argument for pi-condensation is valid.

QUARK DEGREES OF FREEDOM

Until now there is not a satisfactory framework for considering quark degrees of freedom in nuclei. We tried some simple models for it.

MIT Bag Crystal Model For Nuclear Matter

My German collaborators and I started this work in 1984 when I visited Frankfurt University. Since a nucleon is well described by a bag containing quarks and gluons, a nucleus must be an aggregation of bags. For simplicity, also because there are some hints for a crystalization of nuclear matter, we assumed that bags are regularly placed in nuclear matter. They touch each other and open windows between adjacent bags. Quarks and gluons, move freely in bags except a color perturbation between them . They satisfy MIT boundary condition on the bag surface, and periodic conditions on windows. We call these conditions a mixed boundary condition.

Because of the translational symmetry, we need to consider only one bag in the lattice. The periodic condition requirs that the wave function on a pair of windows of this bag at direction s should be the same except a phase factor $\exp(ip_s a_s)$, a_s is the distance between this two windows. The pseudomomentum $\vec{p}$ of the Bloch quark comes in here.

Expand the quark wave function in terms of spherical wave functions:

$$\Psi = \sum_{\kappa\mu} c_{\kappa\mu}\psi_{\kappa\mu}, \qquad \psi_{\kappa\mu} = N_\kappa \begin{pmatrix} j_{l_\kappa}(\varepsilon r)\chi_{\kappa\mu}(\vartheta\varphi) \\ i sign(\kappa) j_{l_{-\kappa}}(\varepsilon r)\chi_{-\kappa\mu}(\vartheta\varphi) \end{pmatrix}. \tag{63}$$

The mixed boundary condition is linear, therefore may be expressed in the form

$$\sum_{\kappa\mu} G_{\kappa\mu}(\vartheta\varphi)c_{\kappa\mu} = 0, \tag{64}$$

each for a direction $\vartheta\varphi$. The scalar production of the rhs of (64) and its Hermite adjoint, integrated over 4π solid angle, gives

$$\sum_{\kappa\mu}\sum_{\kappa'\mu'} c^\star_{\kappa\mu} M_{\kappa\mu,\kappa'\mu'} c_{\kappa'\mu'} = 0, \tag{65}$$

$$M_{\kappa\mu,\kappa'\mu'} = \int G^\dagger_{\kappa\mu}(\vartheta\varphi)G_{\kappa'\mu'}(\vartheta\varphi)d\Omega, \tag{66}$$

$\mathrm{M} = (M_{\kappa\mu,\kappa'\mu'})$ is a semi-positive definite Hermitian matrix. (65) is equivalent to the condition (64). We may solve quark energy ε and wave function $(c_{\kappa\mu})$ from it as functions of pseudomomentum $\vec{p}$. In this way we may obtain the quark energy band and corresponding Bloch wave functions.

The rhs of (65) is non-negative. It means the condition (65) is equivalent to the minimization of its rhs. This in turn is equivalent to the minimum eigenvalue problem of matrix M. In practical solution, we have to truncate the summation in (65) up to a maximum absolute value κ_m of κ. Denote the truncated matrix by $\mathrm{M}(\kappa_{\mathrm{m}})$. Our problem is approximated by the minimum eigenvalue problem of $\mathrm{M}(\kappa_{\mathrm{m}})$. If there is a solution of (65), the minimum eigenvalue M_0 of $\mathrm{M}(\kappa_{\mathrm{m}})$ should monotonically approach zero when κ_m approaches infinite.

We solved the problem up to $\kappa_m = 7$ in 1985 for a simple cubic lattice , and have got reasonable results on quark energy band and its contribution to the energy per-nucleon in nuclear matter[16]. In the numerical calculation we have found a $\Delta\mu = 4$ selection rule for the matrix element of M. We proved this selection rule after my returning China by writing down all irreducible representations of the point group for the simple cubic lattice. My student and I extended the calculation up to $\kappa_m = 15$ by use of this selection rule. We have not seen any essential change of our previous results after this extension.

Because there is no simple way to calculate the Casimir energy in a bag crystal, we reparametrized the bag model. Namely, instead of the Casimir energy, we used the surface energy in our model. For an example, parameters

$$B^{1/4} = 114MeV, \qquad S^{1/3} = -34.1MeV, \qquad and \qquad \alpha_c = 0.77, \tag{67}$$

give exact masses for nucleons(939MeV) and Δ-isobars((1232MeV) . S is the surface tension coefficient of the bag. Applying these parameters to the bag crystal, we calculated the energy per-nucleon as a function of lattice constant a , that is the nuclear equation of state. The calculation of color magnetic energy in nuclear matter has not yet been finished. The result without this contribution shows a saturation at a=2.13fm with the energy per-nucleon equals 805MeV. We hope that the gap between this calculated energy and the corresponding experimental value 923MeV could be filled up by the color magnetic energy, and the calculated equilibrium density will also be improved by this color interaction.

MIT Bag Molecule Model For Finite Nuclei

The basic strategy in our bag crystal model is to reduce the full problem of nuclear matter to one-bag problem by use of its assumed translational symmetry. The same method may be used in considering quark degrees of freedom in finite nuclei. It is to find out the geometric symmetry of the nucleus first, then reduce the problem of the nucleus to its one-bag problem. We have applied this method to two- and three-nucleon system[17], and have got reasonable results.

Quantum Bag Dynamics

Beside the dynamics of quarks and gluons, there must also be dynamics of hadrons in nuclei. According to the bag model of hadrons, it should be a Quantum Bag Dynamics. We realized the MIT boundary condition for dynamic bags in a simple way[18], applied it in considering the collective excitation of hadrons and the two-nucleon problem, and formulated a general framework for its further development.

ACKNOWLEDGEMENT

This work is partially supported by the NSFC of China.

REFERENCES

1. V. R. Pandharipande and R. B. Wiringa, **Rev. Mod. Phys.** 51: 821(1979).
2. B. D. Day, **Phys. Rev. Lett.** 47:226 (1981).
3. A. B. Migdal, **JETP**. 61:2209 (1971) (in Russian).
4. D. P. Sawyer, **Phys. Rev. Lett.** 29:382 (1972).
5. D. J. Scalapino, **Phys. Rev. Lett.** 29:386 (1972).
6. T. D. Lee and G. C. Wick, **Phys. Rev.** D9:2291 (1974).
7. J. D. Walecka, **Ann. Phys.** 83:491 (1974).
8. A. Bohr and B. R. Mottelson, "Nuclear Physics" Benjamin, New York (1969).
9. W. D. Myers and W. J. Swiatecki, **Nucl. Phys.** 81:1 (1966).
10. Q.-R. Zhang, **Phys. Ener. Fort. Phys. Nucl.** 3:75 (1979) (in Chinese).
11. Q.-R. Zhang, **Phys. Ener. Fort. Phys. Nucl.** 4:576 (1980) (in Chinese).
12. J. Boguta, **Nucl. Phys.** A372:386 (1981).
13. Q.-R. Zhang, **Phys. Ener. Fort. Phys. Nucl.** 5:15 (1981) (in Chinese).
14. Q.-R. Zhang, **Phys. Ener. Fort. Phys. Nucl.** 5:314 (1981) (in Chinese).
15. Q.-R. Zhang, **Phys. Lett.** B104:347 (1981).
16. Q.-R. Zhang, C. Derreth, A. Schaefer, and W. Greiner, **J. Phys.** G12:L19 (1986).
17. Q.-R. Zhang, B. Li, L.-S. Yuan, and Z.-X. Li, A Few Nucleon System As A MIT Bag Molecule, in:"Proceedings of International Symposium on Medium Energy Physics", H.-C. Chiang and L.-S. Zheng, ed., World Scientific, Singapore-New Jersey-Hong Kong (1987).
18. Q.-R. Zhang, **J. Phys.** G14:287 (1988).

The Relativistic Mean-Field Model of Nuclear Structure and Dynamics

P.-G. Reinhard and H.G. Döbereiner

Universtät Erlangen, D-8520 Erlangen

and

V. Blum, J. Fink, M. Rufa, J. Maruhn, H. Stöcker, and W. Greiner

Universität Frankfurt, D-6000 Frankfurt

1 Introduction

The relativistic mean-field model has been succesfully applied in various fields of nuclear physics, heavy ion physics and astrophysics, for reviews see [1,2,3]. The model fascinates by its simplicity. It consists of Dirac-nucleons interacting via mesonic mean-fields and the contributions from the Dirac-sea of antinucleons is usually neglected. Yet, the relativistic mean-field model provides an excellent description of nuclear ground-state properties and several other nuclear features. It can very well compete with similar nonrelativistic models. But it has the additional bonus that the spin-orbit force and other spin properties are well determined within the relativistic model.

It is the aim of this contribution to discuss the limits of the relativistic mean-field model which became apparent in some more recent applications. In particluar, we concentrate on one puzzling feature: the very low effective masses which indicate very strong scalar and vector fields. Unfortunately, all attempts to get direct experimental evidence for such strong fields have failed because all seemingly promising observables are found to be masked by strong correlation and polarization effects [3]. Thus one has to procced the tedious way of collecting material from many observables with indirect information on the effective mass or field strength. We want to present here a few steps into that direction.

The contribution is outlined as follows: In section 2, we present the model. In section 3, we discuss some interpretational aspects and show how the model emerges naturally from any relativistic many-body theory. In section 4, we shortly review the successes of the model in describing nuclear ground-state observables. In section 5, we discuss the equation-of-state for neutron matter and its consequences for the stability of neutron stars. In section 6, we discuss the phase transition from spherical to deformed isotopes of Gadolinium. In section 7, we discuss the stability of RPA excitations in symmetric nuclear matter. And in section 8, we present a model study with pseudodata from a relativistic Bethe-Brückner-Goldstone calculation.

The Nuclear Equation of State, Part A
Edited by W. Greiner and H. Stöcker
Plenum Press, New York

2 The Model

The relativistic mean-field model aims at a relativistic selfconsistent description of nuclear structure and dynamics. The internucleon force is mediated by fields which are independent degrees-of-freedom. In practice, we consider:

σ : An isoscalar-scalar field $\Phi(x^\mu)$.
ω : An isoscalar-vector field $V_\nu(x^\mu)$.
ρ : An isovector-vector field $\vec{R}_\nu(x^\mu)$.
γ : A massless vector field $A_\nu(x^\mu)$, the photon.

The σ-field mediates the medium range attraction between the nucleons, the ω-field mediates a short range repulsion, the ρ-field allows to adjust isovector properties, and the photon, of course, accounts for the known electro-magnetic interaction.

The model is formulated on the basis of two approximations, the *mean-field* and the *no-sea* approximation. The *mean-field* approximiation neglects all quantum fluctuations of the meson fields and treats them as classical c-number fields, i.e. we consider only mean values as $\hat{\Phi} \longrightarrow \Phi = < \hat{\Phi} >$ and similarly for the other fields. Its effect is that the nucleons interact only via the mean fields. They accumulate their influence on the fields and the fields push back on the single nucleons independently. The *no-sea* approximation neglects the vacuum polarization effects, i.e. the contribution of the Dirac sea of antinucleons to the various densities. Thus one deals only with valence nucleons and the densities reduce to simple sums over the occupied nucleon states. For example the scalar density becomes in the so called *no-sea* approach $<: \bar{\psi}\psi :> \sim \sum_{\alpha=1}^{F} \bar{\varphi}_\alpha \varphi_\alpha$ and similarly for the other densities.

The relativistic mean-field model is formulated in terms of a covariant action integral which is understood as an effective action to be used in connection with the *mean-field* and *no-sea* approximations. A standard ansatz has evolved in the course of time which employs the simplest linear coupling between nucleons and fields and which is complemented by a cubic and quartic selfcoupling of the scalar field. The coupled field equations are then derived by standard variational techniques. They become for stationary states

$$\epsilon_\alpha \gamma_0 \varphi_\alpha = (-i\vec{\gamma}\cdot\vec{\nabla} + m_B + g_\sigma \Phi + g_\omega V_0 \gamma_0 + \frac{1}{2} g_\rho R_{00} \gamma_0 \tau_0 + e A_0 \gamma_0 \frac{1+\tau_0}{2}) \varphi_\alpha \qquad (1a)$$

$$-\Delta\Phi + \mathcal{U}'(\Phi) = -g_\sigma \rho_s \quad , \quad \rho_s = \sum_{\alpha=1}^{\Omega} w_\alpha \bar{\varphi}_\alpha \varphi_\alpha \qquad (1b)$$

$$(-\Delta + m_\omega^2) V_0 = g_\omega \rho_0 \quad , \quad \rho_\mu = \sum_{\alpha=1}^{\Omega} w_\alpha \bar{\varphi}_\alpha \gamma_\mu \varphi_\alpha \qquad (1c)$$

$$(-\Delta + m_\rho^2) R_{00} = \tfrac{1}{2} g_\rho \rho_{00} \quad , \quad \vec{\rho}_\mu = \sum_{\alpha=1}^{\Omega} w_\alpha \bar{\varphi}_\alpha \vec{\tau} \gamma_\mu \varphi_\alpha \qquad (1d)$$

$$-\Delta A_0 = e \rho_0^{(proton)} \quad , \quad \rho_\mu^{(proton)} = \sum_{\alpha=1}^{\Omega} w_\alpha \bar{\varphi}_\alpha \tfrac{1+\tau_0}{2} \gamma_\mu \varphi_\alpha \qquad (1e)$$

Therebey we have introduced the occupation weights w_α. They are set to one for the occupied states in magic nuclei and they are determined by pairing in the fixed gap approach for non-magic nuclei, where we take for the gap the commonly used $\Delta = 11.2 MeV/\sqrt{A}$.

The scalar field equation employs $\mathcal{U}' = \frac{\partial}{\partial\Phi}\mathcal{U}$ where $\mathcal{U}$ is the nonlinear energy functional for the scalar field. The simplest choice for $\mathcal{U}$ is just a polynomial expansion

$$\mathcal{U}_{standard} = \frac{1}{2} m_\sigma^2 \Phi^2 + \frac{1}{3} b_2 \Phi^3 + \frac{1}{4} b_3 \Phi^4 \qquad (2)$$

This has become the standard ansatz for $\mathcal{U}$ since the introduction by Boguta and Bodmer [4]. The nonlinear selfcoupling of the scalar field is essential to reach a quantitative description of nuclear properties. The problem is, however, that $b_3 < 0$. This makes the scalar meson equation asymptotically unstable, and even worse, the barrier to this instability is practically close to normal nuclear densities. The ansatz can be stabilised by inventing a form of $\mathcal{U}$ which guarantees that $\mathcal{U}'' > 0$ at all field strengths [5]. The best way is to start from a background value $\mathcal{U}'' = m_\infty^2$ and to shape a peak on top. One choice is a $\cosh^{-2}$ peak

$$\mathcal{U}''_{stable}(\Phi) = m_\infty^2 + \Delta m^2 \cosh^{-2}(\frac{\Phi - \Phi_0}{\delta\Phi}) \tag{3}$$

where the parameters of the standard ansatz, m_σ^2, b_2 and b_3 are replaced by the parameters m_∞^2, Δm^2, $\delta\Phi$ and Φ_0. This peak form is positive definite, and it is twice analytically integrable yielding closed expressions for $\mathcal{U}'$ in the meson equation (1b) and for $\mathcal{U}$. The nonlinear meson functional can also be used to account for the effect of the (neglected) Dirac sea within a local density approximation. One can integrate out the sea contributions which yields an additional nonlinear term in the scalar field [1]

$$\delta\mathcal{U}_{eff}^{Sea} = -\frac{m_B^4}{4\pi}\{(1-S)^4\log(1-S) + S - \frac{7}{2}S^2 + \frac{13}{2}S^3 - \frac{25}{12}S^4\} \tag{4}$$

where $S = \frac{g_\sigma\Phi}{m_B}$ Note that this functional adds only contribution ot power S^5 or higher and increases dramatically with S. The combined functional $\mathcal{U}_{standard} + \delta\mathcal{U}_{eff}^{Sea}$ is also unconditionally stable.

3 The concept of an effective Lagrangian

The Lagrangian density of the relativistic mean-field model is considered to be an effective Lagrangian in connection with the *mean-field* and *no-sea* approximation [3]. The meson fields and their couplings are assumed to parametrise the G-matrix for the nucleon-nucleon scattering in the nuclear medium and possibly quantum field effects on the nucleons. The situation is analogous to the nonrelativistic effective Hamiltonians in connection with Hartree-Fock calculations, as e.g. the Skyrme force. They are developed to parametrise the nonrelativistic nucleon-nucleon G-matrix [6]. But things are more involved in the relativistic case because one has to reduce not only the many-body effects (as in the Skyrme forces) but also the quantum field effects. As a consequence, two approximations are involved: the *mean-field* and the *no-sea* approximation. And even worse, there is the fundamental problem in deriving effective Lagrangians for relativistic nuclear models that the microscopic theory of the nucleons and their internal structure, the quantumchromodynamics, is not yet fully understood nor is it very tractable in its present stage [7]. Thus we are yet missing theoretical guidelines for the desirable form of effective Lagrangians. The standard model, as introduced above, was developed with too much emphasis on quantum field theoretical aspects. That holds in particular for the nonlinear selfcoupling of the scalar field in the standard form (2) which was inspired by field theoretical model of a scalar field. We want to point out that quantum field theory is only one, and probably the minor, aspect of the effective theory and that one can very well interpret the form of the relativistic mean-field model as the natural result of any relativistic many-body theory in the low-momentum (q) and low-energy (ω) limit.

We need to make only one assumption to start with: nucleons remain nucleons also in a baryonic environment. Then we can think of a G-matrix for nucleon-nucleon scattering in the medium, whatever the underlying microscopic model of nucleon dynamics might be. This G-matrix will be an integral operator in space and time. The extension of the integral kernel in space and time should not be very different from what we know in conventional nuclear physics. The G-matrix can now be divided into the various interaction channels classified by the exchanged angular momentum and isospin, and it can be expanded, firstly, into a series of derivative couplings for the space extension, and secondly, into a series of various meson masses for the time extension (i.e. a Lehmann spectral representation of the meson fields). This is sketched in fig. 1. The relativistic mean-field approximation consists in considering from the double expansion just the first term in the upper left corner, the strictly local couplings with one meson mass per interaction channel. This is certainly a valid approximation in the low-q- and low-ω-limit. Note that $q \to 0$ and $\omega \to 0$ is realized in a stationary state of nuclear matter. There, the masses of the mesons are completely unimportant because only the fraction g/m is relevant. Thus one can very well replace a spectrum of meson masses by one single meson as a representative for the field. The general

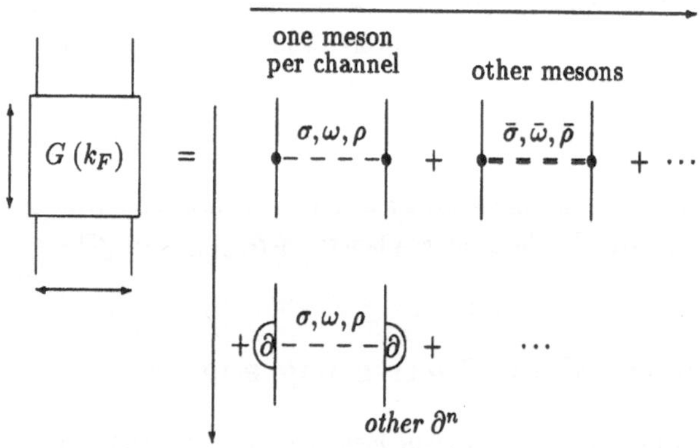

Figure 1.

An illustration of the expansion of the G-matrix in meson fields and its interaction vertices. The arrows on the left-hand-side are to indictes the finite extension of the G-matrix in time and space. The arrows on the right-hand-side indicate the corresponding expansion.

vertex for the nucleon-field coupling can depend on two momenta, the momentum which is transferred via the meson field and the nucleon momentum which "flows through" the vertex. The transferred momentum vanishes in nuclear matter. It remains some uncertainty about the dependence on the through-flowing momentum. But one can very well parametrize this dependence into one effective point-coupling for a given mixture of occupied momenta, i.e. for a certain range of Fermi momenta k_F. Thus the relativistic mean-field approach employing point coupling for one field in each interaction channel is the natural outcome of a low-q-, low-ω-limit independent of the underlying microscopic model. Ground states of finite nuclei challenge the model at finite q because they have finite surfaces and they display shell fluctuations in the spatial densities. Excitations challenge the model at finite q and finite ω. It remains to be seen how far the model reaches in this extended range of applications.

There is furthermore one ingredient which is not visualised in fig. 1: the G-matrix depends on the density of the system (via the Fermi momentum k_F and the related Pauli exclusion operator for which it is computed). Thus all the model parameters, the meson couplings and masses, should also depend on the density. But all these density dependences are parametrised in the standard model simply by the nonlinear selfcouplings (2) of the scalar field. (The stabilized form (3) does not better in that respect.) A theoretical justification for this simplification of the full relativistic G-matrix is presently not yet worked out. A judgement is drawn pragmatically from success or failure of the model. It could very well be that all the problems which arise in an extended range of applications can be traced down to that oversimplistic parametrization.

4 Short review of successes

As we have seen above, the relativistic mean-field model represents an effective theory for computing one-body observables within the mean-field and no-sea approximation. A clear derivation of the effective Lagrangian from a relativistic many-body theory is yet missing.

Table 1.

The meson parameters for a variety of parametrisations. The masses are given in MeV, to get it in units of fm^{-1} one has to divide by $\hbar c = 197.32 MeV fm$. The b_3 is given in fm^{-1}. All other coupling parameters are dimensionless. L-Z is the optimised set for the linear model, NL-Z the optimised set for the standard nonlinear model, the m^*/m are optimised sets in the standard nonlinear model where the effective mass has been kept fixed at the indicated values, and NL-V is an optimised set for the nonlinear model which includes additionally the meson functional from vacuum polarization. The isovector-vector mass is $m_\rho = 763 MeV$ and the nucleon mass is $m_B = 938.9 MeV$ in all sets.

Set	g_σ	g_ω	g_ρ	b_2	b_3	m_σ	m_ω
L-Z	11.1933	13.8256	10.8883	0.0	0.0	551.310	780
NL-Z	10.0553	12.9086	9.69887	-13.5072	-40.2243	488.670	780
$m^*/m = 0.60$	9.99582	12.7285	9.46062	-13.1679	-37.9776	492.363	780
$m^*/m = 0.65$	9.32246	11.7319	9.50398	-11.3003	-25.0851	491.871	780 3
$m^*/m = 0.70$	8.51031	10.7023	9.62038	-10.4491	-13.1752	478.975	780
$m^*/m = 0.75$	7.51141	9.51541	9.85102	-9.57568	1.97056	454.871	780
NL-V	9.67151	12.8395	10.5170	-36.0743	-234.707	441.369	783

Thus the features of the model have to be adjusted and tested phenomenologically. By far the most investigations up to now have been concerned with ground state properties of nuclei and nuclear matter. In order to explore the ability of the model one has to get the optimum possible description within that range of applications. That means one has to find a set of parameters which allows to reproduce nuclear properties as good as possible. To that end we have performed least-squares fits which minimises the sum of the squared deviations from the ground-state observables binding energy E, diffraction radius R, and surface thickness σ in finite nuclei [8,9]. Magic nuclei or near-magic nuclei are most appropriate because correlation effects from low-energy modes are smallest there, and as an extra bonus they are all spherical which keeps computational expense low. A reasonable selection is provided by the eight nuclei: ^{16}O, ^{40}Ca, ^{48}Ca, ^{58}Ni, ^{90}Zr, ^{116}Sn, ^{124}Sn and ^{208}Pb. It covers the range from light to heavy nuclei and they include with Ca and Sn two corners of isotopic chains. The same selection has also been used in comparable nonrelativistic fits [10] which allows a close comparision with those results. We find that the relativistic mean-field model provides an excellent description of nuclear ground state properties which allows to reproduce the energies within 0.26%, the radii within 0.7% and the surfaces within 3.5%. That is as good as comparable fits within the nonrelativistic Skyrme-Hartree-Fock model. But as an extra bonus, the relativistic model predicts the proper spin-orbit force, in contrast to nonrelativistic models which always need an extra adjustment for that. It is to be noted that this excellent description is only possible if nonlinear selfcouplings are included, whereby it does not matter whether one employs eq. (2) or eq. (3).

The extrapolated nuclear matter properties lie in the range of the nonrelativistic predictions and of the commonly accepted values. Onle the effective nucleon mass comes out a bit low with $m^*/m = 0.58$. The small effective mass is due to very strong and counteracting scalar and vector fields. This is still a puzzling problem and the point of most concern in our present investigations. The following sections will discuss some of the attempts to get direct information on the proper magnitude of m^*/m. In order to have parametrizations for further testing, we have produced also a series of fits where in addition to the nuclear ground-state data a given effective mass m^*/m has been fixed. The paramters of these fits and some other parametrizations are given for completeness in table 1.

As discussed in section 2, the relativistic mean-field model simplifies the G-matrix down to a few meson channels with point couplings. The first step to account for a finite extension of the G-matrix is to introduce derivative couplings at the meson-nucleon vertices. The vector mesons have a coupling in first order ∂_μ which survives in the case of stationary states. It is the so called tensor coupling which has already been used in the one-boson-exchange potentials [20]. The effect of tensor couplings in the relativistic mean-field model has been investigated in ref. [9]. We find no significant further improvement of the fits what the nuclear ground-state properties is concerned. On the other hand, there is a large range for a tensor coupling strengths which all provide similar quality of the description. This shows that the model is open for extension to derivative couplings and this freedom may become important if one wants to cover more empirical data within extended relativistic mean-field models. The investigations on such extensions are still in a very preliminary stage. Thus not much more can be said presently about derivative couplings.

5 Neutron matter and neutron stars

In order to explore further the relativistic mean-field model, one needs to consider more and more data, in particular those data which possibly allow to pin down the effective mass problem. In this section, we want to discuss the equation-of-state of infinite nuclear matter as a possible testing ground.

The binding energy per particle as function of the baryon density, i.e. $E/A(\rho_B)$, depends on the incompressibility K for small variations of the density near the equilibrium point. But at higher densities, it depends very sensitively on the effective mass. This is shown in the upper part of fig. 2 where we plot the equation-of-state of neutron matter for the parametrisations with varying m^*/m, see table 1. One sees that the lower effective masses produce the softer equations-of-state and the higher effective masses the stiffer equations-of-state. Note that all parametrisations have almost the same incompressibility of about $K = 220MeV$. Thus the observed trend comes clearly from the effective mass. This can be understood by combining two facts: firstly, low effective mass indicates strong scalar and vector fields whereas high effective mass indicates less strong fields, and secondly, the equation-of-state at high densities is dominated by the vector field. Thus low effective mass correlates to a strong vector field producing a stiff equation-of-state, and vice versa.

The question comes how to access the equation-of-state experimentally. One possibility is to investigate the stability of neutron stars. This is certainly not a direct measurement of the equation-of-state, but it hints on desirable properties and it can help to rule out unreasonable models for which the critical mass for neutron stars comes out lower than observed neutron stars. Thus we calculate properties of cold, spherical symmetric neutron stars, which are in a hydrostatical equilibrium, i.e. the pressure of matter is compensated by the pressure of gravity in each point of the star. The equilibrium condition formulated in the language of general relativity leads to the Oppenheimer-Volkoff-equation [11]

$$\frac{dP}{dr} = \frac{-G[\epsilon(r)/c^2 + P(r)/c^2][m(r) + 4\pi r^3 P(r)/c^2]}{[r - 2Gm(r)/c^2]r} \tag{5}$$

$$dm = \frac{4\pi}{c^2} r^2 \epsilon(r) dr \tag{6}$$

where $m(r)$ is the mass of a spherical shell within a radius R from its center. Therein P is the pressure of matter and ϵ is the total energy density, which is connected with the total mass density by $\rho = \epsilon/c^2$. G is the gravitational constant and c the velocity of light. P and ϵ are related via the equation of state given in the form $P = P(\epsilon)$. For given equation of state and

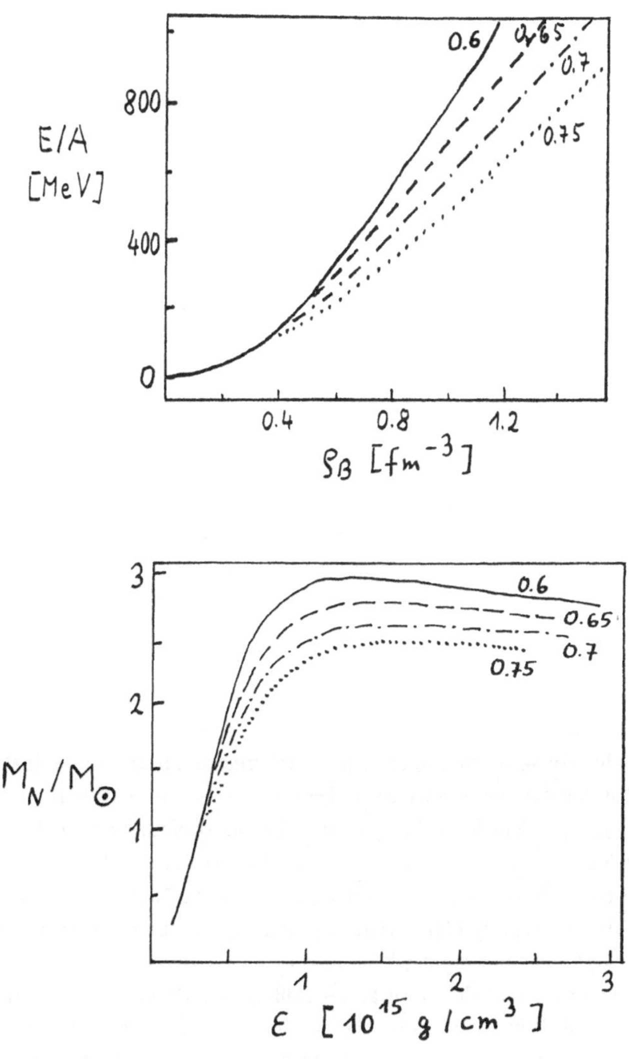

Figure 2.

Equation-of-state for pure neutron matter (upper part) and mass of a neutron star as function of energy density (lower part) for the four parametrisations with systematically varied effective mass, as given in table 1. The star mass M_N is given in units of the solar mass $M_\odot$.

an arbitrary chosen ϵ at $R = 0$ it can be integrated outward until the pressure vanishes, i.e. $P(R) = 0$. In this case $m(R)$ is the total mass of the star and R its radius.

We show in the lower part of fig. 2 the mass of the neutron star as function of the total energy density ϵ for the equations of state from varied effective masses as displayed in the upper part of fig. 2. The mass curves reach a maximum in all cases. This maximum represents the critical mass of a neutron star for the given model and parametrisation. Systems with larger mass are unstable and probably collapse into a black holes. The critical mass varies surprisingly little with the effective nucleon mass, m^*/m, of the parametrisation. That comes because the equilibrium condition for the neutron star does not really explore the high-density part of the equation-of-state but rather concentrates in range of normal densities where all parametrisations behave similar due to construction. A trend with effective mass can nevertheless be read off from fig. 2: the critical mass decreases with softening equation-of-state, i.e. with increasing m^*/m. Finally it is to be noted that all parametrisation give rather large critical masses safely above the haeviest observed neutron star. Thus no parametrisation can be ruled out from that point-of-view. One can, of course, read the above result in the reverse way: all variants of the relativistic mean-field model if fitted properly to nuclear properties give similar properties for neutron stars. Thus the model does have a predictive value for neutron stars.

6 Deformed nuclei

A wide range of available data is accessed if one extends the calculations to deformed systems. One can look for the stable ground-state deformations of strongly deformed nuclei and for the structure of large-amplitude collective motion in nuclei. Deformation properties depend on the strength of the spin-orbit splitting as well as on the effective mass which determines the spectral density. Thus they allow for an indirect test of the single particle spectra and they may contribute information on the effective mass in the relativistic mean-field model. The *Gd* isotopes are a particularly interesting testing ground because they display a phase transition from spherical ^{146}Gd which has a proton subshell closure to manifestly deformed heavier isotopes around ^{158}Gd [12]. Thus we have performed axially deformed calculations of the deformation energy surfaces for the *Gd*-isotopes [13]. For a first survey, we interpret the deformation of the deepest minima as the stable ground-state deformation. In fig. 3 we show this quadrupole deformation as function of the mass number for the *Gd*-isotopes and for various effective masses. We first see that all sets display the transition from spherical to deformed isotopes. This is already quite an achievement and indicates that all sets have a fairly reasonable single particle spectrum. Furthermore, all sets drive finally to the same well developed stable deformation which confirms that this deformation is a merely geometric property rather independent from the dynamics. But most of all we see indeed that there is a sensitive dependence of the transition point on the effective mass of the parametrisation. From fig. 3 we would conclude that an effective mass of $m^*/m \approx 0.63$ is optimal. That is still a low value but already significantly higher than the 0.58 required by the bestfit to spherical data. This result hints at some incompatibilities in the relativistic mean-field model at its present stage. It is to be noted that deformed states explore slightly larger densities than spherical states because the shell fluctuations on the densities are squeezed a little bit in direction of the short axes. Thus the inconsistencies could be related to problems with fitting a wider range of densities. However, it is a bit early to draw the conclusion too far because one needs to solve for the full collective deformation dynamics of the ground-state before concluding on the observed deformation. We are, so to say, in the regime of the dynamical Jahn-Teller effect where every dynamical detail counts [14].

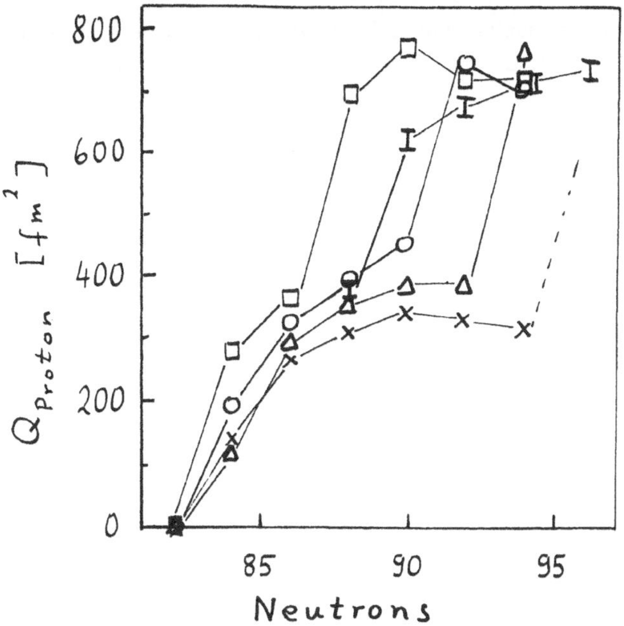

Figure 3.

Ground-state quadrupole deformation for the even isotopes of Gd drawn versus neutron number. The parametrisations with systematically varied effective mass, as given in table 1, have been used. The assignment of the effective mass of the parametrisations in the plot is the following: □ ≡ $m^*/m = 0.6$, ○ ≡ $m^*/m = 0.65$, △ ≡ $m^*/m = 0.7$, *and* × ≡ $\mathrm{m}^*/\mathrm{m} = 0.75$. *Available experimental values are also plotted and indicated by a bar.*

7 RPA instabilities

A wide field of complementing observables is accessed with the collective excitation properties of the nuclei which are describable within the RPA as coherent superposition of 1p1h states. Consequently, the RPA response function of the relativistic mean-field model in nuclear matter is an interesting study object. Much attention has been paid recently to the instabilities, i.e. modes with frequency $\omega = 0$, which occur at higher densities [15,16]. They hint at possible principle limitations of the relativistic mean-field model in its present stage.

We derive the RPA equations by linear response about a given ground-state. For infinite systems we can assume that the mode has definite frequency ω and momentum q. The time dependent wave functions are written as $\varphi_\alpha = \varphi_\alpha^{(0)} + \delta\varphi_\alpha$ and the fields as $\Phi = \Phi^{(0)} + \delta\Phi$ and $V_\mu = V_\mu^{(0)} + \delta V_\mu$ where the $\varphi_\alpha^{(0)}$, $\Phi^{(0)}$ and $V_\mu^{(0)}$ are the ground-state values and $\delta\varphi_\alpha$, $\delta\Phi$ and δV_μ the contributions from the small amplitude RPA vibrations. The Coulomb field and the isovector field do not contribute in symmetric nuclear matter, and only isoscalar vibrations are considered leaving $\delta\underline{R}_\mu = 0$ for the isovector field. We linearize the coupled field equations of the relativistic mean-field and obtain coupled integral equations of finite rank for the variations of wave functions and fields. The effect of the nonlinear scalar selfcouplings on the RPA equations is that an effective scalar mass

$$\overline{m}_\sigma^2 = \left.\frac{\partial^2 \mathcal{U}}{\partial \Phi^2}\right|_{\Phi^{(0)}} \tag{7}$$

occurs in the meson equation and replaces the bare meson mass m_σ^2 of the linear model. After that replacement, all other details can be taken over from the previous investigations. In particular, we are using the explicit analytical expressions for the propagator integrals as given in [16]. The modes separate into two transversal modes from $A = B = \gamma_\mu$ and two coupled longitudinal modes which mix γ_0 and 1. One more longitudinal mode is frozen by the gauge condition $\partial^\mu \delta V_\mu = 0$.

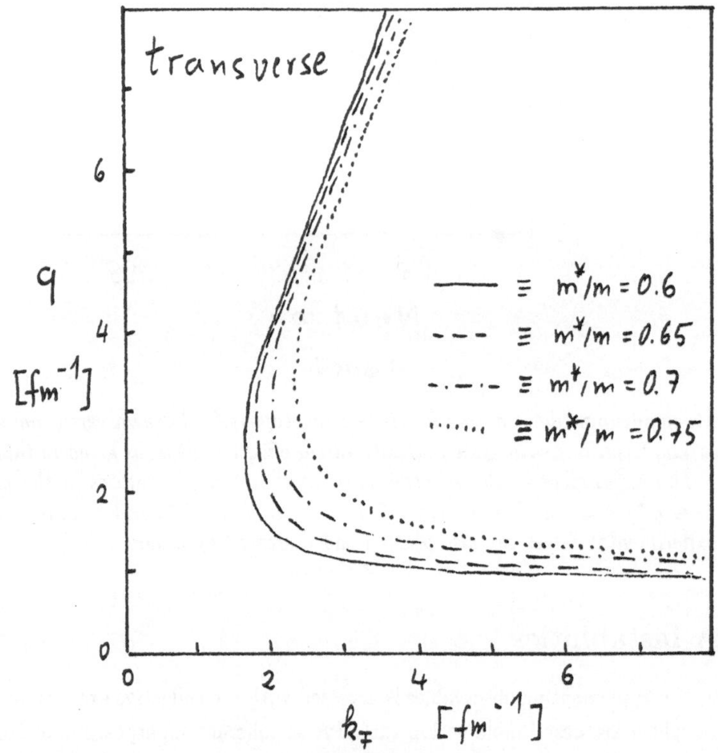

Figure 4.

The instability line, i.e. $\omega = 0$*, in the plane of Fermi momentum* k_F *of the ground-state and the excitation momentum* q *for the transversal mode. Results from various effective masses are drawn, as indicated.*

Here we are interested particularly on the RPA instabilities of the relativistic mean-field model. These are indicated by an excitation mode with zero energy, i.e. $\omega = 0$, but finite momentum q. An $\omega = 0$ mode is the doorway to a spontaneous symmetry breaking of the ground state. In fig. 4 we draw the onset of instability, i.e. the $\omega = 0$ line, in the plane of Fermi momentum k_F and momentum of the mode $q = |\underline{q}|$ for the parametrisations with systematically varied effective nucleon mass m^*/m. We see a strong sensitivity: the instability systematically moves away with rising effective mass. Thus a larger effective mass is highly desirable from that point of view. However, the instabilities may also be cured by other forms of the nonlinear selfcouplings. This is indicated by the fact that the inclusion of vacuum polarization also removes the instabilities [16], and vacuum polarization can be packed into the effective nonlinear functional (4).

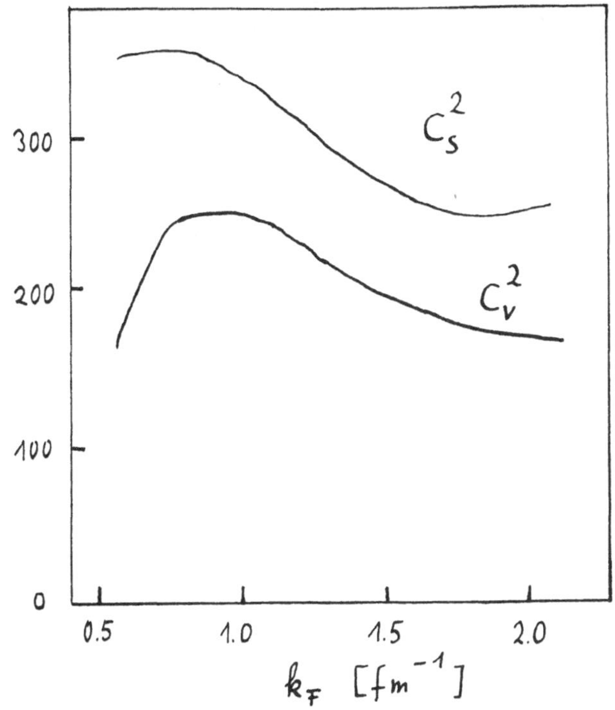

Figure 5.

The scalar and vector coupling parameters of a relativistic mean-field model adjusted to relativistic Bethe-Brückner-Goldstone data at each Fermi momentum k_F separately.

8 Model study with relativistic Bethe-Brückner-Goldstone data

As we have seen in the previous sections, one needs to include more and more observables in the considerations in order to pin down phenomenologically the limits of the relativistic mean-field model and to estimate directions for future extensions of the model. However, the simple observables are already explored and new observables become increasingly complicated to evaluate, as e.g. deformations of excitations. And more deplorable, the most simple nuclear matter is not at all experimentally accessible. On the other hand, nuclear matter is an invaluable tool for first explorations because calculations are simple and most mutual dependences remain fairly transparent. Thus one has to take recourse to pseudodata on nuclear matter.

Here we take as pseudodata the outcome of a deeper lying microscopic theory, namely the relativistic Bethe-Brückner-Goldstone model [18,19] which computes nuclear structure using given One-Boson-Exchange potentials [20]. In particular, we use here the results of Brockmann's calculations [21]. In a previous publication we could show with these data that the relativistic mean-field model is not able to reproduce the equation-of-state in a wide range of densities [22]. Here we want to exploit the fact that these theoretical data have the additional advantage to deliver explicitely the strength of the scalar and vector field which allows to obtain an idea of the possible density dependence of the field parameters. For homogenous systems in the mean-field approach, there is a simple relation between field strength and according density, $m_\sigma^2\Phi = g_\sigma\rho_s$ and similarly for the vector field. The scalar and vector potentials are given by $U_S = g_\sigma\Phi$ and $U_V = g_\omega V_0$. The relativistic Bethe-Brückner-

Goldstone model provides us with $U_S(k_F)$, $U_V(k_F)$, $\rho_\sigma(k_F)$ and $\rho_0(k_F)$ for each given Fermi momentum k_F. Thus we can compute directly the k_F dependent dimensionless couplings as

$$C_S^2 = m_B^2 \frac{U_S}{\rho_s} \quad , \quad C_V^2 = m_B^2 \frac{U_V}{\rho_0} \tag{8}$$

The results are shown in fig. 5. It is obvious that *both* parameters depend on the density, or k_F respectively. That is in clear contradiction to the standard relativistic mean-field model where all nonlinear dependences are cast on the scalar field alone. This approach may be justifiable for a limited range of densities by reshuffling the nonlinear depndences. But all extrapolations to a wider range of densities are questionable.

Thus there is a clear demand for extensions of the relativistic mean-field model such that nonlinear selfcouplings occur in all fields. On the other hand, such extensions need theoretical guidelines in order to avoid proliferation of new model parameters. The fig. 5 suggests that the density-dependence of both parameters goes parallel. This nourishes some hope that one can work out simple rules for the coupled nonlinearities which keep the number of model parameters low.

9 Conclusions

In this contribution, we have discussed a few aspects of the widely used relativistic mean-field model of nuclear structure and dynamcis.

We have first argued that the relativistic mean-field model can be understood as the natural outcome of a low-energy and low-momentum expansion of any relativistic many-body theory, as long as nucleons maintain their identity in the baryonic environment. The only weak point is the parametrisation of the density dependence through nonlinear selfcouplings only in the scalar field.

The further calculations of various observables and the comparison with data concentrated on the problem that good parametrisations of the relativistic mean-field model have a rather low effective nucleon mass hinting at large fields beyond. Thus we have produced a series of parametrisations with systyematically varied effective mass in order to study the influence on the various observables.

The equation-of-state of nuclear and neutron matter at high densities depends uniquely on the effective mass. However, it is hard to measure this region. Neutron stars are found to be not sensitive enough as a test.

Deformation properties can be a sensitive test of the parametrisation. In particular, we find in the *Gd*-isotopes that the transition point from spherical to deformed isotopes varies systematically with the effective mass. Slightly larger effective masses than in spherical nuclei seem to be preferred.

RPA excitations in nuclear matter show an unstable mode at densities somewhat above equilibrium density. The onset of the instability depends strongly on the effective mass. Larger effective masses push it far away whereas the usual low effective masses place the critical point close to equilibrium.

A model study with pseudodata from a relativistic Bethe-Brückner-Goldstone calculation hinted at one possible key to many of the problems: the parameters of all fields should be density-dependent not only those of the scalar field.

Altogether, we see a clear demand for extensions of the relativistic mean-field model, in particular what the implementation of nonlinear selfcouplings is concerned. The problemn which remains is first on the theoretical side: one has to develop flexible forms without too many parameters.

References

[1] B.D. Serot and J.D. Walecka, Adv. in Nucl. Phys., vol 15 (1986)

[2] L.S. Celenza and C.M. Shakin, *Relativistic Nuclear Physics Theories of Structure and Scattering*, World-Scientific, Singapore 1986

[3] P.-G. Reinhard, Rep.Prog.Phys. **52** (1989) 439

[4] J. Boguta and A.R. Bodmer, Nucl.Phys. **A292** (1977) 414

[5] P.-G. Reinhard, Z. Phys. **A329** (1988) 257

[6] J.W. Negele and D. Vautherin, Phys.Rev. **C5** (1972) 1472

[7] C.W. Wong, Phys.Reports **136** (1986) 1

[8] P.-G. Reinhard, M. Rufa, J. Maruhn, W. Greiner and J. Friedrich Z. Phys. **A 323** (1986) 13

[9] M. Rufa, J. Maruhn, W. Greiner, P.-G. Reinhard and M. Strayer, Phys. Rev. **C5** (1988) 390

[10] J. Friedrich and P.-G. Reinhard, Phys. Rev **C33** (1986) 335

[11] J.R. Oppenheimer and G.M. Volkoff, Phys. Rev. **55** (1939) 374

[12] R.F. Casten, D.D. Warner, D.S. Brenner and R.L. Gill, Phys.Rev.Lett. **47** (1981) 1433

[13] J. Fink, V. Blum, P.-G. Reinhard, J.A. Maruhn and W. Greiner, Phys.Lett. **B218** (1989) 277

[14] P.-G. Reinhard and E.W. Otten, Nucl.Phys. **A420** (1984) 173

[15] R.J. Furnstahl and C.J. Horowitz, Nucl.Phys.**A485** (1988) 632

[16] B. Friman and P. Henning, Phys.Lett. **B206** (1988) 579

[17] H.G. Döbereiner and P.-G. Reinhard, Phys.Lett. (1989)

[18] R. Brockmann and R. Machleidt, Phys.Lett. **B149** (1984) 283

[19] B. TerHaar B. and R. Malfliet, Phys.Lett **B172** (1986) 10

[20] K. Holinde, Phys.Rep. **C68** (1981) 122

[21] R. Brockmann, private communication

[22] R. Brockmann and P.-G. Reinhard, Z.Phys. **A331** (1988) 367

THE GIANT MONOPOLE RESONANCE AND THE COMPRESSIBILITY OF NUCLEAR MATTER

A. van der Woude

Kernfysisch Versneller Instituut
9747 AA Groningen, The Netherlands

ABSTRACT

New and precise data on the breathing mode giant monopole resonance (GMR) for ^{208}Pb, the isotopic chains of Sn and Sm and ^{24}Mg have been obtained from inelastic scattering of 120 MeV α-particles at 120 MeV and small scattering angles including 0°. They have been used to determine the nuclear compressibilities as a function of A. In the framework of the scaling model these data have been extrapolated to obtain a value of 300±25 MeV for the compressibility of nuclear matter.

I. INTRODUCTION

The isoscalar giant monopole resonance (GMR) the so-called breathing mode is of particular interest due to its bearing with the compressibility of nuclei and hence, by applying a suitable extrapolation, to the compressibility of nuclear matter. The compressibility K_A is related to the resonance energy E_0 by:

$$E_0 = \hbar\omega_0 = \hbar \sqrt{K_A/(m\langle r_0^2\rangle)} \tag{1}$$

where m is the nucleon mass and $\langle r_0^2\rangle$ is the r.m.s. radius of the nucleus. Using the fact that E_0 is a slowly varying function of A one can make an expansion of K_A in terms of $A^{-1/3}$ in analogy with the semi-empirical mass formula:

$$K_A = K_V + K_s A^{-1/3} + K_C A^{-2/3} + K_\tau \left(\frac{N-Z}{A}\right)^2 + K_{co} Z^2 A^{-4/3} \tag{2}$$

Although this suggests that the various coefficients K are simply related to the corresponding terms in the semi-empirical mass formula , it has been shown by Blaizot[1] that by properly taking into account the

The Nuclear Equation of State, Part A
Edited by W. Greiner and H. Stöcker
Plenum Press, New York

equilibrium condition of the nuclear ground state, the real interpretation of the coefficients is rather complicated. However if the expansion as such is still valid the value of the A-independent terms K_V and K_τ can be determined from a multi-parameter fit to the experimental GMR energies according to eq. (2). It has been shown by Treiner et al.[2] that this is indeed the case in the scaling model of the breathing mode vibration. In the frame work of this model the A expansion of K_A (e.g. eq. (2)) converges rapidly and moreover in that case K_V, the compressibility for $A\to\infty$, equals K_∞, the nuclear matter compressibility.

The curvature term (which goes as $A^{-2/3}$) becomes necessary in eq. (2) if light nuclei are incorporated in the fitting procedure. The associated coefficient K_c has been estimated[2] to be about 300 MeV. Whether or not other A-dependent terms have to be incorporated in eq. (2) has been recently discussed for instance by Sharma et al.[3].

Throughout the last years various data sets on the GMR have been used to extract a value for K_∞.[4,5] These analysises gave a value of about 270 ± 30 MeV, but due to the rather large uncertainties in the experimental input values for the monopole parameters it was not possible to obtain a more precise value while also the other parameters were not very well determined.

The work at the KVI has been performed with the aim to improve the precision with which the various parameters in eq. (2) can be determined from the data on the nuclear compressibilities K_A, by obtaining more accurate data on the GMR for a number of nuclei ranging over a wide range of A-values.

The experimental method used to obtain the data on the GMR will be described in section 2. The results of the analysis in the framework of the scaling model for K_V and the other coefficients in equation 2 are presented in section 3. In section 4 a short discussion will be devoted to the possibility that the isovector GDR is excited simultaneously with the GMR. Some final remarks are made in section 5.

II. EXPERIMENTAL METHOD

The isoscalar GMR is selectively excited by inelastic α-scattering at very small angles including 0^0.[4,5,6] This is demonstrated in figure 1 which shows the calculated excitation cross sections[6] for the various isoscalar giant resonances assuming 100% exhaustion of the corresponding sumrule. Over the angular range from 0^0 to 3^0 the cross section for the

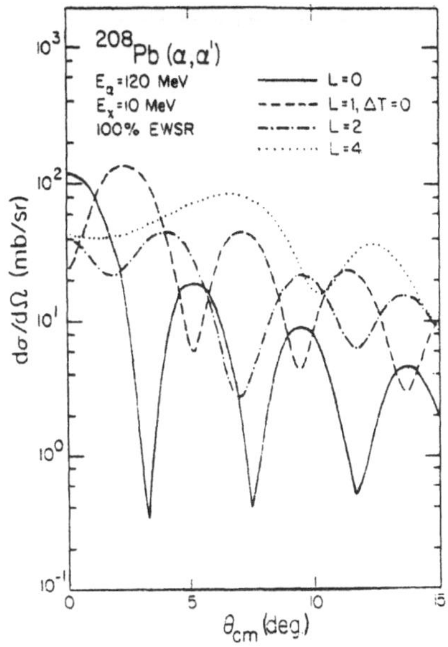

Fig. 1. DWBA prediction for the differential cross sections of various multipoles excited by 120 MeV inelastic α-particle scattering on ^{208}Pb, assuming that 100% of the EWSR is exhausted.

GMR changes very rapidly while for the other multipolarities the cross section is approximately constant. This suggests that by taking simultaneously a spectrum say from 0^0 to 1.5^0 and comparing it to one taken from 1.5^0 to 3^0 the main difference is due to GMR excitation. This effect is even more pronounced if the continuum underlying the giant resonances does also have a weak angular dependence over this angular range.

Small angle scattering including $0^0$is only possible by using a magnetic device, a spectrograph, to separate the beam from the inelastically scattered particles. It requires a very careful adjustment of the beam in order to get rid of beam-halo. Several groups have used this technique.[4,5,6] How well this works is shown in figure 2[6] which shows that by subtracting the spectrum taken over the angular range 1.5^0 to 3^0 from the one of 0^0 to 1.5^0 one indeed is left with a spectrum which is mainly due to GMR excitation. Clearly in this way it is possible to disentangle the GMR from the GQR contributions and thus to determine precisely the energy, width and cross section for the GMR.

We have used this technique to obtain data on ^{208}Pb,[6] $^{112-120}$Sn, and $^{144-152}$Sm.[7] The results for the Sn and Sm isotopes are shown in figure 3. Also shown as squares are older data on the Sn-isotopes,[5] which have an uncertainty of 200-400 keV. Our newer data have an uncertainty of 140-150 keV; systematic errors due to the uncertainty in the energy calibration

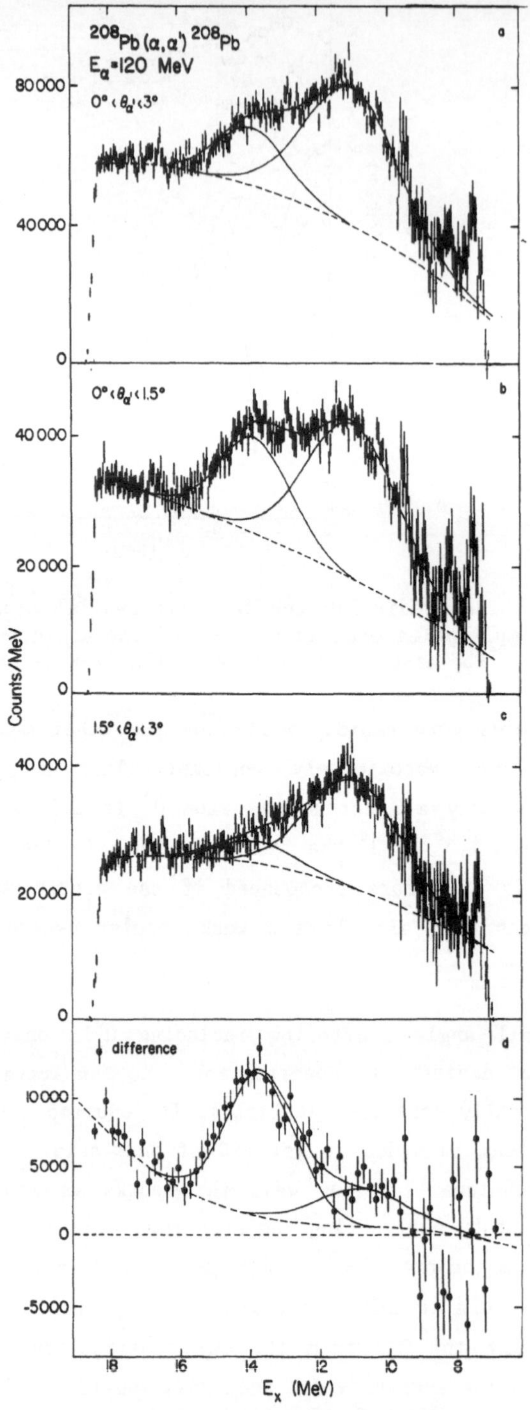

Fig. 2. Singles spectra for ^{208}Pb at E=120 MeV: (a) $0^0<\theta_\alpha<3^0$, (b) $0^0<\theta_\alpha<1.5^0$, (c) $1.5^0<\theta_\alpha<3^0$ and (d) the difference between (b) and (c). The full curves are the contributions of the GQR and the GMR.

and statistical errors contribute about an equal amount. Since these data were taken in the same experiment, the relative uncertainty in the position of the GMR is only about 70 keV. A major limitation is the fact that in these Sn and Sm spectra there is still a considerable background underlying the GMR, which is mainly of instrumental origin.

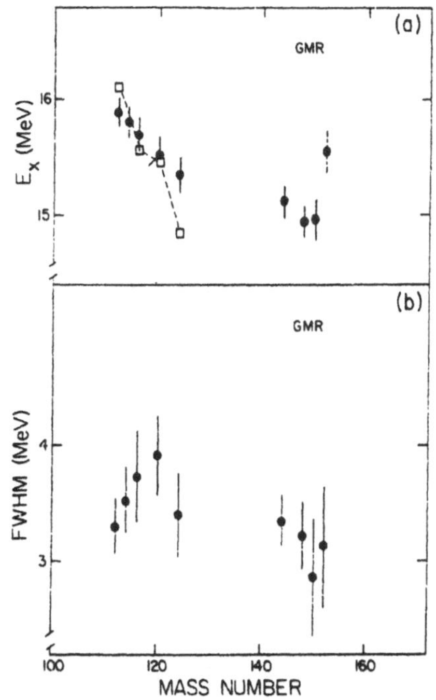

Fig. 3. The centroid energy and the width of the GMR peaks obtained from the fits for Sn and Sm isotopes. The data points for four Sn nuclei from ref. 5 are also shown by empty squares. The centroid energy for ^{152}Sm which is deformed is shown by dashed error bars.

The cross sections obtained from the analysis of these spectra can be converted to fractions of the Energy Weighted Sum Rule (EWSR) using a DWBA analysis with the collective formfactor obtained from the scaling model[2]. In all cases we found that the extracted cross section corresponded to about 100% of the EWSR. This is important since non-detected 0^+ strength will shift the centroid energy of the GMR.

Until a few years ago in light nuclei with $A \lesssim 40$ only a small fraction of the 0^+ EWSR was located. Recently there have been two experiments[8,9] on ^{24}Mg and ^{28}Si where 60-100% of the EWSR was found. This makes it possible to include the results for these nuclei in a multi-parameter fit according to formula 2. However it is than necessary to include in (2) the curvature term.

III. NUCLEAR COMPRESSIBILITIES

Treiner et al.[2] showed that in the scaling model for a GMR with a gaussian strength distribution with a width (Γ) one should define E'_0 by:

$$E'^2_0 = E^2_0 + 3(\Gamma/2.35)^2 \qquad (3)$$

where E'_0 is now related to K_A by the relation

$$E'_0 = \hbar \sqrt{K_A/(m\langle r^2_0\rangle)} \qquad (4)$$

Since we employ the assumptions of the scaling model in the expansion of K_A, we use eqs. (2), (3) and (4) to fit the experimental data. The value of $\langle r^2_0\rangle$ needed to calculate K_A has been calculated using the charge distribution parameters quoted by Bernstein.[10] The parameter K_{co} in eq. (2) can be calculated from:[2]

$$K_{co} = 3/5\ e^2/r_c\ (1 - 27R) \quad \text{with } R = 0.5 - 45/K_\infty.$$

The nuclear-matter compressibility K_∞ and the associated coefficients K_s and K_τ in eq. (2) have been obtained by a three-parameter fit of the experimentally observed energy systematics of the nuclei ^{208}Pb, all Sn isotopes and the two spherical Sm isotopes (^{144}Sm and ^{148}Sm) studied here. All these nuclei exhaust the full E0 EWSR and do not seem to show any anomalous behaviour in the systematics. In these fits we have taken account of the fact that the relative (statistical) uncertainties of ±70 keV in the centroid energies for the GMR for the Sn and Sm nuclei which were measured in the same experiment, are smaller than the total uncertainties of ±140 keV which included statistical and systematic errors. This leads to a slightly better determination of K_τ with the uncertainty of K_τ decreased by ~15%.

The parameters K_∞, K_s and K_τ as obtained from the fits on different sets of nuclei are shown in Table I. It is interesting to notice that including the data for ^{208}Pb hardly changes the parameter values obtained from fitting only the Sn and Sm data. By including light nuclei one might be able to improve the accuracy with which one can resolve the surface and the volume terms. The third set (c) includes in addition to the nuclei of set (b) the datum[8] for ^{24}Mg for which almost the full E0 strength has been located.[8] In this set, the range of the $A^{-1/3}$ variations is considerably increased over that of sets (a) and (b). The results of set (c) show a slight change in the values of the parameters though not outside the error bars. However the uncertainties in the value of the nuclear-matter compressibility and the surface parameters are reduced considerably due to the fact that a much larger range of $A^{-1/3}$ values is now included.

The curvature term, $K_c A^{-2/3}$, is important for light nuclei as it contributes a significant fraction to the compressibility of light nuclei . The curvature coefficient K_c has previously been calculated[2] to be ~300 MeV in the scaling model. Including the curvature term in eq. (2), we have performed fits on the nuclei of set (c). The coefficient K_c is varied between 250 and 400 MeV, with the other three parameters being fitted. We observe that the reduced χ^2 of the fits has a minimum when K_c is taken to be ~375 MeV. This value is slightly higher than the value of K_c calculated as 275 MeV using interaction SIII and comparable to 340 MeV using SkM, in the scaling model.[2] A negative value of this term from our fits is ruled out as the reduced χ^2-value of the fit increases considerably by taking a curvature coefficient as -300 MeV. The results of the fit are shown in set (d). These results are much closer to those of sets (a) and (b) for which the curvature correction is very small. Thus including all important terms in eq. (5) for the nuclei under consideration, we determine the nuclear-matter compressibility and the associated parameters:

$$K_\infty = (\ 300 \pm 25)\ \text{MeV}$$
$$K_s = (-750 \pm 86)\ \text{MeV} \qquad (12)$$
$$K_\tau = (-320 \pm 184)\ \text{MeV}$$

It should be noticed that the results of the fit remain unaffected by the choice of the starting values of the fit, implying that the fit is obtained at a deep minimum of χ^2.

Our values are essentially in agreement with the results quoted in ref. 9, where by using a different set of data including those for ^{28}Si, a value of K_∞=270±13 MeV was obtained without a curvature term. Also in reference 5 a fit is reported to the data of 33 nuclei with 64<A<208: the resulting value of about 270 MeV for the nuclear compressibility without curvature term is again in good agreement with our value if no curvature term is included. This data set included the data for $^{64,66}Zn$ for which only a small amount of the EWSR was found. Thus all analysises using relations 2 to 4 seem to agree with each other in the sense that without a curvature term the value of the compressibility is around 270 MeV. Including the curvature term has the effect of increasing this value to about 300 MeV.

The X^2/N-values shown in table I are very low, while the precision with which the asymmetry term is determined is still not very satisfactory. Both these quantities are very much dependent on the uncertainties of ±70 keV assigned to the relative values of the GMR excitation energies for the Sn- and SM isotopes. If one would for instance decrease this value to ±40 keV the absolute values of the various K-coefficients would barely change but the X^2/N-value would increase to ≃0.7 while the uncertainty in the asymmetry coefficient would decrease to ±100 keV. However we feel that such a small value in the uncertainty of the excitation energy is not justified, especially not in view of the well known problems associated with continuum subtraction.

TABLE I

The nuclear-compressibility parameters derived from fits on the GMR energies on various nuclei. The sets (a), (b) and (c) correspond to the present data and subsequent inclusion of ^{208}Pb and ^{24}Mg, respectively. The numbers in parentheses denote the total number of nuclei included in the fit.

Set of nuclei	Parameters obtained from fits			
	K_v (MeV)	K_s (MeV)	K_τ (MeV)	X^2/N
(a) Sn(5)+Sm(2) = (7)	293±37	-640±155	-341±152	0.074
(b) Sn(5)+Sm(2)+^{208}Pb = (8)	295±48	-647±192	-340±181	0.067
(c) Sn(5)+Sm(2)+^{208}Pb+^{24}Mg = (9)	272±25	-546±89	-295±176	0.277
(d) including curvature term on set (c) with K_c ~ 375 MeV	301±25	-754±89	-323±174	0.104

IV. ISOVECTOR GIANT DIPOLE EXCITATION IN SMALL-ANGLE INELASTIC α-SCATTERING

As will be clear from the foregoing discussion, small-angle inelastic α-scattering has been an essential experimental tool in determining the characteristic parameters of the isoscalar GMR. In a recent paper,[11] however, it was claimed that an appreciable part of the cross section observed in such experiments might be due to the excitation of the isovector GDR instead. If true this would of course seriously effect the method described above for extracting GMR parameters and in fact would put the whole systematic data set obtained for the GMR in doubt. Although various subsequent theoretical and detailed analysises[12,13,14] already indicated that the cross section for isovector GDR excitation should be much smaller than the GMR one, we felt that an experimental study of this problem was called for, especially in view of the fundamental importance of the quantity K_∞.

In our experiment[15] an upper limit for GDR excitation in ^{208}Pb was obtained by measuring in an (α,γ)-coincidence experiment the probability for γ-decay to the ground state of the (bump+continuum) excited in inelastic α-scattering in the GMR excitation energy region, that is in figure 2a the region between 12.5 and 15.5 MeV excitation energy. Figure 3 shows a scatter plot of the relevant coincidence data: here the excitation energy is obtained from the energy of the inelastically scattered α-particle observed in coincidence with the γ-quant. From the ratio of the number of counts observed in the region of interest ($12.5<E_\alpha<15.5$ MeV) to the total number of counts in the (bump+continuum) observed in the singles (α,α') scattering one finds a probability for γ-decay

$$P_\gamma = 2.1*10^{-3}.$$

If one assumes that the observed γ's are due to γ-decay of the bump only and if this bump would have been for 100% due to GDR excitation, one can calculate in a nearly model independent way that

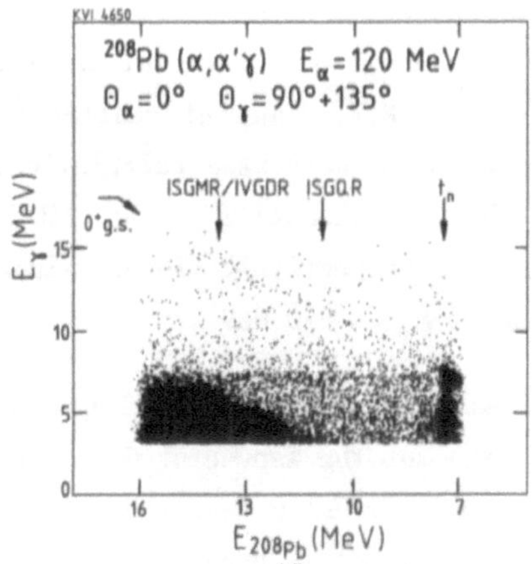

Fig. 4. Two-dimensional scatter plot of the ^{208}Pb excitation energy versus the gamma decay energy. The positions of the GMR, IVGDR and GQR are indicated.

$P_\gamma = 1.7*10^{-2}$.

This implies that at most 12% of the bump between 12.5 and 15.5 MeV can be due to isovector GDR excitation. This corresponds to an upper limit for the isovector GDR excitation cross section of 1.5 mb/str, averaged over the angular range of 0^0 to 3^0. Thus GDR excitation is with respect to GMR excitation only a small effect. It is interesting to note that our upper limit for the GDR excitation cross section indicates that the parameter ξ which is defined as,[13,14]

$$\frac{\rho_p(0)}{\rho_n(0)} = \frac{Z + \xi\,(N-Z)}{N - \xi\,(N-Z)}$$

where $\rho_{p,n}(0)$ are the central proton and neutron densities respectively, should be close to ξ=0.

V. FINAL REMARKS

It has been argued[16] that the value of K_∞ obtained in the way as described here from an extrapolation of the nuclear compressibilities can not be identified with the nuclear matter incompressibility due to the fact that the effective nucleon mass m^* which enters in the Landau definition of the nuclear matter compression modulus K:

$$K = 6\,\frac{k_f^2}{2m^*}\,(1+f_0)$$

is a strong function of the excitation energy around the Fermi surface. This discussion falls outside the scope of the present presentation. The aim of our work has been to show that by using the expansion (2) one can deduce a value for $A\to\infty$, with the result that $K_\infty = 300 \pm 25$ MeV.

It is interesting to note that a recent calculation of Malfliet et al.[17] gives a value K_∞=250 MeV, somewhat smaller but comparable to our experimental value. They also performed calculations for various values of Z/A ratios. They obtain values of K_∞= 250, 240, 210 and 100 MeV for Z/A= 0.5, 0.4, 0.3 and 0.2 respectively. These values are compatible with our experimental value of $K_\tau = -320$ MeV.

All experimental work on the GMR up till now refer to the GMR build on the ground state. Like for the isovector GDR[18] it would be interesting to study the GMR properties build on excited states. For the GDR such a study is experimentally feasible because of the selectivity of the γ-decay mode for this resonance. For the GMR such a specific decay mode would be the decay by internal pair creation, the e^+-e^-mode, which can be estimated to have a probability of $\simeq 10^{-3}$. Moreover by a suitable choice of the e^+-e- opening angle one can effectively discriminate against other processes of pair-creation. Such an experiment to study the GMR properties in hot nuclei is presently being set up[19] using a multi-detector arrangement for e^+-e^- detection based on a previous pair spectrometer design.[20]

The KVI work on the GMR has been a collaborative effort of many people. It is based on the thesis work of S.Brandenburg, W Borghols and T.Poelhekken. M.M.Sharma had the main responsibility for the Sn and Sm work. The close collaboration with M.N.Harakeh is gratefully acknowledged. This work has been performed as part of the research program of the Stichting voor Fundamenteel Onderzoek der Materie, The Netherlands, with financial support of the Nederlandse Organisatie voor Zuiver Wetenschappelijk Onderzoek (ZWO).

REFERENCES

1. J. P. Blaizot, Phys. Rep., 64, 171 (1980).
   J. P. Blaizot, D. Gogny and B. Grammaticos, Nucl. Phys., A265, 315 (1976).
   J. P. Blaizot and B. Grammaticos, Nucl.Phys., A355, 115 (1981).
2. J. Treiner, H. Krivine, O. Bohigas and J. Martorell, Nucl.Phys., A371, 253 (1982).
3. M. M. Sharma, W. Stocker, P. Gleissl and M. Brack, preprint.
   M. M. Sharma, these proceedings.
4. D. H. Youngblood, P. Bogucki, J. D. Bronson, U. Garg, Y.-W. Lui and C. M. Rozsa, Phys. Rev., C23, 1997 (1981).
5. M. Buenerd, in Proc. of Int. Symp. on Highly Excited States and Nuclear Structure, Orsay, France (1983); Jour. de Phys. (Paris) Colloq. 45, C4-115 (1984) and references therein.
6 S. Brandenburg, Ph. D. Thesis, Univ. Groningen (1985), (unpublished).
  S. Brandenburg, W. T. A. Borghols, A. G. Drentje, L. P. Ekström, M. N. Harakeh, A. van der Woude, A. Håkanson, L. Nilsson, N. Olsson, M. Pignanelli and R. De Leo, Nucl. Phys., A466, 29 (1987).
7 M. M. Sharma, W. T. A. Borghols, S. Brandenburg, S. Crona, A. van der Woude and M. N. Harakeh, Phys. Rev., C38, 2562 (1988).
8. H. J. Lu, S. Brandenburg, R. De Leo, M. N. Harakeh, T. D. Poelhekken and A. van der Woude, Phys. Rev., C33, 1116 (1986).
9. Y.-W. Lui, J. D. Bronson, D. H. Youngblood, Y. Toba and U. Garg, Phys. Rev., C31, 1643 (1985).
10. A. M. Bernstein, in Advances in Nucl. Phys., M. Baranger and E. Vogt eds., Plenum Press, New York (1969), Vol. 3, p. 325.
11. R. J. Peterson, Phys. Rev. Lett., 57, 1550, 2771 (1986).
    R. J.Petereson, Phys. Rev. Lett., 59, 1054 (1987).
12. K. Nakayama and G. Bertsch, Phys. Rev. Lett., 59, 1053 (1987).
13. S. Shlomo, Y.-W. Lui, D. H. Youngblood, T. Udagawa and T. Tamura, Phys. Rev., C36, 1317 (1987).
    S. Shlomo, D. H. Youngblood, T. Udagawa and T. Tamura, Phys. Rev. Lett., 59, 1054 (1987).
14. G. R. Satchler, Nucl. Phys., A472, 215 (1987).
15. T. D. Poelhekken, S. K. B. Hesmondhalgh, H. J. Hofmann, H. W. Wilschut and A. van der Woude, Phys. Rev. Lett., 62, 16 (1989).

16. G. E. Brown, Nucl. Phys., A488, 689 (1988).
17. B. ter Haar and R. Malfliet, Phys. Rep., 149, 207 (1987).
R. Malfliet, Progress in Part. and Nucl. Phys., Vol 21, 207 (1988).
18. K. A. Snover, Ann. Rev. Nucl. Part. Sci., 36, (1986).
19. A. Buda, J. Bacelar, J. van Klinken and A. van der Woude, private communication.
20. F. W. N. de Boer et al., Phys. Lett., 180B, 4 (1986).

THE NUCLEAR MATTER COMPRESSIBILITY FROM BREATHING MODE: A SEMI-PHENOMENOLOGICAL APPROACH

Madan M. Sharma+

Department of Physics and Astrophysics
Delhi University, Delhi - 110007, India
and
Sektion Physik, Universität München
D-8046 Garching, West Germany

ABSTRACT

We discuss a semi-phenomenological approach employed to analyse the breathing mode energies. The extraction of the nuclear matter incompressibility from the breathing mode is based upon the liquid-drop-model type expansion of the finite nuclear incompressibility. The validity and applicability of this expansion has been discussed vis-a-vis models of the breathing compression. On the basis of this approach, we have been able to separate the volume and the surface contributions to the nuclear incompressibilty using the recent precision data from Groningen. The data favour a nuclear matter incompressibility $K_\infty \sim (300 \pm 25)$ MeV in contrast with the value (210 ± 30) MeV from the existing microscopic approaches and give a large surface coefficient $K_S \sim (-750 \pm 80)$ MeV.

INTRODUCTION

The equation of state of nuclear matter has been broadly discussed in this conference. We discuss the focal point of this equation of state, i.e., the so called 'incompressibility' of cold nuclear matter at saturation. The question is often asked if the nuclear matter is soft or hard to compress. This is usually answered by quoting the value of the nuclear matter incompressibility K_∞ and has over the last few years been a matter of controversy. This quantity, which is important due to its astrophysical significance, is defined as the second derivative of the binding energy per nucleon as a function of the radial extension of the matter. Since the nuclear matter does not exist in the lab, we have to resort to experimenting only on finite nuclei and consequently we have the nuclear incompressibility of finite nuclei at our disposal, which is also defined analogously (see Eq. (2) below). However, to disentangle K_∞ from K_A is a highly nontrivial problem. We will discuss a semi-phenomenological approach which is employed

The Nuclear Equation of State, Part A
Edited by W. Greiner and H. Stöcker
Plenum Press, New York

to extract K_∞ from finite nuclei. Since K_A is defined at the saturation point, the most direct way to reach this quantity is through the small amplitude motion about the saturation density. Fortunately, there exist breathing mode oscillations in nuclei which correspond to the radial oscillations of the nuclear volume about the equilibrium. We will discuss the breathing mode and the means of realising it in the laboratory in the following.

THE BREATHING MODE

The giant monopole resonance (breathing mode) was first observed in 1975 and since then has been well established in medium and heavy nuclei[1]. The frequency of this compression mode is related to the incompressibility K_A of nuclei as

$$E = \sqrt{\frac{\hbar^2 K_A}{m<r^2>}}. \quad (1)$$

where $< r^2 >$ is the mean square radius of the nucleus and m is the nucleon mass. Thus by observing the giant monopole resonance (GMR) in nuclei in the laboratory, one measures K_A and from this the quantity K_∞ is derived. The procedure of extracting this important parameter from these finite nuclear excitations will be discussed in detail in this talk.

Inelastic Scattering

The breathing mode GMR is excited in inelastic scattering of electrons and hadrons. Since electron scattering excites all electric as well as magnetic resonances,

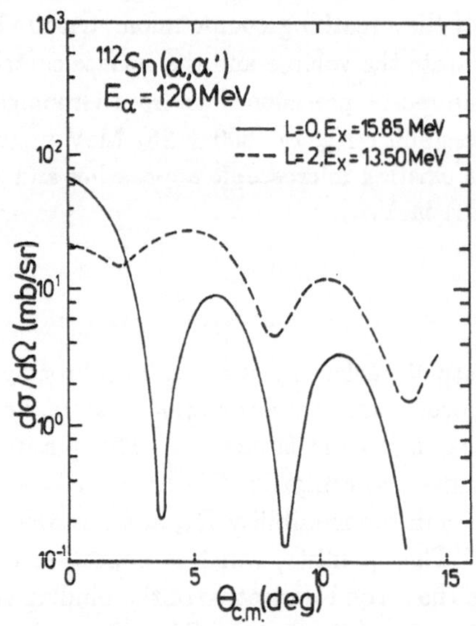

Fig. 1. The angular distribution of the GMR and the GQR.

besides exciting both the isoscalar and isovector type of modes, the extraction and identification of the GMR in electron scattering is very complicated and not completely reliable. In hadron inelastic scattering[2,3] the GMR has been excited using protons, deuterons, ^{3}He and α-particle beams. Since protons excite the isoscalar and isovector electric resonances, the disentangling of the GMR from amongst the mixture of the isovector and other isoscalar resonances is still difficult. Only ^{3}He and ^{4}He beams have been used successfully to excite the GMR whereby only other isoscalar resonance in the neighbourhood of the GMR energies that is also excited is giant quadrupole resonance (GQR). The ^{3}He beam has been used to study GMR at Grenoble and a study has been carried out on a large number of nuclei from light ones as ^{27}Al to heavier ones as ^{208}Pb. At other places namely Texas A & M, Jülich and Groningen the α-particle beams have been used. In these experiments[3], continuum background which accompanies the giant resonances, introduces an uncertainty in determining the resonance parameters. Neither we understand the origin of it properly nor there exist appropriate guidelines to fit this background. Thus, at most of the laboratories the periodic table seems to have been scanned rather fast to observe the GMR without proper consideration of the background.

The angular distribution of the GMR is highly characteristic one peaking strongly at 0°. This requires the use of a Q3D spectrograph which can separate the inelastically scattered particles at 0° from the outgoing beam. The typical angular distribution of the GMR and the GQR are shown in Fig. 1. The GMR peaks strongly at 0° and falls off by about two orders of magnitude in the vicinity of 3°-4°. The GQR and other multipole resonances (not shown in the figure) are approximately flat in the vicinity of 0°. This feature has been exploited in our data analysis to remove the GQR from the spectra.

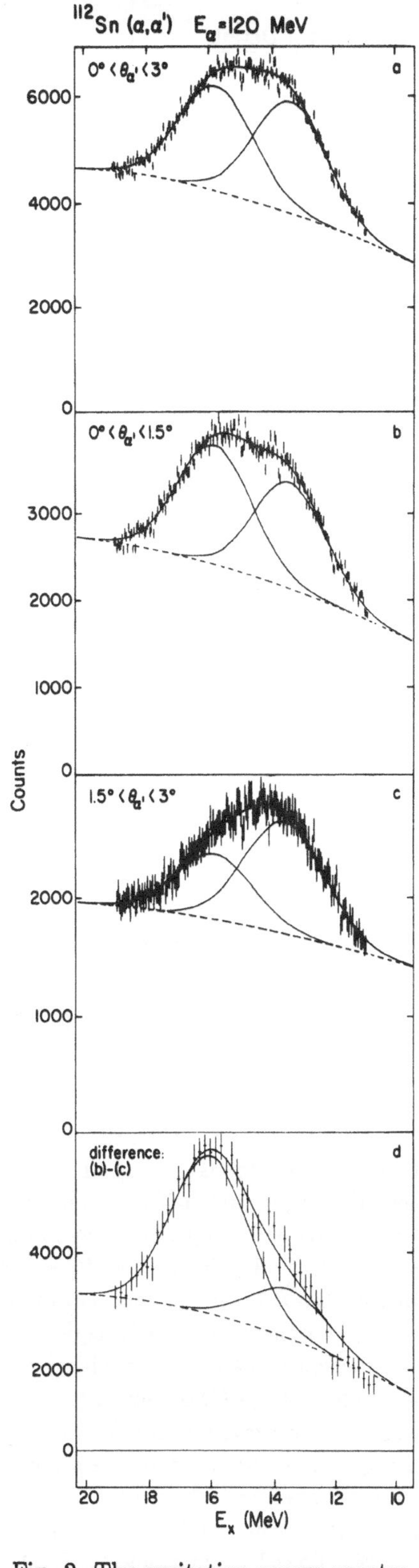

Fig. 2. The excitation-energy spectra

Experimental Details

The inelastic scattering of 120 MeV α-particles has been performed[4] at KVI Groningen on a series of Sn and Sm isotopes, all of them being spherical. The Q3D spectrograph[5] QMG/2 was placed at 0° and the inelastically scattered particles were detected by a 52-cm long two-dimensional position-sensitive detector system[6] at the focal plane of the detector. The spectrograph was set with horizontal opening angle of $\Delta\theta = 6°(-3° \text{ to } +3°)$ and a vertical opening angle of $\Delta\phi = 4°(-2° \text{ to } +2°)$. The angular resolution of the detector system was $\Delta\theta = 0.7°$. The excitation energy range covered was 10-20 MeV covering the GMR and the GQR in all nuclei. The excitation

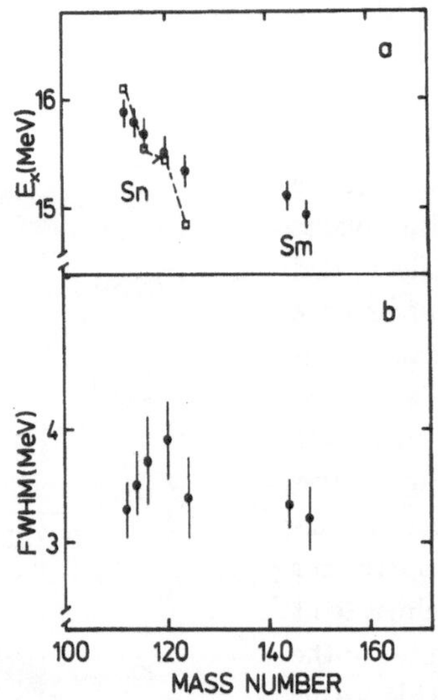

Fig. 3. The energy and width of the GMR

energy spectra for different angular bins for ^{112}Sn are shown in Fig. 2. A broad resonance bump on a underlying continuum background can be seen. The resonance bump was fitted to two Gaussians corresponding to the GMR (higher energy side) and GQR (lower energy side) on top of the continuum background indicated by a dashed curve. In Fig. 2 (a) we have shown the excitation energy spectrum for the full angular range $(0° - 3°)$ where the GQR and the GMR are significantly populated. The spectrum in part (b) for the first half of the bin $(0° - 1.5°)$ shows a GMR contribution which is dominant over the GQR. Similarly in Fig. 2 (c) for the bin $(1.5° - 3°)$ GQR is dominant over the GMR. The intensities of the GMR and the GQR in these figures are consistent with the DWBA calculations shown in Fig. 1. By subtracting the spectra (c) from (b) we have obtained Fig. 2 (d) (squeezed by a factor of 4) which contains a very little contribution of the GQR and whereby the continuum background is also reduced. The parameters of the GMR have been obtained from Fig. 2 (d) which has much better peak to the continuum ratio for the GMR and a very little GQR. All the four spectra in Fig. 2 have been fitted in a self-consistent manner that the GMR and the GQR parameters respectively coming out from the fit of all the spectra were identical. This introduced a constraint on selecting the underlying continuum. This method of data analysis allowed to determine the GMR parameters much more precisely than possible from the full excitation energy spectrum of Fig. 2 (a). Incidentally, at most of the laboratories[3] the systematics of the GMR have been obtained by fitting spectra similar to Fig. 2 (d) where one has too much freedom to select the underlying continuum and thus resulting in ambiguous determinations of the GMR parameters.

The GMR systematics obtained on Sn and Sm nuclei are shown in Fig. 3 (see Refs.[4,7] for details). The centroid energies (shown by error bars) follow a smooth variation with mass number. The error bars are typically about 140-160 keV. The data on Sn nuclei from Grenoble[3] are also shown for comparison by squares. The error bars in these points (not shown in the figure) are roughly ±250 keV. Our data show a contrasting picture with these data though only larger error bars in the Grenoble data[3] make it consistent with ours. The present results[4] for the GMR for Sn nuclei are also in agreement with the previous results[8] from Texas A & M, taking into account the relatively large uncertainties of 300-600 KeV in the centroid energies in the latter. Thus, argument of Blaizot[9] that all the data are consistent with each other is correct only if one takes into account the larger error bars of the other data. This is the *precision* of the Groningen data[4] which allows one to conclude more definitively on nuclear incompressibility than a broader range of it encompassed by the larger uncertainties in other data. We shall later use the Groningen data to extract the nuclear matter incompressibility.

The energy and width of the GQR have also been determined from the spectra. It may be noted that all Sn and Sm nuclei studied exhaust about 100% of the E0 energy-weighted sum-rule (EWSR) and a little more than 100% of EWSR for the GQR. This is presumably due to the presence of the L=4 ($2\hbar\omega$) component at the position of the GQR. We shall compare the GMR and GQR energy systematics with the predictions of a theoretical scaling and 2-dimensional hydrodynamical results using the interaction SkM* in Fig. 6

MODELS OF THE BREATHING MODE

There is no unique definition of the incompressibility K_A of a finite nucleus. It is usually defined[10] as the second derivative of the energy per particle with respect to the radius parameter R at the equilibrium value R_0,

$$K_A = R_0^2 \cdot \frac{\partial^2(E/A)}{\partial R^2}|_{R_0}; \qquad R = < r^2 >^{1/2}. \tag{2}$$

In order to evaluate this expression one has to know the dependence of E on R. Thus the calculation of K_A is model dependent. In particular, two models are discussed[10] leading to two different definitions of the incompressibility:

(i) <u>Constrained Incompressibilty</u>

Putting a constraint on mean square radius R^2 and varying the density of a nucleus (by a time-dependent external field) one obtains the 'constrained' incompressibility:

$$K_A^C = R_0^2 \cdot \frac{\partial^2(E/A)}{\partial R^2}|_{R_0}; \qquad < R^2 >= const. \tag{3}$$

Therefore, when we change the density by constraining R^2, the change in the density will no longer be uniform in the interior of the nucleus.

(ii) Scaling incompressibility

Performing a scaling transformation $\mathbf{r} \rightarrow \lambda \mathbf{r}$ on the single-particle wave function of the HF ground state one gets the 'scaling' incompressibility K_A^{scal}:

$$K_A^{scal} = \frac{\partial^2 (E/A)}{\partial \lambda^2}|_{\lambda=1}. \tag{4}$$

where λ ia a periodic time-dependent collective parameter. On scaling the wavefunctions, the density remains uniform in the interior although both the central density and the surface thickness undergo a change. The breathing-mode energies which are related to the incompressibility K_A can be calculated in terms of the RPA sum rules[11] as $E_k = \sqrt{m_k/m_{k-2}}$. The sum rule m_k can be evaluated with the isoscalar excitation operator $\hat{Q} = r^2$. For $k = 3$, one obtains $K_A = K_A^{scal}$, the scaling incompressibility; and for $k = 1$ one gets $K_A = K_A^C$, the constrained incompressibility of Eq. (3). In the asymptotic limit ($A \rightarrow \infty$) the two incompressibilities, however, yield different results, i.e.

$$\begin{aligned} K_A^{scal} &= K_\infty \\ K_A^C &= \tfrac{7}{10} K_\infty \end{aligned} \tag{5}$$

Thus K_A^C values are ~(3-8)% smaller than K_A^{scal} for real nuclei, so only peak energy considerations do not allow to decide between the two models and a detailed analysis of transition densities is necessary. Only in the scaling mode the asymptotic value of K_A is synonymous with the infinite nuclear matter. On the basis of the presently available data it is difficult to decide which model is correct.

TRANSITION DENSITY

The two models lead to different transition densities. In the constrained model[10] the external field constrains the radius of the nucleus leading to a transition density which may have more than one node. This constraint seems to be quite artificial in the sense that it requires a special restoring force to constrain the *rms* radius in the compression mode, and thus it seems to be far from reality. On the other hand, a simple radial scaling[12] of the ground state density gives

$$\delta\rho(r) = -3\rho_0(r) - r\frac{d}{dr}\rho_0(r). \tag{6}$$

The essential feature of the transition density in the scaling model is the separation of the volume and the surface part. In Fig. 4 we show the transition density of ^{208}Pb nucleus calculated in RPA using the Skyrme interaction SIII and taken from Ref.[13] . The solid curve which is the sum of the contribution of neutrons and protons exhibits a clear separation of the bulk compression and surface decompression. This is close to the scaling transition density, Eq. (6), also shown in the figure. Thus microscopic RPA calculations indeed lead to transition densities in heavy nuclei similar to those obtained in the scaling model. There exists no general theoretical argument for believing that the scaling assumption is correct, yet on the basis of theoretical transition density it may be closer to reality. Only a detailed empirical information about the transition density which is still lacking would allow to determine the validity of the scaling model.

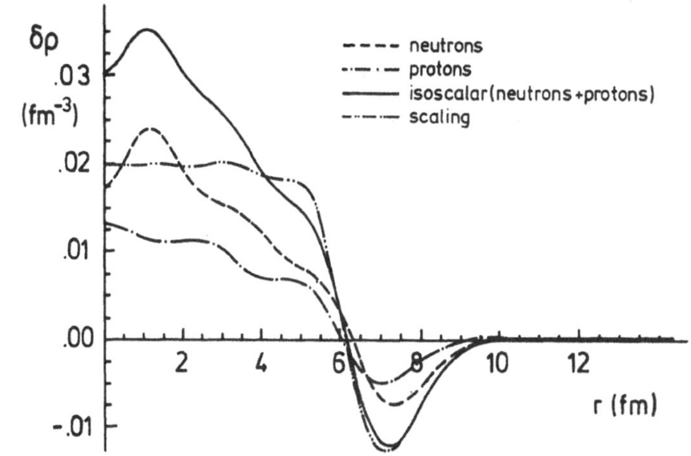

Fig. 4. Transition density of the GMR in ^{208}Pb calculated with the interaction SIII (taken from Ref.[13]).

In a hydrodynamical language, the breathing-mode energy is given by $E = \hbar\sqrt{C/B}$ in terms of a restoring-force parameter C and an inertial parameter B. These quantities depend on the nature of the density vibration which is characterized by the velocity (or displacement) field or the associated transition density as discussed above. They depend in particular on the way in which the nuclear surface is coupled to the nuclear bulk during the breathing-mode vibration[14]. Our approach of analyzing the breathing-mode energies is based upon the assumption that a nucleus, even in a state of compression or decompression during the monopole vibrations displays a division into a homogenous bulk part and a surface skin ('leptodermous' behaviour) similar to the scaling as shown in Fig. 4. Scaling is essentially one special case in this whole class of density oscillations. Only in the scaling case the intertial parameter B can be identified with $m < r^2 >$ leading to Eq. (1) with $K_A = K_A^{scal}$. Using empirical values of $< r^2 >$ one can then extract K_A. In our analysis we have used the assumption of scaling to extract K_A from breathing-mode energies. It may be mentioned that in the adiabatic density-compression mode[10] of the 'constrained model' the leptodermous separation of the transition density into a homogenous bulk and a thin surface region is no more possible.

MICROSCOPIC APPROACH

Blaizot et al.[13,15] have performed microscopic HF + RPA calculations of the breathing-mode energies for several nuclei using different phenomenological effective interactions. The results taken from Ref.[13] are shown in Fig. 5. The incompressibility associated with the interactions varies from 180 MeV to 365 MeV. All these interactions (for details see Ref.[13]) reproduce the ground-state properties of many nuclei well except the interaction B1 which gives lower binding enegies. These interactions, however, differ in the single-particle properties. The Gogny interaction D1 is apparently assumed to reproduce the energies of the GMR and the GQR. The nuclear matter incompressibility K_∞ has been extracted by comparing the then existing experimental data[16] on ^{208}Pb and ^{90}Zr with the resonance energies calculated in RPA. It was found[13] that for being in agreement with the experimental values, the interaction to be used

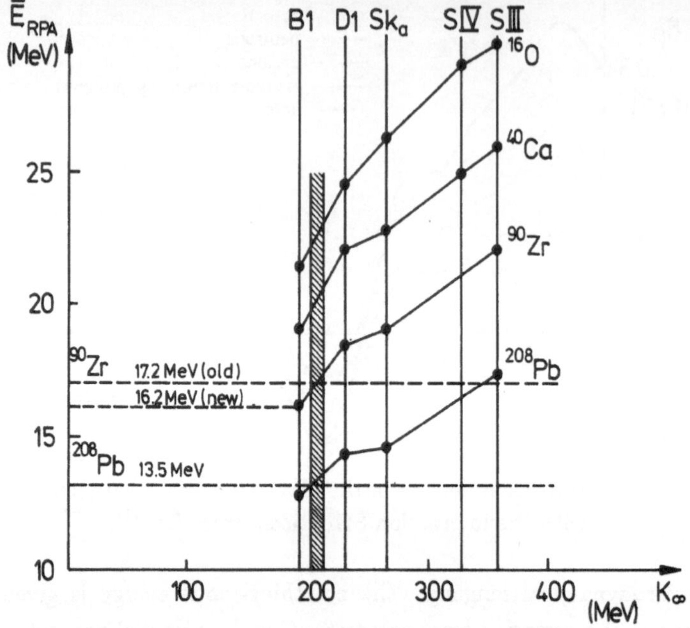

Fig. 5. The energy of the GMR calulated in RPA for different effective interactions (taken from Ref.[13]).

has to be between B1 and D1 (as shown by hatched bar). The conclusion on K_∞ of (210 ± 30) MeV from this approach[13] depends on interpolation between two interactions.

The datum on ^{208}Pb used in this analysis is well established and can be reproduced in this interpolation scheme. The Fig. 5 shows by dashed horizontal line the old datum[16] on ^{90}Zr which is questionable in view of the new datum[17] (also shown by horizontal line) obtained recently in comprehensive measurements at Groningen. The conclusion of Ref.[13], therefore, hinges strongly on the results of ^{208}Pb alone. In order to be able to reproduce the present breathing-mode energy of ^{90}Zr, one will need an interaction (in this approach) having an incompressibility close to that of B1. Thus the results from the microscopic approach[13,15] would be conclusive only if it is possible to reproduce the breathing-mode energy on at least one Sn nucleus in addition to ^{90}Zr and ^{208}Pb. In our approach[18], which we discuss below, it will not be possible to remove the dominant surface effects from the finite nuclear incompressibility only on the basis of one nucleus and determine K_∞. It may be seen in Fig. 7 that slope of the data from ^{208}Pb to ^{90}Zr plays a decisive role on the extraction of K_∞. We will discuss it later in connection with the model calculations of the breathing mode.

LEPTODERMOUS EXPANSION

The approach which is used to extract K_∞ from empirical values of K_A is of using a LDM-type expansion[13,19] of K_A:

$$K_A = K_V + K_S A^{-1/3} + K_\Sigma \left(\frac{N-Z}{N+Z}\right)^2 + K_{Coul} Z^2/A^{-4/3} + K_{curv} A^{-2/3}, \tag{7}$$

where the parameters in Eq. (7) are determined by a least-square-fit of the experimental breathing-mode energies employing scaling assumption Eq. (1). In expansion (7), which was derived by Blaizot[13] taking into account the equilibrium condition of nuclear gound state, the Coulomb term was modified as compared to the original expansion[19]. Thus the coefficients K's are no more simply the second derivative of the respective coefficients in the mass formula. This expansion is not based a priori on the scaling assumption nor on other approximations such as the mean-field theory with Skyrme type interactions. It may therefore supply a semi-phenomenological basis for a fit of its constants to empirical data analogous to the mass formula. It may be mentioned that the expansion converges rapidly[20] in the scaling model and the higher order terms are insignificant. The same is, however, not true for the constrained model in that the expansion converges poorly and the volume term becomes $K_V = \frac{7}{10}K_\infty$. In the following we employ this expansion to determine the coefficients in the framework of scaling.

FIT TO EXPERIMENTAL DATA

The fits of the experimental breathing-mode energies from Groningen, which include new precision data[4] on Sn and Sm nuclei, have been performed using scaling assumption via Eq. (1) on expansion (7). Consequently the volume term corresponds to K_∞. The prescription of Treiner et al.[20] of including the width of the strength distribution of the giant monopole resonances has been employed. The results of fits are shown in Table 1. For details see Refs.[4,18]. In set (i) the breathing-mode energies of five Sn and two Sm isotopes have been fitted. The error bars in the coefficients are large. In set (ii) where ^{208}Pb has been included in the fit, the coefficients do not change. We have also included in the fits a light nucleus ^{24}Mg. Incidentally about full EWSR strength of the GMR has been observed in this nucleus[21]. This provides a confidence to include a light nucleus in the analysis. Including ^{24}Mg, as in set (iii), the error-bars in the volume and surface terms are reduced considerably. This is due to the increase in the range of $A^{-1/3}$ variation which helps resolve the volume and surface terms more accurately. The K_∞ extracted is (300 ± 23) MeV and the surface coefficient is large at a value (-750 ± 80) MeV. The asymmetry term shows a good stability at $\sim (-320 \pm 180)$ MeV. It may be noted that the results on including ^{24}Mg are nearly the same as without this nucleus. This may partly be due to behaviour of this nucleus analogous to the heavier ones during the compression. We have also included in the fits the recently obtained datum[17] on ^{90}Zr from Groningen. It is worth noting from set (iv) that this datum fits very well in the trend of earlier data[4] and that the conclusion about K_∞ remains the same. The curvature term has been kept fixed at a reasonable value of 375 MeV in all these fits. We have also obtained a fit on these data by enforcing a value of the volume coefficient in the fit. By constraining K_∞ to be fixed at 217 MeV, as shown in set (v), we find that the reduced χ^2 is 5-6 times higher than the best-fit one. This seems to rule out a value of K_∞ about 220 MeV. Thus K_∞ extracted from the experimental breathing-mode data is higher than the value of (210 ± 30) MeV obtained from HF + RPA analysis[13,15] of the breathing-mode energies.

The data from Texas A & M and Grenoble have also been analyzed to extract the nuclear compressibility parameters. It has been shown in Ref.[22] that these data do not give consistent fits on different sets of nuclei in contrast with the Groningen data. The values of the parameters fitted depend sensitively on the nuclei included

Table 1.

The nuclear-compressibility parameters obtained from fits on the GMR data from Groningen. The value shown by *) in set (v) is constrained in fit.

Sets of Nuclei	K_V (MeV)	K_S (MeV)	K_Σ (MeV)	χ^2_{rel}
(i) Sn and Sm nuclei	310 ± 35	-797 ± 152	-345 ± 162	.07
(ii) Nuclei of set (i) + ^{208}Pb	311 ± 37	-802 ± 162	-345 ± 177	.06
(iii) Nuclei of set (ii) + ^{24}Mg	301 ± 22	-754 ± 80	-323 ± 182	.10
(iv) Nuclei of set (iii) + ^{90}Zr	300 ± 23	-751 ± 83	-316 ± 192	.17
(v) Constrained fit for nuclei of set (iv)	217*)	-177 ± 85	-206 ± 173	.87

or excluded in the fit. This is due to large error-bars and uncertainties in the data which allow to encompass a broader range in the value of the fitted parameters. This is the reason that previous data have not been able to unambiguously determine these parameters and thus the argument of Blaizot[9] that all data, old and new, lead to the same conclusion of 300 MeV on nuclear matter compressibility is not correct (see Ref.[22] for details). It is the precision of the Groningen data that gives a consistent fit on all sets of nuclei and allows to constrain the value of the fitted parameters.

MODEL CALCULATIONS

In order to check the validity of the leptodermous expansion, incompressibilities K_A and breathing-mode energies are obtained in a density variational approach using the semi-classical local density functional of the extended Thomas-Fermi (ETF) model[14,23]. The interactions used are the phenomenological Skyrme interactions SkM* and SIII. The interaction SkM* was adjusted[24] to reproduce the fission barriers besides giving the correct g.s. properties of nuclei. The spherical neutron and proton density profiles are parametrized by[23]

$$\rho_q(r) = \rho_{0q}\left[1 + exp\left(\frac{r-R_q}{\alpha_q}\right)\right]^{-\gamma_q} , \quad (q = n, p) , \tag{8}$$

The total energy is minimized with respect to the 8 parameters in (8), keeping the nucleon numbers N and Z fixed. The breathing-mode energies are then calculated in an extended hydrodynamical model (for details see Refs.[14,23]), taking the density parameters ρ_{0q} and α_q as time-dependent collective variables (the radii R_q are adjusted at any time to conserve Z and N and γ_q is kept constant). The dynamical coupling of the parameters ρ_{0q} and α_q is described in terms of the coupling parameter β defined by

$$\frac{\alpha_q(t)}{\alpha_q(0)} = \left[\frac{\rho_{oq}(t)}{\rho_{oq}(0)}\right]^{\beta} . \qquad (9)$$

The parameter β represents the coupling between bulk and surface compression and thus defines a whole class of different compression modes. For example, for $\beta = 0$, we have a pure bulk compression; $\beta = -1/3$ defines the scaling model in which case the inertial parameter is $m < r^2 >$; and $\beta = \pm\infty$ gives pure surface modes.

The restoring force parameter $K_A(\beta)$ is determined by numerical differentiation of the total energy with respect to the collective parameters. By solving the continuity equations numerically, the velocity fields corresponding to the bulk density and surface vibrations are determined, and from them the inertial tensor is found. Finally, in a generalized 2-dimensional hydrodynamical model[14,23] (for a given coupling β; β being determined for a given nucleus by minimizing the restoring force), the system of coupled harmonic oscillators (bulk and surface oscillations) is diagonalized. The lower of the two eigenmodes is found to be very close to the experimental breathing-mode energy particularly for nuclei $A > 150$. For heavy nuclei, the lower mode contains almost 100% of EWSR and for the lighter nuclei, less than 50% of the collective r^2 strength is found[23]. It is worth noting that the 2-dim. hydrodynamical model in conjunction with the interaction SkM* is very successful[23] in explaining the lower centroid energies and missing strength of the GMR in light nuclei $A < 90$. In Fig. 6 the results of the calculation in the scaling model (crosses) and the 2-dim. hydrodynamical model (squares) for the GMR (0^+) and the GQR (2^+) energies for Sn nuclei, as taken from Ref.[23], are shown. The experimental data[4,7] from Groningen are shown for comparison by dots with error bars. The scaling overestimates the GMR data by 1.0-1.2 MeV whereas the results of the 2-dim. hydrodynamical model come closer to the experimental 0^+ energies. The scaling, however, reproduces the GQR energies extremely well. This indicates that the interaction SkM* requires some tuning in order to describe both the GMR and the GQR data, and a large surface incompressibility.

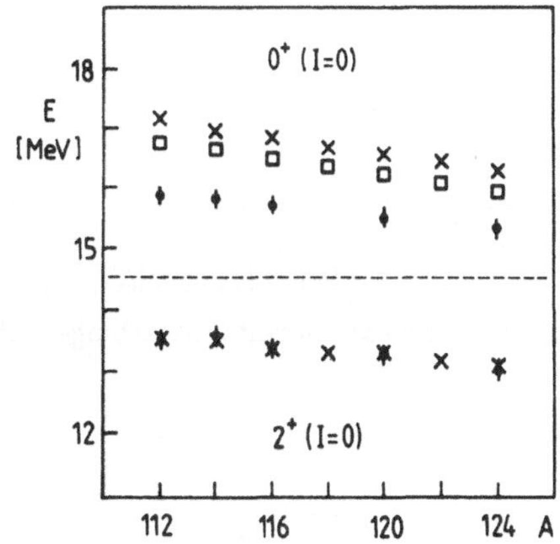

Fig. 6. The GMR (0^+) and the GQR (2^+) energies of Sn isotopes, taken from Ref.[23]. Crosses: scaling, Squares: lower mode of the 2-dim. hydro., Dot with error bars: expt. data[4,7].

Fig. 7 shows the breathing-mode energies[18,23] calculated in the scaling model as well as the 2-dim. hydrodynamical results using the interaction SkM* on several nuclei in addition to the ones in Fig. 6. The experimental data (with error bars) have been taken from the KVI-Groningen work[4,7,17,25]. For ^{208}Pb the calculations are in agreement with the experiment. The dashed lines drawn to guide the eye along the experimental and theoretical data show that going from ^{208}Pb towards ^{90}Zr the difference between theory and experiment is increasing, leading to different slopes of these hypothetical lines[18]. One might naively think that there exists a paradox between the results of SkM* and the experimental data as the latter lead to a higher value of K_∞ (Table 1) as compared to SkM*. This paradox is resolved in Fig. 8, which we shall discuss below. The traingles shown in the figure are the experimental data points used by Blaizot[13] in the analysis of the breathing mode. If one joins these two points by a hypothetical line, it would be closer to the dashed line drawn along the SkM* results. The K_∞ derived from these data points is also very close to that of SkM*. Thus, the conclusion of Blaizot[13] of (210 ± 30) MeV, we emphasize, on the basis of above two points, is not incorrect. It is, however, not compatible[18] with the new precision data on the GMR as shown in Fig. 7. It may be mentioned that the Gogny interaction, which has incompressibility close to the conclusion of Ref.[13], overestimates the GMR data on ^{90}Zr by ~2 MeV and the GQR data on both ^{208}Pb and ^{90}Zr by ~1.5-2 MeV.

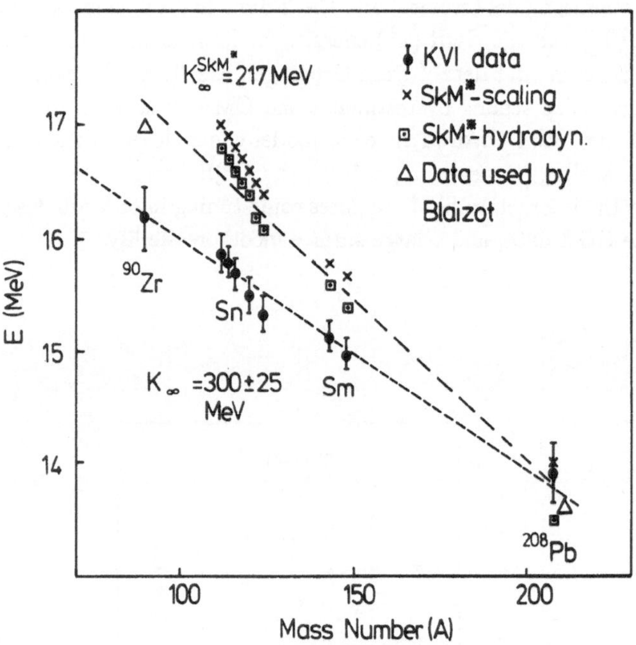

Fig. 7. Theoretical and experimental breathing-mode energies.

FITS TO THEORETICAL DATA

We now present the results of our fitting procedure[18] to determine K_V and K_S employing the theoretical results[18,23] for K_A discussed above. We show the values of the coefficients K_V, K_S, K_Σ and K_{curv} for the scaling and the hydrodynamical model using the interaction SkM* (Table 2) and for the scaling model using the interaction SIII (Table 3). The last column displays the values of the relative χ^2 per degree of freedom. For details of the fits, see Ref.[18]. The exact asymptotic values shown here refer to the determination of the coefficients of expansion (7) in the limit $A \to \infty$ for symmetric system (N=Z) with the Coulomb force switched off. The results of a 2-parameter fit carried out on similar finite nuclei for both the interactions SkM* (Table 2) and SIII (Table 3) are also shown. The volume term in this case has been enforced to its asymptotic values for determining the surface and curvature coefficients. K_S thus obtained is very close to the exact asymptotic one both for SkM* and SIII, whereas K_{curv} is slightly different from its theoretical value reflecting the absence of higher order terms and the use of finite nuclei. The results of a 4-parameter fit on K_A obtained via Eq. (1) from scaling energies (Fig. 7) of finite and realistic nuclei are also given for comparison. It can be noticed that K_V and K_S are practically identical to the exact asymptotic values as well as to the results of the 2-parameter fit. The curvature term compares well with that of a 2-parameter fit on finite nuclei. Thus scaling fulfils the necessary condition for the validity of the LDM expansion (7) and the coefficients K_V and K_S obtained from the fits are practically identical to the asymptotic ones.

The 2-dim hydrodynamical results on K_A obtained for the SkM* have also been fitted to expansion (7) and the results are shown in Table 2. It has been shown in Ref.[23] that the coupling parameter β depends strongly on A and only for very large nuclei it becomes constant. Consequently the LDM coefficients obtained are different from those of the scaling case. Including small nuclei in the fit, one obtains a high relative χ^2-values and quite dramatic variations of the curvature coefficient, reflecting the bad convergence of the expansion. Thus, expansion (7) holds as long as the coupling between bulk and surface remains constant over the range of A values considered in the fit. Nevertheless, fits with 2-dim. hydrodynamical results show a remarkable stability of the volume term K_V, which is our main concern and is in satisfactory agreement with the theoretical value. The 4-parameter fit to these results on finite and realistic nuclei, for which the inertial parameter is different from the scaling case, also reproduces the volume term satisfactorily. Equations (1) and (7) thus allow to extract the parameter K_V correctly even in a theoretical model[23] which goes beyond the scaling assumption and in which case the collective inertia is not proportional to $m < r^2 >$.

The validity and convergence of the LDM expansion (7) can also be shown by Fig. 8, where we have plotted a quantity $\widetilde{K}_A$ defined as

$$\widetilde{K}_A = K_A - K_\Sigma \left(\frac{N-Z}{N+Z}\right)^2 - K_{Coul} Z^2 A^{-4/3} - K_{curv} A^{-2/3} \left(\approx K_\infty + K_S A^{-1/3}\right). \tag{10}$$

If expansion (7) is valid and converges fast, $\widetilde{K}_A$ plotted against $A^{-1/3}$ should give a straight line cutting the $\widetilde{K}_A$-axis at K_V with a slope K_S. The scaling results for SkM* shown by crosses lie on a straight line shown by dashes. It indicates how well expansion (7) converges for scaling. The $\widetilde{K}_A$-points for the best fit of set (iv) of Table 1 are also shown. All data are lying practically on a straight line. This figure also resolves[18] the apparent paradox between SkM* and experimental data as mentioned before. The different slopes of straight lines, bring out the importance of the surface parameter K_S which is empirically found to be much larger in magnitude than in SkM*. This also implies that the assumption of $K_S \simeq -K_V$ and the possibility of one-parameter fit, as

Table 2.

Coefficients of the LDM expansion (7) obtained from fits to theoretical SkM* breathing-mode energies and incompressibilities K_A in the scaling and the 2-dim. hydrodynamical model. See text and Ref.[18] for details.

	K_V (MeV)	K_S (MeV)	K_Σ(MeV)	K_{curv}(MeV)	χ^2_{rel}
<u>scaling</u>					
exact asymptotic values	217	-209 ± 2		-105 ± 10	
2-parameter fit to nuclei 20≤A≤200 (N=Z), without Coulomb	217*)	−217	0.	−57	.006
4-parameter fit to real nuclei 90≤A≤208, with Coulomb	215	− 209	−255	−58	.007
<u>2-dim. hydrodyn.</u>					
3-parameter fit on nuclei N=Z, Coulomb off, with					
(i) 40≤A≤20000	204	0.	0.	−1236	35
(ii) 140≤A≤20000	206	−188	0.	−19	.12
(iii) 200≤A≤20000	208	−224	0.	171	.06
(iv) 600≤A≤20000	211	−311	0.	729	.01
4-parameter fit on real nuclei 90 ≤A≤208. via eq. (1).	212	−352	0.	649	.17

*) Enforced values

Table 3.

Coefficients of the LDM expansion (7) obtained from fits to the scaling energies calculated with interaction SIII.

	K_V (MeV)	K_S (MeV)	K_Σ (MeV)	K_{curv}(MeV)	χ^2_{rel}
<u>scaling</u>					
exact asymptotic values	356	−361±3		−116±10	
2-parameter fit to nuclei 20≤A≤200 (N=Z), without Coulomb	356*)	−371	0.	−72	.004
4-parameter fit to real nuclei 90≤A≤208 with Coulomb	353	− 366	−379	−68	.003

*) Enforced values

discussed by Blaizot[9], is not justified in reality. As can be seen from Tables 2 and 3, this is a property of the Skyrme interactions in the scaling approach. The $\widetilde{K}_A$-points

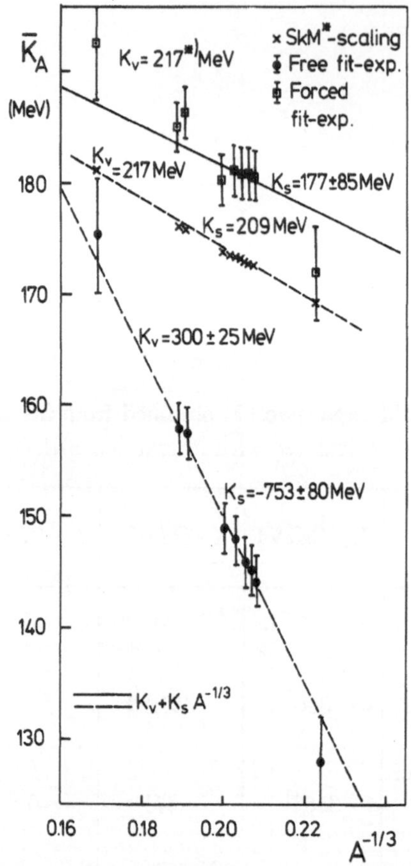

Fig. 8. $\widetilde{K}_A$ extracted from the experimental and the theoretical SkM* data of Fig. 7. K_V and K_S indicated follow from fits shown by lines. The fits include ^{24}Mg, not shown in the figure.

obtained from the experimental data by a constrained fit with $K_V = 217$ MeV imposed (set (v) in Table 1) are also shown for comparison. The badness of this fit (shown by solid line) is reflected by the fact that the points do not lie on a straight line. A value of $K_V \sim 217$ MeV is, therefore, ruled out by these data.

DISCUSSION AND CONCLUSION

We have demonstrated that using the scaling model the LDM expansion of K_A holds well for any interaction used in the model. The expansion converges fast and it is possible to determine the first two coefficients of the expansion from a sample of real nuclei. Using data on a limited number of nuclei, it is indeed possible to disentangle K_A into the volume and the surface parts. This shows the validity of the semi-empirical fitting procedure to determine the nuclear incompressibity from breathing-mode energies. It has been shown that from a fit to a set of new precision data[4,7], one gets $K_\infty \sim (300 \pm 25)$ MeV and a very important surface coefficient $K_S \sim (-750 \pm 80)$ MeV.

Blaizot[9] pointed out that due to a narrow range of $A^{-1/3}$ one measures a quantity $\overline{K}$, which is obtained by averaging the surface, Coulomb and other terms over the $A^{-1/3}$ range used, instead of K_A. This, he argued, is due to large correlation between the volume and the surface terms. This is not quite true in the Groningen data[4]. Looking at the good fit and constrained fit (bad) to the empirical data in Table 1 and Fig. 8, one finds that both these fits have a similar $\overline{K}$ of Blaizot. These, however, differ completely in the quality of fit as evident from the value of relative χ^2 and the $\widetilde{K}_A$ fit, thus disallowing an incompressiblity of $\sim$ 220 MeV. The clear disentangling of the volume and surface terms in the Groningen data becomes possible due to greater precision of data in constrast with the old data which allow a greater range of these parameters due to large error bars.

As a consequence of clear separation of the volume and surface terms, the surface term emerges to be very large. It may be pointed out that there is no Skyrme interaction in vogue with such a large surface coefficient. Correlations in the nuclear surface not describable in the Skyrme/mean-field approximation might be responsible for this large surface term.

It was shown that including the datum[21] on ^{24}Mg in the fits to empirical data, the fit remains as good as without this nucleus. This is in constrast with the fits of the 2-dim. hydrodynamical mode[23]l where including nuclei $A < 90$ results in a very bad fit compared to the scaling case and empirical data. It may therefore be concluded that in reality the coupling of surface to bulk might be less A-dependent than suggested by the theoretical 2-dim. hydrodynamical model.

I thank Adrian van der Woude and collegues of KVI, Groningen and Muhsin Harakeh of Free University, Amsterdam for the experimental collaboration at Groningen. Part of this work has been performed in collaboration with W. Stocker of the University of Munich and Matthias Brack and Peter Gleissl of the University of Regensburg. I am indebted to W. Stocker and Matthias Brack for innumerable discussions. Help and constant support from Jorrit de Boer is gratefully acknowledged. This work was supported by the Alexander von Humboldt Foundation and by BMFT grant No. 06 LM 171.

+Present Address: Dept. of Physics and Astrophysics, Delhi University, Delhi- 110007, India.

REFERENCES

1. A. van der Woude, Prog. Part. Nucl. Phys. 18:217 (1987).

2. J. Speth and A. van der Woude, Rep. Prog. Phys. 44:719 (1981).

3. See for a review, M. Buenerd, in Proc. of the International Symposium on Highly Excited States and Nuclear Structure, Orsay, France, 1983, J. Phys. (Paris) Colloq. 45:C4-115 (1984).

4. M.M. Sharma, W.T.A. Borghols, S. Brandenburg, S. Crona, A. van der Woude and M.N. Harakeh, Phys. Rev. C38:2562 (1988).

5. A.G. Drentje, H.A. Enge and S.B. Kowalski, Nucl. Instrum. Methods 122:485 (1974).

6. J.C. Vermeulen, J. van der Plicht, A.G. Drentje, L.W. Put and J. van Driel, Nucl. Instrum. Methods 180:93 (1981).

7. M.M. Sharma, in Proc. International Winter Meeting on Nuclear Physics, Bormio (Italy), N63:510 (1988), University of Milano.

8. D.H. Youngblood, P. Bogucki, J.D. Bronson, U. Garg, Y.-W. Lui and C.M. Rosza, Phys. Rev. C23:1997 (1981).

9. J.P. Blaizot, these proceedings.

10. B.K. Jennings and A.D. Jackson, Phys. Rep. 66:141 (1980).

11. O. Bohigas, A. M. Lane and J. Martorell, Phys. Rep. 51:276 (1979).

12. G.R. Satchler, Particles and Nuclei 5:105 (1973).

13. J.P. Blaizot, Phys. Rep. 64:171 (1980).

14. M. Brack and W. Stocker, Nucl. Phys. A388:230 (1982);
M. Brack and W. Stocker, Nucl. Phys. A406:413 (1983).

15. J. P. Blaizot, D. Gogny and B. Grammaticos, Nucl. Phys. A265:315 (1976).

16. N. Marty, M. Morlet, A. Willis, V. Comparat and R. Frascaria, Nucl. Phys. A238:93 (1975).

17. W.T.A. Borghols, Ph.D. Thesis (1988) Univ. Groningen (unpublished);
W.T.A. Borghols et al., submitted to Nucl. Phys.A (1988).

18. M.M. Sharma, W. Stocker, P. Gleissl and M. Brack, Nucl. Phys. A (in print).

19. V.R. Pandharipande, Phys. Lett. 31B:635 (1970).

20. J. Treiner, H. Krivine, O. Bohigas and J. Martorell, Nucl. Phys. A371:253 (1981).

21. H.J. Lu, S. Brandenburg, R. De Leo, M.N. Harakeh, T.D. Poelhekken and A. van der Woude, Phys. Rev. C33:1116 (1986).

22. M.M. Sharma, in Proc. Int. Summer School on Nucl. Astrophysics, La Rabida (Spain), 1988 - Nuclear Astrophysics, Research Reports in Physics, M. Lozano et al. (eds.), Springer Verlag, p. 306 (1989).

23. P. Gleissl, M. Brack, J. Meyer and P. Quentin, Preprint NBI-88-62.

24. M. Brack, C. Guet and H.-B. Håkansson, Phys. Rep. 123:275 (1985).

25. S. Brandenburg et al., Nucl. Phys. A446:29 (1987).

NUCLEAR COMPRESSION MODULUS FROM MONOPOLE DATA

J.P. BLAIZOT

Service de Physique Théorique
CEN Saclay
91191 Gif-sur-Yvette Cedex
France

1. INTRODUCTION

The compression modulus K_{NM} is one of the few basic quantities which characterize the nuclear matter equation of state and which may be extracted from empirical data. There are indeed several properties of nuclei which are, to some extent, sensitive to K_{NM}, the most sensitive one being the frequency of the the so-called "breathing mode". About twelve years ago, this special mode of vibration of nuclei was discovered in various experiments, while at the same time microscopic calculations showed that its energy could provide information on K_{NM}. Several analysis [1,2,3], based mostly on microscopic calculations, led then to the value $K_{NM} \simeq 210$MeV.

These analysis have been recently challenged on the basis of new empirical data [4,5]. It is claimed that these new data are more precise than the old ones and allow a much better separation of surface and volume effects in the various contributions to the breathing mode restoring force. The outcome of this new analysis is a larger value of the compression modulus, $K_{NM} \simeq 300$MeV.

This is somewhat puzzling. Indeed, the new data, albeit more precise, are *not* incompatible with the old ones. Furthermore, the type of analysis which is used to extract the value of the compression modulus from the A dependence of the breathing mode frequency has been very much studied in the past [3,6]; it was found then that it led to results similar to those of the microscopic calculations. There is therefore a contradiction, which I shall try to resolve.

In my discussion I shall rely on calculations and analysis which were done about 10 years ago and which are mostly contained in refs.[2] and [3] (with slightly different notations from the ones used in this note). In fact, I shall refer here only to certain aspects of these calculations, i.e. to those which are relevant to the understanding of the present controversy and which, perhaps, have not been fully appreciated.

2. SOME BASIC DEFINITIONS

Let us start by recalling some definitions which pertain to nuclear matter, an idealized uniform system of neutrons and protons with no Coulomb interaction. We consider here nuclear matter at zero temperature. The physical quantity of interest is then the binding energy (per unit volume, or per particle):

$$E = \Omega\,\varepsilon(n) = A\,(E/A) \tag{2.1}$$

where Ω is the volume (assumed to be very large), n is the nucleon density, and A is the total number of nucleons.

The compressibility χ measures the relative change in volume corresponding to a given change in pressure. It is defined by:

$$\chi = -\frac{1}{\Omega}\frac{\partial\Omega}{\partial P} \tag{2.2}$$

where the pressure P is related to the binding energy by the following formulae:

$$P = -\frac{\partial E}{\partial\Omega} = n\frac{d\varepsilon}{dn} - \varepsilon = n^2\,\frac{dE/A}{dn} \tag{2.3}$$

Using these formulae, one can easily express χ^{-1} in terms of the second derivative of E/A with respect to density. However, when discussing nuclear matter equation of state, it is more common to use the compression modulus K_{NM} rather than the compressibility χ. K_{NM} is defined by:

$$K_{NM} = 9\,n_0^2\,\frac{d^2E/A}{dn^2}\bigg|_{n_0} = k_f^2\frac{d^2E/A}{dk_f^2}\bigg|_{k_{f0}} \tag{2.4}$$

where k_f is the Fermi momentum, related to the density by the well known formula: $n=2k_f^3/3\pi^2$. Note that the definition (2.4) holds only at saturation, i.e. at the density n_0 where the energy per particle is minimum.

Some of the properties of nuclear matter at saturation may be obtained rather directly from empirical data. This is the case for example of the binding energy per nucleon which can be deduced from the analysis of nuclear masses; it is approximately equal to 16 MeV. Similarly, the saturation density may be obtained from the measurements of the charge densities of nuclei via electron scattering; it is of the order of 0.17nucleon/fm^3. The value of the compression modulus is somewhat harder to get. Ideally, and according to its definition, it should be deduced from measurements of the binding energy of nuclear matter for several values of the density around the saturation point. Obviously, this is not possible, as we do not have bulk pieces of nuclear matter in the laboratory. Thus, the compression modulus is best determined from dynamical properties in which deviations away from saturation are involved, as it is the case for example in the breathing mode of nuclei. We shall see that a microscopic analysis of breathing mode data suggests a value of K_{NM} around 200 MeV.

Landau's Fermi liquid theory provides a useful expression of the compression modulus:

$$K_{NM} = 6\ \epsilon_F\ (1+F_0) \tag{2.5}$$

This formula isolates two contribution to K_{NM}. One, $6\epsilon_F$, may be viewed as the compression modulus of a gas of nucleons having an effective mass m^* ($\epsilon_F = \hbar^2 k_f^2/2m^*$). The second, $6\epsilon_F F_0$, contains the effect of interactions which are specific to a compression mode. Roughly speaking, the Fermi liquid parameter F_0 represents the change in the potential energy of one nucleon due to the change in the density of the surrounding medium. The formula (2.5) may be used to get an order of magnitude estimate of K_{NM}. Indeed the saturation density corresponds to a Fermi momentum $k_f \simeq 1.35\text{fm}^{-1}$. Taking $m^* \simeq m$, one obtains $\epsilon_F \simeq 38$MeV, i.e. $6\epsilon_F \simeq 228$MeV. It turns out that this value is quite close to the empirical value of K_{NM}; this suggests that the Fermi liquid parameter F_0 is small.

3. THE BREATHING MODE IN NUCLEI

The breathing mode of a heavy nucleus is a monopole isoscalar oscillation of the whole system. The shape of the transition density, i.e that part of the density which oscillates in time, is shown in Fig.1 for the case of the nucleus ^{208}Pb. It can be seen that neutrons and protons are in phase, which confirms the isoscalar character of the mode. Some shell effects can be observed; these are responsible, for example, of the bump at r~2fm in the neutron density. However, for a large nucleus such as ^{208}Pb, these do not affect in a major way the frequency of the mode. In fact, as far as the frequency is concerned, a good estimate may be obtained by assuming for the transition density the so called "scaling form":

$$\delta n(r) = -3\ n_0(r) - r\ \frac{dn_0}{dr} \tag{3.1}$$

also displayed in Fig.1. The transition density (3.1) is obtained by performing a homologous transformation ($r \rightarrow \lambda r$) on the ground state density, keeping the normalisation, i.e. the particle number, constant.

The curves drawn in Fig.1 represent the local amplitudes of the density oscillations, at their maximum. One can then deduce from them that *the breathing mode is a small amplitude vibration*. The relative change in the density $\delta n/n_0$ is at most $0.035/0.17 \sim 20\%$; correspondingly, the relative variation in the r.m.s. radius is less than 5%. This should be kept in mind when trying to extrapolate to large densities.

There are several ways by which one can characterize the "collectivity" of the breathing mode. A microscopic analysis reveals that in heavy nuclei the wave function of the mode has components over many elementary particle-hole excitations. This is not the case for light nuclei where the monopole excitations involve only a few particle-hole configurations. A more global characterization of collectivity relies on the use of sum rules [7,8]. These are defined for moments of the strength function:

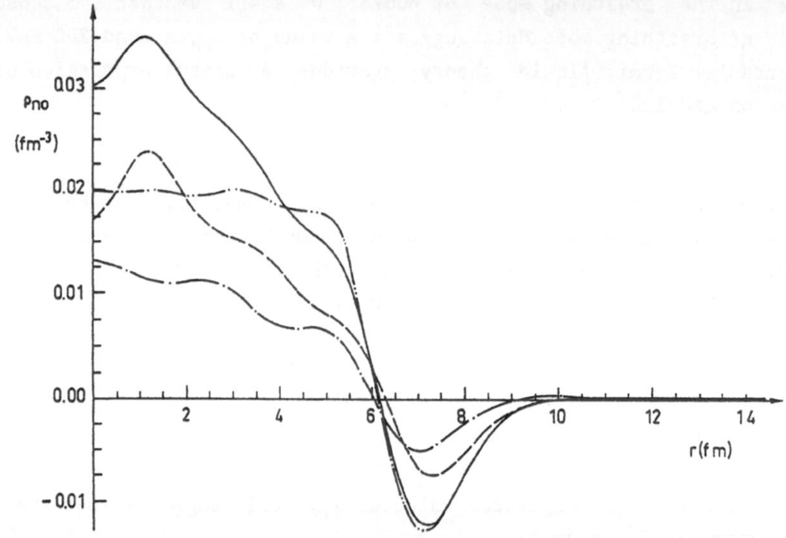

Fig. 1. Transition density of the breathing mode in ^{208}Pb, calculated with the Skyrme interaction SIII. Neutrons:------; protons:-.-.-.; total:———; -..-..-, Eq.(3.1). (From ref.[2])

$$\mathfrak{M}_k = \sum_n (E_n)^k \, |\langle n|r^2|0\rangle|^2 \tag{3.2}$$

some of which can be calculated in close form. A well known example is the *energy weighted sum-rule:*

$$\mathfrak{M}_1 = \sum_n E_n \, |\langle n|r^2|o\rangle|^2 = \frac{2\hbar^2}{m} \langle r^2 \rangle_0 \tag{3.3}$$

where $\langle r^2 \rangle_0$ is a ground state expectation value.

Microscopic calculations indicate that, in heavy nuclei, a single mode exhausts most of the energy weighted sum rule. This is in agreement with the observation of a giant monopole resonance in scattering of light particles on nuclei; this resonance exhausts most of the energy weighted sum rule in nuclei with $A \gtrsim 100$; it occurs at an excitation energy given approximately by $E_0 \sim 80A^{-1/3}$.

Precision measurements of the breathing mode frequency have been performed recently [4,5]. In Table 1, we compare the "new" data with the "old" ones, for nuclei in the range $110 \leqslant A \leqslant 124$. The new measurements are definitely more precise, but, *except for a few isolated cases, they are totally compatible with the old ones;* in Table 1, the exception is ^{124}Sn whose energy is lower in the old data than in the new ones. Similar agreement is found for nuclei other than those of Table1. For example, the recent determination of E_0 in ^{208}Pb is 13.9±0.3MeV [11]; this compares well with the 13.7±0.4MeV of ref.[9], but is not compatible with the 13.2±0.3MeV quoted in ref.[10]. I do not know whether these discrepancies are understood, and whether the new data should be considered as free from the experimental bias from which they may originate.

Table 1. The measured breathing mode energy E_0 (in MeV) in nuclei with $110 \leqslant A \leqslant 124$. The various data are from KVI[4], TAM[9], GR[10].

A	Sn (KVI 88)	Sn (TAM 79)	Sn (GR 79)	Cd (GR 79)
110				15.95±0.25
112	15.88±0.14		16.1±0.25	15.75±0.25
114	15.80±0.14			15.45±0.25
116	15.69±0.16	15.60±0.3	15.55±0.25	15.75±0.25
118		15.5±0.6	15.5±0.6	
120	15.52±0.15	15.4±0.5	15.45±0.25	
124	15.35±0.16	14.8±0.4	14.85±0.25	

I would like to add here a comment concerning the nucleus ^{90}Zr. It has been recently stated [5] that, in the microscopic approach, the determination of the compression modulus relies very much on the value of the breathing mode frequency in this nucleus, for which wrong data were used. It will become clear in the next sections why the first part of this statement is not correct. As for the second part, it is true that in the first paper on the subject [1], we used the only available data at the time, i.e. those coming from the pioneer Orsay experiment [12]: $E_0 \simeq 17.2 \pm 0.5$MeV. Subsequently, other measurements were performed and gave somewhat lower values, 16.2±0.5MeV [9] and 16.4±0.25MeV [10]; these values were certainly not ignored in subsequent analysis, as can be seen in Fig.28 of ref.[2], or in table 7 of ref.[3].

4. BREATHING MODE FREQUENCY AND COMPRESSION MODULUS

We are now coming to the heart of the discussion, i.e. the relation between the frequency $\omega_0 = E_0/\hbar$ of the breathing mode and the nuclear matter compression modulus. It is convenient to write this frequency in the following way:

$$\omega_0 = \sqrt{\frac{K_A}{m \langle r^2 \rangle_0}} \tag{4.1}$$

In some special situations, K_A may be given a transparent physical interpretation as the restoring force of particular modes. However, this is a subtle matter into which we shall not go here. Rather, and for the clarity of the present discussion, we shall view Eq.(4.1) as a convenient parametrization of the empirical data and as the *definition* of K_A. One advantage of this parametrization is that it isolates the dominant A dependence of ω_0 which is contained in $\langle r^2 \rangle_0$; for K_A independent of A, one finds $\omega_0 \sim A^{-1/3}$, in agreement with the main empirical trend. As we shall see, all the information on the compression modulus is contained in K_A.

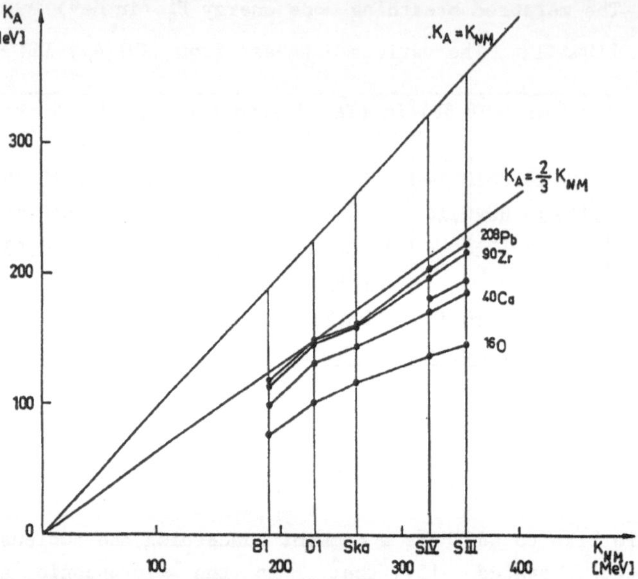

Fig.2. The effective compression modulus K_A calculated from various effective interactions corresponding to different values of the nuclear matter compresion modulus K_{NM}.

One can indeed consider K_A as a function of K_{NM}. *This function is model dependent*, since we need a model to calculate the value of K_A corresponding to a given value of K_{NM}; furthermore, it is only within the framework of a model that we can vary at will the value of the nuclear compression modulus K_{NM}. Microscopic models based on self-consistent mean field approximations allow presently the best determination of such a function. In such models, one calculates separetely, but with the same basic ingredients, namely the same effective interaction, the properties of nuclear matter, e.g. the compression modulus, and those of finite nuclei, e.g. the breathing mode frequency. By changing the parameters of the effective interaction, one can therefore follow the correlated variations of K_A and K_{NM}. The change of parameters, or the choice of effective interactions, is constrained so as to keep constant most properties of nuclei, and in particular $\langle r^2 \rangle_0$. Under such conditions, the only dependence of ω_0 on K_{NM} is contained in K_A.

The relation between K_A and K_{NM} implied by microscopic calculations turns out to be remarkably simple. It is illustrated on Fig.2. One sees there that K_A is roughly proportional to K_{NM}, being approximately equal to 2/3 K_{NM} for the nucleus ^{208}Pb. The variation of K_A with K_{NM} is thus much more important than the variation of K_A with A; as A varies from 40 to 208, K_A varies only by less than 15%. This is reflected in the small A dependence of the quantity $E_0A^{1/3}$ plotted in Fig.3. This figure actually summarizes the main results of microscopic calculations. It shows that the data collected by the year 1980 are compatible with the value of the compression modulus $K_{NM} \simeq 210$MeV. Microscopic calculations would have difficulty to reproduce the empirical value of the breathing mode frequency if K_{NM} deviates from this value by more than 15%. However, as stated in ref.[2] (p.244) one cannot exclude such a possibility.

Let us now consider in more details the A dependence of K_A. First, one ay ask to which limit K_A goes when $A \to \infty$. This limit, which we call K_∞, lepends very much on the assumed structure of the breathing mode. To quote ust but three examples, one has in the liquid drop model, $K_\infty = \frac{\pi^2}{15} K_{NM}$; for :he scaling mode, $K_\infty = K_{NM}$; for the mode defined through a constraint on the :.m.s radius, $K_\infty = \frac{7}{10} K_{NM}$ [13].

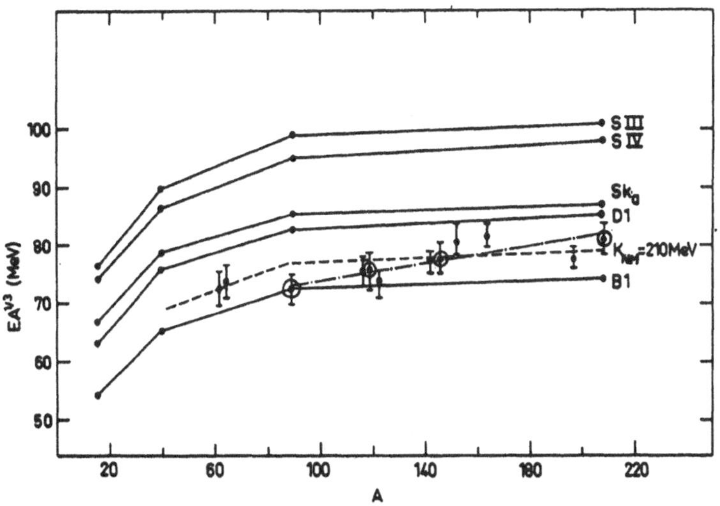

Fig.3. The energy of the breathing mode times $A^{1/3}$ as a function of A (from ref.[2]). Full lines: results of microscopic calculations for various values of K_{NM} (see Fig.2). Dashed line: interpolation corresponding to K_{NM}=210MeV. The dash-dotted line indicates the A-dependence which is implied by Sharma's fit; the circled data points are those taken into account in the fit.

One may try to represent the A dependence of K_A with the following)arametrization:

$$K_A = K_\infty + K_{surf} A^{-1/3} + K_{sym} \left(\frac{N-Z}{A}\right)^2 + K_{coul} \frac{Z^2}{A^{4/3}} \qquad (4.2)$$

rhich isolates contributions with specific N and Z dependences. It is worth :mphasizing that the physical interpretation of the various terms in (4.2) .s by no means staightforward. In particular, more than specific surface)roperties are involved in K_{surf} [2,6]. Also, as we have just seen, the :onnection of the volume term (K_∞) to K_{NM} is ambiguous: it depends strongly)n the structure of the collective mode.

It is tempting to use the formula (4.2) to determine the value of the 1uclear matter compression modulus, by extracting from empirical data the A lependence of K_A. In such an analysis, one usually estimates the Coulomb :erm and fit the other 3 parameters. We should keep in mind that the various coefficients in Eq.(4.2) are, a priori, difficult to determine reliably from a fit. As we have already seen, the A dependence is rather weak; furthermore, in the range $90 \leq A \leq 208$, $A^{-1/3}$ varies only slightly: $0.17 \leq A^{-1/3} \leq 0.22$, and similarly $10^{-2} \leq \left(\frac{N-Z}{A}\right)^2 \leq 4\times10^{-2}$. This situation is to be contrasted with that encountered in the fit of nuclear masses with a mass formula. There, we have many data; besides, the saturation ensures that the ground state density remains roughly constant as A increases, whereas the structure of the breathing mode certainly changes with A.

Table 2. Values of the expansion coefficients in formula (4.2) as deduced from a 3 parameter fit to the data (first 3 columns). $\bar{K}$ is the average defined in Eq.(4.3). The last column gives the value of K_{NM} obtained trough a one parameter fit, assuming a scaling mode (from ref.[3]).

Data	K_∞	K_{surf}	K_{sym}	$\bar{K}$		K_{NM}
GR [10]	300±29	-608±120	-475±176	167		213±2
TAM [9]	357±35	-883±148	-492±210	168		214±5
OR [14]	206±146	-143±571	-148±807	174		231±4
KVI [4]	300±25	-750±86	-320±180			

A thorough analysis along this line has been performed by Treiner et al. [3] who showed that in order to get meaningful results, one has to take into account the known (within a specific model) correlations between the various parameters entering (4.2), reducing this formula to a one parameter formula. Assuming the breathing mode to be a scaling mode, they obtain the following values:

$$K_{NM} = 220\pm20 \text{ MeV}$$
$$K_{surf} = -240\pm70 \text{ MeV}$$
$$K_{sym} = -300\pm100 \text{ MeV}$$

in good agreement with the microscopic calculations. It is worth recalling that a 3 parameter fit to the data available at that time would yield the results displayed in table 2. One notices that the various sets of data do not give identical values of the 3 parameters. In fact, as pointed out by the Orsay group, what can be really determined by the fit is a combination of the 3 parameters, which we call $\bar{K}$:

$$\bar{K} = K_\infty + K_{surf} \langle A^{-1/3} \rangle + K_{sym} \langle \delta^2 \rangle \qquad (4.3)$$

where the brackets denote averages over the families of nuclei considered. The values given in the last column of table 2 were calculated using $\langle A^{-1/3} \rangle = 0.2$ and $\langle \delta^2 \rangle = 0.025$. The values of $\bar{K}$ are indeed remarkably close for the 3 sets of data considered, and so are the values of K_{NM} deduced from a one parameter fit (last column of Table2).

5.DISCUSSION

We are now in position to appreciate the origin of the discrepancy between the "old" analysis and the "new" one presented by Dr.Sharma at this meeting. The fit performed by Sharma et al. (last line in Table 2) looks similar to the 3 parameter fit of the Grenoble data by the Orsay group (first line of Table 2) and, as such, it is subject to the same criticisms [3]. However, in the new fit, emphasis is put on a few nuclei, for which the frequency of the breathing mode is known with great accuracy. One may get a feeling for what it corresponds to by looking at the dash-dotted line in Fig.3. The resulting A dependence of K_A is definitely distinct from that predicted by the microscopic calculations: when the parameters in these calculations are adjusted so that they reproduce the breathing mode energy in ^{208}Pb, they invariably overestimate the breathing mode energy in medium mass nuclei. In fact, this difficulty was already pointed out by C.M. Rozsa et al. in ref.[9] (see Fig.9 in that paper). Certainly, it would be worth trying to understand why this is so. However, as may be inferred from Fig.3, the deviation is rather small. To give an illustrative example, let us consider the case of the nucleus ^{120}Sn (to my knowledge, there are no microscopic calculations of the breathing mode in this nucleus, however the relevant information may be obtained from a reasonable interpolation of known results). One may deduce from Fig.3 that the value K_{NM}=210MeV, which slightly underestimates the breathing mode energy in ^{208}Pb, overestimates it in ^{120}Sn by about 0.5MeV. On the other hand, if instead of 210MeV, one uses a compression modulus of 300MeV, the value obtained by Sharma in its fit of both ^{208}Pb and ^{120}Sn, one overestimates the energy by more than 4 MeV.

Thus, the possible failure of microscopic calculations to accurately reproduce the A dependence of monopole data does not necessarily have the implications suggested by Sharma's analysis. In fact, K_A is a function of two variables, A and K_{NM}. In the range of interest, the dependence on A is relatively mild, while that on K_{NM} is quite appreciable. The analysis based on the A dependence of K_A may, at first sight, look "model independent". However the interpretation of the various terms in the parametrization (4.2) is possible only within a model for the structure of the breathing mode. The analysis based on microscopic calculations relies primarily on the strong correlation between K_A and K_{NM}. Thus, for a given A, the value of K_{NM} needed to fit the breathing mode energy is rather well determined.

As we already emphasized, the formula (4.2) can be justified only within a model. Its validity has been checked in the framework of microscopic calculations or approximations to these (in these respect, the hydrodynamical calculations of ref.[15], for example, should be regarded as approximations to self-consistent mean field calculations). It is therefore difficult to trust conclusions one draws from it, when these are in marked contradiction with those deduced from microscopic calculations.

This being said, ambiguities remain in the determination of the compression modulus, but these are of a totally different nature. As stated earlier, K_{NM} is basically a static concept. We are however deducing its value from the energy of a collective excitation, using a mean field approximation which ignores dynamical effects related to the frequency dependence of various physical quantities. These may affect in a non trivial way the relation between the frequency of the breathing mode and the compression modulus. These effects, which are difficult to estimate in a reliable way, are reviewed in part in ref.[2], and more recently in [16]. Note that such effects may induce a non trivial A dependence in the breathing mode frequency.

REFERENCES

1. J.P. Blaizot, D. Gogny and B. Grammaticos, Nucl.Phys.**A265**:315 (1976).
2. J.P. Blaizot, Phys.Rep. **64**:171 (1980).
3. J. Treiner, H. Krivine, O. Bohigas and J. Martorell, Nucl.Phys.**A371**:253 (1981).
4. M.M. Sharma, W.T.A. Borghols, S. Brandenburg, S.Crona, A. van der Woude and M.N. Harakeh, Phys.Rev. **C38**:2562 (1988).
5. M.M. Sharma, these proceedings.
6. J.P. Blaizot and B. Grammaticos, Nucl.Phys.**A355**:115 (1981).
7. O. Bohigas, A.M. Lane and J. Martorell, Phys.Rep. **51**:276 (1979).
8. E. Lipparini and S. Stringari, Phys.Rep. **175**:103 (1989).
9. D.H. Youngblood, in Proceedings of the "Giant Multiple Resonance" topical conference Oak-Ridge, USA 15-17 Oct.1979.
   D.H. Youngblood, C.M. Rozsa, J.M.Moss, D.R. Brown, and J.D. Bronson, Phys.Rev.Lett. **39**:1188 (1977).
   C.M. Rozsa, D.H. Youngblood, J.D. Bronson, Y.W. Lui and U. Garg, Phys.Rev.**C21**:1252 (1980).
10. M. Buenerd, C. Bonhomme, D. Lebrun, P. Martin, J. Chauvin, G.Duhamel, G. Perrin, and P. de Saintignon, Phys.Lett.**B84**:305 (1979).
    M. Buenerd, D. Lebrun, Ph. Martin, P de Saintignon, and C. Perrin, Phys.Rev.Lett.**45**:1667 (1980).
    M.Buenerd, in Proceedings of the International Symposium on Highly Excited states and Nuclear Structure, Orsay, France, 1983. (J.Phys.Colloq.**45**:C4-115 (1984)).
11. S.Brandenburg et al., Nucl.Phys.**A466**:29 (1987).
12. N. Marty, M. Morlet, A. Willis, V. Comparat and R. Frascaria, Nucl.Phys.**A238**:93 (1975).
13. B.K. Jennings and A.D. Jackson, Phys.Rep.**66**:141 (1980).
14. F.E. Bertrand, G.R. Satchler, D.J. Horen, and A. van der Woude, Phys.Lett.**30B**:198 (1979).
15. M. Brack and W. Stocker, Nucl.Phys.**A406**:413 (1986); MM. Sharma, W. Stocker, P. Gleissl and M. Brack, "Nuclear Matter Incompressibility from a Semi-empirical Analysis of Breathing-mode Energies".
16. G.E. Brown, in Proceedings of the Third International Conference on Nucleus-Nucleus Collisions, Saint-Malo (France) June 6-11, 1988.

Constraints on the Nuclear Equation of State from Type II Supernovae and Newly Born Neutron Stars

Wolfgang Hillebrandt, Ewald Müller and Ralph Mönchmeyer

Max-Planck-Institut für Physik und Astrophysik
Institut für Astrophysik
D–8046 Garching b. München, Fed. Rep. Germany

I. Introduction

Supernovae of type II, i.e. explosions of stars which show strong Balmer lines of hydrogen in their spectra, are now generally believed to originate from massive stars, $M \gtrsim 8M_\odot$, where $M_\odot \simeq 2\ 10^{33}$g is the mass of the sun, at the end of their quiet hydrostatic evolution. If this interpretation is correct they cannot be powered directly by thermonuclear burning and they have to get their energy, at least partially, from another source. It is therefore intriguing to relate the type II phenomenon to the formation of neutron stars, an idea that goes back to Baade and Zwicky (1934) and is supported by the fact that several supernova remnants, including the Crab Nebula, do indeed contain neutron stars. In this picture the energy observed in the explosion must ultimately come from the gain in gravitational binding of the iron core of the star and/or from the binding energy of the newly born neutron star.

Two mechanisms are presently discussed which are potentially able to transform a small fraction of the gravitational energy into outward momentum of the stellar envelope, neither of which, however, so far gives a satisfactory explanation of the entire phenomenon. In one class of models the stellar envelope is ejected by a hydrodynamical shock wave generated by the rebounding core near nuclear matter density, but this mechanism seems to work only for a very limited range of precollapse stellar models. Alternatively, neutrino emission from the hot neutron star may heat some outer layers of the core sufficiently to cause an explosion, but it has still not been confirmed that this mechanism works at all.

The question therefore arises whether the apparent problems of various supernova models are due to our incomplete knowledge of certain physical input data, such as the equation of state of hot and dense matter, or whether something fundamental is missing in the models. In the following sections we will address both aspects in some detail. We will start with a brief review of the observational facts relevant to our discussion, and in particular the information available from Supernova 1987A in the Large Magellanic Cloud will be discussed. We then will present an overview of our present ideas concerning the collapse of non-rotating stellar cores and spherically symmetric supernova explosions. Special emphasis will be on nuclear physics aspects, including the nuclear equation of state, but numerical problems of the models will also be discussed. Because there is now increasing evidence that at least in the case of SN 1987A the progenitor star has been a fast rotator, the influence of rotation on hydrodynamical simulations will be addressed. A summary and conclusions follow.

II. Observations of type II supernovae

According to Zwicky's classificaton, supernovae are called type II if they show strong hydrogen emission lines in their spectra and type I otherwise. Since type I

The Nuclear Equation of State, Part A
Edited by W. Greiner and H. Stöcker
Plenum Press, New York

events are thought to be caused by thermonuclear disruptions of white dwarfs we shall consider only type II supernovae here. At maximum light their luminosity is typically around 10^{44}erg s^{-1}, the total energy in the outburst is found to be of the order of 10^{51}erg, but only a percent or less of this energy is emitted in form of optical radiation. Most of the energy appears to be in the kinetic energy of the ejected matter, which has velocities of the order of 10^4 km s^{-1}. From the spectra one obtains photospheric temperatures near maximum light of about 15 000 K or larger. The light curves stay near maximum for a few days, and in many cases a plateau is observed lasting for about 100 days (see fig. 1). Besides showing strong Balmer lines of hydrogen the overall abundances of elements seem to be approximately solar.

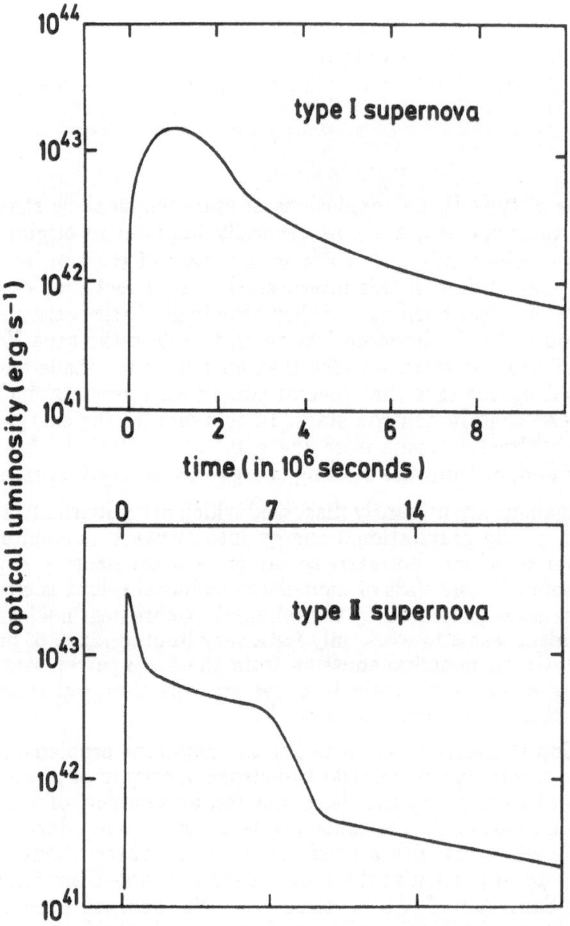

Figure 1. Schematic light curves of type I and type II supernovae. Note the different time scales of both figures. The sharp maximum and the broad plateau after maximum light are indications for an extended hydrogen-rich envelope.

All these findings, together with the fact that type II supernovae are only found in spiral arms of galaxies with young populations of stars, indicate that the progenitor stars are quite massive and have extended hydrogen-rich envelopes. Similar conclusions are reached from X-ray observations of some young supernova remnants, i.e. CasA, Puppis A. Also the observed rate of about one per 50 to 100 years in giant spiral galaxies (Tammann, 1982; van den Bergh et al., 1987) is consistent with the assumption that all stars more massive than about 8 $M_{\odot}$ explode as type II supernovae. Moreover, one can conclude that the progenitors of most type II supernovae will be in the mass range from 8 to 15 $M_{\odot}$, because those are much more frequently born than more massive ones (Salpeter, 1955; Miller and Scalo, 1979).

Next one can ask whether there is observational evidence that our basic hypothesis, namely that type II supernovae are directly related to the formation of neutron stars, is indeed correct. In fact, several young supernova remnants (Crab, Vela, SNR 0540–693, ...) do contain pulsars or X-ray point sources (RCW 103, 3C58, ...), but others don't (CasA, SN 1006, Tycho, ...), and in none of the positive detections it is certain beyond doubt that the explosions were indeed of type II. Moreover, Manchester et al. (1983) have analyzed 85 supernova remnants in the galaxy, the Large Magellanic Cloud (LMC) and the Small Magellanic Cloud (SMC), and found only 5 radio pulsars nearby. Because the frequency of both types of supernovae is approximately equal this result is surprising if one believes that all type II events leave neutron stars behind. Furthermore, not all of the identified pulsars are related to the nearby supernova remnants. On the other hand, there is no obvious contradiction, since many neutron stars may not possess strong magnetic fields or the radio radiation may not be beamed towards us. In this respect only CasA presents a problem because the neutron star should manifest itself as a X-ray point source, but the upper limit obtained for the X-ray flux is far below the values expected from standard neutron star cooling (Tsuruta, 1979).

One may also hope to obtain some information on the explosion mechanism from the observations of element abundances in supernova ejecta, because they are sensitive to the strength of the shock wave and to the position of the mass cut between the newly born neutron star and the ejected material, but again only weak constraints can be obtained. The Crab nebula, for example, has been observed at various wave lengths (Davidson et al., 1982; Henry and McAlpine, 1982; Henry et al., 1984), but with the exception of a large overabundance of helium relative to hydrogen, some possible overabundance of nickel, and a significant underabundance of oxygen, all other elements seem to be rather normal. In CasA, on the other hand, optical data obtained from fast moving knots (Chevalier and Kirshner, 1979) show large abundance inhomogeneities, but virtually no H, He, C or O.

The knots are also depleted in neon and magnesium, but some are enriched in sulphur, argon, and calcium relative to oxygen. These results have been confirmed by X-ray observations (Pravdo and Smith, 1979; Becker et al., 1979; Holt, 1983; Pravdo and Nugent, 1983; Jansen et al., 1985). They indicate that the exploding star has undergone at least oxygen burning, which means that it was a massive star. Moreover, the abundances in the fast moving knots can well be fitted by the products of explosive oxygen burning (El Eid and Langer, 1986), indicating that at least part of the explosion energy came from nuclear burning.

An object similar to CasA has recently been observed (SN 1985f in NGC 4618; Fillipenko and Sargent, 1985). This supernova showed strong oxygen emission lines, indicating more than 5 $M_{\odot}$ of oxygen in the ejecta, very little hydrogen and apparently no helium. The most straightforward interpretation of this event is that we have seen the explosion of a very massive star ($M \simeq 50 M_{\odot}$ on the main sequence) which has lost most of its hydrogen-rich envelope prior to the explosion. An even stranger supernova (SN 1961v) has been observed in NGC 1058. In this case the progenitor star has been observed for about 30 years before the outburst, the velocity of the ejected material was very low, and the optical light curve declined very slowly on timescales longer than about 6 years. The remnant is now a very strong radio source, comparable to CasA, if the estimated distance of about 12 Mpc is correct. Utrobin (1984), by modelling the light curve, has concluded that the exploding star may have had a mass of about 2000 $M_{\odot}$.

Thus it is obvious that type II supernovae do not form a homogeneous class of objects and that observations alone do not give us a clear answer as to what the mass of the progenitor stars was and what the right explosion mechanism is. When on February 23, 1987, a supernova explosion was observed in the LMC, a sattelite of our own galaxy, at a distance of roughly 50 kpc only, this event caused a lot of excitement in the astronomical community, because never since Kepler's supernova in 1604 has a supernova been so close. There was some hope, therefore, that most of the open questions in supernova theory could now be answered. However, it soon became clear that this was not the case. Moreover, SN 1987A raised new problems, which we will discuss in some detail now.

Spectra taken during the second night showed Balmer lines of hydrogen, indicating that the supernova was of type II (Catchpole et al., 1987; Fosbury et al., 1987; Tyson and Boeshaar, 1987). Its position coincided with that of a blue supergiant, Sanduleak (Sk)-69°202 (Walborn et al., 1987; West et al., 1987). When the UV-radiation from the supernova weakened a few weeks after the explosion it became obvious that this star had indeed disappeared (Gilmozzi et al., 1987; Walborn et al., 1987). From its spectral class (B3) and its luminosity class (Ia) (Rousseau et al., 1978; Isserstedt, 1975; West et al., 1987) one could conclude that the main sequence mass of the progenitor had been close to 20 $M_\odot$ (Arnett, 1987; Hillebrandt et al., 1987; Nomoto et al., 1987; Truran and Weiss, 1987).

The first big surprise was that, in contrast to theoretical expectations, it was a blue rather than a red supergiant. Moreover, the presence of low-velocity circumstellar material indicates that the progenitor has been a red supergiant some 7000 years ago (Fransson et al., 1989; Wampler and Richichi, 1989). One possible way to explain this rather complicated evolution is to assume that helium has been mixed more or less homogeneously throughout most of the hydrogen-rich envelope (Nomoto and Hashimoto, 1988; Saio et al., 1988). Additional evidence for strong mixing prior to the explosion comes from the interpretation of the optical spectra (Cassatella, 1987; Blades et al., 1987; Williams, 1987; Höflich, 1988). The reason for these mixing processes is simply not known but may indicate that Sk-69°202 has been a rapidly rotating star (Weiss et al., 1988). In any case, the first nearby supernova has demonstrated that stellar evolution models are still rather uncertain, and one has to keep these uncertainties in mind if attempts are made to model the final explosions.

Also the light curve of SN 1987A was significantly different from "standard" type II supernovae (fig.2). While, however, the very rapid decline during the first few days as well as the slow increase to maximum light can easily be explained from the rather compact structure of the progenitor star (Arnett, 1988a; Shigeyama et al., 1988; Utrobin, 1988; Woosley et al., 1988) if an explosion energy of $(1.5 \pm 0.5)10^{51}$erg is assumed, the flat maximum requires mixing of processed material into the hydrogen-rich layers during the explosion (Arnett, 1988b; Woosley, 1988; Fu, 1988; Nomoto et al., 1989). In fact, this conclusion is not only based on the interpretation of the light curve but also on direct observations. Already in August 1987 γ-ray lines from the radioactive decay of ^{56}Co were discovered (Matz et al., 1988) and at about the same time hard X-rays were detected. The hard X-ray flux remained nearly constant for over one year and then dropped below the detection limit of existing X-ray satellites (Dotani et al., 1987; Sunyaev et al., 1987; Tanaka, 1988). Moreover, strong IR-lines of Ni, Co, and Fe have been seen from November 1987 on with velocities up to 3000 km s^{-1}, corresponding to the matter velocity at the photosphere in April 1987 (Rank et al., 1988; Erickson et al., 1988; Witteborn et al. 1988). However, no evidence for Fe or Co enhancements was found in the early optical spectra. The only reasonable explanation of all these findings is that radioactive Co has been mixed inhomogeneously far out into the H-rich envelope, and has formed clumps there. These effects, which are probably a result of a Rayleigh-Taylor instability caused by the shock propagation through the envelope of the star (see Arnett et al., 1989), can only be modelled by multi-dimensional simulations. As an additional cause for the mixing a Rayleigh-Taylor instability driven by the decay energy of ^{56}Ni located at the edge of the core (see below) has been proposed (Arnett, 1988a; Woosley et al., 1988).

The total amount of radioactive Co produced in SN 1987A can be estimated in two independent ways. Firstly, from figure 2 it is obvious that the late light curve decays on a timescale identical to the lifetime of ^{56}Co. It is therefore reasonable to assume that the late light curve is powered by radioactive decay energy. The initial amount of ^{56}Ni necessary to fit the observations is then about 0.07 $M_\odot$ (Woosley, 1988; Arnett, 1988b; Shigeyama et al., 1988). Secondly, the strength of the IR CoII line at 10.53μ gives 0.0044 $M_\odot$ at day 280 and 0.0023 $M_\odot$ at day 400, in excellent agreement with the estimate obtained from the light curve (Danziger et al., 1989). Moreover, the estimated amount of iron observed at day 280 ($\simeq 0.06 M_\odot$; Danziger et al. (1989)) strengthens the conclusion that the late light curve was indeed powered by radioactive decay. This information sets important constraints on the mass of the neutron star which presumably has formed in SN 1987A. The fact that about 0.1 $M_\odot$ of ^{56}Co have been ejected limits the maximum mass to about 1.7 to 1.8 $M_\odot$, because ^{56}Co is synthesized in the Si-shell and at the inner edge of the O-shell if the explosion energy is around 10^{51}erg as indicated by the early light curve (Nomoto et al., 1989). Moreover, the mass of the final neutron star cannot be below 1.4 $M_\odot$ because otherwise more iron-group elements than observed would have been ejected. Of course, these conclusions rest on the (reasonable) assumption that the main sequence mass of the progenitor star was about 20 $M_\odot$ and that the iron-core masses of $\approx$ 1.4 $M_\odot$ predicted from stellar evolution models for such stars are approximately correct.

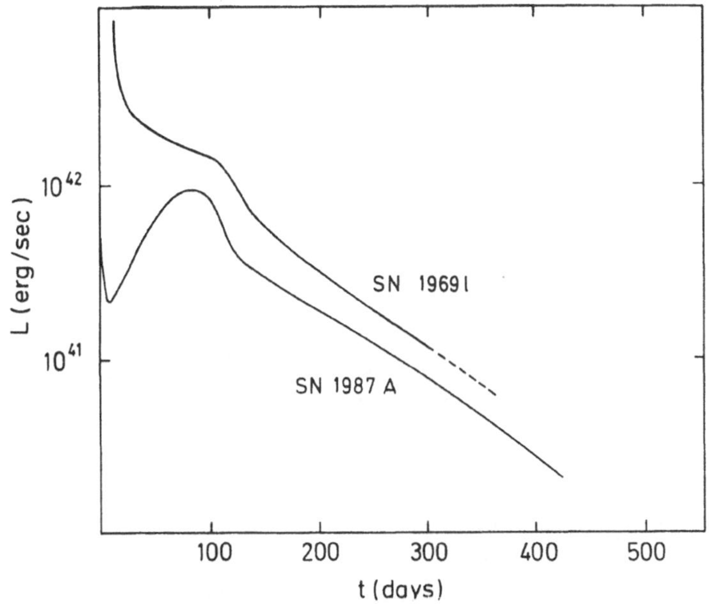

Figure 2. Light curve of SN 1987A in comparison to a typical type II light curve. While the late light curves are similar indicating roughly the same amount of ^{56}Co in both cases, for SN 1987A the lower peak luminosity and the presence of a broad maximum reflect the rather compact nature of the progenitor star.

As will be discussed later, the mass of the newly born neutron star can yield crucial insights into the explosion mechanism. Therefore it is important to ask whether there are additional observational constraints. Of course, the neutrino burst associated with SN 1987A (Bionta et al., 1987; Hirata et al., 1987) has clearly shown that a compact object, presumably a neutron star, has formed during the collapse of the central core of its progenitor star. Because the total energy emitted in form of neutrinos cannot exceed the binding energy of the final cold neutron star, a lower limit on its mass can be obtained. Unfortunately, the total $\bar{\nu}_e$-energies estimated from the KAMIOKANDE and IMB data are only weakly consistent, leaving uncertainties of about a factor of two (see fig.3). Moreover, we do not know whether some of the events were caused by (ν, e)-scattering rather than $(\bar{\nu}_e, p)$-reactions and whether some of the events were due to noise in the detectors. Finally, if the progenitor of SN 1987A was a rapidly spinning star anisotropies of the neutrino emission would add further uncertainties (Janka and Mönchmeyer, 1989a,b). So in conclusion, we can only state that the neutrino data are consistent with the assumption that a neutron star of about 1.5 $M_\odot$ was born in SN 1987A and has radiated away a significant fraction of its binding energy in form of thermal neutrinos during the first few seconds of its life, but neutrino observations do not prove this assumption.

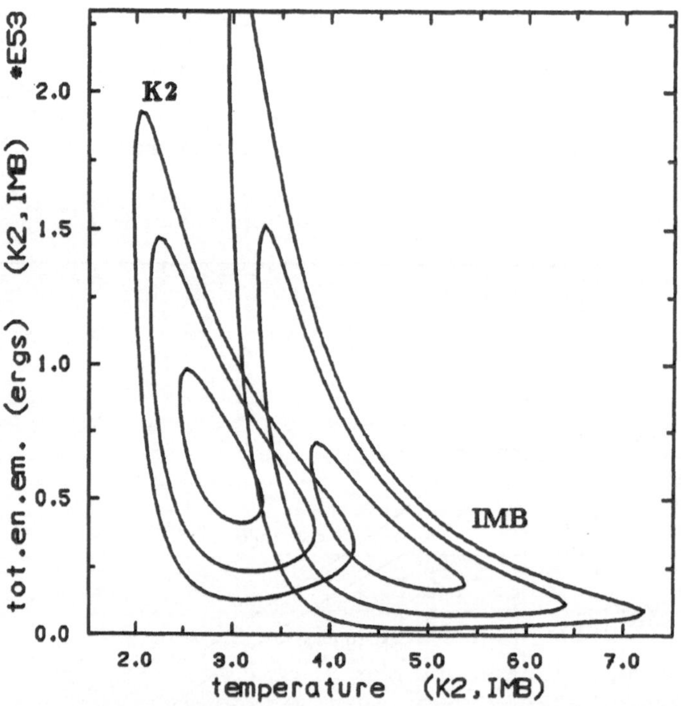

Figure 3. Maximum likelihood regions in the temperature - total energy plane for neutrinos observed by the KAMIOKANDE (K2) and IMB detectors for assumed Fermi-Dirac distributions. The contours mark 68%, 95%, and 99% confidence levels (from Janka and Hillebrandt (1989b)).

Very recently Kristian et al. (1989) reported the discovery of sub-millisecond optical oscillations from SN 1987A. If these are real and should be confirmed they definitely would require the presence of a neutron star. The approximately 2kHz frequency of the oscillations could either be due to rapid rotation near the critical velocity (angular velocity equal to the Keplerian velocity at the equator) or to oscillations in the fundamental radial mode of the neutron star. Moreover, they found sinusoidal frequency modulations of 10^{-6} relative amplitude and about 8 hours period which would require the presence of a very low mass companion ($M \simeq 10^{-3} M_\odot / \sin i$, where i is the inclination angle of the orbit) at a distance of roughly 2 10^{11}cm. One may question this interpretation, however, because it seems to be unlikely that such a low mass companion well inside the hydrogen-rich envelope of the supernova progenitor could have survived the explosion (see, however, Hillebrandt and Meyer (1989) for a scenario which may explain these data). We shall come back to a more detailed discussion of these observations as well as to their interpretation and implications later in section IV.

At the time of writing of this article the observations of Kristian et al. (1989) have not yet been confirmed. This negative result might perhaps be understandable if the rapidly spinning neutron star is surrounded by an accretion disk in which matter that has not escaped from the star gradually settles inward (Hillebrandt and Meyer, 1989). For a sufficiently high rate of accretion, e.g. 10^{18}g/s, one may expect that a shielding electron scattering region is formed around the pulsar. One prediction of this model is that the pulsar should spin up rather than slow down, provided its magnetic field is sufficiently low ($B \lesssim 10^9$ Gauss), and there is, in fact, evidence in the data of Kristian et al. (1989) for such an increase of the pulsation frequency.

We conclude this section with a brief discussion of additional indications for significant anisotropies in SN 1987A. During the first few months after the initial outburst frequency dependent linear polarizations have been observed by several groups (Barrett, 1987; Schwarz, 1987; Mendez et al., 1988). It is not clear yet whether these polarizations were caused by global deformations or local effects like turbulence or clumping, but deviations from spherical symmetry seem to be of the order of 5 to 20% (Höflich et al., 1989). From their recent Speckle observations at optical wavelengths Papaliolios et al. (1989) find that the expanding envelope is rather asymmetric and that the ratio between the minor and major axes is about 2/3. It is difficult to imagine a mechanism for such large asymmetries other than a large deformation of the progenitor's envelope. Therefore it is tempting to conclude that Sk–69°202 was in fact a fast rotator, in which case we would have to revise several of the ideas outlined in the following section.

III. Non-rotating supernova models and neutron stars

Prior to SN 1987A there was no direct evidence that type II supernovae will lead to the formation of neutron stars, although several supernova remnants are known which do contain a neutron star (Crab, Vela, RCW 103, etc.). But in none of those cases do we know beyond doubt that the events were indeed of type II. Moreover, the remnant of an explosion of a rather massive star, CasA, apparently does not contain a neutron star. The assumption, therefore, that massive stars ($M \gtrsim 8M_{\odot}$) are the progenitors of type II supernovae and neutron stars rests mainly on theoretical models of stellar evolution (for recent reviews see, e.g., Hillebrandt (1987a) and Blinnikov et al. (1988)). On the other hand, a significant fraction of all galactic supernovae may escape visual detection, as did CasA some 350 year ago, and may only be discovered by their neutrino emission. It is interesting to note that, if SN 1987A had exploded near the center of our own galaxy, it probably would have been seen as a neutrino source only. We shall now briefly discuss the theoretical models which have been developed over the last 20 years.

Several ways have been suggested which may, under favorable circumstances, convert a few percent or less of the gravitational binding energy of a forming neutron star into outward motion of the stellar envelope, among which at present two scenarios seem to be the most promising ones. Firstly, it has been proposed that the collapse of the iron-core of a massive star leads to the formation of a (proto-) neutron star and a hydrodynamic shock wave. This shock created by the rebounding neutron star and powered by its binding energy may relase enough energy into the stellar envelope to expell it (Arnett, 1977, 1983; Baron et al., 1985ab; Bruenn, 1987; Hillebrandt, 1982; Hillebrandt et al., 1984; Imshennik and Nadyozhin, 1983). However, as will be discussed later, this mechanism requires very special initial conditions and/or special properties of the nuclear equation of state. It may, therefore, work only for a narrow range of stellar masses, and in particular not for stars as massive as Sk-69° 202 (Hillebrandt, 1987b). Secondly, it has been found that neutrinos leaking out of a newly born neutron star may heat up mass zones of the stellar mantle sufficiently to revive a shock that otherwise would not reach the envelope (Wilson, 1985; Bethe and Wilson, 1985; Wilson et al., 1986). This effect, however, is extremely sensitive to the neutrino energies at very small "optical depth", and, therefore, numerical models may not be very reliable. Moreover, the explosion energy found in these simulations is too low to explain the energetic outburst of SN 1987A (see however Wilson, this volume). We shall come back to these questions later.

The outcome of computer simulations of stellar collapse depends on the initial stellar models, the adopted microphysics input data and the numerical methods. Given the fact that the various computations use different initial models, different equations of state, and/or different numerical schemes it is not surprising that the results also differ considerably. A simple estimate can clarify these difficulties. From the observations of SN 1987A we know that almost the entire binding energy of the newly born neutron star ($\gtrsim 10^{53}$ erg) was radiated away by neutrinos. The kinetic energy of the ejecta ($\simeq 10^{51}$ erg; see section II) as well as the energy in electromagnetic radiation ($\lesssim 10^{49}$ erg; see section II) were small corrections at the percent level or less. Another important quantity is the binding energy of the hot proto-neutron star formed at the end of core collapse. Theoretical calculations show that this energy is of the order 10^{52} erg. Therefore, we search for corrections of the order of 10% or less in the energetics of the outburst with numerical models.

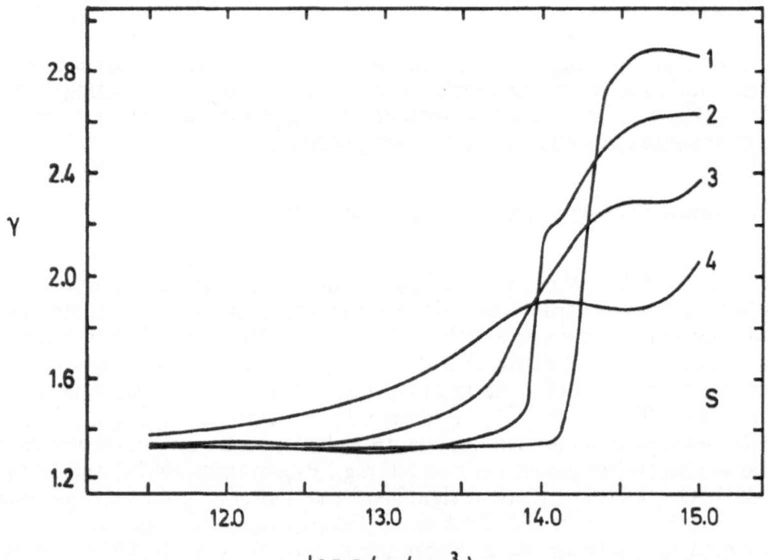

Figure 4. Adiabatic index $\gamma = (d\ ln\ p/d\ ln\ \rho)_s$ for a constant electron concentration of $Y_e = 0.35$ and various entropies per nucleon. The equation of state was calculated by adopting the temperature dependent Hartree-Fock method and a Skyrme-type effective nucleon-nucleon interaction of Köhler (see Hillebrandt et al. (1984) and Hillebrandt and Wolff (1985) for more details). It is apparent that for low entropies and sub-nuclear densities the adiabatic index is very close to the value of a non-interacting relativistic Fermi gas because free nucleons and finite nuclei do not contribute significantly to the total pressure. At densities beyond nuclear saturation density, on the other hand, interactions between the nucleons dominate the equation of state, leading to a steep increase of γ. Note also that temperature effects are large, even at very high densities, as can be seen from a comparison of the $S = 1$ and $S = 2$ curves.

There are, however, several features that are the same in all computations of core collapse using the standard theory of weak interactions and non-rotating stellar models. During the collapse phase the entropy stays low ($S \leq 2k_B$/nucleon) and, therefore, all models collapse to nuclear matter density. The collapsing core furtheron separates into an inner and an outer core. The inner core extends outwards from the center to a (time-dependent) point of maximum infall velocity and collapses homologously, i.e. $v = a(t)r$, where $a(t)$ is independent of r, in agreement with analytical considerations (Goldreich and Weber, 1980; Yahil and Lattimer, 1982). The mass of the inner core is only slightly larger than the Chandrasekhar mass $M_{Ch} \simeq 5.72\, M_{\odot} Y_e^2$, which depends on the electron concentration Y_e. Close to the edge of the inner core exists a sonic point, where the infall velocity exceeds the local sound speed. In the outer core matter falls towards the center with supersonic velocities, which reach 50% to 70% of the free fall velocity. Since M_{Ch} is proportional to Y_e^2 it will decrease during collapse. This decrease depends on the entropy of the initial model, on e^--capture rates, the ν-transport scheme, and the equation of state, and has turned out to be different in most computations.

At core-bounce the central density of the star is in general only slightly higher than nuclear matter density ($\rho_c \gtrsim 3 \times 10^{14}$g cm^{-3}), unless rather soft equations of state are used (Baron et al., 1987). The inner core is stopped on a sound crossing time ($\lesssim 1$ ms) once nuclei in the center of the star have dissolved into a homogeneous fluid of free nucleons and the EOS stiffens (adiabatic index $\gamma \simeq 2.5$ - 3). Because the outer material is still falling with supersonic velocity, a shock must form near the sonic point at M_{CH} (or a radius of about 20 km). The unshocked inner core does not expand against the ram pressure of the supersonic outer layers and achieves hydrostatic equilibrium soon after bounce. From energy conservation one can estimate the energy that is put into the shock (Yahil and Lattimer, 1982) and finds

$$E_{shock} \simeq E_B^{ic} \simeq (4-8) \times 10^{51}\text{erg}, \tag{1}$$

where E_B^{ic} is the binding energy of the unshocked inner core, in agreement with the results of numerical simulations. Here, the main source of uncertainties is the stiffness of the equation of state near nuclear matter density, but also ν-transport can change these numbers considerably. A soft equation of state has the effect that the collapse proceeds to higher densities and, therefore, E_B^{ic} is getting larger. Note, however, that this effect is only pronounced if general relativity is included into the model calculations (Baron et al., 1985a,b). More efficient ν-transport on the other hand, increases the rate of deleptonization and, consequently, reduces the mass of the homologous core and the released binding energy. In addition the stiffness of the equation of state below nuclear saturation density, i.e. the deviation of the adiabatic index from the critical value 4/3, is of importance because it also influences the size of the homologous core (Yahil and Lattimer, 1982). Note that the adibatic index is the second derivative of the energy density with respect to density at constant entropy. Therefore small uncertainties in the energy density may cause significant effects on the dynamics of core collapse.

The outgoing shock wave is heavily damped by energy losses due to nuclear photodissociations, which cost about 8×10^{18} erg/g, and neutrino losses once the shock has passed the neutrino sphere. Neglecting neutrino losses for a moment we find from eq. (1) that at most 0.5 $M_{\odot}$ of heavy nuclei can be dissociated into free nucleons by the shock, and a necessary condition for successful propagation is

$$M_{"Fe"} - M^{ic} \lesssim 0.5 M_{\odot}, \tag{2}$$

where $M_{"Fe"}$ is the mass of the original iron core. Note that the mass difference in eq.(2) can be somewhat larger, if the shocked matter is not completely dissociated into free nucleons, which occurs when the shock gets weaker and the post-shock entropy does no longer exceed a value of $S \approx 5k_B$/nucleon. Numerical models predict $Y_e \lesssim 0.38$ and thus $M^{ic} \lesssim 0.7 M_{\odot}$. It follows immediately that only stellar models with iron-core masses less than about 1.2 $M_{\odot}$ can lead to prompt explosions. Stellar evolution calculations (Hillebrandt et al., 1984; Nomoto, 1984; Woosley, 1986) have shown that this condition limits the mass range of possible progenitor stars to values between 8 and 12 $M_{\odot}$ (or at most $15 M_{\odot}$) on the main sequence, and the only successful collapse computations have indeed been performed with stars in this mass range (Hillebrandt et al., 1984; Baron et al., 1987; see however Cooperstein and Baron, 1989).

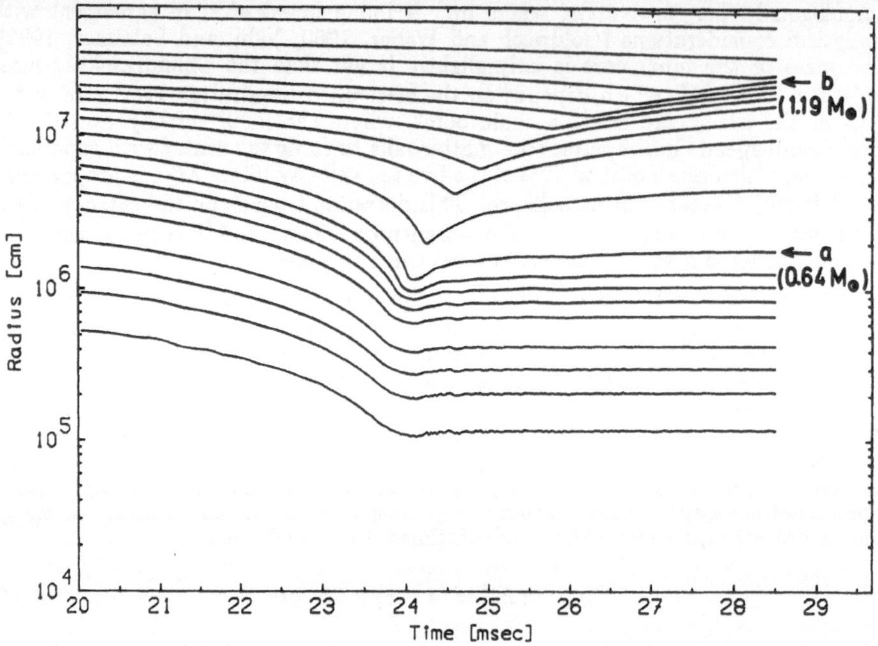

Figure 5. Radius versus time for various mass zones of a collapsing and exploding stellar model of 9 $M_{\odot}$. The unshocked inner core is labelled a, the first zone that reaches escape velocity, b. This computation has been performed with neutrino transport in non-equilibrium flux-limited diffusion approximation, a Hartee-Fock equation of state and general relativistic hydrodynamics. Oxygen burning was included by solving an appropriate nuclear reaction network (from Hillebrandt et al. (1984)).

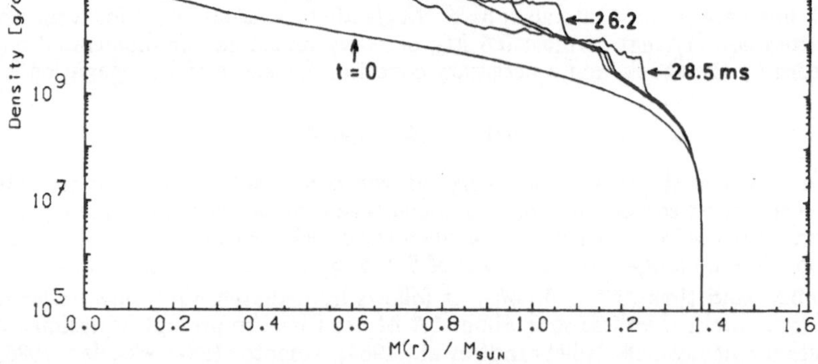

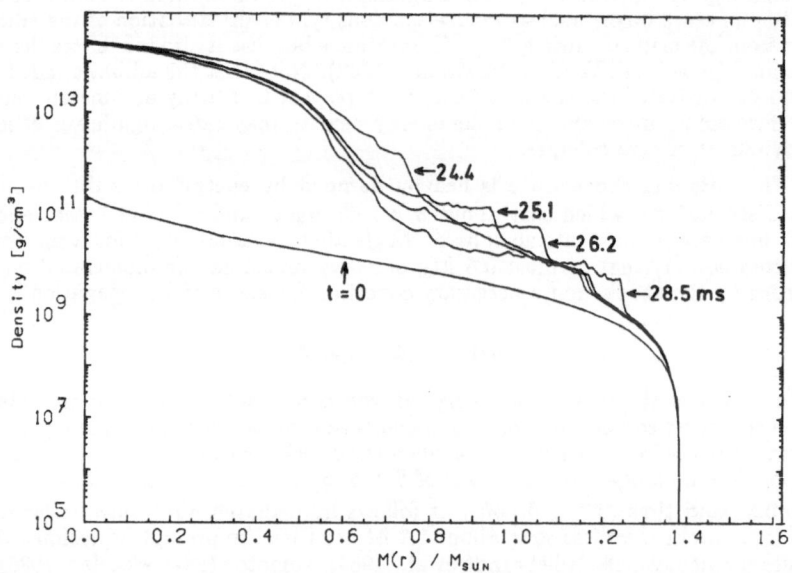

Figure 6. Snapshots of density versus mass in units of solar masses for the supernova model shown in fig.5. The curves are labelled with the time in ms measured from the beginning of the computation. It can be seen that because of the rather stiff equation of state the central density never exceeds nuclear saturation density by much (from Hillebrandt et al. (1984)).

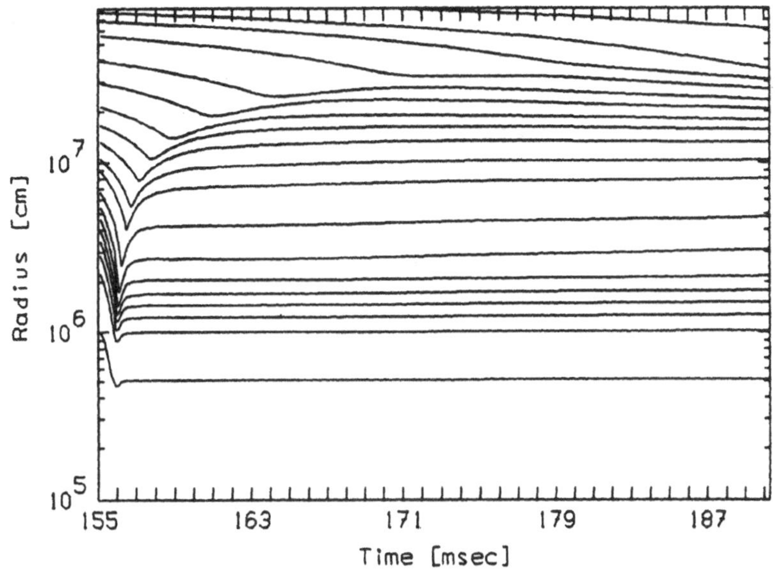

Figure 7. Same as figure 5, but for a stellar model of 20 $M_{\odot}$. It is obvious that this time, because of the larger core mass, the shock stalls at about 300 km (from Hillebrandt (1987a)).

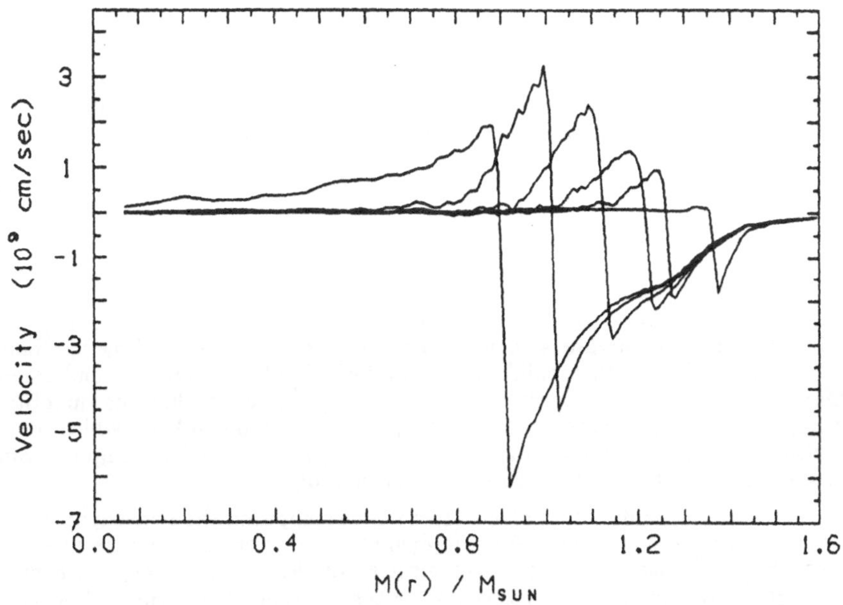

Figure 8. Snapshots of velocity versus mass for the model of figure 7. The curves, from the left, correspond to times 156.2, 156.6, 157.6, 159.9, 163.2, and 202.6 ms, respectively (from Hillebrandt (1987a)).

In figures 5 and 6 some results of such a computation are shown. It should be noted, however, that in other simulations even the most favorable initial conditions do not lead to prompt explosions (Wilson, this volume), probably because different equations of state and different ν-transport schemes were used. Therefore we reach the conclusion that at present we cannot answer the question whether or not stars can explode by the core-bounce mechanism and have to keep this unpleasant situation in mind when we try to interpret the observations of events such as SN 1987A.

Among researchers in the field, however, there is now general agreement that stars with iron cores more massive than about 1.35 $M_\odot$ cannot explode by the core-bounce mechanism, no matter what the right equation of state is. Conventional models of stellar evolution, therefore, exclude main sequence masses above about 15 $M_\odot$. Figures 7 and 8 show a typical example of a simulation where the shock stalled at about 1.2 $M_\odot$.

Because some supernova remnants (e.g., Cas A and Puppis A) show large oxygen overabundances indicating main sequence masses of at least 20 $M_\odot$, even prior to SN 1987A alternative explosion mechanisms had been searched for. A possibility which has been discussed extensively is the so-called delayed explosion model (Wilson, 1985; Bethe and Wilson, 1985). The idea is that a few hundred milliseconds after core-bounce energy transport by neutrinos may revive a stalled shock. At that time the edge of the unshocked inner core ($\simeq 0.7 M_\odot$) will be at a radius of about 20 km, and the density and temperature there are around 10^{12}g cm^{-3} and 2 to 3 MeV, respectively. The shock has changed to an almost standing accretion shock at 1.4 $M_\odot$ and a radius of about 500 km, the temperature and density just behind the shock being 10^7g cm^{-3} and 0.8 MeV, respectively. The shocked matter is irradiated by a neutrino flux originating from the neutrino-sphere at 30 to 70 km. Typical neutrino energies will be around 6 to 10 MeV. The energy gain per gramme of irradiated matter at radius R much larger than R_ν of the neutrino sphere is given by

$$\dot{E}_{gain} = K_{abs} \frac{L_\nu}{4\pi R^2}, \tag{3}$$

where L_ν is the neutrino luminosity and K_{abs} is the absorption opacity. Each gramme of matter will also lose energy by thermal emission of neutrinos, $\dot{E} \sim T^4 K_{emiss}$. Since the neutrino luminosity is proportional to $T^4 R_\nu^2$ and K_{abs} and K_{emiss} are proportional to T_ν^2 and T^2, respectively, the net energy gain can be written as

$$\frac{dQ}{dt} = \frac{7}{16} ac\ K_{abs}\ T_\nu^4 \left[\frac{1}{4} \left(\frac{R_\nu}{R} \right)^2 - \left(\frac{T}{T_\nu} \right)^6 \right], \tag{4}$$

where the temperatures are in MeV, $a = 1.37 \times 10^{26}$erg/(cm^3 MeV^4) and K_{abs} in cgs units is 6×10^{20} $\langle(\epsilon_\nu/m_e c^2)^2\rangle$cm^{2}g^{-1} (Lattimer and Burrows, 1984). The maximum matter temperature that can be obtained from this heating mechanism can be estimated from the equilibrium condition $dQ/dt = 0$, and we find roughly 1 MeV from typical stellar parameters. However, it is clear that the quantity of key importance is the neutrino temperature at the neutrino sphere, which, due to uncertainties in the flux-limited diffusion schemes, is not very well determined. Small changes in T_ν can change the net-heating considerably.

Wilson (1985) and Wilson et al. (1986) have computed the hydrodynamic evolution of several stellar models for approximately 1s after core bounce and found explosions in all cases considered. In most cases, however, the explosion energy ($\lesssim 4 \times 10^{50}$erg) was too low to account for typical type II supernova light curves and in particular for the rather high kinetic energy seen in SN 1987A (see however Wilson, this volume). Only for rather massive stars, $M \gtrsim 25 M_\odot$, did explosive oxygen burning add enough energy to the explosion to explain a typical outburst. This problem can be understood from a simple argument. Neutrino heating proceeds on a time scale much longer than the hydrodynamical time scale. Once, due to neutrino heating, a sufficient overpressure has been built up behind the accretion shock the heated zones will start to expand and further heating will be turned off. The thermal energy needed to build up such an overpressure will be of the order of the binding energy of the overlaying material, i.e., a few times 10^{50}erg. Consequently, one expects that the explosion energy is of the same order.

From numerical experiments Wilson (private communication) has found that small modifications of his neutrino transport scheme can change successful explosions into failures. This, as well as results obtained by Hillebrandt (1985), indicate that the delayed explosion mechanism may not work at all. So we are left with the problem that also (and maybe in particular) for massive stars, $M \simeq 20M_\odot$, the explosion mechanism is not understood. It may well be that we have to invent more complicated scenarios in order to be able to solve these problems. For example, the neglect of rotation, magnetic fields and/or nuclear energy generation may be an oversimplification (see, e.g., Müller and Hillebrandt, 1981; Le Blanc and Wilson, 1970; Symbalisty, 1984; Bodenheimer and Woosley, 1983), or some of our basic assumptions concerning the equation of state (Baron et al., 1987) or the theory of weak interactions are incorrect.

While, therefore, it seems to be difficult to extract information on the nuclear equation of state from supernova models, one may hope to use supernova observations, and in particular the observations of SN 1987A, to place some constraints on the stiffness of nuclear matter.

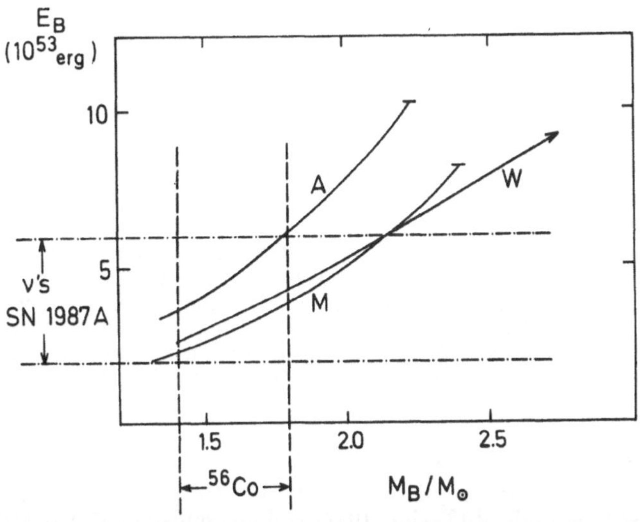

Figure 9. Binding energy versus baryonic mass of neutron star models constructed from various equations of state by Arnett and Bowers (1977). In particular, EOS have been chosen here which fulfill the constraint set by the binary pulsar. The EOS leading to curves A and M are explained in table 3. W is the relativistic mean field model of Walecka (1974). Constraints on the binding energy obtained from the neutrino detections from SN 1987A as well as constraints on the baryonic mass given by the observed amount of ^{56}Co are indicated by horizontal and vertical lines, respectively. It is apparent that the observational limits do not impose strong constraints on the EOS.

As has been discussed before, soft equations of state at densities at and beyond nuclear saturation increase the chance of prompt explosions. In particular Baron et al. (1985a,b) have argued that for successful shock propagation the incompressibility of nuclear matter has to be significantly below the values obtained from fits to the monopole oscillations of lead and other nuclei (Sharma et al., 1988) and from empirically obtained effective nucleon-nucleon interactions. The large explosion energy of SN 1987A, indicating a prompt rather than a delayed explosion, together with the presumably large iron-core mass of the progenitor seems to be in favor of their arguments. However, since it is extremely uncertain by which mechanism massive stars explode, it seems to be dangerous to draw such far-reaching conclusions.

The delayed explosion models, on the other hand, seem to fit very well both the energy spectrum and the duration of the neutrino pulse from SN 1987A (Mayle et al., 1988). This success indicates that the newly born neutron star is rather massive, $M \gtrsim 1.5 M_{\odot}$, which in turn would favor stiff neutron matter equations of state (i.e. stiff for densities larger than a few times nuclear matter density) Again, this conclusion should be regarded with some care because the neutrino fluxes and spectra have been computed by means of the flux limited diffusion approximation and are not very reliable (Janka and Hillebrandt, 1989a). Moreover, as was discussed in section II, the neutron star binding energies estimated from both the KAMIOKANDE and the IMB experiments are only weakly consistent and do not provide enough information to determine accurately the mass of the neutron star.

One also can hope to get better limits on the position of the mass cut (which separates the matter forming the neutron star from the matter being expelled in the explosion) from late time optical and IR spectra of SN 1987A. In the nebular phase matter will be transparent as well in all lines as in the continuum. In this case masses of individual elements can be measured directly from the various line strengths, provided the temperature of the emitting region is sufficiently well known. This method was already successfully applied to determine the total amount of cobalt (Danziger et al., 1989; section II) and will give information of similar quality for elements such as C, O, and Si. A comparison to explosive nucleosynthesis computations based on realistic stellar models should then reveal the desired information, although again the conclusions will not be model independent.

We turn now to a discussion of the properties of slowly rotating neutron stars because, generally speaking, observations of these objects offer a better and more direct way to constrain the equation of state at high densities than supernova observations do. One should keep in mind, however, that supernova cores and neutron stars probe different equations of state. While in supernova cores and newly born neutron stars the electron concentration and thus the proton mass fraction is still around 0.35, during deleptonization Y_e drops to values of about 0.05 or less and the matter becomes very asymmetric with respect to the neutron-proton ratio. Consequently, also the compressibility may change considerably. Moreover, during this cooling process which may take only some 10 s the central density of the neutron star will increase by about one order of magnitude and therefore also a different density regime will be tested by observations of neutron stars.

In the case of the binary pulsar PSR 1913+16 the gravitating masses have very accurately been determined (Taylor, 1987) and one finds $(1.451 \pm 0.007) M_{\odot}$ for the pulsar. The mass of the companion, presumably also a neutron star, is $(1.378 \pm 0.007) M_{\odot}$. If we believe that general relativity is the correct theory of gravity, the maximum mass of a neutron star is only a function of the equation of state. Neutron star models constructed from soft equations of state (i.e. soft for densities larger than a few times nuclear matter density) tend to be rather compact and, therefore, the limiting mass is small. In contrast, if the equation of state is stiff the radius is larger for the same mass, general relativistic effects are reduced and, consequently, also the maximum mass is large. Typical values range from 0.7 $M_{\odot}$ for a non-interacting gas of relativistic fermions to about 2.7 $M_{\odot}$ for equations of state obtained from the non-relativistic mean field theory (Pandharipande and Smith, 1975). Unfortunately, the observed mass of PSR 1913+16 only rules out very soft equations of state, and most realistic equations of state are not in contradiction with this constraint (see, e.g., Arnett and Bowers, 1977; Baym and Pethick, 1979).

Masses of neutron stars in X-ray binaries have also been measured, but these mass estimates depend on the poorly known inclination angle of the orbit. For the best studied case, Her X-1, one finds 1.5 $M_{\odot}$, but the uncertainty is of the order of a few tenths of a solar mass. For some other cases (4U 0900-40, 4U 1538-52, LMC X-4) the mass seems to be somewhat higher, but the data are still consistent with a maximum mass of about 1.6 $M_\odot$ which again only rules out very soft neutron matter equations of state (see, e.g., Rappaport and Joss (1983)). Nevertheless, it appears likely that future observations will reduce the uncertainties of these mass estimates significantly, but one should keep in mind that the absence of very massive neutron stars would not eliminate stiff neutron matter equations of state because the process of formation may favor lower masses.

Certainly, if it should be confirmed that SN 1987A left behind a neutron star spinning with a rotation period of about 0.5 ms, this would rule out most stiff neutron star matter equations of state such as the tensor interaction model or the mean field approach of Pandharipande and Smith (1975), even if effects of rotation on the neutron star's structure are incorporated (Friedman et al., 1986). In the following section, therefore, we shall turn to a discussion of effects of rotation on supernovae and neutron stars.

IV. The effects of rotation on supernovae and neutron stars

Up to now only a few attempts have been made to relax the assumption of spherical symmetry and to perform axisymmetric, i.e. two-dimensional supernova simulations, which allow to study effects due to rotation (LeBlanc and Wilson, 1970; Müller, Różyczka and Hillebrandt, 1980; Tohline, Schombert and Boss, 1980; Müller and Hillebrandt, 1981; Symbalisty, 1984; Mönchmeyer, 1989a,b; Mönchmeyer and Müller, 1989a)

It is obvious that the simulation of rotational core collapse is computationally more difficult and more expensive than a spherical collapse calculation. However, it is still surprising that only a few simulations have been performed in the past, because (i) stars and especially massive stars rotate in general (see e.g. Tassoul, 1978), because (ii) at least some of the compact remnants of type II supernova explosions, the millisecond pulsars, do rotate with significant speed (see e.g. Taylor, 1987), and because (iii) of angular momentum conservation and the increasing centrifugal forces already small initial rotational energies may change the standard collapse picture completely. Therefore the inclusion of rotation may be crucial for the correct modelling of a type II supernova explosion. This argument is supported by the fact that up to now no spherically symmetric collapse simulation for stars with $M \gtrsim 15\ M_\odot$, i.e. with an iron core mass $M_{core} \gtrsim 1.35\ M_\odot$ could convincingly produce a supernova explosion. The need for additional and more detailed rotational core collapse simulations will become even stronger, if the existence of a 0.5 ms pulsar in the remnant of SN 1987A is indeed confirmed (see previous section).

a) Some general remarks on the influence of rotation

From a variety of studies it is known that rotation has a stabilizing influence. A criterion for the dynamic stability of rigid rotators has been given by Ledoux (1945)

$$\gamma > \gamma_{crit} = \frac{4}{3} - \frac{4}{9}\beta \quad , \tag{5}$$

where β is the absolute value of the ratio of rotational to gravitational energy. Computer simulations have shown that this criterion holds for differential rotators, too. The stabilizing effect depends mainly on the amount of rotational energy and not very sensitively on its distribution or on the density stratification (Ostriker and Tassoul, 1969; Ostriker and Bodenheimer, 1973; Durisen and Imamura, 1981). If the (average) adiabatic index in the collapsing core is very close to the value of 4/3, already a small amount of rotation may be enough to stop the collapse and lead to the formation of stable rotating configurations with central densities below nuclear matter density (Tohline 1984; Eriguchi and Müller, 1984; Symbalisty 1984; Mönchmeyer, 1989a,b; Mönchmeyer and Müller, 1989a)

Due to conservation of angular momentum configurations may form during core collapse, which are unstable against tri-axial deformations on secular or even dynamical time scales (Tohline 1984; Eriguchi and Müller, 1984), if $\beta \geq 0.14$ and $\beta \geq 0.27$, respectively (see e.g. Tassoul 1978). Whether these instabilities indeed do occur is a non–trivial question: If the equation of state at sub–nuclear densities is stiff, i.e. if the adiabatic index is very close to 4/3, the core may be stabilized before its rotational energy exceeds the critical value. If on the other hand the initial amount of rotation is small enough for the collapse to proceed to nuclear densities, $\beta > 0.14$ may not be reached before bounce.

Moreover, due to conservation of angular momentum and due to the resulting increase of the centrifugal forces matter will not fall in on radial trajectories. In addition matter in the equatorial plane will not fall towards the center as fast as matter at the polar axis. Especially this last effect leads to a progressively flattening of the core and to a change of the collapse time scale in comparison to a spherically symmetric configuration. As basic properties of the collapsing core depend on the size of the collapse time scale relative to the electron capture time scale and the neutrino diffusion time scale, the kinetic infall energy, the final lepton fraction, and the mass of the inner core may change in the presence of rotation.

After bounce a rotating core will oscillate with a superposition of various axi–symmetric radial and surface modes. The frequency of these modes is determined by the average density of the inner core. In contrast to a spherically symmetric core, which comes to rest soon after bounce, in a rotating core a certain fraction of its kinetic infall energy will be converted into oscillations, which are damped by non–spherical pressure waves. The oscillations will also contribute to the gravitational wave signal of the bouncing core. Due to the asymmetric infall of matter the propagation of the shock wave at the pole will differ from the propagation near the equatorial plane. In particular this effect will be evident in the dissipation rates of kinetic energy and in the maximum entropy values behind the shock front.

In addition it is very well known that rotation can drive meridional circulations in baroclinic regions (see e.g. Tassoul 1978). For type II supernova models the interesting question arises, whether there exist large scale circulation patterns behind the shock, that extend inwards to regions where neutrino energies are high. In the (neutrino) opaque matter those vortices could transport neutrinos towards the shock front much faster than diffusion (Müller and Hillebrandt, 1981).

From 1d–calculations one knows in addition that a negative entropy gradient is established behind the shock after photodisintegration losses have weakened it significantly. This region of decreasing entropy is unstable against convection provided that the stabilizing lepton gradient is not too large (Epstein 1979). Arnett (1985) has first pointed out the possible importance of convection for the type-II supernova mechanism. However, whether convection indeed helps or even harms the propagation of the shock is debated (Burrows 1987; Bethe, Brown and Cooperstein, 1987). Note that for rotating cores mixing of high and low entropy matter may be enhanced, if convectional currents are supported by vortices resulting from rotation in regions, where the deformed surfaces of constant pressure do not coincide with isopycnic surfaces. On the other hand side it has been argued that rotation may have a stabilizing effect on certain types of convective instability modes (see Tassoul 1978). However, especially for centrally condensed and differentially rotating objects the interaction of rotation and convection is not yet understood.

Finally the non–spherical density stratification of a rotating core before and after bounce may modify the neutrino signal in a characteristic way and lead to a directional dependence of the ν–signal (Janka and Mönchmeyer, 1989a,b).

b) Results of recent simulations

Mönchmeyer (1989a) has recently performed two-dimensional (i.e. axisymmetric) simulations of the collapse of rotating cores (see also Hillebrandt et al., 1988; Janka and Mönchmeyer, 1989a,b; Mönchmeyer, 1989b; Mönchmeyer and Müller, 1989a). In these calculations the equation of state of Wolff (see Hillebrandt and Wolff, 1985), which is based on temperature–dependent Hartree–Fock calculations (see Fig. 4) with an effective nucleon–nucleon interaction of Skyrme type, was used to describe the properties of core matter. Contrary to other published equations of state the adiabatic index in Wolff's equation of state is close to 1.32 in the whole density regime $10^{10}\,\mathrm{g/cm^3} \lesssim \rho \lesssim 10^{14}\,\mathrm{g/cm^3}$ and for entropies less than $2\,k_B$/nucleon.

The hydro-code used in the simulations is based on a conservative, second-order accurate, two-dimensional difference scheme in spherical coordinates. (for further details see Mönchmeyer, 1989a, and Mönchmeyer and Müller, 1989b,c). Electron captures on protons were explicitly taken into account and the neutrino transport was simplified by a trapping scheme with a trapping density of 3 10^{11} g/cm^3 during the infall epoch. After bounce the trapping density was reduced to 10^{10} g/cm^3 to take into account the change of the hydrodynamical time scale relative to the diffusion time scale.

The initial radially symmetric core models are based on the 20 $M_\odot$ star of Weaver, Woosley and Fuller (1985), which has a 1.36 $M_\odot$ iron core. The dynamical evolution of the innermost 1.6 $M_\odot$ of this star has been calculated after adding a certain amount of rotation. In particular, Mönchmeyer (1989a) has calculated four model sequences with different amounts and distributions of angular momentum (see Table 1). Note that the initial models were not relaxed with respect to rotation. But due to the initially small rotational velocities and energies (see Table 1) the deviations from equilibrium are relatively small. The results of all four calculations will be published in detail elsewhere (see also Mönchmeyer, 1989b; Mönchmeyer and Müller, 1989a). Therefore after outlining the properties common to all four models we will only discuss the main results of two models (Model A and D) in somewhat more detail here.

Table 1. Parameters defining the initial models. The quantities x, z, r, E_{rot} and Ω are the distance from the core center in the equatorial plane, the distance from the equatorial plane, the corresponding radial distance, the rotational energy and the angular velocity, respectively.

Model		A	D
Ω/Ω_0		$R^2 / (r^2 + R^2)$	$X^2 / (x^2 + X^2) * Z^4 / (z^4 + Z^4)$
R	[10^8 cm]	1.0	–
X	[10^8 cm]	–	1.0
Z	[10^8 cm]	–	1.0
Ω_0	[s^{-1}]	4.0	5.5
E_{rot}	[erg]	$1.92 \cdot 10^{49}$	$3.98 \cdot 10^{49}$
β		0.004	0.009

The choice of the initial angular momentum distribution is one of the most problematic aspects of any 2-d simulation, because there exist no evolutionary calculations of rotating massive stars up to the iron core collapse. Therefore the amount as well as the distribution of the initial angular momentum can be chosen arbitrarily within certain limits imposed by stability considerations (see e.g. Tassoul, 1978).

The initial rotation profile of Model A (see Table 1) is similar to the one used in previous investigations of rotational core collapse (Müller and Hillebrandt, 1981; Symbalisty 1984), where the angular velocity is a function of radius only. The mass interior to the radius R (see Table 1) is about 1.31 $M_\odot$, i.e. a large fraction of the Fe–Ni–core approximately rotates as a rigid rotator. The initial angular momentum distribution of Model D is approximately constant on cylinders (except very close to the axis) for $z \lesssim Z$ (see Table 1). The corresponding rotational energies and the values of β of both initial models are also given in Table 1.

Before discussing the common properties of the rotating models we want to point out that the concept of homology cannot be used for rotating cores. Yet there exists a surface, which separates subsonically from supersonically falling matter. We will speak of the "inner core" when refering to the matter inside a surface of constant density (i.e. inside an isopycnic surface) for which the (absolute) value of the angular averaged radial infall velocity has a maximum. We further define the "sonic mass" as the mass of that region where the matter has subsonic infall velocities. After bounce this inner core corresponds to the unshocked central part of the iron core.

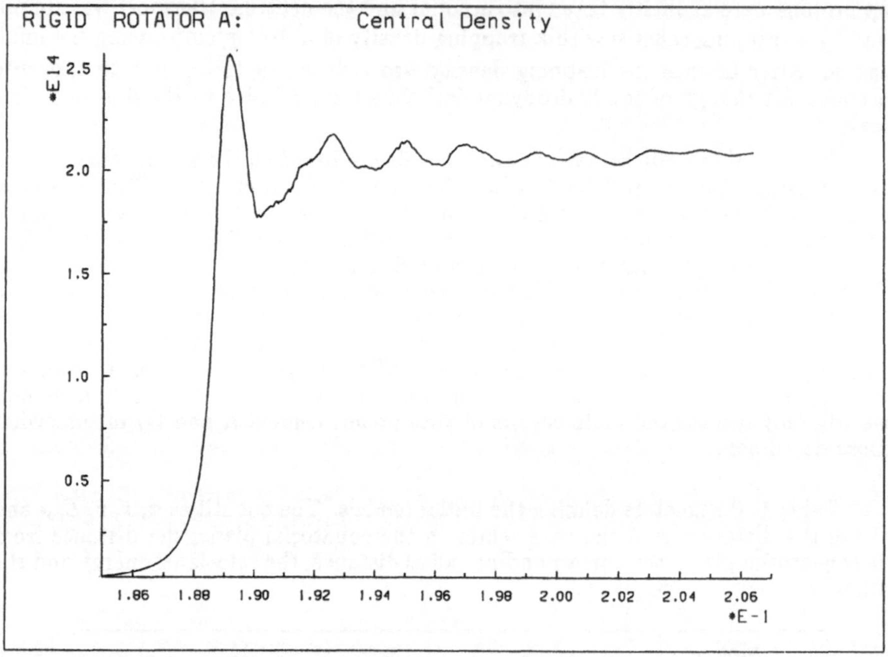

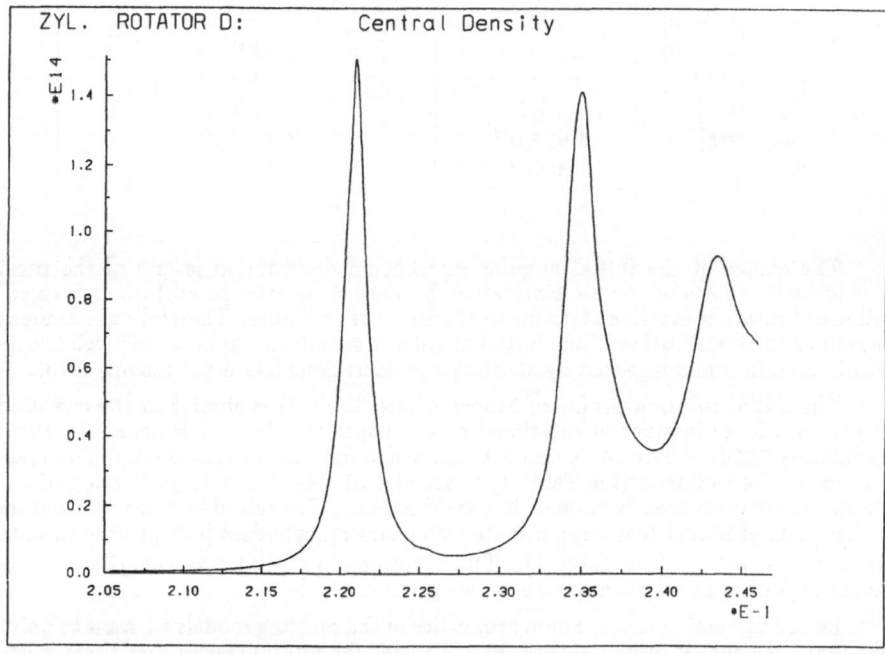

<u>Figure 10.</u> Post-bounce oscillations of the central density ρ_c [gcm^{-3}] of Model A (top) and of Model D (bottom) as a function of time (in msec) since the start of the calculation (from Mönchmeyer, 1989b).

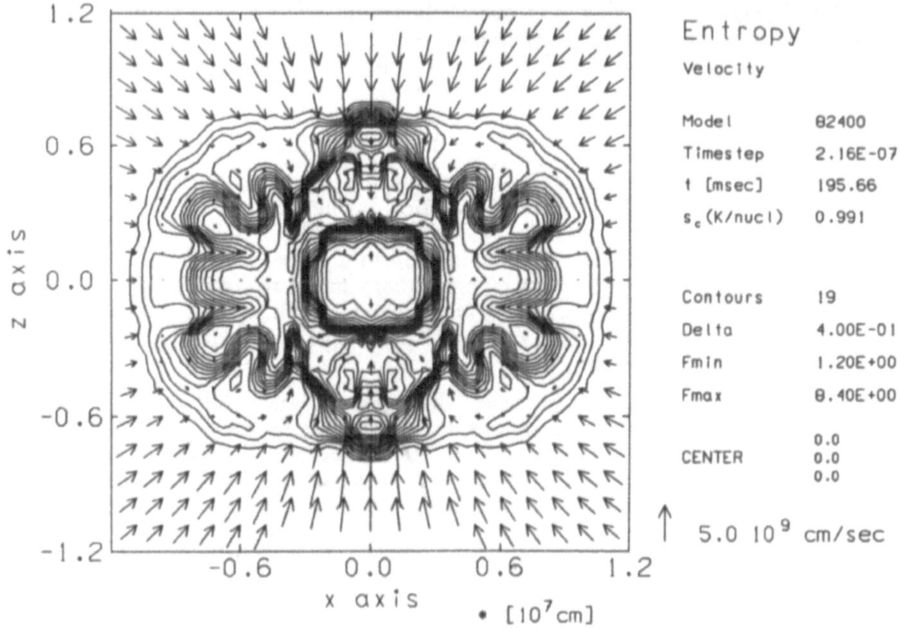

Figure 11. Profiles of the specific entropy and flow pattern of Model A about 6.5 ms after bounce showing the instability at the edge of the high entropy region. The contours cover a range from Fmin to Fmax with a spacing of Delta. The time, the central entropy and the velocity scale are given in the legend of the figure (from Mönchmeyer and Müller, 1989a).

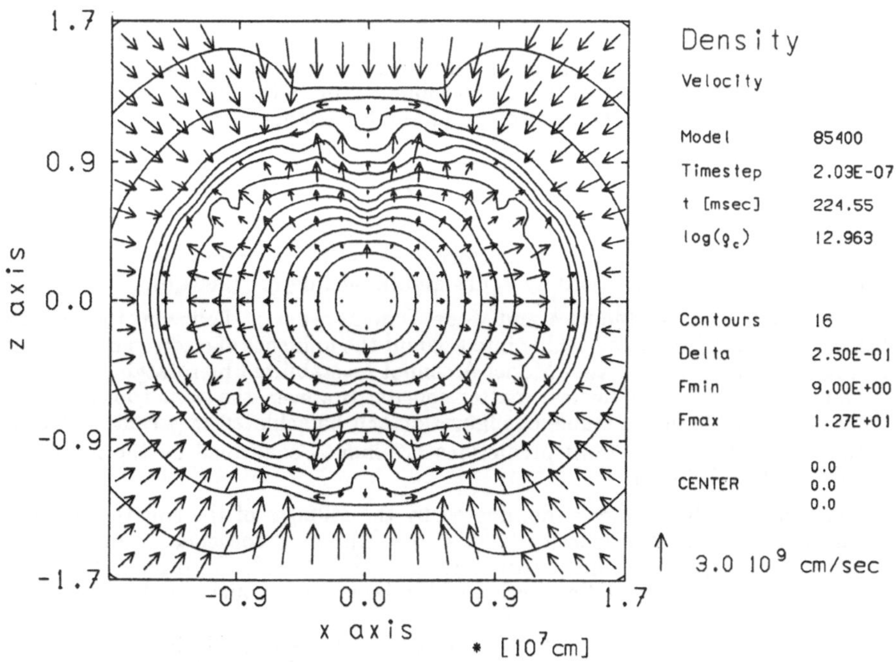

Figure 12. Profiles of the density ($\log_{10} \rho$) and flow pattern for Model D about 3.6 ms after bounce showing the expansion of the inner core behind the shock front (for explanation of legend see Fig. 11; from Mönchmeyer, 1989b).

During a first contraction phase ($\Delta t = 150 - 250$ ms) a subsonically falling inner core forms in each model. Due to an overall flattening of the density stratification the total core quadrupole moment becomes more and more negative. During a subsequent second phase the inner core contracts rapidly for 10 – 20 ms and the quadrupole moment rises until bounce by 5% – 15% over its minimum value. Due to angular momentum conservation the inner core flattens significantly. At bounce the ratio of its polar to equatorial diameter is roughly 0.5 in each model.

In contrast to the oblate deformation of the density contours the contours of constant Ω exhibit a growing prolate deformation during collapse. This effect together with the formation of a very differential Ω–profile in x-direction follows directly from the conservation of the initial angular momentum and from the form of the initial rotation laws given in Table 1. Two off-center maxima of Ω are formed at the polar edges of the inner core. They approximately coincide with the maxima of the infall velocities and cause the formation of typical polar dips in the density stratification of the inner core before bounce.

Due to the centrifugal acceleration the profile of the infall velocity becomes asymmetric during collapse, especially in the supersonic flow region. At bounce the ratio of the maximum polar infall velocity to the maximum equatorial infall velocity typically ranges from 1.8 to 2.5, and the maximum compression of the inner core in polar direction occurs before the equatorial contraction is stopped completely.

The shock front generated at bounce aquires an oblate shape in each of the calculated models. Depending on the values of the centrifugal acceleration behind the shock front two-dimensional flow effects can hinder the propagation of the shock along the axis. The dissipation of the large kinetic energy of the infalling matter at the axis leads to the formation of expanding and (due to buoyancy forces) rising polar entropy blobs.

The amplitudes and the periods of the post bounce volume and surface oscillations of the inner core strongly depend on the amount and distribution of the initial rotational energy, which determines the average inner core density of the models at bounce. The oscillations are damped by the generation of asymmetric pressure waves, which propagate outward through the shock heated matter and are absorbed by the shock.

The rotating equilibrium configuration of the inner core and the distribution of the heated matter left behind the shock wave are strongly deformed. In all models the distribution and the amount of the angular momentum in the final quasi-equilibrium state of the inner core is characterized by (a and b are model dependent parameters)

$$\frac{\partial \Omega}{\partial z} \approx 0 \quad , \qquad \Omega(x) \approx \Omega_0 \cdot \exp(ax + bx^2) \quad , \qquad 0.10 \leq \beta \leq 0.14 \quad , \qquad (6)$$

The consequences resulting from the deformation of the neutrino sphere for the neutrino signal have been discussed by Janka and Mönchmeyer (1989a,b).

We now will discuss Model A and Model D in more detail. Following the considerations of Tohline (1984) and taking into account the influence of electron captures on the pressure, one can show that Model A cannot be stabilized by rotation before the central density ρ_c exceeds nuclear matter density. However, the same considerations imply for Model D that a stable configuration can be achieved at a central density of about $7 \cdot 10^{13}$ gcm^{-3}. Both predictions are indeed confirmed by the numerical simulations (see Table 2).

In Model A the bounce is dominated by the stiffness of the equation of state beyond nuclear saturation density ($2.75 < \gamma < 3$). As a consequence most of the kinetic infall energy of the inner core is transfered to the shock within 0.3 ms after bounce. The average oscillation period of the inner core, which finally achieves an equilibrium central density of $2.6 \cdot 10^{14}$ gcm^{-3} is 2.5 ms. Note that due to the stiffness of the equation of state the amplitudes of the oscillations are relatively small (Fig. 10). Similar to the corresponding non-rotating model the shock stalls at a mass coordinate of 1.3 $M_\odot$ (see also Fig. 8). Neither the evolution of a Rayleigh–Taylor instability (Fig. 11) due to a large negative entropy gradient behind the shock (Mönchmeyer and Müller, 1989a) nor the mixing of low with high entropy matter strengthens the shock (for possible reasons see Bethe et al. (1987)).

Because of the effects of centrifugal forces Model D does bounce at a sub-nuclear central density of $1.5 \cdot 10^{14}$ gcm^{-3}. Therefore the kinetic infall energy of the inner core is only $1.6 \cdot 10^{51}$ erg (see Table 2). But due to the fact that the adiabatic index of matter at sub-nuclear densities is small ($\gamma \approx 4/3$) the inner core can much easier be compressed than in Model A. As a consequence the inner core oscillates around its equilibrium density of $7 \cdot 10^{13}$ gcm^{-3} with a much larger large amplitude than Model A (Fig. 10). For roughly 8ms after bounce the inner core expands (during the first oscillation) behind the shock with an average velocity of $1.0 - 1.5 \cdot 10^9$ cm/s. The polar and equatorial diameters of the inner core increase from 50 km to 215 km and from 100 km to 300 km, respectively. Thus the inner core does compressional work on the surrounding shock heated matter and drives the shock outward like an expanding piston (Fig. 12).

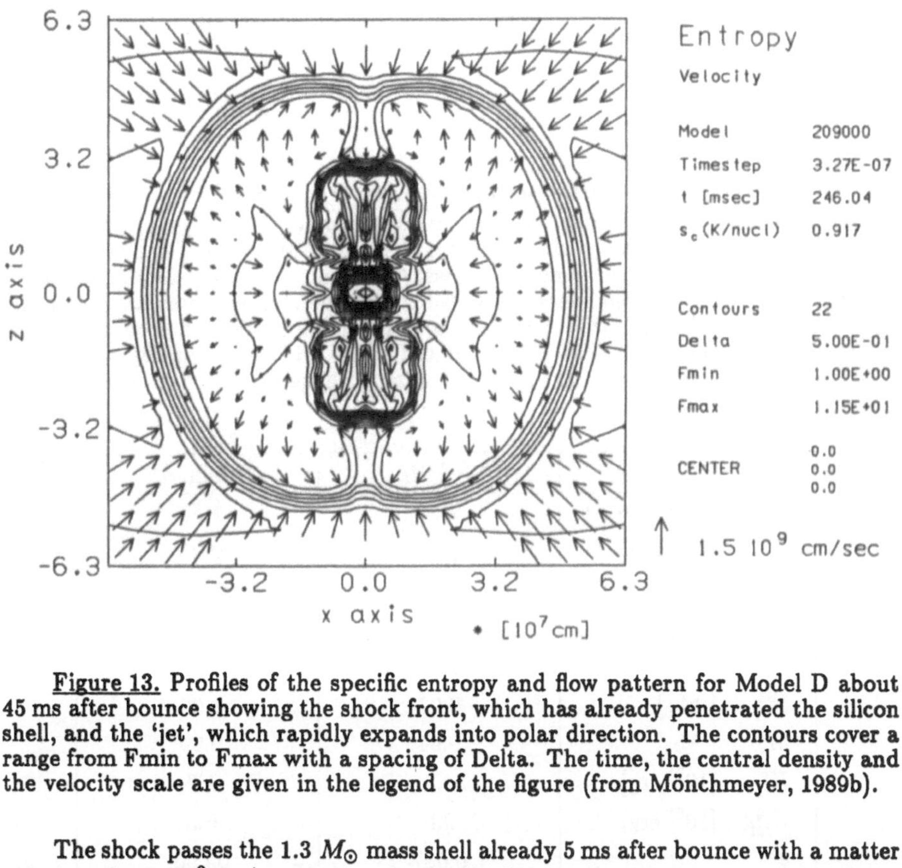

Figure 13. Profiles of the specific entropy and flow pattern for Model D about 45 ms after bounce showing the shock front, which has already penetrated the silicon shell, and the 'jet', which rapidly expands into polar direction. The contours cover a range from Fmin to Fmax with a spacing of Delta. The time, the central density and the velocity scale are given in the legend of the figure (from Mönchmeyer, 1989b).

The shock passes the 1.3 $M_\odot$ mass shell already 5 ms after bounce with a matter velocity of $1.6 \cdot 10^9$ cm/s. Afterwards the shock is weakened due to neutrino losses (especially at the axis; see Janka and Mönchmeyer, 1989b). The polar entropy blobs ($S_{max} \approx 15k_B$/nucleon) cool down, and a region of almost homogeneous specific entropy ($S \approx 5 - 7k_B$/nucleon) is produced behind the shock. A second contraction of the inner core sets in 228 ms after the onset of collapse. The resulting second bounce at 235 ms leads to the formation of a "jet"-like flow pattern confined to a cylindrical diameter of 250 km around the axis. The "jet" propagates through the shock heated and disintegrated matter with velocities up to $2 \cdot 10^9$cm/s (Fig. 13). After a third bounce at 244 ms the calculation was stopped. At this time the mass surrounded by the shock surface is 1.42 $M_\odot$, i.e. the shock has already penetrated the silicon shell. The velocities (see Fig. 13) indicate a weak explosion independent of the evolution of the "jet". The matter inside the "jet" and close to the axis is strongly deleptonized ($Y_{e_{min}} \approx 0.16$) due to neutrino losses.

Table 2 shows that there are large differences in the values of the kinetic energies of the inner cores at bounce. For Model A it can be shown that the value given approximates the energy transferred to the shock at bounce very well, because the inner core reaches equilibrium after some small scale oscillations. However, the results of Model D indicate that the post–bounce evolution and the energetics can become much more complicated if nonlinear, large scale oscillations are excited. Energetically several aspects are important : From the virial theorem it can be shown that among equilibrium configurations with the same potential energy rotating configurations with a large value of β have the largest total (binding) energy if $\gamma < 5/3$. The larger the equilibrium potential energy is, the more energy can be released for a given final value of β, which itself depends on γ and the initial β–value. In this sense Model D is nearly optimal for an explosion, because it bounces with a considerably high value of $\beta \approx 0.12$ just before nuclear densities are reached. In contrast to Model A the inner core of Model D is still far away from equilibrium after bounce. Therefore, the core can transfer energy via expansion to the surrounding matter while achieving its equilibrium stratification with an estimated binding energy of $2 - 3 \cdot 10^{51}$ erg $> K^b_{ic}$.

Table 2. Some characteristic quantities of Model A and D at bounce (marked with superscript 'b') and in the quasi-equilibrium state (marked with superscript 'eq'), respectively. ρ_c, T_c, Ye_c, $Y\nu_c$, M_{ic}, M_s, K_{ic}, E_{rot}, β, T_{os}, r^e_{sh}, r^p_{sh}, M_{sh}, Ω_{axis}, Ω_{ic} and Ω_{sh} are the central density, the central temperature, the central electron–and neutrino concentrations, the inner core mass, the sonic mass, the kinetic energy of the inner core, the rotational energy and β for the inner 1.4 ms, the mean oscillation period of the inner core, the equatorial and polar shock radii at the end of the calculation, the mass inside the shock front, the final mean angular velocity at the axis, the final angular velocities at the equatorial edge of the inner core and the shock, respectively.

		Model A	Model D
ρ^b_c	$[10^{14}\,\mathrm{gcm^{-3}}]$	2.57	1.51
ρ^{eq}_c	$[10^{14}\,\mathrm{gcm^{-3}}]$	2.10	0.75
Ye^{eq}_c	[1/nucleon]	0.324	0.321
$Y\nu^{eq}_c$	[1/nucleon]	0.084	0.085
T^{eq}_c	$[10^{11}\,\mathrm{K}]$	1.55	0.32
M^b_{ic}	$[M_\odot]$	0.97	1.02
M^b_s	$[M_\odot]$	0.94	0.99
K^b_{ic}	$[10^{51}\,\mathrm{erg}]$	3.3	1.7
E^{eq}_{rot}	$[10^{51}\,\mathrm{erg}]$	9.6	6.5
β^{eq}		0.087	0.096
T_{os}	[ms]	2.5	11.5
r^e_{sh}	[km]	120	540
r^p_{sh}	[km]	60	475
M_{sh}	$[M_\odot]$	1.30	1.42
Ω_{axis}	$[10^3\,\mathrm{s^{-1}}]$	3.0	2.5
Ω_{ic}	$[10^3\,\mathrm{s^{-1}}]$	1.5	0.3
Ω_{sh}	$[10^3\,\mathrm{s^{-1}}]$	0.2	0.1

In addition it should be noted that rotation tends to enlarge the mass of the inner core, thereby reducing the total disintegration losses of the shock as compared to 1-d models. Furthermore one has to consider the dissipated kinetic energy of the matter falling into the shock. As rotation helps to stabilize the shock heated matter in the gravitational potential, a much larger part of the dissipated energy can support the expansion of the heated matter behind the shock front until equilibrium is achieved at larger radii and at lower potential energy. The resulting compressional work adds up to the work of the expanding inner core and strengthens the shock. In Model D the effect is enlarged as the adiabatic index of the shock heated matter (up to $m(r) = 1.2\ M_\odot$) is larger than 4/3 due to disintegration of nuclei. In addition centrifugal forces reduce the ram pressure of the supersonic flow considerably in comparison to 1-d models or Model A. All these effects, which should be analyzed in more detail, help the shock to reach the silicon shell in Model D (see Fig. 13).

c) Some important implications

In all models calculated by Mönchmeyer (1989a) the collapsed cores reach large values of the angular velocity ($500\ sec^{-1} \lesssim \Omega \lesssim 1000\ sec^{-1}$), which correspond to periods in the range $6\ msec \lesssim P \lesssim 13\ msec$. These periods are consistent with the idea that the Crab pulsar with its (present!) period of 33 $msec$ was formed by the collapse of a rotating core. Furtheron this demonstrates that at least some (if not all) millisecond pulsars may indeed be born as fast rotating neutron stars, and not be spun up by accretion later in their evolution as has been sugested e.g. by Alpar et al. (1982). Again these arguments will gain tremendous support, if the observation of the 0.5 msec pulsar in the remnant of SN 1987A can be confirmed.

Table 3. Dependence of neutron star properties on various equations of state (adopted from Sato and Suzuki, 1989). M_{NR}, R_{NR}, M_R and Ω_{max} are the maximum mass of a non-rotating neutron star, its corresponding radius, the maximum mass and the maximum angular velocity of a rotating neutron star for a given equation of state, respectively. The table is based on data calculated by Friedman et al. (1986). The equations of state (EOS; ordered with increasing R_{NR}) are: B: Pandharipande (1971b); G: Canuto and Chitre (1974); F: Arponen (1972); A: Pandharipande (1971a); E: Moszkowski (1974); FP: Friedman and Pandharipande (1981); D: Bethe and Johnson V (1974); C: Bethe and Johnson I (1974); N: Relativistic Mean Field, Serot (1979); M: Tensor Interaction, Pandharipande and Smith (1975); L: Mean Field, Pandharipande and Smith (1975).

EOS	$M_{NR}/\ M_\odot$	R_{NR} [km]	$M_R/\ M_\odot$	Ω_{max} [10^4 sec^{-1}]
B	1.36	7.00		
G	1.36	7.01	1.55	1.52
F	1.46	7.97	1.66	1.24
A	1.66	8.43		
E	1.73	9.02		
FP	1.95	9.02	2.30	1.23
D	1.65	9.26		
C	1.85	9.95	2.16	1.11
N	2.60	12.4		
M	1.96	13.6		
L	2.70	13.8	3.18	0.76

The simulations of Mönchmeyer (1989a) do also strongly suggest that the nuclear equation of state is not necessarily probed in rotational core collapse, as the bounce may occur just at or well below nuclear matter density (see also Model B in Mönchmeyer, 1989a,b). Therefore only the subsequent evolution of the collapsed core to the proto-neutron star, which occurs on the neutrino diffusion time scale of a few seconds, may be influenced by the nuclear equation of state. On the other side the simulations have convincingly shown that a very accurate knowledge of the subnuclear equation of state and in particular of the adiabatic index in the density regime $10^{13}\,\mathrm{gcm}^{-3} \lesssim \rho \lesssim 10^{14}\,\mathrm{gcm}^{-3}$ is crucial for the dynamics of rotational core collapse. Although it is often stated in the literature that the equation of state is well known in this density range, one has to keep in mind that this is only meant relative to the knowledge one has about the nuclear equation of state. However, for supernova dynamics a difference in the adiabatic index of 0.01 only (i.e., $\gamma = 1.32$ or $\gamma = 1.31$ for example) can be decisive for the success or failure of the model. As nucleon-nucleon interactions start to become already non-negligible for $\rho \gtrsim 10^{13}\,\mathrm{gcm}^{-3}$, it would be important that nuclear physicist also try to improve the equation of state for densities below nuclear matter density.

d) Rotating neutron star models

Within the framework of general relativity a systematic analysis of the structure and the stability of rotating neutron stars has been performed by Friedman et al. (1986). They have calculated a large set of rigidly rotating neutron star models using a variety of (zero temperature) neutron matter equations of state ranging from very soft ones to extremely stiff ones. For each equation of state Friedman et al. (1986) constructed a sequence of models with increasing angular velocity up to the point, where the (equatorial) surface angular velocity exceeded the Keplerian velocity. Models with even larger angular velocity will shed off mass at the equator. For all equations of state the model sequences do terminate before the neutron star becomes unstable against non-axisymmetric perturbations, which occur when $\beta \gtrsim 0.14$ (see section IV a).

Recently Sato and Suzuki (1989) using the data of Friedman et al. (1986) have investigated the implications on neutron matter equations of state arising from the possible discovery of a 0.5 millisecond pulsar in the remnant of SN 1987A. Some of their results are shown in Fig. 14 and Table 3. In figure 14 the maximum angular velocity (see above) is plotted as a function of the gravitational mass for four equations of state given in Table 3. The figure clearly shows that only a very soft equation of state, such as G, is in accordance with a 0.5 millisecond pulsar. For stiffer equations of state, such as C and L, the maximum angular velocity is too small to account for a period of 0.5 msec. This is because the surface angular velocity of a neutron star constructed with a stiffer equation of state, i.e. with a larger radius, exceeds the Keplerian velocity for a smaller value of Ω than for a softer equation of state.

Sato and Suzuki (1989) have further pointed out that as shown in Table 3, the maximum mass of a non-rotating neutron star M_{NR} is approximately a monotonically increasing function of the radius R_{NR}, if one neglects some exceptions. For example M_{NR} of FP is greater than that of C, although the opposite statement holds for the corresponding radii. This is due to the fact that the pressure of the equation of state FP is lower than that of C for densities $\rho < 10^{15}\,\mathrm{gcm}^{-3}$, but exceeds it for higher densities. Consequently in order to increase the maximum mass without getting into conflict with the existence of a 0.5 msec pulsar, requires a neutron matter equation of state which is soft at low and stiff at high densities, such as FP.

When also taking into account the measured mass of the pulsar PSR 1913+16 ($1.451\,M_\odot$; vertical dashed line in Fig. 14) with a period of 59 msec ($\Omega \approx 0.1 \times 10^4\,\mathrm{sec}^{-1}$) very soft equations of state are excluded, as the corresponding maximum non-rotating neutron star mass (marked by crosses in Fig. 14) is too small (see also Table 3). Therefore from pulsar observations very soft as well as very stiff neutron star matter equations of state seem to be ruled out. Note, however, that if the existence of the sub-millisecond pulsar in SN 1987A cannot be confirmed the presently most stringent limit on the angular velocity arising from the pulsar PSR 1937+214, is too low to rule out any neutron matter equation of state on the basis of observed rotation periods.

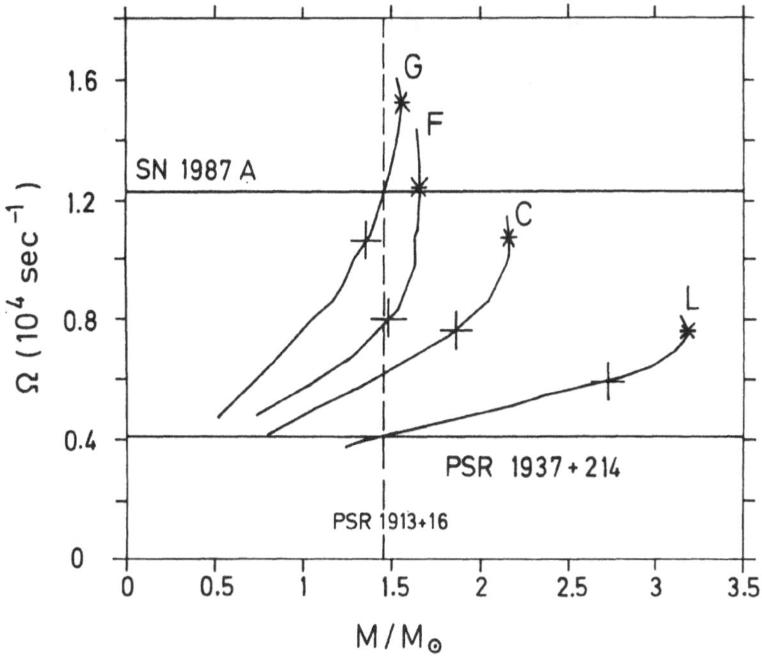

Figure 14. Maximum angular velocity (rotational surface velocity equal to the Keplerian velocity) versus gravitational mass of a rotating neutron star for some of the equations of state given in Table 3 (adopted from Sato and Suzuki, 1989; data taken from Friedman et al. 1986). For each EOS the maximum mass of a rotating and non-rotating neutron star are marked with an asterix and a cross, respectively. In addition the angular velocity of the millisecond pulsar PSR 1937+214 and of the possibly detected pulsar in SN 1987A are marked by horizontal lines. Finally the accurately known mass of the binary pulsar PSR 1913+16 is marked by a dashed vertical line.

V. Summary and Conclusion

Supernova cores and neutron stars are the only cosmic objects in which low entropy matter is compressed to nuclear densities and beyond. It is therefore tempting to use their properties to constrain the equation of state at such high densities. But in doing so one has to be careful not to mingle facts with fiction.

A direct way to obtain the desired information for very asymmetric nuclear matter is, of course, to measure the masses of neutron stars, because large masses ($M > 1.5M_\odot$) would imply rather stiff equations of state. On the other hand, the absence of high mass neutron stars does not imply that the equation of state has to be soft, because the process of formation may select a very limited mass range. Unfortunately, with the exception of the binary pulsar PSR 1913+16, mass determinations of neutron stars are still very uncertain and, therefore, do not impose strong constraints. The mass of PSR 1913+16 ($M = (1.451 \pm 0.007)M_\odot$; Taylor, (1987)) rules out extremely soft equations of state only.

Certainly, if it should be confirmed that SN 1987A left behind a neutron star spinning at a rate of about 2000 revolutions per second, this would rule out most stiff neutron star matter equations of state. Together with the information available for PSR 1913+16 this would restrict the equation of state to a very narrow "allowed" range. However, the observations of Kristian et al. (1989) do not yet rule out radial oscillations as a possible reason for the optical pulsations. Because oscillation frequencies of neutron stars are not very sensitive to both mass and equation of state but are roughly proportional to the inverse square root of the mean density, observations ruling out this explanation are exceedingly important. Therefore, we have to wait for the reappearance of the pulsar before we can draw any conclusions on the properties of the high density equation of state. Of course, if the pulse period should change systematically this would rule out oscillation models.

Observations of type II supernovae as well as supernova models, on the other hand, do not impose strong constraints on the nuclear equation of state. First of all, a surprisingly small number of supernova remnants definitely does contain a neutron star. Moreover, very little observational information on the explosion mechanism is available and even the best studied case, SN 1987A, is no exception to that rule. We only know for sure that some supernovae leave neutron stars behind and that the kinetic energy of the ejecta is typically around 10^{51}erg. Whether or not this energy is supplied by a hydrodynamic shock wave, as suggested by the prompt explosion models, or comes from neutrino energy deposition, nuclear burning, etc., is a completely open question. It may well be that only those stars can explode by the prompt mechanism which possess a significant amount of angular momentum initially. Nevertheless, there is little doubt that soft (supra-nuclear) equations of state favor prompt explosions, at least for non-rotating models, but this should not be regarded as a proof that these equations of state correctly describe the properties of matter at densities beyond nuclear matter density. However, one should keep in mind that very likely in successful 2-d supernova models, which incorporate the effects of rotation, the nuclear equation of state may not be probed at all.

So, in conclusion, there is very little hope to extract information on the equation of state from supernova models unless we know more about the explosion mechanism. In this respect, a galactic type II supernova would certainly help. While in the case of SN 1987A only $\bar{\nu}_e$'s from the early cooling phase of the neutron star have been detected, there is a good chance to see ν_e's from the deleptonization burst in existing neutrino detectors, provided the supernova explodes at a distance of less than a few kpc. Their luminosity and energies are strongly dependent upon the hydrodynamical evolution of the stellar core. Therefore, if the temporal change of both quantities could be measured, this would impose strong constraints on the theoretical models. Finally, there is a fair chance to detect gravity waves from a galactic supernova and, thus, to probe the importance of rotation and asphericities on the dynamics of the stellar core near bounce.

Acknowledgement

The authors would like to thank Peter Höflich and Hans-Thomas Janka for numerous helpful discussions on many aspects of supernova models and observations.

References

Alpar M.A., Cheng A., Ruderman, M., and Shaham J.A., 1982, *Nature* **300** 728

Arnett W.D., 1977 *Astrophys.J.* **218**, 815

Arnett W.D., 1983 *Astrophys.J.* **263**, L55

Arnett W.D., 1985 unpublished

Arnett W.D., 1987 *Astrophys. J.* **319**, 136

Arnett W.D., 1988a *Astrophys.J.* **331**, 377

Arnett W.D., 1988b *Proc. 4th George Mason Astrophysics Workshop: SN1987A in the LMC*, ed. by M. Kafatos and A.G. Michalitsianos, p. 301

Arnett W.D., and Bowers, R.L., 1977 *Astrophys.J.Suppl.* **33**, 415

Arnett W.D., Fryxell B.A., and Müller E., 1989 *Astrophys.J.* bf 341, L63

Arponen J., 1972 *Nucl. Phys.* **A191**, 257

Baade W., and Zwicky F., 1934 *Phys.Rev.* **45**, 138

Baron E., Cooperstein J., and Kahana S., 1985a *Phys.Rev.Lett.* **55**, 126

Baron E., Cooperstein J., and Kahana S., 1985b *Nucl.Phys.* **A440**, 744

Baron E., Bethe H.A., Brown G.E., Cooperstein, J., and Kahana S., 1987 *Phys.Rev. Lett.* **59**, 736

Barrett P., 1987 in *SN 1987A*, ed. by I.J. Danziger, ESO, Garching, p. 173

Baym, G., and Pethick, C., 1975 *Ann.Rev.Astron.Astrophys.* **17** 415

Becker R.H., Holt S.S., Smith B.W., White N.E., Bolt E.A., Mushotzky R.F., and Serlemitsos P.J., 1979 *Astrophys.J.Lett.* **238**, L73

Bethe H.A., Brown G.E., and Cooperstein J., 1987, *Astrophys.J.*, **322**, 201

Bethe H.A., and Johnson M., 1974 *Nucl. Phys.*, **A230**, 1

Bethe H.A., and Wilson J.R., 1985 *Astrophys.J.* **295**, 11

Bionta R.M. *et al.*, 1987 *Phys.Rev.Lett* **58**, 1494

Blades J.C., Wheatley J.M., Panagia N., Grewing M., Pettini M., and Wamsteker W., 1988 *Astrophys.J.* **334**, 308

Blinnikov S.I., Chugai N.N., Golenetskii S.V., Lozinskaya T.A., and Mazets E.P., 1988 *Astrophys. Space Phys. Rev.* **6**, 197

Bodenheimer P., and Woosley S.E., 1983 *Astrophys.J.* **269**, 281

Bruenn S.W., 1987 *Phys.Rev.Lett.* **59**, 938

Bruenn S.W., 1988 *Astrophys.Space Sci.* **143**, 15

Burrows A., 1987, *Astrophys.J.Lett.* **318**, L57

Canuto V., and Chitre S.M., 1974, *Phys. Rev.*, **D9**, 1587

Cassatella A., 1987 in *SN 1987A*, ed. by I.J. Danziger, ESO, Garching, p.101

Catchpole R.M. *et al.*, 1987 *M.N.R.A.S.* **229** , 15

Chevalier R.A., and Kirshner R.P., 1978 *Astrophys.J.* **219**, 931

Cooperstein J., and Baron E., 1989 *Supernovae*, ed. by A. Petschek, Springer, New York, in press

Danziger I.J., Bouchet P., Gouiffes C., and Rufener F. , 1989 *Big Bang, Active Galactic Nuclei an Supernovae*, ed. by S. Hayakawa and K. Sato, Universal Academic Press, Tokyo, p. 429

Davidson K. *et al.*, 1982 *Astrophys.J.* **253**, 696

Dotani T. *et al.*, 1987 *Nature* **330**, 230

Durisen R.H., and Imamura J.N., 1981 *Astrophys. J.* **183**, 215

El Eid M., and Langer N., 1986 *Astron.Astrophys.* **167**, 274

Epstein R.I., 1979 *Mon. Not. Roy. Astron. Soc.* **188**, 305

Erickson E.F., Haas M.R., Colgan S.W.J., Lord S.D., Burton M.G., Wolf J., Hollebach D.J., and Werner M., 1988 *Astrophys.J.* **330**, L39

Eriguchi Y., and Müller E., 1984 *Astron. Astrophys.* **147**, 161

Fillipenko A.V., and Sargent, W.L.W., 1985 *Nature* **316**, 407

Fosbury R.A.E., Danziger I.J., Lucy L.B., Gouiffes C., and Critiani S., 1987 in *SN 1987 A*, ed. by I.J. Danziger, ESO, Garching, p. 139

Fransson C., Cassatella A., Gilmozzi R., Kirshner R.P., Panagia N., 1989 *Astrophys.J.* in press

Friedmann B., and Pandharipande V.R., 1981 *Nucl. Phys.* **A316**, 502

Friedmann J.L., Ipser J.R., and Parker L., 1986 *Astrophys.J.* **304**, 115

Fu A., 1988 *Proc.Astron.Soc.Austrl.* **7**, 505

Gilmozzi R., Cassatella A., Clavel J., Fransson C., Gonzalez R., Gry C., Panagia N., Talavera A., and Wamsteker W., 1987 *Nature* **328**, 318

Goldreich P., and Weber S.V., 1980 *Astrophys.J.* **238**, 991

Henry R.B.C., and McAlpine G.M., 1982 *Astrophys.J.* **258**, 11

Henry R.B.C., McAlpine G.M., and Kishner R.P., 1984 *Astrophys.J.* **278**, 619

Hillebrandt W., 1982 in *Supernovae: A Survey of Current Research*, ed. by M.J. Rees and R.J. Stoneham, NATO-ASI **C90**, Reidel, Dordrecht, p. 123

Hillebrandt W., 1987a in *High Energy Phenomena around Collapsed Stars*, ed. by F. Pacini, NATO-ASI **C195**, Reidel, Dordrecht, p. 73

Hillebrandt W., 1987b in *SN 1987A*, ed. by I.J. Danziger, ESO, Garching, p. 301

Hillebrandt W., Höflich P., Janka H.-T., and Mönchmeyer R., 1988, *Proc. of the 20th Yamada Conf.*, ed. by S. Hayakawa, and K.Sato, Universal Acad. Press, p. 441

Hillebrandt W., and Meyer F., 1989 *Astron.Astrophys.* in press.

Hillebrandt, W., and Wolff, R.G., 1985 in Nucleosynthesis: Challenges and New Developments, ed. by W.D. Arnett, and J.W. Truran, Univ. Chicago Press, Chicago, p. 131

Hillebrandt W., Nomoto K., and Wolff R.G., 1984 *Astron.Astrophys.* **133**, 175

Hillebrandt W., Höflich P., Truran J.W., and Weiss A., 1987 *Nature* **327**, 597

Hirata K. *et al.*, 1987 *Phys.Rev.Lett.* **58**, 1490

Höflich P., 1988 *IAU Symposium 108: Atmospheric Diagnostic of Stellar Evolution*, ed. by K. Nomoto, Springer, Heidelberg, p. 388

Höflich P., Sharp and Ch.M., Zorec J., 1989 *Particle Astrophysics Workshop* Berkley, ed. by C. Pennypacker, in press

Holt S.S., 1983 in *IAU Symposium 101: Supernova Remnants and their X-Ray Emission*, ed. by J. Danziger and P. Gorenstein, Reidel, Dordrecht, p.17

Imshennik V.S., and Nadyozhin D.K., 1983 *Astrophys.Space Phys.Rev.* **2**, 75

Isserstedt J., 1975 *Astron.Astrophys.Suppl.* **19**, 259

Janka H.-T., and Hillebrandt W., 1989a *Astron.Astrophys.Suppl.*, in press

Janka H.-T., and Hillebrandt W., 1989b *Astron.Astrophys.*, in press

Janka H.-T., and Mönchmeyer R., 1989a *Astron.Astrophys.* **209**, L5

Janka H.-T., and Mönchmeyer R., 1989b, *Astron.Astrophys.*, in press

Jansen F.A. *et al.*, 1985 in *Astronomy from Space*, ed. by G.G. Fazio et al., Pergamon Press, Oxford, p. 49

Kristian J.A. *et al.*, 1989 *Nature* **338**, 234

Lattimer J.M., and Burrows A., 1984 in *Problems of Collapse and Numerical Relativity*, ed. by D. Bancel an M. Signore, NATO-ASI **C134**, Reidel, Dordrecht, p. 147

Le Blanc J.M., and Wilson J.R., 1970 *Astrophys.J.* **161**, 541

Ledoux P., 1945 *Astrophys. J.* **102**, 143

Manchester R.N., Tuohy, J.R., and D'Amico, N., 1983 in *IAU Symposium 101: Supernova Remnants and their X-Ray Emission*, ed. by J. Danziger and P. Gorenstein, Reidel, Dordrecht, p. 495

Matz S.M., Share G.H., Leising M.D., Chupp E.L., Vestrand W.T., Purcell W.R., Strickman M.S., and Reppin C., 1988a *Nature* **331**, 416

Mayle R., Wilson J.R., Ellis J., Olive K., Schramm D.N., and Steigman G., 1988 *Phys.Lett.* **B203**, 188

Mendez M., Clocchaiatti A., Benvenuto G., Feinstein C., and Marraco U.G., 1988 *Astrophys.J.* **334**, 295

Miller G.E., and Scalo J.M., 1979 *Astrophys.J.Suppl.* **41**, 513

Moszkowski S., 1974 *Phys. Rev.* **D9**, 1613

Mönchmeyer R., 1989a *Ph.D. thesis*, TU München

Mönchmeyer R., 1989b in *Proc. 5th Workshop on Nuclear Astrophysics*, ed. by W. Hillebrandt, and E. Müller, MPA **P1**, Garching, p.92

Mönchmeyer R. and Müller E., 1989a, in *Timing Neutron Stars*, ed. by H. Ögelman, and E. van den Heuvel, NATO-ASI **C262**, Kluwer, Dordrecht, p.549

Mönchmeyer, R., and Müller, E., 1989b *Astron. Astrophys.*, in press

Mönchmeyer, R., and Müller, E., 1989c in preparation

Müller E., Różyczka M., and Hillebrandt W., 1980 *Astron. Astrophys.* **81**, 288

Müller E., and Hillebrandt W., 1981 *Astron.Astrophys.* **103**, 358

Nomoto K., 1984 *Astrophys.J.* **277**, 791

Nomoto K., and Hashimoto M., 1988 *Physics Reports* **163**, 13

Nomoto K., Shigeyama T., and Hashimoto M., 1987 in *SN 1987A* ed. by I.J. Danziger, ESO, Garching, p. 325

Nomoto K., Hashimoto M., Shigeyama T., Kumagai S., Yamaoka H., and Saio H., 1989 in *Big Bang, Active Galactic Nuclei and Supernovae*, ed. S.Hayakawa and K.Sato, Universal Academy Press, Tokyo, p. 495

Ostriker J.P., and Bodenheimer P., 1973, *Astrophys. J.* **180**, 171

Ostriker J.P., and Tassoul J.L., 1969, *Astrophys. J.* **155**, 987

Pandharipande V.R., 1971a *Nucl.Phys.* **A174**, 641

Pandharipande V.R., 1971b *Nucl.Phys.* **A178**, 123

Pandharipande V.R., and Smith R.A., 1975 *Nucl.Phys.* **A237**, 507

Papaliolios et al., 1989 *Nature* **338**, 565

Pravdo S.H., and Nugent J.J., 1983 in *IAU Symposium 101: Supernova Remnants and their X-Ray emission*, ed. by J. Danziger and P. Gorenstein, Reidel, Dordrecht, p. 29

Pravdo S.H., and Smith B.W., 1979 *Astrophys.J.Lett.* **234**, L195

Rank D.M., Pinto P.A., Woosley S.E., Bregman J.D., Witteborn F.C., Axelrod T.S., and Cohen M., 1988b *Nature* **331**, 505

Rappaport S., and Joss P.C., 1983 in *Accretion Driven Stellar X-Ray Sources*, ed. by W. Lewin and E. Van den Heuvel, Cambridge Univ. Press, Cambridge, p. 1

Rousseau J., Martin N., Prevout L., Reheirot E., Robin A., and Brunet J.P., 1978 *Astron. Astrophys. Suppl.* **31**, 243

Saio H., Kato M., and Nomoto K., 1988 *Astrophys. J.* **331**, 388

Salpeter E.E., 1955 *Astrophys.J.* **121**, 161

Sato K., and Suzuki H., 1989 *UTAP* **88**, preprint

Schwarz H.E., 1987 in *SN 1987A*, ed. by I.J. Danziger, ESO, Garching, p. 167

Serot B.D., 1979 *Phys. Lett.* **86B**, 146 and **87B**, 403

Sharma M.M., Borghols W.T.A., Brandenburg S., Crona S., van der Woude A., and Harakeh M.N., 1988 *Phys.Rev* **C38**, 2562

Shigeyama T., Nomoto K., and Hashimoto M., 1988 *Astron.Astrophys.* **196**, 141

Sunyaev R. *et al.*, 1987 *Nature* **330**, 327

Symbalisty E., 1984 *Astrophys.J.* **285**, 729

Tammann G.A., 1982 in *Supernovae: A Survey of Current Research*, ed. by M.J. Rees and R.J. Stoneham, NATO-ASI **C90**, Reidel, Dordrecht, p. 371

Tanaka Y., 1988 *IAU Colloquium 108: Atmospheric Diagnostic of Stellar Evolution*, ed. by K.Nomoto, Springer, Heidelberg, p. 399

Tassoul, J.-L., 1978 *Theory of Rotating Stars*, Princeton Univ. Press, New Jersey

Taylor J.H., 1987 in *Proc. XIII Texas Symposium on Relativistic Astrophysics*, ed. by P.M. Ulmer, World Scientific, Singapore, p. 467

Tohline J.E., 1984 *Astrophys. J.* **285**, 721

Tohline J.E., Schombert J.M., and Boss A.P., 1980 *Space Sci. Rev.* **27**, 555

Truran J.W., and Weiss A., 1987 in *SN 1987A*, ed. by I.J. Danziger, ESO, Garching, p. 271

Tsuruta S., 1979 *Physics Reports* **56**, 237

Tyson J.A., and Boeshaar P.G., 1987 *PASP* **99**, 905

Utrobin V.P., 1984 *Astrophys. Space Sci.* **98**, 115

Utrobin V.P., 1988 *Atominform*, p.20

Van den Bergh S., Mc Clure R.D., and Evans R., 1987 *Astrophys.J.* **323**, 44

Walborn N.R., Lasker B.M., Laidler V.G., and Chu Y.-H., 1987 *Astrophys.J.* **321**, L41

Walecka J.D., 1974 *Ann. Phys.* **83**, 491

Wampler E.J., and Richichi A., 1989 *Astron.Astrophys.*, in press

Weaver T.A., Woosley S.E., and Fuller G.M., 1985 in *Numerical Astrophysics*, eds. J. Centrella, J. LeBlanc and R. Bowers, Jones and Bartlett, Boston

Weiss A., Hillebrandt W., and Truran J.W., 1988 *Astron.Astrophys.* **197**, L11

West R.M., Lauberts A., Jorgensen H.E., and Schuster H.-E., 1987 *Astron.Astrophys.* **177**, L1

Williams R.E., 1987 *Astrophys. J.* **320**, L117

Wilson J.R., 1985 in *Numerical Astrophysics*, ed. by J. Centrella, J. Le Blanc, and R. Bowers, Jones and Bartlett, p. 422

Wilson J.R. Mayle R.W., Woosley S.E., and Weaver T.A., 1986 *Ann. NY Acad.Sci.* **479**, 267

Witteborn F., Rank D.M., Bregman J.D., Pinto P.A., Wooden D., and Axelrod T.S., 1988 *Astrophys.J.*, in press

Woosley S.E., 1986 in *Nucleosynthesis and Chemical Evolution*, Saas Fee Lecture Notes, ed. by B. Hauck, A, Maeder, and G. Meynet, Geneva Observatory, p. 1

Woosley S.E., 1988 *Astrophys.J.* **330**, 218

Woosley S.E., Pinto P.A., and Weaver T.A., 1988 *Proc.Astron.Soc.Austrl.* **7**, 355

Yahil A., and Lattimer J.M., 1982 in *Supernovae: A Survey of Current Research*, ed. by M.J. Rees and R.J. Stoneham, NATO-ASI **C90**, Reidel, Dordrecht, p. 53

THE HADRONIC EQUATION OF STATE AND SUPERNOVAE

S.H. Kahana

Department of Physics
Brookhaven National Laboratory
Upton, New York 11973

ABSTRACT

The mechanism for type II supernovae is still not fully understood. A direct explosion, driven by the shock wave created after core collapse, is still a strong likelihood. The conditions favourable to this simplest and most natural mechanism are discussed and recent calculations presented which highlight our uncertainties in the conditions just before collapse. Details of the pre-collapse density and entropy profiles, and not just the "iron" core mass, matter very much.

1. INTRODUCTION

The subject of this school, the nuclear equation of state, involves concepts easy enough to define theoretically but hard to confront in nature. Generally one can formulate an equation of state only for material in the bulk and in or near equilibrium. Clearly, most terrestrial examples of nuclear systems do not have these attributes. Nuclear matter in macroscopic amounts was present in the early universe, can be found in stellar interiors and above normal saturation density only in highly condensed objects. Unfortunately such realizations of nuclear matter are not easily accessed and our knowledge of such material is culled from theory or from observations of supernovae or pulsars; stellar evolution leading to gravitationally collapsed objects is critically dependent on the properties of dense nuclear matter. A most helpful recent event has been the sighting of SN1987a,[1] with the detection of neutrinos[2] yielding a direct look at the very densest material in the collapsed core of 1987a. One can now be certain[3] that the gravitational collapse scenario for Type II supernovae is substantially correct. It was also possible that SN1987a might have been used to constrain the detailed mechanism responsible for the explosion. A high explosion energy would indicate the presence of direct explosion due to a prompt shock, and rule out explosion by a stalled shock, resuscitated through deposition of neutrino energy in material behind the shock. The total explosion energy, estimated from optical analysis of the SN1987a remnant, is in the range 0.5 to 1.5×10^{51} ergs[4,5], too low to settle this ambiguity outright.

The Nuclear Equation of State, Part A
Edited by W. Greiner and H. Stöcker
Plenum Press, New York

The above two mechanisms arising within the same general framework, seem the only likely candidates for generators of type II supernovae. Other possibilities such as rotation induced detonation of the oxygen shell, are far less natural. The observations[6] in January of this year claiming to have seen a pulsar with rotational period 1/2 has not been confirmed in many subsequent sightings. Should, however, Pennypacker and co-workers[6] prove correct the spherically symmetric collapse models must be questioned and rapid rotation taken seriously.

My task in this talk is to consider in some detail the conditions favourable for direct explosion by a prompt shock. My collaborators Ed Baron and Jerry Cooperstein[7,8] are still valiantly completing the program of extending hydrodynamic calculations some several seconds beyond collapse and any results of theirs quoted here are only preliminary. Their results are both encouraging and discouraging. Introduction of all species of neutrinos into the transport and of neutrino–electron scattering in the allowed weak processes has considerably weakened the shocks generated in simulations based on the current initial models of Weaver and Woosley[4] or Nomoto and Hashimoto[5]. Variants of these "realistic" initial models have been created[7,8], however, which do lead to vigorous explosion. The crucial feature which generates direct explosions is the use, at high density, of a relatively soft hadronic equation of state[9,10]. A recent re-examination of neutron star masses[11] suggests that an even softer equation of state than previously thought, is possible in the range of densities relevant to supernovae. The laboratory evidence favours such a state of affairs. The breathing mode in heavy nuclei is consistent[12] with an incompressibility $K_0 = 210 \pm 30$ MeV at saturation density while the most recent analyses[13] of relativistic ion collisions, contrary to earlier presentations, tends to confirm such a picture, finding $K_0 \approx 215$ MeV. The direct mechanism is alive and well on Long Island and likely elsewhere in the universe.

2. INITIAL MODELS

The staged burning of increasingly heavier elements in the stellar core must eventually cease with ^{28}Si burning leading to ^{56}Co, and by various weak process other elements near iron in atomic mass. The scene is set for hydrodynamic simulation of collapse by conditions reached at the end of this lengthy period of quasi-static evolution. Fig. 1 shows the density profiles from initial modeling by Weaver and Woosley[4] for a main sequence star masses $M_{ms} = 15M_{\odot}$ and iron core mass $M_C = 1.27M_{\odot}$ or by Hashimoto and Nomoto[5] for $M_{ms} = 13M_{\odot}$ and $M_C = 1.18M_{\odot}$. Over the past decade the core mass has evolved downwards from $M_C = 1.55M\odot$ for Weaver-Woosley-Zimmerman[14], to $M_C = 1.36M\odot$ for Weaver-Woosley-Fuller[15] to the lower values given above. Since the shock must traverse the core from its birth near the edge of the homologous core, at mass $\approx 0.6 - 0.7M_{\odot}$, to the edge of the entire iron core, a smaller *total* core mass is more favourable. Dissociation of $0.1M_{\odot}$ from heavy nuclei to nuclear matter costs $\sim 1.6 \times 10^{51}$ ergs, and is a principle reason for failure of the direct mechanism. Neutrino production losses are a second cause of failure. The artificial models created by Baron and Cooperstein[8] are intermediate between the models displayed in Figure 1 and possess somewhat lower temperatures.

The mass of the iron core is of course closely related to the Chandresakhar mass

$$M_{Ch} = 5.6Y_e^2 \tag{1}$$

where $Y_e \approx Z/A$ is the electron fraction per baryon. The electron captures and β-decays taking place during the quasi-static burning, especially during silicon burning,

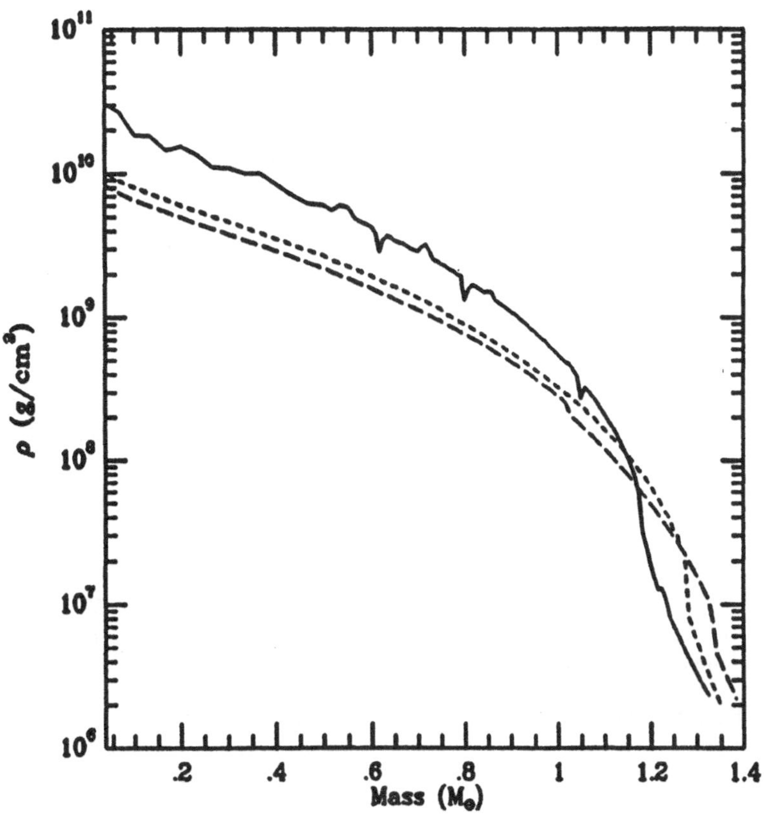

Fig. 1. A comparison of the density structure of three initial models. The solid line corresponds to $13\,M_\odot$ ($3.3\,M_\odot$ helium core) model of Nomoto and Hashimoto (Ref. (5)), while the short and long dashed lines correspond to $15\,M_\odot$ and $18\,M_\odot$ models of Woosley and Weaver (Ref. (4)). One might think that because the Nomoto-Hashimoto model falls off more rapidly than the Woosley-Weaver models in the mass regions $1.10\,M_\odot < M < 1.4\,M_\odot$, important to final propagation of the shock out of the core, that this model is easier to explode directly. In fact the $15\,M_\odot$ model of Woosley and Weaver leads to more vigorous shocks.

determines Y_e and hence M_C. Aufderheide and Brown[16] are re-examining these critical weak processes and are finding, in contrast to earlier studies, that β-decays are frequently faster than capture. This is a natural consequence of large β-decay Q values and produces a slightly higher Y_e, but lower temperatures in the pre-collapse core. This cooling, and a consequential further drop in core mass, is what our collaboration[8] has tried to represent in constructed initial models. In any case the general trend in realistic models has been towards cores highly favourable to direct explosion. Perhaps the most fruitful future direction to pursue would be to greatly improve these initial simulations.

3. HYDRODYNAMICS SIMULATION OF COLLAPSE

A self-consistent hydrostatic core using the low density equation of state of Cooperstein[17] is constructed from the realistic models or variations thereof. This model is destablized in a variety of ways, in earlier simulations by a drop in pressure throughout the core and presently by constructing the model with a low value for Newton's constant G which is increased to initiate collapse. It is instructive to keep in mind at least the Newtonian equation for hydrodynamics

$$P\frac{dU}{dt}(R) = -\rho\frac{GM(R)}{R^2} - \frac{dP}{dR} \tag{2}$$

with R the radial coordinate, $\rho(R)$ the density, $M(R)$ the Lagrangian mass, and $U(R)$ the velocity of the mass element $dM(r) = 4\pi R^2\rho(R)\,dR$. The pressure is partitioned into matter and leptonic components

$$P = P_{\text{hadronic}} + P_{\text{leptonic}} \tag{3}$$

The scheme employed is spherically symmetric[9]. Although the period for the SN1987a pulsar is uncertain, and perhaps short[6], it is likely most type II progenitors contain slowly rotating cores. These must also explode and the one dimensional mechanism should adequately describe them.

The additional physics input for collapse mainly invokes:

1. Transport for neutrinos plus a description of the weak leptonic interactions.

2. An hadronic equation of state.

The former has been improved in stages[7]. We began with a simplified treatment[9], allowing for trapping or free streaming above or below a critical density $P_{\text{trapping}} \approx 10^{12} g/cm^3$. Neutrinos with energies of 10 MeV interacting with matter at this density possess mean free paths of approximately 1 km, i.e. short in comparison to the 20 km scale of the collapsed core at maximum density. With trapping taking place in the central core, electron capture ceases, and a lowered value $Z/A = 0.31 to 0.32$ is achieved. The transport now[7] includes all neutrino species, neutrino absorption on free protons, and allows for neutrino-electron scattering. The latter degrades neutrino energies, allowing them to escape the core more easily and lowers the post-collapse core Y_e and mass. The shock will then find it more difficult to survive its outward passage in the presence of such scattering.

The second element of imput physics, the equation of state (*eos*), is most influential. Simply put, if the *eos* is sufficiently soft and general relativity is accounted for, the prompt shock will produce viable explosions. For a pre-collapse core supported by the electron pressure

$$P_{\text{lepton}} \approx \rho^{\Gamma} \tag{4}$$

and with adiabatic index $\Gamma \approx 4/3$, the total energy is given by the non-relativistic virial theorem

$$U_{\text{gravitation}} + U_{\text{internal}} = \frac{\Gamma - 4/3}{\Gamma - 1} U_{\text{gravitation}} \approx 0 \tag{5}$$

This energy balance remains throughout the almost adiabatic collapse but $U_{\text{gravitation}}$ and U_{internal} rise in magnitude to 1 to 2×10^{53} ergs, whereas a healthy shock possesses an energy near $1 \times ^{51}$ ergs, hence the delicacy of the simulation.

To transfer positive energy to the shock the collapsed core must be left gravitationally bound. A softer equation of state increases gravitational binding in the core and greatly enhances the initial shock energy.

4. PHENOMENOLOGICAL EQUATION OF STATE

I will concentrate on the *eos* at suprasaturation density where our knowledge is limited. Theoretical derivations of the *eos* for the highest densities reached in simulations cannot be taken seriously[9–11]. I introduce therefore, for the cold hadronic pressure at $\rho \gg \rho_0(\Delta)$:

$$\begin{aligned} P_{\text{hadronic}}(\rho) =& \frac{K_0(\Delta)\,\rho_0(\Delta)}{9\gamma}\left(u^{\gamma} - 1\right) \\ u =& [\rho/\rho_0(\Delta)] \\ \Delta =& \frac{N - Z}{A} \end{aligned} \tag{6}$$

Use of this phenomenological form, with the symmetric saturation incompressibility $K_0(0)$ and the limiting adiabatic index, $\Gamma \to \gamma$ for $\rho \gg \rho_0$, as parameters, allows one to bypass a theoretical derivation and to proceed with the collapse simulation. Two other parameters occur in the energy at finite temperature, the effective mass at the fermi surface and the symmetry energy coefficient W_s[9–11].

Near saturation $P(\rho)$ is purely nuclear and $K_0(0)$ is safely deduced from properties of just nuclear rather than hadronic matter. At higher densities other hadrons, π, Δ's and hyperons play a role. I assume however that Eq. (6) gives the *total* hadronic pressure. Some authors misunderstand this point, suggesting that these other hadron species soften the equation of state further. Such effects are included here a priori.

Constraints on the *eos* are produced by:

1. Laboratory measurements of the symmetric bulk incompressibility $K_0(0)$.
2. Relativistic heavy ion collisions.
3. Neutron star masses.

Let me take a didactic point of view since again many misconceptions have been propagated recently. The only way $K_0(0)$ can be extracted reliably from the breathing mode in the heaviest finite nuclei is to perform two separate calculations with a single well defined two body interaction:

1. One of the breathing mode in a self-consistent conserving approximation, i.e. using the RPA.

2. A second for infinite nuclear matter to determine $K_0(0)$.

Blaizot had performed just such a calculation a decade ago and discusses this in these proceedings[12]. His result is $K_0(0) = 210 \pm 30$ MeV.

The work of Sharma *et al.*[18], also presented in this school, is essentially without merit. There is no way one can ignore shell effects in a particular nucleus and accurately treat the energy of finite nuclei as an expansion in powers of $A^{1/3}$. The so–called (negative) surface energy in the computations of Ref. 18 is too large and too important in the calculation to be extracted by fitting a series of nuclei. Shell properties cannot be treated sufficiently smoothly to validate the analyses of Sharma *et al.*. A warning is provided by the latter authors themselves who ascribe Blaizot's lower values of $K_0(0)$ to an incorrect value for the breathing mode energy in ^{90}Zr. In fact Blaizot has used virtually the same Zr energy[12]. The A-dependent series used in Ref. 18 (originally suggested by Blaizot[12]) requires a high incompressibility for large A to compensate for an overly negative "surface" energy in Zr. Finally, to ascribe a statistically significant role to nuclei as light as magnesium makes little sense.

It is of great interest that almost all present analyses of the sideways flow[13] in relativistic ion collisions also yield a low value $K_0(0) \approx 215$ MeV, when momentum dependence is properly included in the mean field. I remind this audience that recent history[19] tended to favour much stiffer equations of state. All of this change is favourable for the prompt-shock mechanism.

Neutron star masses have been discussed elsewhere[9,20,21], but alas do not strongly constrain supernova modeling. This follows simply from the small overlap between the density ranges $\rho \lesssim 4\rho_0$ for supernovae and $3\rho_0 \lesssim \rho \lesssim 10\rho_0$ for neutron stars. To this orthogonality one must add an uncertainty in symmetry behaviour, i.e., for supernovae $\Delta \approx 1/3$ is most critical but for neutron stars $\Delta = 0$.

Symmetry Properties

I have indicated in the past[9-11] that a considerable softening in the eos results from the extrapolation from symmetric ($\Delta = 0$) to asymmetric ($\Delta \approx 1/3$) matter, required because the collapsed core has experienced considerable β-capture. The softening is approximately described by

$$K_0(\Delta) = K_0(0)\ \left[1 - a\Delta^2\right] \tag{7}$$

$$\rho_0(\Delta) = \rho_0(0)\ \left[1 - b\Delta^2\right] \tag{8}$$

with $a = 2.0$, $b = 0.75$ for one Skyrme model[22], but is a general feature of the trend from bound symmetric matter to unbound neutron matter. For the neutron star one must use an unbound equation of state

$$P(\rho) = \frac{K_N}{9\gamma_{NS}} u^{\gamma_{NS}} \tag{9}$$

with the incompressibility

$$K_N = K(\rho_0(0), \Delta = 1) \tag{10}$$

perhaps considerably larger than the value 180 MeV used by Baron, Cooperstein and myself[9]. In a recent calculation[11] I found that it is just possible in the lower supernova density range to use a softer eos than that embodied in the standard BCK parametrisation, $K_0(0) = 180$ MeV, and $\gamma = 2.5$, while simultaneously keeping the maximum neutron star mass at $1.58 M_{\odot}$. This should help in the direct mechanism simulation and shows up in simulations using recent initial modeling[8]. Once again I pray that people who see danger to the direct mechanism from neutron star masses would try to understand these simple points.

5. RESULTS

Table 1. Equation of state parameters $(K_0(0), \gamma)$ are the incompressibility at saturation for symmetric matter and the high density adiabatic index, as given in Eq. (6). All calculations take full account of general relativity except for model #38*, which is Newtonian. The maximum central density reached in the calculation ρ^c_{max} is in units of the saturation density appropriate to the asymmetric matter $(\Delta = 1/3)$ relevant to bounce, $(\rho_0(1/3) = 2.4 \times 10^{14}$ g/cm^3). The precollapse models for #40-45 are the main sequence $M = 12$, $15 M_{\odot}$ models of Ref. (15), while #61-63 are from the $13 M_{\odot}$ model of Ref. (4). The explosion energy E_{expl} was obtained from the estimated shock energy by correcting for oxygen burning in the mantle and gravitational binding of mantle and envelope. Models are further distinguished by the symmetry energy W_s which we believe is experimentally closer to the higher values in the table and by a trapping density which is set at 0.4×10^{12} g/cm^3 in the first six models in the table, and at the more realistic 1×10^{12} g/cm^3 for #62, 63.

Model #	Mass $M_{\odot}$	$K_0(0)$ MeV	Γ	W_s MeV	$\frac{\rho^c_{max}}{\rho_0(0.33)}$	E_{expl} 10^{51} ergs
38*	12	180	2	29.3	2.3	0.1
40	12	180	2	29.3	12.0	3.2
41	12	180	3	29.3	3.1	0.8
43	15	180	2.5	29.3	4.1	1.7
45	15	90	3	29.3	4.0	0.8
61	13	180	2.5	29.3	4.1	2.4
62	13	180	2.5	36.0	4.1	2.6
63	13	180	2.5	34.0	4.1	1.9

I turn briefly to the numerical simulations. Table 1 displays some earlier calculations by the Brookhaven-StonyBrook-Cornell collaboration[9,23] and Table 2 some recent modeling by Baron and Cooperstein[8] using the new initial model described above. Bruenn[24] has explored the parameter space for the prompt shock mechanism more completely. Not only the core mass of the initial model (hard to define but near $1.15 M_{\odot}$ for the model in Table 2) but also the density and entropy profiles are crucial in determining the fate of the shock. With the standard BCK parametrisation[9] $K_0(0) = 180$ MeV, $\gamma = 2.5$ the earlier calculations excluding some elements of neutrino transport, some species and electron-neutrino scattering, led to vigorous explosion. The symmetry energy W_s plays a key role[23], higher values near 36 MeV suppress the free proton fraction and hence limit β-capture. This keeps the post collapse value of Y_e high and prevents the shock from forming at too low a value of the Lagrangian mass parameter, whence propagation out of the core would be impossible.

The explosions in Table 2 are also achieved with standard parameters but now allow for all neutrino species and neutrino–electron scattering. Even more vigorous explosions would obtain for say reductions in γ below 2.5 for the densities appropriate to supernova, reductions which I now believe possible for densities less than say $4\rho_0$.

Table 2. Using the models of references 4 and 5 as guides it is possible to create artificial models which do lead to vigorous explosions, even after inclusion of all neutrino transport refinements. The models are constructed to be hydrostatically self consistent[8], and to include the cooling suggested by Aufderheide and Brown[16]. The shock energy is as given in Ref. 9. $U_{\max}$, R_S, M_S are the maximum velocity, mass radius and included mass for the shock in the last snapshot in the simulation, i.e., just as the shock is leaving the mathematical grid defining the initial core. The iron core has a "mass" $1.15\,M_{\odot}$, i.e., to the point where some shell *Si* burning is still taking place.

Parameters					
$K_0\,(0)$ MeV	γ	E_S	$U_{\max}$	R_S(km)	M_S ($M_{\odot}$)
180	5.0	0.5	1.6	2680	1.2
180	2.5	1.3	2.0	3160	1.2

6. THEORETICAL DERIVATION OF THE EQUATION OF STATE

It is difficult to summarize the three or so decades of nuclear matter calculations, at varying densities and using an array of approximation schemes. To first approximation purely Brueckner-like treatments cannot be believed much beyond normal saturation, since they are hole-line expansions, i.e., expansions in powers of the density. The variational calculations of Pandharipande and collaborators[25] can in principle handle higher densities but are every bit as phenomenological as my Equation (6). These authors fit their three body interaction to the observed properties of nuclear matter at saturation and further presuppose that $K_0\,(0) \approx 240$ MeV, the latter without any empirical justification.

The mean field theories based on work by Walecka and Serot[26] or by Celenza and Shakin[27] are attractive theoretical alternatives, incorporating as they do relativity in a fundamental fashion. However, in their simplest form these theories predict excessive density dependence in the nucleon effective rest mass and in any case do not account for $SU\,(2) \times SU\,(2)$ chiral invariance; they ignore the π-meson. Recent work[28,29] on the extension of Nambu, Jona-Lasinio theories to quarks (and hence to hadrons[29]) presents a promising direction to pursue.

7. CONCLUSIONS

A final eventual resolution of the problem of which supernova mechanism (for type II) is used by nature must await considerable improvements in both input physics and techniques in the pre-collapse modeling. It is clearly here that the greatest uncertainties now lie. The collapse and short time post-collapse simulations seem better understood, although there is room for further refining of neutrino transport schemes. Recent Monte-Carlo calculations by the Munich group are most promising in this regard[30]. In any case the direct mechanism as I have said is still very much in the picture. Completion of long time calculations by Baron and Cooperstein should yield an independent view of the delayed mechanism of Wilson and collaborators[31].

ACKNOWLEDGMENT

Work supported under contract number DE-AC02-76CH00016 with the U.S. Department of Energy.

REFERENCES

1. Shelton, I., *International Astronomical Circular* No. 4316, 24 February 1987.
2. Hirata, K. *et al.* , Phys. Rev. Lett. 58, 1490 (1987);
   Bionta, R.M. *et al.* , Phys. Rev. Lett. 58, 1494 (1987).
3. Kahana, S., Cooperstein, J., Baron, E., Phys. Lett. B196, 259 (1987).
4. Woosley, S.E., Weaver, T.A., Bull. Amer. Astr. Soc. 16, 971 (1984);
   Woosley, S.E., Weaver, T.A., Ann. Rev. Astron. Astrophys. 24, 205 (1986);
   Woosley, S.E., Weaver, T.A., Phys. Rep. 163, 79 (1988), and private communication.
5. Nomoto, K., Hasimoto, M., Phys. Rep. 163, 13 (1988);
   Nomoto, K., Hashimoto, M., Prog. in Part. & Nucl. Phys. 17, 2670 (1986), and private communication.
6. Middleditch, J. *et al.* , IAU Circular No. 4735 (1989).
7. Baron, E., Cooperstein, J., Review Article, to be published.
8. Baron, E., Cooperstein, J., Brookhaven-Stony Brook preprint 1989.
9. Baron, E., Cooperstein, J., Kahana, S., Nucl. Phys. A440, 744 (1985);
   Baron, E., Cooperstein, J., Kahana, S., Phys. Rev. Lett. 55, 126 (1985).
10. Kahana, S., *Windsurfing the Fermi Sea*, Proc. Int. Conf. and Symp. on Unified Concepts of Many-Body Problems, September 1986, eds. T.T.S. Kuo and J. Speth, Elsevier Science Pub. (1987).
11. Kahana, S., to appear in *Supernova and the Hadronic Equation of State*, Ann. Rev. of Nucl. & Particle Physics (1989).
12. Blaizot, J.P, Phys. Rep. 64, 171 (1980);
    Blaizot, J.P., Goguy, B., Grammaticos, B., Nucl. Phys. A265, 305 (1976); see also present proceedings.
13. Gale, C., Bertsch, G., Das Gupta, S., Phys. Rev. C (1987);
    Gale, C., Welke, G.M., Prakash, M., Lee, S.Y., Das Gupta, S., preprint (1989).
14. Weaver, T.A., Zimmerman, B., Woosley, S.E., Astrophys. J. 225, 1021 (1978).
15. Weaver, T.A., Woosley, C.E. and Fuller, G.M., Bull. Am. Astron. Soc. 14, No. 4, 957 (1982).
16. Aufderheide, M., Brown, G.E., private communication.
17. Cooperstein, J., PhD. Thesis, SUNY at Stony Brook, unpublished (1983).
18. Sharma, M.M., Borghois, W.T.A., Brandenburg, S., Crona, S., van der Woude, A., Harakeh, M.N., Phys. Rev. C38, 2562 (1988; see also present proceedings.
19. Gustafsson, H.A. *et al.* , Phys. Rev. Lett. 52, 1590 (1984);
    Stöcker, H., Greiner, W., Phys. Rep. 137, 277 (1986);
    Aichelin, J., Rosenhauer, A., Peilert, G., Stöcker, H., Greiner, W., Phys. Rev. Lett. 58, 1926 (1987).
20. Cooperstein, J., Phys. Rev. C37, 786 (1988);
    Kahana, S.H., Cooperstein, J., Baron, E.A. and Gerdes, D., Phys. Rev. Lett. 60, 68 (1988);
    Prakash, M., Ainsworth, T.L., Lattimer, J.M., Phys. Rev. Lett. 61, 2518 (1988.
21. Glendenning, N.K., Phys. Rev. Lett. 57, 1120 (1986).
22. Kohlemainen, K., Prakash, M., Lattimer, J., Treiner, J., Nucl. Phys. A439, 535 (1985).
23. Baron, E., Bethe, H.A., Brown, G.E., Cooperstein, J., Kahana, S.H., Phys. Rev. Lett. 59, 726 (1987).
24. Bruenn, S., Phys. Rev. Lett. 59, 938 (1987), Astro. J. 340, 955 (1989).
25. Pandharipande, V.R., Wiringa, R.B., Rev. Mod. Phys. 51, 821 (1979);
    Friedman, B., Pandharipande, V.R., Nucl. Phys. A361, 502 (1981).
26. Serot, B.D., Walecka, J.D., Adv. Nucl. Phys. 16, 1-321 (1985).
27. Celenza, J.S., Shakin, C.M. in *Relativistic Nuclear Physics*, World Scientific Pub., Singapore (1985).

28. Bernard, V., Meissner, U., Zahed, I., Phys. Rev. D36, 819 (1987).
29. Kahana, D.E., University of Regensburg preprint (1989).
30. Hillebrandt, W., private communication.
31. Wilson, J.R, in *Numerical Astrophysics*, Eds. J. Centrella, J. Leblanc, R. Bowes, Boston: Jones Bartlett (1985);
Bethe, H.A., Wilson, J.R., Astrophys. J. 295, 14 (1985);
Mayle, R., PhD. Thesis, University of California at Berkeley (1984); unpublished.

SUPERNOVAE AND STELLAR COLLAPSE

James R. Wilson and Ronald W. Mayle

Lawrence Livermore National Laboratory, Livermore, California

Abstract: In this paper model calculations of the collapse of the iron core of a star at the end of its thermonuclear life are presented. The neutrino processes that lead to an explosion are described. Comparisons are made between the model calculations and the observations of SN1987a.

I. Physical and Numerical Model: The numerical model is based on a fully general relativistic treatment of hydrodynamics and neutrino flow. The neutrino time evolution is approximated by a flux limited diffusion equation. The neutrinos are described by three functions, one representing electron neutrinos, another electron antineutrinos and a third to represent muon and tauon neutrinos and their antiparticles; these functions depend on time, space and neutrino energy. All neutrinomatter interactions thought to be important in the collapse and explosion process are included. The equation of state for matter below nuclear density is represented by the species: photons, electrons, positrons, neutrons, protons, helium nuclei and heavy nuclei. The heavy nuclei have properties dependent on the chemical potentials of the other constituents; Saha equations are solved to determine the abundance of all particles. A simple nuclear burn model consisting of He, C, O, Ne, Si, and Ni is used to carry matter up "iron" (nuclear statistical equilibrium); after matter has burned to "iron" it is assumed to be in statistical equilibrium thereafter. The above nuclear density equation of state is discussed in Section V of this paper. In our calculation the densities range from 10^2 to 10^{15} gm/cc and the temperatures range from 10^{-2} to 10^2 MeV. When the star becomes convectively unstable by either the Le Doux or the salt finger criteria, convection is treated in the mixing length approximation.

II. Collapse and Bounce: After the inner part (1.2 – 1.5 solar masses) of a massive star burns to "iron", the core cools by neutrino emission and consequently increases in density due to slow contraction. When densities reach about 10^9 gm/cc electron capture on heavy nuclei and the accompanying energy loss becomes fast enough that the core begins to collapse dynamically. The central part of the iron core contracts homologously with the in fall velocity proportional to the distance from the center of the star. The size of this

The Nuclear Equation of State, Part A
Edited by W. Greiner and H. Stöcker
Plenum Press, New York

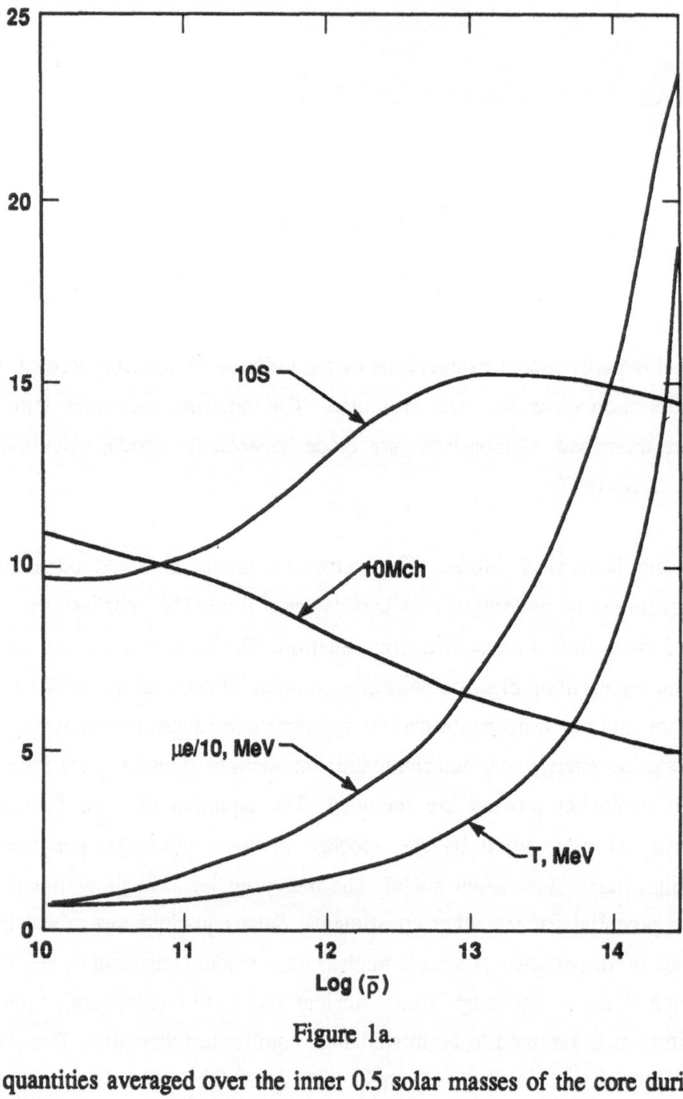

Figure 1a.

Various quantities averaged over the inner 0.5 solar masses of the core during collapse versus the mean density in the inner 0.5 solar masses. a) temperature (T), electron chemical potential (μ_e), Chandrasekhar mass (M_{ch}), and entropy per baryon (S). b) mass fraction of nucleons in heavy nuclei, mass fraction of nucleons in helium, ratio of the number of electrons minus positrons to the number of baryons (Y_e), and the ratio of pressure to internal energy per volume ($P/\rho\varepsilon$). Heavy nuclei and helium vanish at high densities since much of the matter is above nuclear density.

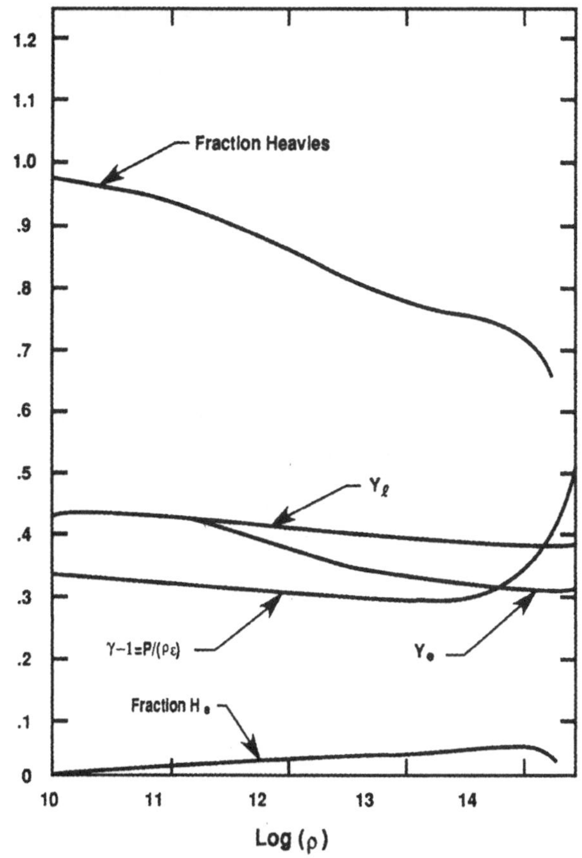

Figure 1b

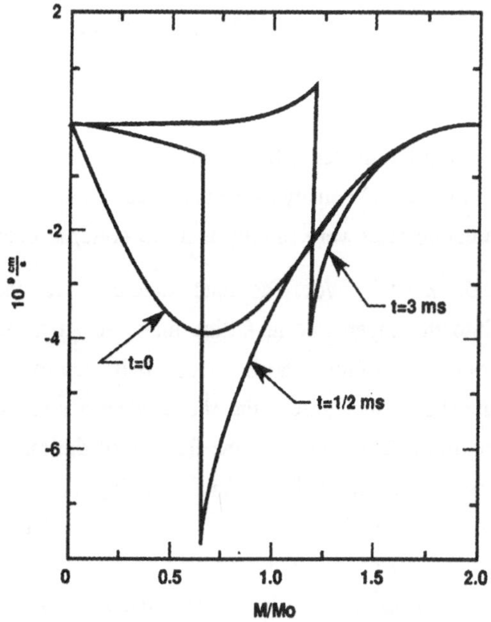

Figure 2

Velocity versus mass for three times near bounce time = 0.0ms.

homologous core is about the Chandrasekhar mass (M_{ch}) = 5.8 Y_e^2 solar masses, where Y_e is the number of electrons per baryon. In what follows we will give results of a collapse calculation using as initial data the evolved core of a 20 solar mass stellar model of SN1987a supplied by T.A. Weaver and S. Woosley (1988). This model had an iron core mass of about 1.45 solar masses.

At the start of the collapse $Y_e \approx .43$ giving M_{ch} = 1.12 solar masses; by the time nuclear density is reached at the center of the core $Y_e \approx .30$ and M_{ch} = 0.53 solar masses. In Figs 1ab various physical quantities averaged over the inner 0.50 solar masses are presented as a function of the mean density in the inner 0.50 solar masses. The entropy per baryon starts at about 1.0 and rises to 1.5 as the density passes from 10^{11} to 10^{12} gm/cc, as can be seen in Figure 1a. The star passes from being transparent to neutrinos for $\rho < 10^{11}$ gm/cc to being diffusive for $\rho > 10^{12}$ gm/cc. In the range $10^{11} < \rho < 10^{12}$ gm/cc the neutrinos are very interactive with the matrbut not able to stay in equilibrium; this leads to the entropy increase. After the density rises above 10^{14} gm/cc the entropy slowly falls due to neutrino cooling. In Figure 1b it is seen that $Y_e \approx Y_l$ (Y_l is the lepton number per baryon) for $\rho < 10^{11}$ gm/cc and ($Y_l - Y_e$) is constant for $\rho > 3 \times 10^{12}$ gm/cc reflecting again the transition of the state of the neutrinos from complete non interaction to complete equilibrium with the matter. The fraction of baryon matter contained in heavy nuclei is .98 at the start of collapse. Due to the rise in temperature and density during the core contraction, this fraction drops to about .75, keeping the pressure lower than it would have been without the break up of the heavy nuclei and allowing the collapse to accelerate.

In Figure 2 the collapse velocity versus radius is given for 3 times. At time = 0.0 ms, the core has attained the maximum in fall kinetic energy of 1.1 x 10^{52} ergs; below the sonic point (the position in the star where the sound speed equals the in fall velocity) the matter is collapsing homologously. The position of the sonic point is close in mass to $M_{ch} \approx .50$ solar masses. Outside the sonic point the matter is falling in supersonically with more than half the free fall velocity. After the central density exceeds nuclear matter density, $\approx$ 2.4 x 10^{14} gm/cc at Y_l = .37, the pressure rises very rapidly and the collapse decelerates; the peak central density reached is 5.6 x 10^{14} gm/cc. The inner sonic core rebounds and starts a pressure wave moving out into the supersonic in falling material. At a mass of about 0.60 solar masses the pressure wave turns into a shock wave; this is seen in Figure 2 at the time of 0.5ms. At the third time shown in Figure 2 the shock wave has proceeded to 1.2 solar masses and is also moving outward in radius. Some fraction of the maximum in fall kinetic energy of 1.1 x 10^{52} ergs should be available for energizing the outward moving shock wave. However, the shock has to proceed through nearly 0.9 solar masses of in falling iron; to dissociate one solar mass of iron requires about 1.65 x 10^{52} ergs of energy. In addition to the energy expended dissociating iron the neutrino luminosity just outside the shock front minus the luminosity at the neutrinosphere (the point in the star inside of which the neutrino mean free path is less than the neutrinosphere radius) is 3.2 x 10^{51} ergs. The combination of energy losses to iron dissociation and neutrino emission weakens the shock so that after a few tens of milliseconds it turns into an almost stationary accretion shock at a radius of about 4 x 10^7 cm; see Figures 3 and 5.

III. Late Time Neutrino Heating

After a few tenths of seconds the first phase of neutrino heating occurs. Figure 4 illustrates the various neutrino luminosities as functions of time. Well outside the neutrinosphere we can approximate the neutrino energy exchange with the matter in the following manner. The heating rate is given by

$$\dot{E}_{+} = K(T_p)\, L / 4\pi R^2 \qquad 1$$

where L is the total luminosity in electron neutrinos and electron antineutrinos. The opacity, $K(T_p)$, is evaluated at the neutrinosphere temperature T_p and is proportional to the square of the temperature. Since the major source of opacity in this phase of neutrino heating is the emission and absorption of electron neutrinos and antineutrinos on free baryons, the opacity is also proportional to the free baryon mass fraction (this is not explicitly shown in the notation). The cooling rate is given by

$$\dot{E}_{-} = - K(T_m)\, a'c\ T_m^4 \qquad 2$$

where T_m is the matter temperature at the point of interest, a' is the radiation constant for Fermi particles (7/8 the photon radiation constant 'a'), and c is the speed of light. If we let $L = 4\pi R_p^2 a'c\ T_p^2/4$ then the net heating rate becomes

$$\dot{E}_{net} = a'\, c\, K(T_p)\, T_p^4 \left[\ (R_p/ 2R_m)^2 - (T_m / T_p)^6\ \right] . \qquad 3$$

Thus if the matter temperature is low enough compared to the neutrinosphere temperature a positive net heating can occur.

At the times under consideration the matter falling through the shock wave is iron; the kinetic energy dissipated in the shock is approximately the gravitational energy since the shock position is nearly constant. So we may write

$$\frac{GM}{R} \approx \left(\frac{3}{2} f + 3\, Y_e \right) kT + f\ I \qquad 4$$

Where the first two terms in the left are the thermal energy of free baryons and electrons; in the last term I is the dissociation energy and f is the degree of dissociation. For our example calculation M ≈ 1.5 solar masses, $Y_e \approx 0.5$ so we have with R_7 equal to the radius in units of 10^7 cm and temperatures and energies measured in MeV

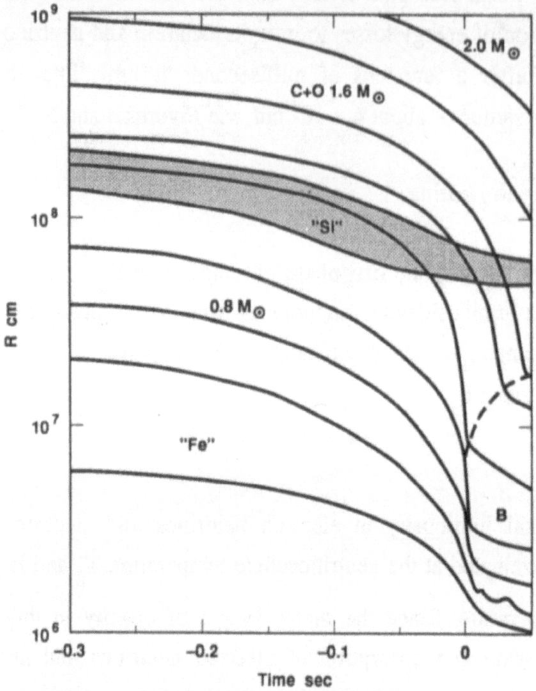

Figure 3

Radius versus time for several mass points. The inner 1.45 solar mass is composed of iron like nuclei. The silicon region is cross hatched. Carbon an Oxygene are above the silicon. The proto–neutron star that will form at about 0.5 sec has a baryon mass of 1.64 solar masses.

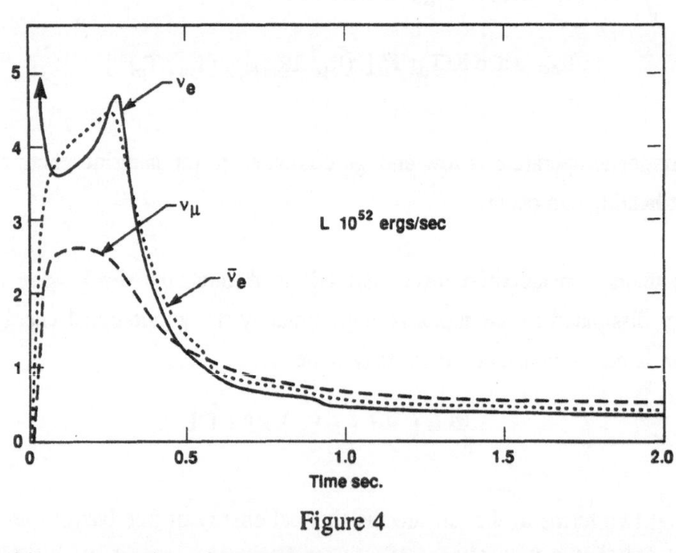

Figure 4

Luminosities versus time for electron neutrinos, electron antineutrinos, and muon neutrinos. Muon antineutrinos tauon neutrinos and tauon antineutrinos have the same luminosity as muon neutrinos.

$$\frac{20}{R_7} \approx 1.5\,(\,1.+f\,)\,T + 8.4\,f\ . \qquad 5$$

At the densities just below the shock (10^8 to 10^9 gm/cc) the decomposition temperature is about 1 to 2 MeV. Above a radius of about 4 x 10^7 cm the material is predominately undecomposed iron and helium which have smaller opacities for neutrino interactions than free baryons; heating will not occur in this region. The heating as a maximum at about 2 x 10^7 cm.

This phase of neutrino heating does not occur immediately after bounce. As time progresses the density of in falling matter decreases which lowers the temperature at which decomposition occurs. Also the proto–neutron star contracts with time and the neutrinosphere temperature increases. These two effects lead to the heating phase being delayed for several tenths of a second after bounce. When neutrino heating occurs the heated matter expands; the amount of material in the heating region decreases due to the expansion thus resulting in a decrease of the heating rate. The heating rate is high from about 0.3 to 0.5 seconds after core bounce; see Figure 5. From Figure 5 we can see that the shock wave begins moving outward rapidly again after 0.5 seconds. The shock wave passes through the iron and into the predominately oxygen region; in this region the shock raises the temperature sufficiently to burn some of the oxygen to nickel. After the shock passes a radius of about 2 x 10^8 cm the density becomes too low and the shock too weak to induce nuclear burn. The amount of Ni^{56}, which is an observable, produced depends directly on the energy deposited in this first phase of neutrino heating. The energy contained in the hot bubble between the neutrinosphere and the shock is still much less than the gravitational binding energy of the matter above the shock. Further energy deposition is needed.

After the hot bubble has expanded a different kind of neutrino heating can occur, neutrino–antineutrino annihilation. This was discussed as an explosion mechanism by Goodman, Dar and Nussinov (1987); Cooperstein , van den Horn and Baron (1987) dismissed neutrino–antineutrino annihilation as being a small effect since the annihilation can only occur near the neutrinosphere and if the density in this region is too high the deposited energy is imediately re–radiated by electron–positron capture on protons and neutrons. However, after the first phase of neutrino heating the density gradient outside the neutrinosphere becomes progressively steeper and eventually becomes sufficiently so that matter on the surface of the proto–neutron star is heated and blown off.

The cross section for neutrino antineutrino annihilation depends on the neutrino energy , ε_ν, and the angle, Θ_ν, between the neutrino and antineutrino in the combination $\varepsilon_\nu^2\,(1 - \cos\,\Theta_\nu)^2$. That is the cross section is proportional to the square of the collision energy in the center of mass frame of the colliding neutrino and antineutrino. In the numerical computer model we carry only the energy density of

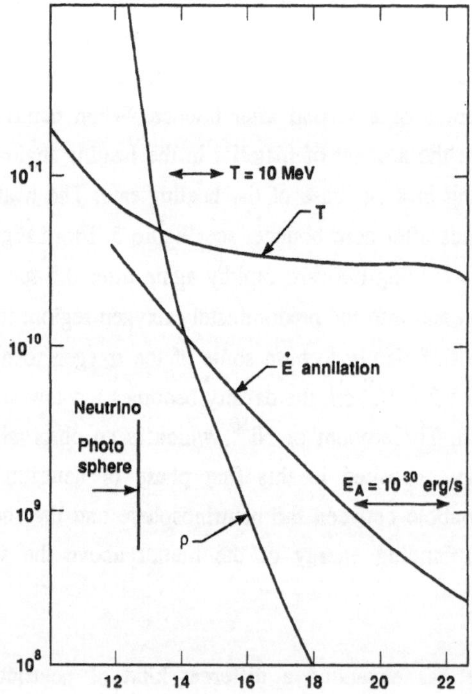

Figure 6

Profile of the proto–neutron star at 2.0 sec after bounce. The annihilation heating rate becomes greater than the electron–positron capture cooling rate at about 14 km.

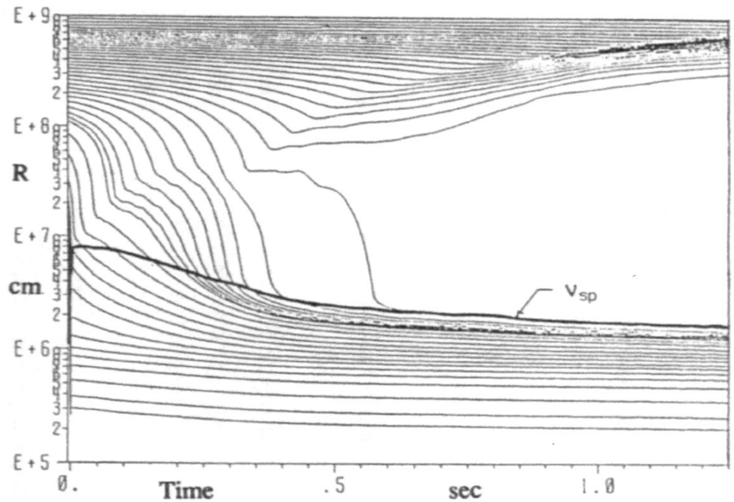

Figure 5

Radius versus time for several mass points post bounce. The neutrinosphere is indicated by ν_{SP}. Accretion of in falling material continues up to .35 sec after bounce. The first neutrino heating phase occurs during the interval of .35 to 55 sec. The second neutrino heating phase begins near 1.0 sec.

neutrinos, F_0, as a function of the neutrino energy. We must infer the angular distribution indirectly. In the diffuse limit we approximate the angular distribution by

$$F(R,\varepsilon,\mu) = F_0(R,\varepsilon) + \frac{3D\mu}{c}\frac{\partial F_0(R,\varepsilon)}{\partial R} \qquad 6$$

where D, the flux limited diffusion coefficient, depends on $\lambda\left|\frac{\partial \log F_0}{\partial R}\right|$ with λ being the neutrino mean free path (see Bowers and Wilson 1982) and μ the cosine of the angle of the neutrinos with respect to the radial direction. Integration of the cross section over μ gives the angular factor

$$Q_1 = 1 - \frac{3}{2}\frac{D\bar{D}}{c^2}\left|\frac{\partial \log F_0}{\partial R}\right|\left|\frac{\partial \log \bar{F}_0}{\partial R}\right| \qquad 7$$

where the bars over D and F_0 denote the flux limited diffusion coefficient and the distribution function for the antineutrinos. In the limit of an infinitely sharp neutron star boundary the angular integration gives a factor

$$Q_2 = (1 - x)^2(5 + 4x + x^2)/8 \qquad 8$$

with $x = \sqrt{1 - (R^*/R)^2}$ where R^* is the radius of the neutron star which we will take as the neutrinosphere radius. We use the larger of Q_1 and Q_2 for our angular factor; in the limit where Q_2 dominates the net heating per unit volume goes as $(1/R)^8$ for large R.

In Figure 6 the structure near the neutrinosphere is shown for the star at a time of 0.9 sec after bounce. The second phase of neutrino heating is just getting underway. At a density a little under 10^{10} gm/cc the neutrino–antineutrino annihilation energy deposition becomes greater than the electron–positron annihilation energy loss to neutrino production. The energy deposition rat is not high, only a few times 10^{50} ergs/sec, but it will continue as long as the proto–neutron star is emitting energy in neutrinos.

An additional neutrino process also occurs at late times when matter in the bubble region has heated sufficiently that most of its internal energy is in photons and electron and positron pairs. As soon as this happens we may write

$$E = \frac{11}{4}\frac{aT^4}{\rho} \qquad 9$$

$$n_p = \frac{7}{4}\frac{aT^3}{3k} \qquad 10$$

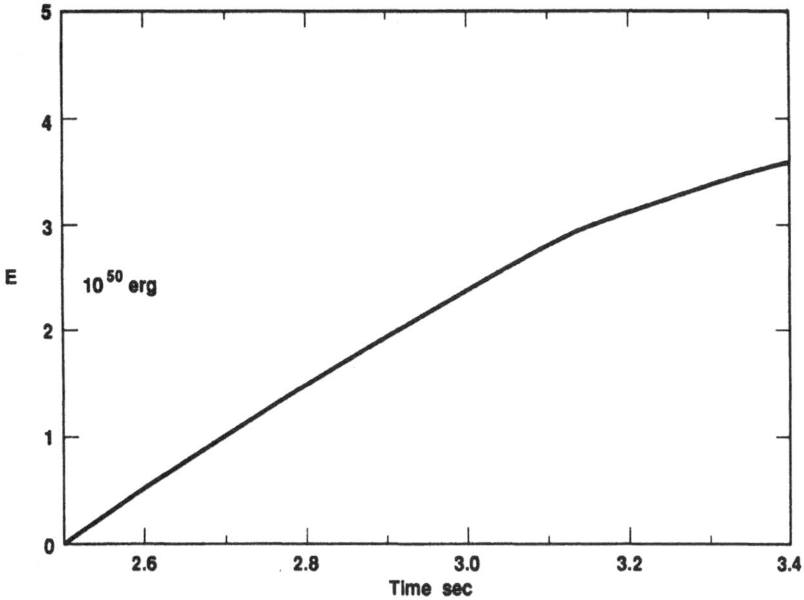

Figure 7

Net energy versus time of all the matter above the lowest radius where the matter's internal energy is greater than its gravitational binding energy.

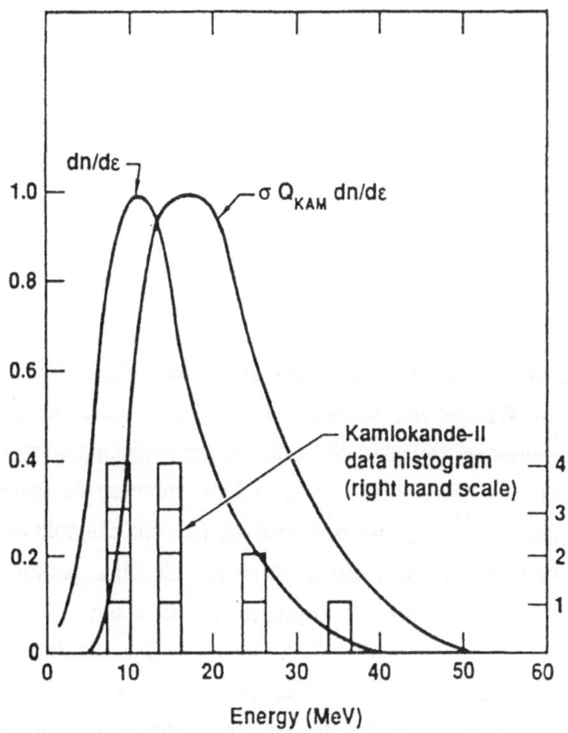

Figure 8a.

$$\sigma_e = \frac{5}{4}\sigma_0 \frac{\varepsilon_\nu T}{(m_e c^2)^2} \qquad 11$$

$$\dot{E} = \sigma_e n_p L / 4\pi R^2 = \frac{11}{4} a\dot{T}^4 \qquad 12$$

$$\frac{\dot{T}^4}{T^4} = \frac{5}{12}\frac{\sigma_0}{k}\frac{\varepsilon_\nu}{(m_e c^2)^2}\frac{L}{4\pi R^2} \equiv \frac{1}{\tau} \qquad 13$$

where a is the photon radiation constant, k is Boltzmanns constant, c is the speed of light, m_e is the mass of an electron, $\sigma_0 = 1.7 \times 10^{-44}\ (\mathrm{cm})^2$ is a fundamental weak interaction cross section (see Tubbs and Schramm 1975), ε_ν is the neutrino energy, n_p is the number of electron–positron pairs, L is the neutrino luminosity, E is the internal energy of matter per gram and R is the radius. At a time of 2.5 sec after bounce at a radius of 10^7 cm $\tau = .5$ sec. This pair heating does not produce much total energy deposition but it takes matter heated by the other two neutrino heating processes and raises the entropy per baryon to a few thousand. The calculation was carried out to a time of 3.6 sec. after bounce. At that time the density near 10^7 cm had fallen to about 100 gm/cc.

The calculation becomes very expensive in computer time so it was not completed. We extrapolate the energy production rate to estimate the final explosion energy. In Figure 7 the net energy above the neutrinosphere is plotted versus time. Note that not until 2.5 sec after bounce is there a net explosion energy. At 3.6 sec about two thirds of the final binding energy of the neutron star has been emitted in neutrinos that have escaped the star. If we assume that the energy deposition efficiency for the remaining energy to be emitted in neutrinos is the same as it is in the interval from 2.5 to 3.6 seconds then the final explosion energy should be 1.4×10^{51} ergs.

IV. Comparison of Calculations and Observations

As stated earlier the calculation described above was based on a stellar evolution calculation made by Weaver and Woosley for a 20 solar mass star that is thougt to be similar to the progenitor of SN1987a. We estimate the explosion energy to be about 1.4×10^{51} ergs. From the immediate post explosion light curve an estimate of the explosion energy of 0.6 to 2.0×10^{51} ergs has been derived (see for example Shigeyama, Nomoto and Hashimoto 1988 and Woosley 1988). From the late time photon luminosity of the suppernovae the Ni^{56} production is estimated to be about .075 solar masses. Our calculation gave .065 solar masses of Ni^{56}. At the end of our calculation the baryonic mass of the neutron star is 1.63 solar masses and then calculated the neutrino emission assuming the star is in quasi static equilibrium. To check the mean energy of the electron antineutrinos we folded our energy spectrum with the detector efficiencies of the IMB detector (see Bionta et al. 1987) and the Kamiokande–II detector (see Hirata et al. 1987). Since each detector only saw a few events we grouped the observed events into histograms. Our mean neutrino energy weighted by the detector efficiencies falls midway (see Figures 8ab) between the average event energies of the two detectors. To examine the temporal emission of the neutrinos we combined the events of the two detectors with equal weight (we counted an IMB event as 11/8 of a Kamiokande–II event since IMB saw 8 events and Kamiokande–II saw 11). In Figure 9 we compare the calculated emission folded with an average detector efficiency (to take into account the two different detector efficiencies) and the combined observational emission. Thus we have good overall agreement of observation and calculation. The final neutron star gravitational mass we estimate to be 1.45 solar masses; however, this is not observable.

V. Equation of State Considerations

In Section I of this paper the subnuclear density equation of state we use is described. Above nuclear density we use a zero temperature equation of state plus a thermal component. The zero temperature nuclear EOS was suggested by H. A. Bethe 1987) who used results described in Muther, Prakash and Ainsworth (1987) (and references therein). The internal energy per baryon, E_0 (in units of MeV), of the zero temperature component of our EOS is taken to be

$$E_0 = -16 + \frac{1}{9} K_0 \left(\eta^{\gamma} - 1 + \gamma(\eta - 1)\right) / \left(\eta\, \gamma (\gamma - 1)\right) + E_{SYM} \qquad 14$$

$$E_{SYM} = 16\,(1 - 2\,Y_e)^2\, \eta\, (1 + 72/(1 + 4\,\eta)) \qquad 15$$

where K_0 = 200 MeV, γ = 2.75, and $\eta = \rho/\rho_N$ with $\rho_N = 2.656 \times 10^{14}$ gm/cc.

Using 14 and 15 to compute the pressure, $P_0 = \rho^2\, \partial E_0/\partial\rho)$, a cold neutron star can be constructed for any central density (also assuming $Y_e \approx 0$.) We find that 14 and 15 produce a maximum baryon mass for a stable neutron star of 1.93 solar masses (the gravitational mass is 1.68 solar masses). For a cold neutron star of total baryon mass 1.63 solar masses 14 and 15 give a gravitational mass of 1.46 solar masses, a central density of 1.45×10^{15} gm/cc and a binding energy of 3.04×10^{53} ergs. The zero temperature component of our supranuclear EOS does have a superluminal sound speed for high enough density, but for the highest density reached in the model calculation we present in this paper, the sound speed is subluminal.

The free energy of the thermal part of the supranuclear EOS is taken to be

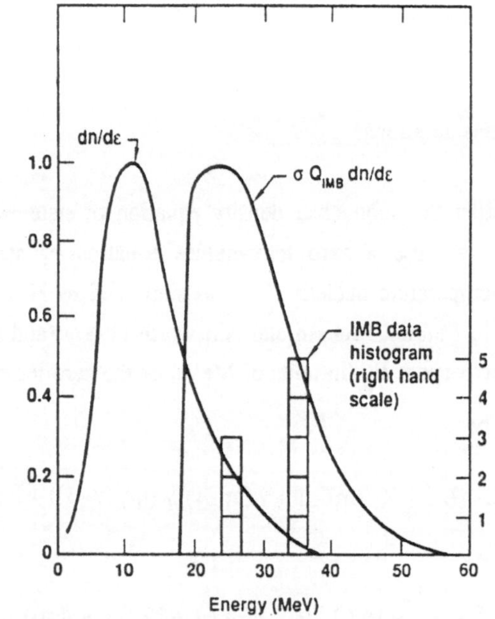

Figure 8b

Relative antineutrino spectra. dn/dε is the electron antineutrino spectrum from the computer model normalized so that the maximum is unity. σ is the antineutrino capture cross section for electron antineutrino capture on protons. Q is the detector efficiency. The observed detected events are put into a histogram for visualization. a) Kamiokande. b) IMB.

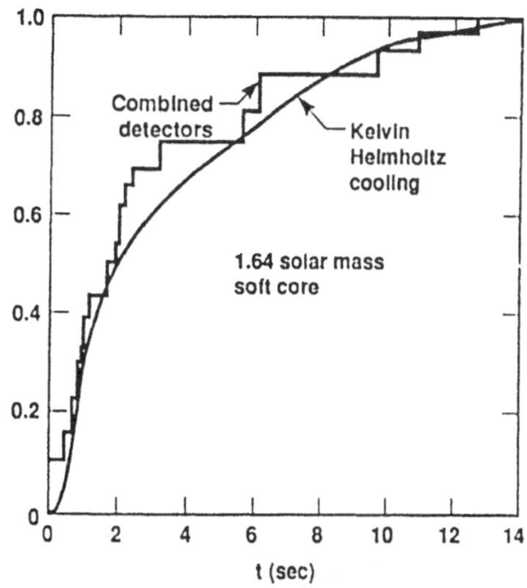

Figure 9

Cumulative counts versus time. Stepped line is formed by combining the events from the Kamiokande and IMB detectors weighted inversely by their respective number of detections. Smooth line is from a kelvin–Helmholtz cooling calculation of a 1.63 solar mass proto–neutron star.

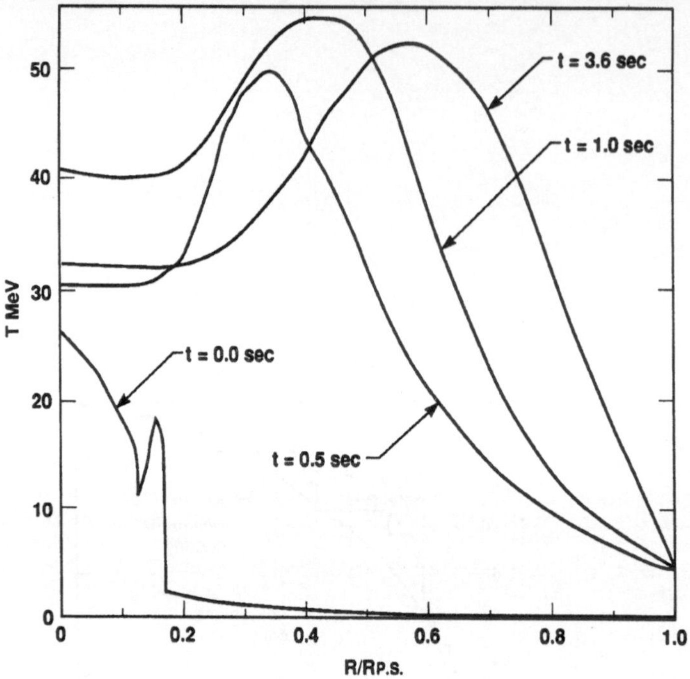

Figure 10a

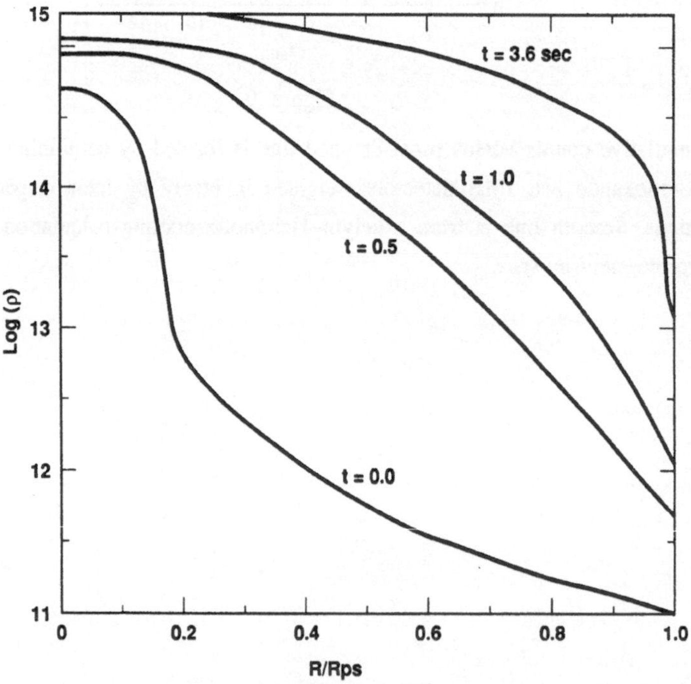

Figure 10b

$$F = \frac{3}{2} kT \left[\sqrt{1+Y^2} - Y - \log\left((1 + \sqrt{1+Y^2}) / (.93Y) \right) \right] \qquad 16$$

$$Y = \left(\frac{3}{20} \frac{1}{m_B^{5/3}} \left(\frac{3}{8\pi} h^3 \right)^{2/3} \right) \frac{\rho^{2/3}}{kT} \qquad 17$$

where m_B is the baryon mass, k is Boltzmanns constant and h is Plancks constant. The form of the free energy is an approximation to a noninteracting gas of nonrelativistic fermions. The contributions of electrons, positron and photons to the free energy are added to the nuclear component.

In Figures 10a–f the density ρ, temperature T, entropy per baryon S, the ratio of the number of electrons minus positrons to the number of baryons Y_e, electron chemical potential μ_e, and the neutron minus proton chemical potentials $\hat{\mu}$ are presented for several times after bounce. At the later times the thermodynamic conditions are not far from those inferred from hydrodynamic modeling of heavy ion collisions at the Bevalac energies.
The presence of pions, which is ignored in the present calculations, will make an appreciable change in the EOS. Estimates using a model similar to that described in the work of Friedman, Pandharipande and Usmani (1981) show that a lowering of the peak temperature seen in our model calculations by 10 MeV could be expected. The neutrino opacity would increase by as much as a factor of two if pions were included in the EOS. The biggest effect expected is from the reduction of the electron density by the conversion of electrons to negative pions. The effect of this latter process on the cooling of the proto–neutron star is hard to estimate at present. We have made only one calculation spanning a long time after the core bounce, but we think the neutrino heating process will work for equations of state not greatly different from the one described in this paper.

Acknowledgments
This work was performed under the auspices of the USDOE at Lawrence Livermore National Laboratory under contract no. W–7405–ENG–48

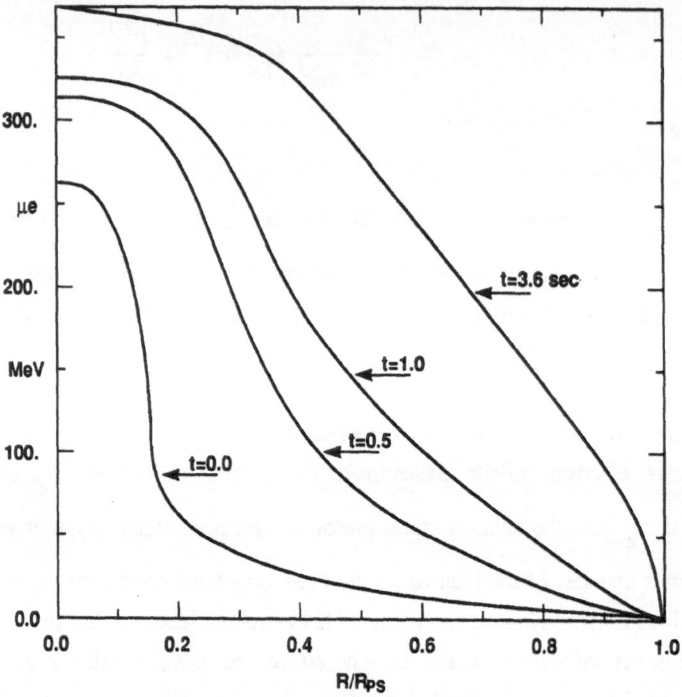

Figure 10c

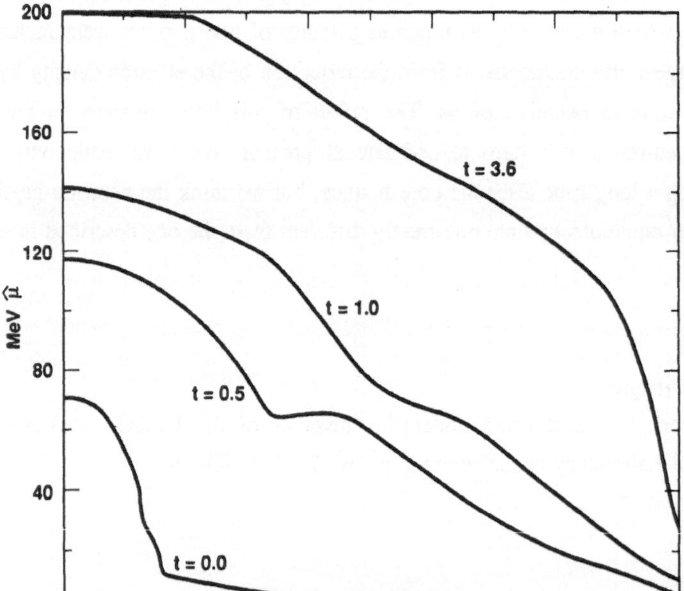

Figure 10d

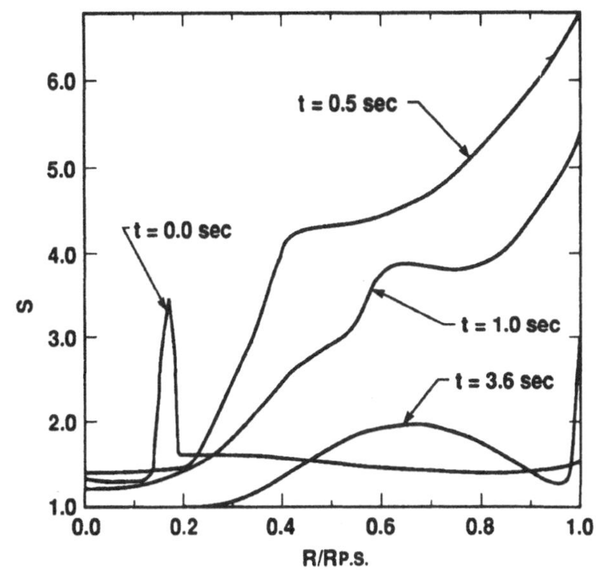

Figure 10e

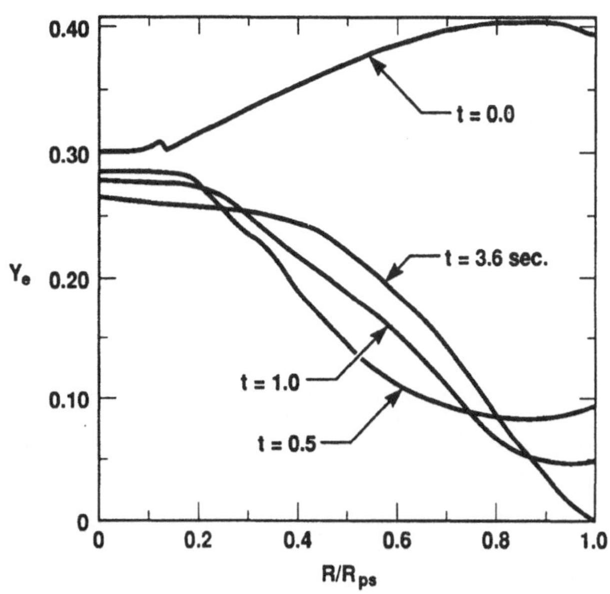

Figure 10f

Various physical quantities versus the ratio of radius to the neutrinosphere radius at four times. The mass out to the neutrinosphere radius is 1.0 solar masses at t=0 sec, increases to 1.63 solar masses at 0.5 sec and then decreases to 1.61 solar masses at 3.6 sec. a) density. b) temperature. c) electron chemical potential. d) neutron minus proton chemical potential. e) entropy. f) electron to baryon ratio.

References

Bethe (1987) private communication

Bionta et al. 1987 Phys. Rev. Lettr. 58, 1494

Bowers and Wilson 1982, A. J. Suppl. 50,115

Cooperstein, van den Horn and Baron 1987, Ap. J., 309,653

Friedman, Pandharipande and Usmani 1981, Nuc. Phys. A 372, 483

Goodman, Dar and Nussinov 1987, Ap. J., 314, L10

Hirata et al. 1987, Phys. Rev. Lettr. 60, 1999

Muther, Prakash and Ainsworth 1987. Phys. Lettr. B, 199, 469

Shigeyama, Nomoto and Hashimoto 1988, Astron. Astrophys. 196, 141

Tubbs and Schramm 1975, Ap. J., 201, 467

Weaver and Woosley 1988, private communication

Neutron Stars, Fast Pulsars, Supernovae and the Equation of State of Dense Matter

Norman K. Glendenning

Nuclear Science Division
Lawrence Berkeley Laboratory
1 Cyclotron Road
Berkeley, California 94720

1 Preliminaries

Astrophysical constraints on the equation of state are imposed by three sources, (1) masses of neutron stars, (2) rotational frequencies of very fast pulsars (neutron stars), and (3) supernovae. Of the three, the theoretical and observational status of the first is most secure. For the second, a half millisecond pulse over an eight hour period was reported in January but not seen again two weeks later at the next turn at the telescope, nor at any time since. If substantiated it would be the fastest known pulsar. If it can be interpreted as neutron star rotation, then theory seems to indicate that it provides very stringent conditions on the equation of state when combined with the mass constraint. The physics of supernova involves so many factors of comparable importance but high uncertainty, that they cannot be said to provide any constraint at the present time. Indeed the consensus is that it is not known whether supernovae can explode by the prompt bounce mechanism [1, 2]. no simulations whose physics has been agreed upon have been successful and earlier claims to success of the prompt explosion of $1.35M_\odot$ iron core progenitors have been later withdrawn [3, 4]. The delayed explosion mechanism produces only small explosions so far, and may be the means by which some stars explode, but not evidently the one by which all type II stars explode.

I shall talk mostly about slow neutron stars, ones for which the rotational energy is very small compared to the binding energy and can be ignored. This is the case for almost all pulsars, but I will report the current situation with respect to fast pulsars.

Nuclear and astrophysics are connected through Einstein's theory of general relativity. The field equations can be obtained through a variation of a generalized action involving the metric of space-time and the matter fields. They are,

$$G_{\mu\nu} = -8\pi T_{\mu\nu}$$
$$\frac{\delta\mathcal{L}_m}{\delta\phi} - \partial_\mu \frac{\delta\mathcal{L}_m}{\delta(\partial_\mu\phi)} = 0, \quad \text{each matter field } \phi \tag{1}$$

Here $G_{\mu\nu}$ is Einstein's curvature tensor, a function of the space-time metric functions $g_{\mu\nu}$ and $T_{\mu\nu}$ is the matter stress-energy tensor, derivable from the matter Lagrangian $\mathcal{L}_m$. We don't usually think of the matter Lagrangian and field equations as depending on the metric

The Nuclear Equation of State, Part A
Edited by W. Greiner and H. Stöcker
Plenum Press, New York

because we usually ignore gravity and write $g_{\mu\nu} = (1, -1, -1, -1)\delta_{\mu\nu}$. In strong gravitational fields this is not true, and we need to justify it for nuclear astrophysics. This we now do. In the process we become familiar with gravitational units, and get a qualitative description of neutron stars.

In empty space outside a static spherical star of radius R and mass M, Schwarzschild showed that the solution of Einstein's equations have a simple form. All but the diagonal components of the metric vanish, and they are simple. The line element is

$$ds^2 = \left(1 - \frac{2M}{r}\right)dt^2 - \left(1 - \frac{2M}{r}\right)^{-1} dr^2 - r^2 d\theta^2 d\theta^2 - r^2 \sin^2\theta d\phi^2, \quad (r > R) \tag{2}$$

I want to show you that the metric functions, g_{tt}, g_{rr} in front of dt^2 and dr^2 change by an infinitesimal amount over the distance between nucleons in a star that is near the limit of collapse to a black hole. We will see that such a star is what is referred to as a neutron star. To do this I need to show you how to compute in gravitational units $G = 1 = c$.

$$\begin{aligned} 1 = c &= 3 \times 10^{10} \text{ cm/s} \\ 1 = G &= 6.7 \times 10^{-8} \text{ cm}^3 \text{ g}^{-1} \text{ s}^{-2} \end{aligned} \tag{3}$$

These can be treated as equations so that for example,

$$\begin{aligned} 1\, s &= 3 \times 10^{10} \text{ cm} \\ 1\, g &= 7.4 \times 10^{-29} \text{ cm} \\ 1\, s^{-2} &= 1.5 \times 10^{7} \text{ g/cm}^3 \\ 1 \text{ erg} = 1 \text{ g cm}^2 \text{ s}^{-2} &= 8.2 \times 10^{-50} \text{ cm} \end{aligned} \tag{4}$$

Next we estimate the mass and radius of a star near the limit. Notice that the metric becomes singular at $r = 2M$. For actual stars, this radius is interior to the star itself where the Schwarzschild solution does not hold, but in the special case where the star lies within the "gravitational radius", it must be a black hole. Let us estimate the properties of a star near the limit, $R = 2M$. Assume that gravity packs nucleons up to their hard cores, say $r_0 \approx 0.5 \times 10^{-13}$ cm. Then

$$R \approx r_0 A^{1/3}, \quad M \approx Am \tag{5}$$

where A is the number of baryons in the star and m is their mass

$$m = 940 \text{ MeV} = 1.7 \times 10^{-24} \text{ g} = 1.3 \times 10^{-52} \text{ cm} \tag{6}$$

Hence substituting eq.(5) into the equation, $R = 2M$ we find,

$$A^{2/3} = r_0/(2m) = 1.9 \times 10^{38} \tag{7}$$

Putting this answer back into the expressions for radius and mass we have

$$\begin{aligned} A &= 2.6 \times 10^{57} \\ R &= r_0 A^{1/3} = 7 \text{ km} \\ M = R/2 &= 3.5 \text{ km} = 2.3 M_\odot \end{aligned} \tag{8}$$

where I used the solar mass,

$$M_\odot = 2 \times 10^{33} \text{ g} = 1.5 \text{ km} \tag{9}$$

So here we have an estimate of the baryon number, radius and mass of a star at the limit. We expect a slightly smaller mass and larger radius than the the values given by the Schwarzschild relation, say,

$$M \approx 2M_\odot, \quad R \approx 10 \text{ km} \tag{10}$$

The average density of such an object is

$$\bar{\rho} = 9.7 \times 10^{14}\ \text{g/cm}^3 \approx 4\rho_0 \tag{11}$$

where $\rho_0 = 2.4 \times 10^{14}$ g/cm^3 is the density of symmetric matter at saturation. Since the density of the star near the limit is supernuclear and since it must be charge neutral else the repulsive Coulomb force will overwhelm gravity, it will be dominated by neutrons and is called a neutron star. Putting the mass and radius into the metric we have,

$$\frac{g_{rr}(R)}{g_{rr}(0)} = \left(1 - \frac{2M}{R}\right)^{-1} = \left(1 - \frac{6}{10}\right)^{-1} = 2.5 \tag{12}$$

So the metric changes by a small amount over the dimension of the star. It changes by $2r_0/R = 2A^{-1/3} \approx 10^{-19}$ of this over the spacing of nucleons in the star. Later we shall also be interested in pulsars with very high angular velocity. Since the gravitational attraction must exceed the centrifugal repulsion else the star would fly apart, we are assured that the curvature of space-time due to rotation must be even less than that due to the mass, which we just saw is completely negligible in any local frame in a region spanning the distance between many nucleons. So in solving the field equations for matter, we make negligible error for neutron stars by solving them in the absence of gravity and then using the resulting stress-energy tensor, which is diagonal in a co-moving frame,

$$T_{\mu\nu} = (\epsilon, p, p, p)\delta_{\mu\nu} \tag{13}$$

in Einstein's field equation to find how matter is compacted under the influence of gravity.

In the special case of a static star Einstein's equations take a special form first written down by Oppenheimer and Volkoff.

$$4\pi r^2 dp(r) = \\ \frac{GM(r)dM(r)}{r^2}\left(1 + \frac{p(r)}{\epsilon(r)}\right)\left(1 + \frac{4\pi r^3 p(r)}{M(r)}\right)\left(1 - \frac{2GM(r)}{r}\right)^{-1} \tag{14}$$

$$dM(r) = 4\pi r^2 \epsilon(r)\, dr \tag{15}$$

The interpretation is very simple. Think of a shell of matter in the star of radius r and thickness dr. The second equation gives the mass energy in this shell. The pressure of matter exterior to the shell is $p(r)$ and interior to it $p(r) + dp(r)$. The left side of the first equation is the net force acting outward on the surface of the shell by the pressure, and the first term on the right side is the attractive force of gravity acting on the shell by the mass interior to it in Newton's theory. The remaining three factors are the exact corrections for general relativity. So these equations express the balance of internal pressure and gravity. The equation of state $p = p(\epsilon)$ is the manner in which matter enters the equations of star structure. Otherwise they are completely specified and their correctness is confirmed by the observational tests of general relativity.

They can be integrated from the origin with the initial conditions that $M(0) = 0$ and an arbitrary value for the central energy density $\epsilon(0)$, until the pressure, $p(r)$, becomes zero. That point, R, defines the radius of the star, and $M(R)$ its mass. For the given equation of state, there is a unique relationship between the mass and central density, $\epsilon(0)$. So for each possible equation of state there is a unique family of stars, parameterized by, say, the central density. Several such families are shown in Fig.1 for different values of the nuclear compression. It will be noted that each family has a maximum mass star, called the limiting mass and that the central density of the limiting mass star is higher the softer the equation of state. The part of the curve for which the slope is positive corresponds to stable configurations. For negative slope, one can readily verify that the star is unstable to radial perturbations. In fact those beyond the maximum are unstable to collapse to black holes. It is in the limiting mass that a constraint on the equation of state arises. Obviously an acceptable equation of state must have a limiting mass at least as large as the largest observed mass.

2 Why pulsars are neutron stars

About 400 pulsars have been found since the first discovery in 1967 in the pulsed signals of a radio-telescope. The period of the pulses range from milliseconds to seconds, and is interpreted as the period of a rotation. Why? Ordinary stars have magnetic fields (~ 100 gauss) and rotate. When they collapse from a radius of 10^6 km to 10 km, both the rotation frequency and field are scaled up by the conservation laws of angular momentum and magnetic flux. The field is typically scaled to 10^{12} gauss. There is other evidence of such strong fields. The remnant of the crab pulsar is still accelerating with an apparent energy input of $\sim 10^{38}$ ergs/s, and the most likely source of input energy is the absorption of magnetic dipole radiation from the fast pulsar within it. Energy balance implies about the same strength for the magnetic field as quoted above,

$$10^{38}\text{ergs/s} = -\frac{dE}{dt} = -\frac{d}{dt}(\frac{1}{2}I\omega^2) = \frac{1}{6}R^6B^2\omega^4\sin^2\alpha \tag{16}$$

Given the observed period and rate of change of period,

$$T \sim \frac{1}{30}\text{ s}, \quad \dot{T} \sim 4 \times 10^{-13}\text{ s/s} \tag{17}$$

we find (taking $\sin\alpha = 1$),

$$B \sim 4 \times 10^{12}\text{ gauss}, \qquad I \sim 2 \times 10^{44}\text{ g cm}^2$$

$$E_{rot} \sim \frac{1}{2}I\omega^2 \sim 4 \times 10^{48}\text{ ergs} \tag{18}$$

where I use 3.5×10^{24} gauss cm $= 1$ in gravitational units. The field will in general be oriented in a different direction than the rotation axis, say by an angle α. It is believed, but not understood, that radiation over a broad band of frequencies is emitted within some angular spread along the magnetic axis. Given the rotation, one has a beacon which we see as pulses as the star rotates.

We can use the period of rotation to estimate an average density of a millisecond pulsar. For the star to hold together under the opposing forces of gravity and centrifuge, we must have,

$$\frac{GmM}{R^2} > m\omega^2R \tag{19}$$

Hence for the average density,

$$\bar{\rho} = M/\left(\frac{4\pi}{3}R^3\right) > \frac{3\pi}{T^2} = 1.4 \times 10^{14}\text{ g/cm}^3 \tag{20}$$

where the last equality holds for a millisecond pulsar. Since $\rho_0 \approx 2.4 \times 10^{14}\ g/cm^3$, we learn that the average density of a pulsar has to be about as large or larger than nuclear density and so pulsars must be neutron stars. In actual models that I have studied, $\bar{\rho}/\rho_0$ ranges from 3 for stiff equations of state to 5 for the standard compression found a decade ago in the analysis of the giant monopole resonance in nuclei ($K = 210$ MeV) to 6 or more for softer equations of state. Clearly the stiffer the equation of state the more sensitive the neutron star limiting mass will be to properties near saturation.

3 Nuclear and neutron star matter

The idealized matter of the interior of nuclei and the matter of neutron stars have similarities and differences which need to be understood. The similarities include the fact that they are composed of hadrons and the densities are the same within an order of magnitude. The differences arise from two facts. (1) Nuclei are bound by the charge symmetric nuclear force, but neutron stars are bound by gravity. Hence in nuclei, $N \approx Z$. However since the repulsive Coulomb force is so much stronger than the gravitational force that binds stars, the net charge in a star must be very small,

$$\frac{(Z_{net}e)e}{R^2} < \frac{G(Am)m}{R^2} \longrightarrow Z_{net} = Z_p + Z_e < 10^{-36}A \tag{21}$$

(where we use $\hbar c/e^2 = 137$ and $\hbar = 1.1 \times 10^{-27}$ g cm^2 s^{-1} = 2.6×10^{-66} cm^2.) Although Z is not nearly so small as Z_{net}, it is considerably smaller than $A/2$. So nuclei are symmetric and neutron stars are asymmetric. (2) There is another and profound difference arising from the weak interaction time scale $\tau_w \sim 10^{-10}$ seconds. Because of the high density of neutron stars and the fact that baryons obey the Pauli principle, it is energetically favorable for nucleons at the top of the Fermi sea to convert to other baryons, including strange ones (hyperons). This is possible because strangeness is conserved only on the strong interaction time scale, not on the weak. Even the time scale of supernova is long compared to the weak. So strangeness is not conserved in astrophysical objects. It wouldn't be conserved in stable nuclei either, but energetically it is not favorable to have hyperons in the ground state. Nuclear reactions on the other hand are so fast that strangeness is conserved. So the matter studied in nuclei or their reactions has zero net strangeness, whereas neutron stars can and almost certainly do contain hyperons.

These are the differences. Of course the properties of such systems as the hot symmetric non-strange matter produced in relativistic nuclear collisions and the cold asymmetric charge neutral and strangeness carrying matter of neutron stars are related in any comprehensive theory of matter. It is through relativistic nuclear field theory that I shall make the connection between them. This theory can be generalized to incorporate nucleons and higher mass baryon states, interacting through exchange of mesons[5]. Its coupling constants can be fixed by properties of symmetric nuclear matter. It describes numerous properties of finite nuclei[6]. It can be extended to finite temperature[7]. It can be extrapolated to hot dense matter and its composition (nucleons, hyperons, deltas, mesons)[5]. *One and the same theory* with fixed coupling constants describes:

1. Symmetric nuclear matter and the matter produced in high energy collisions when the field equations are solved subject to the constraints of *isospin symmetry and strangeness conservation.*

2. Neutron star matter when the field equations are solved subject to the constraints of *charge neutrality and generalized equilibrium.*

Therefore we are able to conveniently characterize the neutron star matter equation of state by the compression modulus of the corresponding symmetric matter, and shall always do so.

Neutron stars are not pure in neutron as their name implies, and as they were first thought of. Charge neutrality is automatically respected by pure neutron matter, but this is not the lowest energy state of dense neutral matter. The reason is that as the density of neutron matter is increased, the Fermi energy of the neutron would soon equal the energy of a proton, electron and neutrino. At this point beta decay occurs. The neutrino and any gamma ray leaks out of the star, thus lowering its energy. As the density is further increased, other thresholds are reached. In a Fermi gas model the thresholds are found from the masses of the particles. In general they depend on the interactions as well. The isospin symmetry energy arising from the coupling of baryon isospin to the neutral rho meson is very important in this respect. Obviously it favors conversion of neutrons to baryons of opposite isospin projection, consistent of course with charge neutrality. For these reasons neutron star matter is very complex and the Lagrangian used in nuclear field theory has to be generalized to include these complications[9]. Fig.2 shows the result of such a general calculation for the populations of neutron star matter. For low density, the charge neutral uniform matter is

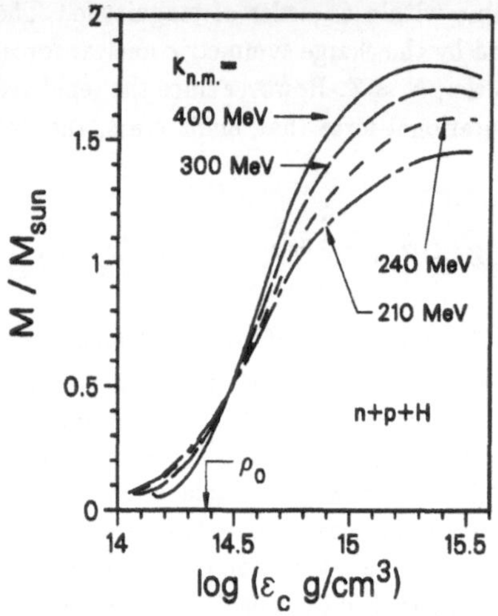

Figure 1. Families of neutron stars corresponding to equations of state with different compression modulus.

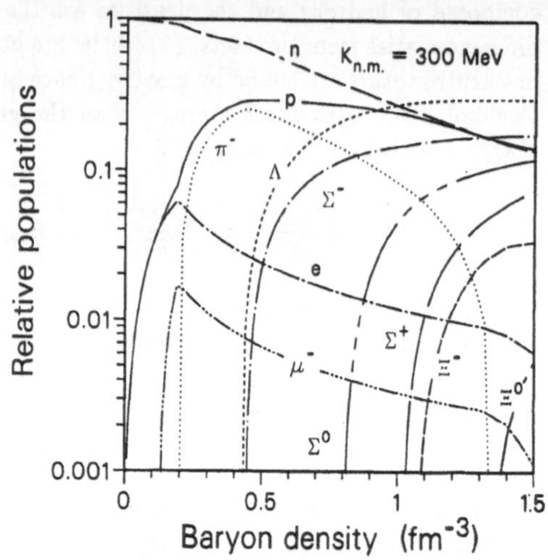

Figure 2. Relative populations in neutron star matter as a function of density [8].

pure in neutron, but with increasing density the proton and electron in equal numbers are populated; when the electron Fermi energy increases to the muon mass then the muon as well will be populated. The pion, as we discuss later, may also condense and then at densities beginning at about three times nuclear, hyperon thresholds are reached, and with further increase in density become important components of neutron star matter. The Lagrangian employed was,

$$\begin{aligned} \mathcal{L} &= \sum_B \overline{\psi}_B(i\gamma_\mu\partial^\mu - m_B + g_{\sigma B}\sigma - g_{\omega B}\gamma_\mu\omega^\mu - \tfrac{1}{2}g_{\rho B}\gamma_\mu\tau_3\rho_3^\mu)\psi_B \\ &+ \tfrac{1}{2}(\partial_\mu\sigma\partial^\mu\sigma - m_\sigma^2\sigma^2) - \tfrac{1}{4}\omega_{\mu\nu}\omega^{\mu\nu} + \tfrac{1}{2}m_\omega^2\omega_\mu\omega^\mu \\ &- \tfrac{1}{4}\rho_{\mu\nu}\cdot\rho^{\mu\nu} + \tfrac{1}{2}m_\rho^2\rho_\mu\cdot\rho^\mu - \tfrac{1}{3}bm_n(g_\sigma\sigma_0)^3 - \tfrac{1}{4}c(g_\sigma\sigma_0)^4 + \cdots \\ &+ \sum_\lambda \bar{\psi}_\lambda(i\gamma_\mu\partial^\mu - m_\lambda)\psi_\lambda \end{aligned} \tag{22}$$

The first line is the sum of baryon Lagrangians and the interactions with the scalar, vector and vector-isovector mesons (σ, ω, ρ). The second line contains the Lagrangians of the scalar and vector mesons, whose interactions with the baryons give rise respectively to attraction and short range repulsion. The third line contains the Lagrangian for the isovector meson which couples to the isospin of baryons and gives rise to the charge symmetry energy. This line also contains self-interaction terms of the scalar field. The last line contains the Lagrangians for the leptons (electrons and muons) which are important agents in the charge neutrality of neutron star matter. The sum over baryons is over the charge states of nucleons, deltas and hyperons. It is of course essential to keep track of the individual charge states so that charge neutrality can be enforced. The five coupling constants appearing in the Lagrangian are fixed by properties at saturation, namely the binding energy $B = 16.3$ MeV, density, $\rho_0 = 0.153$ fm^{-3}, symmetry energy coefficient $a_{sym} = 32.5$ MeV, the effective mass at saturation, $m_{eff} \approx 0.8m$ and $K = 200 - 300$ MeV. The last quantity seems to be the main uncertainty.

I do not write down the field equations. This has been done in detail elsewhere[9]. Instead I describe what has to be done. A description of neutron star matter is obtained as the self-consistent solution to a system of coupled non-linear equations in 7+N unknowns, as follows:

(1) three field equations for the meson fields (σ, ω, ρ)

(2) equation for electrical neutrality (μ_e)

(3) equation for baryon density (μ_n)

(4) two equations for lepton Fermi momenta, (k_e, k_μ)

(5) N equations for the Fermi momenta of N baryon species in chemical equilibrium (k_n, k_p, $k_\Lambda, \ldots, k_\Xi, \ldots$).

Such a system of non-linear equations is generally difficult to solve, which probably accounts for the fact that most applications of equations of state to neutron star structure approximate the star as either pure in neutron, or else as involving beta equilibrium in the restricted sense of equilibrium among only neutrons, protons and electrons. This is not really very useful in connection with the mass constraint on the equation of state, because both are gross approximations. Two early studies, one by Pandharipande [10], and one by Bethe and Johnson [11] did include hyperons. These works are non-relativistic so the equation of state explicitly violates causality at high density, and of course must be wrong even at a lower density than the point where causality is first violated. The work of Bethe and Johnson suggests a smaller fraction of hyperons than in the work of Pandharipande. However it suffers from the fact that no control was placed on the compression and the symmetry energy, even though neutron stars are highly compressed and isospin asymmetric. The hyperon fraction that I find is of similar importance as found by Pandharipande. The advantage of the present approach is that it is relativistically covariant, and the coupling constants of the theory are related to nuclear matter properties in a way that allows one to investigate the dependance of neutron star structure on nuclear matter properties. Is there such a dependance? Although the density at the center of a neutron star may be quite high it would be incorrect to assume that properties near saturation density are not important. This is so because the center contributes little to the mass on account of the volume element. I find that the average density is only three to five or so times nuclear density, depending on how stiff or soft the equation of state is. (See Fig.3). Moreover the equation of state at higher density is connected to that near saturation through continuity and causality.

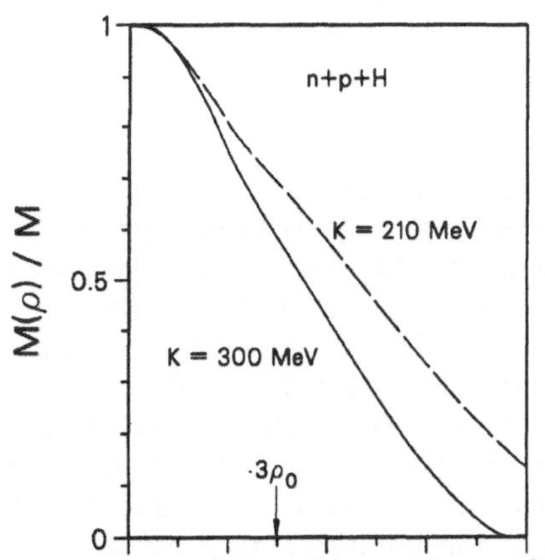

Figure 3. Fraction of mass of neutron star resident in matter at at densities greater than ρ.

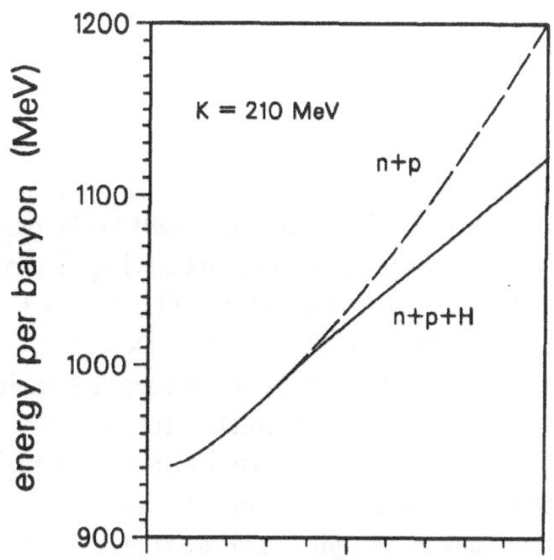

Figure 4. Compares equation of state of neutron star matter with and without hyperons included in the equilibrium composition.

In terms of the solution for the field amplitudes, chemical potentials and Fermi momenta, the equation of state is given by,

$$\begin{aligned}
p &= -\tfrac{1}{3}bm_n(g_\sigma\sigma_0)^3 - \tfrac{1}{4}c(g_\sigma\sigma_0)^4 - \tfrac{1}{2}m_{\sigma_0}^2\sigma_0^2 + \tfrac{1}{2}m_{\omega_0}^2\omega_0^2 + \tfrac{1}{2}m_{\rho_{03}}^2\rho_{03}^2 \\
&\quad + \frac{1}{3}\sum_B \frac{2J_B+1}{2\pi^2}\int_0^{k_B} \frac{p^4}{\sqrt{k^2+(m_B-g_{\sigma B}\sigma_0)^2}}dp \\
&\quad + \frac{1}{3}\sum_\lambda \frac{1}{\pi^2}\int_0^{k_\lambda} \frac{p^4}{\sqrt{k^2+m_\lambda^2}}dp \\
\epsilon &= \tfrac{1}{3}bm_n(g_\sigma\sigma_0)^3 + \tfrac{1}{4}c(g_\sigma\sigma_0)^4 + \tfrac{1}{2}m_{\sigma_0}^2\sigma_0^2 + \tfrac{1}{2}m_{\omega_0}^2\omega_0^2 + \tfrac{1}{2}m_{\rho_{03}}^2\rho_{03}^2 \\
&\quad + \sum_B \frac{2J_B+1}{2\pi^2}\int_0^{k_B} \sqrt{k^2+(m_B-g_{\sigma B}\sigma_0)^2}\,p^2dp \\
&\quad + \sum_\lambda \frac{1}{\pi^2}\int_0^{k_\lambda} \sqrt{k^2+m_\lambda^2}\,p^2dp
\end{aligned} \tag{23}$$

We illustrate in Fig.4 the equation of state in the form of $E/A \equiv \epsilon/\rho$ vs ρ for two cases, one in which only beta equilibrium between neutrons, protons and electrons is taken into account, and one in which full equilibrium between all particle species to convergence is taken into account. The latter is considerably softer than the former, for the reason that the Fermi pressure of neutrons and protons near the top of the Fermi sea is relieved by allowing them to hyperonize. The corresponding results for neutron star masses is shown in Fig.5 and we see that gravity very effectively exploits the softening at higher density introduced by hyperonization. A similar reduction in limiting mass can be found between pure neutron matter and n+p matter.

4 Pion and Kaon condensation

Pion condensation was a subject of much investigation a few years ago, especially for symmetric nuclear matter and nuclei. Pions are more likely to condense in neutron star matter. Condensation occurs in nuclear matter if the pion energy becomes degenerate with the normal state, which might happen if the interaction is attractive and strong enough. However in neutron stars, charge neutrality favors pion condensation. This is so because as a function of increasing density, neutrons at the top of the Fermi sea will decay to proton plus electron. The electrons are fermions and their Fermi level increases with further increase in density. When the electron chemical potential (Fermi energy) becomes equal to the *effective* pion mass in the medium, it will be favorable thereafter for negative pions to play the role that the electrons had in preserving charge neutrality because they are bosons and can all condense in the lowest state. Thus while $\mu_e = \mu_n - \mu_p$ is essentially zero in symmetric matter, it is positive in neutron star matter, thus favoring the π^- since $\mu_{\pi^-} = \mu_e$. On the other hand the other charge states of the pion are excluded since $\mu_{\pi^+} = -\mu_e$ and $\mu_{\pi^0} = 0$.

When pions condense the growth of the electron chemical potential with further increase in density is arrested at a value equal to the effective pion mass. This means that the electron chemical potential cannot approach the mass of the negative kaon, which is greater than that of the pion, and so they cannot condense. Condensation of the other kaon states is even less likely for the same reason as given above for the other charge states of the pion.

Since the hyperons have charges of both signs and carry the conserved baryon charge, charge neutrality can be achieved mainly among baryons at sufficiently high density. This means that the electron chemical will initially be an increasing function of density, will saturate if and when it becomes equal to the pion effective mass, will remain essentially saturated through a range of density, and then will decrease as the hyperon populations grow. The pions at this point will be reabsorbed. So pion condensation is an intermediate density phenomenon in neutron stars. This can be seen in Fig.2. I do not find that pion condensation has a large effect on neutron star structure, an example of which can be seen in Fig.6. In this calculation the effective pion mass was taken to be the vacuum mass. This probably overestimates the role of the pion. In neutron star matter, the pion experiences a repulsive s-wave interaction which would increase its effective mass. It experiences also an attractive p-wave interaction which however it must pay for by having a finite momentum. Our earlier estimate of the effective mass due to the latter is $\sim 200 MeV$[12].

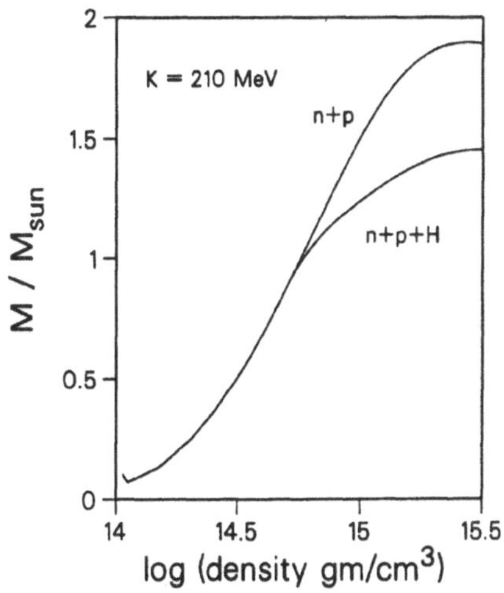

Figure 5. Neutron star masses compared with and without inclusion of hyperons in the equilibrium composition.

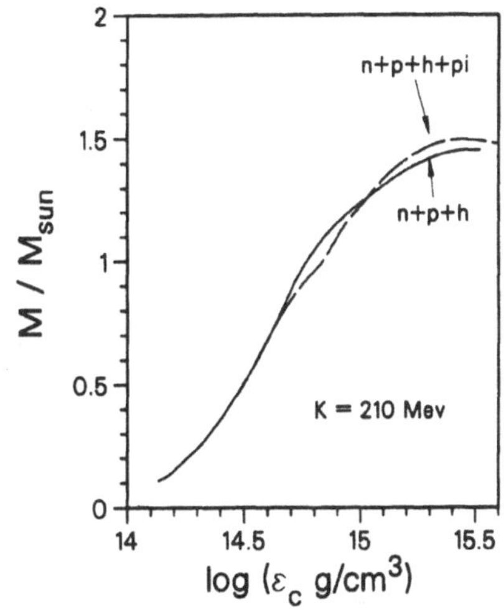

Figure 6. Neutron star masses compared with and without pion condensation.

5 Limiting neutron star mass and the equation of state

I have shown you that neutron stars are complex in composition. Models that treat them as pure in neutron or treat only beta equilibrium among neutrons, protons and electrons are inadequate since they overestimate the limiting mass by as much as a half a solar mass. This is very large in comparison with the range in which the limiting mass can lie. For a non-interacting neutron gas, which provides a lower limit because there is no repulsive force, the limiting mass is $\sim .75 M_{\odot}$. At the other extreme the hardest equations of state without full equilibrium among species give $\sim 2.75 M_{\odot}$. So the effective range in which it can fall is about 2 solar masses, and taking into account more realistic models is less than this.

Do we know the limiting mass in nature? We do know of the existence of about 400 pulsars, but we know the masses of only seven. This is because mass measurements are possible only for binaries. Of these the most massive is 4U0900-40 for which $M = 1.85 \pm 0.3 M_{\odot}$. The most accurately measured is PSR1913+16 for which $M = 1.442 \pm .003$. With so few known masses it is unlikely that we know the most massive one. Therefore we can know only a lower limit on the maximum mass which translates to a lower limit on the stiffness of the equation of state. In Fig.7 I show calculated limiting masses from the above described theory as a function of compression K. There is some uncertainty owing to the fact that one of the saturation properties used to fix the coupling constants is the nucleon effective mass at saturation. While this is apparently known in a narrow range, there remains some uncertainty. To account for PSR1913+16 we need $K > 200$ MeV. To account for 4U0900-40 at the most probable mass we need $K > 300$ MeV or at the lower limit $K > 200$ MeV.

Of course it is possible to criticize the above results as being model dependent. Nonetheless the model contains a lot of physics, much more than parameterizations, or models based on potential interactions of nucleons which become acausal at high density, and which omit the large effect of equilibrium in the star. Recall that our theory is constrained at saturation by five nuclear properties and at all densities, but most importantly as a constraint at high densities, it is causal. These are strong constraints and should be contrasted with theories or parameterizations with fewer constraints. To emphasize this we contrast results for the linear and non-linear versions of nuclear field theory which allow respectively three and five properties to be constrained at saturation (two and four respectively for the symmetric matter equation of state). We show in Fig.8 that in the first case the theory computed in mean field approximation can be made to agree at saturation, but no where else, with the theory which includes vacuum renormalization. In contrast, with the additional constraints (K and m^*), the mean field theory agrees within three percent up to ten times nuclear density, although the constraints apply only at saturation density! [8].

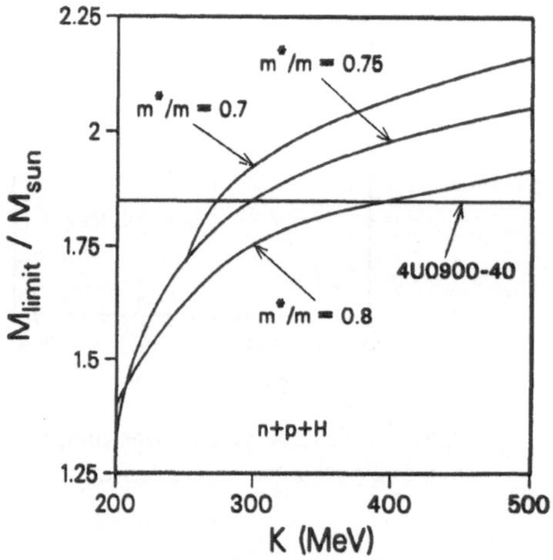

Figure 7. Limiting neutron star masses as a function of K for several choices of the value of m^* at saturation.

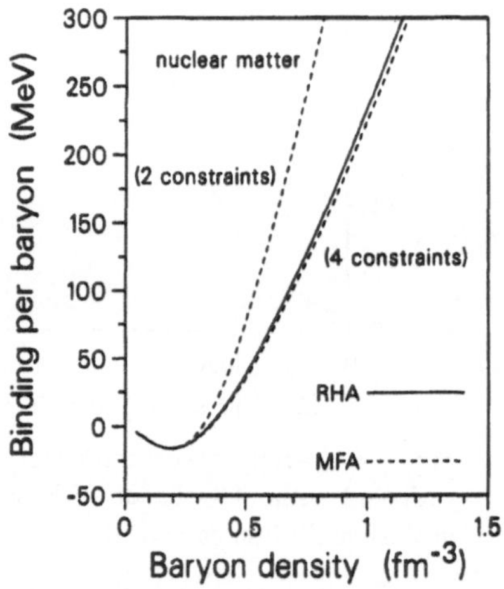

Figure 8. Equation of state with vacuum polarization (RHA, solid line) approximated in mean field theory with 2 parameter constraint at saturation, (B and ρ_0), and with four, (K and m^* in addition) [8].

Let us compare results with a recent work that purports to explain neutron star masses with much softer equations of state[13]. In Fig.9, results from that work are shown for the limiting mass of pure neutron stars and stars with only neutrons, protons and electrons. Essentially I agree with these results as far as they go. However the authors claim on the basis of their calculations that neutron stars can be accounted for with K as low as 120 MeV. What there results clearly show is that there is a substantial reduction in limiting mass in going from pure neutron to beta equilibrium. This is well known. There is a further reduction due to hyperonization as I have frequently emphasized which is shown by my calculation by the solid line. When all of this is taken account of, K has to be much larger than the low value they quote. This illustration is rather typical of calculations that are incomplete with respect to the space of baryon types or which are constrained by only several saturation properties.

Nonetheless no matter that we constrain the theory as best as can be presently done, both at saturation and at high density by causality, it is a fact that the connection between nuclear matter and neutron star matter can be made only through theory. Moreover the use of K as a characterization of the stiffness of the equation of state at high density is model dependent. Within nuclear field theory K characterizes the stiffness at high density if the effective mass of the nucleon at saturation is well known, since the latter uniquely specifies the coupling constant of the vector meson which dominates the energy at high density. To facilitate the comparison of different models and parameterizations we propose reference to a standard form for the equation of state. Define a standard stiffness by an effective K^{eff} of symmetric matter which is chosen so that the standard form equals the binding energy per nucleon of the equation of state in question at $4\rho_0$. This measures the stiffness of the compression energy from saturation to $4\rho_0$ relative to the standard form,

$$E(\rho) = \frac{K^{eff}}{9\rho_0(\rho+\rho_0)}(\rho-\rho_0)^2 - 16 \quad \text{MeV} \tag{24}$$

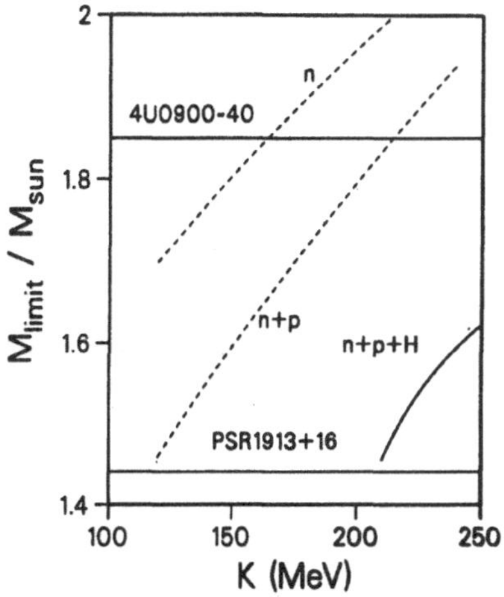

Figure 9. Limiting neutron star mass as function of K for pure neutron matter and matter with neutron and proton from [13] compared to results when hyperons are taken account of (this work).

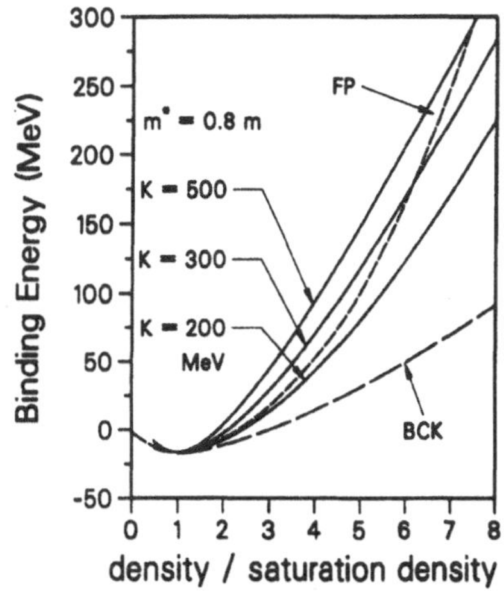

Figure 10. Comparison of various equations of state for symmetric nuclear matter. See Table I for further identification.

This will be recognized as the average of two commonly employed parameterizations, one of which is asymptotically linear in ρ and the other quadratic. The suggested form here is asymptotically linear and hence causal at high density. At $4\rho_0$ it has the value

$$E(4\rho_0) = K^{eff}/5 - 16 \quad \text{MeV} \tag{25}$$

which by equating to any equation of state at $4\rho_0$ is solved for the effective stiffness. In Table I, we list the effective stiffness for several equations of state, and in Fig.10 we plot several of them. It should be emphasized that this effective stiffness is not any longer related to the

Table 1. Effective stiffness for several equations of state

Description	K^{eff} (MeV)
This work $K = 500$ MeV, $m^*/m = 0.8$	540
This work $K = 300$ MeV, $m^*/m = 0.8$	440
This work $K = 200$ MeV, $m^*/m = 0.8$	290
This work $K = 300$ MeV, $m^*/m = 0.75$	500
This work $K = 300$ MeV, $m^*/m = 0.85$	340
Friedman and Pandharipande [14]	330
Ainsworth et al. [15]	130
ter Haar and Malfliet [16]	650
BCK [17, 18] $K = 180$ MeV $\gamma = 2.5$	154

curvature at saturation, but is rather a more global curvature spanning the density range up to $4\rho_0$. By this measure we can see that the equation of state Ainsworth et al. and of BCK (the one that is referred to in [19] as the preferred one) are very soft, and that of ter Harr and Malfliet, calculated in relativistic Dirac-Brueckner approach is very stiff.

6 Rapidly rotating neutron stars

Pulsed signals from the location of the supernova explosion of 1987 were reported in January of this year. A period of 1/2 millisecond was measured and the pulses were frequency modulated with a seven hour period, that is presumed to be due to the doppler shift caused by orbital motion of the pulsar with a companion of about a Jupiter mass. The signals, which have been attributed to a pulsar, were not seen two weeks later at next search, nor have they been seen since, either by the reporting group nor any other [20, 21]. Nonetheless, because of the high quality of the data it cannot be lightly dismissed. In Fig.11 the amplitude of the frequency modulation is shown over the eight hours of observation in the laboratory frame. A skeptic who asserts that the signal is due to instrumental error, or to a signal in the observatory, would be very hard pressed to explain how it is that by taking into account the earth's rotation about its axis and its orbital motion about the sun, the laboratory data is transformed into a sinusoidal curve as would be characteristic of an emitting pulsar in a seven hour orbit. If the pulses represent rotation, it would be the fastest known pulsar. Friedman, Ipser and Parker[22] have studied the fast rotation of neutron stars based on a number of equations of state from various models, assuming uniform rotation. Very rapidly rotating stars provide an additional constraint. This can be understood from eq.(19), which says that the maximum frequency of a star of mass M and radius R is

$$\omega \approx \sqrt{\frac{M}{R^3}} = 3.7 \times 10^5 \sqrt{\frac{M/M_\odot}{(R/km)^3}} \quad s^{-1} \tag{26}$$

Of course this is only a rough indication of the way the mass and radius determine the limiting frequency. The relativistic equations have to be solved for the rapidly rotating star, and these equations are much more complicated than those for the static stars that we have studied up to now. Friedman et al.[23] find from their numerical studies that as a rough rule of thumb, the above formula gives the limiting frequency of the limiting mass star when the factor 3.7 in the above formulae is replaced by 2.4. and M, R are those of the non-rotating star.

The limiting mass of a rapidly rotating star is somewhat (<20 %) larger than for a static star, because of the additional resistance to collapse provided by the centrifugal force. If a star near the mass limit at high frequency slows down with time due to radiation damping which it will do, it will collapse to a black hole. Because of the very short period of the pulsar, so much data was collected in the eight hours of observation that a measurement of the rate of change of the period was possible. It appears that $dT/dt = 3.3 \times 10^{-15}$ s/s. At this very slow rate of change the pulsar will not collapse in our lifetime. (1 yr = 3×10^7 s)

Friedman et al.[23] have studied rotating stars based on an old compilation of equations of state [24] that range from soft to stiff in the sense of how large the limiting mass star is for the eos. Their findings are summarized in Fig.12. Most models fail by a large margin to support the very high spin implied by the signals seen in January. There are two cases (B and G) that do support a high spin, but the same equations of state would not support slowly rotating pulsars at their observed mass, not even PSR1913+16, let alone the more massive star I referred to earlier and so are unsatisfactory. There are several cases for which the star barely reaches the observed frequency. Stars near the limiting frequency are expected to be unstable to non-axial deformations which would damp the star's rotation by gravitational radiation causing it to spin down rapidly. So these too, for a different reason are unsatisfactory. It may be remarked that most of the models that come close to supporting the alleged spin, violate causality at densities not much above the termination point. All of

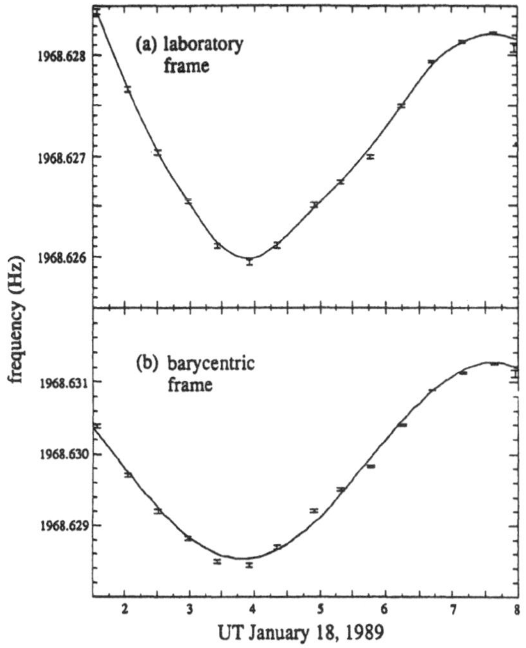

Figure 11. Shows frequency modulation of the new pulsar in SN1987A as seen in the laboratory frame and when corrected for the earth's motion (barycentric frame) [20]

those that come close or exceed the spin of the new pulsar have central densities of 15 to 20 times ρ_0. They are therefore being sampled in a density region where a description in terms of hadrons is no longer possible, since they would be strongly overlapping.

So as it stands, if the observed signals in January are those of a rapidly spinning star, it appears that a very stringent condition may be placed on the equation of state. However none of the models included in the study by Friedman et al are satisfactory. It also appears that none of my models will account for the high spin, though only one has been tested so far. It remains to be found what attributes of the equation of state will allow it at one and the same time to support the observed masses of slowly rotating stars and the high spin of this fast one. One possibility is discussed next.

6.1 Pure quark matter star

I mention here that I have computed a pure quark matter star consisting of u,d,s massless quarks, and a bag constant $B^{1/4} = 170$ MeV, which is half way between the value used in the MIT bag model of baryon resonances and the value which yields a phase transition in hot matter with zero baryon chemical potential at the temperature favored by lattice QCD simulations. This star has zero charge with no need for leptons. According to Witten, such a star may be absolutely stable [25]. Bethe et al. argue to the contrary[26]. Both arguments, pro and con are schematic since only perturbative models with unsatisfactory convergence are available. Maybe fast pulsars can give a hint. Using the rule of thumb quoted above for fast pulsars, the quark star has a limiting value of $\omega = 1.3 \times 10^4\ \mathrm{s}^{-1}$ compared to the observed value of 1.237 in the above units so that it appears to be able to support slightly higher spin than that seen in the fast pulsar. It can also support masses observed for slowly rotating neutron stars, since its limiting mass is $1.5 M_\odot$. The window for which the bag constant can produce a star of sufficiently high mass but small radius so that the double constraint can be satisfied is very small. For example $B^{1/4} = 145$ MeV produces a star with mass and radius of the limiting star of $M \approx 2M_\odot, R \approx 11$ km which does not satisfy the rule of thumb for high spin. On the otherhand $B^{1/4} = 200$ MeV produces a limiting star of $M \approx 1.1 M_\odot$ so that it cannot satisfy the mass constraint. It remains to be seen whether with the obvious refinements of the model and when tested in the relativistic fast rotating star calculation, it

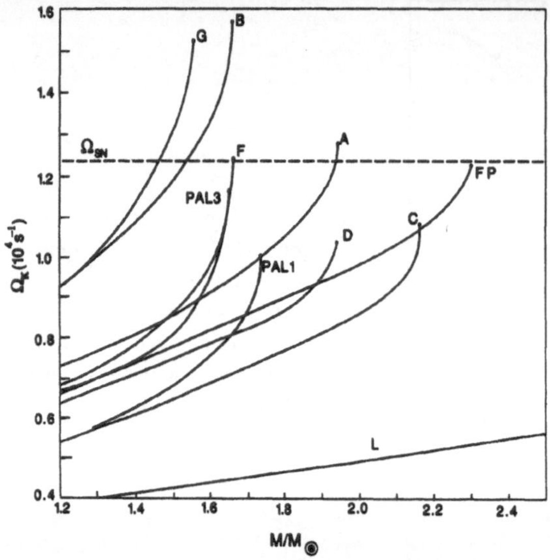

Figure 12. Maximum angular velocity, Ω_K, vs M computed for a number of equation of state. The final dot in each case represents the maximum mass and angular velocity consistent with stability. [23] The angular velocity of the 1/2 millisecond pulsar is shown by the dashed line.

really does support the observed spin. Also of great importance, it is not known at this time how unique this solution is to the double constraint of high spin of the fast pulsar and the observed masses of a slow pulsars.

There is an altogether different possibility however, namely that a pure quark matter star is stable against mass loss at any frequency of rotation. However it might still be unstable to non-axisymmetric modes which would damp the spin rapidly until the axisymmetric oblate deformation of the star becomes stable. Any halo of hadronic matter may have already been ejected by the rotation in the case of a fast pulsar. Stability against mass loss at high spin would then depend upon whether or how fast quarks at the surface could recombine to form hadrons. Cooling could occur by means of the fission of chromo-electric flux tubes formed as quarks occasionally cross the surface of the star in the course of their random motion [27, 28]. This preferentially leads to pion emission, which however must be very slow even at temperatures of a few MeV because the escape probability is appreciable only for leading quarks with momentum greater than about 1 GeV. [27]. Emission of baryons by this process is much further inhibited by the necessity of double pair-creation of quarks in the color field. The loss of baryon number of a high-spin pure quark star is likely to be very slow therefore.

In any case it is interesting to contrast a pure quark star with a neutron star. In the latter, the hadrons are bound only by gravity and for low mass neutrons stars, near the lower limit of stability, the outer regions of the star become very diffuse because of the small gravitational attraction of the low mass. As the limiting mass is approached neutron stars become very compact, and the outer crust and atmosphere become thin. A quark matter star on the other hand is first and foremost bound by the QCD confinement, which in the simple model is represented by the bag pressure. Gravity is what makes it dense, and therefore what possibly transforms a neutron star into a quark matter star. As a consequence the behavior of the radius with mass is very different for neutron stars and pure quark stars, as shown in Fig.13.

7 The Disappearance of the fast pulsar

We noted above the high quality of the pulsed data coming from the center of the supernova 1987A, and in particular how the transformation to the barycentric frame brought the data on the frequency modulation into sinusoidal form as would be the case for modulation due to orbital motion with a companion. However the pulsed signals were not seen again two weeks later at the time of the next search, nor have they been seen since on fourteen subsequent searches reported as of this date. The disappearance can be interpreted in a number of ways:

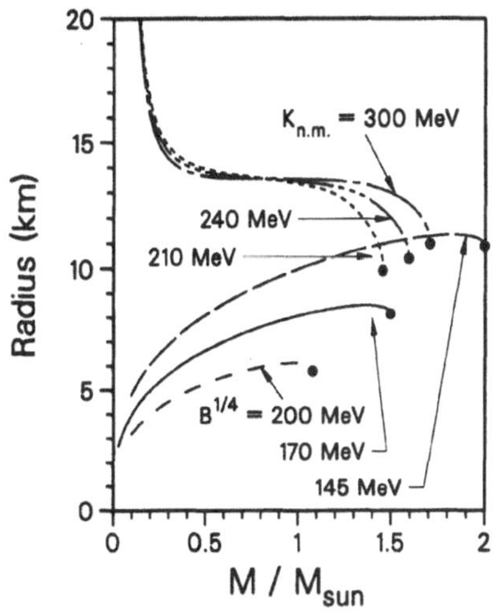

Figure 13. Radius vs mass for neutron stars (top three curves) and pure quark matter stars (bottom three) labeled by the compression and bag constant respectively.

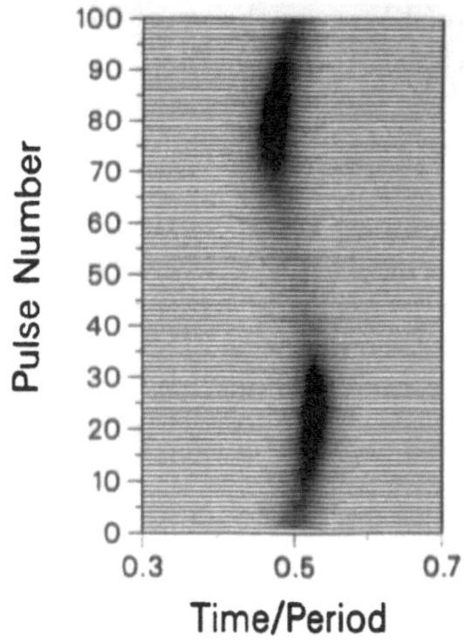

Figure 14. Simulation of pulses plotted modulo the period for a central slice of the period and stacked by pulse number, for $e = -0.2$, $\Omega/n = 100$, $\gamma = 0.05\pi$. Observer's orientation given by polar angle $\theta = -.35\pi$. [29]

1. The signals were of a different origin, instrumental or other earth bound source. This seems to be refuted by the order brought into the frequency modulation by the transformation to the barycentric frame.

2. The signals are from a pulsar which has temporarily 'disappeared'. There are several possibilities:

   (a) The pulsar has been obscured by debris from the supernova. The debris would likely share the same angular momentum of the progenitor star as the fast pulsar, and clouds of it could have moved into the line of sight following the discovery. However with time it should grow optically thinner, and the longer the elapsed time the less likely this explanation.

   (b) The pulsar may have a slight non-axial symmetric deformation which would cause precession, possibly rotating the magnet axis in such a way that the cone of radiation emitted around its direction no longer subtends our line of sight. We examine details of such a possibility[29].

3. The pulsar has collapsed to a black hole. According to the measured rate of change of the period, it was slowing down extremely slowly at the time of observation, and spin down to the point of loss of the additional stability provided by the high spin is unlikely in such a short time scale. We will examine another scenario which leads to rapid spindown and collapse to a black hole[30].

7.1 Cyclic appearances and disappearances of pulsars

We discuss how a very small eccentricity whose symmetry axis is inclined at an angle to the direction of the angular velocity vector will slowly rotate a pulsar about that axis, causing cyclic disappearance and reappearance of radiation beamed along a magnetic axis fixed in the star [29]. Using the observed rate of change of the millisecond period as a constraint on the eccentricity, we find that the period of the cycle for disappearances could range from several hours to the age of the age of the universe, if no subsequent adjustment of the shape occurred.

Precession in planets and stars has been discussed previously assuming that the body is neither perfectly rigid nor fluid [31, 32]. In such a case the motion is described relative to a reference system in which the elastic energy vanishes[32]. Otherwise the precession is like that of a rigid body. In the following what we refer to as the body symmetry axis should be understood more generally as the symmetry axis of the above reference system, which coincide for rigid bodies, and similarly other variables should be understood as effective variables in the reference system. From the mechanics of a rigid body possessing an axis of symmetry that makes an angle β with the angular momentum axis which is fixed in space, the body will rotate about its own symmetry axis with an angular velocity of [33],

$$n = \Omega \frac{e}{1+e} \cos \beta, \qquad e \equiv \frac{I_3}{I_1} - 1 \tag{27}$$

where I_1, I_2, I_3 are the principal moments of inertia and Ω is the precession frequency of the symmetry axis about the fixed direction of the angular momentum vector. For illustration we have taken $I_1 = I_2$. For very small 'eccentricity', e, Ω is almost equal to the angular velocity, ω, of the body. The precise relation is given by,

$$\Omega \cos \beta = \omega \frac{(1+e)}{\sqrt{1 + (1+e)^2 \tan^2 \beta}} \tag{28}$$

From eq.(27) we have the inequality,

$$\left| \frac{n}{\Omega} \right| < \left| \frac{e}{1+e} \right| \tag{29}$$

If the pulsar in SN1987A, in addition to the oblate deformation caused by its high spin, has an additional non-axially deformation, there will be a time varying mass quadrupole field, and the eccentricity of the latter must be very small because of the measured $\dot{T} \approx 3.3 \times 10^{-15}$ and the short period, $T \approx 5 \times 10^{-4}$ seconds. Gravitational radiation yields a rate of change of the period according to,

$$\dot{T} = (2\pi)^4 \frac{64}{25} \frac{1}{T^3} M R^2 e^2 \tag{30}$$

in gravitational units ($G = c = 1$). For a rapidly rotating neutron star near the limits of stability, we take $M = 1.7 M_\odot$, $R = 11$ km [22]. We obtain the constraint $e < 10^{-7}$. According to eq.(29) this would yield a period for the rotation of the star around its own axis of $> 6 \times 10^3$ seconds. If instead we assumed a period of a week for the rotation about the symmetry axis, then we find $e \geq 8 \times 10^{-10}$, which would make gravitational radiation completely negligible at the lower bound.

We tentatively propose that the subsequent failed searches for the signals in Sn1987A have occurred during the portions of the above cycle when the magnetic axis has been rotated too far out of our line of sight by the slow rotation about the symmetry axis. If this explanation is correct, the pulsar should reappear and disappear at cyclic intervals. The duration of the appearance need not be the same as the duration of the disappearance. This depends on the angle between the symmetry axis and the angular momentum axis, on the angle between the magnetic axis and the symmetry axis, and on the beam width of the radiation about the magnetic axis. As shown above, a period of the reappearance cycle anywhere from two hours to infinity is compatible with the observed rate of change of the precession associated with the millisecond period. (An eccentricity associated with the mass of a fly (taken as 1/10 g) situated on the surface of the star would yield a period for the rotation about the symmetry axis of $\sim 10^{23}$ years, assuming a solar mass for the star and assuming no readjustment in the shape of the star. Because of the eccentricity of the earth ($e \approx -0.003$), it rotates about its polar axis with a period of a little more than 400 days.) With respect to very long disappearance cycles we should mention the caveat that the shape is likely to make adjustments on a shorter time scale that brings the star to a different precessional motion. Starquakes (with periods of months to years) are an example of an adjustment. The new motion might never bring the beamed radiation back into view.

We show in Fig.14 an example of how the frequency is modulated by a precession such as discussed above and how the signal seen by an observer fixed in space will vary in intensity. We use parameters suitable for graphical display which leads to a ratio of periods of 100 rather than 10^9. The intensity variation is more clearly illustrated in Fig.15 The signals from the model pulsar were computed assuming a beam of radiation whose intensity is Gaussian about the direction of the magnetic axis. This axis is taken to be inclined at an angle γ from the symmetry axis. The polar angles of the magnetic axis referred to the angular momentum direction as the z-axis, with origin at the center of the star are given by,

$$\Theta(t) = \arctan\left(\frac{\sqrt{(\sin\beta + \cos\beta\tan\gamma\cos nt)^2 + \tan^2\gamma\sin^2 nt}}{\cos\beta - \sin\beta\tan\gamma\cos nt}\right) \tag{31}$$

$$\Phi(t) = \Omega t + \arctan\left(\frac{\tan\gamma\sin nt}{\sin\beta + \cos\beta\tan\gamma\cos nt}\right) \tag{32}$$

The first term of eq.(32) shows that the *precession* frequency, and not ω, is the pulsar frequency and the second term exhibits a frequency modulation.

The frequency modulation exhibited in Fig.14 raises the question of whether the observed seven hour modulation of the millisecond pulses could be related to precession. The frequency modulation in the second term of the equation for $\Phi(t)$ is sinusoidal to high accuracy under the condition $\tan\gamma << \tan\beta$,

$$\dot{\Phi} \approx \Omega + \left\{1 + \left(\frac{\tan\gamma\sin nt}{\sin\beta}\right)^2\right\}^{-1} n\frac{\tan\gamma}{\sin\beta}\cos nt \tag{33}$$

as for a pulsar whose signal is modulated by circular orbital motion. We check the constraints imposed on such an hypothesis for the pulsar discovered in the remnant of SN1987A. The seven hour modulation is closely sinusoidal and its amplitude is $\Delta\Omega/\Omega \approx 7.5\times10^{-7}$. The seven hour period requires $e > 10^{-7}/6$, which together with the constraint calculated above for the rate of change of the short period due to gravitational radiation, imposes the compatible conditions, $\frac{1}{6} < e/10^{-7} < 1$. However from eq.(7), the amplitude of the frequency modulation is given by,

$$\frac{\Delta\Omega}{\Omega} << \left|\frac{e}{1+e}\right| < 1\times10^{-7} \tag{34}$$

whereas the observed amplitude of the frequency modulation is 7 times larger than the right side. Although the general form of the modulation is difficult to analyze, it seems unlikely that the observed modulation is due to precession.

The proposition seems viable that the disappearance of the pulsar is due to the slow rotation of the magnetic axis, fixed in the star, about an axis established by a very small eccentricity offset from the angular momentum axis. It can be tested if the slow period is not too long by making observations over a sequence of nights that would not be synchronous with a periodic phenomenon. Of course the disappearance could have been occasioned by debris moving into the line of sight. However this hypothesis becomes less tenable as time passes, because such supernova debris is expanding. Elsewhere we have discussed the possible extinction of the pulsar by collapse to a black hole [30].

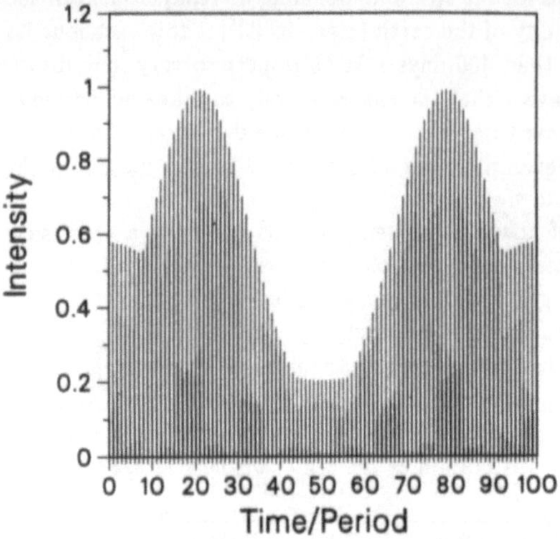

Figure 15. For same case as previous figure, intensity of individual pulses is plotted for a hundred of the fast pulses over one period of the rotation of the star about the axis containing the non-axial deformation. [29]

We mention that the above model of an axially deformed pulsar gains some support from the fact that it can qualitatively simulate several features observed in the signals of many other pulsars, namely subpulse drifting, nulling, and mode switching, which is the subject of another manuscript [34].

7.2 Possible gravitational collapse into a black hole

We shall explore a scenario in which a rapidly spinning neutron star will subside into a black hole within the two week interval between first sighting and next search, due to the recapture of a small mass object ejected at the early stage of formation of the pulsar and the consequent braking of the rotation to a sub-critical value due to the gravitational radiation caused by the aspherical transport of the captured object [30]. The constraints found for the object are that it is a black hole of sub-Jupiter mass up to $\frac{1}{20}M_{Jupiter}$. Interestingly enough, since this suggestion was first made the authors of the discovery paper have improved their analysis of the data and find evidence of a second companion of mass $\approx \frac{1}{60}M_{Jupiter}$, which is close enough for our purpose[35].

The key to this scenario is the observation that a rapidly spinning neutron star can be stabilized at a mass that is greater than the limiting mass of a star with the same baryon number but lower frequency [22]. This is almost certainly the case for this millisecond pulsar as the study of Friedman, Ipser and Parker[22] suggests. Assume that following the main supernova explosion the collapsing core had angular velocity that exceeded the limit for its mass and that it shed matter in the equatorial plane, of which the observed Jupiter-mass companion is one result. This is plausible since the companion is unlikely to have survived the explosion had it been formed earlier. Because the seven hour frequency modulation of the millisecond signals from the pulsar is rather closely sinusoidal, with only small deviations, the Jupiter-mass companion must form the bulk of material in orbit with the pulsar but it is unlikely that all of the matter expelled was so coherent as to form a single companion because of the turbulent conditions. We suppose that one additional small object, its orbit damped by gravitational radiation and perturbed by the other companion and the pulsar, has fallen back onto the surface of the neutron star near the equatorial plane from which it was first ejected, creating an aspherical transport of matter in the neutron star. Since the neutron star will have cooled substantially in the intervening two years since its birth to $\sim$ 100 KeV temperature the dense matter is highly degenerate and the viscosity is expected to be high. The captured object is expected to remain localized for tens of years from the estimate of the viscous damping time of neutron star matter by Comins [36] as interpreted by Friedman et al. [22]. In this event the pulsar now has a time-varying mass quadrupole moment which will produce gravitational waves and damp the pulsar's rotation. We calculate how massive the recaptured object must have been so that the resulting gravitational radiation will have damped the rotation sufficiently in two weeks to bring it below the critical angular velocity that provides the marginal stability against collapse to a black hole.

For a spherical rotating star of radius R and mass M with a lump of mass m attached at its equator, the angular velocity, ω, is damped by gravity waves according to

$$\dot{\omega} = -16\frac{m^2}{M}R^2\omega^5 \equiv -A\omega^5 \tag{35}$$

This equation integrates to

$$\frac{1}{\omega^4} - \frac{1}{\omega_0^4} = 4A(t - t_0) \tag{36}$$

From this we find the period doubling time,

$$t_{(2)} = \frac{15}{4A}\frac{1}{\omega_0^4} \approx 2.7 \times 10^{10}\frac{M}{M_\odot}\left(\frac{M_\odot}{m}\right)^2\frac{T_s^4}{R_{10}^2} \quad s \tag{37}$$

where T_s is the period of the pulsar in seconds and R_{10} is its radius in units of 10 km. Assuming that the pulsar's observed frequency is near the limit for its mass we estimate from the tables of Friedman, Ipser and Parker [22] that the mass is $\approx 1.66M_\odot$ with a radius of $R \approx 11$ km. We find from Eq.(37) that the mass of the object must have been $m \approx 4.4 \times 10^{-5}M_\odot \approx \frac{1}{20}M_{Jupiter}$ so as to double the period in two weeks. This appears not to be unreasonably large, its mass being constrained from above qualitatively by the small degree of perturbation of the pulsar's orbit as inferred from its frequency modulation. The effects of period doubling of pulsars near the limiting frequency is drastic as can be inferred from Fig. 1 of Ref.[22] so that the above mass is *more* than sufficient to destabilize the pulsar. We used the period doubling only as a rough but conservative estimate of the destabilization effects of spindown.

For the proposed scenario to work, the captured object must satisfy the two mass constraints stated above (sufficiently large to destabilize the pulsar within two weeks but sufficiently small compared to the Jupiter-companion as not to appreciably alter the sinusoidal frequency modulation). Also it must not be torn apart by tidal forces and it must be small compared to the neutron star so that its mass is localized after capture. There are no known astrophysical objects of $M \approx \frac{1}{20}M_{Jupiter}$ which are small compared to expected neutron star dimensions ($R = 10 - 15$ km) excepting for a small black hole. Planets and white dwarfs are much larger in size, and in any case would be destroyed by tidal forces and accreted as dust at a rate consistent with the Eddington limit. Exotic stars have been conjectured (see refs. [37, 38, 39] and cited references) that can fall in the desired size and mass range, but they are more likely to have been created in the early universe, if any of them exist, than in the collapse of a star. Black holes accrete background radiation (3°K) or evaporate according to whether their mass is greater or less than $\sim 10^{-7}M_\odot$, so that we have no concern that the conjectured small black hole originating in the turbulence of the supernova will have evaporated in the intervening two years.

Of course the most plausible reason for a temporary disappearance of the pulsar is that it has been obscured by a cloud of debris from the supernova which meanwhile has moved into the line of sight because of its angular velocity. This conjecture becomes less plausible with time because if nothing else happens to the pulsar itself, then it will reappear as the cloud becomes less opaque due to its expansion or as the cloud is carried away by its angular velocity. In any case the scenario proposed above shows that whether or not the newly born pulsar has subsided into a black hole by now, it is in peril of doing so promptly following the aspherical capture of a object having mass of the above order.

Whether "last gasp" signals emitted just prior to disappearance of a neutron star into a black hole could be observed depends to a large degree on the brevity of the final collapse, since a concentrated pulse of neutrinos or gravitational radiation is more easily detected than a long one carrying the same energy.

If this scenario does indeed describe the fate of the young pulsar so fleetingly glimpsed in January, then we have witnessed a remarkable sequence of events that is not likely to occur again for many generations, a spectacular supernova display whose equal has not occurred in this part of the universe since the Crab supernova 900 years ago, the birth of a neutron star at its center, followed soon after by its disappearance into a black hole.

8 Supernovae

Supernovae involve a number of factors of comparable importance but high uncertainty. Among the factors in this connection we note:

(1). Stars in the range $8 < M/M_{\odot} < 100$ which evolve to be the progenitors for type II supernovae, cook for 10^7 years, evolving from a hydrogen gas to layers of heavier elements and eventually an iron core. The mass and entropy of the core are crucial but highly uncertain because of the large network of nuclear reactions and their rates that must be simulated, some of them known and others highly controversial. The results of the simulation of such an evolution provide the initial conditions for the simulation of the collapse. The presupernova evolution calculations yield iron core masses in the range $\sim 1.3 - 2.5M_{\odot}$[40].

(2). When stability is lost at a point corresponding to approximately a Chandrasekhar mass of iron core, collapse proceeds on the millisecond time scale.

(3). When the core reaches supernuclear density a shock originates at a radius that is *interior* to the iron core at a point that includes about $1/2M_{\odot}$. It must propagate outward through the remaining iron core and expel most of the stars mass beyond the core, $> 8M_{\odot}$. Otherwise no supernova and neutron star; instead a black hole.

(4). The shock suffers severe energy loss as it propagates through the infalling overlaying core to the mantle that it must expel. The losses are due to the fact that the shock front dissociates nuclei. The energy loss is easy to calculate. For each $1/10M_{\odot}$ of nuclei dissociated ($B \sim 10$ MeV/nucleon) the energy dissipated is $\sim A/10 \times B = 10^{57}/10 \times 10$ MeV $=1.6 \times 10^{51}$ ergs. This is typically of the same order as the entire energy of the supernova explosion. For example SN1987A is believed to have exploded with an energy $1-3\times 10^{51} ergs$.

(5). The lower limit on the iron core mass found in published stellar evolution calculations is $\sim 1.27M_{\odot}$[41]. This is marginally within the domain of possible success of the explosion. Obviously from the above points, the smaller this mass the more favorable for the success of the prompt shock mechanism. Recent work has therefore focussed on attempts to make plausible a lower core mass.

(6). Neutrino physics is as yet highly uncertain and is treated differently by the various groups doing supernova simulations. Some early claims to success were later understood to be mistaken, by failing to account for neutrino losses of the core. The excess of electron neutrinos provided an unrealistic pressure boost to the shock [1].

(7). It had been claimed by the Stony Brook - Brookhaven group that if the equation of state is sufficiently soft then supernovae with iron cores as massive as $1.35M_{\odot}$ can be successfully simulated [18, 19]. This has been widely cited as evidence that the equation of state must be soft. The claim has floundered on two important issues: (i) The equation of state that was used is too soft to support the masses of several known neutron stars [42] (Ref. [18] seems to suggest, erroneously as it turns out, that it will do so). (ii) Corrections to the handling of the neutrino transport, whose importance was emphasized by Bludman [1] is now conceded to have played such an important role in the previously claimed success for the prompt explosion of $1.35M_{\odot}$ iron core models that it is now agreed that no matter how soft the equation of state, such models cannot be made to explode promptly [3, 4].

We show in Fig.16 that the equation of state used in the Stony Brook-Brookhaven simulations is too soft to support known neutron star masses. In the one case we use a Z/A ratio of 1/3 which is on the average appropriate for the matter of the collapsing material. Later as the neutron star is formed further neutronization occurs because it leads to a lower energy state and hence softer equation of state. Therefore we show also the neutron star masses for a Z/A ratio more appropriate to the evolved neutron star matter, and of course the limiting mass is even less. The appropriate lepton contributions to pressure and energy are included in both cases. In neither case are the observed neutron star masses supported by this equation of state. Effectively the explosion energy in the supernova simulations has been bought at the cost of neutron star mass. It might be claimed that the BCK equation of state is intended for use only at lower densities than in the core of a neutron star. However in one of the models (# 40) reported in ref. [18] the maximum central density achieved just before bounce is $12\rho_0$ so in fact it was used at very high density by its authors. It might also be claimed that it stiffens at high density for some reason or other. However new physical effects will come into play at high density only if they are energetically favorable because physical systems arrange themselves so as to find the lowest possible energy. Of course any processes that lower the energy also soften the equation of state. Examples of this are the neutronization and hyperonization whose softening effect on the equation of state were discussed earlier. I recently read another formulation of the same idea by Bethe in discussing dense matter. "The results of these considerations are that the repulsive nuclear loops saturate, but the attractive forces increase rapidly in magnitude.... These behaviors follow the general rule that repulsive interactions tend to screen themselves so as to cut down the repulsion, while attractive interactions do not."[43]

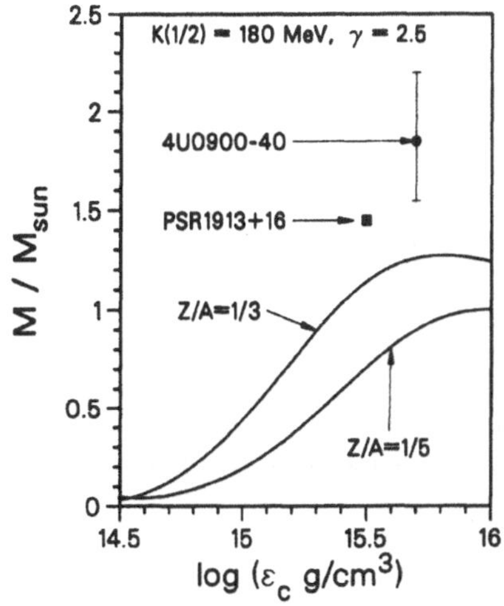

Figure 16. For the equation of state used in the supernova simulations neutron star masses are shown for two values of Z/A. The smaller corresponds more closely to that prevalent in neutron stars. In neither case can the BCK equation of state support stars of observed mass. [42]

The foregoing is the story as regards supernovae and the equation of state as of the published literature to this date. Recent developments for the reasons discussed above have focussed on making a case for smaller iron cores in the lighter presupernova stars. This will reduce the energy losses of the shock wave and presumably will also relieve the need for very soft equations of state. Reaction rates, some not well known, are crucial to the presupernova evolution scenarios. Of particular importance is the $C^{12} + \alpha \rightarrow \gamma + O^{16}$ reaction which has been measured at Muenster and Cal. Tech. They are difficult experiments and they are not in complete agreement. Moreover the accelerator experiments have to be extrapolated to thermal energies, a highly uncertain matter. A particular extrapolation of Brown and Woosley [44] yields iron cores $1 - 1.1 M_\odot$ down from the $\sim 1.3 - 1.4 M_\odot$ of the earlier evolution scenarios[40]. If there is a class of progenitors that have such low iron core masses, they clearly become possible candidates for the prompt explosion mechanism. However it should be noted that the neutron stars that would be produced are of very low mass. Although it may be argued that neutron stars with masses as large as that of 4U0900-40 with $M = 1.85 \pm 0.3 M_\odot$ may have acquired their high mass after initial formation through accretion from a companion, this cannot be claimed for the neutron star in PSR1913+16, since this system, studied over years, is very well understood in terms of general relativity without mass loss or accretion[45]. It is unlikely therefore that such a scenario with such light iron cores can account for anything but a subclass of supernova.

In connection with the last point we mention the late-time neutrino reheating mechanism of Wilson [46, 47, 48]. The shock typically stalls at ~ 100 km because of severe energy losses as discussed above and turns into an accretion shock. Wilson found that it is possible for such shocks to be revived on a long time scale by reheating due to absorption of neutrino's diffusing from the proto-neutron star as it further neutronizes. The scenario appears to work for a broad range of iron core masses and with no strong constraints on the equation of state.

Conclusions for supernovae:

1. It has not been established that supernovae provide a constraint on the equation of state as was frequently quoted. There are many other factors involved of comparable importance but high uncertainty.

2. It is not presently understood why stars explode.

9 Summary

We have reviewed the prospects for obtaining constraints on the equation of state from astrophysical sources. Neutron star masses although few are known at present, provide a very direct constraint in as much as the connection to the equation of state involves only the assumption that Einstein's general theory of relativity is correct at the macroscopic scale. This is almost as secure as Maxwell's equations. If the millisecond pulses briefly observed in the remnant of SN1987A can be attributed to uniform rotation of a pulsar, then a very severe constraint is placed on the equation of state. The theory again is very secure. The precise nature of the constraint is not yet understood, but it appears that the equation of state must be neither too soft nor stiff, and it may be that there is information not only on the stiffness of the equation of state but on its shape. Supernovae simulations involve such a plethora of physical processes including those involved in the evolution of the precollapse configuration, not all of them known or understood, that they provide no constraint at the present time. Not even the broad category of mechanism for the explosion is agreed upon (prompt shock, delayed shock, or nuclear explosion).

Acknowledgements: This work was supported by the Director, Office of Energy Research, Office of High Energy and Nuclear Physics, Division of Nuclear Physics, of the U.S. Department of Energy under Contract DE-AC03-76SF00098.

References

[1] S. A. Bludman, Physics Reports **163** (1988) 47.

[2] W. Hillebrandt, elsewhere in this volume.

[3] E. Baron in paper presented at *Particle and Astrophysics Workshop*, Berkeley, December, 1988.

[4] J. Cooperstein and E. Baron, in *Supernovae* ed. by A. Petschek (Springer-Verlag, to appear, 1989).

[5] S. I. A. Garpman, N. K. Glendenning and Y. J. Karant, Nuc. Phys. **A322** (1979) 382.

[6] P.-G. Reinhard, M. Rufa, J. Maruhn, W. Greiner and J. Friedrich, Z. Phys. **A323** (1986) 13.

[7] B. D. Serot and J. D. Walecka, *The Relativistic Nuclear Many-Body Problem*, in "Advances in Nuclear Physics", eds. J. W. Negele and E. Vogt, (Plenum Press N. Y.), 1986.

[8] N. K. Glendenning, Nucl. Phys. **A493** (1989) 521.

[9] N. K. Glendenning, Phys. Lett. **114B** (1982) 392;
N. K. Glendenning, Astrophys. J. **293** (1985) 470;
N. K. Glendenning, Z. Phys. **A 326** (1987) 57;
N. K. Glendenning, Z. Phys. A, **327** (1987) 295.

[10] V. R. Pandharipande, Nucl. Phys. **A178** (1971) 123.

[11] H. A. Bethe and M. Johnson, Nucl. Phys. **A230** (1974) 1974.

[12] N. K. Glendenning, P. Hecking and V. Ruck, Ann. Phys. (N. Y.) **149** (1983) 22.

[13] M. Prakash, T. L. Ainsworth and J. M. Lattimer, Phys. Rev. Lett. **61** (1988) 2518.

[14] B.Friedman, V. R. Pandharipande and Q. N. Usmani, Nucl. Phys. **A372** (1981) 483.

[15] T. L. Ainsworth, E. Baron, G. E. Brown, J. Cooperstein, M. Prakash, Nucl. Phys. **A464** (1987) 740.

[16] B. ter Haar and R. Malfliet, Phys. Reports **149** (1987) 208.

[17] E. Baron, J. Cooperstein and S. Kahana, Nucl. Phys. **A440** (1985) 744.

[18] E. Baron, J. Cooperstein and S. Kahana, Phys. Rev. Lett. **55** (1985) 126.

[19] E. Baron, H. A. Bethe, G. E. Brown, J. Cooperstein and S. Kahana, Phys. Rev. Lett. **59** (1987) 736.

[20] J. Kristian, C. R. Pennypacker, J. Middleditch, M. A. Hamuy, J. N. Imamura, W. E. Kunkel, R. Lucinio, D. E. Morris, R. A. Muller, S. Perlmutter, S. J. Rawlings, T. P. Sasseen, I. K. Shelton, T. Y. Steinman-Cameron and I. R. Tuohy, NATURE **385** (1989) 234.

[21] J. Middleditch et. al. *Further Analysis of the PSR1987A Discovery Data*, preprint, Aug, 1989.

[22] J. L. Friedman, J. A. Ipser and L. Parker, Astrophys. J. **304** (1986) 115.

[23] J. L. Friedman, J. A. Ipser and L. Parker, Phys. Rev. Let. **62** (1989) 3015.

[24] W. D. Arnett and R. L. Bowers, Astrophys. J. Suppl. **33** (1977) 415.

[25] E. Witten, Phys. Rev. D **30** (1984) 272.

[26] H. A. Bethe, G. E. Brown and J. Cooperstein, Nucl. Phys. **A462** (1987) 791.

[27] B. Banerjee, N. K. Glendenning and T Matsui, Phys. Lett. **127B** (1983) 453.

[28] N. K. Glendenning and T. Matsui, Phys. Rev. D **28** (1983) 28.

[29] N. K. Glendenning, Phys. Rev. Let. **63** (1989) 1443.

[30] N. K. Glendenning, *Possible Gravitational Collapse into a Black Hole of the Pulsar in Supernova 1987A*, LBL-26978, March 1989.

[31] W. H. Munk and G. J. F. MacDonald, *The Rotation of the Earth* (Cambridge University Press, Cambridge, 1960).

[32] D. Pines and J. Shaham, Nature Phys. Sci. **235** (1972) 43.

[33] J. L. Synge and B. A. Griffith *Principles of Mechanics* McGraw-Hill, N.Y. (1949) 425.

[34] N. K. Glendenning, *Deformed Pulsar Model of Subpulse Drift, Nulling and Mode Switching*, LBL-27026, April 1989.

[35] J. Middleditch et. al. *The Pulse Profile and Planetary Companions of the Sub-millisecond Pulsar in SN1987A*, preliminary preprint, May, 1989.

[36] N. Comins, Mon. Not. Royal Astron. Soc. **189** (1979) 233.

[37] T. D. Lee, Phys. Rev. D **35** (1987) 3637;
R. Friedberg, T. D. Lee and Y. Pang, Phys. Rev. D **35** (1987) 3640; ibid. 3658;
T. D. Lee and Y. Pang, Phys. Rev. D **35** (1987) 3678.

[38] J. J. Van der Bij and M. Gleiser, Phys. Lett. B **194** (1987) 482.

[39] N. K. Glendenning, T. Kodama and F. R. Klinkhammer, Phys. Rev. D **38** (1988) 3226.

[40] S. E. Woosley and T. A. Weaver, Ann. Rev. Astron. Astrophys. **24** (1986) 205.

[41] S. E. Woolsey, P. A. Pinto, L. Ensman, Astrophys. J. **324** (1988) 466.

[42] N. K. Glendenning, Phys. Rev. C **37** (1988) 2733.

[43] H. A. Bethe, Ann. Rev. Nucl. and Part. Sc. **38** (1988) 1.

[44] G. E. Brown, (private communication, January, 1989).

[45] J. M. Weisenberg and J. H. Taylor, Phys. Rev. Lett. **52** (1984) 1348;
J. H. Taylor (private communication, April 1987).

[46] J. R. Wilson, in *Numerical Astrophysics* ed. by J. Centrella, J. LeBlanc and R. Bowers (Jones and Bartlett, Boston 1985) p422; J. R. Wilson, R. Mayle, S. E. Woosley and T. Weaver, Ann. New York Acadamy of Sciences **470** (1986) 267.

[47] J. R. Wilson, R. Mayle, S. E. Woosley and T. Weaver, Ann. New York Acadamy of Sciences **470** (1986) 267.

[48] J. R. Wilson, see elsewhere in this volume.

Constraints on the equation of state for neutron stars from fast rotating pulsars

Xuejun Wu, Michael Soffel and Hans Ruder

Lehrstuhl für Theoretische Astrophysik
Universität Tübingen

Béla Waldhauser*, Horst Stöcker and Walter Greiner

Institut für Theoretische Physik der
Johann Wolfgang Goethe-Universität,
6000 Frankfurt am Main

Abstract

We analyze the restriction on the equation of state for neutron stars given by the more or less known masses ($1.445 \pm 0.007 M_{\odot}$ and $1.85 \pm 0.30 M_{\odot}$) of neutron stars and the also more or less known frequencies of fast rotating pulsars like PSR1957+20 ($\Omega = 4033\,\mathrm{Hz}$) and perhaps by a newborne pulsar in the supernova SN1987A ($\Omega = 12369\,\mathrm{Hz}$), where the latter one is seen only by one group for 8 hours and then never again. Softer equations of state generally yield stable rotating pulsars of lower mass, while stiffer ones are leading to heavier neutron stars.

1 Introduction

The equation of state of nuclear matter reflects on a variety of physical topics, such as the liquid vapor phase transition, the monopol resonances, bulk properties of nuclei, high energy heavy ion collisions, supernovae, neutron stars and the phase transition to the quark gluon plasma ([1] and references therein).

There have been a lot of attempts to pin down the nuclear equation of state in all of these areas of nuclear physics, high energy physics and astrophysics up till now. For example Stock et al. [2] proposed a rather stiff equation of state by using the measured pion yields to fix the behaviour of nuclear matter at higher densities. But as there are a lot of uncertainties, like the functional form of the equation of state [3], the thermal behaviour or even the inclusion of higher baryon resonances [4] it is not possible to

*invited speaker

deduce the stiffness of the equation of state via the pion yields only.

Also the collective flow in high energy heavy ion collisions was used in this context. The transverse momentum transfer turned out to be quite sensitive to the nuclear equation of state and seemed to predict also a stiff equation of state [5], but later on it turns out that the inclusion of momentum-dependent interactions simulates the effect of a stiff equation of state with a soft one [6]. Furthermore detailed investigations have revealed that the fragment flow depends also on the not well known in medium NN cross sections [7]. Therefore, it is premature to draw any conclusion from heavy ion experiments.

Another attempt was done by Baron et al. [8], namely to investigate the influence of the equation of state on the prombt explosion of supernovae. Their conclusion tends to a soft equation of state. But as there are a lot of ambiguities in the composition of supernovae, as well as for the explosion mechanism it is not possible to find any conclusionary result from supernova explosion up till now [9].

It was Glendenning [10] who tried to combine all those works to deduce the equation of state from nuclear and astrophysical evidence. But even this work combines the deduction of the compression constant with that of the stiffness of the nuclear equation of state, which is absolutely meaningless. The stiffness of the equation of state depends even more on the parametrization of the compression part as well as on the specific thermal part which is used than on the compression constant at ground state. Prakash, Ainsworth and Lattimer investigated exactly this question for non-rotating neutron stars and confirmed the above statement [11].

Even the phase transition to the quark gluon plasma can not help us for this job. Besides the poor experimental information about the phase transition, the influence of the nuclear equation of state is as large as that of the quark gluon plasma one [12].

Therefore we are trying here not to pin down the exact value of the compression constant, but instead to give some constraints for the stiffness of the equation of state for neutron stars with some "hard" and some "soft" values of the mass and the frequency of rotating pulsars.

Meanwhile it is widely appreciated that the existence of fast rotating pulsars such as PSR 1957+20 ($\Omega = 4033\,\mathrm{Hz}$) or PSR 1937+214 ($\Omega = 3910\,\mathrm{Hz}$) [13] places important constraints on the equation of state for neutron star matter [14]. If the existence of a newborne pulsar in SN1987A with $\Omega \simeq 12369\,\mathrm{Hz}$ [15] will be confirmed then it will provide one of the best testing grounds for the equation of state of condensed matter above nuclear density.

For a given equation of state and pulsar rotation frequency our model calculations for the structure of the neutron star in the framework of General Relativity usually yield a certain mass range for stable configurations.

Now, observationally determined neutron star masses are quite uncertain with one exception: the pulsar mass in the binary pulsar system PSR 1913+16 has been determined to 1.445 ± 0.007 solar masses [16]. However, there are neutron stars with a higher mass, namely the 4U0900-40 with $1.85 \pm 0.30 M_{\odot}$ [17], where the lower bound of the mass measurement coincides more or less with the mass of the PSR 1913+16.

Under this hypothesis the neutron star structure calculation shows: in order to allow for neutron star masses around $1.5 M_{\odot}$, the neutron star equation of state should be neither too soft nor even too hard to allow a fast rotation. Softer equations of state generally yield stable rotating configurations of lower mass, stiffer ones those of higher mass.

2 The equations for a rotating neutron star

To compute the structure of rotating neutron stars one has to integrate Einstein's field equations for a given equation of state and energy-momentum tensor. For the non-rotating (spherically symmetric) case one has to deal with the well known Tolman-Oppenheimer-Volkoff equations [18]

$$\frac{dm}{dr} = 4\pi r^2 \rho(r); \qquad m(r) = \frac{c^2 r}{2G}\left(1 - e^{-\lambda(r)}\right)$$

$$\frac{dp}{dr} = -G\frac{(\rho + p/c^2)(m + 4\pi r^3 p/c^2)}{r^2(1 - 2Gm/c^2 r)} \tag{1}$$

and

$$\frac{d\nu}{dr} = -\frac{2}{(\rho + p/c^2)}\frac{dp}{dr} \quad . \tag{2}$$

Here, ρ is the mass energy density, p is the pressure and $-e^{\nu}$ (e^{λ}) is the g_{tt} (g_{rr}) component of the metric tensor. (Please be aware that "*density*" or "ρ" does not stand for the baryon density, but instead for the energy density like usually used in astrophysics.) For a given equation of state and a fixed central density the numerical integration outwards to the neutron star surface, where $p = 0$, gives all information for the non-rotating case.

For the rotating neutron star we have employed Hartle's approximation [19] for the Einstein equations, which starts from the spherically symmetric, non-rotating case and involves an expansion of all interesting quantities in terms of Ω/Ω_c up and including the second order. Here, the critical angular velocity Ω_c might be taken as the Kepler frequency

$$\Omega_c = \sqrt{\frac{GM_0}{R_0^3}} \simeq (M/M_\odot)^{1/2}(R/10km)^{-3/2}10^4\text{Hz} \quad . \tag{3}$$

The (Hartle-) equation linear in Ω yields the angular velocity of frame dragging; the equations quadratic in Ω yield corrections for radius, mass and quadrupole moment (oblateness, ellipticity) of the neutron star.

In order to have a feeling of the validity of our expansion We have compared our calculations with the exact numerical integrations performed by Friedmann, Ipser, Parker [20].

In Figure 1 we compare for the equation of state "G" cited in reference [20] (originating from the Arnett and Bowers collection [21]) the radius as well as the eccentricity ($e = \sqrt{(radius\ at\ equator/radius\ at\ pole)^2 - 1}$) of the rotating neutron star dependent on the angular velocity Ω (full line are our results). One sees that even close to the critical frequency (where the Hartle approximation breaks down) we are able to reproduce all results from the exact integration to within $\leq 10\%$ (the same accuracy holds for the mass). So we are encouraged to compare the influence of different neutron star matter equations of state on the maximum angular velocity, mass, radius etc.

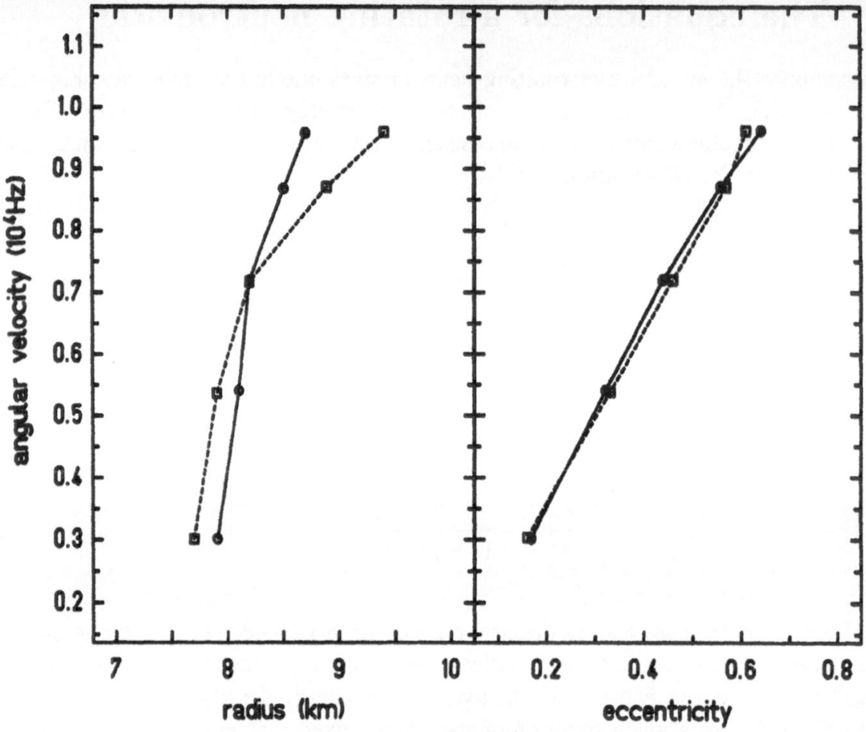

Figure 1. **Comparison with Friedman et al.**

3 Equation of state for neutron stars

For a first try, we will use some equations of state often used in astrophysics. The first one was listed up by Baym, Pethick and Sutherland (BPS) [22] using a low density part (up to two times ground state density) from Baym, Bethe and Pethick together with a high density part for hyperonic matter by Pandharipande for densities larger than three times ground state density. Unfortunately the connection between the two parts is not continous (compare figure 2), but this does not alter the results. For comparison we are using a stiffer equation of state by Tsuruta and Cameron [23].

Our aim is now, to investigate an equation of state model normally used in nuclear and high energy physics [4], [10], [12], [24], [25], namely the mean field model [26] with scalar self-interactions up to fourth order [27], in order to have the same input like in those calculations. For the details of the mean field model and the special sets of coupling constants we refer to [28] and references therein.

For sake of simplicity we have choosen two different sets of coupling constants which satisfy the known bulk properties, like binding energy per nucleon $E/A = -16MeV$ at the nuclear ground state density $n_0 = 0.15/fm^3$ with the compression constant $K = 300MeV$. The only difference is the effective mass at ground state, which is $m^* = 0.85m$ for the softer equation of state and $m^* = 0.55m$ for the harder one, because it is the effective nucleon mass at the ground state, which determines the

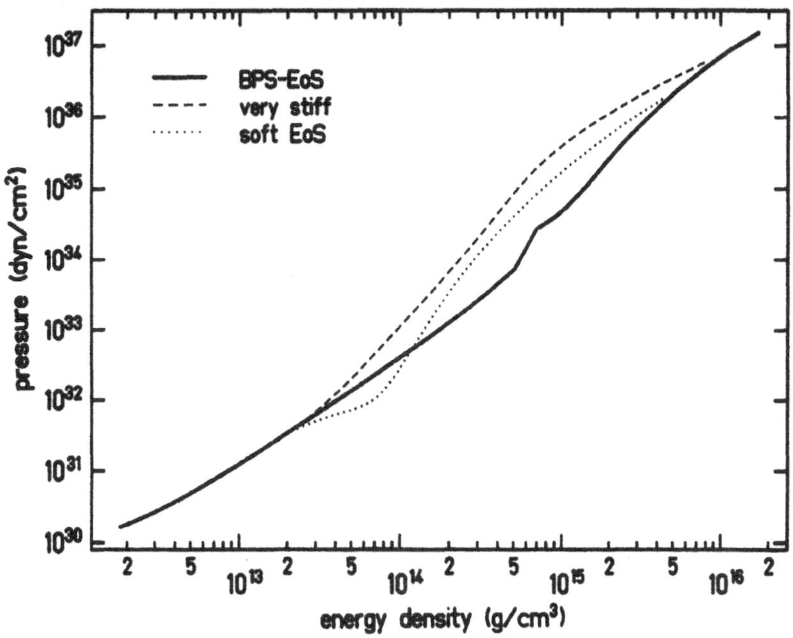

Figure 2. **Equations of state for cold neutron stars**

stiffness of the equation of state at high densities [28], [12].

Our choice for a soft equation of state is, for symmetric nuclear matter, comparable with the soft equation of state used in the QMD-model, whereas our stiff equation of state is even stiffer than their stiff choice [6].

In figure 2 they are plotted together with the BPS equation of state in a logarithmic plot with pressure versus energy density. One can observe, that we have fitted the mean field equations of state to the BPS equation of state at one tenth of normal ground state density.

We had to point out that the mean field equations of state used here, are for pure neutron matter. So it is clear that the inclusion of a percentage of protons will soften the equation of state and an even larger effect will be due to the inclusion of hyperons. They will push the equation of state down and therefore leading to a decrease of the maximum mass of the neutron star by one half solar masses [29]. However, our calculations are an upper estimate for the mass of a neutron star as well as a lower estimate of his rotation frequency.

4 Results

For a whole set of equations of state series of model calculations for different values of central density were performed for static neutron stars as well as for $\Omega = 4033$ and 12 369 Hz.

One example is provided by the following Table 1 for $\Omega = 12369\,Hz$. The first part refers to the BPS equation of state mentioned above. One finds that $\Omega < \Omega_c$ only for central densities larger than about $2 \times 10^{15}\,\mathrm{g/cm^3}$; for central densities larger than about 5×10^{15} the configuration becomes unstable with respect to perturbations. (Compare Figure 3, where the stable and unstable regime is shown for the BPS equa-

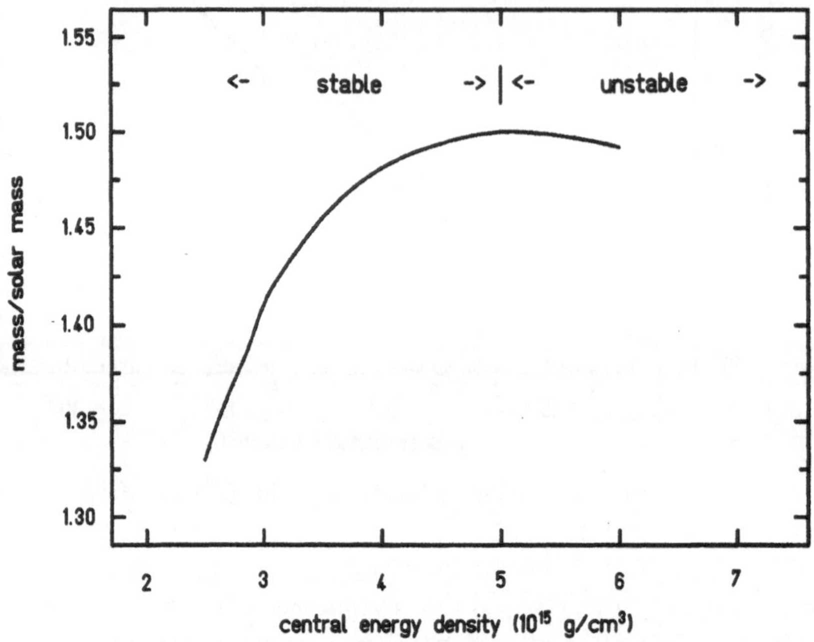

Figure 3. **Stable and unstable regions for neutron stars**

Table 1. **SN1987A pulsar:** $\Omega = 12369\,Hz$

BPS EOS for hyperonic matter			
$E_c(10^{15}g/cm^3)$	$M_r/M_\odot$	$\Omega_c(10^4\ Hz)$	stable?
2.50	1.33	1.60	yes
3.37	1.45	1.86	yes
5.00	1.50	2.19	yes
5.50	1.50	2.26	no
V_γ - EOS (Tsuruta and Cameron)			
$E_c(10^{15}g/cm^3)$	$M_r/M_\odot$	$\Omega_c(10^4\ Hz)$	stable?
1.50	2.22	1.26	yes
2.00	2.24	1.42	yes
3.00	2.18	1.63	no

tion of state. The configuration with maximum mass marks the onset of instability to collapse [30]. The instability is secular, leading to gravitational collapse on a time scale associated with redistribution of angular momentum.)

These two conditions restrict the mass range of stable neutron stars configurations for the BPS equation of state to 1.3 to 1.5 solar masses, including the favoured 1.445 $M_\odot$ value, but not consistent with the mass of the 4U0900-40 ($1.85 \pm 0.30 M_\odot$), even, if one considers only the lower bound as a serious value. A counterexample is shown in the lower part of Table 1, where the V_γ equation of state from Tsuruta and Cameron [23] was used. Here, only in a very small mass regime around 2.2 $M_\odot$ a neutron star could rotate as fast as the SN1987A pulsar. On the other hand such an equation of state includes for the non-rotating case the measured masses of all known neutron stars. Please note that Tables 2 and 3 contain the masses for the rotating (M_r), as well as for the non-rotating (M_{nr}) case.

The results for the soft as well as for the very stiff mean field equation of state are summarized in Tables 2 and 3. We see that for the PSR 1957+20 pulsar the stable mass regime lies between 1.45 and 3.2 for the very stiff equation of state and between 0.73 and 2.1 for the soft one. So there is no contradiction between those equations of state, neither a soft one nor a very stiff one, and confirmed results of rotating and non-rotating neutron stars. Even the soft equation of state has a mass regime large enough to include the 1.85 $M_\odot$ of the 4U0900-40. The radii of these two very different equations of state are ranging from 12 to approximately 19 km (for the very stiff case), whereas the eccentricity has more or less similar values for the different equations of state varying with the central energy density and the angular velocity (from 0.35 to 0.85).

However, if the new pulsar in SN1987A exists, then the very stiff equation of state seems to conflict with nature. No stable configuration with $\Omega < \Omega_c$ was found and even for the soft one only a small mass regime around 2.5 solar masses would be stable. On the other hand, we mentioned already that these two mean field equations of state are calculated for pure neutron matter. So we expect a considerable decrease in the stiffness of both and therefore an increase in the stable rotation frequency. Furthermore we investigated only two extreme cases for the equation of state. We think that it is more likely to find a proper equation of state, which satisfies all conditions, in between. However, we expect the range of stiffness (for a proper equation of state) to be rather narrow.

After the end of the International Advanced Courses on

THE NUCLEAR EQUATION OF STATE

two papers appeared which give similar conclusions reached here [31], [32]. Especially the numerically exact solutions of Ipser and Lindblom [31], for the normal modes of pulsation of rapidly rotating Newtonian configurations, lead to the prediction that the critical angular velocity, at which nonaxisymmetric instability sets in is given by $0.9\,\Omega_c$ for a neutron star of the age of that in SN1987A. So this condition is even more stringent and decreases the range of acceptable equations of state [32].

Table 2. **PSR1957+20:** $\Omega = 4033 Hz$

$E_c(10^{15} g/cm^3)$	$M_r/M_\odot$	$M_{nr}/M_\odot$	e	R(km)	$\Omega_c(10^4\ Hz)$	stable?
soft mean field EOS						
0.3	0.73	0.61	0.86	15.4	0.54	yes
0.6	1.56	1.43	0.57	14.6	0.82	yes
1.0	1.97	1.88	0.44	13.7	1.01	yes
2.0	2.12	2.08	0.33	12.1	1.26	no
very stiff mean field EOS						
$E_c(10^{15} g/cm^3)$	$M_r/M_\odot$	$M_{nr}/M_\odot$	e	R(km)	$\Omega_c(10^4\ Hz)$	stable?
0.3	1.45	1.23	0.83	18.8	0.56	yes
0.6	2.80	2.63	0.50	17.0	0.87	yes
1.0	3.17	3.06	0.39	15.6	1.04	yes
2.0	3.06	3.00	0.32	14.0	1.21	no

Table 3. **SN1987A pulsar:** $\Omega = 12369 Hz$

$E_c(10^{15} g/cm^3)$	$M_r/M_\odot$	$M_{nr}/M_\odot$	e	R(km)	$\Omega_c(10^4\ Hz)$	stable?
soft mean field EOS						
2.0	2.53	2.08	–	12.75	1.26	yes
2.5	2.44	2.07	0.94	12.07	1.34	no
very stiff mean field EOS						
$E_c(10^{15} g/cm^3)$	$M_r/M_\odot$	$M_{nr}/M_\odot$	e	R(km)	$\Omega_c(10^4\ Hz)$	stable?
3.0	3.34	2.87	0.92	13.44	1.29	no
4.0	3.17	2.76	0.89	12.99	1.33	no

References

[1] J. Cleymans, R.V. Gavai and E. Suhonen, Phys.Rep. **130** (1986), 217
R. Stock, Phys.Rep. **135** (1986), 259
H. Stöcker and W. Greiner, Phys.Rep. **137** (1986), 277
T.L. Ainsworth, E. Baron, G.E. Brown, J.Cooperstein and M. Prakash, Nucl.Phys. **A464** (1987), 740

[2] R. Stock et al., Phys.Rev.Lett. **49** (1982), 1236

[3] J.A. Maruhn and H. Stöcker, Z.Phys.A **327** (1987), 75

[4] B.M. Waldhauser, J.A. Maruhn, H. Stöcker and W. Greiner, Z.Phys.A **328** (1987), 19 and Proceedings of the Eighth Balaton Conference on Nuclear Physics 1987, page 93

[5] Joseph J. Molitoris and Horst Stöcker, Phys.Rev.C **32** (1985), 346

[6] J. Aichelin, A. Rosenhauer, G. Peilert, H. Stöcker and W. Greiner, Phys.Rev.Lett. **58** (1987), 1926

[7] G. Peilert, H. Stöcker, W. Greiner, A. Rosenhauer, A. Bohnet and J. Aichelin, Phys.Rev.C **39** (1989), 1402

[8] E. Baron, H.A. Bethe, G.E. Brown, J. Cooperstein and S.Kahana, Phys.Rev.Lett. **59** (1987), 736

[9] K. Hillebrandt, these proceedings

[10] Norman K. Glendenning, Phys.Rev.C **37** (1988), 2733 and these proceedings

[11] M. Prakash, T.L. Ainsworth and J.M. Lattimer, Phys.Rev.Lett. **61** (1988), 2518

[12] B.M. Waldhauser, D.H. Rischke, J.A. Maruhn, H. Stöcker and W. Greiner Z.Phys.C **43** (1989), 411

[13] A.S. Fruchter, D.R. Stinebring and J.H. Taylor, Nature (London) **333** (1988), 237

[14] John L. Friedman, James N. Imamura, Richard H. Durisen and Leonard Parker, Nature (London) **336** (1988), 560

[15] J. Kristian et al., Nature (London) **338** (1989), 234

[16] J.M. Weisberg and J.H. Taylor, Phys.Rev.Lett. **52** (1984), 1348

[17] P.C. Joss and S.A. Rappaport, Ann.Rev.Astr.Astrophys. **22** (1984), 537

[18] Richard C. Tolman, Phys.Rev. **55** (1939), 364
J.R. Oppenheimer and G.M. Volkoff, Phys.Rev. **55** (1939), 374

[19] James B. Hartle, Ap.J. **150** (1967), 1005
James B. Hartle and Kip S. Thorne, Ap.J. **153** (1968), 807

[20] John L. Friedman, James R. Ipser and Leonard Parker, Ap.J. **304** (1986), 115

[21] W. David Arnett and Richard L. Bowers, Ap.J.Suppl. **33** (1977), 415

[22] Gordon Baym, Cristopher Pethick and Peter Sutherland, Ap.J. **170** (1971), 299

[23] S. Tsuruta and A. Cameron, Can.J.Phys. **44** (1966), 1895

[24] J. Boguta, Phys.Lett. **106B** (1981), 245
J. Boguta and H. Stöcker, Phys.Lett. **120B** (1983), 289

[25] S.I.A. Garpman, N.K. Glendenning and Y.J. Karant, Nucl.Phys. **A322** (1979), 382
B.M. Waldhauser, J. Theis, J.A. Maruhn, H. Stöcker and W. Greiner, Phys.Rev.C **36** (1987), 1019

[26] M.H. Johnson and E. Teller, Phys.Rev. **98** (1955), 783
H.-P. Duerr, Phys.Rev. **103** (1956), 469
J.D. Walecka, Ann.Phys.(New York) **83** (1974), 491

[27] J. Boguta and A.R. Bodmer, Nucl.Phys. **A292** (1977), 413

[28] B.M. Waldhauser, J.A. Maruhn, H. Stöcker and W. Greiner, Phys.Rev.C **38** (1988), 1003

[29] N.K. Glendenning, Z.Phys.A **327** (1987), 295

[30] J.L. Friedman, J.R. Ipser and R. Sorkin, Ap.J. **325** (1988), 722

[31] James R. Ipser and Lee Lindblom, Phys.Rev.Lett. **62** (1989), 2777

[32] John L. Friedman, James R. Ipser and Leonard Parker, Phys.Rev.Lett. **62** (1989), 3015

NEUTRON STAR CALCULATIONS IN THE RELATIVISTIC HARTREE- AND HARTREE-FOCK APPROXIMATION†

F. Weber and M. K. Weigel

Sektion Physik der Ludwig-Maximilians
Universität München
Am Coulombwall 1
D-8046 Garching, Fed. Rep. of Germany

ABSTRACT

The baryon composition and gross structural parameters (radius, gravitational mass, moment of inertia, redshift) of static, spherically symmetric neutron stars are calculated in the Hartree- and Hartree-Fock theory, respectively. We find for the maximum stable neutron star mass $M/M_\odot = 1.98$ for the Hartree approximation and values of 2.18 and 2.31 for the latter treatment. The baryon composition calculated for both of these approximations differ considerably from each other.

1. INTRODUCTION

If one assumes that general relativity is the correct theory of gravity, then the Oppenheimer-Volkoff (OV) equations[1] determine the masses and radii of non-rotating, spherically symmetric neutron stars (NS). The unknown therein is the equation of state (EOS), i.e. $P(\epsilon)$, of matter whose particle density ranges from zero up to about $10\varrho_0$ ($\varrho_0 \approx .17\ \mathrm{fm}^{-3}$) in the star's core. Equations of state for the outer and inner surface of neutron stars have been calculated by Harrison-Wheeler (HW)[2] and Negele-Vautherin (NV),[3] respectively. A reliable theoretical determination of the EOS, which describes the matter in the core of a neutron star, demands for a relativistic fieldtheoretic treatment. A relatively simple approach is Walecka's scalar-vector (SV) model - which describes the nuclear forces by the exchange of scalar ($\sigma-$) and vector ($\omega-$) mesons - treated in the so-called mean-field approximation (MFT).[4] Our aim is to calculate the bulk parameters of neutron stars from a more refined Lagrangian density in which the nuclear forces are mediated by the exchange of $\sigma-, \omega-, \pi-$, and $\varrho-$mesons (SVI theory).[5] By making use of the relativistic

† Supported by the German Bundesminister of Forschung and Technology under the contract number 06LM117/VII.

The Nuclear Equation of State, Part A
Edited by W. Greiner and H. Stöcker
Plenum Press, New York

Green's function formalism the EOS is calculated in the Hartree (H) and Hartree-Fock (HF) approximation for charge neutral neutron star matter whose fundamental constituents are charged hyperons and baryons.[6]

2. NEUTRON STARS

The bulk parameters of neutron stars are determined by the OV equations: [1,5]

$$\frac{dP(r)}{dr} = -G\,\frac{[\epsilon(r)+P(r)]\ [m(r)+4\pi r^3 P(r)]}{r^2\left[1-\frac{2Gm(r)}{r}\right]}, \quad m(r) = 4\pi \int_0^r dr'\, r'^2\, \epsilon(r'). \tag{2.1,2}$$

Once the EOS has been calculated the OV equations (2.1,2) can be solved for the star's radius and gravitational mass. The moment of inertia (I) is given by[5,6]

$$I = \frac{8\pi}{3}\int_0^R dr'\, r'^4\, \frac{[\epsilon(r')+P(r')]\, e^{-\Phi(r')/2}}{\sqrt{1-\frac{2Gm(r')}{r'}}}, \quad \frac{d\Phi(r)}{dr} = 2\,G\,\frac{[m(r)+4\pi r^3 P(r)]}{r^2\left[1-\frac{2Gm(r)}{r}\right]}, \tag{2.3,4}$$

where $\Phi(r)$ denotes the metric function. (Eq. (2.3) neglects the dragging of local inertial frames.[7]) The fractional redshift of a photon emitted at the surface of a NS and received very far away is defined by:[6]

$$z = e^{-\Phi(R)/2} - 1 = \frac{1}{\sqrt{1-\frac{2MG}{R}}} - 1\,. \tag{2.5,6}$$

3. LAGRANGIAN DENSITY AND GREEN'S FUNCTIONS

Our description of relativistic many-baryon/lepton matter is based on the following SVI-Lagrangian:[5,6]

$$\mathcal{L}(x) = \sum_B \mathcal{L}^0_B(x) + \sum_{M=\sigma,\omega,\pi,\varrho}\left[\mathcal{L}^0_M(x) + \mathcal{L}_{BM}(x)\right] + \mathcal{L}^{(\sigma^4)}(x) + \mathcal{L}^0_{Lept}(x). \tag{3.1}$$

The sum in (3.1) extends over baryon fields of type $B = p, n, \Sigma^{\pm,0}, \Lambda, \Xi^{0,-}, \Delta^{++,+,0,-}$. A well-known tool to study the properties of many-body systems are the so-called Green's functions. Of special interest are the relativistic two-point propagators ($\lambda = e^-, \mu^-$):[5,6]

$$g^B_{\zeta\zeta'}(p) = \int_{-\infty}^{+\infty} d\omega \frac{a^B_{\zeta\zeta'}(\omega,\vec{p})}{\omega - (p^0 - \mu^B)(1+i\eta)}, \quad G^\lambda_{\alpha\alpha'}(p) = \int_{-\infty}^{+\infty} d\omega \frac{\Gamma^\lambda_{\alpha\alpha'}(\omega,\vec{p})}{\omega - (p^0 - \mu^\lambda)(1+i\eta)}. \tag{3.2, 3}$$

Each baryon propagator $g^B(p)$ of (3.2) obeys a Dyson equation of the following type:

$$\Big\{ (1)_{\zeta\zeta''} \Big[m_B + \Sigma^B_S(p^0,\vec{p}) \Big] + (\vec{\gamma}\cdot\hat{p})_{\zeta\zeta''} \Big[|\vec{p}| + \Sigma^B_V(p^0,\vec{p}) \Big] + (\gamma^0)_{\zeta\zeta''} \Big[\Sigma^B_0(p^0,\vec{p}) - p^0 \Big] \Big\} g^B_{\zeta''\zeta'}(p^0,\vec{p}) = (1)_{\zeta\zeta'}, \tag{3.4}$$

where the general decomposition

$$\Sigma^B(p) \equiv \Sigma^B_S(p) + \vec{\gamma}\cdot(\vec{p}/|\vec{p}|)\, \Sigma^B_V(p) + \gamma^0\, \Sigma^B_0(p) \tag{3.5}$$

of the baryon self-energy in infinite, homogeneous matter has been used. For Σ^B_i ($i = S, V, 0$) of (3.4) one obtains in the Hartree-Fock approximation ($\Sigma^{H,B}$ denotes the Hartree self-energy of baryon B, $\Sigma^{F,B}$ is the corresponding Fock (exchange-) contribution):

$$\Sigma^{H,B}_{\zeta_1\zeta_1'}(p^\mu)\,|_\sigma = -i\,\delta_{\zeta_1\zeta_1'}\, \Delta^0(0)\, g_{\sigma B} \sum_{B'} g_{\sigma B'} \int \frac{d^4q}{(2\pi)^4}\, e^{i\eta q^0}\, Tr\, g^{B'}(q), \tag{3.6}$$

$$\Sigma^{F,B}_{\zeta_1\zeta_1'}(p^\mu)\,|_\sigma = i\, g^2_{\sigma B} \int \frac{d^4q}{(2\pi)^4}\, e^{i\eta q^0}\, \Delta^0(p-q) \Big(1 \otimes g^B(q) \otimes 1\Big)_{\zeta_1\zeta_1'}, \tag{3.7}$$

$$\Sigma^{H,B}_{\zeta_1\zeta_1'}(p^\mu)\,|_\omega = i\,\gamma^\mu_{\zeta_1\zeta_1'}\, \Delta^0_{\mu\nu}(0)\, g_{\omega B} \sum_{B'} g_{\sigma B'} \int \frac{d^4q}{(2\pi)^4}\, e^{i\eta q^0}\, Tr\, \Big(\gamma^\nu g^{B'}(q)\Big), \tag{3.8}$$

$$\Sigma^{F,B}_{\zeta_1\zeta_1'}(p^\mu)\,|_\omega = -i\, g^2_{\omega B} \int \frac{d^4q}{(2\pi)^4}\, e^{i\eta q^0}\, \gamma^\mu_{\zeta_1\zeta_3}\, \Delta^0(p-q)_{\mu\nu}\, g^B_{\zeta_3\zeta_2}(q)\, \gamma^\nu_{\zeta_2\zeta_1'}, \tag{3.9}$$

$$\Sigma^{F,B}_{\zeta_1\zeta_1'}(p^\mu)\,|_\pi = i\Big(\frac{f_{\pi B}}{m_\pi}\Big)^2 \int \frac{d^4q}{(2\pi)^4}\, e^{i\eta q^0} (\gamma_5\gamma_\nu)_{\zeta_1\zeta_3} \times (p-q)^\mu\, (p-q)^\nu\, \Delta^0(p-q)\, g^B_{\zeta_3\zeta_2}(q)\, (\gamma_5\gamma_\mu)_{\zeta_2\zeta_1'}, \tag{3.10}$$

$$\Sigma^{H,B}_{\zeta_1\zeta_1'}(p^\mu)\,|_\varrho = i \sum_{B'} \int \frac{d^4q}{(2\pi)^4}\, e^{i\eta q^0} \Big(g_{\varrho B}\gamma_\mu - i\big(\frac{f_{\varrho B}}{2m_B}\big)(p-q)^\lambda\, \sigma_{\lambda\mu} \Big)_{\zeta_1\zeta_1'} \times \Delta^0(0)^{\mu\nu}\, Tr \Big[\Big((g_{\varrho B'}\gamma_\nu + i\big(\frac{f_{\varrho B'}}{2m_{B'}}\big)(p-q)^\kappa \sigma_{\kappa\nu} \Big)\, g^{B'}(q) \Big] \tag{3.11}$$

$$\Sigma^{F,B}_{\zeta_1\zeta_1'}(p^\mu)\mid_\varrho = -i\int\frac{d^4q}{(2\pi)^4}\,e^{i\eta q^0}\Big(g_{\varrho B}\gamma_\mu - i(\frac{f_{\varrho B}}{2m_B})\,(p-q)^\lambda\,\sigma_{\lambda\mu}\Big)_{\zeta_1\zeta_3}$$
$$\times\,\Delta^0(p-q)^{\mu\nu}g^B_{\zeta_3\zeta_2}(q)\Big(g_{\varrho B}\gamma_\nu + i(\frac{f_{\varrho B}}{2m_B})(p-q)^\kappa\sigma_{\kappa\nu}\Big)_{\zeta_2\zeta_1'}. \qquad (3.12)$$

By using the techniques described in Refs. 5,6 g^B in (3.6)-(3.12) can be eliminated in favour of the spectral functions a^B. For instance, the contribution to the self-energy for the σ-interaction reads as follows ($\Theta^B(\vec{q}) \equiv \Theta(p_{F,B}-\mid\vec{q}\mid)$, $\hat{p}\equiv\vec{p}/\mid\vec{p}\mid$):[5,6]

$$\Sigma^{H,B}_{\zeta_1\zeta_1'}(p)\mid_\sigma = -2\,\delta_{\zeta_1\zeta_1'}\left(\frac{g_{\sigma B}}{m_\sigma}\right)^2\sum_{B'}(2J_{B'}+1)\Big(\frac{g_{\sigma B'}}{g_{\sigma B}}\Big)\int\frac{d^3\vec{q}}{(2\pi)^3}a^{B'}_S(\vec{q})\,\Theta^{B'}(\vec{q})\,, \qquad (3.13)$$

$$\Sigma^{F,B}_{\zeta_1\zeta_1'}(p)\mid_\sigma = g^2_{\sigma B}\int\frac{d^3\vec{q}}{(2\pi)^3}\Big\{\delta_{\zeta_1\zeta_1'}\,a^B_S(\omega^B(\vec{q}),\vec{q}) + (\vec{\gamma}\cdot\hat{p})_{\zeta_1\zeta_1'}\,\hat{p}\cdot\hat{q}\,a^B_V(\omega^B(\vec{q});\vec{q})$$
$$+\,\gamma^0_{\zeta_1\zeta_1'}\,a^B_0(\omega^B(\vec{q}),\vec{q})\Big\}\Delta^0(p^0-\omega^B(\vec{q}),\vec{p}-\vec{q})\,\Theta^B(\vec{q})\,. \qquad (3.14)$$

The spectral functions $a^B_i(\vec{q})$ ($\equiv a^B_i(\omega^B(\vec{q}),\vec{q})$) of (3.13), (3.14) are given by:[6]

$$a^B_S(\omega^B(\vec{p}),\vec{p}) = \Big[m_B + \Sigma^B_S(\omega^B(\vec{p}),\vec{p})\Big]\left[\frac{\partial D^B(\omega,\vec{p})}{\partial\omega}\mid_{\omega^B(\vec{p})}\right]^{-1}, \qquad (3.15)$$

$$a^B_V(\omega^B(\vec{p}),\vec{p}) = \Big[\mid\vec{p}\mid + \Sigma^B_V(\omega^B(\vec{p}),\vec{p})\Big]\left[\frac{\partial D^B(\omega,\vec{p})}{\partial\omega}\mid_{\omega^B(\vec{p})}\right]^{-1}, \qquad (3.16)$$

$$a^B_0(\omega^B(\vec{p}),\vec{p}) = \Big[\omega^B(\vec{p}) - \Sigma^B_0(\omega^B(\vec{p}),\vec{p})\Big]\left[\frac{\partial D^B(\omega,\vec{p})}{\partial\omega}\mid_{\omega^B(\vec{p})}\right]^{-1}, \qquad (3.17)$$

$$\frac{\partial D^B(\omega,\vec{p})}{\partial\omega} = 2\left\{\Big[m_B+\Sigma^B_S(\omega,\vec{p})\Big]\frac{\partial\Sigma^B_S(\omega,\vec{p})}{\partial\omega} + \Big[\mid\vec{p}\mid+\Sigma^B_V(\omega,\vec{p})\Big]\frac{\partial\Sigma^B_V(\omega,\vec{p})}{\partial\omega}\right.$$
$$\left.+\Big[\Sigma^B_0(\omega,\vec{p})-\omega\Big]\Big[1-\frac{\partial\Sigma^B_0(\omega,\vec{p})}{\partial\omega}\Big]\right\}. \qquad (3.18)$$

The energy-momentum relation $\omega^B(\vec{p})$ in (3.15-17) has the form (I_{3B} denotes the third component of isospin of baryon B)

$$\omega^B(\vec{p}) = \Sigma^B_0(\vec{p}) + I_{3B}\,\Sigma^B_{03}(\vec{p}) + \sqrt{\Big[m_B+\Sigma^B_S(\vec{p})\Big]^2 + \Big[\mid\vec{p}\mid+\Sigma^B_V(\vec{p})\Big]^2}. \qquad (3.19)$$

The chemical potential of a baryon of type B is then defined by $\mu^B = \omega^B(p_{F,B})$. The leptons of (3.1) are treated as free fermions. Hence their energy-momentum relations (chemical potentials) are given by

$$\omega^\lambda(\vec{p}) = \sqrt{m_\lambda^2 + \vec{p}^2}, \qquad \mu^\lambda = \omega^\lambda(p_{F,\lambda}) . \tag{3.20 – 21}$$

For the constraint of charge neutrality of neutron star matter one obtains[6,8]

$$\varrho_{tot}^{el} \equiv \varrho_{Bary}^{el} + \varrho_{Lept}^{el} \equiv 0 . \tag{3.22}$$

In summary, equations (3.6)-(3.12) (cf. (3.13),(3.14)) must be solved in combination with (3.15)-(3.22) for the following unknowns of the baryon-lepton system:[6]

$$\Sigma^{H,B}\,|_{\sigma,\omega,\varrho}\,, \qquad \Sigma^{F,B}\,|_{\sigma,\omega,\pi,\varrho}\,, \qquad B = p,\ n,\ \Sigma^{\pm,0},\ \Lambda,\ \Xi^{0,-},\ \Delta^{++,+,0,-}\,;$$

$$\mu^n\,, \qquad \mu^e\,;$$

$$p_{F,e},\ p_{F,\mu},\ p_{F,p},\ p_{F,n},\ p_{F,\Sigma^{\pm,0}},\ p_{F,\Lambda},\ p_{F,\Xi^{0,-}},\ p_{F,\Delta^{++,+,0,-}}\,.$$

For the purpose of briefty we do not give the electric charge densities of baryons ϱ_{Bary}^{el} and leptons ϱ_{Lept}^{el}, respectively, which can be taken from Ref. 6.
The EOS follows from the stress-energy density tensor $\mathcal{T}_{\mu\nu}$. The energy denstiy (ϵ) and pressure (P) of the system are given by $\epsilon \equiv < \mathcal{T}_{00} >$ and $P \equiv \frac{1}{3}\sum_{i=1}^{3} < \mathcal{T}_{ii} >$. Explicit expressions for ϵ and P are given in Ref. 6 and will not be repeated here. The basic point is that in the Green's functions method the EOS can be expressed solely in terms of Σ^B and a^B. Thus, once (3.6)-(3.12) and (3.15)-(3.22) have been solved self-consistently the EOS is known up to simple mathematical operations.[5,6]

4. RESULTS AND DISCUSSION

Our calculations are based on one Hartree parameter set[8] (denoted by HV) and two Hartree-Fock parameter sets[5,6] (denoted by HFIII and HFV) with coupling constants adjusted to the bulk parameters of nuclear matter. For the coupling strengths of the hyperons ($\Sigma^{\pm,0}, \Lambda, \Xi^{0,-}$) and baryons ($\Delta^{++,+,0,-}$) the prescription of universal coupling has been chosen. As a special feature the parameter set HFIII contains cubic and quartic self-interactions of the scalar σ-field. In Fig. 1 we exhibit the hadronic equation of state, i.e., pressure P versus energy density ϵ for the Hartree and the Hartree-Fock calculations. At low ϵ values only p, n, e^-, and μ^- are present. The

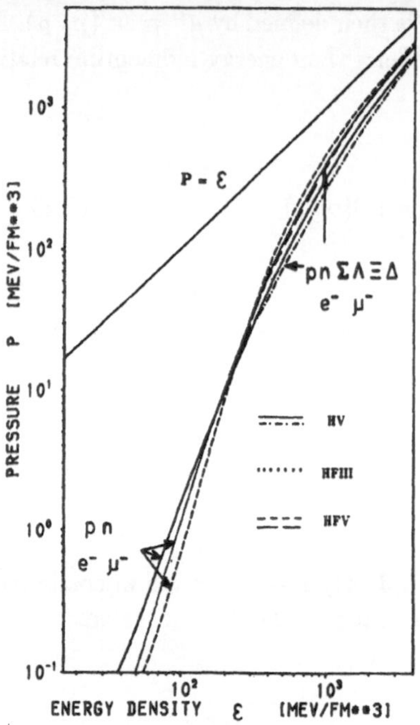

Figure 1

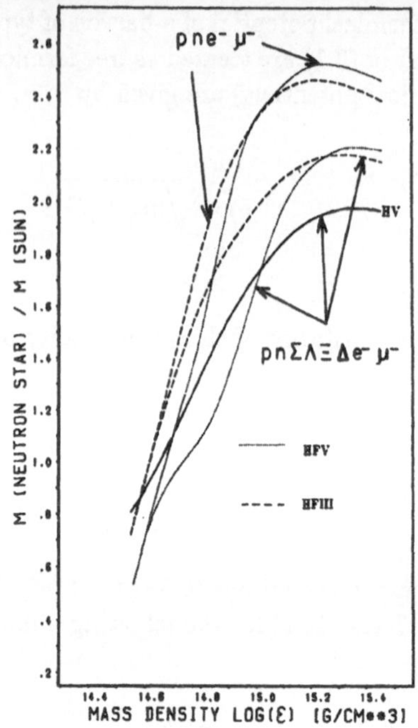

Figure 2

more massive baryons become populated at about $\epsilon > 320$ MeV/fm^3, which leads to a softening of the EOS. This is shown for the parameter sets HV and HFV.

The calculated gravitational neutron star masses as a function of central energy density are exhibited in Fig. 2 for the parameter sets HV, HFIII, and HFV, respectively. The underlying equations of state have been joined with the ones calculated by HW (for the outer surface of the star, i.e., $4.4 \cdot 10^{-12} < \epsilon/(\mathrm{MeV/fm^3}) < 5.6 \cdot 10^{-2}$) and NV (for the inner surface, i.e., $5.6 \cdot 10^{-2} < \epsilon/(\mathrm{MeV/fm^3}) < 5.6$). The influence of baryon population on the neutron star mass is shown for the two chosen HF parameter sets. The bulk parameters of the maximum stable NS are summarized in Table 1. The maximum stable masses are consistent with values estimated for the upper bound of mass of NS 4U 0900-40. The corresponding fractional redshifts, defined by (2.6), are $z = .4488$ (HV), .5057 (HFIII), and .6135 (HFV). (This results refer to calculations in which charged hyperon and baryon states (cf. (3.1)) are allowed to become populated.)

Table 1. Central energy density ϵ_c, maximum stable neutron star mass in solar units, radius, and moment of inertia of neutron stars (universal coupling).

	$\log(\epsilon_c)$ [g/cm^3]	$\frac{M}{M_\odot}\|_{\max}$	R [km]	log(I) [g cm^2]
HV	15.37	1.98	11.16	45.56
HFIII	15.31	2.18	11.51	45.69
HFV	15.37	2.21	10.59	45.72

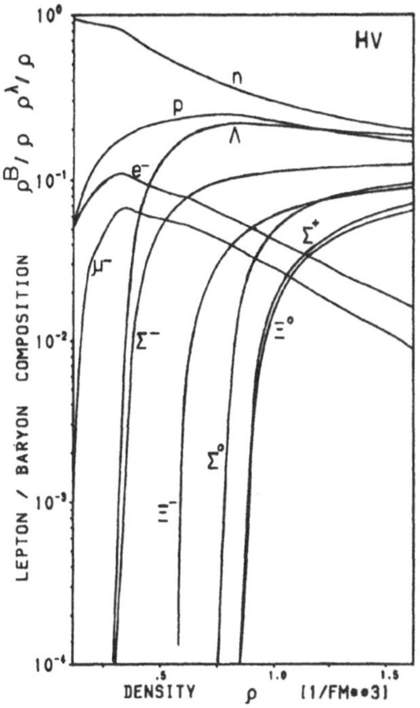

Figure 3

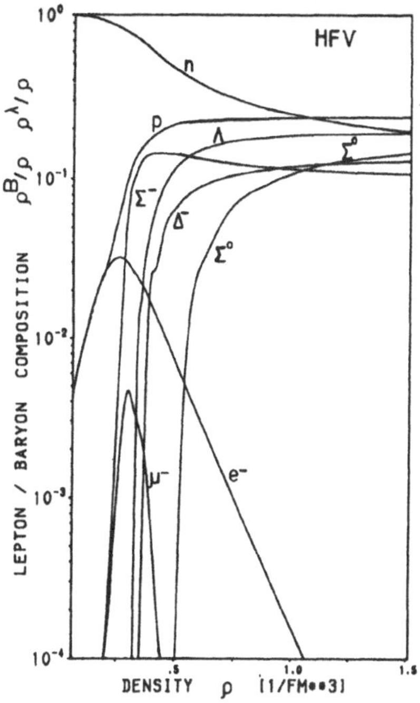

Figure 4

In Figs. 3,4 the lepton-baryon composition calculated for HV and HFV for charge neutral neutron star matter is shown. Here ϱ^B/ϱ (ϱ^λ/ϱ) denotes the ratio of the baryon (lepton) density to the total baryon density.[6] As can be seen the lepton-baryon composition calculated for the Hartree parameter set (Fig. 3) differs drastically from the Hartree-Fock outcome (Fig. 4). A detailed discussion of the baryon threshold is beyond the scope of this presentation. It can qualitatively be understood in terms of charge-, isospin-, and mass favouring arguments.[6,8] The star matter at low densities consists for both cases of nearly 100% neutrons and a small admixture of protons and electrons. At larger densities ($\varrho > 1.5\,\varrho_0$) the population of the Λ-hyperon (HV) respectively Σ^- (HFV) sets in. Due to the small value of the vector coupling constant of the ϱ-meson in the case of HFV (denoted by g_ϱ^{HFV}), which is only approximately one tenth of g_ϱ^{HV} (cf. Ref. 6), the Δ-state is no longer prevented from becoming populated by reasons of isospin, and the isospin-unfavoured but charge-favoured Δ^- can become populated even at rather low nuclear densities. (This trend is even reinforce by the smaller effective mass of the Δ-particle in the HFV case.) Furthermore one sees in Figs. 3,4 that the relative lepton concentrations saturate. This is due to the occurence of baryon population and vanishes for matter which consists of only p, n, e^-, μ^-. The Δ^- with its large statistical weight of spin leads to large contributions for ϱ^Δ/ϱ. Consequently less electrons and muons are needed in order to fulfill the constraint of charge neutrality. For this reason ϱ^e/ϱ and ϱ^μ/ϱ calculated for HFV (Fig. 4) are significantly reduced as compared to the

Hartree case (Fig. 3). We found for the chemical potential of the electron $\mu^e \leq 222$ MeV for the Hartree treatment and $\mu^e \leq 155$ MeV for the HF theory, respectively. The rather low HF value can eventually prevent pion condensation, depending on the effective pion mass in the nuclear medium (which is however not very accurately known).

REFERENCES

1. J. R. Oppenheimer and G. M. Volkoff, On Massive Neutron Cores, Phys. Rev. 55:374 (1939).
2. B. K. Harrison and J. A. Wheeler, chapter 10, in: "Gravitation Theory and Gravitational Collapse," B. K. Harrison et al., University of Chicago Press, Chicago (1965).
3. J. W. Negele and D. Vautherin, Neutron Star Matter at Sub-nuclear densities, Nucl. Phys. A 207:298 (1973).
4. B. D. Serot and J. D. Walecka, The Relativistic Nuclear Many-Body Problem, in: "Advances in Nuclear Physics," Vol. 16, J. W. Negele, E. Vogt, ed., Plenum Press, New York (1986).
5. F. Weber and M. K. Weigel, Neutron Star Properties and the Relativistic Nuclear Equation of State of Many-Baryon Matter, Nucl. Phys. A 493:549 (1989); Equation of State of Dense Baryonic Matter, J. Phys. G 15:765 (1989).
6. F. Weber and M. K. Weigel, Baryon Composition and Macroscopic Properties of Neutron Stars, submitted to Nucl. Phys. A.
7. B. D. Serot and H. Uechi, Neutron Stars in Relativistic Hadron-Quark Models, Ann. Phys. 179:272 (1987).
8. N. K. Glendenning, The Hyperon Composition of Neutron Stars, Phys. Lett. B 114:392 (1982); Neutron Stars are Giant Hypernuclei?, Astrophys. J. 293:470 (1985); Neutron Star Masses as a Constraint on the Nuclear Compression Modulus, Phys. Rev. Lett. 57:1120 (1986).

THE EQUATION OF STATE AND PROPERTIES OF NEUTRON STARS

T. L. Ainsworth

Institut für Kernphysik, Kernforschungsanlage

D-5170 Jülich, West Germany

One of the most intriguing possibilities to determine properties of the nuclear equation of state (EOS) is the astrophysical data of dense stars. Type II supernova and neutron stars offer two independent means of studying the nuclear EOS. Bethe[1] has recently reviewed the nuclear physics of supernova. Neutron star EOS's sample a different range of densities and isospin asymmetry than supernova. Thus these two situations yield related but not identical information about the EOS. It was hoped that the structure of neutron stars and their observed masses would constrain the wide variety of neutron star EOS's.[2] The only "hard" constraint is that the minimum maximum neutron star mass, M_{max}, that an EOS must support is $\sim 1.44\ M_{\odot}$,[3,4] where $M_{\odot}$ is the solar mass. Recently several authors have concentrated on this aspect of the neutron star EOS.[5-9] The common conclusion is that too "soft" of an EOS does not support a massive enough neutron star.

The existence of millisecond pulsars[10,11] places an upper limit on the "stiffness" of the EOS. In practice, this limit is easily met, except for the recently reported 0.5 msec. period of the SN1987A pulsar.[12] A 0.5 msec. rotational period places the upper limit below the lower limit, thus eliminating "all" EOS's.[13,14] If this observation is confirmed and results from rigid-body rotation of the neutron star then there are strong restrictions on the stiffness of the neutron star EOS.

Setting aside for the moment the problem of rapidly rotating stars: Can the non-rotating star constrain the nuclear EOS? Can model-independent conclusions be drawn?

There are several general many-body methods used to calculate the EOS of interacting hadrons: (i) non-relativistic potentials models[15,16] which employ realistic two-body potentials fit to phase-shift data, and often are augmented by a three-nucleon interaction; (ii) field theoretic models[17] with and without scalar self-interactions which are inherently relativistic calculations; (iii) hybrids of these two approaches.[18,19] The energy per particle of asymmetric matter is well approximated[9,15,16,20] by

$$E(n,\alpha) = E(n,\alpha=0) + \alpha^2 E_{sym}(n), \tag{1}$$

where $\alpha = 1 - 2\,Z/A$ and $E_{sym}(n) = E(n,\alpha=1) - E(n,\alpha=0)$, the difference between pure neutron matter and symmetric nuclear matter. Neutron star matter is in equilibrium with respect to weak interactions. Therefore the total energy per particle is the sum of

The Nuclear Equation of State, Part A
Edited by W. Greiner and H. Stöcker
Plenum Press, New York

$E(n,\alpha)$ and the lepton energy. The asymmetry of matter in β-equilibrium is the value of α which minimizes the total energy. Once the total energy per particle is provided, the neutron star radius, surface red shift, moment of inertia and binding energy as a function of mass are given by the TOV equations.[2]

To investigate the role of the components of neutron star EOS's a simple parameterization of the nuclear EOS is assumed,[8]

$$E/A(u,\alpha=0) = \frac{3}{5}E_F^{(0)}u^{2/3} + \frac{1}{2}Au + \frac{Bu^\sigma}{1+B'u^{\sigma-1}} + 3\sum_{i=1,2} C_i\left(\frac{\Lambda_i}{p_F^{(0)}}\right)^3\left(\frac{p_F}{\Lambda_i} - \tan^{-1}\frac{p_F}{\Lambda_i}\right), \tag{2}$$

where $u = n/n_0$, and $E_F^{(0)}$ and $p_F^{(0)}$ are the Fermi energy and momentum at $n_0 = 0.16\ \mathrm{fm}^{-3}$. The energy per particle is the sum of kinetic energy (1st term), and potential energy terms arising from local zero-range (2nd & 3rd) and finite-range interactions (last term). The form is similar to that used by Gale, Bertsch and Das Gupta[21] in the study of heavy-ion collisions. The differences are the presence of the second finite-range term and the denominator, $1 + B'u^{\sigma-1}$, of the third term. The denominator serves to keep the EOS causal when $\sigma > 1$ and the second finite-range term allows the parameterization to fit a wider range of incompressibilities. Similar parameterizations have reproduced properties of finite nuclei[22] and flow observables in heavy-ion collisions.[23] The coefficients are adjusted to reproduce the properties of symmetric nuclear matter at n_0, *i.e.* the binding energy, ~ -16 MeV; effective mass, ~ 0.7; single-particle potential depth, ~ -76 MeV. The values of the coefficients and further details are given in Ref. 8.

The symmetry energy is separated into kinetic and potential terms, the density dependence of the potential contributions being parameterized by the function $F(u)$,

$$E_{sym}(n) = (2^{2/3}-1)\frac{3}{5}E_F^{(0)}\{u^{2/3} - F(u)\} + E_{sym}^{(0)}F(u), \tag{3}$$

where $F(1) = 1$ and $E_{sym}^{(0)} = 30$ MeV. The forms chosen for $F(u)$ mimic results of microscopic calculations. Non-relativistic potential models[15,16] have an E_{sym} which initially increases with density and then saturates at high density, whereas the field theoretic[5,9,24] and relativistic Brueckner[19,20] calculations have E_{sym} increasing linearly at high density.

A suite of models are considered with a range of compression moduli and three functional forms of $F(u)$ for the symmetry energy. The results for M_{max} neutron stars with the tabulated values of K_0 and assumed forms of $F(u)$ are shown in Table 1. For a given parameterization M_{max} scales approximately with $K_0^{1/2}$.[7,8,25] An EOS with K_0 as low as 120 MeV can support $\sim 1.5\ M_\odot$ stars. This contrasts with field theoretic models[9,24] where $K_0 > 225$ MeV was required for $1.5\ M_\odot$ stars. The symmetry energy, especially its behavior with density, determines the proton fraction of neutron star matter. E_{sym} becomes less important for the stiffer EOS's.[8,9,20] The difference in M_{max} between the pure neutron and β-stable EOS's is greatest for the most rapidly increasing E_{sym}, *i.e.* $F(u) = 2u^2/(1+u)$. A hefty, linearly increasing symmetry energy significantly softens the β-stable EOS relative to the pure neutron EOS.

Four different EOS's, each fitting nuclear matter properties, show that the variations in M_{max} arising from effects other than K_0 are large. These EOS's[5,9,16,20] all yield $K_0 \sim 200$ MeV with the resulting M_{max} stars of 1.3, 1.36, 2.0 and 2.4 $M_\odot$, respectively.

Table 1. Results for spherical β-stable (pure neutron) neutron stars are shown. See text for additional discussion.

$F(u)$	K_0 [MeV]	$M_{max}/M_\odot$	R [km]	Ω_K [rad/sec]
	120	1.458 (1.70)	9.114	15,980
u	180	1.722 (1.90)	9.879	15,390
	240	1.935 (2.07)	10.57	14,740
	120	1.470 (1.95)	9.895	14,180
$2u^2/(1+u)$	180	1.738 (2.10)	10.32	14,480
	240	1.952 (2.24)	10.93	14,070
	120	1.404 (1.45)	8.435	17,610
$\sqrt{u}$	180	1.679 (1.71)	9.324	16,570
	240	1.895 (1.92)	10.11	15,590

(The present model, with $K_0 = 200$ MeV and $F(u) = u$, yields $M_{max} = 1.8\ M_\odot$.) The primary reason for this wide variation is that different models are used to extrapolate from n_0 to the higher densities, $\sim 6n_0$. All methods have drawbacks: Field theoretic models have large loop corrections which do not converge, even weakly, at 2nd-order.[26] Non-relativistic variational calculations become acausal at very high density. And phenomenological models are not microscopic. The possible presence of hyperons, pion condensates and quark/gluon phase transitions further obfuscates analysis of the high density EOS. Thus there are no unique, model-independent means of constraining the high density EOS from properties determined near saturation density.

Does it matter? Do static neutron stars depend critically upon the high density EOS? Yes.[2,8,9] In Fig. 1, the mass fraction at or below a given density is plotted for the M_{max} stars of Table 1, with $F(u) = u$. Most of the star mass lies above $3\rho_0$. Only when the EOS is very stiff is this conclusion modified.

There exist quantitative measures of the "stiffness" or "softness" an EOS. Commonly used are the adiabatic index, $\Gamma = d\ln P/d\ln\rho$, the sound speed, $c_s^2 = c^2\,dP/d\epsilon$, and the incompressibility, $K = 9\ dP/d\rho$. The structure of neutron stars depends, via the TOV equations, on the pressure and energy density integrated over a wide range of densities, typically between ~ 8 and $\sim 2 \times 10^{15}$ gm/cm^3.[2] Therefore the value of Γ, c_s or K at any single density does not reflect gross properties of neutron stars. The sound speed may be an appropriate gauge of neutron star properties because it directly relates the pressure and energy density, the two variables upon which the TOV equations depend. Additionally, causality requires that $c_s < c$ for all densities setting an absolute upper limit on the stiffness of an EOS. Detailed comparison of the relative stiffness of EOS's thus requires the juxtaposition of, say, c_s/c for each EOS as a function of density. However, an empirical, simple and straightforward measure of the *average* stiffness of a neutron star EOS is the value of M_{max} determined by the TOV equations.

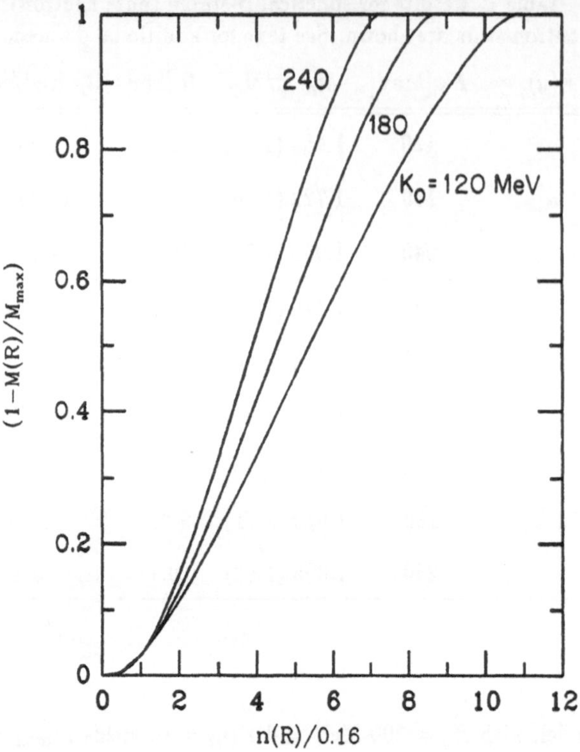

Figure 1. The mass fraction at or below a given density is plotted for maximum mass stars. See text for details.

The recent observation of the 0.5 msec. period of the SN1987a pulsar[12] provides another constraint on neutron star EOS's.[13,14] Detailed analysis of rotating neutron stars is beyond the scope of this paper, however using models of slowly rotating stars general conclusions can be drawn. The Keplerian angular velocity, the critical angular velocity for equatorial mass from a spherical star, is $\Omega_K = (GM_G/R^3)^{1/2}$. The critical angular velocity for stability of a rapidly rotating star, Ω_{crit}, is smaller than Ω_K by $\sim$ 25% for stiff EOS's and up to $\sim$ 45% for soft EOS's because the rapidly rotating star is ellipsoidal rather than spherical.[27] An additional 6 – 8% reduction results from the analysis of non-axial symmetric deformations.[14] Thus the value of $\Omega_{crit} \sim .6 \pm .1\ \Omega_K$. These results are model dependent, however they should provide reasonable estimates. The maximum Ω_K, and Ω_{crit}, values are obtained from the maximum mass stars, *i.e.* largest M_G and smallest R. The observed Ω is $\sim 12,370$ rad./sec. which implies that Ω_K for the spherical star should be greater than $\sim 20,600$ rad./sec.

Neutron stars from stiff EOS's have relatively large radii and thus smaller Ω_{crit}, whereas too soft an EOS does not give $M_{max} > 1.44\ M_\odot$. The interesting point is that *no* published neutron star EOS appears able to simultaneously give a non-rotating 1.44 $M_\odot$ star and a stable rotating star with a 0.5 msec. period.[13,14] The basic requirements of an EOS which meets both criterion are a stiff high-density ($n > 4\,n_0$) EOS coupled to a very soft low-density ($1 - 2\,n_0$) EOS.[13] The hard core is needed to support a massive enough star, and the soft surroundings provides a density profile with a relatively small radius.

One definite conclusion is that the properties of static neutron stars do not provide constraints on symmetric matter at or near to saturation density. However, the recent observation of the SN1987A pulsar in conjunction with the required minimum M_{max} appears to constrain the neutron star matter EOS for densities at or above n_0. In particular, detailed knowledge of the stiffness of neutron star EOS's is required for two distinct density regimes: one above $\sim 4\,n_0$ and the other near $1 - 2\,n_0$. One reasonable measure of stiffness is suggested: the speed of sound. The symmetry energy, especially at high density, deserves further consideration. Hopefully, more detailed observations will be able to provide guidance for future investigations of the high density equation of state.

ACKNOWLEDGMENTS

This paper is based on work done in collaboration with M. Prakash and J. M. Lattimer. Discussions with them on these and related topics have been stimulating and fruitful.

REFERENCES

1. H. A. Bethe, *Ann. Rev. Nuc. Part. Sci.*, **38**:1, (1988).
2. W. D. Arnett and R. L. Bowers, *Astrophys. J. Suppl.*, **33**:415, (1977).
3. S. Rappaport and P. C. Joss, *Ann. Rev. Astron. Astrophys.*, **22**:537, (1984).
4. J. M. Weisberg and J. H. Taylor, *Phys. Rev. Lett.*, **52**:1348, (1984).
5. N. K. Glendenning, *Nucl. Phys.*, **A493**:521, (1989); and references therein.
6. F. Weber and M. K. Weigel, *Nucl. Phys.*, **A493**:549, (1989).
7. J. Cooperstein, *Phys. Rev. C*, **37**:786, (1988).
8. M. Prakash, T. L. Ainsworth and J. M. Lattimer, *Phys. Rev. Lett.*, **61**:2518, (1988).
9. M. Prakash and T. L. Ainsworth, *Phys. Rev. C*, **36**:346, (1987).
10. D. C. Backer *et al*, *Nature*, **300**:615, (1982).
11. A. S. Fruchter, D. R. Stinebring and J. H. Taylor, *Nature*, **333**:237, (1988).
12. J. A. Kristian *et al*, *Nature*, **338**:234, (1989).
13. J. M. Lattimer and M. Prakash, private communication.
14. J. R. Ipser and L. Lindblom, *Phys. Rev. Lett.*, **62**:2777, (1989).
15. I. E. Lagaris and V. R. Pandharipande, *Nucl. Phys.*, **A369**:470, (1981).
16. R. B. Wiringa, V. Fiks and A. Fabrocini, *Phys. Rev. C*, **38**:1010, (1988).
17. B. D. Serot and J. D. Walecka, in *The Relativistic Nuclear Many-Body Problem*, J. W. Negele and E. Vogt, eds., Plenum, New York, (1985).
18. R. Brockman and R. Machleidt, *Phys. Lett. B*, **149**:283, (1984).
19. B. ter Haar and R. Malfliet, *Phys. Rev. Lett.*, **56**:1237, (1986).
20. H. Müther, M. Prakash and T. L. Ainsworth, *Phys. Lett. B*, **199**:469, (1987).
21. C. Gale, G. Bertsch and S. Das Gupta, *Phys. Rev. C*, **35**:1666, (1987).
22. J. P. Blaizot, *Phys. Rept.*, **64**:171, (1980).
23. G. Bertsch and S. Das Gupta, *Phys. Rept.*, **160**:189, (1988).
24. N. K. Glendenning, *Phys. Rev. Lett.*, **57**:1120, (1985).
25. S. A. Bludman, *Astrophys. J.*, **183**:649, (1973).
26. R. J. Furnstahl, R. J. Perry and B. D. Serot, preprint, (1988).
27. J. L. Friedman, J. R. Ipser and L. Parker, *Astrophys. J.*, **304**:115, (1986).

INDEX